McGRAW-HILL CONCISE ENCYCLOPEDIA OF ENVIRONMENTAL SCIENCE

McGraw-Hill

New York Chicago San Francisco Lisbon London Madrid Mexico City
Milan New Delhi San Juan Seoul Singapore Sydney Toronto

The McGraw-Hill Companies

Library of Congress Cataloging in Publication Data

McGraw-Hill concise encyclopedia of environmental science.
p. cm.
Includes bibliographical references and index.
ISBN 0-07-143951-X
1. Environmental sciences—Encyclopedias. I. Title: Concise encyclopedia of environmental science.

GE10.M377 2005
363.7′003—dc22 2005043792

This material was extracted from the *McGraw-Hill Encyclopedia of Science & Technology*, Ninth Edition, © 2002 by The McGraw-Hill Companies, Inc. All rights reserved.

McGRAW-HILL CONCISE ENCYCLOPEDIA OF ENVIRONMENTAL SCIENCE, copyright © 2005 by The McGraw-Hill Companies, Inc. All rights reserved. Printed in the United States of America. Except as permitted under the United States Copyright Act of 1976, no part of this publication may be reproduced or distributed in any form or by any means, or stored in a database or retrieval system, without the prior written permission of the publisher.

1 2 3 4 5 6 7 8 9 0 DOC/DOC 0 10 9 8 7 6 5

0-07-143951-X

This book was printed on acid-free paper.

It was set in Helvetica Black and Souvenir by TechBooks, Fairfax, Virginia.

The book was printed and bound by RR Donnelley, The Lakeside Press.

McGraw-Hill books are available at special quantity discounts to use as premiums and sales promotions, or for use in corporate training programs. For more information, please write to the Director of Special Sales, McGraw-Hill Professional, Two Penn Plaza, New York, NY 10121-2298. Or contact your local bookstore.

CONTENTS

EDITORIAL STAFF

Mark D. Licker, Publisher

Elizabeth Geller, Managing Editor
Jonathan Weil, Senior Staff Editor
David Blumel, Editor
Alyssa Rappaport, Editor
Charles Wagner, Manager, Digital Content
Renee Taylor, Editorial Assistant

EDITING, DESIGN, AND PRODUCTION STAFF

Roger Kasunic, Vice President—Editing, Design, and Production

Joe Faulk, Editing Manager
Frank Kotowski, Jr., Senior Editing Supervisor
Ron Lane, Art Director
Vincent Piazza, Assistant Art Director
Thomas G. Kowalczyk, Production Manager
Pamela A. Pelton, Senior Production Supervisor

CONSULTING EDITORS

Dr. Mark Chase. *Molecular Systematics Section, Jodrell Laboratory, Royal Botanic Gardens, Kew, Richmond, Surrey, United Kingdom.* PLANT TAXONOMY.

Dr. Peter J. Davies. *Professor, Department of Plant Biology, Cornell University, Ithaca, New York.* PLANT PHYSIOLOGY.

Dr. John P. Harley. *Department of Biological Sciences, Eastern Kentucky University, Richmond.* MICROBIOLOGY.

Dr. S. C. Jong. *Senior Staff Scientist and Program Director, Mycology and Protistology Program, American Type Culture Collection, Manassas, Virginia.* MYCOLOGY.

Dr. Peter M. Kareiva. *Director of Conservation and Policy Projects, Environmental Studies Institute, Santa Clara University, Santa Clara, California.* ECOLOGY AND CONSERVATION.

Dr. John Knauss. *Scripps Institution of Oceanography, La Jolla, California.* OCEANOGRAPHY.

Prof. Robert E. Knowlton. *Department of Biological Sciences, George Washington University, Washington, DC.* INVERTEBRATE ZOOLOGY.

Prof. Konrad B. Krauskopf. *Formerly, School of Earth Sciences, Stanford University, Stanford, California.* GEOCHEMISTRY.

Dr. Donald W. Linzey. *Wytheville Community College, Wytheville, Virginia.* VERTEBRATE ZOOLOGY.

Prof. Scott M. McLennan. *Chair, Department of Geosciences, State University of New York at Stony Brook.* GEOLOGY (PHYSICAL, HISTORICAL, AND SEDIMENTARY).

Dr. Henry F. Mayland. *Soil Scientist, Northwest Irrigation and Soils Research Laboratory, USDA-ARS, Kimberly, Idaho.* SOILS.

Dr. Orlando J. Miller. *Professor Emeritus, Center for Molecular Medicine and Genetics, Wayne State University School of Medicine, Detroit, Michigan.* GENETICS.

Prof. J. Jeffrey Peirce. *Department of Civil and Environmental Engineering, Edmund T. Pratt Jr. School of Engineering, Duke University, Durham, North Carolina.* ENVIRONMENTAL ENGINEERING.

Dr. Kenneth P. H. Pritzker. *Pathologist-in-Chief and Director, Head, Connective Tissue Research Group, and Professor, Laboratory Medicine and Pathobiology, University of Toronto, Mount Sinai Hospital, Toronto, Ontario, Canada.* MEDICINE AND PATHOLOGY.

Dr. Roger M. Rowell. *USDA-Forest Service, Forest Products Laboratory, Madison, Wisconsin.* FORESTRY.

Dr. Steven A. Slack. *Associate Vice President for Agricultural Administration, Director, Ohio Agricultural Research and Development Center, and Associate Dean for Research, College of Food, Agricultural, and Environmental Sciences, Ohio State University, Wooster.* PLANT PATHOLOGY.

Prof. John F. Timoney. *Department of Veterinary Science, University of Kentucky, Lexington.* VETERINARY MEDICINE.

Dr. Bruce A. Voyles. *Professor, Department of Biological Chemistry, Grinnell College, Grinnell, Iowa.* VIROLOGY.

Prof. Pao K. Wang. *Department of Atmospheric and Oceanic Sciences, University of Wisconsin-Madison.* METEOROLOGY AND CLIMATOLOGY.

Prof. Thomas A. Wikle. *Head, Department of Geography, Oklahoma State University, Stillwater.* PHYSICAL GEOGRAPHY.

PREFACE

For more than four decades, the *McGraw-Hill Encyclopedia of Science & Technology* has been an indispensable scientific reference work for a broad range of readers, from students to professionals and interested general readers. Found in many thousands of libraries around the world, its 20 volumes authoritatively cover every major field of science. However, the needs of many readers will also be served by a concise work covering a specific scientific or technical discipline in a handy, portable format. For this reason, the editors of the *Encyclopedia* have produced this series of paperback editions, each devoted to a major field of science or engineering.

The articles in this *McGraw-Hill Concise Encyclopedia of Environmental Science* cover all the principal topics of this field. Each one is a condensed version of the parent article that retains its authoritativeness and clarity of presentation, providing the reader with essential knowledge in environmental science without extensive detail. The initials of the authors are at the end of the articles; their full names and affiliations are listed in the back of the book.

The reader will find over 800 alphabetically arranged entries, many illustrated with images or diagrams. Most include cross references to other articles for background reading or further study. Dual measurement units (U.S. Customary and International System) are used throughout. The Appendix includes useful information complementing the articles. Finally, the Index provides quick access to specific information in the articles.

This concise reference will fill the need for accurate, current scientific and technical information in a convenient, economical format. It can serve as the starting point for research by anyone seriously interested in environmental science, even professionals seeking information outside their own specialty. It should prove to be a much used and much trusted addition to the reader's bookshelf.

MARK D. LICKER
Publisher

ORGANIZATION OF THE ENCYCLOPEDIA

Alphabetization. The more than 800 article titles are sequenced on a word-by-word basis, not letter by letter. Hyphenated words are treated as separate words. In occasional inverted article titles, the comma provides a full stop. The index is alphabetized on the same principles. Readers can turn directly to the pages for much of their research. Examples of sequencing are:

Plant-water relations
Plants, life forms of
Precipitation (meteorology)
Precipitation measurement
Rain shadow
Rainforest
Water-borne disease
Water treatment

Cross references. Virtually every article has cross references set in CAPITALS AND SMALL CAPITALS. These references offer the user the option of turning to other articles in the volume for related information.

Measurement units. Since some readers prefer the U.S. Customary System while others require the International System of Units (SI), measurements in the Encyclopedia are given in dual units.

Contributors. The authorship of each article is specified at its conclusion, in the form of the contributor's initials for brevity. The contributor's full name and affiliation may be found in the "Contributor" section at the back of the volume.

Appendix. Every user should explore the variety of succinct information supplied by the Appendix, which includes conversion factors, measurement tables, fundamental constants, and a biographical listing of scientists. Users wishing to go beyond the scope of this Encyclopedia will find recommended books and journals listed in the "Bibliographies" section; the titles are grouped by subject area.

Index. The 7000-entry index offers the reader the time-saving convenience of being able to quickly locate specific information in the text, rather than approaching the Encyclopedia via article titles only. This elaborate breakdown of the volume's contents assures both the general reader and the professional of efficient use of the *McGraw-Hill Concise Encyclopedia of Environmental Science*.

Abscission The process whereby a plant sheds one of its parts. Leaves, flowers, seeds, and fruits are parts commonly abscised. Almost any plant part, from very small buds and bracts to branches several inches in diameter, may be abscised by some species. However, other species, including many annual plants, may show little abscission, especially of leaves.

Abscission may be of value to the plant in several ways. It can be a process of self-pruning, removing injured, diseased, or senescent parts. It permits the dispersal of seeds and other reproductive structures. It facilitates the recycling of mineral nutrients to the soil. It functions to maintain homeostasis in the plant, keeping in balance leaves and roots, and vegetative and reproductive parts.

In most plants the process of abscission is restricted to an abscission zone at the base of an organ (see illustration); here separation is brought about by the disintegration of the walls of a special layer of cells, the separation layer. The portion of the abscission zone which remains on the plant commonly develops into a corky protective layer that becomes continuous with the cork of the stem.

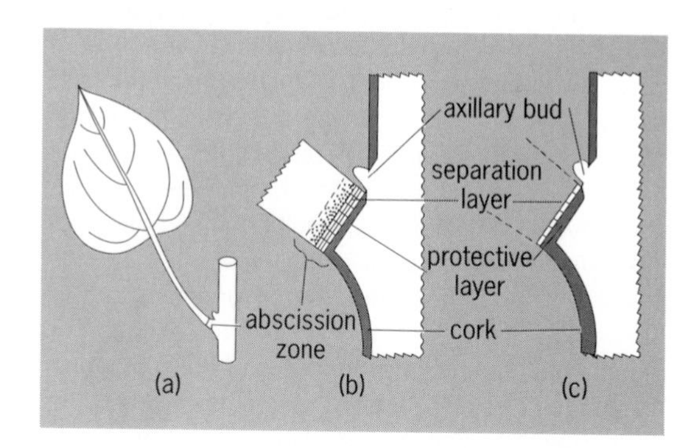

Diagrams of the abscission zone of a leaf. (*a*) A leaf with the abscission zone indicated at the base of the petiole. (*b*) The abscission zone layers shortly before abscission and (*c*) the layers after abscission.

Auxin applied experimentally to the distal (organ) side of an abscission zone retards abscission, while auxin applied to the proximal (stem) side accelerates abscission. The gibberellins are growth hormones which influence abscission. When applied to young fruits or to leaves, they tend to promote growth, delay maturation, and thereby indirectly prevent or delay abscission. Abscisic acid has the ability to promote abscission and senescence and to retard growth. Small amounts of ethylene have profound effects on the growth of plants and can distort and reduce growth and promote senescence and abscission.

[F.T.A.]

Acari A subclass of Arachnida, the mites and ticks; also called Acarina. All are small (0.004–1.2 in. or 0.1–30 mm, most less than 0.08 in. or 2 mm in length), have lost most traces of external body segmentation, and have the mouthparts borne on a discrete body region, the gnathosoma. They are apparently most closely related to Opiliones and Ricinulei.

The chelicerae may be chelate or needlelike. Pedipalps are generally smaller than the walking legs, and are simple or sometimes weakly chelate. Legs may be modified by projections, enlarged claws, heavy spines, or bladelike rows of long setae for crawling, clinging to hosts, transferring spermatophores, or swimming. Mites frequently have one or two simple eyes (ocelli) anterolaterally on the idiosoma (occasionally one anteromedially as well). The ganglia of the central nervous system are coalesced into a single "brain" lying around the esophagus. A simple dorsal heart is present in a few larger forms.

Mites are ubiquitous, occurring from oceanic trenches below 13,200 ft (4000 m) to over 20,800 ft (6300 m) in the Himalayas and suspended above 3300 ft (1000 m) in the atmosphere; there are species in Antarctica. Soil mites show the least specialized adaptations to habitat and are frequently well-sclerotized predators. Inhabitants of stored grains, cheese, and house dust are also relatively unspecialized. The dominant forms inhabiting mosses are heavily sclerotized beetle mites (Oribatidae). Flowering plant associates include spider mites (Tetranychidae) and gall mites (Eriophyidae). Fungivores are usually weakly sclerotized inhabitants of moist or semiaquatic habitats, while the characteristic fresh-water mites include sluggish crawlers, rapid swimmers, and planktonic drifters. Mites associated with invertebrate animals include internal, external, and social parasites as well as inactive phoretic stages on Insecta, Crustacea, Myriapoda, Chelicerata, Mollusca, and Parazoa. A similar diversity of species are parasites or commensals of vertebrates. Some parasitic mites are disease vectors. Over 30,000 species have been described, and it is estimated that as many as 500,000 may exist. [D.B.]

Acid rain Precipitation that incorporates anthropogenic acids and acidic materials. The deposition of acidic materials on the Earth's surface occurs in both wet and dry forms as rain, snow, fog, dry particles, and gases. Although 30% or more of the total deposition may be dry, very little information that is specific to this dry form is available. In contrast, there is a large and expanding body of information related to the wet form: acid rain or acid precipitation. Acid precipitation, strictly defined, contains a greater concentration of hydrogen (H^+) than of hydroxyl (OH^-) ions, resulting in a solution pH less than 7. Under this definition, nearly all precipitation is acidic. The phenomenon of acid deposition, however, is generally regarded as being anthropogenic, that is, resulting from human activity.

Theoretically, the natural acidity of precipitation corresponds to a pH of 5.6, which represents the pH of pure water in equilibrium with atmospheric concentrations of carbon dioxide. Atmospheric moisture, however, is not pure, and its interaction with ammonia, oxides of nitrogen and sulfur, and windblown dust results in a pH between 4.9 and 6.5 for most "natural" precipitation. The distribution and magnitude of precipitation pH in the United States (see illustration) suggest the impact of anthropogenic rather than natural causes. The areas of highest precipitation acidity (lowest pH) correspond to areas within and downwind of heavy industrialization and urbanization where emissions of sulfur and nitrogen oxides are high. It is with these emissions that the most acidic precipitation is thought to originate.

The transport of acidic substances and their precursors, chemical reactions, and deposition are controlled by atmospheric processes. In general, it is convenient to distinguish between physical and chemical processes, but it must be realized that both

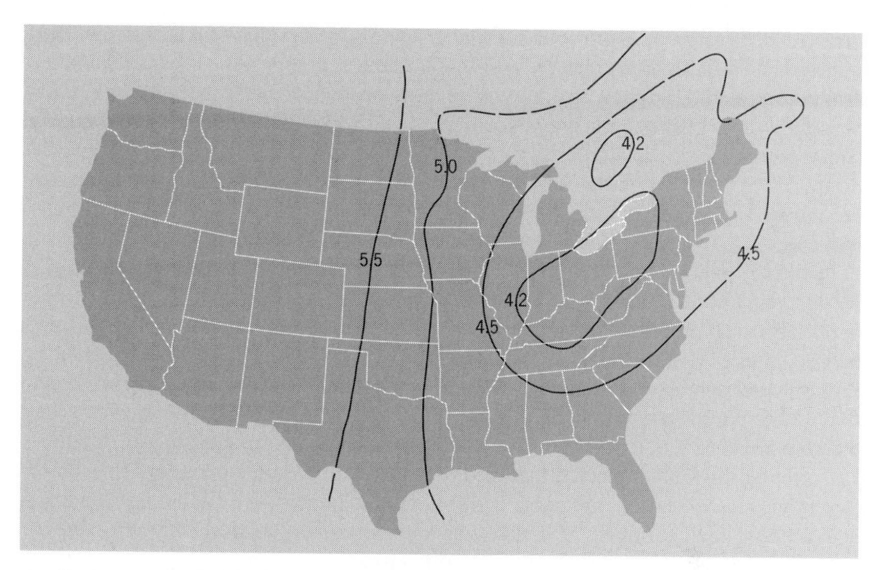

Distribution of rainfall pH in the eastern United States.

types may be operating simultaneously in complicated and interdependent ways. The physical processes of transport by atmospheric winds and the formation of clouds and precipitation strongly influence the patterns and rates of acidic deposition, while chemical reactions govern the forms of the compounds deposited.

There are a number of chemical pathways by which the primary pollutants, sulfur dioxide (SO_2) from industry, nitric oxide (NO) from both industry and automobiles, and reactive hydrocarbons mostly from trees, are transformed into acid-producing compounds. Some of these pathways exist solely in the gas phase, while others involve the aqueous phase afforded by the cloud and precipitation. As a general rule, the volatile primary pollutants must first be oxidized to more stable compounds before they are efficiently removed from the atmosphere. Ironically, the most effective oxidizing agents, hydrogen peroxide (H_2O_2) and ozone (O_3), arise from photochemical reactions involving the primary pollutants themselves. *See* AIR POLLUTION.

The effect of acid deposition on a particular ecosystem depends largely on its acid sensitivity, its acid neutralization capability, the concentration and composition of acid reaction products, and the amount of acid added to the system. As an example, the major factors influencing the impact of acidic deposition on lakes and streams are (1) the amount of acid deposited; (2) the pathway and travel time from the point of deposition to the lake or stream; (3) the buffering characteristics of the soil through which the acidic solution moves; (4) the nature and amount of acid reaction products in soil drainage and from sediments; and (5) the buffering capacity of the lake or stream.

Acid precipitation may injure trees directly or indirectly through the soil. Foliar effects have been studied extensively, and it is generally accepted that visible damage occurs only after prolonged exposure to precipitation of pH 3 or less (for example, acid fog or clouds). Measurable effects on forest ecosystems will then more likely result indirectly through soil processes than directly through exposure of the forest canopy. Many important declines in the condition of forest trees have been reported in Europe and North America during the period of increasing precipitation acidity. These cases include injury to white pine in the eastern United States, red spruce in the Appalachian Mountains of eastern North America, and many economically important species in central

Europe. Since forest trees are continuously stressed by competition for light, water, and nutrients; by disease organisms; by extremes in climate; and by atmospheric pollutants, establishing acid deposition as the cause of these declines is made more difficult. Each of these sources of stress, singly or in combination, produces similar injury. However, a large body of information indicates that accelerated soil acidification resulting from acid deposition is an important predisposing stress that in combination with other stresses has resulted in increased decline and mortality of sensitive tree species and widespread reduction in tree growth. *See* FOREST ECOSYSTEM; TERRESTRIAL ECOSYSTEM.

Acidic deposition impacts aquatic ecosystems by harming individual organisms and by disrupting flows of energy and materials through the ecosystem. The effect of acid deposition is commonly assessed by studying aquatic invertebrates and fish. Aquatic invertebrates live in the sediments of lakes and streams and are vitally important to the cycling of energy and material in aquatic ecosystems. These small organisms break down large particulate organic matter for further degradation by microorganisms, and they are an important food source for fish, aquatic birds, and predatory invertebrates.

Currently, there are concerns that acid deposition is causing the loss of fish species, through physiological damage and by reproductive impairment. While fish die from acidification, their numbers and diversity are more likely to decline from a failure to reproduce.The effects of acid deposition on individuals in turn elicit changes in the composition and abundance of communities of aquatic organisms. The degree of change depends on the severity of acidification, and the interaction of other factors, such as metal concentrations and the buffering capacity of the water. The pattern most characteristic of aquatic communities in acidified waters is a loss of species diversity, and an increase in the abundance of a few, acid-tolerant taxa.

Community-level effects may occur indirectly, as a result of changes in the food supply and in predator-prey relations. Reduction in the quality and amount of periphyton may decrease the number of herbivorous invertebrates, which may in turn reduce the number of organisms (predatory invertebrates and fish) that feed upon herbivorous invertebrates. The disappearance of fish may result in profound changes in plant and invertebrate communities. Dominant fish species function as keystone predators, controlling the size distribution, diversity, and numbers of invertebrates. Their reduction alters the interaction within and among different levels of the food web and the stability of the ecosystem as a whole.

The impact of acid deposition on terrestrial and aquatic ecosystems is not uniform. While increases in acid deposition may stress some ecosystems and reduce their stability and productivity, others may be unaffected. The degree and nature of the impact depend on the acid input load, organismal susceptibility, and buffering capacity of the particular ecosystem. *See* BIOGEOCHEMISTRY. [R.R.Sc./D.Lam./H.B.Pi./D.Ge.]

Acquired immune deficiency syndrome (AIDS) A viral disease of humans caused by the human immunodeficiency virus (HIV), which attacks and compromises the body's immune system. Individuals infected with HIV proceed through a spectrum of stages that ultimately lead to the critical end point, acquired immune deficiency syndrome. The disease is characterized by a profound progressive irreversible depletion of T-helper-inducer lymphocytes (CD4+ lymphocytes), which leads to the onset of multiple and recurrent opportunistic infections by other viruses, fungi, bacteria, and protozoa, as well as various tumors (Kaposi's sarcoma, lymphomas). HIV infection is transmitted by sexual intercourse (heterosexual and homosexual), by blood and blood products, and perinatally from infected mother to child (prepartum, intrapartum, and postpartum via breast milk).

Since retroviruses such as HIV-1 integrate their genetic material into that of the host cell, infection is generally lifelong and cannot be eliminated easily. Therefore, medical efforts have been directed toward preventing the spread of virus from infected individuals. *See* RETROVIRUS.

Approximately 50–70% of individuals with HIV infection experience an acute mononucleosis-like syndrome approximately 3–6 weeks following primary infection. In the acute HIV syndrome, symptoms include fever, pharyngitis, lymphadenopathy, headache, arthralgias, myalgias, lethargy, anorexia, nausea, and erythematous maculopapular rash. These symptoms usually persist for 1–2 weeks and gradually subside as an immune response to HIV is generated.

Although the length of time from initial infection to development of the clinical disease varies greatly from individual to individual, a median time of approximately 10 years has been documented for homosexual or bisexual men, depending somewhat on the mode of infection. Intravenous drug users experience a more aggressive course than homosexual men and hemophiliacs because their immune systems have already been compromised.

As HIV replication continues, the immunologic function of the HIV-infected individual declines throughout the period of clinical latency. At some point during that decline (usually after the CD4+ lymphocyte count has fallen below 500 cells per microliter), the individual begins to develop signs and symptoms of clinical illness, and sometimes may demonstrate generalized symptoms of lymphadenopathy, oral lesions, herpes zoster, and thrombocytopenia.

Secondary opportunistic infections are a late complication of HIV infection, usually occurring in individuals with less than 200 CD4+ lymphocytes per microliter. They are characteristically caused by opportunistic organisms such as *Pneumocystis carinii* and cytomegalovirus that do not ordinarily cause disease in individuals with a normally functioning immune system. However, the spectrum of serious secondary infections that may be associated with HIV infection also includes common bacterial pathogens, such as *Streptococcus pneumoniae*. Secondary opportunistic infections are the leading cause of morbidity and mortality in persons with HIV infection. Tuberculosis has also become a major problem for HIV-infected individuals. Therefore, HIV-infected individuals are administered protective vaccines (pneumococcal) as well as prophylactic regimens for the prevention of infections with *P. carinii, Mycobacterium tuberculosis*, and *M. avium* complex. *See* MYCOBACTERIAL DISEASES; OPPORTUNISTIC INFECTIONS; PNEUMOCOCCUS; STREPTOCOCCUS; TUBERCULOSIS.

Antiretroviral treatment with deoxyribonucleic acid (DNA) precursor analogs—for example, azidothymidine (AZT), dideoxyinosine (ddI), and dideoxycytidine (ddC)—has been shown to inhibit HIV infection by misincorporating the DNA precursor analogs into viral DNA by the viral DNA polymerase. Nevertheless, these agents are not curative and do not completely eradicate the HIV infection. [A.M.Ma.; E.P.Go.]

Actinobacillus A genus of gram-negative, immotile and nonspore-forming, oval to rod-shaped, often pleomorphic bacteria which occur as parasites or pathogens in mammals (including humans), birds, and reptiles. They are facultatively aerobic, capable of fermenting carbohydrates (without production of gas) and of reducing nitrates. The genomic DNA contains between 40 and 47 mol % guanine plus cytosine. The actinobacillus group shares many biological properties with the genus *Pasteurella*.

Actinobacillus (*Pasteurella*) *ureae* and *A. hominis* occur in the respiratory tract of healthy humans and may be involved in the pathogenesis of sinusitis, bronchopneumonia, pleural empyema, and meningitis. *Actinobacillus actinomycetemcomitans* occurs in the human oral microflora, and together with anaerobic or capnophilic organisms may cause endocarditis and suppurative lesions in the upper alimentary tract. Actinobacilli are susceptible to most antibiotics of the β-lactam family, aminoglycosides, tetracyclines, chloramphenicol, and many other antibacterial chemotherapeutics. *See* MEDICAL BACTERIOLOGY. [W.Ma.]

Adaptation (biology) A characteristic of an organism that makes it fit for its environment or for its particular way of life. For example, the Arctic fox (*Alopex lagopus*) is well adapted for living in a very cold climate. Appropriately, it has much thicker fur than similar-sized mammals from warmer places; measurement of heat flow through fur samples demonstrates that the Arctic fox and other arctic mammals have much better heat insulation than tropical species. Consequently, Arctic foxes do not have to raise their metabolic rates as much as tropical mammals do at low temperatures. The insulation is so effective that Arctic foxes can maintain their normal deep body temperatures of 100°F (38°C) even when the temperature of the environment falls to −112°F (−80°C). Thus, thick fur is obviously an adaptation to life in a cold environment.

In contrast to that clear example, it is often hard to be sure of the effectiveness of what seems to be an adaptation. For example, the scombrid fishes (tunnies and mackerel) seem to be adapted to fast, economical swimming. The body has an almost ideal streamlined shape. However, some other less streamlined-looking fishes are equally fast for their sizes. There are no measurements of the energy cost of scombrid swimming, but measurements on other species show no clear relationship between energy cost and streamlining.

Evolution by natural selection tends to increase fitness, making organisms better adapted to their environment and way of life. It might be inferred that this would ultimately lead to perfect adaptation, but this is not so. It must be remembered that evolution proceeds by small steps. For example, squids do not swim as well as fish. The squid would be better adapted for swimming if it evolved a fishlike tail instead of its jet propulsion mechanism, but evolution cannot make that change because it would involve moving down from the lesser adaptive summit before climbing the higher one. [R.M.Al.]

Adaptive management An approach to management of natural resources that emphasizes how little is known about the dynamics of ecosystems and that as more is learned management will evolve and improve. Natural systems are very complex and dynamic, and human observations about natural processes are fragmentary and inaccurate. As a result, the best way to use the available resources in a sustainable manner remains to be determined. Furthermore, much of the variability that affects natural populations is unpredictable and beyond human control. This combination of ignorance and unpredictability means that the ways in which ecosystems respond to human interventions are unknown and can be described only in probabilistic terms. Nonetheless, management decisions need to be made. Adaptive management proceeds despite this uncertainty by treating human interventions in natural systems as large-scale experiments from which more may be learned, leading to improved management in the future.

A key first step in the development of an adaptive management program is the assessment of the problem. During this stage, existing knowledge and interdisciplinary experience is synthesized and formally integrated by developing a dynamic model of the system. This modeling exercise helps to identify key information gaps and to postulate hypotheses about possible system responses to human intervention consistent with available information. Different management policies have to be screened in order to narrow down the alternatives to a few plausible candidates.

The second stage involves the formal design of a management and monitoring program. To the extent that new information can result in improved future management, adaptive management programs may include large-scale experiments deliberately designed to accelerate learning. Some management actions may be more effective than others at filling the relevant information gaps. In cases where spatial replication is possible (such as small lakes, patches of forest, and reefs), policies that provide contrasts between different management units will be much more informative about the system dynamics than those that apply the same rule everywhere. There are other barriers to the implementation of large-scale management experiments. Experiments usually have associated costs; thus, in order to be worthwhile, benefits derived from learning must overcompensate short-term sacrifices. Choices may be also restricted by social concerns or biological constraints, or they may have unacceptably high associated risks.

Once a plan for action has been chosen, the next stage is to implement the program in the field. This is one of the most difficult steps, because it involves a concerted and sustained effort from all sectors involved in the use, assessment, and management of the natural resources. Beyond the implementation of specific initial actions, putting in place an adaptive management program involves a long-term commitment to monitoring the compliance of the plan, evaluating the effects of management interventions, and adjusting management accordingly.

No matter how thorough and complete the initial assessment and design may have been, systems may always respond in manners that could not be foreseen at the planning stage. Ecosystems exhibit long-term, persistent changes at the scale of decades and centuries; thus, recent experience is not necessarily a good basis for predicting future behavior. The effects of global climatic change on the dynamics of ecosystems, which are to a large extent unpredictable, will pose many such management challenges. Adaptive management programs have to include a stage of evaluation and adjustment. Outcomes of past management decisions must be compared with initial forecasts, models have to be refined to reflect new understanding, and management programs have to be revised accordingly. New information may suggest new uncertainties and innovative management approaches, leading to another cycle of assessment, design, implementation, and evaluation. [A.M.Pa.]

Adenoviridae A family of viral agents associated with pharyngoconjunctival fever, acute respiratory disease, epidemic keratoconjunctivitis, and febrile pharyngitis in children. A number of types have been isolated from tonsils and adenoids removed from surgical patients. Although most of the illnesses caused by adenoviruses are respiratory, adenoviruses are frequently excreted in stools, and certain adenoviruses have been isolated from sewage. Distinct serotypes of mammalian and avian species are known. These genera contain 87 and 14 species, respectively. *See* ANIMAL VIRUS.

Infective virus particles, 70 nanometers in diameter, are icosahedrons with shells (capsids) composed of 252 subunits (capsomeres). No outer envelope is known. The genome is double-stranded deoxyribonucleic acid (DNA), with a molecular weight of $20–25 \times 10^6$. Three major soluble antigens are separable from the infectious particle by differential centrifugation. These antigens—a group-specifc antigen common to all adenovirus types, a type-specific antigen unique for each type, and a toxinlike material which also possesses group specificity—represent virus structural protein subunits that are produced in large excess of the amount utilized for synthesis of infectious virus.

The known types of adenoviruses of humans total at least 33, and previously unrecognized types continue to be isolated. The serotypes are antigenically distinct in neutralization tests, but they share a complement-fixing antigen, which is probably a smaller soluble portion of the virus.

The virus does not commonly produce acute disease in laboratory animals but is cytopathogenic, that is, destroys cells, in cultures of human tissue. Certain human adenovirus serotypes produce cancer when injected into newborn hamsters.

Base ratio determinations have revealed three distinct groups of adenoviruses: those with a low guanine plus cytosine (G + C) content (48–49%); those with an intermediate G + C content (50–53%); and those with a high G + C content (56–60%). The strongly oncogenic adenovirus types 12, 18, and 31 are the only members of the group with low G + C, and certain adenoviruses in the intermediate group (types 3, 7, 14, 16, and 21) are mildly oncogenic. The adenovirus mRNA observed in transformed and tumor cells has a G + C content of 50–52% in the DNA. This suggests that viral DNA regions containing 47–48% G + C are integrated into the tumor cells or that such regions are preferentially transcribed. However, the mRNA from tumor cells induced by one subgroup such as the highly oncogenic adenoviruses (types 12 and 18) do not hybridize with DNA from the other two subgroups. Apparently, different viralcoded information is involved in carcinogenesis by the three different groups of adenoviruses.

With simian adenovirus 7 (SA7), the intact genome, as well as the heavy and light halves of the viral DNA, is capable of inducing tumors when injected into newborn hamsters. Extensive studies have failed to demonstrate adenovirus DNA or viral-specific mRNA in human tumors.

Live virus vaccines against type 4 and type 7 have been developed and used extensively in military populations. When both are administered simultaneously, vaccine recipients respond with neutralizing antibodies against both virus types. *See* DEPENDOVIRUS. [J.L.Me.; M.E.Re.]

Aeromonas A bacterial genus in the family Vibrionaceae comprising oxidase-positive, facultatively anaerobic, monotrichously flagellated gram-negative rods. The mesophilic species are *A. hydrophila*, *A. caviae*, and *A. sobria*; the psychrophilic one is *A. salmonicida*. Aeromonads are of aquatic origin and are found in surface and waste water but not in seawater. They infect chiefly cold-blooded animals such as fishes, reptiles, and amphibians and only occasionally warm-blooded animals and humans. Human wound infections may occur following contact with contaminated water. Septicemia has been observed mostly in patients with abnormally low white blood counts or liver disease. There is evidence of intestinal carriers. The three mesophilic species are also associated with diarrheal disease (enteritis and colitis) worldwide. *See* DIARRHEA.

A related lophotrichous genus, *Plesiomonas* (single species, *P. shigelloides*), is also known as an aquatic bacterium and is associated with diarrhea chiefly in subtropical and tropical areas. It is also found in many warm-blooded animals. Systemic disease in humans is rare. *See* MEDICAL BACTERIOLOGY. [A.W.C.V.G.]

Aeronautical meteorology The branch of meteorology that deals with atmospheric effects on the operation of vehicles in the atmosphere, including winged aircraft, lighter-than-air devices such as dirigibles, rockets, missiles, and projectiles. The air which supports flight or is traversed on the way to outer space contains many potential hazards.

Poor visibility caused by fog, snow, dust, and rain is a major cause of aircraft accidents and the principal cause of flight cancellations or delays.

The weather conditions of ceiling and visibility required by regulations for crewed aircraft during landing or takeoff are determined by electronic and visual aids operated by the airport and installed in the aircraft. The accurate forecasting of terminal conditions is critical to flight economy, and to safety where sophisticated landing aids are not available. Improved prediction methods are under continuing investigation and development, and are based on mesoscale and microscale meteorological analyses, electronic computer calculations, radar observations of precipitation areas, and observations of fog trends. *See* MICROMETEOROLOGY; RADAR METEOROLOGY.

Atmospheric turbulence is principally represented in vertical currents and their departures from steady, horizontal airflow. When encountered by an aircraft, turbulence produces abrupt excursions in aircraft position, sometimes resulting in discomfort or injury to passengers, and sometimes even structural damage or failure. Major origins of turbulence are (1) mechanical, caused by irregular terrain below the flow of air; (2) thermal, associated with vertical currents produced by heating of air in contact with the Earth's surface; (3) thunderstorms and other convective clouds; (4) mountain wave, a regime of disturbed airflow leeward of mountains or hills, often comprising both smooth and breaking waves formed when stable air is forced to ascend over the mountains; and (5) wind shear, usually variations of horizontal wind in the vertical direction, occurring along air-mass boundaries, temperature inversions (including the tropopause), and in and near the jet stream.

While encounters with strong turbulence anywhere in the atmosphere represent substantial inconvenience, encounters with rapid changes in wind speed and direction at low altitude can be catastrophic. Generally, wind shear is most dangerous when encountered below 1000 ft (300 m) above the ground, where it is identified as low-altitude wind shear. Intense convective microbursts, downdrafts usually associated with thunderstorms, have caused many aircraft accidents often resulting in a great loss of life. The downdraft emanating from convective clouds, when nearing the Earth's surface, spreads horizontally as outrushing rain-cooled air. When entering a microburst outflow, an aircraft first meets a headwind that produces increased performance by way of increased airspeed over the wings. Then within about 5 s, the aircraft encounters a downdraft and then a tailwind with decreased performance. A large proportion of microburst accidents, both after takeoff and on approach to landing, are caused by this performance decrease, which can result in rapid descent. *See* THUNDERSTORM.

Turbulence and low-altitude wind shear can readily be detected by a special type of weather radar, termed Doppler radar. By measuring the phase shift of radiation backscattered by hydrometeors and other targets in the atmosphere, both turbulence and wind shear can be clearly identified. It is anticipated that Doppler radars located at airports, combined with more thorough pilot training regarding the need to avoid microburst wind shear, will provide desired protection from this dangerous aviation weather phenomenon.

Since an aircraft's speed is given by a propulsive component plus the speed of the air current bearing the aircraft, there are aiding or retarding effects depending on wind direction in relation to the track flown. Wind direction and speed vary only moderately from day to day and from winter to summer in certain parts of the world, but fluctuations of the vector wind at middle and high latitudes in the troposphere and lower stratosphere can exceed 200 knots (100 mph). The role of the aeronautical meteorologist is to provide accurate forecasts of the wind and temperature field, in space and time, through the operational ranges of each aircraft involved. For civil jet-powered aircraft, the optimum flight plan must always represent a compromise among wind, temperature, and turbulence conditions. *See* WIND.

The jet stream is a meandering, shifting current of relatively swift wind flow which is embedded in the general westerly circulation at upper levels. Sometimes girdling the globe at middle and subtropical latitudes, where the strongest jets are found, this band of strong winds, generally 180–300 mi (300–500 km) in width, has great operational significance for aircraft flying at cruising levels of 4–9 mi (6–15 km). The jet stream challenges the forecaster and the flight planner to utilize tailwinds to the greatest extent possible on downwind flights and to avoid retarding headwinds as much as practicable on upwind flights. As with considerations of wind and temperature, altitude and horizontal coordinates are considered in flight planning for jet-stream conditions. Turbulence in the vicinity of the jet stream is also a forecasting problem. *See* JET STREAM.

An electrical discharge or lightning strike to or from an aircraft is experienced as a blinding flash and a muffled explosive sound. Atmospheric conditions favorable for lightning strikes follow a consistent pattern, characterized by solid clouds or enough clouds for the aircraft to be flying intermittently on instruments; active precipitation of an icy character; and ambient air temperature near or below 32°F (0°C). Saint Elmo's fire, radio static, and choppy air often precede the strike. However, the charge separation processes necessary for the production of strong electrical fields is destroyed by strong turbulence. Thus turbulence and lightning usually do not coexist in the same space. *See* LIGHTNING.

Modern aircraft operation finds icing to be a major factor in the safe flight. Icing usually occurs when the air temperature is near or below freezing (32°F or 0°C) and the relative humidity is 80% or more. Clear ice is most likely to form when the air temperature is between 32 and −4°F (0 and −20°C) and the liquid water content of the air is high (large drops or many small drops). As these drops impinge on the skin of an aircraft, the surface temperature of which is 32°F (0°C) or less, the water freezes into a hard, high-density solid. When the liquid water content is small and when snow or ice pellets may also be present, the resulting rime ice formation is composed of drops and encapsulated air, producing an ice that is less dense and opaque in appearance. Accurate forecasts and accurate delineation of freezing conditions are essential for safe aircraft operations. [J.T.Le.; J.M.]

Aeronomy The study of the chemistry and physics of the regions above the tropopause or upper part of the atmosphere. The region of the atmosphere below the tropopause is the site of most of the weather phenomena that so directly affect all life on the planet; this region has primarily been the domain of meteorology.

The chemical and physical properties of the atmosphere and the changes that result from external and internal forces impact all altitudes and global distributions of atoms, molecules, ions, and electrons, both in composition and in density. Dynamical effects are seen in vertical and horizontal atmospheric motion, and energy is transferred through radiation, chemistry, conduction, convection, and wave propagation.

The atmosphere of the Earth is separated into regions defined by the variation of temperature with height. In the middle atmosphere, that region of the atmosphere between the tropopause and the mesopause (10–100 km or 6–60 mi), the temperature varies from 250 K (−9.7°F) at the tropopause to 300 K (80°F) at the stratopause and back down to 200 K (−100°F) at the mesopause (see illustration). These temperatures are average values, and they vary with season and heat and winds due to the effect of the Sun on the atmosphere. Over this same height interval the atmospheric density varies by over five orders of magnitude. Although there is a constant mean molecular weight over this region, that is, a constant relative abundance of the major atmospheric constituents of molecular oxygen and nitrogen, there are a number of minor constituents that have a profound influence on the biosphere and an increasing influence on the change in the atmosphere below the tropopause associated with the general topic of global change. These constituents, called the greenhouse gases (water

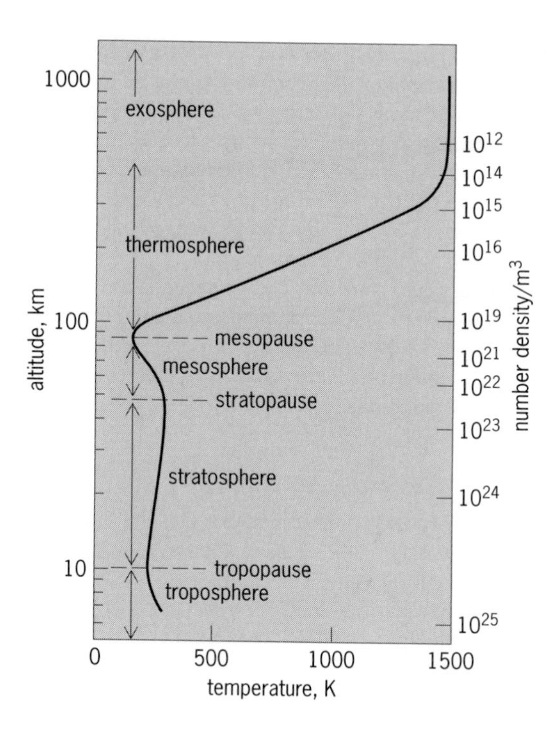

Typical atmospheric temperature variation with altitude and constituent number density (number of atoms and molecules per cubic meter) at high solar activity. °F = (K × 1.8) − 459.67. 1 km = 0.6 mi.

vapor, ozone, carbon dioxide, methane, chlorine compounds, nitrogen oxides, chlorofluorocarbons, and others), are all within the middle atmosphere. *See* GREENHOUSE EFFECT; MESOSPHERE; STRATOSPHERE; TROPOSPHERE.

Understanding the aeronomy of the middle atmosphere requires the study of the physical motion of the atmosphere. The particular composition and chemistry at any given time, location, or altitude depends on how the various constituents are transported from one region to another. Thus, global circulation models for both horizontal and vertical motions are needed to completely specify the chemical state of the atmosphere. In understanding the dynamics of the middle atmosphere, internal gravity waves and acoustic gravity waves play significant roles at different altitudes, depending on whether the particle motion associated with the wave is purely transverse to the direction of propagation or has some longitudinal component. These waves originate primarily in meteorological events such as wind shears, turbulent storms, and weather fronts; and their magnitude can also depend on orographic features on the Earth's surface. *See* ATMOSPHERIC CHEMISTRY.

The upper atmosphere is that region above the middle atmosphere that extends from roughly 100 km (60 mi) to the limit of the detectable atmosphere of the planet. This region is characterized by an increasing temperature until it reaches a constant exospheric temperature. There is a slow transition from the region of constant mean molecular weight associated with the middle atmosphere to that of almost pure atomic hydrogen at high altitudes of the exosphere. This is also the region of transition between transport dominated by collision and diffusion, and transport influenced by plasma convection in the magnetic field. The neutral density varies by over ten orders of magnitude from one end to the other and is dominated by molecular processes in the high-density region and an increasing importance of atomic, electron, and ion processes as the density decreases with altitude.

As the Sun sets on the atmosphere, dramatic changes due to the loss of the solar radiation occur. The excitation of the atmosphere declines, and chemical reactions between the constituents become more and more dominant as the night progresses. When observed with very sensitive instruments, the night sky appears to glow at various colors of light. Prominent green and red atomic emissions due to oxygen at 555.7 and 630 nm, respectively, and yellow sodium light at 589 nm appear, while molecular bands of the hydroxyl radical and molecular oxygen even further in the red collectively contribute most of the total intensity of the nightglow spectrum.

The aurora that appears in the southern and northern polar regions is the optical manifestation of the energy loss of energetic particles precipitating into the atmosphere. The region of highest probability of occurrence is called the auroral oval. At high altitudes, electrons and ions present in the magnetosphere are accelerated along magnetic field lines into the atmosphere at high polar latitudes. [G.J.R.]

Aflatoxin Any of a group of secondary metabolites produced by the common molds *Aspergillus flavus* and *A. parasiticus* that cause a toxic response in vertebrates when introduced in low concentration by a natural route. The group constitutes a type of mycotoxin. The naturally occurring aflatoxins are identified in physicochemical assays as intensely blue (aflatoxins B_1 and B_2) or blue-green (aflatoxins G_1 and G_2) fluorescent compounds under long-wave ultraviolet light. The common structural feature of the four major aflatoxins is a dihydrodifurano or tetrahydrodifurano group fused to a substituted coumarin group (see illustration). The relative proportions of the four major aflatoxins synthesized by *Aspergillus* reflect the genetic constitution of the producing strain and the parameters associated with fungal growth. In addition, derivative aflatoxins are produced as metabolic or environmental products. *See* TOXIN.

Aflatoxins are formed through a polyketide pathway involving a series of enzymatically catalyzed reactions. In laboratory cultures, aflatoxins are biosynthesized after active growth has ceased, as is typical for secondary metabolites. By using blocked mutants and metabolic inhibitors, many of the intermediates have been identified as brightly colored anthraquinones.

Aflatoxins are potent molecules with many biological effects. They are toxigenic, carcinogenic, mutagenic, and teratogenic in various animal species. Aflatoxin B_1 is usually the most abundant naturally occurring member of the family, and most studies on the pharmacological activity of aflatoxin have been conducted with this congener. Aflatoxin B_1 is the most potent hepatocarcinogenic agent known, although the liver by no means is the only organ susceptible to aflatoxin carcinogenesis. Aflatoxin is listed as a probable human carcinogen by the International Agency for Research on Cancer. *See* MUTAGENS AND CARCINOGENS; PLANT PATHOLOGY.

Aflatoxins are a major agricultural problem. Contamination can occur in the field, during harvest, or in storage and processing. Corn, rice, cottonseed, and peanuts are the major crops regularly displaying high levels of aflatoxin contamination. Since *A. flavus* and *A. parasiticus* are nearly ubiquitous in the natural environment, numerous other grain, legume, nut, and spice crops, as well as coffee and cocoa, have been reported to contain aflatoxins. Given the potential of aflatoxins as human carcinogens and their known activity as toxins in animal feeds, many international regulatory agencies monitor aflatoxin levels in susceptible crops. Prevention is the main line of defense against aflatoxins entering the food chain. Moisture, temperature, and composition of the substrate are the chief factors affecting fungal growth and toxin production. In the field, insect damage is often involved. Detoxification is a last line of

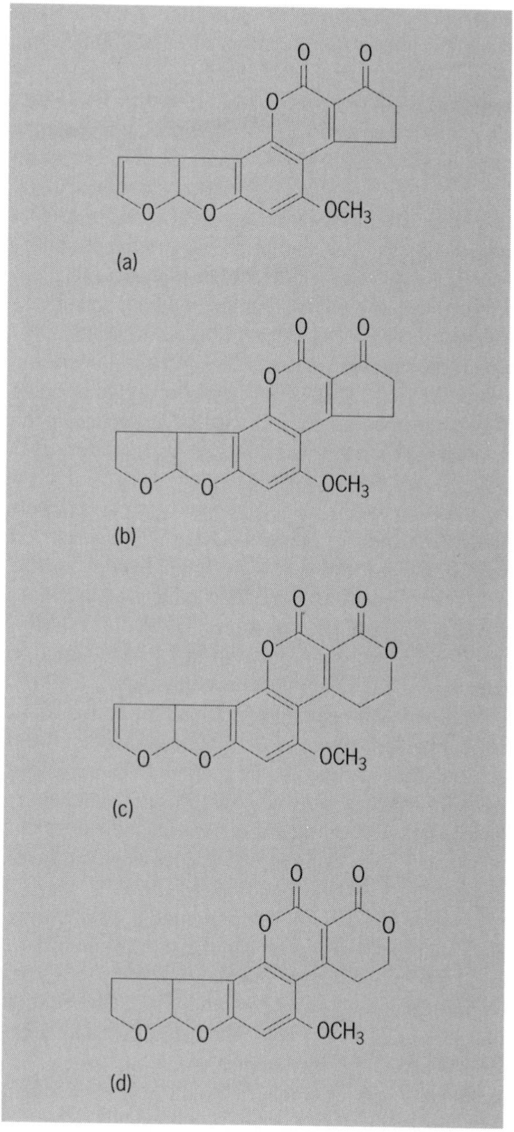

Structures of major naturally occurring aflatoxins. (*a*) B_1. (*b*) B_2. (*c*) G_1. (*d*) G_2.

defense. Several commercially feasible methods of ammoniation have been developed for reducing levels of aflatoxin contamination in animal feeds. *See* AGRONOMY; MYCOTOXIN.

[J.W.Be.]

Africa A continent that straddles the Equator, extending between 37°N and 35°S. It is the second largest continent, exceeded by Eurasia. The area, shared by 55 countries, is 11,700,000 mi^2 (30,300,00 km^2), approximately 20% of the world's total land area. Despite its large area, it has a simple geological structure, a compact shape with a smooth outline, and a symmetrical distribution of climate and vegetation.

Africa has few inlets or natural harbors and a small number of offshore islands that are largely volcanic in origin. Madagascar is the largest island, with an area of 250,000 mi^2 (650,000 km^2).

Africa is primarily a high interior plateau bounded by steep escarpments. These features show evidence of the giant faults created during the drift of neighboring continents. The surface of the plateau ranges from 4000–5000 ft (1200–1500 m) in the south to about 1000 ft (300 m) in the Sahara. These differences in elevation are particularly apparent in the Great Escarpment region in southern Africa, where the land suddenly drops from 5000 ft (1500 m) to a narrow coastal belt. Although most of the continent is classified as plateau, not all of its surface is flat. Rather, most of its physiographic features have been differentially shaped by processes such as folding, faulting, volcanism, erosion, and deposition. *See* ESCARPMENT; FAULT AND FAULT STRUCTURES; PLATEAU.

The rift valley system is one of the most striking features of the African landscape. Sliding blocks have created wide valleys 20–50 mi (30–80 km) wide bounded by steep walls of variable depth and height. Within the eastern and western branches of the system, there is a large but shallow depression occupied by Lake Victoria. *See* RIFT VALLEY.

Several volcanic features are associated with the rift valley system. The most extensive of these are the great basalt highlands that bound either side of the rift system in Ethiopia. These mountains rise over 10,000 ft (3000 m), with the highest peak, Ras Dashan, reaching 15,158 ft (4500 m). There are also several volcanic cones, including the most renowned at Mount Elgon (14,175 ft; 4321 m); Mount Kenya (17,040 ft; 5194 m); and Mount Kilimanjaro, reaching its highest point at Mount Kibo (19,320 ft; 5889 m). Mounts Kenya and Kilimanjaro are permanently snowcapped.

Since the Equator transects the continent, the climatic conditions in the Northern Hemisphere are mirrored in the Southern Hemisphere. Nearly three-quarters of the continent lies within the tropics and therefore has high temperatures throughout the year. Frost is uncommon except in mountainous areas or some desert areas where nighttime temperatures occasionally drop below freezing. These desert areas also record some of the world's highest daytime temperatures, including an unconfirmed record of 136.4°F (58°C) at Azizia, Tripoli. *See* EQUATOR.

Africa can be classified into broad regions based on the climatic conditions and their associated vegetation and soil types. The tropical rainforest climate starts at the Equator and extends toward western Africa. The region has rainfall up to 200 in. (500 cm) per year and continuously high temperatures averaging 79°F (26°C). The eastern equatorial region does not experience these conditions because of the highlands and the presence of strong seasonal winds that originate from southern Asia.

The areal extent of the African rainforest region (originally 18%) has dwindled to less than 7% as a result of high rates of deforestation. Despite these reductions, the region is still one of the most diverse ecological zones in the continent. *See* RAINFOREST.

Extensive savanna grasslands are found along the Sudanian zone of West Africa, within the Zambezian region and the Somalia-Masai plains. Large areas such as the Serengeti plains in the Somalia-Masai plains are home to a diverse range of wild animals.

The tropical steppe forms a transition zone between the humid areas and the deserts. This includes the area bordering the south of the Sahara that is known as the Sahel, the margins of the Kalahari basin, and the Karoo grasslands in the south.

The structural evolution of the continent has much to do with the drainage patterns. Originally, most of the rivers did not drain into the oceans, and many flowed into the large structural basins of the continent. However, as the continental drift occurred and coasts became more defined, the rivers were forced to change courses, and flow over the escarpments in order to reach the sea. Several outlets were formed, including deep canyons, waterfalls, cataracts, and rapids as the rivers carved out new drainage patterns across the landscape. Most of the rivers continue to flow through or receive some of their drainage from the basins, but about 48% of them now have a direct access into the surrounding oceans. The major rivers are the Nile, Congo (Zaire), Niger, and Zambezi. *See* RIVER.

The tremendous diversity in wildlife continues to be one of the primary attractions of this continent. Africa is one of the few remaining places where one can view game fauna in a natural setting. There is a tremendous diversity in species, including birds, reptiles, and large mammals. Wildlife are concentrated in central and eastern Africa because of the different types of vegetation which provide a wide range of habitats.

Africa is not a densely populated continent. With an estimated population of 743,000,000, its average density is 64 per square mile (26 per square kilometer). However, some areas have large concentrations, including the Nile valley, the coastal areas of northern and western Africa, the highland and volcanic regions of eastern Africa, and parts of southern Africa. These are mostly areas of economic or political significance.

[F.L.M.]

African horsesickness A highly fatal insect-borne viral disease of horses and mules, and a mild subclinical disease in donkeys and zebras. It normally occurs in sub-Saharan Africa but occasionally spreads to North Africa, the Iberian Peninsula, and Asia Minor.

The African horsesickness virus is an orbivirus (family Reoviridae) measuring 68–70 nanometers in diameter. The outer layer of the double-layered protein shell is ill defined and diffuse and is formed by two polypeptides. The highly structured core consists of five structural proteins arranged in icosa-hedral symmetry. The viral genome is composed of 10 double-stranded ribonucleic acid (RNA) segments (genes) ranging in size from 240,000 to 2,530,000 daltons. Nine distinct serotypes which can be distinguished by neutralization tests are known. The virus can be cultivated in various cell cultures, in the brains of newborn mice, and in embryonated hen eggs by intravascular inoculation. *See* ANIMAL VIRUS.

African horsesickness is a noncontagious disease that can readily be transmitted by the injection of infective blood or organ suspensions. In nature, the virus is biologically transmitted by midges of the genus *Culicoides*, such as *C. imicola*. The disease has a seasonal incidence in temperate regions (late summer to autumn), and its prevalence is influenced by climatic conditions favoring insect breeding (for example, warm, moist conditions in low-lying areas). Mechanical transmission by large biting flies is possible, but plays a much smaller role than biological transmission in the epidemiology of this disease.

There is no specific treatment for this disease. Infected animals should be given complete rest as the slightest exertion may result in death. Stabling of horses at night during the African horsesickness season reduces exposure to insect vectors and hence the risk of disease transmission. Prophylactic vaccination is the most practical and effective control measure. In epidemic situations outside Africa, the causal virus should be serotyped as soon as possible, allowing the use of a monovalent vaccine. However, in endemic regions it is imperative to use a polyvalent vaccine, which should render protection against all nine serotypes of African horsesickness virus. *See* DISEASE; EPIDEMIC.

[B.E.]

Agricultural chemistry The science of chemical compositions and changes involved in the production, protection, and use of crops and livestock. As a basic science, it embraces, in addition to test-tube chemistry, all the life processes through which humans obtain food and fiber for themselves and feed for their animals. As an applied science or technology, it is directed toward control of those processes to increase yields, improve quality, and reduce costs. One important branch of it, chemurgy, is concerned chiefly with utilization of agricultural products as chemical raw materials.

The goals of agricultural chemistry are to expand understanding of the causes and effects of biochemical reactions related to plant and animal growth, to reveal opportunities for controlling those reactions, and to develop chemical products that will provide the desired assistance or control. Every scientific discipline that contributes to agricultural progress depends in some way on chemistry. Hence agricultural chemistry is not a distinct discipline, but a common thread that ties together genetics, physiology, microbiology, entomology, and numerous other sciences that impinge on agriculture.

Chemical materials developed to assist in the production of food, feed, and fiber include scores of herbicides, insecticides, fungicides, and other pesticides, plant growth regulators, fertilizers, and animal feed supplements. Chief among these groups from the commercial point of view are manufactured fertilizers, synthetic pesticides (including herbicides), and supplements for feeds. The latter include both nutritional supplements (for example, minerals) and medicinal compounds for the prevention or control of disease. *See* AGRICULTURE; FERTILIZER; HERBICIDE; PESTICIDE; SOIL. [R.N.H.]

Agricultural engineering A discipline concerned with solving the engineering problems of providing food and fiber for the people of the world. These problems include designing improved tools to work the soil and harvest the crops, as well as developing water supplies for agriculture and systems for irrigating and draining the land where necessary. Agricultural engineers design buildings in which to house animals or store grains. They also work on myriad problems of processing, packaging, transporting, and distributing the food and fiber products. Agricultural engineering combines the disciplines of mechanical, civil, electrical, and chemical engineering with a basic understanding of biological sciences and agricultural practices. Some agricultural engineers work directly with farmers. Most, however, work with the companies that manufacture and supply equipment, feeds, fertilizers, and pesticides. Others work for companies that provide services to farmers, such as developing irrigation and drainage systems or erecting buildings and facilities. Still others work with food-processing companies. *See* AGRICULTURE. [R.E.Ga.]

Agricultural meteorology A branch of meteorology that examines the effects and impacts of weather and climate on crops, rangeland, livestock, and various agricultural operations. The branch of agricultural meteorology dealing with atmospheric-biospheric processes occurring at small spatial scales and over relatively short time periods is known as micrometeorology, sometimes called crop micrometeorology for managed vegetative ecosystems and animal biometeorology for livestock operations. The branch that studies the processes and impacts of climatic factors over larger time and spatial scales is often referred to as agricultural climatology. *See* CLIMATOLOGY; MICROMETEOROLOGY.

Agricultural meteorology, or agrometeorology, addresses topics that often require an understanding of biological, physical, and social sciences. It studies processes that occur from the soil depths where the deepest plant roots grow to the atmospheric levels where seeds, spores, pollen, and insects may be found. Agricultural meteorologists characteristically interact with scientists from many disciplines.

Agricultural meteorologists collect and interpret weather and climate data needed to understand the interactions between vegetation and animals and their atmospheric environments. The climatic information developed by agricultural meteorologists is valuable in making proper decisions for managing resources consumed by agriculture, for optimizing agricultural production, and for adopting farming practices to minimize any adverse effects of agriculture on the environment. Such information is vital to ensure the economic and environmental sustainability of agriculture now and in the future. *See* WEATHER OBSERVATIONS.

Agricultural meteorologists also quantify, evaluate, and provide information on the impact and consequences of climate variability and change on agriculture. Increasingly, agricultural meteorologists assist policy makers in developing strategies to deal with climatic events such as floods, hail, or droughts and climatic changes such as global warming and climate variability.

Agricultural meteorologists are involved in many aspects of agriculture, ranging from the production of agronomic and horticultural crops, trees, and livestock to the final delivery of agricultural products to market. They study the energy and mass exchange processes of heat, carbon dioxide, water vapor, and trace gases such as methane, nitrous oxide, and ammonia, within the biosphere on spatial scales ranging from a leaf to a watershed and even to a continent. They study, for example, the photosynthesis, productivity, and water use of individual leaves, whole plants, and fields. They also examine climatic processes at time scales ranging from less than a second to more than a decade.

[B.L.B.]

Agriculture The art and science of crop and livestock production. In its broadest sense, agriculture comprises the entire range of technologies associated with the production of useful products from plants and animals, including soil cultivation, crop and livestock management, and the activities of processing and marketing. The term agribusiness has been coined to include all the technologies that mesh in the total inputs and outputs of the farming sector. In this light, agriculture encompasses the whole range of economic activities involved in manufacturing and distributing the industrial inputs used in farming; the farm production of crops, animals, and animal products; the processing of these materials into finished products; and the provision of products at a time and place demanded by consumers.

Many different factors influence the kind of agriculture practiced in a particular area. Among these are climate, soil, water availability, topography, nearness to markets, transportation facilities, land costs, and general economic level. Climate, soil, water availability, and topography vary widely throughout the world. This variation brings about a wide range in agricultural production enterprises. Certain areas tend toward a specialized agriculture, whereas other areas engage in a more diversified agriculture. As new technology is introduced and adopted, environmental factors are less important in influencing agricultural production patterns. Continued growth in the world's population makes critical the continuing ability of agriculture to provide needed food and fiber.

The primary agricultural products consist of crop plants for human food and animal feed and livestock products. The crop plants can be divided into 10 categories: grain crops (wheat, for flour to make bread, many bakery products, and breakfast cereals; rice, for food; maize, for livestock feed, syrup, meal, and oil; sorghum grain, for livestock feed; and oats, barley, and rye, for food and livestock feed); food grain legumes (beans, peas, lima beans, and cowpeas, for food; and peanuts, for food and oil); oil seed crops (soybeans, for oil and high-protein meal; and linseed, for oil and high-protein meal); root and tuber crops (principally potatoes and sweet potatoes); sugar crops (sugarbeets and sugarcane); fiber crops (principally cotton, for fiber to make textiles and for seed to produce oil and high-protein meal); tree and small fruits; nut crops; vegetables; and forages (for support of livestock pastures and range grazing lands and for hay and silage crops). The forages are dominated by a wide range of grasses and legumes, suited to different conditions of soil and climate.

Livestock products include cattle, for beef, tallow, and hides; dairy cattle, for milk, butter, cheese, ice cream, and other products; sheep, for mutton (lamb) and wool; pigs, for pork and lard; poultry (chiefly chickens but also turkeys and ducks) for meat and eggs; and horses, primarily for recreation. [J.J.]

Agroecosystem A model for the functionings of an agricultural system, with all inputs and outputs. An ecosystem may be as small as a set of microbial interactions that take place on the surface of roots, or as large as the globe. An agroecosystem may be at the level of the individual plant-soil-microorganism system, at the level of crops or herds of domesticated animals, at the level of farms or agricultural landscapes, or at the level of entire agricultural economies.

Characteristics. Agroecosystems differ from natural ecosystems in several fundamental ways. First, the energy that drives all autotrophic ecosystems, including agroecosystems, is either directly or indirectly derived from solar energy. However, the energy input to agroecosystems includes not only natural energy (sunlight) but also processed energy (fossil fuels) as well as human and animal labor. Second, biodiversity in agroecosystems is generally reduced by human management in order to channel as much energy and nutrient flow as possible into a few domesticated species. Finally, evolution is largely, but not entirely, through artificial selection where commercially desirable phenotypic traits are increased through breeding programs and genetic engineering. Agroecosystems are usually examined from a range of perspectives including energy flux, exchange of materials, nutrient budgets, and population and community dynamics.

Solar energy influences agroecosystem productivity directly by providing the energy for photosynthesis and indirectly through heat energy that influences respiration, rates of water loss, and the heat balance of plants and animals. *See* BIOLOGICAL PRODUCTIVITY; ECOLOGICAL ENERGETICS; PHOTOSYNTHESIS.

Nutrient uptake from soil by crop plants or weeds is primarily mediated by microbial processes. Some soil bacteria fix atmospheric nitrogen into forms that plants can assimilate. Other organisms influence soil structure and the exchange of nutrients, and still other microorganisms may excrete ammonia and other metabolic by-products that are useful plant nutrients. There are many complex ways that microorganisms influence nutrient cycling and uptake by plants. Some microorganisms are plant pathogens that reduce nutrient uptake in diseased plants. Larger organisms may influence nutrient uptake indirectly by modifying soil structure or directly by damaging plants. *See* NITROGEN CYCLE; SOIL MICROBIOLOGY.

Although agroecosystems may be greatly simplified compared to natural ecosystems, they can still foster a rich array of population and community processes such as herbivory, predation, parasitization, competition, and mutualism. Crop plants may compete among themselves or with weeds for sunlight, soil nutrients, or water. Cattle overstocked in a pasture may compete for forage and thereby change competitive interactions among pasture plants, resulting in selection for unpalatable or even toxic plants. Indeed, one important goal of farming is to find the optimal densities for crops and livestock. *See* HERBIVORY; POPULATION ECOLOGY; WEEDS.

Widespread use of synthetic chemical pesticides has bolstered farm production worldwide, primarily by reducing or eliminating herbivorous insect pests. Traditional broad-spectrum pesticides such as DDT, however, can have far-ranging impacts on agroecosystems. For instance, secondary pest outbreaks associated with the use of many traditional pesticides are not uncommon due to the elimination of natural enemies or resistance of pests to chemical control. Growers and pesticide developers in temperate regions have begun to focus on alternative means of control. Pesticide developers have begun producing selective pesticides, which are designed to target only pest species and to spare natural enemies, leaving the rest of the agroecosystem community intact. Many growers are now implementing integrated pest management programs that incorporate the new breed of biorational chemicals with cultural and other types of controls. *See* FOREST PEST CONTROL; INSECT CONTROL, BIOLOGICAL; PESTICIDE.

Genetic engineering. The last few decades have seen tremendous advances in molecular approaches to engineering desirable phenotypic traits in crop plants. Although artificially modifying crop plants is nothing new, the techniques used in genetic engineering allow developers to generate new varieties an order-of-magnitude faster than traditional plant breeding. In addition, genetic engineering differs from traditional breeding in that the transfer of traits is no longer limited to same-species organisms. Scientists are still assessing the effects that the widespread deployment of these traits may have on agroecosystems and natural ecosystems. There is some concern, for instance, that engineered traits may escape, via genes in pollen transferred by pollinators, and become established in weedy populations of plants in natural ecosystems, in some cases creating conservation management problems and new breeds of superweeds. As with pesticides, there is evidence that insects are already becoming resistant to some more widespread traits used in transgenic plants, such as the antiherbivore toxin produced by the bacterium *Bacillus thuringiensis*. *See* BIOTECHNOLOGY; GENETIC ENGINEERING. [C.R.Ca.; C.A.H.; J.E.Ba.]

Agronomy The science and study of crops and soils. Agronomy is the umbrella term for a number of technical research and teaching activities: crop physiology and management, soil science, plant breeding, and weed management frequently are included in agronomy; soil science may be treated separately; and vegetable and fruit crops generally are not included. Thus, agronomy refers to extensive field cultivation of plant species for human food, livestock and poultry feed, fibers, oils, and certain industrial products. *See* AGRICULTURE.

Agronomic studies include some basic research, but the specialists in this field concentrate on applying information from the more basic disciplines, among them botany, chemistry, genetics, mathematics, microbiology, and physiology. Agronomists also interact closely with specialists in other applied areas such as ecology, entomology, plant pathology, and weed science. The findings of these collaborative efforts are tested and recommended to farmers through agricultural extension agents or commercial channels to bring this knowledge into practice. This critical area is now focused on the efficiency of resource use, profitability of management practices, and minimization of the impact of farming on the immediate and the off-farm environment. *See* AGROECOSYSTEM; CONSERVATION OF RESOURCES; SOIL CONSERVATION. [C.A.F.]

Air A predominantly mechanical mixture of a variety of individual gases enveloping the terrestrial globe to form the Earth's atmosphere. In this sense air is one of the three basic components, air, water, and land (atmosphere, hydrosphere, and lithosphere), that interblend to form the life zone at the face of the Earth. *See* ATMOSPHERE. [C.V.C.]

Air mass In meteorology, an extensive body of the atmosphere which is relatively homogeneous horizontally. An air mass may be followed on the weather map as an entity in its day-to-day movement in the general circulation of the atmosphere. The expressions air mass analysis and frontal analysis are applied to the analysis of weather maps in terms of the prevailing air masses and of the zones of transition and interaction (fronts) which separate them.

The relative horizontal homogeneity of an air mass stands in contrast to sharper horizontal changes in a frontal zone. The horizontal extent of important air masses is reckoned in millions of square miles. In the vertical dimension an air mass extends at most to the top of the troposphere, and frequently is restricted to the lower half or less of the troposphere. *See* FRONT; METEOROLOGY; WEATHER MAP.

The occurrence of air masses as they appear on the daily weather maps depends upon the existence of air-mass source regions, areas of the Earth's surface which are sufficiently uniform that the overlying atmosphere acquires similar characteristics throughout the region. *See* ATMOSPHERIC GENERAL CIRCULATION.

The thermodynamic properties of air mass determine not only the general character of the weather in the extensive area that it covers, but also to some extent the severity of the weather activity in the frontal zone of interaction between air masses. Those properties which determine the primary weather characteristics of an air mass are defined by the vertical distribution of water vapor and heat (temperature). On the vertical distribution of water vapor depend the presence or absence of condensation forms and, if present, the elevation and thickness of fog or cloud layers. On the vertical distribution of temperature depend the relative warmth or coldness of the air mass and, more importantly, the vertical gradient of temperature, known as the lapse rate. The lapse rate determines the stability or instability of the air mass for thermal convection and consequently, the stratiform or convective cellular structure of the cloud forms and precipitation. The most unstable moist air mass is characterized by severe turbulence and heavy showers or thundershowers. In the most stable air mass there is observed an actual increase (inversion) of temperature with increase of height at low elevations. *See* TEMPERATURE INVERSION. [H.C.Wi.; E.Ke.]

Air pollution The presence in the atmospheric environment of natural and artificial substances that affect human health or well-being, or the well-being of any other specific organism. Pragmatically, air pollution also applies to situations where contaminants impact structures and artifacts or esthetic sensibilities (such as visibility or smell). Most artificial impurities are injected into the atmosphere at or near the Earth's surface. The lower atmosphere (troposphere) cleanses itself of some of these pollutants in a few hours or days as the larger particles settle to the surface and soluble gases and particles encounter precipitation or are removed through contact with surface objects. Unfortunately, removal of some pollutants (for example, sulfates and nitrates) by precipitation and dry deposition results in acid deposition, which may cause serious environmental damage. Also, mixing of the pollutants into the upper atmosphere may dilute the concentrations near the Earth's surface, but can cause long-term changes in the chemistry of the upper atmosphere, including the ozone layer. *See* ATMOSPHERE; TROPOSPHERE.

Types of sources. Sources may be characterized in a number of ways. First, a distinction may be made between natural and anthropogenic sources. Another frequent classification is in terms of stationary (power plants, incinerators, industrial operations, and space heating) and moving (motor vehicles, ships, aircraft, and rockets) sources. Another classification describes sources as point (a single stack), line (a line of stacks), or area (city).

Different types of pollution are conveniently specified in various ways: gaseous, such as carbon monoxide, or particulate, such as smoke, pesticides, and aerosol sprays; inorganic, such as hydrogen fluoride, or organic, such as mercaptans; oxidizing substances, such as ozone, or reducing substances, such as oxides of sulfur and oxide s of nitrogen; radioactive substances, such as iodine-131; inert substances, such as pollen or fly ash; or thermal pollution, such as the heat produced by nuclear power plants.

Air contaminants are produced in many ways and come from many sources; it is difficult to identify all the various producers. Also, for some pollutants such as carbon dioxide and methane, the natural emissions sometimes far exceed the anthropogenic emissions.

Both anthropogenic and natural emissions are variable from year to year, depending on fuel usage, industrial development, and climate. In some countries where pollution control regulations have been implemented, emissions have been significantly reduced. For example, in the United States sulfur dioxide emissions dropped by about 30% between 1970 and 1992, and carbon monoxide (CO) emissions were cut by over 30% in the same period. However, in some developing countries emissions continually rise as more cars are put on the road and more industrial facilities and power plants are constructed. In dry regions, natural emissions of nitrogen oxides (NO_x), carbon dioxide (CO_2), and hydrocarbons can be greatly increased during a season with high rainfall and above-average vegetation growth.

The anthropogenic component of most estimates of the methane budget is about two-thirds. Ruminant production and emissions from rice paddies are regarded as anthropogenic because they result from human agricultural activities. The perturbations to carbon dioxide since the industrial revolution are also principally the result of human activities. These emissions have not yet equilibrated with the rest of the carbon cycle and so have had a profound effect on atmospheric levels, even though emissions from fossil fuel combustion are dwarfed by natural emissions.

Effects. The major concern with air pollution relates to its effects on humans. Since most people spend most of their time indoors, there has been increased interest in air-pollution concentrations in homes, workplaces, and shopping areas. Much of the early information on health effects came from occupational health studies completed prior to the implementation of general air-quality standards.

Air pollution principally injures the respiratory system, and health effects can be studied through three approaches, clinical, epidemiological, and toxicological. Clinical studies use human subjects in controlled laboratory conditions, epidemiological studies assess human subjects (health records) in real-world conditions, and toxicological studies are conducted on animals or simple cellular systems. Of course, epidemiological studies are the most closely related to actual conditions, but they are the most difficult to interpret because of the lack of control and the subsequent problems with statistical analysis. Another difficulty arises because of differences in response among different people. For example, elderly asthmatics are likely to be more strongly affected by sulfur dioxide than the teenage members of a hiking club. *See* EPIDEMIOLOGY.

Damage to vegetation by air pollution is of many kinds. Sulfur dioxide may damage field crops such as alfalfa and trees such as pines, especially during the growing season. Both hydrogen fluoride (HF) and nitrogen dioxide (NO_2) in high concentrations have been shown to be harmful to citrus trees and ornamental plants, which are of economic, importance in central Florida. Ozone and ethylene are other contaminants that cause damage to certain kinds of vegetation.

Air pollution can affect the dynamics of the atmosphere through changes in longwave and shortwave radiation processes. Particles can absorb or reflect incoming short-wave solar radiation, keeping it from the Earth's surface during the day. Greenhouse gases can absorb long-wave radiation emitted by the Earth's surface and atmosphere.

Carbon dioxide, methane, fluorocarbons, nitrous oxides, ozone, and water vapor are important greenhouse gases. These represent a class of gases that selectively absorb long-wave radiation. This effect warms the temperature of the Earth's atmosphere and surface higher than would be found in the absence of an atmosphere (the greenhouse effect). Because the amount of greenhouse gases in the atmosphere is rising, there is a possibility that the temperature of the atmosphere will gradually rise, possibly resulting in a general warming of the global climate over a time period of several generations. *See* GREENHOUSE EFFECT.

Researchers are also concerned with pollution of the stratosphere (10–50 km or 6–30 mi above the Earth's surface) by aircraft and by broad surface sources. The stratosphere is important, because it contains the ozone layer, which absorbs part of the Sun's short-wave radiation and keeps it from reaching the surface. If the ozone layer is significantly depleted, an increase in skin cancer in humans is expected. Each 1% loss of ozone is estimated to increase the skin cancer rate 3–6%. *See* STRATOSPHERE.

Visibility is reduced as concentrations of aerosols or particles increase. The particles do not just affect visibility by themselves but also act as condensation nuclei for cloud or haze formation. In each of the three serious air-pollution episodes discussed above, smog (smoke and fog) were present with greatly reduced visibility.

Chemistry. Air pollution can be divided into primary and secondary compounds, where primary pollutants are emitted directly from sources (for example, carbon monoxide, sulfur dioxide) and secondary pollutants are produced by chemical reactions between other pollutants and atmospheric gases and particles (for example, sulfates, ozone). Most of the chemical transformations are best described as oxidation processes. In many cases these secondary pollutants can have significant environmental effects, such as acid rain and smog.

Smog is the best-known example of secondary pollutants formed by photochemical processes, as a result of primary emissions of nitric oxide (NO) and reactive hydrocarbons from anthropogenic sources such as transportation and industry as well as natural sources. Energy from the Sun causes the formation of nitrogen dioxide, ozone (O_3), and peroxyacetalnitrate, which cause eye irritation and plant damage.

It has been shown that when emissions of sulfur dioxide and nitrogen oxide from tall power plant and other industrial stacks are carried over great distances and combined with emissions from other areas, acidic compounds can be formed by complex chemical reactions. In the absence of anthropogenic pollution sources, the average pH of rain is around 5.6 (slightly acidic). In the eastern United States, acid rain with a pH less than 5.0 has been measured and consists of about 65% dilute sulfuric acid, 30% dilute nitric acid, and 5% other acids. [S.R.H.; P.J.S.]

Air pressure The force per unit area that the air exerts on any surface in contact with it, arising from the collisions of the air molecules with the surface. It is equal and opposite to the pressure of the surface against the air, which for atmospheric air in normal motion approximately balances the weight of the atmosphere above, about 15 pounds per square inch (psi) at sea level. It is the same in all directions and is the force that balances the weight of the column of mercury in the Torricellian barometer, commonly used for its measurement.

The units of pressure traditionally used in meteorology are based on the bar, defined as equal to 1,000,000 dynes/cm^2. One bar equals 1000 millibars or 100 centibars.

In the meter-kilogram-second or International System of Units (SI), the unit of force, the pascal (Pa), is equal to 1 newton/m^2. One millibar equals 100 pascals. The normal pressure at sea level is 1013.25 millibars or 101.325 kilopascals.

Also widely used in practice are units based on the height of the mercury barometer under standard conditions, expressed commonly in millimeters or in inches. The standard atmosphere (760 mmHg) is also used as a unit, mainly in engineering, where large pressures are encountered. The following equivalents show the conversions between the commonly used units of pressure, where $(\text{mmHg})_n$ and $(\text{in. Hg})_n$ denote the millimeter and inch of mercury, respectively, under standard (normal) conditions, and where $(\text{kg})_n$ and $(\text{lb})_n$ denote the weight of a standard kilogram and pound mass, respectively, under standard gravity.

$$
\begin{aligned}
1\text{ kPa} &= 10\text{ millibars} = 1000\text{ N/m}^2 \\
&= 7.50062\,(\text{mmHg})_n \\
&= 0.295300\,(\text{in. Hg})_n \\
1\text{ millibar} &= 100\text{ Pa} = 1000\text{ dynes/cm}^2 \\
&= 0.750062\,(\text{mmHg})_n \\
&= 0.0295300\,(\text{in. Hg})_n \\
1\text{ atm} &= 101.325\text{ kPa} = 1013.25\text{ millibars} \\
&= 760\,(\text{mmHg})_n = 29.9213\,(\text{in. Hg})_n \\
&= 14.6959\,(\text{lb})_n/\text{in.}^2 \\
&= 1.03323\,(\text{kg})_n/\text{cm}^2 \\
1\,(\text{mmHg})_n &= 1\text{ torr} = 0.03937008\,(\text{in. Hg})_n \\
&= 1.333224\text{ millibars} \\
&= 133.3224\text{ Pa} \\
1\,(\text{in. Hg})_n &= 33.8639\text{ millibars} \\
&= 25.4\,(\text{mmHg})_n \\
&= 3.38639\text{ kPa}
\end{aligned}
$$

Because of the almost exact balancing of the weight of the overlying atmosphere by the air pressure, the latter decreases with height. A standard equation is used in practice to calculate the vertical distribution of pressure with height above sea level. The temperature distribution in a standard atmosphere, based on mean values in middle latitudes, has been defined by international agreement. The use of the standard atmosphere yields a definite relation between pressure and height. This relation is used in all altimeters which are basically barometers of the aneroid type. The difference between the height estimated from the pressure and the actual height is often considerable; but since the same standard relationship is used in all altimeters, the difference is the same for all altimeters at the same location, and so causes no difficulty in determining the relative position of aircraft. Mountains, however, have a fixed height, and accidents have been caused by the difference between the actual and standard atmosphere.

In addition to the large variation with height, atmospheric pressure varies in the horizontal and with time. The variations of air pressure at sea level, estimated in the case of observations over land by correcting for the height of the ground surface, are routinely plotted on a map and analyzed, resulting in the familiar weather map representation with its isobars showing highs and lows. The movement of the main features of the sea-level pressure distribution, typically from west to east, produces characteristic fluctuations of the pressure at a fixed point, varying by a few percent within a few days. Smaller-scale variations of sea-level pressure, too small to appear on the ordinary weather map, are also present. These are associated with various forms of atmospheric motion, such as small-scale wave motion and turbulence. Relatively large variations are found in and near thunderstorms, the most intense being the low-pressure region in a tornado. The pressure drop within a tornado can be a large fraction of an atmosphere, and is the principal cause of the explosion of buildings over which a tornado passes. *See* ISOBAR (METEOROLOGY); WEATHER MAP.

It is a general rule that in middle latitudes at localities below 1000 m (3280 ft) in height above sea level, the air pressure on the continents tends to be slightly higher in winter than in spring, summer, and autumn; whereas at considerably greater heights on the continents and on the ocean surface, the reverse is true.

The practical importance of air pressure lies in its relation to the wind and weather. It is because of these relationships that pressure is a basic parameter in weather forecasting, as is evident from its appearance on the ordinary weather map.

The large-scale variations of pressure at sea level shown on a weather map are associated with characteristic patterns of vertical motion of the air, which in turn affect the weather. Descent of air in a high heats the air and dries it by adiabatic compression, giving clear skies, while the ascent of air in a low cools it and causes it to condense and produce cloudy and rainy weather. These processes at low levels, accompanied by others at higher levels, usually combine to justify the clear-cloudy-rainy marking on the household barometer. [R.J.D.; E.Ke.]

Air temperature The temperature of the atmosphere represents the average kinetic energy of the molecular motion in a small region, defined in terms of a standard or calibrated thermometer in thermal equilibrium with the air. Many different types of thermometer are used for the measurement of air temperature, the most common depending on the expansion of mercury with temperature, the variation of electrical resistance with temperature, or the thermoelectric effect (thermocouple).

The temperature of a given small mass of air varies with time because of heat added or subtracted from it, and also because of work done during changes of volume.

The rate at which the temperature changes at a particular point, that is, as measured by a fixed thermometer, depends on the movement of air as well as physical processes such as absorption and emission of radiation, heat conduction, and changes of phase of water involving latent heat of condensation and freezing. The large changes of air temperature from day to day are mainly due to the horizontal movement of air, bringing relatively cold or warm air masses to a particular point, as the large-scale pressure-wind systems move across the weather map. *See* AIR MASS; AIR PRESSURE.

Temperatures near the surface are read at one or more fixed times daily, and the day's extremes are obtained from special maximum and minimum thermometers, or from the trace (thermogram) of a continuously recording instrument (thermograph). The average of these two extremes, technically the midrange, is considered in the United States to be the day's average temperature. The true daily mean, obtained from a thermogram, is closely approximated by the mean of 24 hourly readings, but may differ from the midrange by 1 or 2°F (0.6 or 1°C), on the average. In many countries temperatures are read daily at three or four fixed times, so that their weighted mean closely approximates the true daily mean.

Averages of daily maximum and minimum temperature for a single month for many years give mean daily maximum and minimum temperatures for that month. The average of these values is the mean monthly temperature, while their difference is the mean daily range for that month. Monthly means, averaged through the year, give the mean annual temperature; the mean annual range is the difference between the hottest and coldest mean monthly values. The hottest and coldest temperatures in a month are the monthly extremes; their averages over a period of years give the mean monthly maximum and minimum (used extensively in Canada), while the absolute extremes for the month (or year) are the hottest and coldest temperatures ever observed. The interdiurnal range or variability for a month is the average of the successive differences, regardless of sign, in daily temperatures.

Over the oceans the mean daily, interdiurnal, and annual ranges are slight, because water absorbs the insolation and distributes the heat through a thick layer. In tropical regions the interdiurnal and annual ranges over the land are small also, because the annual variation in insolation is relatively small. The daily range also is small in humid tropical regions, but may be large (up to 40°F or 22°C) in deserts. Interdiurnal and annual ranges increase generally with latitude, and also with distance from the ocean; the mean annual range defines continentality. The daily range depends on aridity, altitude, and noon Sun elevation. *See* ATMOSPHERE; INSOLATION. [R.J.D.]

Alder A deciduous tree, *Alnus rubra*, which grows from Alaska to northern California and eastern Idaho. It is recognized by its stalked buds, simple leaves, and dry, conelike, ellipsoid fruit. With the big-leaf maple it shares the role of principal hardwood tree in the Pacific Northwest, where most of the commercially important trees are conifers. The wood is used in furniture. [A.H.G./K.P.D.]

Algae An informal assemblage of predominantly aquatic organisms that carry out oxygen-evolving photosynthesis but lack specialized water-conducting and food-conducting tissues. They may be either prokaryotic (lacking an organized nucleus) and therefore members of the kingdom Monera, or eukaryotic (with an organized nucleus) and therefore members of the kingdom Plantae, constituting with fungi the subkingdom Thallobionta. They differ from the next most advanced group of plants, Bryophyta, by their lack of multicellular sex organs sheathed with sterile cells and by their failure to retain an embryo within the female organ. Many colorless organisms are referable to the algae on the basis of their similarity to photosynthetic forms with respect to structure, life history, cell wall composition, and storage products. The study of algae is called algology (from the Latin *alga*, meaning sea wrack) or phycology (from the Greek *phykos*, seaweed). *See* PLANT KINGDOM; THALLOBIONTA.

General form and structure. Algae range from unicells 1–2 micrometers in diameter to huge thalli [for example, kelps often 100 ft (30 m) long] with functionally and structurally distinctive tissues and organs. Unicells may be solitary or colonial, attached or free-living, with or without a protective cover, and motile or nonmotile. Colonies may be irregular or with a distinctive pattern, the latter type being flagellate or nonmotile. Multicellular algae form packets, branched or unbranched filaments, sheets one or two cells thick, or complex thalli, some with organs resembling roots, stems, and leaves (as in the brown algal orders Fucales and Laminariales). Coenocytic algae, in which the protoplast is not divided into cells, range from microscopic spheres to thalli 33 ft (10 m) long with a complex structure of intertwined siphons (as in the green algal order Bryopsidales).

Classification. Sixteen major phyletic lines (classes) are distinguished on the basis of differences in pigmentation, storage products, cell wall composition, flagellation of motile cells, and structure of such organelles as the nucleus, chloroplast, pyrenoid, and eyespot. These classes are interrelated to varying degrees, the interrelationships being expressed by the arrangement of classes into divisions (the next-higher category). Among phycologists there is far greater agreement on the number of major phyletic lines than on their arrangement into divisions.

Superkingdom Prokaryotae
Kingdom Monera
Division Cyanophycota (= Cyanophyta, Cyanochloronta)
Class Cyanophyceae, blue-green algae
Division Prochlorophycota (= Prochlorophyta)
Class Prochlorophyceae
Superkingdom Eukaryotae
Kingdom Plantae
Subkingdom Thallobionta
Division Rhodophycota (= Rhodophyta, Rhodophycophyta)
Class Rhodophyceae, red algae
Division Chromophycota (= Chromophyta)
Class: Chrysophyceae, golden or golden-brown algae
Prymnesiophyceae (= Haptophyceae)
Xanthophyceae (= Tribophyceae), yellow-green algae
Eustigmatophyceae
Bacillariophyceae, diatoms
Dinophyceae, dinoflagellates
Phaeophyceae, brown algae
Raphidophyceae, chloromonads
Cryptophyceae, cryptomonads
Division Euglenophycota (= Euglenophyta, Euglenophycophyta)
Class Euglenophyceae
Division Chlorophycota (= Chlorophyta, Chlorophycophyta)
Class: Chlorophyceae, green algae
Charophyceae, charophytes
Prasinophyceae

Placing more taxonomic importance on motility than on photosynthesis, zoologists traditionally have considered flagellate unicellular and colonial algae as protozoa, assigning each phyletic line the rank of order. *See* PROKARYOTAE; PROTOZOA; THALLOBIONTA.

Although some unicellular algae are naked or sheathed by mucilage or scales, most are invested with a covering (wall, pellicle, or lorica) of diverse composition and construction. These coverings consist of at least one layer of polysaccharide (cellulose, alginate, agar, carrageenan, mannan, or xylan), protein, or peptidoglycan that may be impregnated or encrusted with calcium carbonate, iron, manganese, or silica. They are often perforated and externally ornamented. Diatoms have a complex wall composed almost entirely of silica. In multicellular and coenocytic algae, most reproductive cells are naked, but vegetative cells have walls whose composition varies from class to class.

Characteristics. Prokaryotic algae lack membrane-bounded organelles. Eukaryotic algae have an intracellular architecture comparable to that of higher plants but more varied. Among cell structures unique to algae are contractile vacuoles in some freshwater unicells, gas vacuoles in some planktonic blue-green algae, ejectile organelles in dinoflagellates and cryptophytes, and eyespots in motile unicells and reproductive cells of many classes. Chromosome numbers vary from $n = 2$ in some red and green algae to $n \geq 300$ in some dinoflagellates. The dinoflagellate nucleus is in some respects intermediate between the chromatin region of prokaryotes and the nucleus of eukaryotes and is termed mesokaryotic. Some algal cells characteristically are multinucleate, while others are uninucleate. Chloroplasts, which always originate by division of preexisting chloroplasts, have the form of plates, ribbons, disks, networks, spirals, or stars and may be positioned centrally or along the cell wall. Photosynthetic membranes (thylakoids) are arranged in distinctive patterns and contain pigments diagnostic of individual classes. *See* CELL PLASTIDS; PHOTOSYNTHESIS; PLANT CELL.

In all classes of algae except Prochlorophyceae, there are cells that are capable of movement. The slow, gliding movement of certain blue-green algae, diatoms, and reproductive cells of red algae presumably results from extracellular secretion of mucilage. Ameboid movement, involving pseudopodia, is found in certain Chrysophyceae and Xanthophyceae. An undulatory or peristaltic movement occurs in some Euglenophyceae. The fastest movement is produced by flagella, which are borne by unicellular algae and reproductive cells of multicellular algae representing all classes except Cyanophyceae, Prochlorophyceae, and Rhodophyceae.

Internal movement also occurs in algae in the form of cytoplasmic streaming and light-induced orientation of chloroplasts.

Sexual reproduction is unknown in prokaryotic algae and in three classes of eukaryotic unicells (Eustigmatophyceae, Cryptophyceae, and Euglenophyceae), in which the production of new individuals is by binary fission. In sexual reproduction, which is found in all remaining classes, the members of a copulating pair of gametes may be morphologically indistinguishable (isogamous), morphologically distinguishable but with both gametes motile (anisogamous), or differentiated into a motile sperm and a relatively large nonmotile egg (oogamous). Gametes may be formed in undifferentiated cells or in special organs (gametangia), male (antheridia) and female (oogonia). Sexual reproduction may be replaced or supplemented by asexual reproduction, in which special cells (spores) capable of developing directly into a new alga are formed in undifferentiated cells or in distinctive organs (sporangia). *See* REPRODUCTION (PLANT).

Most algae are autotrophic, obtaining energy and carbon through photosynthesis. All photosynthetic algae liberate oxygen and use chlorophyll *a* as the primary photosynthetic pigment. Secondary (accessory) photosynthetic pigments, which capture light energy and transfer it to chlorophyll *a*, include chlorophyll *b* (Prochlorophyceae, Euglenophyceae, Chlorophycota), chlorophyll *c* (Chromophycota), fucoxanthin among other xanthophylls (Chromophycota), and phycobiliproteins (Cyanophyceae, Rhodophyceae, Cryptophyceae). Other carotenoids, especially β-carotene, protect the photosynthetic pigments from oxidative bleaching. Except for different complements of accessory pigments (resulting in different action spectra), photosynthesis in algae is identical to that in higher plants. Carbon is predominantly fixed through the C_3 pathway. *See* CHLOROPHYLL.

The source of carbon for most photosynthetic algae is carbon dioxide (CO_2), but some can use bicarbonate. Many photosynthetic algae are also able to use organic substances (such as hexose sugars and fatty acids) and thus can grow in the dark or in the absence of CO_2. Colorless algae obtain both energy and carbon from a wide variety of organic compounds in a process called oxidative assimilation.

Numerous substances are liberated into water by living algae, often with marked ecological effects. These extracellular products include simple sugars and sugar alcohols, wall polysaccharides, glycolic acid, phenolic substances, and aromatic compounds. Some secreted substances inhibit the growth of other algae and even that of the secreting alga. Some are toxic to fishes and terrestrial animals that drink the water.

Occurrence. Algae are predominantly aquatic, inhabiting fresh, brackish, and marine waters without respect to size or degree of permanence of the habitat. They may be planktonic (free-floating or motile) or benthic (attached). Benthic marine algae are commonly called seaweeds. Substrates include rocks (outcrops, boulders, cobbles, pebbles), plants (including other algae), animals, boat bottoms, piers, debris, and less frequently sand and mud. Some species occur on a wide variety of living organisms, suggesting that the hosts are providing only space. Many species, however, have a restricted range of hosts and have been shown to be (or are suspected of being) at least partially parasitic. All reef-building corals contain dinoflagellates, without which their calcification ability is greatly reduced. Different phases in a life history may have different substrate preferences. Many fresh-water algae have become adapted to a nonaquatic habitat, living on moist soil, masonry and wooden structures, and trees. A few parasitize higher plants (expecially in the tropics), producing diseases in such crops as tea, coffee, and citrus. Thermophilic algae (again, chiefly blue-greens) live in hot springs at temperatures up to 163°F (73°C), forming a calcareous deposit known as tufa. One of the most remarkable adaptations of certain algae (blue-greens and greens) is their coevolution with fungi to form a compound organism, the lichen. *See* PHYTOPLANKTON.

Geographic distribution. Fresh-water algae, which are distributed by spores or fragments borne by the wind or by birds, tend to be widespread if not cosmopolitan, their distribution being limited by the availability of suitable habitats. Certain species, however, are characteristic of one or another general climatic zone, such as cold-temperate regions or the tropics. Marine algae, which are spread chiefly by water-borne propagules or reproductive cells, often have distinctive geographic patterns. Many taxonomic groups are widely distributed, but others are characteristic of particular climatic zones or geographic areas. *See* PLANT GEOGRAPHY.

Economic importance. Numerous red, brown, and green seaweeds as well as a few species of fresh-water algae are consumed by the peoples of eastern Asia, Indonesia, Polynesia, and the North Atlantic. Large brown seaweeds may be chopped and added to poultry and livestock feed or applied whole as fertilizer for crop plants. The purified cell-wall polysaccharides of brown and red algae (alginate, agar, carrageenan) are used as gelling, suspending, and emulsifying agents in numerous industries. Some seaweeds have specific medicinal properties, such as effectiveness against worms. Petroleum is generally believed to result from bacterial degradation of organic matter derived primarily from planktonic algae.

Planktonic algae, as the primary producers in oceans and lakes, support the entire aquatic trophic pyramid and thus are the basis of the fisheries industry. Concomitantly, their production of oxygen counteracts its uptake in animal respiration. The ability of certain planktonic algae to assimilate organic nutrients makes them important in the treatment of sewage. *See* FOOD WEB; SEWAGE TREATMENT.

On the negative side, algae can be a nuisance by imparting tastes and odors to drinking water, clogging filters, and making swimming pools, lakes, and beaches unattractive. Sudden growths (blooms) of planktonic algae can produce toxins of varying potency. In small bodies of fresh water, the toxin (usually from blue-green algae) can kill fishes and livestock that drink the water. In the ocean, toxins produced by dinoflagellate blooms (red tides) can kill fishes and render shellfish poisonous to humans.

Fossil algae. At least half of the classes of algae are represented in the fossil record, usually abundantly, in the form of siliceous, calcareous, or organic remains, impressions, or indications. Blue-green algae were among the first inhabitants of the Earth, appearing in rocks at least as old as 2.3 billion years. Their predominance in shallow Precambrian seas is indicated by the extensive development of stromatolites.

All three classes of seaweeds (reds, browns, and greens) were well established by the close of the Precambrian, 600 million years ago (mya). By far the greatest number of fossil taxa belong to classes whose members are wholly or in large part planktonic. Siliceous frustules of diatoms and endoskeletons of silicoflagellates, calcareous scales of coccolithophorids, and highly resistant organic cysts of dinoflagellates contribute slowly but steadily to sediments blanketing ocean floors, as they have for tens of millions of years. Cores obtained in the Deep Sea Drilling Project have revealed an astounding chronology of the appearance, rise, decline, and extinction of a succession of species and genera. From this chronology, much can be deduced about the climate, hydrography, and ecology of particular geological periods. [P.C.Si.; R.L.Moe.]

Allelopathy The biochemical interactions among all types of plants, including microorganisms. The term is usually interpreted as the detrimental influence of one plant upon another but is used more and more, as intended originally, to encompass both detrimental and beneficial interactions. At least two forms of allelopathy are distinguished: (1) the production and release of an allelochemical by one species inhibiting the growth of only other adjacent species, which may confer competitive advantage for the allelopathic species; and (2) autoallelopathy, in which both the species producing the allelochemical and unrelated species are indiscriminately affected. The term allelopathy, frequently restricted to interactions among higher plants, is now applied to interactions among plants from all divisions, including algae. Even interactions between plants and herbivorous insects or nematodes in which plant substances attract, repel, deter, or retard the growth of attacking insects or nematodes are considered to be allelopathic. Interactions between soil microorganisms and plants are important in allelopathy. Fungi and bacteria may produce and release inhibitors or promoters. Some bacteria enhance plant growth through fixing nitrogen, others through providing phosphorus. The activity of nitrogen-fixing bacteria may be affected by allelochemicals, and this effect in turn may influence ecological patterns. The rhizosphere must be considered the main site for allelopathic interactions. *See* NITROGEN FIXATION; RHIZOSPHERE.

Allelopathy is clearly distinguished from competition: In allelopathy a chemical is introduced by the plant into the environment, whereas in competition the plant removes or reduces such environmental components as minerals, water, space, gas exchange, and light. In the field, both allelopathy and competition usually act simultaneously. [M.Ru.]

Allergy Altered reactivity in humans and animals to allergens (substances foreign to the body that cause allergy) induced by exposure through injection, inhalation, ingestion, or skin contact. The most common clinical manifestations of allergy are hay fever, asthma, hives, atopic (endogenous) eczema, and eczematous skin lesions caused by direct contact with allergens such as poison ivy or certain chemicals.

A large variety of substances may cause allergies: pollens, animal proteins, molds, foods, insect venoms, foreign serum proteins, industrial chemicals, and drugs. Most natural allergens are proteins or polysaccharides of moderate molecular size (molecular weights of 10,000 to 200,000). Chemicals or drugs of lower molecular weight (haptens) have first to bind to the body's own proteins (carriers) in order to become fully effective allergens.

For the development of the hypersensitivity state underlying clinical allergies, repeated contact with the allergen is required. Duration of the sensitization period is usually dependent upon the sensitizing strength of the allergen and the intensity of exposure. Some allergens (for example, saliva, urine, and hair proteins of domestic animals) are more sensitizing than others. In most instances, repeated contact with minute amounts of allergen is required; several annual seasonal exposures to grass pollens or ragweed pollen usually occur before an overt manifestation of hay fever. On the other hand, allergy to cow milk proteins in infants can develop within a few weeks. When previous contacts with allergens have not been apparent (for example, antibiotics in food), an allergy may become clinically manifest even upon the first conscious encounter with the offending substance.

Besides the intrinsic sensitizing properties of allergens, individual predisposition of the allergic person to become sensitized also plays an important role. Clinical manifestations, such as hay fever, allergic asthma, and atopic (endogenous) dermatitis, occur more frequently in some families. In other clinical forms of allergy, genetic predisposition, though possibly present as well, is not as evident.

Exposure to sensitizing allergens may induce several types of immune response, and the diversity of immunological mechanisms involved is responsible for the various clinical forms of allergic reactions which are encountered in practice. Three principal types of immune responses are encountered: the production of IgE antibodies, IgG or IgM antibodies, and sensitized lymphocytes.

Diagnosis of allergic diseases encompasses several facets. Since many clinical manifestations of allergy are mimicked by nonallergic mechanisms, it is usually necessary to use additional diagnostic procedures to ascertain whether the person has developed an immune response toward the incriminated allergen. Such procedures primarily consist of skin tests, in which a small amount of allergen is applied on or injected into the skin. If the individual is sensitized, a local immediate reaction ensues, taking the form of a wheal (for IgE-mediated reactions), or swelling and redness occurs after several hours (for delayed hypersensitivity reactions). The blood may also be analyzed for IgE and IgG antibodies by serological assays, and sensitized lymphocytes are investigated by culturing them with the allergen.

Since the discovery of the responsible allergens markedly influences therapy and facilitates prediction of the allergy's outcome, it is important to achieve as precise a diagnosis as possible. Most tests indicate whether the individual is sensitized to a given allergen, but not whether the allergen is in fact still causing the disease. Since in most cases the hypersensitive state persists for many years, it may well happen that sensitization is detected for an allergen to which the individual is no longer exposed and which therefore no longer causes symptoms. In such cases, exposition tests, consisting of close observation of the individual after deliberate exposure to the putative allergen, may yield useful information.

The most efficient treatment, following identification of the offending allergen, remains elimination of allergen from the person's environment and avoidance of further exposure. This form of treatment is essential for allergies caused by most household and workplace allergens. [A.L.deW.]

Alpine vegetation Plant growth forms characteristic of upper reaches of forests on mountain slopes. In such an environment, trees undergo gradual changes that, though subtle at first, may become dramatic beyond the dense forest as the zone of transition leads into the nonforested zone of the alpine tundra. In varying degrees, depending on the particular mountain setting, the forest is transformed from a closed-canopy forest to one of deformed and dwarfed trees interspersed with alpine tundra species. This zone of transition is referred to as the forest-alpine tundra ecotone. The trees within the ecotone are stunted, often shrublike, and do not have the symmetrical shape of most trees within the forest interior. *See* PLANT GEOGRAPHY.

The forest-alpine tundra ecotone is a mosaic of both tree and alpine tundra species; and it extends from timberline (the upper limit of the closed-canopy forest of symmetrically shaped, usually evergreen trees) to treeline (the uppermost limit of tree species) and the exposed alpine tundra. With elevational increases, tree deformation is magnified, tree height is reduced, and the total area occupied by trees becomes smaller as the alpine shrub, grass, and herbaceous perennials become more dominant.

The environment in which these tenacious individuals survive is harsh and involves a complex interaction of many factors, with the major controlling factor often being climate. The climate is characterized by a short growing season, low air temperatures, frozen soils, drought, high levels of ultraviolet radiation, irregular accumulation of snow, and strong winds. The interaction of all these factors produces varying levels of stress within the trees. *See* WIND.

The ultimate cause of the tree deformations and of the eventual complete cessation of tree growth lies in the inability of the tissues of the shoots and the needles to mature and prepare for the harsh environmental conditions. As the length of the growing season decreases with elevation, new needles often do not mature; they have thinner cuticles (the waxlike covering on the needles that protects against desiccation and wind abrasion), and they are less acclimated against low air temperatures. Factors that particularly affect the length of the growing season include air and soil temperatures, and the depth and distribution of snow. *See* AIR TEMPERATURE. [K.J.H.]

Altitudinal vegetation zones Intergrading regions on mountain slopes characterized by specific plant life forms or species composition, and determined by complex environmental gradients. Along an altitudinal transect of a mountain, there are sequential changes in the physiognomy (growth form) of the plants and in the species composition of the communities.

Such life zones are associated with temperature gradients present along mountain slopes. Research on patterns of altitudinal zonation has centered on the response of species and groups of species to a complex of environmental gradients. Measurements of a species along a gradient, for example, the number of individuals, biomass, or ground coverage, generally form a bell-shaped curve. Peak response of a species occurs under optimum conditions and falls off at both ends of the gradient. The unique response of each species is determined by its physiological, reproductive, growth, and genetic characteristics. Zones of vegetation along mountain slopes are formed by intergrading combinations of species that differ in their tolerance to environmental conditions. Zones are usually indistinct entities rather than discrete groupings of species. However, under some conditions of localized disjunctions, very steep sections of gradients, or competitive exclusion, discontinuities in the vegetation can create discrete communities. Vegetation zones are often defined by the distributions of species having the dominant growth form, most frequently trees. *See* ECOLOGICAL COMMUNITIES.

Altitudinal vegetation zonation, therefore, is an expression of the response of individual species to environmental conditions. Plants along an altitudinal transect are exposed, not to a single environmental gradient, but to a complex of gradients, the most important of which are solar radiation, temperature, and precipitation. Although these major environmental gradients exist in most mountain ranges of the world, the gradients along a single altitudinal transect are not always smooth because of topographic and climatic variability.

The solar energy received by mountain surfaces increases with altitude, associated with decreases in air density and the amount of dust and water vapor. An overcast sky is more efficient at reducing short-wave energy reaching low elevations and can increase the difference in energy input to 160%. However, more frequent clouds over high elevations relative to sunnier lower slopes commonly reduces this difference. Vegetation patterns are also strongly influenced by the decline in air temperature with increasing altitude, called the adiabatic lapse rate. Lapse rates are generally between 1.8°F to 3.6°F per 1000 ft (1°C to 2°C per 300 m), but vary with the amount of moisture present; wet air has a lower lapse rate. Thus, plants occurring at higher elevations generally experience cooler temperatures and shorter growing periods than low-elevation plants. Variation in the temperature gradient can be caused by differences in slope, aspect, radiation input, clouds, and air drainage patterns. The precipitation gradient in most mountains is the reverse of the temperature gradient: precipitation increases with altitude. *See* AIR TEMPERATURE; PRECIPITATION (METEOROLOGY).

General changes in vegetation with increases in altitude include reduction in plant size, slower growth rates, lower production, communities composed of fewer species, and less interspecific competition. However, many regional exceptions to these trends exist.

Characteristics of vegetation zones also vary with latitude. Mountains at higher latitudes have predominantly seasonal climates, with major temperature and radiation extremes between summer and winter. Equatorial and tropical mountains have a strong diurnal pattern of temperature and radiation input with little seasonal variation. The upper altitudinal limit of trees, and the maximum elevation of plant growth generally, decreases with distance from the Equator, with the exception of a depression near the Equator. [J.S.C.]

Ameba Any protozoon moving by means of protoplasmic flow. In their entirety, the ameboid protozoa include naked amebas, those enclosed within a shell or test, as well as more highly developed representatives such as the heliozoians, radiolarians, and foraminiferans. Ameboid movement is accomplished by pseudopods—cellular extensions which channel the flow of protoplasm. Pseudopods take varied forms and help distinguish among the different groups. A lobe-shaped extension or lobopod is perhaps the simplest type of pseudopod. The shapelessness and plasticity of these locomotory organelles impart an asymmetric, continually changing aspect to the organism. Other, more developed, representatives have pseudopodial extensions containing fibrous supporting elements (axopods) or forming an extensive network of anastomosing channels (reticulopods). Though involved in locomotion, these organelles are also functional in phagocytosis—the trapping and ingesting of food organisms (usually bacteria, algae, or other protozoa) or detritus.

Amebas range from small soil organisms, such as *Acanthamoeba* (20 micrometers), to the large fresh-water forms *Amoeba proteus* (600 μm; see illustration) and *Pelomyxa* (1 mm, or more). Some types, such as *Amoeba,* are uninucleate; others are multinucleate. Reproduction is by mitosis with nuclear division preceding cytoplasmic division to produce two daughters. Multinucleate forms have more unusual patterns of division, since nuclear division is not immediately or necessarily followed by cytoplasmic division. Transformation of the actively feeding ameba into a dormant cyst occurs in many species, particularly those found in soil or as symbionts. The resting stages allow survival over periods of desiccation, food scarcity, or transmission between hosts.

Amebas are found in a variety of habitats, including fresh-water and marine environments, soil, and as symbionts and parasites in body cavities and tissues of vertebrates and invertebrates. Because of their manner of locomotion, amebas typically occur on surfaces, such as the bottom of a pond, on submerged vegetation, or floating debris. In soil, they are a significant component of the microfauna, feeding extensively on bacteria and small fungi. Amebas in marine habitats may be found as planktonic forms adapted for floating at the surface (having oil droplets to increase bouyancy and projections to increase surface area), where they feed upon bacteria, algae, and other protozoa. Several species of amebas may be found in the human intestinal tract as harmless

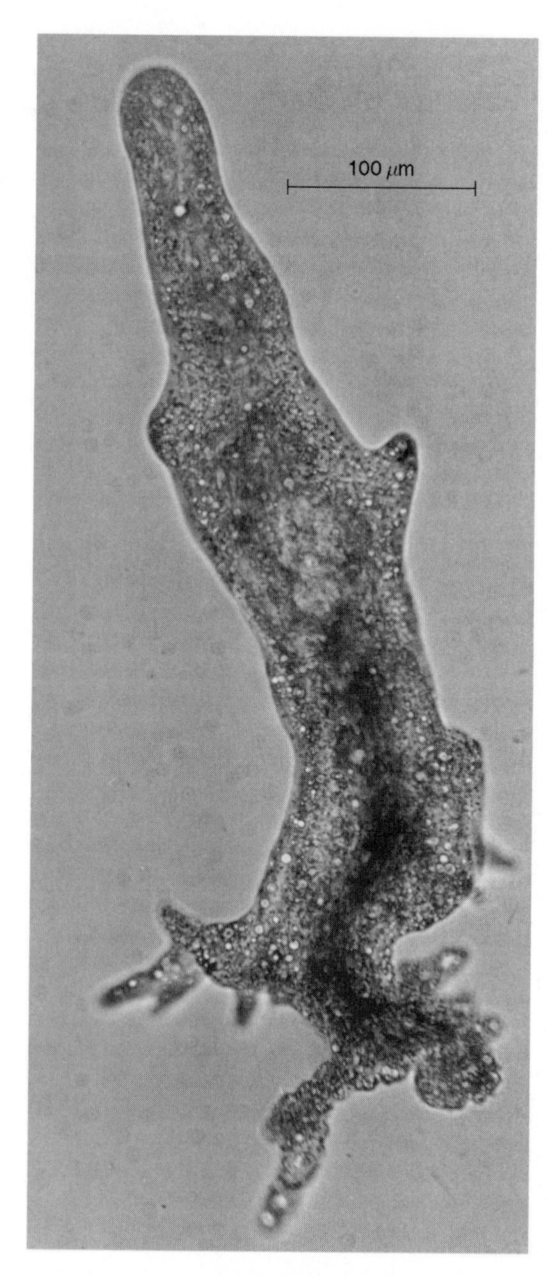

Phase-contrast photomicrograph of *Amoeba proteus*, a large fresh-water ameba. The organism is seen moving by means of a single lobose pseudopod.

commensals (for example, *Entamoeba coli*) or as important parasites responsible for amebic dysentery (*E. histolytica*).
[F.L.Sc.]

Amino acid dating Determination of the relative or absolute age of materials or objects by measurement of the degree of racemization of the amino acids present. With the exception of glycine, the amino acids found in proteins can exist in two isomeric forms called D- and L-enantiomers. Although the enantiomers of an amino acid rotate plane-polarized light in equal but opposite directions (the D form rotates it to the right and the L form to the left), their other chemical and physical properties are identical. It was discovered by L. Pasteur around 1850 that only L-amino acids are generally found in living organisms, but scientists still have not formulated a convincing reason to explain why life is based on only L-amino acids.

Under conditions of chemical equilibrium, equal amounts of both enantiomers are present (D/L = 1.0); this is called a racemic mixture. Living organisms maintain a state of disequilibrium through a system of enzymes that selectively utilize only the L-enantiomers. Once a protein has been synthesized and isolated from active metabolic processes, the L-amino acids are subject to a racemization reaction that converts them into a racemic mixture. Since racemization is a chemical process, the extent of racemization is dependent not only on the time that has elapsed since the L-amino acids were synthesized but also on the exposure temperature: the higher the temperature, the faster the rate of racemization. The rate of racemization is also different for most of the various amino acids.

A variety of analytical procedures can be used to separate amino acid enantiomers; gas chromatography and high-performance liquid chromatography are the most widely used. Since these techniques have sensitivities in the parts per billion range, only a few hundred milligrams of sample material are normally required. Samples are first hydrolyzed in hydrochloric acid to break down the proteins into free amino acids, which are then isolated by cation-exchange chromatography.

Since the late 1960s, the geochemical and biological significance of amino acid racemization has been extensively investigated. Geochemical uses of amino acid racemization include the dating of fossils or, in the case of known age specimens, the determination of their temperature history. Fossil types such as bones, teeth, and shells have been studied, and racemization has been found to be particularly useful for dating specimens that were difficult to date by other methods. Racemization has also been observed in the metabolically inert tissues of living mammals. Racemization can be studied in certain organisms and used to assess the biological age of a variety of mammalian species; in addition, it may be important in determining the biological lifetime of certain proteins. Fossils have been found to contain both D- and L-amino acids, and the extent of racemization generally increases with geologic age. *See* GEOCHRONOMETRY; RADIOCARBON DATING. [J.L.Ba.]

Anaerobic infection An infection caused by anaerobic bacteria (organisms that are intolerant of oxygen). Most such infections are mixed, involving more than one anaerobe and often aerobic or facultative bacteria as well. Anaerobes are prevalent throughout the body as indigenous flora, and virtually all anaerobic infections arise endogenously, the principal exception being *Clostridium difficile* colitis. Factors predisposing to anaerobic infection include those disrupting mucosal or other surfaces (trauma, surgery, and malignancy or other disease), those lowering redox potential (impaired blood supply, tissue necrosis, and growth of nonanaerobic bacteria), drugs inactive against anaerobes (such as aminoglycosides), and virulence factors produced by the anaerobes (toxins, capsules, and collagenase, hyaluronidase, and other enzymes). Anaerobic gram-negative bacilli (*Bacteroides, Prevotella, Porphyromonas, Fusobacterium*) and anaerobic gram-positive cocci (*Peptostreptococcus*) are the most common anaerobic pathogens. *Clostridium* (spore formers) may cause serious infection. The prime pathogen among gram-positive nonsporulating anaerobic bacilli is *Actinomyces*. Of the infections commonly involving anaerobes, the oral and dental pleuropulmonary, intraabdominal, obstetric-gynecologic, and skin and soft tissue infections are most important in terms of frequency of occurrence. To document anaerobic infection properly, specimens for culture must be obtained so as to exclude normal flora and must be transported under anaerobic conditions. Therapy includes surgery and antimicrobial agents. *See* INFECTION. [S.M.F.]

Anaplasmosis A disease of ruminants caused by a specialized group of gram-negative bacteria of the order Rickettsiales, family Anaplasmataceae, genus *Anaplasma*. *Anaplasma* is an obligate intracellular parasite infecting erythrocytes of cattle, sheep, goats, and wild ruminants in most of the tropical and subtropical world. The most important species is *A. marginale*, which causes anemia and sometimes death in cattle. Anaplasmosis is one of the most important diseases of cattle and results in significant economic losses. *See* RICKETTSIOSES.

Transmission of *Anaplasma* occurs biologically by species of hard ticks. Rickettsiae are ingested with a blood meal and undergo complex development beginning in the tick gut. *Anaplasma* is transmitted from the salivary glands while ticks feed. Mechanical transmission occurs when the mouthparts of biting flies become contaminated while feeding on infected cattle and then quickly move to uninfected animals. Contaminated needles and instruments for dehorning, castration, and tagging may also transmit the organism throughout a herd.

The average prepatent period is 21 days postinfection, but animals may not exhibit clinical disease for as long as 60 days after infection. In acute anaplasmosis, cattle exhibit depression, loss of appetite, increased temperature, labored breathing, dehydration, jaundice, and a decrease in milk production. Death may result from severe anemia, and abortions may occur. Recovered animals gradually regain condition but remain chronically infected and are subject to periodic relapses.

The clinical manifestations of anaplasmosis can be halted or prevented if tetracycline antibiotics are administered early. The drug inhibits rickettsial protein synthesis but does not kill the organism. Tetracyclines may also be added to feeds to prevent clinical symptoms. Controlling fly and tick populations by chemical spraying or dipping may be used to reduce the spread of disease. Vaccination with killed *A. marginale* from erythrocytes provides protection against the acute disease but does not prevent infection. Boosters must be given annually. The less virulent species, *A. centrale*, is used as a live vaccine in some countries and provides some protection against *A. marginale*, but it can revert to a virulent form.

Once cattle are infected with *Anaplasma*, they remain persistently infected and may serve as reservoirs of infection for other cattle or tick vectors. During this time the parasitemia fluctuates and at times may be undetectable. The spread of anaplasmosis may occur when these carrier cattle are moved to nonendemic areas. Ticks have been shown to transmit new infections after feeding on animals with undetectable parasitemias. Carrier animals may also have relapse infections, producing new herd infections.

Cattle which recover from clinical anaplasmosis are protected from subsequent exposure to the same geographic strain of *Anaplasma*. Calves have a natural immunity to clinical disease if exposed within their first year. Protection is attributed to a complex combination of antibody and cell-mediated responses. [E.F.B.]

Animal communication A discipline within the field of animal behavior that focuses upon the reception and use of signals. Animal communication could well include all of animal behavior, since a liberal definition of the term signal could include all stimuli perceived by an animal. However, most research in animal communication deals only with those cases in which a signal, defined as a structured stimulus generated by one member of a species, is subsequently used by and influences the behavior of another member of the same species in a predictable way (intraspecific communication). In this context, communication occurs in virtually all animal species.

The field of animal communication includes an analysis of the physical characteristics of those signals believed to be responsible in any given case of information transfer. A large part of this interest is due to technological improvements in signal detection, coupled with analysis of the signals obtained with such devices.

Information transmission between two individuals can pass in four channels: acoustic, visual, chemical, and electrical. An individual animal may require information from two or more channels simultaneously before responding appropriately to reception of a signal. Furthermore, a stimulus may evoke a response under one circumstance but be ignored in a different context.

Acoustic signals have characteristics that make them particularly suitable for communication, and virtually all animal groups have some forms which communicate by means of sound. Sound can travel relatively long distances in air or water, and obstacles between the source and the recipient interfere little with an animal's ability to locate the source. Sounds are essentially instantaneous and can be altered in important ways. Both amplitude and frequency modulation can be found in sounds emitted by animals; in some species sound signals have discrete patterns due to frequency and timing of utterances. Since a wide variety of sound signals are possible, each species can have a unique set of signals in its repertoire.

Sound signals are produced and received primarily during sexual attraction, including mating and competition. They may also be important in adult–young interactions, in the coordination of movements of a group, in alarm and distress calls, and in intraspecific signaling during foraging behavior.

Visual signaling between animals can be an obvious component of communication. Besides the normal range of human vision (visible light), visual signals include additional frequencies in the infrared and ultraviolet ranges. The quality of light that is often considered is color, but other characteristics are important in visual communication. Alterations of brightness, pattern, and timing also provide versatility in signal composition. The visual channel suffers from the important limitation that all visual signals must be line of sight. Information transfer is therefore largely restricted to the daytime (except for animals such as fireflies) and to rather close-range situations.

Intraspecific visual signaling appears to occur primarily during mate attraction. The color dimorphism of birds, the patterns of butterfly wings, the posturing of some fish, and firefly flashing are examples. Some parent–young interactions involve visual signaling. A young bird in the nest may open its mouth when it sees the underside of its parent's beak. Other examples are the synchronized behavior observed in schooling fish and flocking birds.

Chemical signals, like visual and sound signals, can travel long distances, but with an important distinction. Distant transmission of chemical signals requires a movement of air or water. Therefore, an animal cannot perceive an odor from a distance; it can only perceive molecules brought to it by a current of air or water. Animals do not hunt for an odor source by moving other than upwind or upcurrent in water because chemical signals do not travel in still air or water since diffusion is far too slow.

The fact that chemical signals comprise molecules means that, unlike acoustical or visual signals, chemical signals have a time lag. Chemical signals have to be of an appropriate concentration if they are to be effective. A chemical normally considered to be an attractant can serve as a repellent if it is too strong. Chemical signals may persist for a while, and time must pass before the concentration drops below the threshold level for reception by a searching animal. Since molecules of different sizes and shapes have varying degrees of persistence in the environment, the chemical channel is often involved in territorial marking, odor trail formation, and mate attraction. This channel is particularly suitable where acoustical or visual signals might betray the location of a signaler to a potential predator.

The array of molecular structure is essentially limitless, permitting a species-specific nature for chemical signals. Unfortunately, that specificity can make interception and analysis of chemical signals a difficult matter for research.

Pheromones are chemical signals that are produced by an animal and are exuded to influence the behavior of other members of the same species. If pheromones are incorporated into a recipient's body (by ingestion or absorption), they may chemically alter the behavior of such an individual for a considerable period of time. *See* CHEMICAL ECOLOGY.

Some electric fish and electric eels live in murky water and have electric generating organs that are really modified muscle bundles. Communication by electric signaling is rapid; signals can travel throughout the medium (even murky water), and rather complex signals can be generated, permitting species-specific communication during sexual attraction. However, the electrical mode is apparently restricted to those species that have electric generating organs.

Animal communication is one of the most difficult areas of study in science for several reasons. First, experiments must be designed and executed in such a manner that extraneous cues (artifacts) are eliminated as potential causes of the observed results. Second, once supportive evidence has been obtained, each hypothesis must be tested. In animal communication studies, adequate tests often rely upon direct evidence—that is, evidence obtained by artificially generating the signal presumed responsible for a given behavioral act, providing that signal to a receptive animal, and actually evoking a specific behavioral act in a predictable manner. *See* ETHOLOGY. [A.M.We.]

Animal evolution The theory that modern animals are the modified descendants of animals that formerly existed and that these earlier forms descended from still earlier and different organisms.

Animals are multicellular organisms that feed by ingestion of other organisms or their products, being unable to derive energy through photosynthesis or chemosynthesis. Animals are currently classed into about 30 to 35 phyla, each of which has evolved a distinctive body plan or architecture.

All phyla began as invertebrates, but lineages of the phylum Chordata developed the internal skeletal armature, with spinal column, which was exploited in numerous fish groups and which eventually gave rise to terrestrial vertebrates. The number of phyla is uncertain partly because most of the branching patterns and the ancestral body plans from which putative phyla have arisen are not yet known. For example, arthropods (including crustaceans and insects) may have all diversified from a common ancestor that was a primitive arthropod, in which case they may be grouped into a single phylum; or several arthropod groups may have evolved independently from nonarthropod ancestors, in which case each such group must be considered a separate phylum. So far as known, all animal phyla began in the sea.

Some features of the cells of primitive animals resemble those of the single-celled Protozoa, especially the flagellates, which have long been believed to be animal ancestors. Molecular phylogenies have supported this idea and also suggest that the phylum Coelenterata arose separately from all other phyla that have been studied by this technique. Thus animals may have evolved at least twice from organisms that are not themselves animals, and represent a grade of evolution and not a single branch (clade) of the tree of life. Sponges have also been suspected of an independent origin, and it is possible that some of the extinct fossil phyla arose independently or branched from sponges or cnidarians. *See* PROTOZOA.

The earliest undoubted animal fossils (the Ediacaran fauna) are soft-bodied, and first appear in marine sediments nearly 650 million years (m.y.) old. This fauna lasted about 50 m.y. and consisted chiefly of cnidarians or cnidarian-grade forms, though it contains a few enigmatic fossils that may represent groups that gave rise to more advanced phyla. Then, nearly 570 m.y. ago, just before and during earliest Cambrian time, a diversification of body architecture began that produced most of the living phyla as well as many extinct groups. The body plans of some of these groups involved mineralized skeletons which, as these are more easily preserved than soft tissues, created for the first time an extensive fossil record. The soft-bodied groups were markedly diversified, though their record is so spotty that their history cannot be traced in detail. A single, exceptionally preserved soft-bodied fauna from the Burgess Shale of British Columbia that is about 530 m.y. old contains not only living soft-bodied worm phyla, but extinct groups that cannot be placed in living phyla and do not seem to be ancestral to them.

Following the early phase of rampant diversification and of some concurrent extinction of phyla and their major branches, the subsequent history of the durably skeletonized groups can be followed in a general way in the marine fossil record. The composition of the fauna changed continually, but three major associations can be seen: one dominated by the arthropodlike trilobites during the early Paleozoic, one dominated by articulate brachiopods and crinoids (Echinodermata) in the remaining Paleozoic, and one dominated by gastropod (snail) and bivalve (clam) mollusks during the Mesozoic and Cenozoic. The mass extinction at the close of the Paleozoic that caused the contractions in so many groups may have extirpated over 90% of marine species and led to a reorganization of marine community structure and composition into a modern mode. Resistance to this and other extinctions seems to have been a major factor in the rise of successive groups to dominance. Annelids, arthropods, and mollusks are the more important invertebrate groups that made the transition to land. The outstanding feature of terrestrial fauna is the importance of the insects, which appeared in the late Paleozoic and later radiated to produce the several million living species, surpassing all other life forms combined in this respect. *See* ANNELIDA; ARTHROPODA; INSECTA. [J.W.V.]

The phylum Chordata consists largely of animals with a backbone, the Vertebrata, including humans. The group, however, includes some primitive nonvertebrates, the protochordates: lancelets, tunicates, acorn worms, pterobranchs, and possibly the extinct graptolites and conodonts. The interrelationships of these forms are not well understood. With the exception of the colonial graptolites, they are soft-bodied and have only a very limited fossil record. They suggest possible links to the Echinodermata in developmental, biochemical, and morphological features. In addition, some early Paleozoic fossils, the carpoids, have been classified alternatively as chordates and as echinoderms, again suggesting a link. In spite of these various leads, the origin of the chordates remains basically unclear.

Chordates are characterized by a hollow, dorsal, axial nerve chord, a ventral heart, a system of slits in the larynx that serves variously the functions of feeding and respiration, a postanal swimming tail, and a notochord that is an elongate supporting structure lying immediately below the nerve chord. The protochordates were segmented, although sessile forms such as the tunicates show this only in the swimming, larval phase.

The first vertebrates were fishlike animals in which the pharyngeal slits formed a series of pouches that functioned as respiratory gills. An anterior specialized mouth permitted ingestion of food items large in comparison with those of the filter-feeding protochordates. Vertebrates are first known from bone fragments found in rocks of

Cambrian age, but more complete remains have come from the Middle Ordovician. Innovations, related to greater musculoskeletal activity, included the origin of a supporting skeleton of cartilage and bone, a larger brain, and three pairs of cranial sense organs (nose, eyes, and ears). At first the osseous skeleton served as protective scales in the skin, as a supplement to the notochord, and as a casing around the brain. In later vertebrates the adult notochord is largely or wholly replaced by bone, which encloses the nerve chord to form a true backbone. All vertebrates have a heart which pumps blood through capillaries, where exchanges of gases with the external media take place. The blood contains hemoglobin in special cells which carry oxygen and carbon dioxide. In most fishes the blood passes from the heart to the gills and thence to the brain and other parts of the body. In most tetrapods, and in some fishes, blood passes to the lungs, is returned to the heart after oxygenation, and is then pumped to the various parts of the body.

The jawless fish, known as Agnatha, had a sucking-rasping mouth apparatus rather than true jaws. They enjoyed great success from the Late Cambrian until the end of the Devonian. Most were heavily armored, although a few naked forms are known. They were weak swimmers and lived mostly on the bottom. The modern parasitic lampreys and deep-sea scavenging hagfish are the only surviving descendants of these early fish radiations.

In the Middle to Late Silurian arose a new type of vertebrate, the Gnathostomata, characterized by true jaws and teeth. They constitute the great majority of fishes and all tetrapod vertebrates. The jaws are modified elements of the front parts of the gill apparatus, and the teeth are modified bony scales from the skin of the mouth. With the development of jaws, a whole new set of ecological opportunities was open to the vertebrates. Along with this, new swimming patterns appeared, made possible by the origin of paired fins, forerunners of which occur in some agnathans.

Four groups of fishes quickly diversified. Of these, the Placodermi and Acanthodii are extinct. The Placodermi were heavily armored fishes, the dominant marine carnivores of the Silurian and Devonian. The Acanthodii were filter-feeders mostly of small size. They are possibly related to the dominant groups of modern fishes, the largely cartilaginous Chondrichthyes (including sharks, rays, and chimaeras) and the Osteichthyes (the higher bony fishes). These also arose in the Late Silurian but diversified later.

The first land vertebrates, the Amphibia, appeared in the Late Devonian and were derived from an early group of osteichthyans called lobe-finned fishes, of which two kinds survive today, the Dipnoi or lungfishes, and the crossopterygian coelacanth *Latimeria*. They were lung-breathing fishes that lived in shallow marine waters and in swamps and marshes. The first amphibians fed and reproduced in or near the water. True land vertebrates, Reptilia, with a modified (amniote) egg that could survive on land, probably arose in the Mississippian.

By the Middle Pennsylvanian a massive radiation of reptiles was in process. The most prominent reptiles belong in the Diapsida: dinosaurs, lizards and snakes, and pterosaurs (flying reptiles). The birds, Aves, which diverged from the dinosaur radiation in the Late Triassic or Early Jurassic, are considered to be feathered dinosaurs, and thus members of the Diapsida, whereas older authorities prefer to treat them as a separate case. In addition, there were several Mesozoic radiations of marine reptiles such as ichthyosaurs and plesiosaurs. Turtles (Chelonia) first appeared in the Triassic and have been highly successful ever since.

The line leading to mammals can be traced to primitive Pennsylvanian reptiles, Synapsida, which diversified and spread worldwide during the Permian and Triassic. The first true mammals, based on characteristics of jaw, tooth, and ear structure, arose in the Late Triassic. Derived mammals, marsupials (Metatheria) and placentals (Eutheria), are known from the Late Cretaceous, but mammalian radiations began only in the early Cenozoic. By the end of the Eocene, all the major lines of modern mammals had become established. Molecular analyses (blood proteins, deoxyribonucleic acid, ribonucleic acid) of living mammals show that the most primitive group of placentals is the edentates (sloths, armadillos, and anteaters). An early large radiation included the rodents, primates (including monkeys, apes, and humans), and bats, possibly all closely related to the insectivores and carnivores. The newest radiations of mammals are of elephants and sea cows, while the whales are related to the artiodactyls (cattle, camels). [K.S.Th.]

Animal virus A small infectious agent that is unable to replicate outside a living animal cell. Unlike other intracellular obligatory parasites (for example, chlamydiae and rickettsiae), they contain only one kind of nucleic acid, either deoxyribonucleic acid (DNA) or ribonucleic acid (RNA). They do not replicate by binary fission. Instead, they divert the host cell's metabolism into synthesizing viral building blocks, which then self-assemble into new virus particles that are released into the environment. During the process of this synthesis, viruses utilize cellular metabolic energy, many cellular enzymes, and organelles which they themselves are unable to produce. Animal viruses are not susceptible to the action of antibiotics. The extracellular virus particle is called a virion, while the name virus is reserved for various phases of the intracellular development.

Morphology. Virions are small, 20–300 nanometers in diameter, and pass through filters which retain most bacteria. However, large virions (for example, vaccinia, which is 300 nm in diameter) exceed in size some of the smaller bacteria. The major structural components of the virion are proteins and nucleic acid, but some virions also possess a lipid-containing membranous envelope. The protein molecules are arranged in a symmetrical shell, the capsid, around the DNA or RNA. The shell and the nucleic acid constitute the nucleocapsid.

In electron micrographs of low resolution, virions appear to possess two basic shapes: spherical and cylindrical. High-resolution electron microscopy and x-ray diffraction studies of crystallized virions reveal that the "spherical" viruses are in fact polyhedral in their morphology, while the "cylindrical" virions display helical symmetry. The polyhedron most commonly encountered in virion structures is the icosahedron, in which the protein molecules are arranged on the surface of 20 equilateral triangles. Based on these morphological features, viruses are classified as helical or icosahedral. Certain groups of viruses do not exhibit any discernible features of symmetry and are classified as complex virions. Further distinction is made between virions containing RNA or DNA as their genomes and between those with naked or enveloped nucleocapsids.

Viral nucleic acid. The outer protein shell of the virion furnishes protection to the most important component, the viral genome, shielding it from destructive enzymes (ribonucleases or deoxyribonucleases). The viral genome carries information which specifies all viral structural and functional components required for the initiation and establishment of the infectious cycle and for the generation of new virions. This information may be contained in a double-stranded or single-stranded (parvoviruses) DNA, or double-stranded (reoviruses) or single-stranded RNA. The viral DNA may be linear or circular, and the viral RNA may be a single long chain or a number of shorter chains (fragmented genomes), each of which contains different genetic information. Furthermore, some RNA viruses have the genetic information expressed as a complementary nucleotide sequence. These are classified as negative-strand RNA viruses. Finally, the RNA tumor viruses have an intracellular DNA phase, during which the genetic information contained in the virion RNA is transcribed into a DNA and integrated into the host cell's genome. The discovery of this process came as a surprise, since it was believed that the flow of genetic information was unidirectional from DNA to RNA to protein and could not take place in the opposite direction. The transcription of RNA to DNA was termed reverse transcription, and the RNA tumor viruses are sometimes referred to as retroviruses. *See* GENETIC CODE; TUMOR VIRUSES.

When introduced into a susceptible cell by either chemical or mechanical means, the naked viral nucleic acid is in most cases itself infectious. Two exceptions are the negative-strand RNA viruses and the RNA tumor viruses. In these cases the RNA has to be first transcribed and reverse-transcribed, respectively, into the proper form of genetic information before the infectious process can take place. This task is carried out by means of an enzyme which is contained in the protein shell of the virion nucleocapsid. The whole nucleocapsid is therefore required for infectivity.

Viral infection is composed of several steps: adsorption, penetration, uncoating and eclipse, and maturation and release. Adsorption takes place on specific receptors in the membrane of an animal cell. The presence or absence of these receptors determines the tissue or species susceptibility to infection by a virus. Enveloped viruses exhibit surface spikes which are involved in adsorption; however, most animal viruses do not possess obvious attachment structures. Penetration takes place through invagination and ingestion of the virion by the cell membrane (phagocytosis or viropexis). Penetration is followed by uncoating of the nucleic acid, or in some cases by uncoating of the nucleocapsid. At this stage, the identity of the virion has disappeared, and viral infectivity cannot be recovered from disrupted cells.

The absence of infectious particles in cell extracts is characteristic of the eclipse period. During the eclipse the biochemical processes of the cell are manipulated to synthesize viral proteins and nucleic acids. The eclipse period in infections with DNA viruses starts with the transcription of the genetic information in the nucleus of the cell (except poxviruses), processing into mRNAs, and their translation into proteins (in the cytoplasm). This process is divided into early and late transcription. The early proteins are virus-encoded functional proteins which will participate in the synthesis of viral DNA and of intermediate and late viral proteins, as well as in the shutoff of various cellular functions which might be detrimental to viral synthesis. The major late products are the structural proteins of the nucleocapsid. Almost as soon as these proteins are synthesized, they assemble with newly synthesized DNA molecules into virion nucleocapsids.

The events of the eclipse period in infections with RNA viruses are similar, except that they take place in the cytoplasm (influenza virus excepted), and a division into early and late transcription cannot be made. In the case of positive-strand RNA viruses, the viral RNA is itself the mRNA. In infections with negative-strand RNA viruses, the virion RNA in the nucleocapsid is first transcribed into positive mRNAs. Intracellular nucleocapsids are present throughout the entire infectious cycle, and the eclipse period cannot be defined in the classical sense. RNA tumor viruses reverse-transcribe their RNA into DNA, which enters the cell nucleus and becomes integrated into the cellular DNA. All viral mRNAs and genomic RNAs are generated by transcription of the integrated DNA.

The event characteristic of the maturation step is virion assembly and release. In many cases the protein shell is assembled first (procapsid) and the nucleic acid is inserted into it. During this insertion, processing of some shell proteins by cleavage takes place and is accompanied by a modification of the structure to accommodate the nucleic acid. Unenveloped viruses which mature in the cytoplasm (for example, poliovirus) often exit the cell rapidly by a reverse-phagocytosis process, even before the breakdown of the cell. In some cases, however, a large number of virus particles may accumulate inside the cell in crystalline arrays called inclusion bodies. Viruses

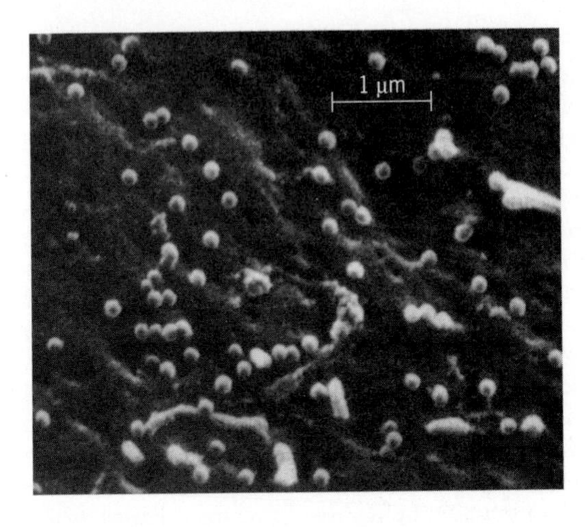

Scanning electron micrograph of the surface of a mouse cell infected with murine leukemia virus. A large number of virus particles are shown in the process of budding. (*Courtesy of R. MacLeod*)

that mature in the nucleus are usually released slowly, and the damage to the cell is extensive. Enveloped viruses exit the cell by a process of budding. Viral envelope proteins (glycoproteins) become inserted at various sites into the cell membrane, where they also interact with matrix proteins and with nucleocapsids. The cellular membrane then curves around the complex and forms a bud which detaches from the rest of the cell (see illustration).

Effect of viral infections. Two extreme types of effects are identified with viral infections: lytic infections, which cause cell death by a variety of mechanisms with cell lysis as the most common outcome, and persistent infections, accompanied either by no apparent change in the host cell or by some interference with normal growth control, as in transformation of normal to cancer cells. In animals, extensive destruction of tissue may accompany an infection by a lytic virus.

As a defense to certain conditions of infection, animal cells generate a group of substances called interferons which, by a complex mechanism, inhibit replication of viruses. They are specific to the cell species from which they were derived but not to the virus which elicited their generation. (Mouse interferon will protect mouse but not human cells from any viral infection.)

Pathology. Virus infections spread in several ways: through aerosols and dust, by direct contact with carriers or their excretions, and by bites or stings of animal and insect vectors. At the point of entry, infected cells undergo viremia. From there, the virus becomes disseminated by secretions. It is carried through the lymphatic system and bloodstream to other target organs, where secondary viremias occur (except in localized infections like warts). In most cases viral infections are of short duration and great severity. However, persistent infections are not uncommon (herpes, adeno, various paramyxoviruses such as measles).

The afflicted organism mounts a variety of defenses, the most important of which is the immune response. Circulating antibodies against viral proteins are generated. Those interacting with virion surface proteins neutralize the infectious potential of the virus. Although the antibodies are specific against the virus which has elicited them, they will cross-react with closely related virus strains. The specificity of neutralizing antibodies obtained from experimentally injected animals is utilized for diagnostic purposes or in quantitative assays. In addition to the circulating antibodies, cell-mediated immune responses also take place. The most important of these is the production of cytotoxic thymus-derived lymphocytes, found in the lymph nodes, spleen, and blood. They destroy all cells which harbor in their membrane viral glycoproteins. The cell-mediated immunity has been demonstrated to be more important to the process of recovery than circulating antibodies. In spite of their beneficial role, immune responses often seriously contribute to the pathology of the disease. Circulating antigen-antibody complexes can lodge in organs and cause inflammation; cell-mediated responses have been known to produce severe shock syndromes in patients with a history of previous exposures to the virus.

Control. Viruses are resistant to the antibiotics commonly used against bacterial infections. The use of chemotherapeutic agents with antiviral activity is plagued by their toxicity to the animal host. However, the application of vaccines has been successful in the control of several viruses. The vaccines elicit immune responses and provide sometimes life-long protection. Two types of vaccines have been applied: inactivated virus and live attentuated virus. Various inactivation procedures are available. An attenuated laboratory strain of smallpox has been applied so successfully that the disease is considered to be eradicated. A small probability of back mutations of the attenuated virus to a virulent strain makes applications of live vaccines somewhat riskier. On the other hand, protection is longer-lasting and, by virtue of spread to nonvaccinated individuals, more beneficial to the population group (herd effect). *See* POLIOMYELITIS.

In order to achieve full protection, it is important that the vaccine contain all the distinct antigenic types of the virus. Development of monoclonal antibodies led to a better characterization of these types in naturally occurring viruses. This information will undoubtedly lead to better vaccines. Moreover, monoclonal antibodies have aided investigations into the molecular structure of viral antigenic groups and brightened future prospects for synthetic vaccines. *See* VIRUS. [M.E.Re.]

Annelida The phylum comprising the multisegmented, invertebrate wormlike animals, of which the most numerous are the marine bristle worms and the most familiar the terrestrial earthworms. The Annelida (meaning little annuli or rings) include the Polychaeta (meaning many setae); the earthworms and fresh-water worms, or Oligochaeta (meaning few setae); the marine and fresh-water leeches or Hirudinea; and two other marine classes having affinities with the Polychaeta: the Archiannelida (meaning primitive annelids), small heteromorphic marine worms, and the Myzostomaria (meaning sucker mouths), parasites of crinoid echinoderms. These five groups share few common characters and little resemblance except that most have a wormlike body. Typically they are bilaterally symmetrical, lack a skeleton, and have a short to long linear body divided into rings or segments, which are separated from one another by transverse walls or septa. The mouth is an anteroventral or anterior vent at the forward end of the alimentary tract, and the anus posterodorsal or posterior at the hind end of the gut. *See* HIRUDINEA; MYZOSTOMARIA; OLIGOCHAETA; POLYCHAETA.

The linear series of segments, or metameres, from anterior to posterior ends constitute the annelid body. These segments may be similar throughout, resulting in an annulated cylinder, as in earthworms and *Lumbrineris*. More frequently the successive segments are dissimilar, resulting in regions modified for particular functions. Each segment may be simple (uniannular) corresponding to a metamere, or it may be divided (multiannulate). The total number of segments varies from five to several hundred. Segments may have lateral fleshy outgrowths called parapodia (meaning side feet), armed with special secreted bristles or rods, called setae and acicula; they provide protection and aid in locomotion. Setae are lacking in Hirudinea and some polychaetes. The body is covered by a thin to thick epithelium which is never shed.

Annelids have sense organs of many kinds. Most conspicuous are those on the anterior and on parapodia. Eyes which function as photoreceptors may be variously developed. Feelers or tactoreceptors are frequently on the prostomium (a lobe of skin projecting from the first body segment) as antennae; on anterior segments as long filiform or thick fleshy tentacles; and on parapodia as cirri, papillae, scales, or tactile hairs. Cilia in bands or clusters or in grooves occur in specific patterns. The most conspicuous receptors are the large nuchal organs of amphinomid polychaetes, fleshy, folded, paired organs surrounding the cephalic structures.

Setae detect changes in the environment through their basal attachments. Shallow-water species often have short, strong, resistant setae, whereas abyssal species have long, slender, simple setae. Each organ is unique and well adapted to its role in the development, growth, protection, and reproduction of the species involved.

Chromatophores are cell clusters which change their shape and size to conform to the shadows of the animal' s background, responding to changes in light intensities, and therefore are generally protective. They are well developed in translucent pelagic larvae which exist at the surface of the sea; they screen damaging intensities of light from delicate tissues. Oligochaetes and hirudineans, which generally lack eyes or special light receptors, are sensitive to light changes through the surface epithelium.

The alimentary tract of annelids is a straight or sinuous tube, consisting of mouth, pharynx, esophagus, stomach, gut, and pygidium or anus. The mouth may be a simple anterior (oligochaetes) or anteroventral (many polychaetes) pore provided with highly complex organs or accessory parts. Accessory organs include the grooved palpi of many polychaetes, which direct food to the mouth or also select and propel nutrients along ciliated tracts. The building organs and cementing glands of some tubicolous annelids are associated with the mouth; they select inert particles for shape and size and attach them in specific patterns to a basic secreted mucoid membrane, resulting in tubes which are highly characteristic.

The mouth is followed by the buccal cavity which may be modified as a phoboscis or saclike eversible pouch. The buccal cavity may be followed by a short to long muscularized eversible proboscis which captures and breaks up or compresses food particles. Its inner walls may be fortified by papillae or hard gnaths or jaws. A short esophagus leads to the muscular stomach or digestive region. Lateral ceca or pouches may be present, along esophageal and stomach portions, to increase the amount of surface for secretion and digestion, especially in short-bodied worms. Peristaltic (clasping and compressing) movements are rhythmic and result in the food bolus being digested, and wastes separated and pushed into the gut. In nonselective-feeding annelids the gut may be distended with great amounts of inert materials; in selective feeders there are few remains but those of living animals. The proctodaeum, or region preceding the anus, expels the wastes as fecal pellets of characteristic form.

The nervous system consists of a dorsal, bilaterally symmetrical, ganglionic mass or brain within or behind the preoral region. The brain is connected to the ventral cord through the circumesophageal connectives which extend about the oral cavity. The ventral cord may be single or paired, nearly smooth or nodular, or ganglionated according to the segmental pattern of the body. The brain sends out lateral branches to the eyes, palpi, antennae, or other structures; the ventral cord has lateral branches to all fleshy parts which receive stimuli. A giant axon is an enlarged part of the ventral cord, present in many long, muscular, or actively moving oligochaetes and polychaetes. It permits rapid transfer of stimuli and muscular response, resulting in abrupt response.

The circulatory system consists of dorsal and ventral longitudinal, median vessels located above and below the alimentary tract. Lateral branches extend to all parts of the body. Pulsating or propelling contractile portions, sometimes called hearts, are in the anterior dorsal vessel or also at intervals in the ventral vessels. In oligochaetes these are segmental vessels of varying number connecting the dorsal and ventral vessels and surrounding some portion of the alimentary tract. The contained blood may be red, through the presence of a hemoglobinlike substance, or green, through the presence of chlorocruorin. These colors when diluted are yellow or colorless, as in many small annelids.

Some annelids lack a closed circulatory system so that blood and coelomic fluids mix freely, resulting in a hemocoel; it may be partial or complete. Many annelids have a special organ called a cardiac or heart body surrounding the pulsating vessel and sometimes visible as a thick brown or red body of spongy tissue; its function is to dispose of circulatory wastes. The circulatory and coelomic fluids aid in maintaining turgidity of the annelid body, and with musculature control they act as a kind of skeleton.

Special organs for excretion are called nephridia. In their simplest form they are protonephridia, consisting of a strand of cells connecting the coelom to the body wall, usually a pair to a segment. More complex organs, or metanephridia, have a ciliated nephrostome or funnel opening into the coelom and continued to the surface as a complex organ. They function to transport wastes and at sexual maturity may serve to release gonadial products. Complex nephridia are present in all oligochaetes and many polychaetes.

The muscular system consists of an outer circular and an inner longitudinal system of muscles, each varying in extent and density according to species. In addition, an oblique series, between outer and inner layers, is well developed in annelids performing complex lateral movements. Long, very active burrowers or crawlers have an extensive musculature, whereas short, sluggish forms may have diminished musculature development. Movements are achieved mainly by coordinated muscular contractions and expansions of the laterally projecting parapodia or setae, resulting in an undulating or meandering movement. Swimming species may move from side to side or by successive forward darts and stops. Some annelids with reduced parapodia and a long proboscis use the latter in progression by extension and withdrawal of the eversible part of the alimentary tract.

The ability to replace lost parts is highly developed in annelids. Most frequent is the replacement of tail, parapodia, and setae. The anterior end may be replaced provided the break is postpharyngeal. The torn end is first covered over with scar tissue, then differentiated into epithelial cells and all other tissues characteristic of the whole animal.

Depending on the species, reproduction may be sexual, asexual, or both. Sexual reproduction may be dioecious, in which male and female are similar, rarely dissimilar. Individuals may be hermaphroditic, both male and female, but with cross fertilization. Some annelids are protandric hermaphrodites, in which the sexual stages alternate. [O.H.]

Fossil annelids, or segmented worms, are mostly soft-bodied, yet sufficiently common to indicate that this large and varied group of invertebrates has been abundant for more than 500 million years. The bulk of the annelidan fossil record is represented by the polychaetes. The earliest definite occurrences are from the Lower Cambrian of Greenland (Sirius Passet fauna).

The fossil record of oligochaetes is poor, in part because their predominantly terrestrial and fresh-water habitats have a relatively poor rock record. Nevertheless, the reproductive cocoons characteristic of the clitellates (oligochaetes and leeches) have been recognized as far back as the Triassic and may be more common as fossils than is generally realized. [S.C.M.]

Anoxic zones Oxygen-depleted regions in marine environments. The dynamic steady state between oxygen supply and consumption determines the oxygen concentration. In regions where the rate of consumption equals the rate of supply, seawater becomes devoid of oxygen and thus anoxic. In the open ocean, the only large regions which approach anoxic conditions are between 165 and 3300 ft (50 and 1000 m) deep in the equatorial Pacific and between 330 and 3300 ft (100 and 1000 m) in the northern Arabian Sea and the Bay of Bengal in the Indian Ocean. The Pacific region consists of vast tongues extending from Central America and Peru nearly to the middle of the ocean in some places. In parts of this zone, oxygen concentrations become very low, 15 μmol/liter (atmospheric saturation is 200–300 μmol/liter). Pore waters of marine sediments are sometimes anoxic a short distance below the sediment-water interface. The degree of oxygen consumption in sediment pore waters depends upon the amount of organic matter reaching the sediments and the rate of bioturbation (mixing of the surface sediment by benthic animals). In shallow regions (continental shelf and slope), pore waters are anoxic immediately below the sediment-water interface; in relatively rapid sedimentation-rate areas of the deep sea, the pore waters are usually anoxic within a few centimeters of the interface; and in pore waters of slowly accumulating deep-sea sediments, oxygen may never become totally depleted. *See* MARINE SEDIMENTS.

Restricted basins (areas where water becomes temporarily trapped) are often either permanently or intermittently anoxic. Classic examples are the Black Sea, the Carioca Trench off the coast of Venezuela, and fiords which occupy the Norwegian and British Columbia coasts. Lakes which receive a large amount of nutrient inflow (either from natural or human-produced sources) are often anoxic during the period of summer stratification. *See* BLACK SEA; FIORD.

The chemistry of many elements dissolved in seawater (particularly the trace elements) is vastly changed by the presence or absence of oxygen. Since large areas of the ocean water mass are in contact with oxygen-depleted pore waters, the potential exists for anoxic conditions to have a marked effect on the chemistry of the sea. *See* SEAWATER; SEAWATER FERTILITY. [S.R.E.]

Antarctic Circle An imaginary line that delimits the northern boundary of Antarctica. It is a distinctive parallel of latitude at approximately 66°30' south. Thus it is located about 4590 mi (7345 km) south of the Equator and about 1630 mi (2620 km) north of the south geographic pole.

All of Earth's surface south of the Antarctic Circle experiences one or more days when the Sun remains above the horizon for at least 24 h. The Sun is at its most southerly position on or about December 21 (slightly variable from year to year). This date is known as the summer solstice in the Southern Hemisphere and as the winter solstice in the Northern Hemisphere. At this time, because Earth is tilted on its axis, the circle of illumination reaches 23.50° to the far side of the South Pole and stops short 23.50° to the near side of the North Pole.

The longest period of continuous sunshine at the Antarctic Circle is 24 h, and the highest altitude of the noon Sun is 47° above the horizon at the time of the summer solstice. The long days preceding and following the solstice allow a season of about 5 months of almost continuous daylight.

Six months after the summer solstice, the winter solstice (Southern Hemisphere terminology) occurs on or about June 21 (slightly variable from year to year). On this date the Sun remains below the horizon for 24 h everywhere south of the Antarctic Circle; thus the circle of illumination reaches 23.50° to the far side of the North Pole and stops short 23.50° to the near side of the South Pole. *See* ARCTIC OCEAN. [T.L.M.]

Antarctic Ocean The Antarctic Ocean, sometimes called the Southern Ocean, is the watery belt surrounding Antarctica. It includes the great polar embayments of the

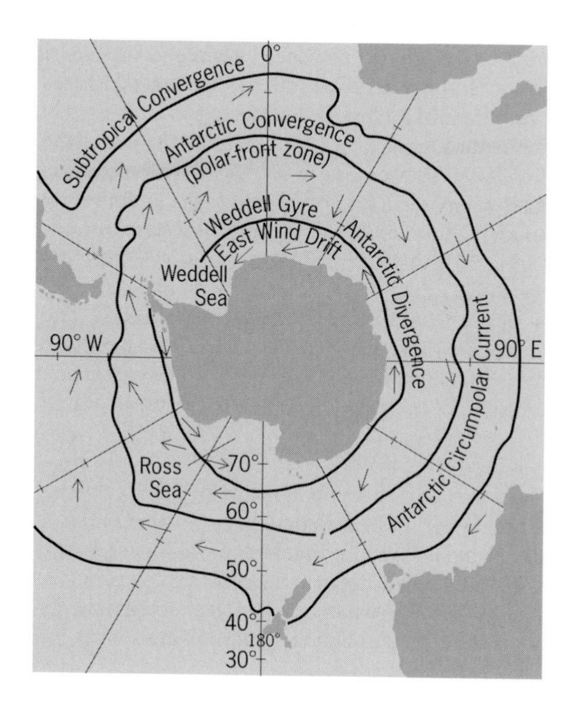

Direction of the surface circulation and major surface boundaries of the Antarctic Ocean.

Weddell Sea and Ross Sea, and the deep circumpolar belt of ocean between 50 and 60°S and the southern fringes of the warmer oceans to the north. Its northern boundary is often taken as 30°S (see illustration). The Antarctic is a cold ocean, covered by sea ice during the winter from Antarctica's coast northward to approximately 60°S.

The remoteness of the Antarctic Ocean severely hampers the ability to observe its full character. The sparse data collected and the more recent addition of data obtained from satellite-borne sensors have led to an appreciation of the unique role that this ocean plays in the Earth's ocean and climate. Between 50 and 60°S there is the greatest of all ocean currents, the Antarctic Circumpolar Current sweeping seawater from west to east, blending waters of the Pacific, Atlantic, and Indian oceans. Observed within this current is the sinking of cool (approximately 4°C; 39.2°F), low-salinity waters to depths of near 1 km (0.6 mi), which then spreads along the base of the warm upper ocean waters or thermocline of more hospitable ocean environments. The cold polar atmosphere spreading northward from Antarctica removes great amount of heat from the ocean, heat which is carried to the sea surface from ocean depths, brought into the Antarctic Ocean from warmer parts of the ocean. At some sites along the margins of Antarctica, there is rapid descent of cold (near the freezing point of seawater, −1.9°C; 28.6°F) dense water, within thin convective plumes. This water reaches the sea floor, where it spreads northward, chilling the lower 2 km (1.2 mi) of the global ocean, even well north of the Equator.

The major flow is the Antarctic Circumpolar Current, or West Wind Drift (see illustration). Along the Antarctic coast is the westward-flowing East Wind Drift. The strongest currents are in the vicinity of the polar front zone and restricted passages such as the Drake Passage, and over deep breaks in the meridionally oriented submarine ridge systems.

The extreme cold of the polar regions causes an extensive ice field to form over the southern regions of the Antarctic Ocean. The extent of the ice is seasonal in that during the October-to-March period the area decreases, and it increases during the remaining months. The seasonal difference in the volume of sea ice is estimated as 2.3×10^{19} grams (8.1×10^{17} oz). Satellite photographs reveal that the sea ice field is not uniform, but has many large polynyas (areas of water). The sea ice plays an important role in the heat balance since it reflects much more solar radiation (and therefore heat) into space than would be the case for a water surface. The polynyas would therefore be of special interest in radiation and heat-balance studies. In addition to the ice formed at sea, the ice calving at the coast of Antarctica introduces icebergs into the ocean at a rate of approximately 1×10^{18} g/year (3.5×10^{12} oz/year). *See* HEAT BALANCE, TERRESTRIAL ATMOSPHERIC; ICEBERG; SEA ICE.

Glacial (fresh-water) ice and the ocean meet along the shores of Antarctica. This occurs not only at the northern face of the ice sheet but also at hundreds of meters depth along the bases of floating ice shelves. Ocean-glacial ice interaction is believed to be a major factor in controlling Antarctica's glacial ice mass balance and stability. [A.L.G.]

Antarctica The coldest, windiest, and driest continent, overlying the South Pole. The lowest temperature ever measured on Earth was recorded at the Russian Antarctic station of Vostok at −89.2°C (−128.5°F) in July 1983. Katabatic (cold, gravitational) winds with velocities up to 50 km/h (30 mi/h) sweep down to the coast and occasionally turn into blizzards with 150 km/h (nearly 100 mi/h) wind velocities. Antarctica's interior is a cold desert with only a few centimeters of water-equivalent precipitation, while the coastal areas average 30 cm (12 in.).

Antarctica's area is about 14 million square kilometers (5.4 million square miles), which is larger than the contiguous 48 United States and Mexico together. It is the third smallest continent, after Australia and Europe. About 98% of it is buried under a thick ice sheet, which in places is 4 km (13,000 ft) thick, making it the highest continent, with an average elevation of over 2 km (6500 ft).

Although most of Antarctica is covered by ice, some mountains rise more than 3 km (almost 10,000 ft) above the ice sheet. The largest of these ranges is the Transantarctic Mountains separating east from west Antarctica, and the highest peak in Antarctica is Mount Vinson, 5140 m (16,850 ft), in the Ellsworth Mountains. Other mountain ranges, such as the Gamburtsev Mountains in East Antarctica, are completely buried, but isolated peaks called nunataks frequently thrust through the ice around the coast.

The Antarctic ice sheet is the largest remnant of previous ice age glaciations. It has probably been in place for the last 20 million years and perhaps up to 50 million years. It is the largest reservoir of fresh water on Earth, with a volume of about 25 million cubic kilometers (6 million cubic miles). Glaciers flow out from this ice sheet and feed into floating ice shelves along 30% of the Antarctic coastline. The two biggest ice shelves are the Ross and Filchner-Ronne. These shelves may calve off numerous large tabular icebergs, with thicknesses of several hundred meters, towering as high as 70–80 m (250 ft) above the sea surface. *See* GLACIOLOGY.

Year-round life on land in Antarctica is sparse and primitive. North of the Antarctic Peninsula a complete cover of vegetation, including moss carpets and only two species of native vascular plants, may occur in some places. For the rest of Antarctica, only lichen, patches of algae in melting snow, and occasional microorganisms occur. In summer, however, numerous migrating birds nest and breed in rocks and cliffs on the continental margins, to disappear north again at the beginning of winter. South of the Antarctic Convergence, 43 species of flying birds breed annually. They include petrels, skuas, and terns, cormorants, and gulls. Several species of land birds occur on the subantarctic islands. The largest and best-known of the Antarctic petrels are the albatrosses, which breed in tussock grass on islands north of the pack ice. With a wing span of 3 m (10 ft), they roam freely over the westerly wind belt of the Southern Ocean.

[G.We.]

Anthrax An acute infectious zoonotic disease caused by the bacterium *Bacillus anthracis* and primarily associated with herbivorous mammals. Carnivorous mammals, birds, reptiles, amphibians, fish, and insects are generally resistant to anthrax infection. However, carnivorous and omnivorous mammals often succumb after ingestion of infected meat containing the anthrax toxins, which can cause swelling in the throat and suffocation. Humans primarily present with cutaneous lesions, appearing as black scabs or eschars, after contact with infected animals, carcasses, or animal products. *See* ZOONOSES.

Anthrax is responsible for the deaths of thousands of domesticated and wild herbivorous animals annually. Parts of Africa, Asia, southern Europe, and North and South America are subject to repeated outbreaks. In the Western Hemisphere, anthrax is well controlled in livestock.

Bacillus anthracis is a gram-positive, rod-shaped, endospore-forming bacterium, approximately 1.0–1.2 micrometers in diameter and 3–8 μm long. The spores resist drying, cold, heat, and disinfectants, and can remain viable for many years in soil, water, and animal hides and products. *Bacillus anthracis* possesses three virulence factors: lethal toxin, edema toxin, and a poly-D-glutamic acid capsule. Lethal toxin is composed of two proteins, lethal factor and protective antigen. The protective antigen is produced by the anthrax bacillus at a molecular weight of 83 kDa, but must be cleaved by either serum or target cell surface proteases to 63 kDa before it complexes with lethal factor to form lethal toxin. The edema toxin is composed of edema factor and protective antigen, and it is believed to complex in a manner similar to that seen for lethal toxin. Protective antigen plays a central role in that it is required for transport of lethal factor and edema factor into host target cells. The macrophage appears to be the primary host target cell for lethal toxin, whereas the neutrophil appears to be the target cell for edema toxin in addition to other cells involved in edema formation. The third virulence factor is the capsule, which inhibits phagocytosis through its negatively charged poly-D-glutamic acid composition. All three toxin components are encoded by a plasmid, pXO1, whereas the enzymes required for capsule synthesis are encoded for by the pXO2 plasmid. Strains lacking either or both plasmids are avirulent, such as the veterinary vaccine Sterne strain, which lacks the pXO2 plasmid.

Anthrax consists of two clinical forms, cutaneous and septicemic. The cutaneous form begins as a blisterlike lesion that eventually becomes an intensely dark, relatively painless, edematous lesion forming a black eschar. The lesions rapidly become sterile after antibiotic therapy and take several weeks to resolve, even with treatment. The cutaneous form is reported only in humans, rabbits, swine, and horses.

The septicemic form arises from various initial sites of infection, including cutaneous, oropharyngeal, gastrointestinal, or inhalational exposures. The course of septicemic disease depends on the exposure route and the susceptibility of the animal host. The vast majority of systemic anthrax cases in herbivorus animals occur from trauma to mucosal linings of the mouth and upper alimentary canal caused by ingested fibrous foods. Inhalation anthrax is believed to be initiated by phagocytosis of spores within the lungs by alveolar macrophages. Spore-laden macrophages pass through lymphatic channels to the sinuses of regional lymph nodes or migrate to the spleen, where the spores germinate within the macrophages, multiply, and overwhelm and escape the macrophages to invade the efferent lymphatics. For other portals of entry, mesenteric lymph nodes become involved. The bacilli move to the spleen, where they induce pronounced splenomegaly (enlargement of the spleen), and finally enter the bloodstream, where they induce secondary sites of infection, massive bacillemia, toxemia, and sudden death. Failure of the blood to clot, hemorrhages of skin, hemorrhagic meningitis, and reduced rigor mortis are frequently found in anthrax-infected carcasses. Exposure of contaminated body fluids to the lower atmospheric levels of carbon dioxide results in sporulation of the bacilli. Therefore, opening of infected carcasses should be avoided.

Besides its central role for binding the lethal and edema toxins to target cells, protective antigen plays an important role in the host's protective immune response against anthrax, hence the term protective antigen. Vaccines lacking protective antigen are not protective. For United States and United Kingdom human anthrax vaccines, protective antigen bound to aluminum salts is the principal immunogen. However, veterinary vaccines are composed of viable spores of *B. anthracis* Sterne strain, a nonencapsulated toxigenic variant. Full protection against anthrax with the veterinary vaccine is afforded by primary and annual booster vaccinations. *See* INFECTIOUS DISEASE; MEDICAL BACTERIOLOGY. [J.W.Ez.]

Antimicrobial agents Chemical compounds biosynthetically or synthetically produced which either destroy or usefully suppress the growth or metabolism of a variety of microscopic or submicroscopic forms of life. On the basis of their primary activity, they are more specifically called antibacterial, antifungal, antiprotozoal, antiparasitic, or antiviral agents. Antibacterials which destroy are bactericides or germicides; those which merely suppress growth are bacteriostatic agents. *See* FUNGISTAT AND FUNGICIDE.

Of the thousands of antimicrobial agents, only a small number are safe chemotherapeutic agents, effective in controlling infectious diseases in plants, animals, and humans. A much larger number are used in almost every phase of human activity: in

Common antimicrobial agents and their uses

Use	Agents
Chemotherapeutics (animals and humans)	
Antibacterials	Sulfonamides, isoniazid, *p*-aminosalicylic acid, penicillin, streptomycin, tetracyclines, chloramphenicol, erythromycin, novobiocin, neomycin, bacitracin, polymyxin
Antiparasitics (humans)	Emetine, quinine
Antiparasitics (animal)	Hygromycin, phenothiazine, piperazine
Antifungals	Griseofulvin, nystatin
Chemotherapeutics (plants)	Captan (*N*-trichlorothio-tetrahydrophthalimide), maneb (manganese ethylene bisdithiocarbamate), thiram (tetramethylthiuram disulfide)
Skin disinfectants	Alcohols, iodine, mercurials, silver compounds, quaternary ammonium compounds, neomycin
Water disinfectants	Chlorine, sodium hypochlorite
Air disinfectants	Propylene glycol, lactic acid, glycolic acid, levulinic acid
Gaseous disinfectants	Ethylene oxide, β-propiolactone, formaldehyde
Clothing disinfectants	Neomycin
Animal-growth stimulants	Penicillin, streptomycin, bacitracin, tetracyclines, hygromycin
Food preservatives	Sodium benzoate, tetracycline

agriculture, food preservation, and water, skin, and air disinfection. A compilation of some common uses for antimicrobials is shown in the table.

The most important antimicrobial discovery of all time, that of the chemotherapeutic value of penicillin, was made in 1938. In the next 20 years, more than a score of new and useful microbially produced antimicrobials entered into daily use. New synthetic antimicrobials are found today by synthesis of a wide variety of compounds, followed by broad screening against many microorganisms. Biosynthetic antimicrobials, although first found in bacteria, fungi, and plants, are now being discovered primarily in actinomycetes.

Antimicrobial agents contain various functional groups. No particular structural type seems to favor antimicrobial activity. The search for correlation of structure with biological activity goes on, but no rules have yet appeared with which to forecast activity from contemplated structural changes. On the contrary, minor modifications may lead to unexpected loss of activity. [G.M.S.]

Aphid One of a group of mostly soft-bodied plant-feeding insects of the suborder Homoptera, superfamily Aphidoidea. The worldwide fauna of over 4000 species is most abundant in north temperate regions. Aphids feed on phloem sap from vascular plants, tapping it through a feeding tube formed from modified mandibles and maxillae called stylets. In so doing they may transmit viruses from plant to plant, spreading serious disease in crops such as potatoes, cereals, sugarbeet, and citrus. Plants sometimes react to aphid feeding by forming galls in which the aphids live protected from drought and enemies. The so-called Chinese gall is valued in commerce for its high tannin content.

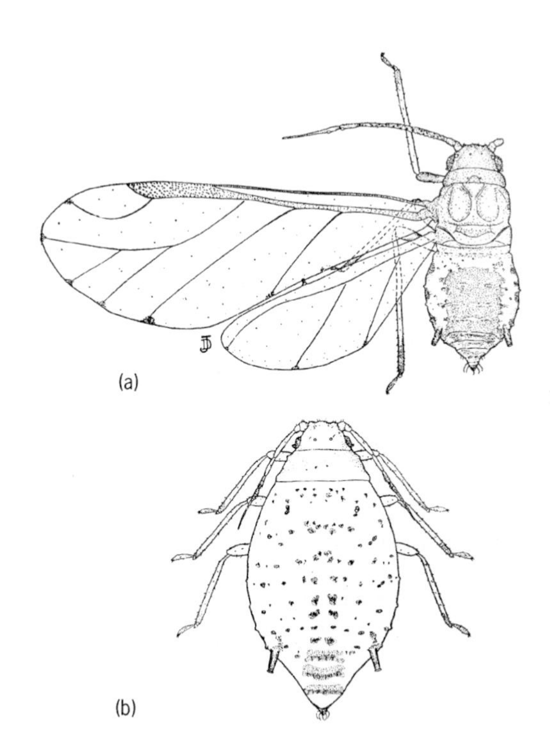

Viviparae of the tulip bulb aphid: (*a*) the winged form and (*b*) the wingless form. (*After J. Davidson, On some aphids infesting tulips, Bull. Entomol. Res., 18:51–62, 1927*)

Aphids have evolved complex life styles to exploit the changing growth phases of plants. Many divide their yearly cycle by flying between a primary host, on which sexual forms mate and lay winter eggs, and a secondary host, where only parthenogenetic females multiply. Only one generation of males and sexual oviparous females occurs each year, usually in autumn. Most parthenogenetic females are also viviparous, and reproduce very rapidly under favorable conditions. Viviparae are winged or wingless (see illustration). Development of young aphids can be switched toward either wingedness or sexuality by outside factors, such as crowding, decreasing temperature, or shortening days. Aphids in the tropics often remain wholly parthenogenetic. [H.L.G.S.]

Apostomatida A group of ciliates comprising an order of the Holotrichia. The majority occur as commensals on marine crustaceans. Their life histories may become exceedingly complicated, and they appear to bear a direct relationship to the molting cycles of their hosts. Apostomes are particularly characterized by the presence of a

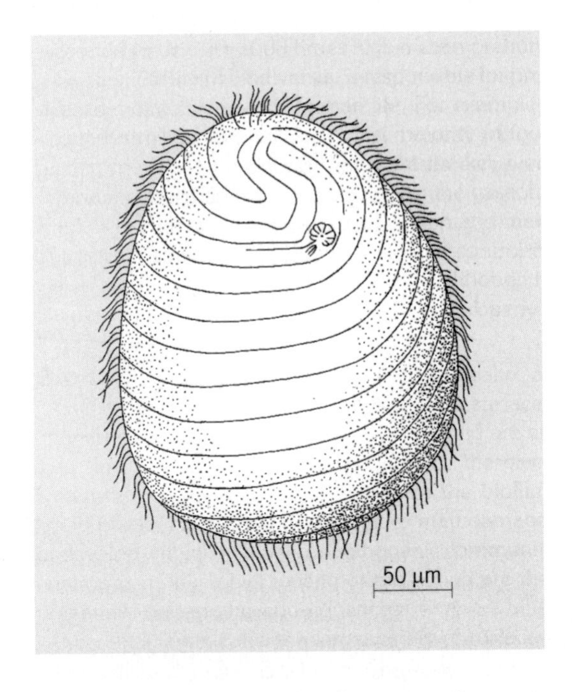

***Foettingeria*, an example of an apostomatid.**

unique rosette in the vicinity of an inconspicuous mouth opening and the possession of only a small number of ciliary rows wound around the body in a spiral fashion (see illustration). *See* PROTOZOA. [J.O.C.]

Aquaculture The cultivation of fresh-water and marine species (the latter type is often referred to as mariculture). Aquacultural ventures occur worldwide. China grows macroalgae (seaweeds) and carp. Japan cultures a wide range of marine organisms, including yellowtail, sea bream, salmonids, tuna, penaeid shrimp, oysters, scallops, abalone, and algae. Russia concentrates on the culture of fish such as sturgeon, salmon, and carp. North America grows catfish, trout, salmon, oysters, and penaeid shrimp. Europe cultures flatfish, trout, oysters, mussels, and eels. Presently, plant aquaculture is almost exclusively restricted to Japan, China, and Korea, where the national diets include substantial amounts of macroalgae.

The worldwide practice of aquaculture runs the gamut from low-technology extensive methods to highly intensive systems. At one extreme, extensive aquaculture can be little more than contained stock replenishment, using natural bodies of water such as coastal embayments, where few if any alterations of the environment are made. Such culture usually requires a low degree of management and low investment and operating costs; it generally results in low yields per unit area. At the other extreme, intensive aquaculture, animals are grown in systems such as tanks and raceways, where the support parameters are carefully controlled and dependence on the natural environment is minimal. Such systems require a high degree of management and usually involve substantial investment and operating costs, resulting in high yields per unit area.

A unique combination of highly intensive and extensive aquaculture occurs in ocean ranching, as commonly employed with anadromous fish (which return from the ocean to rivers at the time of spawning). The two most notable examples are the ranching of salmon and sturgeon. In both instances, highly sophisticated hatchery systems are used to rear young fish, which are then released to forage and grow in their natural environment. The animals are harvested upon return to their native rivers.

Intensive aquaculture brings with it high energy costs, necessitating the design of energy-efficient systems. As this trend continues, aquaculture will shift more to a year-round, mass-production industry using the least amount of land and water possible. With this change to high technology and dense culturing, considerable knowledge and manipulation of the life cycles and requirements of each species are necessary. Specifically, industrialized aquaculture has mandated the development of reproductive control, hatchery technology, feeds technology, disease control, and systems engineering.

Regardless of the type of system used, aquacultural products are marketed as are fisheries products (which are caught in the ocean), except for some advantages. For one, fisheries products often must be transported on boats and may experience spoilage; whereas cultured products, which are land-based, can be delivered fresh to the various nearby markets. Also, intensively cultured products through genetic selection can result in a more desirable food than those caught in the wild, with uniform size and improved taste resulting from controlled feeding and rearing in pollution-free water. *See* AGRICULTURE; MARINE FISHERIES. [W.H.C.; A.B.McG.]

Aquifer A subsurface zone that yields economically important amounts of water to wells. The term is synonymous with water-bearing formation. An aquifer may be porous rock, unconsolidated gravel, fractured rock, or cavernous limestone.

Aquifers are important reservoirs storing large amounts of water relatively free from evaporation loss or pollution. If the annual withdrawal from an aquifer regularly exceeds the replenishment from rainfall or seepage from streams, the water stored in the aquifer will be depleted. This mining of groundwater results in increased pumping costs and sometimes pollution from sea water or adjacent saline aquifers. Lowering the piezometric pressure in an unconsolidated artesian aquifer by overpumping may cause the aquifer and confining layers of silt or clay to be compressed under the weight of the overlying material. The resulting subsidence of the ground surface may cause structural damage to buildings, altered drainage paths, increased flooding, damage to wells, and other problems. *See* ARTESIAN SYSTEMS. [R.K.Li.]

Araeolaimida An order of nematodes in which the amphids are simple spirals that appear as elongate loops, shepherd's crooks, question marks, or circular forms. The cephalic sensilla are often separated into three circlets: the first two are papilliform or the second coniform, and the third is usually setiform; rarely are the second and third whorls combined. Body annulation is simple. The stoma is anteriorly funnel shaped and posteriorly tubular; rarely is it armed. Usually the esophagus ends in a bulb that may be valved. In all but a few taxa the females have paired gonads. Male preanal supplements are generally tubular, rarely papilloid.

There are three araeolaimid superfamilies: Araeolaimoidea, Axonolaimoidea, and Plectoidea. The distinguishing characteristics of the Araeolaimoidea are in the amphids (sensory receptors), stoma, and esophagus. The amphids are in the form of simple spirals, elongate loops, or hooks. Although araeolaimoids are chiefly found in the marine environment, many species have been collected from fresh water and soil.

The amphids in the Axonolaimoidea are generally prominent features of the anterior end, visible as a single-turn loop of a wide sausage shape. Feeding habits are unknown. All known species occur in marine or brackish-water environments.

Plectoidea comprise small free-living nematodes, found mainly in terrestrial habitats, frequently in moss; some are fresh-water, and a few are marine. Those that inhabit moss cushions can withstand lengthy desiccation. For most, the feeding habit is unconfirmed; where it is known, they are microbivorous. Many are easily raised on agar cultures that support bacteria. *See* NEMATA. [A.R.M.]

Arboretum An area set aside for the cultivation of trees and shrubs for educational and scientific purposes. An arboretum differs from a botanical garden in emphasizing woody plants, whereas a botanical garden includes investigation of the growth and development of herbaceous plants as well as trees and shrubs. The largest of the arboretums in the United States is the Arnold Arboretum of Harvard University, founded in 1872. *See* BOTANICAL GARDENS. [E.L.C.]

Arboriculture A branch of horticulture concerned with the selection, planting, and care of woody perennial plants. Knowing the potential form and size of plants is essential to effective landscape planning as well as to the care needed for plants. Arborists are concerned primarily with trees since they become large, are long-lived, and dominate landscapes both visually and functionally.

Plants can provide privacy, define space, and progressively reveal vistas; they can be used to reduce glare, direct traffic, reduce soil erosion, filter air, and attenuate noise; and they can be positioned so as to modify the intensity and direction of wind. They also influence the microclimate by evaporative cooling and interception of the Sun's rays, as well as by reflection and reradiation. Certain plants, however, can cause human irritations with their pollen, leaf pubescence, toxic sap, and strong fragrances from flowers and fruit. Additionally, trees can be dangerous and costly: branches can fall, and roots can clog sewers and break paving. [R.W.Ha.]

Arborvitae A plant, sometimes called the tree of life, belonging to the genus *Thuja* of the order Pinales (Coniferales). It is characterized by flattened branchlets with two types of scalelike leaves. At the edges of the branchlets the leaves may be keeled or rounded; on the upper and lower surfaces they are flat, and often have resin glands. The cones, about 1/2 in. (1.25 cm) long, have the scales attached to a central axis.

The tree is valued both for its wood and as an ornamental. *Thuja occidentalis*, of the eastern United States, is known as the northern white cedar. It occurs in moist or swampy soil from Nova Scotia to Manitoba and in adjacent areas of the United States, and extends south in the Appalachians to North Carolina and Tennessee. Other important species include the giant arborvitae (*T. plicata*); oriental arborvitae (*T. orientalis*); and Japanese arborvitae (*T. standishii*). Among the horticultural forms are the dwarf pendulous and juvenile varieties. *See* FOREST AND FORESTRY; TREE. [A.H.G./K.P.D.]

Arboviral encephalitides A number of diseases, such as St. Louis, Japanese B, and equine encephalitis, which are caused by arthropod-borne viruses (abbreviated "arboviruses"). In their most severe human forms, the diseases invade the central nervous system and produce brain damage, with mental confusion, convulsions, and coma; death or serious aftereffects are frequent in severe cases. Inapparent infections are common.

The arbovirus "group" comprises more than 250 different viruses, many of them differing fundamentally from each other except in their ecological property of being transmitted through the bite of an arthropod. A large number of arboviruses of antigenic groups A and B are placed in the family Togaviridae, in two genera, alphavirus (serological group A) and flavivirus (serological group B). Still other arboviruses, related structurally and antigenically to one another but unrelated to Togaviridae, are included in the family Bunyaviridae, consisting chiefly of the numerous members of the Bunyamwera supergroup—a large assemblage of arboviruses in several antigenic groups which are cross-linked by subtle interrelationships between individual members. The nucleic acid genomes of all arboviruses studied thus far have been found to be RNA.

Members of serological group A include western equine encephalitis, eastern equine encephalitis, and Venezuelan equine encephalitis viruses; and Mayaro, Semliki Forest, Chikungunya, and Sindbis viruses, which have nonencephalitic syndromes. Group A viruses are chiefly mosquito-borne. Serological group B viruses include Japanese B, St. Louis, and Murray Valley encephalitis viruses (mosquito-borne), and the viruses of the Russian tick-borne complex, some of which produce encephalitis (Russian spring-summer), whereas others cause hemorrhagic fevers (Omsk, Kyasanur Forest) or other syndromes, such as louping ill. Also in group B are the nonneurotropic viruses of West Nile fever, yellow fever, dengue, and other diseases. *See* LOUPING ILL; YELLOW FEVER.

There is no proved specific treatment. In animals, hyperimmune serum given early may prevent death. Killed virus vaccines have been used in animals and in persons occupationally subjected to high risk. A live, attenuated vaccine against Japanese B encephalitis virus, developed in Japan, has been used experimentally with some success, not only in pigs to reduce amplification of the virus in this important vertebrate reservoir but also in limited trials in humans. In general, however, control of these diseases continues to be chiefly dependent upon elimination of the arthropod vector. *See* ANIMAL VIRUS; VIRUS.

[J.L.Me.]

Archaebacteria A group of prokaryotic organisms that are more closely related to eukaryotes than bacteria. Based on comparative analyses of small subunit ribosomal ribonucleic acid (rRNA) sequences and selected protein sequences, the three primary lines of descent from the common ancestor are the Archaea (archaebacteria), the Bacteria, and the Eucarya (eukaryotes). Although the Archaea look like Bacteria cytologically (they are both prokaryotes), they are not closely related to them. *See* CLASSIFICATION, BIOLOGICAL.

The Archaea can be divided into two evolutionary lineages on the basis of rRNA sequence comparisons, the Crenarchaeotae and the Euryarchaeotae. The crenarchaeotes are organisms that grow at high temperatures (thermophiles) and metabolize elemental sulfur. Most are strict anaerobes that reduce sulfur to hydrogen sulfide (sulfidogens), but a few can grow aerobically and oxidize sulfur to sulfuric acid. The euryarchaeotes have a number of different phenotypes. *Thermococcus* and *Pyrococcus* are sulfidogens like many crenarchaeotes. *Archaeoglobus* reduces sulfate to sulfide. *Thermoplasma* grows under acidic conditions aerobically or anaerobically (as a sulfidogen). Many euryarchaeotes are methane-producing anaerobes (methanogens) and some grow aerobically in the presence of very high concentrations of salt (halophiles). *See* METHANOGENESIS (BACTERIA).

The thermophilic archaea are found in high-temperature environments around the world. They have been isolated from soils and shallow marine sediments heated by nearby volcanoes and from deep-sea hydrothermal vents. Some are used as a source for heat-stable enzymes useful for industrial applications. The methanogenic archaea inhabit the digestive tracts of animals (especially ruminants like cows), sewage sludge digesters, swamps (where they produce marsh gas), and sediments of marine and fresh-water environments. They are of interest commercially because of their ability to produce methane from municipal garbage and some industrial wastes. Halophilic archaea live in the Great Salt Lake, the Dead Sea, alkaline salt lakes of Africa, and salt-preserved fish and animal hides. They are also commonly found in pools used to evaporate seawater to obtain salt.

The discovery of the Archaea caused a major revision in the understanding of evolutionary history. It had previously been thought that all prokaryotes belonged to one evolutionary lineage. Since their cellular organization is simpler, prokaryotes were assumed to be ancestors of eukaryotes. The discovery of the relationship of the Archaea to the Eucarya revealed that prokaryotes do not comprise a monophyletic group since they can be divided into two distinct lineages. Although the three descended from a common ancestor, modern eukaryotes may have arisen from fusions of bacterial and archaeal endosymbionts with ancestral eukaryotes. Chloroplasts and mitochondria arose from free-living bacteria which became endosymbionts. The discovery of the Archaea has also given microbiologists a better picture of the common ancestor. The deepest-branching eukaryotes (like *Giardia*) are strict anaerobes that lack mitochondria, and they diverged much later than the deepest-branching bacteria and archaea. The earliest archaea and bacteria (*Thermotoga* and *Aquifex*) are also anaerobes and are also extreme thermophiles. Therefore the common ancestor of these groups was probably also an extremely thermophilic anaerobe. Therefore, it is possible that life may have arisen in a relatively hot environment, perhaps like that found in deep-sea hydrothermal vents. *See* BACTERIA; PROKARYOTAE. [K.M.N.]

Arctic and subarctic islands Defined primarily by climatic rather than latitudinal criteria, arctic islands are those in the Norhern Hemisphere where the mean temperature of the warmest month does not exceed 50°F (10°C) and that of the coldest is not above 32°F (0°C). Subarctic islands are those in the Northern Hemisphere where the mean temperature of the warmest month is over 50°F (10°C) for less than 4 months and that of the coldest is less than 32°F (0°C). Such islands generally are in high latitudes. Distribution of land and sea masses, ocean currents, and atmospheric circulation greatly modifies the effect of latitude so that it is often misleading to use location relative to the Arctic Circle as a significant criterion of arctic or subarctic. The largest proportion by area of the islands lies in the Western Hemisphere, primarily in Greenland and in the Canadian Arctic Archipelago. Within this general description, individual islands vary considerably (see table).

Physiographically, the islands include all the varied major landforms found elsewhere in the world, from rugged mountains over 8000 ft (2500 m) high, through plateaus and hills, to level plains only recently emerged from the sea. All have been glaciated except Sakhalin and some of the islands in the Bering Sea sector. Removal of the weight of ice sheets and the resultant crustal rebound has exposed prominent marine beaches and wave-cut cliffs on many of the islands. These now commonly occur at elevations of over 300 ft (150 m) above sea level.

Size of larger arctic and subarctic Islands*

Name	Area	
	mi^2	km^2
Aleutian Is.		
Unimak I.	15,500	40,100
Unalaska I.	10,800	28,000
St. Lawrence I.	18,200	47,100
Nunivak I.	16,000	41,400
Kodiak I.	37,400	96,900
Canadian Arctic Archipelago	500,000	1,295,000
Baffin I.	196,000	507,000
Ellesmere I.	76,000	197,000
Victoria	84,000	217,000
Banks	27,000	70,000
Devon	21,000	55,000
Axel Heiberg	17,000	43,000
Melville	16,000	42,000
Southhampton	16,000	42,000
Prince of Wales	13,000	33,033
Newfoundland	42,734	109,000
Greenland	840,000	2,176,000
Iceland	39,961	102,000
Svalbard (archipelago)	24,100	62,000
Vest-Spitsbergen	15,250	39,000
Franz Josef Land (archipelago)	7,000	18,000
Novaya Zemlya (archipelago)	36,000	93,000
Severny I.	21,000	54,000
Yughny I.	15,000	39,000
Severnaya Zemlya (archipelago)	14,000	36,000
New Siberian Is.	12,000	31,000
Wrangel I.	2,000	5,000
Sakhalin I.	27,000	70,000
Kuriloe Is.	6,000	16,000

*Approximate only in some cases because of incomplete mapping.

The general climatic pattern of these islands is set by their location relative to the two semipermanent centers of low pressure over the Aleutian Islands and over Iceland. Most of the precipitation is cyclonic in origin. Because they are marine areas, the islands receive more precipitation than they otherwise would, yet even so this is very light for most of the arctic islands removed from the zone of cyclonic activity. Also, because they are marine areas, the islands, regions of low temperatures by definition, are not regions of extreme low temperatures. In general, the larger the island and the closer its proximity to a continental landmass, the higher are the summer temperatures and the lower its winter temperature. *See* POLAR METEOROLOGY.

The climatic differences between arctic and subarctic islands are reflected in their natural vegetation. The arctic islands are treeless. Natural vegetation consists of the tundra—mosses, sedges, lichens, grasses, and creeping shrubs. Bare ground is often exposed and in some places plant growth may be lacking completely except for a few rock-encrusting lichens. In such places the ground surface may consist of frost-shattered rock fragments, tidal mud flats, boulder-strewn fell fields, or snow patches and ice. Permafrost (permanently frozen ground) occurs throughout the Arctic (and in parts of the subarctic) and is reflected in impeded drainage and patterned ground. *See* PERMAFROST; TUNDRA.

The natural vegetation of subarctic islands characteristically is the boreal forest or taiga, composed predominantly of conifers such as spruce, fir, pine, and larch with deciduous trees such as birch, aspen, and willow; the latter are especially common in regrowth of clearings in the forest. Impeded drainage because of permafrost or glaciation gives rise to numerous ponds and muskeg areas. A transitional type of vegetation, the forest-tundra, is recognized on some subarctic islands in sectors where smaller trees are widely spaced and abundant mosses cover the ground. *See* MUSKEG; TAIGA.

The typical soils of the subarctic islands are podzols—the grayish-white surface soil beneath the raw humus layer and highly acidic in nature. The tundra soils of the arctic islands really consist only of a dark-brown peaty surface layer over poorly defined thin horizons, and much of the ground cannot properly be termed soil. [W.C.Wo.]

Arctic Circle The parallel of latitude approximately $66^1/_2{}^\circ$ (66.55°) north of the Equator, or $23^1/_2{}^\circ$ from the North Pole. The Arctic Circle has the same angular distance from the Equator as the inclination of the Earth's axis from the plane of the ecliptic. Thus, when the Earth in its orbit is at the Northern Hemisphere summer solstice, June 21, and the North Pole is tilted $23^1/_2{}^\circ$ toward the Sun, the Sun's rays extend beyond the pole $23^1/_2{}^\circ$ to the Arctic Circle, giving that parallel 24 h of sunlight. On this same date the Sun's rays at noon will just reach the horizon at the Antarctic Circle, $66^1/_2{}^\circ$ south. The highest altitude of the noon Sun at the Arctic Circle is on June 21, when it is 47° above the horizon.

At the Arctic Circle the Sun remains above the horizon continuously only 24 h at the longest period. However, with twilight considered, it remains daylight or twilight continuously for about 5 months. Twilight can be considered to last until the Sun drops 18° below the horizon. *See* MATHEMATICAL GEOGRAPHY. [V.H.E.]

Arctic Ocean The north polar ocean lying between North Armerica and Asia, extending over about 386,000 mi² (10^6 km²). It is nearly completely covered by 6–9 ft (2–3 m) of ice in winter, and in summer it becomes substantially open only at its peripheries. Its extent has been variably defined, but it is oceanogaphically appropriate to consider it bounded on the south by a line running from northern Greenland through Smith, Jones, and Lancaster sounds, along northwestern Baffin Island to the Canadian mainland, thence to the Alaskan coast, across Bering Strait, along the Siberian coast to Novaya Zemlya, across to Franz Josef Land and Spitsbergen, and over to northern Greenland. This definition omits the Barents, Norwegian, and Greenland seas and Baffin Bay, which have a pronounced North Atlantic character.

The central polar basin, somewhat triangular in shape, is surrounded by continental shelves which are interrupted only by the deep passage running through Fram Strait. The upper 650 ft (200 m) of the Arctic Ocean, referred to as Surface Water or Arctic Water, is characterized by a significant density stratification produced by the strong increase in salinity downward from the surface. This density stratification is of considerable importance, for it prevents a deep-reaching convection from developing within the Arctic Ocean and also prevents the heat of the underlying warm Atlantic Water from reaching the surface. The relatively low salinity at the surface is maintained against the upward diffusion of salt by the addition of fresh water, principally through river outflow. The upper 100–160 ft (30–50 m) of Surface Water tends to be relatively uniform vertically in temperature and salinity. Except for areas which become ice-free in summer, the water will be near the freezing point. Currents in the upper waters tend to be relatively slow (4 in./s or 10 cm/s or less), and they are similar in both speed and direction to the ice motion. The overall circulation in the upper waters has its ultimate cause in the prevailing wind pattern over the Arctic Ocean.

As in other oceans, the current at any instant can vary greatly from the mean condition. The most spectacular example observed in the Arctic Ocean occurs on an occasional basis in the Canadian Basin, consisting of a high-speed current core. *See* OCEAN CIRCULATION; SEAWATER.

Below the Surface Water, the temperature increases to a maximum, which over most of the region is about 33°F (0.5°C) and lies between 1000 and 1500 ft (300 and 500 m). The salinity is nearly uniform, and since at low temperatures the density of seawater depends almost solely on salinity, there is virtually no density stratification beneath the upper waters. Significant deviations from the stated temperature occur only in the southern Eurasian Basin closest to Spitsbergen, for it is there that the warm and saline water (called Atlantic Water) which maintains the temperature maximum throughout the Arctic Ocean first enters. This water has its origin in the North Atlantic. Once into the Arctic Ocean it sinks because of its high salinity and moves eastward along the Eurasian continental slope. Beneath the Atlantic Water lies cold, nearly uniform Bottom Water. These two water masses together constitute over 90% of the volume of the Arctic Ocean. The Bottom Water is formed in the Greenland Sea. [K.A.]

Artesian systems Groundwater conditions formed by water-bearing rocks (aquifers) in which the water is confined above and below by impermeable beds. Because the water table in the intake area of an artesian system is higher than the top of the aquifer in its artesian portion, the water is under sufficient head to cause it to rise in a well above the top of the aquifer. Many of the systems have sufficient head to cause the water to overflow at the surface, at least where the land surface is relatively low. Flowing artesian wells were extremely important during the early days of the development of groundwater from drilled wells, because there was no need for pumping. Their importance has diminished with the decline of head that has occurred in many artesian systems and with the development of efficient pumps and cheap power with which to operate the pumps. [A.N.S./R.K.Li.]

Arthropoda A phylum that includes the well-known insects, spiders, ticks, and crustaceans, as well as many smaller groups, some of which are known only as fossils. Arthropodous animals make up about 75% of all animals that have been described. The estimated number of known species exceeds 780,000. Of this number the class Insecta alone contains about 700,000 described species. Arthropods vary in size from the microscopic mites to the giant decapod crustaceans, such as the Japanese crab with an appendage span of 5 ft (1.5 m) or more.

The adult arthropod typically has a body composed of a series of ringlike segments, muscularly movable on each other. The integument is sclerotized by the formation of hardening substances in the cuticle, and the segmental limbs are many-jointed. These characteristics, taken together, distinguish the arthropods from all other animals. Young stages may be quite different from the adults, and some parasitic species differ very radically from their relatives.

Arthropod evolution is no longer the clear-cut subdivision of a single phylum, Arthropoda, into three structurally divergent subphyla. Advances in functional morphology, comparative embryology, spermatology, serology, and paleontology have brought an array of new hypotheses about relationships of arthropodous animals. At the center of debate is the question of monophyly versus polyphyly: Did all arthropodous animals evolve from a common ancestor or did several distinct lineages evolve along similar pathways? Two opposing classification schemes are presented; numerous variations on these schemes can be found in the literature. The first pair of classifications is as follows:

Phylum Uniramia
  Subphylum: Onychophora
    Myriapoda
    Hexapoda (Insecta)
Phylum Trilobita (Trilobitomorpha)
Phylum Crustacea
Phylum Chelicerata

Versus

Phylum Arthropoda
  Subphylum Arachnata
    Superclass: Trilobita
      Chelicerata
  Subphylum Mandibulata
    Superclass: Crustacea
      Myriapoda
      Insecta
Phylum Onychophora

Alternatively, a slightly different and expanded pair of classifications is as follows:

Phylum Uniramia
  Subphylum Onychophora
  Subphylum Myriapoda
    Class: Chilopoda
      Diplopoda
      Symphyla
      Pauropoda
      Arthropleurida
  Subphylum Hexapoda
    Class: Protura
      Collembola
      Diplura
      Thysanura
      Pterygota (Insecta)
Phylum Crustacea
    Class: Cephalocarida
      Remipedia
      Branchiopoda
      Ostracoda
      Tantulocarida
      Maxillopoda
      Malacostraca
Phylum Cheliceriformes
  Subphylum Pycnogonida
  Subphylum Chelicerata
    Class: Merostomata
      Arachnida

Phylum Trilobitomorpha
Class: Trilobitoidea
Trilobita

Versus

Phylum Onychophora
Phylum Arthropoda
Subphylum Cheliceromorpha
Infraphylum: Pycnogonida
Chelicerata
Superclass: Xiphosurida
Cryptopneustida
Class: Eurypterida
Archnida
Subphylum Gnathomorpha
Infraphylum: Trilobitomorpha
Class: Trilobita
Trilobitodea
Infraphylum: Mandibulata
Class: Cheloniellida
Crustacea
Myriapoda
Insecta

Body segmentation, or metamerism, is the most fundamental character of the arthropods, but it is shared by the annelid worms, so there can be little doubt that these two groups of animals are related. The limbs of all modern arthropods develop in the embryo from small lateroventral outgrowths of the body segments that lengthen and become jointed. Hence it may be inferred that the arthropods originated from some segmented worm that acquired similar lobelike limb rudiments and thus, as a crawling or walking animal, became distinguished from its swimming relatives. Then, with sclerotization of the integument, the limbs could lengthen and finally become jointed, providing greater locomotor efficiency. In their later evolution, some of these limbs became modified for many other purposes, such as feeding, grasping, swimming, respiration, silk spinning, egg laying, and sperm transfer. The body segments, corresponding to specialized sets of appendages, tend to become consolidated or united in groups, or tagmata, forming differentiated body regions, such as head, thorax, and abdomen. Annelida; Metameres.

Sclerotization of the cuticle may be continuous around the segments. More usually, it forms discrete segmental plates, or sclerites. A back plate of a segment is a tergum, or notum; a ventral plate is a sternum; and lateral plates are pleura. The consecutive tergal and sternal plates, unless secondarily united, are connected by infolded membranes, and are thus movable on each other by longitudinal muscles attached on anterior marginal ridges of the plates. Since nearly all the body and limb muscles are attached on integumental sclerites, there is little limit to the development of skeletomuscular mechanisms.

All arthropods have all the internal organs essential to any complex animal. An alimentary canal extends either straight or coiled from the subapical ventral mouth to the terminal anus. Its primary part is the endodermal stomach, or mesenteron, but there are added ectodermal ingrowths that form a stomodeum anteriorly and a proctodeum posteriorly. The nervous system includes a brain and a subesophageal ganglion in the head, united by connectives around the stomodeum, and a ventral nerve cord of interconnected ganglia. Some of the successive ganglia, however, may be condensed into composite ganglionic masses. Nerves proceed from the ganglia. Internal proprioceptors and surface sense organs of numerous kinds are present, chiefly tactile, olfactory, and optic. A usually tubular pulsatory heart lies along the dorsal side of the body and keeps the blood in circulation. In some arthropods arteries distribute the blood from the heart; in others it is discharged from the anterior end of the tube directly into the body cavity. The blood reenters the heart through openings along its sides.

Aquatic arthropods breathe by means of gills. Most terrestrial species have either flat air pouches or tubular tracheae opening from the outside surface; some have both. A few small, soft-bodied forms respire through the skin. Excretory organs open either at the bases of some of the appendages or into the alimentary canal. Most arthropods have separate sexes, but some are hermaphroditic, and parthenogenesis is of common occurrence. The genital openings differ in position in different groups and are not always on the same body segment in the two sexes. *See* INSECTA. [J.C.Ro.]

Ascaridida An order of nematodes in which the oral opening is generally surrounded by three or six labia; in some taxa labia are absent, but the cephalic sensilla are always evident. Usually there are eight cephalic or labial sensilla; the submedians may be fused and then only four sensilla are seen. The stoma varies from being completely reduced to spacious or globose. The esophagus varies from club shaped to nearly cylindrical, never rhabditoid. There may be posterior esophageal or anterior intestinal ceca. The collecting tubules of the excretory system may extend posteriorly and anteriorly. Males generally have two spicules; however, in some taxa there may be none or only one. The gubernaculum may also be present or absent. Though females generally have two ovaries, multiple ovaries do occur. The number of uteri is also variable: two, three, four, or six. Phasmids are sometimes large and pocketlike. Reportedly, the larvae lack a stomatal hook or barb.

The order probably comprises seven superfamilies: Ascaridoidea, Seuratoidea, Camallanoidea, Dracunculoidea, Subuluroidea, Dioctophymatoidea, and Muspiceoidea (*incertae sedis*).

The Ascaridoidea include about 65 genera which comprise large parasitic roundworms whose adult stages usually occur in the stomach or small intestine of terrestrial and aquatic mammals, birds, reptiles, and fishes; the parasitic larval stages of many species occur, either temporarily or indefinitely, in other parts of the host's body.

Many species have a direct life cycle. Others, mainly species with marine mammals, birds, and fishes as definitive hosts, require an intermediate host, such as a fish, amphibian, insect, crustacean, or small mammal. Infestation is typically characerized by pulmonary damage and distress initially, and digestive disturbances later. Damage may also occur during larval migration to other parts of the body, including the liver and brain.

Dracunculoidea, the superfamily of parasitic nematodes, comprise obligate tissue parasites of fishes, reptiles, and mammals. All known species require an intermediate host in order to complete their life cycle, and that host is always a water flea (*Cyclops*). The most widely known example is *Dracunculus medinensis*, the guinea worm.

Ingestion of water containing infective *Cyclops* is the only known source of infection. The encysted nematode larvae are released from *Cyclops* by the digestive juices of the duodenum. Then the larvae burrow through the intestinal wall, and upon reaching the loose connective tissue, they develop to adulthood in 8 months to 1 year.

The gravid females, 28–48 in. (70–120 cm) long, migrate from the site of development to the surface of the skin, and a papule is formed, then a blister, usually on the lower extremities. When the blister comes in contact with fresh water, the uterus bursts through the anterior part of the nematode's body, and the worm also bursts, releasing cloudlike swarms of motile larvae. These larvae are then filtered from the water by *Cyclops* and subsequently ingested.

The formation of the blister and subsequent rupturing of the female produce a profound allergic reaction. This reaction results from the release of large amounts of toxic by-products from the worm. Upon discharge of larvae in fresh water, much of the allergic reaction abates. The reactions and systemic prodromes include erythema, urticarial rash, pruritus, vomiting, diarrhea, and giddiness. Septicemia, suppurating cysts, and chronic abscesses are not uncommonly associated with these infections. The worms can be removed surgically, or in the native manner of winding upon a stick. Chemotherapy is also available.

Control in endemic areas includes keeping infected persons from wading or bathing in water used for drinking purposes, and the education to avoid drinking suspect water. *See* NEMATA. [A.R.M.]

Ascomycota A phylum in the kingdom Fungi, representing the largest of the major groups of fungi, and distinguished by the presence of the ascus, a specialized saclike cell in which fusion of nuclei and reduction division occur and the resulting nuclei form ascospores. In most ascomycetes, each ascus contains eight ascospores, but the number may vary from one to several hundred. In the simplest ascomycetes (yeasts), the vegetative body (thallus) is unicellular; however, in the majority of ascomycetes, the thallus is more complex and consists of a tubular, threadlike hypha with cross walls which grows in or on the substrate. These hyphae eventually form structures called ascomata (ascocarps), on or in which the asci are formed. In addition to their sexual reproduction, most ascomycetes reproduce asexually by means of conidia.

Traditionally, the structure of the ascoma and ascus has served as the basis for subdividing the Ascomycota into five classes: Hemiascomycetes, Plectomycetes, Pyrenomycetes, Discomycetes, Loculoascomycetes. The introduction of molecular data, however, is changing concepts of the relationships of different groups of ascomycetes and will eventually lead to a much-revised classification scheme.

The ascomycetes occur throughout the world in all types of habitats and on both living and dead substrates. An estimated 33,000 species are arranged in about 3300 genera, with new species being described regularly. Ecologically ascomycetes function as primary decomposers of plant materials, but they also are important as plant and human pathogens; in baking, brewing, and winemaking; in enzyme and acid production; and as sources of antibiotics and other drugs. *See* EUMYCOTA; FUNGI; PLANT PATHOLOGY; YEAST. [R.T.Ha.]

Ash A genus, *Fraxinus*, of deciduous trees of the olive family Oleaceae, order Scrophulariales, which have opposite, pinnate leaflets, except in one species, *F. anomala*, which has only a single leaflet. There are about 65 species in the Northern Hemisphere. This tree occurs in America south to Mexico, in Asia south to Java, and in Europe.

The white ash (*F. americana*), of the eastern United States, has stalked leaflets, rusty-colored winter buds, and an erect trunk that is valuable for lumber. The wood is light, strong, but flexible, and is used for oars, baseball bats, furniture, motor vehicle parts, boxes, baskets, and crates. The black ash (*F. nigra*) grows in wet soils in the northeastern United States and Canada and has sessile leaflets and friable outer bark. The wood of black ash is used for the same purposes as that of white ash. The red ash (*F. pennsylvanica*), also of the eastern United States and adjacent Canada, has pubescent (hairy) twigs and leafstalks. The uses of the wood of this species are also similar to those of white ash. Some species of ash are ornamental trees, such as the flowering ash (*F. ornus*) with gray winter buds and white flowers, and the European ash (*F. excelsior*) with black buds and sessile leaflets. *See* FOREST AND FORESTRY; TREE.

[A.H.G./K.P.D.]

Asia The largest of the world's continents. With its peninsular extension, commonly called the continent of Europe, it is the major portion of the broad east-west extent of the Northern Hemisphere land masses. In many ways Asia is more a cultural concept than a physical entity. There is no logical physical separation between Asia and Europe, and even Africa is separated from Asia merely by the width of the Suez Canal. For convenience, however, the Eurasian land mass is considered to be divided by the Ural Mountains into Europe in the west and Asia in the east. Thus restricted, Asia has an area of about 17,700,000 mi^2 (45,800,000 km^2), about one-third of the land area of the Earth. In the north, Siberia reaches past the 80th latitude. Southward, India and Sri Lanka (Ceylon) reach nearer than 10°N of the Equator, while the Indonesian islands extend more than 10°S of the Equator. The continental heart of Asia is more than 2000 mi (3200 km) from the nearest ocean. *See* CONTINENT; EUROPE.

Topography. In the topographic framework of Asia, the great mountain systems are the most impressive features. From the central knot of the mighty Pamirs and Kopet Dagh in the heart of the continent originate chains radiating in several directions. In the Peter the First Range there are such heights as Qullai Ismoili Somoni, 24,584 ft (7493 m), and Lenin Peak, 23,377 ft (7125 m), above sea level. Running westward through Afghanistan is the Hindu Kush, reaching elevations over 20,000 ft (6100 m). The mountain trendline continues, after a jog northwestward, in the Elburz of northern Iran and thence in the Armenian highlands and the Caucasus, each with elevations reaching 18,000 ft (5500 m), decreasing thereafter to the Pontus and Taurus ranges of northern and southern Turkey. In western and southern Iran are the massive Zagros and Makran ranges.

Southeastward from the Pamir knot run the three most imposing mountain chains on Earth: the Karakorum, which continues the line of the Hindu Kush eastward in an arc convex to the north; the Himalaya in an arc convex to the south; and the shorter Trans-Himalaya, or Nyen-chen Tangla, north of the Himalaya, with higher average elevations but peaks of lesser height. In all of these, the average elevations exceed 4 mi (6400 m), with several scores of peaks reaching a height in excess of 25,000 ft (7600 m) above sea level. Everest, 29,141 ft (8882 m), and Kinchinjunga, 28,146 ft (8579 m), lie in the Himalaya, while the peak designated as K2, 28,250 ft (8611 m), rises in the Karakorum.

In eastern Tibet the Himalaya and Nyen-chen Tangla bend sharply toward the south, and the former is cut through by the gorge of the Brahmaputra River. From the bend zone, great ridges divided by deep gorges run south to form the Burma-China frontiers and the mountain backbones of the Malay peninsula and Vietnam. The Nan-ling system of south China diverges eastward to divide the Yang-tzu (Yangtze) from the Hsi (Si) drainage.

From the western Himalaya, the 11,000-ft (3400-m) Sulaiman Range runs south and, together with the Kirthar Range, divides West Pakistan from Afghanistan.

Beginning at heights over 20,000 ft (6100 m) and branching off from the Karakorum south of Kashgar, the Kuen-lun Mountains run eastward across western China. Genetically they form the longest mountain system of China. With their eastward extensions in the 12,000 ft (3700 m) Ch'in-ling and the lesser Ta-pieh mountains and Huai-yang hills, they reach almost to the Pacific. Together with the northeastward arc of the Altyn Tagh and the Nan Shan branching from it, the Kuen-lun forms the northern wall of the Tibetan plateau. Near the eastern end of the Kuen-lun proper lie the Amne Machin Mountains, with peaks up to 25,000 ft (7600 m) in elevation.

Northeastward of the Pamir knot runs the east-west oriented Tien Shan, over 1000 mi (1600 km) long and maintaining heights of 18,000–20,000 ft (5500–6100 m) over much of its length. Roughly parallel and trending east and west is a series of great ranges to its north, with mutual connections in the west. These include the Altai-Sayan, the Tannu Ola, and the Kentei, which form natural boundaries for Outer Mongolia. They continue the systems of young mountains crossing central Asia; farther northeast, they extend further in the Stanovoi Mountains of Eastern Siberia.

The Asian plateaus are in various stages of erosion and thus present a great variety of landscapes. The Tibetan plateau is a prime example. The western half, because of little rainfall, exhibits a rolling topography with relatively slight local relief except where mountain chains cross it; it is a land of internal drainage basins. Average elevations are over 16,000 ft (4900 m). The eastern half is humid or subhumid and is cut by numerous rivers, producing deep canyons and great ridges. In contrast to this is the Mongolian plateau. This plateau consists mostly of vast, rather level plains 3000–5000 ft (900–1500 m) high, surmounted in places by mountains, and containing broad, shallow basins divided by land swells of low elevation.

Other major topographic units of Asia are blocs of hill lands. Most of southern China and much of southeastern Asia comprise hills which may be roughly defined as slope lands with local relief under 1000–1500 ft (300–450 m) although in absolute elevation they may rise many thousands of feet above sea level. Hilly lands are found to predominate in the northern part of the Indian peninsula and along both flanks of the Indian plateau, where they are called ghats. In southern India are the Nilgiri and Cardomom hills, rising to mountainous elevations of 8000 ft (2400 m). Many parts of different plateaus have hilly regions where erosion has produced uneven local relief, as in the Shan or North Vietnam plateau. Hills are prominent features of southwestern Asia, including eastern Mediterranean regions, such as Israel, Syria, and Lebanon.

The most significant topographic units of Asia are the great alluvial plains and river deltas. The gross drainage pattern of Asia is radial; the rivers flow from the highlands in the heart of the continent and run outward in all directions. Only in the south, east, and north sectors of the continent do the rivers reach the sea. Flowing into the peripheral seas of the Pacific are such mighty rivers as the Mekong, the Hsi, the Yang-tzu, the Huai, the Yellow, and the Amur, each building large, heavily populated plains and, with the exception of the Amur, densely settled deltas. The Yellow Plain (North China Plain), with some 125,000 mi^2 (324,000 km^2) of area, and the Yangtzu Plain, with about 75,000 mi^2 (194,000 km^2), are among the most extensive alluvial plains of the Earth. In the shallow South China, East China, and Yellow seas, the deltas of the first five rivers mentioned above are pushing steadily seaward.

Important sectors of Asia, containing some 200,000,000 people, are completely insular. The most important are the Japanese, Philippine, and Indonesian islands and Taiwan. Almost all of Asia's islands lie in great volcanic arcs bounding large seas off the continent's Pacific coast. At least 160 active volcanoes are found here and in Kamchatka. Few islands lie along the Asiatic coasts of the Indian Ocean, although the Sunda chain of Indonesia has perhaps more of a claim to Indian Ocean frontage than to Pacific frontage. Sri Lanka is the only significant island in the northern part of the Indian Ocean west of Sumatra. In the Persian Gulf off the north coast of Arabia lies the small island Bahrein.

Few islands lie off the alluviated coastlands of northern Siberia. Some moderately large ones are included in the barren and rocky Severnaya Zemlya group, the New Siberian Islands, and Wrangel Island. The Commander Islands and Karaginski Island lie in the Bering Sea only a short distance from the Aleutians.

Climates. Five major climatic types may be distinguished in the Asian region: (1) the monsoonal system of eastern Asia, (2) the monsoonal system of southern Asia, (3) the equatorial regions of southeastern Asia and their extension into the Southern Hemisphere as they are influenced by the Australian monsoon, (4) the winter rainfall areas of southwestern Asia, and (5) the cyclonic and convectional storm systems of central and northern Asia.

Fundamental to understanding the climates of Asia are the vastness of the unbroken landmass and the long latitudinal stretch from the polar realm to south of the Equator. These are responsible for the great temperature and humidity extremes that occur. The greatest ranges of temperatures in the world have been recorded in interior Asia. Continentality, therefore, is the outstanding feature of climates of interior Asia. In coastal and insular areas of east Asia, however, winds moving over the warm, northward-flowing Japan Current and the western Pacific waters moderate the coastland and island climates. *See* MARITIME METEOROLOGY; MONSOON METEOROLOGY.

The driest portions of Asia include the vast areas of southern Mongolia, Hsin-chiang, former Soviet Central Asia, and southwestern Asia. Except for small, favored mountain areas, most of this region from the Gobi to the Red Sea gets less than 10 in. (25 cm) of precipitation per year. With the exception of southern Arabia, which is subtropical desert, these are mid-latitude desert and dry steppe regions. Favored with higher rainfall are the Yemen Mountains and the coastal mountains of Turkey, together with Lebanon, Syria, and northern Israel. The highlands of Armenia and the Elburz of Iran are favored also with more abundant rainfall, which may range from 25 to 50 in. (64 to 127 cm) or more per year.

The northeastern Siberian mountains and the Arctic coastal lands also receive meager rainfall, less than 8 in. (20 cm), but are not dry because evaporation is low and the water table is high. Most of Siberia has permafrost below a few feet of surface soil, so that rainwater does not filter far down into the earth. Between the arid belt of central Asia and the northeast Siberian low-precipitation zone, the annual rainfall ranges between 10 and 18 in. (25 and 45 cm).

In eastern Asia the precipitation increases in a southeasterly direction from interior Asia to the coast. The annual maximum seldom exceeds 80 in. (203 cm) in the wetter southeast coastal regions, whereas this drops to less than 30 in. (76 cm) in the North China Plain and less than 15 in. (38 cm) at the Great Wall. In some mountainous parts of Japan and Taiwan, the yearly average may be more than 100 in. (254 cm).

In the Indian subcontinent rainfall is heaviest along the western plateau fringe and in East Bengal, where it may average over 100 in. (254 cm) per year. The interior of the peninsula is relatively dry. Northwestern India and Pakistan share the drought of southwestern Asia. With the exception of the extreme north, Sri Lanka generally has abundant rainfall.

Southeastern Asia has the heaviest rainfall of the entire Asiatic region. The mainland mountains facing the southwest summer monsoon crossing the Bay of Bengal, and parts of the Vietnamese and Laotian cordilleras facing the humidified northeast winter monsoons of eastern Asia, regularly get average rainfalls of 120–150 in. (305–381 cm) or even more. Equally heavy rainfalls occur in the southwestern half of Sumatra, southwestern Java, the northwestern half of Borneo, and the Pacific fringe of the Philippine Islands. With a few small exceptions, southeastern Asia has no areas that are subject to severe drought.

Vegetation. Asia's vegetation belts and zones follow, in general, the climatic patterns from desert lands through tropical to Arctic margins.

A wide belt of tundra made irregular by topography occupies the entire Arctic lowland of Siberia with widths varying from 250 to 500 mi (400 to 800 km) north and south. It is widest in the extreme northeast and it extends southward and inland with higher elevations. The frozen subsoil permits the growth of little more than mosses, lichens, dwarfed trees, and scrub. *See* PERMAFROST; TUNDRA.

The largest unbroken expanse of forest in the world is the Siberian taiga, a dominantly coniferous forest of larches, spruce, fir, and pines, with such deciduous trees as birch and aspen occurring intermixed with the conifers or taking over as a secondary growth in burnt-over areas. The width of this belt in Siberia is more than 1000 mi (1600 km) and it stretches about 4000 mi (6400 km) from the Sea of Okhotsk to the Urals. *See* TAIGA.

Various admixtures of coniferous and deciduous trees compose the vegetation of mid-latitude mixed forests. In the west Siberian plain there is a narrow zone of mixed taiga and deciduous forests including oaks, maples, ash, and lindens. This zone, with a width of 50–100 mi (80–160 km), lies somewhat south of the parallel of 60°N and fades into the steppelands that form the great spring-wheat region of Siberia. Mixed midlatitude deciduous and coniferous forest areas of a similar type occupy most of Korea, the northern half of Honshu in Japan, and the hill lands surrounding the Yellow Plain, as well as the Ch'in-ling Mountains. In southern Asia these forests are found chiefly in a narrow belt of mountain land in the outer ranges of the Himalaya. The remaining areas of these mixed forests run from the Elburz Mountains through the Armenian highlands and the Black Sea fringe of Turkey to the Aegean coast, and in southwestern Asia in the Elburz of northern Iran.

From the mixed and deciduous forests of the west Siberian plain southward, an increasingly dry steppeland is encountered. It extends for 400–500 mi (640–800 km) in a belt about 1000 mi (1600 km) long between the Urals and the Altai-Sayan and associated uplands. The northern half of this belt with its higher annual precipitation of 12–16 in. (30–40 cm) is the agricultural heart of the plain. The southern part gradually changes to desert steppe and then to desert along about the 50th parallel. Eastward of Lake Baikal a broadened steppe zone occupies the Trans-Baikal region extending southward to the Gobi Desert of southern Mongolia and eastward to the Great Hsing-an Mountains, where the zone, about 200 mi (320 km) wide, runs southward in Inner Mongolia. The steppe zone in Inner Mongolia widens with the increasing moisture south of the Great Wall to include most of China's loess plateau. Grasses also form the natural vegetation of the Manchurian plain, with tall grass in the eastern portion thinning out to short-grass steppe in the Hsing-an Mountain flanks. The Gobi Desert is flanked by steppelands to its north, east, and south, as well as by mountain steppe zones in the eastern Altai and eastern T'ien Shan.

Mixed evergreen forests appear to be limited mostly to interior southern China and to Japan from the Kwanto Plain southward, South of the Yang-tzu Valley, this forest type extends from the coast at Shanghai to the gorge lands of eastern Tibet. In Asia the characteristic trees of the mixed forest include broad-leafed evergreen trees such as banyans and camphor, and coniferous trees such as pines, cedars, and cypresses, as well as varieties of bamboo.

Tropical and subtropical rainforest is restricted to warm or hot regions of southern and southeastern Asia which get ample rainfall the year round or get so much rain during a large part of the year that a high groundwater table is maintained during the short dry season. The subtropical sectors are found along the southeastern China coast, in Taiwan, and in northern Burma; they merge with the tropical rainforest farther south, where rainfall and temperature increase. *See* RAINFOREST.

Monsoon tropical deciduous forests comprise the tropical parts of Asia which have a moderately high rainfall but a long dry season (usually in the low-sun period or winter). These forests consist mostly of mixed species, but sometimes a single species becomes dominant as a result of selection from frequent burnings.

A large region of savanna grassland surrounds the Thar Desert of northwestern India and occupies most of the Indus Valley, the Punjab, and the Kathiawar peninsula. Much of the drier interior peninsular Deccan of India also has this as a natural vegetation. Other Asian regions with similar cover are found in Yemen and the region in southeastern Arabia from Oman as far westward as the Qatar peninsula; and similar vegetation extends over the Korat plateau of Thailand, lower Thailand west of Bangkok, southern Cambodia, and small areas in interior Borneo and the Philippines. *See* SAVANNA.

Immense areas of central and southwestern Asia have little or no vegetative cover, and bare rock alternates with sand veneering. In places shifting sand dunes are formed. Although the deserts are not necessarily lifeless, the vegetation is so widely spaced that much bare ground is exposed. The tropical desert areas generally receive their meager rainfall in torrential downpours on rare occasions. After such rains numerous herbs may spring to life and flower, while the bunch grass here and there may become green for a short season. [H.J.Wi.]

Asthma An allergic inflammatory disease of the airways, involving mast cells, eosinophils, macrophages, fibroblasts, and neutrophils. Such inflammatory changes are associated with widespread airflow obstruction, which is variable and improves (reverses) spontaneously or with appropriate therapy. Inflammation progresses to increased airway irritability (hyperresponsiveness) induced by the inhalation of allergens, cold air, and occupational factors. Although bronchospasm can be induced immediately after exposure to a specific allergen in an appropriately sensitized recipient, it is the late allergic response that most resembles the inflammatory reaction occurring in asthma. Central to this reaction is the release from mast cells, eosinophils, and lymphocytes of chemical mediators such as histamine, leukotrienes (potent bronchoconstricting agents), and various cytokines which perpetuate the response. Potent neurohumoral agents derived from neural pathways contribute further to the bronchospasm.

Wheezing, nocturnal breathlessness, coughing, and chest tightness often relieved by expectoration are highly suggestive of asthma. Episodes of breathlessness which result from exposure to an irritant (such as cold air) or an allergen (such as dust mites) following exercise or a viral infection and which are reversed spontaneously or with therapy are diagnostic of asthma. Eczema and edema in the folds of the nasal chambers are suggestive of a hereditary allergy, the major predictor of asthma. Objective measures of airflow obstruction which improved spontaneously or with therapy are also central to establishing an asthma diagnosis. Atopy, the genetic predisposition for developing an immunoglobulin-E (IgE) mediated (allergic) response to inhaled environmental allergens, is the strongest predisposing factor for developing asthma. Asthma may be classified, therefore, according to severity, etiology, or pattern of airflow obstruction. It is helpful to differentiate those factors that induce inflammation from those that incite acute bronchospasm in susceptible individuals. The association of an elevated serum IgE and the occurrence of asthma in all age groups, including those who are not atopic, makes antigenic stimulation causal in all instances of asthma. The severity of asthma can best be defined in terms of peak-flow monitoring (monitoring the severity of the allergy). Such evaluations as mild, moderate, and severe are useful in applying therapy in a stepwise manner contingent on severity.

Successful management of asthma requires education of the sick individual coupled with the development of a partnership with an asthma management health-care team;

assessing and monitoring the severity of asthma, with utilization of objective parameters of assessment (for example, the peak-flow meter, a device that measures the amount of air that enters and leaves the lungs); environmental management to avoid asthma triggers; and establishment of a drug regimen that controls asthma (medications include bronchodilators, which act as relievers, and bronchodilators, which act as preventers), as well as a written plan to prevent the condition from becoming worse. Adequate management of asthma should control the symptoms, prevent asthma attacks, return and maintain pulmonary function as close to normal as possible, maintain normal activity levels including exercise, avoid adverse side effects from the drugs, reduce and prevent irreversible airway changes, and prevent mortality. *See* ALLERGY. [A.L.She.]

Astomatida An order of protozoans, subclass Holotrichia, in which all species are mouthless. All species are parasitic in other animals, typically oligochaete annelids. Many astomatids possess an elaborate holdfast organelle. *Anoplophyra* is a typical example. *See* OLIGOCHAETA. [J.O.C.]

Atlantic Ocean The large body of sea water separating the continents of North and South America in the west from Europe and Africa in the east and extending south from the Arctic Ocean to the continent of Antarctica. The Atlantic is the second largest ocean water body and in area covers nearly one-fifth of the Earth's surface. The two major divisions, North and South Atlantic oceans, have the Equator as the common boundary. The North Atlantic, because of projecting land areas and island arcs, has numerous subdivisions. These include three large mediterranean-type seas, the Mediterranean Sea, the Gulf of Mexico plus Caribbean Sea, and the Arctic Ocean; two small mediterranean-type seas, the Baltic Sea and Hudson Bay; and four marginal seas, the North Sea, English Channel, Irish Sea, and Gulf of St. Lawrence. Parts of the Atlantic are given special names but lack precise boundaries, such as the Bahama Sea, Irminger Sea, Labrador Sea, and Sargasso Sea.

The mean depth of the Atlantic Ocean is 12,960 ft (3868 m), and its volume is 76,300,000 mi^3 (318,000,000 km^3). Broad shelves with depths less than 660 ft (200 m) are found in the region of the North Sea and the British Isles, on the Grand Banks of Newfoundland, and off the coasts of northeastern South America and Patagonia. The Mid-Atlantic Ridge, which extends from the Arctic Ocean to 55°S, is less than 9800 ft (3000 m) beneath the surface and is characterized by a pronounced relief. It separates the east and west Atlantic troughs, both of which have relatively uniform relief.

Three marked east-west ridges—the Greenland-Scotland Ridge in the North Atlantic and the Walvis and Rio Grande Ridges in the South Atlantic—and several less-conspicuous east-west rises separate the two Atlantic troughs into a series of basins including the West European, Canary, and Angola in the eastern Atlantic and the North American, Brazilian, and Argentine basins in the western Atlantic.

Islands in the Atlantic are mostly of volcanic origin. The Bermudas are the northernmost coral reefs, rising from an old submarine volcanic cone. Some islands, such as the British Isles, are continental in character. *See* OCEANIC ISLANDS.

The primary circulation of surface winds over the Atlantic Ocean is characterized by a zonal distribution pattern oriented in an east-west direction. The greatest storm frequency, more than 30% in winter, is in the zone of the prevailing westerlies. Air temperatures also follow a zonal pattern of distribution. They are lower in the South Atlantic than in the North Atlantic, and lower in the tropics and subtropics over the eastern Atlantic, than they are in the same latitudes over the western Atlantic. Maximum precipitation occurs in the doldrum zone (80 in. or 2000 mm/year). Precipitation also is relatively great in the zone of westerlies but is low in the trade-wind zones.

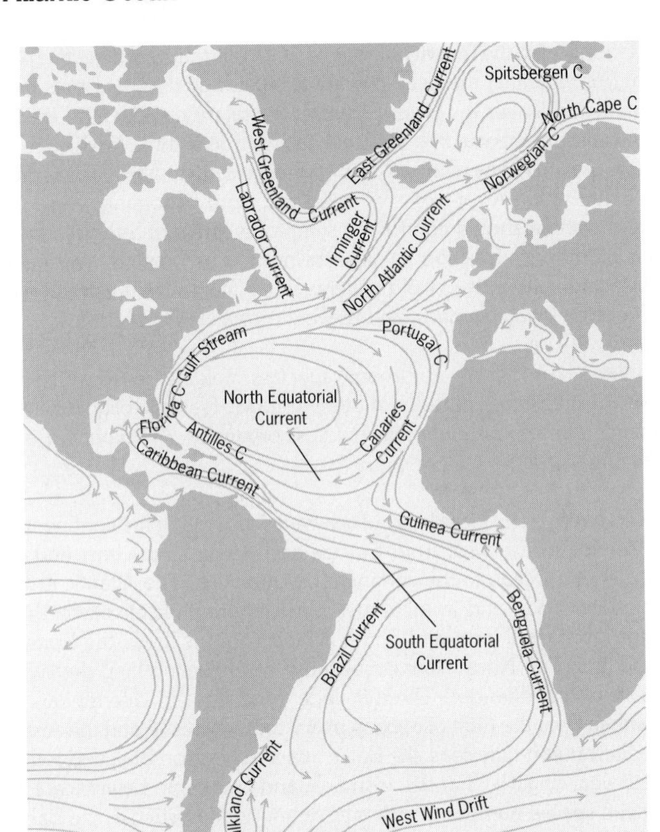

Currents of the Atlantic Ocean. (*Adapted from J. Bartholomew, Advanced Atlas of Modern Geography, McGraw-Hill, 3d ed., 1957*)

Sea ice is formed in the northernmost and southernmost parts of the Atlantic Ocean. From these areas drift ice moves equatorward into neighboring regions where it becomes a hazard to sea traffic and limits fishing. Many icebergs drift southward into the sea lanes of the North Atlantic. Most of these have their origin in the valley glaciers of western Greenland. Icebergs generally drift south of the Grand Banks, and some are known to have drifted southeast of Bermuda. In the South Atlantic large, tabular icebergs separate from the Antarctic ice shelf and drift northward. *See* ICEBERG; SEA ICE.

Surface currents in the Atlantic Ocean flow in much the same direction as the prevailing surface winds (see illustration). Deflections from these directions are caused by the bottom topography and the latitude or increased effect of Coriolis forces. The fairly constant flow of the North and South Equatorial currents is sustained largely by the trade winds. As a result, warm water is piled up along the poleward borders of these currents and on the western sides of the Atlantic Ocean. *See* OCEAN; OCEAN CIRCULATION.

The surface water in certain areas takes on a particularly high density in winter under the influence of climatic conditions. These water masses sink to a depth where the surrounding waters have a corresponding density and then spread out at that level. At the same time they are constantly mixing with the surrounding waters. In this way a multistoried stratification arises. Compared with that of the Indian and Pacific oceans, the deep circulation in the Atlantic Ocean is very vigorous, and the deeper water is therefore rich in oxygen. The abundance of nutrients permits a greater rate of organic production where the nutrient-rich waters nearly reach the surface, as in the Antarctic waters. *See* SEAWATER; SEAWATER FERTILITY.

The semidiurnal tidal form predominates in the Atlantic Ocean. The mean tidal range is about 3.3 ft (1 m) in the open ocean, but it decreases to 6.3 in. (16 cm) off Rio Grande do Sul in southern Brazil and to 3.5 in. (9 cm) off Puerto Rico. Tidal ranges increase beyond broad shelves under favorable physical conditions. The tides of the mediterranean and marginal seas are cooscillations of the tides of the Atlantic Ocean.

The Atlantic Ocean, especially the North Atlantic, is by far the most important bearer of the world'ssea traffic. Favorable trend include increased transportation capacities for handling bulk goods, regular weather observations for the safety of air and sea traffic by weather ships in selected positions, and the observation and reporting of drifting icebergs by the International Ice Patrol. Communication facilities, including telegraph and telephone cables and radio stations, have been improved and increased in number. [G.O.D.]

Atmosphere A gaseous layer that envelops the Earth and most other planets in the solar system. Earth, Venus, Mars, Jupiter, Saturn, Uranus, Neptune, and Titan (Saturn's largest satellite) are all known to possess substantial atmospheres that are held by the force of gravity. The structure and properties of the various atmospheres are determined by the interplay of physical and chemical processes. Structural features of Earth's atmosphere detailed below can often be identified in the atmospheres of other planetary bodies.

The composition of the Earth's atmosphere is primarily nitrogen (N_2), oxygen (O_2), and argon (Ar) [see table]. The concentration of water vapor (H_2O) is highly variable, especially near the surface, where volume fractions can vary from nearly 0% to as high as 4% in the tropics. There are many minor constituents or trace gases, such as neon (Ne), helium (He), krypton (Kr), and xenon (Xe), that are inert, and active species such as carbon dioxide (CO_2), methane (CH_4), hydrogen (H_2), nitrous oxide (NO), carbon monoxide (CO), ozone (O_3), and sulfur dioxide (SO_2), that play an important role in radiative and biological processes.

In addition to the gaseous component, the atmosphere suspends many solid and liquid particles. Aerosols are particulates usually less than 1 micrometer in diameter that are created by gas-to-particle reactions or are lifted from the surface by the wind. A portion of these aerosols can become centers of condensation or deposition in the growth of water and ice clouds. Cloud droplets and ice crystals are made primarily of water with some trace amounts of particles and dissolved gases. Their diameters range from a few micrometers to about 100 μm. Water or ice particles larger than about 100 μm begin to fall because of gravity and may result in precipitation at the surface. *See* CLOUD PHYSICS; PRECIPITATION (METEOROLOGY).

One of the remarkable properties of the Earth's atmosphere is the large amount of free molecular oxygen in the presence of gases such as nitrogen, methane, water vapor, hydrogen, and others that are capable of being oxidized. The atmosphere is in a highly oxidizing state that is far from chemical equilibrium. This is in sharp contrast to the atmospheres of Venus and Mars, the planets closest to the Earth, which are composed almost entirely of the more oxidized state, carbon dioxide. The chemical disequilibrium on the Earth is maintained by a continuous source of reactive gases derived from biological

Composition of the atmosphere*

Molecule	Fraction volume near surface	Vertical distribution
Major constituents		
N_2	7.8084×10^{-1}	Mixed in homosphere; photochemical dissociation high in thermosphere
O_2	2.0946×10^{-1}	Mixed in homosphere; photochemically dissociated in thermosphere, with some dissociation in mesosphere and stratosphere
Ar	9.34×10^{-3}	Mixed in homosphere with diffusive separation increasing above
Important radiative constituents		
CO_2	3.5×10^{-4}	Mixed in homosphere; photochemical dissociation in thermosphere
H_2O	Highly variable	Forms clouds in troposphere; little in stratosphere; photochemical dissociation above mesosphere
O_3	Variable	Small amounts, 10^{-8}, in troposphere; important layer, 10^{-6} to 10^{-5}, in stratosphere; dissociated above
Other constituents		
Ne	1.82×10^{-5}	
He	5.24×10^{-6}	Mixed in homosphere with diffusive separation increasing above
Kr	1.14×10^{-6}	
CH_4	1.15×10^{-6}	Mixed in troposphere; dissociated in upper stratosphere and above
H_2	5×10^{-7}	Mixed in homosphere; product of H_2O photochemical reactions in lower thermosphere, and dissociated above
NO	$\sim 10^{-8}$	Photochemically produced in stratosphere and mesosphere

*Other gases, for example, CO, N_2O, NO_2, and many by-products of atmospheric pollution also exist in small amounts.

processes. Life plays a vital role in maintaining the present atmospheric composition. *See* ATMOSPHERIC CHEMISTRY.

The total mass of the Earth's atmosphere is about 5.8×10^{15} tons (5.3×10^{15} metric tons). The vertical distribution of gaseous mass is maintained by a balance between the downward force of gravity and the upward pressure gradient force. The balance is known as the hydrostatic balance or the barometric law. Hence, the declining atmospheric pressure that is measured while ascending in the atmosphere is a result of gravity. The globally averaged pressure at mean sea level is 1013.25 millibars (101,325 pascals).

Below about 60 mi (100 km) in altitude, the atmosphere's composition of major constituents is very uniform. This region is known as the homosphere to distinguish it from the heterosphere above 60 mi (100 km), where the relative amounts of the major constituents change with height. In the homosphere there are sufficient atmospheric motions and a short enough molecular free path to maintain uniformity in composition. Above the boundary between the homosphere and the heterosphere, known as the homopause or turbopause, the mean free path of the individual molecules becomes long enough that gravity is able to partially separate the lighter molecules from the heavier ones. The mean free path is the average distance that a particle will travel before encountering a collision. Hence the average molecular weight of the heterosphere decreases with height as the lighter atoms dominate the composition.

The vertical structure of the atmosphere is in large part determined by the transfer properties of the solar and terrestrial radiation streams. The energy of the smallest unit of radiation, the photon, is directly proportional to its frequency. The type of interaction that occurs between photons and the atmosphere depends on the energy of the photons.

The most energetic of the photons are x-rays and extreme ultraviolet radiation of the eletromagnetic spectrum, which are capable of dissociating and ionizing the gaseous molecules. The less energetic near-ultraviolet photons are able to excite molecules and atoms into higher electronic levels. As a result, most of the ultraviolet and x-ray radiation is attenuated by the upper atmosphere. A cloudless atmosphere, however, is relatively transparent to visible light, where most of the solar energy resides. At the opposite end of the spectrum toward the lower frequencies of radiation is the infrared part, which is capable of inducing various vibrational and rotational motions in triatomic and polyatomic molecules.

In order to maintain an energy balance, the Earth must emit about the same amount of radiation as it absorbs from the Sun. The terrestrial radiation occurs in the infrared part of the spectrum and hence is strongly affected by water vapor, clouds, carbon dioxide, and ozone and other trace gases. The ability of these gases to absorb and emit in the infrared allows them to effectively trap some of the outgoing radiation that is emitted by the surface, creating the so-called greenhouse effect. *See* GREENHOUSE EFFECT; INSOLATION; TERRESTRIAL RADIATION.

The atmospheric layer that extends from the surface to about 7 mi (11 km) is called the troposphere. The tropopause, which is the top of the troposphere, has an average altitude that varies from about 11 mi (18 km) near the Equator to about 5 mi (8 km) near the Poles. The actual tropopause height varies considerably on time scales from a few days to an entire year. The troposphere contains about 80% of the atmospheric mass and exhibits most of the day-to-day weather fluctuations that are observed from the ground. Temperatures generally decrease with increasing altitude at an average lapse rate of about 17°F/mi (6°C/km), although this rate varies considerably, depending on time and location. *See* TROPOPAUSE; TROPOSPHERE.

The stratosphere is the atmospheric layer that extends from the tropopause up to the stratopause at about 30 mi (50 km) above the surface. It is characterized by a nearly isothermal layer in the first 6 mi (10 km) overlaid by a layer in which the temperature increases with height to a maximum of about 32°F (0°C) at the stratopause. The reversal in the temperature lapse rate is a result of direct absorption of solar radiation, mainly by ozone and oxygen at the ultraviolet frequencies. *See* STRATOSPHERE.

The reversal of the temperature lapse rate makes the stratosphere vertically stable. This stability limits the amount of vertical mixing and results in molecular residence times of many months to years. Another consequence of a stable stratosphere is that it acts as a lid on the troposphere, confining the strong vertical overturning and hence most of the surface-based weather phenomena. *See* WEATHER.

The mesosphere is the atmospheric layer extending from the stratopause up to the mesopause at an altitude of about 53 mi (85 km). The mesosphere is characterized by temperatures decreasing with height at a rate of about 12°F/mi (4°C/km). Although the mesosphere has less vertical stability than the stratosphere, it is still more stable than the troposphere and does not experience rapid overturning. The coldest temperatures of the entire atmosphere are encountered at the mesopause, with values as low as −150°F (−100°C). The temperature lapse rate found in the mesosphere is a result of the gradual weakening with height of the direct absorption of solar radiation by ozone. The radiative infrared cooling to space by the carbon dioxide molecules is responsible for the low temperatures near the mesopause. *See* MESOSPHERE.

The thermosphere is found above the mesopause. The thermosphere is characterized by rising temperatures with height up to an altitude of about 190 mi (300 km) and then is nearly isothermal above that. Although there is no clear upper limit to the thermosphere, it is convenient to consider it extending several thousand kilometers. Embedded within the thermosphere is the ionosphere, comprising those atmospheric layers in which the ionized molecules and atoms are dominating the processes.

Molecular species dominate the lower thermosphere, while atomic species are dominant above 190 mi (300 km). The distribution of the constituents is controlled by diffusive equilibrium in which the concentration of each constituent decreases exponentially with height according to its molecular weight. Hence the concentration of the heavier constituents such as nitrogen, oxygen, and carbon dioxide will decrease with height faster than the lighter constituents such as helium and hydrogen. At an altitude of 560 mi (900 km) helium becomes the dominant constituent while hydrogen dominates above 1900 mi (3000 km).

The ionosphere can be defined operationally as that part of the atmosphere that is sufficiently ionized to affect the propagation of radio waves. In the ionosphere, the dominant negative ion is the electron, and the main positive ions include O^+, NO^+, and O_2^+. The ionosphere is classified into four subregions. The D region extends from 40 to 60 mi (60 to 90 km) and contains complex ionic chemistry; most of the ionization is caused by ultraviolet ionization of NO and by galactic cosmic rays. This region is responsible for the daytime absorption of radio waves, which prevents distant propagation of certain frequencies. The E region extends from 60 to 90 mi (90 to 150 km) and is caused primarily by the x-rays from the Sun. The F1 region from 90 to 125 mi (150 to 200 km) is caused by the extreme ultraviolet radiation from the Sun and disappears at night. Finally, the F2 region includes all the ionized particles above 125 mi (200 km), with the peak ion concentrations occurring near 190 mi (300 km).

The exosphere is the atmosphere above 300 mi (500 km) where the probability of interatomic collisions is so low that some of the atoms traveling upward with sufficient velocity can escape the Earth's gravitational field. The dominant escaping atom is hydrogen since it is the lightest constituent. Calculations of the thermal escape of hydrogen (also known as the Jeans escape) yield a value of about 3×10^8 atoms $\cdot$ $cm^{-2} \cdot s^{-1}$. This is a very small amount since at this rate less than 0.5% of the oceans would disappear over the current age of the Earth.

The magnetosphere is the region surrounding the Earth where the movement of ionized gases is dominated by the geomagnetic field. The lower boundary of the magnetosphere, which occurs at an altitude of nearly 75 mi (120 km), can be roughly defined as the height where there are enough neutral atoms that the ion-neutral particle collisions dominate the ion motion. The dynamics of the magnetosphere is dictated in part by its interaction with the plasma of ionized gases that blows away from the Sun, the solar wind. The solar wind interacts with the Earth's magnetic field and severely deforms it, producing a magnetosphere around the Earth. It extends about 40,000 mi (60,000 km) toward the Sun but extends beyond the orbit of the Moon away from the Sun. [G.B.L.]

Atmospheric chemistry A scientific discipline concerned with the chemical composition of the Earth's atmosphere. Topics include the emission, transport, and deposition of atmospheric chemical species; the rates and mechanisms of chemical reactions taking place in the atmosphere; and the effects of atmospheric species on human health, the biosphere, and climate.

A useful quantity in atmospheric chemistry is the atmospheric lifetime, defined as the mean time that a molecule resides in the atmosphere before it is removed by chemical reaction or deposition. The atmospheric lifetime measures the time scale on which changes in the production or loss rates of a species may be expected to translate into changes in the species concentration. The atmospheric lifetime can also be compared to the time scales for atmospheric transport to infer the spatial variability of a species in the atmosphere; species with lifetimes longer than a decade tend to be uniformly mixed, while species with shorter lifetimes may have significant gradients reflecting the distributions of their sources and sinks. *See* ATMOSPHERIC GENERAL CIRCULATION.

The principal constituents of dry air are nitrogen (N_2; 78% by volume), oxygen (O_2; 21%), and argon (Ar; 1%). The atmospheric concentrations of N_2 and Ar are largely determined by the total amounts of N and Ar released from the Earth's interior since the origin of the Earth. The atmospheric concentration of O_2 is regulated by a slow atmosphere-lithosphere cycle involving principally the conversion of O_2 to carbon dioxide (CO_2) by oxidation of organic carbon in sedimentary rocks (weathering), and the photosynthetic conversion of CO_2 to O_2 by marine organisms which precipitate to the bottom of the ocean to form new sediment. This cycle leads to an atmospheric lifetime for O_2 of about 4 million years. *See* BIOSPHERE; PHOTOSYNTHESIS.

Water vapor concentrations in the atmosphere range from 3% by volume in wet tropical areas to a few parts per million by volume (ppmv) in the stratosphere. Water vapor, with a mean atmospheric lifetime of 10 days, is supplied to the troposphere by evaporation from the Earth's surface, and it is removed by precipitation. Because of this short lifetime, water vapor concentrations decrease rapidly with altitude, and little water vapor enters the stratosphere. Oxidation of methane represents a major source of water vapor in the stratosphere, comparable to the source contributed by transport from the troposphere.

The most abundant carbon species in the atmosphere is CO_2. It is produced by oxidation of organic carbon in the biosphere and in sediments. The atmospheric concentration of CO_2 is rising, and there is concern that this may cause significant warming of the Earth's surface because of the ability of CO_2 to absorb infrared radiation emitted by the Earth (the greenhouse effect). The total amount of carbon present in the atmosphere is small compared to that present in the other geochemical reservoirs, and therefore it is controlled by exchange with these reservoirs. Equilibration of carbon between the atmosphere, biosphere, soil, and surface ocean reservoirs takes place on a time scale of decades. *See* GREENHOUSE EFFECT.

Methane is the second most abundant carbon species in the atmosphere and an important greenhouse gas. It is emitted by anaerobic decay of biological carbon (for example, in wetlands, landfills, and stomachs of ruminants), by exploitation of natural gas and coal, and by combustion. It has a mean lifetime of 12 years against atmospheric oxidation by the hydroxyl (OH) radical, its principal sink.

Many hydrocarbons other than methane are emitted to the atmosphere from vegetation, soils, combustion, and industrial activities. The emission of isoprene [$H_2C{=}C(CH_3){-}CH{=}CH_2$] from deciduous vegetation is particularly significant. Nonmethane hydrocarbons have generally short lifetimes against oxidation by OH (a few hours for isoprene), so that their atmospheric concentrations are low. They are most important in atmospheric chemistry as sinks for OH and as precursors of tropospheric ozone, organic nitrates, and organic aerosols.

Carbon monoxide (CO) is emitted to the atmosphere by combustion, and it is also produced within the atmosphere by oxidation of methane and other hydrocarbons. It is removed from the atmosphere by oxidation by OH, with a mean lifetime of 2 months. Carbon monoxide is the principal sink of OH and hence plays a major role in regulating the oxidizing power of the atmosphere.

Nitrous oxide (N_2O) is of environmental importance as a greenhouse gas and as the stratospheric precursor for the radicals NO and NO_2. The principal sources of N_2O to the atmosphere are microbial processes in soils and the oceans; the main sinks are photolysis and oxidation in the stratosphere, resulting in an atmospheric lifetime for N_2O of about 130 years.

About 90% of total atmospheric ozone (O_3) resides in the stratosphere, where it is produced by photolysis of O_2. The ultraviolet photons ($\lambda < 240$ nm) needed to photolyze O_2 are totally absorbed by ozone and O_2 as solar radiation travels through the stratosphere. As a result, ozone concentrations in the troposphere are much lower than in the stratosphere. *See* STRATOSPHERE; TROPOSPHERE.

Tropospheric ozone plays a central role in atmospheric chemistry by providing the primary source of the strong oxidant OH. It is also an important greenhouse gas. In surface air, ozone is of great concern because of its toxicity to humans and vegetation. Ozone is supplied to the troposphere by slow transport from the stratosphere, and it is also produced within the troposphere by a chain reaction involving oxidation of CO and hydrocarbons by OH in the presence of NO_x. Ozone production by this mechanism is particularly rapid in urban areas, where emissions of NO_x and of reactive hydrocarbons are high.

Sulfuric acid produced in the atmosphere by oxidation of sulfur dioxide (SO_2) is a major component of aerosols in the atmosphere and an important contributor to acid deposition. Sources of SO_2 to the atmosphere include emission from combustion, smelters, and volcanoes, and oxidation of oceanic dimethylsulfide [$(CH_3)_2S$] emitted by phytoplankton. It is estimated that about 75% of total sulfur emission to the atmosphere is anthropogenic. *See* AIR POLLUTION. [D.J.J.]

Atmospheric general circulation The statistical description of atmospheric motions over the Earth, their role in transporting energy, and the transformations among different forms of energy. Through their influence on the pressure distributions that drive the winds, spatial variations of heating and cooling generate air circulations, but these are continually dissipated by friction. While large day-to-day and seasonal changes occur, the mean circulation during a given season tends to be much the same from year to year. Thus, in the long run and for the global atmosphere as a whole, the generation of motions nearly balances the dissipation. The same is true of the long-term balance between solar radiation absorbed and infrared radiation emitted by the Earth-atmosphere system, as evidenced by its relatively constant temperature. Both air and ocean currents, which are mainly driven by the winds, transport heat. Hence the atmospheric and oceanic general circulations form cooperative systems. *See* MARITIME METEOROLOGY; OCEAN CIRCULATION.

Owing to the more direct incidence of solar radiation in low latitudes and to reflection from clouds, snow, and ice, which are more extensive at high latitudes, the solar radiation absorbed by the Earth-atmosphere system is about three times as great in the equatorial belt as at the poles, on the annual average. Infrared emission is, however, only about 20% greater at low than at high latitudes. Thus in low latitudes (between about 35°N and 35°S) the Earth-atmosphere system is, on the average, heated by radiation, and in higher latitudes cooled by radiation. The Earth's surface absorbs more radiative heat than it emits, whereas the reverse is true for the atmosphere. Therefore, heat must be transferred generally poleward and upward through processes other than radiation. At the Earth-atmosphere interface, this transfer occurs in the form of turbulent flux of sensible heat and through evapotranspiration (flux of latent heat). In the atmosphere the latent heat is released in connection with condensation of water vapor. *See* CLIMATOLOGY; HEAT BALANCE, TERRESTRIAL ATMOSPHERIC.

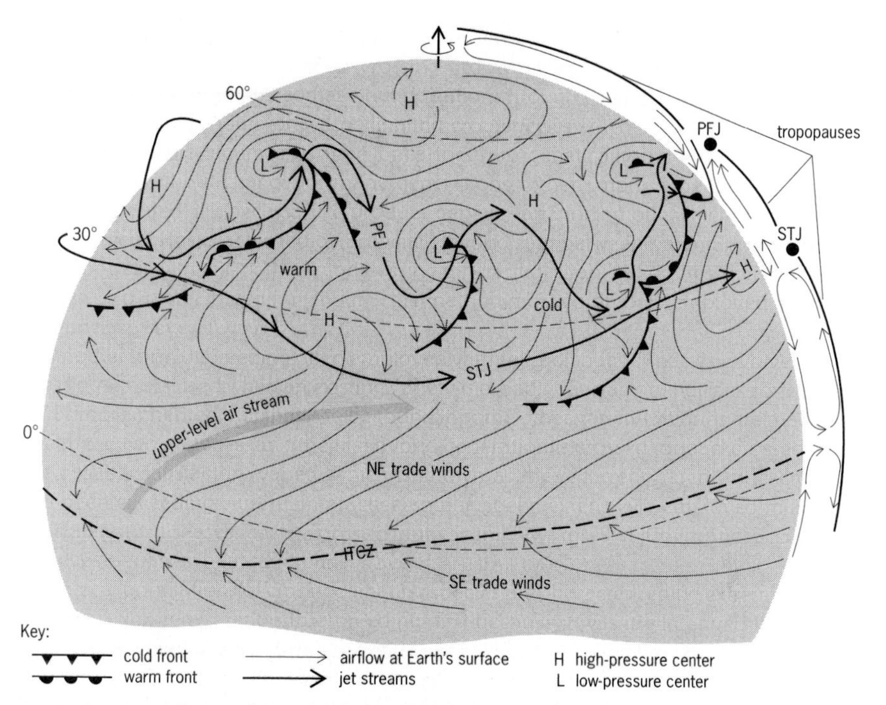

Schematic circulations over the Northern Hemisphere in winter. The intertropical convergence zone (ITCZ) lies entirely north of the Equator in the summer. Eastward acceleration in the upper-level tropical airstream is due to Earth rotation and generates the subtropical jet stream (STJ). The vertical section (right) shows the dominant meridional circulation in the tropics and shows airstreams relative to the polar front in middle latitudes.

Considering the atmosphere alone, the heat gain by condensation and the heat transfer from the Earth's surface exceed the net radiative heat loss in low latitudes. The reverse is true in higher latitudes. The meridional transfer of energy, necessary to balance these heat gains and losses, is accomplished by air currents. These take the form of organized circulations, whose dominant features are notably different in the tropical belt (roughly the half of the Earth between latitudes 30°N and 30°S) and in extratropical latitudes.

Characteristic circulations over the Northern Hemisphere are shown in the illustration. In the upper troposphere, there are two principal jet-stream systems: the subtropical jet (STJ) near latitude 30°, and the polar-front jet (PFJ), with large-amplitude long waves and superimposed shorter waves associated with cyclone-scale disturbances. The long waves on the polar-front jet move slowly eastward, and the shorter waves move rapidly. At the Earth's surface, northeast and southeast trade winds of the two hemispheres meet at the intertropical convergence zone (ITCZ), in the vicinity of which extensive lines and large clusters of convective clouds are concentrated. Westward-moving waves and vortices form near the intertropical convergence zone and, in summer, within the trades. Heat released by condensation in convective clouds of the intertropical convergence zone, and the mass of air conveyed upward in them, drive meridional circulations (right side of the illustration), whose upper-level poleward branches generate the subtropical jet stream at their poleward boundaries. *See* TROPICAL METEOROLOGY.

In extratropical latitudes, the circulation is dominated by cyclones and anticyclones. Cyclones develop mainly on the polar front, where the temperature contrast between polar and tropical air masses is concentrated, in association with upper-level waves on the polar-front jet stream. In winter, cold outbreaks of polar air from the east coasts of continents over the warmer oceans result in intense transfer of heat and water vapor into the atmosphere. Outbreaks penetrating the tropics also represent a sporadic exchange in which polar air becomes transformed into tropical air. Tropical airstreams, poleward on the west sides of the subtropical highs, then supply heat and water vapor to the extratropical disturbances. *See* CYCLONE; FRONT.

The characteristic flow in cyclones takes the form of slantwise descending motions on their west sides and ascent to their east in which extensive clouds and precipitation form. Heat that is released in condensation drives the ascending branch, and the descending branch consists of polar air that has been cooled by radiation in higher latitudes. When viewed relative to the meandering polar-front zone (right side of the illustration), the combined sinking of cold air and ascent of warm air represents a conversion of potential energy into kinetic energy. This process maintains the polar jet stream. The branches of the circulation transfer heat both upward, to balance the radiative heat loss by the atmosphere, and poleward, to balance the radiative heat deficit in high latitudes. [C.W.N.]

Atoll An annular coral reef, with or without small islets, that surrounds a lagoon without projecting land area. Most atolls are isolated reefs rising from the deep sea, and vary considerably in size. Small rings, usually without islets, may be less than a mile in diameter, but many atolls have a diameter of about 20 mi (32 km) and bear numerous islets.

The reefs of the atoll ring are flat, pavementlike areas, large parts of which, particularly along the seaward margin, may be exposed at times of low tide. The reefs vary in width from narrow ribbons to broad bulging areas more than a mile (1.6 km) across. The structures form a most effective baffle that robs the incoming waves of much of their destructive power, and at the same time brings a constant supply of refreshing sea water with oxygen, food, and nutrient salts to wide expanses of the reef.

Atolls, like other types of coral reefs, require strong light and warm waters and are limited in the existing seas to tropical and near-tropical latitudes. A large percentage of the world's atolls are contained in an area known as the former Darwin Rise that covers much of the central and southwestern Pacific. Atolls are also numerous in parts of the Indian Ocean and a number are found, mostly on continental shelves, in the Caribbean area. *See* OCEANIC ISLANDS; REEF. [H.S.L.]

Australia An island continent in the Southern Hemisphere with a total area of 2,941,526 mi^2 (7,618,552 km^2). It is bounded on the west by the Indian Ocean and on the east by the Pacific Ocean and the Tasman Sea. Numerous small and several large islands lie off the coast, including Tasmania and New Zealand. Australia is generally of remarkably low elevation and moderate relief. Three-fourths of the land mass lies between 600 and 1500 ft (180 and 450 m) in the form of a huge plateau. A cross section from east to west shows first a narrow belt of coastal plain, then the steep escarpments of the eastern face of the Great Dividing Range, stretching 1200 mi (1900 km) from the north of Queensland to the south of Victoria. The descent on the western slope of the Dividing Range is gradual until often elevation in the inland basins is below sea level, rising gradually again across the great plateau until the low ranges of western Australia fringing the plateau are reached, and beyond these lies another coastal plain. With the exception of the Gulf of Carpentaria and Cape York peninsula in the north and the

Great Australian Bight in the south, there are few striking features in the configuration of the coast. Australia may conveniently be divided into three great structural and landform regions.

The region called East Australian Highlands consists of a narrow plain extending north and south along the eastern coast. Flanking the plain are the series of ranges and tablelands making up the Great Dividing Range. The East Australian Highlands is the best-watered region in Australia, and some of the river systems are of considerable size. On the flanks of the East Australian Highlands are Australia's principal coal deposits—in the vicinity of Sydney and Newcastle and in the Bowen and Ipswich fields in Queensland. Petroliferous basins at Surat (Roma), flanking the divide in Queensland and off the coast of Victoria in Bass Strait, are Australia's most promising deposits of petroleum and natural gas.

The region known as the Interior Lowland Basins comprises a region of sedimentary rocks that occupy one-third of the continent between the western slope of the eastern highlands and the inner eastern margin of the ancient shield which forms the Western Plateau. Little land is over 500 ft (150 m), and some is below sea level. The rivers of the Murray-Darling Basin, draining the western slopes of the Great Dividing Range, have a marked seasonal variation in flow but never dry up in the lower reaches. South Australia's shallow lakes are more often dry expanses of encrusted white salt than bodies of water—the result of low rainfall and high evaporation. In most parts of the region water from deep artesian wells is available.

The region known as the Western Plateau is the largest area, occupying almost three-fifths of the continent, and is a great shield of ancient rocks standing 750–1500 ft (225–450 m) high. Much of it is buried in desert sand, and only a few ridges of ancient mountains (such as the Macdonnel and Musgrave ranges) break the monotony of the plateau surface. Only in the southwestern corner of the continent and along the northwestern coast is rainfall sufficient to support a sclerophyll forest of eucalypts and a monsoon woodland, respectively. In the north, coastal rivers are of considerable size but change from flooded torrents after rains to a succession of water holes in dry seasons.

Tasmania is a small mountainous island lying 150 mi (240 km) southeast of Australia across Bass Strait, with a total area of 26,383 mi^2 (68,332 km^2). The island is structurally similar to the East Australian Highlands. The dominant feature is the central plateau, falling from a general level of 3500 ft (1070 m) in the northwest toward the southeast. A dense eucalyptus forest covers most of the island except along the wetter west coast, where beech forest predominates. The rivers have short, rapid courses with little seasonal variation in flow. *See* NEW ZEALAND. [K.B.C.]

Avalanche In general, a large mass of snow, ice, rock, earth, or mud in rapid motion down a slope or over a precipice. In the English language, the term avalanche is reserved almost exclusively for snow avalanche. Minimal requirements for the occurrence of an avalanche are snow and an inclined surface, usually a mountainside. Most avalanches occur on slopes between 30 and 45°.

Two basic types of avalanches are recognized according to snow cover conditions at the point of origin. A loose-snow avalanche originates at a point and propagates downhill by successively dislodging increasing numbers of poorly cohering snow grains, typically gaining width as movement continues downslope. This type of avalanche commonly involves only those snow layers near the surface. The mechanism is analogous to dry sand. The second type, the slab avalanche, occurs when a distinct cohesive snow layer breaks away as a unit and slides because it is poorly anchored to the snow or ground below. A clearly defined gliding surface as well as a lubricating layer may be identifiable at the base of the slab, but the meteorological conditions which create these layers are complex.

In the case of the loose avalanche, release mechanisms are primarily controlled by the angle of repose, while slab releases involve complex strength-stress problems. A release may occur simply as a result of the overloading of a slope during a single snowstorm and involve only snow which accumulated during that specific storm, or it may result from a sequence of meteorological events and involve snow layers comprising numerous precipitation episodes. Most large snow slides are believed to be caused by an unstable layer of ice grains that develop deep in mountain snow. Called depth hoar by students of avalanche dynamics, these crystals owe their formation to heat from earth and rock which are buried by the snow, and which in late autumn are warmer than the surrounding air. Snow nearest the ground vaporizes, causing growth of angular ice grains that exhibit poor bonding qualities. Gravity combining with the weakness of the depth hoar crystals loosens the upper stable layers. Once the stable layers begin to slide, the depth hoar acts in a manner similar to ball beatings to speed the descent of the slide.

Where snow avalanches constitute a hazard, that is, where they directly threaten human activities, various defense methods have evolved. Attempts are made to prevent the avalanche from occurring by artificial supporting structures or reforestation in the zone of origin. The direct impact of an avalanche can be avoided by construction of diversion structures, dams, sheds, or tunnels. Hazardous zones may be temporarily evacuated while avalanches are released artificially, most commonly by explosives. Finally, attempts are made to predict the occurrence of avalanches by studying relationships between meteorological and snow cover factors. [R.L.A.]

Avian leukosis A complex of several related and unrelated viruses (both C-type retroviruses and herpesviruses) that are collectively responsible for a variety of benign and malignant neoplasms in chickens and, to a lesser extent, in other avian species. Although most neoplasms observed in avian species are induced by viruses, there are some of unknown etiology.

The neoplastic diseases induced by the leukosis-sarcoma group of retroviruses include lymphoid leukosis, myeloid or erythroid leukemias or solid tumors, tumors of connective tissue origin (for example, sarcomas, fibromas, and chondromas), epithelial carcinomas, and endothelial tumors. The many viral strains involved have similar physical and chemical characteristics and share a group-specific antigen; some can cause more than one type of neoplasm. The viruses are about 100 nanometers in diameter; have a core composed of ribonucleic acid; contain a reverse transcriptase; mature by budding from the cell membrane; and are divided into subgroups based on envelope glycoproteins. Some strains carry their own specific oncogenes that induce neoplasms within days or weeks. Others lack an oncogene and cause neoplasms less frequently and only after several months, probably by activating a specific cellular oncogene. *See* CANCER (MEDICINE); ONCOLOGY.

Lymphoid leukosis is the most important of the leukosis sarcoma diseases. The lymphoid leukosis virus is transmitted vertically from hen to chick through the egg. Infection can result in leukotic neoplasms in various visceral organs following metastasis from primary tumors in the bursa of Fabricius. Large-scale transmission of the lymphoid leukosis virus can be eradicated by eliminating individual infected breeders.

Reticuloendotheliosis virus strains constitute another retrovirus group, unrelated to the lymphoid-sarcoma group, and also may carry a specific oncogene. They can cause a chronic neoplastic form of reticuloendotheliosis or other neoplasms in turkeys, chickens, ducks, geese, quail, and pheasants, and a runting disease has been seen in chickens after accidental contamination of vaccines with reticuloendotheliosis virus.

Marek's disease in chickens is caused by an oncogenic, cell-associated, lymphotrophic, highly contagious herpesvirus. Inhalation of the virus causes an active infection in lymphoid organs; after about 1 week, a latent infection develops in lymphocytes. T-cell lymphomas may develop within a few weeks or months, depending on age, genetic makeup, virus virulence, and other factors. Degenerative, inflammatory and lymphoproliferative lesions occur principally in the peripheral nerves (causing paralysis), lymphoid tissues, visceral organs, muscle, and skin. Eye involvement (gray eye) can cause blindness. The disease is of great economic importance in chickens, and several vaccines, injected at 1 day of age, have been in worldwide use since about 1970. *See* ANIMAL VIRUS; TUMOR VIRUSES. [B.W.C.]

B

Bacillary dysentery A highly contagious intestinal disease caused by rod-shaped bacteria of the genus *Shigella*. Bacillary dysentery is a significant infection of children in the developing world, where it is transmitted by the fecal-oral route. The global disease burden is estimated as 165 million episodes and 1.3 million deaths annually. Common-source outbreaks occasionally occur in developed countries, usually as a result of contaminated food. The most common species isolated in developed countries is *S. sonnei*, while *S. flexneri* serotypes predominate in endemic areas. Epidemics of *S. dysenteriae* 1 occur in equatorial regions, and these outbreaks can involve adults as well as children.

When ingested even in very small numbers, shigellae multiply in the intestine and invade the epithelial lining of the colon. Infection of this tissue elicits an acute inflammatory response (colitis) that is manifested as diarrhea or bloody, mucoid stools (dysentery). The virulence of all *Shigella* species, and *Shigella*-like enteroinvasive *Escherichia coli*, depends on an extrachromosomal genetic element (virulence plasmid) that encodes four invasion plasmid antigen (Ipa) proteins and a secretory system (Type III) for these proteins. Secreted Ipa proteins help shigellae to initiate colonic invasion through specialized endocytic intestinal cells (M cells). After shigellae pass through these M cells, they are phagocytized by tissue macrophages in the underlying lymphoid tissue. Ipa proteins then induce apoptosis (programmed cell death) in infected macrophages, releasing cytokines (primarily IL-1) that initiate an acute, localized inflammatory infiltrate. This infiltrate of polymorphonuclear leukocytes destabilizes tight junctions between absorptive epithelial cells (enterocytes), making the tissue more susceptible to additional *Shigella* invasion. Secreted Ipa proteins induce uptake of shigellae by the colonic enterocytes. The virulence plasmid also encodes an intercellular spread protein (IcsA) that recruits mammalian cytoskeletal elements (primarily actin) to the bacterial surface. This actin is organized into a cytoplasmic motor that facilitates spread of shigellae to adjacent enterocytes. *See* DIARRHEA; ESCHERICHIA.

In otherwise healthy individuals, bacillary dysentery is typically a short-term disease lasting less than a week. The symptoms can be truncated by appropriate antibiotic therapy (such as oral ampicillin or cyprofloxacin) that rapidly eliminates shigellae from the intestinal lumen and tissues. When *S. dysenteriae* 1 is the etiologic agent, however, hemolytic uremic syndrome can be manifested as a serious consequence of disease. This species produces a cytotoxin (Shiga toxin or Stx) that is functionally identical to the toxin of enterohemorrhagic *E. coli* (for example, O157:H7). Stx inhibits protein synthesis, damaging endothelial cells of the intestinal capillary bed; the toxin may also damage renal tubules, causing acute renal failure with chronic sequela in up to one-third of hemolytic uremic syndrome patients. *See* MEDICAL BACTERIOLOGY.

[T.L.Ha.]

Bacteria Extremely small—usually 0.3 to 2.0 micrometers in diameter—and relatively simple microorganisms possessing the prokaryotic type of cell construction. Although traditionally classified within the fungi as Schizomycetes, they show no phylogenetic affinities with the fungi, which are eukaryotic organisms. The only group that is clearly related to the bacteria are the blue-green algae. Bacteria are found almost everywhere, being abundant, for example, in soil, water, and the alimentary tracts of animals. Each kind of bacterium is fitted physiologically to survive in one of the innumerable habitats created by various combinations of space, food, moisture, light, air, temperature, inhibitory substances, and accompanying organisms. Dried but often still living bacteria can be carried into the air. Bacteria have a practical significance for humans. Some cause disease in humans and domestic animals, thereby affecting health and the economy. Some bacteria are useful in industry, while others, particularly in the food, petroleum, and textile industries, are harmful. Some bacteria improve soil fertility. As in higher forms of life, each bacterial cell arises either by division of a preexisting cell with similar characteristics or through a combination of elements from two such cells in a sexual process. *See* MEDICAL BACTERIOLOGY; PETROLEUM MICROBIOLOGY; SOIL MICROBIOLOGY.

Descriptions of bacteria are preferably based on the studies of pure cultures, since in mixed cultures it is uncertain which bacterium is responsible for observed effects. Pure cultures are sometimes called axenic, a term denoting that all cells had a common origin in being descendants of the same cell, without implying exact similarity in all characteristics. Pure cultures can be obtained by selecting single cells, but indirect methods achieving the same result are more common.

If conditions are suitable, each bacterium grows and divides, using food diffused through the gel, and produces a mass of cells called a colony. Colonies always develop until visible to the naked eye unless toxic products or deficient nutrients limit them to microscopic dimensions.

The morphology, that is, the shape, size, arrangement, and internal structures, of bacteria can be distinguished microscopically and provides the basis for classifying the bacteria into major groups. Three principal shapes of bacteria exist, spherical (coccus), rod (bacillus), and twisted rod (spirillum). The coccus may be arranged in chains of cocci as in *Streptococcus*, or in tetrads of cocci as in *Sarcina*. The rods may be single or in filaments. Stains are used to visualize bacterial structures otherwise not seen, and the stain reaction with Gram's stain provides a characteristic used in classifying bacteria.

Many bacteria are not motile. Of the motile bacteria, however, some move by means of tiny whirling hairlike flagella extending from within the cell. Others are motile without flagella and have a creeping or gliding motion. Many bacteria are enveloped in a capsule, a transparent gelatinous or mucoid layer outside the cell wall. Some form within the cell a heat- and drought-resistant spore, called an endospore. Cytoplasmic structures such as reserve fat, protein, and volutin are occasionally visible within the bacterial cell.

The nucleus of bacteria is prokaryotic, that is, not separated from the rest of the cell by a membrane. It contains the pattern material for forming new cells. This material, deoxyribonucleic acid (DNA), carrying the information for synthesis of cell parts, composes a filament with the ends joined to form a circle. The filament consists of two DNA strands joined throughout their length. The joining imparts a helical form to the double strand. The double-stranded DNA consists of linearly arranged hereditary units, analogous and probably homologous with the "genes" of higher forms of life. During cell division and sexual reproduction, these units are duplicated and a complete set is distributed to each new cell by an orderly mechanism.

The submicroscopic differences that distinguish many bacterial genera and species are due to structures such as enzymes and genes that cannot be seen. The nature of these structures is determined by studying the metabolic activities of the bacteria. Data are accumulated on the temperatures and oxygen conditions under which the bacteria grow, their response in fermentation tests, their pathogenicity, and their serological reactions. There are also modern methods for determining directly the similarity in deoxyribonucleic acids between different bacteria. *See* FERMENTATION; PATHOGEN.

Bacteria are said to be aerobic if they require oxygen and grow best at a high oxygen tension, usually 20% or more. Microaerophilic bacteria need oxygen, but grow best at, or may even require, reduced oxygen tensions, that is, less than 10%. Anaerobic bacteria do not require oxygen for growth. Obligatorily anaerobic bacteria can grow only in the complete absence of oxygen. Some bacteria obtain energy from the oxidation of reduced substances with compounds other than oxygen (O_2). The sulfate reducers use sulfate, the denitrifiers nitrate or nitrite, and the methanogenic bacteria carbon dioxide as the oxidizing agents, producing H_2S, nitrogen (N_2), and methane (CH_4), respectively, as reduction products.

Interrelationships may be close and may involve particular species. Examples are the parasitic association of many bacteria with plant and animal hosts, and the mutualistic association of nitrogen-fixing bacteria with leguminous plants, of cellulolytic bacteria with grazing animals, and of luminous bacteria with certain deep-sea fishes. *See* NITROGEN FIXATION; POPULATION ECOLOGY. [R.E.H.]

Endospores are resistant and metabolically dormant bodies produced by the gram-positive rods of *Bacillus* (aerobic or facultatively aerobic), *Clostridia* (strictly anaerobic), by the coccus *Sporosarcina*, and by certain other bacteria. Sporeforming bacteria are found mainly in the soil and water and also in the intestines of humans and animals. Some sporeformers are found as pathogens in insects; others are pathogenic to animals and humans. Endospores seem to be able to survive indefinitely. Spores kept for more than 50 years have shown little loss of their capacity to germinate and propagate by cell division. The mature spore has a complex structure which contains a number of layers. The unique properties of bacterial spores are their extreme resistance to heat, radiation from ultraviolet light and x-rays, organic solvents, chemicals, and desiccation. The capacity of a bacterial cell to form a spore is under genetic control, although the total number of genes specific for sporulation is not known. The actual phenotypic expression of the spore genome depends upon a number of external factors. For each species of sporeforming bacteria, there exist optimum conditions for sporogenesis which differ from the optimal conditions for vegetative growth. These conditions include pH, degree of aeration, temperature, metals, and nutrients. The three processes involved in the conversion of the spore into a vegetative cell are (1) activation (usually by heat or aging), which conditions the spore to germinate in a suitable environment; (2) germination, an irreversible process which results in the loss of the typical characteristics of a dormant spore; and (3) outgrowth, in which new classes of proteins and structures are synthesized so that the spore is converted into a new vegetative cell. [H.O.H.; K.Hu.; C.O.]

Bacterial genetics The study of gene structure and function in bacteria. Genetics itself is concerned with determining the number, location, and character of the genes of an organism. The classical way to investigate genes is to mate two organisms with different genotypes and compare the observable properties (phenotypes) of the parents with those of the progeny. Bacteria do not mate (in the usual way), so there is no way of getting all the chromosomes of two different bacteria into the same cell. However, there are a number of ways in which a part of the chromosome or genome from one bacterium can be inserted into another bacterium so that the outcome can be studied. *See* GENETICS.

All organisms have diverged from a common ancestral prokaryote whose precise location in the evolutionary tree is unclear. This has resulted in three primary kingdoms, the Archaebacteria, the Eubacteria, and the Eukaryotae. All bacteria are prokaryotes, that is, the "nucleus" or nucleoid is a single circular chromosome, without a nuclear membrane. Bacteria also lack other membrane-bounded organelles such as mitochondria or chloroplasts, but they all possess a cytoplasmic membrane. Most bacteria have a cell wall that surrounds the cytoplasmic membrane, and some bacteria also contain an outer membrane which encompasses the cell wall. Duplication occurs by a process of binary fission, in which two identical daughter cells arise from a single parent cell. Every cell in a homogeneous population of bacterial cells retains the potential for duplication. Bacteria do not possess the potential for differentiation (other than spore formation) or for forming multicellular organisms. *See* ARCHAEBACTERIA; BACTERIA; PROKARYOTAE.

One of the most frequently used organisms in the study of bacterial genetics is the rod-shaped bacillus *Escherichia coli*, whose normal habitat is the colon. Conditions have been found for growing *E. coli* in the laboratory, and it is by far the best understood of all microorganisms. The single circular chromosome of *E. coli* contains about 4.5×10^6 base pairs, which is enough to make about 4500 average-size genes (1000 base pairs each). In regions where mapping studies are reasonably complete, the impression is obtained of an efficiently organized genome. Protein coding regions are located adjacent to regulatory regions. There is no evidence for significant stretches of nonfunctional deoxyribonucleic acid (DNA), and there is no evidence for introns [regions that are removed by splicing the messenger RNA (mRNA) before it is translated into protein] in the coding regions. Very little repetitive DNA exists in the *E. coli* chromosome other than the seven sequence-related rRNA genes that are dispersed at different locations on the chromosome. *See* GENETIC CODE.

The first step in performing genetic research on bacteria is to select mutants that differ from wild-type cells in one or more genes. Then crosses are made between mutants and wild types, or between two different mutants, to determine dominance-recessive relationships, chromosomal location, and other properties. Various genetic methods are used to select bacterial mutants, antibiotic-resistant cells, cells with specific growth requirements, and so on.

Certain genes that have the function of modulating the expression of other genes are known as regulatory genes. Mutations that affect the action of regulatory proteins are of two types: those that occur in the genes that encode the regulatory proteins, and those that affect the genetic loci where the regulatory protein interacts to modulate the level of gene expression. Some regulatory gene mutations cause overproduction and some cause underproduction of gene products. This is the hallmark of a mutation that influences the functioning of a regulatory protein or regulatory factor-binding site; it affects the quantity but not the quality of other gene products. Furthermore, regulatory gene mutations are frequently pleiotropic, that is, they influence the rate of synthesis of several gene products simultaneously.

Frequently, geneticists want to increase the number or types of mutants that can be obtained as a result of spontaneous mutagenesis. In such instances, they treat a bacterial population with a mutagenic agent to increase the mutation frequency. This is called induced mutagenesis. The simplest techniques of induced mutagenesis involve measured exposure of the bacteria to a mutagenic agent, such as x-rays or chemical mutagenic agents. Such procedures have a general effect on the increase in the mutation rate. More sophisticated procedures involve isolating the gene of interest and making a change in the desired location. This is called site-directed mutagenesis. The goal is usually to determine the effects of a change at a specific gene locus. The gene in question is isolated, modified, and reinserted into the organism. Discrete alterations can be made in a variety of ways on any DNA in cell-free culture, and the effect of such alterations can be subsequently tested in the organism. *See* GENETIC ENGINEERING; MUTAGENS AND CARCINOGENS.

Bacteria do not mate to form true zygotes, but they are able to exchange genetic information by a variety of processes in which partial zygotes (merozygotes) are formed. The first type of genetic exchange between bacteria to be observed was transformation. Naturally occurring transformation involves the uptake of DNA. This phenomenon is observed only for a limited number of bacterial species and is a relatively difficult technique to use for gene manipulation. In 1946 direct chromosomal exchange by conjugation between *E. coli* cells was discovered by J. Lederberg and E. Tatum, and in 1951 transduction, the virus-mediated transfer of bacterial genes, was discovered. Both conjugation and transduction provide facile, generally applicable methods for moving part of the bacterial chromosome from one cell to another. The discovery of bacterial transposons (a class of mobile genetic elements commonly found in bacterial populations) in the 1970s has been useful in marking and mobilizing genes of interest. The purely genetic approaches to mapping have been supplemented by the biochemical approaches of hybrid plasmid construction and DNA sequence analysis.

At any given time, only a small percentage of the *E. coli* genome is being actively transcribed. The remainder of the genome is either silent or being transcribed at a very low rate. When growth conditions change, some active genes are turned off and other, inactive genes are turned on. The cell always retains its totipotency, so that within a short time (seconds to minutes), and given appropriate circumstances, any gene can be fully turned on. The maximal activity for transcription varies from gene to gene. For example, a β-galactosidase gene makes about one copy per minute, and a fully turned-on biotin synthase gene makes about one copy per 10 min. In the maximally repressed state, both of these genes express less than one transcript per 10 min. The level of transcription for any particular gene usually results from a complex series of control elements organized into a hierarchy that coordinates all the metabolic activities of the cell. For example, when the rRNA genes are highly active, so are the genes for ribosomal proteins, and the latter are regulated in such a way that stoichiometric amounts of most of the ribosomal proteins are produced. When glucose is abundant, most genes involved in processing more complex carbon sources are turned off in a process called catabolite repression. If the glucose supply is depleted and lactose is present, the genes involved in lactose breakdown (catabolism) are expressed. In *E. coli* the production of most RNAs and proteins is regulated exclusively at the transcriptional level, although there are notable exceptions. [G.Z.]

Bacterial growth The processes of both the increase in number and the increase in mass of bacteria. Growth has three distinct aspects: biomass production, cell production, and cell survival. Biomass production depends on the physical aspects of the environment (water content, pH, temperature), the availability of resources (carbon and energy, nitrogen, sulfur, phosphorus, minor elements), and the enzymatic machinery for catabolism (energy trapping), anabolism (biosynthesis of amino acid, purines, pyrimidines, and so forth), and macromolecular synthesis [proteins, ribonucleic acid (RNA), and deoxyribonucleic acid (DNA)]. Cell production is contingent on biomass production and involves, in addition, the triggering of chromosome replication and subsequent cell division. The cells may or may not separate from each other, and the division may partition the cell evenly or unevenly. Alternatively, growth may occur by budding (unequal division). Most cells so produced are themselves capable of growing and dividing; consequently, viability is usually very high when growth conditions are favorable. Moreover, in many cases the incidence of death is surprisingly low in the absence of needed nutrients. Many bacteria differentiate into resistant resting forms (such as spores); others may simply reduce their rate of metabolism and persist in the vegetative state for long times. [A.L.Ko.]

Bacterial luminescence The production of visible light by bacteria; with very few exceptions this light is blue-green. The phenomenon is seen in many species of several genera, including *Vibrio, Photobacterium, Alteromonas*, and *Xenorhabdus*. Luminous bacteria are primarily marine, but there are some genera with terrestrial (*Xenorhabdus*) and fresh-water (*Vibrio*) species. In the marine environment the bacteria are found in various habitats, including planktonic (free-floating), saprophytic (on a variety of marine proteinaceous materials), parasitic (on a number of marine invertebrates), and symbiotic. The symbiotic habitat can take one of several forms. The symbiotic bacteria may be loosely associated as gut symbionts in many different marine organisms; they may be specifically and more tightly associated in the light organs of marine fishes and squids; or they may be very tightly associated as intracellular symbionts in luminescent pyrosomes (light-omitting organelles). When associated as light-organ symbionts, the bacteria are used by the host fish or squid as a biological light bulb. Under these conditions the bacteria are maintained in specialized organs where they are cultured by the host organism, kept free from contaminants, and continuously emit light. The actual light emission is then controlled physically by the host's use of shutters, chromatophores, or other mechanisms. These symbiotic relationships are probably the most common habitats in which bacterial luminescence is observed in the ocean.

The chemistry of bacterial luminescence is unique among luminous organisms. The enzyme that catalyzes light emission is luciferase; it combines with a riboflavinlike substance called flavin mononucleotide ($FMNH_2$). This complex then reacts with a long-chain aldehyde, and with molecular oxygen to form an excited state capable of emitting light. The molecule that actually emits the light is an altered form of the flavin. This unique biochemistry has been used as an indicator of the presence of luminous bacteria in cases where the symbiotic bacteria could not be obtained in pure culture or could not be grown free from their host. *See* BACTERIAL PHYSIOLOGY AND METABOLISM.

[K.H.N.]

Bacterial physiology and metabolism The biochemical reactions that together enable bacteria to live, grow, and reproduce. Strictly speaking, metabolism describes the total chemical reactions that take place in a cell, while physiology describes the role of metabolic reactions in the life processes of a bacterium. The study of bacteria has significance beyond the understanding of bacteria themselves. Since bacteria are abundant, easily grown, and relatively simple in cellular organization, they have been used extensively in biological research. Functional analyses of bacterial systems have provided a foundation for much of the current detailed knowledge about molecular biology and genetics. Bacteria are prokaryotes, lacking the complicated cellular organization found in higher organisms; they have no nuclear envelope and no specialized organelles. Yet they engage in all the basic life processes—transport of materials into and out of the cell, catabolism and anabolism of complex organic molecules, and the maintenance of structural integrity. To accomplish this, bacteria must obtain nutrients and convert them into a form of energy that is useful to the cell. [M.R.J.S.]

Enzymes. A list of bacterial enzymes (organic catalysts) includes many of the enzymes found in mammalian tissues, as well as many enzymes not found in higher

Enzymes excreted by microorganisms of medical importance

Organism	Enzyme	Substrate	End products
Clostridium	Lecithinase	Lecithin	Diglyceride, phosphoryl choline
Clostridium	Collagenase	Collagen	?
Streptococcus	Hyaluronidase	Hyaluronic acid polymer	Hyaluronic acid
Streptococcus	Streptodornase	Deoxyribonucleic acid	Nucleotides
Streptococcus	Streptokinase	Activates plasminogen to plasmin	Results in lysis of fibrin clots
Staphylococcus	Coagulase	Coagulase reacting factor	Results in coagulation of plasma
Proteus	Urease	Urea	Ammonia and carbon dioxide
Corynebacterium diphtheriae	Diphtheria toxin	Nicotinamide dinucleotide (NAD)	Splits NAD and adds ADP-ribose to elongation factor 2 to prevent protein synthesis by freezing ribosome movement

forms of life. By combining with such enzymes, many antibiotics are able to exert a selective killing or inhibition of bacterial growth without causing toxic reactions in the mammalian host. The great capability of the bacterial cell to metabolize a wide variety of substances, as well as to control to some extent the environment in which the cell lives, is reflected in its ability to form inducible enzymes. The majority of bacterial enzymes require cofactors for activity. These cofactors may be inorganic cations of organic molecules called coenzymes.

Bacterial enzymes may be classified in numerous ways, for example, on the basis of (1) whether they are inducible or constitutive (constitutive enzymes are defined as those enzymes formed by the bacterial cell under any or all conditions of growth, whereas inducible enzymes are formed by the bacterial cell only in response to an inducer); (2) whether they are degradative (catabolic; resulting in the release of energy) or synthetic (anabolic; using energy to catalyze the formation of macromolecules); or (3) whether they are exoenzymes (enzymes secreted from the cell to hydrolyze insoluble polymers—wood, starch, protein, and so on—into smaller, soluble compounds which can be taken into the cytoplasm of the bacterium).

In addition, bacterial enzymes are involved in the transport of substrates across the cell wall, in the oxidation of inorganic molecules to provide energy for the cell, and in the destruction of a large number of antibiotics.

Many pathogenic microorganisms excrete enzymes which may play an important role in pathogenesis in some cases (see table). The α-toxin (lecithinase) of *Clostridium perfringens* illustrates a highly active enzyme which is responsible for the necrotizing action associated with gas gangrene infections due to this microorganism. *Streptococcus pyogenes* excretes hyaluronidase which degrades ground substance (polymer of hyaluronic acid), and streptokinase, which activates plasmin resulting in a system that lyses fibrin. Other examples include coagulase of the *Staphylococcus*, which activates clotting of plasma, urease of *Proteus vulgaris*, which splits urea to ammonia and carbon dioxide, and collagenase of *Clostridium*, which hydrolyzes collagen. *See* DIPHTHERIA; STAPHYLOCOCCUS.

Many bacteria are able to synthesize enzymes which will hydrolyze or modify an antibiotic so that it is no longer effective. Essentially all of these enzymes are coded by DNA that exists in bacterial plasmids. As a result, the ability to produce enzymes which destroy antibiotics can be rapidly passed from one organism to another either by conjugation in gram-negative organisms or by transduction in both gram-negative and gram-positive bacteria.

[W.A.V.]

Bacterial catabolism. Bacterial catabolism comprises the biochemical activities concerned with the net breakdown of complex substances to simpler substances by living cells. Substances with a high energy level are converted to substances of low energy content, and the organism utilizes a portion of the released energy for cellular processes. Endogenous catabolism relates to the slow breakdown of nonvital intracellular constituents to secure energy and replacement building blocks for the maintenance of the structural and functional integrity of the cell. This ordinarily occurs in the absence of an external supply of food. Exogenous catabolism refers to the degradation of externally available food. The principal reactions employed are dehydrogenation or oxygenation (either represents biological oxidation), hydrolysis, hydration, decarboxylation, and intermolecular transfer and substitution. The complete catabolism of organic substances results in the formation of carbon dioxide, water, and other inorganic compounds and is known as mineralization. Catabolic processes may degrade a substance only part way. The resulting intermediate compounds may be reutilized in biosynthetic processes, or they may accumulate intra- or extracellularly. Catabolism also implies a conversion of the chemical energy into a relatively few energy-rich compounds or "bonds," in which form it is biologically useful; also, part of the chemical energy is lost as heat.

Bacterial intermediary metabolism relates to the chemical steps involved in metabolism between the starting substrates and the final product. Normally these intermediates, or precursors of subsequent products, do not accumulate inside or outside the bacterial cell in significant amounts, being transformed serially as rapidly as they are formed. The identification of such compounds, the establishment of the coenzymes and enzymes catalyzing the individual reaction steps, the identity of active forms of the intermediates, and other details of the reaction mechanisms are the objectives of a study of bacterial intermediary metabolism. [J.W.Fo./R.E.K.]

Many bacteria are able to decompose organic compounds and to grow in the absence of oxygen gas. Such anaerobic bacteria obtain energy and certain organic compounds needed for growth by a process of fermentation. This consists of an oxidation of a suitable organic compound, using another organic compound as an oxidizing agent in place of molecular oxygen. In most fermentations both the compounds oxidized and the compounds reduced (used as an oxidizing agent) are derived from a single fermentable substrate. In other fermentations, one substrate is oxidized and another is reduced. Different bacteria ferment different substrates. Many bacteria are able to ferment carbohydrates such as glucose and sucrose, polyalchohols such as mannitol, and salts of organic acids such as pyruvate and lactate. Other compounds, such as cellulose, amino acids, and purines, are fermented by some bacteria. [H.A.B.]

Bacterial anabolism. Bacterial anabolism comprises the physiological and biochemical activities concerned with the acquisition, synthesis, and organization of the numerous and varied chemical constituents of a bacterial cell. Clearly, when a cell grows and divides to form two cells, there exists twice the amount of cellular components that existed previously. These components are drawn, directly or indirectly, from the environment around the cell, and (usually) modified extensively in the growth processes when new cell material is formed (biosynthesis). This build-up, or synthesis, begins with a relatively small number of low-molecular-weight building blocks which are either assimilated directly from the environment or produced by catabolism. By sequential and interrelated reactions, they are fashioned into different molecules (mostly of high molecular weight, and hence called macromolecules), for example, lipids, polysaccharides, proteins, and nucleic acids, and many of these molecules are in turn arranged into more complex arrays such as ribosomes, membranes, cell walls, and flagella. Other typical anabolic products, of lower molecular weight, include pigments, vitamins, antibiotics, and coenzymes. The enzymes responsible for the sequential reactions in any one biosynthetic pathway or assembly sequence are often located on or in cellular structures and thus in physical proximity to the preceding and succeeding enzymes, and their products, and to the site(s) where cellular structures are to be formed. Anabolism also includes the transport of molecules into cells, of building blocks to reaction sites, energetic activations, and the transfer and incorporation of the finished products to their ultimate sites in or outside the cell. [E.R.L.]

Bacteriology The science and study of bacteria, and hence a specialized branch of microbiology. It deals with the nature and properties of the bacteria as living entities, their morphology and developmental history, ecology, physiology and biochemistry, genetics, and classification.

The major subjects that have consecutively occupied the forefront of bacteriological research have been the origin of bacteria, the constancy or variability of their properties, their role as causative agents of disease and of spoilage of foods, their significance in the cycle of matter, their classification, and their physiological, biochemical, and genetic features. *See* BACTERIA; MICROBIOLOGY. [C.B.V.N.]

Bacteriophage Any of the viruses that infect bacterial cells. They are discrete particles with dimensions from about 20 to about 200 nanometers. A given bacterial virus can infect only one or a few related species of bacteria; these constitute its host range. Bacteriophages consist of two essential components: nucleic acid, in which genetic information is encoded (this may be either ribonucleic acid or deoxyribonucleic acid), and a protein coat (capsid), which serves as a protective shell containing the nucleic acid and is involved in the efficiency of infection and the host range of the virus.

The description of a bacterial virus involves a study of its shape and dimensions by electron microscopy (see illustration), its host range, the serological properties of its capsid, the kind of nucleic acid it contains, and the characters of the plaques it

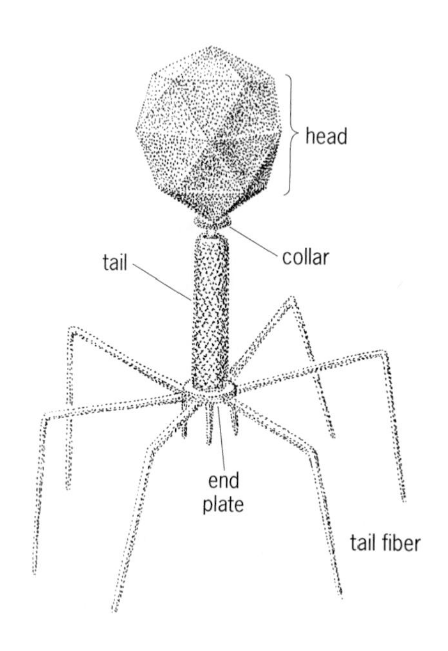

Diagram of a T4 bacteriophage.

forms on a given host. Both the nucleic acid and the capsid proteins are specific to the individual virus; in the case of the capsid proteins this specificity is the basis for serological identification of the virus.

The most striking form of phage infection is that in which all of the infected bacteria are destroyed in the process of the formation of new phage particles. This results in the clearing of a turbid liquid culture as the infected cells lyse. When lysis occurs in cells fixed as a lawn of bacteria growing on a solid medium, it produces holes, or areas of clearing, called plaques. These represent colonies of bacteriophage. The size and other properties of the plaque vary with individual viruses and host cells. *See* COLIPHAGE; VIRUS. [L.B.]

Balsa A fast-growing tree, *Ochroma lagopus*, widely distributed in tropical America, especially in Ecuador. The leaves are simple, angled, or lobed, and the flowers are large and yellowish-white or brownish, and they are terminal on the branches.

With plenty of room for growth in a rich, well-drained soil at low elevations, the wood is very light and soft. However, under adverse conditions, the wood is heavier. Culture is important, for if the trees are injured only slightly, the wood develops a hard and fibrous texture, thereby losing its commercial value. To secure a uniform product the trees must be grown in plantations.

The wood decays easily in contact with the soil and is subject to sap stain if not promptly dried. Seasoned lumber absorbs water quickly, but this can be largely overcome by waterproofing.

Balsa owes most of its present commercial applications to its insulating properties. Balsa also has sound-deadening qualities, and is also used under heavy machinery to prevent transmission of vibrations. The heartwood of balsa is pale brown or reddish, whereas the sapwood is nearly white, often with a yellowish or pinkish hue. Luster is usually rather high, and the wood is odorless and tasteless. [A.H.G./K.P.D.]

Baltic Sea A semienclosed brackish sea located in a humic zone, with a positive water balance relative to the adjacent ocean (the North Sea and the North Atlantic). The Baltic is connected to the North Sea by the Great Belt (70% of the water exchange), the Øresund (20% of the water exchange), and the Little Belt. The total area of the Baltic is 147,414 mi^2 (381,705 km^2), its total volume 4982 mi^3 (20,764 km^3), and its average depth 181 ft (55.2 m). The greatest depth is 1510 ft (459 m), in the Landsort Deep.

The topography of the Baltic is characterized by a sequence of basins separated by sills and by two large gulfs, the Gulf of Bothnia (40,100 mi^2 or 104,000 km^2) and the Gulf of Finland (11,400 mi^2 or 29,500 km^2). More than 200 rivers discharge an average of 104 mi^3 (433 km^3) annually from a watershed area of 637,056 mi^2 (1,649,550 km^2). The largest river is the Newa, with 18.5% of the total fresh-water discharge. From December to May, the northern and eastern parts of the Baltic are frequently covered with ice. On the average, the area of maximum ice coverage is 82,646 km^2 (214,000 km^2). The mean maximum surface-water temperature in summer is between 59 and 63°F (15 and 17°C).

As the Baltic stretches from the boreal to the arctic continental climatic zone, there are large differences between summer and winter temperature in the surface waters, ranging from about 68 to 30°F (20 to −1°C) in the Western Baltic and 57 to 32°F (14 to −0.2°C) in the Gulf of Bothnia and the Gulf of Finland.

The salt content of the Baltic waters is characterized by two major water bodies; the brackish surface water and the more saline deep water. Salinities for the surface water range from 8 to 6‰ in the Western and Central Baltic and 6 to 2000 in the Gulf of

Bothnia and the Gulf of Finland; salinities for the deep water range from 18 to 13‰ in the Western and Central Baltic and 10 to 4‰ in the Gulf of Bothnia and the Gulf of Finland.

The surface currents of the Baltic are dominated by a general counterclockwise movement and by local and regional wind-driven circulations. A complex system of small- and medium-scale gyres develops especially in the central parts of the Baltic. The currents in the Belt Sea are dominated by the topography; they are due to sea-level differences between the Baltic proper and the North Sea. Tides are of minor importance, ranging between 0.8 and 4.7 in. (2 and 12 cm). Water-level changes of more than 6 ft (2 m) occur occasionally as a result of onshore or offshore winds and the passage of cyclones over the Baltic Sea area. The frequency of longitudinal sea-level oscillations is about 13.5 h. *See* OCEAN CIRCULATION.

The flora and fauna of the Baltic are those of a typical brackish-water community, with considerably reduced numbers of species compared to an oceanic community. The productivity is relatively low compared to other shelf seas. The major commercially exploited species are cod, herring, sprat, flounder, eel, and salmon, and some fresh-water species such as whitefish, pike, perch, and trout. The total annual catch amounts to about 880,000 tons (800,000 metric tons). The Baltic is completely divided into fishery zones, with exclusive fishing rights belonging to the respective countries.

Other than fish the only major resources that have been exploited are sand and gravel in the Western Baltic Sea. It is believed that the deeper layer under the Gotland Basin contains mineral oil, but so far only exploratory drilling has been carried out in the near-coastal regions. Limited amounts of mineral oil have also been located in the Gulf of Kiel.

[K.G.]

Bamboo The common name of various perennial, ornamental grasses (Gramineae). There are five genera with approximately 280 species. They have a wide distribution, but occur mainly in tropical and subtropical parts of Asia, Africa, and America, extending from sea level to an elevation of 15,000 ft (4600 m). Their greatest development occurs in the monsoon regions of Asia. Most plants are woody; a few are herbaceous or climbing. The economic uses of bamboo are numerous and varied. The seeds and young shoots are used as food, and the leaves make excellent fodder for cattle. In varying sizes, the stems are used for pipes, timber, masts, bows, furniture, bridges, cooking vessels, buckets, wickerwork, paper pulp, cordage, and weaving. Entire houses are made of bamboo stems. Certain bamboos have been naturalized in California, Louisiana, and Florida.

[P.D.St./E.L.C.]

Bark A word generally referring to the surface region of a stem or a root. Sometimes part or all of the bark is called rind. Occasionally the word bark is used as a substitute for periderm or for cork only. Most commonly, however, it refers to all tissues external to the cambium. If this definition is applied to stems having only primary tissues, it includes phloem, cortex, and epidermis; the bark of roots of corresponding age would contain cortex and epidermis. In the more general usage the term bark is restricted to woody plants with secondary growth.

In most roots with secondary growth, the bark consists of phloem and periderm since the cortex and epidermis are sloughed off with the first cork formation in the pericycle that is beneath the cortex. In stems, the first cork cambium may be formed in any living tissue outside the vascular cambium, and the young bark may include any or all of the cortex in addition to the phloem and periderm. The region composed of the successive layers of periderm and the enclosed dead tissues is called outer bark. The outer bark composed of dead phloem alternating with bands of cork is called, technically, rhytidome. Both stems and roots may have rhytidome. The inner bark is living, and consists of phloem only. *See* CORTEX (PLANT); EPIDERMIS (PLANT); PARENCHYMA; PERICYCLE; PERIDERM; PHLOEM; ROOT (BOTANY); STEM. [H.W.Bl.]

Barrier islands Elongate, narrow accumulations of sediment which have formed in the shallow coastal zone and are separated from the mainland by some combination of coastal bays and marshes. They are typically several times longer than their width and are interrupted by tidal inlets. Although their origin has been widely discussed, at least three possibilities exist: longshore spit development and subsequent cutting of inlets; drowning of old coastal ridges; and upward shoaling of subtidal sediment accumulations. All three may have occurred; however, the last seems most likely and most prevalent.

Barrier islands must be considered in terms of the adjacent and closely related environments within the coastal system. Beginning offshore and proceeding landward, the sequence of environments crossed is shoreface, beach, dunes, back-island flats or marsh, coastal bay, marsh, and mainland. The barrier island proper consists of the beach, dunes, and back-island flats or marsh; however, of the remaining environments, at least the shoreface is closely integrated with the barrier island in terms of morphology, processes, and sediments. *See* DUNE.

A variety of physical processes exists along the coast. These processes act to shape and maintain the barrier-island system and also to enable the barrier to migrate landward as sea level continues to rise. The most important process in the barrier-island system is the waves, which also give rise to longshore currents. Waves and longshore currents dominate the outer portion of the barrier system, whereas tidal currents are dominant landward of the barrier, although small waves may also be present. Tidal currents are most prominent in and adjacent to the inlets. On the supratidal portion of the barrier island, the wind is the most dominant physical process.

Barrier-island sands represent one of the best sources of oil and gas, with the tight organic-rich source rocks being in the form of the bay and shelf muds and the barrier itself being the reservoir rock. These elongate sand bodies have been sought by exploration geologists for decades. The Tertiary sequences of the Texas Gulf coasts are an example of such barrier systems which have been very productive. *See* COASTAL LANDFORMS. [R.A.D.]

Basidiomycota A phylum in the kingdom fungi; commonly known as basidiomycetes. Basidiomycetes traditionally included four artificial classes: Hymenomycetes, Gasteromycetes, Urediniomycetes, and Ustilaginomycetes. They are mostly filamentous fungi characterized by the production of basidia. These are microscopic, often club-shaped end cells in which nuclear fusion and meiosis usually take place prior to the maturation of external, typically haploid basidiospores, which are then disseminated. Common basidiomycetes are the rusts and smuts, which cause severe plant diseases, mushrooms (edible and poisonous), boletes, puffballs, stinkhorns, chanterelles, false truffles, jelly fungi, bird's-nest fungi, and conk or bracket fungi. Basidiomycetes are the most important decayers of wood, living or dead, in forests or buildings, causing either brown rot (for example, dry rot) or white rot. Many, especially mushrooms and boletes, are the primary fungal partners in symbiotic ectomycorrhizal associations with tree roots. Plant litter and soil are other major habitats. A few basidiomycetes are cultivated for food. Some are luminescent, hallucinogenic, lichenized, nematophagous, or aquatic. Some are cultivated by ants or termites, or are symbiotic with captured scale insects. Some can convert to a yeast (or single-cell) phase, one of which causes cryptococcosis in humans and animals. *See* MUSHROOM; MYCORRHIZAE; RUST (MICROBIOLOGY); SMUT (MICROBIOLOGY); WOOD DEGRADATION. [S.A.R.]

Basin A low-lying area which is wholly or largely surrounded by higher land. An example is Hudson Bay in northeastern Canada, which was formed by depression beneath the center of a continental ice sheet 18,000 years ago. Another example, the Qattara depression, is 150 mi (240 km) long and the largest of several wind-excavated basins of northern Egypt. Depressions in the ocean floor are also basins, such as the Canary Basin, west of northern Africa, or the Argentine Basin, east of Argentina. These basins occur in regions where cold, dense oceanic crust lies between the topographically elevated ocean ridges and the continental margins. *See* CONTINENTAL MARGIN; MARINE GEOLOGY.

A drainage basin is the entire area drained by a river and its tributaries. Thus, the Mississippi Basin occupies most of the United States between the Rocky Mountains and the Appalachians. Interior drainage basins consist of depressions that drain entirely inward, without outlet to the sea. Examples may be quite small, such as the Salton Sea of southern California or the Dead Sea of central Asia. One of the most remarkable examples of an interior drainage basin is the Chad Basin in northern Africa, the center of which is occupied by Lake Chad. The fresh waters of the lake drain underground to feed oases in the lowlands 450 mi (720 km) to the northeast.

In the geologic sense, a basin is an area in which the continental crust has subsided and the depression has been filled with sediments. Such basins were interior drainage basins at the time of sediment deposition but need not be so today. As these basins subside, the layers of sediment are tilted toward the axis of maximum subsidence. Consequently, when the sedimentary layers are observed in cross section, their geometry is a record of the subsidence of the basin through time and contains clues about the origin of the basin.

The origin of geologic basins is a topic of continuing interest in both applied and basic geological studies. They contain most of the world's hydrocarbon reserves, and they are regarded as some of the best natural laboratories in which to understand the thermal and mechanical processes that operate deep in the interior of the Earth and that shape the Earth's surface. [G.Bo.; M.Ko.]

Basswood A member of the linden family in the order Malvales. One species, known as the American linden (*Tilia americana*), is a timber tree of the northeastern quarter of the United States and the adjacent area of Canada. *Tilia* is also an ornamental tree. *Tilia europea*, or lime tree of Europe, is often cultivated along the streets. The lindens are also important as bee trees. The leaves are heart-shaped, coarsely toothed, long, pointed, and alternate. All species of *Tilia* can be recognized by the winter buds, which have a large outer scale that gives the bud a humped, asymmetrical appearance, and by the small, spherical, nutlike fruits borne in clusters.

The wood of basswood, also known as whitewood, is white and soft and is used for boxes, venetian blinds, millwork, furniture, and woodenware. There are about 30 species in the temperate regions of the Northern Hemisphere in North America south to Mexico, but none in the western part. In Asia lindens grow south to central China and southern Japan. [A.H.G./K.P.D.]

Beech A genus, *Fagus*, of deciduous trees of the beech family Fagaceae, order Fagales. They can best be distinguished by their long (often more than 1 in. or 2.5 cm), slender, scaly winter buds; their thin, gray bark, smooth even in old trees; and their simple, toothed, ovate or ovate-oblong, deciduous leaves.

The American beech (*F. grandifolia*) is native in the United States east of the Mississippi River and in the lower Mississippi Valley. The hard, strong wood is used for furniture, handles, woodenware, cooperage, and veneer. The small, edible, three-sided nuts, called beechnuts, are valuable food for wildlife. The European beech (*F. sylvatica*) is more popular as an ornamental tree than the American species. Its leaves are smaller, with 5–9 pairs of primary veins compared with 9–14 pairs in the American beech. The leaf teeth are also shorter. Important ornamental varieties are *F. sylvatica purpurea*, the copper or purple beech; var. *incisa*, the cut-leaved or fern-leaved beech; and *F. pendula*, the weeping European beech. [A.H.G./K.P.D.]

Behavioral ecology The branch of ecology that focuses on the evolutionary causes of variation in behavior among populations and species. Thus it is concerned with the adaptiveness of behavior, the ultimate questions of why animals behave as they do, rather than the proximate questions of how they behave. The principles of natural selection are applied to behavior with the underlying assumption that, within the constraints of their evolutionary histories, animals behave optimally by maximizing their genetic contribution to future generations. For example, animals must maintain their internal physiological conditions within certain limits in order to function properly, and often they do this by behavior. Small organisms may avoid desiccation by living under logs or by burrowing. Many insects must raise body temperatures to 86–95°F (30–35°C) for effective flight, and achieve this by muscular activity such as the shivering of butterflies in the early morning or by orienting to the Sun. Other adaptive behaviors that are studied may fall in the categories of habitat selection, foraging, territoriality, and reproduction. *See* ETHOLOGY; MIGRATORY BEHAVIOR. [H.Di.; P.Fr.]

Behavioral toxicology The study of behavioral abnormalities induced by exogenous agents such as drugs, chemicals in the general environment, and chemicals encountered in the workplace. Just as some substances are hazardous to the skin or liver, some are hazardous to the function of the nervous system. In the case of permanent effects, changes in sensation, mood, intellectual function, or motor coordination would obviously be undesirable, but even transient alterations of behavior are considered toxic in some situations. For example, operating room personnel accidentally exposed to very small doses of anesthetic do not exhibit altered performance on an intelligence test, a dexterity test, or a vigilance task. However, significant decrements in performance occur in recognizing and recording visual displays, detecting changes in audiovisual displays, and recalling series of digits.

By comparing the behavior of exposed subjects and control subjects, behavioral toxicologists seek to identify agents capable of altering behavior and to determine the level of exposure at which undesirable effects occur. When the agent under study is one in common use, and there is no evidence of its being hazardous to health, experiments may be carried out on human volunteers, or comparisons may be made from epidemiologic data. More frequently, safety considerations dictate the use of laboratory animals in toxicology research.

Perhaps the best-known example of toxicity in humans is methyl mercury poisoning (Minimata disease), which occurred in epidemic proportions in a Japanese coastal town where the inhabitants ate fish contaminated with mercury from industrial pollution. Although mercury affects a variety of behaviors, the most obvious symptoms are tremors and involuntary movements.

A different set of functional problems is exemplified by the effects of ethyl alcohol, a single agent with direct and indirect, short- and long-term consequences. The short-term, low-dose effects of alcohol include sensory disturbances, motor problems, and difficulties with processing information. Neurologically, alcohol is usually described as a central nervous system depressant which is general, in the sense that it disrupts many functions.

In some individuals, large quantities of alcohol consumed over a long period lead to permanent damage to the nervous system. Behaviorally, individuals with Korsakoff's syndrome exhibit severe memory deficits. Anatomically, their brains are found to have degenerative changes in the thalamus. This syndrome is not just an extension of the short-term effects. In fact, it is thought to arise from alcohol-induced malnutrition rather than as a direct effect of alcohol itself.

Lasting injuries to the nervous system have been reported to occur in children exposed to alcohol before birth. The behavioral problems associated with fetal alcohol syndrome do not appear to be related to either Korsakoff's syndrome or the immediate effects of alcohol. Rather, they constitute a third set of effects, including learning deficits, problems with inhibiting inappropriate behavior, and fine motor dysfunction, along with some visible physical abnormalities. While malnutrition may play a role in this congenital syndrome, the mechanism and locus of damage are not known.

When toxicity is considered only in terms of direct risk to survival, behavioral toxicity may seem to be of minor importance. However, survival is not the only criterion of good health. In a complex society that places heavy demands on an individual's educability, alertness, and emotional stability, even small deviations in behavior are potentially hazardous. Severe disabilities, as in the gross motor malfunctions of Minimata disease, have drawn attention to behavioral toxicology. Such incidents represent failures of control of toxic substances. Successes are difficult to measure, for they can be seen only in reduction of risk—the ultimate goal of toxicology. *See* TOXICOLOGY. [P.M.R.]

Bering Sea A water body north of the Pacific Ocean, 875,000 mi^2 (2,268,000 km^2) in area, bounded by Siberia, Alaska, and the Aleutian Islands. The Bering Sea is a biologically productive area, with large populations of marine birds and mammals. An active pollock fishery and a developing bottom-fish industry are evidence of its rich biological resources.

The Bering Sea consists of a large, deep basin in the southwest portion, where depths as great as 9900 ft (3000 m) are encountered. To the north and east, an extremely wide, shallow continental shelf extends north to the Bering Strait. The two major regions are separated by a shelf break, the position of which coincides with the southernmost extent of sea ice in a cold season. Ice is a prominent feature of the Bering Sea shelf during the cold months. Coastal ice begins to form in late October, and by February coastal ice is found in the Aleutians. The sea ice may extend as far south as 58°N. Thus, the ice edge in the eastern Bering Sea advances and retreats seasonally over a distance as great as 600 mi (1000 km). Ice-free conditions can be expected throughout the entire region by early July. *See* SEA ICE.

The main water connections with the Pacific are in the west of the Aleutian Islands, the 6600-ft-deep (2000-m) pass between Attu and Komandorskiye Islands and the 14,000-ft-deep (4400-m) pass between the Komandorskiyes and Kamchatka. Aleutian passes also serve to exchange water. The Bering Sea connection with the Arctic Ocean (Chukchi Sea) is the Bering Strait, 53 mi (85 km) wide and 30 mi (45 m) deep.

Tides in the Bering Sea are semidiurnal, with a strong diurnal inequality typical of North Pacific tides. Three water masses are associated with Bering sea water—Western Subarctic, Bering Sea, and the Alaskan Stream. The general circulation of the Bering Sea is counterclockwise, with many small eddies superimposed on the large-scale pattern. The currents in the Bering Sea are generally a few centimeters per second except along the continental slope, the coast of Kamchatka, and in certain eddies, where somewhat higher values have been found. See OCEAN CIRCULATION. [V.A.]

Biochemical engineering The application of engineering principles to conceive, design, develop, operate, or use processes and products based on biological and biochemical phenomena. Biochemical engineering, a subset of chemical engineering, impacts a broad range of industries, including health care, agriculture, food, enzymes, chemicals, waste treatment, and energy. Historically, biochemical engineering has been distinguished from biomedical engineering by its emphasis on biochemistry and microbiology and by the lack of a health care focus. However, now there is increasing participation of biochemical engineers in the direct development of health care products. Biochemical engineering has been central to the development of the biotechnology industry, especially with the need to generate prospective products (often using genetically engineered microorganisms) on scales sufficient for testing, regulatory evaluation, and subsequent sale. See BIOTECHNOLOGY.

In the discipline's initial stages, biochemical engineers were chiefly concerned with optimizing the growth of microorganisms under aerobic conditions at scales of up to thousands of liters. While the scope of the discipline has expanded, this focus remains. Often the aim is the development of an economical process to maximize biomass production (and hence a particular chemical, biochemical, or protein), taking into consideration raw-material and other operating costs. The elemental constituents of biomass (carbon, nitrogen, oxygen, hydrogen, and to a lesser extent phosphorus, sulfur, mineral salts, and trace amounts of certain metals) are added to the biological reactor (often called a fermentor) and consumed by the bacteria as they reproduce and carry out metabolic processes. Sufficient amounts of oxygen (usually supplied as sterile air) are added to the fermentor in such a way as to promote its availability to the growing culture. See BIOMASS; FERMENTATION.

In some situations, microorganisms may be cultivated whose activity is adversely affected by the presence of dissolved oxygen. Anaerobic cultures are typical of fermentations in which organic acids and solvents are produced; these systems are usually characterized by slower growth rates and lower biomass yields. The largest application of anaerobic microorganisms is in waste treatment, where anaerobic digesters containing mixed communities of anaerobic microorganisms are used to reduce the quantity of solids in industrial and municipal wastes. See SEWAGE TREATMENT.

While the operation and optimization of large-scale, aerobic cultures of microorganisms is still of major importance in biochemical engineering, the capability of cultivating a wide range of cell types has become important also. Biochemical engineers are often involved in the culture of plant cells, insect cells, and mammalian cells, as well as the genetically engineered versions of these cell types. Metabolic engineering uses the tools of molecular genetics, often coupled with quantitative models of metabolic pathways and bioreactor operation, to optimize cellular function for the production of specific metabolites and proteins. Enzyme engineering focuses on the identification, design, and use of biocatalysts for the production of useful chemicals and biochemicals. Tissue engineering involves material, biochemical, and medical aspects related to the transplant of living cells to treat diseases. Biochemical engineers are also actively involved in many aspects of bioremediation, immunotechnology, vaccine development, and the use of cells and enzymes capable of functioning in extreme environments. [R.M.Ke.]

Biodegradation The destruction of organic compounds by microorganisms. Microorganisms, particularly bacteria, are responsible for the decomposition of both natural and synthetic organic compounds in nature. Mineralization results in complete conversion of a compound to its inorganic mineral constituents (for example, carbon dioxide from carbon, sulfate or sulfide from organic sulfur, nitrate or ammonium from organic nitrogen, phosphate from organophosphates, or chloride from organochlorine). Since carbon comprises the greatest mass of organic compounds, mineralization can be considered in terms of CO_2 evolution. Radioactive carbon-14 (^{14}C) isotopes enable scientists to distinguish between mineralization arising from contaminants and soil organic matter. However, mineralization of any compound is never 100% because some of it (10–40% of the total amount degraded) is incorporated into the cell mass or products that become part of the amorphous soil organic matter, commonly referred to as humus. Thus, biodegradation comprises mineralization and conversion to innocuous products, namely biomass and humus. Primary biodegradation is more limited in scope and refers to the disappearance of the compound as a result of its biotransformation to another product. *See* HUMUS.

Compounds that are readily biodegradable are generally utilized as growth substrates by single microorganisms. Many of the components of petroleum products (and frequent ground-water contaminants), such as benzene, toluene, ethylbenzene, and xylene, are utilized by many genera of bacteria as sole carbon sources for growth and energy.

The process whereby compounds not utilized for growth or energy are nevertheless transformed to other products by microorganisms is referred to as cometabolism. Chlorinated aromatic hydrocarbons, such as diphenyldichloroethane (DDT) and polychlorinated biphenyls (PCBs), are among the most persistent environmental contaminants; yet they are cometabolized by several genera of bacteria, notably *Pseudomonas*, *Alcaligenes*, *Rhodococcus*, *Acinetobacter*, *Arthrobacter*, and *Corynebacterium*. Cometabolism is caused by enzymes that have very broad substrate specificity. *See* BACTERIAL GROWTH.

The use of microorganisms to remediate the environment of contaminants is referred to as bioremediation. This process is most successful in contained systems such as surface soil or ground water where nutrients, mainly inorganic nitrogen and phosphorus, are added to enhance growth of microorganisms and thereby increase the rate of biodegradation. The process has little, if any, applicability to a large open system such as a bay or lake because the nutrient level (that is, the microbial density) is too low to effect substantive biodegradation and the system's size and distribution preclude addition of nutrients.

Remediation of petroleum products from ground waters is harder to achieve than surface soil because of the greater difficulty in distributing the nutrients throughout the zone of contamination, and because of oxygen (O_2) limitations. [D.D.Fo.]

Biodiversity The variety of all living things; a contraction of biological diversity. Biodiversity can be measured on many biological levels ranging from genetic diversity within a species to the variety of ecosystems on Earth, but the term most commonly refers to the number of different species in a defined area.

Recent estimates of the total number of species range from 7 to 20 million, of which only about 1.75 million species have been scientifically described. The best-studied groups include plants and vertebrates (phylum Chordata), whereas poorly described groups include fungi, nematodes, and arthropods (see table). Species that live in the ocean and in soils remain poorly known. For most groups of species, there is a gradient of increasing diversity from the Poles to the Equator, and the vast majority of species are concentrated in the tropical and subtropical regions.

Human activities, such as direct harvesting of species, introduction of alien species, habitat destruction, and various forms of habitat degradation (including environmental

Numbers of extant species for selected taxonomic groups

Kingdom	Phylum	Number of species described	Estimated number of species	Percent described
Protista		100,000	250,000	40.0
Fungi	Eumycota	80,000	1,500,000	5.3
Plantae	Bryophyta	14,000	30,000	46.7
	Tracheophyta	250,000	500,000	50.0
Animalia	Nematoda	20,000	1,000,000	2.0
	Arthropoda	1,250,000	20,000,000	5.0
	Mollusca	100,000	200,000	50.0
	Chordata	40,000	50,000	80.0

*With permission, modified from G. K. Meffe and C. R. Carroll, *Principles of Conservation Biology*, 1997.

pollution), have caused dramatic losses of biodiversity; current extinction rates are estimated to be 100–1000 times higher than prehuman extinction rates.

Some measure of biodiversity is responsible for providing essential functions and services that directly improve human life. For example, many medicines, clothing fibers, and industrial products and the vast majority of foods are derived from naturally occurring species. In addition, species are the key working parts of natural ecosystems. They are responsible for maintenance of the gaseous composition of the atmosphere, regulation of the global climate, generation and maintenance of soils, recycling of nutrients and waste products, and biological control of pest species. Ecosystems surely would not function if all species were lost, although it is unclear just how many species are necessary for an ecosystem to function properly. [M.A.Ma.]

Biofilm An adhesive substance, the glycocalyx, and the bacterial community which it envelops at the interface of a liquid and a surface. When a liquid is in contact with an inert surface, any bacteria within the liquid are attracted to the surface and adhere to it. In this process the bacteria produce the glycocalyx. The bacterial inhabitants within this microenvironment benefit as the biofilm concentrates nutrients from the liquid phase. However, these activities may damage the surface, impair its efficiency, or develop within the biofilm a pathogenic community that may damage the associated environment. Microbial fouling or biofouling are the terms applied to these actual or potentially undesirable consequences.

Microbial fouling affects a large variety of surfaces under various conditions. Microbial biofilms may form wherever bacteria can survive; familiar examples are dental plaque and tooth decay. Dental plaque is an accumulation of bacteria, mainly streptococci, from saliva. The process of tooth decay begins with the bacteria colonizing fissures in and contact points between the teeth. Dietary sucrose is utilized by the bacteria to form extracellular glucans that make up the glycocalyx and assist adhesion to the tooth. Within this microbial biofilm or plaque the metabolic by-products of the bacterial inhabitants are trapped; these include acids that destroy the tooth enamel, dentin, or cementum. [H.L.Sc.; J.W.C.]

Biogeochemistry The study of the cycling of chemicals between organisms and the surface environment of the Earth. The chemicals either can be taken up by organisms and used for growth and synthesis of living matter or can be processed to obtain energy. The chemical composition of plants and animals indicates which elements, known as nutrient elements, are necessary for life. The most abundant nutrient elements, carbon (C), hydrogen (H), and oxygen (O), supplied by the environment in the form of carbon dioxide (CO_2) and water (H_2O), are usually present in excess. The other nutrient elements, which are also needed for growth, may sometimes be in short supply; in this case they are referred to as limiting nutrients. The two most commonly recognized limiting nutrients are nitrogen (N) and phosphorus (P).

Biogeochemistry is concerned with both the biological uptake and release of nutrients, and the transformation of the chemical state of these biologically active substances, usually by means of energy-supplying oxidation-reduction reactions, at the Earth's surface. Emphasis is on how the activities of organisms affect the chemical composition of natural waters, the atmosphere, rocks, soils, and sediments. Thus, biogeochemistry is complementary to the science of ecology, which includes a concern with how the chemical composition of the atmosphere, waters, and so forth affects life. *See* ECOLOGY.

The two major processes of biogeochemistry are photosynthesis and respiration. Photosynthesis involves the uptake, under the influence of sunlight, of carbon dioxide, water, and other nutrients by plants to form organic matter and oxygen. Respiration is the reverse of photosynthesis and involves the oxidation and breakdown of organic matter and the return of nitrogen, phosphorus, and other elements, as well as carbon dioxide and water, to the environment. *See* PHOTOSYNTHESIS; PLANT RESPIRATION.

Biogeochemistry is usually studied in terms of biogeochemical cycles of individual elements. There are short-term cycles ranging from days to centuries and long-term (geological) cycles ranging from thousands to millions of years.

There has been increasing interest in biogeochemistry because the human influence on short-term biogeochemical cycling has become evident. Perhaps the best-known example is the changes in the biogeochemical cycling of carbon due to the burning of fossil fuels and the cutting and burning of tropical rainforests. The cycles of nitrogen and phosphorus have been altered because of the use of fertilizer and the addition of wastes to lakes, rivers, estuaries, and the oceans. Acid rain, which results from the addition of sulfur and nitrogen compounds to the atmosphere by humans, affects biological systems in certain areas.

Carbon cycle. Carbon is the basic biogeochemical element. The atmosphere contains carbon in the form of carbon dioxide gas. There is a large annual flux of atmospheric carbon dioxide to and from forests and terrestrial biota, amounting to nearly 7% of total atmospheric carbon dioxide. This is because carbon dioxide is used by plants to produce organic matter through photosynthesis, and when the organic matter is broken down through respiration, carbon dioxide is released to the atmosphere. The concentration of atmospheric carbon dioxide shows a yearly oscillation because there is a strong seasonal annual cycle of photosynthesis and respiration in the Northern Hemisphere.

Photosynthesis and respiration in the carbon cycle can be represented by the reaction below. Breakdown of organic matter via respiration is accomplished mainly by

$$CO_2 + H_2O \underset{\text{respiration}}{\overset{\text{photosynthesis}}{\rightleftharpoons}} CH_2O + O_2$$

bacteria that live in soils, sediments, and natural waters. There is a very large reservoir of terrestrial carbon in carbonate rocks, which contain calcium carbonate ($CaCO_3$), and in rocks such as shales which contain organic carbon. Major exchange of carbon between rocks and the atmosphere is very slow, on the scale of thousands to millions of years, compared to exchange between plants and the atmosphere, which can even be seasonal. *See* MICROBIAL ECOLOGY; SOIL MICROBIOLOGY.

The oceans taken as a whole represent a major reservoir of carbon. Carbon in the oceans occurs primarily as dissolved $(HCO_3)^-$ and to a lesser extent as dissolved carbon dioxide gas and carbonate ion $[(CO_3)^{2-}]$. The well-mixed surface ocean (the top 250 ft or 75 m) rapidly exchanges carbon dioxide with the atmosphere. However, the deep oceans are cut off from the atmosphere and mix with it on a long-term time scale of about 1000–2000 years. Most of the biological activity in the oceans occurs in the surface (or shallow) water where there is light and photosynthesis can occur. *See* MARITIME METEOROLOGY.

The main biological process in seawater is photosynthetic production of organic matter by phytoplankton. Some of this organic matter is eaten by animals, which are in turn eaten by larger animals farther up in the food chain. Almost all of the organic matter along the food chain is ultimately broken down by bacterial respiration, which occurs primarily in shallow water, and the carbon dioxide is quickly recycled to the atmosphere. *See* FOOD WEB; PHYTOPLANKTON; SEAWATER.

Another major biological process is the secretion of shells and other hard structures by marine organisms. A biogeochemical cycle of calcium and bicarbonate exists within the oceans, linking the deep and shallow water areas. Bottom dwellers in shallow water, such as corals, mollusks, and algae, provide calcium carbonate skeletal debris. Since the shallow waters are saturated with respect to calcium carbonate, this debris accumulates on the bottom and is buried, providing the minerals that form carbonate rocks such as limestone and dolomite. Calcium carbonate is also derived from the shells of organisms inhabiting surface waters of the deep ocean; these are tiny, floating plankton such as foraminiferans, pteropods, and coccoliths. Much of the calcium carbonate from this source dissolves as it sinks into the deeper ocean waters, which are undersaturated with respect to calcium carbonate. The undissolved calcium carbonate accumulates on the bottom to form deep-sea limestone. The calcium and the bicarbonate ions [Ca^{2+} and $(HCO_3)^-$] dissolved in the deep ocean water eventually are carried to surface and shallow water, where they are removed by planktonic and bottom-dwelling organisms to form their skeletons.

The long-term biogeochemical carbon cycle occurs over millions of years when the calcium carbonate and organic matter that are buried in sediments are returned to the Earth's surface. There, weathering occurs which involves the reaction of oxygen with sedimentary organic matter with the release of carbon dioxide and water (analogous to respiration), and the reaction of water and carbon dioxide with carbonate rocks with the release of calcium and bicarbonate ions. *See* WEATHERING PROCESSES.

Fossil fuels (coal and oil) represent a large reservoir of carbon. Burning of fossil fuels releases carbon dioxide to the atmosphere, and an increase in the atmospheric concentration of carbon dioxide has been observed since the mid-1950s. While much of the increase is attributed to fossil fuels, deforestation by humans accompanied by the decay or burning of trees is another possible contributor to the problem.

When estimates are made of the amount of fossil fuels burned from 1959 to 1980, only about 60% of the carbon dioxide released can be accounted for in the atmospheric increase in carbon dioxide. The remaining 40% is known as excess carbon dioxide. The surface oceans are an obvious candidate for storage of most of the excess carbon dioxide by the reaction of carbon dioxide with dissolved carbonate to form bicarbonate. Because the increase in bicarbonate concentration in surface waters due to excess carbon dioxide uptake would be small, it is difficult to detect whether such a change has occurred. Greater quantities of excess carbon dioxide could be stored as bicarbonate in the deeper oceans, but this process takes a long time because of the slow rate of mixing between surface and deep oceans.

An increase in atmospheric carbon dioxide is of concern because of the greenhouse effect. The carbon dioxide traps heat in the atmosphere; notable increases in atmospheric carbon dioxide should cause an increase in the Earth's surface temperature by as much as several degrees. This temperature increase would be greater at the poles, and the effects could include melting of polar ice, a rise in sea level, and changes in rainfall distribution, with droughts in interior continental areas such as the Great Plains of the United States. *See* DROUGHT; GREENHOUSE EFFECT.

Nitrogen cycle. Nitrogen is dominantly a biogenic element and has no important mineral forms. It is a major atmospheric constituent with a number of gaseous forms, including molecular nitrogen gas (N_2), nitrogen dioxide (NO_2), nitric oxide (NO), ammonia (NH_3), and nitrous oxide (N_2O). As an essential component of plant and animal matter, it is extensively involved in biogeochemical cycling. On a global basis, the nitrogen cycle is greatly affected by human activities.

Nitrogen gas (N_2) makes up 80% of the atmosphere by volume; however, nitrogen is unreactive in this form. In order to be available for biogeochemical cycling by organisms, nitrogen gas must be fixed, that is, combined with oxygen, carbon, or hydrogen. There are three major sources of terrestrial fixed nitrogen: biological nitrogen fixation by plants, nitrogen fertilizer application, and rain and particulate dry deposition of previously fixed nitrogen. Biological fixation occurs in plants such as legumes (peas and beans) and lichens in trees, which incorporate nitrogen from the atmosphere into their living matter; about 30% of worldwide biological fixation is due to human cultivation of these plants. Nitrogen fertilizers contain industrially fixed nitrogen as both nitrate and ammonium. *See* FERTILIZER.

Fixed nitrogen in rain is in the forms of nitrate $[(NO_3)^-]$ and ammonium $[(NH_4)^+]$ ions. Major sources of nitrate, which is derived from gaseous atmospheric nitrogen dioxide (and nitric oxide), include (in order of importance) combustion of fossil fuel, especially by automobiles; forest fires (mostly caused by humans); and lightning. Nitrate in rain, in addition to providing soluble fixed nitrogen for photosynthesis, contributes nitric acid (HNO_3), a major component of acid rain. Sources of ammonium, which is derived from atmospheric ammonia gas (NH_3), include animal and human wastes, soil loss from decomposition of organic matter, and fertilizer release. *See* ACID RAIN.

The basic land nitrogen cycle involves the photosynthetic conversion of the nitrate and ammonium ions dissolved in soil water into plant organic material. Once formed, the organic matter may be stored or broken down. Bacterial decomposition of organic matter (ammonification) produces soluble ammonium ion which can then be either taken up again in photosynthesis, released to the atmosphere as ammonia gas, or oxidized by bacteria to nitrate ion (nitrification).

Nitrate ion is also soluble, and may be used in photosynthesis. However, part of the nitrate may undergo reduction (denitrification) by soil bacteria to nitrogen gas or to nitrous oxide which are then lost to the atmosphere. Compared to the land carbon cycle, the land nitrogen cycle is considerably more complex, and because of the large input of fixed nitrogen by humans, it is possible that nitrogen is building up on land. However, this is difficult to determine since the amount of nitrogen gas recycled to the atmosphere is not known and any changes in the atmospheric nitrogen concentration would be too small to detect. *See* NITROGEN CYCLE.

The oceans are another major site of nitrogen cycling: the amount of nitrogen cycled biogenically, through net primary photosynthetic production, is about 13 times that on land. The main links between the terrestrial and the oceanic nitrogen cycles are the atmosphere and rivers. Nitrogen gases carried in the atmosphere eventually fall as dissolved inorganic (mainly nitrate) and organic nitrogen and particulate organic nitrogen in rain on the oceans. The flux of river nitrogen lost from the land is only about 9% of the total nitrogen recycled biogeochemically on land each year and only about 25% of the terrestrial nitrogen flux from the biosphere to the atmosphere.

River nitrogen is an important nitrogen source to the oceans; however, the greatest amount of nitrogen going into ocean surface waters comes from the upwelling of deeper waters, which are enriched in dissolved nitrate from organic recycling at depth. Dissolved nitrate is used extensively for photosynthesis by marine organisms, mainly plankton. Bacterial decomposition of the organic matter formed in photosynthesis results in the release of dissolved ammonium, some of which is used directly in photosynthesis. However, most undergoes nitrification to form nitrate, and much of the nitrate may undergo denitrification to nitrogen gas which is released to the atmosphere. A small amount of organic-matter nitrogen is buried in ocean sediments, but this accounts for a very small amount of the nitrogen recycled each year. There are no important inorganic nitrogen minerals such as those that exist for carbon and phosphorus, and thus there is no mineral precipitation and dissolution. *See* UPWELLING.

Phosphorus cycle. Phosphorus, an important component of organic matter, is taken up and released in the form of dissolved inorganic and organic phosphate. Phosphorus differs from nitrogen and carbon in that it does not form stable atmospheric gases and therefore cannot be obtained from the atmosphere. It does form minerals, most prominently apatite (calcium phosphate), and insoluble iron (Fe) and aluminum (Al) phosphate minerals, or it is adsorbed on clay minerals. The amount of phosphorus used in photosynthesis on land is large compared to phosphorus inputs to the land. The major sources of phosphorus are weathering of rocks containing apatite and mining of phosphate rock for fertilizer and industry. A small amount comes from precipitation and dry deposition.

Phosphorus is lost from the land principally by river transport, which amounts to only 7% of the amount of phosphorus recycled by the terrestrial biosphere; overall, the terrestrial biosphere conserves phosphorus. Humans have greatly affected terrestrial phosphorus: deforestation and agriculture have doubled the amount of phosphorus weathering; phosphorus is added to the land as fertilizers and from industrial wastes, sewage, and detergents. Thus, about 75% of the terrestrial input is anthropogenic; in fact, phosphorus may be building up on the land.

In the oceans, phosphorus occurs predominantly as dissolved orthophosphates [PO_4^{3-}, $(HPO_4)^{2-}$ and $(H_2PO_4)^{-}$]. Since it follows the same cycle as do carbon and nitrogen, dissolved orthophosphate is depleted in surface ocean waters where both photosynthesis and respiration occur, and the concentration builds up in deeper water where organic matter is decomposed by bacterial respiration. The major phosphorus input to the oceans is from rivers, with about 5% coming from rain. However, 75% of the river phosphorus load is due to anthropogenic pollutants; humans have changed the ocean balance of phosphorus. Most of the dissolved oceanic orthophosphate is derived from recycled organic matter. The output of phosphorus from the ocean is predominantly biogenic: organic phosphorus is buried in sediments; a smaller amount is removed by adsorption on volcanic iron oxides. In the geologic past, there was a much greater inorganic precipitation of phosphorite (apatite) from seawater than at present, and this has resulted in the formation of huge deposits which are now mined.

Nutrients in lakes. Biogeochemical cycling of phosphorus and nitrogen in lakes follows a pattern that is similar to oceanic cycling: there is nutrient depletion in surface waters and enrichment in deeper waters. Oxygen consumption by respiration in deep water sometimes leads to extensive oxygen depletion with adverse effects on fish and other biota. In lakes, phosphorus is usually the limiting nutrient.

Many lakes have experienced greatly increased nutrient (nitrogen and phosphorus) input due to human activities. This stimulates a destructive cycle of biological activity:

very high organic productivity, a greater concentration of plankton, and more photosynthesis. The result is more organic matter falling into deep water with increased depletion of oxygen and greater accumulation of organic matter on the lake bottom. This process, eutrophication, can lead to adverse water quality and even to the filling up of small lakes with organic matter. *See* EUTROPHICATION; LIMNOLOGY.

Biogeochemical sulfur cycle. A dominant flux in the global sulfur cycle is the release of 65–70 teragrams of sulfur per year to the atmosphere from burning of fossil fuels. Sulfur contaminants in these fuels are released to the atmosphere as sulfur dioxide (SO_2) which is rapidly converted to aerosols of sulfuric acid (H_2SO_4), the primary contributor to acid rain. Forest burning results in an additional release of sulfur dioxide. Overall, the broad range of human activities contribute 75% of sulfur released into the atmosphere. Natural sulfur sources over land are predominantly the release of reduced biogenic sulfur gases [mainly hydrogen sulfide (H_2S) and dimethyl sulfide] from marine tidal flats and inland waterlogged soils and, to much lesser extent, the release of volcanic sulfur. The atmosphere does not have an appreciable reservoir of sulfur because most sulfur gases are rapidly returned (within days) to the land in rain and dry deposition. There is a small net flux of sulfur from the atmosphere over land to the atmosphere over the oceans.

Ocean water constitutes a large reservoir of dissolved sulfur in the form of sulfate ions [$(SO_4)^{2-}$]. Some of this sulfate is thrown into the oceanic atmosphere as sea salt from evaporated sea spray, but most of this is rapidly returned to the oceans. Another major sulfur source in the oceanic atmosphere is the release of oceanic biogenic sulfur gases (such as dimethyl sulfide) from the metabolic activities of oceanic organisms and organic matter decay. Marine organic matter contains a small amount of sulfur, but sulfur is not a limiting element in the oceans.

Another large flux in the sulfur cycle is the transport of dissolved sulfate in rivers. However, as much as 43% of this sulfur may be due to human activities, both from burning of fossil fuels and from fertilizers and industrial wastes. The weathering of sulfur minerals, such as pyrite (FeS_2) in shales, and the evaporite minerals, gypsum and anhydrite, make an important contribution to river sulfate. The major mechanism for removing sulfate from ocean water is the formation and burial of pyrite in oceanic sediments, primarily nearshore sediments. (The sulfur fluxes of sea salt and biogenic sulfur gases do not constitute net removal from the oceans since the sulfur is recycled to the oceans.)

Biogeochemical cycles and atmospheric oxygen. The main processes affecting atmospheric oxygen are photosynthesis and respiration; however, these processes are almost perfectly balanced against one another and, thus, do not exert a simple effect on oxygen levels. Only the very small excess of photosynthesis over respiration, manifested by the burial of organic matter in sediments, is important in raising the level of oxygen. This excess is so small, and the reservoir of oxygen so large, that if the present rate of organic carbon burial were doubled and the other rates remained constant, it would take 5–10 million years for the amount of atmospheric oxygen to double. Nevertheless, this is a relatively short time from a geological perspective. *See* ATMOSPHERE; ATMOSPHERIC CHEMISTRY; BIOSPHERE; GEOCHEMISTRY; HYDROSPHERE; MARINE SEDIMENTS.

[E.K.B.; R.A.Ber.]

Biogeography A synthetic discipline that describes the distributions of living and fossil species of plants and animals across the Earth's surface as consequences of ecological and evolutionary processes. Biogeography overlaps and complements many biological disciplines, especially community ecology, systematics, paleontology, and evolutionary biology. *See* PLANT GEOGRAPHY; ZOOGEOGRAPHY.

Biogeographic realms

Realm	Continental areas included	Examples of distinctive or endemic taxa
Palearctic	Temperate Eurasia and northern Africa	Hynobiid salamanders
Oriental	Tropical Asia	Lower apes
Ethiopian	Sub-Saharan Africa	Great apes
Australian	Australia, New Guinea, and New Zealand	Marsupials
Nearctic	Temperate North America	Pronghorn antelope, ambystomatid salamanders
Neotropic	Subtropical Central America and South America	Hummingbirds, antbirds, marmosets

Based on relatively complete compilations of species within well-studied groups, such as birds and mammals, biogeographers identified six different realms within which species tend to be closely related and between which turnovers in major groups of species are observed (see table). The boundaries between biogeographic realms are less distinct than was initially thought, and the distribution of distinctive groups such as parrots, marsupials, and southern beeches (*Nothofagus* spp.) implies that modern-day biogeographic realms have been considerably mixed in the past. *See* ANIMAL EVOLUTION; PLANT EVOLUTION; SPECIATION.

Two patterns of species diversity have stimulated a great deal of progress in developing ecological explanations for geographic patterns of species richness. The first is that the number of species increases in a regular fashion with the size of the geographic area being considered. The second is the nearly universal observation that there are more species of plants and animals in tropical regions than in temperate and polar regions.

In order to answer questions about why there are a certain number of species in a particular geographic region, biogeography has incorporated many insights from community ecology. Species number at any particular place depends on the amount of resources available there (ultimately derived from the amount of primary productivity), the number of ways those resources can be apportioned among species, and the different kinds of ecological requirements of the species that can colonize the region. The equilibrium theory of island biogeography arose as an application of these insights to the distribution of species within a specified taxon across an island archipelago. This theory generated specific predictions about the relationships among island size and distance from a colonization source with the number and rate of turnover of species. Large islands are predicted to have higher equilibrium numbers of species than smaller islands; hence, the species area relationship can be predicted in principle from the ecological attributes of species. Experimental and observational studies have confirmed many predictions made by this theory. *See* ECOLOGICAL COMMUNITIES; ISLAND BIOGEOGRAPHY.

The latitudinal gradient in species richness has generated a number of explanations, none of which has been totally satisfactory. One explanation is based on the observation that species with more temperate and polar distributions tend to have larger geographic ranges than species from tropical regions. It is thought that since species with large geographic ranges tend to withstand a wider range of physical and biotic conditions, this allows them to penetrate farther into regions with more variable climates at higher latitudes. If this were true, then species with smaller geographic ranges would tend to concentrate in tropical regions where conditions are less variable. While this might be generally true, there are many examples of species living in high-latitude regions that have small geographic regions. *See* ALTITUDINAL VEGETATION ZONES.

Biogeography is entering a phase where data on the spatial patterns of abundance and distribution of species of plants and animals are being analyzed with sophisticated mathematical and technological tools. Geographic information systems and remote sensing technology have provided a way to catalog and map spatial variation in biological processes with a striking degree of detail and accuracy. These newer technologies have stimulated research on appropriate methods for modeling and analyzing biogeographic patterns. Modern techniques of spatial modeling are being applied to geographic information systems data to test mechanistic explanations for biogeographic patterns that could not have been attempted without the advent of the appropriate technology. *See* GEOGRAPHIC INFORMATION SYSTEMS. [B.A.M.]

Bioherm A lenslike to moundlike structure of strictly organic origin. This term involves two concepts: shape and organic internal composition.

The term shape denotes original topographic relief above the sea floor as well as a three-dimensional quality: crudely conical (sugar loaf–shaped) or ellipsoidal (bread loaf–shaped). Such forms are massive or unbedded, their upbuilding resulting from the very rapid rate of accretion of organic carbonate once it starts in a favorable locality. There are size limitations: bioherms a meter or so in diameter are known, and some rise 300 ft (100 m) or more above the sea floor.

The second concept, organic internal composition, not only embraces sessile, bottom-dwelling organisms forming frame-building reefy bondstone but also includes piles of organically derived debris replete with organisms which encrust it and cement it in place. Even inorganic cement precipitated from marine and meteoric water is known to play a role in a buildup of massive, moundlike structures. When the internal material is coarse and identifiable, no problem is encountered in applying the term bioherm in its original sense. In some buildups, however, an appreciable amount of lime mud is present, and relatively few organisms are identifiable which could have secreted, bound, encrusted, or trapped carbonate mud (as in some early Carboniferous mounds). In such cases, problems arise in applying that part of the definition based on internal composition. In fact, as a field term, bioherm can hardly ever be completely diagnostic because careful petrographic study is commonly necessary for details of internal composition to be ascertained.

Bioherms may occur on shelves (where they are normally lens-shaped) or in shallow basins, often at the basin margin. In the latter position, they have been called reef knolls or pinnacle reef. *See* BIOSTROME; REEF. [J.L.Wi.]

Biological productivity The amount and rate of production which occur in a given ecosystem over a given time period. It may apply to a single organism, a population, or entire communities and ecosystems. Productivity can be expressed in terms of dry matter produced per area per time (net production), or in terms of energy produced per area per time (gross production = respiration + heat losses + net production). In aquatic systems, productivity is often measured in volume instead of area. *See* BIOMASS.

Ecologists distinguish between primary productivity (by autotrophs) and secondary productivity (by heterotrophs). Plants have the ability to use the energy from sunlight to convert carbon dioxide and water into glucose and oxygen, producing biomass through photosynthesis. Primary productivity of a community is the rate at which biomass is produced per unit area by plants, expressed in either units of energy [joules/(m^2)(day)] or dry organic matter [kg/(m^2)(year)]. The following definitions are useful in calculating production: Gross primary production (GPP) is the total energy fixed by photosynthesis per unit time. Net primary production (NPP) is the gross production minus losses due to plant respiration per unit time, and it represents the actual new biomass that is available for consumption by heterotrophic organisms. Secondary production is the rate of production of biomass by heterotrophs (animals, microorganisms), which feed on plant products or other heterotrophs. *See* PHOTOSYNTHESIS.

Productivity is not spread evenly across the planet. For instance, although oceans cover two-thirds of Earth's surface, they account for only one-third of the Earth's productivity. Furthermore, the factors that limit productivity in the ocean differ from those limiting productivity on land, producing differences in geographic patterns of productivity in the two systems. In terrestrial ecosystems, productivity shows a latitudinal trend, with highest productivity in the tropics and decreasing progressively toward the Poles; but in the ocean there is no latitudinal trend, and the highest values of net primary production are found along coastal regions. [D.C.C.; E.Gry.]

Biology A natural science concerned with the study of all living organisms. Although living organisms share some unifying themes, such as their origin from the same basic cellular structure and their molecular basis of inheritance, they are diverse in many other aspects. The diversity of life leads to many divisions in biological science involved with studying all aspects of living organisms. The primary divisions of study in biology consist of zoology (animals), botany (plants), and protistology (one-celled organisms), and are aimed at examining such topics as origins, structure, function, reproduction, growth and development, behavior, and evolution of the different organisms. In addition, biologists consider how living organisms interact with each other and the environment on an individual as well as group basis. Therefore, within these divisions are many subdivisions such as molecular and cellular biology, microbiology (the study of microbes such as bacteria and viruses), taxonomy (the classification of organisms into special groups), physiology (the study of function of the organism at any level), immunology (the investigation of the immune system), genetics (the study of inheritance), and ecology and evolution (the study of the interaction of an organism with its environment and how that interaction changes over time).

The study of living organisms is an ongoing process that allows observation of the natural world and the acquisition of new knowledge. Biologists accomplish their studies through a process of inquiry known as the scientific method, which approaches a problem or question in a well-defined orderly sequence of steps so as to reach conclusions. The first step involves making systematic observations, either directly through the sense of sight, smell, taste, sound, or touch, or indirectly through the use of special equipment such as the microscope. Next, questions are asked regarding the observations. Then a hypothesis—a tentative explanation or educated guess—is formulated, and predictions about what will occur are made. At the core of any scientific study is testing of the hypothesis. Tests or experiments are designed so as to help substantiate or refute the basic assumptions set forth in the hypothesis. Therefore, experiments are repeated many times. Once they have been completed, data are collected and organized in the form of graphs or tables and the results are analyzed. Also, statistical tests may be performed to help determine whether the data are significant enough to support or disprove the hypothesis. Finally, conclusions are drawn that provide explanations or insights about the original problem. By employing the scientific method, biologists aim to be objective rather than subjective when interpreting the results of their experiments. Biology is not absolute: it is a science that deals with theories or relative truths. Thus, biological conclusions are always subject to change when new evidence is presented.

As living organisms continue to evolve and change, the science of biology also will evolve. *See* BOTANY; ECOLOGY; GENETICS; MICROBIOLOGY; PLANT; TAXONOMY; ZOOLOGY. [L.Co.]

Biomass The organic materials produced by plants, such as leaves, roots, seeds, and stalks. In some cases, microbial and animal metabolic wastes are also considered biomass. The term "biomass" is intended to refer to materials that do not directly go into foods or consumer products but may have alternative industrial uses. Common sources of biomass are (1) agricultural wastes, such as corn stalks, straw, seed hulls, sugarcane leavings, bagasse, nutshells, and manure from cattle, poultry, and hogs; (2) wood materials, such as wood or bark, sawdust, timber slash, and mill scrap; (3) municipal waste, such as waste paper and yard clippings; and (4) energy crops, such as poplars, willows, switchgrass, alfalfa, prairie bluestem, corn (starch), and soybean (oil). *See* BIOLOGICAL PRODUCTIVITY.

Biomass is a complex mixture of organic materials, such as carbohydrates, fats, and proteins, along with small amounts of minerals, such as sodium, phosphorus, calcium, and iron. The main components of plant biomass are carbohydrates (approximately 75%, dry weight) and lignin (approximately 25%), which can vary with plant type. The carbohydrates are mainly cellulose or hemicellulose fibers, which impart strength to the plant structure, and lignin, which holds the fibers together. Some plants also store starch (another carbohydrate polymer) and fats as sources of energy, mainly in seeds and roots (such as corn, soybeans, and potatoes).

A major advantage of using biomass as a source of fuels or chemicals is its renewability. Utilizing sunlight energy in photosynthesis, plants metabolize atmospheric carbon dioxide to synthesize biomass. An estimated 140 billion metric tons of biomass are produced annually.

Major limitations of solid biomass fuels are difficulty of handling and lack of portability for mobile engines. To address these issues, research is being conducted to convert solid biomass into liquid and gaseous fuels. Both biological means (fermentation) and chemical means (pyrolysis, gasification) can be used to produce fluid biomass fuels. For example, methane gas is produced in China for local energy needs by anaerobic microbial digestion of human and animal wastes. Ethanol for automotive fuels is currently produced from starch biomass in a two-step process: starch is enzymatically hydrolyzed into glucose; then yeast is used to convert the glucose into ethanol. About 1.5 billion gallons of ethanol are produced from starch each year in the United States. [B.Y.Ta.]

Biome A major community of plants and animals having similar life forms or morphological features and existing under similar environmental conditions. The biome, which may be used at the scale of entire continents, is the largest useful biological community unit. In Europe the equivalent term for biome is major life zone, and throughout the world, if only plants are considered, the term used is formation. *See* ECOLOGICAL COMMUNITIES.

Each biome may contain several different types of ecosystems. For example, the grassland biome may contain the dense tallgrass prairie with deep, rich soil, while the desert grassland has a sparse plant canopy and a thin soil. However, both ecosystems have grasses as the predominant plant life form, grazers as the principal animals, and a climate with at least one dry season. Additionally, each biome may contain several successional stages. A forest successional sequence may include grass dominants at an early stage, but some forest animals may require the grass stage for their habitat, and all successional stages constitute the climax forest biome. *See* DESERT; ECOLOGICAL SUCCESSION; ECOSYSTEM; GRASSLAND ECOSYSTEM.

Distributions of animals are more difficult to map than those of plants. The life form of vegetation reflects major features of the climate and determines the structural nature of habitats for animals. Therefore, the life form of vegetation provides a sound basis for ecologically classifying biological communities. Terrestrial biomes are usually identified by the dominant plant component, such as the temperate deciduous forest. Marine biomes are mostly named for physical features, for example, for marine upwelling, and for relative locations, such as littoral. Many biome classifications have been proposed, but a typical one might include several terrestrial biomes such as desert, tundra, grassland, savanna, coniferous forest, deciduous forest, and tropical forest. Aquatic biome examples are fresh-water lotic (streams and rivers), fresh-water lentic (lakes and ponds), and marine littoral, neritic, upwelling, coral reef, and pelagic. *See* FRESH-WATER ECOSYSTEM; MARINE ECOLOGY; PLANTS, LIFE FORMS OF; TERRESTRIAL ECOSYSTEM. [P.Ri.]

Biomedical chemical engineering The application of chemical engineering principles to the solution of medical problems due to physiological impairment. A knowledge of organic chemistry is required of all chemical engineers, and many also study biochemistry and molecular biology. This training at the molecular level gives chemical engineers a unique advantage over other engineering disciplines in communication with life scientists and clinicians in medicine. Practical applications include the development of tissue culture systems, the construction of three-dimensional scaffolds of biodegradable polymers for cell growth in the laboratory, and the design of artificial organs.

Cell transplantation is explored as a means of restoring tissue function. With this approach, individual cells are harvested from a healthy section of donor tissue, isolated, expanded in culture, and implanted at the desired site of the functioning tissue. Isolated cells cannot form new tissues on their own and require specific environments that often include the presence of supporting material to act as a template for growth. Three-dimensional scaffolds can be used to mimic their natural counterparts, the extracellular matrices of the body. These scaffolds serve as both a physical support and an adhesive substrate for isolated parenchymal cells during cell culture and subsequent implantation. The scaffold must be made of biocompatible materials. As the transplanted cell population grows and the cells function normally, they will begin to secrete their own extracellular matrix support. The need for an artificial support will gradually diminish; and thus if the implant is biodegradable, it will be eliminated as its function is replaced. The development of processing methods to fabricate reproducibly three-dimensional scaffolds of biodegradable polymers that will provide temporary scaffolding to transplanted cells will be instrumental in engineering tissues.

Chemical engineers have made significant contributions to the design and optimization of many commonly used devices for both short-term and long-term organ replacement. Examples include the artificial kidney for hemodialysis and the heart-lung machine employed in open heart surgery. The artificial kidney removes waste metabolites (such as urea and creatinine) from blood across a polymeric membrane that separates the flowing blood from the dialysis fluid. The mass transport properties and biocompatibility of these membranes are crucial to the functioning of hemodialysis equipment. The heart-lung machine replaces both the pumping function of the heart and the gas exchange function of the lung in one fairly complex device. While often life saving, both types of artificial organs only partially replace real organ function. Long-term use often leads to problems with control of blood coagulation mechanisms to avoid both excessive clotting initiated by blood contact with artificial surfaces and excessive bleeding due to platelet consumption or overuse of anticoagulants.

Other chemical engineering applications include methodology for development of artificial bloods, utilizing fluorocarbon emulsions or encapsulated or polymerized hemoglobin, and controlled delivery devices for release of drugs or of specific molecules (such as insulin) missing in the body because of disease or genetic alteration. [L.V.M.; A.G.M.]

Biometeorology A branch of meteorology and ecology that deals with the effects of weather and climate on plants, animals, and humans.

The principal problem for living organisms is maintaining an acceptable thermal equilibrium with their environment. Organisms have natural techniques for adapting to adverse conditions. These techniques include acclimatization, dormancy, and hibernation, or in some cases an organism can move to a more favorable environment or microenvironment. Humans often establish a favorable environment through the use of technology. *See* DORMANCY; MICROMETEOROLOGY.

Homeotherms, that is, humans and other warm-blooded animals, maintain relatively constant body temperatures under a wide range of ambient thermal and radiative conditions through physiological and metabolic mechanisms. Poikilotherms, that is, cold-blooded animals, have a wide range in body temperature that is modified almost exclusively by behavioral responses. Plants also experience a wide range of temperatures, but because of their immobility they have less ability than animals to adapt to extreme changes in environment.

Humans are physically adapted to a narrow range of temperature, with the metabolic mechanism functioning best at air temperatures around 77°F (25°C). There is a narrow range above and below this temperature where survival is possible. To regulate heat loss, warm-blooded animals developed hair, fur, and feathers. Humans invented clothing and shelter. The amount of insulation required to maintain thermal equilibrium is governed by the conditions in the atmospheric environment. There are a limited number of physiological mechanisms, controlled by the hypothalamus, that regulate body heat.

Clothing, shelter, and heat-producing objects can largely compensate for environmental cold, but with extensive exposure to cold, vasoconstriction in the peripheral organs can lead to chilblains and frostbite on the nose, ears, cheeks, and toes. This exposure is expressed quantitatively as a wind chill equivalent temperature that is a function of air temperature and wind speed. The wind chill equivalent temperature is a measure of convective heat loss and describes a thermal sensation equivalent to a lower-than-ambient temperature under calm conditions, that is, for wind speeds below 4 mi/h (1.8 m/s). Persons exposed to extreme cold develop hypothermia, which may be irreversible when the core temperature drops below 91°F (33°C.)

The combination of high temperature with high humidity leads to a very stressful thermal environment. The combination of high temperature with low humidity leads to a relatively comfortable thermal environment, but such conditions create an environment that has a very high demand for water.

Conditions of low humidity exist principally in subtropical deserts, which have the highest daytime temperatures observed at the Earth's surface. Human and animal bodies are also exposed to strong solar radiation and radiation reflected from the surface of the sand. This combination makes extraordinary demands on the sweat mechanism. Water losses of 1.06 quarts (1 liter) per hour are common in humans and may be even greater with exertion. Unless the water is promptly replaced by fluid intake, dehydration sets in. *See* DESERT; HUMIDITY.

When humans are exposed to warm environments, the first physiological response is dilation of blood vessels, which increases the flow of blood near the skin. The next response occurs through sweating, panting, and evaporative cooling. Since individuals differ in their physiological responses to environmental stimuli, it is difficult to develop a heat stress index based solely on meteorological variables. Nevertheless, several useful indices have been developed.

Since wind moves body heat away and increases the evaporation from a person, it should be accounted for in developing comfort indices describing the outdoor environment. One such index, used for many years by heating and ventilating engineers, is the effective temperature. People will feel uncomfortable at effective temperatures above 81°F (27°C) or below 57°F (15°C); between 63°F (17°C) and 77°F (25°C) they will feel comfortable.

Both physiological and psychological responses to weather changes (meteorotropisms) are widespread and generally have their origin in some bodily impairment. Reactions to weather changes commonly occur in anomalous skin tissue such as scars and corns; changes in atmospheric moisture cause differential hygroscopic expansion and contraction between healthy and abnormal skin, leading to pain. Sufferers from rheumatoid arthritis are commonly affected by weather changes; both pain and swelling of affected joints have been noted with increased atmospheric humidity. Sudden cooling can also trigger such symptoms. Clinical tests have shown that in these individuals the heat regulatory mechanism does not function well, but the underlying cause is not understood.

Weather is a significant factor in asthma attacks. Asthma as an allergic reaction may, in rare cases, be directly provoked by sudden changes in temperature that occur after passage of a cold front. Often, however, the weather effect is indirect, and attacks are caused by airborne allergens, such as air pollutants and pollen. An even more indirect relationship exists for asthma attacks in autumn, which often seem to be related to an early outbreak of cold air. This cold air initiates home or office heating, and dormant dust or fungi from registers and radiators are convected into rooms, irritating allergic persons. *See* ALLERGY; ASTHMA.

A variety of psychological effects have also been attributed to heat. They are vaguely described as lassitude, decrease in men-j tal and physical performance, and increased irritability. Similar reactions to weather have been described for domestic animals, particularly dogs; hence hot, humid days are sometimes known as dog days.

Meteorological and seasonal changes in natural illumination have a major influence on animals. Photoperiodicity is widespread. The daily cycle of illumination triggers the feeding cycle in many species, especially birds. In insectivores the feeding cycle may result from the activities of the insects, which themselves show temperature-influenced cycles of animation. Bird migration may be initiated by light changes, but temperature changes and availability of food are also involved. In the process of migration, especially over long distances, birds have learned to take advantage of prevailing wind patterns. In humans, light deprivation, as is common in the cold weather season in higher latitudes, is suspected as a cause of depression. Exposure to high-intensity light for several hours has been found to be an effective means of treating this depression. *See* MIGRATORY BEHAVIOR.

Humans and animals often exhibit a remarkable ability to adapt to harsh or rapidly changing environmental conditions. An obvious means of adaptation is to move to areas where environmental conditions are less severe; examples are birds and certain animals that migrate seasonally, animals that burrow into the ground, and animals that move to shade or sunshine depending on weather conditions. Animals can acclimatize to heat and cold. The acclimatization process is generally complete within

2–3 weeks of exposure to the stressful conditions. For example, in hot climates heat regulation is improved by the induction of sweating at a lower internal body temperature and by the increase of sweating rates. The acclimatization to cold climates is accomplished by increase in the metabolic rate, by improvement in the insulating properties of the skin, and by the constriction of blood vessels to reduce the flow of blood to the surface.

Unlike humans and animals, plants cannot move from one location to another; therefore, they must adapt genetically to their atmospheric environment. Plants are often characteristic for their climatic zone, such as palms in the subtropics and birches or firs in regions with cold winters. Whole systems of climatic classification are based on the native floras. *See* ALTITUDINAL VEGETATION ZONES; ECOLOGY; METEOROLOGY; PLANT GEOGRAPHY.

[B.L.B.; H.E.L.]

Biosphere All living organisms and their environments at the surface of the Earth. Included in the biosphere are all environments capable of sustaining life above, on, and beneath the Earth's surface as well as in the oceans. Consequently, the biosphere overlaps virtually the entire hydrosphere and portions of the atmosphere and outer lithosphere. *See* ATMOSPHERE; HYDROSPHERE.

Neither the upper nor lower limits of the biosphere are sharp. Spores of microorganisms can be carried to considerable heights in the atmosphere, but these are resting stages that are not actively metabolizing. A variety of organisms inhabit the ocean depths, including the giant tubeworms and other creatures that were discovered living around hydrothermal vents. Evidence exists for the presence of bacteria in oil reservoirs at depths of about 6600 ft (2000 m) within the Earth. The bacteria are apparently metabolically active, utilizing the paraffinic hydrocarbons of the oils as an energy source. These are extreme limits to the biosphere; most of the mass of living matter and the greatest diversity of organisms are within the upper 330 ft (100 m) of the lithosphere and hydrosphere, although there are places even within this zone that are too dry or too cold to support much life. Most of the biosphere is within the zone which is reached by sunlight and where liquid water exists.

The biosphere is characterized by the interrelationship of living things and their environments. Communities are interacting systems of organisms tied to their environments by the transfer of energy and matter. Such a coupling of living organisms and the nonliving matter with which they interact defines an ecosystem. An ecosystem may range in size from a small pond, to a tropical forest, to the entire biosphere. Ecologists group the terrestrial parts of the biosphere into about 12 large units called biomes. Examples of biomes include tundra, desert, grassland, and boreal forest. *See* BIOME; ECOLOGICAL COMMUNITIES; ECOSYSTEM.

Human beings are part of the biosphere, and some of their activities have an adverse impact on many ecosystems and on themselves. As a consequence of deforestation, urban sprawl, spread of pollutants, and overharvesting, both terrestrial and marine ecosystems are being destroyed or diminished, populations are shrinking, and many species are dying out. In addition to causing extinctions of some species, humans are expanding the habitats of other organisms, sometimes across oceanic barriers, through inadvertent transport and introduction into new regions. Humans also add toxic or harmful substances to the outer lithosphere, hydrosphere, and atmosphere. Many of these materials are eventually incorporated into or otherwise affect the biosphere, and water and air supplies in some regions are seriously fouled. *See* HAZARDOUS WASTE.

[R.M.M.]

Biostrome An evenly bedded and generally horizontally layered stratum composed mostly of organic remains, normally considered to be those of sedentary organisms which lived, died, and were buried essentially in place.

The criterion of formation by in-place growth of organisms is subject to some interpretation. Crinoidal limestones obviously resulted from the accumulation of decayed pieces of millions of these stalked echinoderms, often with accompanying detritus of associated fenestrate bryozoans. Usually it is impossible to ascertain whether such debris dropped vertically a few centimeters or meters through the water column as the organisms died and collapsed (an essentially in-place deposit) or whether the layers of crinoidal grainstone were piled mechanically by currents. The same is true of coquinas of many other thin-shelled calcareous tests, such as those of brachiopods, bryozoans, and trilobites.

Biostromal layers need not have been horizontally deposited when they occur as flanking beds around organic buildups. Dips of up to 25 or 30° are possible here. Such biostromes are probably veneers of sessile organisms which lived somewhat above the realm of deposition and were buried as sediment cascaded down the flank of a mound. [J.L.Wi.]

Biotechnology Generally, any technique that is used to make or modify the products of living organisms in order to improve plants or animals, or to develop useful microorganisms. In modern terms, biotechnology has come to mean the use of cell and tissue culture, cell fusion, molecular biology, and in particular, recombinant deoxyribonucleic acid (DNA) technology to generate unique organisms with new traits or organisms that have the potential to produce specific products. Some examples of products in a number of important disciplines are described below.

Recombinant DNA technology has opened new horizons in the study of gene function and the regulation of gene action. In particular, the ability to insert genes and their controlling nucleic acid sequences into new recipient organisms allows for the manipulation of these genes in order to examine their activity in unique environments, away from the constraints posed in their normal host. Genetic transformation normally is achieved easily with microorganisms; new genetic material may be inserted into them, either into their chromosomes or into extrachromosomal elements, the plasmids. Thus, bacteria and yeast can be created to metabolize specific products or to produce new products. *See* GENE.

Genetic engineering has allowed for significant advances in the understanding of the structure and mode of action of antibody molecules. Practical use of immunological techniques is pervasive in biotechnology.

Few commercial products have been marketed for use in plant agriculture, but many have been tested. Interest has centered on producing plants that are resistant to specific herbicides. This resistance would allow crops to be sprayed with the particular herbicide, and only the weeds would be killed, not the genetically engineered crop species. Resistances to plant virus diseases have been induced in a number of crop species by transforming plants with portions of the viral genome, in particular the virus's coat protein.

Biotechnology also holds great promise in the production of vaccines for use in maintaining the health of animals. Interferons are also being tested for their use in the management of specific diseases.

Animals may be transformed to carry genes from other species including humans and are being used to produce valuable drugs. For example, goats are being used to produce tissue plasminogen activator, which has been effective in dissolving blood clots.

Plant scientists have been amazed at the ease with which plants can be transformed to enable them to express foreign genes. This field has developed very rapidly since the first transformation of a plant was reported in 1982, and a number of transformation procedures are available.

Genetic engineering has enabled the large-scale production of proteins which have great potential for treatment of heart attacks. Many human gene products, produced with genetic engineering technology, are being investigated for their potential use as commercial drugs. Recombinant technology has been employed to produce vaccines from subunits of viruses, so that the use of either live or inactivated viruses as immunizing agents is avoided. Cloned genes and specific, defined nucleic acid sequences can be used as a means of diagnosing infectious diseases or in identifying individuals with the potential for genetic disease. The specific nucleic acids used as probes are normally tagged with radioisotopes, and the DNAs of candidate individuals are tested by hybridization to the labeled probe. The technique has been used to detect latent viruses such as herpes, bacteria, mycoplasmas, and plasmodia, and to identify Huntington's disease, cystic fibrosis, and Duchenne muscular dystrophy. It is now also possible to put foreign genes into cells and to target them to specific regions of the recipient genome. This presents the possibility of developing specific therapies for hereditary diseases, exemplified by sickle-cell anemia.

Modified microorganisms are being developed with abilities to degrade hazardous wastes. Genes have been identified that are involved in the pathway known to degrade polychlorinated biphenyls, and some have been cloned and inserted into selected bacteria to degrade this compound in contaminated soil and water. Other organisms are being sought to degrade phenols, petroleum products, and other chlorinated compounds. *See* GENETIC ENGINEERING. [M.Z.]

Biotelemetry The use of telemetry methods for sending signals from a living organism over some distance to a receiver. Usually, biotelemetry is used for gathering data about the physiology, behavior, or location of the organism. Generally, the signals are carried by radio, light, or sound waves. Consequently, biotelemetry implies the absence of wires between the subject and receiver.

Generally, biotelemetry techniques are necessary in situations when wires running from a subject to a recorder would inhibit the subject's activity; when the proximity of an investigator to a subject might alter the subject's behavior; and when the movements of the subject and the duration of the monitoring make it impractical for the investigator to remain within sight of the subject. Biotelemetry is widely used in medical fields to monitor patients and research subjects, and now even to operate devices such as drug delivery systems and prosthetics. Sensors and transmitters placed on or implanted in animals are used to study physiology and behavior in the laboratory and to study the movements, behavior, and physiology of wildlife species in their natural environments.

Biotelemetry is an important technique for biomedical research and clinical medicine. Perhaps cardiovascular research and treatment have benefited the most from biotelemetry. Heart rate, blood flow, and blood pressure can be measured in ambulatory subjects and transmitted to a remote receiver-recorder. Telemetry also has been used to obtain data about local oxygen pressure on the surface of organs (for example, liver and myocardium) and for studies of capillary exchange (that is, oxygen supply and discharge). Biomedical research with telemetry includes measuring cardiovascular performance during the weightlessness of space flight and portable monitoring of radioactive indicators as they are dispersed through the body by the blood vessels.

Telemetry has been applied widely to animal research, for example, to record electroencephalograms, heart rates, heart muscle contractions, and respiration, even from sleeping mammals and birds. Telemetry and video recording have been combined in research of the relationships between neural and cardiac activity and behavior. Using miniature electrodes and transmitters, ethologists have studied the influence of one bird's song on the heart rate and behavior of a nearby bird.

Many species of wildlife are difficult to find and observe because they are secretive, nocturnal, wide-ranging, or move rapidly. Most commonly, a transmitter is placed on a wild animal so that biologists can track or locate it by homing toward the transmitted signal or by estimating the location by plotting the intersection of two or more bearings from the receiver toward the signal. For some purposes, after homing to a transmitter-marked animal, the biologists observe its behavior. For other studies, successive estimates of location are plotted on a map to describe movement patterns, to delineate the amount of area the animal requires, or to determine dispersal or migration paths. Ecologists can associate the vegetation or other features of the environment with the locations of the animal.

There are usually two concerns associated with the use of biotelemetry: the distance over which the signal can be received, and the size of the transmitter package. Often, both of these concerns depend on the power source for the transmitter. Integrated circuits and surface mount technology allow production of very small electronic circuitry in transmitters, making batteries the largest part of the transmitter package. However, the more powerful transmitters with their larger batteries are more difficult to place on or implant in a subject without affecting the subject's behavior or energetics. [M.R.F.]

Birch A deciduous tree of the genus *Betula* which is distributed over much of North America, in Asia south to the Himalaya, and in Europe. About 40 species are known. The birches comprise the family Betulaceae in the order Fagales. The sweet birch, *B. lenta*, the yellow birch, *B. alleghaniensis*, and the paper birch, *B. papyrifera*, are all important timber trees of eastern United States. The yellow and the paper species extend into Canada. The gray birch, *B. populifolia*, is a smaller tree of the extreme northeastern United States and adjacent Canada. Both sweet (black) and yellow birches can be recognized by the wintergreen taste of the bark of the young twigs. A flavoring similar to oil of wintergreen is obtained from the bark of the sweet birch. The paper and gray birches can be easily identified by their white bark. The bark of the paper birch peels off in thin papery sheets, a characteristic not true of the gray birch. The river birch, *B. nigra*, is a less common tree of wet soils and banks of streams and is important as an ornamental and for erosion control. The hard, strong wood of the yellow and the sweet birches is used for furniture, boxes, baskets, crates, and woodenware. The European birches, *B. pubescens* and *B. pendula*, are the counterparts of the paper and gray birches in the United States. European birches are also cultivated in America. [A.H.G./K.P.D.]

Black Sea A semienclosed marginal sea with an area of 420,000 km^2 (160,000 mi^2) bounded by Turkey to the south, Georgia to the east, Russia and Ukraine to the north, and Romania and Bulgaria to the west. The physical and chemical structure of the Black Sea is critically dependent on its hydrological balance. As a result, it is the world's largest anoxic basin. It has recently experienced numerous types of environmental stress.

The Black Sea consists of a large basin with a depth of about 2200 m (7200 ft). The continental shelf is mostly narrow except for the broad shelf in the northwest region. Fresh-water input from rivers, especially the Danube, Dniester, and Don, and precipitation exceeds evaporation. Low-salinity surface waters are transported to the Mediterranean as a surface outflow. High-salinity seawater from the Mediterranean enters the Black Sea as a subsurface inflow through the Bosporus. This estuarine circulation (seawater inflow at depth and fresh-water outflow at the surface) results in an unusually strong vertical density gradient determined mainly by the salinity. Thus the Black Sea has a two-layered structure with a lower-salinity surface layer and a higher-salinity deep layer.

The vertical stratification has a strong effect on the chemistry of the sea. Respiration of particulate organic carbon sinking into the deep water has used up all the dissolved oxygen. Thus, conditions favor bacterial sulfate reduction and high sulfide concentrations. As a result, the Black Sea is the world's largest anoxic basin and is commonly used as a modern analog of an environment favoring the formation of organic-rich black shales observed in the geological sedimentary record.

Before the 1970s the Black Sea had a highly diverse and healthy biological population. Its species composition was similar to that of the Mediterranean but with less quantity. The phytoplankton community was characterized by a large diatom bloom in May-June followed by a smaller dinoflagellate bloom. The primary zooplankton were copepods, and there were 170 species of fish, including large commercial populations of mackerel, bonito, anchovies, herring, carp, and sturgeon.

Since about 1970 there have been dramatic changes in the food web due to anthropogenic effects and invasions of new species. It is now characterized as a nonequilibrium, low-diversity, eutrophic state. The large increase in input of nitrogen due to eutrophication and decrease in silicate due to dam construction have increased the frequency of noxious algal blooms and resulted in dramatic shifts in phytoplankton from diatoms (siliceous) to coccolithophores and flagellates (nonsiliceous). The most dramatic changes have been observed in the northwestern shelf and the western coastal regions, which have the largest anthropogenic effects. The water overlying the sediments in these shallow areas frequently go anoxic due to this eutrophication. In the early 1980s the grazer community experienced major increases of previously minor indigenous species such as the omnivorous dinoflagellate *Noctilluca scintillans* and the medusa *Aurelia aurita*. The ctenophore *Mnemopsis leidyi* was imported at the end of the 1980s from the east coast of the United States as ballast water in tankers and experienced an explosive unregulated growth. These changes plus overfishing resulted in a collapse of commercial fish stocks during the 1990s. [J.W.M.]

Blackleg An acute, usually fatal, disease of cattle and occasionally of sheep, goats, and swine. The infection is caused by *Clostridium chauvoei* (*C. feseri*), a strictly anaerobic, sporeforming bacillus of the soil. The disease is also called symptomatic anthrax or quarter-evil. The characteristic lesions in the natural infection consist of crepitant swellings in involved muscles, which at necropsy are dark red, dark brown, or blue black. Artificial immunization is possible; animals surviving an attack of blackleg are permanently immune to recurrence of the disease. [L.S.McC.]

Blastomycetes A class of the subdivision Deuteromycotina comprising anamorphic (asexual or imperfect) yeast fungi that lack fruit bodies (conidiomata), have no dikaryophase, and are usually unicellular rather than filamentous. The thallus consists of individual cells. Approximately 80 genera comprising about 600 species are recognized.

The Blastomycetes, like other groups of deuteromycetes, are artificial, composed entirely of anamorphic fungi of ascomycete or basidiomycete affinity. Taxa are referred to as form genera and form species because the absence of sexual, perfect, or meiotic states forces classification and identification by artificial rather than phylogenetic means. Black yeasts are distinguished from anamorphic yeasts by the presence of melanin in the cell walls, abundant production of septate mycelium (filamentous), and aerial dispersal of conidia. Unlike other deuteromycetes, the number of morphological and developmental features for classification of Blastomycetes, although useful, is limited. The emphasis in yeast systematics has therefore been on physiological and biochemical tests, supplemented extensively by serological, electrophoretic, and molecular techniques.

Anamorphic yeasts can be recovered from most ecological niches—animals, plants and their surfaces, fresh and marine water, soils, and environments such as manufacturing plants, tanning fluids, and mineral oils. Blastomycetes are of great economic importance in two respects: the production of products and the spoilage of raw materials and products. Selected strains of *Saccharomyces cerevisiae* are used in the baking, brewing, distilling, and wine industries.

Blastomycetes are also recognized pathogens in medicine. Both *Candida*, causing candidiasis or candidosis, and *Cryptococcus*, causing cryptococcosis, are opportunistic pathogens that cause systemic infections only in individuals with lowered resistance. Esophageal candidiasis and cryptococcosis of the central nervous system are both regarded as being particularly strong indicators of AIDS. *See* EUMYCOTA; FUNGI; YEAST.

[B.C.S.]

Bluetongue An arthropod-borne disease of ruminant species. Its geographic distribution is dependent upon a susceptible ruminant population and climatic conditions that favor breeding of the primary vector, a mosquito (*Culicoides* species).

Bluetongue virus is the prototype of the genus *Orbivirus* (family Reoviridae). The viral genome exists as 10 segments of the double-stranded ribonucleic acid (RNA) that encode for seven structural and three nonstructural proteins. The viral particle has a double capsid, with the outer coat (morphologically poorly defined) being composed of two proteins. Twenty-four serotypes of bluetongue virus have been defined, and their distribution throughout the world is varied. *See* ANIMAL VIRUS.

While multiple ruminant species can become infected, only sheep and deer typically display clinical bluetongue disease. Severity of the disease is dependent upon multiple factors, including virus strain, animal breed, and environmental conditions. Upon infection by a gnat bite, the virus apparently replicates in the local lymphatic system prior to the viral particles moving into the blood (viremia). Viral replication occurs in the endothelial cells of small vessels, resulting in narrowing of the vessel, release of proteinaceous material into the surrounding tissues, and possibly hemorrhage, with the respiratory tract, mucous membranes, cardiac and skeletal musculature, and skin being most affected. Animals experiencing acute clinical symptoms typically die from pneumonia or pulmonary failure; hemorrhage at the base of the pulmonary artery indicates the presence of a vascular lesion.

Control of bluetongue disease requires the application of vaccines and modulation of the farm environment. While blue-tongue virus vaccines are available, efficacy is often incomplete and variable, in part because of the multiplicity of serotypes active throughout the world and limited cross-serotype protection. Furthermore, use of polyvalent (multiple-serotype) vaccines in the United States has been discouraged because of potential genetic reassortment between vaccine viruses and wild-type viruses, a process that could possibly lead to pathogenic variants. Relative to environment, elimination of vector breeding sites can also facilitate control of virus transmission. With the multiplicity of serotypes typically active in an endemic area, and the minimal cross-serotype protection observed, administration of vaccine in the face of an outbreak may be of limited value.

[J.L.Sto.]

Bog Nutrient-poor, acid peatlands with a vegetation in which peat mosses (*Sphagnum* spp.), ericaceous dwarf shrubs, and to a lesser extent, various sedges (Cyperaceae) play a prominent role. The terms muskeg, moor, heath, and moss are used locally to indicate these sites. *See* MUSKEG.

Bogs are most abundant in the Northern Hemisphere, especially in a broad belt including the northern part of the deciduous forest zone and the central and southern parts of the boreal forest zone. Farther south, and in drier climates farther inland, they become sporadic and restricted to specialized habitats. To the north, peatlands controlled by mineral soil water (aapa mires) replace them as the dominant wetlands.

Bogs are much less extensive in the Southern Hemisphere because there is little land in cold temperate latitudes. In these Southern Hemisphere peatlands, *Sphagnum* is much less important, and Epacridaceae and Restionaceae replace the Ericaceae and Cyperaceae of the Northern Hemisphere.

Bogs have a fibric, poorly decomposed peat consisting primarily of the remains of *Sphagnum*. Peat accumulation is the result of an excess of production over decomposition. Obviously, the very presence of bogs shows that production exceeded decay over the entire period of bog formation. However, in any given bog present production can exceed, equal, or be less than decomposition, depending on whether it is actively developing, in equilibrium, or eroding. In most bogs, production and decomposition appear to be in equilibrium at present.

Slow decay rather than high productivity causes the accumulation of peat. Decomposition of organic matter in peat bogs is slow due to the high water table, which causes the absence of oxygen in most of the peat mass, and to the low fertility of the peat. Bogs, in contrast to other peatlands, can accumulate organic matter far above the groundwater table.

Bogs show large geographic differences in floristic composition, surface morphology, and development. Blanket bogs, plateau bogs, domed bogs, and flat bogs represent a series of bog types with decreasing climatic humidity. Concentric patterns of pools and strings (peat dams) become more common and better developed northward. Continental bogs are often forest-covered, whereas oceanic bogs are dominated by dwarf shrub heaths and sedge lawns, with forests restricted to the bog slope if the climate is not too severe.

Bogs have long been used as a source of fuel. In Ireland and other parts of western Europe, the harvesting of peat for domestic fuel and reclamation for agriculture and forestry have affected most of the peatlands, and few undisturbed bogs are left. Other uses are for horticultural peat, air layering in greenhouses, litter for poultry and livestock, and various chemical and pharmaceutical purposes. Mechanical extraction of peat for horticultural purposes has affected large bog areas worldwide. *See* BIOMASS; SWAMP, MARSH, AND BOG.

[A.W.H.D.]

Bordetella A genus of gram-negative bacteria which are coccobacilli and obligate aerobes, and fail to ferment carbohydrates. These bacteria are respiratory pathogens. *Bordetella pertussis*, *B. parapertussis*, and *B. bronchiseptica* share greater than 90% of their deoxyribonucleic acid (DNA) sequences and would not warrant separate species designations except that the distinctions are useful for clinical purposes. *Bordetella pertussis* is an obligate human pathogen and is the causative agent of whooping cough (pertussis). *Bordetella parapertussis* causes a milder form of disease in humans and also causes respiratory infections in sheep. *Bordetella bronchiseptica* has the broadest host range, causing disease in many mammalian species, but kennel cough in dogs and atrophic rhinitis, in which infected piglets develop deformed nasal passages, have the biggest economic impact. *Bordetella avium* is more distantly related to the other species. A pathogen of birds, it is of major economic importance to the poultry industry.

Infection by all four species is characterized by bacterial adherence to the ciliated cells that line the windpipe (trachea), *B. pertussis* releases massive amounts of peptidoglycan, causing an exaggerated immune response that is ultimately deleterious, resulting in self-induced death of the ciliated cells. *Bordetella* also produces protein toxins. The best-characterized is pertussis toxin, made only by *B. pertussis*. This toxin interferes with the mechanisms used by host cells to communicate with one another.

Bordetella pertussis is spread by coughing and has no environmental reservoir other than infected humans. Culturing the organism is difficult. Erythromycin is the antibiotic used most frequently to treat whooping cough. Unfortunately, antibiotic treatment improves the patient's condition only if given early, when the disease is most difficult to diagnose, and does not help after whooping has begun. This is consistent with the concept that the early symptoms of the disease result from bacterial damage to the respiratory tract and the later symptoms are due to toxins released by the bacteria. Antibiotics can eradicate the microorganisms but cannot reverse the effects of toxins, which can cause damage far from the site of bacterial growth.

Vaccines have been developed for whooping cough and kennel cough. Multicomponent pertussis vaccines consisting of inactivated pertussis toxin and various combinations of filamentous hemagglutinin, pertactin, and fimbriae are now replacing the older whole-cell vaccines consisting of killed bacteria, which were suspected but not proven to cause rare but serious side effects. Vaccination programs have greatly reduced the incidence of whooping cough in affluent nations, but worldwide nearly half a million deaths occur each year, most of which are vaccine-preventable. *See* MEDICAL BACTERIOLOGY. [A.We.]

Borrelia A genus of spirochetes that have a unique genome composed of a linear chromosome and numerous linear and circular plasmids. Borreliae are motile, helical organisms with 4–30 uneven, irregular coils, and are 5–25 micrometers long and 0.2–0.5 μm wide. All borreliae are arthropod-borne. Of the 24 recognized species, 21 cause relapsing fever and similar diseases in human and rodent hosts; two are responsible for infections in ruminants and horses; and the remaining one, for borreliosis in birds. *See* BACTERIA.

The borreliae of human relapsing fevers are transmitted by the body louse or by a large variety of soft-shelled ticks of the genus *Ornithodoros*. The species *B. burgdorferi*, the etiologic agent of Lyme disease and related disorders, is transmitted by ticks of the genus *Ixodes*. *Borrelia anserina*, which causes spirochetosis in chickens and other birds, is propagated by ticks of the genus *Argas*. Various species of ixodid ticks are responsible for transmitting *B. theileri* among cattle, horses, and sheep. *Borrelia coriaceae*, isolated from *O. coriaceus*, is the putative cause of epizootic bovine abortion in the western United States. *See* RELAPSING FEVER.

Polyacrylamide gel electrophoresis of spirochetes has shown that the outer surface of the microorganisms contains numerous variable lipoproteins of which at least two are abundant. The antigenic variability is well known for the relapsing fever borreliae. A switch in the major outer-surface proteins leads to recurrent spirochetemias. Tetracyclines, penicillins, and doxycycline are the most effective antibiotics for treatment of spirochetes. Two vaccines consisting of recombinant *B. burgdorferi* have been evaluated in subjects of risk for Lyme disease. Both proved safe and effective in the prevention of this disease. *See* MEDICAL BACTERIOLOGY. [W.Bu.; P.Ro.]

Botanical gardens A garden for the culture of plants collected chiefly for scientific and educational purposes. Such a garden is more properly called a botanical institution, in which the outdoor garden is but one portion of an organization including the greenhouse, the herbarium, the library, and the research laboratory. *See* HERBARIUM.

It was only in modern Europe, after the foundation of the great medieval universities, that botanical gardens for educational purposes began to be established in connection with the schools. The oldest gardens are those in Padua (established 1533) and at Pisa (1543). The botanical garden of the University of Leiden was begun in 1587, and the first greenhouse is said to have been constructed there in 1599. The Royal Botanical Gardens at Kew, England, were officially opened in 1841. This institution came to be known as the botanical capital of the world.

The first of the great tropical gardens was founded at Calcutta in 1787. The original name, Royal Botanic Garden, was changed in 1947 to Indian Botanic Garden. Another great tropical garden, the Jardin Botanico of Rio de Janeiro, was founded in 1808. The great tropical botanical garden of Buitenzorg (Bogor), Java, which originated in 1817, has an area of 205 acres (83 hectares) with an additional 150 acres (61 hectares) in the Mountain Garden.

The first great garden of the United States was founded by Henry Shaw at St. Louis in 1859, and is now known as the Missouri Botanical Garden. The New York Botanical Garden was chartered in 1891 and the Brooklyn Botanic Garden in 1910. The Jardin Botanique of Montreal, the leading garden of Canada, was opened in 1936. *See* ARBORETUM. [E.L.C.]

Botany That branch of biological science which embraces the study of plants and plant life. Botanical studies may range from microscopic observations of the smallest and obscurest plants to the study of the trees of the forest. One botanist may be interested mainly in the relationships among plants and in their geographic distribution, whereas another may be primarily concerned with structure or with the study of the life processes taking place in plants.

Botany may be divided by subject matter into several specialties, such as plant anatomy, plant chemistry, plant cytology, plant ecology (including autecology and synecology), plant embryology, plant genetics, plant morphology, plant physiology, plant taxonomy, ethnobotany, and paleobotany. It may also be divided according to the group of plants being studied; for example, agostology, the study of grasses; algology (phycology), the study of algae; bryology, the study of mosses; mycology, the study of fungi; and pteridology, the study of ferns. Bacteriology and virology are also parts of botany in a broad sense. Furthermore, a number of agricultural subjects have botany as their foundation. Among these are agronomy, floriculture, forestry, horticulture, landscape architecture, and plant breeding. *See* AGRICULTURE; AGRONOMY; BACTERIOLOGY; ECOLOGY; FLORICULTURE; GENETICS; PLANT ANATOMY; PLANT GROWTH; PLANT MORPHOGENESIS; PLANT PATHOLOGY; PLANT PHYSIOLOGY; PLANT TAXONOMY. [A.Cr.]

Botulism An illness produced by the exotoxin of *Clostridium botulinum* and occasionally other clostridia, and characterized by paralysis and other neurological abnormalities. There are seven principal toxin types involved (A–G); only types A, B, E, and F have been implicated in human disease. Types C and D produce illness in birds and mammals. Strains of *C. barati* and *C. butyricum* have been found to produce toxins E and F and have been implicated in infant botulism. *See* ANAEROBIC INFECTION; VIRULENCE.

The three clinical forms of botulism are classic botulism, infant botulism, and wound botulism. Classic botulism is typically due to ingestion of preformed toxin, infant botulism involves ingestion of *C. botulinum* spores with subsequent germination and toxin production in the gastrointestinal tract, and wound botulism involves production of toxin by the organism's infecting or colonizing a wound. The incubation period is from a few hours to more than a week (but usually 1–2 days), depending primarily on the amount of toxin ingested or absorbed.

There is classically acute onset of bilateral cranial nerve impairment and subsequent symmetrical descending paralysis or weakness. Commonly noted are dysphagia (difficulty in swallowing), dry mouth, diplopia (double vision), dysarthria (a neuromuscular disorder affecting speech), and blurred vision. Nausea, vomiting, and fatigue are common as well. Ileus (impaired intestinal motility) and constipation are much more typical than diarrhea; there may also be urinary retention and dry mucous membranes. Central nervous system function and sensation remain intact, and fever does not occur in the absence of complications. Fever may even be absent in wound botulism. *See* TOXIN.

In food-borne botulism, home-canned or home-processed foods (particularly vegetables) are commonly implicated, with commercially canned foods involved infrequently. Outbreaks usually involve only one or two people, but may affect dozens. In infant botulism, honey and corn syrup have been implicated as vehicles. Therapy involves measures to rid the body of unabsorbed toxin, neutralization of unfixed toxin by antitoxin, and adequate intensive care support. *See* FOOD POISONING; MEDICAL BACTERIOLOGY; POISON. [S.M.F.]

Bovine virus diarrhea A viral disease of cattle. The bovine viral diarrhea virus is common worldwide, infects cattle of all ages, and causes a variety of disease processes. In addition to cattle, the virus infects most even-toed ungulates, including sheep, swine, goats, deer, bison, llama, and antelope. *See* ANIMAL VIRUS.

The mature virion is spherical and 40–60 nanometers in diameter. It is composed of a spherical core that is surrounded by a lipid envelope. The viral genome consists of a single strand of positive-sense ribonucleic acid (RNA) about 12,300 nucleotides in length. Although all bovine viral diarrhea viruses are related serologically, variation is common among viral isolates in nucleotide sequence, antigenic sites on viral proteins, and biologic properties expressed in vitro or in vivo.

Cattle infected with bovine viral diarrhea virus usually develop a clinically mild, acute disease that is characterized by fever, low white blood cell count, mild depression, and transient loss of appetite. A clinically severe, acute disease occurs in cattle infected with some viruses. The clinical signs of severe disease include prolonged fever, pronounced depression, rapid respiration, ulceration of the mouth, esophagus, and intestines, hemorrhaging, diarrhea, dehydration, and death. The virus replicates in lymphoid cells and may suppress their immune function, potentially lowering resistance of the host to other infectious agents. Adverse effects on the fetus are common following infection of pregnant cattle with bovine viral diarrhea virus. Embryonic resorption, abortion, stillbirth, and congenital anomalies may result when the virus crosses the placenta and infects the fetus. A clinically severe disease, termed mucosal disease, occurs only in persistently infected cattle. The signs of mucosal disease are fever, low white blood cell count, diarrhea that is often bloody, mucosal ulcerations of the alimentary tract, inappetence, and death.

Oral or nasal exposure with virus is the primary mode of transmission of acute disease. Persistently infected cattle are important in viral transmission because they shed large quantities of virus in their saliva and nasal secretions. Contact with other cattle undergoing an acute infection is another important method of viral transmission. Biting insects and artificial insemination with semen contaminated with virus have been reported to transmit the virus. Because wild ruminants may be infected with bovine viral diarrhea virus, there is a possibility that disease may spread from farm to farm with movement of wildlife.

Treatment for bovine virus diarrhea primarily is based on supportive care. Prevention of disease can be achieved through use of vaccination. The vaccines usually are given first when calves are about 6 months of age. Vaccination is often repeated at regular intervals during the life of the animal. Identification and elimination of persistently infected cattle also is important for control of viral spread and for prevention of disease. [S.R.Bo.]

Breeding (animal) The application of genetic principles to improving heredity for economically important traits in domestic animals. Examples are improvement of milk production in dairy cattle, meatiness in pigs, feed requirements or growth rate in beef cattle, and egg production in chickens. Selection permits the best parents to leave more offspring in the next generation than do poor parents.

Selection is the primary tool for generating directed genetic changes in animals. It may be concentrated on one characteristic, may be directed independently on several traits, or may be conducted on an index or total score which includes information on several traits. In general, the third method is preferable when several important heritable traits need attention. In practice, selection is likely to be a mixture of the second and third methods.

Heritability, the fraction of the total variation in a trait that is due to additive genetic differences, is a key parameter in making decisions in selection. Most traits are strongly to moderately influenced by environmental or managemental differences. Therefore, managing animals to equalize environmental influences on them, or statistically adjusting for environmental differences among animals, is necessary to accurately choose those with the best inheritance for various traits.

The improvement achieved by selection is directly related to the accuracy with which the breeding values of the subjects can be recognized. Accuracy, in turn, depends upon the heritabilities of the traits and upon whether they can be measured directly upon the subjects for selection (mass selection), upon their parents (pedigree selection), upon their brothers and sisters (family selection), or upon their progeny (progeny testing). For traits of medium heritability, the following sources of information are about equally accurate for predicting breeding values of subjects: (1) one record measured on the subject; (2) one record on each ancestor for three previous generations; (3) one record each on five brothers or sisters where there is no environmental correlation between family members; and (4) one record each on five progeny having no environmental correlations, each from a different mate.

Propagation of improved animal stocks is achieved primarily with purebred strains descended from imported or locally developed groups or breeds of animals which have been selected and interbred for a long enough period to be reasonably uniform for certain trademark characteristics, such as coat color. Because the number of breeding animals is finite and because breeders tend to prefer certain bloodlines and sires, some inbreeding occurs within the pure breeds, but this has not limited productivity in most of these breeds. Crossbreeding makes use of the genetic phenomenon of heterosis. Heterosis is improved performance of crossbred progeny, exceeding that of the average performance of their parents. Most commercial pigs, sheep, and beef cattle are produced by crossbreeding. *See* GENETICS.

Advances in a variety of technologies have application for improvement of domestic animals, including quantitative genetics, reproductive physiology, and molecular genetics. Quantitative geneticists use statistical and genetic information to improve domestic animals. Typically a statistical procedure is used to rank animals based on their estimated breeding values for traits of economic importance. The statistical procedures used allow ranking animals across herds or flocks, provided the animals in different herds or flocks have relatives in common. The primary contribution of reproductive physiology to genetic improvement is to reduce the generation interval. If genetic improvement is increasing at the same rate per generation, more generations can be produced for a fixed time, and thus more gain per unit of time. The most important development was artificial insemination, which allows extensive use of superior males. Another development was embryo transplantation, which allows more extensive use of females. Cloning is a relatively new technique, by which whole and healthy animals have been produced that have the same DNA as the animal from which the cells were taken.

Due to advances in molecular genetics, knowledge is increasing regarding the location of genes on chromosomes and the distance between the genes. In domestic animals, polymorphisms (changes in the order of the four bases) that are discovered in the DNA may be associated with economic traits. When the polymorphisms are associated with or code for economic traits, they are called quantitative trait loci (QTL). When a few or several quantitative trait loci are known that control a portion of the variability in a trait, increasing the frequencies of favorable alleles can enhance the accuracy of selection and augment production. Another use of molecular genetics is to detect the genes that code for genetically predetermined diseases. An example is the bovine leukocyte deficiency gene, which does not allow white blood cells to migrate out of the blood supply into the tissues to fight infection. The calves perish at a young age. Screening all sires that enter artificial breeding organizations and not using sires that transmit the defect has effectively controlled this condition. [A.E.Fr.]

Breeding (plant) The application of genetic principles to improve cultivated plants. New varieties of cultivated plants can result only from genetic reorganization that gives rise to improvements over the existing varieties in particular characteristics or in combinations of characteristics. Thus, plant breeding can be regarded as a branch of applied genetics, but it also makes use of the knowledge and techniques of many aspects of plant science, especially physiology and pathology. Related disciplines, like biochemistry and entomology, are also important, and the application of mathematical statistics in the design and analysis of experiments is essential. *See* GENETICS.

The cornerstone of all plant breeding is selection, or the picking out of plants with the best combinations of agricultural and quality characteristics from populations of plants with a variety of genetic constitutions. Seeds from the selected plants are used to produce the next generation, from which a further cycle of selection may be carried out if there are still differences. Conventional breeding is divided into three categories on the basis of ways in which the species are propagated. First come the species that set seeds by self-pollination; that is, fertilization usually follows the germination of pollen on the stigmas of the same plant on which it was produced. The second category of species sets seeds by cross-pollination; that is, fertilization usually follows the germination of pollen on the stigmas of different plants from those on which it was produced. The third category comprises the species that are asexually propagated; that is, the commercial crop results from planting vegetative parts or by grafting. The procedures used in breeding differ according to the pattern of propagation of the species. Several innovative techniques have been explored to enhance the scope, speed, and efficiency of producing new, superior cultivars. Advances have been made in extending conventional sexual crossing procedures by laboratory culture of plant organs and tissues and by somatic hybridization through protoplast fusion.

The essential attribute of self-pollinating crop species, such as wheat, barley, oats, and many edible legumes, is that, once they are genetically pure, varieties can be maintained without change for many generations. When improvement of an existing variety is desired, it is necessary to produce genetic variation among which selection can be practiced. This is achieved by artificially hybridizing between parental varieties that may contrast with each other in possessing different desirable attributes. This system is known as pedigree breeding, and it is the method most commonly employed, and can be varied in several ways.

Another form of breeding often employed with self-pollinating species involves backcrossing. This is used when an existing variety is broadly satisfactory but lacks one useful and simply inherited trait that is to be found in some other variety. Hybrids are made between the two varieties, and the first hybrid generation is crossed, or backcrossed, with the broadly satisfactory variety which is known as the recurrent parent. Backcrossing has been exceedingly useful in practice and has been extensively employed in adding resistance to diseases, such as rust, smut, or mildew, to established and acceptable varieties of oats, wheat, and barley. *See* PLANT PATHOLOGY.

Natural populations of cross-pollinating species are characterized by extreme genetic diversity. No seed parent is true-breeding, first because it was itself derived from a fertilization in which genetically different parents participated, and second because of the genetic diversity of the pollen it will have received. In dealing with cultivated plants with this breeding structure, the essential concern in seed production is to employ systems in which hybrid vigor is exploited, the range of variation in the crop is diminished, and only parents likely to give rise to superior offspring are retained.

Plant breeders have made use either of inbreeding followed by hybridization or of some form of recurrent selection. During inbreeding programs normally cross-pollinated species, such as corn, are compelled to self-pollinate by artificial means. Inbreeding is continued for a number of generations until genetically pure, true-breeding, and uniform inbred lines are produced. During the production of the inbred lines, rigorous selection is practiced for general vigor and yield and disease resistance, as well as for other important characteristics. To estimate the value of inbred lines as the parents of hybrids, it is necessary to make tests of their combining ability. The test that is used depends upon the crop and on the ease with which controlled cross-pollination can be effected.

Breeding procedures designated as recurrent selection are coming into limited use with open-pollinated species. In theory, this method visualizes a controlled approach to homozygosity, with selection and evaluation in each cycle to permit the desired stepwise changes in gene frequency. Experimental evaluation of the procedure indicates that it has real possibilities. Four types of recurrent selection have been suggested: on the basis of phenotype, for general combining ability, for specific combining ability, and reciprocal selection. The methods are similar in the procedures involved, but vary in the type of tester parent chosen, and therefore in the efficiency with which different types of gene action (additive and nonadditive) are measured.

Varieties of asexually propagated crops consist of large assemblages of genetically identical plants, and there are only two ways of introducing new and improved varieties: by sexual reproduction and by the isolation of somatic mutations. (A very few asexually propagated crop species are sexually sterile, like the banana, but the majority have some sexual fertility.) The latter method has often been used successfully with decorative plants, such as chrysanthemum, and new forms of potato have occasionally arisen in this way. When sexual reproduction is used, hybrids are produced on a large scale between existing varieties; the small number that have useful arrays of characters are propagated vegetatively until sufficient numbers can be planted to allow agronomic evaluation. [R.Ri.]

Cell technologies have been used to extend the range and efficiency of asexual plant propagation. For example, plant cell culture involves the regeneration of entire mature plants from single cells or tissues excised from a source plant and cultured in a nutrient medium. In micropropagation and cloning, tissues are excised from root, stem, petiole, or seedling and induced to regenerate plantlets. All regenerants from tissues of one source plant constitute a clone. Microspore or anther culture is the generation of plants from individual cells with but one set of chromosomes, haploid cells, as occurs in the development of pollen. Microspores are isolated from anthers and cultured on nutrient media, or entire anthers are cultured in this manner. Doubling of chromosomes that may occur spontaneously or can be induced by treatment with colchicine leads to the formation of homozygous dihaploid plants. *See* PLANT PROPAGATION; POLLEN.

Breeding for new, improved varieties of crop plants is most often based on cross-pollination and hybrid production. Such breeding is limited to compatible plants, and compatibility lessens with increasing distance in the relationship between plants. Breeding would benefit from access to traits inherent in sexually noncompatible plants. Biotechnological techniques such as in vitro fertilization and embryo rescue (the excision and culture of embryos on nutrient media) have been employed to overcome incompatibility barriers, as have somatic hybridization and DNA technologies. Somatic hybridization involves enzymatic removal of walls from cells of leaves and seedlings to furnish individual naked cells, that is, protoplasts, which can then be fused to produce hybrids. Similarity of membrane structure throughout the plant kingdom permits the fusion of distantly related protoplasts. Cell fusion may lead to nuclear fusion, resulting in amphi-diploid somatic hybrid cells. Fusion products of closely related yet sexually incompatible plants have been grown to flowering plants; the most famous example is the potato + tomato hybrid = pomato (*Solanum tuberosum* + *Lycopersicon esculentum*). DNA technologies enable the isolation of desirable genes from bacteria, plants, and animals (genes that confer herbicide resistance or tolerance to environmental stress, or encode enzymes and proteins of value to the processing industry) and the insertion of such genes into cells and tissues of target plants by direct or indirect uptake has led to the genetic transformation of plant cells. The regeneration of transformed plant cells and tissues results in new and novel genotypes (transgenic plants). Contrary to hybrids obtained by cross-pollination, such plants are different from their parent by only one or two single, defined traits. *See* GENETIC ENGINEERING. [F.Co.]

Brucellosis An infectious, zoonotic disease of various animals and humans caused by *Brucella* species. Each species tends to preferentially infect a particular animal, but several types can infect humans. *Brucella melitensis* (preferentially infects goats and sheep), *B. suis* (infects pigs), and *B. abortus* (infects cattle) are the most common causes of human brucellosis. *Brucella melitensis* is the most virulent for humans, followed by *B. suis* and *B. abortus*. *Brucella canis* and *B. ovis*, which infect dogs and sheep respectively, rarely infect humans. Although brucellosis is found all over the world, in many countries the disease has been eradicated. The brucellae are small, gram-negative coccobacilli which are defined as facultative intracellular parasites since they are able to replicate within specialized cells of the host.

In animals the brucellae often localize in the reproductive tract, mammary gland, and lymph node. They have a particular affinity for the pregnant uterus, leading to abortion and reduced milk production with resultant economic loss to the farmer.

Wildlife, including elk, feral pigs, bison, and reindeer, can become infected and can spread the disease to domestic livestock.

Brucellosis in humans is characterized by undulant fever, cold sweats, chills, muscular pain, and severe weakness. Some individuals may have recurrent bouts of the disease in which a variety of organs may be affected, sometimes resulting in death. The disease can be contracted by consuming unpasteurized milk or cheese, or via the introduction of organisms through small skin lesions or as an aerosol through the conjunctiva and the respiratory system. Treatment with tetracycline and other antibiotics is most successful if started early after symptoms occur. Development of the disease can be prevented if treatment is initiated immediately after contact with potentially infected material.

At present there are no effective vaccines for humans. The disease can be eliminated only by eradicating it in animals. A major source of brucellosis in humans is the consumption of *B. melitensis*–infected milk and cheese from goats. Incidence can be reduced by pasteurizing milk. Animals can be vaccinated to increase their immunity against brucellosis and therefore reduce abortions and disease transmission. *See* EPIDEMIOLOGY; MEDICAL BACTERIOLOGY. [W.W.Sp.]

Buckeye A genus, *Aesculus*, of deciduous trees or shrubs belonging to the plant order Sapindales, buckeyes grow in North America, southeast Europe, and eastern Asia to India. The distinctive features are opposite, palmately compound leaves and a large fruit having a firm outer coat and containing usually one large seed with a conspicuous hilum.

The Ohio buckeye (*A. glabra*) is found mainly in the Ohio valley and in the southern Appalachians. It can be recognized by the glabrous winter buds, prickly fruits, and compound leaves having five leaflets. Another important species, the yellow buckeye (*A. octandra*), is native in the Central states, has five leaflets and smooth buds, but differs in its smooth, larger fruit. The horse chestnut (*A. hippocastanum*), which usually has seven leaflets and resinous buds, is a native of the Balkan Peninsula. It is planted throughout the United States and is a beautiful ornamental tree bearing cone-shaped flower clusters in early summer.

The seeds of all species contain a bitter and narcotic principle. The wood of the native tree species is used for furniture, boxes, crates, baskets, and artificial legs. [A.H.G./K.P.D.]

Bud An embryonic shoot containing the growing stem tip surrounded by young leaves or flowers or both, and the whole frequently enclosed by special protective leaves, the bud scales.

The bud at the apex of the stem is called a terminal bud (illus. *a*). Any bud that develops on the side of a stem is a lateral bud. The lateral bud borne in the axil (angle between base of leaf and stem) of a leaf is the axillary bud (illus. *a* and *d*). It develops concurrently with the leaf which subtends it, but usually such buds do not unfold and grow rapidly until the next season. Because of the inhibitory influence of the apical or other buds, many axillary buds never develop actively or may not do so for many years. These are known as latent or dormant buds. Above or beside the axillary buds, some plants regularly produce additional buds called accessory, or supernumerary, buds. Accessory buds which occur above the axillary bud are called superposed buds (illus. *c*), and those beside it collateral buds (illus. *d*). Under certain conditions, such as removal of terminal and axillary buds, other buds may arise at almost any point on the stem, or even on roots or leaves. Such buds are known as adventitious buds. *See* PLANT GROWTH.

Buds that give rise to flowers only are termed flower buds, or in some cases, fruit buds. If a bud grows into a leafy shoot, it is called a leaf bud, or more accurately, a

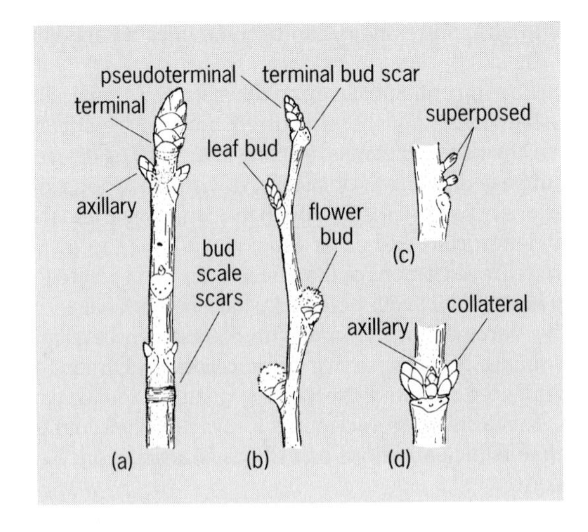

Bud positions. (*a*) Terminal and axillary (buckeye). (*b*) Pseudoterminal (elm). (*c*) Superposed (butternut). (*d*) Collateral (red maple).

branch bud. A bud which contains both young leaves and flowers is called a mixed bud.

Buds of herbaceous plants and of some woody plants are covered by rudimentary foliage leaves only. Such buds are called naked buds. In most woody plants, however, the buds are covered with modified protective leaves in the form of scales. These buds are called scaly buds or winter buds. In the different species of plants, the bud scales differ markedly. They may be covered with hairs or with water-repellent secretions of resin, gum, or wax. Ordinarily when a bud opens, the scales fall off, leaving characteristic markings on the stem (bud scale scars). *See* LEAF. [N.A.]

C

Cancer (medicine) The common name for a malignant neoplasm or tumor. Neoplasms are new growths and can be divided into benign and malignant types, although in some instances the distinction is unclear. The most important differentiating feature is that a malignant tumor will invade surrounding structures and metastasize (spread) to distant sites whereas a benign tumor will not. Other distinctions between benign and malignant growth include the following: malignancies but not benign types are composed of highly atypical cells; malignancies tend to show more rapid growth than benign neoplasms, and are composed, in part, of cells showing frequent mitotic activity; and malignant tumors tend to grow progressively without self-limitation. *See* TUMOR.

Malignant neoplasms that arise from cells of mesenchymal origin (for example, bone muscle, connective tissue) are called sarcomas. Those that develop from epithelial cells and tissues (for example, skin, mucosal membranes, and glandular tissues) are termed carcinomas. Carcinomas usually metastasize initially by way of lymphatic channels, whereas sarcomas spread to distant organs through the bloodstream.

The cause of most types of human cancers is unknown. However, a number of factors are thought to be operative in the development of some malignant neoplasms. Genetic factors are thought to be causally related to some human malignancies such as lung cancer in that the incidence of cancer among persons with a positive family history of cancer may be three times as high as in those who do not have a family history. A number of different neoplasms are known to be genetically related and may be due to damage or changes in chromosome structure. Radiation in various forms is thought to be responsible for up to 3% of all cancers. In the United States the carcinogens in tobacco account for up to one-third of all cancer deaths in men and 5–10% in women. The increasing incidence and death rate from cancer of the lung in women is alarming, and is directly related to the increasing prevalence of cigarette smoking by women. Cigarette smoking and the heavy consumption of ethyl alcohol appear to act synergistically in the development of oral, esophageal, and gastric cancers. There are several carcinogens to which people are exposed occupationally that result in the development of cancer, although the mechanisms by which they cause neoplasms are sometimes poorly understood. For example, arsenic is associated with lung, skin, and liver cancer and asbestos causes mesotheliomas (cancer of the pleural, peritoneal, and pericardial cavities). Certain drugs and hormones have been found to cause certain types of neoplasms. Postmenopausal women taking estrogen hormones have a much higher incidence of endometrial cancer (cancer of the lining of the uterine cavity). The role of diet and nutrition in the development of malignant tumors is controversial and still under investigation. Some epidemiologic studies have shown that certain diets, such as those high in saturated fats, are associated with an increased incidence of certain types of neoplasm, such as colon cancer. The role of viruses in the development of human cancers is being studied. *See* MUTAGENS AND CARCINOGENS; TUMOR VIRUSES.

It is generally accepted that the neoplastic condition is caused by alterations in genetic mechanisms involved in cellular differentiation. In malignant cells, normal cellular processes are bypassed due to the actions of a select group of genes called oncogenes which regulate cellular activities. A group of these highly conserved genes exist in normal cells and are called proto-oncogenes. These genes appear to be important in regulating cellular growth during embryonic development. It is thought that in carcinogenesis these proto-oncogenes become unmasked or changed during the breakage or translocation of chromosomes. These genes that were previously suppressed in the cell then become functional, and in some instances lead to the excessive production of growth factors which could be important in the neoplastic state.

The physical changes that cancer produces in the body vary considerably, depending on the type of tumor, location, rate of growth, and whether it has metastasized. The American Cancer Society has widely publicized cancer's seven warning signals: (1) a change in bowel or bladder habits; (2) a sore that does not heal; (3) unusual bleeding or discharge; (4) a thickening or lump in the breast or elsewhere; (5) indigestion or difficulty in swallowing; (6) an obvious change in a wart or mole; and (7) a nagging cough or hoarseness. In current medical practice, most cancers are staged according to tumor size, metastases to lymph nodes, and distant metastases. This type of staging is useful in determining the most effective therapy and the prognosis.

The progression, or lack thereof, of a given cancer is highly variable and depends on the type of neoplasm and the response to treatment. Treatment modalities include surgery, chemotherapy, radiation therapy, hormonal manipulation, and immunotherapy. In general, each type of cancer is treated very specifically, and often a combination of the various modalities is used, for example, surgery preceded or followed by radiation therapy. The response to treatment depends on the type of tumor, its size, and whether it has spread. *See* ONCOLOGY. [S.P.H.]

Canine distemper A fatal viral disease of dogs and other carnivores, with a worldwide distribution. Canine distemper virus has a wide host range; most terrestrial carnivores are susceptible to natural canine distemper virus infection. All animals in the families Canidae (such as dog, dingo, fox, coyote, wolf, jackal), Mustelidae (such as weasel, ferret, mink, skunk, badger, stoat, marten, otter), and Procyonidae (such as kinkajou, coati, bassariscus, raccoon, panda) may succumb to canine distemper virus infection. Members of other Carnivora families, including domestic cats and swine, may become subclinically infected. The virus has also been isolated from large cats (lions, tigers, leopards) that have died in zoological parks in North America, from wild lions in the Serengeti National Park (Tanzania), and from wild javelinas (collared peccaries).

Canine distemper virus is classified as a morbillivirus within the Paramyxoviridae family, closely related to measles virus and rinderpest virus of cattle and the phocine (seal) and dolphin distemper virus. The virus is enveloped with a negative-sense ribonucleic acid and consists of six structural proteins: the nucleoprotein and two enzymes in the nucleocapsid, the membrane protein on the inside, and the hemagglutinating and fusion proteins on the outside of the lipoprotein envelope. *See* ANIMAL VIRUS; PARAMYXOVIRUS.

Canine distemper is enzootic worldwide. Aerosol transmission in respiratory secretions is the main route of transmission. Virus shedding begins approximately 7 days after the initial infection. Acutely infected dogs and other carnivores shed virus in all body excretions, regardless of whether they show clinical signs or not.

Great variations occur in the duration and severity of canine distemper, which may range from no visible signs to severe disease, often with central nervous system involvement, with approximately 50% mortality in dogs. The first fever 3–6 days after infection may pass unnoticed; the second peak (several days later and intermittent thereafter) is usually associated with nasal and ocular discharge, depression, and anorexia. A low lymphocyte count is always present during the early stages of infection. Gastrointestinal and respiratory signs may follow, often enhanced by secondary infection.

A specific antiviral drug having an effect on canine distemper virus in dogs is not presently available. Treatment of canine distemper, therefore, is nonspecific and supportive. Antibiotic therapy is recommended because of the common occurrence of secondary bacterial infections of the respiratory and alimentary tracts. Administration of fluids and electrolytes may be the most important therapy for canine distemper because diseased dogs with diarrhea are often dehydrated. [M.J.G.A.; B.A.S.]

Canine parvovirus infection Severe enteritis caused by a small nonenveloped single-stranded deoxyribonucleic acid (DNA) virus that is resistant to inactivation and remains infectious in the environment for 5–7 months. First observed in dogs in 1976, canine parvovirus may have originated by mutation of a closely related parvovirus of cats or wildlife. The original virus was designated as canine parvovirus, type 2 (CPV-2); however, since its discovery the virus has undergone two minor genetic alterations, designated CPV-2a and CPV-2b. These alterations may have enabled the virus to adapt to its new host, replicate, and spread more effectively. *See* ANIMAL VIRUS.

Canine parvovirus is transmitted between dogs by the fecaloral route. The incubation period is 3–7 days. Virus is first shed in the feces on day 3, and shedding continues for an additional 10 days. Chronically infected dogs that shed virus intermittently are rare. Most naturally occurring infections in dogs are subclinical or result in mild signs of the disease. Dogs with subclinical infections play an important role in the spread of the disease by shedding large amounts of virus into the environment. This shedding, along with the ability of the virus to persist in the environment, contributes to the endemicity of the disease. The development of disease following infection ranges from 20 to 90%, and mortality ranges 0 to 50%.

The goal of treatment is to support the animal until the infection runs its course. There are no specific antiviral therapies available. The intensity of treatment depends on the severity of signs. Dehydrated pups require intensive intravenous fluid therapy. Antimicrobial drugs are useful because of the risk of secondary bacterial infections, and antiemetic drugs help control vomiting and nausea. Good nursing care is essential. All food and water should be withheld until the pup is no longer vomiting, and the pup should be kept warm, clean, and dry. Because of the infectious nature of the disease, pups should be isolated from other dogs. [M.S.Co.]

Cat scratch disease In humans, typically a benign, subacute regional disease of the lymph nodes (lymphadenopathy) resulting from dermal inoculation of the causative agent, the bacterium *Bartonella henselae*. The domestic cat is the major reservoir of *B. henselae*, and the cat flea, *Ctenocephalides felis*, is the main vector of transmission from cat to cat. *Bartonella clarridgeiae* has been isolated from domestic cats. Recently, a new bacterium, *B. koehlerae*, has also been isolated from the blood of domestic cats.

Cat scratch disease occurs in immunocompetent patients of all ages, with 55–80% being less than 20 years of age. More than 90% of cases have a history of contact with cats, and 57–83% recall being scratched by a cat. Incidence varies by season; most cases occur in the fall and winter. More cases are observed in males than females.

In humans, 1–3 weeks may elapse between the scratch (or bite) and the appearance of clinical signs. In 50% of the cases, a small skin lesion, often resembling an insect bite, appears at the inoculation site (usually on the hand or forearm) and evolves from a pimple (papule) to a skin blister to partially healed ulcers. These lesions resolve within a few days to a few weeks. Inflammation of lymph nodes develops approximately 3 weeks after exposure. Swelling of the lymph node is usually painful and persists for several weeks or months. In 25% of the cases, a discharge of pus occurs. A large majority of the cases show signs of systemic infection, such as fever, chills, malaise, anorexia, or headaches. In general, the disease is benign and heals spontaneously without aftereffects.

No major clinical signs of cat scratch disease have been reported in cats, although enlargement of the lymph nodes caused by a cat scratch disease–like organism has been reported.

Most individuals with cat scratch disease experience mild illness and require minimal treatment. In severe forms, antibiotics such as ciprofloxacin, rifampin, or gentamicin have been recommended. Use of oral azithromycin for 5 days has shown significant clinical benefit in typical cat scratch disease. [B.B.Ch.]

Cave A natural cavity located underground or in the side of a hill or cliff, generally of a size to admit a human. Caves occur in all types of rocks and topographic situations. They may be formed by many different erosion processes. The most important are created by ground waters that dissolve the common soluble rocks—limestone, dolomite, gypsum, and salt. Limestone caves are the most frequent, longest, and deepest. Lava-tube caves, sea caves created by wave action, and caves caused by piping in unconsolidated rocks are the other important types. The science of caves is known as speleology.

Caves are important sediment traps, preserving evidences of past erosional, botanic, and other phases that may be obliterated aboveground. Chemical deposits are very important. More than 100 different minerals are known to precipitate in caves. Most abundant and significant are stalactites, stalagmites, and flowstones of calcite. These may be dated with uranium series methods, thus establishing minimum ages for the host caves. They contain paleomagnetic records. Their oxygen and carbon isotope ratios and trapped organic materials may record long-term changes of climate and vegetation aboveground that can be dated with great precision. As a consequence, cave deposits are proving to be among the most valuable paleoenvironmental records preserved on the continents. *See* STALACTITES AND STALAGMITES. [D.C.F.]

Cedar Any of a large number of evergreen trees having fragrant wood of great durability. Arborvitae is sometimes called northern white cedar. *See* ARBORVITAE.

Chamaecyparis thyoides, the southern white cedar, grows only in swamps near the eastern coast of North America, where it is also known as Atlantic white cedar. The wood is soft, fragrant, and durable in the soil and is used for boxes, crates, small boats, tanks, woodenware, poles, and shingles. The Port Orford cedar (*C. lawsoniana*), also known as Lawson cypress, is native to southwestern Oregon and northwestern California. It is the principal wood for storage battery separators, but is also used for venetian blinds and construction purposes. Alaska cedar (*C. nootkatensis*) is found from Oregon to Alaska. The wood is used for interior finish, cabinetwork, small boats, and furniture. It is also grown as an ornamental tree. Incense cedar (*Libocedrus decurrens*) is found from Oregon to western Nevada and Lower California. Incense cedar is one of the chief woods for pencils, and is also used for venetian blinds, rough construction, and fence posts and as an ornamental and shade tree.

Eastern red cedar (*Juniperus virginiana*) is distributed over the eastern United States and adjacent Canada. The very fragrant wood is durable in the soil and is used for fence posts, chests, wardrobes, flooring, and pencils. Cedarwood oil is used in medicine and perfumes. Cedar of Lebanon (*Cedrus libani*) and Atlas cedar (*C. atlantica*) resemble the larch, but the leaves are evergreen and the cones are much larger and erect on the branches. The cedar of Lebanon is a native of Asia Minor. *See* LARCH.

The cigarbox cedar (*Cedrela odorata*), also known as the West Indian cedar, belongs to the mahogany family, is a broadleaved tree with pinnate, deciduous leaves, and is related to the *Ailanthus* and sumac. The wood is very durable and fragrant and is valued in the West Indies for the manufacture of cabinets, furniture, and canoes. [A.H.G./K.P.D.]

Cell plastids Specialized structures found in the cytoplasm of plant cells, diverse in distribution, size, shape, composition, structure, function, and mode of development. A number of different types are recognized. Chloroplasts occur in the green parts of plants and are responsible for the green coloration, for they contain the chlorophyll pigments. These pigments, along with certain others, absorb the light energy that drives the processes of photosynthesis, by which sugars, starch, and other organic materials are synthesized. Amyloplasts, nearly or entirely colorless, are packed with starch grains and occur in cells of storage tissue. Proteoplasts are less common and contain crystalline, fibrillar, or amorphous masses of protein, sometimes along with starch grains. In chromoplasts the green pigment is masked or replaced by others, notably carotenoids, as in the cells of carrot roots and many flowers and fruits. *See* CHLOROPHYLL.

All types of plastids have one structural feature in common, a double envelope consisting of two concentric sheets of membrane. The outer of these is in contact with the cytoplasmic ground substance; the inner with the plastid matrix, or stroma. They are separated by a narrow space of about 10 nanometers.

Another system of membranes generally occupies the main body of the plastid. This internal membrane system is especially well developed in chloroplasts, where the unit of construction is known as a thylakoid. In its simplest form this is a sac such as would be obtained if a balloon-shaped, membrane-limited sphere were to be flattened until the internal space was not much thicker than the membrane itself. It is usual, however, for thylakoids to be lobed, branched, or fenestrated.

The surface area of thylakoids is very large in relation to the volume of the chloroplast. This is functionally significant, for chlorophyll molecules and other components of the light-reaction systems of photosynthesis are associated with these membranes. A chloroplast, however, is much more than a device for carrying out photosynthesis. It can use light energy for uptake and exchange of ions and to drive conformational changes. The stroma contains the elements of a protein-synthesizing system—as much deoxyribonucleic acid (DNA) as a small bacterium, various types of ribonucleic acid (RNA), distinctive ribosomes, and polyribosomes. There is evidence to indicate that much of the protein synthesis of a leaf takes place within the chloroplasts. *See* PHOTOSYNTHESIS.

One of the most challenging problems in cell biology concerns the autonomy of organelles, such as the plastids. Chloroplasts, for instance, have their own DNA, DNA-polymerase, and RNA-polymerase; can make proteins; and, significantly, can mutate. All this suggests a measure of independence. It is known, however, that some nuclear genes can influence the production of molecules that are normally found only in chloroplasts, so their autonomy cannot be complete. It remains to be seen whether they control and regulate their own morphogenetic processes. [B.E.S.G.]

Cereal Any member of the grass family (Gramineae) which produced edible grains usable as food by humans and livestock. Common cereals are rice, wheat, barley, oats, maize (corn), sorghum, rye, and certain millets, with corn, rice, and wheat being the most important. Developed by scientists, triticale is a new cereal derived from crossing wheat and rye and then doubling the number of chromosomes in the hybrid. Occasionally, grains from other grasses (for example, teff) are used for food. Cereals provide more food for human consumption than any other crops.

Four general groups of foods are prepared from the cereal grains. (1) Baked products, made from flour or meal, include breads, pastries, pancakes, cookies, and cakes. (2) Milled grain products, made by removing the bran and usually the germ (or embryo of the seed), include polished rice, farina, wheat flour, cornmeal, hominy, corn grits, pearled barley, semolina (for macaroni products), prepared breakfast cereals, and soup, gravy, and other thickenings. (3) Beverages such as beer and whiskey, made from fermented grain products (distilled or undistilled) and from boiled, roasted grains. (4) Whole-grain products include rolled oats, brown rice, popcorn, shredded and puffed gains, and breakfast foods.

All cereal grains have high energy value, mainly from the starch fraction but also from the fat and protein. In general, the cereals are low in protein content, although oats and certain millets are exceptions. *See* GRAIN CROPS. [L.P.R.]

Chaparral A vegetation formation characterized by woody plants of low stature (3.3–10 ft or 1–3 m tall), impenetrable because of tough, rigid, interlacing branches, with small, simple, waxy, evergreen, thick leaves. The term refers to evergreen oak, Spanish *chapparo*, and therefore is uniquely southwestern North American. This type of vegetation has its center in California and occurs continuously over wide areas of mountainous to sloping topography. The Old World Mediterranean equivalent is called maquis or macchie, with nomenclatural and ecological variants in the countries from Spain to the Balkans. Physiognomically similar vegetation occurs also in South Africa, Chile, and southwestern Australia in areas of Mediterranean climates, that is, with very warm, dry summers and maximum precipitation during the cool season. The floras of these five areas with Mediterranean climates are altogether different.

The characteristic species of the true chaparral of California include *Adenostema fasciculatum*, *Ceanothus cuneatus*, *Quercus dumosa*, *Heteromeles arbutifolia*, *Rhamnus californica*, *R. crocea*, and *Cercocarpus betuloides*, plus a host of endemic species of *Arctostaphylos* and *Ceanothus* and other Californian endemics, both shrubby and herbaceous. These plants determine the formation's physiognomy. It is a dense, uniform-appearing, evergreen, shrubby cover with sclerophyllous leaves and deep-penetrating roots.

Ecologically, chaparral occurs in a climate which is hot and dry in summer, cool but not much below freezing in winter, with little or no snow, and with excessive winter precipitation that leaches the soil of nutrients. The need for water and its supply are exactly out of phase.

Chaparral soils are generally rocky, often shallow, or of extreme chemistry such as those derived from serpentine, and are always low in fertility. In the very precipitous southern Californian mountains, soil erosion rates may be 0.04 in. (1 mm) per year over large watershed areas. [J.Ma.]

Charcoal A porous solid product containing 85–98% carbon produced by heating carbonaceous materials such as cellulose, wood, peat, and coals of bituminous or lower rank at 930–1100°F (500–600°C) in the absence of air.

Chars or charcoals from cellulose or wood are soft and friable. They are used chiefly for decolorizing solutions of sugar and other foodstuffs and for removing objectionable tastes and odors from water. Chars from nutshells and coal are dense, hard carbons.

They are used in gas masks and in chemical manufacturing for many mixture separations. Another use is for the tertiary treatment of waste water. Residual organic matter is adsorbed effectively to improve the water quality. [J.H.Fi.]

Chemical ecology The study of ecological interactions mediated by the chemicals that organisms produce. These substances, known as allelochemicals, serve a variety of functions. They influence or regulate interspecific and intraspecific interactions of microorganisms, plants, and animals, and operate within and between all trophic levels—producers, consumers, and decomposers—and in terrestrial, fresh-water, and marine ecosystems.

Function is an important criterion for the classification of allelochemicals. Allelochemicals beneficial to the emitter are called allomones; those beneficial to the recipient are called kairomones. An allomone to one organism can be a kairomone to another. For example, floral scents benefit the plant (allomones) by encouraging pollinators, but also benefit the insect (kairomones) by providing a cue for the location of nectar.

The chemicals involved are diverse in structure and are often of low molecular weight ($<10,000$). They may be volatile or nonvolatile; water-soluble or fat-soluble. Proteins, polypeptides, and amino acids are also found to play an important role.

Plant allelochemicals are often called secondary compounds or metabolites to distinguish them from those chemicals involved in primary metabolism, although this distinction is not always clear.

Chemical defense in plants. Perhaps to compensate for their immobility, plants have made wide use of chemicals for protection against competitors, pathogens, herbivores, and abiotic stresses. A chemically mediated competitive interaction between higher plants is referred to as allelopathy. Allelopathy appears to occur in many plants, may involve phenolics or terpenoids that are modified in the soil by microorganisms, and is at least partly responsible for the organization of some plant communities. *See* ALLELOPATHY.

Chemicals that are mobilized in response to stress or attack are referred to as active or inducible chemicals, while those that are always present in the plant are referred to as passive or constitutive. In many plants, fungus attack induces the production of defensive compounds called phytoalexins, a diverse chemical group that includes isoflavonoids, terpenoids, polyacetylenes, and furanocoumarins. *See* PHYTOALEXINS.

Defensive chemicals can be induced by herbivore attack. There has been increasing evidence that inducible defenses, such as phenolics, are important in plant-insect interactions.

Constitutive defenses include the chemical hydrogen cyanide. Trefoil, clover, and ferns have been found to exist in two genetically different forms, one containing cyanide (cyanogenic) and one lacking it (acyanogenic); acyanogenic forms are often preferred by several herbivores.

Chemical defenses frequently occur together with certain structures which act as physical defenses, such as spines and hairs. While many chemicals protect plants by deterring herbivore feeding or by direct toxic effects, other defenses may act more indirectly. Chemicals that mimic juvenile hormones, the antijuvenile hormone substances found in some plants, either arrest development or cause premature development in certain susceptible insect species.

Plant chemicals potentially affect not only the herbivores that feed directly on the plant, but also the microorganisms, predators, or parasites of the herbivore. For example, the tomato plant contains an alkaloid, tomatine, that is effective against certain insect herbivores. The tomato hornworm, however, is capable of detoxifying this alkaloid and can thus use the plant successfully—but a wasp parasite of the hornworm cannot detoxify tomatine, and its effectiveness in parasitizing the hornworm is reduced. Therefore, one indirect effect of the chemical in the plant may be to reduce the effectiveness of natural enemies of the plant pest, thereby actually working to the disadvantage of the plant.

Most plant chemicals can affect a wide variety of herbivores and microorganisms, because the modes of action of the chemicals they manufacture are based on a similarity of biochemical reaction in most target organisms (for example, cyanide is toxic to most organisms). In addition, many plant chemicals may serve multiple roles: resins in the creosote bush serve to defend against herbivores and pathogens, conserve water, and protect against ultraviolet radiation.

It is argued that there are two different types of defensive chemicals in plants. The first type occurs in relatively small amounts, is often toxic in small doses, and poisons the herbivore. These compounds may also change in concentration in response to plant damage; that is, they are inducible. These kinds of qualitative defensive compounds are the most common in short-lived or weedy species that are often referred to as unapparent. They are also characteristic of fast-growing species with short-lived leaves. In contrast, the second type of defensive chemicals often occurs in high concentrations, is not very toxic, but may inhibit digestion by herbivores and is not very inducible. These quantitative defenses are most common in long-lived, so-called apparent plants such as trees that have slow growth rates and long-lived leaves. Some plants may use both types of defenses.

There is accumulating evidence that marine plants may be protected against grazing by similar classes of chemicals to those found in terrestrial plants. One interesting difference in the marine environment is the large number of halogenated organic compounds that are rare in terrestrial and fresh-water systems.

Through evolution, as plants accumulate defenses, herbivores that are able to bypass the defense in some way are selected for and leave more offspring than others. This in turn selects for new defenses on the part of the plant in a continuing process called coevolution.

Animals that can exploit many plant taxa are called generalists, while those that are restricted to one or a few taxa are called specialists. Specialists often have particular detoxification mechanisms to deal with specific defenses. Some generalists possess powerful, inducible detoxification enzymes, while others exhibit morphological adaptations of the gut which prevent absorption of compounds such as tannins, or provide reservoirs for microorganisms that accomplish the detoxification. Animals may avoid eating plants, or parts of plants, with toxins.

Some herbivores that have completely surmounted the plant toxin barrier use the toxin itself as a cue to aid in locating plants. The common white butterfly, *Pieris rapae*, for example, uses mustard oil glycosides, which are a deterrent and toxic to many organisms, to find its mustard family hosts.

Chemical defense in animals. Many animals make their own defensive chemicals—such as all of the venoms produced by social insects (bees, wasps, ants), as well as snakes and mites. These venoms are usually proteins, acids or bases, alkaloids, or combinations of chemicals. They are generally injected by biting or stinging, while other defenses are produced as sprays, froths, or droplets from glands.

Animals frequently make the same types of toxins as plants, presumably because their function as protective agents is similar. Other organisms, particularly insects, use plant chemicals to defend themselves. Sequestration may be a low-cost defense mechanism and probably arises when insects specialize on particular plants.

Microbial defenses. Competitive microbial interactions are regulated by many chemical exchanges involving toxins. They include compounds such as aflatoxin, botulinus toxin, odors of rotting food, hallucinogens, and a variety of antibiotics. *See* TOXIN.

Microorganisms also play a role in chemical interaction with plants and animals that range from the production of toxins that kill insects, such as those produced by the common biological pest control agent *Bacillus thuringiensis*, to cooperative biochemical detoxification of plant toxins by animal symbionts.

Information exchange. A large area of chemical ecology concerns the isolation and identification of chemicals used for communication. Pheromones, substances produced by an organism that induce a behavioral or physiological response in an individual of the same species, have been studied particularly well in insects. These signals are compounds that are mutually beneficial to the emitter and sender, such as sex attractants, trail markers, and alarm and aggregation signals. Sex pheromones are volatile substances, usually produced by the female to attract males. Each species has a characteristic compound that may differ from that of other species by as little as a few atoms.

Pheromones are typically synthesized directly by the animal and are usually derived from fatty acids. In a few cases the pheromone or its immediate precursors may be derived from plants, as in danaid butterflies.

Very little work has been done in identifying specific pheromones in vertebrates, particularly mammals. It is known, however, that they are important in marking territory, in individual recognition, and in mating and warning signals. Chemical communication may also occur among plants and microorganisms, although it is rarer and less obvious than in animals. *See* TERRITORIALITY. [C.G.J.; A.C.L.]

Chemical engineering The application of engineering principles to conceive, design, develop, operate, or use processes and products based on chemical and physical phenomena. The chemical engineer is considered an engineering generalist because of a unique ability (among engineers) to understand and exploit chemical change. Drawing on the principles of mathematics, physics, and chemistry and familiar with all forms of matter and energy and their manipulation, the chemical engineer is well suited for working in a wide range of technologies.

Although chemical engineering was conceived primarily in England, it underwent its main development in America, propelled at first by the petroleum and heavy-chemical industries, and later by the petrochemical industry with its production of plastics, synthetic rubber, and synthetic fibers from petroleum and natural-gas starting materials. In the early twentieth century, chemical engineering developed the physical separations such as distillation, absorption, and extraction, in which the principles of mass transfer, fluid dynamics, and heat transfer were combined in equipment design. The chemical and physical aspects of chemical engineering are known as unit processes and unit operations, respectively.

Chemical engineering now is applied in biotechnology, energy, environmental, food processing, microelectronics, and pharmaceutical industries, to name a few. In such industries, chemical engineers work in production, research, design, process and product development, marketing, data processing, sales, and, almost invariably, throughout top management. *See* BIOCHEMICAL ENGINEERING; BIOMEDICAL CHEMICAL ENGINEERING; BIOTECHNOLOGY. [W.F.F.]

Chemostratigraphy A subdiscipline of stratigraphy and geochemistry that involves correlation and dating of marine sediments and sedimentary rocks through the use of trace-element concentrations, molecular fossils, and certain isotopic ratios that can be measured on components of the rocks. The isotopes used in chemostratigraphy can be divided into three classes: radiogenic (strontium, neodymium, osmium), radioactive (radiocarbon, uranium, thorium, lead), and stable (oxygen, carbon, sulfur). Trace-element concentrations (that is, metals such as nickel, copper, molybdenum, and vanadium) and certain organic molecules (called biological markers or bio-markers) are also employed in chemostratigraphy. *See* DATING METHODS.

Radiogenic isotopes are formed by the radioactive decay of a parent isotope to a stable daughter isotope. The application of these isotopes in stratigraphy is based on natural cycles of the isotopic composition of elements dissolved in ocean water, cycles which are recorded in the sedimentary rocks.

The elements hydrogen, carbon, nitrogen, oxygen, and sulfur owe their isotopic distributions to physical and biological processes that discriminate between the isotopes because of their different atomic mass. The use of these isotopes in stratigraphy is also facilitated by cycles of the isotopic composition of seawater, but the isotopic ratios in marine minerals are also dependent on water temperature and the mineral-forming processes. *See* SEAWATER.

Certain organic molecules that can be linked with a particular source (called biomarkers) have become useful in stratigraphy. The sedimentary distributions of biomarkers reflect the biological sources and inputs of organic matter (such as that from algae, bacteria, and vascular higher plants), and the depositional environment.

Certain trace metals, such as nickel, copper, vanadium, magnesium, iron, uranium, and molybdenum, are concentrated in organic-rich sediments in proportion to the amount of organic carbon. Although the processes controlling their enrichment are complex, they generally form in an oxygen-poor environment (such as the Black Sea) or at the time of global oceanic anoxic events, during which entire ocean basins become oxygen poor, resulting in the death of many organisms; hence large amounts of organic carbon are preserved in marine sediments. The trace-metal composition of individual stratigraphic units may be used as a stratigraphic marker, or "fingerprint." *See* GEOCHEMISTRY; MARINE SEDIMENTS. [B.L.I.; D.J.DeP.]

Chemotaxonomy The use of biochemistry in taxonomic studies. Living organisms produce many types of natural products in varying amounts, and quite often the biosynthetic pathways responsible for these compounds also differ from one taxonomic group to another. The distribution of these compounds and their biosynthetic pathways correspond well with existing taxonomic arrangements based on more traditional criteria such as morphology. In some cases, chemical data have contradicted existing hypotheses, which necessitates a reexamination of the problem or, more positively, chemical data have provided decisive information in situations where other forms of data are insufficiently discriminatory. *See* PLANT TAXONOMY.

Modern chemotaxonomists often divide natural products into two major classes: (1) micromolecules, that is, those compounds with a molecular weight of 1000 or less, such as alkaloids, terpenoids, amino acids, fatty acids, flavonoid pigments and other phenolic compounds, mustard oils, and simple carbohydrates; and (2) macromolecules, that is, those compounds (often polymers) with a molecular weight over 1000, including complex polysaccharides, proteins, and the basis of life itself, deoxyribonucleic acid (DNA).

A crude extract of a plant can be separated into its individual components, especially in the case of micromolecules, by using one or more techniques of chromatography, including paper, thin-layer, gas, or high-pressure liquid chromatography. The resulting chromatogram provides a visual display or "fingerprint" characteristic of a plant species for the particular class of compounds under study.

The individual, separated spots can be further purified and then subjected to one or more types of spectroscopy, such as ultraviolet, infrared, or nuclear magnetic resonance or mass spectroscopy (or both), which may provide information about the structure of the compound. Thus, for taxonomic purposes, both visual patterns and structural knowledge of the compounds can be compared from species to species.

Because of their large, polymeric, and often crystalline nature, macromolecules (for example, proteins, carbohydrates, DNA) can be subjected to x-ray crystallography, which gives some idea of their three-dimensional structure. These large molecules can then be broken down into smaller individual components and analyzed by using techniques employed for micromolecules. In fact, the specific amino acid sequence of portions or all of a cellular respiratory enzyme, cytochrome *c*, has been elucidated and used successfully for chemotaxonomic comparisons in plants and especially animals.

Cyctochrome *c* is a small protein or polypeptide chain consisting of approximately 103–112 amino acids, depending on the animal or plant under study. About 35 of the amino acids do not vary in type or position within the chain, and are probably necessary to maintain the structure and function of the enzyme. Several other amino acid positions vary occasionally, and always with the same amino acid substitution at a particular position. Among the remaining 50 positions scattered throughout the chain, considerable substitution occurs, the number of such differences between organisms indicating how closely they are related to one another. When such substitutional patterns were subjected to computer analysis, an evolutionary tree was obtained showing the degree of relatedness among the 36 plants and animals examined. This evolutionary tree is remarkably similar to evolutionary trees or phylogenies constructed on the basis of the actual fossil record for these organisms. Thus, the internal biochemistry of living organisms reflects a measure of the evolutionary changes which have occurred over time in these plants and animals. Since each amino acid in a protein is the ultimate product of a specific portion of the DNA code, the substitutional differences in this and other proteins in various organisms also reflect a change in the nucleotide sequences of DNA itself. *See* GENETIC CODE; PHYLOGENY.

In the case of proteins, it is often not necessary to know the specific amino acid sequence of a protein, but, rather, to observe how many different proteins, or forms of a single protein, are present in different plant or animal species. The technique of electrophoresis is used to obtain a pattern of protein bands of spots much like the chemical fingerprint of micromolecules. Because each amino acid in a protein carries a positive, negative, or neutral ionic charge, the total sum of charges of the amino acids constituting the protein will give the whole protein a net positive, negative, or neutral charge.

By using other techniques of molecular biology, such as DNA hybridization and genetic cloning, the specific gene function of individual fragments may be identified. Their nucleotide sequences can be determined and then compared for different taxa. Such data may prove useful at several different taxonomic levels. *See* GENETIC ENGINEERING.

While the organellar DNA does not contain the number of genetic messages of the organism that nuclear DNA does, and its transmission from parent to offspring may vary somewhat depending on the organism, the convenient size of organellar DNA and its potential for direct examination of the genetic code suggest that it is a potent macromolecular approach to chemosystematics. *See* GENETIC CODE. [D.E.G.]

Chestnut Any of seven species of deciduous, nut-bearing trees of the genus *Castanea* Corden Fagales native to the Northern Hemisphere and introduced throughout the world. The nuts are actually fruits, with the shells enclosing cotyledons. Trees bear both male and female flowers in late spring but must be cross-pollinated for nut production. Nuts are borne in a spiny involucre or bur that opens to release the nuts in late fall.

Japanese chestnuts (*C. crenata*) and Chinese chestnuts (*C. mollissima*) are grown in Asia and the United States for their nuts, and many cultivars have been selected. European chestnuts (*C. sativa*) are an important food source, both cooked whole and ground into flour. They are native to the Caucasus mountains, and distributed throughout southern Europe. American chestnuts (*C. dentata*) have smaller nuts than Asian or European species and are usually sweeter. Only American trees served as an important source of lumber, because of the length of their unbranched trunks; all chestnut species have been used as a source of tannin for the leather-tanning industry. American and Chinese chinquapins (*C. pumila* and *C. henryi*) have very small nuts that are an important source of food for wildlife. All of the species can be crossed, and hybrids have been selected primarily as orchard cultivars. [S.L.A.]

Chinook A mild, dry, extremely turbulent westerly wind on the eastern slopes of the Rocky Mountains and closely adjoining plains. The term is an Indian word which means "snow-eater," appropriately applied because of the great effectiveness with which this wind reduces a snow cover by melting or by sublimation. The chinook is a particular instance of a type of wind known as a foehn wind. Foehn winds, initially studied in the Alps, refer to relatively warm, rather dry currents descending the lee slope of any substantial mountain barrier. The dryness is an indirect result of the condensation and precipitation of water from the air during its previous ascent of the windward slope of the mountain range. The warmth is attributable to adiabatic compression, turbulent mixing with potentially warmer air, and the previous release of latent heat of condensation in the air mass and to the turbulent mixing of the surface air with the air of greater heat content aloft. In winter the chinook wind sometimes impinges upon much colder stagnant polar air along a sharp front located in the foothills of the Rocky Mountains or on the adjacent plain. Small horizontal oscillations of this front have been known to produce several abrupt temperature rises and falls of as much as 45–54°F (25–30°C) at a given location over a period of a few hours. Damaging winds sometimes occur as gravity waves, which are triggered along the interface between the two air masses. *See* FRONT; PRECIPITATION (METEOROLOGY); WIND. [F.S.; H.B.B.]

Chlamydia A genus of bacteria with a growth cycle differing from that of all other microorganisms. Chlamydiae grow only in living cells and cannot be cultured on artificial media. Although capable of synthesizing macromolecules, they have no system for generating energy; the host cell's energy system fuels the chlamydial metabolic processes. The genome is relatively small; the genomes of *C. pneumoniae* and *C. trachomatis* have been completely sequenced.

The chlamydial infectious particle, called the elementary body, is round and about 350–450 nanometers in diameter. It enters a susceptible host cell and changes to a metabolically active and larger (approximately 800–1000 nm in diameter) reticulate body that divides by binary fission. The entire growth cycle occurs within a vacuole that segregates the chlamydia from the cytoplasm of the host cell. The reticulate bodies change back to elementary bodies, and then the cell lyses and the infectious particles are released. The growth cycle takes about 48 h.

Human diseases are caused by three species of *Chlamydia*. *Chlamydia trachomatis* is almost exclusively a human pathogen, and one of the most common. Infections occur in two distinct epidemiologic patterns. In many developing countries, *C. trachomatis* causes trachoma, a chronic follicular keratoconjunctivitis. It is the world's leading cause of preventable blindness, affecting approximately 500 million people. In areas where this condition is highly endemic, virtually the entire population is infected within the first few years of life. Most active infections are found in childhood. By age 60, more than 20% of a population can be blinded as a result of trachoma.

Chlamydia trachomatis is the most common sexually transmitted bacterial pathogen; an estimated 3–4 million cases occur each year in the United States, and there are close to 90 million worldwide. The most common manifestation is nongonococcal urethritis in males. The cervix is the most commonly infected site in women. Ascending infections can occur in either sex, resulting in epididymitis in males or endometritis and salpingitis in females. Chlamydial infection of the fallopian tube can cause late consequences such as infertility and ectopic pregnancy, even though the earlier infection is asymptomatic. The infant passing through the infected birth canal can acquire the infection and may develop either conjunctivitis or pneumonia. A more invasive form of *C. trachomatis* causes a systemic sexually transmitted disease called lymphogranuloma venereum. *See* SEXUALLY TRANSMITTED DISEASES.

Chlamydia psittaci is virtually ubiquitous among avian species and is a common pathogen among lower mammals. It is economically important in many countries as a cause of abortion in sheep, cattle, and goats. It causes considerable morbidity and mortality in poultry. *Chlamydia psittaci* can infect humans, causing the disease psittacosis. Psittacosis can occur as pneumonia or a febrile toxic disease without respiratory symptoms.

Chlamydia pneumoniae appears to be a human pathogen with no animal reservoir. It is of worldwide distribution and may be the most common human chlamydial infection. It appears to be an important cause of respiratory disease.

Azithromycin is the drug of choice for uncomplicated chlamydial infection of the genital tract. Two therapeutic agents require longer treatment regimens: doxycycline, a tetracycline antibiotic, is the first alternate treatment; erythromycin may be used for those who are tetracycline-intolerant, as well as for pregnant women or young children. *See* MEDICAL BACTERIOLOGY. [J.S.]

Chlorophyll The generic name for the intensely colored green pigments which are the photoreceptors of light energy in photosynthesis. These pigments belong to the tetrapyrrole family of organic compounds.

Five closely related chlorophylls, designated *a* through *e*, occur in higher plants and algae. The principal chlorophyll (Chl) is Chl *a*, found in all oxygen-evolving organisms; photosynthetic bacteria, which do not evolve O_2, contain instead bacteriochlorophyll (Bchl). Higher plants and green algae contain Chl *b*, the ratio of Chl *b* to Chl *a* being 1:3. Chlorophyll *c* (of two or more types) is present in diatoms and brown algae. Chlorophyll *d*, isolated from marine red algae, has not been shown to be present in the living cell in large enough quantities to be observed in the absorption spectrum of these algae. Chlorophyll *e* has been isolated from cultures of two algae, *Tribonema bombycinum* and *Vaucheria hamata*. In higher plants the chlorophylls and the above-mentioned pigments are contained in lipoprotein bodies, the plastids. *See* CELL PLASTIDS; PHOTOSYNTHESIS.

Chlorophyll molecules have three functions: They serve as antennae to absorb light quanta; they transmit this energy from one chlorophyll to another by a process of "resonance transfer;" and finally, this chlorophyll molecule, in close association with enzymes, undergoes a chemical oxidation (that is, an electron of high potential is ejected from the molecule and can then be used to reduce another compound). In this way the energy of light quanta is converted into chemical energy.

The chlorophylls are cyclic tetrapyrroles in which four 5-membered pyrrole rings join to form a giant macrocycle. Chlorophylls are members of the porphyrin family, which

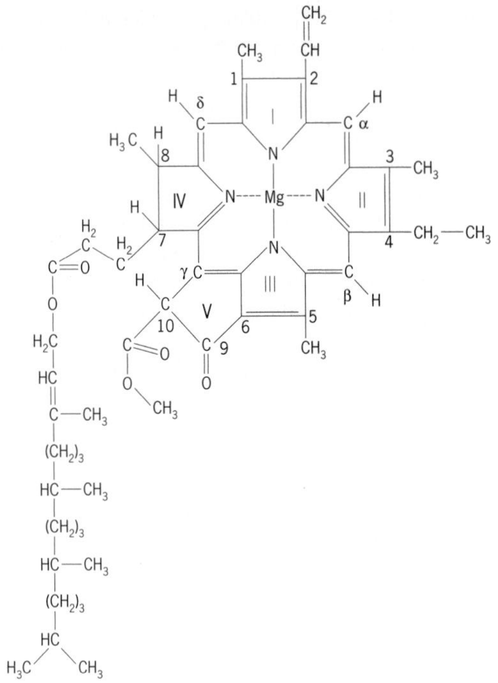

Structure of chlorophyll *a* ($C_{55}H_{72}O_5N_4Mg$).

plays important roles in respiratory pigments, electron transport carriers, and oxidative enzymes.

It now appears that the chlorophyll *a* group may be made up of several chemically distinct Chl *a* species. The structure of monovinyl cholorophyll *a*, the most abundant of the Chl *a* species, is shown in the illustration.

The two major pigments of protoplasm, green chlorophyll and red heme, are synthesized from ALA (δ-aminolevulinic acid) along the same biosynthetic pathway to protoporphyrin. ALA is converted in a series of enzymic steps, identical in plants and animals, to protoporphyrin. Here the pathway branches to form (1) a series of porphyrins chelated with iron, as heme and related cytochrome pigments; and (2) a series of porphyrins chelated with magnesium which are precursors of chlorophyll.

Chlorophylls reemit a fraction of the light energy they absorb as fluorescence. Irrespective of the wavelength of the absorbed light, the emitted fluorescence is always on the long-wavelength side of the lowest energy absorption band, in the red or infrared region of the spectrum.

The fluorescent properties of a particular chlorophyll are functions of the structure of the molecule and its immediate environment. Thus, the fluorescence spectrum of chlorophyll in the living plant is always shifted to longer wavelengths relative to the fluorescence spectrum of a solution of the same pigment. This red shift is characteristic of aggregated chlorophyll. [G.; S.Gr.; G.P.]

Cholera A severe diarrheal disease caused by infection of the small bowel of humans with *Vibrio cholerae*, a facultatively anaerobic, gram-negative, rod-shaped bacterium. Cholera is transmitted by the fecal-oral route. Cholera has swept the world in seven pandemic waves. These involved the Western Hemisphere several times in the 1800s, and again in Peru in 1991. Whereas previous cholera outbreaks were associated with high mortality rates, through understanding of its pathophysiology it can now be said that no one should die of cholera who receives appropriate treatment soon enough.

Cholera produces a secretory diarrhea caused by the protein cholera enterotoxin (CTX). The toxin causes hypersecretion of chloride and bicarbonate and inhibition of sodium absorption in host membranes leading to the secretion of the large volumes of isotonic fluid which constitute the diarrhea of severe cholera. Treatment consists of replacing the fluids and electrolytes lost in the voluminous cholera stool. This can be done intravenously or orally. Appropriate antibiotics can also be used. The incubation period may be less than one day or up to several days; properly treated, the patient should recover in 4 or 5 days. The disease produces immunization, and convalescents rarely get cholera again. *See* DIARRHEA.

Despite the fact that the cholera bacteria were first discovered by Robert Koch in 1883 and a cholera vaccine was introduced 3 years later, there is still no effective, economical, and nonreactogenic vaccine. Use of a killed whole-cell vaccine administered parenterally (via injection) was eliminated because of expense, reactogenicity, and lack of efficacy. Experimental vaccines currently being evaluated include genetically engineered living attentuated preparations administered orally (or intranasally), killed whole-cell vaccines administered orally, and conjugated vaccines (polysaccharide and toxin antigens) administered parenterally. Efforts are also being made to include cholera antigens transgenically in edible plants.

A complicating feature is the fact that of approximately 150 recognized serogroups of *V. cholerae*, until 1992 only two, classical (first described by Koch) and El Tor (recognized later), of serogroup O1 have been responsible for all epidemic cholera. In 1992 a recently recognized serogroup, O139, caused epidemic cholera in India and Bangladesh and, for a time, replaced the resident El Tor vibrios. O139 and El Tor are antigenically distinct, so a new vaccine will be required for O139. The emergence of O139 raises the specter that other serogroups of *V. cholerae* may acquire virulence and epidemicity.

The best ways to avoid cholera are by chlorination of water, sanitary disposal of sewage, and avoidance of raw or improperly cooked seafood, which may have become infected by ingesting infected plankton in epidemic areas. [R.A.Fin.]

Chromadorida An order of nematodes in which the amphid manifestation is variable but within superfamilies some constancy is apparent. The various amphids are reniform, transverse elongate loops, simple spirals, or multiple spirals not seen in any other orders or subclasses. The cephalic sensilla are in one or two whorls at the extreme anterior. In all taxa the cuticle shows some form of ornamentation, usually punctations that are apparent whether the cuticle is smooth or annulated. When developed, the stoma is primarily esophastome and is usually armed with a dorsal tooth, jaws, or protrusible rugae. The corpus of the esophagus is cylindrical; the isthmus is not seen; and the postcorpus, in which the heavily cuticularized lumen forms the cresentic valve, is distinctly expanded. The esophagointestinal valve is triradiate or flattened. The females usually have paired reflexed ovaries.

There are four chromadorid superfamilies: Choanolaimoidea, Chromadoroidea, Comesomatoidea, and Cyatholaimoidea. Choanolaimoidea are distinguished by a complex stoma in two parts. The group occupies marine habitats; some species are predaceous, but for many the feeding habits are unknown. Chromadoroidea comprise small to moderate-sized free-living forms that are mainly marine but are also found in fresh water and soil. Known species either are associated with algal substrates or are nonselective deposit feeders in softbottom sediments. Comesomatoidea, containing only the family Comesomatidae, are found in marine habitats, but the feeding habits are unknown. Cyatholaimoidea are found in marine, terrestrial, and fresh-water environments. *See* NEMATA. [A.R.M.]

Ciliatea The single class of the subphylum Ciliophora. This group has the characteristics of those defined for the subphylum. This protozoan class is divided into the subclasses Holotrichia, Peritrichia, Suctoria, and Spirotrichia. [J.O.C.]

Classification, biological A human construct for grouping organisms into hierarchical categories. The most inclusive categories of any classification scheme are called kingdoms, which are delimited so that organisms within a single kingdom are more related to each other than to organisms grouped in the other kingdoms. Classification (grouping) is a part of biological systematics or taxonomy, science that involves naming and sorting organisms into groups.

Historically, organisms have been arranged into kingdoms based on practical characteristics, such as motility, medicinal properties, and economic value for food or fiber. In the nineteenth and twentieth centuries, advances such as electron and light microscopy, biochemistry, genetics, ethology, and greater knowledge of the fossil record provided new evidence upon which to construct more sophisticated classification schemes. Proposals were made to assign organisms into four, five, and even thirteen kingdoms. Earlier two-kingdom and three-kingdom systems were devised without awareness of the profound distinction between prokaryotes and eukaryotes, which Edouard Chatton (1937) recognized as a fundamental evolutionary discontinuity. Prokaryotic cells do not have either a nucleus or any other internal membrane-bounded structures, the so-called organelles. By contrast, eukaryotic cells have both a nucleus and organelles. The deoxyribonucleic acid (DNA) of eukaryotes is combined with protein to form chromosomes. All bacteria are prokaryotes. Plants, fungi, animals, and protoctists are eukaryotes. *See* PROKARYOTAE.

By the late 1990s, biologists widely accepted a system that classifies all organisms into five kingdoms based on key characteristics of function and structure. These are Superkingdom Prokarya [Kingdom Bacteria (Monera) with two subkingdoms, Archaea and Eubacteria] and Superkingdom Eukarya with four kingdoms, Protoctista (or Protozoa in some classification schemes), Fungi, Animalia, and Plantae. Key charactistics used to classify organisms into these categories include the mode by which the organism obtains its nutrients; presence or absence of an embryo; and whether and how the organism achieves motility.

Cladistics may be defined as an approach to grouping organisms that classifies them according to the time at which branch points occur along a phylogenetic tree. Such a phylogenetic tree is represented by a diagram called a cladogram. In this approach, classification is based on a sequence of phylogenetic branching. In a cladogram, the phylogenetic tree branches dichotomously and repeatedly, reflecting cladogenesis (production of biological diversity by evolution of new species from parental species).

Modern classification is based on evidence derived from developmental pattern, biochemistry, molecular biology, genetics, and detailed morphology of extant organisms and their fossils. Because information is drawn from such diverse sources, and because 10–30 million (possibly 100 million) species are probably alive today, informed judgments must be made to integrate the information into classification hierarchies. Only about 1.7 million species have been formally classified in the taxonomic literature of biology to date. Thus, as new evidence about the evolutionary relationships of organisms is weighed, it must be anticipated that biological classification schemes will continue to be revised. *See* PLANT TAXONOMY. [K.V.S.]

Clay The finest-grain particles in a sediment, soil, or rock. Clay is finer than silt, characterized by a grain size of less than approximately 4 micrometers. However, the term clay can also refer to a rock or a deposit containing a large component of clay-size material. Thus clay can be composed of any inorganic materials, such as clay minerals, allophane, quartz, feldspar, zeolites, and iron hydroxides, that possess a sufficiently fine grain size. Most clays, however, are composed primarily of clay minerals.

Although the composition of clays can vary, clays can share several properties that result from their fine particle size. These properties include plasticity when wet, the ability to form colloidal suspensions when dispersed in water, and the tendency to flocculate (clump together) and settle out in saline water.

Clays, together with organic matter, water, and air, are one of the four main components of soil. Clays can form directly in a soil by precipitation from solution (neoformed clays); they can form from the partial alteration of clays already present in the soil (transformed clays); or they can be inherited from the underlying bedrock or from sediments transported into the soil by wind, water, or ice (inherited clays). *See also* SOIL.

The type of clays neoformed in a soil depends on the composition of the soil solution, which in turn is a function of climate, drainage, original rock type, vegetation, and time. Generally, neoformed clays that have undergone intense leaching, such as soils formed under wet, tropical climates, are composed of the least soluble elements, such as ferric iron, aluminum, and silicon. These soils contain clays such as gibbsite, kaolinite, goethite, and amorphous oxides and hydroxides of aluminum and iron. Clays formed in soils that are found in dry climates or in soils that are poorly drained can contain more soluble elements, such as sodium, potassium, calcium, and magnesium, in addition to the least soluble elements. These soils contain clays such as smectite, chlorite, and illite, and generally are more fertile than those formed under intense leaching conditions.

Examples of clays formed by the transformation of other clays in a soil include soil chlorite and soil vermiculite, the first formed by the precipitation of aluminum hydroxide in smectite interlayers, and the second formed by the leaching of interlayer potassium from illite. Examples of inherited clays in a soil are illite and chlorite-containing soils formed on shales composed of these minerals.

Clays also occur abundantly in sediments and sedimentary rocks. For example, clays are a major component of many marine sediments. These clays generally are inherited from adjacent continents, and are carried to the ocean by rivers and wind, although some clays (such as smectite and glauconite) are neoformed abundantly in the ocean. Hydrothermal clays can form abundantly where rock has been in contact with hot water or steam. Illite and chlorite, for example, form during the deep burial of sediments, and smectite and chlorite form by the reaction of hot, circulating waters at ocean ridges. *See* MARINE SEDIMENTS; SEDIMENTARY ROCKS.

Various clays possess special properties which make them important industrially. For example, bentonite, a smectite formed primarily from the alteration of volcanic ash, swells; is readily dispersible in water; and possesses strong absorptive powers, including a high cation exchange capacity. These properties lead to uses in drilling muds, as catalysts and ion exchangers, as fillers and absorbents in food and cosmetics, and as binders for taxonite and fertilizers. Other important uses for clays include the manufacture of brick, ceramics, molding sands, decolorizers, detergents and soaps, medicines, adhesives, liners for ponds and landfills, lightweight aggregate, desiccants, molecular sieves, pigments, greases, paints, plasticizing agents, emulsifying, suspending, and stabilizing agents, and many other products. [D.D.E.]

Climate history The long-term records of precipitation, temperature, wind, and all other aspects of the Earth's climate. The climate, like the Earth itself, has a history extending over several billion years. Climatic changes have occurred at time scales ranging from hundreds of millions of years to centuries and decades. Processes in the atmosphere, oceans, cryosphere (snow cover, sea ice, continental ice sheets), biosphere, and lithosphere (such as plate tectonics and volcanic activity) and certain extraterrestrial factors (such as the Sun) have caused these changes of climate.

The present climate can be described as an ice age climate, since large land surfaces are covered with ice sheets (for example, Antartica and Greenland). The origins of the present ice age may be traced, at least in part, to the movement of the continental plates. With the gradual movement of Antarctica toward its present isolated polar position, ice sheets began to develop there about 30 million years ago. For the past several million years, the Antarctic ice sheet reached approximately its present size, and ice sheets appeared on the lands bordering the northern Atlantic Ocean. During the past million years of the current ice age, about 10 glacial-interglacial cycles have been documented. Changes in the Earth's orbital parameters, eccentricity, obliquity, and longitude of perihelion are thought to have initiated, or paced, these cycles through the associated small changes in the seasonal and latitudinal distribution of solar radiation. The most recent glacial period ended between about 15,000 and 6000 years ago with the rapid melting of the North American and European ice sheets and an associated rise in sea level, and the atmospheric concentration of carbon dioxide.

The climates of the distant geologic past were strongly influenced by the size and location of continents and by large changes in the composition of the atmosphere. For example, around 250 million years ago the continents were assembled into one supercontinent, Pangaea, producing significantly different climatic patterns than are seen today with widely distributed continents. In addition, based upon models of stellar evolution, it is hypothesized that the Sun's radiation has gradually increased by 10–20% over the past several billion years and, if so, this has contributed to a significant warming of the Earth. *See* PALEOCLIMATOLOGY.

Instrumental records of climatic variables such as temperature and precipitation exist for the past 100 years in many locations and for as long as 200 years in a few locations. These records provide evidence of year-to-year and decade-to-decade variability, but they are completely inadequate for the study of century-to-century and longer-term variability. Even for the study of short-term climatic fluctuations, instrumental records are incomplete, because most observations are made from the continents (covering only 29% of the Earth's surface area). Aerological observations, which permit the study of atmospheric mass, momentum and energy budgets, and the statistical structure of the large-scale circulation, are available only since about the mid-1960s. Again there is a bias toward observations over the continents. It is only with the advent of satellites that global monitoring of the components of the Earth's radiation budget (clouds;

planetary albedo, from which the net incoming solar radiation can be estimated; and the outgoing terrestrial radiation) became possible.

Evidence of climatic changes prior to instrumental records comes from a wide variety of sources. Tree rings, banded corals, and pollen and trace minerals retrieved from laminated lake sediments and ice sheets yield environmental records for past centuries and millennia. Advanced drilling techniques have made it possible to obtain long cores from ocean sediments that provide geologic records of climatic conditions going back hundreds of millions of years.

Many extraterrestrial and terrestrial processes have been hypothesized to be possible causes of climatic fluctuations. These include solar irradiance, variations in orbital parameters, motions of the lithosphere, volcanic activity, internal variations of the climate system, and human activities. It is likely that all of the natural processes have played a role in past climatic changes. Also, the climatic response to some particular causal process may depend on the initial climatic state, which in turn depends upon previous climatic states because of the long time constants of lithosphere, oceans, and cryosphere. True equilibrium climates may not exist, and the climate system may be in a continual state of adjustment.

Because of the complexity of the real climate system, simplified numerical models of climate are being used to study particular processes and interactions. Some models treat only the global-average conditions, whereas others, particularly the dynamical atmosphere and ocean models, simulate detailed patterns of climate. These models will undoubtedly be of great importance in attempts to understand climatic processes and to assess the possible effects of human activities on climate. *See* CLIMATE MODELING; CLIMATOLOGY. [J.E.K.]

Climate modeling Construction of a mathematical model of the climate system of the Earth capable of simulating its behavior under present and altered conditions. The Earth's climate is continually changing over time scales ranging from millions of years to a few years. Since the climate is determined by the laws of classical physics, it should be possible in principle to construct such a model. The advent of a worldwide weather observing system capable of gathering data for validation and the development and widespread routine use of digital computers have made this undertaking possible.

The Earth's average temperature is determined mainly by the balance of radiant energy absorbed from sunlight and the radiant energy emitted by the Earth system. About 30% of the incoming radiation is reflected directly to space, and 72% of the remainder is absorbed at the surface. The radiation is absorbed unevenly over the Earth, which sets up thermal contrasts that in turn induce convective circulations in the atmosphere and oceans. Climate models attempt to calculate from mathematical algorithms the effects of these contrasts and the resulting motions in order to understand better and perhaps predict future climates in some probabilistic sense. *See* TERRESTRIAL RADIATION.

Climate models differ in complexity, depending upon the application. The simplest models are intended for describing only the surface thermal field at a fairly coarse resolution. These mainly thermodynamical formulations are successful at describing the seasonal cycle of the present climate, and have been used in some simulations of past climates, for example, for different continental arrangements millions of years ago. At the other end of the spectrum are the most complex climate models, which are extensions of the models in weather forecasts. These models aim at simulating seasonal and even monthly averages just shortly into the future, based upon conditions such as the temperatures of the tropical-sea surfaces. Intermediate to these extremes are models that attempt to model climate on a decadal basis, and these are used mainly in studies of the impact of hypothesized anthropogenically induced climate change. *See* WEATHER FORECASTING AND PREDICTION.

Attempts at modeling climate have demonstrated the extreme complexity and subtlety of the problem. This is due largely to the many feedbacks in the system. One of the simplest and yet most important feedbacks is that due to water vapor. If the Earth is perturbed by an increase in the solar radiation, for example, the first-order response of the system is to increase its temperature. But an increase in air temperature leads to more water vapor evaporating into the air; this in turn leads to increased absorption of space-bound long-wave radiation from the ground (greenhouse effect), which leads to an increased equilibrium temperature. Water vapor feedback is not the only amplifier in the system. Another important one is snowcover: a cooler planet leads to more snow and hence more solar radiation reflected to space, since snow is more reflecting of sunlight than soil or vegetation. Other, more subtle mechanisms that are not yet well understood include those involving clouds, oceans, and the biosphere.

While water vapor and snowcover feedback are fairly straightforward to model, the less understood feedbacks differ in their implementations from one climate model to another. These differences as well as the details of their different numerical formulations have led to slight differences in the sensitivity of the various models to such standard experimental perturbations as doubling carbon dioxide in the atmosphere. All models agree that the planetary average temperature should increase if carbon dioxide concentrations are doubled. However, the predicted response in planetary temperatures ranges from 4.5 to 9°F (2.5 to 5.0°C). Regional predictions of temperature or precipitation are not reliable enough for detailed response policy formulation. Many of the discrepancies are expected to decrease as model resolution increases (more grid points), since it is easier to include such complicated phenomena as clouds in finer-scale formulations and coupling with dynamic models of the ocean. Similarly, it is anticipated that some observational data (such as rainfall over the oceans) that are needed for validation of the models will soon be available from satellite sensors. [R.E.D.; E.S.S.]

Climate modification Alteration of the Earth's climate by human activities. This can occur on various scales. For example, conventional agriculture alters the microclimate in the lowest few meters of air, causing changes in the evapotranspiration and local heating characteristics of the air-surface interface. These changes lead to different degrees of air turbulence over the plants and to different moisture and temperature distributions in the local air. An example at a larger scale is that the innermost parts of cities are several degrees warmer than the surrounding countryside, and have slightly more rainfall. These changes are brought about by the differing surface features of urban land versus natural countryside and the ways that cities dispose of water (for example, storm sewers). The urban environment prevents evaporation cooling of surfaces in the city. The modified surface texture of cities (horizontal and vertical planes of buildings and streets versus gently rolling surfaces over natural forest or grassland) leads to a more efficient trapping of solar heating of the near-surface air. The scales of buildings and other structures also lead to a different pattern of atmospheric boundary-layer turbulence, modifying the stirring efficiency of the atmosphere. *See* MICROMETEOROLOGY; URBAN CLIMATOLOGY.

At the next larger scale, human alteration of regional climates is caused by changes in the Earth's average reflectivity to sunlight. For example, the activities of building roads and highways and deforestation alter the amount of sunshine that is reflected to space as opposed to being absorbed by the surface and thereby heating the air through contact. Such contact heating leads to temperature increases and evaporation of liquid water at the surface. Vapor wakes from jet airplanes are known to block direct solar radiation near busy airports by up to 20%. Human activities also inject dust, smoke, and other aerosols into the air, causing sunlight to be scattered back to space. Dust particles screen out sunlight before it can enter the lower atmosphere and warm the near-surface air. *See* AIR POLLUTION; SMOG. [G.R.N.]

One of the most important and best understood features of the atmosphere is the process that keeps the Earth's surface much warmer than it would be with no atmosphere. This process involves several gases in the air that trap infrared radiation, or heat, emitted by the surface and reradiate it in all directions, including back to the surface. The heat-trapping gases include water vapor, carbon dioxide (CO_2), methane (CH_4), and nitrous oxide (N_2O). These gases constitute only a small fraction of the atmosphere, but their heat-trapping properties raise the surface temperature of the Earth by a large amount, estimated to be more than 55°F (30°C). Human activities, however, are increasing the concentrations of CO_2, CH_4, and N_2O in the atmosphere, and in addition industrially synthesized chemicals—chlorofluorocarbons (CFCs) and related compounds—are being released to the air, where they add to the trapping of infrared radiation. Carbon dioxide is released to the air mainly from fossil fuel use, which also contributes to the emissions of CH_4 and N_2O. Agricultural and industrial processes add to the emissions of these gases. These concentration increases add to the already powerful heat-trapping capability of the atmosphere, raising the possibility that the surface will warm above its past temperatures, which have remained roughly constant, within about 3°F (1.7°C) for the past 10,000 years. Treaties are in existence that control internationally the production and use of many of the CFCs and related compounds, so their rate of growth has slowed, and for some a small decrease in atmospheric concentration has been observed. Large emissions of sulfur dioxide in industrial regions are thought to result in airborne sulfate particles that reflect sunlight and decrease the amount of heating in the Northern Hemisphere. *See* ATMOSPHERE; GREENHOUSE EFFECT. [J.Fi.]

Climatology The scientific study of climate. Climate is the expected mean and variability of the weather conditions for a particular location, season, and time of day. The climate is often described in terms of the mean values of meteorological variables such as temperature, precipitation, wind, humidity, and cloud cover. A complete description also includes the variability of these quantities, and their extreme values. The climate of a region often has regular seasonal and diurnal variations, with the climate for January being very different from that for July at most locations. Climate also exhibits significant year-to-year variability and longer-term changes on both a regional and global basis.

The goals of climatology are to provide a comprehensive description of the Earth's climate over the range of geographic scales, to understand its features in terms of fundamental physical principles, and to develop models of the Earth's climate for sensitivity studies and for the prediction of future changes that may result from natural and human causes. *See* CLIMATE HISTORY; CLIMATE MODELING; CLIMATE MODIFICATION; WEATHER. [D.L.Ha.]

Clinical microbiology The adaptation of microbiological techniques to the study of the etiological agents of infectious disease. Clinical microbiologists determine the nature of infectious disease and test the ability of various antibiotics to inhibit or kill the isolated microorganisms. In addition to bacteriology, a contemporary clinical microbiologist is responsible for a wide range of microscopic and cultural studies in mycology, parasitology, and virology. The clinical microbiologist is often the most competent person available to determine the nature and extent of hospital-acquired infections, as well as public-health problems that affect both the hospital and the community. *See* ANIMAL VIRUS; HOSPITAL INFECTIONS; MEDICAL BACTERIOLOGY; MEDICAL MYCOLOGY; MEDICAL PARASITOLOGY; VIRUS.

Bacteriology. Historically, the diagnosis of bacterial disease has been the primary job of clinical microbiology laboratories. Many of the common ailments of humans are bacterial in nature, such as streptococcal sore throat, diphtheria, and pneumococcal pneumonia. The bacteriology laboratory accepts specimens of body fluids, such as sputum, urine, blood, and respiratory or genital secretions, and inoculates the specimens onto various solid and liquid growth media. Following incubation at body temperature, the microbiologist examines these agar plates and tubes and makes a determination as to the relative numbers of organisms growing from the specimen and their importance in the disease process. The microbiologist then identifies these alleged causes of disease and determines their pattern of antibiotic susceptibility to a few chosen agents.

Clinical microbiologists also microscopically examine these body fluids. They report on the presence of bacteria in body fluids and the cellular response to infection, such as the numbers or types of white blood cells observed in the specimen. [R.C.T.]

Nonculture methods. While direct microscopy and culture continue to be methodological mainstays in diagnostic microbiology laboratories, nonculture methods are growing in the variety of applications and the sophistication of the technology. For example, polyclonal antibodies raised in animals such as mice, sheep, goats, and rabbits, and monoclonal antibodies produced by hybridization technology are used to detect bacteria, fungi, parasites, or virus-infected cells by using direct or indirect fluorescent techniques. Additional methods include latex agglutination tests to detect particulate antigens and enzyme immunoassays to detect soluble antigens.

Probes for deoxyribonucleic acid (DNA) or messenger ribonucleic acid (mRNA) are available for various applications. Probes are used for direct detection of organisms in clinical material and for culture confirmation.

Further increases in analytical sensitivity have been achieved by nucleic acid amplification techniques. In polymerase chain reaction (PCR), double-stranded DNA is denatured; oligonucleotide probes bind to homologous strands of single-stranded DNA, and the enzyme polymerase extends the probes using deoxyribonucleotides in the milieu. In ligase chain reaction (LCR), the enzyme ligase fills the 3-nucleotide gap between two probes that attach to homologous, target, single-stranded DNA. In nucleic acid sequence–based amplification (NASBA), reverse transcriptase is used to make double-stranded complementary DNA (cDNA) and the target RNA is digested by ribonuclease H.

Analysis of lipopolysaccharides and proteins by sodium dodecyl sulfate–polyacrylamide gel electrophoresis (SDS-PAGE) and cellular fatty acid analysis by gas-liquid chromatography have given way to nucleic acid–based methods. Restriction enzymes, which cut DNA at a constant position within a specific recognition site usually composed of four to six base pairs, are used to cut chromosomal DNA; the resulting fragments are compared by pulse field gel electrophoresis (PFGE) or ribotyping. Electrophoresis of isolated plasmid DNA is another method for comparing organisms. DNA sequencing can also compare segments of the DNA of organisms from the same genus and species.

DNA chip or microarray technology is expected to have a greater effect on medicine than either DNA sequencing or PCR. Over 30,000 small cDNA clones of expressed fragments of individual genes are spotted onto a thumbnail-sized glass chip. Fluorescein-labeled genomic or cDNA from the sample being evaluated is passed over the chip to allow hybridization. A laser measures the fluorescent emissions and a computer analyzes the data. [C.A.Sp.]

Clostridium A genus of bacteria comprising large anaerobic spore-forming rods that usually stain gram-positive. Most species are anaerobes, but a few will grow minimally in air at atmospheric pressure.

The clostridia are widely distributed in nature, and are present in the soil and in the intestinal tracts of humans and animals. They usually live a saprophytic existence, and play a major role in the degradation of organic material in the soil and other nature environments. A number of clostridia release potent exotoxins and are pathogenic for humans and animals. Among the human pathogens are the causative agents of botulism (*Clostridium botulinum*), tetanus (*C. tetani*), gas gangrene (*C. perfringens*), and an antibiotic-associated enterocolitis (*C. difficile*). *See* ANAEROBIC INFECTION; BOTULISM; TETANUS; TOXIN.

Clostridial cells are straight or slightly curved rods, 0.3–1.6 micrometers wide and 1–14 μm long. They may occur singly, in pairs, in short or long chains, or in helical coils. The length of the cells of the individual species varies according to the stage of growth and growth conditions. Most clostridia are motile with a uniform arrangement of flagella.

The endospores produced by clostridia are dormant structures capable of surviving for prolonged periods of time, and have the ability to reestablish vegetative growth when appropriate environmental conditions are provided. The spores of clostridia are oval or spherical and are wider than the vegetative bacterial cell. Among the distinctive forms are spindle-shaped organisms, club-shaped forms, and tennis racket-shaped structures:

Clostridia are obligate anaerobes: they are unable to use molecular oxygen as a final electron acceptor and generate their energy solely by fermentation. Clostridia exhibit varying degrees of intolerance of oxygen. Some species are sensitive to oxygen concentrations as low as 0.5%, but most species can tolerate concentrations of 3–5%. The sensitivity of clostridia to oxygen restricts their habitat to anaerobic environments; habitats that contain large amounts of organic matter provide optimal conditions for their growth and survival.

A primary property of all species of *Clostridium* is their inability to carry out a dissimilatory reduction of sulfate. Most species are chemoorganotrophic. The substrate spectrum for the genus as a whole is very broad and includes a wide range of naturally occurring compounds. Extracellular enzymes are secreted by many species, enabling the organism to utilize a wide variety of complex natural substrates in the environment. [H.P.W.]

Cloud Suspensions of minute droplets or ice crystals produced by the condensation of water vapor (the ordinary atmospheric cloud). Other clouds, less commonly seen, are composed of smokes or dusts. *See* AIR POLLUTION; DUST STORM.

If water vapor is cooled sufficiently, it becomes saturated, that is, in equilibrium with a plane surface of liquid water (or ice) at the same temperature. Further cooling in the presence of such a surface causes condensation upon it. In the atmosphere, even in the apparent absence of any surfaces, there are invisible motes upon which the condensation proceeds at barely appreciable cooling beyond the state of saturation. Consequently, when atmospheric water vapor is chilled sufficiently, such motes, or condensation nuclei, swell into minute waterdroplets and form a visible cloud.

The World Meteorological Organization (WMO) uses a classification which divides clouds into low-level (base below about 1.2 mi or 2 km), middle-level (about 1.2–4 mi or 2–7 km), and high-level (4–8 mi or 7–14 km) forms within the middle latitudes. The names of the three basic forms of clouds are used in combination to define 10 main characteristic forms, or "genera."

1. Cirrus are high white clouds with a silken or fibrous appearance.
2. Cumulus are detached dense clouds which rise in domes or towers from a level low base.
3. Stratus are extensive layers or flat patches of low clouds without detail.
4. Cirrostratus is cirrus so abundant as to fuse into a layer.
5. Cirrocumulus is formed of high clouds broken into a delicate wavy or dappled pattern.
6. Stratocumulus is a low-level layer cloud having a dappled, lumpy, or wavy structure.
7. Altocumulus is similar to stratocumulus but lies at intermediate levels.
8. Altostratus is a thick, extensive, layer cloud at intermediate levels.
9. Nimbostratus is a dark, widespread cloud with a low base from which prolonged rain or snow falls.
10. Cumulonimbus is a large cumulus which produces a rain or snow shower.

See CLOUD PHYSICS. [F.H.Lu.]

Cloud physics The study of the physical and dynamical processes governing the structure and development of clouds and the release from them of snow, rain, and hail (collectively known as precipitation).

The factors of prime importance are the motion of the air, its water-vapor content, and the numbers and properties of the particles in the air which act as centers of condensation and freezing. Because of the complexity of atmospheric motions and the enormous variability in vapor and particle content of the air, it seems impossible to construct a detailed, general theory of the manner in which clouds and precipitation develop. However, calculations based on the present conception of laws governing the growth and aggregation of cloud particles and on simple models of air motion provide reasonable explanations for the observed formation of precipitation in different kinds of clouds.

Clouds are formed by the lifting of damp air which cools by expansion under continuously falling pressure. The relative humidity increases until the air approaches saturation. Then condensation occurs on some of the wide variety of aerosol particles present; these exist in concentrations ranging from less than 2000 particles/in.^{3} (100/cm^3) in clean, maritime air to perhaps 10^7/in.^{3} (10^6/cm^3) in the highly polluted air of an industrial city. A portion of these particles are hygroscopic and promote condensation at relative humidities below 100%; but for continued condensation leading to the formation of cloud droplets, the air must be slightly supersaturated. Among the highly efficient condensation nuclei are the salt particles produced by the evaporation of sea spray, but it appears that particles produced by human-made fires and by natural combustion (for example, forest fires) also make a major contribution. Condensation onto the nuclei continues as rapidly as the water vapor is made available by cooling of the air and gives rise to droplets of the order of 0.0004 in. (0.01 mm) in diameter. These droplets, usually present in concentrations of several thousand per cubic inch, constitute a nonprecipitating water cloud.

Cloud droplets are seldom of uniform size. Droplets arise on nuclei of various sizes and grow under slightly different conditions of temperature and supersaturation in different parts of the cloud. A droplet appreciably larger than average will fall faster than the smaller ones, and so will collide and fuse (coalesce) with some of those which it overtakes.

The second method of releasing precipitation can operate only if the cloud top reaches elevations where temperatures are below 32°F (0°C) and the droplets in the upper cloud regions become supercooled. At temperatures below −40°F (−40°C) the droplets freeze automatically or spontaneously; at higher temperatures they can freeze only if they are infected with special, minute particles called ice nuclei. As the temperature falls below 32°F (0°C), more and more ice nuclei become active, and ice crystals appear in increasing numbers among the supercooled droplets. Such a mixture of supercooled droplets and ice crystals is unstable. After several minutes the growing crystals will acquire definite falling speeds, and several of them may become joined together to form a snowflake. In falling into the warmer regions of the cloud, however, the snowflake may melt and reach the ground as a raindrop.

The deep, extensive, multilayer-cloud systems, from which precipitation of a usually widespread, persistent character falls, are generally formed in cyclonic depressions (lows) and near fronts. Although the structure of these great raincloud systems, which are being explored by aircraft and radar, is not yet well understood, radar signals from these clouds usually take a characteristic form which has been clearly identified with the melting of snowflakes.

Precipitation from shower clouds and thunderstorms, whether in the form of raindrops, pellets of soft hail, or true hailstones, is generally of greater intensity and shorter duration than that from layer clouds and is usually composed of larger particles. The clouds themselves are characterized by their large vertical depth, strong vertical air currents, and high concentrations of liquid water, all these factors favoring the rapid growth of precipitation elements by accretion.

The development of precipitation in convective clouds is accompanied by electrical effects culminating in lightning. The mechanism by which the electric charge dissipated in lightning flashes is generated and separated within the thunderstorm has been debated for more than 200 years, but there is still no universally accepted theory. However, the majority opinion holds that lightning is closely associated with the appearance of the ice phase, and the most promising theory suggests that the charge is produced by the rebound of ice crystals or a small fraction of the cloud droplets that collide with the falling hail pellets. *See* LIGHTNING.

The various stages of the precipitation mechanisms raise a number of interesting and fundamental problems in classical physics. Worthy of mention are the supercooling and freezing of water; the nature, origin, and mode of action of the ice nuclei; and the mechanism of ice-crystal growth which produces the various snow crystal forms.

The maximum degree to which a sample of water may be supercooled depends on its purity, volume, and rate of cooling. The freezing temperatures of waterdrops containing foreign particles vary linearly as the logarithm of the droplet volumes for a constant rate of cooling. This relationship, which has been established for drops varying between 10 micrometers and 1 centimeter in diameter, characterizes the heterogeneous nucleation of waterdrops and is probably a consequence of the fact that the ice-nucleating ability of atmospheric aerosol increases logarithmically with decreasing temperature.

Measurements made with large cloud chambers on aircraft indicate that the most efficient nuclei, active at temperatures above 14°F (−10°C), are present in concentrations of only about 10 in a cubic meter of air, but as the temperature is lowered, the numbers of ice crystals increase logarithmically to reach concentrations of about 1 per liter at −4°F (−20°C) and 100 per liter at −22°F (−30°C). Since these measured concentrations of nuclei are less than one-hundredth of the numbers that apparently are consumed in the production of snow, it seems that there must exist processes by which the original number of ice crystals are rapidly multiplied, Laboratory experiments suggest the fragmentation of the delicate snow crystals and the ejection of ice splinters from freezing droplets as probable mechanisms.

The most likely source of atmospheric ice nuclei is provided by the soil and mineral-dust particles carried aloft by the wind. Laboratory tests have shown that, although most common minerals are relatively inactive, a number of silicate minerals of the clay family produce ice crystals in a supercooled cloud at temperatures above −4°F (−18°C). A major constituent of some clays, kaolinite, which is active below 16°F (−9°C), is probably the main source of highly efficient nuclei.

The fact that there may often be a deficiency of efficient ice nuclei in the atmosphere has led to a search for artificial nuclei which might be introduced into supercooled clouds in large numbers. In general, the most effective ice-nucleating substances, both natural and artificial, are hexagonal crystals in which spacings between adjacent rows of atoms differ from those of ice by less than 16%. The detailed surface structure of the nucleus, which is determined only in part by the crystal geometry, is of even greater importance.

Collection of snow crystals from clouds at different temperatures has revealed their great variety of shape and form. This multiple change of habit over such a small temperature range is remarkable and is thought to be associated with the fact that water molecules apparently migrate between neighboring faces on an ice crystal in a manner which is very sensitive to the temperature. The temperature rather than the supersaturation of the environment is primarily responsible for determining the basic shape of the crystal, though the supersaturation governs the growth rates of the crystals, the ratio of their linear dimensions, and the development of dendritic forms.

The presence of either ice crystals or some comparatively large waterdroplets (to initiate the coalescence mechanism) appears essential to the natural release of precipitation. Rainmaking experiments are conducted on the assumption that some clouds precipitate inefficiently, or not at all, because they are deficient in natural nuclei; and that this deficiency can be remedied by "seeding" the clouds artificially with dry ice or silver iodide to produce ice crystals, or by introducing waterdroplets or large hygroscopic nuclei. *See* PRECIPITATION (METEOROLOGY); WEATHER MODIFICATION.

[B.J.M.]

Coastal landforms The characteristic features and morphology of the land in the coastal zone. They are subject to processes of erosion and deposition as produced by winds, waves, tides, and river discharge. The interactions of these processes and the coastal environments produce a wide variety of landforms. Processes directed seaward from the land are dominated by the transport of sediment by rivers, but also include gravity processes such as landslides, rockfalls, and slumping. The dominant processes on the seaward side are wind, waves, and wave-generated currents. Mixed among these locations are tidal currents which also carry large volumes of sediment.

Subcontinental- to continental-scale coastal landform patterns are related to plate tectonics. The three major tectonic coastal types are leading-edge, trailing-edge, and marginal sea coasts. Leading-edge coasts are associated with colliding plate boundaries where there is considerable tectonic activity. Trailing-edge coasts are on stable continental margins, and marginal seas have fairly stable coasts with plate margins, commonly characterized by island arcs and volcanoes, that form their seaward boundaries. *See* CONTINENTAL MARGIN.

The primary characteristics of leading-edge coasts is the rugged and irregular topography, commonly displaying cliffs or bluffs right up to the shoreline. Seaward, the topography reflects this with an irregular bottom and deep water near the shoreline. The geology is generally complex with numerous faults and folds in the strata of the coastal zone. These coasts tend to be dominated by erosion with only local areas of deposition, typically in the form of small beaches or spits between headlands. Waves tend to be large because of the deep nearshore water, and form wave-cut platforms, terraces, notches, sea stacks, and caves.

The most diverse suite of coastal landforms develops along trailing-edge coasts. These coasts are generally developed along the margin of a coastal plain, they are fed by well-developed and large river systems, and they are subjected to a low-to-modest wave energy because of the gently sloping adjacent continental shelf. The overall appearance is little topographic relief dominated by deposition of mud and sand. The spectrum of environments and their associated landforms includes deltas, estuaries, barrier islands, tidal inlets, tidal flats, and salt marshes. *See* BARRIER ISLANDS; COASTAL PLAIN; DELTA; FLOODPLAIN; SALT MARSH.

Marginal coastal settings are along a stable continental mass and are protected from open ocean processes, commonly an island arc system or other form of a plate boundary. The consequence is a coastal zone that tends to be subjected to small waves and where considerable mud is allowed to accumulate in the coastal zone. Many coastal landforms are present in much the same fashion as on the trailing-edge coasts. Marginal sea coasts tend to have large river deltas. Examples are the eastern margin of Asia along the Gulf of Korea and the China Sea, the entire Gulf of Mexico, and the Mediterranean Sea. *See* DEPOSITIONAL SYSTEMS AND ENVIRONMENTS. [R.A.D.]

Coastal plain An extensive, low-relief area that is bounded by the sea on one side and by some type of relatively high-relief province on the landward side. The geologic province of the coastal plain actually extends beyond the shoreline across the continental shelf. It is only during times of glacial melting and high sea level that much of the coastal plain is drowned. *See* CONTINENTAL MARGIN.

The coastal plain is a geologic province that is linked to the stable part of a continent on the trailing edge of a plate. The extent and nature of the coastal plains of the world range widely. Some are very large and old, whereas others are small and geologically young. For example, the Atlantic and Gulf coasts of the United States are among the largest in the world. (see illustration). In some areas the coastal plain is hundreds of kilometers wide and extends back about 100 million years. By contrast, local coastal plains in places like the east coast of Australia and New Zealand are only 1 or 2 million years old and extend only tens of kilometers landward from the shoreline.

The typical character of a coastal plain is one of strata that dip gently and uniformly toward the sea. There may be low ridges that are essentially parallel to the coast and that have developed from erosion of alternating resistant and nonresistant strata.

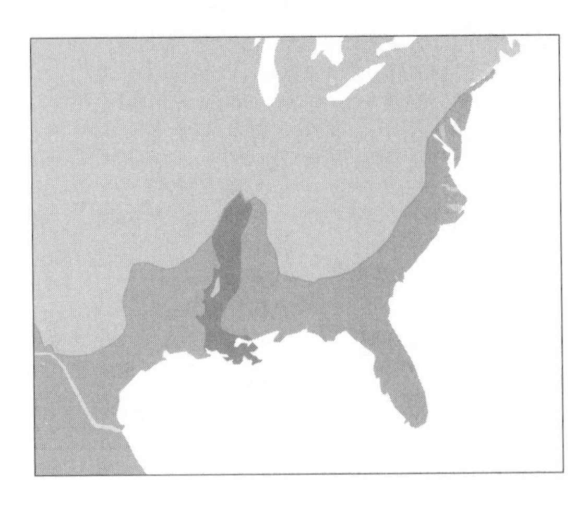

Coastal plains of the United States.

These strata are commonly a combination of mudstone, sandstone, and limestone, although the latter is typically a subordinate amount of the total. These strata resulted from deposition in fluvial, deltaic, and shelf environments as sea level advanced and retreated over this area. Coastal plain strata have been a source of considerable oil and gas as well as various economic minerals. Although the coastal plain province is typically stable tectonically, there may be numerous normal faults and salt dome intrusions. *See* COASTAL LANDFORMS; DELTA; PLAINS. [R.A.D.]

Cold hardiness (plant) The ability of temperate zone plants to survive subzero temperatures. This characteristic is a predominant factor that determines the geographical distribution of native plant species and the northern limits of cultivation of many important agronomic and horticultural crops. Further, freezing injury is a major cause of crop loss resulting from early fall frosts, low midwinter temperatures, or late spring frosts. Problems of cold hardiness are of concern to farmers in diverse areas of agriculture. As a result, the development of varieties of cultivated plants with improved cold hardiness is of long-standing concern.

Within the plant kingdom there is a wide range of diversity in low-temperature tolerance—from low levels of hardiness in herbaceous species such as potatoes, to intermediate levels of hardiness for winter annuals such as wheat and rye, to extremely hardy deciduous trees and shrubs such as black locust and red osier dogwood that can withstand temperatures of liquid nitrogen. Within a given species, the range in hardiness can be substantial. Within a given plant there is a wide range in the cold hardiness of different tissues and organs. For example, roots are much less tolerant of subzero temperatures than shoots; flower buds are more sensitive than vegetative buds.

The cold hardiness of a given species is an inherent genetic trait that requires certain environmental cues for its expression. With the shorter days and cooler nights of autumn, temperate zone plants become dormant and increase their cold hardiness. This process is referred to as cold acclimation. In the spring, increasing daylength and warmer temperatures result in the resumed growth and development of the plant and a corresponding decrease in cold hardiness. Cold hardiness may be influenced by radiation, temperature, photoperiod, precipitation, and stage of development of the plant, with different optimum conditions for different species or cultivars and ecotypes within a species. The various environmental cues serve to synchronize plant development with the environment. This synchronization has taken centuries to evolve, and freezing injury in cultivated species can result from any factor that disrupts this synchrony.

Temperature is the key environmental parameter for increasing a plant's capacity to withstand freezing temperatures. Low, above-freezing temperatures are conducive to an increase in hardiness in the fall, and warm temperatures are responsible for the decrease in the spring. Generally, it is considered that most plants will acclimate as temperatures are gradually lowered below 50°F (10°C). However, during acclimation, the progressive decline in temperatures is extremely important. The development of cold hardiness may take 4 to 6 weeks.

Photoperiod is the second major factor influencing cold acclimation, but only in those species that are photoperiodically responsive in relation to growth cessation or induction of dormancy (a true physiological rest period). In other species, light is important only in providing sufficient photosynthetic reserves required for the cold acclimation process. In some cases (for example, germinating seeds), sufficient energy reserves are already present and acclimation can occur in the dark. *See* PHOTOPERIODISM.

There are conflicting reports on the role of moisture in relation to cold hardiness. High soil moisture may reduce the degree of cold acclimation; however, severe winter injury of evergreens will occur if soil moisture levels are too low. Most often tissue moisture levels will influence the survival to a given freeze-thaw cycle rather than directly influencing the process of cold acclimation. Thus, whereas temperature and light effects on hardiness are probably mediated through the development of hardiness (cold acclimation), tissue moisture content directly affects the stresses that are incurred during a freeze-thaw cycle. In addition, various cultural practices can influence the cold hardiness of a given plant. For example, late fall applications of fertilizer or improper pruning practices may stimulate flushes of growth that do not have sufficient time to acclimate. Conversely, insufficient mineral nutrition can also impair the development of maximum cold hardiness.

The process of cold acclimation results in numerous biochemical changes within the plant. These include increases in growth inhibitors and decreases in growth promoters; changes in nucleic acid metabolism; alterations in cellular pigments such as carotenoids and anthocyanins; the accumulation of carbohydrates, amino acids, and water-soluble proteins; increases in fatty acid unsaturation; changes in lipid composition; and the proliferation of cellular membrane systems. Some of these are merely changes in response to slower growth rates and decreased photosynthate utilization; others are changes associated with growth at low, above-zero temperatures; and still others are associated with other developmental phenomena, such as vernalization or the induction of dormancy, that also occur during the period of cold acclimation.

Large increases in cellular solute concentrations are one of the most universal manifestations of cold acclimation. A doubling of the intracellular solute concentration, most notably sugars, is not uncommon. Such increases have several beneficial effects. First, they serve to depress the freezing point of the intracellular solution. More important, a doubling of the initial intracellular solute concentration will decrease the extent of cell dehydration at any subzero temperature by 50%. An increase in intracellular solutes will also decrease the concentration of toxic solutes at temperatures below 32°F (0°C), because less water will be removed. Following cold acclimation there are also substantial changes in the lipid composition of the plasma membrane. This includes an increase in free sterols with corresponding decreases in steryl glucosides and acylated steryl glucosides, a decrease in the glucocerebroside content, and an increase in the phospholipid content. The complexity of the plasma membrane lipid composition and the numerous changes that occur during cold acclimation preclude the possibility of any simple correlative analysis; however, studies have demonstrated that differential behavior of the plasma membrane observed in protoplasts isolated from nonacclimated and cold-acclimated leaves is a consequence of alterations in the lipid composition. *See* ALTITUDINAL VEGETATION ZONES; PLANT PHYSIOLOGY; PLANT-WATER RELATIONS. [P.L.S.]

Coliphage Any bacteriophage able to infect the bacterium *Escherichia coli*; many are able to attack more than one strain of this organism. The T series of phage (T1–T7), propagated on a special culture of *E. coli*, strain B, have been used in extensive studies, from which most of the knowledge of phages is derived. *See* BACTERIOPHAGE. [P.B.C.]

Common cold An acute infectious disorder characterized by nasal obstruction and discharge that may be accompanied by sneezing, sore throat, headache, malaise, cough, and fever. The disorder involves all human populations, age groups, and geographic regions; it is more common in winter than in summer in temperate climates. Most people in the United States experience at least one disabling cold (causing loss of time from work or school or a physician visit) per year. Frequencies are highest in children and are reduced with increasing age.

Most, or possibly all, infectious colds are caused by viruses. More than 200 different viruses can induce the illness, but rhinoviruses, in the picornavirus family, are predominant. Rhinoviruses are small ribonucleic acid-containing viruses with properties similar to polioviruses. Other viruses commonly causing colds include corona, parainfluenza, influenza, respiratory syncytial, entero, and adeno. *See* ADENOVIRIDAE; ENTEROVIRUS; PARAINFLUENZA VIRUS; RHINOVIRUS.

Cold viruses are spread from one person to another in either of two ways: by inhalation of infectious aerosols produced by the sneezing or coughing of ill individuals, or by inoculation with virus-containing secretions through direct contact with a person or a contaminated surface. Controlled experiments have not shown that chilling produces or increases susceptibility to colds. Infection in the nasopharynx induces symptoms, with the severity of the illness relating directly to the extent of the infection. Recovery after a few days of symptoms is likely, but some individuals may develop a complicating secondary bacterial infection of the sinuses, ear, or lung (pneumonia).

Colds are treated with medications designed to suppress major symptoms until natural defense mechanisms terminate the infection. Immunity to reinfection follows recovery and is most effective in relation to antibody in respiratory secretions. There is no established method for prevention of colds; however, personal hygiene is recommended to reduce contamination of environmental air and surfaces with virus that may be in respiratory secretions. *See* PNEUMONIA. [R.B.C.]

Conservation of resources Management of the human use of natural resources to provide the maximum benefit to current generations while maintaining capacity to meet the needs of future generations. Conservation includes both the protection and rational use of natural resources.

Earth's natural resources are either nonrenewable, such as minerals, oil, gas, and coal, or renewable, such as water, timber, fisheries, and agricultural crops. The combination of growing populations and increasing levels of resource consumption is degrading and depleting the natural resource base. The world's population stood at 850 million at the onset of the industrial age. The global population has grown to nearly seven times as large (6 billion), and the level of consumption of resources is far greater. This human pressure now exceeds the carrying capacity of many natural resources.

Nonrenewable resources, such as fossil fuels, are replaced over geologic time scales of tens of millions of years. Human societies will eventually use up all of the economically available stock of many nonrenewable resources, such as oil. Conservation entails actions to use these resources most efficiently and thereby extend their life as long as possible. By recycling aluminum, for example, the same piece of material is reused in a series of products, reducing the amount of aluminum ore that must be mined. Similarly, energy-efficient products help to conserve fossil fuels since the same energy services, such as lighting or transportation, can be attained with smaller amounts of fuel. *See* HUMAN ECOLOGY.

It may be expected that the biggest challenge of resource conservation would involve nonrenewable resources, since renewable resources can replenish themselves after harvesting. In fact, the opposite is the case. Historically, when nonrenewable resources have been depleted, new technologies have been developed that effectively substitute for the depleted resources. Indeed, new technologies have often reduced pressure on these resources even before they are fully depleted. Fiber optics, for example, has substituted for copper in many electrical applications, and it is anticipated that renewable sources of energy, such as photovoltaic cells, wind power, and hydropower, will ultimately take the place of fossil fuels when stocks are depleted. Renewable resources, in contrast, can be seriously depleted if they are subjected to excessive harvest or otherwise degraded, and no substitutes are available for, say, clean water or food products such as fish or agricultural crops. Moreover, when the misuse of biological resources causes the complete extinction of a species or the loss of a particular habitat, there can be no substitute for that diversity of life.

"Conservation" is sometimes used synonymously with "protection." More appropriately, however, it refers to the protection and sustainable use of resources. Critical elements of the effective conservation of natural resources include sustainable resource management, establishment of protected areas, and ex situ (off-site) conservation.

Resource management. Some of the most pressing resource conservation problems stem directly from the mismanagement of important biological resources. Many marine fisheries are being depleted, for example, because of significant overcapacity of fishing vessels and a failure of resource managers to closely regulate the harvest. In theory, a renewable resource stock could be harvested at its maximum sustainable yield and maintain constant average annual productivity in perpetuity. In practice, however, fishery harvest levels are often set too high and, in many regions, enforcement is weak, with the result that fish stocks are driven to low levels. A similar problem occurs in relation to the management of timber resources. Short-term economic incentives encourage cutting as many trees as quickly as possible. *See* FISHERIES ECOLOGY; FOREST MANAGEMENT.

A number of steps are being taken to improve resource conservation in managed ecosystems. (1) Considerable scientific research has been undertaken to better understand the natural variability and productivity of economically important resources. (2) Many national and local governments have enacted regulations for resource management practices on public and private lands. (3) In some of regions, programs recently have been established either to involve local communities who have a greater incentive to manage for long-term production more directly in resource management decisions or to return to them resource ownership rights. (4) Efforts are under way to manage resources on a regional or ecosystem scale using methods that have come to be known as ecosystem management or bioregional management. Since the actions taken in one location often influence species and processes in other locations, traditional resource conservation strategies were often focused too narrowly to succeed.

Protected areas. One of the most effective strategies to protect species from extinction is the establishment of protected areas designed to maintain populations of a significant fraction of the native species in a region. Worldwide, 9832 protected areas, totaling more than 9.25 million square kilometers (24 million square miles), cover about 8% of land on Earth. Although these sites are not all managed exclusively for the conservation of species, they play an essential role in protecting species from extinction.

Many problems remain, however, in ensuring effective protected-area conservation networks. For example, several regions with important biodiversity still lack effective protected-area networks. In addition, where protected areas have been designated, human and financial resources are not always available to effectively manage the areas. Particularly in developing countries, the establishment of protected areas has resulted in conflicts with local communities that had been dependent upon the areas for their livelihood. These challenges are now being addressed through international efforts, such as the International Convention on Biological Diversity, which aims to increase the financing available for protected areas and to integrate conservation and development needs.

Ex situ conservation. The most effective and efficient means for conserving biological resources is to prevent the loss of important habitats and to manage resources for their long-term productivity of goods and services. In many cases, effective conservation in the field is no longer possible. For example, some species have been so depleted that only a few individuals remain in their natural habitat. In these cases, there is no alternative to the ex situ conservation of species and genetic resources in zoos, botanical gardens, and seed banks. Ex situ collections play important conservation roles as well as serving in public education and research. Worldwide, zoos contain more than 3000 species of birds, 1000 species of mammals, and 1200 species of reptiles, and botanic gardens are believed to hold nearly 80,000 species of plants. These collections hold many endangered species, some of which have breeding populations and thus could potentially be returned to the wild. Genebanks hold an important collection of the genetic diversity of crops and livestock. *See* LAND-USE PLANNING; SOIL CONSERVATION; WATER CONSERVATION. [W.Re.]

Continent A protuberance of the Earth's crustal shell with an area of several million square miles and with sufficient elevation above neighboring depressions (the ocean basins) so that much of it is above sea level.

The great majority of maps now in use imply that the boundaries of continents are their shorelines. From the geological point of view, however, the line of demarcation between a continent and an adjacent ocean basin lies offshore, at distances ranging from a few to several hundred miles, where the gentle slope of the continental shelf changes somewhat abruptly to a steeper declivity. This change occurs at depths ranging from a few to several hundred fathoms (1 fathom = 6 feet) at different places around the periphery of various continents. *See* CONTINENTAL MARGIN.

On such a basis, numerous offshore islands, including the British Isles, Greenland, Borneo, Sumatra, Java, New Guinea, Tasmania, Taiwan, Japan, and Sri Lanka, are parts of the nearby continent. Thus, there are six continents: Eurasia (Europe, China, and India are parts of this largest continent), Africa, North America, South America, Australasia (including Australia, Tasmania, and New Guinea), and Antarctica.

All continents have similar structural features but display great variety in detail. Each includes a basement complex (shield) of metamorphosed sedimentary and volcanic rocks of Precambrian age, with associated igneous rocks, mainly granite. Originally formed at considerable depths below the surface, this shield was later exposed by extensive erosion, then largely covered by sediments of Paleozoic, Mesozoic, and Cenozoic age, chiefly marine limestones, shales, and sandstones. In at least one area on each continent, these basement rocks are now at the surface (an example is the Canadian Shield of North America). In some places they have disappeared beneath a sedimentary platform occupying a large fraction of the area of each continent, such as the area in the broad lowland drained by the Mississippi River in the United States. In each continent there are long belts of mountains in which thick masses of sedimentary rocks have been compressed into folds and broken by faults.

Continents are the less dense, subaerially exposed portion of the plates that make up the Earth's lithosphere, or outer shell of rigid rock material. As such, continents together with part of the ocean's floor are intimately joined portions of the lithospheric plates. As plates rip apart and migrate horizontally over the Earth's surface, so too do continents rip apart and migrate, sometimes colliding with other continental segments scores of millions of years later. Mountain systems, such as the Appalachian-Ouachita, the Arbuckle-Wichita, and the Urals systems, are now believed to represent the sutures of former continents attached to their respective plates which collided long ago. The Red Sea and the linear rift-volcano-lakes district of Africa are also believed by many to manifest continental ripping and early continental drifting. Such continental collision and accretion are believed to have occurred throughout most of the Earth's history. *See* AFRICA; ANTARCTICA; ASIA; EUROPE; NORTH AMERICA; SOUTH AMERICA. [D.L.J.]

Continental margin The submerged portions of the continental masses on crustal plates, including the continental shelf, the continental slope, and the continental rise. All continental masses have some continental margin, but there is great variety in the size, shape, and geology depending upon the tectonic setting.

The most common settings are the trailing-edge margin and the leading-edge margin. The former is associated with tectonic stability, as exemplified by the Atlantic side of the North American landmass (see illustration). Here the margin is wide and geologically relatively uncomplicated, with thick sequences of coastal plain to shallow marine strata dipping slightly toward the ocean basin. By contrast, the leading-edge margin (for example, the Pacific side of the United States) is narrow, rugged, and geologically complicated. The global distribution of these widespread continental margin types is controlled by the plate tectonic setting in which the landmass resides. Some major

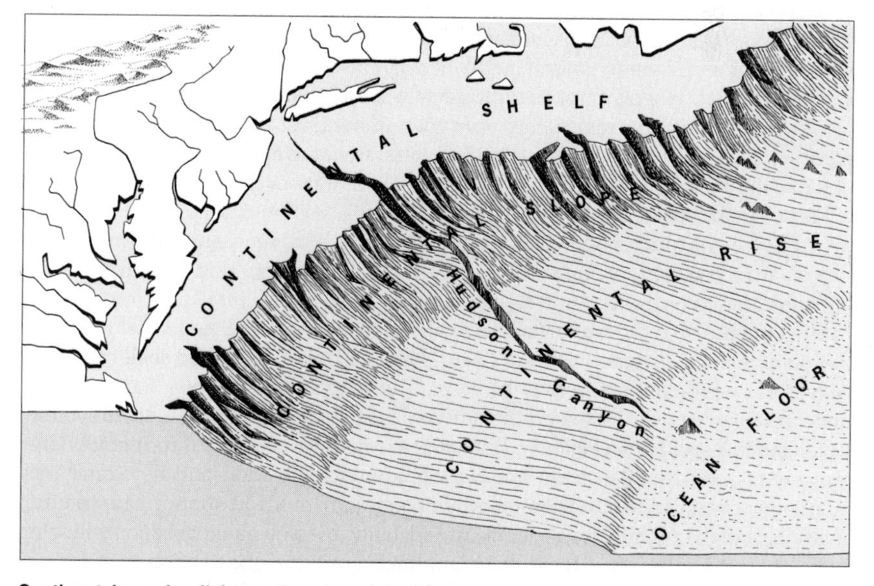

Continental margin off the northeastern United States.

landmasses, such as Australia, are surrounded by wide margins, but most, such as North and South America, have some of both types.

Any consideration of the continental margin must include a general understanding of global seal-level history over the past few million years. As glaciers expanded greatly just over 2 million years ago, sea level was lowered more than 300 ft (100 m). The cyclic growth and decay of glaciers during this period caused the shoreline to move from near its present position to near the edge of the present continental shelf on multiple occasions. These sea-level changes had a profound effect on the entire continental margin, particularly the shelf and the rise. During times of glacial advance, the coast was near the shelf edge, causing large volumes of river-borne sediment to flow down the continental slope and pile up on the rise; deltas were poorly developed for lack of place for sediment to accumulate. During times of high sea-level stand similar to the present time, little sediment crossed the shelf and large volumes of riverine sediment accumulated in large fluvial deltas. *See* DELTA.

The continental shelf is simply an extension of the adjacent landmass. It is characterized by a gentle slope and little relief except for shelf valleys (see illustration), which are old rivers that were active during times of low sea level. The outer limit of the shelf shows a distinct change in gradient to the much steeper slope.

The continental slope and rise of the outer continental margin includes the relatively steep slope and the rise that accumulates at the base of the slope. This continental material has the same general composition as the landmass.

The leading-edge continental margin that is commonly associated with a crustal plate boundary displays a very different geology, geomorphology, and bathymetry than the outer continental margin. In this type there is no distinct shelf, slope, and rise. Like the trailing-edge margin, the leading-edge margin exhibits the same characteristics as the adjacent landmass, in this case a structurally complex geology with numerous fault basins and high relief. The borderland is narrow and overall steep. Its geomorphology consists of numerous local basins that receive sediment through numerous submarine canyons. The canyons commonly extend nearly to the beach; there is no shelf as such.

The continental margin contains a vast amount and array of natural resources, most of which are being harvested. The primary fishing grounds around the globe are in shelf waters. The Grand Banks off northeastern North America and the North Sea adjacent to Europe are among the most heavily fished. There are also many mineral resources that are taken from shelf sediments, including heavy minerals that are sources of titanium, phosphate, and even placer gold. Important commodities such as sand, gravel, and shell are also taken in large quantities from the inner shelf. Salt domes that underlie the shelf, especially in the northern Gulf of Mexico, provide salt and sulfur.

Probably the most important resource obtained from the continental margin is petroleum, in the form of both oil and gas. Production is extremely high is some places, ranging from the deltas at the coast across the entire shelf and onto the outer margin, and reserves are high. *See* MARINE GEOLOGY. [R.A.D.]

Shelf circulation is the pattern of flow over continental shelves. An important part of this pattern is any exchange of water with the deep ocean across the shelf-break and with estuaries or marginal seas at the coast. The circulation transports and distributes materials dissolved or suspended in the water, such as nutrients for marine life, freshwater and fine sediments originating in rivers, and domestic and industrial waste. Water movements over continental shelves include tidal motions, wind-driven currents, and long-term mean circulation. The inflow of fresh water from land also contributes to shelf circulation, because such water would tend to spread out on the surface on account of its low density. Rapid nearshore mixing reduces the density contrast, and the Earth's rotation deflects the offshore flow into a shore-parallel direction, leaving the coast to the right. A compensating shoreward flow at depth is deflected in the opposite direction, adding to the complexity of shelf circulation. *See* OCEAN CIRCULATION. [G.T.C.]

Contour The locus of points of equal elevation used in topographic mapping. Contour lines represent a uniform series of elevations, the difference in elevation between adjacent lines being the contour interval of the given map. Thus, contours represent the shape of terrain on the flat map surface (*see* illustration). Closely spaced contours

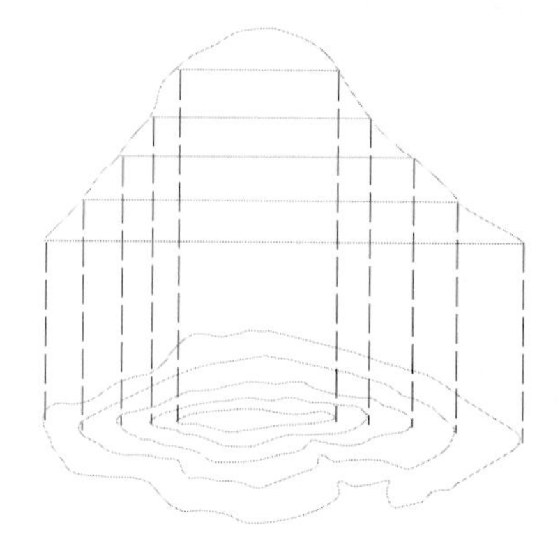

Contour representation.

indicate steep ground; sparse-ness or absence of contours indicates gentle slope or flat ground. Contours do not cross each other unless there is an overhang. [R.H.Do.]

Corn *Zea mays* occupies a larger area than any other grain crop in the United States, where 60% of the world production is grown. Although corn is grown in the United States primarily for livestock feed, about 10% is used for the manufacture of starch, sugar, corn meal, breakfast cereals, oil, alcohol, and several other specialized products. In many tropical countries, corn is used primarily for human consumption.

As a crop. The origin of corn is still unsettled, but the most widely held hypothesis assumes that corn developed from its wild relative teosinte (*Z. mexicana*) through a combination of favorable mutations, recognized and selectively propagated by early humans. Corn migrated from its center of origin, presumed to be Mexico or Central America, and was being cultivated by the Indians as far north as New England upon the arrival of the first European colonists, whose survival was due largely to the use of corn as food.

Botanically, corn is a member of the grass family. Each form (botanical variety) is conditioned by fairly few genetic differences, and each may exhibit the full range of differences in color, plant type, maturity, and so on, characteristic of the species. All types have the same number of chromosomes (10 pairs), and all may be intercrossed to produce fertile progeny. Dent corns are the most important in the United States. Sweet corn is grown more extensively in the United States than in any other country. It is eaten as fresh corn or canned or frozen. In other countries, flint, dent, or flour corns may be eaten fresh, but at a much more mature stage than the sweet corn eaten in the United States. The commercial production of popcorn is almost exclusively American. *See* GENETICS; REPRODUCTION (PLANT).

Corn is a cross-pollinated plant; the staminate (male) and pistillate (female) inflorescences (flower clusters) are borne on separate parts of the same plant (see illustration). Plants of this type are called monoecious. The staminate inflorescence is the tassel; it produces pollen that is carried by the wind to the silks produced on the ears.

A corn plant in full tassel and silk. The tassel produces pollen that is blown by wind to the silks. (*Courtesy of J. W. McManigal*)

The development of varieties and strains of corn made possible the extension of its culture under diverse soil and climatic conditions. However, modern research methods led to the present widespread use of hybrid corn. Hybrid corn is the first generation of cross involving inbred lines. Inbred lines are developed by controlled self-pollination. When continued for several generations, self-pollination leads to reduction in vigor but permits the isolation of types which are genetically pure or homozygous. Intense selection is practiced during the inbreeding phase to identify and maintain genotypes having the desired plant and ear type and maturity characteristics, and relative freedom from insect and disease attacks. Crosses involving any two unrelated lines will exhibit heterosis, that is, yields above the means of the two parents. *See* BREEDING (PLANT).

Planting dates depend upon temperature and soil conditions. Germination is very slow at soil temperatures of 50°F (10°C), and seedling growth is limited at temperatures of 60°F (16°C) or below. Planting rates are influenced by water supply, soil type, and fertility and by the maturity characteristics of the hybrid grown. With planting rates above 16,000 plants per acre (40,000 per hectare), drilling in rows 24–36 in. (60–90 cm) apart has become common practice. The use of nitrogen fertilizer has increased greatly; lesser amounts of phosphorus and potash are applied as needed.

In the 1930s most corn was husked by hand, and the ears were stored in slatted cribs. The mechanical picker supplanted hand harvesting. The mechanical picker, in turn, has been replaced by the picker-sheller or corn combine, which harvests the crop as shelled grain. When harvested as shelled grain, at a relatively high moisture content (20–30%), the grain must be dried artificially for safe storage. High-moisture corn to be used for livestock feed may be stored in airtight silos or may be treated with certain chemical preservatives such as propionic acid. Corn stored under either of these systems is not suitable either for industrial processing or for seed. [G.F.S.]

Corn is highly productive largely because it can use solar energy so efficiently. The corn plant grows vegetatively until about silking, after which all weight increase is in the form of grain. Almost the entire grain yield results from photosynthesis during the grain growth period, which runs from silking to maturity. Contrary to much popular opinion, grain yields are highest under cool conditions, when the lengthened grain growth period more than compensates for the slower growth rate. Relationships among solar radiation, temperature, growing-season length, soil moisture, day length, soil fertility, and corn genotype in producing grain yields are complex and not well understood. Attempts to study the system as a whole, using simulation models on digital computers, may add considerably to knowledge of the subject. [W.G.D.]

Processing. Corn kernels (seeds) are subjected to both wet and dry milling. The goal of both processes is to separate the germ, the endosperm, and the pericarp (hull).

Wet milling separates the chemical constituents of corn into starch, protein, oil, and fiber fractions, the primary objective being to produce refined corn starch. Worldwide, the production of nutritive sweeteners is the largest use for the starch obtained from corn. The manufacture of corn sweeteners begins with the wet milling process. The starch is first cooked, or pasted. Then, the starch polymers are hydrolyzed (depolymerized) using an acid, an enzyme, a combination of enzymes, or an acid-enzyme combination. The resulting solutions are refined and concentrated to 70–80% solids. These syrups are known worldwide as glucose syrups, but in the United States are often called corn syrups. When starch is completely hydrolyzed, that is, converted into its monomer units, the only product is D-glucose (dextrose), which can be crystallized from concentrated solutions. Isomerization of some of the D-glucose in a high-glucose hydrolyzate to D-fructose produces high-fructose corn syrups (HFCS), which are known simply as high-fructose syrups (HFS) outside the United States. Fructose is approximately 20% sweeter than sucrose on an equal weight basis.

Corn starch is less extensively depolymerized to make products other than sweeteners. Very slight hydrolysis makes products known as acid-modified or thin-boiling starches. A little more modification with an acid produces dextrins. One application is as remoistenable adhesives on envelopes. Hydrolysis catalyzed by acid or enzymes produces starch oligomers, which are known as maltooligosaccharides or maltodextrins. Maltodextrins are used extensively in foods for their bulking and binding properties and the protection they give to frozen foods. Hydrolysis gives mixtures of breakdown products that, when dried, are known as corn syrup solids. Corn syrup solids dissolve rapidly, are mildly sweet, and are used as bulking materials in food.

Most of the processing of dry-milled corn is done by tempering-degerming systems. Cleaned kernels are transferred to a tempering bin, where they are held for various times at various temperatures depending on the miller and the desired product. Tempered kernels are passed through a degerminator, which removes the bran (pericarp) and germ while leaving the endosperm intact. The endosperm may be converted into as many as 16 different fractions. The main products are regular grits, coarse grits, flaking grits, and corn flour. Other products are corn cones and corn meal.

Nixtamalization is the process of cooking and soaking corn kernels in water containing calcium hydroxide (lime) to soften the pericarp and hydrate the protein matrix and starch of the endosperm. The cooked, steeped product, called nixtamal, is then ground, using stone attrition mills. The product, masa, is sheeted, cut into pieces, and baked, producing tortillas, tortilla chips, taco shells, and corn chips. [J.N.BeM.]

Cortex (plant) The mass of primary tissue in roots and stems extending inward from the epidermis to the phloem. The cortex may consist of one or a combination of three major tissues: parenchyma, collenchyma, and sclerenchyma. In roots the cortex almost always consists of parenchyma, and is bounded, more or less distinctly, by the hypodermis (exodermis) on the periphery and by the endodermis on the inside.

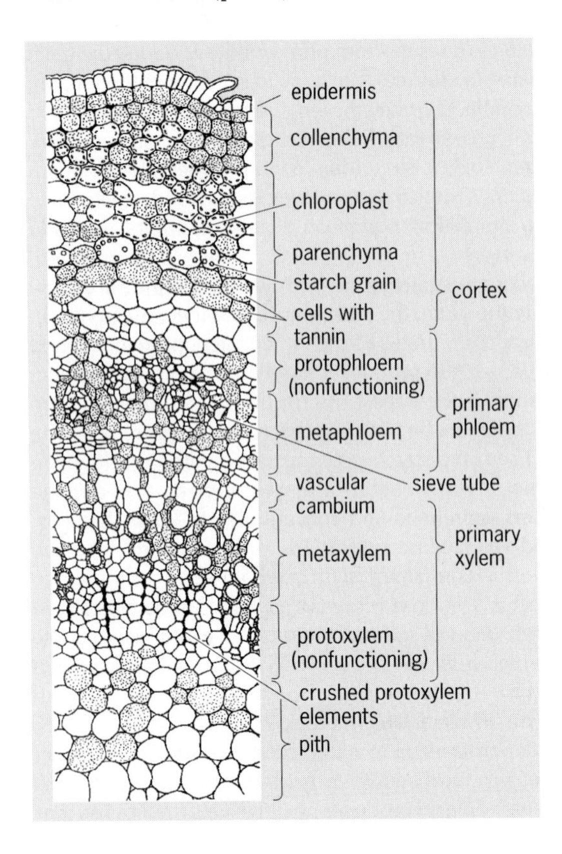

Transverse section of the *Prunus* stem showing the cortex which is composed of collenchyma and parenchyma. (*After K. Esau, Plant Anatomy, 2d ed., 1967*)

Cortical parenchyma is composed of loosely arranged thin-walled living cells. Prominent intercellular spaces usually occur in this tissue. In stems the cells of the outer parenchyma may appear green due to the presence of chloroplasts in the cells (see illustration). This green tissue is sometimes called chlorenchyma, and it is probable that photosynthesis takes place in it.

In some species the cells of the outer cortex are modified in aerial stems by deposition of hemicellulose as an additional wall substance, especially in the corners or angles of the cells. This tissue is called collenchyma, and the thickening of the cell walls gives mechanical support to the shoot.

The cortex makes up a considerable proportion of the volume of the root, particularly in young roots, where it functions in the transport of water and ions from the epidermis to the vascular (xylem and phloem) tissues. In older roots it functions primarily as a storage tissue.

In addition to being supportive and protective, the cortex functions in the synthesis and localization of many chemical substances; it is one of the most fundamental storage tissues in the plant. The kinds of cortical cells specialized with regard to storage and synthesis are numerous.

Because the living protoplasts of the cortex are so highly specialized, patterns and gradients of many substances occur within the cortex, including starch, tannins, glucosides, organic acids, crystals of many kinds, and alkaloids. Oil cavities, resin ducts, and laticifers (latex ducts) are also common in the midcortex of many plants. [D.S.V.F.]

Cover crops Unharvested crops grown to improve soil quality or enhance pest management. Increasing numbers of farmers plant cover crops as a means of conserving soil, enhancing production, and reducing off-farm inputs. The sustainable agriculture movement has been a driving force for the increased use of cover crops. Cover cropping can accomplish a wide range of desired benefits, although there can be some drawbacks.

Cover crops play a vital role in controlling erosion by (1) shielding the soil surface from the impact of falling raindrops; (2) holding soil particles in place; (3) preventing crust formation; (4) improving the soil's capacity to absorb water; (5) slowing the velocity of runoff; and (6) removing subsurface water between storms through transpiration. Cover crops also improve soil structure by adding organic matter. Legume cover crops add nitrogen to the soil, which can then be used by crops. Insect, weed, and nematode management can be affected by cover cropping, as cover crops provide a food source to many beneficial insect species. In turn, these beneficial insects may feed on adjacent crop pests.

Cover crops require knowledge and management to attain the desired benefit. If they are not properly selected or managed, there are drawbacks to their use, including depletion of soil moisture, competition with the adjacent crop (when present) for soil moisture and nutrients, increased frost hazard in orchards and vineyards, increased insect, nematode, and weed pests, and added costs to purchase and plant seeds.

Cover crops can be readily grown in humid climates and in arid and semiarid climates where irrigation provides sufficient water that cover crops do not rob the soil of needed moisture. Cover crops that reseed themselves are often sown on rangeland. Long-lived grasses have been widely seeded on many different soil types in the Great Plains.

In cold climates, many common spring-sown crops can be advantageously used as cover crops. Buckwheat (*Fagopyrum esculentum*) can be used as a cover crop to protect the soil and as a smother crop to control weeds; it also improves the soil upon incorporation.

Cover cropping in orchard and vineyard middles is often easier than in annual crops because they are grown between the crop spatially rather than temporally. A wide range of species, mixes, and management systems are used in orchards and vineyards. *See* AGRICULTURAL SOIL AND CROP PRACTICES; NITROGEN FIXATION; SOIL CONSERVATION.

[C.In.; P.J.Z.]

Coxsackievirus A large subgroup of the genus *Enterovirus* in the family Picornaviridae. The coxsackieviruses produce various human illnesses, including aseptic meningitis, herpangina, pleurodynia, and encephalomyocarditis of newborn infants. *See* ENTEROVIRUS; PICORNAVIRIDAE.

Coxsackieviruses measure about 28 nanometers in diameter; they resemble other enteroviruses in many biological properties, but differ in their high pathogenicity for newborn mice. At least 23 antigenically distinct types in group A are now recognized, and 6 in group B.

After incubation for 2–9 days, during which the virus multiplies in the enteric tract, clinical manifestations appear which vary widely. Diagnosis is by isolation of virus in tissue culture or infant mice. Stools are the richest source of virus. Neutralizing and complement-fixing antibodies form during convalescence and are also useful in diagnosis.

The coxsackieviruses have worldwide distribution. Infections occur chiefly during summer and early fall, often in epidemic proportions. Spread of virus, like that of other enteroviruses, is associated with family contact and contacts among young children. *See* ANIMAL VIRUS.

[J.L.Me.; M.E.Re.]

Crown gall A neoplastic disease of primarily woody plants, although the disease can be reproduced in species representing more than 90 plant families. The disease results from infection of wounds by the free-living soil bacterium *Agrobacterium tumefaciens* which is commonly associated with the roots of plants.

The first step in the infection process is the site-specific attachment of the bacteria to the plant host. Up to half of the bacteria become attached to host cells after 2 h. At 1 or 2 weeks after infection, swellings and overgrowths take place in tissue surrounding the site of infection, and with time these tissues proliferate into large tumors (see illustration). If infection takes place around the main stem or trunk of woody hosts, continued tumor proliferation will cause girdling and may eventually kill the host. Crown gall is therefore economically important, particularly in nurseries where plant material for commercial use is propagated and disseminated.

Unlike healthy normal cells, crown gall tumor cells do not require an exogenous source of phytohormones (auxins and cytokinin) for growth in culture because they readily synthesize more than sufficient quantities for their own growth. They also synthesize basic amino acids, each conjugated with an organic acid, called opines. The tumor cells also grow about four times faster and are more permeable to metabolities than normal cells.

These cellular alterations, such as the synthesis of opines and phytohormone regulation, result from bacterial genes introduced into host plant cells by *A. tumefaciens* during infection. Although it is not understood how these genes are introduced into the plant cell, the genes for the utilization of these opines and for regulating phytohormone

Crown gall on peach.

production have been found to be situated on an extrachromosomal element called the pTi plasmid. This plasmid, harbored in all tumor-causing *Agrobacterium* species, also carries the necessary genetic information for conferring the tumor-inducing and host-recognition properties of the bacterium.

Crown gall is consequently a result of this unique bacteria-plant interaction, whereby *A. tumefaciens* genetically engineers its host to produce undifferentiated growth in the form of a large tumor, in which there is the synthesis of a unique food source in the form of an opine for specific use by the bacterial pathogen. *See* BACTERIAL GENETICS; GENETIC ENGINEERING; PLANT PATHOLOGY. [C.J.Ka.]

Cyanobacteria A large and heterogeneous group of photosynthetic microorganisms, formerly referred to as blue-green algae. They had been classified with the algae because their mechanism of photosynthesis is similar to that of algal and plant chloroplasts; however, the cells are prokaryotic, whereas the cells of algae and plants are eukaryotic. The name cyanobacteria is now used to emphasize the similarity in cell structure to other prokaryotic organisms. *See* ALGAE; CELL PLASTIDS.

All cyanobacteria can grow with light as an energy source through oxygen-evolving photosynthesis; carbon dioxide (CO_2) is fixed into organic compounds via the Calvin cycle, the same mechanism used in green plants. Thus, all species will grow in the absence of organic nutrients. However, some species will assimilate organic compounds into cell material if light is available, and a few isolates are capable of growth in the dark by using organic compounds as carbon and energy sources. Some cyanobacteria can shift to a different mode of photosynthesis, in which hydrogen sulfide rather than water serves as the electron donor. Molecular oxygen is not evolved during this process, which is similar to that in purple and green photosynthetic sulfur bacteria. The photosynthetic pigments of cyanobacteria include chlorophyll *a* (also found in algae and plants) and phycobiliproteins. *See* CHLOROPHYLL; PHOTOSYNTHESIS.

Cyanobacteria are extremely diverse morphologically. Species may be unicellular or filamentous. Both types may aggregate to form macroscopically visible colonies. The cells range in size from those typical of bacteria (0.5–1 micrometer in diameter) to 60 μm.

When examined by electron microscopy, the cells of cyanobacteria appear similar to those of gram-negative bacteria. Many species produce extracellular mucilage or sheaths that promote the aggregation of cells or filaments into colonies.

The photosynthetic machinery is located on internal membrane foldings called thylakoids. Chlorophyll *a* and the electron transport proteins necessary for photosynthesis are located in these lipid membranes, whereas the water-soluble phycobiliprotein pigments are arranged in particles called phycobilisomes which are attached to the lipid membrane.

Several other types of intracellular structures are found in some cyanobacteria. Gas vesicles, which may confer buoyancy on the organisms, are often found in cyanobacteria that grow in the open waters of lakes. Polyhedral bodies, also known as carboxysomes, contain large amounts of ribulose bisphosphate carboxylase, the key enzyme of CO_2 fixation via the Calvin cycle. Several types of storage granules may be found.

Cyanobacteria can be found in a wide variety of fresh-water, marine, and soil environments. They are more tolerant of environmental extremes than are eukaryotic algae. For example, they are the dominant oxygenic phototrophs in hot springs (at temperatures up to 72°C or 176°F) and in hypersaline habitats such as may occur in marine intertidal zones.

Cyanobacteria are often the dominant members of the phytoplankton in fresh-water lakes that have been enriched with inorganic nutrients such as phosphate. It is now known that high population densities of small, single-celled cyanobacteria occur in the oceans, and that these are responsible for 30–50% of the CO_2 fixed into organic matter in these environments. About 8% of the lichens involve a cyanobacterium, which can provide both fixed nitrogen and fixed carbon to the fungal partner. *See* PHYTOPLANKTON.

Cyanobacteria are thought to be the first oxygen-evolving photosynthetic organisms to develop on the Earth, and hence responsible for the conversion of the Earth's atmosphere from anaerobic to aerobic about 2 billion years ago. This development permitted the evolution of aerobic bacteria, plants, and animals. *See* BACTERIA. [A.Ko.]

Cyclone An atmospheric circulation system in which the sense of rotation of the wind about the local vertical is the same as that of the Earth's rotation. Thus, a cyclone rotates clockwise in the Southern Hemisphere and counterclockwise in the Northern Hemisphere. In meteorology the term cyclone is reserved for circulation systems with horizontal dimensions of hundreds (tropical cyclones) or thousands (extratropical cyclones) of kilometers. For such systems the Coriolis force due to the Earth's rotation, which is directed to the right of the flow in the Northern Hemisphere, and the pressure gradient force, which is directed toward low pressure, are in opposite directions. Thus, there must be a pressure minimum at the center of the cyclone, and cyclones are sometimes simply called lows. *See* AIR PRESSURE.

Extratropical cyclones are the common weather disturbances which travel around the world from west to east in mid-latitudes. They are generally associated with fronts, which are zones of rapid transition in temperature. Extratropical cyclones arise due to the hydrodynamic instability of the upper-level jet stream flow. *See* FRONT; JET STREAM.

Tropical cyclones, by contrast, derive their energy from the release of latent heat of condensation in precipitating cumulus clouds. Over the tropical oceans, where moisture is plentiful, tropical cyclones can develop into intense vortical storms (hurricanes and typhoons), which can have wind speeds in excess of 200 mi/h ($100 \text{ m} \cdot \text{s}^{-1}$). *See* HURRICANE; WIND. [J.R.H.]

Cypress The true cypress (*Cupressus*), which is very close botanically to the cedars (*Chamaecyparis*). All of the species of *Cupressus* in the United States are western and are found from Oregon to Mexico. The Arizona cypress (*Cupressus arizonica*) of the southwestern United States and the Monterey cypress (*Cupressus macrocarpa*) of California are medium-sized trees and are chiefly of ornamental value. The Italian cypress (*Cupressus sempervirens*) and its varieties are handsome ornamentals, but usually do well only in the southern parts of the United States. Other trees are also called cypress, such as the Port Orford cedar (*Chamaecyparis lawsoniana*) known also as the Lawson cypress, and the Alaska cedar (*Chamaecyparis nootkatensis*), known also as the Nootka cypress or cedar.

The bald cypress (*Taxodium distichum*) is an entirely different tree that is found in the swamps of the South Atlantic and Gulf coastal plains and in the lower Mississippi Valley. The soft needlelike leaves and short branches are deciduous; hence, they drop off in winter and give the tree its common name. Also known as the southern or red cypress, this tree yields a valuable decay-resistant wood used principally for building construction, especially for exposed parts or where a high degree of resistance to decay is required as in ships, boats, greenhouses, railway cars, and railroad ties. *See* CEDAR. [A.H.G./K.P.D.]

Cytomegalovirus infection A common asymptomatic infection caused by cytomegalovirus, which can produce life-threatening illnesses in the immature fetus and in immunologically deficient subjects.

Cytomegalovirus is a member of the herpesvirus group, which asymptomatically infects 50–100% of the normal adult population. Such infections usually take place during the newborn period when the virus can be transmitted from the mother to the baby if the virus is present in the birth canal or in breast milk. Toddlers may also acquire the infection in nurseries. Later in life, the virus may be transmitted by heterosexual or male homosexual activity. After infection, cytomegalovirus remains latent in the body because it cannot be completely eradicated even by a competent immune system. It may be activated and cause illnesses when there is a breakdown of the immune system.

Congenital or transplacental cytomegalovirus infection is also a fairly common event. With rare exceptions, it too is usually asymptomatic. Congenital cytomegalovirus disease results from transplacental transmission of the virus, usually from a mother undergoing initial or primary cytomegalovirus infection, during pregnancy. Its manifestations range from subtle sensory neural hearing loss detectable only later in life, to a fulminating multisystem infection and eventual death of the newborn. This important congenital disease occurs in about 1 in 1000 pregnancies.

The only cytomegalovirus illness clearly described in mature, immunologically normal subjects is cytomegalovirus mononucleosis. This is a self-limited illness like infectious mononucleosis, the main manifestation of which is fever. *See* INFECTIOUS MONONUCLEOSIS.

Otherwise, cytomegalovirus illnesses are usually seen only when cellular immunity is deficient. They constitute the most important infection problem after bone marrow and organ transplantations. Manifestations vary from the self-limited cytomegalovirus mononucleosis to more serious organ involvement such as pneumonia, hepatitis, gastrointestinal ulcerations, and widespread dissemination. The virus causing these illnesses may come from activation of the patient's own latent infection, or it may be transmitted from an outside source, usually from latent cytomegalovirus infecting the graft from a donor.

Cytomegalovirus illnesses are also serious, fairly frequent complications of the acquired immunodeficiency syndrome (AIDS). One reason is that most individuals with human immunodeficiency virus (HIV) infection are already infected with cytomegalovirus. Disease manifestations are similar to what is seen in transplant cases, except they may be more severe. Cytomegalovirus retinitis is a typical problem associated with advanced AIDS. Without treatment, the retina is progressively destroyed such that blindness of one or both eyes is inevitable. *See* ACQUIRED IMMUNE DEFICIENCY SYNDROME (AIDS).

Cytomegalovirus diseases can be treated with two antivirals, ganciclovir or foscarnet, with varying degrees of success. Cytomegalovirus pneumonia in the bone marrow transplant recipient cannot be cured by antivirals alone because it probably has an immunopathologic component. Cytomegalovirus diseases in persons with AIDS can be contained but not cured by specific treatment. For example, ganciclovir treatment of cytomegalovirus retinitis is effective only as long as maintenance therapy is continued. *See* ANIMAL VIRUS. [Mo.H.]

D

Dating methods Relative and quantitative techniques used to arrange events in time and to determine the numerical age of events in history, geology, paleontology, archeology, paleoanthropology, and astronomy. Relative techniques allow the order of events to be determined, whereas quantitative techniques allow numerical estimates of the ages of the events. Most numerical techniques are based on decay of naturally occurring radioactive nuclides, but a few are based on chemical changes through time, and others are based on variations in the Earth's orbit. Once calibrated, some relative techniques also allow numerical estimates of age.

Relative dating methods rely on understanding the way in which physical processes in nature leave a record that can be ordered. Once the record of events is ordered, each event is known to be older or younger than each other event. In most cases the record is contained within a geological context, such as a stratigraphic sequence; in other cases the record may be contained within a single fossil or in the arrangement of astronomical bodies in space and time. The most important relative dating methods are stratigraphic dating and paleontologic dating. Other relative dating methods include paleomagnetic dating, dendrochronology, and tephrostratigraphy.

Several chemical processes occur slowly, producing changes over times of geological interest; among these are the hydration of obsidian, and the conversion of L- to D-amino acids (racemization or epimerization). Determination of age requires measurement of a rate constant for the process, knowledge of the temperature history of the material under study, and (particularly for amino acid racemization) knowledge of the chemical environment of the materials. *See* AMINO ACID DATING.

Unlike chemical methods, in which changes depend both on time and on environmental conditions, isotopic methods which are based on radioactive decay depend only on time. A parent nuclide may decay to one stable daughter in a single step by simple decay [for example, rubidium decays to strontium plus a beta particle ($^{87}Rb \rightarrow {}^{87}Sr + \beta$)]; to two daughters by branched decay through different processes [for example, potassium captures an electron to form argon, or loses a beta particle to form calcium ($^{40}K + e^{-} \rightarrow {}^{40}Ar$; $^{40}K \rightarrow {}^{40}Ca + \beta$)]; to one stable daughter through a series of steps (chain decay); or into two unequal-sized fragments by fission. In all cases, the number of parent atoms decreases as the number of daughter atoms increases, so that for each method there is an age-sensitive isotopic ratio of daughter to parent that increases with time. Many different isotopes have been exploited for measuring the age of geological and archeological materials. For example, the table shows parent isotopes, their half-lives, and the resulting daughter products.

Astronomers have estimated the age of the universe, and of the Milky Way Galaxy, by various methods. It is well known that the universe is expanding equally from all points, and that the velocity of recession of galaxies observed from Earth increases with distance. The rate of increase of recession velocity with distance is called the Hubble

Principal parent and daughter isotopes used in radiometric dating

Radioactive parent isotope	Stable daughter isotope	Half-life, years
Carbon-14	Nitrogen-14	5730
Potassium-40	Argon-40	1.25×10^9
Rubidium-87	Strontium-87	4.88×10^{10}
Samarium-147	Neodymium-143	1.06×10^{11}
Lutetium-176	Hafnium-176	3.5×10^{10}
Rhenium-187	Osmium-187	4.3×10^{10}
Thorium-232	Lead-208	1.4×10^{10}
Uranium-235	Lead-207	7.04×10^8
Uranium-238	Lead-206	4.47×10^9

constant; and knowing the recession rate and distance of galaxies at some distance, it is simple to find how long it took them to get there. Initial estimates for the age of the universe were approximately 20 billion years; but as the rate of expansion decreases with time, revised estimates are nearer 13 billion years. By contrast, the Earth and other bodies in the solar system are only about 4.5 billion years old. Comparison of present-day osmium isotope ratios with theoretically estimated initial ratios yields estimates of 8.6–15.7 billion years for the age of the Galaxy. [F.S.B.]

Deciduous plants Plants that regularly lose their leaves at the end of each growing season. Dropping of the leaves occurs at the inception of an unfavorable season characterized by either cold or drought or both. Most woody plants of temperate climates have the deciduous habit, and it may also occur in those of tropical regions having alternating wet and dry seasons. Many deciduous trees and shrubs of regions with cold winters become evergreen when grown in a warm climate. Conversely, such trees as magnolias, evergreen in warm areas, become deciduous when grown in colder climates. *See* LEAF; PLANT PHYSIOLOGY; PLANT TAXONOMY. [N.A.]

Deep-marine sediments The term “deep marine” refers to bathyal sedimentary environments occurring in water deeper than 200 m (656 ft), seaward of the continental shelf break, on the continental slope and the basin (see illustration). The continental rise, which represents that part of the continental margin between continental slope and abyssal plain, is included under the broad term “basin.” On the slope and basin environments, sediment-gravity processes (slides, slumps, debris flows, and turbidity currents) and bottom currents are the dominant depositional mechanisms, although pelagic and hemipelagic deposition is also important. *See* BASIN; CONTINENTAL MARGIN; GULF OF MEXICO; MARINE SEDIMENTS.

Types of processes. The mechanics of deep-marine processes is critical in understanding the nature of transport and deposition of sand and mud in the deep sea. In deep-marine environments, gravity plays the most important role in transporting and depositing sediments. Sediment failure under gravity near the shelf edge commonly initiates gravity-driven deep-marine processes, such as slides, slumps, debris flows, and turbidity currents (see illustration). Sedimentary deposits reflect only depositional mechanisms, not transportational mechanisms. *See* SEDIMENTOLOGY.

A slide is a coherent mass of sediment that moves along a planar glide plane and shows no internal deformation. Slides represent translational movement. Submarine slides can travel hundreds of kilometers. For example, the runout distance of Nuuanu Slide in offshore Hawaii is 230 km (143 mi). Long runout distances of 50–100 km (31–62 mi) of slides are common.

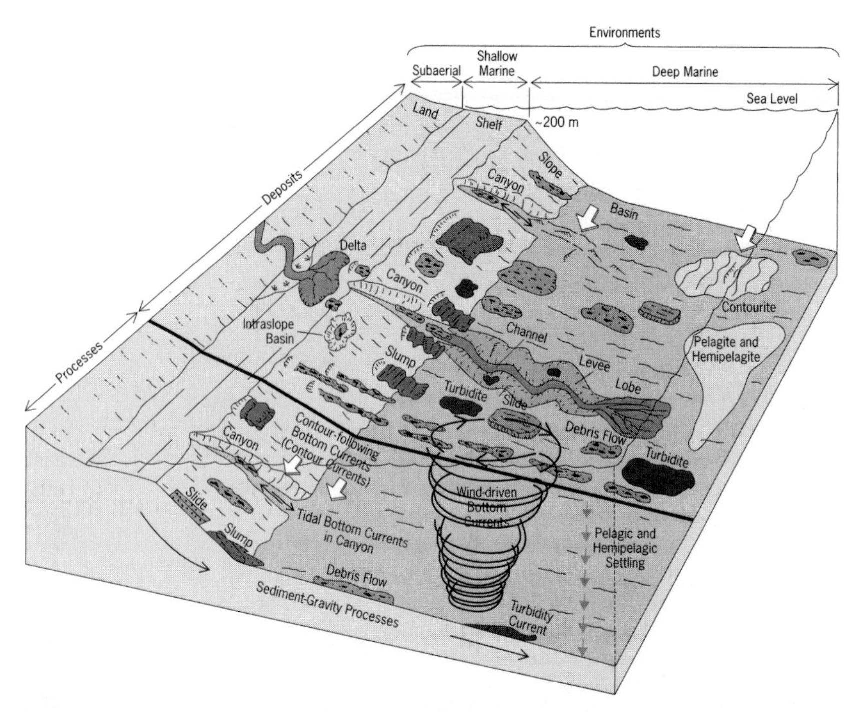

Slope and basinal deep-marine sedimentary environments occurring at water depths greater than 200 m (656 ft). Slides, slumps, debris flows, turbidity currents, and various bottom currents are important processes in transporting and depositing sediment in the deep sea. Note the complex distribution of deep-marine deposits.

A slump is a coherent mass of sediment that moves on a concave-up glide plane and undergoes rotational movements causing internal deformation.

A downslope increase in mass disaggregation results in the transformation of slumps into debris flows. Sediment is now transported as an incoherent viscous mass, as opposed to a coherent mass in slides and slumps. A debris flow is a sediment-gravity flow with plastic rheology (that is, fluids with yield strength) and laminar state. Deposition from debris flows occurs through freezing. The term "debris flow" is used here for both the process and the deposit of that process. The terms "debris flow" and "mass flow" are used interchangeably because each exhibits plastic flow behavior with shear stress distributed throughout the mass. Although only muddy debris flows (debris flows with mud matrix) received attention in the past, recent experimental and field studies show that sandy debris (debris flows with sand matrix) flows are equally important. Rheology is more important than grain-size distribution in controlling sandy debris flows, and the flows can develop in slurries of any grain size (very fine sand to gravel), any sorting (poor to well), any clay content (low to high), and any modality (unimodal and bimodal).

With increasing fluid content, plastic debris flows tend to become turbidity currents. Turbidity currents can occur in any part of the system (proximal and distal), and can also occur above debris flows due to flow transformation in density-stratified flows. A turbidity current is a sediment-gravity flow with newtonian rheology (that is, fluids without yield strength) and turbulent state. Deposition from turbidity currents occurs through suspension settling. Deposits of turbidity currents are called turbidites. Although turbidity currents have received a lot of emphasis in the past, other processes are equally important in the deep sea (see illustration). In terms of transporting coarse-grained sediment into the deep sea, sandy debris flows and other mass flows appear to play a greater role than turbidity currents.

Bottom currents. In large modern ocean basins, such as the Atlantic, thermohaline-induced geostrophic bottom currents within the deep and bottom water masses commonly flow approximately parallel to bathymetric contours (that is, along the slope (see illustration). They are generally referred to as contour currents. However, because not all bottom currents follow regional bathymetric contours, it is preferred that the term "contour current" be applied only to currents flowing parallel to bathymetric contours, and other currents be termed bottom currents. For example, wind-driven surface currents may flow in a circular motion (see illustration) and form eddies that reach the deep-sea floor, such as the Loop Current in the Gulf of Mexico, and the Gulf Stream in the North Atlantic. Local bottom currents that move up- and downslope can be generated by tides and internal waves, especially in submarine canyons. These currents are quite capable of erosion, transportation, and redeposition of fine-to-coarse sand in the deep sea. *See* GULF STREAM; OCEAN CIRCULATION.

Pelagic and hemipelagic settling. Pelagic and hemipelagic processes generally refer to settling of mud fractions derived from the continents and shells of microfauna down through the water column throughout the entire deep-ocean floor (see illustration). Hemipelagites are deposits of hemipelagic settling of deep-sea mud in which more than 25% of the fraction coarser than 5 micrometers is of terrigenous, volcanogenic, or neritic origin. Although pelagic mud and hemipelagic mud accumulate throughout the entire deep-ocean floor, they are better preserved in parts of abyssal plains (see illustration). Rates of sedmentation vary from millimeters to greater than 50 cm (20 in.) per 1000 years, with the highest rates on the upper continental margin.

Submarine slope environments. Submarine slopes are considered to be of the sea floor between the shelf-slope break and the basin floor (see illustration). Modern continental slopes around the world average 4°, but slopes range from less than 1° to greater than 40°. Slopes of active margins (for example, California and Oregon, about 2°) are relatively steeper than those of passive margins (for example, Louisiana, about 0.5°). On constructive continental margins with high sediment input, gravity tectonics involving salt and shale mobility and diapirism forms intraslope basins of various sizes and shapes (for example, Gulf of Mexico). Erosional features, such as canyons and gullies, characterize intraslope basins. Deposition of sand and mud occurs in intraslope basins. Slope morphology plays a major role in controlling deep-marine deposition through (1) steep versus gentle gradients, (2) presence or absence of canyons and gullies, (3) presence or absence of intraslope basins, and (4) influence of salt tectonics. *See* EROSION.

Submarine canyon and gully environments. Submarine canyons and gullies are erosional features that tend to occur on the slope. Although canyons are larger than gullies, there are no standardized size criteria to distinguish between them. Submarine canyons are steep-sided valleys that incise the continental slope and shelf. They serve as major conduits for transporting sediment from land and the continental shelf to the basin floor. Modern canyons are relatively narrow, deeply incised, steeply walled, often sinuous valleys with predominantly V-shaped cross sections. Most canyons originate near the continental shelf break and generally extend to the base of the continental slope. Canyons commonly occur off the mouths of large rivers such as the Hudson and Mississippi, although many others, such as the Bering Canyon in the southern Bering Sea, have developed along structural trends. *See* BERING SEA.

Modern submarine canyons vary considerably in their dimensions. Their average length of canyons has been estimated to be about 55 km (34 mi), although the Bering

Canyon, the world's longest, is nearly 1100 km (684 mi). The shortest canyons are those off the Hawaiian Islands, with average lengths of about 10 km (6.2 mi). [G.Sh.]

Deep-sea fauna The deep sea may be regarded as that part of the ocean below the upper limit of the continental slopes (see illustration). Its waters fill the deep ocean basins, cover about two-thirds of the Earth's surface, have an average depth of about 12,000 ft (4000 m), and provide living space for communities of animals that are quite different from those inhabiting the land-fringing waters which overlie the continental shelves (neritic zone). *See* ECOLOGICAL COMMUNITIES.

The deep-sea fauna consists of pelagic animals (swimming and floating forms between the surface and deep-sea floor) and below these the benthos, or bottom dwellers, which live on or near the ocean bottom. Pelagic animals can be divided into the usually smaller forms that tend to drift with the currents (zooplankton) and the larger and more active nekton, such as squids, fishes, and cetaceans. Pelagic, deep-sea animals are frequently termed bathypelagic in contrast to the epipelagic organisms of the surface waters (see illustration). *See* ZOOPLANKTON.

All animal life in the sea, pelagic and benthic, depends on the growth of microscopic plants (phytoplankton). From the surface down to a maximum depth of about 300 ft (100 m) there is sufficient light for photosynthesis and vigorous phytoplanktonic growth. This layer is known as the photic zone. *See* PHYTOPLANKTON.

Bathypelagic fauna. The typical bathypelagic animals begin to appear below depths of about 600 ft (200 m). The bathypelagic fauna is most diverse in the tropical and temperate parts of the ocean. Numerous species are found in all three temperature zones, but many appear to have a more limited distribution. Each species also has a definite vertical occurrence. Findings suggest that there are three main vertical zones, each with a characteristic community. Here the term bathypelagic is used for the fauna between about 3000 and 6000 ft (1000 and 2000 m), that above (between 600 and 3000 ft or 200 and 1000 m) being called mesopelagic and that below 6000 ft (2000 m) abyssopelagic (see illustration). The typical forms of the mesopelagic fauna (stomiatoids and lantern fishes) live in the twilight zone of the deep sea (between the 68 and 50°F or 20 and 10°C isotherms), while the bathypelagic species (ceratioid angler fishes and *Vampyroteuthis*) occur in the dark, cooler parts below the 50°F (10°C) isotherm.

Perhaps the most conspicuous feature of pelagic deep-sea life is the widespread occurrence of luminescent species bearing definite light organs (photophores). Many of the squids and fishes have definite patterns of such lights, as do some of the larger crustaceans (hoplophorid and sergestid prawns and euphausiids). Investigations have

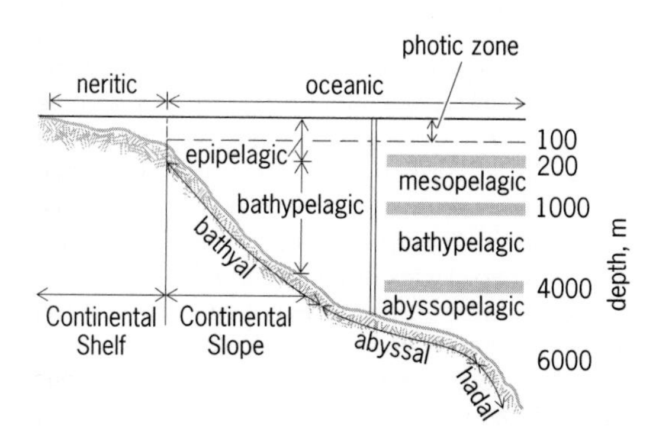

Classification of marine environments. Right side of diagram illustrates the proposal to divide the bathypelagic zone into mesopelagic, bathypelagic, and abyssopelagic zones. Division of benthic region into bathyal, abyssal, and hadal zones also is shown. 1 m = 3.3 ft.

shown that flashes from luminescent organisms could be detected down to depths of 12,300 ft (3750 m).

Benthic fauna. There are two main ecological groups of bottom-living animals in the ocean: organisms that attach to the bottom and those that freely move over the bottom. The benthic fauna is most diverse in the temperate and tropical ocean, although the arctic and antarctic areas have their characteristic species. As in the pelagic fauna, certain species occur in all three oceanic zones, while others appear to have a more restricted occurrence. While a number of species—particularly among the polychaete worms, gastropod mollusks, and the brittle stars (Ophiuroidea)—range from littoral to abyssal regions, most forms tend to live within smaller ranges of depth. Data suggest that there are typical communities of animals over the continental slopes (see illustration) extending down to about 9000 ft (3000 m; bathyal zone); others occur below this in the abyssal zone. *See* MARINE ECOLOGY. [N.B.M.]

Deep-sea trench A long, narrow, very deep, and asymmetrical depression of the sea floor, with relatively steep sides. Oceanic trenches characterize active margins at the ocean-basin–continent or ocean-basin–island-arc boundaries. They contain the greatest oceanic depths and are associated with the most active volcanism, largest negative gravity anomalies, most frequent shallow seismicity, and almost all of the intermediate and deep-focus earthquake activity. As the surface expression of the widely accepted process of subduction by which oceanic crustal material is returned to the upper mantle, they are key elements in current models of plate tectonic evolution on Earth and possibly on Venus. *See* VOLCANOLOGY.

Deep-sea trenches are the signature relief form of the Pacific; in a counterclockwise direction, they occur from southern Chile to just northeast of North Island, New Zealand. A secondary or outer branch trends southward from near Tokyo Bay in a festoon of arcs to south of Palau. The principal gaps in the circum-Pacific chain are from Baja California to south-eastern Alaska, and off the northern coast of New Guinea. From eastern New Guinea to southern Vanuatu the trenches lie southwest, or "inside," the island chains; otherwise their characteristics are like those facing the Pacific. The Indian Ocean contains only the very long contorted Sunda Trench that appears near the northwestern end of Sumatra and extends southeast and east past Timor, to curve north and west near Aru and end adjacent to Buru. In the Atlantic, the Puerto Rico–Antillean trench system extends outside the island arc from eastern Hispaniola around to Trinidad; but south of 14°N, off Barbados, the trench is filled with sediment. In the far South Atlantic a typical island-arc–trench complex extends from near South Georgia through the South Sandwich Archipelago.

A series of pioneering gravity observations with pendulum instruments on Dutch submarines during the 1920s and 1930s established that the East Indian trenches, and several others, were characterized by a belt of negative gravity anomalies of 150–200+ milligals, that is, values 150–200 parts per million less than normal, interpretable as deficiency of mass near and at their axes.

It was established that oceanic crust is thin, that crust under island arcs is thicker, and its layers display different sound transmission velocities, indicating different composition. Shipboard studies in the Middle America, Tonga, Cedros, Aleutian, Peru-Chile, and Sunda trenches established that the characteristic oceanic crustal layer [that is, 6.8–7.0 km/s (4.1–4.2 mi/s) compressional wave velocity] does not end or thin under the trench; rather, it may thicken slightly but does deepen steeply as it passes beneath the island arc or continental slope by the process of subduction. *See* EARTH CRUST; FAULT AND FAULT STRUCTURES; OCEANIC ISLANDS. [R.L.Fi.]

Defoliant and desiccant Defoliants are chemicals that cause leaves to drop from plants; defoliation facilitates harvesting. Desiccants are chemicals that kill leaves of plants; the leaves may either drop off or remain attached; in the harvesting process the leaves are usually shattered and blown away from the harvested material. Defoliants are desirable for use on cotton plants because dry leaves are difficult to remove from the cotton fibers. Desiccants are used on many seed crops to hasten harvest; the leaves are cleaned from the seed in harvesting. Defoliants and desiccants have also been used during war to destroy vegetation. [A.S.C.]

Delta A deposit of sediment at the mouth of a river or tidal inlet. It is also used for storm washovers of barrier islands and for sediment accumulations at the mouths of submarine canyons. *See* FLOODPLAIN.

The shape and internal structure of a delta depend on the nature and interaction of two forces: the sediment-carrying stream from a river, tidal inlet, or submarine canyon, and the current and wave action of the water body in which the delta is building. This interaction ranges from complete dominance of the sediment-carrying stream (still-water deltas) to complete dominance of currents and waves, resulting in redistribution of the sediment over a wide area (no deltas). This interaction has a large effect on the shape and structure of the delta body.

Most of the sediment carried into the basin is deposited when the inflowing stream decelerates. If there is little density contrast, this deceleration is sudden and most sediment is deposited near the mouth of the river. If the inflowing water is much lighter than the basin water, for example, fresh water flowing into a colder sea, the outflow spreads at the surface over a large distance away from the outlet. If the inflow is very dense, for instance, cold muddy water in a warm lake, it may form a density flow on or near the bottom, and the principal deposition may occur at great distance from the outlet.

Three principal components make up the bodies of most deltas in varying proportions: topset, foreset, and bottomset beds (see illustration). As defined for most deltas, the topset beds comprise the sediments formed on the subaerial delta: channel deposits, natural levees, floodplains, marshes, and swamp and bay sediments. The foreset beds are those formed in shallow water, mostly as a broad platform fronting the delta shore, and the bottomset beds are the deep-water deposits beyond the deltaic bulge. In marine deltas the fluviatile influence decreases and the marine influence increases from the topset to the bottomset beds.

In a different way, deltas can be viewed as being composed of three structural elements: (1) a framework of elongate coarse bodies (channels, river-mouth bars, levee

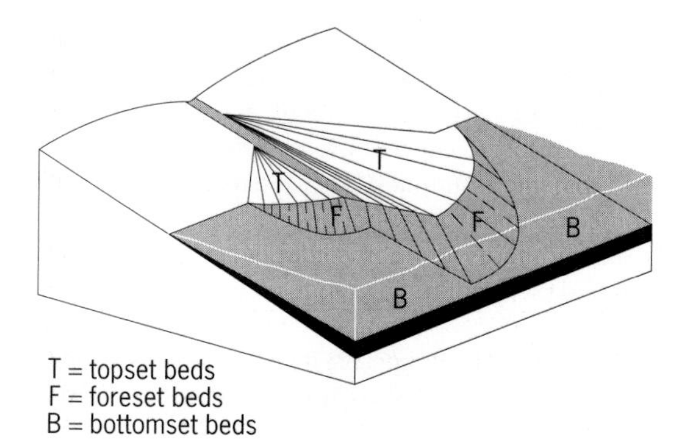

Schematic diagram showing two stages of growth and a Gilbert-type delta. (*After P. H. Kuenen, Marine Geology, John Wiley, 1950*).

deposits), which radiate from the apex to the distributary mouths (sand fingers); (2) a matrix of fine-grained floodplain, marsh, and bay sediments; and (3) a littoral zone, usually of beach and dune sands which result from sorting and longshore transport of river-mouth deposits by waves, currents, tides, and wind. The relative proportions of these components vary widely. The Mississippi delta consists almost entirely of framework and matrix; its rapid seaward growth is the result of deposition of river-mouth bars and extension of levees, and the areas in between are filled later with matrix. This gives the delta its characteristic bird-foot outline. A different makeup is presented by the Rhone delta, where the supply of coarse material at the distributary mouths is slow, and dispersal by wave action and longshore drift fairly efficient, so that nearly all material is evenly redistributed as a series of coastal bars and dunes across a large part of the delta front. This delta advances as a broad lobate front, while the present Mississippi delta grows at several localized and sharply defined points.

Despite difficult engineering problems, many cities, such as Calcutta, Shanghai, Venice, Alexandria (Egypt), and New Orleans, were constructed on deltas. These problems include shifting and extending shipping channels; lack of firm footing for construction except on levees; steady subsidence; poor drainage; and extensive flood danger. Moreover, in certain deltas the tendency of the main flow to shift away to entirely different areas, with resulting disappearance of the main channels for water traffic, is a constant problem that is difficult and costly to counter. *See* ESTUARINE OCEANOGRAPHY.

[T.H.V.A.]

Dendrology The division of forestry concerned with taxonomy of trees and other woody plants. Dendrology, called forest botany in some countries, usually is limited to taxonomy of trees but may also include shrubs and woody vines. This basic subject in the training of foresters teaches how trees are named (nomenclature), described (morphology), and grouped (classification); how to find the name of an unknown tree and recognize important forest species (identification); and where trees occur both by geographic ranges of species and by forest types (distribution). Forest stands of similar composition, appearance, and structure are grouped together into areas characterized by major forest types or formation, and are named from the predominant or characteristic species. *See* FOREST AND FORESTRY; PLANT TAXONOMY. [E.L.L.]

Dependovirus Any of a group (genus) of defective viruses which seem unable to reproduce without help from adenoviruses. They were formerly known as adeno-associated viruses and belong to the family Parvoviridae. These 20-nanometer virus particles were found during electron microscopic studies (see illustration) in several preparations of adenovirus from both human and monkey sources, and have since been observed in many adenovirus stocks. *See* ADENOVIRIDAE.

Like adenovirus, the dependovirus has a deoxyribonucleic acid (DNA) core. However, the dependovirus, although dependent upon adenovirus for its growth, does not appear to be structurally related to its "helper." Its genetic content, as well as its size, is much smaller than that of adenovirus, and its protein coat is completely different from that of adenovirus. These particles contain single-stranded DNA and a protein coat with icosahedral symmetry. The single-stranded DNA has been shown to be present within the dependovirus virion as either plus or minus complementary strands in separate particles. Upon extraction, the minus and plus strands unite to form a double-stranded helix.

The genetic material of dependovirus not only can persist in cultured cells for long periods without giving evidence of its presence but also can survive within human beings in a latent state, becoming detectable only in specimens taken when the person is

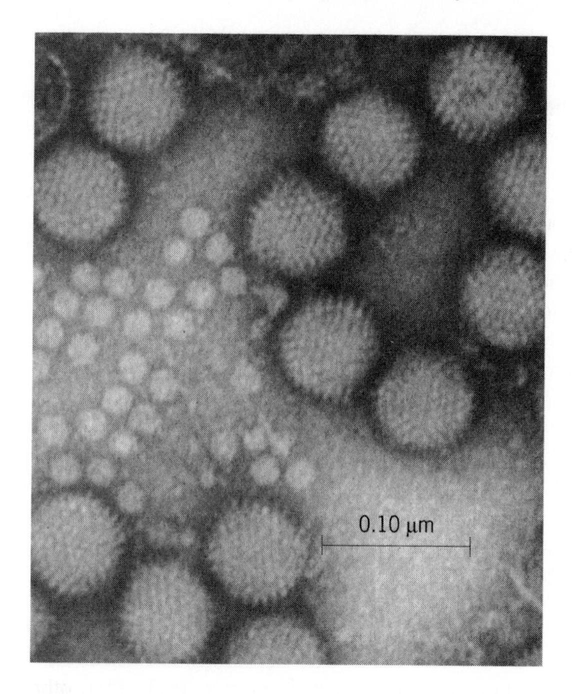

Electron micrograph showing adenovirus particles (70-nm diameter; 0.10 μm = 100 nm) and dependoviruses (20-nm diameter).

concurrently infected with adenovirus. This characteristic permits survival of a defective virus in nature, even though it cannot regularly replicate or be passed in infective form from host to host. [J.L.Me.; M.E.Re.]

Depositional systems and environments Depositional systems are descriptions of the interrelationships of form and the physical, chemical, or biological processes involved in the development of stratigraphic sequences. Depositional environments are the locations where accumulations of sediment have been deposited by either mechanical or chemical processes.

Depositional systems. Traditional stratigraphic analysis, which emphasized the physically descriptive aspects of strata, has changed; a critical genetic dimension has been added. Where once, for example, a formation may have been described in physical terms as a fine- to medium-grained quartzose sandstone, overlying a thick dark-gray shale sequence and underlying a coal-bearing sequence with discontinuous sands, the same sandstone may now be recognized as a delta-front sandstone, the underlying shale as a prodelta facies, and the overlying coal-bearing formation as the product of deposition on a delta plain. In this interpretation, the three distinct stratigraphic units become part of a genetically related sequence, each a component facies of a prograding delta system.

In a modern setting, a particular system of deposition is directly observable and known, whether it is a major delta, an alluvial fan, a meandering river, a barrier bar, a carbonate platform, or the like. Through specific observation, description, and delineation, it may be determined that a variety of depositional processes are active. At the terminus of rivers, sands may be deposited as bars in river mouths, creating the delta front; muds may be carried by suspension into the oceanic waters and deposited through flocculation, creating the prodelta; on the delta plain, a distributary channel may be depositing bed-load sands and, during flooding, may be carrying suspended muds to flanking flood basins or breaching the natural levee to form crevasse splays. Each process or combination of processes gives rise to distinct, specific environments of deposition, with each resulting in a deposit which can be characterized by such features as lithologic composition, texture, sedimentary structures, geometry, size, and relationship to other deposits. *See* DELTA; FLOODPLAIN; RIVER.

Distinct physical, or in some cases biologic or chemical, products of deposition can be related directly to definable, operative processes. Such data permit the development of models of modern deposition in which processes and resulting deposits or facies are linked. By recognizing comparable physical, chemical, and biologic attributes of ancient strata, modern depositional analogs can be applied, and the original processes forming the ancient deposit can be inferred. Such an ancient deposit is called a genetic facies. It contains the sedimentary record and constitutes a three-dimensional stratigraphic (ancient) depositional system. *See* FACIES (GEOLOGY).

An ancient depositional system is a three-dimensional, genetically defined, physical stratigraphic unit that consists of a contiguous set of process-related sedimentary facies. Several corollaries have evolved from the application of this concept. Depositional systems, such as delta, fluvial, and shelf systems, (1) are the stratigraphic equivalents of major physical geographic units; (2) form the principal building blocks of the sedimentary basin fill; and (3) can be applied where principal boundaries of the systems are preserved and where the geometry of the framework facies can be mapped.

The major realms of deposition may be classed broadly as terrigenous clastic depositional systems and biogenic-chemical depositional systems. Each of these major systems is subdivided according to particular systems of deposition, and within each of the subdivisions is an assemblage of genetic facies, which are the fundamental units of depositional systems.

Terrigenous clastic systems, composed chiefly of sands and shales, embrace eight major systems: (1) fluvial or river systems; (2) delta systems; (3) strike coastal systems; (4) fan or clastic wedge systems; (5) lacustrine systems; (6) continental eolian systems; (7) shelf systems; and (8) slope and abyssal systems.

The biogenic-chemical systems consist of three major systems: (1) carbonate systems; (2) glauconitic and authigenic shelf systems; and (3) evaporite systems. [W.L.Fi.]

Depositional environments. Depositional environments may be distinguished from erosional environments, in which erosion of the Earth's surface is taking place. Both depositional and erosional environments are of interest to geomorphologists. However, most attention to depositional environments has come from sedimentologists, particularly in order to understand the origin of sedimentary rocks. *See* EROSION.

Sediment is derived mainly from source areas that are actively undergoing uplift and erosion, and is deposited mainly in areas that are undergoing subsidence. Location of the source and basin of deposition is mainly controlled by large-scale geophysical processes acting within the Earth's mantle, so a major factor affecting the nature and distribution of sedimentary environments is the overall structural development, or tectonics, of the area.

Tectonics determines the major geological structure or setting of an environment of deposition, including the location and nature of the main areas undergoing uplift or subsidence. Areas with high relief, such as mountains and volcanoes, suffer rapid erosion and supply much more sediment to basins of deposition than larger areas of low relief. One investigation, for example, found that 82% of the suspended solids (mud) discharged by the Amazon River were supplied by the 12% of the drainage basin located within the Andes Mountains.

A second important major control is climate. This includes the average temperature, the range of temperature variation, the aridity or humidity (ratio of evaporation to precipitation), and the magnitude and frequency of floods and storms. Climate in turn has an important influence on such physical factors as the salinity and energy of the environment (wind and water speeds and degree of turbulence, for example), as well as on the abundance and types of plants and animals.

In areas of subsidence and sedimentation, topography results from and controls sedimentary environments. Along a coastline of low relief, for example, spits and barrier islands are produced by waves generated in the open sea. Shallow lagoons on the landward side of barrier islands are protected from wave action by the islands themselves. The distinctive features of the lagoon environment are a result of a topography which has been produced by the accumulation of sediment in another sedimentary environment (the barrier island). *See* BARRIER ISLANDS.

Sedimentary environments can be classified into three categories: terrestrial, including alluvial fans, fluvial plains, sandy deserts, lakes, and glacial regions; mixed (shore-related), including deltas, estuaries, barrier island complexes, and glacial marine environments; and marine, including terrigenous shelves or shallow seas, carbonate shelves or platforms, continental slopes, continental rises, basin plains, ocean ridges, and ocean trenches.

Although the importance of tectonics and climate in controlling sedimentary environments is widely recognized, most classifications are based mainly on topography. Almost all distinguish terrestrial (sub-aerial or fresh-water) from marine environments, and also recognize an important group of mixed or shore-related environments.

A number of major processes operate within environments and determine the types of sediment deposited in the environment, including water depth, energy (waves and current), temperature, and salinity. Biological factors also exert a very strong influence. There would be little or no oxygen in the atmosphere if it were not for the photosynthetic activity of plants. Deposition of calcium carbonate and silica in lakes and the oceans takes place largely through the action of plants and animals, and organic matter deposited along with mineral particles is largely responsible for the development of reducing conditions within sediments after deposition. Vegetable material accumulates in swamps to form peat and coal, and fine organic detritus settles with marine muds and is the ultimate source of oil and gas. Both terrestrial and aquatic plants exert a trapping and binding action that tends to immobilize sedimentary particles, as, for example, when coastal sand dunes or tidal flats become stabilized by the growth of salt-tolerant grasses. Terrestrial vegetation plays an important role in rock weathering. [G.V.M]

Desert No precise definition of a desert exists. From an ecological viewpoint the scarcity of rainfall is all important, as it directly affects plant productivity which in turn affects the abundance, diversity, and activity of animals. It has become customary to describe deserts as extremely arid where the mean precipitation is less than 2.5–4 in. (60–100 mm), arid where it is 2.5–4 to 6–10 in. (60–100 to 150–250 mm), and semiarid where it is 6–10 to 10–20 in. (150–250 to 250–500 mm). However, mean figures tend to distort the true state of affairs because precipitation in deserts is unreliable and variable. In some areas, such as the Atacama in Chile and the Arabian Desert, there may be no rainfall for several years. It is the biological effectiveness of rainfall that matters and this may vary with wind and temperature, which affect evaporation rates. The vegetation cover also alters the evaporation rate and increases the effectiveness of rainfall. Rainfall, then, is the chief limiting factor to biological processes, but intense solar radiation, high temperatures, and a paucity of nutrients (especially of nitrogen) may also limit plant productivity, and hence animal abundance. Of the main desert regions of the world, most lie within the tropics and hence are hot as well as arid.

The Namib and Atacama coastal deserts are kept coot by the Benguela and Humboldt ocean currents, and many desert areas of central Asia are cool because of high latitude and altitude.

The diversity of species of animals in a desert is generally correlated with the diversity of plant species, which to a considerable degree is correlated with the predictability and amount of rainfall. There is a rather weak latitudinal gradient of diversity with relatively more species nearer the Equator than at higher latitudes. This gradient is much more conspicuous in wetter ecosystems, such as forests, and in deserts appears to be overridden by the manifold effects of rainfall. Animals, too, may affect plant diversity: the burrowing activities of rodents create niches for plants which could not otherwise survive, and mound-building termites tend to concentrate decomposition and hence nutrients, which provide opportunities for plants to colonize.

Each desert has its own community of species, and these communities are repeated in different parts of the world. Very often the organisms that occupy similar niches in different deserts belong to unrelated taxa. The overall structural similarity between American cactus species and African euphorbias is an example of convergent evolution, in which separate and unrelated groups have evolved almost identical adaptations under similar environmental conditions in widely separated parts of the world. Convergent structural modification occurs in many organisms in all environments, but is especially noticeable in deserts where possibly the small number of ecological niches has necessitated greater specialization and restriction of way of life. The face and especially the large ears of desert foxes of the Sahara and of North America are remarkably similar, and there is an extraordinary resemblance between North American sidewinding rattlesnakes and Namib sidewinding adders. *See* ECOLOGY; PHYSIOLOGICAL ECOLOGY (PLANT); PRECIPITATION (METEOROLOGY). [D.F.Ow.]

Desert erosion features A distinctive topography carved by erosion in regions of low rainfall and high evaporation where vegetation is scanty or absent. Although rainfall is low, it is the most important climatic factor in the formation of desert erosion features. Desert rains commonly occur as torrential downpours of short duration with a consequent high percentage of runoff. As a result of the dryness, wind and mechanical weathering also play an important part in desert erosion. *See* SEDIMENTOLOGY; WEATHERING PROCESSES.

When storms of the so-called cloudburst type occur in the desert, sudden rushes of water, or flash floods, sweep down the normally dry washes or the narrow canyons in the mountains bordering the basins. The comparatively large volume of water combined with a high velocity due to the steepness of the slopes give the short-lived streams power to carry large amounts of fine and coarse rock fragments. As a result, the streams have great erosive power.

When intermittent streams leave the canyons and spread out at the foot of a desert mountain, they lose velocity and quickly drop the coarsest of the transported material to build an alluvial fan. Some of the water sinks into the fan, and some evaporates, but whatever remains may follow one of the channels on the fan or spread out in the form of a sheetflood, in either case carrying coarse sand, silt, and clay, and perhaps rolling some larger rock fragments along.

When the water reaches the toe of the fan, it spreads still more, dropping all but the finest silt and clay. Any excess water follows shallow washes to the lowest part of the basin, where it may form a playa lake. This evaporates in a few hours or a few days, depositing the silt and clay, mixed perhaps with soluble salts. The flat-surfaced area resulting from the silt and clay deposition is a playa. *See* PLAYA.

The lack of moisture during most of the year and the scanty vegetation make the wind a more potent agent of erosion in deserts than in humid lands. The finest material is blown high in the air and may be carried entirely out of the area, a process known as deflation. The larger sand grains are rolled along the surface, bouncing into the air when they strike an obstacle, knocking more grains into the air as they hit the ground again, until eventually a sheet of sand is moving along in the 3 or 4 ft (1 or 1.3 m) above the surface. This moving sand abrades rocks and other objects with which it comes in contact; at the same time the grains themselves become rounded and frosted. If movement is impeded by vegetation or other obstacles, sand accumulates to form dunes. *See* DUNE.

Desert landscapes evolve in three stages. In the early, or youthful, stage, alluvial fans are built, washes develop, playas form, and the basins slowly fill with detritus. As this stage progresses, some alluvial fans coalesce to form bajadas or piedmont alluvial plains along the mountain fronts, and individual basins may become deeply filled with waste to form bolsons. Desert flats develop between alluvial fans (or bajadas) and playas, and isolated dunes accumulate on the lee sides of the latter. If the original highlands are flat-topped rather than tilted mountain blocks, mesas develop. As the mountain fronts slowly retreat under the attack of the atmosphere and running water, small bare rock surfaces or pediments form at the canyon mouths, the result of lateral cutting by the intermittent streams. The general tendency during youth is for relief to decrease.

The middle, or mature, stage is initiated by the development of exterior drainage or the capture of higher basins by lower ones as drainage channels erode headward through low divides. The fill deposited during youth undergoes erosion, and pediments become more widely developed. The mountains are worn still lower, and more and more channels extend completely through them, cut by the streams engaged in draining and dissecting the higher basins. Playa deposits or other easily eroded sediments are cut into badlands before being entirely removed, and mesas are reduced to buttes. Undissected remnants of older deposits become covered with desert pavement (flat-lying, interlocking, angular stones left after finer particles are removed by deflation). Where winds are turbulent and large supplies of sand are available, complex dune areas develop. Relief shows some net increase during maturity.

At the late, or old-age, stage of desert evolution, the original mountains are so reduced in elevation that the winds sweep over them with little or no condensation of moisture, and rains become still more infrequent. Great expanses of wind-scoured bare rock, or hammada, are exposed, with here and there a more resistant remnant standing above the general level as an inselberg. Buttes are reduced to smaller bornhardts and finally disappear.Those parts of the flat surface floored by earlier deposits are covered and protected by extensive areas of desert pavement. The rock fragments may be colored brown to black by desert varnish, a coating of manganese and iron oxides. Sand blown from the bare rock surfaces and from the sediments may form large dune areas. If there are no obstacles to obstruct movement or cause wind turbulence, the sand may move as a sheet, forming large expanses of flat or gently undulating sand surfaces. Relief slowly decreases in old age. *See* DESERT. [T.C.]

Desertification Land degradation in low-rainfall and seasonally dry areas of the Earth. It can be viewed as both a process and the resulting condition. Desertification involves the impoverishment of vegetation and soil resources. Key characteristics include the degradation of natural vegetation cover and undesirable changes in the composition of forage species, deterioration in soil quality, decreasing water availability, and increased soil erosion from wind and water. Various stages of desertification can be seen in most of the world's drylands. In rare cases, desertification leads to abandoned, desertlike landscapes.

It is generally agreed that human activities, particularly excessive resource use and abusive land-use practices, are the primary cause of desertification. Specific activities leading to desertification include clearing and cultivation of low-rainfall areas where such cultivation is not sustainable, overgrazing of rangelands, clearing of woody plant species for fuelwood and building materials, and mismanagement of irrigated cropland leading to the buildup of mineral salts in the soil (salinization). Drought is often cited as a basic cause of desertification; however, it merely accelerates or accentuates land degradation processes already under way. *See* DROUGHT.

Consequences of desertification include reduced biological productivity, reduction of biodiversity, a gradual loss of agricultural potential and resource value, loss of food security, reduced carrying capacity for humans and livestock, increased risks from drought and flooding, and in extreme cases, barren lands that are effectively beyond restoration. Paleostudies, supported by model simulations, have shown that the intensity of Northern Hemisphere desert conditions has waxed and waned over the past 9000 years in response to the precession of the Earth's orbit about the Sun. Thus, it may be that the causal factors of desertification, whether climate change or human activities, depend on the time scale being addressed. *See* CLIMATE MODIFICATION; DESERT. [W.Swe.]

Dew The deposit of liquid water resulting from condensation of atmospheric water vapor to exposed surfaces that cool during the night. Dewfall is noticeable in the early morning after a calm, cool, clear night, usually as beads of liquid water on the outside and upward-facing surfaces of trees, buildings, and so forth. If the ground is moist, some of the condensed water can be evaporated surface moisture. Dew forms when the surface temperature drops sufficiently to saturate air in contact with the surface (that is, when the surface drops to below atmospheric dew-point temperature); when the surface cools to below freezing temperature, frost occurs. *See* DEW POINT; FROST.

Hygroscopic particles on surfaces can act as sites for condensation at temperatures higher than the atmospheric dew-point temperature. Thus, if a surface is not clean, the first deposit of moisture from the air can occur well before the surface cools to dew-point temperature. For some chemicals, such as common salt, condensation can start to occur when the local relative humidity reaches 80%; the humidity must be 100% in the case of a clean surface. Some desert plants exude hygroscopic salts from the interior of leaves which provide preferred sites for condensation and thereby create a supply of water for the plant. *See* HUMIDITY. [B.Hi.]

Dew point The temperature at which air becomes saturated when cooled without addition of moisture or change of pressure. Any further cooling causes condensation; fog and dew are formed in this way.

Frost point is the corresponding temperature of saturation with respect to ice. At temperatures below freezing, both frost point and dew point may be defined because water is often liquid (especially in clouds) at temperatures well below freezing; at freezing (more exactly, at the triple point, $+.01^\circ$C) they are the same, but below freezing the frost point is higher. For example, if the dew point is -9°C, the frost point is -8°C. Both dew point and frost point are single-valued functions of vapor pressure. *See* DEW; FOG; HUMIDITY. [J.R.F.]

Diarrhea The passage of loose or watery stools, usually at more frequent than normal intervals. Diarrhea is a symptom of many diseases and may be accompanied by nausea, vomiting, griping, tenesmus, and other general or specific indications of a disease.

The more common specific disorders which may produce diarrhea include intestinal infections, such as dysentery, cholera, typhoid fever, food poisonings, and parasitic infestations; food sensitivities; drug and chemical irritation; and vitamin deficiency states.

Emotional and psychic disturbances frequently produce diarrhea and other visceral derangements. The poorly understood entities of regional enteritis and ulcerative colitis are perhaps related to these disturbances, as are other psychosomatic disorders.

Diarrhea is a common symptom in gastrointestinal obstruction or in inflammations from local infections or tumor invasion. *See* BACILLARY DYSENTERY; FOOD POISONING; MEDICAL PARASITOLOGY. [E.G.St./N.K.M.]

Diphtheria An acute infectious disease of humans caused by *Corynebacterium diphtheriae*. Classically, the disease is characterized by low-grade fever, sore throat, and a pseudomembrane covering the tonsils and pharynx. Complications such as inflammation of the heart, paralysis, and even death may occur due to exotoxins elaborated by toxigenic strains of the bacteria. The upper respiratory tract is the most common portal of entry for *C. diphtheriae*. It can also invade the skin and, more rarely, the genitalia, eye, or middle ear. The disease has an insidious onset after a usual incubation period of 2–5 days.

The only specific therapy is diphtheria antitoxin, administered in doses proportional to the severity of the disease. Antitoxin is produced by hyperimmunizing horses with diphtheria toxoid and toxin. It is effective only if administered prior to the binding of circulating toxin to target cells. Antibiotics do not alter the course, the incidence of complications, or the outcome of diphtheria, but are used to eliminate the organism from the patient.

Persons with protective antitoxin titers may become infected with diphtheria but do not develop severe disease. Since the 1920s, active immunization with diphtheria toxoid has proved safe and effective in preventing diphtheria in many countries. Diphtheria toxoid is produced by incubating the toxin with formalin. Active immunization requires a primary series of four doses, usually at 2, 4, 6, and 18 months of age, followed by a booster at school entry. *See* MEDICAL BACTERIOLOGY; TOXIN. [R.Ch.; C.V.]

Diplogasterida An order of nematodes in which the labia are seldom well developed; however, a hexaradiate symmetry is distinct. The external circle of labial sensilla may appear setose, but they are always short, never long or hair-like. The stoma may be slender and elongate, or spacious, or any gradation between. The stoma may be armed or unarmed; the armature may be movable teeth, fossores, or a pseudostylet. The corpus is always muscled and distinct from the postcorpus, which is divisible into an isthmus and glandular posterior bulb. The metacorpus is almost always valved. The female reproductive system may have one or two ovaries, and males may or may not have caudal alae; however, a gubernaculum is always present. The male tail commonly has nine pairs of caudal papillae; three are preanal and six are caudal.

There are two superfamilies: The members of Diplogasteroidea are predators, bacterial feeders, and omnivores; they are often found in association with insects or the fecal matter of herbivores. Cylindrocorporoidea include both free-living forms and intestinal parasites of amphibians, reptiles, and certain mammals. *See* NEMATA. [A.R.M.]

Disease A deleterious set of responses which occurs at the subcellular level, stimulated by some injury, and which is often manifested in altered structure or functioning of the affected organism. With advances in understanding and the development of sensitive probes, it has become clear that the fundamental causes of diseases are based on biochemical and biophysical responses within the cell. These responses are now being categorized and, slowly, the mechanisms are being understood.

The term homeostasis refers to functional equilibrium in an organism and to the processes that maintain it. There is a range of responses that is considered normal. If cells are pushed to respond beyond these limits, there may be an increase, a decrease, or a loss of normal structure or function. These changes may be reversible or irreversible. If irreversible, the cells may die. Thus, subcellular changes may be reflected in altered tissues, organs, and consequently organisms, and result in a condition described as diseased.

Lesions are the chemical and structural manifestations of disease. Subjective manifestations of a disease process such as weakness, pain, and fatigue are called symptoms. The objective measurable manifestations such as temperature, blood pressure, and respiratory rate changes are called signs or physical findings. Changes in the chemical or cellular makeup of an organ, tissue, or fluid of the body or its excretory products are called laboratory findings. To make a diagnosis is to determine the nature of the pathologic process by synthesizing information from these sources evaluated in the light of the patient's history and compared with known patterns of signs and symptoms. In common usage, the term disease indicates a constellation of specific signs and symptoms attributable to altered reactions in the individual which are produced by agents that affect the body or its parts.

Etiology is the study of the cause or causes of a disease process. Although a disease may have one principal etiologic agent, it is becoming increasingly apparent that there are several factors involved in the initiation of a disease process. Susceptibility of the individual is an ever present variable. The etiologic factors can conveniently be divided into two categories (see table). One group consists of endogenous (internal; within the body) factors, and may originate from errors in the genetic material. The other category of etiologic factors is exogenous (environmental). These account for the majority of disease reactions. Exogenous factors include physical, chemical, and biotic agents.

Pathogenesis refers to the mechanisms by which the cell, and consequently the body, responds to an etiologic agent. It involves biochemical and physiological responses which are reflected in ultrastructural, microscopic, or gross anatomic lesions. There are a limited number of ways in which cells respond to injury. The nature of the response is modified by the nature of the agent, dose, portal of entry, and duration of exposure, as well as many host factors such as age, sex, nutritional state, and species and individual susceptibility.

The diseases which are important in causing human death have changed in the last 80 years. In 1900 six of the ten leading causes of death in the United States were infectious (biotic) agents. At present, only one of the ten leading causes of death in the United States, influenza and pneumonia, is due to biotic agents. While most of the biotic causes of diseases were being brought under control, continued population growth (in large part, a consequence of the control of infectious disease) and the remarkable growth of industrialization have been associated with an increased prevalence of diseases caused by physical and chemical agents. These include cancer, cirrhosis, and cardiovascular disease.

General principles of the organism's response to toxic substances, some of which occur naturally in the environment, have evolved from a great number of investigations of agent-host interaction. They are: (1) All substances entering the organism are toxic; none is harmless. Dose rate of exposure and route of entry into the body determine whether a toxic response will occur or not. (2) All agents evoke multiple responses.

Common exogenous and endogenous causes of disease

Causative agent	Disease
EXOGENOUS FACTOR	
Physical	
Mechanical injury	Abrasion, laceration, fracture
Nonionizing energy	Thermal bums, electric shock, frostbite, sunburn
Ionizing radiation	Radiation syndrome
Chemical	
Metallic poisons	Intoxication from methanol, ethanol, glycol
Nonmetallic inorganic poisons	Intoxication, from phosphorous, borate, nitrogen dioxide
Alcohols	Intoxication from methanol, ethanol, glycol
Asphyxiants	Intoxication from carbon monoxide, cyanide
Corrosives	Burns from acids, alkalies, phenols
Pesticides	Poisoning
Medicinals	Barbiturism, salicylism
Warfare agents	Burns from phosgene, mustard gas
Hydrocarbons (some)	Cancer
Nutritional deficiency	
Metals (iron, copper, zinc)	Some anemias
Nonmetals (iodine, fluorine)	Goiter, dental caries
Protein	Kwashiorkor
Vitamins:	
A	Epithelial metaplasia
D	Rickets, osteomalacia
K	Hemorrhage
Thiamine	Beriberi
Niacin	Pellagra
Folic acid	Macrocytic anemia
B_{12}	Pemicious anemia
Ascorbic acid	Scurvy
Biological	
Plants (mushroom, fava beans, marijuana, poison ivy, tobacco, opium)	Contact dermatitis, systemic toxins, cancer, hemorrhage
Bacteria	Abscess, scarlet fever, pneumonia, meningitis, typhoid, gonorrhea, food poisoning, cholera, whooping cough, undulant fever, plague, tuberculosis, leprosy, diphtheria, gas gangrene, botulism, anthrax
Spirochetes	Syphilis, yaws, relapsing fever, rat bite fever
Virus	Warts, measles, German measles, smallpox, chickenpox, herpes, roseola, influenza, psittacosis, mumps, viral hepatitis, poliomyelitis, rabies, encephalitis, trachoma
Rickettsia	Rocky Mountain spotted fever, typhus
Fungus	Ringworm, thrush, actinomycosis, histoplasmosis, coccidiomycosis
Parasites (animal)	
Protozoa	Amebic dysentery, malaria, toxoplasmosis, trichomonas vaginitis
Helminths (worms)	Hookworm, trichinosis, tapeworm, filariasis, ascariasis
ENDOGENOUS FACTOR	
Hereditary	Phenylketonuria, alcaptonuria, glycogen storage disease, Down syndrome (trisomy 21), Turner's syndrome, Klinefelter's syndrome, diabetes, familial polyposis
Hypersensitivity	Asthma, serum sickness, eczema drug idiosyncrasy

(3) Most of the biological responses are undesirable, leading to the development of pathological changes. (4) A given dose of an agent does not produce the same degree of response in all individuals. Thus, when disease is viewed as interaction between the environment and the individual, the control of disease is largely the management of the environmental causes of disease. [N.K.M.; C.Qu]

Disease ecology The interaction of the behavior and ecology of hosts with the biology of pathogens, as it relates to the impact of diseases on populations.

Threshold theorem. For a disease to spread, on average it must be successfully transmitted to a new host before its current host dies or recovers. This observation lies at the core of the most important idea in epidemiology: the threshold theorem. The threshold theorem states that if the density of susceptible hosts is below some critical value, then on average the transmission of a disease will not occur rapidly enough to cause the number of infected individuals to increase. In other words, the reproductive rate of a disease must be greater than 1 for there to be an epidemic, with the reproductive rate being defined as the average number of new infections created per infected individual. Human immunization programs are based on applying the threshold theorem of epidemiology to public health; specifically, if enough individuals in a population can be vaccinated, then the density of susceptible individuals will be sufficiently lowered that epidemics are prevented. *See* EPIDEMIOLOGY.

In general, the rate of reproduction for diseases is proportional to their transmissibility and to the length of time that an individual is infectious. For this reason, extremely deadly diseases that kill their hosts too rapidly may require extremely high densities of hosts before they can spread. All diseases do not behave as simply as hypothesized by the threshold theorem, the most notable exceptions being sexually transmitted diseases. Because organisms actively seek reproduction, the rate at which a sexually transmitted disease is passed among hosts is generally much less dependent on host density.

Population effects. Cycles in many animal populations are thought to be driven by diseases. For example, the fluctuations of larch bud moths in Europe are hypothesized to be driven by a virus that infects and kills the caterpillars of this species. Cycles of red grouse in northern England are also thought to be driven by disease, in this case by parasitic nematodes. It is only when grouse are laden with heavy worm burdens that effects are seen, and those effects take the form of reduced breeding success or higher mortality during the winter. This example highlights a common feature of diseases: their effects may be obvious only when their hosts are assaulted by other stresses as well (such as harsh winters and starvation).

The introduction of novel diseases to wild populations has created massive disruptions of natural ecosystems. For example, the introduction of rinderpest virus into African buffalo and wildebeest populations decimated them in the Serengeti. African wild ungulates have recovered in recent years only because a massive vaccination program eliminated rinderpest from the primary reservoir for the disease, domestic cattle. But the consequences of the rinderpest epidemic among wild ungulates extended well beyond the ungulate populations. For example, human sleeping sickness increased following the rinderpest epidemic because the tsetse flies that transmit sleeping sickness suffered a shortage of game animals (the normal hosts for tsetse flies) and increasingly switched to humans to obtain meals.

It is widely appreciated that crop plants are attacked by a tremendous diversity of diseases, some of which may ruin an entire year's production. Diseases are equally prevalent among wild populations of plants, but their toll seems to be reduced because natural plant populations are so genetically variable that it is unlikely that any given pathogen strain can sweep through and kill all of the plants—there are always some resistant genotypes. But when agronomists have bred plants for uniformity, they have often depleted genetic diversity and created a situation in which a plant pathogen that evolves to attack the crop encounters plants with no resistance (all the plants are the same). For example, when leaf blight devastated the United States corn crop, 70% of that crop shared genetically identical cytoplasm, and the genetic uniformity of the host exacerbated the severity of the epidemic. *See* PLANT PATHOLOGY.

Disease emergence. Humans are dramatically altering habitats and ecosystems. Sometimes these changes can influence disease interactions in surprising ways. Lyme disease in the eastern United States provides a good example of the interplay of human habitat modifications and diseases. Lyme disease involves a spirochete bacterium transmitted to humans by ticks. However, humans are not the normal hosts for this disease; instead, both the ticks and the bacterium are maintained primarily on deer and mouse populations. Human activities influence both deer and mice populations, and in turn tick populations, affecting potential exposure of humans to the disease. Much less certain are the impacts of anticipated global warming on diseases. There is some cause for concern about the expansion of tropical diseases into what are now temperate regions in those cases where temperature sets limits to the activity or distribution of major disease vectors. *See* LYME DISEASE; POPULATION ECOLOGY. [P.Ka.]

Diving Skin diving, scuba diving, saturation diving, and "hard hat" diving are techniques used by scientists to investigate the underwater environment. Skin diving is usually without breathing apparatus and is done with fins and faceplate. The diver's underwater observation is limited to the time that breath can be held (1–2 min). Diving with scuba (self-contained underwater breathing apparatus) and "hard hat" provide the diver with a breathable gas, thus expanding the submerged time and the depth range of underwater observations. This type of diving is limited by human physiology and the diver's reaction to the pressure and nature of the breathing gas. Saturation diving permits almost unlimited time down to depths of 100 ft (30 m).

Scuba diving. Scuba is used by trained personnel as a tool for direct observation in marine research and underwater engineering. This equipment is designed to deliver through a demand-type regulator a breathable gas mixture at the same pressure as that exerted on the diver by the overlying water column. The gas which is breathed is carried in high-pressure cylinders (at starting pressures of 2000–3000 psi or 14–20 megapascals) worn on the back.

Scuba can be divided into three types: closed-circuit, semiclosed-circuit, and open-circuit. In the first two, which use pure oxygen or various combinations of oxygen, helium, and nitrogen, exhaled gas is retained and passed through a canister containing a carbon dioxide absorbent for purification and then recirculated to a bag worn by the diver. During inhalation additional gas is supplied to the bag by various automatic devices from the high-pressure cylinders. These two types of equipment are much more efficient than the open-circuit system, in which the exhaled gas is discharged directly into the water after breathing. Most open-circuit systems use compressed air because it is relatively inexpensive and easy to obtain. Although open-circuit scuba is not as efficient as the other types, it is preferred because of its safety, the ease in learning its use, and its relatively low cost.

For physiological reasons scuba diving is limited to about 165 ft (50 m) of water depth. Below this depth when using compressed air as a breathing gas, the diver is limited, not by equipment, but by the complex temporary changes which take place in the body chemistry while breathing gas (air) under high pressure.

Saturation diving. This type of diving permits long periods of submergence (1–2 weeks). It allows the diver to take advantage of the fact that at a given depth the body will become fully saturated with the breathing gas and then, no matter how long the submergence period, the decompression time needed to return to the surface will not be increased. Using this method, the diver can live on the bottom and make detailed measurements and observations, and work with no ill effects. This type of diving requires longer periods of decompression in specially designed chambers to free the diver's body of the high concentration of breathing gas. Decompression times of days or weeks (the time increases with depth) are common on deep dives of over 200 ft (60 m). [R.F.Di.]

Physiology. Environmental effects on the submerged diver are quite different from those experienced at sea level. Two elements are very evident during the dive. As the depth of surrounding water increases, pressure on the air the diver breathes also increases. In addition, as the pressure increases, the solubility of the gases in the diver's tissues increases. The tissues, therefore, accumulate certain gases which are not metabolized. The increased presence of certain gases causes specific and often dangerous physiological effects.

The total pressure of the atmosphere at sea level is approximately 760 mmHg or 30.4 in. Hg. Pressure increases underwater at the rate of 1 atm (10 kilopascals) for each 33 ft (10 m) that the diver descends. The total pressure applied to the body and to the breathing gas increases proportionately with depth. As pressure of the gas increases, the amount of gas that is absorbed by the body increases. This is particularly evident if the diver is breathing air within a caisson since the percentage of nitrogen in the breathing gas increases proportionately to the amount of oxygen that is removed. Likewise, the amount of carbon dioxide in the body increases during the dive, particularly if the exhaled air is not separated from the inhaled air.

One of the more obvious effects of gases on divers is caused by nitrogen. This gas makes up about 78% of the air that is normally breathed, and its solubility in the tissues increases as atmospheric pressure increases. When nitrogen is dissolved in the body, more than 50% is contained in the fatty tissues; this includes the myelin sheaths which surround many nerve cells. When divers undergo increased pressure, amounts of nitrogen in nerve tissue increase and lead to nitrogen narcosis or "rapture of the deep." Nitrogen narcosis can occur at 415 ft (130 m) or 5 atm (500 kPa), and increases in severity as the diver descends below this depth. The irrational behavior and euphoria often seen in nitrogen narcosis can result in serious, even deadly mistakes during a dive. The maximum time that divers can remain underwater without showing symptoms decreases with increasing depth.

One of the earlier recognized problems associated with human diving is known as decompression sickness, the bends, or caisson disease. If a diver is allowed to stay beneath the surface for long periods of time, the volume of dissolved gases in the tissues will increase. This is particularly true of nitrogen in the case of air breathing. When nitrogen accumulates in the tissues, it remains in solution as long as the pressure remains constant. However, when pressure decreases during the ascent, bubbles form in the tissues. Nitrogen bubbles can occur in nerves or muscles and cause pain, or they can occur in the spinal cord or brain and result in paralysis, dizziness, blindness, or even unconsciousness. Bubbles forming in the circulatory system result in air embolism. If the embolism occurs in the circulation of the lungs, a condition known as the chokes occurs.

A method of prevention of decompression sickness was suggested by J. S. Haldane in 1907. Haldane introduced the method of stage decompression, in which the diver is allowed to ascend a few feet and then remain at this level until the gases in the tissues have been allowed to reequilibrate at the new pressure. This stepwise ascent is continued until the diver finally reaches the surface. A modern variation of this method consists of placing the diver in a decompression chamber after the surface is reached, to allow for periods of decompression which simulate ocean depths. [J.H.F.]

Dogwood A tree, *Cornus florida*, also known as flowering dogwood, which may reach a height of 40 ft (12 m) and is found in the eastern half of the United States and in southern Ontario, Canada. It has opposite, simple, deciduous leaves with entire margins. When this tree is in full flower, the four large, white, notched bracts or petallike growths surrounding the small head of flowers give an ornamental effect. Pink, rose, and cream-colored varieties are commonly planted. The wood is very hard and is used for roller skates, carpenters' planes, and other articles in which hardness is desired. The Pacific dogwood (*C. nuttalli*), which grows in Idaho and from southwestern British Columbia to southern California, is similar to the eastern dogwood, but has rounded bracts. The Japanese dogwood (*C. kousa*) is a similar small tree with pointed bracts and blooms in June. Other shrubby species of dogwood are used as ornamentals. *See* FOREST AND FORESTRY; TREE. [A.H.G.; K.P.D.]

Dormancy In the broadest sense, the state in which a living plant organ (seed, bud, tuber, bulb) fails to exhibit growth, even when environmental conditions are considered favorable. In a stricter context, dormancy pertains to a condition where the inhibition of growth is internally controlled by factors restricting water and nutrient absorption, gas exchange, cell division, and other metabolic processes necessary for growth. By utilizing the latter definition, dormancy can be distinguished from other terms such as rest and quiescence which reflect states of inhibited development due to an unfavorable environment.

Physically induced dormancy can be separated into two distinct classes, based on external conditions imposed by the environment (light, temperature, photoperiod) and restraints induced by structural morphology (seed-coat composition and embryo development).

The physical environment plays a key role in dormancy induction, maintenance, and release in several plant species.

1. *Temperature*. The onset of dormancy in many temperate-zone woody species coincides with decreasing temperature in the fall. However, it is the chilling temperature of the oncoming winter which is more crucial, particularly in regard to spring budbreak.

2. *Light duration and quality*. Possibly the single most important environmental variable affecting dormancy is day length or photoperiod. *See* PHOTOPERIODISM.

3. *Water and nutrient status*. Dormancy is affected by the availability of water and nutrients as demonstrated by many grasses, desert species, and subtropical fruits which go into dormancy when confronted by drought or lack of soil fertility. *See* FERTILIZER; PLANT MINERAL NUTRITION; PLANT-WATER RELATIONS.

4. *Environmental interactions*. Several of the factors previously discussed do not simply act independently, but combine to influence dormancy.

Examples of dormancy imposed by physical restrictions are most evident in the structural morphology of dormant seeds. These restrictions specifically pertain to the physical properties of the seed coat and developmental status of the embryo.

1. *Seed-coat factors*. The seed-coat material surrounding embryos of many plants consists of several layers of tissue, termed integuments, which are infiltrated with waxes and oils. In effect these waterproofing agents enable the seed coat to inhibit water absorption by the embryo. This results in a type of seed dormancy very characteristic of legume crops (clover and alfalfa). The environment itself can break this type of seed-coat dormancy through alternating temperature extremes of freezing and thawing. The extreme heat induced by forest fires is especially effective.

Seed-coat-induced dormancy can also result from mechanical resistance due to extremely hard, rigid integuments commonly found in conifer seeds and other tree species with hard nuts.

2. *Embryonic factors*. The morphological state of the embryo is yet another physical factor affecting dormancy. Often the embryo is in a rudimentary stage when the seed is shed from the maternal plant; dormancy will usually cease in these plants as the embryos reach an adequate state of maturation.

Studies dealing with dormancy have resulted in searches for endogenous plant hormones which regulate the process. Studies involving dormant buds of ash (*Fraxinus americana*) and birch (*Betula pubescens*) revealed the presence of high concentrations of a growth inhibitor or dormancy-inducing and -maintaining compound. This compound was later identified as abscisic acid. As buds of these trees began to grow and elongate, the levels of abscisic acid fell appreciably, supporting the role for abscisic acid in the regulation of dormancy. Abscisic acid is also important in the regulation of seed dormancy, as exemplified by seeds of ash in which abscisic acid levels are high during the phase of growth inhibition, but then decline rapidly during stratification, resulting in germination.

In conjunction with decreased levels of abscisic acid, the endogenous supply of many growth promoters, such as gibberellins, cytokinins, and auxins, have been reported to rise during budbreak in sycamore (*Acer pseudoplatanus*) as well as in Douglas fir (*Pseudotsuga menziesii*). Levels of these dormancy-releasing compounds also correlate well with the breaking of seed dormancy. The hormonal regulation of dormancy can best be perceived as a balance between dormancy inducers or maintainers and dormancy-releasing agents.

In addition to endogenous hormones, there are a variety of compounds that can break dormancy in plant species when they are applied exogenously. Many of these substances are synthetic derivatives or analogs of naturally occurring, dormancy-releasing agents.

The physical environment exerts a marked influence on dormancy. The plant, however, needs a receptor system to perceive changes in the environment so it can translate them into physiological responses which in most cases are under hormonal control. In the case of changing day length or photoperiod, phytochrome may serve as a receptor pigment. Phytochrome essentially favors the production of either abscisic acid (short days) or gibberellic acid (long days). Stress conditions, such as limited water or nutrient availability, favor the production of abscisic acid, whereas a period of chilling often promotes synthesis of gibberellic acid and other compounds generally considered as growth promoters. *See* PHYTOCHROME.

The mode of action of endogenous growth regulators can only be postulated at this time. Whatever the specific mechanism, it probably involves the regulation of gene action at the level of deoxyribonucleic acid (DNA) and ribonucleic acid (RNA), which subsequently controls protein synthesis. In this framework, abscisic acid is believed to repress the functioning of nucleic acids responsible for triggering enzyme and protein synthesis needed for growth. Gibberellic acid, on the other hand, promotes synthesis of enzymes essential for germination as in the case of α-amylase production that is crucial for barley seed growth. *See* BUD; PLANT GROWTH; SEED. [C.S.M.]

Dorylaimida An order of nematodes in which the labia are generally well developed; however, many taxa exhibit a smoothly rounded anterior. The labial region is often set off from the general body contour by a constriction. The cephalic sensilla are all located on the labial region. When there is no constriction, the labial region is defined as that region anterior to the amphids. The amphidial pouch is shaped like an inverted stirrup, and the aperture is ellipsoidal or a transverse slit. The stoma is armed with a movable mural tooth or a hollow axial spear. The anterior portion of the tooth or spear is produced by a special cell in the anterior esophagus. The esophagus is divided into a slender, muscular anterior region and an elongated or pyriform glandular/muscular posterior region. There are generally five esophageal glands with orifices posterior to the nerve ring. In some taxa there are three glands, and in others seven have been reported. The esophagointestinal valve is well developed. The mesenteron is often clearly divided into an anterior intestine and a prerectum. Females have one or two reflexed ovaries; when there is only one, the vulva may shift anteriorly. Males have paired equal spicules that are rarely accompanied by a gubernaculum. The males often have the ventromedial preanal supplements preceded by paired adanal supplements.

There are seven dorylaimid superfamilies: Actinolaimoidea, Belondiroidea, Diphtherophoroidea, Dorylaimoidea, Encholaimoidea, Nygolaimoidea, and Trichodoroidea. *See* NEMATA. [A.R.M.]

Double diffusion A type of convective transport in fluids that depends on the difference in diffusion rates of at least two density-affecting components. This phenomenon was discovered in 1960 in an oceanographic context, where the two components are heat and dissolved salts. Besides different diffusivities, it is necessary to have an unstable or top-heavy distribution of one component.

In the oceanographic context, if the unstable component is the slower-diffusing one (salt), with the overall gravitational stability maintained by the faster-diffusing component (heat), then "salt fingers" will form. Since warm, salty tropical waters generally overlie colder, fresher waters from polar regions, this is a very common stratification in the mid- to low-latitude ocean. Salt fingers arise spontaneously when small parcels of warm, salty water are displaced into the underlying cold, fresh water. Thermal conduction then removes the temperature difference much quicker than salt diffusion can take effect. The resulting cold, salty water parcel continues to sink because of its greater density. Conversely, a parcel of cold, fresh water displaced upward gains heat but not salt, becoming buoyant and continuing to rise. The fully developed flow has intermingled columns of up- and downgoing fluid, with lateral exchange of heat but not salt, carrying advective vertical fluxes of salt and to a lesser extent heat.

Another form of double-diffusive convection occurs when the faster-diffusing component has an unstable distribution. In the ocean, this happens when cold, fresh water sits above warmer, saltier and denser water. Such stratifications are common in polar regions and in local areas above hot springs at the bottom of the deep sea. *See* OCEAN CIRCULATION; OCEANOGRAPHY; SEAWATER.

The importance of double diffusion lies in its ability to affect water mass structure with its differential transport rates for heat and salt. This is believed to play a significant role in producing certain oceanic water types with well-defined relationships between temperature and salinity. *See* OCEAN CIRCULATION; SEAWATER. [R.W.Sc.]

Douglas-fir A large coniferous tree, *Pseudotsuga menziesii* (formerly *P. taxifolia*), known also as red fir, belonging to the pine family (Pinaceae). It is one of the most widespread and most valuable tree species of western North America and ranks among the world's most important. In the United States, this species is first in total stand volume, lumber production, and veneer for plywood. It is the most common Christmas tree in the West and is the state tree of Oregon.

Douglas-fir is a large to very large evergreen tree with narrow pointed crown of slightly drooping branches; at maturity it reaches a height of 80–200 ft (24–60 m) and a trunk diameter of 2–5 ft (0.6–1.5 m). The bark is dark or reddish brown, very thick, deeply furrowed, and often corky. The needlelike leaves spreading mostly in two rows are flat, and flexible, with a very short leafstalk. Buds are distinctive, conical pointed, scaly, and dark red. The cones are elliptical and light brown, with many thin rounded cone-scales, each above a longer distinctive three-pointed bract.

The natural distribution of Douglas-fir extends from southwestern Canada (British Columbia and Alberta) south through the western United States (Washington, Oregon, the Sierra Nevada, and the Rocky Mountains), and south to central Mexico. This species has been introduced into the eastern United States, Europe, and elsewhere.

Two varieties are distinguished: coast Douglas-fir of the Pacific region and Rocky Mountain or inland Douglas-fir. The latter has shorter cones with bracts bent backward. Though not so large, it is hardier and grows better in the East. *See* FOREST AND FORESTRY; TREE. [E.L.L.]

Drought A general term implying a deficiency of precipitation of sufficient magnitude to interfere with some phase of the economy. Agricultural drought, occurring when crops are threatened by lack of rain, is the most common. Hydrologic drought, when reservoirs are depleted, is another common form. The Palmer index is used by agriculturalists to express the intensity of drought as a function of rainfall and hydrologic variables.

The meteorological causes of drought are usually associated with slow, prevailing, subsiding motions of air masses from continental source regions. These descending air motions, of the order of 660–1000 ft (200 or 300 m) per day, result in compressional warming of the air and therefore reduction in the relative humidity. Since the air usually starts out dry, and the relative humidity declines as the air descends, cloud formation is inhibited—or if clouds are formed, they are soon dissipated. [J.N.]

Drug resistance The ability of an organism to resist the action of an inhibitory molecule or compound. Examples of drug resistance include disease-causing bacteria evading the activity of antibiotics, the human immunodeficiency virus resisting antiviral agents, and human cancer cells replicating despite the presence of chemotherapy agents. There are many ways in which cells or organisms become resistant to drugs, and some organisms have developed many resistance mechanisms, each specific to a different drug. Drug resistance is best understood as it applies to bacteria, and the increasing resistance of many common disease-causing bacteria to antibiotics is a global crisis.

Genetic basis. Some organisms or cells are innately or inherently resistant to the action of specific drugs. In other cases, the development of drug resistance involves a change in the genetic makeup of the organism. This change can be either a mutation in a chromosomal gene or the acquisition of new genetic material from another cell or the environment.

Organisms may acquire deoxyribonucleic acid (DNA) that codes for drug resistance by a number of mechanisms. Transformation involves the uptake of DNA from the environment. Once DNA is taken up into the bacterial cell, it can recombine with the recipient organism's chromosomal DNA. This process plays a role in the development and spread of antibiotic resistance, which can occur both within and between species.

Transduction, another mechanism by which new DNA is acquired by bacteria, is mediated by viruses that infect bacteria (bacteriophages). Bacteriophages can integrate their DNA into the bacterial chromosome.

Conjugation is the most common mechanism of acquisition and spread of resistance genes among bacteria. This process, which requires cell-to-cell contact, involves direct transfer of DNA from the donor cell to a recipient cell. While conjugation can

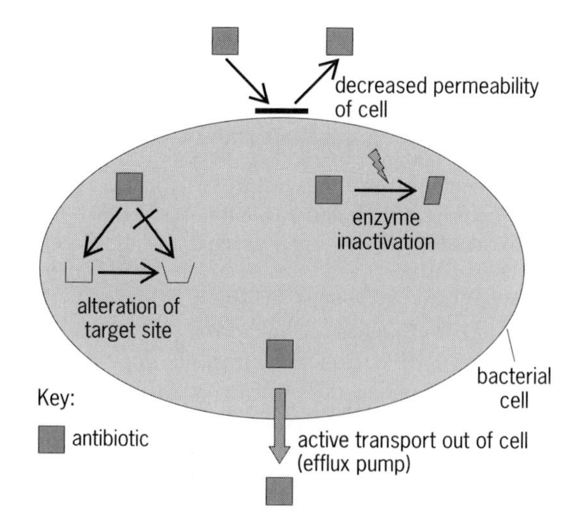

Four common mechanisms of antibiotics resistance.

involve cell-to-cell transfer of chromosomal genes, bacterial resistance genes are more commonly transferred on nonchromosomal genetic elements known as plasmids or transposons.

Mechanisms of resistance. The four most important antibiotic resistance mechanisms are alteration of the target site of the antibiotic, enzyme inactivation of the antibiotic, active transport of the antibiotic out of the bacterial cell, and decreased permeability of the bacterial cell wall to the antibiotic (see illustration).

By altering the target site to which an antibiotic must bind, an organism may decrease or eliminate the activity of the antibiotic. Alteration of the target site is the mechanism for one of the most problematic antibiotic resistances worldwide, methicillin resistance among *Staphylococcus aureus*. *See* BACTERIAL GENETICS.

The most common mechanism by which bacteria are resistant to antibiotics is by producing enzymes that inactivate the drugs. For example, β-lactam antibiotics (penicillins and cephalosporins) can be inactivated by enzymes known as β-lactamases.

Active transport systems (efflux pumps) have been described for the removal of some antibiotics (such as tetracyclines, macrolides, and quinolones) from bacterial cells. In these situations, even though the drug can enter the bacterial cell, active efflux of the agent prevents it from accumulating and interfering with bacterial metabolism or replication.

Bacteria are intrinsically resistant to many drugs based solely on the fact that the drugs cannot penetrate the bacterial cell wall or cell membrane. In addition, bacteria can acquire resistance to a drug by an alteration in the porin proteins that form channels in the cell membrane. The resistance that *Pseudomonas aeruginosa* exhibits to a variety of penicillins and cephalosporins is mediated by an alteration in porin proteins.

Promoters. In the hospital environment, many factors combine to promote the development of drug resistance among bacteria. Increasing use of powerful new antibiotics gives selective advantage to the most resistant bacteria. In addition, advances in medical technology allow for the survival of sicker patients who undergo frequent invasive procedures. Finally, poor infection control practices in hospitals allow for the unchecked spread of already resistant strains of bacteria.

Outside the hospital environment, other important factors promote antibiotic resistance. The overuse of antibiotics in outpatient medicine and the use of antibiotics in agriculture exert selective pressure for the emergence of resistant bacterial strains. The spread of these resistant strains is facilitated by increasing numbers of children in close contact at day care centers, and by more national and international travel.

Control. A multifaceted worldwide effort will be required to control drug resistance among disease-causing microorganisms. Ongoing programs to decrease the use of antibiotics, both in the clinics and in agriculture, will be necessary. The increased use of vaccines to prevent infection can help limit the need for antibiotics. Finally, the development of novel classes of antibiotics to fight emerging resistant bacteria will be required. *See* BACTERIA. [D.J.D.]

Drumlin A streamlined, oval-shaped hill which has been shaped by flowing glacial ice. The long axis is parallel to the direction of ice flow, the up-glacier slope is usually steeper than the lee slope, and composition includes a variety or combination of materials—till, outwash, or bedrock. Drumlins are highly localized, but where present, they occur in large numbers. Some drumlins are clearly erosional in origin, but in others till deposition appears to have been synchronous with drumlin formation. Thus, one or both processes must be operative at some time in the subglacial environment where drumlins form. [W.H.J.]

Dune Mobile accumulation of sand-sized material that occurs along shorelines and in deserts because of wind action. Dunes are typically located in areas where winds decelerate and undergo decreases in sand-carrying capacity. Dunefields are composed of rhythmically spaced mounds of sand that range from about 3 ft (1 m) to more than 650 ft (200 m) in height and may be spaced as much as 5000 ft (1.5 km) apart. Smaller accumulations of windblown sand, typically ranging in height from 0.25 to 0.6; in. (5 to 15 mm) and in wavelength from 3 to 5 in. (7 to 12 cm), are known as wind ripples. Dunes and ripples are two distinctly different features. The lack of intermediate forms shows that ripples do not grow into dunes. Ripples commonly are superimposed upon dunes, typically covering the entire upwind (stoss) surface and much of the downwind (leeward) surface as well.

Virtually any kind of sand-sized material can accumulate as dunes. The majority of dunes are composed of quartz, an abundant and durable mineral released during weathering of granite or sandstone. Dunes along subtropical shorelines, however, are commonly composed of grains of calcium carbonate derived in part from the breakdown of shells and coral. Along the margins of seasonally dry lakes, dunes may be composed of gypsum (White Sands, New Mexico) or sand-sized aggregates of clay minerals (Laguna Madre, Texas).

The leeward side of most dunes is partly composed of a slip face, that is, a slope at the angle of repose. For dry sand, this angle is approximately 33°. When additional sand is deposited at the top of such a slope, tonguelike masses of sand avalanche to the base of the slope. The dune migrates downwind as material is removed from the gently sloping stoss side of the dune and deposited by avalanches along the slip face. Much of the sand on the leeward side of the dune is later reworked by side winds into wind ripple deposits. Because the coarsest grains preferentially accumulate at the crests of wind ripples, the layering in wind ripple deposits is distinctive and relatively easily recognized—each thin layer is coarser at its top than at its base.

Dunes can be classified on the basis of their overall shape and number of slip faces. Three kinds of dunes exist with a single slip face; each forms in areas with a single dominant wind direction. Barchans are crescent-shaped dunes; their arms point downwind. They develop in areas in which sand is in small supply. If more sand is available, barchans coalesce to form sinuous-crested dunes called barchanoid ridges. Transverse

Linear dune, Imperial County, California.

dunes with straight crests develop in areas of abundant sand supply. The axis of each of these dune types is oriented at right angles to the dominant wind, and the dunes migrate rapidly relative to other dune types. The migration rate of individual dunes is quite variable, but in general, the larger the dune, the slower the migration rate. In the Mojave Desert of southern California, barchans having slip faces 30 ft (10 m) long migrate about 50 ft (15 m) per year.

Dunes having more than one slip face develop in areas with more complex wind regimes. Linear dunes, sometimes called longitudinal or self dunes, possess two slip faces which meet along a greatly elongated, sharp crest (see illustration). Some linear dunes in Saudi Arabia reach lengths of 120 mi (190 km). Experimental evidence has shown that linear dunes are the result of bidirectional winds that differ in direction by more than 90°. The trend of these dunes is controlled by wind direction, strength, and duration, but the nature of the wind regime cannot be deduced from a knowledge of dune trends. Star dunes bear many slip faces and consist of a central, peaked mound from which several ridges radiate. Because they do not migrate appreciably, they grow in height as sand is delivered to them, some reaching 1000 ft (300 m).

Plant growth appears to be important to the growth and maintenance of two types of dunes. Coppice dunes are small mounds of sand that are formed by the wind-baffling and sand-trapping action of desert plants. The crescentic shape of parabolic dunes gives them a superficial resemblance to barchan dunes, but their arms point upward. Plants commonly colonize and anchor only the edges of a dune, leaving the body of the dune free to migrate. The retarded migration rate of the dune margin leads to the formation of the trailing arms of a parabolic dune. *See* DUNE VEGETATION.

Other dune types are dependent on special topographic situations for their formation. Climbing dunes develop on the upwind side of mountains or cliffs; falling dunes are formed at the sheltered, downward margin of similar features. [D.B.Lo.]

Dune vegetation Plants occupying sand dunes and the slacks, or swales, and flats between them. The density and diversity of dune vegetation are greater on coastal dunes than on desert dunes. *See* DUNE.

A zonation pattern is evident in the vegetation of the coastal dunes (see illustration). A wrack, or debris, line occurs at the upper limit of the beach. Seeds caught in decaying plant material and other debris washed in on the high tides germinate here and trap windblown sand, initiating the formation of a dune. The foredunes, also called the primary dunes, are those closest to the water and lie behind the wrack line. The plants on these dunes, mostly grasses, are tolerant to sea spray, high winds, and sand accretion. Behind the primary dunes are the secondary dunes, sometimes called the dune field.

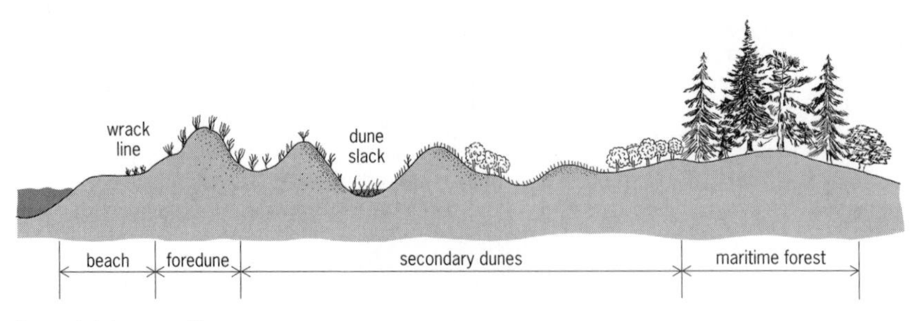

A coastal dune profile.

In this more favorable environment, the vegetation is denser and more diverse; the foredunes block sea spray and reduce wind velocity. The dune slacks are the low areas between dunes and are frequently a result of a blowout, an area where sand has been blown away down to where the sand is moist and close to the water table. Plants typical of wetlands often vegetate these areas. Shrub communities also inhabit the dune field and often form dense patches of vegetation. A maritime forest may be found behind the secondary dunes. In coastal barrier beach or island locations, a salt marsh adjacent to a bay or sound may lie behind the forest.

Although the plant species occupying the sand dunes of the United States vary from coast to coast, their functions and adaptations are essentially the same. Plants growing on sand dunes are adapted to the environment. The plants closest to the sea are usually the most tolerant of salt spray. The plants in the wrack line must be tolerant to salinity, wind, and burial by sand.

The foredune plants must be tolerant of sand burial, sea spray, and a nutrient-poor substrate. By a system of underground stems called rhizomes, they overcome burial by sand and spread throughout the dune with new shoots arising from buds on the rhizomes. The roots and rhizomes of these dune grasses are important in stabilizing the dune sand and preventing wind erosion. The foredune plants participate in dune formation; by slowing the wind, they favor sand deposition. Furthermore, some of these plants have specialized bacterium named *Azotobacter* associated with their roots, and these bacteria fix atmospheric nitrogen into a form usable by the plant.

The plants in the dune slacks have morphological and physiological adaptations for growth in flooded areas. For example, the sedge American three-square contains large air spaces (aerenchyma) in its stems and roots, which provide an oxygen pathway from the shoots above the water to the oxygen-deprived roots in the flooded soil. When these plants are flooded, they increase their production of the hormone ethylene, which may stimulate the production of aerenchymatous tissue. [D.M.Se.]

Dust storm A strong, turbulent wind carrying large clouds of dust. In a large storm, clouds of fine dust may be raised to heights well over 10,000 ft (3000 m) and carried for hundreds or thousands of miles (1 mi = 1.6 km).

Sandstorms differ by the larger mass, more rapid setting speeds of the particles involved, and the stronger transporting winds required. The sand cloud seldom rises above 3.3–6.6 ft (1–2 m) and is not carried far from the place where it was raised.

Dust storms cause enormous erosion of the soil, as in the dust bowl disasters of 1933–1937 in the Great Plains of the United States. Besides causing acute physical discomfort, they present a severe hazard to transportation by reducing the visibility to very low ranges. Conditions required are an ample supply of fine dust or loose soil, surface winds strong enough to stir up the dust, and sufficient atmospheric instability for marked vertical turbulence to occur.

Small dust particles increase scattering of light, mainly in short (blue) wavelengths. The Sun often appears a deep orange or red when seen through a dust cloud; however, optical effects are variable. Large particles are effective reflectors, and an observer in an aircraft above a dust storm may see a solid sheet with an apparent dust horizon.

[C.W.N.]

E

Earth The third planet from the Sun and the largest of the four inner, or terrestrial, planets. The Sun is an average-sized, middle-aged star situated toward the outer edge of one of the spiral arms of the Milky Way Galaxy. So far as is known, Earth is unique in the solar system in having life. Whether life exists in the universe beyond the solar system is unknown.

Earth has one natural satellite, the Moon. Otherwise, Earth's nearest neighbors in space are Venus, which is about 108×10^6 km (67×10^6 mi) from the Sun, and Mars, about 228×10^6 km (141×10^6 mi) from the Sun. Earth is about 150×10^6 km (93×10^6 mi) from the Sun.

Earth completes an orbit around the Sun in 365 days, 5 h, 48 min, 46 s; the orbit defines the length of the year. The length of the day is determined by the period of Earth's rotation about its axis. The fact that the year is not a whole number of days has affected the development of the calendar.

Earth rotates on its axis once each day. The axis of rotation is perpendicular to the Equator, and the Equator is inclined at about 23.5° to the plane of Earth's orbit around the Sun. As Earth moves in its orbit, the north spin axis, or north geographic pole, points in the direction of the star Polaris, making it the North Star or polestar. One result of the tilt of the Equator relative to the orbital plane is that different parts of Earth receive differing amounts of sunlight through the year; this is the primary cause of seasons. *See* EQUATOR.

Earth is an oblate spheroid. The mean equatorial radius is 6378.139 km (3963.37 mi), and the polar radius is 6356.779 km (3950.10 mi), the difference being 21.360 km (13.27 mi).

Earth's mass is 5.976×10^{27} g (0.2108×10^{27} oz), being the sum of 5.974×10^{27} g (0.2107×10^{27} oz) for solid Earth, 1.4×10^{24} g (0.049×10^{24} oz) for the ocean, and 5.1×10^{21} g (0.18×10^{21} oz) for the atmosphere. Earth's average density is 5.518 g/cm^3, which is just about double the density of the common rocks that form at Earth's surface, indicating that Earth's interior is more dense than the surface. Seismic studies have confirmed that Earth is layered both compositionally and mechanically (see illustration). *See* ATMOSPHERE; OCEANOGRAPHY.

The deepest compositional layer is the core, which is divided into a solid inner core and a liquid outer core. Both the inner and outer core have the same composition, believed to be nickel-iron plus a small amount of lighter elements such as sulfur and silicon. Electric currents moving in the molten metal outer core are believed to be the origin of Earth's magnetic field. Above the core is the mantle which, on the basis of density of rare rock samples brought up from deep in the mantle in kimberlite pipes, and other evidence, is believed to be composed of silicate minerals, and in particular olivine and pyroxene. A rock composed largely of olivine and pyroxene is called a peridotite.

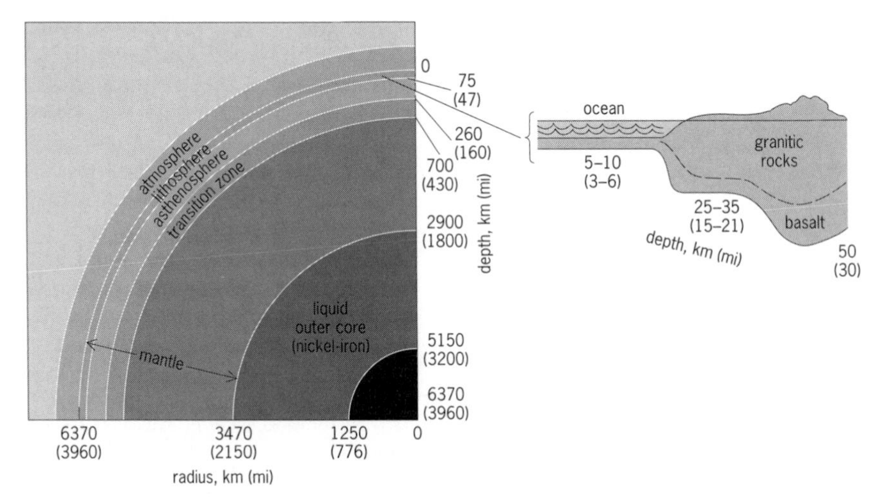

Principal layers of Earth.

Above the mantle is Earth's crust, and between the crust and the mantle there is a pronounced seismic discontinuity known as the Mohorovičić discontinuity, or Moho. The crust is of two kinds, both of which are less dense and compositionally different from the peridotitic mantle below. Beneath the ocean the crust is basaltic in composition and about 8 km (5 mi) thick. The crust beneath the continents is granitic in composition and averages 35 km (21.7 mi) in thickness but ranges up to 80 km (49.7 mi), as beneath Tibet. The oceanic crust is geologically young because it is continually created and destroyed through the process of plate tectonics. No part of the oceanic crust that is older than about 180×10^6 years has yet been discovered. The continental crust is much older than the oceanic crust. Continental rocks as old as 4×10^9 years have been discovered in Canada, and the fact that they are highly deformed indicates a long and eventful history. *See* EARTH CRUST.

The surface of solid Earth has a bimodal distribution of elevations. If the water from the ocean could be removed, it would be apparent that continents stand high (average elevation is 840 m or 2755 ft above sea level), while the ocean floor sits low (average elevation is 3800 m or 12,464 ft below sea level). This difference in elevation arises because rigid lithosphere floats on the weak asthenosphere, and because the density of oceanic lithosphere (that is, lithosphere capped by oceanic crust) is greater than the density of continental lithosphere.

On the continents, mountain belts are the most dramatic features. They range in elevation from Mount Everest, 8848 m (29,030 ft), in the Himalaya Mountains to older, deeply eroded ranges that are now barely above sea level. Granitic and metamorphic rocks are generally exposed in the cores of mountain ranges. The overlying rocks that cover most of the Earth's surface are sedimentary, mainly of shallow marine origin, that may or may not have been deformed. The deformation is the result of compression and tension that causes folding and faulting, and may be accompanied by intrusion and metamorphism. Movements and collisions of tectonic plates are the principal cause of mountain building. Mountains generally are formed over several tens of millions of years. The rocks deformed in the process are generally marine sedimentary rocks formed along the margins of continents. *See* DEEP-MARINE SEDIMENTS; MARINE SEDIMENTS; SEDIMENTARY ROCKS.

The topographic features underlying the oceans are similarly diverse and reveal more evidence of a dynamic Earth. The continental shelf, an area covered by shallow water, generally less than 150 m (500 ft) deep, surrounds the continents at most places. Such areas are generally underlain by continental, granitic rocks, and are submerged parts of the continents. Continental slopes are the transition between the continental shelf and the ocean floors. Their tops are generally less than 150 m below sea level, and they slope down to about 4400 m (14,000 ft). They are narrow, steep features, with slopes generally between 2 and 6°, but some are up to 45°. They are generally underlain by thick accumulations of sedimentary rocks. *See* SEA-FLOOR IMAGING.

Submarine trenches and their associated volcanic island arcs are formed as a result of a tectonic plate of lithosphere sinking into the mantle beneath the edge of an overriding plate. The deepest place on Earth is in the Mariana Trench, 11,022 m (36,152 ft) below sea level.

The ocean floor is the most widespread surface feature of Earth. Beneath an average of 4.4 km (2.75 mi) of seawater are about 2.3 km (1.4 mi) of sedimentary rocks with some intercalated basalt, and below that is the oceanic crust, consisting of 4–6 km (2.5–3.7 mi) of basaltic rocks. Interrupting the ocean floor at many places are submarine mountains formed by basaltic volcanoes. Some of these volcanoes are very large and form oceanic islands such as the Hawaiian Islands.

New oceanic lithosphere capped by basaltic crust is created at the mid-ocean ridges, and this newly formed plate moves away from the ridges. The tectonic plates formed in this way may carry continents on them, and are the mechanism of continental drift. Paleomagnetic data from the continents indicate that the continents have moved relative to each other. The tectonic plates capped by basaltic crust plates are consumed at the trench-volcanic island arc areas.

As well as the ridge and the trench, a third type of plate boundary occurs where two plates slide past each other at a transform fault. Such collisions account for the deformed rocks found in the crust.

The evidence for continental drift in the geological past includes matching of rock types, ages, fossils, climates, and structures (mountain ranges), as well as the paleomagnetic data. Evidence showing or suggesting present movements consists of shallow earthquakes along mid-ocean ridges and transform faults that offset them; deep earthquakes associated with deep-sea trench-volcanic island arc areas; direct measurement of movement; volcanic activity at mid-ocean ridges; and volcanic activity at trench-island arc areas.

Earth's temperature and gravitation are such that an atmosphere is present. The major constituents are nitrogen and oxygen. A thin ozone layer in the atmosphere shields the Earth from lethal ultraviolet radiation from the Sun. The atmosphere, especially oxygen, and the presence of water, both at the surface and in the atmosphere, make life possible. Precipitation, mainly rain, results in running water such as streams and rivers on the continents. Running water is the main cause of erosion of the continents, and most of the landscapes have been eroded by water, although some are eroded by wind or ice (glaciers). *See* EROSION; GLACIOLOGY.

Earth, along with the rest of the solar system, is believed to have formed about 4.55×10^9 years ago. This age is determined by dating radioactive isotopes in meteorities. Meteorites are believed to be fragments produced by collisions among small bodies formed by the same process that created the solar system. Theoretical studies of the Sun and other studies of radioactive isotopes also suggest a similar age. [B.J.S.]

Earth crust The low-density outermost layer of the Earth above the Mohorovičić discontinuity (the Moho), a global boundary that is defined as the depth in the Earth where the compressional-wave seismic velocity increases rapidly or discontinuously to a value in excess of 4.7 mi/s (7.6 km/s; the upper mantle). The crust is also the cold, upper portion of the Earth's lithosphere, which in terms of plate tectonics is the mobile, outer layer that is underlain by the hot, convecting asthenosphere.

Continental crust. The Earth's continental crust has evolved over the past 4 billion years, and is highly variable in geologic composition and internal structure. The world-wide mean thickness of continental crust is 24 mi (40 km), with a standard deviation of 5.4 mi (9 km). The thinnest continental crust (found in the Afar Triangle, northeast Africa) is about 9 mi (15 km) thick, and the thickest crust (the Himalayan Mountains in China) is about 47 mi (75 km) thick. Ninety-five percent of all continental crust has a thickness within two standard deviations of the mean thickness, between 13 mi (22 km) and 37 mi (58 km). The Antarctic continent has a crustal thickness of 24 mi (40 km) in the ancient, stable (cratonic) region of East Antarctica, and about 12 mi (20 km) in the recently stretched (extended) crust of West Antarctica. Continental margins, which mark the transition from oceanic to continental crust, range in thickness from about 9 mi (15 km) to 18 mi (30 km). *See* CONTINENTAL MARGIN.

Despite its geologic complexity, the continental crust may generally be divided into four layers: an uppermost sedimentary layer, and an upper, middle, and lower crust composed of crystalline rocks. The sedimentary cover of the continental crust is an important source of natural resources. This cover averages 0.6 mi (1 km) in thickness, and varies in thickness from zero (for example, on shields) to more than 9 mi (15 km) in deep basins. In stable continental crust of average thickness (25 mi or 40 km), the crystalline upper crust is commonly 6–9 mi (10–15 km) thick and has an average composition equivalent to a granite. The middle crust is 3–9 mi (5–15 km) thick and has a composition equivalent to a diorite; and the lower crust is 3–12 mi (5–20 km) thick and has a composition equivalent to a gabbro. Due to increasing temperature and pressure with depth, the metamorphic grade of rocks increases with depth, and the rocks within the deep continental crust generally are metamorphic rocks, even if they originated as sedimentary or igneous rocks.

Crustal properties vary systematically with geologic setting, which may be divided into six groups: orogens (mountain belts), shields and platforms, island arcs (volcanic arcs), continental magmatic arcs, rifts, extended (stretched) crust, and forearcs. Orogens are typified by thick crust [average thickness is 29 mi (46 km), but the maximum thickness is as much as 47 mi (75 km) in the Himalayas]. Shields and platforms, such as the Canadian Shield and the Russian Platform, commonly have an approximately 26-mi-thick (42-km) crust, including a 3–6 mi-thick (5–10 km) lower crust. In comparison with shields, island arcs (such as Japan) have thinner crusts and significantly shallower middle and lower crustal layers due to the intrusion of mafic (that is, low silica content) plutons. Continental magmatic arcs, such as the Cascades volcanoes of the northwestern United States, intrude preexisting continental crust, and therefore they are generally 3–9 mi (10–15 km) thicker than island arcs. Continental rifts, such as the East African and Rio Grande rifts, have an average crustal thickness of about 22 mi (36 km). Extended continental crust, such as the Basin and Range Province of the western United States, averages 18 mi (30 km) in thickness. Forearcs are regions that were formed oceanward of volcanic arcs, such as much of the west coast of North America. They typically have thin crust, about 15 mi (25 km), and have a thick (9 mi or 15 km) upper crustal section that consists of relatively low-density metasedimentary rocks. *See* NORTH AMERICA; OCEANIC ISLANDS; RIFT VALLEY; SEDIMENTARY ROCKS; VOLCANO.

At least three processes provide new continental crust. The first is the accretion and consolidation of island arcs, such as Japan or the Aleutian Islands, onto a continental margin. The second process is the tectonic underplating of oceanic crust at active subduction zones. In this process, the continental crust grows from below as oceanic crust is welded to the base of the continental margin, either when subduction stops or when subduction steps oceanward and a new trench is formed. This process has been identified in western Canada and southern Alaska. The third process is the magmatic inflation of the crust at continental arcs, rifts, and regions of crustal extension. This process has been identified in many regions. [W.D.Mo.]

Oceanic crust. The surface of the ocean crust, except for some locally high volcanoes and plateaus, resides some 1–3 mi (2–5 km) below sea level, and about another kilometer below the average level of the continents. The ocean crust represents the youngest and geologically most dynamic portion of the Earth's surface. Most of it was produced at mid-ocean ridges during the process of sea-floor spreading. The ridges define the trailing edges, or accreting boundaries, of the major lithospheric plates that are moving about the surface of the Earth at present. Thus, the oldest rocks of the ocean crust date back no earlier than the rifting episodes that created most of these plates and initiated the most recent phase of continental drift, the Pangaean breakup, in Late Jurassic times. *See* MID-OCEANIC RIDGE.

There are fault slices of types of ocean crust on land, known as ophiolites, where nearly or entirely complete cross sections through the crust can be mapped and sampled. These strongly indicate that the ocean crust consists in downward sequence of submarine extrusives (usually pillow basalts), feeder dikes (often vertically sheeted), or sills, gabbros, and peridotites. There is much uncertainty, however, about the extent to which typical ophiolites, most of which formed in island-arc or backarc environments, can represent abyssal ocean crust, which is produced at the major accreting plate boundaries. Moreover, the physical correspondence of the rocks in ophiolites to ocean crust is often complicated by their complex structure and extent of alteration and metamorphism, particularly in the ultramafic sections. [J.H.Na.]

Earth sciences Sciences that involve attempts to understand the nature, origin, evolution, and behavior of the Earth or of its parts and to comprehend its place in the universe, especially in the solar system. Understanding has advanced primarily through improved appreciation of the complex, usually cyclical interactions that take place among distinct parts of the Earth such as the lithosphere, atmosphere, hydrosphere, and biosphere. Geophysics is the study of the physics of the Earth, emphasizing its physical structure and dynamics. Geochemistry is the study of the chemistry of the Earth, dealing with its composition and chemical change. Geology is the study of the solid Earth and of the processes that have formed and modified it throughout its 4.5-billion-year history. *See* GEOCHEMISTRY.

Many branches of geology are considered separate sciences. Mineralogy is the study of the composition, structure, and properties of minerals. Petrology involves understanding how rocks originate and evolve, as well as rock description and classification. Specialties related to petrology include sedimentology and volcanology. Stratigraphy is the study of the origin, age, and development of layered, generally sedimentary rocks. Paleontology is the study of ancient (fossil) life. Historical geology is the study of the evolution of the Earth and its life. Geomorphology is the study of landscapes and their evolution. Seismology is the study of earthquakes and their effects. Structural geology is the study of deformed rocks. Engineering geology relates to the support of human constructions by underlying rock. *See* ENGINEERING GEOLOGY; HYDROLOGY; VOLCANOLOGY.

Oceanography is the study of the oceans; limnology, the study of lakes; hydrology, the study of underground and surface water; and glaciology, the study of glaciers, ice caps, and ice sheets. These disciplines address the study of water in and on the Earth. The gaseous outer parts of the planet are the province of the atmospheric sciences, including meteorology, which is concerned with the weather and weather forecasting; climatology, which deals with longer-term and regional variations; and aeronomy which, because it deals with the outermost ionized region of the atmosphere, is much concerned with solar terrestrial interactions, including the aurora borealis and aurora australis. The biosphere embodies all life on Earth, and its study includes molecular biology, zoology, botany, and ecology. Geography, the study of all that happens at the Earth's surface, has been distinct insofar as it has encompassed not only physical and biological sciences but also the social sciences, including aspects of political science and economics. This distinction is fading rapidly as other earth sciences become more involved with social considerations. [K.Bu.]

Earthquake The sudden movement of the Earth caused by the abrupt release of accumulated strain along a fault in the interior. The released energy passes through the Earth as seismic waves (low-frequency sound waves), which cause the shaking. Seismic waves continue to travel through the Earth after the fault motion has stopped. Recordings of earthquakes, called seismograms, illustrate that such motion is recorded all over the Earth for hours, and even days, after an earthquake.

Earthquakes are not distributed randomly over the globe but tend to occur in narrow, continuous belts of activity. Approximately 90% of all earthquakes occur in these belts, which define the boundaries of the Earth's plates. The plates are in continuous motion with respect to one another at rates on the order of centimeters per year; this plate motion is responsible for most geological activity.

Plate motion occurs because the outer cold, hard skin of the Earth, the lithosphere, overlies a hotter, soft layer known as the asthenosphere. Heat from decay of radioactive minerals in the Earth's interior sets the asthenosphere into thermal convection. This convection has broken the lithosphere into plates which move about in response to the convective motion. As the plates move past each other, little of the motion at their boundaries occurs by continuous slippage; most of the motion occurs in a series of rapid jerks. Each jerk is an earthquake. This happens because, under the pressure and temperature conditions of the shallow part of the Earth's lithosphere, the frictional sliding of rock exhibits a property known as stick-slip, in which frictional sliding occurs in a series of jerky movements, interspersed with periods of no motion—or sticking. In the geologic time frame, then, the lithospheric plates chatter at their boundaries, and at any one place the time between chatters may be hundreds of years.

The periods between major earthquakes is thus one during which strain slowly builds up near the plate boundary in response to the continuous movement of the plates. The strain is ultimately released by an earthquake when the frictional strength of the plate boundary is exceeded. *See* FAULT AND FAULT STRUCTURES.

Most great earthquakes occur on the boundaries between lithospheric plates and arise directly from the motions between the plates. These may be called plate boundary earthquakes. There are many earthquakes, sometimes of substantial size, that cannot be related so simply to the movements of the plates. At many plate boundaries, earthquakes occur over a broad zone—often several hundred miles wide—adjacent to the plate boundary. These earthquakes, which may be called plate boundary-related earthquakes, are secondarily caused by the stresses set up at the plate boundary. Some earthquakes also occur, although infrequently, within plates. These earthquakes, which are not related to plate boundaries, are called intraplate earthquakes. The immediate cause of intraplate earthquakes is not understood.

In addition to the tectonic types of earthquakes described above, some earthquakes are directly associated with volcanic activity. These volcanic earthquakes result from the motion of undergound magma that leads to volcanic eruptions.

Earthquakes often occur in well-defined sequences in time. Tectonic earthquakes are often preceded, by a few days to weeks, by several smaller shocks (foreshocks), and are nearly always followed by large numbers of aftershocks. Foreshocks and aftershocks are usually much smaller than the main shock. Volcanic earthquakes often occur in flurries of activity, with no discernible main shock. This type of sequence is called a swarm.

Earthquakes range enormously in size, from tremors in which slippage of a few tenths of an inch occurs on a few feet of fault, to the greatest events, which may involve a rupture many hundreds of miles long, with tens of feet of slip.

The size of an earthquake is given by its moment: average slip times the fault area that slipped times the elastic constant of the Earth. The units of seismic moment are dyne-centimeters. An older measure of earthquake size is magnitude, which is proportional to the logarithm of moment. Magnitude 2.0 is about the smallest tremor that can be felt. Most destructive earthquakes are greater than magnitude 6; the largest shock known was the 1960 Chile earthquake, with a moment of 10^{30} dyne-centimeters (10^{23} newton-meters) or magnitude 9.5. It involved a fault 600 mi (1000 km) long slipping 30 ft (10 m).

The intensity of an earthquake is a measure of the severity of shaking and its attendant damage at a point on the surface of the Earth. The same earthquake may therefore have different intensities at different places. The intensity usually decreases away from the epicenter (the point on the surface directly above the onset of the earthquake), but its value depends on many factors and generally increases with moment. Intensity is usually higher in areas with thick alluvial cover or landfill than in areas of shallow soil or bare rock. Poor building construction leads to high intensity ratings because the damage to structures is high. Intensity is therefore more a measure of the earthquake's effect on humans than an innate property of the earthquake.

Many additional effects may be produced by earthquake shaking, including landslides and tsunamis. *See* LANDSLIDE; TSUNAMI.

Earthquake prediction research has been going on for nearly a century. Unfortunately, successful earthquake predictions are extremely rare. There are two basic categories of earthquake predictions: forecasts (months to years in advance) and short-term predictions (hours or days in advance). Forecasts are based a variety of research, including the history of earthquakes in a specific region, the identification of fault characteristics (including length, depth, and segmentation), and the identification of strain accumulation. Data from these studies are used to provide rough estimates of earthquake sizes and recurrence intervals. [C.H.S.; K.M.S.]

Ebola virus Ebola viruses are a group of exotic viral agents that cause a severe hemorrhagic fever disease in humans and other primates. The four known subtypes or species of Ebola viruses are Zaire, Sudan, Reston, and Côte d'Ivoire (Ivory Coast), named for the geographic locations where these viruses were first determined to cause outbreaks of disease. Ebola viruses are very closely related to, but distinct from, Marburg viruses. Collectively, these pathogenic agents make up a family of viruses known as the Filoviridae.

Filoviruses have an unusual morphology, with the virus particle, or virion, appearing as long thin rods. A filovirus virion is composed of a single species of ribonucleic acid (RNA) molecule that is bound together with special viral proteins, and this RNA–protein complex is surrounded by a membrane derived from the outer membrane of infected cells. Infectious virions are formed when the virus buds from the surface of infected cells and is released. Spiked structures on the surface of virions project from the virion and serve to recognize and attach to specific receptor molecules on the surface of susceptible cells, allowing the virion to penetrate the cell. The genetic information contained in the RNA molecule directs production of new virus particles by using the cellular machinery to drive synthesis of new viral proteins and RNA. *See* VIRUS.

Although much is known about the agents of Ebola hemorrhagic fever disease, the ecology of Ebola viruses remains a mystery. The natural hosts of filoviruses remain unknown, and there has been little progress at unraveling the events leading to outbreaks or identifying sources of filoviruses in the wild. Fortunately, the incidence of human disease is relatively rare and has been limited to persons living in equatorial Africa or working with the infectious viruses. The virus is spread primarily through close contact with the body of an infected individual, his or her body fluids, or some other source of infectious material.

Ebola virus hemorrhagic fever disease in humans begins with an incubation period of 4–10 days, which is followed by abrupt onset of illness. Fever, headache, weakness, and other flulike symptoms lead to a rapid deterioration in the condition of the individual. In severe cases, bleeding and the appearance of small red spots or rashes over the body indicate that the disease has affected the integrity of the circulatory system. Individuals with Ebola virus die as a result of a shock syndrome that usually occurs 6–9 days after the onset of symptoms. This shock is due to the inability to control vascular functions and the massive injury to body tissues.

It appears that the immune response is impaired and that a strong cellular immune response is key to surviving infections. This immunosuppression may also be a factor in death, especially if secondary infections by normal bacterial flora ensue.

Outbreaks of Ebola virus disease in humans are controlled by the identification and isolation of infected individuals, implementation of barrier nursing techniques, and rapid disinfection of contaminated material. Diagnosis of Ebola virus cases is made by detecting virus proteins or RNA in blood or tissue specimens, or by detecting antibodies to the virus in the blood.

Dilute hypochlorite solutions (bleach), 3% phenolic solutions, or simple detergents (laundry or dish soap) can be used to destroy infectious virions. No known drugs have been shown to be effective in treating Ebola virus (or Marburg virus) infections, and protective vaccines against filoviruses have not been developed. [A.San.]

Ebony A genus, *Diospyros*, of the ebony family, containing more than 250 species. Some species are important for their succulent fruits, such as date plum, kaki plum, and persimmon, and several for their timber, particularly the heartwood, which is the true ebony of commerce.

Although it is popularly supposed to be a black wood, most species have a heartwood that is only streaked and mottled with black. The heartwood is very brittle, and is difficult to work, but it has long been in demand. The sapwood is white, becoming bluish or reddish when cut.

Black ebony is used for knife handles, piano keys, finger boards of violins, hairbrush backs, inlays, and marquetry. Some of the woods called ebony, however, belong to different families, especially the pulse family, Leguminosae.

Persimmon (*D. virginiana*), of the southeastern United States, is one of numerous tropical or subtropical species. The species in tropical America are too small or rare to be of economic value, although several of them have black heartwood used locally for making walking sticks, inlays, and miscellaneous articles of turnery and carving. [A.H.G./K.P.D.]

Echovirus One of the divisions of the enterovirus subgroup, within the picornavirus group of viruses. The name is derived from the term enteric cytopathogenic human orphan virus. More than 34 antigenic types exist. Only certain types have been associated with human illnesses, particularly with aseptic meningitis and febrile disease. Their epidemiology is similar to that of other enteroviruses. Echoviruses resemble polioviruses and coxsackieviruses in size (about 28 nanometers) and in many other properties. Diagnosis is made by isolation and typing of the viruses in tissue culture. Antibodies form during convalescence. *See* ANIMAL VIRUS; ENTEROVIRUS; PICORNAVIRIDAE.

[J.L.Me.]

Ecological communities Assemblages of living organisms that occur together in an area. The nature of the forces that knit these assemblages into organized systems and those properties of assemblages that manifest this organization have been topics of intense debate among ecologists since the beginning of the twentieth century. On the one hand, there are those who view a community as simply consisting of species with similar physical requirements, such as temperature, soil type, or light regime. The similarity of requirements dictates that these species be found together, but interactions between the species are of secondary importance and the level of organization is low. On the other hand, there are those who conceive of the community as a highly organized, holistic entity, with species inextricably and complexly linked to one another and to the physical environment, so that characteristic patterns recur, and properties arise that one can neither understand nor predict from a knowledge of the component species. In this view, the ecosystem (physical environment plus its community) is as well organized as a living organism, and constitutes a superorganism. Between these extremes are those who perceive some community organization but not nearly enough to invoke images of holistic superorganisms. *See* ECOSYSTEM.

Every community comprises a given group of species, and their number and identities are distinguishing traits. Most communities are so large that it is not possible to enumerate all species; microorganisms and small invertebrates are especially difficult to census. However, particularly in small, well-bounded sites such as lakes or islands, one can find all the most common species and estimate their relative abundances. The number of species is known as species richness, while species diversity refers to various statistics based on the relative numbers of individuals of each species in addition to the number of species. The rationale for such a diversity measure is that some communities have many species, but most species are rare and almost all the individuals (or biomass) in such a community can be attributed to just a few species. Such a community is not diverse in the usual sense of the word. Patterns of species diversity abound in the ecological literature; for example, pollution often effects a decrease in species diversity.

The main patterns of species richness that have been detected are area and isolation effects, successional gradients, and latitudinal gradients. Larger sites tend to have more species than do small ones, and isolated communities (such as those on oceanic islands) tend to have fewer species than do less isolated ones of equal size. Later communities in a temporal succession tend to have more species than do earlier ones, except that the last (climax) community often has fewer species than the immediately preceding one. Tropical communities tend to be very species-rich, while those in arctic climates tend to be species-poor. This observation conforms to a larger but less precise rule that communities in particularly stressful environments tend to have few species.

Communities are usually denoted by the presence of species, known as dominants, that contain a large fraction of the community's biomass, or account for a large fraction of a community's productivity. Dominants are usually plants. Determining whether communities at two sites are truly representatives of the "same" community requires knowledge of more than just the dominants, however. "Characteristic" species, which are always found in combination with certain other species, are useful in deciding whether two communities are of the same type, though the designation of "same" is arbitrary, just as is the designation of "dominant" or "characteristic."

Communities often do not have clear spatial boundaries. Occasionally, very sharp limits to a physical environmental condition impose similarly sharp limits on a community. For example, serpentine soils are found sharply delimited from adjacent soils in many areas, and have mineral concentrations strikingly different from those of the neighboring soils. Thus they support plant species that are very different from those found in nearby nonserpentine areas, and these different plant species support animal species partially different from those of adjacent areas.

Here two different communities are sharply bounded from each other. Usually, however, communities grade into one another more gradually, through a broad intermediate region (an ecotone) that includes elements of both of the adjacent communities, and sometimes other species as well that are not found in either adjacent community. *See* ECOTONE.

The environment created by the dominant species, by their effects on temperature, light, humidity, and other physical factors, and by their biotic effects, such as allelopathy and competition, may entrain some other species so that these other species' spatial boundaries coincide with those of the dominants. *See* PHYSIOLOGICAL ECOLOGY (PLANT); POPULATION ECOLOGY.

More or less distinct communities tend to follow one another in rather stylized order. As with recognition of spatial boundaries, recognition of temporal boundaries of adjacent communities within a sere (a temporary community during a successional sequence at a site) is partly a function of the expectations that an observer brings to the endeavor. Those who view communities as superorganisms are inclined to see sharp temporal and spatial boundaries, and the perception that one community does not gradually become another community over an extended period of time confirms the impression that communities are highly organized entities, not random collections of species that happen to share physical requirements. However, this superorganismic conception of succession has been replaced by an individualistic succession. Data on which species are present at different times during a succession show that there is not abrupt wholesale extinction of most members of a community and concurrent simultaneous colonization by most species of the next community. Rather, most species within a community colonize at different times, and as the community is replaced most species drop out at different times. That succession is primarily an individualistic process does not mean that there are not characteristic changes in community properties as most successions proceed. Species richness usually increases through most of the succession, for example, and stratification becomes more highly organized and well defined. A number of patterns are manifest in aspects of energy flow and nutrient cycling. *See* ECOLOGICAL SUCCESSION.

Living organisms are characterized not only by spatial and temporal structure but by an apparent purpose or activity termed teleonomy. In the first place, the various species within a community have different trophic relationships with one another. One species may eat another, or be eaten by another. A species may be a decomposer, living on dead tissue of one or more other species. Some species are omnivores, eating many kinds of food; others are more specialized, eating only plants or only animals, or even just one other species. These trophic relationships unite the species in a community into a common endeavor, the transmission of energy through the community. This energy flow is analogous to an organism's mobilization and transmission of energy from the food it eats.

By virtue of differing rates of photosynthesis by the dominant plants, different communities have different primary productivities. Tropical forests are generally most productive, while extreme environments such as desert or alpine conditions harbor rather unproductive communities. Agricultural communities are intermediate. Algal communities in estuaries are the most productive marine communities, while open ocean communities are usually far less productive. The efficiency with which various animals ingest and assimilate the plants and the structure of the trophic web determine the secondary productivity (production of organic matter by animals) of a community. Marine secondary productivity generally exceeds that of terrestrial communities. *See* AGROECOSYSTEM; BIOLOGICAL PRODUCTIVITY.

A final property that any organism must have is the ability to reproduce itself. Communities may be seen as possessing this property, though the sense in which they do so does not support the superorganism metaphor. A climax community reproduces itself through time simply by virtue of the reproduction of its constituent species, and may also be seen as reproducing itself in space by virtue of the propagules that its species transmit to less mature communities. For example, when a climax forest abuts a cutover field, if no disturbance ensues, the field undergoes succession and eventually becomes a replica of the adjacent forest. Both temporally and spatially, then, community reproduction is a collective rather than an emergent property, deriving directly from the reproductive activities of the component species. *See* ALTITUDINAL VEGETATION ZONES; BOG; CHAPARRAL; DESERT; ECOLOGY; GRASSLAND ECOSYSTEM; MANGROVE; MUSKEG; PARAMO; PUNA. [D.Sim.]

Ecological competition The interaction of two (or more) organisms (or species) such that, for each, the birth or growth rate is depressed and the death rate increased by the presence of the other organisms (or species). Competition is recognized as one of the more important forces structuring ecological communities, and interest in competition led to one of the first axioms of modern ecology, the competitive exclusion principle. The principle suggests that in situations where the growth and reproduction of two species are resource-limited, only one species can survive per resource.

The competitive exclusion principle was originally derived by mathematicians using the Lotka-Volterra competition equations. This model of competition predicts that if species differ substantially in competitive ability, the weaker competitor will be eliminated by the stronger competitor. However, a competitive equilibrium can occur if the negative effect of each species on itself (intraspecific competition) is greater than the negative effect of each species on the other species (interspecific competition). Because the competitive exclusion principle implies that competing species cannot coexist, it follows that high species diversity depends upon mechanisms through which species avoid competition. *See* MATHEMATICAL ECOLOGY.

In general, competitive exclusion can be prevented if the relative competitive abilities of species vary through time and space. Such variation occurs in two ways. First, dispersal rates into particular patches may fluctuate, causing fluctuations in the numerical advantage of a species in a particular patch. Second, competitive abilities of species may be environmentally dependent and, therefore, fluctuate with local environmental changes. Competitive exclusion can also be avoided if fluctuations in environmental factors reduce the densities of potentially competing species to levels where competition is weak and population growth is for a time insensitive to density.

Coexistence is not merely a result of environmental harshness or fluctuations but also involves the critical element of niche differentiation (that is, species must differ from one another if they are to coexist). However, the focus is not how species coexist by partitioning resources, but how species can coexist on the same resources by differing sufficiently in their responses to environmental conditions and fluctuations. *See* ECOLOGICAL COMMUNITIES; ECOLOGICAL SUCCESSION.

Competition theory has been applied to human-manipulated ecosystems used to produce food, fiber, and forage crops as well as in forestry and rangeland management. Although many characteristics of agricultural systems are similar to those of natural ecosystems, agricultural communities are unique because they are often managed for single-species (sometimes multispecies) production and they are usually characterized by frequent and intense disturbance. Studies of competition in agriculture have primarily examined crop loss from weed abundance under current cropping practices, and have evaluated various weed control tactics and intercropping systems. Factors that influence competition in agroecosystems include the timing of plant emergence, growth rates, spatial arrangements among neighbors, plant–plant-environment interactions, and herbivory. *See* ECOLOGY; ECOLOGY, APPLIED. [P.C.M.]

Ecological energetics The study of the flow of energy within an ecological system from the time the energy enters the living system until it is ultimately degraded to heat and irretrievably lost from the system. It is also referred to as production ecology, because ecologists use the word production to describe the process of energy input and storage in ecosystems.

Ecological energetics provides information on the energetic interdependence of organisms within ecological systems and the efficiency of energy transfer within and between organisms and trophic levels. Nearly all energy enters the biota by green plants' transformation of light energy into chemical energy through photosynthesis; this is referred to as primary production. This accumulation of potential energy is used by plants, and by the animals which eat them, for growth, reproduction, and the work necessary to sustain life. The energy put into growth and reproduction is termed secondary production. As energy passes along the food chain to higher trophic levels (from plants to herbivores to carnivores), the potential energy is used to do work and in the process is degraded to heat. The laws of thermodynamics require the light energy fixed by plants to equal the energy degraded to heat, assuming the system is closed with respect to matter. An energy budget quantifies the energy pools, the directions of energy flow, and the rates of energy transformations within ecological systems. *See* BIOLOGICAL PRODUCTIVITY; FOOD WEB; PHOTOSYNTHESIS.

The essentials of ecological energetics can be most readily appreciated by considering energy flowing through an individual; it is equally applicable to populations, communities, and ecosystems. Of the food energy available, only part is harvested in the process of foraging. Some is wasted, for example, by messy eaters, and the rest consumed. Part of the consumed food is transformed but is not utilized by the body, leaving as fecal material or as nitrogenous waste, the by-product of protein metabolism. The remaining energy is assimilated into the body, part of which is used to sustain the life functions and to do work—this is manifest as oxygen consumption. The remainder of the assimilated energy is used to produce new tissue, either as growth of the individual or as development of offspring. Hence production is also the potential energy (proteins, fats, and carbohydrates) on which other organisms feed. Production leads to an increase in biomass or is eliminated through death, migration, predation, or the shedding of, for example, hair, skin, and antlers.

Energy flows through the consumer food chain (from plants to herbivores to carnivores) or through the detritus food chain. The latter is fueled by the waste products of the consumer food chain, such as feces, shed skin, cadavers, and nitrogenous waste. Most detritus is consumed by microorganisms, although this food chain includes conspicuous carrion feeders like beetles and vultures. In terrestrial systems, more than 90% of all primary production may be consumed by detritus feeders. In aquatic systems, where the plants do not require tough supporting tissues, harvesting by herbivores may be efficient with little of the primary production passing to the detrivores.

Traditionally the calorie, a unit of heat energy, has been used in ecological energetics, but this has been largely replaced by the joule. Production is measured from individual growth rates and the reproductive rate of the population to determine the turnover time. The energy equivalent of food consumed, feces, and production can be determined by measuring the heat evolved on burning a sample in an oxygen bomb calorimeter, or by chemical analysis—determining the amount of carbon or of protein, carbohydrate, and lipid and applying empirically determined caloric equivalents to the values. The latter three contain, respectively, 16.3, 23.7, and 39.2 kilojoules per gram of dry weight. Maintenance costs are usually measured indirectly as respiration (normally the oxygen consumed) in the laboratory and extrapolated to the field conditions. Error is introduced by the fact that animals have different levels of activity in the field and are subject to different temperatures, and so uncertainty has surrounded these extrapolations. Oxygen consumption has been measured in animals living in the wild by using the turnover rates of doubly labeled water (D_2O).

Due to the loss of usable energy with each transformation, in an area more energy can be diverted into production by plants than by consumer populations. For humans this means that utilizing plants for food directly is energetically much more efficient than converting them to eggs or meat. *See* BIOMASS; ECOLOGICAL COMMUNITIES; ECOSYSTEM.

[W.F.H.]

Ecological modeling The use of computer simulations or mathematical equations to address questions that cannot be answered solely by experiments or observations. Ecological models have two major aims: to provide general insight into how ecological systems or ecological interactions work; and to provide specific predictions about the likely futures of particular populations, communities, or ecosystems.

Models can be used to indicate general possibilities or to forecast the most likely outcomes of particular populations or ecosystems. Models differ in whether they are "basic" or are intended to address management decisions. As ecology has grown in its sophistication, models are increasingly used as decision support tools for policymakers. Models of virtually every possible type of ecological interaction have been developed (competition, parasitism, disease, mutualism, plant-herbivore interactions, and so forth). The models vary in their level of detail. Some models simply keep track of the density of organisms, treating all organisms of any species as identical (mass action models). At the other extreme, the movement and fate of each individual organism may be tracked in an elaborate computer simulation (individual behavior models). *See* POPULATION ECOLOGY.

Simple algebraic models are very useful for indicating general principles and possibilities. In order to be a management tool, the model must be more complicated and detailed to reflect the specific situation under examination. For example, instead of a few equations, ecologists have modeled spotted owl populations and old growth forests in Washington using a detailed computer simulation that keeps track of habitat in real maps at the scale of hectares. In these simulation models, owls are moved as individuals from one hectare to another, and their fate (survival, death, or reproduction) is recorded in the computer's memory. By tracking hundreds or even thousands of owls moving around in this computer world, different forestry practices corresponding to different logging scenarios can be examined. *See* ECOLOGY, APPLIED; MATHEMATICAL ECOLOGY; SYSTEMS ECOLOGY.

A model is a formal way of examining the consequences of a series of assumptions about how nature works. Such models refine thinking and clarify what results are implied by any set of assumptions. As models become more complicated and specific, they can also be used to conduct experiments that are too expensive or impractical in the field.

One danger of ecological modeling is the uncertainty of the models and the shortage of supporting data. Properly used, models allow exploration of a wide range of uncertainty, pointing out the limits of current knowledge and identifying critical information required prior to management decision making. However, it would not be prudent to rely solely on the output of any model. *See* ECOLOGY. [P.Ka.]

Ecological succession A directional change in an ecological community. Populations of animals and plants are in a dynamic state. Through the continual turnover of individuals, a population may expand or decline depending on the success of its members in survival and reproduction. As a consequence, the species composition of communities typically does not remain static with time. Apart from the regular fluctuations in species abundance related to seasonal changes, a community may develop progressively with time through a recognizable sequence known as the sere. Pioneer populations are replaced by successive colonists along a more or less predictable path toward a relatively stable community. This process of succession results from interactions between different species, and between species and their environment, which govern the sequence and the rate with which species replace each other. The rate at which succession proceeds depends on the time scale of species' life histories as well as on the effects species may have on each other and on the environment which supports them. In some cases, seres may take hundreds of years to complete, and direct observation at a given site is not possible. Adjacent sites may be identified as successively older stages of the same sere, if it is assumed that conditions were similar when each seral stage was initiated. *See* ECOLOGICAL COMMUNITIES; POPULATION ECOLOGY.

The course of ecological succession depends on initial environmental conditions. Primary succession occurs on novel areas such as volcanic ash, glacial deposits, or bare rock, areas which have not previously supported a community. In such harsh, unstable environments, pioneer colonizing organisms must have wide ranges of ecological tolerance to survive. In contrast, secondary succession is initiated by disturbance such as fire, which removes a previous community from an area. Pioneer species are here constrained not by the physical environment but by their ability to enter and exploit the vacant area rapidly.

As succession proceeds, many environmental factors may change through the influence of the community. Especially in primary succession, this leads to more stable, less severe environments. At the same time interactions between species of plant tend to intensify competition for basic resources such as water, light, space, and nutrients. Successional change results from the normal complex interactions between organism and environment which lead to changes in overall species composition. Whether succession is promoted by changing environmental factors or competitive interactions, species composition alters in response to availability of niches. Populations occurring in the community at a point in succession are those able to provide propagules (such as seeds) to invade the area, being sufficiently tolerant of current environmental conditions, and able to withstand competition from members of other populations present at the same stage. Species lacking these qualities either become locally extinct or are unable to enter and survive in the community.

Early stages of succession tend to be relatively rapid, whereas the rates of species turnover and soil changes become slower as the community matures. Eventually an approximation to the steady state is established with a relatively stable community, the nature of which has aroused considerable debate. Earlier, the so-called climax vegetation was believed to be determined ultimately by regional climate and, given sufficient time, any community in a region would attain this universal condition. This unified concept of succession, the monoclimax hypothesis, implies the ability of organisms progressively to modify their environment until it can support the climatic climax community. Although plants and animals do sometimes ameliorate environmental conditions, evidence suggests overwhelmingly that succession has a variety of stable end points. This hypothesis, known as the polyclimax hypothesis, suggests that the end point of a succession depends on a complex of environmental factors that characterize the site, such as parent material, topography, local climate, and human influences.

Actions of the community on the environment, termed autogenic, provide an important driving force promoting successional change, and are typical of primary succession where initial environments are inhospitable. Alternatively, changes in species composition of a community may result from influences external to the community called allogenic.

Whereas intrinsic factors often result in progressive successional changes, that is, changes leading from simple to more complex communities, external (allogenic) forces may induce retrogressive succession, that is, toward a less mature community. For example, if a grassland is severely overgrazed by cattle, the most palatable species will disappear. As grazing continues, the grass cover is reduced, and in the open areas weeds characteristic of initial stages of succession may become established.

In some instances of succession, the food web is based on photosynthetic organisms, and there is a slow accumulation of organic matter, both living and dead. This is termed autotrophic succession. In other instances, however, addition of organic matter to an ecosystem initiates a succession of decomposer organisms which invade and degrade it. Such a succession is called heterotrophic. *See* EUTROPHICATION; FOOD WEB.

Observed changes in the structure and function of seral communities result from natural selection of individuals within their current environment. Three mechanisms by which species may replace each other have been proposed; the relative importance of each apparently depends on the nature of the sere and stage of development.

1. The facilitation hypothesis states that invasion of later species depends on conditions created by earlier colonists. Earlier species modify the environment so as to increase the competitive ability of species which are then able to displace them. Succession thus proceeds because of the effects of species on their environment.

2. The tolerance hypothesis suggests that later successional species tolerate lower levels of resources than earlier occupants and can invade and replace them by reducing resource levels below those tolerated by earlier occupants. Succession proceeds despite the resistance of earlier colonists.

3. The inhibition hypothesis is that all species resist invasion of competitors and are displaced only by death or by damage from factors other than competition. Succession proceeds toward dominance by longer-lived species.

None of these models of succession is solely applicable in all instances; indeed most examples of succession appear to show elements of all three replacement mechanisms.

Succession has traditionally been regarded as following an orderly progression of changes toward a predictable end point, the climax community, in equilibrium with the prevailing environment. This essentially deterministic view implies that succession will always follow the same course from a given starting point and will pass through a recognizable series of intermediate states. In contrast, a more recent view of succession is based on adaptations of independent species. It is argued that succession is disorderly and unpredictable, resulting from probabilistic processes such as invasion of propagules and survival of individuals which make up the community. Such a stochastic view reflects the inherent variability observed in nature and the uncertainty of environmental conditions. In particular, it allows for succession to take alternative pathways and end points dependent on the chance outcome of interactions among species and between species and their environment.

Consideration of community properties such as energy flow supports the view of succession as an orderly process. The rate of gross primary productivity typically becomes limited also by the availability of nutrients, now incorporated within the community biomass, and declines to a level sustainable by release from decomposer organisms. Species diversity tends to rise rapidly at first as successive invasions occur, but declines again with the elimination of the pioneer species by the climax community.

Stochastic aspects of succession can be represented in the form of models which allow for transitions between a series of different "states." Such models, termed Markovian models, can apply at various levels: plant-by-plant replacement, changes in tree size categories, or transitions between whole communities. A matrix of replacement probabilities defines the direction, pathway, and likelihood of change, and the model can be used to predict the future composition of the community from its initial state. [P.Ran.]

Ecology The subdiscipline of biology that concentrates on the relationships between organisms and their environments; it is also called environmental biology. Ecology is concerned with patterns of distribution (where organisms occur) and with patterns of abundance (how many organisms occur) in space and time. It seeks to explain the factors that determine the range of environments that organisms occupy and that determine how abundant organisms are within those ranges. It also emphasizes functional interactions between co-occurring organisms. In addition to being a unique component of the biological sciences, ecology is both a synthetic and an integrative science since it often draws upon information and concepts in other sciences, ranging from physiology to meteorology, to explain the complex organization of nature.

Environment is all of those factors external to an organism that affect its survival, growth, development, and reproduction. It can be subdivided into physical, or abiotic, factors, and biological, or biotic, factors. The physical components of the environment include all nonbiological constituents, such as temperature, wind, inorganic chemicals, and radiation. The biological components of the environment include the organisms. A somewhat more general term is habitat, which refers in a general way to where an organism occurs and the environmental factors present there. *See* ENVIRONMENT.

A recognition of the unitary coupling of an organism and its environment is fundamental to ecology; in fact, the definitions of organism and environment are not separate. Environment is organism-centered since the environmental properties of a habitat are determined by the requirements of the organisms that occupy that habitat. For example, the amount of inorganic nitrogen dissolved in lake water is of little immediate significance to zooplankton in the lake because they are incapable of utilizing inorganic nitrogen directly. However, because phytoplankton are capable of utilizing inorganic nitrogen directly, it is a component of their environment. Any effect of inorganic nitrogen upon the zooplankton, then, will occur indirectly through its effect on the abundance of the phytoplankton that the zooplankton feed upon. *See* PHYTOPLANKTON; ZOOPLANKTON.

Just as the environment affects the organism, so the organism affects its environment. Growth of phytoplankton may be nitrogen-limited if the number of individuals has become so great that there is no more nitrogen available in the environment. Zooplankton, not limited by inorganic nitrogen themselves, can promote the growth of additional phytoplankton by consuming some individuals, digesting them, and returning part of the nitrogen to the environment.

Ecology is concerned with the processes involved in the interactions between organisms and their environments, with the mechanisms responsible for those processes, and with the origin, through evolution, of those mechanisms. It is distinguished from such closely related biological subdisciplines as physiology and morphology because it is not intrinsically concerned with the operation of a physiological process or the function of a structure, but with how a process or structure interacts with the environment to influence survival, growth, development, and reproduction.

Major subdivisions of ecology by organism include plant ecology, animal ecology, and microbial ecology. Subdivisions by habitat include terrestrial ecology, the study of organisms on land; limnology, the study of fresh-water organisms and habitats; and oceanography, the study of marine organisms and habitats.

The levels of organization studied range from the individual organism to the whole complex of organisms in a large area. Autecology is the study of individuals, population ecology is the study of groups of individuals of a single species or a limited number of species, synecology is the study of communities of several populations, and ecosystem, or simply systems, ecology is the study of communities of organisms and their environments in a specific time and place. *See* POPULATION ECOLOGY; SYSTEMS ECOLOGY.

Higher levels of organization include biomes and the biosphere. Biomes are collections of ecosystems with similar organisms and environments and, therefore, similar ecological properties. All of Earth's coniferous forests are elements in the coniferous forest biome. Although united by similar dynamic relationships and structural properties, the biome itself is more abstract than a specific ecosystem. The biosphere is the most inclusive category possible, including all regions of Earth inhabited by living things. It extends from the lower reaches of the atmosphere to the depths of the oceans. *See* BIOME; BIOSPHERE.

The principal methodological approaches to ecology are descriptive, experimental, and theoretical. Descriptive ecology concentrates on the variety of populations, communities, and habitats throughout Earth. Experimental ecology involves manipulating organisms or their environments to discover the underlying mechanisms governing distribution and abundance. Theoretical ecology uses mathematical equations based on assumptions about the properties of organisms and environments to make predictions about patterns of distribution and abundance. *See* THEORETICAL ECOLOGY. [S.J.McN.]

Ecology, applied The application of ecological principles to the solution of human problems and the maintenance of a quality life. It is assumed that humans are an integral part of ecological systems and that they depend upon healthy, well-operating, and productive systems for their continued well-being. For these reasons, applied ecology is based on a knowledge of ecosystems and populations, and the principles and techniques of ecology are used to interpret and solve specific environmental problems and to plan new management systems in the biosphere. Although a variety of management fields, such as forestry, agriculture, wildlife management, environmental engineering, and environmental design, are concerned with specific parts of the environment, applied ecology is unique in taking a view of whole systems, and attempting to account for all inputs to and outputs from the systems—and all impacts. In the past, applied ecology has been considered as being synonymous with the above applied sciences.

The objective of applied ecology management is to maintain the system while altering its inputs or outputs. Often, ecology management is designed to maximize a particular output or the quantity of a specific component. Since outputs and inputs are related, maximization of an output may not be desirable; rather, the management objective may be the optimum level. Optimization of systems can be accomplished through the use of systems ecology methods which consider all parts of the system rather than a specific set of components. In this way, a series of strategies or scenarios can be evaluated, and the strategy producing the largest gain for the least cost can be chosen for implementation.

A variety of general environmental problems within the scope of applied ecology relate to the major components of the Earth: the atmosphere, water, land, and the biota. Applied ecology also is concerned with the size of the human population, since many of the impacts of human activities on the environment are a function of the number and concentration of people. *See* ECOLOGY; ECOSYSTEM; HUMAN ECOLOGY; SYSTEMS ECOLOGY.

[F.B.Go.]

Ecosystem A functional system that includes an ecological community of organisms together with the physical environment, interacting as a unit. Ecosystems are characterized by flow of energy through food webs, production and degradation of organic matter, and transformation and cycling of nutrient elements. This production of organic molecules serves as the energy base for all biological activity within ecosystems. The consumption of plants by herbivores (organisms that consume living plants or algae) and detritivores (organisms that consume dead organic matter) serves to transfer energy stored in photosynthetically produced organic molecules to other organisms. Coupled to the production of organic matter and flow of energy is the cycling of elements. *See* ECOLOGICAL COMMUNITIES; ENVIRONMENT.

All biological activity within ecosystems is supported by the production of organic matter by autotrophs (organisms that can produce organic molecules such as glucose from inorganic carbon dioxide; see illustration). More than 99% of autotrophic production on Earth is through photosynthesis by plants, algae, and certain types of bacteria. Collectively these organisms are termed photoautotrophs (autotrophs that use energy from light to produce organic molecules). In addition to photosynthesis, some production is conducted by chemoautotrophic bacteria (autotrophs that use energy stored in the chemical bonds of inorganic molecules such as hydrogen sulfide to produce organic molecules). The organic molecules produced by autotrophs are used to support the organism's metabolism and reproduction, and to build new tissue. This new tissue

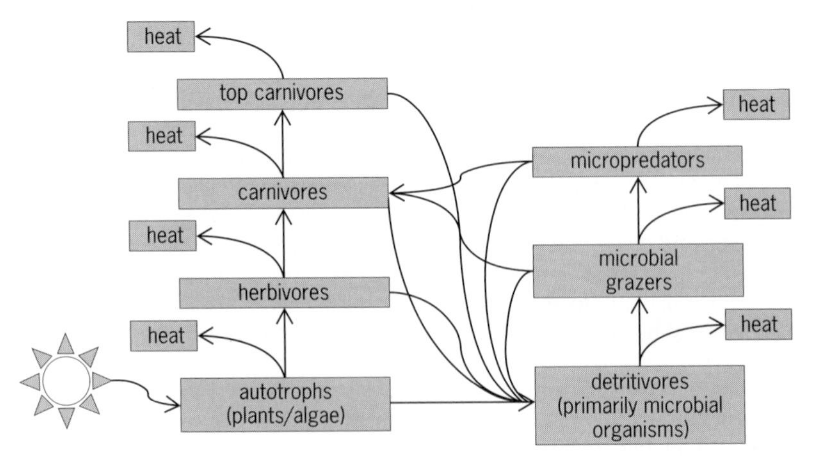

General model of energy flow through ecosystems.

is consumed by herbivores or detritivores, which in turn are ultimately consumed by predators or other detritivores.

Terrestrial ecosystems, which cover 30% of the Earth's surface, contribute a little over one-half of the total global photosynthetic production of organic matter—approximately 60×10^{15} grams of carbon per year. Oceans, which cover 70% of the Earth's surface, produce approximately 51×10^{15} g C y^{-1} of organic matter. *See* BIOMASS.

Food webs. Organisms are classified based upon the number of energy transfers through a food web (see illustration). Photoautotrophic production of organic matter represents the first energy transfer in ecosystems and is classified as primary production. Consumption of a plant by a herbivore is the second energy transfer, and thus herbivores occupy the second trophic level, also known as secondary production. Consumer organisms that are one, two, or three transfers from photoautotrophs are classified as primary, secondary, and tertiary consumers. Moving through a food web, energy is lost during each transfer as heat, as described by the second law of thermodynamics. Consequently, the total number of energy transfers rarely exceeds four or five; with energy loss during each transfer, little energy is available to support organisms at the highest levels of a food web. *See* ECOLOGICAL ENERGETICS; FOOD WEB.

Biogeochemical cycles. In contrast to energy, which is lost from ecosystems as heat, chemical elements (or nutrients) that compose molecules within organisms are not altered and may repeatedly cycle between organisms and their environment. Approximately 40 elements compose the bodies of organisms, with carbon, oxygen, hydrogen, nitrogen, and phosphorus being the most abundant. If one of these elements is in short supply in the environment, the growth of organisms can be limited, even if sufficient energy is available. In particular, nitrogen and phosphorus are the elements most commonly limiting organism growth. This limitation is illustrated by the widespread use of fertilizers, which are applied to agricultural fields to alleviate nutrient limitation. *See* BIOGEOCHEMISTRY; NITROGEN CYCLE.

Carbon cycles between the atmosphere and terrestrial and oceanic ecosystems. This cycling results, in part, from primary production and decomposition of organic matter. Rates of primary production and decomposition, in turn, are regulated by the supply of nitrogen, phosphorus, and iron. The combustion of fossil fuels is a recent change in the global cycle that releases carbon that has long been buried within the Earth's crust to the atmosphere. Carbon dioxide in the atmosphere traps heat on the Earth's surface and is a major factor regulating the climate. This alteration of the global carbon cycle along with the resulting impact on the climate is a major issue under investigation by ecosystem ecologists. *See* AIR POLLUTION; CONSERVATION OF RESOURCES; ECOLOGY, APPLIED; HUMAN ECOLOGY; WATER POLLUTION. [J.B.Jo.]

Ecotone A geographic boundary or transition zone between two different groups of plant or animal distributions. The term has been used to denote transitions at different spatial scales or levels of analysis, and may refer to any one of several attributes of the organisms involved. For example, an ecotone could refer to physiognomy (roughly, the morphology or appearance of the relevant organisms), such as between the boreal forest and grassland biomes; or it could refer to composition, such as between oak-hickory and maple-basswood forest associations; or it could refer to both. Ecotones are generally distinguished from other geographic transitions of biota by their relative sharpness. The ecotone between boreal forest and prairie in central Saskatchewan occurs over a hundred kilometers or so, in contrast to the transition from tropical forest to savanna in South America or Africa that is associated with increasing aridity and is dispersed over hundreds of kilometers. The "tension zone" between broadleaf deciduous forests in south-cental Michigan and mixed forests to the north is similarly sharp. Ecotones are thought to reflect concentrated long-term gradients of one or more current environmental (rather than historical or human) factors. Though often climatic, these factors can also be due to substrate materials, such as glacial sediments or soils. Regardless of their specific environmental basis, most ecotones are thought to be relatively stable.

Ecotones are often reflected in the distributions of many biota besides the biota used to define them. The prairie-forest ecotone, for example, is defined not only by the dominant vegetation components but also by many faunal members of the associated ecosystems, such as insects, reptiles and amphibians, mammals, and birds, that reach their geographic limits there. *See* ALPINE VEGETATION; BIOME; ECOLOGICAL COMMUNITIES; ECOSYSTEM; FOREST ECOSYSTEM; GRASSLAND ECOSYSTEM; SAVANNA; ZOOGEOGRAPHY. [J.R.Har.]

Ehrlichiosis A tick-borne infection that often is asymptomatic but also can produce an illness ranging from a few mild symptoms to an overwhelming multisystem disease. Ehrlichiosis is included with those infections that are said to be emerging, either because they have been recognized only recently or because they were previously well known but now are occurring more frequently.

Human ehrlichiosis is caused by two distinct species: *E. chaffeensis* and an unnamed ehrlichial species. In the United States, *Ehrlichia chaffeensis* infects primarily mononuclear blood cells; the disease produced by this species is referred to as human monocytic ehrlichiosis. The other ehrlichial species invades granulocytic blood cells, causing human granulocytic ehrlichiosis. The latter organism closely resembles *E. equi*, a species that infects horses.

Both of these ehrlichia species are transmitted to humans by the bite of infected ticks. *Ehrlichia chaffeensis* occurs most commonly in the south-central and southeastern states, where it is associated primarily with the Lone Star tick (*Amblyomma americanum*); it is also transmitted by the common dog tick (*Dermacentor variabilis*). The agent of human granulocytic ehrlichiosis is found in the upper midwestern states of Wisconsin and Minnesota, as well as in several northeastern states. This agent seems to be transmitted principally by the deer tick (*Ixodes scapularis*). Although ticks (the vector) are the mode of transmission of ehrlichial infections to humans, the ticks must acquire the ehrlichial organisms from animal sources (the reservoir hosts).

The forms of the disease caused by the two ehrlichial species are indistinguishable. Illness occurs most often during April–September, corresponding to the period when ticks are most active and humans are pursuing outdoor activities. In ehrlichiosis, the incubation period can last from 1 to 3 weeks after exposure to the infected tick. Thereafter, individuals develop fever, chills, headache, and muscle pains. Gastrointestinal symptoms such as nausea, vomiting, and loss of appetite also are common. Laboratory abnormalities regularly include anemia, low white blood cell and platelet counts, and abnormal liver function. More severely ill individuals also may manifest abnormalities of the central nervous system, lungs, and kidneys. Because the clinical presentation is nonspecific, the diagnosis of ehrlichiosis may not be immediately apparent. Prolonged intervals between the onset of illness and the administration of appropriate therapy can lead to more severe disease symptoms and a greater risk of fatality. *See* CLINICAL MICROBIOLOGY; INFECTION.

An important clue to the diagnosis of human granulocytic ehrlichiosis is the recognition of cytoplasmic vacuoles filled with ehrlichiae (morulae) in circulating neutrophils. Careful examination of stained smears of peripheral blood often yields such findings in human granulocytic ehrlichiosis. In disease caused by *E. chaffeensis*, laboratory diagnosis usually is made by detecting an increase in species-specific antibodies in serum specimens obtained during the acute and convalescent phases of the illness. However, such serologic testing is of no use in establishing the diagnosis before treatment is initiated. Therefore, therapy must be initiated on clinical suspicion.

Ehrlichiosis closely resembles another tick-borne illness, Rocky Mountain spotted fever, except that the rash, characteristic of spotted fever, is usually absent or modest. Hence, ehrlichiosis has been referred to as spotless fever. Fortunately, both diseases can be treated with tetracycline antibiotics. Most individuals respond to tetracycline therapy within 48–72 h. *See* RICKETTSIOSES.

The avoidance of tick bites is fundamental to preventing ehrlichiosis. [W.Scha.; S.M.St.]

El Niño In general, an invasion of warm water into the central and eastern equatorial Pacific Ocean off the coast of Peru and Ecuador, with a return period of 4–7 years. El Niño events come in various strengths: weak, moderate, strong, very strong, and extraordinary. The size of an El Niño event can be determined using various criteria: the amount of warming of sea surface temperatures in the central and eastern Pacific from their average condition; the areal extent of that warm water anomaly; and the length of time that the warm water lingers before being replaced by colder-than-average sea surface temperatures in this tropical Pacific region.

Under normal conditions the winds blow up the west coast of South America and then near the Equator turn westward to Asia. The surface water is piled up in the western Pacific, and the sea level there is several tens of centimeters above average while the sea level in the eastern Pacific is below average. As the water is pushed toward the west, cold water from the deeper part of the ocean along the Peruvian coast wells up to the surface to replace it. This cold water is rich with nutrients, making the coastal upwelling region along western South America among the most productive fisheries in the world. *See* UPWELLING.

Every 4–7 years those winds tend to die down and sometimes reverse, allowing the warm surface waters that piled up in the west to move back toward the eastern part of the Pacific Basin. With reduced westward winds the surface water also heats up. Sea level drops in the western Pacific and increases in the eastern part of the basin. El Niño condition can last for 12–18 months, sometimes longer, before the westward flowing winds start to pick up again. Occasionally, the opposite also occurs: the eastern Pacific becomes cooler than normal, rainfall decreases still more, atmospheric surface pressure increases, and the westward winds become stronger. This irregular cyclic swing of warm and cold phases in the tropical Pacific is referred to as ENSO (El Niño Southern Oscillation).

El Niño is considered to be the second biggest climate-related influence on human activities, after the natural flow of the seasons. Although the phenomenon is at least thousands of years old, its impacts on global climate have only recently been recognized. Due to improved scientific understanding and forecasting of El Niño's interannual process, societies can prepare for and reduce its impacts considerably. *See* CLIMATOLOGY; MARITIME METEOROLOGY; TROPICAL METEOROLOGY. [M.H.G.]

Numerical models that couple the atmosphere to the ocean have been used to successfully predict the sea surface temperature of the tropical Pacific a year or so in advance. The basic reason that the cycle is predictable is that ENSO evolves slowly and regularly. If the initial state of the atmosphere-ocean system can be characterized accurately, the classification of this state in the ENSO sequence is made (even if it is not completely recognizable in each system separately), and the future evolution of the cycle can be predicted. *See* CLIMATE MODELING. [E.S.S.]

Elements, geochemical distribution of The distribution of the chemical elements within the Earth in space and time. Knowledge of the geochemical distribution of the elements in the Earth, particularly in the Earth's crust, and of the processes that lead to the observed distributions make it possible to locate and use efficiently essential elements and minerals and to predict their dispersal patterns when they reenter the natural environment after use.

To understand the present-day distribution of the elements in the Earth, it is necessary to go back to the time of Earth formation approximately 4.5 billion years ago. It is generally believed that the Earth and the other planets in the solar system formed by agglomeration of smaller fragments of solid material orbiting around the Sun. This material had precipitated from a cooling hot gas cloud (the solar nebula), with the most refractory materials condensing out first, the most volatile last. The distribution of elements in the solar system in this early phase thus had much to do with volatility, and the solid material that aggregated to form the planets was a mix of volatile and nonvolatile materials.

Although the Earth may have been an approximately homogeneous mixture of accreted materials at the time of its formation, it is now made of many chemically distinct parts. At the fundamental level, these are the core, the mantle, and the crust. While chemical fractionation in the solar nebula depended upon volatility, chemical differentiation within the Earth took place by the separation of molten material from unmelted residue under the influence of gravity. Because large amounts of energy were released from accreting fragments, the early Earth was very hot, and during the accretion stage itself, temperatures in some parts exceeded the melting point of iron metal. Pools of dense molten iron, with dissolved nickel and other elements, aggregated and sank through the Earth under gravity to form the core, leaving behind a mantle of silicate and oxide minerals. The present core constitutes about 32.4% of the Earth's mass. The distinct parts of the Earth possess unique overall compositions (see table).

The large-scale distribution of the elements in the Earth depends on the affinity of each element for specific compounds or phases. Those elements that alloy easily with iron, for example, are mostly sequestered in the Earth's core; those which form

Element distribution among some of the major subdivisions of the Earth*

Element	Continental crust	Oceanic crust	Upper mantle	Core†
Oxygen	45.3	43.6	44.2	—
Silicon	26.7	23.1	21.0	—
Aluminium	8.39	8.47	1.75	—
Iron	7.04	8.16	6.22	85.5
Calcium	5.27	8.08	1.86	—
Magnesium	3.19	4.64	24.0	—
Sodium	2.29	2.08	0.25	—
Potassium	0.91	0.13	0.02	—
Titanium	0.68	1.12	0.11	—
Nickel	0.011	0.014	0.20	5.5
Sulfur	NA	NA	NA	9.0

*Estimates of element abundances are in percent by weight and are arranged in order of decreasing abundance in the continental crust. Sulfur contents are not well known and are designated "not applicable."

†The estimate for the core is just one of several models. Others substitute light elements such as oxygen, carbon, or silicon for some or most of the sulfur shown here.

oxides and silicate minerals tend to be concentrated in the Earth's crust and mantle. Although many elements display multiple characteristics depending on the chemical environment, a classification according to geochemical affinity is nevertheless useful. The categories in this classification include atmophile (elements that are gases and concentrate in the atmosphere), lithophile (elements that form silicates or oxides and are concentrated in the minerals of the Earth's crust), siderophile (elements that alloy easily with iron and are concentrated in the core), and chalcophile (elements such as copper which commonly form sulfide minerals if sufficient sulfur is available).

Although geochemists have a good general knowledge of the overall distribution of elements in the core and mantle, much more detailed information is available about the chemical composition of the crust, which is accessible. The crust is actually composed of two major parts with quite different compositions, thickness, and average age: the continental crust and the oceanic crust.

Although all elements are present, the crust is made almost entirely of just nine chemical elements: oxygen, silicon, aluminum, iron, magnesium, calcium, sodium, potassium, and titanium. Oxygen and silicon are by far the most abundant. The most common minerals in the crust are those of the silicate family, in which the basic building block is a silicon atom surrounded by four oxygen atoms in the form of a tetrahedron. The crust is essentially a framework of oxygen atoms bound together by the common cations.

A variety of processes act to make the crust chemically heterogeneous on many scales. Many of these processes involve liquid water. Running water physically sorts particles depending on size and density, which are ultimately related to chemical composition. It is also a superb solvent, carrying many elements in solution under different conditions of temperature and pressure, and depositing them when these conditions change. Processes involving water account for many ore deposits, in which extreme concentrations of some elements occur relative to their average abundance in the crust. One example is the circulating hydrothermal solutions in volcanically active parts of the crust, which can leach metals from their normally dispersed state in large volumes of volcanic rocks and deposit them in concentrated zones as the solutions cool and encounter different rock types. Another example is the action of weathering in tropical regions with high rainfall, which can leach away all but the least soluble components from large volumes of rock, leaving behind mineral deposits rich in aluminum or, depending on the original composition of the rocks being weathered, metals such as iron and nickel. *See* GEOCHEMISTRY; WEATHERING PROCESSES. [G.Fau.; J.D.MacD.]

Elm Any species of *Ulmus*, a genus of hardwood trees in the Northern Hemisphere, with simple, serrate, deciduous leaves. The American or white elm (*U. americana*) is the most important species. It ranges from the eastern half of the United States westward as far as the base of the northern Rockies and southward through central Texas to the Gulf of Mexico. The tree is also found in southern Canada.

The tree was very popular as a shade tree, perhaps better adapted as a street tree than any other species because the upper branches spread, joining with elms across the street to form an arch. Once abundant, the elm has been severely attacked by the lethal Dutch elm disease imported from Europe. The future of the species is uncertain. [A.H.G./K.P.D.]

Endangered species A species that is in danger of extinction throughout all or a significant portion of its range. "Threatened species" is a related term, referring to a species likely to become endangered within the foreseeable future. The main factors that cause species to become endangered are habitat destruction, invasive species, pollution, and overexploitation.

Habitat destruction is the single greatest threat to species around the globe. Natural habitat includes the breeding sites, nutrients, physical features, and processes such as periodic flooding or periodic fires that species need to survive. Humans have altered, degraded, and destroyed habitat in many different ways. Logging around the world has destroyed forests that are habitat to many species. This has a great impact in tropical areas, where species diversity is highest. Although cut forests often regrow, many species depend upon old-growth forests that are over 200 years old; these forests are destroyed much faster than they can regenerate. Agriculture has also resulted in habitat destruction. In the United States, tallgrass prairies that once were home to a variety of unique species have been almost entirely converted to agriculture. Housing development and human settlement have cleared large areas of natural habitat. Mining has destroyed habitat because the landscape often must be altered in order to access the minerals. Finally, water development, especially in arid regions, has fundamentally altered habitat for many species. Dams change the flow and temperature of rivers and block the movements of species up and down the river. Also, the depletion of water for human use (usually agriculture) has dried up vegetation along rivers and left many aquatic species with insufficient water.

The invasion of nonnative species is another major threat to species worldwide. Invasive species establish themselves and take over space and nutrients from native species; they are especially problematic for island species, which often do not have defensive mechanisms for the new predators or competitors. Habitat destruction and invasion of nonnative species can be connected in a positive feedback loop: when habitat is degraded or changed, the altered conditions which are no longer suitable for native species can be advantageous for invasive species. In the United States, approximately half of all endangered species are adversely affected by invasive species.

Pollution directly and indirectly causes species to become endangered. In some cases, pesticides and other harmful chemicals are ingested by animals low on the food chain. When these animals are eaten by others, the pollutants become more and more concentrated, until the concentration reaches dangerous levels in predators and omnivores. These high levels cause reproductive problems and sometimes death. In addition, direct harm often occurs when pollutants make water uninhabitable. Agriculture and industrial production cause chemicals such as fertilizers and pesticides to reach waterways. Lakes have become too acidic from acid rain. Other human activities such as logging, grazing, agriculture, and housing development cause siltation in waterways. Largely because of this water pollution, two out of three fresh-water mussel species in the United States are at risk of extinction. *See* ACID RAIN; WATER POLLUTION.

Many species have become endangered or extinct from killing by humans throughout their ranges. For example, the passenger pigeon, formerly one of the most abundant birds in the United States, became extinct largely because of overexploitation. This overexploitation is especially a threat for species that reproduce slowly, such as large mammals and some bird species. Overfishing by large commercial fisheries is a threat to numerous marine and fresh-water species.

Efforts to save species focus on ending exploitation, halting habitat destruction, restoring habitats, and breeding populations in captivity. In the United States, the Endangered Species Act of 1973 protects endangered species and the ecosystems upon which they depend. Internationally, endangered species are protected from trade which depletes populations in the wild, through the Convention on International Trade in Endangered Species (CITES). Over 140 member countries act by banning commercial international trade of endangered species and by regulating and monitoring trade of other species that might become endangered. For example, the international ivory trade was halted in order to protect elephant populations from further depletion.

Typically, the first step is identifying which species are in danger of extinction throughout all or part of their range and adding them to an endangered species list. In the United States, species are placed on the endangered species list if one or more factors puts it at risk, including habitat destruction or degradation, overutilization, disease, and predation. Florida and California contain the most endangered species of all the contiguous 48 states. Hawaii has more endangered species than any other state. Hawaii, like other islands, has a diversity of unique species that occur nowhere else in the world. These species are also highly susceptible to endangerment because they tend to have small population sizes, and because they are particularly vulnerable to introduced competitors, predators, and disease.

For many endangered species, a significant captive population exists in zoos and other facilities around the world. By breeding individuals in captivity, genetic variation of a species can be more easily sustained, even when the species' natural habitat is being destroyed. Some species exist only in captivity because the wild population became extinct. For a few species, captive individuals have been reintroduced into natural habitat in order to establish a population where it is missing or to augment a small population. Depending on the species, reintroduction can be very difficult and costly, because individual animals may not forage well or protect themselves from predators. *See* ECOLOGY; EXTINCTION (BIOLOGY). [L.H.W.]

Endodermis The single layer of plant cells that is located between the cortex and the vascular (xylem and phloem) tissues. It has its most obvious development in roots and subaerial stems. The endodermis has many apparent functions: absorption of water, selection of solutes and ions, and production of oils, antibiotic phenols, and acetylenic acids.

The endodermis has been found to have extra sets of chromosomes as compared with cortical and other cells in the plant. In some plants the chromosome numbers may be so high in the endodermis that four sets of chromosomes may occur in each endodermal cell. The larger amount of nuclear material and nucleic acid in the cells of the endodermis may in part account for the great capacity of endodermal cells to produce large amounts of chemical substances, such as acetylenic oils, high in caloric energy. *See* CORTEX (PLANT); HYPODERMIS. [D.S.V.F.]

Endotoxin A biologically active substance produced by bacteria and consisting of lipopolysaccharide, a complex macromolecule containing a polysaccharide covalently linked to a unique lipid structure, termed lipid A. All gram-negative bacteria synthesize lipopolysaccharide, which is a major constituent of their outer cell membrane. One major function of lipopolysaccharide is to serve as a selectively permeable barrier for organic molecules in the external environment. Different types of gram-negative bacteria synthesize lipopolysaccharide with very different polysaccharide structures. The biological activity of endotoxic lipopolysaccharide resides almost entirely in the lipid A component.

When lipopolysaccharides are released from the outer membrane of the microorganism, significant host responses are initiated in humans and other mammals. It is generally accepted that lipopolysaccharides are among the most potent microbial products, known for their ability to induce pathophysiological changes, in particular fever and changes in circulating white blood cells. In humans as little as 4 nanograms of purified lipopolysaccharide per kilogram of body weight is sufficient to produce a rise in temperature of about 3.6°F (2°C) in several hours. This profound ability of the host to recognize endotoxin is thought to serve as an early warning system to signal the presence of gram-negative bacteria.

Unlike most microbial protein toxins (which have been termed bacterial exotoxins), endotoxin is unique in that its recognized mode of action does not result from direct damage to host cells and tissues. Rather, endotoxin stimulates cells of the immune system, particularly macrophages, and of the vascular system, primarily endothelial cells, to become activated and to synthesize and secrete a variety of effector molecules that cause an inflammatory response at the site of bacterial invasion. These mediator molecules promote the host response which results in elimination of the invading microbe. Thus, under these circumstances lipopolysaccharide is not a toxin at all, but serves an important function by helping to mobilize the host immune system to fight infection.

Even though endotoxin stimulation of host cells is important to host defense against infection, overstimulation due to excess production of endotoxin can lead to serious consequences. Endotoxin-induced multiple-organ failure continues to be a major health problem, particularly in intensive care; it has been estimated that as many as 50,000 deaths annually occur in the United States as the result of endotoxin-induced shock.

Immunization of humans with endotoxin vaccines to protect against endotoxin shock has not been considered practical. Efforts to provide immunologic protection against endotoxin-related diseases have focused upon development of antibodies that recognize the conserved lipid A structure of endotoxin as a means of passive protection against the lethal effects of this microbial product. *See* BACTERIA; MEDICAL BACTERIOLOGY. [D.C.M.]

Energy sources Sources from which energy can be obtained to provide heat, light, and power. Sources of energy have evolved from human and animal power to fossil fuels, uranium, water power, wind, and the Sun.

The principal fossil fuels are coal, lignite, peat, petroleum, and natural gas; other potential sources of fossil fuels include oil shale and tar sands. As fossil fuels become depleted, nonfuel sources and fission and fusion sources will become of greater importance since they are renewable. Nuclear power is based on the fission of uranium, thorium, and plutonium, and the fusion power is based on the forcing together of the nuclei of two light atoms such as deuterium, tritium, or helium-3. *See* NATURAL GAS; PETROLEUM.

Nonfuel sources of energy include wastes, water, wind, geothermal deposits, biomass, and solar heat. *See* BIOMASS; GEOTHERMAL POWER; SOLAR ENERGY; WIND POWER.

Fuels which do not exist in nature are known as synthetic fuels. They are synthesized or manufactured from varieties of fossil fuels which cannot be used conveniently in their original forms. Substitute natural gas is manufactured from coal, peat, or oil shale. Synthetic liquid fuels can be produced from coal, oil shale, or tar sands. Both gaseous and liquid fuels can be synthesized from renewable resources, collectively called biomass. These carbon sources are trees, grasses, algae, plants, and organic waste. Production of synthetic fuels, particularly from renewable resources, increases the scope of available energy sources.

Energy management includes not only the procurement of fuels on the most economical basis, but the conservation of energy by every conceivable means. Whether this is done by squeezing out every Btu through heat exchangers, or by room-temperature processes instead of high-temperature processes, or by greater insulation to retain heat which has been generated, each has a role to play in requiring less energy to produce the same amount of goods and materials. [G.C.G.]

Engineering geology The application of education and experience in geology and other geosciences to solve geological problems posed by civil engineering works. The branches of the geosciences most applicable are surficial geology, petrofabrics, rock and soil mechanics, geohydrology, and geophysics, particularly exploration geophysics and earthquake seismology.

The terms engineering geology and environmental geology often seem to be used interchangeably. Specifically, environmental geology is the application of engineering geology in the solution of urban problems; in the prediction and mitigation of natural hazards such as earthquakes, landslides, and subsidence; and in solving problems inherent in disposal of dangerous wastes and in reclaiming mined lands.

Another relevant term is geotechnics, the combination of pertinent geoscience elements with civil engineering elements to formulate the civil engineering system that has the optimal interaction with the natural environment. [W.R.J.]

Enterovirus One of the two subgroups of human picornaviruses. Enteroviruses include the polioviruses, the coxsackieviruses, and the echoviruses. They are small (17–28 nanometers in diameter), contain ribonucleic acid (RNA), and are resistant to ether. The enteroviruses multiply chiefly in the alimentary tract. The polioviruses, most echoviruses, and a number of the coxsackieviruses can be grown in cell cultures of monkey origin, as well as in human cells.

Enteroviruses are widespread during summer and fall in temperate climates, but may circulate throughout the year in tropical areas. The majority of enterovirus infections are benign and inapparent. However, when these viruses invade tissues other than the enteric tract, serious diseases may result, as when poliovirus invades the spinal cord or when some of the coxsackievirus types invade the heart muscle. *See* ANIMAL VIRUS; COXSACKIEVIRUS; ECHOVIRUS; PICORNAVIRIDAE; POLIOMYELITIS; RHINOVIRUS. [J.L.Me.]

Entodiniomorphida An order of the Spirotrichia. These are strikingly different-looking ciliates, covered with a smooth, firm pellicle. They are devoid of external ciliature except for the adoral zone of membranelles and, occasionally, one or two other tufts or zones of other specialized cilia (see illustration). Internal organization of the body is

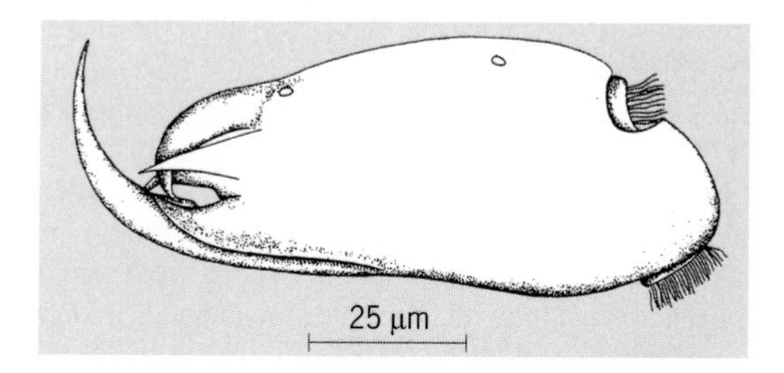

***Ophryoscolex*, an entodinomorphid.**

very specialized and complex. These organisms are considered to be highly evolved. Entodiniomorphids occur exclusively as endocommensals of herbivorous mammals, either in the rumen and reticulum of ruminants or in the colon of certain higher mammals. *See* PROTOZOA. [J.O.C.]

Entomology, economic The study of insects that have a direct influence on humanity. Though this includes beneficial as well as harmful species, most attention is devoted to the latter and how they become pests and are controlled. The emphasis on managing harmful insects reflects the immediacy and seriousness of pest problems, particularly the destruction of food and the transmission of disease. These are highly visible problems, whereas the benefits gained from useful insects are not so clearly understood, nor so well documented economically.

Central to the definition of a pest is determination of the economic threshold. Any insect population, when introduced into a favorable environment, increases numbers until reaching an environmental carrying capacity. In pest insects, there exists a density above which the insect population interferes with human health, comfort, convenience, or profits. When this economic threshold is reached, a decision must be made to utilize some control measure to prevent further increase in numbers. Often, the presence of even a single insect is sufficient to warrant control measures, for instance, when that insect is a flea harboring the plague bacillus, or a mosquito capable of transmitting malaria. Also, consumer expectations in most markets are for insect-free produce, so that the economic threshold can be very low on items that people eat. However, economic thresholds may be higher for insects that damage only the inedible portions of crop plants such as the leaves of beans, tomatoes, and apple trees. In any case, knowledge of the amount of injury which is due to different densities of insects is an important prerequisite for efficient management.

Pest management. Management of insect pests begins with prevention. Many of the United States' most noxious insects have been imported from overseas: most domestic cockroaches, the gypsy moth, Japanese beetle, corn borer, housefly, cabbageworm, and codling moth are just a few. Some North American insects have spread elsewhere too—the Colorado potato beetle to Europe and the fall webworm to Japan. To stem the flow of insect invasions, the federal government's Animal and Plant Health Inspection Service maintains inspection facilities for the examination of all incoming shipments of plant or other material that may harbor pests.

Once a pest is established, its spread can sometimes be slowed by an efficient system of local quarantines, early detection, and local eradication. A widely used eradication method is the application of synthetic chemical insecticides. Insecticides were once regarded as a panacea for pest problems, but the development of resistant strains of major insect pests, together with the rising cost of materials and application, has led to recognition that insecticides are more efficiently utilized in a program integrating them with other techniques in a framework of total crop management. *See* INSECTICIDE.

Another method is biological control, in which insects have had their numbers checked by natural enemies. Economic entomologists have effectively reduced densities of several pests by releasing parasites or predators. Natural enemies that are mobile and relatively restricted in diet do the best job of biological control. *See* INSECT CONTROL, BIOLOGICAL.

Crop rotation is a standard agronomic practice that often reduces damage due to insects. Rotation of alfalfa or soybeans with corn reduces populations of corn rootworms, wireworms, and white grubs. The physical disruption of autumnal plowing and disking destroys many insects that could overwinter in stubble or on the soil surface.

The cleanup of breeding and gathering sites is useful, especially in management of medically important insects, many of which have evolved resistance to the commoner insecticides. In general, the most successful programs of insect pest management rely on integrated control or the use of several methods in concert to control a complex of pests.

Beneficial insects. It has been estimated that the dollar value gained from a single insect, the honeybee, equals the loss from damage plus cost of control for all pests combined. Honeybees are managed for their honey and beeswax, but their most valued service is pollination of crop plants. Bees of many species are the chief pollinators, though wasps, flies, moths, butterflies, and beetles pollinate as well. *See* POLLINATION.

Silk is produced by larvae of the silkworm, an insect so thoroughly domesticated that it cannot climb its food plant, mulberry, with its degenerated legs. The silkworm apparently no longer survives in the wild. Many uses of silk have been taken over by less expensive synthetic materials.

Other insects may be equally beneficial, but their value is not so easily calculated. Foremost among these are predatory insects of several orders. These predators may prevent other insects from ever reaching an economic threshold and thus becoming pests. Innumerable insect species are scavengers, quietly but efficiently breaking down the remains of dead plants and animals. A lack of scavenging insects would, however, result in a great increase of decomposing organic material lying about. Plant-eating insects have been set to beneficial use when their diets consist mainly of unwanted weeds.

Certain rare and showy butterflies and beetles are sought after so that they have considerable economic worth. Conservation of rare and endangered insects incurs some expense as well. Habitat management to conserve rare insects is a valid and growing concern of economic entomologists.

Finally, insects have rendered invaluable service to science, and thus to humanity, as easily reared experimental animals for investigation of basic principles of genetics, biochemistry, development, and behavior. *See* INSECTA. [D.J.Hor.]

Environment The sum of all external factors, both biotic (living) and abiotic (nonliving), to which an organism is exposed. Biotic factors include influences by members of the same and other species on the development and survival of the individual. Primary abiotic factors are light, temperature, water, atmospheric gases, and ionizing radiation, influencing the form and function of the individual.

For each environmental factor, an organism has a tolerance range, in which it is able to survive. The intercept of these ranges constitutes the ecological niche of the organism. Different individuals or species have different tolerance ranges for particular environmental factors—this variation represents the adaptation of the organism to its environment. The ability of an organism to modify its tolerance of certain environmental factors in response to a change in them represents the plasticity of that organism. Alterations in environmental tolerance are termed acclimation. Exposure to environmental conditions at the limit of an individual's tolerance range represents environmental stress. *See* ADAPTATION (BIOLOGY); ECOLOGY; PHYSIOLOGICAL ECOLOGY (ANIMAL); PHYSIOLOGICAL ECOLOGY (PLANT).

Abiotic factors. The spectrum of electromagnetic radiation reaching the Earth's surface is determined by the absorptive properties of the atmosphere. Biologically, the most important spectral range is 300–800 nanometers, incorporating ultraviolet, visible, and infrared radiation. Visible light provides the energy source for most forms of life. Light absorbed by pigment molecules (chlorophylls, carotenoids, and phycobilins) is converted into chemical energy through photosynthesis. Light availability is especially important in determining the distribution of plants. Photosynthetic organisms can exist within a wide range of light intensities. Full sunlight in the tropics is around 2000 μmol photons $\cdot$ $m^{-2} \cdot s^{-1}$. Photosynthetic organisms have survived in locations where the mean light is as low as 0.005% of this value. *See* INSOLATION; PHOTOSYNTHESIS.

In addition to providing energy, light is important in providing an organism with information about its surroundings. The human eye, for example, is able to respond to wavelengths of light between 400 and 700 nm—the visible range. Within this range, sensitivity is greatest in the green part of the spectrum. This is the portion of the spectrum that plants absorb least, and so is the principal part of the spectrum to be reflected.

Temporal variation in light also provides an important stimulus. Life forms from bacteria upward are able to detect and respond to daily light fluctuations. Such a response may be directly controlled by the presence or absence of light (diurnal rhythms) or may persist when the variation in light is removed (circadian rhythms). In the latter case, regulation is through an internal molecular clock, which is able to predict the daily cycle. Such circadian clocks are normally reset by light on a daily basis. Processes controlled by circadian clocks range from the molecular (gene expression) to the behavioral (for example, sleep patterns in animals or leaf movements in plants). *See* PHOTOPERIODISM.

Ultraviolet radiation has the ability to break chemical bonds and so may lead to damage to proteins, lipids, and nucleic acids. Damage to DNA may result in genetic mutations. The ozone layer in the stratosphere is responsible for absorbing a large proportion of ultraviolet radiation reaching the outer atmosphere. As ozone is destroyed by the action of pollutants such as chloroflurocarbons, the proportion of ultraviolet radiation reaching the surface of the Earth rises.

Water is ubiquitous in living systems, as the universal solvent for life, and is essential for biological activity. Many organisms have evolved the ability to survive prolonged periods in the total absence of water, but this is achieved only through the maintenance of an inactive state. Water availability remains a primary environmental factor limiting survival on land. Primitive land organisms possess little or no ability to conserve water within their cells and are termed poikilohydric. Examples include amphibians and primitive plants such as most mosses and liverworts. These are confined to places where water is in plentiful supply or they must be able to tolerate periods of desiccation. Lichens can survive total water loss and rapidly regain activity upon rewetting. Such organisms must be able to minimize the damage caused to cellular structures when water is lost. Dehydration causes irreversible damage to membranes and proteins. This damage can be prevented by the accumulation of protective molecules termed compatible solutes.

Homeohydric organisms possess a waterproof layer that restricts the loss of water from the cells. Such waterproofing is never absolute, as there is still a requirement to exchange gas molecules and to absorb organic or mineral nutrients through a water phase. Water conservation allows organisms to live in environments in which the water supply is extremely low. In extremely arid environments, behavioral adaptations may allow the water loss to be minimized. Animals may be nocturnal, emerging when temperatures are lower and hence evaporation minimized. Cacti possess a form of photosynthesis, crassulacean acid metabolism (CAM), that allows them to separate gas exchange and light capture. *See* GROUND-WATER HYDROLOGY; PLANT-WATER RELATIONS.

Temperature is a determinant of survival in two ways: (1) as temperatures decrease, the movement of molecules slows and the rate of chemical reactions declines; (2) temperature determines the physical state of water.

The slowing of metabolic activity at low temperatures is illustrated in reptiles. Such poikilothermic animals, unable to maintain their internal temperature, are typically inactive in the cold of morning. They bask in the sun to increase their body temperature and so become active. High temperatures will cause the three-dimensional structure of proteins to break down, preventing the organisms from functioning. Organisms adapted to extremely high temperatures need more rigid proteins that maintain their structure. Temperature also affects the behavior of cell membranes, made up of lipids and proteins in a liquid crystalline state. At low temperatures, the membrane structure becomes rigid and liable to break. At high temperatures, it becomes too fluid and again liable to disintegrate. In adapting to different temperatures, organisms alter the composition of the lipids in their membranes, whose melting temperature is thereby changed. This outcome also applies to storage lipids. Hence, cold-water fish are a useful source of oils, whereas mammals, with their higher body temperature, contain fats. The effect of temperature on membranes is thought to be a key factor determining the temperature range that an organism is able to survive.

The effect of temperature on the physical state of water is essential to determining the availability of that water to organisms. Poikilothermic organisms may find that the water in their cells begins to freeze at low temperatures. Certain species can survive total freezing through the prevention of ice crystal formation altogether which would otherwise damage cellular structures. To survive low temperatures, cells must be able to survive desiccation, and so low-temperature tolerance involves the formation of compatible solutes. High temperatures increase the rate of evaporation of water. Hence, where water supply is limiting, an organism's ability to survive high temperatures is impaired.

Mammals and birds, homeothermic organisms, are able to regulate their internal temperature, limiting the effects of external temperature variations. Temperature still acts as an environmental constraint in such organisms, however. Cooling is achieved through sweating and hence loss of water. Heat is produced through the metabolism of food, and hence survival in cold climates requires a high metabolic rate.

The atmosphere on Earth is thought to be determined to a large extent by the presence of life. At the same time, organisms have evolved to survive in the atmosphere as it is. The atmospheric constitutents with the most direct biological importance are oxygen (O_2) and carbon dioxide (CO_2). Oxygen makes up approximately 20% of the atmosphere and is due to the occurrence of oxygenic photosynthesis. This process involves the simultaneous uptake of CO_2 to make sugars. Aerobic respiration involves the reverse of this process, the release of CO_2 and the uptake of O_2 to form water. Hence, the current atmosphere represents the balance of previous biological activity. For most terrestrial organisms, neither CO_2 nor O_2 is limiting in the atmosphere; however, the need to get either or both of these gases to cells may represent a limitation on size or on the ability to tolerate water stress. Limitation of either gas may be important in aquatic environments, where the concentration of each is significantly lower.

Nitrogen is also required by all organisms but cannot be used by most in the gaseous form. Nitrogen fixation, the conversion of N_2 gas into a biologically useful form, occurs in some species of bacteria and cyanobacteria or may be caused by lightning.

Atmospheric gases are important in determining the climate and the light environment. Absorption of electromagnetic radiation by the atmosphere determines the spectrum of light reaching the Earth's surface. Absorption and reflectance of infrared radiation by greenhouse gases such as CO_2 and water vapor regulate temperature. *See* PHOTORESPIRATION.

Among other environmental factors determining the range and distribution and form of organisms are mechanical stimuli such as wind or water movement, and the presence of metals, inorganic nutrients, and toxins in the air, soil, or food.

Biotic factors. The biotic environment of an individual is made up of members of the same or other species. Intraspecific interactions involve the need to breed with other individuals, to gain protection through living in a group, and to compete for resources such as food, light, nutrients, and space. The optimal population density depends on the availability of resources and on the behavior, size, and structure of the organism. Interspecific interactions may also be positive or negative. For example, symbiotic relationships involve the mutual benefit of the individuals involved, whereas competition for resources is deleterious to both. Although predation exerts a negative influence on the population as a whole, the success of an individual may be enhanced if a predator removes one of its conspecific competitors.

Humans alter their environment in ways that exceed the impact of all other organisms. For example, the release of greenhouse gases into the atmosphere contributes to climate alterations over the entire planet. This in turn has impacts on the distribution of all other species. The release of pollutants into the environment brings organisms into contact with stresses to which they were not previously exposed. This causes the evolution of new varieties, eventually perhaps new species, adapted to the polluted environments. *See* AIR POLLUTION; BIOSPHERE; HUMAN ECOLOGY; WATER POLLUTION.

For any given organism, it is often possible to identify a factor in the environment that limits survival and growth. The limiting factor may change through time. Such a change may cause the organism to be at the limit of or outside its tolerance range for that or another environmental factor. In such cases, the organism is said to suffer stress. If the stress to which an individual is exposed is extreme, it may result in irreversible damage and death. Exposure to moderate stress, however, results in a period of acclimation within the organism that allows it to adjust to the new conditions. Organisms exposed gradually to new conditions usually have a higher chance of survival than those exposed suddenly. *See* POPULATION ECOLOGY.

Where a particular environmental factor (or combination of factors) dominates the growth and development of organisms, it is often found that the adaptations and gross features of the landscape will be the same, even when the actual species are different. Thus, mediterranean vegetation is found not only around the Mediterranean Sea but also in California and South Africa, where the conditions of hot dry summers and warm wet winters occur. Regions with similar environmental conditions are classed as biomes. The occurrence of such global vegetation types clearly illustrates the role played by the environment in determining the form and function of individual species. [G.J.]

Environmental engineering The division of engineering concerned with the environment and management of natural resources. The environmental engineer places special attention on the biological, chemical, and physical reactions in the air, land, and water environments and on improved technology for integrated management systems, including reuse, recycling, and recovery measures.

Environmental engineering began with consideration of the need for acceptable drinking water and for management of liquid and solid wastes. Abatement of air and land contamination became new challenges for the environmental engineer, followed by toxic-waste and hazardous-waste concerns. The environmental engineer is also instrumental in the mitigation and protection of wildlife habitat, preservation of species, and the overall well-being of ecosystems.

The principal environmental engineering specialties are air-quality control, water supply, wastewater disposal, stormwater management, solid-waste management, and hazardous-waste management. Other specialties include industrial hygiene, noise control, oceanography, and radiology. *See* AIR POLLUTION; HAZARDOUS WASTE; WATER POLLUTION. [R.A.Cor.]

Environmental geology The branch of geology that deals with the ways in which geology affects people. Examples of the effect of geology on human civilizations include (1) the ways that fertile soils develop from rocks and how these soils can become polluted by human activities; (2) how rocks and soils move down-slope to destroy roads, houses, and other human constructions; (3) sources of surface and subsurface water supplies and how they become polluted; (4) why floods occur where they do and how human activities affect floods; (5) locations of earthquakes and volcanic eruptions and the dangers they pose; (6) location of mineral resources such as copper, oil and gas, and uranium, and how mining these resources can pollute the environment; (7) how human activities can pollute the atmosphere and cause global warming, sea-level rise, and ozone depletion. *See* AIR POLLUTION; EARTHQUAKE; SOIL CONSERVATION; VOLCANO; WATER POLLUTION. [H.Bl.]

Environmental management The development of strategies to allocate and conserve resources, with the ultimate goal of regulating the impact of human activities on the surrounding environment. "Environment"here usually means the natural surroundings, both living and inanimate, of human lives and activities. However, it can also mean the artificial landscape of cities, or occasionally even the conceptual field of the noosphere, the realm of communicating human minds.

Environmental management is a mixture of science, policy, and socioeconomic applications. It focuses on the solution of the practical problems that humans encounter in cohabitation with nature, exploitation of resources, and production of waste. In a purely anthropocentric sense, the central problem is how to permit technology to evolve continuously while limiting the degree to which this process alters natural ecosystems. Environmental management is thus intimately intertwined with questions regarding economic growth, equitable distribution of consumable goods, and conserving resources for future generations. Environmental managers fall within a broad spectrum, from those who would limit human interference in nature to those who would increase it in order to guide natural processes along benign paths. Participants in the process of environmental management fall into seven main groups: (1) governmental organizations at the local, regional, national, and international levels, including world bodies such as the United Nations Environment Programme and the U.N. Conference on Environment and Development; (2) research institutions, such as universities, academies, and national laboratories; (3) bodies charged with the enforcement of regulations, such as the U.S. Environmental Protection Agency; (4) businesses of all sizes and multinational corporations; (5) international financial institutions, such as the World Bank and International Monetary Fund; (6) environmental nongovernmental organizations, such as the World Wildlife Fund for Nature; and (7) representatives of the users of the environment, including tribes, fishermen, and hunters. The agents of environmental management include foresters, soil conservationists, policy-makers, engineers, and resource planners.

Some common themes of environmental management are bilateral and multilateral environmental treaties; design and use of decision-support systems; environmental policy formulation, enactment, and policing of compliance; estimation, analysis, and management of environmental risk; management of recreation and tourism; natural resource evaluation and conservation; positive environmental economics; promotion of positive environmental values by education, debate, and information dissemination; and strategies for the rehabilitation of damaged environments.

The need to improve management of the environment has given rise to several new techniques. There is environmental impact analysis, which was first formulated in California and is codified in the U.S. National Environmental Policy Act (NEPA). Through the environmental impact statement, it prescribes the investigatory and remedial measures that must be taken in order to mitigate the adverse effects of new development. It is intended to act in favor of both prudent conservation and participatory democracy. Another technique is environmental auditing, which uses the model of the financial audit to examine the processes and outcomes of environmental impacts. It requires value judgments, which are usually set by public preference, ideology, and policy, to define what are regarded as acceptable outcomes. Audits use techniques such as life-cycle analysis and environmental burden analysis to assess the impact of, for example, manufacturing processes that consume resources and create waste. [D.Ale.]

Environmental toxicology A broad field of study encompassing the production, fate, and effects of natural and synthetic pollutants in the environment. The breadth of this field depends on the definition of environment. It can be defined as narrowly as the home and workplace or as broadly as the entire Earth and its biosphere. Environmental toxicology is truly an interdisciplinary science. The effects of a pollutant on the environment depend on the amount released (the dose) and its chemical and physical properties. Pollutants can be grouped according to their origin and effects.

Pollution from nutrients is generally a problem of aquatic systems. Carbon, nitrogen, and phosphorus are essential nutrients and, when present in excess, can result in an overstimulation of microbial and plant growth. Nutrients enter the environment in runoff from fertilized agricultural areas, and in effluents from municipal and industrial waste and decaying plant material. *See* EUTROPHICATION.

Pathogenic bacteria and protozoa can be a major source of pollution in areas that receive untreated sewage, items from ocean dumping, and improperly discarded hospital waste. Toxic metabolites of fungal origin (mycotoxins) are also potential pollutants. *See* AFLATOXIN.

Forest fires, volcanic eruptions, and dust storms can be major sources of suspended materials. These materials can also originate in runoff from agricultural areas, construction and mining sites, and roads and other paved areas. Truck and automobile exhaust and industrial discharge to the atmosphere are also sources of suspended solids.

Metabolic processes and natural combustion and thermal activity (such as forest fires and volcanoes) can release large amounts of gaseous by-products to the atmosphere. However, natural inputs are minor compared to atmospheric pollutants due to human activity. Although most anthropogenic air pollution is produced by the various forms of transportation, emission from stationary sources of fuel combustion (for example, factories and power plants) are responsible for the greatest amount of hazardous materials released. *See* AIR POLLUTION; GREENHOUSE EFFECT.

All living organisms require certain metals for physiological processes. These elements, when present at concentrations above the level of homeostatic regulation, can be toxic. In addition, there are metals that are chemically similar to, but higher in molecular weight than, the essential metals (heavy metals).

Organic solvents are used widely and in large amounts in industries, laboratories, and homes. They are released to the atmosphere as vapor and can pose a significant inhalation hazard. Improper storage, use, and disposal have resulted in the contamination of surface and ground waters and drinking water. *See* WATER POLLUTION.

The pesticides represent an important group of materials that can enter the environment as pollutants. They are highly toxic, and many nontarget organisms can suffer harmful effects if misuse or unintended release occurs. *See* PESTICIDE.

Coal and petroleum-derived materials and by-products are major environmental pollutants. Widespread use has led to enormous releases to the environment of distillate fuels, crude oils, runoff from coal piles, exhaust from internal combustion-fired power plants, industrial emissions, and emissions from municipal incinerators. The toxicity of polycyclic aromatic hydrocarbons is perhaps one of the most serious long-term problems associated with the use of petroleum. They accumulate in soil, sediment, and biota, and at high concentrations can be acutely toxic. *See* FOSSIL FUEL.

Polychlorinated biphenyls (PCBs) are produced by the chlorination of biphenyl, giving rise to mixtures of up to 210 possible products. They have been used worldwide in electrical equipment, vacuum pumps, hydraulic fluids, heat-transfer systems, lubricants, and inks. The related polybrominated biphenyls (PBBs) have been used as fire retardants. Major sources of polychlorinated biphenyls have included leaks from waste disposal facilities, vaporization during combustion, and disposal of industrial fluids. Their use has been largely restricted or eliminated. Environmental concentrations are decreasing, but with their persistence they remain significant pollutants. Chlorinated dibenzo-*p*-dioxins and dibenzofurans are formed during the heating of chlorophenols, and have been identified as potential contaminants in the herbicide 2,4,5-T. They can be formed during the incineration of municipal wastes, polychlorinated biphenyls, or plant materials treated with chlorophenols. *See* ENVIRONMENTAL ENGINEERING; TOXICOLOGY. [J.T.O.]

Eolian landforms Topographic features generated by the wind. The most commonly seen eolian landforms are sand dunes created by transportation and accumulation of windblown sand. Blankets of wind-deposited loess, consisting of fine-grained silt, are less obvious than dunes, but cover extensive areas in some part of the world.

Where abundant loose sand is available for the wind to carry, sand dunes develop. As soon as enough sand accumulates in one place, it interferes with the movement of air and a wind shadow is produced which contributes to the shaping of the pile of sand. Dunes advance downwind by erosion of sand on the windward side and redeposition on the slip face. Dunes may have a variety of shapes, depending on wind conditions, vegetation, and sand supply. The fine silt and clay winnowed out from coarser sand is often blown longer distances before coming to rest as a blanket of loess mantling the preexisting topography. Thick deposits of loess are most often found in regions downwind from glacial outwash plains or alluvial valleys. *See* DUNE; LOESS; SAND. [D.J.E.]

Epidemic The occurrence of cases of disease in excess of what is usually expected for a given period of time. Epidemics are commonly thought to involve outbreaks of acute infectious disease, such as measles, polio, or streptococcal sore throat. More recently, other types of health-related events such as homicide, drownings, and even hysteria have been considered to occur as "epidemics."

Confusion sometimes arises because of overlap between the terms epidemic, outbreak, and cluster. Although they are closely related, epidemic may be used to suggest problems that are geographically widespread, while outbreak and cluster are reserved for problems that involve smaller numbers of people or are more sharply defined in terms of the area of occurrence. For example, an epidemic of influenza could involve an entire state or region, whereas an outbreak of gastroenteritis might be restricted to a nursing home, school, or day-care center. The term cluster may be used to refer to noncommunicable disease states.

In contrast to epidemics, endemic problems are distinguished by their consistently high levels over a long period of time. Lung cancer in males has been endemic in the United States, whereas the surge of lung cancer cases in women in the United States represents an epidemic problem that has resulted from increase in cigarette smoking among women in general. A pandemic is closely related to an epidemic, but it is a problem that has spread over a considerably larger geographic area; influenza pandemics are often global.

Disease and epidemics occur as a result of the interaction of three factors, agent, host, and environment. Agents cause the disease, hosts are susceptible to it, and environmental conditions permit host exposure to the agent. An understanding of the interaction between agent, host, and environment is crucial for the selection of the best approach to prevent or control the continuing spread of an epidemic.

For infectious diseases, epidemics can occur when large numbers of susceptible persons are exposed to infectious agents in settings or under circumstances that permit the spread of the agent. Spread of an infectious disease depends primarily on the chain of transmission of an agent: a source of the agent, a route of exit from the host, a suitable mode of transmission between the susceptible host and the source, and a route of entry into another susceptible host. Modes of spread may involve direct physical contact between the infected host and the new host, or airborne spread, such as coughing or sneezing. Indirect transmission takes place through vehicles such as contaminated water, food, or intravenous fluids; inanimate objects such as bedding, clothes, or surgical instruments; or a biological vector such as a mosquito or flea. *See* EPIDEMIOLOGY; INFECTIOUS DISEASE; PUBLIC HEALTH. [R.A.Go.]

Epidemic viral gastroenteritis A clinical syndrome characterized by acute infectious gastroenteritis with watery diarrhea, vomiting, malaise, and abdominal cramps with a relatively short incubation period (12–36 h) and duration (24–48 h). A viral etiology is suspected when bacterial and parasitic agents are not found. In the United States, no etiologic agent can be found in 70% of the outbreaks of gastroenteritis. Most of these may be due to viral agents, such as the Norwalk, Snow Mountain, and Hawaii agents, astroviruses, caliciviruses, adenoviruses, nongroup A rotaviruses, and paroviruses. Epidemics are common worldwide and have occurred following the consumption of fecally contaminated raw shellfish, food, or water, although the virus may be spread by airborne droplets as well. Epidemics are most frequent in residential homes, camps, institutions, and cruise ships. Many individual cases of mild diarrhea may in fact occur in epidemics for which the source of the infection cannot be found. Epidemic viral gastroenteritis is distinct from rotavirus diarrhea, a seasonal disease in winter that is the most common cause of diarrhea in young children, and affects virtually all children in the first 4 years of life. Since the diarrhea is often mild and of short duration, attention should be given to rehydration therapy and prevention by identification of the source. Fatalities have been associated with severe dehydration and loss of fluids and electrolytes in the stool. *See* ANIMAL VIRUS; DIARRHEA; INFANT DIARRHEA. [R.G.]

Epidemiology The study of the distribution of diseases in populations and of factors that influence the occurrence of disease. Epidemiology examines epidemic (excess) and endemic (always present) diseases; it is based on the observation that most diseases do not occur randomly, but are related to environmental and personal characteristics that vary by place, time, and subgroup of the population. The epidemiologist attempts to determine who is prone to a particular disease; where risk of the disease is highest; when the disease is most likely to occur and its trends over time; what exposure its victims have in common; how much the risk is increased through exposure; and how many cases of the disease could be avoided by eliminating the exposure.

In the course of history, the epidemiologic approach has helped to explain the transmission of communicable diseases, such as cholera and measles, by discovering what exposures or host factors were shared by individuals who became sick. Modern epidemiologists have contributed to an understanding of factors that influence the risk of chronic diseases, particularly cardiovascular diseases and cancer, which account for most deaths in developed countries today. Epidemiology has established the causal association of cigarette smoking with heart disease; shown that acquired immune deficiency syndrome (AIDS) is associated with certain sexual practices; linked menopausal estrogen use to increased risk of endometrial cancer but to decreased risk of osteoporosis; and demonstrated the value of mammography in reducing breast cancer mortality.

By identifying personal characteristics and environmental exposures that increase the risk of disease, epidemiologists provide crucial input to risk assessments and contribute to the formulation of public health policy.

Epidemiologic studies, based mainly on human subjects, have the advantage of producing results relevant to people, but the disadvantage of not always allowing perfect control of study conditions. For ethical and practical reasons, many questions cannot be addressed by experimental studies in humans and for which observational studies (or experimental studies using laboratory animals or biomedical models) must suffice. Still, there are circumstances in which experimental studies on human subjects are appropriate, for example, when a new drug or surgical procedure appears promising and the potential benefits outweigh known or suspected risks. *See* DISEASE; EPIDEMIC.

Descriptive epidemiologic studies provide information about the occurrence of disease in a population or its subgroups and trends in the frequency of disease over time. Data sources include death certificates, special disease registries, surveys, and population censuses; the most common measures of disease occurrence are (1) mortality (number of deaths yearly per 1000 of population at risk); (2) incidence (number of new cases yearly per 100,000 of population at risk); and (3) prevalence (number of existing cases at a given time per 100 of population at risk). Descriptive measures are useful for identifying populations and subgroups at high and low risk of disease and for monitoring time trends for specific diseases. They provide the leads for analytic studies designed to investigate factors responsible for such disease profiles.

Analytic epidemiologic studies seek to identify specific factors that increase or decrease the risk of disease and to quantify the associated risk. In observational studies, the researcher does not alter the behavior or exposure of the study subjects, but observes them to learn whether those exposed to different factors differ in disease rates. Alternatively, the researcher attempts to learn what factors distinguish people who have developed a particular disease from those who have not. In experimental studies, the investigator alters the behavior, exposure, or treatment of people to determine the impact of the intervention on the disease. Usually two groups are studied, one that experiences the intervention (the experimental group) and one that does not (the control group). Outcome measures include incidence, mortality, and survival rates in both the intervention and control groups. [V.L.E.]

Epidermis (plant) The outermost layer (occasionally several layers) of cells on the primary plant body. Its structure is variable; this article singles out five structural components of the tissue: (1) cuticle; (2) stomatal apparatus (including guard cells and subsidiary cells); (3) bulliform (motor) cells; (4) trichomes; and (5) root hairs.

Leaves, herbaceous stems, and floral organs usually retain the epidermis through life. Most woody stems retain it for one to many years, after which it is replaced. In roots it is usually short-lived. *See* LEAF; PERIDERM.

Cutin is a mixture of fatty substances characteristically found in epidermal cells. It impregnates the outer cell walls and occurs as a continuous layer (cuticle) on the outer surface. The cuticle covers the surfaces of young stems, leaves, floral organs, and even apical meristems. Waxes appear as a deposit on the outside of the cuticle in many plants; the bloom on purple grapes and plums is an example. Most often the waxes are present in small quantity, but the leaves of some plants may be almost white with wax (*Echeveria subrigida*). The waxes of a few species are of great commercial value in the manufacture of polishes for floors, furniture, automobiles, and shoes. Other substances, such as gums, resins, and salts, usually in crystalline form, may be deposited on the outside of the cuticle.

The apertures in the epidermis which are surrounded by two specialized cells, the guard cells, are known as stomata. The singular form, stoma, is derived from the Greek word for mouth. However, some authorities prefer to include both aperture and guard cells within the concept of stoma. The apertures of stomata are contiguous with the intercellular space system of underlying tissues and thus permit gas exchange between internal cells and the external environment. The opening and closing of the stomatal

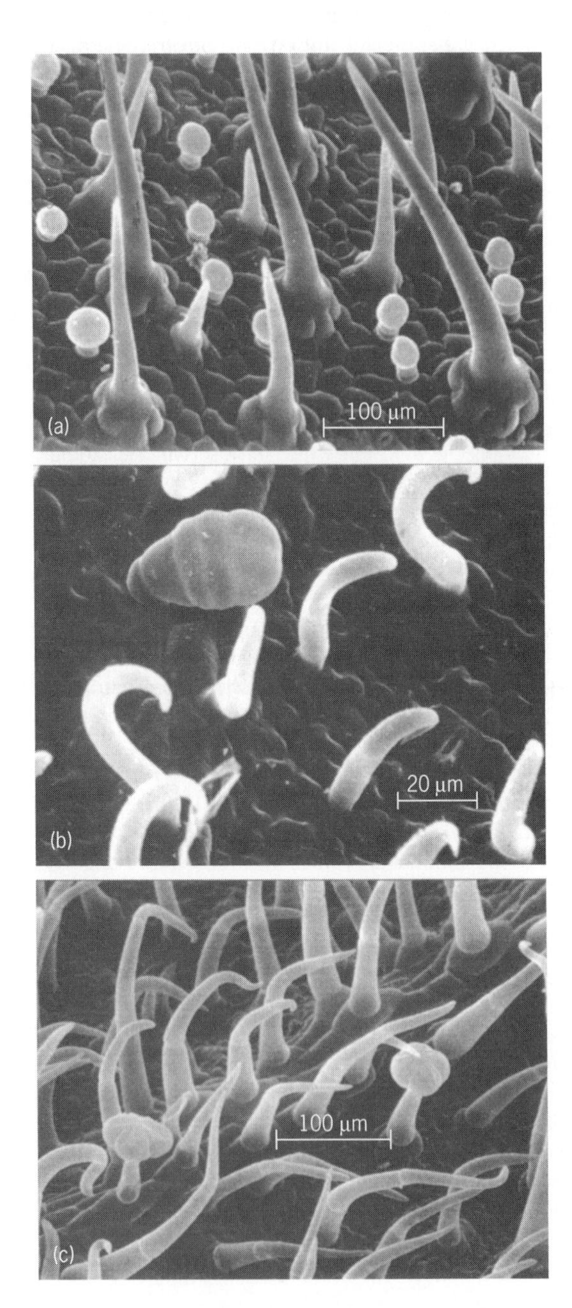

Trichomes. (*a*) Unicellular and glandular (colleters) hairs of the geranium (*Pelargonium*), (*b*) Unicellular-hooked and uniseriate, club-shaped hairs of the bean (*Phaseolus*). (*c*) Uniseriate and glandular hairs of the tomato (*Lycopersicon*).

aperture is caused by relative changes in turgor between the guard cells and surrounding epidermal cells.

Bulliform (motor) cells are large, highly vacuolated cells that occur on the leaves of many monocotyledons but are probably best known in grasses. They are thought to play a role in the unfolding of developing leaves and in the rolling and unrolling of mature leaves in response to alternating wet and dry periods.

Appendages derived from the protoderm are known as trichomes; the simplest are protrusions from single epidermal cells. Included in the concept, however, are such diverse structures as uniseriate hairs, multiseriate hairs (*Begonia, Saxifraga*), anchor hairs, stellate hairs, branched (candelabra) hairs, peltate scales, stinging hairs, and glandular hairs (see illustration). Cotton and kapok fibers are unicellular epidermal hairs.

Root hairs are thin-walled extensions of certain root epidermal cells. They develop only on growing root tips and may arise from any epidermal cell, or from specialized cells known as trichoblasts. The life of a given root hair is usually numbered in days. *See* ROOT (BOTANY). [N.H.B.]

Equator The great circle around the Earth, equally distant from the North and South poles, which divides the Earth into Northern and Southern hemispheres. It is the greatest circumference of the Earth because of centrifugal force from rotation, and resultant flattening of the polar areas.

The Earth's rotational axis is vertical to the plane of the Equator, and because the inclination of the axis is $66^1/_2{}^\circ$ from the plane of the ecliptic, the plane of the Equator is always inclined $23^1/_2{}^\circ$ from the ecliptic.

The celestial equator in astronomy is equally distant from the celestial poles and is the great circle in which the plane of the terrestrial Equator intersects the celestial sphere. *See* MATHEMATICAL GEOGRAPHY. [V.H.E.]

Equatorial currents Ocean currents near the Equator. The westward trade winds that prevail over the tropical Atlantic and Pacific oceans drive complex oceanic circulations characterized by alternating bands of eastward and westward currents. The intense currents are confined to the surface layers of the ocean; below a depth of approximately 100 m (330 ft) the temperature is much lower, and the speed of ocean currents is much slower. The westward surface currents tend to be divergent—they are associated with a parting of the surface waters—and therefore entrain cold water from below. The water temperature rises as the currents flow westward, so that temperatures are low in the east and high in the west, except between 3 and 10°N where eastward surface currents create a band of warm water across the Pacific and Atlantic oceans. The distinctive sea surface temperature pattern in which surface waters are warm in the west and cold in the east, except for the warm band just north of the Equator, reflects the oceanic circulation. A dramatic change in this pattern every few years during El Niño episodes, when the temperature of the eastern tropical Pacific Ocean rises, is associated with an intensification of the eastward currents and a weakening (sometimes reversal) of the westward currents.

The South Equatorial Current flows westward in the upper ocean, has its northern boundary at approximately 3°N, and attains speeds in excess of 1 m/s (3.3 ft/s) near the Equator. It is directly driven by the westward trade winds and has its origins in the cold, northwestward-flowing Peruvian coastal current. Because the Coriolis force deflects water parcels to their right in the Northern Hemisphere and to their left in the Southern Hemisphere, this current is divergent at the Equator. As a consequence, cold water from below wells up along the Equator. *See* UPWELLING.

The North Equatorial Countercurrent flows eastward immediately to the north of the South Equatorial Current. The boundary between these two currents is a sharp thermal front that is clearly evident in satellite photographs. The front can literally be a green line, hundreds of yards wide, because of the abundance of phytoplankton. This current, which is counter to the wind, is driven by the torque (curl) that the wind exerts on the ocean. To its north is a colder westward current known as the North Equatorial Current. *See* OCEAN WAVES; PHYTOPLANKTON.

The Equatorial Undercurrent, which in the Pacific Ocean was originally known as the Cromwell Current, is an intense, narrow, eastward, subsurface jet that flows precisely along the Equator across the width of the Pacific. Its core, where speeds can be in excess of 1.5 m/s (5 ft/s), is at an approximate depth of 100 m (330 ft); its width is approximately 200 km (120 mi). A similar current exists in the Atlantic Ocean. In the Indian Ocean it is often present along the Equator, in the western part of the basin during March and April when westward winds prevail over that region. Such winds (including the trade winds over the Pacific and Atlantic oceans) pile up warm surface waters in the west while exposing cold waters to the surface in the east. *See* ATLANTIC OCEAN; INDIAN OCEAN; PACIFIC OCEAN. [S.G.P.]

Equine infectious anemia A lentivirus-induced disease of the horse family with an almost worldwide distribution. It is characterized by recurring fever, platelet reduction, weight loss, edema, and anemia. Although death can occur, there is usually an eventual cessation of clinical signs. However, host defenses are unable to completely eliminate the virus, and the animal remains a persistently infected inapparent carrier.

The equine infectious anemia virus is in the *Lentivirus* genus of the family Retroviridae. Although it is closely related to the human immunodeficiency viruses (HIV-1, HIV-2), its genetic organization is the least complex of this group of viruses.

Equine infectious anemia virus infects only members of the horse family. The mechanical transfer of blood between animals by large blood-feeding insects (mainly horse flies and deer flies) is probably the most important mechanism of natural transmission. However, the virus can be efficiently transmitted by humans if sterile techniques are not observed during veterinary procedures.

Clinical responses to equine infectious anemia virus infection can range from an extremely severe disease resulting in death to an absence of obvious signs. However, when disease occurs it is usually observed shortly after exposure to the virus (incubation periods of 10–45 days are common) and consists of fever, platelet reduction (thrombocytopenia), lethargy, loss of appetite, and petechial hemorrhages. Most horses survive this acute episode but in many cases progress to the chronic form of the disease, characterized by recurring fever, thrombocytopenia, anemia, weight loss, edema, and hemorrhage. Each fever episode is associated with massive amounts of viral replication that occurs predominantly in the mature tissue macrophages of the liver and spleen and results in the release of millions of viral particles into the bloodstream. This extensive replication triggers the release of powerful molecules called cytokines, such as tumor necrosis factor alpha (TNFα), that are normally associated with inflammatory reactions.

Equine infectious anemia virus employs numerous mechanisms to avoid removal by host defenses. These mechanisms range from infecting and possibly compromising the efficiency of a key cell type in the immune system to possessing a surface unit protein with a fundamental structure that confers partial resistance to neutralizing antibodies. Another property of this protein is its ability to withstand considerable alterations in amino acid content without loss of function. This, in combination with a high mutation rate, facilitates the emergence of variants that are completely resistant to the strain-specific neutralizing antibodies produced within the first months of infection. Although additional strain-specific neutralizing antibodies may be generated against these variants, the mutational capabilities of this virus permit still more resistant types to arise.

A safe and effective vaccine that rapidly induces protection against all strains of the equine infectious anemia virus is not yet available; this virus is controlled in most areas of the world by serological testing that detects the presence of antibodies against the major core protein of the virus. If a horse tests positive, further transmission of the equine infectious anemia virus can usually be prevented by segregation or quarantine. *See* RETROVIRUS. [R.F.Co.; C.J.I.]

Ergot and ergotism Ergot is the seedlike body of fungi (molds) of the genus *Claviceps*; ergotism is a complex disease of humans and certain domestic animals caused by ingestion of grains and cereals infested with ergot. Ingestion of these long, hard, purplish-black structures called sclerotia may lead to convulsions, abortion, hallucinations, or death. During the Middle Ages, hundreds of thousands of people are believed to have died from this disease, often referred to as holy fire, St. Anthony's fire, or St. Vitus' dance. Epidemics in humans, although less prevalent in modern times, last occurred in 1951, and the potential danger is always present, as shown by annual livestock losses due to ergot poisoning. There are 32 recognized species of *Claviceps*, most of which infect members of the grass family. Only three species are parasitic on the rushes and sedges. *See* PLANT PATHOLOGY; POISON.

Sclerotia have an unusual chemical makeup. They carry only 10% water by weight, and 50% of the dry weight is composed of fatty acids, sugars, and sugar alcohols, which make the ergot a storehouse of energy. Unfortunately, they also contain the poisonous alkaloids, ranging from 0 to 1.2% of the dry weight.

There are three types of ergotism (gangrenous, convulsive, and hallucinogenic). Their symptoms often overlap; the hallucinogenic form is usually observed in combination with one of the other two. The unusual combination of gangrenous and convulsive symptoms is sometimes observed in the Balkans and areas near the Rhine River.

The hallucinogenic form often includes symptoms of one of the other types. In its more pure form, it is referred to as choreomania, St. Vitus' dance, or St. John's dance. Vivid hallucinations are accompanied by psychic intoxication reminiscent of the effects of many of the modern psychedelic drugs. Early reports state that the disease usually manifested itself in the form of strange public dances that might last for days or weeks on end. Dancers made stiff jerky movements accompanied by wild hopping, leaping, screaming, and shouting. They were often heard conversing with devils or gods, and danced compulsively, as if possessed, until exhaustion caused them to fall unconscious or to lie twitching on the ground. High mortality rates were associated with severe epidemics involving any of the three forms of ergotism. The success with which the disease is controlled in humans has been brought about by (1) agricultural inspection, (2) use of wheat, potatoes, and maize instead of rye, (3) limited control of ergot, (4) reserves of sound grain, and (5) forecasting severe ergot years. The most recent and best-recorded epidemic was in southern France in 1951 when an unscrupulous miller used moldy grain to make flour.

In the early twentieth century two ergot alkaloids (ergotoxine and ergotamine) were isolated. Unfortunately, they caused significant side effects and were not as specific or active as some of the crude aqueous preparations. Shortly after its discovery, ergotamine was found to be effective in the treatment of migraines. Both ergotoxine and ergotamine cause vasoconstriction that can lead to gangrene with chronic use.

In 1935 a new water-soluble ergot alkaloid, ergonovine, was synthesized. Ergonovine is used to facilitate childbirth by stimulating uterine contractions. Many other important lysergic acid derivatives have been produced by means of semisynthesis. Of these derivatives, LSD-25 (*d*-lysergic acid diethylamide) is the most famous. LSD has been used experimentally, mainly in psychiatry and neurophysiology.

Through extensive research, many other uses for ergot alkaloids has been found. Ergotamine and dihydroergotamine are used to treat migraines, and methysergine is used in migraine prophylaxis. Dihydroergotoxine is prescribed for hypertension, cerebral diffuse sclerosis, and peripheral vascular disorders. Ergocorine and the less toxic agroclavine have been reported as unusual experimental birth-control agents. The drugs appear to inhibit implantation of the ovum. Several semisynthetic alkaloids are also active implantation inhibitors. [R.L.M.]

Erosion The result of processes that entrain and transport earth materials along coastlines, in streams, and on hillslopes. Wind and water are common agents through which forces are applied to resistant rocks, soils, or other unconsolidated materials. Erosion types often are designated on the basis of the agent: wind erosion, fluvial (water) erosion, and glacial erosion. Fluvial erosion usually has been regarded as the most effective type in shaping the land surface during recent geologic time. Under certain environmental conditions, however, wind erosion moves considerable quantities of earth materials, as demonstrated during the "dust bowl" years in the United States. Glacial erosion shaped much of the land surface during the Quaternary Period of geologic time. Each type of erosion produces distinctive landforms, contributing to the diversity of terrestrial landscapes. *See* DESERT EROSION FEATURES; EOLIAN LANDFORMS; FLUVIAL EROSION LANDFORMS; GLACIOLOGY; MASS WASTING; QUATERNARY; STREAM TRANSPORT AND DEPOSITION.

Forces exerted by erosion processes must exceed resistances of earth materials for entrainment and transportation to occur. Environmental conditions determine the magnitude of the forces, the resistances, and the relations among them. Erosion rates are highly variable in time and space due to changing relations between forces and resistances. The major factors governing wind-erosion rates are wind velocity, topography, surface roughness, soil properties and soil moisture, vegetation cover, and land use. The major factors governing fluvial-erosion rates on hillslopes are rainfall energy, topography, soil properties, vegetation cover, and land use. The major factors governing fluvial-erosion rates in stream channels are depth and velocity of water flow, together with the size and cohesiveness of the bed and bank materials. The major factors governing glacial-erosion rates are the depth and velocity of ice flow, together with the hardness of the bed and side-wall materials.

Accelerated erosion by fluvial processes may be the most important environmental problem worldwide because of its spatial and temporal ubiquity. Erosion rates commonly exceed soil-formation rates, causing depletion of soil resources. The effects of erosion are insidious due to the removal of the fertile topsoil horizon, compromising food production. Sediment frequently is transported well beyond the source area to degrade water quality in streams and lakes, harm aquatic life, reduce the water-storage capacity of reservoirs, and increase channel-maintenance costs. *See* SOIL; SOIL CONSERVATION. [T.J.T.]

Escarpment A long line of cliffs or steep slopes that break the general continuity of the land by separating it into two level or sloping surfaces. Some very high escarpments, or scarps, may form by vertical movement along faults. Often a whole block of land may be forced upward while the adjacent block is downfaulted. *See* FAULT AND FAULT STRUCTURES.

Other types of escarpments form by differential weathering and erosion of contrasted rock types. Less resistant rocks, such as clay or shale, are often eroded from beneath resistant cap rocks, such as sandstone and limestone. With support removed from below, the cap rock fails and the escarpment retreats. Escarpments are often very prominent in arid regions, where hardened weathering products may form extensive cap rocks known as duricrusts.

Some of the largest known escarpments occur on the planet Mars, where erosion has presumably been much slower than on the Earth in reducing primary structural relief. [V.R.B.]

Escherichia A genus of bacteria named for Theodor Escherich, an Austrian pediatrician and bacteriologist, who first published on these bacteria in 1885. *Escherichia coli* is the most important of the six species which presently make up this genus, and it is among the most extensively scientifically characterized living organisms. *Escherichia coli* are gram-negative rod-shaped bacteria approximately 0.5 × 1–3 micrometers in size. Molecular taxonomic analysis based on the nucleotide sequences of ribosomal ribonucleic acid (RNA) has revealed that *Shigella*, a bacterial genus of medical importance previously thought to be distinct from *E. coli*, is actually the same species.

The natural habitat of *E. coli* is the colon of mammals, reptiles, and birds. In humans, *E. coli* is the predominant bacterial species inhabiting the colon that is capable of growing in the presence of oxygen. The presence of *E. coli* in the environment is taken to be an indication of fecal contamination.

Most strains of *E. coli* are harmless to the humans and other animals they colonize, but some strains can cause disease when given access to extraintestinal sites or the intestines of noncommensal hosts. *Escherichia coli* is the most important cause of urinary tract infections. Women are more susceptible than men; four out of ten women experience at least one urinary tract infection in their lifetime. Urinary tract infections may extend into the bloodstream, especially in hospitalized patients whose defenses are compromised by the underlying illness. This may lead to a type of whole-body inflammatory response known as sepsis, which is frequently fatal. Certain *E. coli* strains can invade the intestine of the newborn and cause sepsis and meningitis. These strains are acquired at birth from *E. coli* which have colonized the vagina of the mother.

Several different strains of *E. coli* cause intestinal infections. In the developing world, the most important of these are the enterotoxigenic *E. coli*, which produce enterotoxins that act on the epithelial cells lining the small intestine, causing the small intestine to reverse its normal absorptive function and secrete fluid. This leads to a dehydrating diarrhea which can be fatal, especially in poorly nourished infants. Therapy consists of oral or, in serious cases, intravenous rehydration. Enterotoxigenic *E. coli* are transmitted by ingestion of fecally contaminated water and food, and are a common cause of diarrheal disease in travelers in developing countries.

An important group of pathogenic *E. coli* in developed countries are the enterohemorrhagic strains, especially the serotype known as *E. coli* O157:H7. These strains are normal in cattle but cause bloody diarrhea in humans. A complication of approximately 10% of cases is a potentially fatal disease known as hemolytic uremic syndrome. The virulence of these strains involves the close attachment of bacteria to epithelial cells lining the colon, resulting in alteration of the epithelial cell structure, and the production of Shiga toxin. The toxin enters the bloodstream after being absorbed in the colon and damages the endothelial cells lining the blood vessels of the colon, resulting in bloody diarrhea. In cases of hemolytic uremic syndrome, the toxin circulating in the blood damages blood vessels in the kidney, resulting in kidney failure and anemia. Enterohemorrhagic *E. coli* are acquired by the ingestion of undercooked beef, uncooked vegetables, or unpasteurized juices from fruits which have been contaminated with the feces of infected cattle. An infection can also be acquired from contact with a human infected with the organism and from contaminated water. Children and the elderly are at greatest risk of developing hemolytic uremic syndrome.

Other strains which are pathogenic in the human colon include the enteroinvasive *E. coli* (including *Shigella*) and the enteropathogenic *E. coli*. Enteroinvasive *E. coli* cause a disease called bacillary dysentery characterized by bloody diarrhea. Enteropathogenic *E. coli* have been associated with protracted diarrhea in infants and can occasionally cause severe wasting. *See* DIARRHEA; TOXIN. [S.L.M.]

Esker A sinuous ridge composed predominantly of sand and gravel deposited by glacial meltwater. Eskers vary in degree of continuity, and range in size from a few meters (1 m = 3.3 ft) to tens of meters high and from a few meters to a hundred or more kilometers long. They have steep ice-contact slopes and were deposited in channels confined by ice. Most eskers generally parallel the direction of ice flow, and while most follow valleys and have a normal down-drainage slope, some trend up a regional or local slope. [W.H.J.]

Estuarine oceanography The study of the physical, chemical, biological, and geological characteristics of estuaries. An estuary is a semienclosed coastal body of water which has a free connection with the sea and within which the sea water is measurably diluted by fresh water derived from land drainage. Many characteristic features of estuaries extend into the coastal areas beyond their mouths, and because the techniques of measurement and analysis are similar, the field of estuarine oceanography is often considered to include the study of some coastal waters which are not strictly, by the above definition, estuaries. Also, semienclosed bays and lagoons exist in which evaporation is equal to or exceeds freshwater inflow, so that the salt content is either equal to that of the sea or exceeds it. Hypersaline lagoons have been termed negative estuaries, whereas those with precipitation and river inflow equaling evaporation have been called neutral estuaries. Positive estuaries, in which river inflow and precipitation exceed evaporation, form the majority, however.

Within estuaries, the river discharge interacts with the sea water, and river water and sea water are mixed by the action of tidal motion, by wind stress on the surface, and by the river discharge forcing its way toward the sea. There is a small difference in salinity between river water and sea water, but it is sufficient to cause horizontal pressure gradients within the water which affect the way it flows. Salinity is consequently a good indicator of estuarine mixing and the patterns of water circulation.

Estuarine ecological environments are complex and highly variable when compared with other marine environments. They are richly productive, however. Because of the variability, fewer species can exist as permanent residents in this environment than in some other marine environments, and many of these species are shellfish that can easily tolerate short periods of extreme conditions. Motile species can escape the extremes. A number of commercially important marine torms are indigenous to the estuary, and the environment serves as a spawning or nursery ground for many other species. *See* MARINE ECOLOGY.

The patterns of sediment distribution and movement depend on the type of estuary and on the estuarine topography. The type of sediment brought into the estuary by the rivers, by erosion of the banks, and from the sea is also important; and the relative importance of each of these sources may change along the estuary. Fine-grained material will move in suspension and will follow the residual water flow, although there may be deposition and re-erosion during times of locally low velocities. The coarser-grained material will travel along the bed and will be affected most by high velocities and, consequently, in estuarine areas, will normally tend to move in the direction of the maximum current. [K.R.D.]

Ethology The study of animal behavior. Modern ethology includes many different approaches, but the original emphasis, as expounded by Konrad Lorenz and Niko Tinbergen, was placed on the natural behavior of animals. This contrasted with the focus of comparative psychologists on behavior in artificial laboratory situations such as mazes and puzzle boxes. Ethologists view the naturalistic approach as crucial because it reveals the environmental and social circumstances in which the behavior originally evolved, and prepares the way for more realistically designed laboratory experiments. The approach goes back to the stress that Charles Darwin placed on hereditary contributions to behavior in all species, including humans. Viewing behavior as a product of evolutionary history has helped to elucidate many otherwise puzzling aspects of its biology and has paved the way for the new science of neuroethology, concerned with how the structure and functioning of the brain controls behavior and makes learning possible.

A central concept in classical ethology is that of the innate release mechanism. If a species has had a long history of experience with certain stimuli, especially those involving survival and reproduction, then to the extent that genes affect the ability to attend closely to such stimuli, natural selection leads to adaptations enhancing responsiveness to them. A common first step in the study of these adaptations was investigation of the development of responsiveness to such stimuli in infancy, focusing on situations that the ethologist knew to be especially relevant to survival. The term innate releasing mechanism, set forth by Tinbergen and Lorenz, eschews notions of innate mental imagery and has proved fertile in understanding how genes influence behavioral development, and in focusing attention of neuroethologists on inborn physiological mechanisms that permit learning while encouraging the infant to attend closely to very specific stimuli, the nature of which varies from species to species according to differences in ecology and social organization.

In birds, such as the herring gull, innate release mechanisms are also better thought of as having evolved to guide processes of perceptual learning, rather than to design animals as though they were automata. Learning is as important in the development of the behavior of many animals as in the human species. Yet as the young organism interacts with social and physical environments with which it has evolved adaptive relationships, the course that learning takes in nature is guided and profoundly influenced by the innate predispositions that the organism brings to bear on dealing with the situation.

Perceptions of the external world provide a basis for both thought and action. It is a fundamental axiom of ethology that each organism's brain is armed with genetically determined programs of action which, in their own way, are as predictable and controlled as the genetic programs for developing anatomical structures such as a brain or a face. Ethologists have shown that it is possible to reconcile the need to modify patterns of action on the basis of experience with the possession of basic patterns of action that are coordinated by the brain, innately controlled, and often distinct from species to species. The concept of the fixed action pattern remains fundamental to understanding the development of the ability to act. Innate "motor programs," generated by the brain, are the natural units out of which behavior emerges during development. Each of these programs designates not a single completely stereotyped action, but a range of options which are limited but sufficient that selection among them allows for adjustments through experience. Close study reveals that sometimes the modifiability of actions lies in the potential flexibility of orientation, timing, and sequencing of actions rather than in the basic patterns or coordinations from which complex actions are built up. Thus nervous systems make some behavioral adjustments promptly and easily and others only with much greater difficulty, in harmony with species differences in the requirements for patterns of action as dictated by the species' structure and mode of life. Ethologists have found repeatedly that while social experience is vital in many animals for normal development of actions and responses, animals reared in restricted environments may still develop many units of action that are normal. The animal has to learn, however, how to put them together in an adaptive sequence.

Modern research on the ethology of learning began when Lorenz discovered imprinting in geese. He found that if he led a flock of newly hatched goslings himself they became imprinted on him. When mature, they would court people as though confused about their own species identity. Learning occurred very rapidly and tended to be restricted to a short sensitive phase early in life. The learning is highly focused by genetically determined preferences both to follow a parent-object with particular appearance and emitting species-specific calls, and also to learn most quickly and accurately at a particular stage of development. The interplay between nature (genetic predisposition) and nurture (environmental influence) in learning is displayed especially clearly in imprinting, hence its special interest to biologists and psychiatrists. Indications are that it is not concerned so much with learning about species as with learning to recognize individual parents and kin, both to ensure mating with one's own kind and to avoid incestuous inbreeding.

There are many forms of imprinting. So-called filial imprinting, ensuring that ducklings and goslings follow only their parent, is distinct from sexual imprinting, affecting mate choice in adulthood; the sensitive phases for learning are different in each case. Imprinting-like processes also shape the development of food preferences and abilities to use the Sun and stars in navigation.

Unlike psychological studies of animal learning in the laboratory, which have tended to favor the "blank-slate" view of the brain's contribution to learning, ethology emphasizes the need to understand all aspects of the biology of a species under study before one can hope to understand how the animal learns to cope with the many complexities of individual existence and social living. Thus ethology may lead not only to an understanding of how natural behavior evolves, but also to new insights into how brains help organisms learn to cope with social and environmental problems confronting them as individuals. *See* ANIMAL COMMUNICATION. [P.R.M.]

Eucalyptus A large and important genus of Australian forest trees; includes about 500 species in the family Myrtaceae. Only two species occur naturally outside Australia in the adjacent islands. Eucalypts occur throughout Australia except in coastal tropical and subtropical rainforests in Queensland and New South Wales and in temperate rainforests in Victoria and Tasmania. They are confined to water courses in the extensive arid zones of central and northwest Australia. Eucalypts grow from sea level to tree line (6600 ft or 2000 m). Because of its large geographic range the genus exhibits many habits, from tall trees to multistemmed, shrubby species called mallees.

Eucalyptus is an evergreen genus with four different leaf types—seedling, juvenile, intermediate, and adult—depending on plant maturity. Juvenile leaves of some species, particularly those that are silvery blue and oval, are extensively used for floral decorations. Most species have white or cream flowers. Some species, particularly those from Western Australia, are planted widely as ornamentals.

Eucalypts have been planted widely for commercial use in Brazil and other South American countries, Africa, the Indian subcontinent, and the Middle East. They are used extensively for fuel and construction and are an important component of third world economies. Foliage of some species yields essential oils for medicines and perfumes. Tannins are extracted from the bark of certain species. *See* TREE. [R.L.E.]

Eumycota True fungi, a group of heterotrophic organisms with absorptive nutrition, capable of utilizing insoluble food from outside the cell by secretion of digestive enzymes and absorption. Glycogen is the primary storage product of fungi. Most fungi have a well-defined cell wall that is composed of chitin and glucans. Spindle pole bodies, rather than centrioles, are associated with the nuclear envelope during cell division in most species. Typically, the fungal body (thallus) is haploid and consists of microscopic, branched, threadlike hyphae (collectively called the mycelium), which develop into radiating, macroscopic colonies within a substrate or host. The filamentous hypha may be divided by cross walls (septa) into compartments. Hyphal growth is apical. Some species are coenocytic (without cross walls); others, including yeasts, are unicellular. Reproductive bodies are highly variable in morphology and size, and may be asexual or sexual.

Great changes in understanding the phylogenetic relationships of fungi have been brought about by the use of characters derived from deoxyribonucleic acid (DNA) sequences and the use of computer-assisted phylogenetic analysis; these changes are reflected in current classification schemes. A modern classification follows:

Eumycota
- Phylum: Chytridiomycota
- Phylum: Zygomycota
  - Class: Zygomycetes
    - Trichomycetes
- Phylum: Ascomycota
  - Class: Archiascomycetes
    - Hemiascomycetes
    - Euascomycetes
- Phylum: Basidiomycota
  - Class: Hymenomycetes
    - Urediniomycetes
    - Ustilaginomycetes

Fungi are found in practically every type of habitat. Most are strictly aerobic, although a few are anaerobes that live in the gut of herbivores. Some species are thermophilic. Many fungi form saprobic (including parasitic) relationships with animals and plants; the majority are saprobes. As now recognized, the Eumycota are a monophyletic group of the crown eukaryotes, presumed to have been present in the fossil record 900–570 million years ago. *See* FUNGI. [M.Bl.]

Europe Although long called a continent, in many physical ways Europe is but a great western peninsula of the Eurasian landmass. Its eastern limits are arbitrary and are conventionally drawn along the water divide of the Ural Mountains, the Ural River, the Caspian Sea, and the Caucasus watershed to the Black Sea. On all other sides Europe is surrounded by salt water. Of the oceanic islands of Franz Josef Land, Spitsbergen (Svalbard), Iceland, and the Azores, only Iceland is regarded as an integral part of Europe; thus the northwestern boundary is drawn along the Danish Strait.

Europe is not only peninsular but has a large ratio of shoreline to land area reflecting a notable interfingering of land and sea. Excluding Iceland, the maximum north-south distance is (3529 mi) (5680 km); and the greatest east-west extent is 2398 mi (3860 km). Of Europe's area of 3,881,000 mi^2 (10,050,000 km^2) 73% is mainland, 19% peninsulas, and 8% islands. Also, 51% of the land is less than 155 mi (250 km) from shores and another 23% lies closer than 310 mi (500 km). This situation is caused by the inland seas that enter, like arms of the ocean, deep into the northern and southern regions of Europe, which thus becomes a peninsula of peninsulas. The most notable of these branching arms of salt water are the White Sea, the North Sea, the Baltic Sea with the Gulf of Bothnia, the English Channel (La Manche), the Mediterranean Sea with its secondary branches, and finally, the Black Sea. Even the Caspian Sea, presently the largest saltwater lake of the world, formed part of the southern seas before the folding of the Caucasus. The penetration of the landmass by these seas brings marine influences deep into the continent and provides Europe with a balanced climate favorable for human evolution and settlement.

Europe has a unique diversity of land forms and natural resources. The relief, as varied as that of other continents, has an average elevation of 980 ft (300 m) as compared with North America's 1440 ft (440 m). The shape and the overall physiographic aspect of the great peninsula are controlled by geologic structure which delimits the major regional units.

Climate is determined by a number of factors. Probably the most important are a favorable location between 35° and 71°N latitudes on the western or more maritime side of the world's largest continental mass; the west-to-east trend (rather than north-south) of the lofty southern ranges and the Central Lowlands, as well as of the inland seas, which permit the prevailing westerly winds of these latitudes to carry marine influences deep into the continent; the beneficial influence of the North Atlantic Drift, which makes possible ice-free coasts far within the Arctic Circle; and the low elevation of the northwestern mountain ranges and the Urals, which allows the free shifting of air masses over their crests.

The intricate relief and the climates of Europe are well reflected in the drainage system. Extensive drainage basins with large slow-flowing rivers are developed only in the Central Lowlands, especially in the eastern part. Streams with the greatest discharge empty into the Black Sea and the North Sea, although Europe's longest river, the Volga, feeds the Caspian Sea. Second in dimension is the Danube, which crosses the Carpathian Basin and cuts its way twice through mountain ranges at the Gate of Bratislava and at the Iron Gate. The Rhine and Rhone are the two major Alpine rivers with headwater sources close to each other but feeding the North Sea and the Western Mediterranean Basin, respectively. Abundant precipitation throughout the year, as well as the permeable soils and the dense vegetation which temporarily store the water, provides the streams of Europe north of the Southern Highlands with ample water throughout the seasons. The combined effects of poor vegetation, rocky and desolate limestone karstlands, and slight annual precipitation result in intermittent flow of the rivers along the Mediterranean coast, especially on the eastern side of peninsulas. Only the Alpine rivers carry enough water, and if it were not for the Danube and Rhone, both originating in regions north of the Alps, the only major river of the Mediterranean basin would be the Po. *See* ATLANTIC OCEAN; BALTIC SEA; BLACK SEA; CONTINENT; MEDITERRANEAN SEA. [G.T.]

Eutrophication The deterioration of the esthetic and life-supporting qualities of lakes and estuaries, caused by excessive fertilization from effluents high in phosphorus, nitrogen, and organic growth substances. Algae and aquatic plants become excessive, and, when they decompose, a sequence of objectionable features arise. Water supplies drawn from such lakes must be filtered and treated. Diversion of sewage, better utilization of manure, erosion control, improved sewage treatment and harvesting of the surplus aquatic crops alleviate the symptoms. Prompt public action is essential. *See* WATER CONSERVATION; WATER POLLUTION. [A.D.H.]

Evergreen plants Plants that retain their green foliage throughout the year. Popularly, needle-leaved trees (pine, fir, juniper, spruce) and certain broad-leaved shrubs (rhododendron, laurel) are called evergreens. In warm regions many broad-leaved trees (magnolia, live oak) are evergreen, and in the tropics most trees are evergreen and nearly all have broad leaves. Many herbaceous biennials and perennials have basal rosettes with leaves close to the ground that remain green throughout the winter. *See* FOREST AND FORESTRY; LEAF; PLANT TAXONOMY. [N.A.]

Exotic viral diseases Viral diseases that occur only rarely in human populations of developed countries. However, many of these diseases cause significant human morbidity and mortality in underdeveloped areas, and have the proven capacity to be transported to population centers in developed countries and to cause explosive outbreaks or epidemics. Most of the exotic viruses are zoonotic, that is, they are transmitted to humans from an ongoing life cycle in animals or arthropods; the exception is smallpox.

Important diseases caused by exotic viruses include yellow fever, Venezuelan equine encephalitis, Rift Valley fever, tick-borne encephalitis, Crimean hemorrhagic fever, rabies, Lassa fever, hemorrhagic fever, and Marburg and Ebola hemorrhagic fevers. Control of these diseases is often very difficult because of the lack of detailed knowledge about the natural history of the viruses in their natural animal hosts, and because of the difficulty of controlling natural populations of alternative hosts such as insects or rodents. *See* ANIMAL VIRUS. [F.A.M.]

Extinction (biology) The death and disappearance of a species. The fossil record shows that extinctions have been frequent in the history of life. Mass extinctions refer to the loss of a large number of species in a relatively short period of time. Episodes of mass extinction occur at times of rapid global environmental change; five such events are known from the fossil record of the past 600 million years. Human activity is causing extinctions on a scale comparable to the mass extinctions in the fossil record.

Record. An extinction may be of two types; phyletic or terminal. Phyletic extinction occurs when one species evolves into another with time; in this case, the ancestral species can be called extinct. However, because the evolutionary lineage has continued, such extinctions are really pseudoextinctions. In contrast, terminal extinction marks the end of an evolutionary lineage, termination of a species without any descendants. Most extinctions recorded in the fossil record and those occurring today are terminal. It has been estimated that 99% of all species that have ever lived are now extinct. *See* ORGANIC EVOLUTION.

The fossil record is best known for marine organisms. The mass extinctions of the marine fossil record occurred during the Late Ordovician, Late Devonian, Late Permian, Late Triassic, and Late Cretaceous. These mass extinctions affected a variety of organisms in many different ecological settings. Terrestrial and marine mass extinctions seem to occur at about the same time. The Late Permian, Late Triassic, and Late Cretaceous are also times of extinction for terrestrial vertebrates; the most dramatic extinction of terrestrial vertebrates took place at the end of the Cretaceous, when the last dinosaurs died off.

The best record of terrestrial vertebrate extinction is that of the Pleistocene. Late Pleistocene extinctions in North America are especially well known—33 genera of mammals vanished during the last 100,000 years. These extinctions were concentrated among the large mammals—those over 100 lb (44 kg) in weight—and most occurred during a short time interval approximately 11,000 years ago.

Causes. Ever since the work of Georges Cuvier, the French naturalist who demonstrated the reality of extinction, explanations have fascinated both scientists and the general public. Cuvier invoked sudden catastrophic events, whereas his contemporaries favored more gradual processes. These two themes, catastrophism and gradualism, are still debated.

In 1980 high concentrations of iridium were reported precisely at the Cretaceous-Tertiary boundary. Iridium is rare in most rocks but more abundant in meteorites. It was proposed, therefore, that an asteroid struck the Earth 65 million years ago. The impact darkened the atmosphere with dust, caused a catastrophic short-term cooling of the climate, and thus led to the extinction of dinosaurs and many other Cretaceous species. The iridium-rich layer at the boundary marks this terminal Cretaceous event.

Astronomical theories have been put forward to explain the Late Cretacous extinctions as well as the 26-million-year periodicity. In one theory, the Sun has a distant companion star that would pass in orbit near the solar system's cloud of comets every 26 million years. This might perturb many comets, sending a few into the Earth. A comet would produce the same effects as an asteroid.

Other explanations for mass extinctions include lowered sea level, climatic cooling, and changes in oceanic circulation. Biotic processes such as disease, predation, and competition may also cause the extinction of species but are difficult to prove from the fossil record because they leave little evidence. Biotic factors usually affect only one or a few interdependent species. Predation and competition are important causes of more recent extinctions, which continue today. Human activities such as hunting and fishing (predation), habitat alteration (competition for space), and pollution have probably destroyed thousands of species. These activities, together with continued tropical deforestation and resulting changes in climate, are likely to cause extinctions that will be comparable to the mass extinctions seen in the fossil record. [K.W.F.]

F

Facies (geology) Any observable attribute of rocks, such as overall appearance, composition, or conditions of formation, and changes that may occur in these attributes over a geographic area. The term facies is widely used in connection with sedimentary rock bodies, but is not restricted to them. In general, facies are not defined for sedimentary rocks by features produced during weathering, metamorphism, or structural disturbance. In metamorphic rocks specifically, however, facies may be identified by the presence of minerals that denote degrees of metamorphic change.

Sedimentary facies. The term sedimentary facies is applied to bodies of sedimentary rock on the basis of descriptive or interpretive characteristics. Descriptive facies are based on lithologic features such as composition, grain size, bedding characteristics, and sedimentary structures (lithofacies) or on biological (fossil) components (biofacies), or on both. Individual lithofacies or biofacies may be single beds a few millimeters thick or a succession of beds tens to hundreds of meters thick. For example, a river deposit may consist of decimeters-thick beds of a conglomerate lithofacies interbedded with a cross-bedded sandstone lithofacies. The fill of certain major Paleozoic basins may be divided into units hundreds of meters thick comprising a shelly facies, containing such fossils as brachiopods and trilobites, and graptolitic facies.

The term facies can be used also in an interpretive sense for groups of rocks that are thought to have been formed under similar conditions. This usage may emphasize specific depositional processes, such as a turbidite facies, or a particular depositional environment such as a shelf carbonate facies, encompassing a range of depositional processes.

Groups of facies (usually lithofacies) that are commonly found together in the sedimentary record are known as facies assemblages or facies associations. These groupings provide the basis for defining broader, interpretive facies for the purpose of paleogeographic reconstruction. [A.D.M.]

A metamorphic facies is a collection of rocks containing characteristic mineral assemblages developed in response to burial and heating to similar depths and temperatures. It can represent either the diagnostic mineral assemblages that indicate the physical conditions of metamorphism or the pressure-temperature conditions that produce a particular assemblage in a rock of a specific composition.

The metamorphic facies to which a rock belongs can be identified from the mineral assemblage present in the rock; the pressure and temperature conditions represented by each facies are broadly known from experimental laboratory work on mineral stabilities. These facies names are based on the mineral assemblages that develop during metamorphism of a rock with the composition of a basalt, which is a volcanic rock rich in iron and magnesium and with relatively little silica. For example, the dominant mineral of the blueschist facies (in a rock of basaltic composition) is a sodium- and magnesium-bearing silicate called glaucophane, which is dark blue in outcrop and blue or violet when viewed under the microscope. Characteristic minerals of the greenschist facies include chlorite and actinolite, both of which are green in outcrop and under the microscope. Basaltic rocks metamorphosed in the amphibolite facies are largely composed of an amphibole called hornblende. The granulite facies takes its name from a texture rather than a specific mineral: the pyroxenes and plagioclase that are common minerals in granulite facies rocks typically form rounded crystals of similar size that give the rock a granular fabric. [J.Sel.]

Farm crops Farm crops may be roughly classed as follows: (1) food crops—the bread grains (wheat and rye), rice, sugar crops (sugarbeets and sugarcane), potatoes, and dry legume seeds (peanuts, beans, and peas); (2) feed crops—corn, sorghum grain, oats, barley, and all hay and silage; and (3) industrial crops—cotton (lint and seed), soybeans, flax, and tobacco. See separate articles on these topics.

Crop production is regionalized in the United States in response to the combination of soil and climatic conditions and to the land topography, which favors certain kinds of crop management. In general, commercial farm crops are confined to land in humid and subhumid climates that can be managed to minimize soil and water erosion damage, where soil productivity can be kept at a relatively high level, and where lands are smooth enough to permit large-scale mechanized farm operations. In less well-watered regions, cropping is practiced efficiently on fairly level, permeable soils, where irrigation water can be supplied. The tilled crops, such as corn, sorghums, cotton, potatoes, and sugar crops, are more exacting in soil requirements than the close-seeded crops, such as wheat, oats, barley, rye, and flax. The crops planted in soil stands, mostly hay crops (as well as pastures), are efficient crops for lands that are susceptible to soil and water erosion.

The major farming regions of the United States are named from the predominant kinds of crops grown, even though there is tremendous diversity within each region. The Corn Belt includes a great central area extending from Nebraska and South Dakota east across much of Iowa, Missouri, Illinois, and Indiana to central Ohio. To the north and east of this region is the Hay and Dairy Region, which actually grows large quantities of feed grains. To the south of the Corn Belt is the Corn and Winter-Wheat Belt. The southern states, once the Cotton Belt, now concentrate on hay, pasture, and livestock, with considerable acreages of soybeans and peanuts. The cultivated portions of the Great Plains, extending from Canada to Mexcio, are divided into a spring-wheat region in the Dakotas and a winter-wheat region from Texas to Nebraska. The Intermountain Region, between the Rocky Mountains and the Cascade-Sierra Nevada mountain ranges, is cropped only where irrigation is feasible, and a wide range of farm crops is grown. In the three states of the Pacific Region, a great diversity of crops is grown. Cotton is now concentrated in the irrigated regions from Texas to California.

The United States is known most widely for its capacity to produce and export wheat. However, this nation has become a major producer of soybeans for export, and rice exports have become important. Most of the other farm crops are consumed within the United States. Tobacco and sugar crops are high acre-value crops, as are potatoes, peanuts, and dry beans. The production of these high acre-value crops is concentrated in localized areas where soils and climate are particularly favorable, rather than in broad acreages. [H.B.S.]

Fault and fault structures Products of fracturing and differential movements along fractures in continental and oceanic crustal rocks. Faults range in length and magnitude of displacement from small structures visible in hand specimens, displaying offsets of a centimeter or less, to long, continuous crustal breaks, extending

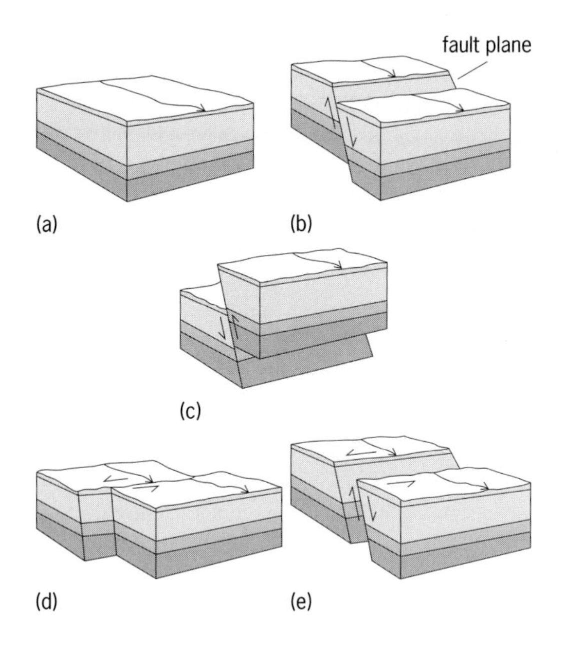

Slip on faults. (*a*) Block before faulting; (*b*) normal-slip; (*c*) reverse-slip; (*d*) strike-slip; (*e*) oblique-slip. (*After F. Press and R. Siever, Earth, 2d ed., 1978*)

hundreds of kilometers in length and accommodating displacements of tens or hundreds of kilometers. Faults exist in deformed rocks at the microscopic scale, but these are generally ignored or go unrecognized in most geological studies. Alternatively, where microfaults systematically pervade rock bodies as sets of very closely spaced subparallel, planar fractures, they are recognized and interpreted as a type of cleavage which permitted flow of the rock body. Fractures along which there is no visible displacement are known as joints. Large fractures which have accommodated major dilational opening (a meter or more) perpendicular to the fracture surfaces are known as fissures. Formation of fissures is restricted to near-surface conditions, for example, in areas of crustal stretching of subsidence.

In addition to describing the physical and geometric nature of faults and interpreting time of formation, it has been found to be especially important to determine the orientations of minor fault structures (such as striae and drag folds) which record the sense of relative movement. Evaluating the movement of faulting can be difficult, for the apparent relative movement (separation) of fault blocks as seen in map or outcrop may bear little or no relation to the actual relative movement (slip). The slip of the fault is the actual relative movement between two points or two markers in the rock that were coincident before faulting (see illustration). Strike-slip faults have resulted in horizontal movements between adjacent blocks; dip-slip faults are marked by translations directly up or down the dip of the fault surface; in oblique-slip faults the path of actual relative movement is inclined somewhere between horizontal and dip slip.

Recognizing even the simplest translational fault movements in nature is often enormously difficult because of complicated and deceptive patterns created by the interference of structure and topography, and by the absence of specific fault structures which define the slip path. While mapping, the geologist mainly documents apparent relative movement (separation) along a fault, based on what is observed in plan-view or cross-sectional exposures. [G.H.D.]

Feline infectious peritonitis A fatal disease of both domestic and exotic cats (particularly cheetahs) caused by feline infectious peritonitis virus, a member of the Coronaviridae family. There are multiple strains of the virus which vary in virulence. Feline infectious peritonitis virus is closely related morphologically, genetically, and antigenically to other members of the Coronaviridae and arises as mutants from feline enteric coronaviruses, which infect cats but generally induce very mild or inapparent gastroenteritis. All coronaviruses are single-stranded ribonucleic acid (RNA) viruses with poor or no error correction during replication, resulting in relatively high mutation rates.

Feline coronaviruses are contracted primarily during exposure to infectious cat feces in the environment, but also via ingestion or inhalation during cat-to-cat contact. Feline infectious peritonitis virus is relatively labile once outside the cat's body but may be able to survive for as long as 7 weeks if protected from heat, light, and desiccation. It is readily inactivated by most disinfectants.

Signs of infection with feline infectious peritonitis virus depend upon the severity of infection, the relative ability of the immune system to minimize some of the characteristic inflammatory lesions, and the organ systems affected. There are two clinical forms of the disease. Wet feline infectious peritonitis, which is characterized by protein and fluid leakage into the abdominal cavity (or less frequently thoracic and scrotal spaces), occurs in cats with overwhelming infection, with poor immunity, or during late stages of other forms of the virus. In contrast, if a cat has a moderately competent (but ultimately ineffectual) immune response to the virus infection, granulomas may arise in infected tissues. This form of the disease, known as dry feline infectious peritonitis, commonly affects kidneys, liver, lymph nodes, mesentery, diaphragm, the outer surface of the intestine, and the neurological system. Wet or effusive feline infectious peritonitis is characterized by abdominal swelling, jaundice, and typically difficulty in breathing.

As with most viral infections, there is no specific antiviral drug of proven efficacy in the treatment of feline infectious peritonitis. Clinical management rests upon palliative treatment of the specific signs exhibited by each cat, and upon antibiotics, when indicated, to reduce secondary bacterial infections. The most important therapeutic approach involves the administration of immunosuppressive doses of corticosteroids to reduce the cat's immune response to the virus. *See* ANIMAL VIRUS; VIRUS. [J.E.F.]

Feline leukemia A type of cancer caused by the feline leukemia virus, a retrovirus which affects only a small percentage of freely roaming or domestic cats. The feline leukemia virus is genetically and morphologically similar to murine leukemia virus, from which it presumably evolved several million years ago.

About 1–5% of healthy-appearing wild or freely roaming domestic cats have lifelong (persistent) infections. These carrier cats shed the virus in urine, feces, and saliva. The principal route of infection is oral. Infections occurring in nature are usually inapparent or mild, and 95% of such cats recover without any signs of illness. Mortality due to a feline leukemia virus infection occurs mainly among persistently infected cats and at a rate of around 50% per year.

There is no treatment that eliminates the virus. Supportive or symptomatic treatment may prolong life for weeks or months, depending on the particular disease manifestation. All cats should be tested for the presence of the virus prior to putting them in contact with feline leukemia virus-free animals. Healthy-appearing or ill infected cats should not be in intimate contact with noninfected cats, even if the latter have been vaccinated. Feline leukemia virus vaccines are available and should be administered annually, although they should not be considered a substitute for testing, elimination, and quarantine procedures. *See* RETROVIRUS. [N.C.P.]

Feline panleukopenia An acute viral infection of cats, also called feline viral enteritis and (erroneously) feline distemper. The virus infects all members of the cat family (Felidae) as well as some mink, ferrets, and skunks (Mustelidae); raccoons and coatimundi (Procyonidae); and the binturong (Viverridae). Panleukopenia is the most important infectious disease of cats. This disease occurs worldwide, and nearly all cats are exposed by their first year because the virus is stable and ubiquitous; the disease is rarely seen in older cats. Without treatment, this disease is often fatal.

Feline panleukopenia virus is classified as a parvovirus, and is one of the smallest known viruses. It is antigenically identical to the mink enteritis virus, and only minor antigenic differences exist between feline panleukopenia virus and canine parvovirus. It is believed that canine parvovirus originated as a mutation from feline panleukopenia virus.

The disease is severe and life threatening in 20–50% of cases. The cat is depressed and may refuse food or water; vomiting and diarrhea are common, resulting in severe dehydration. The cat may have a fever or a subnormal temperature. A low white blood cell count confirms the diagnosis as panleukopenia. Diagnosis can be made by autopsy and evidence of the destruction of the intestinal crypts and villus shortening.

Highly effective and safe vaccines are available for the prevention of panleukopenia. Premises contaminated by feline panleukopenia virus are extremely difficult to disinfect; chlorine bleach, formaldehyde, or a certain quaternary ammonium disinfectant will destroy the virus. A cat should be successfully immunized before being introduced to premises where a pan-leukopenia-infected cat previously lived. *See* ANIMAL VIRUS.

[J.H.Car.]

Fermentation Decomposition of foodstuffs generally accompanied by the evolution of gas. The best-known example is alcoholic fermentation, in which sugar is converted into alcohol and carbon dioxide. During fermentation organic matter is decomposed in the absence of air (oxygen); hence, there is always an accumulation of reduction products, or incomplete oxidation products. Some of these products (for example, alcohol and lactic acid) are of importance to humans, and fermentation has therefore been used for their manufacture on an industrial scale. There are also many microbiological processes that go on in the presence of air while yielding incomplete oxidation products. Good examples are the formation of acetic acid (vinegar) from alcohol by vinegar bacteria, and of citric acid from sugar by certain molds (for example, *Aspergillus niger*). These microbial processes, too, have gained industrial importance, and are often referred to as fermentations, even though they do not conform to L. Pasteur's concept of fermentation as a decomposition in the absence of air. [C.B.V.N.]

Fertilizer Materials added to the soil, or applied directly to crop foliage, to supply elements needed for plant nutrition. These materials may be in the form of solids, semisolids, slurry suspensions, pure liquids, aqueous solutions, or gases.

The chemical elements nitrogren, phosphorus, and potassium are the macronutrients, or primary fertilizer elements, which are required in greatest quantity. Sulfur, calcium, and magnesium, called secondary elements, are also necessary to the health and growth of vegetation, but they are required in lesser amounts compared to the macronutrients. The other elements of agronomic importance, called micronutrients and provided for plant ingestion in small (or trace) amounts, include boron, cobalt, copper, iron, manganese, molybdenum, and zinc. All these fertilizer elements, along with other chemical elements, occur naturally in agricultural soils in varying concentrations and mineral compositions which may or may not be in forms readily accessible to root systems of plants. The addition of fertilizer to soils used for the production of commercial crops is necessary to correct natural deficiencies and to replace the components absorbed by the crops in their growth.

Crop requirements of fertilizer components could be satisfied by the spreading of individual materials for each element deficient in the soil. However, economy favors the single application of a balanced mixture that satisfies all nutritional needs of a crop. Many commercial fertilizers therefore contain more than one of the primary fertilizer elements.

The compositions of fertilizer mixtures, in terms of the primary fertilizer elements, are identified by an N-P-K code: N denotes elemental nitrogen; P denotes the anhydride of phosphoric acid (P_2O_5); K denotes the oxide of potassium (K_2O). All are expressed numerically in percentage composition, or units of 20 lb each per short ton (10 kg per metric ton) of finished fertilizer as packaged. Formula 8-32-16 thus contains a mixture aggregating 8 wt % N in some form of nitrogen compounds, 32 wt % P_2O_5 in some form of phosphates, and 16 wt % K_2O in some form of potassium compounds, to give a product with a total of 56 fertilizer units. The commercial N-P-K formulas are generally in whole numbers. None of the N-P-K formulas totals 100% plant nutrients because the formulas indicate only the nutrient portions of the primary-element compounds and do not account for any other materials present.

Aqueous solutions of urea, ammonia, and ammonium nitrate (UAN solutions) are used directly by the farmers as well as in the preparation of granular N-P-K products by mixing with other materials. UAN solutions are also spread directly by field application or used to prepare complete N-P-K fertilizer solutions or suspensions. Suspension fertilizers consist of aqueous slurries of fine crystals in saturated solutions that are stabilized by small amounts of gelling materials, such as attapulgite clay. Suspensions can be maintained in uniform composition during spreading on the fields, and give better dispersion than granular material. *See* FERTILIZING. [A.Lo.]

Organic fertilizers are organic materials of vegetable and animal origin which contain certain macro, secondary, or micro nutrients that can be utilized by plants after application to agricultural soils. The primary nutrient sources of vegetable origin are crop residues, green manures, oilseed cakes, seaweeds, and miscellaneous food processing and distillery wastes. Also included in this category is biologically fixed nitrogen from legumes in association with root-nodulating bacteria of the genus *Rhizobium*. Animal sources include animal manures and urine, sewage sludge, septage, latrine wastes, and to a lesser extent materials such as blood meal, bone meal, and fish scraps. Often organic fertilizers are of mixed animal and vegetable origin, such as most farmyard manures, rural and urban composts, and sewage effluents and sludges. *See* NITROGEN FIXATION; SOIL MICROBIOLOGY. [R.I.P.; J.F.P.]

Fertilizing Addition of elements or other materials to the soil to increase or maintain plant yields. Fertilizers may be organic or inorganic. Organic fertilizers are usually manures and waste materials which in addition to providing small amounts of growth elements also serve as conditioners for the soil. Commercial fertilizers are most often inorganic. *See* FERTILIZER.

Methods of applying fertilizers vary widely and depend on such factors as kind of crop and stage of growth, application rates, physical and chemical properties of the fertilizer, and soil type. Two basic application methods are used, bulk spreading and precision placement. Time and labor are saved by the practice of bulk spreading, in which the fertilizer is broadcast over the entire area by using large machines which cover many acres in a short time. Precision placement, in which the fertilizer is applied in one or more bands in a definite relationship to the seed or plants, requires more equipment and time, but usually smaller amounts of fertilizer are needed to produce a given yield increase.

For some deep-rooted plants, subsoil fertilization to depths of 12–20 in. (30–50 cm) is advantageous. This is usually a separate operation from planting, and uses a modified subsoil plow followed by equipment to bed soil over the plow furrow, thereby eliminating rough soil conditions unfavorable for good seed germination. Top-dressings are usually applied by broadcasting over the soil surface for closely spaced crops such as small grains.

Since solid fertilizers range from dense heavy materials to light powders and liquid fertilizers range from high pressure to zero pressure, a variety of equipment is required for accurate metering and placement. In addition, application rates may be as low as 50 lb/acre (56 kg/hectare) or as high as 6 tons/acre (13.3 metric tons/hectare). Large bulk spreaders usually use drag chains or augers to force the material through a gate or opening whose size is varied to regulate the amount passing through and falling on the spreader.

Liquid fertilizer of the high-pressure type (for example, anhydrous ammonia) is usually regulated by valves or positive displacement pumps. The size of the orifice may be controlled manually or automatically by pressure-regulating valves. Low-pressure solutions may be metered by gravity flow through orifices, but greater accuracy is obtained by using compressed air or other gases to maintain a constant pressure in the tank. This method eliminates the effect of temperature and volume changes. Nonpressure solutions may be metered by gravity or by gear, roller, piston, centrifugal, or hose pumps. The accuracy of the gravity type can be improved by the use of a constant head device by which all air is introduced into the tank at the bottom.

Nonvolatile fertilizer solutions are often pumped into the supply lines of irrigation systems to allow simultaneous fertilization and irrigation. With the exception of bulk spreaders and other broadcasters, most fertilizer application devices are built as attachments which can be mounted in conjunction with planters, cultivators, and herbicide applicators. Often the tanks, pumps, and controls used for liquid fertilizers are also used for applying other chemicals such as insecticides. [J.G.F.]

Fescue A group of approximately 100 species of grass; more than 30 are represented in the United States. Tall fescue (*Festuca arundinacea*), a perennial cool-season plant introduced from Europe, occupies about 35×10^6 acres (15×10^6 hectares), primarily in the humid south-central region of the United States. It is popular because of its ease of establishment, vigor, wide range of adaptation, long grazing season, tolerance to abuse, sufferance of drought and poor soils, pest resistance, good seed production, and esthetic value when used for turf, ground cover, and conservation purposes. It is used primarily as pasture and hay for beef cattle, with lesser use for dairy cows or replacement heifers, sheep, and horses. The leafy and vigorous plants can grow to 3–4 ft (0.9–1.2 m) if undisturbed; under grazing or clipping, they can form a dense sod when sufficient water and fertility are available.

Other important fescues include meadow fescue (*F. elatior*), red fescue (*F. rubra*), Chewings fescue (*F. rubra* var. *commutata*), Idaho fescue (*F. idahoensis*), and sheep fescue (*F. ovina*). *See* GRASS CROPS. [H.A.Fr.]

Fiber crops Crops that are grown because of their content or yield of fibrous material which is used for many commercial purposes and for home industry. Fibers may be extracted from various parts of different plants.

Long, multiple-celled fibers can be subdivided into hard, or leaf, fibers that traditionally are used for cordage, such as sisal for binder and baler twine and abaca or manila hemp for ropes; soft, or bast (stem), fibers that are used for textiles, for example, flax for linen, hemp for small twines and canvases, and jute and kenaf for industrial textiles such as burlap; and miscellaneous fibers that may come from the roots, such as "broom" root for brushes, or stems, as Spanish moss for upholstery, or fruits, as coir from coconut husks for cordage and floor coverings. See separate articles on these topics.

Short, one-celled fibers come from the seeds or seed pods of plants such as cotton and kapok. Cotton is the world's most widely grown and used textile fiber. [E.G.N.]

Filtration The separation of solid particles from a fluidsolids suspension of which they are a part by passage of most of the fluid through a septum or membrane that retains most of the solids on or within itself. The septum is called a filter medium, and the equipment assembly that holds the medium and provides space for the accumulated solids is called a filter. The fluid may be a gas or a liquid. The solid particles may be coarse or very fine, and their concentration in the suspension may be extremely low (a few parts per million) or quite high (>50%).

The object of filtration may be to purify the fluid by clarification or to recover clean, fluid-free particles, or both. In most filtrations the solids-fluid separation is not perfect. In general, the closer the approach to perfection, the more costly the filtration; thus the operator of the process cannot justify a more thorough separation than is required.

Gas filtration involves removal of solids (called dust) from a gas-solids mixture because: (1) the dust is a contaminant rendering the gas unsafe or unfit for its intended use; (2) the dust particles will ultimately separate themselves from the suspension and create a nuisance; or (3) the solids are themselves a valuable product that in the course of its manufacture has been mixed with the gas.

Three kinds of gas filters are in common use. Granular-bed separators consist of beds of sand, carbon, or other particles which will trap the solids in a gas suspension that is passed through the bed. Bag fitters are bags of woven fabric, felt, or paper through which the gas is forced; the solids are deposited on the wall of the bag. Air filters are light webs of fibers, often coated with a viscous liquid, through which air containing a low concentration of dust can be passed to cause entrapment of the dust particles.

Liquid filtration is used for liquid-solids separations in the manufacture of chemicals, polymer products, medicinals, beverages, and foods; in mineral processing; in water purification; in sewage disposal; in the chemistry laboratory; and in the operation of machines such as internal combustion engines.

Liquid filters are of two major classes, cake filters and clarifying filters. The former are so called because they separate slurries carrying relatively large amounts of solids. They build up on the filter medium as a visible, removable cake which normally is discharged "dry" (that is, as a moist mass), frequently after being washed in the filter. It is on the surface of this cake that filtration takes place after the first layer is formed on the medium. The feed to cake filters normally contains at least 1% solids. Clarifying filters, on the other hand, normally receive suspensions containing less than 0.1% solids, which they remove by entrapment on or within the filter medium without any visible formation of cake. The solids are normally discharged by backwash or by being discarded with the medium when it is replaced. [S.A.M.]

Fiord A segment of a troughlike glaciated valley partly filled by an arm of the sea. It differs from other glaciated valleys only in the fact of submergence. The floors of many fiords are elongate basins excavated in bedrock, and in consequence are shallower at the fiord mouths than in the inland direction. The seaward rims of such basins represent lessening of glacial erosion at the coastline, where the former glacier ceased to be confined by valley walls and could spread laterally. Some rims are heightened by glacial drift deposited upon them in the form of an end moraine.

Fiords occur conspicuously in British Columbia and southern Alaska, Greenland, Arctic islands, Norway, Chile, New Zealand, and Antarctica—all of which are areas of rock resistant to erosion, with deep valleys, and with strong former glaciation. [R.F.Fl.]

Fir Any tree of the genus *Abies*, of the pine family, characterized by erect cones, by the absence of resin canals in the wood but with many in the bark, and by flattened needlelike leaves which lack definite stalks. The leaves usually have two white lines on the underside and leave a circular scar when they fall.

The native fir of the northeastern United States and adjacent Canada is *A. balsamea*. Its principal uses are for paper pulp, lumber, boxes and crates, and as a source of the liquid resin called Canada balsam. In the eastern United States the fir is commonly used as a Christmas tree.

The Fraser fir (*A. fraseri*) is a similar species found in the southern Appalachians.

Several species of *Abies* grow in the Rocky Mountains region and westward to the Pacific Coast. The most important commercially is the white fir (*A. concolor*), also known as silver fir. Other western species of commercial importance are the subalpine fir (*A. lasiocarpa*), grand fir (*A. grandis*), Pacific silver fir (*A. amabilis*), California red fir (*A. magnifica*), and noble fir (*A. procera*). [A.H.G./K.P.D.]

Fisheries ecology The study of ecological processes that affect exploited aquatic organisms, in both marine and fresh-water environments. Because this field is primarily motivated by an attempt to harvest populations, special attention has been given to understanding the regulation of aquatic populations by nature and humans. The foundations of fisheries ecology lie in population and community ecology, with ideas and methods from physiology, genetics, molecular biology, and epidemiology being increasingly relevant. Because of issues associated with harvesting biological resources, fisheries ecology must also go beyond biology and ecology into sociology and economics. *See* ECOLOGY.

The problem of regulating the exploitation of aquatic organisms in order to ensure sustained harvest lies at the core of fisheries ecology. Most experts agree that the harvest of marine resources has peaked and increased yields are likely to come only from fine-tuning of regulations on stocks that are fully exploited.

The typical unit at which management efforts are directed is the exploited population, customarily termed a stock. A central aspect of assessing fisheries resources is to identify these stocks and determine how isolated they are from other stocks of the same species. For example, there are 20 major recognized stocks of cod in the North Atlantic Ocean, each on the whole isolated from every other and distinct with regard to several biological characteristics that determine the potential for harvest. The three main population processes that govern the size and productivity of given stock are somatic growth, mortality, and recruitment (the incorporation of new individuals into the population through birth). A typical assessment of an exploited stock includes the study of these three processes, as well as some protocol to estimate abundance. From this information, and aided by mathematical models and statistical tools, fisheries biologists produce recommendations on how many individuals to harvest, of what size or age, when, and where.

The early history of modern fishing was characterized by attempts to ensure an increasing supply of aquatic organisms at all costs. In the last two decades, that view has yielded to a more realistic perspective, increasingly heedful to the natural limits of aquatic resources and to many environmental aspects of harvested systems. These include habitat degradation, overfishing, incidental mortality of nontarget species, and the indirect effects of species removal at the ecosystem level. *See* MARINE CONSERVATION; MARINE FISHERIES. [M.Pa.; P.D.]

Floodplain The relatively broad and smooth valley floor that is constructed by an active river and periodically covered with floodwater from that river during intervals of overbank flow. Engineers consider the floodplain to be any part of the valley floor subject to occasional floods that threaten life and property. Various channel improvements or impoundments may be used to restrict the natural process of overbank flow. Geomorphologist consider the floodplain to be a surface that develops by the active erosional and depositional processes of a river. Floodplains are underlain by a variety of sediments, reflecting the fluvial history of the valley.

Most floodplains consist of the following types of deposits: colluvium—slope wash and mass-wasting products from the valley sides, as is common in small, narrow floodplains; channel lag—coarse debris marking the bottoms of former channels; lateral accretion deposits—sand and gravel deposited as the meandering river migrates laterally; vertical accretion deposits—clay and silt deposited by overbank flooding of the river; crevassesplay deposits—relatively coarse sediment carried through breaks in the natural river levees and deposited in areas that usually receive overbank deposition; and channel-fill deposits—fills of former river channels. Channel fills may be coarse for sandy rivers. The noncohesive character of coarse sediments allows these rivers to easily erode laterally. *See* STREAM TRANSPORT AND DEPOSITION. [V.R.B.]

Floriculture The segment of horticulture concerned with commercial production, marketing, and sale of bedding plants, cut flowers, potted flowering plants, foliage plants, flower arrangements, and noncommercial home gardening.

Commercial crops are grown either in the field or under protected cultivation, such as in glass or plastic structures. Field production is confined to warm climates or to summer months in colder areas. Typical field crops are gladiolus, peonies, stock, gypsophila, asters, and chrysanthemums. Greenhouse production is not as confined by climate or season, but most greenhouses are located in areas that have advantages such as high light intensity, cool night temperatures, or ready access to market. Jet air transportation resulted in major changes in international crop production.

Pronounced improvements in cultivars have been realized because of excellent breeding programs conducted by commercial propagators and by some horticulture departments. Modern cultivars have traits such as more attractive flower colors and forms, longer-lasting flowers, better growth habit, increased resistance to insects and disease organisms, or ability to grow and flower at cooler night temperatures. *See* BREEDING (PLANT). [R.A.L.]

Flower A higher plant's sexual apparatus in the aggregate, including the parts that produce sex cells and closely associated attractive and protective parts (Fig. 1). "Flower" as used in this article will be limited, as is usual, to the angiosperms, plants with enclosed seeds and the unique reproductive process called double fertilization. In its most familiar form a flower is made up of four kinds of units arranged concentrically. The green sepals (collectively termed the calyx) are outermost, showy petals (the corolla) next, then the pollen-bearing units (stamens, androecium), and finally the centrally placed seed-bearing units (carpels, gynoecium). This is the "complete" flower of early botanists, but it is only one of an almost overwhelming array of floral forms. One or more kinds of units may be lacking or hard to recognize depending on the species,

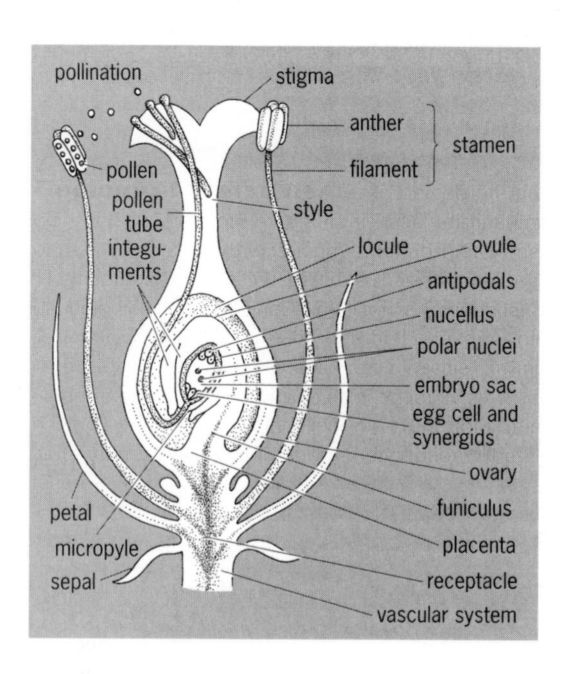

Fig. 1. Flower structure, median longitudinal section.

and evolutionary modification has been so great in some groups of angiosperms that a flower cluster (inflorescence) can took like a single flower.

Flora diversity. Most botanical terms are descriptive, and a botanist must have a large store of them to impart the multiformity of flowers. The examples that follow are only a smattering. An extra series of appendages alternating with the sepals, as in purple loosestrife, is an epicalyx. A petal with a broad distal region and a narrow proximal region is said to have a blade and claw: the crape myrtle has such petals. The term perianth, which embraces calyx and corolla and avoids the need to distinguish between them, is especially useful for a flower like the tulip, where the perianth parts are in two series but are alike in size, shape, and color. The members of such an undifferentiated perianth are tepals. When the perianth has only one series of parts, however, they are customarily called sepals even if they are petallike, as in the windflower.

A stamen commonly consists of a slender filament topped by a four-lobed anther, each lobe housing a pollen sac. In some plants one or more of the androecial parts are sterile rudiments called staminodes: a foxglove flower has four fertile stamens and a staminode. Carpellode is the corresponding term for an imperfectly formed gynoecial unit.

A gynoecium is apocarpous if the carpels are separate (magnolia, blackberry) and syncarpous if they are connate (tulip, poppy). Or the gynoecium may regularly consist of only one carpel (bean, cherry). A solitary carpel or a syncarpous gynoecium can often be divided into three regions: a terminal, pollen-receptive stigma; a swollen basal ovary enclosing the undeveloped seeds (ovules); and a constricted, elongate style between the two. The gynoecium can be apocarpous above and syncarpous below; that is, there can be separate styles and stigmas on one ovary (wood sorrel).

Every flower cited so far has a superior ovary: perianth and androecium diverge beneath it (hypogyny). If perianth and androecium diverge from the ovary's summit, the ovary is inferior and the flower is epigynous (apple, banana, pumpkin). A flower is perigynous if the ovary is superior within a cup and the other floral parts diverge from

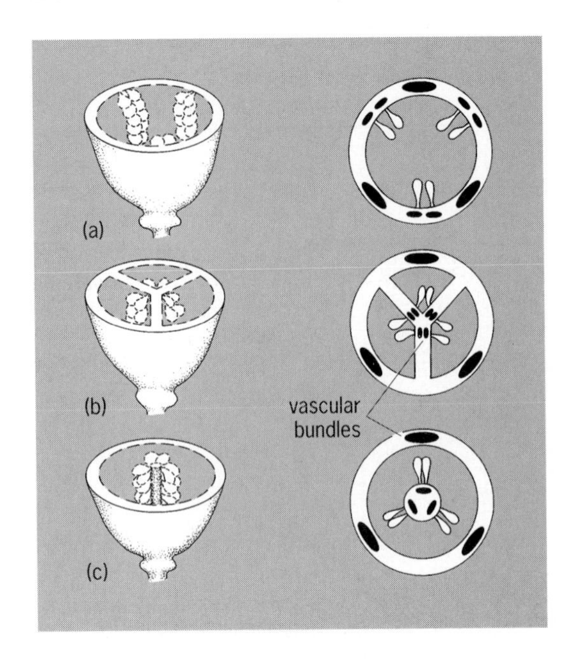

Fig. 2. Placentation: (*a*) parietal, (*b*) axile, and (*c*) free central. (*After P. H. Raven, R. F. Evert, and H. Curtis, Biology of Plants, Worth Publishers, 1976*)

the cup's rim (cherry). A syncarpous ovary is unilocular if it has only one seed chamber, plurilocular if septa divide it into more than one. The ovules of a plurilocular ovary are usually attached to the angles where the septa meet; this is axile placentation, a placenta being a region of ovular attachment. There are other ways in which the ovules can be attached—apically, basally, parietally, or on a free-standing central placenta—each characteristic of certain plant groups (Fig. 2).

The term bract can be applied to any leaflike part associated with one or more flowers but not part of a flower. Floral bracts are frequently small, even scalelike, but the flowering dogwood has four big petallike bracts below each flower cluster. The broad end of a flower stalk where the floral parts are attached is the receptacle. The same term is used, rather inconsistently, for the broad base bearing the many individual flowers (florets) that make up a composite flower like a dandelion or a sunflower.

Sexuality. A plant species is diclinous if its stamens and carpels are in separate flowers. A diclinous species is monoecious if each plant bears staminate and carpellate (pistillate) flowers, dioecious if the staminate and carpellate flowers are on different plants. The corn plant, with staminate inflorescences (tassels) on top and carpellate inflorescences (ears) along the stalk, is monoecious. Hemp is a well-known dioecious plant.

Nectaries. Flowers pollinated by insects or other animals commonly have one or more nectaries, regions that secrete a sugar solution. A nectary can be nothing more than a layer of tissue lining part of a floral tube or cup (cherry), or it can be as conspicuous as the secretory spur of a nasturtium or a larkspur. It can be a cushionlike outgrowth at the base of a superior ovary (orange blossom) or atop an inferior ovary (parsley family). Gladiolus and a number of other monocotyledons have septal nectaries, deep secretory crevices where the carpels come together. Substances that give off floral odors—essential oils for the most part—ordinarily originate close to the nectar-producing region but are not coincident with it. Production by the epidermis of perianth parts is most common, but in some species the odor emanates from a more restricted region and may even come from a special flap or brush. Most insect-pollinated plants have visual cues, some of them outside the human spectral range, as well as odor to bring the pollinators to the flowers and guide them to the nectar.

Inflorescence. Inflorescence structure, the way the flowers are clustered or arranged on a flowering branch, is almost as diverse as floral structure. To appreciate this, one need only contrast the drooping inflorescences (catkins) of a birch tree with the coiled flowers of a forget-me-not or with the solitary flower of a tulip. In some cases one kind of inflorescence characterizes a whole plant family. Queen Anne's lace and other members of the parsley family (Umbelliferae) have umbrellalike inflorescences with the flower stalks radiating from almost the same point in a cluster. The stalkless flowers (florets) of the grass family are grouped into clusters called spikelets, and these in turn are variously arranged in different grasses.

Flowers of the arum family (calla lily, jack-in-the-pulpit), also stalkless, are crowded on a thick, fleshy, elongate axis. In the composite family, florets are joined in a tight head at the end of the axis; the heads of some composites contain two kinds, centrally placed florets with small tubular corollas and peripheral ray florets with showy, strap-shaped corollas (the "petals" one plucks from a daisy). *See* INFLORESCENCE.

Anatomy. Some of the general anatomical features of leaves can be found in the floral appendages. A cuticle-covered epidermis overlies a core of parenchyma cells in which there are branching vascular bundles (solitary bundles in most stamens). Sepal parenchyma and petal parenchyma are often spongy, but palisade parenchyma occurs only rarely in flowers and then only in sepals. As in other parts of the plant, color comes mostly from plastids in the cytoplasm and from flavonoids in the cell sap. Cells of the petal epidermis may have folded side walls that interlock so as to strengthen the tissue. In some species the outer walls of the epidermis are raised as papillae; apparently, this is part of the means of attracting pollinators, for the papillae are light reflectors.

Stamen. As a stamen develops, periclinal divisions in the second cell layer of each of its four lobes start a sequence that will end with the shedding of pollen. The first division makes two cell layers. The outer daughter cells give rise to the wall of the pollen sac, and the inner ones are destined to become pollen after further divisions. When mature, a pollen sac typically has a prominent cell layer just below a less distinctive epidermis. The inner wall and the side walls of an endothecium cell carry marked thickenings, but the outer wall does not. Splitting of the ripe anther is due partly to the way in which these differentially thickened walls react to drying and shrinking and partly to the smaller size of the cells along the line of splitting. *See* POLLEN.

Carpel. Like other floral parts, a carpel is made up of epidermis, parenchyma, and vascular tissue. In addition, a carpel commonly has a special tissue system on which pollen germinates and through which, or along which, pollen tubes are transmitted to the ovules. Most angiosperms have solid styles, and the transmitting tissue is a column of elongate cells whose softened walls are the medium for tubal growth. The epidermis at the stigmatic end of a carpel usually changes to a dense covering of papillae or hairs; the hairs can be unicellular or pluricellular, branched or unbranched. In taxa with hollow styles, the transmitting tissue is a modified epidermis running down the stylar canal. There are two kinds of receptive surfaces, and they are distributed among the monocotyledons and the dicotyledons with taxonomic regularity. One kind has a fluid medium for germinating the pollen, and the other has a dry proteinaceous layer over the cuticle. The proteins of the dry stigmas have a role in the incompatibility reactions that encourage outbreeding.

Ovule. Ovule development usually takes place as the gynoecium forms, but it may be retarded when there is a long interval between pollination and fertilization (oaks, orchids). A typical ovule has a stalk (funiculus), a central bulbous body (nucellus), and one or two integuments (precursors of seed coats), which cover the nucellus except for a terminal pore (micropyle). Orientation of the ovule varies from group to group. It can be erect on its stalk or bent one way or another to differing degrees. There are also taxonomic differences in the extent to which the ovule is vascularized by branches from the gynoecial vascular system. *See* FRUIT; REPRODUCTION (PLANT). [R.H.E.]

Fluvial erosion landforms Landforms that result from erosion by water flowing on land surfaces. This water may concentrate in channels as streams and rivers or flow in thin sheets and rills down slopes. Essentially all land surfaces are subjected to modification by running water, and it is among the most important surface processes. Valleys are cut, areas become dissected, and sediment is moved from land areas to ocean basins. With increasing dissection and lowering of the landscape, the land area may pass through a series of stages known as the fluvial erosion cycle.

The most distinctive fluvial landform is the stream valley. Valleys range greatly in size and shape, as do the streams that flow in them. They enlarge both through down and lateral cutting by the stream and mass wasting processes acting on the valley sides.

Waterfalls occur where there is a sudden drop in the stream bed. This is often the case where a resistant rock unit crosses the channel and the stream is not able to erode through it at the same rate as the adjacent less resistant rock. Waterfalls also occur where a main valley has eroded down at a faster rate than its tributary valleys which are left hanging above the main stream. With time, waterfalls migrate upstream and are reduced to rapids.

Many streams flow in a sinuous or meandering channel, and stream velocity is greatest around the outside of meander bends. Erosion is concentrated in this area, and a steep, cut bank forms. If the river meander impinges against a valley wall, the valley will be widened actively.

A stream terrace represents a former floodplain which has been abandoned as a result of rejuvenation or downcutting by the stream. It is a relatively flat surface with a scarp slope that separates it from the current floodplain or from a lower terrace. Terraces are common features in valleys and are the result of significant changes in the stream system through time. *See* FLOODPLAIN.

Fluvial erosion also has regional effects. Streams and their valleys form a drainage network which reflects the original topography and geologic conditions in the drainage basin. A dendritic drainage pattern, like that of a branching tree, is the most common and reflects little or no control by underlying earth materials. Where the underlying earth materials are not uniform in resistance, streams develop in the least resistant areas, and the drainage pattern reflects the geology. If the rocks contain a rectangular joint pattern, a rectangular drainage pattern develops; if the rock units are tilted or folded, a trellis pattern of drainage is common. Topography also controls drainage development; parallel and subparallel patterns are common on steep slopes, and a radial pattern develops when streams radiate from a central high area. *See* EROSION; RIVER; STREAM TRANSPORT AND DEPOSITION. [W.H.J.]

Fluvial sediments Deposits formed by rivers. A river accumulates deposits because its capacity to carry sediment has been exceeded, and some of the sediment load is deposited. Such accumulations range from temporary bars deposited on the insides of meander bends as a result of a loss of transport energy within a local eddy, to deposits tens to hundreds of meters thick formed within major valleys or on coastal plains as a result of the response of rivers to a long-term rise in base level or to the uplift of sediment source areas relative to the alluvial plain. The same processes control the style of rivers and the range of deposits that are formed, so that a study of the deposits may enable the geologist to reconstruct the changes in controlling factors during the accumulation of the deposits. *See* DEPOSITIONAL SYSTEMS AND ENVIRONMENTS; RIVER; STREAM TRANSPORT AND DEPOSITION.

Coarse debris generated by mechanical weathering, including boulders, pebbles, and sand, is rolled or bounced along the river bed and is called bedload. The larger particles may be moved only infrequently during major floods. Finer material, of silt and clay grade, is transported as a suspended load, and there may also be a dissolved load generated by chemical weathering. Whereas the volume of sediment tends to increase downstream within a drainage system, as tributaries run together, the grain size generally decreases as a result of abrasion and selective transport. This downstream grain-size decrease may assist in the reconstruction of transport directions in ancient deposits where other evidence of paleogeography has been obscured by erosion or tectonic change. *See* MASS WASTING.

River deposits of sediment occur as four main types. (1) Channel-floor sediments consist of the coarsest bedload, such as gravel, waterlogged vegetation, or fragments of caved bank material. (2) Bar sediments are accumulations of gravel, sand, or silt which occur along river banks and are deposited within channels, forming bars that may be of temporary duration, or may last for many years, eventually becoming vegetated and semipermanent. (3) Channel-top and bar-top sediments are typically composed of fine-grained sand and silt, and are formed in the shallow-water regions on top of bars, in the shallows at the edges of channels, and in abandoned channels. (4) Floodplain deposits are formed when the water level rises above the confines of the channel and overflows the banks. Much of the coarser floodplain sediment is deposited close to the channel, in the form of levees; silt and mud may be carried considerable distances from the channel, forming blanketlike deposits. *See* FLOODPLAIN.

The thickest (up to 6 mi or 10 km) and most extensive fluvial deposits occur in convergent plate-tectonic settings, including regions of plate collision, because this is where the highest surface relief and consequently the most energetic rivers and most abundant debris are present. Some of the most important accumulations occur in foreland basins, which are formed where the continental margin is depressed by the mass of thickened crust formed by convergent tectonism. *See* BASIN.

Thick fluvial deposits also occur in rift basins, where continents are undergoing stretching and separation. The famous hominid-bearing sediments of Olduvai Gorge and Lake Rudolf are fluvial and lacustrine deposits formed in the East Africa Rift System. Fluvial deposits are also common in wrench-fault basins, such as those in California.

Significant volumes of oil and gas are trapped in fluvial sandstones. Placer gold, uranium, and diamond deposits of considerable economic importance occur in the ancient rock record in South Africa and Ontario, Canada, and in Quaternary deposits in California and Yukon Territory. Fluvial deposits are also essential aquifers, especially the postglacial valley-fill complexes of urban Europe and North America. [A.D.M.]

Fly A member of the insect order Diptera. Insects of other orders are popularly called flies; mayflies, stoneflies, dragonflies, dobsonflies, and caddis flies all have four wings and therefore are not true flies. Mosquitoes, gnats, and midges all have two wings and are therefore also true flies, order Diptera. Hindwings of all Diptera are greatly reduced balancing organs called halteres.

Flies are a numerous and diverse lot, with over 85,000 described species worldwide. North America has 16,000 species belonging to 107 families. Flies are old, their fossils dating from the Triassic, over 200,000,000 years ago.

The flies of greatest importance to humanity are those that suck blood from people or domestic animals. Females of many mosquitoes, deerflies, horseflies, blackflies, and gnats require a meal of blood before producing eggs. When biting, they may introduce pathogenic microorganisms. Mosquitoes may transmit malaria, yellow fever, viral encephalitis, or parasitic roundworms. Tropical blackflies transmit onchocerciasis (river blindness), and sandflies transmit leishmaniasis, a debilitating protozoan infection. Tsetse flies, in which both sexes bite, transmit African sleeping sickness.

Some flies that do not bite may become a nuisance because of their sheer numbers and association with human habitations. The ubiquitous housefly is bothersome and, in unsanitary situations, may contaminate food with the pathogens of hepatitis, polio, cholera, typhoid, or tuberculosis.

Many thousands of flies are predatory, and they doubtless help to suppress populations of insect pests. Especially important are larvae of the family Syrphidae that eat up to 50 aphids per day. Scavenging Diptera are quite important in aiding the quick breakdown of dead animals and plants. *See* ENTOMOLOGY, ECONOMIC. [D.J.Hor.]

Fog A cloud comprising waterdroplets or (less commonly) ice crystals formed near the ground and resulting in a reduction in visibility to below 0.6 mi (1 km). This is lower than that occurring in mist, comprising lower concentration of waterdroplets, and haze, comprising smaller-diameter aerosol particles.

Fog results from the cooling of moist air below its saturation (dew) point. Droplets form on hygroscopic nuclei originating from ocean spray, combustion, or reactions involving trace chemicals in the atmosphere. Visibility is reduced even more when such nuclei are present in high concentrations and faster cooling rates activate a larger fraction of such nuclei. Thus, polluted fog, with more numerous smaller droplets, results in lower visibility for a given water content. *See* DEW POINT.

Haze, the precursor to fog and mist, forms at relative humidity below 100% to about 80%. It is composed of hygroscopic aerosol particles grown by absorption of water vapor to a diameter of about 0.5 micrometer, concentration 1000 to 10,000 per cubic centimeter. Fog and mist form as the relative humidity increases just beyond saturation (100%), so that larger haze particles grow into cloud droplets with a diameter of 10 μm and a concentration of several hundred per cubic centimeter. Fog and mist are a mix of lower-concentration cloud droplets and higher-concentration haze particles. By contrast, smog is formed of particles of 0.5–1-μm diameter, produced by photochemical reactions with organic vapors from automobile exhaust. *See* ATMOSPHERIC CHEMISTRY; HUMIDITY; SMOG. [J.Hal.]

Food poisoning An acute gastrointestinal or neurologic disorder caused by bacteria or their toxic products, by viruses, or by harmful chemicals in foods.

Bacteria may produce food poisoning by three means: (1) they infect the individual following consumption of the contaminated food; (2) they produce a toxin in food before it is consumed; or (3) they produce toxin in the gastrointestinal tract after the individual consumes the contaminated food.

Infectious bacteria associated with food poisoning include *Brucella*, *Campylobacter jejuni*, enteroinvasive *Escherichia coli*, enterohemorrhagic *E. coli*, *Listeria monocytogenes*, *Salmonella*, *Shigella*, *Vibrio parahaemolyticus*, *V. vulnificus*, and *Yersinia enterocolitica*. These organisms must be ingested for poisoning to occur, and in many instances only a few cells need be consumed to initiate a gastrointestinal infection. *Salmonella* and *C. jejuni* are the most prevalent causes of food-borne bacterial infections. *See* YERSINIA.

Staphylococcus aureus and *Clostridium botulinum* are bacteria responsible for food poisonings resulting from ingestion of preformed toxin. *Staphylococcus aureus* produces heat-stable toxins that remain active in foods after cooking. *Clostridium botulinum* produces one of the most potent toxins known. Botulinal toxin causes neuromuscular paralysis, often resulting in respiratory failure and death. *See* BOTULISM; STAPHYLOCOCCUS; TOXIN.

Food-poisoning bacteria that produce toxin in the gastrointestinal tract following their ingestion include *Bacillus cereus*, *Clostridium perfringens*, enterotoxigenic *E. coli*, and *V. cholerae*. *Bacillus cereus* and *C. perfringens* are spore-forming bacteria that often survive cooking and grow to large numbers in improperly refrigerated foods. Following ingestion, their cells release enterotoxins in the intestinal tract. Enterotoxigenic *E. coli* is a leading cause of travelers' diarrhea. *See* DIARRHEA; ESCHERICHIA.

Viruses that cause food-borne disease generally emanate from the human intestine and contaminate food through mishandling by an infected individual, or by way of water or sewage contaminated with human feces. Hepatitis A virus and Norwalk-like virus are the preeminent viruses associated with food-borne illness. *See* HEPATITIS.

Chemical-induced food poisoning is generally characterized by a rapid onset of symptoms which include nausea and vomiting. Foods contaminated with high levels of heavy metals, insecticides, or pesticides have caused illness following ingestion. *See* MEDICAL BACTERIOLOGY; TOXICOLOGY. [M.Do.]

Food web A diagram depicting those organisms that eat other organisms in the same ecosystem. In some cases, the organisms may already be dead. Thus, a food web is a network of energy flows in and out of the ecosystem of interest. Such flows can be very large, and some ecosystems depend almost entirely on energy that is imported. A food chain is one particular route through a food web.

A food web helps depict how an ecosystem is structured and functions. Most published food webs omit predation on minor species, the quantities of food consumed, the temporal variation of the flows, and many other details.

Along a simple food chain, A eats B, B eats C, and so on. For example, the energy that plants capture from the sun during photosynthesis may end up in the tissues of a hawk. It gets there via a bird that the hawk has eaten, the insects that were eaten by the bird, and the plants on which the insects fed. Each stage of the food chain is called a trophic level. More generally, the trophic levels are separated into producers (the plants), herbivores or primary consumers (the insects), carnivores or secondary consumers (the bird), and top carnivores or tertiary consumers (the hawk).

Food chains may involve parasites as well as predators. The lice feeding in the feathers of the hawk are yet another trophic level. When decaying vegetation, dead animals, or both are the energy sources, the food chains are described as detrital. Food chains are usually short; the shortest have two levels. One way to describe and simplify various food chains is to count the most common number of levels from the top to the bottom of the web. Most food chains are three or four trophic levels long (if parasites are excluded), though there are longer ones.

There are several possible explanations for why food chains are generally short. Between each trophic level, much of the energy is lost as heat. As the energy passes up the food chain, there is less and less to go around. There may not be enough energy to support a viable population of a species at trophic level five or higher.

This energy flow hypothesis is widely supported, but it is also criticized because it predicts that food chains should be shorter in energetically poor ecosystems such as a bleak arctic tundra or extreme deserts. These systems often have food chains similar in length to energetically more productive systems. *See* ECOLOGICAL ENERGETICS.

Another hypothesis about the shortness of food chains has to do with how quickly particular species recover from environmental disasters. For example, in a lake with phytoplankton, zooplankton, and fish, when the phytoplankton decline the zooplankton will also decline, followed by the fish. The phytoplankton may recover but will remain at low levels, kept there by the zooplankton. At least transiently, the zooplankton may reach higher than normal levels because the fish, their predators, are still scarce. The phytoplankton will not completely recover until all the species in the food chain have recovered. Mathematical models can show that the longer the food chain, the longer it will take its constituent species to recover from perturbations. Species atop very long food chains may not recover before the next disaster. Such arguments predict that food chains will be longer when environmental disasters are rare, short when they are common, and will not necessarily be related to the amount of energy entering the system.

The number of trophic levels a food web contains will determine what happens when an ecosystem is subjected to a short, sharp shock—for example, when a large number of individuals of one species are killed by a natural disaster or an incident of human-made pollution and how quickly the system will recover. The food web will also influence what happens if the abundance of a species is permanently reduced (perhaps because of harvesting) or increased (perhaps by increasing an essential nutrient for a plant).

Some species have redundant roles in an ecosystem so that their loss will not seriously impair the system's dynamics. Therefore, the loss of such species from an ecosystem will not have a substantial effect on ecosystem function. The alternative hypothesis is that more diverse ecosystems could have a greater chance of containing species that survive or that can even thrive during a disturbance that kills off other species. Highly connected and simple food webs differ in their responses to disturbances, so once again the structure of food webs makes a difference. *See* ECOLOGICAL COMMUNITIES; ECOSYSTEM; POPULATION ECOLOGY. [S.Pi.]

Foot-and-mouth disease A highly contagious viral disease of domesticated and wild cloven-hoofed animals with the potential to cause enormous economic losses. It is characterized by the formation of vesicles on the feet, in and around the mouth, and on the mammary gland. At the acute stage there is high fever, depression, lameness, and reduced appetite. Milking animals show a sudden reduction in production. The mortality in adult animals is usually less than 3%, but in young animals it can exceed 90%.

The causative virus is a member of the *Aphthovirus* genus of the family Picornaviridae. The virus can infect by different routes and mechanisms. During the acute phase of disease, large amounts of virus occur throughout the tissues and organs of the infected animal and in its excretions and secretions. The movement of infected animals and the transmission of virus by contact to susceptible animals in a new herd or flock is by far the most common mechanism of spread. Next in frequency is the movement of contaminated animal products, such as meat and milk—most likely to infect pigs through ingestion.

Immunity to the disease is primarily mediated through neutralizing antibodies directed against the structural proteins of the virus. Specific antibodies are detectable by bioassay 4–5 days after infection, at which time viremia ceases and there is a progressive reduction in virus excretion as the lesions heal. The detection of the viral antigen is sufficient for a positive diagnosis. Normally, viral antigen is detected and serotyped by enzyme-linked immunosorbent assay (ELISA).

Control in countries where the disease is endemic or sporadic is mainly by vaccination. The vaccines contain inactivated viral antigens of one or more serotypes, depending on the prevailing disease situation. Immunity following primary and booster doses of vaccine lasts for around 6 months. *See* ANIMAL VIRUS; EPIDEMIOLOGY. [A.I.D.]

Forage crops Grasses and legumes that make up grasslands and are used as forages for livestock. The grasslands represent an ancient renewable natural resource. They form 25% of the world's vegetation and occupy the largest area of any single plant type. They benefit humanity indirectly by providing food for both wild animals and domesticated livestock, some of which are ruminants that, because their digestive systems contain microorganisms, are able to digest fibrous forage material. Thus, the prime value of grassland areas lies in the meat, milk, or work produced by the livestock that graze on them. *See* GRASSLAND ECOSYSTEM.

Grasslands have other attributes as well. In order to withstand the tug and pull of the grazing animal, a forage plant must have an extensive root system, and this makes a contribution to soil fertility. When the plant is grazed or cut, the photosynthetic area that remains is not large enough to provide sugars to maintain root respiration, and so part of the root system dies and adds to the organic material in the soil and to soil structure. Such is the basis of many highly productive crop rotations that maintain soil fertility without expensive fertilizer applications. Most forage plants are long-lived perennials that can be defoliated repeatedly during a growing season.

Forages also protect the soil from erosion by both wind and water. In fact, where row crop farming has led to soil deterioration, as it did when dust bowl conditions prevailed in North America, forages are used to restore and stabilize the land. *See* EROSION. [P.D.W.]

Forest and forestry A plant community consisting predominantly of trees and other woody vegetation, growing closely together, is a forest. Forests cover about one-fourth of the land area on Earth. The trees can be large and densely packed, as they are in the coastal forests of the Pacific Northwest, or they can be relatively small and sparsely scattered, as they are in the dry tropical forests of sub-Saharan Africa.

Forests are complex ecosystems that also include soils and decaying organic matter, fungi and bacteria, herbs and shrubs, vines and lichens, ferns and mosses, insects and spiders, reptiles and amphibians, birds and mammals, and many other organisms. All of these components constitute an intricate web with many interconnections. *See* FOREST ECOSYSTEM.

Forests have important functions, such as cleansing the air, moderating the climate, filtering water, cycling nutrients, providing habitat, and performing a number of other vital environmental services. They also supply a variety of valuable products ranging from pharmaceuticals and greenery to lumber and paper products. *See* LUMBER.

Forest types. There are many ways to classify forests, as by (1) location (for example, temperate zone forests, tropical zone forests); (2) ownership (for example, public forests, private forests); (3) age or origin (for example, old-growth forests, second-growth forests, plantation forests); (4) important species (such as Douglas-fir forests, redwood forests); (5) economic and social importance (for example, commercial forests, noncommercial forests, urban forests, wilderness); (6) wood properties (for example, hardwood forests, softwood forests); (7) botanical makeup (for example, broadleaf forests, evergreen forests); or (8) a combination of features (such as moist temperate coniferous forests, dry tropical deciduous forests). The last approach tends to be the most descriptive because it often integrates several dominant characteristics related to climate, geography, and botanical features.

Some examples of the major forest types are: Northern coniferous forests which span the cold, northern latitudes of Canada and Europe; Temperate mixed forests which occupy the eastern United States, southeastern Canada, central Europe, Japan, and East Asia, and parts of the Southern Hemisphere in Chile, Argentina, Australia, and New Zealand; Temperate rainforests which are situated along moist, coastal regions of the Pacific Northwest, southern Chile, southeastern Australia, and Tasmania; Tropical rainforests which are found in the equatorial regions of Central and South America (for example, Costa Rica, Brazil, and Ecuador); on the west coast of Africa (for example, Congo, Ivory Coast, and Nigeria); and Southeast Asia (for example, Thailand, Malaysia, and Indonesia); Dry forests which occur in the southwestern United States, the Mediterranean region, sub-Saharan Africa, and semiarid regions of Mexico, India, and Central and South America; and mountain forests which are characteristic of mountainous regions throughout the world.

Characteristics. Although forests take a variety of forms, they have several features in common that allow them to develop in their respective environments. Forests generally contain a broad array of species, each of which is well adapted to the environmental conditions of the region. This biodiversity and adaptability help the forests cope with natural (and in some cases human-caused) forces of destruction, including wildfire, windstorms, floods, and pests. This built-in resiliency also allows the periodic extraction of wood and other products without jeopardizing the long-term health and productivity of the ecosystems—provided such harvesting operations are performed with care. Forests are dynamic—they are constantly changing at both landscape and smaller scales of resolution. This natural propensity to change and develop over time is called forest succession. Forests have a mitigating influence on the environment. This characteristic not only facilitates their own survival and development but also moderates the surrounding climate.

Ecological processes and hydrologic cycle. Forests play a vital role in ecological processes. From a global perspective, they help convert carbon dioxide in the atmosphere to oxygen, thereby facilitating life for aerobic organisms. Forests can also capture, store, convert, and recycle a variety of nutrients such as nitrogen, phosphorus, and sulfur. Forests also play a critical role in the hydrologic cycle. Finally, forests play a crucial ecological role in the habitat that they provide for countless organisms. *See* AIR POLLUTION; ECOLOGICAL SUCCESSION; FORESTRY, URBAN.

Forestry and forest management. The Society of American Foresters defines forestry as the science, the art, and the practice of managing and using for human benefit the natural resources that occur on and in association with forest lands. Natural resources have traditionally entailed major commodities such as wood, forage, water, wildlife, and recreation. However, the concept of forestry has expanded to encompass consideration of the entire forest ecosystem, ranging from mushrooms to landscapes. The practice of forestry requires in-depth knowledge of the complex biological nature of the forest. It also requires an understanding of geology and soils, climate and weather, fish and wildlife, forest growth and development, and social and economic factors. Foresters, wildlife managers, park rangers, and other natural resource specialists are trained in biology, physics, chemistry, mathematics, statistics, computer science, communications, economics, and sociology. *See* FOREST MANAGEMENT.

Silviculture is the art, science, and practice of controlling the establishment, composition, and growth of a forest. It entails the use of both natural and induced processes to foster forest development. For example, reforestation of a harvested or burned-over area can be accomplished by natural seeding from nearby trees or by planting seedlings. *See* REFORESTATION; SILVICULTURE.

Laws and policies. The management of forest land in the United States is regulated by numerous laws and policies. Federal agencies must comply with laws such as the National Forest Management Act (1976), the Forest and Rangeland Renewable Resources Planning Act (1974), and the National Environmental Policy Act (1969). Other public and private forest landowners generally must comply with state regulations or guidelines designed to promote sound forest stewardship in their respective regions.

Policy makers in government, industry, environmental organizations, and the private sector strive to balance the multitude of interests surrounding forest resources. Input from the public as well as resource managers and specialists is a crucial ingredient in the process.

Utilization of forest resources. Forests are often focused on particular uses. For example, plantation forests are generally designed to produce wood and fiber products. Conversely, public forests are increasingly devoted to nonconsumptive purposes such as the preservation of biodiversity, natural conditions, and scenic vistas. However, all forests can provide multiple benefits, including harvestable products, watershed protection, recreation opportunities, wildlife habitat, and ecological services. *See* FOREST GENETICS; FOREST HARVEST AND ENGINEERING; FOREST MANAGEMENT. [M.J.Cou.]

Forest ecosystem The entire assemblage of organisms (trees, shrubs, herbs, bacteria, fungi, and animals, including people) together with their environmental substrate (the surrounding air, soil, water, organic debris, and rocks), interacting inside a defined boundary. Forests and woodlands occupy about 38% of the Earth's surface, and they are more productive and have greater biodiversity than other types of terrestrial vegetation. Forests grow in a wide variety of climates, from steamy tropical rainforests to frigid arctic mountain slopes, and from arid interior mountains to windy rain-drenched coastlines. The type of forest in a given place results from a complex of factors, including frequency and type of disturbances, seed sources, soils, slope and aspect, climate, seasonal patterns of rainfall, insects and pathogens, and history of human influence.

Ecosystem concept. Often forest ecosystems are studied in watersheds draining to a monitored stream: the structure is then defined in vertical and horizontal dimensions. Usually the canopy of the tallest trees forms the upper ecosystem boundary, and plants with the deepest roots form the lower boundary. The horizontal structure is usually described by how individual trees, shrubs, herbs, and openings or gaps are distributed. Wildlife ecologists study the relation of stand and landscape patterns to habitat conditions for animals.

Woody trees and shrubs are unique in their ability to extend their branches and foliage skyward and to capture carbon dioxide and most of the incoming photosynthetically active solar radiation. Some light is reflected back to the atmosphere and some passes through leaves to the ground (infrared light). High rates of photosynthesis require lots of water, and many woody plants have deep and extensive root systems that tap stored ground water between rain storms. Root systems of most plants are greatly extended through a relation between plants and fungi, called mycorrhizal symbiosis. *See* PHOTOSYNTHESIS; ROOT (BOTANY).

The biomass of a forest is defined here as the mass of living plants, normally expressed as dry weight per unit area. Biomass production is the rate at which biomass is accrued per unit area over a fixed interval, usually one year. If the forest is used to grow timber crops, production measures focus on the biomass or volume of commercial trees. Likewise, if wildlife populations are the focus of management, managers may choose to measure biomass or numbers of individual animals. Ecologists interested in the general responses of forest ecosystems, however, try to measure net primary production (Npp), usually expressed as gross primary production (Gpp) minus the respiration of autotrophs (Ra).

Another response commonly of interest is net ecosystem production (NEP),

$$\mathrm{NEP} = \mathrm{Gpp} - (\mathrm{Ra} + \mathrm{Rh}) = \mathrm{Npp} - \mathrm{Rh}$$

usually expressed as where Rh is respiration of heterotrophs. *See* BIOMASS.

Productivity is the change in production over multiple years. Monitoring productivity is especially important in managed forests. Changes in forest productivity can be detected only over very long periods. *See* BIOLOGICAL PRODUCTIVITY.

Forested ecosystems have great effect on the cycling of carbon, water, and nutrients, and these effects are important in understanding long-term productivity. Cycling of carbon, oxygen, and hydrogen are dominated by photosynthesis, respiration, and decomposition, but they are also affected by other processes. Forests control the hydrologic cycle in important ways. Photosynthesis requires much more water than is required in its products. Water is lost back to the atmosphere (transpiration), and water on leaf and branch surfaces also evaporates under warm and windy conditions. Water not taken up or evaporated flows into the soil and eventually appears in streams, rivers, and oceans where it can be reevaporated and moved back over land, completing the cycle.

Forest plants and animals alter soil characteristics, for example, by adding organic matter, which generally increases the rate at which water infiltrates and is retained. Nutrient elements cycle differently from water and from each other.

Elements such as phosphorus, calcium, and magnesium are released from primary minerals in rocks through chemical weathering. Elements incorporated into biomass are returned to the soil with litterfall and root death; these elements become part of soil organic matter and are mineralized by decomposers or become a component of secondary minerals. All elements can leave ecosystems through erosion of particles and then be transported to the oceans and deposited as sediment. Deeply buried sediments undergo intense pressure and heat that reforms primary minerals. Volcanoes and plate tectonic movements eventually distribute these new minerals back to land.

Nitrogen is the most common gas in the atmosphere. Only certain bacteria can form a special enzyme (nitrogenase) which breaks apart N_2 and combines with photosynthates to form amino acids and proteins. In nature, free-living N_2-fixing microbes and a few plants that can harbor N_2-fixing bacteria in root nodules play important roles controlling the long-term productivity of forests limited by nitrogen supply. Bacteria that convert ammonium ion ($NH^{+}{}_{4}$) to $NO^{-}{}_{3}$ (nitrifiers) and bacteria that convert $NO^{-}{}_{3}$ back to N_2 (denitrifiers) are important in nitrogen cycling as well. *See* HYDROLOGY; NITROGEN CYCLE.

Changes in the plant species of a forest over 10 to 100 years or more are referred to as succession. Changes in forest structure are called stand development; changes in composition, structure, and function are called ecosystem development. Simplified models of succession and development have been created and largely abandoned because the inherent complexity of the interacting forces makes model predictions inaccurate. *See* ECOLOGICAL SUCCESSION; FOREST FIRE. [B.T.B.]

Streams. One of the products of an undisturbed forest is water of high quality flowing in streams. The ecological integrity of the stream is a reflection of the forested watershed that it drains. When the forest is disturbed (for example, by cutting or fire), the stream ecosystem will also be altered. Forest streams are altered by any practices or chemical input that alter forest vegetation, by the introduction of exotic species, and by the construction of roads that increase sediment delivery to streams. *See* ACID RAIN; STREAM TRANSPORT AND DEPOSITION. [J.L.Mey.]

Vertebrates. Forest animals are the consumers in forest ecosystems. They influence the flow of energy and cycling of nutrients through systems, as well as the structure and composition of forests, through their feeding behavior and the disturbances that they create. In turn, their abundance and diversity is influenced by the structure and composition of the forest and the intensity, frequency, size, and pattern of disturbances that occur in forests. Forest vertebrates make up less than 1% of the biomass in most forests, yet they can play important functional roles in forest systems.

Invertebrates. Invertebrates are major components of forest ecosystems, affecting virtually all forest processes and uses. Many species are recognized as important pollinators and seed dispersers that ensure plant reproduction. Even so-called pests may be instrumental in maintaining ecosystem processes critical to soil fertility, plant productivity, and forest health.

Invertebrates affect forests primarily through the processes of herbivory and decomposition. They are also involved in the regulation of plant growth, survival, and reproduction; forest diversity; and nutrient cycling. Typically, invertebrate effects on ecosystem structure and function are modest compared to the more conspicuous effects of plants and fungi. However, invertebrates can have effects disproportionate to their numbers or biomass.

Changes in population size also affect the ecological roles of invertebrates. For example, small populations of invertebrates that feed on plants may maintain low rates of foliage turnover and nutrient cycling, with little effect on plant growth or survival, whereas large populations can defoliate entire trees, alter forest structure, and contribute a large amount of plant material and nutrients to the forest floor. Different life stages also may represent different roles. Immature butterflies and moths are defoliators, whereas the adults often are important pollinators. [T.Sc.]

Microorganisms. Microorganisms, including bacteria, fungi, and protists, are the most numerous and the most diverse of the life forms that make up any forest ecosystem. The structure and functioning of forests are dependent on microbial interactions. Four processes are particularly important: nitrogen fixation, decomposition and nutrient cycling, pathogenesis, and mutualistic symbiosis.

Nitrogen fixation is crucial to forest function. While atmospheric nitrogen is abundant, it is unavailable to trees or other plants unless fixed, that is, converted to ammonia (NH_4), by either symbiotic or free-living soil bacteria. *See* AIR POLLUTION; NITROGEN FIXATION.

Most microorganisms are saprophytic decomposers, gaining carbon from the dead remains of other plants or animals. In the process of their growth and death, they release nutrients from the forest litter, making them available once again for the growth of plants. Their roles in carbon, nitrogen, and phosphorus cycling are particularly important. Fungi are generally most important in acid soils beneath conifer forests, while bacteria are more important in soils with a higher pH. Bacteria often are the last scavengers in the food web and in turn serve as food to a host of microarthropods.

Microorganisms reduce the mass of forest litter and, in the process, contribute significantly to the structure and fertility of soils as the organic residue is incorporated.

Some bacteria and many fungi are plant pathogens, obtaining their nutrients from living plants. Some are opportunists, successful as saprophytes, but capable of killing weakened or wounded plant tissues. Others require a living host, often preferring the most vigorous trees in the forest. Pathogenic fungi usually specialize on roots or stems or leaves, on one species or genus of trees.

Pathogenic fungi are important parts of all natural forest ecosystems. The forest trees evolved with the fungi, and have effective means of defense and escape, reducing the frequency of infection and slowing the rates of tissue death and tree mortality. However, trees are killed, and the composition and structure of the forest is shaped in large part by pathogens.

Pathogens remove weak or poorly adapted organisms from the forest, thus maintaining the fitness of the population. Decay fungi that kill parts of trees or rot the heartwood of living trees create an essential habitat for cavity-nesting birds and the other animals dependent on hollow trees.

By killing trees, pathogens create light gaps in the forest canopy. The size and rate of light gap formation and the relative susceptibility of the tree species present on the site determine the ecological consequences of mortality. Forest succession is often advanced as shade-tolerant trees are released in small gaps. Gaps allow the growth of herbaceous plants in the island of light, creating habitat and food diversity for animals within the forest. In many forests, pathogens are the most important gap formers and the principal determinants of structure and succession in the long intervals between stand-replacing disturbances such as wildfires or hurricanes. *See* ECOLOGICAL SUCCESSION; PLANT PATHOLOGY.

The fungus roots of trees, and indeed most plants, represent an intimate physical and physiological association of particular fungi and their hosts. Mycorrhizae are the products of long coevolution between fungus and plant, resulting in mutual dependency. Mycorrhizae are particularly important to trees because they enhance the uptake of phosphorus from soils. Mycorrhizal fungi greatly extend the absorptive surface of roots through the network of external hyphae. *See* ECOSYSTEM; FOREST AND FORESTRY; MYCORRHIZAE. [E.Ha.]

Forest fire The term wildfire refers to all uncontrolled fires that burn surface vegetation (grass, weeds, grainfields, brush, chaparral, tundra, and forest and woodland); often these fires also involve structures. In addition to the wildfires, several million acres of forest land are intentionally burned each year under controlled conditions to accomplish some silvicultural or other land-use objective or for hazard reduction.

Most wildfires are caused by human beings, directly or indirectly. In the United States less than 10% of all such fires are caused by lightning, the only truly natural cause. In the West (the 17 Pacific and Rocky Mountain states) lightning is the primary cause, with smoking (cigarettes, matches, and such) the second most frequent. Combined they account for 50 to 75% of all wildfires. In the 13 southern states (Virginia to Texas) the primary cause is incendiary. This combined with smoking and debris burning make up 75% of the causes. The 20 eastern states have smoking and debris burning as causing close to 50% of all wildfires. Miscellaneous causes of wildfires are next in importance in most regions. The other causes of wildfires are machine use and campfires. Machine use includes railroads, logging, sawmills, and other operations using equipment.

The manner in which fuel ignites, flame develops, and fire spreads and exhibits other phenomena constitutes the field of fire behavior. Factors determining forest fire behavior may be considered under four headings: attributes of the fuel, the atmosphere, topography, and ignition. A forest fire may burn primarily in the crowns of trees and shrubs (a crown fire); primarily in the surface litter and loose debris of the forest floor and small vegetation (a surface fire); or in the organic material beneath the surface litter (a ground fire). The most common type is a surface fire.

The U.S. Forest Service has developed a National Fire Danger Rating System (NFDRS) to provide fire-control personnel with numerical ratings to help them with the tasks of fire-control planning and the suppression of specific fires. The system includes three basic indexes: an occurrence index, a burning index, and a fire load index. Each of these is related to a specific part of the fire-control job. These indexes are used by dispatchers in making decisions on setting up firefighting forces, lookout systems, and so forth. [W.S.Br.]

Forest genetics The subdiscipline of genetics concerned with genetic variation and inheritance in forest trees. The study of forest genetics is important because of the unique biological nature of forest trees (large, long-lived plants covering 30% of the Earth's surface) and because of the trees' social and economic importance. Forest genetics is the basis for conservation, maintenance, and management of healthy forest ecosystems; and development of programs which breed high-yielding varieties of commercially important tree species.

Variation in natural forests. The outward appearance of a tree is called its phenotype. The phenotype is any characteristic of the tree that can be measured or observed such as its height or leaf color. The phenotype is influenced by (1) the tree's genetic potential (its genotype); and (2) the environment in which the tree grows as determined by climate, soil, diseases, pests, and competition with other plants.

No two trees of the same species have exactly the same phenotype, and in most forests there is tremendous phenotypic variation among trees of the same species. Forest geneticists often question whether the observed phenotypic variation among trees in forests is caused mostly by genetic differences or by differences in environmental effects. Common garden tests are often used to hold the environment constant and therefore isolate the genetic and environmental effects on phenotypic variability.

Geographic variation. The term "provenance" refers to a specific geographical location within the natural range of a tree species. Natural selection during the course of evolution has adapted each provenance to its particular local environment. This means that there are large genetic differences among provenances growing in different environments. Provenances originating from colder regions, for example, tend to have narrower crowns with flatter branches better adapted to the dry snow and types of frosts in colder climates. To demonstrate that these differences are genetic in origin, common garden tests called provenance tests have been planted. That is, seed has been collected from several provenances and planted for comparison in randomized, replicated studies in various forest locations. The study of geographic variation through provenance tests should be a first step in the genetic research of any tree species.

Genetic variation. In addition to genetic differences among provenances, there is usually substantial genetic variation among trees within the same provenance and even within the same forest stand. There are two reasons for this genetic diversity: (1) Different trees have different genotypes in most natural stands. (2) Each tree is heterozygous for many genes, meaning that a given tree has multiple forms (different alleles) of many genes. Population genetics studies patterns of genetic diversity in populations (such as forest stands). Results of many studies have shown that most forest tree species maintain very high levels of genetic diversity within populations. *See* POPULATION GENETICS.

Forest tree breeding. Beginning with the natural genetic variation that exists in an undomesticated tree species, tree breeding programs use selection, breeding, and other techniques to change gene frequencies for a few key traits of the chosen species. As with agricultural crops, tree breeders produce genetically improved, commercial varieties that are healthier, grow faster, and yield better wood products. After an existing forest stand is harvested, a new stand of trees is planted to replace the previous stand in the process called reforestation. Use of a genetically improved variety for this reforestation means that the new plantation will grow faster and produce wood products sooner than did the previous stand.

Several laboratory techniques promise to make major contributions to forest genetics and tree breeding: (1) Somatic embryogenesis is a technique to duplicate (or propagate) selected trees asexually from their vegetative (somatic) cells, and this allows the best trees to be immediately propagated commercially as clones without sexual reproduction (that is, no seed is involved). (2) Genetic mapping of some important tree species is well under way, and these maps will be useful in many ways to understand the genetic control of important traits, such as disease resistance. (3) Marker-assisted selection is the use of some kinds of genetic maps to help select excellent trees at very early ages based on their deoxyribonucleic acid (DNA) genotype as assessed in a laboratory (instead of growing trees in the field and selecting based on performance in the forest). (4) Functional genomic analysis is an exciting new field of genetics that aims to understand the function, controlled expression, and interaction of genes in complex traits such as tree growth. (5) Genetic engineering or genetic modification is the insertion of new genes into trees from other species.

Conservation of genetic resources. For commercially important species of forest trees, gene conservation is practiced by tree breeding programs to sustain the genetic diversity needed by the program. However, conservation of genetic diversity is a major global concern and is important for all forest species to maintain the health and function of forest ecosystems, and to sustain the genetic diversity of noncommercial species that may eventually have economic value. There are two broad categories of gene conservation programs. In-situ programs conserve entire forest ecosystems in forest reserves, national parks, wilderness areas, or other areas set aside for conservation purposes. Ex-situ programs obtain a sample of the genotypes from different provenances of a single tree species and collect seed or vegetative plant material from each genotype to store in a separate location (such as a seed bank in a refrigerated room). Both types of programs are important for conserving the world's forest resources. *See* BREEDING (PLANT); FOREST AND FORESTRY; FOREST ECOSYSTEM; GENETIC ENGINEERING; PLANT PROPAGATION. [T.L.W.]

Forest harvest and engineering Application of engineering principles to the solution of forestry problems, such as those dealing with harvesting, forest transportation, materials handling, and mechanical silviculture, with regard to long-range environmental and economic effects.

The work that forest engineers perform varies widely throughout the United States. In the Northeast and Southeast, tasks include mechanization of harvesting, site preparation, planting, and product handling. Development and modification of machinery is an important part of this job, as is the improvement of the ability of workers to use machines in the woods, an activity known as work science. In the West, the terrain changes the job, as does the size of the trees usually harvested. Planning, design, and construction of road systems are major operational and environmental challenges in the West. Harvesting with mechanical fellers or yarding with ground skidders is often limited by tree size and terrain. Cable systems are commonly used to overcome these constraints, and the design and layout of these cable logging units require a great deal of engineering skill if logging is to be done in an economic and environmentally acceptable manner. In short, the skills possessed by forest engineers in the eastern United States parallel most closely those of the mechanical or agricultural engineer. In the West, the skills are more closely aligned with civil engineering. *See* AGRICULTURAL ENGINEERING; FOREST AND FORESTRY; SILVICULTURE. [J.E.O'L.]

Forest management The planning and implementation of sustainable production of forest crops and other forest resources and uses. Key decisions include land allocation to different uses or combination of uses, silvicultural method and practices, intensity of management, timber harvest scheduling, and environmental protection.

Nearly three-fourths (360 million acres or 140 million hectares) of all commercial forest land in the United States is privately owned by farmers, forest investment groups, other types of nonindustrial owners, or industrial firms engaged in the business of growing and harvesting timber for conversion to wood products. The objectives and practices of private owners are extremely diverse. Many states and local governments have enacted laws that regulate the practice of private forestry. Therefore, management planning for a specific property requires a detailed review of the owner's objectives, resources, and any legal constraints regarding land uses or choice of management practices in the local area.

One-fourth (about 120 million acres or 49 million hectares) of the commercial forest in the United States is administered by federal and state agencies. The National Forest System, managed by the U.S. Forest Service, is particularly important in the western United States, where it includes nearly one-half of the commercial forest. Each national forest is required by federal law to develop a long-term land management plan, involve the public in evaluating alternatives, project future practices and outputs, and identify methods to mitigate adverse environmental impacts. These plans provide for timber harvesting, wilderness management, watershed protection, wildlife habitat, and other services in combinations that vary from forest to forest. The recovery plans of many endangered species occur primarily on public forest lands.

Forest land-use planning and project implementation requires information about the physical, vegetative, and developmental characteristics of forest resources within the management unit. Aerial photography, satellite imagery, and statistically designed ground surveys are commonly used to obtain the necessary information. The resource assessment should normally include estimates of timber volume classified according to species, age or size class, quality, location, and other attributes that affect value in the local market or have relevance to decision-making. Typically, statistically designed sampling procedures using plots or strips are employed by specially trained technicians to estimate volume. Information about nontimber resources such as wildlife, streams and lakes, fisheries, and historical and archeological sites may also be required, depending upon the owner's objectives and local forest practice regulations. Assessment methods normally utilize both ground and aerial surveys, and professional specialists employed by the owner or by outside consultants. *See* FOREST MEASUREMENT; GEOGRAPHIC INFORMATION SYSTEMS; LAND-USE PLANNING.

A forest typically will have complex structure. There may be a range of soil types, slopes, and aspects that differ in potential productivity. To facilitate planning for such complex situations, optimization methods may be used to schedule management activities over time, determine the timber harvest level, allocate the land base among alternative uses, and calculate benefits and costs. A common method, linear programming, requires that the manager specify both a linear objective function to be maximized and a set of linear equations that describe management constraints. *See* SILVICULTURE.

Modern forest management usually involves the production of multiple services of value to the owner or to society. Determining the best mix of services requires technical information on trade-offs between the different outputs, costs and values, and pertinent legal constraints. Subjective evaluations may be required in the case of unpriced services such as wildlife, water, and scenery. Forest structure has a major impact on the mix of outputs. Forest outputs or services can generally be classified as complementary or competitive. Services are complementary if an increase in the output of one is accompanied by an increase in the output of the other. They are competitive if effort to increase the output of one results in a reduction in the output of the other. In response to concerns about landscapes, such as maximum opening sizes, spatial diversity, and wildlife habitat, new scheduling methods are being developed and used (for example, tabu search, simulated annealing, and heuristics). Unlike linear programming, these methods do not guarantee optimal solutions. However, they do provide quality solutions that consider the complexity of modern forest management and can be implemented on the ground. *See* FOREST AND FORESTRY. [D.E.Te.]

Forest measurement The science and practice of measuring the volume, growth, and development of trees individually and collectively and estimating the products obtainable from them. Foresters use quantitative sciences such as mathematics and statistics for these measurements.

Regardless of the land management objectives—timber, wildlife, recreation, watershed, or a combination of these resources—the tree overstory (the forest canopy) must be quantified for informed decision making. Forest cover is an important part of wildlife habitat, and the understory component is related to the overstory characteristics. The recreation potential of wildland is a function of many variables, including the size and number of trees present. Water yields are related to the composition and density of the tree canopy. The measurement principles discussed here are applicable to all forest resource management situations that require quantitative information about the tree component of the land base.

Standing trees are commonly measured for diameter, height, and age. The diameter and height measurements are used to estimate the volume (or weight) and value of individual trees; ages are used in assessing site quality and predicting growth. *See* DENDROLOGY; WOOD ANATOMY.

In addition to inventories aimed at determining current conditions, land managers need trend data. Monitoring consists of collecting information over time, generally on a sample basis by measuring change in key indicator variables, in order to determine the effects of management treatments in the long term. These data, along with research results, can be used to modify management on a continuing basis to ensure that objectives are being met. The sampling design for monitoring generally involves repeated measurements on the same sample plots or individuals.

Forest inventory information is commonly stored, updated, and retrieved through geographic information systems (GISs). A GIS is a computerized database for storing, manipulating, and displaying map (spatial) data and tabular (attribute) information. In a GIS, forest inventory information can be stored in a computer and directly linked to associated forest maps, making it easier and faster to analyze and graphically display the results of forest inventories. GISs can make forest inventory information more powerful by allowing forest resource managers to integrate it with other data commonly needed to make management decisions.

Foresters estimate site quality to assess present and future forest productivity and to provide a frame of reference for land management. Many parameters that affect productivity are difficult or impossible to measure directly, and as a consequence site quality is determined indirectly. Most commonly, site quality is evaluated from tree height in relation to age. Theoretically, height growth is sensitive to differences in site quality, is little affected by varying stand-density levels, is relatively stable under varying thinning intensities, and is strongly correlated with volume. Thus height has been found to be a practical, consistent, and useful indication of site quality.

Quantitative measures of stand density are used when deriving silvicultural prescriptions and predicting growth and yield. The two most commonly used measures of stand density are tree basal area per unit area and number of trees per unit area. Stand basal area is the cross-sectional area at diameter at breast height of all stems, or some specified portion of the stand, expressed on a per-acre (or per-hectare) basis. Similarly, trees per acre may be determined for all stems or for some specified portion of the stand. *See* SILVICULTURE.

Growth is the increase (increment) over a given period of time. Yield is the total amount available for harvest at a given time, that is, the summation of the annual increments. The factors most closely related to growth and yield of forest stands of a given species composition are the point in time in stand development, the site quality, and the degree to which the site is occupied. *See* FOREST AND FORESTRY; FOREST MANAGEMENT. [H.E.B.]

Forest pest control Forest pest control or forest protection refers to the approaches and tactics for protecting forests from insects and pathogens. The traditional view is that plant-feeding insects and pathogens are destructive agents that must be controlled to protect forest resources. Pest activity generally is triggered by specific changes in host-tree condition and density that often result from forest management practices. Integrated forest pest management represents the current approach to optimize accomplishment of forest management goals by evaluating the costs and benefits of various forest species for production of multiple resources. A number of pest management tools are available, including computerized models that facilitate evaluation and decision-making, and a variety of chemical, biological, and silvicultural techniques for manipulating pest abundances. *See* PLANT PATHOLOGY.

A variety of organisms can interfere with forest management objectives. Most of these are insects, fungal pathogens, and nematodes. Insects are responsible for vectoring some microbial pathogens, and pathogens frequently increase the vulnerability of infected trees to insects.

A critical first step in integrated forest pest management is identification of the forest management goals. If not justified by contribution to forest management goals, that is, optimized production of forest resources, suppression represents unnecessary costs in terms of time, money, and environmental quality.

Forest managers must be aware of potential impediments, including effects of insect and pathogens, in the accomplishment of their goals. Effective management requires information on which species can affect forest resources and at what densities. The relative threats of potential pests to particular management goals must be weighed carefully to determine tolerable or optimal abundances. Substantial data are needed to evaluate pest status. Examples of information needed for assessment include current and projected abundances of potential pests, action thresholds (abundance at which loss of resources exceeds costs of control efforts), environmental conditions favorable to various pests, and factors that influence the effectiveness of control tactics. This information can be used to project losses or gains for various forest resources as a result of specific insect or pathogen species. Such projections can be improved greatly by use of computerized models that synthesize available data and permit simulation and prediction of resource production under various environmental conditions or pest management scenarios.

The objective of pest suppression should be maintenance of pest populations below their action thresholds. Elimination of native species is impractical and would interfere with their natural ecological functions. Reducing abundances to levels that no longer interfere with management goals is sufficient. However, preventing the establishment of exotic species may be critical to sustainability of forest resources.

A variety of control options are available, but many have limited utility against particular pest species. Pesticides can be applied as aerial or ground aerosols or as fumigants. Fungicides are relatively ineffective against fungal pathogens that generally are protected from exposure. Microbial pathogens and antibiotics can be delivered as aerosols or applied to surfaces exposed to infectious agents. Other biological control options include augmentation of natural enemy populations. Biological control is most effective when the predator, parasite, or pathogen selectively and efficiently preys on the pest species. Pheromones are chemicals produced by animals, most commonly to attract potential mates. In some species, especially of bark beetles, a combination of attractive and repellent pheromones limits population density and reduces competition for resources. Silvicultural options include thinning, prescribed understory burning, and fertilization to reduce competition among trees for light, water, and nutrient resources. Thinning also slows spread of insects and pathogens between trees.

A goal of integrated forest pest management is variation in control tactics over time and across landscapes to minimize development of resistance to particular control options. *See* FUNGISTAT AND FUNGICIDE; INSECT CONTROL, BIOLOGICAL; PESTICIDE; SILVICULTURE. [T.Sc.]

Forest recreation Recreation involving direct contact with forests in various activities ranging from walking in the woods to wilderness backpacking. The primary suppliers of forest recreation opportunities in the United States are federal land management agencies such as the U.S. Department of Agriculture (USDA) Forest Service, the National Park Service, and the Bureau of Land Management; state recreation, park, wildlife, and forest departments; and private landowners and corporations (ski areas, industrial forests, resorts).

In the early decades of the twentieth century, recreation management was mostly custodial management, keeping areas free from litter and pollutants and providing fire protection. As recreation use increased dramatically in the 1950s, a concept of activity-based management emerged with a focus on the numbers of users, the activities in which they participated, and the sites necessary for participation. In 1958 the U.S. Congress empaneled the Outdoor Recreation Resources Review Commission to assess the situation and make recommendations for the future of outdoor recreation. *See* FOREST MANAGEMENT.

As outdoor recreation use of forests grew and as research began to focus on behavior of participants, the activity-based management concept evolved into a more sophisticated concept of experience-based management that focused on the achievement of satisfying experiences from recreation engagement. Three important concepts that have come from experience-based management are a definition of recreation that is behaviorally based, the Recreation Opportunity Spectrum (ROS) approach to recreation classification and management, and the Limits of Acceptable Change (LAC) planning system.

Using specific criteria to define different types of outdoor recreation, the Recreation Opportunity Spectrum allows managers to characterize the kind of recreation to be offered and to guide management decisions. It also is used to assess the impacts of recreation and proposed recreation on other uses such as timber management.

The Limits of Acceptable Change incorporates the Recreation Opportunity Spectrum and explicitly makes monitoring and evaluation of use and impacts a part of planning and management. The Limits of Acceptable Change is important in integrating forest recreation issues and concerns into multiple-use land management.

Recreation management has evolved to focus on the potential benefits that recreationists, society, and communities might realize from recreation opportunities and participation. This has become known as benefits-based management, and expected benefits are used to guide design of recreation sites and their access systems, to develop recreation information programs, and to integrate recreation with other forest uses.

The educational, physical, and mental health effects on individuals engaged in recreation on forest lands can have positive effects on sustainability of environmental and natural resources. Likewise, recreation can be detrimental. For example, wildlife can be adversely affected by off-road vehicle users or traffic on recreational roads. A major challenge for forestland managers is to help people achieve their recreational goals, but in ways that minimize negative impacts on ecosystems. *See* FOREST AND FORESTRY; FOREST ECOSYSTEM. [P.J.Bro.]

Forestry, urban The planning and implementation of actions to establish, protect, restore, and maintain trees and forests in cities and smaller communities. When it began in the 1960s, the field of urban forestry focused on individual trees along streets and adjacent to homes and buildings, and on groups of trees in specific spatial areas such as parks. Since then, it has evolved toward more holistic consideration of the structure, processes, and functions of urban ecosystems at a larger scale. Scientists have come to realize that trees and forests play a critical role in maintaining healthy urban ecosystems, providing ecological services such as filtering water and air pollution, reducing stormwater flows, sequestering carbon emissions, conserving energy, and reducing soil erosion, as well as providing human health benefits, recreation, esthetics, and fish and wildlife habitat. These ecological services can be translated into economic values worth billions of dollars by comparing them to the costs of achieving the same benefits with technology and human-made infrastructure. Urban trees, forests, and related vegetation are often referred to as green infrastructure in order to compare and contrast them with human-made or hard infrastructure, such as buildings and roads. *See* ECOLOGICAL COMMUNITIES; ECOLOGICAL SUCCESSION.

New satellite remote sensing and geographic information system tools are enabling communities to assess changes in their green infrastructure and to develop plans for restoring and maintaining urban trees and forests. City planners and policymakers use the same tools to address other urban infrastructure needs, such as transportation and housing, which allows them to integrate green infrastructure information into plans for other infrastructure development.

As urban forestry has expanded to include larger ecosystems, spatial boundaries between urban and rural areas have begun to blur. Their ecological, social, and economic links have come into view. Watershed connections have become a key consideration, while other social and economic links related to movements of people, businesses, goods, and services have also become clearer. Rural lands, both public and private, provide an array of ecological services to nearby cities, including drinking water, recreational opportunities, fish and wildlife habitat, and agricultural and forest products.

Urban forests are owned by a diverse array of public and private entities, including individual homeowners, private businesses, and federal, state, and local governments. Their management and use are closely tied to the homes, buildings, transportation systems, utilities, and other infrastructure in an urban area. A key challenge is coordinating the efforts of the large number of individuals, groups, and organizations involved in urban forestry, each with its own information, resources, and objectives. *See* LAND-USE PLANNING. [G.J.Ga.; G.A.Mo.]

Formation A fundamental geological unit used in the description and interpretation of layered sediments, sedimentary rocks, and extrusive igneous rocks. A formation is defined on the basis of lithic characteristics and position within a stratigraphic succession. It is usually tabular or sheetlike, and is mappable at the Earth's surface or traceable in the subsurface (for example, between boreholes or in mines). Examples are readily recognized in the walls of the Grand Canyon of northern Arizona. Each formation is referred to a section or locality where it is well developed (a type section), and assigned an appropriate geographic name combined with the word formation or a descriptive lithic term such as limestone, sandstone, or shale (for example, Temple Butte Formation, Hermit Shale). This usage of "formation" by geologists differs from its informal lay usage for stalactites, stalagmites, and other mineral buildups in caves.

Distinctive lithic characteristics used to designate formations include chemical and mineralogical composition, particle size and other textural features, primary sedimentary or volcanic structures related to processes of accumulation, fossils or other organic content, and color. Contacts or boundaries between formations are chosen at surfaces of abrupt lithic change or within zones of gradational lithic character. Commonly, these contacts correspond with recognizable changes in topographic expression, related to variations in resistance to weathering. *See* SEDIMENTARY ROCKS; SEDIMENTOLOGY.

Mappability is an essential characteristic of a formation because such units are used to delineate geological structure (faults and folds), and it is useful to be able to recognize individual formations in isolated outcrops or areas of poor exposure. Well-established formations are commonly divisible into two or more smaller-scale units termed members and beds (for example, subdivisions of the Redwall Limestone). In other cases, formations of similar lithic character or related genesis are combined into composite units termed groups and supergroups (for example, Supai Group). The rank of a named unit may vary from one area to another (for example, from group to formation) according to whether or not subunits are readily mappable. Changes in rank are also justified in the light of new geological knowledge.

Although commonly used as a framework for interpreting geological history, formations and related units are conceptually independent of geological time. They may represent either comparatively short or comparatively long intervals of time. Accumulation of a particular unit may have begun earlier in some places than in others, and the time span represented by a unit may be influenced by later erosion. In some cases, a formation cropping out at one locality may be entirely older or younger than the same lithic unit at another locality. Although the concept of time plays no role in the definition of a formation, evidence of age is useful in the recognition of lithologically similar units far from their type localities. [N.C.B.]

Fossil fuel Any naturally occurring carbon-containing material which when burned with air (or oxygen) produces (directly) heat or (indirectly) energy. Fossil fuels can be classified according to their respective forms at ambient conditions. Thus, there are solid fuels (coals); liquid fuels (petroleum, heavy oils, bitumens); and gaseous fuels (natural gas, which is usually a mixture of methane, CH_4, with lesser amounts of ethane, C_2H_6, hydrogen sulfide, H_2S, and numerous other constituents in small proportions).

One important aspect of the fossil fuels is the heating value of the fuel, which is measured as the amount of heat energy produced by the complete combustion of a unit quantity of the fuel. For solid fuels and usually for liquid fuels the heating value is quoted for mass, whereas for gaseous fuels the heating value is quoted for volume. The heating values are commonly expressed as British thermal units per pound (Btu/lb). In SI units the heating values are quoted in megajoules per

Heating values of representative fuels

Fossil fuel	Btu/lb	Btu/ft^3	MJ/k	MJ/m^3
Natural gas		900		33.5
Petroleum	19,000		44.1	
Heavy oil	18,000		41.8	
Tar-sand bitumen	17,800		41.3	
Coal				
Lignite	8,000*		18.6	
Subbituminous	10,500*		24.4	
Bituminous	15,500*		36.0	
Anthracite	15,000*		34.8	

*Representative values are given because of the spread of subgroups with various heating values.

kilogram (MJ/kg). For gases, the heating values are expressed as Btu per cubic foot (Btu/ft^3) or as megajoules per cubic meter (MJ/m^3). The table gives heating values of representative fuels. *See* ENERGY SOURCES; NATURAL GAS; PETROLEUM. [J.G.S.]

Fresh-water ecosystem Fresh water is best defined, in contrast to the oceans, as water that contains a relatively small amount of dissolved chemical compounds. Some studies of fresh-water ecosystems focus on water bodies themselves, while others include the surrounding land that interacts with a lake or stream. *See* ECOLOGY; ECOSYSTEM.

Fresh-water ecosystems are often categorized by two basic criteria: water movement and size. In lotic or flowing-water ecosystems the water moves steadily in a uniform direction, while in lentic or standing-water systems the water tends to remain in the same general area for a longer period of time. Size varies dramatically in each category. Lotic systems range from a tiny rivulet dripping off a rock to large rivers. Lentic systems range from the water borne within a cup formed by small plants or tree holes to very large water bodies such as the Laurentian Great Lakes. Fresh-water studies also consider the interactions of the geological, physical, and chemical features along with the biota, the organisms that occur in an area.

Physical environment. The quantity and spectral quality of light have major influences on the distribution of the biota and also play a central role in the thermal structure of lakes. The light that reaches the surface of a lake or stream is controlled by latitude, season, time of day, weather, and the conditions that surround a water body. Light penetration is controlled by the nature of water itself and by dissolved and particulate material in a water column.

Water exhibits a number of unusual thermal properties, including its existence in liquid state at normal earth surface temperatures, a remarkable ability to absorb heat, and a maximum density at 39.09°F (3.94°C), which leads to a complex annual cycle in the temperature structure of fresh-water ecosystems.

As water is warmed at the surface of a lake, a stable condition is reached in which a physically distinct upper layer of water, the epilimnion, is maintained over a deeper, cooler stratum, the hypolimnion. The region of sharp temperature changes between these two layers is called the metalimnion. The characteristic establishment of two layers is of major importance in the chemical cycling within lakes and consequently for the biota.

As the surface waters of a lake cool, the density of epilimnetic waters increases, which decreases their resistance to mixing with the hypolimnion. If cooling continues, the entire water column will mix, an event known as turnover. At temperatures below 39.09°F (3.94°C), water again becomes less dense; ice and very cold water float above slightly warmer water, maintaining liquid water below ice cover even in lakes in the Antarctic. Many lakes in the temperate zones undergo two distinct periods of mixing annually, one in the spring and the other in the fall, that separate periods of stratification in the summer and winter.

Water movement is more extensive in lotic than in standing-water ecosystems, but water motion has important effects in both types. Turbulence occurs ubiquitously and affects the distribution of organisms, particles, dissolved substances, and heat. Turbulence increases with the velocity of flowing water, and the amount of material transported by water increases with turbulence. Flowing-water ecosystems are characterized by large fluctuations in the velocity and amount of water. Aside from surface waves on large lakes, most water movement in lentic systems is not conspicuous. *See* LAKE; LIMNOLOGY; RIVER.

Chemical environment. For an element, three basic parameters are of importance: the forms in which it occurs, its source, and its concentration in water relative to its biological demand or effect. Most elements are derived from dissolved gases in the atmosphere or from minerals in geological materials surrounding a lake. In some cases the presence of elements is strongly mediated by biological activities. *See* BIOGEOCHEMISTRY.

Oxygen occurs as dissolved O_2 and in combination with other elements resulting from chemical or biological reactions. It enters water primarily from the atmosphere through a combination of diffusion and turbulent mixing. When biological demands for oxygen exceed supply rates, it can be depleted from fresh-water ecosystems. Anoxic conditions occur in hypolimnia during summer and under ice cover in winter when lake strata are isolated from the atmosphere. Oxygen depletion may also occur in rivers that receive heavy organic loading. Aside from specialized bacteria, few organisms can occur under anoxic conditions.

Carbon dioxide is derived primarily from the atmosphere, with additional sources from plant and animal respiration and carbonate minerals. Its chemical species exert a major control on the hydrogen ion concentration of water (the acidity or pH).

Phosphorus occurs primarily as a phosphate ion or in a number of complex organic forms. It is the element which is most commonly in the shortest supply relative to biological demand. Phosphorus is thus a limiting nutrient, and its addition to fresh-water ecosystems through human activities can lead to major problems due to increased growth of aquatic plants.

Nitrogen occurs in water as N_2, NO_2, NO_3, NH_4, and in diverse organic forms. It may be derived from precipitation and soils, but its availability is usually regulated by bacterial processes. Nitrogen occurs in relatively short supply relative to biological demand. It may also limit growth in some fresh-water systems, particularly when phosphorus levels have been increased because of human activity. *See* NITROGEN FIXATION.

A variety of other elements also help determine the occurrence of fresh-water organisms either directly or by the elements' effects on water chemistry.

Biota. In addition to taxonomy, fresh-water organisms are classified by the areas in which they occur, the manner in which they move, and the roles that they occupy in trophic webs. Major distinctions are made between organisms that occur in bottom areas and those within the water column, the limnetic zone. Production is the most difficult variable to measure, but it provides the greatest information on the role of organisms in an ecosystem. *See* BIOLOGICAL PRODUCTIVITY; BIOMASS.

Plankton organisms occur in open water and move primarily with general water motion. Planktonic communities occur in all lentic ecosystems. In lotic systems they are important only in slow-moving areas.

Phytoplankton (plant plankton) comprise at least eight major taxonomic groups of algae, most of which are microscopic. They exhibit a diversity of forms ranging from one-celled organisms to complex colonies. *See* ALGAE; PHYTOPLANKTON.

Zooplankton (animal plankton) comprise protozoans and three major groups of eukaryotic organisms: rotifers, cladocerans, and copepods. Most are microscopic but some are clearly visible to the naked eye. *See* POPULATION ECOLOGY; ZOOPLANKTON.

Animals, such as fishes and swimming insects, that occur in the water column and can control their position independently of water movement are termed nekton. In addition to their importance as a human food source, fishes may affect zooplankton, benthic invertebrates, vegetation, and lake sediments.

Benthic organisms are a diverse group associated with the bottoms of lakes and streams. The phytobenthos ranges from microscopic algae to higher plants. Benthic animals range from microscopic protozoans and crustaceans to large aquatic insects and fishes. *See* FOOD WEB.

Bacteria occur throughout fresh-water ecosystems in planktonic and benthic areas and play a major role in biogeochemical cycling. Most bacteria are heterotrophic, using reduced carbon as an energy source; others are photosynthetic or derive energy from reduced compounds other than carbon. *See* BACTERIAL PHYSIOLOGY AND METABOLISM.

Interactions. Ultimately the conditions in a fresh-water ecosystem are controlled by numerous interactions among biotic and abiotic components. Primary production in a fresh-water ecosystem is controlled by light and nutrient availability. As light diminishes with depth in a column or water, a point is reached where energy for photosynthesis balances respiratory energy demands. In benthic areas, the region where light is sufficient for plant growth is termed the littoral zone; deeper areas are labeled profundal.

Nutrient availability generally controls the total amount of primary production that occurs in fresh-water ecosystems. One classification scheme for lakes ranks them according to total production, ranging from oligotrophic lakes, where water is clear and production is low, to eutrophic systems, characterized by high nutrient concentrations, high standing algal biomass, high production, low water clarity, and low concentrations of oxygen in the hypolimnion. Eutrophic conditions are more likely to occur as a lake ages. This aging process, termed eutrophication, occurs naturally but can be greatly accelerated by anthropogenic additions of nutrients. A third major lake category, termed dystrophy, occurs when large amounts of organic materials that are resistant to decomposition wash into a lake basin. These organic materials stain the lake water and have a major influence on water chemistry which results in low production. *See* BOG; EUTROPHICATION. [T.M.F.]

Front An elongated, sloping zone in the troposphere, within which changes of temperature and wind velocity are large compared to changes outside the zone. Thus the passage of a front at a fixed location is marked by rather sudden changes in temperature and wind and also by rapid variations in other weather elements such as moisture and sky condition.

In its idealized sense, a front can be regarded as a sloping surface of discontinuity separating air masses of different density or temperature. In practice, the temperature change from warm to cold air occurs mainly within a zone of finite width, called a transition or frontal zone. The three-dimensional structure of the frontal zone is shown in the illustration. In typical cases, the zone is about 1 km (0.6 mi) in depth and 100–200 km (60–120 mi) in width, with a slope of approximately 1/100. The cold air lies

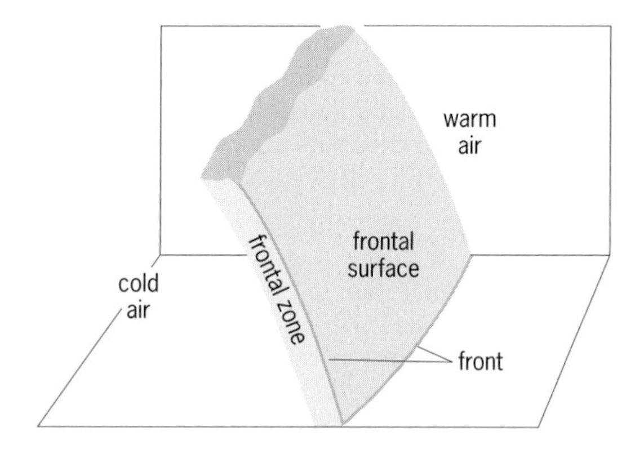

Schematic diagram of the frontal zone, angle with Earth's surface much exaggerated

beneath the warm in the form of a shallow wedge. Temperature contrasts generally are strongest at or near the earth's surface, the frontal zone usually being narrowest near the ground and becoming wider and more diffuse with height.

The surface separating the frontal zone from the adjacent warm air mass is referred to as the frontal surface, and it is the line of intersection of this surface with a second surface, usually horizontal or vertical, that strictly speaking constitutes the front. According to this more precise definition, the front represents a discontinuity in temperature gradient rather than in temperature itself. The boundary on the cold air side is often ill-defined, especially near the earth's surface, and for this reason is not represented in routine analysis of weather maps. *See* WEATHER MAP.

The wind gradient, or shear, like the temperature gradient, is larger within the frontal zone than on either size of it. In well-developed fronts the shift in wind direction often is concentrated along the frontal surface, while a more gradual change in wind speed may occur throughout the frontal zone. An upper-level jet stream normally is situated above the frontal zone. *See* JET STREAM. [S.E.M.]

Frost A covering of ice in one of several forms produced by the freezing of supercooled water droplets on objects colder than 32°F (0°C). The partial or complete killing of vegetation, by freezing or by temperatures somewhat above freezing for certain sensitive plants, also is called frost. Air temperatures below 32°F (0°C) sometimes are reported as "degrees of frost"; thus, 10°F (−12°C) is 22 degrees of frost (this usage is confined to the Fahrenheit scale and is not applied to Celsius temperatures).

Frost forms in exactly the same manner as dew except that the individual droplets that condense in the air a fraction of an inch from a subfreezing object are themselves supercooled, that is, colder than 32°F (0°C). When the droplets touch the cold object, they freeze immediately into individual crystals. When additional droplets freeze as soon as the previous ones are frozen, and hence are still close to the melting point because all the heat of fusion has not been dissipated, amorphous frost or rime results.

At more rapid rates of condensation, the drops form a thin layer of liquid before freezing, and glaze or glazed frost ("window ice" on house windows, "clear ice" on aircraft) generally follows. Glaze formation on plants, buildings and other structures, and especially on wires sometimes is called an ice storm, a silver frost storm, or thaw.

At slower deposition rates, such that each crystal cools well below the melting point before the next joins it, true crystalline or hoar frosts form. These include fernlike assemblages on snow surfaces, called surface hoar; similar feathery plumes in cold buildings, caves, and crevasses, called depth hoar; and the common window frost or ice flowers on house windows.

Killing frosts or freezes damage or kill vegetation depending on their duration and their intensity, that is, how far the plant temperatures go below 32°F (0°C). Such conditions result from advection of much colder air, which then cools the plants, as in the infamous cold waves of the north-central United States; or from radiational cooling of the plants themselves, by long-wave radiation to clear skies at night. In either case, the extent to which plant fluids freeze determines the severity of the frost. *See* AIR TEMPERATURE; DEW; DEW POINT. [A.Cou.]

Fruit A matured carpel or group of carpels (the basic units of the gynoecium or female part of the flower) with or without seeds, and with or without other floral or shoot parts (accessory structures) united to the carpel or carpels. Carpology is the study of the morphology and anatomy of fruits. The ovary develops into a fruit after fertilization and usually contains one or more seeds, which have developed from the fertilized ovules. Parthenocarpic fruits usually lack seeds. Fruitlets are the small fruits or subunits of aggregate or multiple fruits. Flowers, carpels, ovaries, and fruits are, by definition, restricted to the flowering plants (angiosperms), although fruitlike structures may enclose seeds in certain other groups of seed plants. The fruit is of ecological significance because of seed dispersal. *See* SEED.

Morphology. A fruit develops from one or more carpels. Usually only part of the gynoecium, the ovary, develops into a fruit; the style and stigma wither. Accessory (extracarpellary or noncarpellary) structures may be closely associated with the carpel or carpels and display various degrees of adnation (fusion) to them, thus becoming part of the fruit. Such accessory parts include sepals (as in the mulberry), the bases of sepals, petals, and stamens united into a floral tube (apple, banana, pear, and other species with inferior ovaries), the receptacle (strawberry), the pedicel and receptacle (cashew), the peduncle (fleshy part of the fig), the involucre composed of bracts and bracteoles (walnut and pineapple), and the inflorescence axis (pineapple). *See* FLOWER.

A fruit derived from only carpellary structures is called a true fruit, or, because it develops from a superior ovary (one inserted above the other floral parts), a superior fruit (corn, date, grape, plum, and tomato). Fruits with accessory structures are called accessory (or inaptly, false or spurious) fruits (pseudocarps), or, because of their frequent derivation from inferior ovaries (inserted below the other floral parts), inferior fruits (banana, pear, squash, and walnut).

Fruits can be characterized by the number of ovaries and flowers forming the fruit. A simple fruit is derived form one ovary, an aggregate fruit from several ovaries of one flower (magnolia, rose, and strawberry). A multiple (collective) fruit is derived from the ovaries and accessory structures of several flowers consolidated into one mass (fig, pandan, pineapple, and sweet gum).

The fruit wall at maturity may be fleshy or, more commonly, dry. Fleshy fruits range from soft and juicy to hard and tough. Dry fruits may be dehiscent, opening to release seeds, or indehiscent, remaining closed and containing usually one seed per fruit. Fleshy fruits are rarely dehiscent.

The pericarp is the fruit wall developed from the ovary. In true fruits, the fruit wall and pericarp are synonymous, but in accessory fruits the fruit wall includes the pericarp plus one or more accessory tissues of various derivation. Besides the fruit wall, a fruit contains one or more seed-bearing regions (placentae) and often partitions (septa).

Anatomy. Anatomically or histologically, a fruit consists of dermal, ground (fundamental), and vascular systems and, if present, one or more seeds. After fertilization the ovary and sometimes accessory parts develop into the fruit; parthenocarpy is fruit production without fertilization. The fruit generally increases in size and undergoes various anatomical changes that usually relate to its manner of dehiscence, its mode of dispersal, or protection of its seeds. The economically important, mainly fleshy fruits have received the most histological and developmental study.

Size increase of fruits is hormonally controlled and results from cell division and especially from cell enlargement. Cell number, volume, and weight thus control fruit weight. Cell division generally is more pronounced before anthesis (full bloom); cell enlargement is more pronounced after.

Functional aspects. Large fruits generally require additional anatomical modifications for nutrition or support or both. The extra phloem in fruit vascular bundles and the often increased amount of vascular tissue in the fruit wall and septa supply nutrients to the developing seeds and, especially in fleshy fruits, to the developing walls. Large, especially fleshy fruits (apple, gourd, and kiwi) usually contain proportionally more vascular tissue than small fruits. Vascular tissue also serves for support and in lightweight fruits may be the chief means of support.

Crystals, tannins, and oils commonly occur in fruits and may protect against pathogens and predators. The astringency of tannins, for example, may be repellent to organisms. With fruit maturation, tannin content ordinarily decreases, so the tannin repellency operative in early stages is superseded in fleshy fruits by features (tenderness, succulence, sweetness through odor and increased sugar content, and so on) attractive to animal dispersal agents. Many fruits are dispersed by hairs, hooks, barbs, spines, and sticky mucilage adhering the fruit to the surface of the dispersal agent. Lightweight fruits with many air spaces or with wings or plumes may be dispersed by wind or water. Gravity is always a factor in dispersal of fruits and seeds. [R.S.]

Fuel gas A fuel in the gaseous state whose potential heat energy can be readily transmitted and distributed through pipes from the point of origin directly to the place of consumption. The types of fuel gases are natural gas, LP gas, refinery gas, coke oven gas, and blast-furnace gas. The last two are used in steel mill complexes.

Most fuel gases are composed in whole or in part of the combustibles hydrogen, carbon monoxide, methane, ethane, propane, butane, and oil vapors and, in some instances, of mixtures containing the inerts nitrogen, carbon dioxide, and water vapor. *See* NATURAL GAS. [J.Hu.]

Fungal ecology The subdiscipline in mycology and ecology that examines community composition and structure; responses, activities, and interactions of single species; and the functions of fungi in ecosystems. These organisms display an extraordinary diversity of ecological interactions and life history strategies, but are alike in being efficient heterotrophs. Fungi, along with bacteria, are the primary decomposers, facilitating the flow of energy and the cycling of materials through ecosystems. *See* ECOLOGICAL ENERGETICS; FUNGI.

Fungi occur in many different habitat types—on plant surfaces; inside plant tissues; in decaying plant foliage, bark and wood; and in soil—generally changing in abundance and species composition through successional stages of decomposition. Fungi are also found in marine and aquatic habitats; in association with other fungi, lichens, bacteria, and algae; and in the digestive tracts and waste of animals. Some fungi grow in extreme environments: rock can harbor free-living endolithic fungi or the fungal mutualists of lichens; thermotolerant and thermophilic fungi can grow at temperatures above 45°C (113°F); psychrophilic fungi can grow at temperatures to below −3°C (27°F). Xerotolerant fungi are able to grow in extremely dry habitats, and osmotolerant fungi grow on subtrates with high solute concentrations. Most fungi are strict aerobes, but species of the chytrid *Neocallimastix*, which inhabit the rumen of herbivorous mammals, are obligate anaerobes. Several aquatic fungi are facultative anaerobes. Many fungi occur as free-living saprobes, but fungi are particularly successful as mutualistic, commensal, or antagonistic symbionts with other organisms. *See* POPULATION ECOLOGY.

Fungi possess unique features that affect their capacity to adapt and to function in ecosystems:

1. Fungi are composed of a vegetative body (hyphae or single cells) capable of rapid growth. Hyphae are linear strands composed of tubular cells that are in direct contact with the substrate. The cells secrete extracellular enzymes that degrade complex polymers, such as cellulose, into low-molecular-weight units that are then absorbed and catabolized. Many fungi also produce secondary metabolites such as mycotoxins and plant growth regulators that affect the outcomes of their interactions with other organisms.

2. Filamentous fungi are able to mechanically penetrate and permeate the substrate.

3. Fungi have an enormous capacity for metabolic variety. Fungal enzymes are able to decompose highly complex organic substances such as lignin, and to synthesize structurally diverse, biologically active secondary metabolites. Saprotrophic fungi are very versatile; some are able to grow on tree resins and even in jet fuel.

4. Structural and physiological features of fungi facilitate absorption and accumulation of mineral nutrients as well as toxic elements. The capacity of fungi to absorb, accumulate, and translocate is especially significant ecologically where hyphal networks permeate soil and function as a link between microhabitats.

5. Fungi have the capacity for indeterminate growth, longevity, resilience, and asexual reproduction. The vegetative cells of Eumycota are often multinucleate, containing dissimilar haploid nuclei. This combination of features gives the fungi an unparalleled capacity for adaptation to varying physiological and ecological circumstances and ensures a high level of genetic diversity.

6. Many species of fungi have a capacity to shift their mode of nutrition. The principal modes are saprotrophy (the utilization of dead organic matter) and biotrophy, which is characteristic of parasitic, predacious, and mutualistic fungi (including mycorrhizae and lichen fungi). *See* BIODEGRADATION; FUNGAL GENETICS; MYCORRHIZAE.

Fungi interact with all organisms in ecosystems, directly or indirectly, and are key components in ecosystem processes. As decomposers, fungi are crucial in the process of nutrient cycling, including carbon cycling as well as the mineralization or immobilization of other elemental constituents. As parasites, pathogens, predators, mutualists, or food sources, fungi can directly influence the species composition and population dynamics of other organisms with which they coexist. Fungi may act both as agents of successional change or as factors contributing to resilience and stability. Mycorrhizal fungi function as an interface between plant and soil, and are essential to the survival of most plants in natural habitats.

There are several economically important areas that benefit from application of the principles of fungal ecology: biotechnology, biological control, bioremediation, agriculture, forestry, and land reclamation. With only a small fraction of the total species known, the fungi offer a rich potential for bioprospecting, the search for novel genetic resources with unique, useful biochemical properties. *See* ECOLOGY; ECOSYSTEM; FUNGI.

[M.Ch.; J.K.St.]

Fungal genetics The study of gene structure and function in fungi. Genetic research has provided important knowledge about genes, heredity, genetic mechanisms, metabolism, physiology, and development in fungi, and in higher organisms in general, because in certain respects the fungal life cycle and cellular attributes are ideally suited to both mendelian and molecular genetic analysis.

Fungal nuclei are predominantly haploid; that is, they contain only one set of chromosomes. This characteristic is useful in the study of mutations, which are usually recessive and therefore masked in diploid organisms. Mutational dissection is an important technique for the study of biological processes, and the use of haploid organisms conveniently allows for the immediate expression of mutant genes. *See* MUTATION.

Reproduction in fungi can be asexual, sexual, or parasexual (see illustration). Asexual reproduction involves mitotic nuclear division during the growth of hyphae, cell division, or the production of asexual spores. Sexual reproduction is based on meiotic nuclear divisions fairly typical of eukaryotes in general. In ascomycetes and basidiomycetes, the spores, containing nuclei that are the four products of a single meiosis, remain together in a group called a tetrad. The isolation and testing of the phenotypes of cultures arising from the members of a tetrad (tetrad analysis) permit the study of the genetic events occurring in individual meioses; this possibility is offered by virtually no

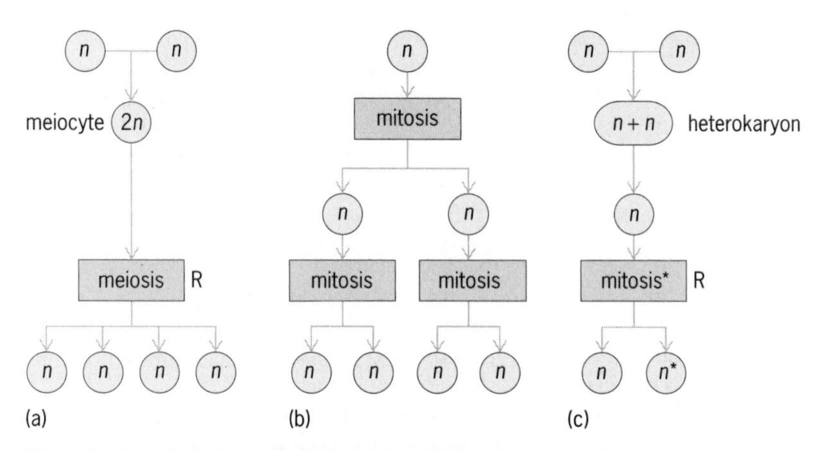

Three different kinds of reproduction occurring in fungi, each of which provides opportunities for genetic analysis. (*a*) Sexual reproduction leads to recombination (R) of genes at meiosis. (*b*) Asexual reproduction (shown here in a typical haploid fungal cell) usually reproduces the gene set faithfully. (*c*) Parasexual reproduction derives from an atypical mitotic division of an unstable cell that produces haploid cells and other aneuploid (deviating from normal chromosome complement) unstable intermediates.

other eukaryotic group. In other groups, genetic analysis is limited to products recovered randomly from different meioses. Since a great deal of genetic analysis is based on meiosis, fungal tetrads have proved to be pivotal in shaping current ideas on this key process of eukaryotic biology.

Because their preparation in large numbers is simple, fungal cells are useful in the study of rare events (such as mutations and recombinations) with frequencies as little as one in a million or less. In such cases, selective procedures must be used to identify cells derived from the rare events. The concepts and techniques of fungal asexual and parasexual genetics have been applied to the genetic manipulation of cultured cells of higher eukaryotes such as humans and green plants. However, the techniques remain much easier to perform with fungi.

The fact that each enzyme is coded by its own specific gene was first recognized in fungi and was of paramount importance because it showed how the many chemical reactions that take place in a living cell could be controlled by the genetic apparatus. The discovery arose from a biochemical study of nutritional mutants in *Neurospora*.

In genetically transformed organisms, the genome has been modified by the addition of DNA, a key technique in genetic engineering. The cell wall is temporarily removed; exogenous DNA is then taken up by cells and the cell wall is restored. The incorporation of DNA must be detected by a suitable novel genetic marker included on the assimilated molecule in order to distinguish transformed from nontransformed cells. The fate of the DNA inside the cell depends largely on the nature of the vector or carrier. Some vectors can insert randomly throughout the genome. Others can be directed to specific sites, either inactivating a gene for some purpose or replacing a resident gene with an engineered version present on the vector. A third kind of vector remains uninserted as an autonomously replicating plasmid. The ability to transform fungal cells has permitted the engineering of fungi with modified metabolic properties for making products of utility in industry. *See* GENETIC ENGINEERING.

A surprising development in the molecular biology of eukaryotes was the discovery of transposons, pieces of DNA that can move to new locations in the chromosomes. Although transposons were once known only in bacteria, they are now recognized in many eukaryotes. The transposons found in fungi mobilize by either of two processes: one type via a ribonucleic acid (RNA) intermediate that is subsequently reverse-transcribed to DNA, and the other type via DNA directly. In either case, a DNA copy of the transposed segment is inserted into the new site and may contain, in addition to the transposon itself, segments of contagious DNA mobilized from the original chromosomal site. Because of the rearrangements which transposons may produce, they have been important in the evolution of the eukaryotic genome. *See* FUNGI; GENETICS. [A.J.F.G.; R.U.]

Fungal infections Several thousand species of fungi have been described, but fewer than 100 are routinely associated with invasive diseases of humans. In general, healthy humans have a very high level of natural immunity to fungi, and most fungal infections are mild and self-limiting. Intact skin and mucosal surfaces and a functional immune system serve as the primary barriers to colonization by these ubiquitous organisms, but these barriers are sometimes breached.

Unlike viruses, protozoan parasites, and some bacterial species, fungi do not require human or animal tissues to perpetuate or preserve the species. Virtually all fungi that have been implicated in human disease are free-living in nature. However, there are exceptions, including various *Candida* spp., which are frequently found on mucosal surfaces of the body such as the mouth and vagina, and *Malassezia furfur*, which is usually found on skin surfaces that are rich in sebaceous glands. These organisms are often cultured from healthy tissues, but under certain conditions they cause disease. Only a handful of fungi cause significant disease in healthy individuals. Once established, these diseases can be classified according to the tissues that are initially colonized.

Superficial mycoses. Four infections are classified in the superficial mycoses. Black piedra, caused by *Piedraia hortai*, and white piedra, caused by *Trichosporn beigleii*, are infections of the hair. The skin infections include tinea nigra, caused by *Exophiala werneckii*, and tinea versicolor, caused by *M. furfur*. Where the skin is involved, the infections are limited to the outermost layers of the stratum corneum; in the case of hairs, the infection is limited to the cuticle. In general, these infections cause no physical discomfort to the patient, and the disease is brought to the attention of the physician for cosmetic reasons.

Cutaneous mycoses. The cutaneous mycoses are caused by a homogeneous group of keratinophilic fungi termed the dermatophytes. Species within this group are capable of colonizing the integument and its appendages (the hair and the nails). In general, the infections are limited to the nonliving keratinized layers of skin, hair, and nails, but a variety of pathologic changes can occur depending on the etiologic agent, site of infection, and immune status of the host. The diseases are collectively called the dermatophytoses, ringworms, or tineas. They account for most of the fungal infections of humans.

Subcutaneous mycoses. The subcutaneous mycoses include a wide spectrum of infections caused by a heterogeneous group of fungi. The infections are characterized by the development of lesions at sites of inoculation, commonly as a result of traumatic implantation of the etiologic agent. The infections initially involve the deeper layers of the dermis and subcutaneous tissues, but they eventually extend into the epidermis. The lesions usually remain localized or spread slowly by direct extension via the lymphatics, for example, subcutaneous sporotrichosis.

Systemic mycoses. The initial focus of the systemic mycoses is the lung. The vast majority of cases in healthy, immunologically competent individuals are asymptomatic or of short duration and resolve rapidly, accompanied in the host by a high degree of specific resistance. However, in immunosuppressed patients the infection can lead to life-threatening disease. *See* FUNGI; MEDICAL MYCOLOGY. [G.S.K.]

Fungal virus Any of the viruses that infect fungi (mycoviruses). In general these viruses are spheres of 30–45-nanometer diameter composed of multiple units of a single protein arranged in an icosahedral structure enclosing a genome of segmented double-stranded ribonucleic acid (dsRNA). Viruses are found in most species of fungi, where they usually multiply without apparent harm to the host. Most fungal viruses are confined to closely related species in which they are transmitted only through sexual or asexual spores to progeny or by fusion of fungal hyphae (filamentous cells). Some fungal strains are infected with multiple virus species. Although hundreds of virus-containing fungi have been reported, very few have been studied in significant detail. Three families of mycoviruses are recognized by the International Committee on Taxonomy of Viruses. The most thoroughly studied mycoviruses are in the family Totiviridae. *See* FUNGI; MYCOLOGY; PLANT PATHOLOGY; VIRUS. [R.F.Bo.]

Fungi Nucleated, usually filamentous, sporebearing organisms devoid of chlorophyll; typically reproducing both sexually and asexually; living as parasites in plants, animals, or other fungi, or as saprobes on plant or animal remains, in aquatic, marine, terrestrial, or subaerial habitats. Yeasts, mildews, rusts, mushrooms, and truffles are examples of fungi.

Some fungal classifications were constructed to facilitate identification, whereas others emphasize phylogeny. The more widely used classifications reflect a series of compromises between identification and phylogeny, and tend to conserve the vocabulary and nomenclature familiar to broad groups of users. The following is a conventional classification, in which all organisms are treated as members of the kingdom Fungi:

Division: Eumycota
  Subdivision: Mastigomycotina
    Class: Chytridiomycetes
      Hyphochytriomycetes
      Oomycetes
  Subdivision: Zygomycotina
    Class: Zygomycetes
      Trichomycetes
  Subdivision: Ascomycotina
    Class: Hemiascomycetes
      Plectomycetes
      Pyrenomycetes
      Discomycetes
      Loculoascomycetes
  Subdivision: Basidiomycotina
    Class: Hymenomycetes
      Gasteromycetes
      Urediniomycetes
      Ustilaginomycetes
  Subdivision: Deuteromycotina
    Class: Blastomycetes
      Hyphomycetes
      Coelomycetes
      Agonomycetes
Division: Myxomycota

[F.M.D.]

Organisms in the kingdom Fungi are mostly haploid, use chitin as a structural cell-wall polysaccharide, and synthesize lysine by the alpha amino adipic acid pathway; and their body is made of branching filaments (hyphae). The fungi arose about 1 billion years ago along with plants (including green algae), animals plus choanoflagellates, red algae, and stramenopiles. Ribosomal comparison indicates that the closest relatives to the fungi are the animals plus choanoflagellates.

Ascomycetes are the most numerous fungi (75% of all described species), and include lichen-forming symbionts. The group has traditionally been divided into unicellular yeasts and allies with naked asci, and hyphal forms with protected asci. However, ribosomal gene sequences indicate that some traditional yeasts and allied forms diverged early (early ascomycetes), at about the time ascomycetes were diverging from basidiomycetes. Hyphal ascomycetes protect their asci with a variety of fruiting bodies; the earliest fruiting bodies may have been open cups (Discomycetes), while in more recent groups they are flask shaped (Pyrenomycetes and Loculoascomycetes) or are completely closed (Plectomycetes). Ascomycetes lacking sexual structures have been classified in the Fungi Imperfecti, but molecular comparisons now allow their integration with the ascomycetes. *See* ASCOMYCOTA. [J.W.T.]

The mycelium, generally the vegetative body of fungi, is extremely variable. Unicellular forms, thought to be primitive or derived, grade into restricted mycelial forms; in most species, however, the mycelium is extensive and capable of indefinite growth. Some are typically perennial though most are ephemeral. The mycelium may be nonseptate, that is, coenocytic, with myriad scattered nuclei lying in a common cytoplasm, or septate, with each cell containing one to a very few nuclei or an indefinite number of nuclei. Septa may be either perforate or solid. Cell walls are composed largely of chitinlike materials except in one group of aquatic forms that have cellulose walls. Most mycelia are white, but a wide variety of pigments can be synthesized by specific forms and may be secreted into the medium or deposited in cell walls and protoplasm. Mycelial consistency varies from loose, soft wefts of hyphae to compact, hardened masses that resemble leather. Each cell is usually able to regenerate the entire mycelium, and vegetative propagation commonly results from mechanical fragmentation of the mycelium.

Asexual reproduction, propagation by specialized elements that originate without sexual fusion, occurs in most species and is extremely diverse. The most common and important means of asexual reproduction are unicellular or multicellular spores of various types that swim, fall, blow, or are forcibly discharged from the parent mycelium.

Sexual reproduction occurs in a majority of species of all classes. Juxtaposition and fusion of compatible sexual cells are achieved by four distinct sexual mechanisms, involving various combinations of differentiated sexual cells (gametes), undifferentiated sexual cells (gametangia), and undifferentiated vegetative cells. [J.R.Ra.]

Fungi obtain organic substances (food) from their environment which have been produced through the (photosynthetic) activities of green plants, since fungi do not contain chlorophyll and are unable to manufacture their own food. Fungi are able to digest food externally by releasing enzymes into their environment. These smaller molecules can be absorbed into the fungal body and transported to various locations where they can be used for energy or converted into different chemicals to make new cells or to serve other purposes. Some of the by-products of fungal metabolism may be useful to humans. Most fungi use nonliving plant material for food, but a few use nonliving animal material and therefore are called saprophytic organisms. In nature the decomposition of dead plant material is an important function of fungi, as the process releases nutrients back into the surrounding ecosystem where they can be reused by other organisms, including humans. *See* BIODEGRADATION; FUNGAL ECOLOGY.

A few fungi have the physiological capability to grow on living plants and may cause diseases such as wheat rust or corn smut on these economically important plants. Some fungi can grow on grains and may produce substances known as aflatoxins which can be detrimental to animals or humans. A few species of fungi have the ability to grow and acquire their food from skin or hair on living animals such as cats, horses, and humans. The disease known as ringworm may result. It is not caused by a worm but by an expanding circular growth of a fungus which has the physiological capability to use the components of skin or hair as the food source. The most frequently encountered fungal disease in humans is candidiasis, which is caused by one of the few fungi that is normally found associated with humans (*Candida albicans*). *See* AFLATOXIN; MEDICAL MYCOLOGY; PLANT PATHOLOGY; YEAST INFECTION.

A number of fungal species are able to enter plant roots and develop an association that may be beneficial to the plant under natural field conditions. This association of a higher plant root and a fungus that does not produce a disease is called a mycorrhiza. This fungal association with the plant root may permit the plant to live under soil conditions where it may not otherwise survive because of an excess of acid in the soil or a lack or excess of certain nutrients. *See* MYCORRHIZAE.

Certain species of fungi have been used by humans since early times in the preparation of foods such as leavened bread, cheeses, and beverages. Additional by-products of fungal physiology are used in industrial applications such as antibiotics, solvents, and pharmaceuticals. *See* YEAST. [M.C.W.]

Fungistat and fungicide Synthetic or biosynthetic compounds used to control fungal diseases in animals and plants. A fungistat prevents the spread of a fungus, whereas a fungicide kills the fungus.

Seeds and seedlings are protected against fungi in the soil by treating the seeds and the soil with fungicides. Seed-treating materials must be safe for seeds and must resist degradation by soil and soil microorganisms. Some soil fungicides are safe to use on living plants. Others are injurious to seeds and living plants. These compounds are useful because they are volatile. Used before planting, they have a chance to kill soil fungi and then escape from the soil.

Formic acid, acetic acid, and propionic acid up through pelargonic acid and capric acid (the C_1-C_{10} volatile fatty acids) possess significant fungicidal activity. Many of them are present in natural foodstuffs that are resistant to fungal attack. Use of the volatile fatty acids and their salts in bread to prevent ropy mold is widespread. A. I. Virtanen was awarded the Nobel prize in 1945 for his discovery and development of these lower volatile fatty acids to prevent fungal growth and so preserve the nutritious quality of cattle fodders. Since these volatile fatty acids stop fungal growth, they prevent mycotoxin generation and lessen the risk of cancer from exposure to mycotoxins. [J.G.H.; S.Ri.]

Two types of fungicides are used to control plant diseases: (1) Surface protectants remain on the plant surface and exert their toxic action on fungi before they have penetrated into plant tissue. (2) Systemic fungicides move into plant tissue and exert their toxic action on fungi which have already penetrated internally. These fungicides can also provide surface protection by acting on fungi before they have penetrated the plant.

Most agricultural fungicides are systemic compounds that act at a single target site in fungal cells, such as cell membranes, microtubules, or ribosomes.

There have been very few instances of fungal resistance to surface protectant fungicides which is attributed to their action at multiple sites within fungal cells. However, most systemic fungicides that act at a single site have generated serious problems with fungal resistance. A single gene mutation in a fungus can lead to loss of effectiveness of all fungicides in a particular mode-of-action group. Experience has shown that frequent, uninterrupted use of a fungicide increases the risk for development of a resistant strain of the target organism. That risk can be reduced by alternating use of fungicides with different modes of action or by using them in mixtures. *See* FUNGI. [H.D.S.]

G

Gene The basic unit in inheritance. There is no general agreement as to the exact usage of the term, since several criteria that have been used for its definition have been shown not to be equivalent.

The facts of mendelian inheritance indicate the presence of discrete hereditary units that replicate at each cell division, producing remarkably exact copies of themselves, and that in some highly specific way determine the characteristics of the individuals that bear them. The evidence also shows that each of these units may at times mutate to give a new equally stable unit (called an allele), which has more or less similar but not identical effects on the characters of its bearers. These hereditary units are the genes, and the criteria for the recognition that certain genes are alleles have been that they (1) arise from one another by a single mutation, (2) have similar effects on the characters of the organism, and (3) occupy the same locus in the chromosome. It has long been known that there were a few cases where these criteria did not give consistent results, but these were explained by special hypotheses in the individual cases. However, such cases have been found to be so numerous that they appear to be the rule rather than the exception. *See* MENDELISM; MUTATION.

The term gene, or cistron, may be used to indicate a unit of function. The term is used to designate an area in a chromosome made up of subunits present in an unbroken unit to give their characteristic effect.

Every gene consists of a linear sequence of bases in a nucleic acid molecule. Genes are specified by the sequence of bases in DNA in prokaryotic, archaeal, and eukaryotic cells, and in DNA or ribonucleic acid (RNA) in prokaryotic or eukaryotic viruses. The ultimate expressions of gene function are the formation of structural and regulatory RNA molecules and proteins. These macromolecules carry out the biochemical reactions and provide the structural elements that make up cells. *See* VIRUS.

One goal of molecular biology is to understand the function, expression, and regulation of a gene in terms of its DNA or RNA sequence. The genetic information in genes that encode proteins is first transcribed from one strand of DNA into a complementary messenger RNA (mRNA) molecule by the action of the RNA polymerase enzyme. Many kinds of eukaryotic and a limited number of prokaryotic mRNA molecules are further processed by splicing, which removes intervening sequences called introns. In some eukaryotic mRNA molecules, certain bases are also changed posttranscriptionally by a process called RNA editing. The genetic code in the resulting mRNA molecules is translated into proteins with specific amino acid sequences by the action of the translation apparatus, consisting of transfer RNA (tRNA) molecules, ribosomes, and many other proteins. The genetic code in an mRNA molecule is the correspondence of three contiguous (triplet) bases, called a codon, to the common amino acids and translation stop signals; the bases are adenine (A), uracil (U), guanine (G), and cytosine (C). There are 61 codons that specify the 20 common amino acids, and 3 codons that lead to translation stopping. *See* GENETIC CODE. [A.H.St.]

In many cases, the genes that mediate a specific cellular or viral function can be isolated. The recombinant DNA methods used to isolate a gene vary widely depending on the experimental system, and genes from RNA genomes must be converted into a corresponding DNA molecule by biochemical manipulation using the enzyme reverse transcriptase. The isolation of the gene is referred to as cloning, and allows large quantities of DNA corresponding to a gene of interest to be isolated and manipulated.

After the gene is isolated, the sequence of the nucleotide bases can be determined. The goal of the large-scale Human Genome Project is to sequence all the genes of several model organisms and humans. The sequence of the region containing the gene can reveal numerous features. If a gene is thought to encode a protein molecule, the genetic code can be applied to the sequence of bases determined from the cloned DNA. The application of the genetic code is done automatically by computer programs, which can identify the sequence of contiguous amino acids of the protein molecule encoded by the gene. If the function of a gene is unknown, comparisons of its nucleic acid or predicted amino acid sequence with the contents of huge international databases can often identify genes or proteins with analogous or related functions. These databases contain all the known sequences from many prokaryotic, archaeal, and eukaryotic organisms. Putative regulatory and transcript-processing sites can also be identified by computer. These putative sites, called consensus sequences, have been shown to play roles in the regulation and expression of groups of prokaryotic, archaeal, or eukaryotic genes. However, computer predictions are just a guide and not a substitute for analyzing expression and regulation by direct experimentation. *See* GENETIC ENGINEERING. [M.E.Wi.]

Genetic code The rules by which the base sequences of deoxyribonucleic acid (DNA) are translated into the amino acid sequences of proteins. Each sequence of DNA that codes for a protein is transcribed or copied into messenger ribonucleic acid (mRNA). Following the rules of the code, discrete elements in the mRNA, known as codons, specify each of the 20 different amino acids that are the constituents of proteins. During translation, another class of RNAs, called transfer RNAs (tRNAs), are coupled to amino acids, bind to the mRNA, and, in a step-by-step fashion provide the amino acids that are linked together in the order called for by the mRNA sequence. The specific attachment of each amino acid to the appropriate tRNA, and the precise pairing of tRNAs via their anticodons to the correct codons in the mRNA, form the basis of the genetic code.

The genetic information in DNA is found in the sequence or order of four bases that are linked together to form each strand of the two-stranded DNA molecule. The bases of DNA are adenine, guanine, thymine, and cytosine, which are abbreviated as A, G, T, and C. Chemically, A and G are purines, and C and T are pyrimidines. The two strands of DNA are wound about each other in a double helix that looks like a twisted ladder. Each rung of the ladder is formed by two bases, one from each strand, that pair with each other by means of hydrogen bonds. For a good fit, a pyrimidine must pair with a purine; in DNA, A bonds with T, and G bonds with C.

Ribonucleic acids such as mRNA or tRNA also comprise four bases, except that in RNA the pyrimidine uracil (U) replaces thymine. During transcription a single-stranded mRNA copy of one strand of the DNA is made.

If two bases at a time are grouped together, then only 4×4 or 16 different combinations are possible, a number that is insufficient to code for all 20 amino acids that are found in proteins. However, if the four bases are grouped together in threes, then there

	U		C		A		G	
U	UUU, UUC	Phe	UCU, UCC, UCA, UCG	Ser	UAU, UAC	Tyr	UGU, UGC	Cys
	UUA, UUG	Leu			UAA, UAG	*Stop*	UGA	*Stop*
							UGG	Trp
C	CUU, CUC, CUA, CUG	Leu	CCU, CCC, CCA, CCG	Pro	CAU, CAC	His	CGU, CGC, CGA, CGG	Arg
					CAA, CAG	Gln		
A	AUU, AUC, AUA	Ile	ACU, ACC, ACA, ACG	Thr	AAU, AAC	Asn	AGU, AGC	Ser
	AUG	Met			AAA, AAG	Lys	AGA, AGG	Arg
G	GUU, GUC, GUA, GUG	Val	GCU, GCC, GCA, GCG	Ala	GAU, GAC	Asp	GGU, GGC, GGA, GGG	Gly
					GAA, GAG	Glu		

Universal (standard) genetic code. Each of the 64 codons found in mRNA specifies an amino acid (indicated by the common three-letter abbreviation) or the end of the protein chain (*stop*). The amino acids are phenylalanine (Phe), leucine (Leu), isoleucine (Ile), methionine (Met), valine (Val), serine (Ser), proline (Pro), threonine (Thr), alanine (Ala), tyrosine (Tyr), histidine (His), glutamine (Gln), asparagine (Asn), lysine (Lys), aspartic acid (Asp), glutamic acid (Glu), cysteine (Cys), tryptophan (Trp), arginine (Arg), and glycine (Gly).

are $4 \times 4 \times 4$ or 64 different combinations. Read sequentially without overlapping, those groups of three bases constitute a codon, the unit that codes for a single amino acid.

The 64 codons can be divided into 16 families of four (see illustration), in which each codon begins with the same two bases. With the number of codons exceeding the number of amino acids, several codons can code for the same amino acid. Thus, the code is degenerate. In eight instances, all four codons in a family specify the same amino acid. In the remaining families, the two codons that end with the pyrimidines U and C often specify one amino acid, whereas the two codons that end with the purines A and G specify another. Furthermore, three of the codons, UAA, UAG, and UGA, do not code for any amino acid but instead signal the end of the protein chain.

On the ribosome, the nucleic acid code of an mRNA is converted into an amino acid sequence with the aid of tRNAs. These RNAs are relatively small nucleic acids, varying from 75 to 93 bases in length, that are folded in three dimensions to form an L-shaped molecule to which an amino acid can be attached. At the other end of the tRNA molecule, three bases are free to pair with a codon in the mRNA. These three bases of a tRNA constitute the anticodon. Each amino acid has one or more tRNAs, and because of the degeneracy of the code, many of the tRNAs for a specific amino acid have different anticodon sequences. However, the tRNAs for one amino acid are capable of pairing their anticodons only with the codon or codons in the mRNA that specify that amino acid. The tRNAs act as interpreters of the code, providing the correct amino acid in response to each codon by virtue of precise codon-anticodon pairing. The tRNAs pair with the codons and sequentially insert their amino acids in the exact order specified by the sequence of codons in the mRNA.

The rules of the genetic code are virtually the same for all organisms, but there are some interesting exceptions. In the microorganism *Mycoplasma capricolum*, UGA is not a stop codon; instead it codes for tryptophan. This alteration in the code is also found in the mitochondria of some organisms. In addition to changes in the meanings of codons, a modified system for reading codons that requires fewer tRNAs is found in mitochondria. *See* GENE; GENETICS. [P.Sc.; H.E.We.]

Genetic engineering The artificial recombination of nucleic acid molecules in the test tube, their insertion into a virus, bacterial plasmid, or other vector system, and the subsequent incorporation of the chimeric molecules into a host organism in which they are capable of continued propagation. The construction of such molecules has also been termed gene manipulation because it usually involves the production of novel genetic combinations by biochemical means.

Genetic engineering provides the ability to propagate and grow in bulk a line of genetically identical organisms, all containing the same artificially recombinant molecule. Any genetic segment as well as the gene product encoded by it can therefore potentially be amplified. For these reasons the process has also been termed molecular cloning or gene cloning. *See* GENE.

Basic techniques. The central techniques of such gene manipulation involve (1) the isolation of a specific deoxyribonucleic acid (DNA) molecule or molecules to be replicated (the passenger DNA); (2) the joining of this DNA with a DNA vector (also known as a vehicle or a replicon) capable of autonomous replication in a living cell after foreign DNA has been inserted into it; and (3) the transfer, via transformation or transfection, of the recombinant molecule into a suitable host.

Isolation of passenger DNA. Passenger DNA may be isolated in a number of ways; the most common of these involves DNA restriction. Restriction endonucleases make possible the cleavage of high-molecular-weight DNA. Although three different classes of these enzymes have been described, only type II restriction endonucleases have been used extensively in the manipulation of DNA. Type II restriction endonucleases are DNAases that recognize specific short nucleotide sequences (usually 4 to 6 base pairs in length), and then cleave both strands of the DNA duplex, generating discrete DNA fragments of defined length and sequence. A number of restriction enzymes make staggered cuts in the two DNA strands, generating single-stranded termini.

The various fragments generated when a specific DNA is cut by a restriction enzyme can be easily resolved as bands of distinct molecular weights by agarose gel electrophoresis. Specific sequences of these bands can be identified by a technique known as Southern blotting. In this technique, DNA restriction fragments resolved on a gel are denatured and blotted onto a nitrocellulose filter. The filter is incubated together with a radioactively labeled DNA or RNA probe specific for the gene under study. The labeled probe hybridizes to its complement in the restricted DNA, and the regions of hybridization are detected autoradiographically. Fragments of interest can then be eluted out of these gels and used for cloning. Purification of particular DNA segments prior to cloning reduces the number of recombinants that must later be screened.

Another method that has been used to generate small DNA fragments is mechanical shearing. Intense sonification of high-molecular-weight DNA with ultrasound, or high-speed stirring in a blender, can both be used to produce DNA fragments of a certain size range. Shearing results in random breakage of DNA, producing termini consisting of short, single-stranded regions. Other sources include DNA complementary to poly(A) RNA, or cDNA, which is synthesized in the test tube, and short oligonucleotides that are synthesized chemically.

Joining DNA molecules. Once the proper DNA fragments have been obtained, they must be joined. When cleavage with a restriction endonuclease creates cohesive ends, these can be annealed with a similarly cleaved DNA from another source, including a vector molecule. When such molecules associate, the joint has nicks a few base pairs apart in opposite strands. The enzyme DNA ligase can then repair these nicks to form an intact, duplex recombinant molecule, which can be used for transformation and the subsequent selection of cells containing the recombinant molecule. Cohesive ends can also be created by the addition of synthetic DNA linkers to blunt-ended DNA molecules.

Another method for joining DNA molecules involves the addition of homopolymer extensions to different DNA populations followed by an annealing of complementary homopolymer sequences. For example, short nucleotide sequences of pure adenine can be added to the 3′ ends of one population of DNA molecules and short thymine blocks to the 3′ ends of another population. The two types of molecules can then anneal to form mixed dimeric circles that can be used directly for transformation.

The enzyme T4 DNA ligase carries out the intermolecular joining of DNA substrates at completely base-paired ends; such blunt ends can be produced by cleavage with a restriction enzyme or by mechanical shearing followed by enzyme treatment.

Transformation. The desired DNA sequence, once attached to a DNA vector, must be transferred to a suitable host. Transformation is defined as the introduction of foreign DNA into a recipient cell. Transformation of a cell with DNA from a virus is usually referred to as transfection.

Transformation in any organism involves (1) a method that allows the introduction of DNA into the cell and (2) the stable integration of DNA into a chromosome, or maintenance of the DNA as a self-replicating entity.

Escherichia coli is usually the host of choice for cloning experiments, and transformation of *E. coli* is an essential step in these experiments. *Escherichia coli* treated with calcium chloride are able to take up DNA from bacteriophage lambda as well as plasmid DNA. Calcium chloride is thought to effect some structural alterations in the bacterial cell wall. An efficient method for transformation in *Bacillus* species involves polyethylene glycol-induced DNA uptake in bacterial protoplasts and subsequent regeneration of the bacterial cell wall. Actinomycetes can be similarly transformed. Transformation can also be achieved by first entrapping the DNA with liposomes followed by their fusion with the host cell membrane. Similar transformation methods have been developed for lower eukaryotes such as the yeast *Saccharomyces cerevisiae* and the filamentous fungus *Neurospora crassa*.

Several methods are available for the transfer of DNA into cells of higher eukaryotes. Specific genes or entire viral genomes can be introduced into cultured mammalian cells in the form of a coprecipitate with calcium phosphate. DNA complexed with calcium phosphate is readily taken up and expressed by mammalian cells. DNA complexed with diethylamino-ethyl-dextran (DEAE-dextran) or DNA trapped in liposomes or erythrocyte ghosts may also be used in mammalian transformation. Alternatively, bacterial protoplasts containing plasmids can be fused to intact animal cells with the aid of chemical agents such as polyethylene glycol (PEG). Finally, DNA can be directly introduced into cells by microinjection. The efficiency of transfer by each of these methods is quite variable.

Introduction of DNA sequences by insertion into the transforming (T)-DNA region of the tumor-inducing (Ti) plasmid of *Agrobacterium tumefaciens* is a method of introducing DNA into plant cells and ensuring its integration. Because of the limitations of the host range of *A. tumefaciens*, however, alternative transformation systems are being developed for gene transfer in plants. They include the use of liposomes, as well as induction of DNA uptake in plant protoplasts. Foreign DNA has been introduced into plant cells by a technique called electroporation. This technique involves the use of electric pulses to make plant plasma membranes permeable to plasmid DNA molecules. Plasmid DNA taken up in this way has been shown to be stably inherited and expressed.

Cloning vectors. There is a large variety of potential vectors for cloned genes. The vectors differ in different classes of organisms.

Prokaryotes and lower eukaryotes. Three types of vectors have been used in these organisms: plasmids, bacteriophages, and cosmids. Plasmids are extrachromosomal

DNA sequences that are stably inherited. *Escherichia coli* and its plasmids constitute the most versatile type of host-vector system known for DNA cloning. Several natural plasmids, such as ColE1, have been used as cloning vehicles in *E. coli*. In addition, a variety of derivatives of natural plasmids have been constructed by combining DNA segments and desirable qualities of older cloning vehicles. The most versatile and widely used of these plasmids is pBR322. Transformation in yeast has been demonstrated using a number of plasmids, including vectors derived from the naturally occurring 2μ plasmid of yeast.

Bacteriophage lambda is a virus of *E. coli*. Several lambda-derived vectors have been developed for cloning in *E. coli*, and for the isolation of particular genes from eukaryotic genomes. These lambda derivatives have several advantages over plasmids: (1) Thousands of recombinant phage plaques can easily be screened for a particular DNA sequence on a single petri dish by molecular hybridization. (2) Packaging of recombinant DNA in laboratory cultures provides a very efficient means of DNA uptake by the bacteria. (3) Thousands of independently packaged recombinant phages can be easily replicated and stored in a single solution as a "library" of genomic sequences. *See* BACTERIOPHAGE.

Plasmids have also been constructed that contain the phage *cos* DNA site, required for packaging into the phage particles, and ColE1 DNA segments, required for plasmid replication. These plasmids have been termed cosmids. The recombinant cosmid DNA is injected into a host and circularizes like phage DNA but replicates as a plasmid. Transformed cells are selected on the basis of a vector drug resistance marker.

Animal cells. In contrast to the wide variety of plasmid and phage vectors available for cloning in prokaryotic cells, relatively few vectors are available for introducing foreign genes into animal cells. The most commonly used are derived from simian virus 40 (SV40). Normal SV40 cannot be used as a vector, since there is a physical limit to the amount of DNA that can be packaged into the virus capsid, and the addition of foreign DNA would generate a DNA molecule too large to be packaged. However, SV40 mutants lacking portions of the genome can be propagated in mixed infections in which a "helper" virus supplies the missing function.

Plant cells. Two systems for the delivery and integration of foreign genes into the plant genome are the Ti plasmid of the soil bacterium *Agrobacterium* and the DNA plant virion cauliflower mosaic virus. The Ti plasmid is a natural gene transfer vector carried by *A. tumefaciens*, a pathogenic bacterium that causes crown gall tumor formation in dicotyledonous plants. A T-DNA segment present in the Ti plasmid becomes stably integrated into the plant cell genome during infection. This property of the Ti plasmid has been exploited to show that DNA segments inserted in the T-DNA region can be cotransferred to plant DNA. *See* CROWN GALL.

Applications. Recombinant DNA technology has permitted the isolation and detailed structural analysis of a large number of prokaryotic and eukaryotic genes. This contribution is especially significant in the eukaryotes because of their large genomes. The methods outlined above provide a means of fractionating and isolating individual genes, since each clone contains a single sequence or a few DNA sequences from a very large genome. Isolation of a particular sequence of interest has been facilitated by the ability to generate a large number of clones and to screen them with the appropriate "probe" (radioactively labeled RNA or DNA) molecules.

Genetic engineering techniques provide pure DNAs in amounts sufficient for mapping, sequencing, and direct structural analyses. Furthermore, gene structure-function relationships can be studied by reintroducing the cloned gene into a eukaryotic nucleus and assaying for transcriptional and translational activities. The DNA sequences can be altered by mutagenesis before their reintroduction in order to define precise functional regions.

Genetic engineering methodology has provided means for the large-scale production of polypeptides and proteins. It is now possible to produce a wide variety of foreign proteins in *E. coli*. These range from enzymes useful in molecular biology to a vast range of polypeptides with potential human therapeutic applications, such as insulin, interferon, growth hormone, immunoglobins, and enzymes involved in the dynamics of blood coagulation. *See* BIOTECHNOLOGY.

Finally, experiments showing the successful transfer and expression of foreign DNA in plant cells using the Ti plasmid, as well as the demonstration that whole plants can be regenerated from cells containing mutated regions of T-DNA, indicate that the Ti plasmid system may be an important tool in the genetic engineering of plants. Such a system will help in the identification and characterization of plant genes as well as provide basic knowledge about gene organization and regulation in higher plants. Once genes useful for crop improvement have been identified, cloned, and stably inserted into the plant genome, it will be possible to engineer plants to be resistant to environmental stress, to pests, and to pathogens. *See* BREEDING (PLANT); GENE. [P.K.M.]

Genetics The science of biological inheritance, that is, the causes of the resemblances and differences among related individuals.

Genetics occupies a central position in biology, for essentially the same principles apply to all animals and plants, and understanding of inheritance is basic for the study of evolution and for the improvement of cultivated plants and domestic animals. It has also been found that genetics has much to contribute to the study of embryology, biochemistry, pathology, anthropology, and other subjects.

Genetics may also be defined as the science that deals with the nature and behavior of the genes, the fundamental hereditary units. From this point of view, evolution is seen as the study of changes in the gene composition of populations, whereas embryology is the study of the effects of the genes on the development of the organism. *See* POPULATION GENETICS. [A.H.St.]

The field of molecular genetics describes the basis of inheritance at the molecular level. It focuses on two general questions: how do genes specify the structure and function of organisms, and how are genes replicated and transmitted to successive generations? Both questions have been answered. Genes specify organismal structure and function according to a process described by the central dogma of molecular biology: DNA is made into messenger ribonucleic acid (mRNA), which specifies the structure of a protein; the mRNA molecule then serves as a template for protein synthesis, which is carried out by complex machinery that comprises a particle called a ribosome and special adapter RNA molecules called transfer RNA.

The structure of DNA provides a simple mechanism for genes to be faithfully reproduced: the specific interaction between the nucleotides means that each strand of the double helix carries the information for producing the other strand. *See* GENETIC CODE; GENETIC ENGINEERING; MUTATION. [M.J.]

Geochemical prospecting The use of chemical properties of naturally occurring substances (including rocks, glacial debris, soils, stream sediments, waters, vegetation, and air) as aids in a search for economic deposits of metallic minerals or hydrocarbons. In exploration programs, geochemical techniques are generally integrated with geological and geophysical surveys. *See* GEOCHEMISTRY.

General principles. Mineral deposits represent anomalous concentrations of specific elements, usually within a relatively confined volume of the Earth's crust. Most mineral deposits include a central zone, or core, in which the valuable elements or minerals are concentrated, often in percentage quantities, to a degree sufficient to permit economic exploitation. The valuable elements surrounding this core generally decrease in concentration until they reach levels, measured in parts per million (ppm) or parts per billion (ppb), which appreciably exceed the normal background level of the enclosing rocks. These zones or halos afford means by which mineral deposits can be detected and traced; they are the geochemical anomalies being sought by all geochemical prospectors.

The zone surrounding the core deposit is known as a primary halo or anomaly, and it represents the distribution patterns of elements which formed as a result of primary dispersion. Primary dispersion halos vary greatly in size and shape as a result of the numerous physical and chemical variables that affect fluid movements in rocks. Some halos can be detected at distances of hundreds of meters from their related ore bodies; others are no more than a few centimeters in width.

Abnormal chemical concentrations in weathering products are known as secondary dispersion halos or anomalies and are more widespread. They are sometimes referred to as dispersion trains. The shape and extent of secondary dispersion trains depend on a host of factors, of which topography and groundwater movement are perhaps most important. Groundwaters frequently dissolve some of the constituents of mineralized bodies and may transport these for considerable distances before eventually emerging in springs or streams. Further dispersion may ensue in stream sediments when soil or weathering debris that has anomalous metal content becomes incorporated through erosion in stream sediment. Analysis of the fine sand arid silt of stream sediment can be a particularly effective method for detection of mineralized bodies within the area drained by the stream.

Survey design. The degree of success of a geochemical survey in a mineral exploration program is often a reflection of the amount of care taken with initial planning and survey design. This phase of activity is often referred to as an orientation survey; its practical importance cannot be overstressed.

When a geochemical prospecting survey is contemplated, four basic considerations must be addressed: the nature of the mineral deposits being sought; the geochemical properties of the elements likely to be present in the target mineral deposit; geological factors likely to cause variations in geochemical background; and environmental, or landscape, factors likely to influence the geochemical expression of the target mineral deposit. Elucidation of these factors in an orientation survey will permit design of a geochemical prospecting survey that is most likely to prove effective under the prevailing conditions.

Geochemical prospecting surveys fall into two broad categories, strategic or tactical, which may be further subdivided according to the material sampled. Strategic surveys imply coverage of a large area (generally several thousands of square kilometers) where the primary objective is to identify districts of enhanced mineral potential; tactical surveys comprise the more detailed follow-up to strategic reconnaissance. Typically the area covered by a tactical survey is divided into discrete areas of high mineral potential within the general anomalous district.

Soil and glacial till surveys have been used extensively in geochemical prospecting and have resulted in the discovery of a number of ore bodies. Generally, such surveys are of a detailed nature and are run over a closely spaced grid.

Biogeochemical surveys are of two types. One type utilizes the trace-element content of plants to outline dispersion halos, trains, and fans related to mineralization; the other uses specific plants or the deleterious effects of an excess of elements in soils on plants as indicators of mineralization. The latter type of survey is often referred to as a geobotanical survey.

Rock geochemical surveys are reconnaissance surveys carried out on a grid or on traverses of an area, with samples taken of all available rock outcrops or at some specific interval. One or several rock types may be selected for sampling and analyzed for various elements. Geochemical maps are compiled from the analyses, and contours of equal elemental values are drawn. These are then interpreted, often by using statistical methods. Under favorable conditions, mineralized zones or belts may be outlined in which more detailed work can be concentrated. If the survey is executed over a large expanse of territory, geochemical provinces may be outlined.

Isotopic surveys are applicable to elements which exist in two or more isotopic forms. They employ the ratios between isotopes such as ^{204}Pb, ^{206}Pb, ^{207}Pb, ^{208}Pb, or ^{32}S and ^{34}S to "fingerprint" or indicate certain types of mineral deposits which may share a common origin. Isotopic ratios may also be used to determine the ages of minerals or given rock types and may, thus, assist in elucidating questions of ore formation.

Geochemistry applied to hydrocarbon exploration differs from that in the search for metallic mineral deposits; the former chiefly involves detection and study of organic substances found during drilling; the latter, detection and study of inorganic substances at the surface. Once hydrocarbon accumulations have been discovered, their classification into geochemical families is important. The final stages of detailed exploration may involve complex multivariate computer-aided modeling of all available geological, geochemical, geophysical, and hydrological data—to determine the ultimate hydrocarbon potential of a given basin. [R.F.H.]

Geochemistry A field that encompasses the investigation of the chemical composition of the Earth, other planets, and the solar system and universe as a whole, as well as the chemical processes that occur within them. The discipline is large and very important because basic knowledge about the chemical processes involved is critical for understanding subjects as diverse as the formation of economically valuable ore deposits, safe disposal of toxic wastes, and variations in the Earth's climate.

Isotope geochemistry is based on the fact that the isotopic compositions of various chemical elements may reveal information about the age, history, and origin of terrestrial and extraterrestrial materials. Isotopes of an element share the same chemical properties but have slightly different nuclear makeups and therefore different masses. Some naturally occurring isotopes are radioactive and decay at known rates to form daughter isotopes of another element; for example, radioactive uranium isotopes decay to stable isotopes of lead. Radioactive decay is the basis of geochronology, or age determination: the age of a sample can be found by measuring its content of the daughter isotope. Both radioactive decay and the processes that enrich or deplete materials in certain isotopes cause different parts of the Earth and solar system to have different, characteristic isotopic compositions for some elements. These differences serve as fingerprints for tracing the origins of, and characterizing the interactions between, various geochemical reservoirs. *See* DATING METHODS; ELEMENTS, GEOCHEMICAL DISTRIBUTION OF; GEOCHRONOMETRY.

Cosmochemistry deals with nonearthly materials. Typically, cosmochemists use the same kinds of analytical and theoretical approaches as other geochemists but apply them to problems involving the origin and history of meteorites, the formation of the solar system, the chemical processes on other planets, and the ultimate origin of the elements themselves in stars.

Organic geochemistry deals with carbon-containing compounds, largely those produced by living organisms. These are widely dispersed in the outer part of the Earth—in the oceans, the atmosphere, soil, and sedimentary rocks. Organic geochemistry is important for understanding many of the chemical cycles that occur on Earth because biology often plays a major role. Organic geochemists are also active in investigating such areas as the origin of life, the formation of some types of ore deposits that may be biologically mediated, and the origin of coal, petroleum, and natural gas. *See* BIOGEOCHEMISTRY; NATURAL GAS; ORGANIC GEOCHEMISTRY; PETROLEUM.

In recent years there has been widespread application of geochemical techniques to problems in paleoclimatology and paleoceanography. In this approach, ocean sediments, sedimentary rocks on land, ice cores, and other continuous records of the Earth's history are analyzed for fossil chemical evidence of past climates or seawater composition. As in most areas of geochemistry, precise and accurate analytical methods for determining the isotopic and elemental composition of the samples are critical. *See* PALEOCLIMATOLOGY. [J.D.MacD.]

Geochronometry The measurement of the age of rocks, minerals, water, and biological materials. Measurements are based primarily on the radioactive decay or fission of such naturally occurring isotopes as ^{238}U, ^{235}U, ^{232}Th, ^{187}Re, ^{176}Lu, ^{147}Sm, ^{87}Rb, ^{40}K, ^{129}I, ^{36}Cl, ^{26}Al, ^{14}C, and ^{10}Be. These radioactive isotopes can be divided into two groups: primordial isotopes that are residual from early nucleosynthesis, and cosmogenic isotopes that are continuously produced by cosmic-ray-induced spallation reactions primarily within the Earth's atmosphere or on the surfaces of meteorites. For the first group, the relative amounts of the radioactive parent and radiogenic daughter are used as a measure of age. Age is determined for the second group by the amount of radioactive isotope remaining after the object is isolated from further intake—for example, by death of an organism participating in the carbon-oxygen cycle, or by trapping of the cosmogenic isotope in sediment or ice. Tree-ring dating (dendrochronology), which is based on the counting of annual rings, may also be used and provides a very precise measure of age of the last eight millennia. *See* RADIOCARBON DATING.

There are also methods of establishing the relative sequence of events in time, most importantly, the use of unidirectional biologic evolution upon which the boundaries of the Phanerozoic time scale are based (5.5×10^8 years to the present). The virtues of isotopic dating are its applicability to the full range of geologic time, including the Precambrian for which an adequate paleontologic time scale does not exist; better resolution of events during the Cenozoic (6.5×10^7 years to present); and provision of the fourth physical dimension of astronomic time to quantify rates and energies involved in geologic processes. These isotopic chronometers have been used to measure the age of the Earth, Moon, and meteorites (4.5×10^9 years), the age of the oldest datable rocks (3.7×10^9 years), and many other significant geologic events such as the advance and retreat of continental glaciers. They have also been used to establish a Precambrian time scale, to calibrate the Phanerozoic time scale in solar years, and to provide a chronology for significant biologic, cultural, and environmental events related to the evolution of the human race. On a much shorter time scale, these methods have been used to determine rates of flow of water through aquifers and rates of material (aerosols) transport through the atmosphere. *See* AMINO ACID DATING; DATING METHODS; GEOLOGIC TIME SCALE. [P.E.Da.]

Geographic information systems Computer-based technologies for the storage, manipulation, and analysis of geographically referenced information. Attribute and spatial information is integrated in geographic information systems (GIS) through the notion of a data layer, which is realized in two basic data models: raster and vector. The major categories of applications comprise urban and environmental inventory and management, policy decision support, and planning; engineering and defense applications; and scientific analysis and modeling.

A geographic information system differs from other computerized information systems in two major respects. First, the information in this type of system is geographically referenced (geocoded). Second, a geographic information system has considerable capabilities for data analysis and scientific modeling, in addition to the usual data input, storage, retrieval, and output functions.

A geographic information system is composed of software, hardware, and data. The notion of data layer (or coverage) and overlay operation lies at the heart of most software designed for geographic information systems.

Two fundamental data models, the vector and raster models, embody the overlay idea in geographic information systems. In a vector geographic information system, the geometrical configuration of a coverage is stored in the form of points, arcs (line segments), and polygons, which constitute identifiable objects in the database. In a raster geographic information system, a layer is composed of an array of elementary cells of pixels, each holding an attribute value without explicit reference to the geographic feature of which the pixel is a part.

A data layer or coverage integrates two kinds of information: attribute and spatial (geographic). The functionality of a geographic information system consists of the ways in which that information may be captured, stored, manipulated, analyzed, and presented to the user. Spatial data capture (input) may be from primary sources such as remote sensing scanners, radars, or global positioning systems, or from scanning or digitizing images and maps derived from remote sensing. Output (whether as a display on a cathode-ray tube or as hard copy) is usually in map or graph form, accompanied by tables and reports linking spatial and attribute data. The critical data management and analysis functions fall into four categories: retrieval, classification, and measurement; overlay functions; neighborhood operations; and connectivity functions. [H.Co.]

Business applications of geographic information systems are increasingly widespread and include market analysis, store location, and agribusiness (for example, determining the correct amount of fertilizers or pesticides needed at each point of a cultivated field). Engineers use geographic information systems when modeling terrain, building roads and bridges, maintaining cadastral maps, routing vehicles, drilling for water, determining what is visible from any point on the terrain, integrating intelligence information on enemy targets, and so forth. Such applications have been facilitated through the integration of geographic information systems with global positioning systems.

Among the earliest and still most widespread applications of the technology are land information and resource management systems (for example, forest and utility management). Other common uses of geographic information systems in an urban policy context include emergency planning, determination of optimal locations for fire stations and other public services, assistance in crime control and documentation, and electoral and school redistricting. Uses of geographic information systems have spread well beyond geography, the source discipline, and now involve most applied sciences, both social and physical, that deal with spatial data. The nature of the applications of geographic information systems in these areas ranges from simple thematic mapping for illustration purposes to complex statistical and mathematical modeling for the exploration of hypotheses or the representation of dynamic processes. [H.Co.]

Geography The study of physical and human landscapes, the processes that affect them, how and why they change over time, and how and why they vary spatially. Geographers consider, to varying degrees, both natural and human influences on the landscape, although a common division separates human and physical geography. Physical geographers may study landforms (geomorphology), water (hydrology), climate and meteorology (climatology), the biotic environment (biogeography), or soils (pedology). Human geographers include urban, regional, and environmental planners; cultural geographers; regional and area specialists; economic geographers; political geographers; transportation analysts; location analysts; and specialists in the spatial nature of ethnic or gender issues. *See* BIOGEOGRAPHY; CLIMATOLOGY; HYDROLOGY; PEDOLOGY.

Many geographers are involved with the development of techniques and applications that support spatial analytical studies or the display of spatial information and data. Maps, whether printed, digital, or conceptual, are the basic tools of geography. Geographers are involved in map interpretation and use, as well as map production and design. Cartographers supervise the compilation, design, and development of maps, globes, and other graphic representations.

A geographic information system (GIS) is a relatively new technology that combines the advantages of computer-assisted cartography with those of spatial database management. It facilitates the storage, retrieval, and analysis of spatial information in the form of digital map "overlays," each representing a different landscape component (terrain, hydrologic features, roads, vegetation, soil types, or any mappable factor). Each of these data layers can be fitted digitally to the same map scale and map projection—in any combination—permitting the analysis of relationships among any combination of environmental variables for which data have been input into the geographic information systems. *See* GEOGRAPHIC INFORMATION SYSTEMS.

Many geographers are applied practitioners, solving problems using a variety of tools, including computer-assisted cartography, statistical methods, remotely sensed imagery, the Global Positioning System (GPS), and geographic information systems. Today, nearly all geographers, regardless of their subdisciplinary emphases, employ some or all of these techniques in their professional endeavors. *See* PHYSICAL GEOGRAPHY; TERRAIN AREAS. [J.F.P.]

Geologic thermometry The measurement or estimation of temperatures at which geologic processes take place. Methods used can be divided into two groups, nonisotopic and isotopic. Nonisotopic methods involve measurements of earth temperatures either directly by surface and near-surface features or indirectly from various properties of minerals and fossils. The isotopic methods involve the determination of distribution of isotopes of the lighter elements between pairs of compounds in equilibrium at various temperatures, and application of these data to problems

of the temperature at which these compounds (commonly minerals) form in nature.
[E.I.]

Geologic time scale An ordered, internally consistent, internationally recognized sequence of time intervals, each distinct in its own history and record of life on Earth, including the assignment of absolute time in years to each geologic period. The geologic time scale (see table) has a relative scale, consisting of named intervals of geologic history arranged in historical sequence; and a numerical (or absolute) time scale, providing absolute ages for the boundaries of these intervals.

In order to establish a geologic time scale, an independent means of dating rocks is required. Before the discovery of radioactivity, crude estimates of the length of a geologic history were made based on the total thicknesses of sedimentary rock and assumed rates of erosion and sedimentation. These estimates varied by as much as a factor of 10.

Geologic time scale

Eon / Era / Period [system] / Epoch [series]	Age at beginning of interval, 10^6 years	Interval length, 10^6 years
Phanerozoic		
Cenozoic		65
Quaternary (Q)*		1.8
Recent	0.01	0.01
Pleistocene	1.8	1.79
Tertiary (T)	65	63.2
Pliocene (Tpl)	5.3	3.5
Miocene	23.8	18.5
Oligocene (To)	33.7	9.9
Eocene (Te)	54.8	21.1
Paleocene (Tp)	65	10.2
Mesozoic	250	185
Cretaceous (K)	144	79
Jurassic (J)	206	62
Triassic (Tr)	250	44
Paleozoic	543	297
Permian (P)	290	40
Carboniferous (M, P)	354	64
Devonian	417	63
Silurian	443	26
Ordovician	490	47
Cambrian	543	53
Precambrian		
Proterozoic	2500	1957
Late (Z)† (Neoproterozoic)	900	357
Middle (Y) (Mesoproterozoic)	1600	700
Early (X) (Paleoproterozoic)	~2500	900
Archean	3800	
Late (W)	3000	500
Middle (U)	3400	400
Early (V)	>3800	>400

*In parentheses are the symbols for the periods and epochs used on geologic maps and figures in North America, as well as other parts of the world.

†Letter designations of Precambrian age intervals are used by the U.S. Geological Survey.

The modern geologic time scale is based on many measurements of various rock types by quantitative isotopic chronometers such as uranium-lead (U-Pb) and potassium-argon (K-Ar). [A.R.P.; J.W.Gei.; J.L.K.]

Geophysical fluid dynamics The branch of physics that studies the dynamics of naturally occurring large-scale flows in the atmosphere and oceans. Examples of such flows are weather patterns, atmospheric fronts, and ocean currents. The fluids are either air or water in a moderate range of temperatures and pressures.

Because of their large scale (from tens of kilometers up to the size of the planet), geophysical flows are strongly influenced by the diurnal rotation of the Earth, which is manifested in the equations of motion as the Coriolis force. Another fundamental characteristic is stratification, that is, density heterogeneity within the fluid in the presence of the Earth's gravitational field, which is responsible for buoyancy forces. Thus, geophysical fluid dynamics may be considered to be the study of rotating and stratified fluids. It is the common denominator of dynamical meteorology and physical oceanography. *See* METEOROLOGY; OCEANOGRAPHY.

The first of the two distinguishing attributes of geophysical fluid dynamics is the effect of the Earth's rotation. Because geophysical flows are relatively slow and spread over long distances, the time taken by a fluid particle (be it a parcel of air in the atmosphere or water in the ocean) to traverse the region occupied by a certain flow structure is comparable to, and often longer than, a day. Thus, the Earth rotates significantly during the travel time of the fluid, and rotational effects enter the dynamics. Fluid flows viewed in a rotating framework of reference are subject to two additional types of forces, namely the centrifugal force and the Coriolis force. (Properly speaking, these originate not as actual forces but as acceleration terms to correct for the fact that viewing the flow from a rotating frame—the rotating Earth in the case of geophysical fluid dynamics—demands a special transformation of coordinates.) Contrary to intuition, the centrifugal force plays no role on fluid motion because it is statically compensated by the tilting of the gravitational force caused by the departure of the Earth's shape from sphericity. Thus, of the two, only the Coriolis force acts on fluid parcels.

Variations of moisture in the atmosphere, of salinity in the ocean, and of temperature in either can modify the density of the fluid to such an extent that buoyancy forces become comparable to other existing forces. The fluid then has a strong tendency to arrange itself vertically so that the denser fluid sinks under the lighter fluid. The resulting arrangement is called stratification, the second distinguishing attribute of geophysical fluid dynamics. The greater the stratification in the fluid, the greater the resistance to vertical motions, and the more potential energy can affect the amount of kinetic energy available to the horizontal flow.

A quantity central to the understanding of geophysical flows, which are simultaneously rotating and stratified, is the potential vorticity, q. This quantity incorporates both rotation and stratification. Geophysical flows are replete with vortices, resulting from baroclinic instability. Their interactions generate highly complex flows not unlike those commonly associated with turbulence. Unlike classical fluid turbulence, however, geophysical flows are wide and thin (with, furthermore, a high degree of vertical rigidity as a result of rotational effects), and their turbulence is nearly two-dimensional.

In meteorology, geophysical fluid dynamics has been the key to understanding the essential properties of midlatitude weather systems, including the formation of cyclones and fronts. Geophysical fluid dynamics also explains the dynamical features of hurricanes and tornadoes, sea and land breezes, the seasonal formation and break-up of the polar vortex that is associated with high-latitude stratospheric ozone holes, and a host of other wind-related phenomena in the lower atmosphere. *See* CYCLONE; HURRICANE; TORNADO.

In oceanography, successes of geophysical fluid dynamics include the explanation of major oceanic currents, such as the Gulf Stream. Coastal river plumes, coastal upwelling, shelf-break fronts, and open-ocean variability on scales ranging from tens of kilometers to the size of the basin are among the many other marine applications. The El Niño phenomenon in the tropical Pacific is rooted in processes that fall under the scope of geophysical fluid dynamics. *See* GULF STREAM. [B.C.R.]

Geothermal power Thermal or electrical power produced from the thermal energy contained in the Earth (geothermal energy). Use of geothermal energy is based thermodynamically on the temperature difference between a mass of subsurface rock and water and a mass of water or air at the Earth's surface. This temperature difference allows production of thermal energy that can be either used directly or converted to mechanical or electrical energy.

Commercial exploration and development of geothermal energy to date have focused on natural geothermal reservoirs—volumes of rock at high temperatures (up to 662°F or 350°C) and with both high porosity (pore space, usually filled with water) and high permeability (ability to transmit fluid). The thermal energy is tapped by drilling wells into the reservoirs. The thermal energy in the rock is transferred by conduction to the fluid, which subsequently flows to the well and then to the Earth's surface.

There are several types of natural geothermal reservoirs. All the reservoirs developed to date for electrical energy are termed hydrothermal convection systems and are characterized by circulation of meteoric (surface) water to depth. The driving force of the convection systems is gravity, effective because of the density difference between cold, downward-moving, recharge water and heated, upward-moving, thermal water. A hydrothermal convection system can be driven either by an underlying young igneous intrusion or by merely deep circulation of water along faults and fractures. Depending on the physical state of the pore fluid, there are two kinds of hydrothermal convection systems: liquid-dominated, in which all the pores and fractures are filled with liquid water that exists at temperatures well above boiling at atmospheric pressure, owing to the pressure of overlying water; and vapor-dominated, in which the larger pores and fractures are filled with steam. Liquid-dominated reservoirs produce either water or a mixture of water and steam, whereas vapor-dominated reservoirs produce only steam, in most cases superheated.

Although geothermal energy is present everywhere beneath the Earth's surface, its use is possible only when certain conditions are met: (1) The energy must be accessible to drilling, usually at depths of less than 2 mi (3 km) but possibly at depths of 4 mi (6–7 km) in particularly favorable environments (such as in the northern Gulf of Mexico Basin of the United States). (2) Pending demonstration of the technology and economics for fracturing and producing energy from rock of low permeability, the reservoir porosity and permeability must be sufficiently high to allow production of large quantities of thermal water. (3) Since a major cost in geothermal development is drilling and since costs per meter increase with increasing depth, the shallower the concentration of geothermal energy the better. (4) Geothermal fluids can be transported economically by pipeline on the Earth's surface only a few tens of kilometers, and thus any generating or direct-use facility must be located at or near the geothermal anomaly.

Equally important worldwide is the direct use of geothermal energy, often at reservoir temperatures less than 212°F (100°C). Geothermal energy is used directly in a number of ways: to heat buildings (individual houses, apartment complexes, and even whole communities); to cool buildings (using lithium bromide absorption units); to heat greenhouses and soil; and to provide hot or warm water for domestic use, for product processing (for example, the production of paper), for the culture of shellfish and fish, for swimming pools, and for therapeutic (healing) purposes. [J.P.M.]

The use of geothermal energy for electric power generation has become widespread because of several factors. Countries where geothermal resources are prevalent have desired to develop their own resources in contrast to importing fuel for power generation. In countries where many resource alternatives are available for power generation, including geothermal, geothermal has been a preferred resource because it cannot be transported for sale, and the use of geothermal energy enables fossil fuels to be used for higher and better purposes than power generation. Also, geothermal steam has become an attractive power generation alternative because of environmental benefits and because the unit sizes are small (normally less than 100 MW). Moreover, geothermal plants can be built much more rapidly than plants using fossil fuel and nuclear resources, which, for economic purposes, have to be very large in size. Electrical utility systems are also more reliable if their power sources are not concentrated in a small number of large units.

The most common process is the steam flash process, which incorporates steam separators to take the steam from a flashing geothermal well and passes the steam through a turbine that drives an electric generator. A more efficient utilization of the resource can be obtained by using the binary process on resources with a temperature less than 360°F (180°C). This process is normally used when wells are pumped. The pressurized geothermal brine yields its heat energy to a second fluid in heat exchangers and is reinjected into the reservoir. The second fluid (commonly referred to as the power fluid) has a lower boiling temperature than the geothermal brine and therefore becomes a vapor on the exit of the heat exchangers. It is separately pumped as a liquid before going through the heat exchangers. The vaporized, high-pressure gas then passes through a turbine that drives an electric generator. [T.C.Hi.]

Geyser A natural spring or fountain which discharges a column of water or steam into the air at more or less regular intervals. Perhaps the best-known area of geysers is in Yellowstone Park, Wyoming, where there are more than 100 active geysers and more than 3000 noneruptive hot springs.

The eruptive action of geysers is believed to result from the existence of very hot rock not far below the surface. The neck of the geyser is usually an irregularly shaped tube partly filled with water which has seeped in from the surrounding rock. Far down the pipe the water is at a temperature much above the boiling point at the surface, because of the pressure of the column of water above it. Its temperature is constantly increasing, because of the volcanic heat source below. Eventually the superheated water changes into steam, lifting the column of water out of the hole. [A.N.S./R.K.Li.]

Giardiasis A disease caused by the protozoan parasite *Giardia lamblia*, characterized by chronic diarrhea that usually lasts 1 or more weeks. The diarrhea may be accompanied by one or more of the following: abdominal cramps, bloating, flatulence, fatigue, or weight loss. The stools are malodorous and have a pale greasy appearance. Infection without symptoms is also common.

Giardiasis occurs worldwide. In community epidemics caused by contaminated drinking water, as many as 50 to 70% of the residents have become infected. Outbreaks also occur among backpackers and campers who drink untreated stream water. Both human and animal (beaver) fecal contamination of stream water has been implicated as the source of *Giardia* cysts in waterborne outbreaks. *Giardia* species in dogs and possibly other animals are also considered infectious for humans. Epidemics resulting from person-to-person transmission occur in day-care centers for preschool-age children and institutions for the mentally retarded. Infants and toddlers in day-care centers are more commonly infected than older children who have been toilet-trained. *See* EPIDEMIOLOGY. [D.D.J.]

Glacial geology The scientific study of the effects of glaciers on the broad land areas, on the oceans, and on climate, of their erosion and deposition, and of their modification of the Earth's surface in detail. Included in the realm of glacial geology is the history of glacial theory, consideration of the origin of glacial ages, extent and times of past glaciations, erosion and sculpturing of plains and mountains, deposition of ice-contact and meltwater sediments, and the consequences of glaciers on worldwide climate, and also on local climate around their edges. Quite distinct from glacial geology, however, is the separate, growing subscience of glaciology, the study of glaciers themselves. *See* GLACIOLOGY.

Features on the Earth's surface explained by former worldwide glaciation are numerous, embracing, for example, glacially eroded and molded valleys and mountains; ice-transported and deposited sediments and nonglacial sediments; abandoned stream channels with associated floodwater deposits; elevated silts and clays that collected around continental edges when sea level was higher; valleys eroded across and into continental shelves and slopes when sea level was much lower; communities of plants and animals similar to each other but separated by shallow seaways where land bridges once existed; fossil shells and microorganisms in deep-sea sediments reflecting colder or warmer water temperatures than today; vegetated sand dunes aligned to wind systems no longer operating; ancient shorelines and beach ridges ringing dry empty lake basins far inland; and orderly patterns of stones and fine sediments next to glacier margins in polar regions and high mountains. *See* DRUMLIN; MORAINE. [S.E.Wh.]

Glaciology A broad field encompassing all aspects of the study of ice, specifically glaciers, the largest ice masses on Earth.

Glaciers are classified principally on the basis of size, shape, and temperature. Cirque glaciers occupy spectacular steep-walled, overdeepened basins a few square kilometers ($1\ km^2 = 0.36\ mi^2$) in area, called cirques. Most cirques are in high mountain areas that have been repeatedly inundated by ice. The cirques and the deep valleys leading away from them were, in fact, eroded by larger glaciers over the past 3 million years.

As a cirque glacier expands, it is usually constrained, at least initially, to move down such a valley. It then becomes a valley glacier (see illustration). Where such a valley ends in a deep fiord in the sea, the glacier is called a tidewater glacier. *See* FIORD.

In contrast, some glaciers are situated on relatively flat topography. Such glaciers can spread out in all directions from a central dome. When small, on the order of a few tens of kilometers across, these are called ice caps. Large ones, like those in Antarctica and Greenland, are ice sheets. *See* ANTARCTICA; ICE FIELD.

Thermally, glaciers are usually classified as either temperate or polar. In the simplest terms, a temperate glacier is one that is at the melting point throughout. The term melting point is used in this context rather than 0°C (32°F), because the temperature at which ice melts decreases as the pressure increases. Thus, the temperature at the base of a temperate glacier that is 500 m (1700 ft) thick will be about −0.4°C (31.2°F), but if heat energy is added to the ice, it can melt without an increase in temperature. Most valley glaciers are temperate.

In polar glaciers the temperature is below the melting point nearly everywhere. The temperature of a polar glacier increases with depth, however, because the deeper ice is warmed by heat escaping from within the Earth and by frictional heat generated by

Storglaciären, a small valley glacier in northern Sweden.

deformation of the ice. Thus, at its base, a polar glacier may be frozen to the substrate or may be at the melting point. Ice caps and ice sheets are normally polar, as are some valley glaciers in high latitudes.

As was the case with the classification based on size and shape, there is a continuum of thermal regimes in glaciers. The most common intermediate type has a surficial layer of cold ice, a few tens of meters (1 m = 3.28 ft) thick in its lower reaches, but is temperate elsewhere. Such glaciers are sometimes called subpolar or polythermal.

Glaciers exist because there are places where the climate is so cold that some or all of the winter snow does not melt during the following summer. The next winter's snow then buries that remaining from the previous winter, and over a period of years a thick snow pack or snowfield develops. Deep in such a snow pack the snow is compacted by the weight of the overlying snow. In addition, evaporation of water molecules from the tips of snowflakes and condensation of this water in intervening hollows results in rounding of grains. These processes of compaction and metamorphism gradually transform the deeper snow, normally known as firn, into ice. Melt water percolating downward into this firn may refreeze, accelerating the transformation.

When, during a given year, the mass of snow added in the accumulation area of a glacier exceeds the mass of ice lost from the ablation area, the glacier is said to have had a positive mass balance. If such a situation persists for several years, the glacier will advance to lower elevations or more temperate latitudes, thus increasing the size of its ablation area and the mass loss. Conversely, persistent negative mass balances lead to retreat. Contrary to one implication of the word retreat, a retreating glacier does not flow backward. Rather, a glacier retreats when the ice flow toward the terminus is less than the melt rate at the terminus.

Under certain rather rare circumstances, the high subglacial water pressures that develop in the spring do not dissipate quickly but persist for weeks. This occurs under glaciers that have been thickening for several years or decades but have not advanced appreciably as a result of the thickening. On these occasions, the increase in sliding speed resulting from the increased water pressure inhibits development of an integrated subglacial conduit system, so water pressures remain high. The glacier then may advance at speeds of meters to tens of meters per day, in what is known as a surge.

Ice stream flow occurs in some parts of the Antarctic Ice Sheet. These high flow rates occur in streams of ice tens of kilometers wide and hundreds of kilometers long, and are sustained for centuries. These streams are bounded not by valley sides but by ice that is moving much more slowly. Ice Stream B, for example, which drains to the Ross Ice Shelf in West Antarctica, has a maximum speed of 825 m/yr, while ice on either side of it is moving at only 10–20 m/yr. The high speeds of these ice streams are attributed to slippery conditions at the bed, where high water pressures reduce friction between the ice and the bed. Changes in paths of water flow at the bed are believed to be responsible for the changes in ice stream activity.

Among the hazards associated with glaciers are jökulhlaups, or sudden releases of water from lakes dammed by glaciers. Jökulhlaup is an Icelandic word; it has entered the vocabulary of geology because such floods are common in Iceland where localized volcanic heat is responsible for the presence of deep lakes surrounded by ice. In other regions, the lakes are more commonly formed where a glacier in a trunk valley extends across the open mouth of a tributary valley. *See* GLACIAL GEOLOGY. [R.LeB.H.]

Glanders A contagious zoonosis affecting primarily horses, mules, and donkeys and caused by the bacterium *Burkholderia (Pseudomonas) mallei*. Glanders (farcy) was once common throughout the world but is now found only in parts of Africa, Russia, and Asia. *Burkholderia mallei* is a gram-negative, non-acid-fast, nonsporulating, nonmotile, unencapsulated bacillus occasionally showing bipolar staining; it is obligately aerobic and oxidase-positive. *Burkholderia mallei* is highly infectious for humans, who may acquire it by handling or treating glanderous animals or during laboratory investigations. *See* ZOONOSES.

Glanders is usually contracted by ingestion of contaminated food or water, by contact, and by inhalation of infectious droplets. All equids are highly susceptible. The disease is usually acute and often fatal in donkeys and mules, and chronic in horses, some of which may ultimately recover but continue to carry *B. mallei*. It is characterized by formation of nodules and ulcerations of the skin and respiratory membranes and by granulomatous nodules in the lungs, lymphatic channels, and lymph nodes.

Although *B. mallei* is sensitive to sulfonamides and tetracyclines, affected horses are not usually treated since destruction of cases has been found to be extremely effective in control and eradication. Essential components of diagnosis include clinical examinations at frequent intervals to detect the cutaneous and nasal forms; immunological tests to detect serum antibody; and skin and intradermopalpebral (within the skin of the eyelid) injection of mallein, a glycoprotein of *B. mallei*, to detect hypersensitivity. [J.F.T.]

Global climate change The periodic fluctuations in global temperatures and precipitation, such as the glacial (cold) and interglacial (warm) cycles of the Pleistocene (a geological period from 1.8 million to 10,000 years ago). Presently, the increase in global temperatures since 1900 is of great interest. Many atmospheric scientists and meteorologists believe it is linked to human-produced carbon dioxide (CO_2) in the atmosphere.

Greenhouse effect. The greenhouse effect is a process by which certain gases (water vapor, carbon dioxide, methane, nitrous oxide) trap heat within the Earth's atmosphere and thereby produce warmer air temperatures. These gases act like the glass of a greenhouse: they allow short (ultraviolet; UV) energy waves from the Sun to penetrate into the atmosphere, but prevent the escape of long (infrared) energy waves that are emitted from the Earth's surface. *See* ATMOSPHERE; GREENHOUSE EFFECT.

Human-induced changes in global climate caused by release of greenhouse gases into the atmosphere, largely from the burning of fossil fuels, have been correlated with global warming. Since 1900, the amount of two main greenhouse gases (carbon dioxide and methane) in the Earth's atmosphere has increased by 25%. Over the same period, mean global temperatures have increased by about 0.5°C (0.9°F). The most concern centers on carbon dioxide. Not only is carbon dioxide produced in much greater quantities than any other pollutant, but it remains stable in the atmosphere for over 100 years. Methane, produced in the low-oxygen conditions of rice fields and as a by-product of coal mining and natural gas use, is 100 times stronger than carbon dioxide in its greenhouse effects but is broken down within 10 years.

Chloroflurocarbon (CFC) pollution, from aerosol propellants and coolant systems, affects the Earth's climate because CFCs act as greenhouse gases and they break down the protective ozone (O_3) layer. Other pollutants released into the atmosphere are also likely to influence global climate. Sulfur dioxide (SO_2) from car exhaust and industrial processes, such as electrical generation from coal, cool the Earth's surface air temperatures and counteract the effect of greenhouse gases. Nevertheless, there have been attempts in industrialized nations to reduce sulfur dioxide pollution because it also causes acid rain. *See* AIR POLLUTION.

Possible impact. A rise in mean global temperatures is expected to cause changes in global air and ocean circulation patterns, which in turn will alter climates in different regions. Changes in temperature and precipitation have already been detected. In the United States, total precipitation has increased, but it is being delivered in fewer, more extreme events, making floods (and possibly droughts) more likely. *See* OCEAN CIRCULATION.

Global warming has caused changes in the distribution of a species throughout the world. By analyzing preserved remains of plants, insects, mammals, and other organisms which were deposited during the most recent glacial and interglacial cycles, scientists have been able to track where different species lived at times when global temperatures were either much warmer or much cooler than today's climate. Several studies have documented poleward and upward shifts of many plant and insect species during the current warming trend.

Changes in the timing of growth and breeding events in the life of an individual organism, called phenological shifts, have resulted from global warming. For example, almost one-third of British birds are nesting earlier (by 9 days) than they did 25 years ago, and five out of six species of British frog are laying eggs 2–3 weeks earlier.

Community reassembly, changes in the species composition of communities, has resulted from climate change because not all species have the same response to environmental change.

To date, there have been no extinctions of species directly attributable to climate change. However, there is mounting evidence for drastic regional declines. For example, the abundance of zooplankton (microscopic animals and immature stages of many species) has declined by 80% off the California coast. This decline has been related to gradual warming of sea surface temperatures. *See* CLIMATE HISTORY; CLIMATE MODIFICATION; EXTINCTION (BIOLOGY). [C.Pa.]

Globe (Earth) A sphere on the surface of which is a map of the world. The map may be drawn, engraved, or painted directly on the surface but is more commonly prepared as a series of gores, or segments in other designs, to be affixed to the globe ball (see illustration).

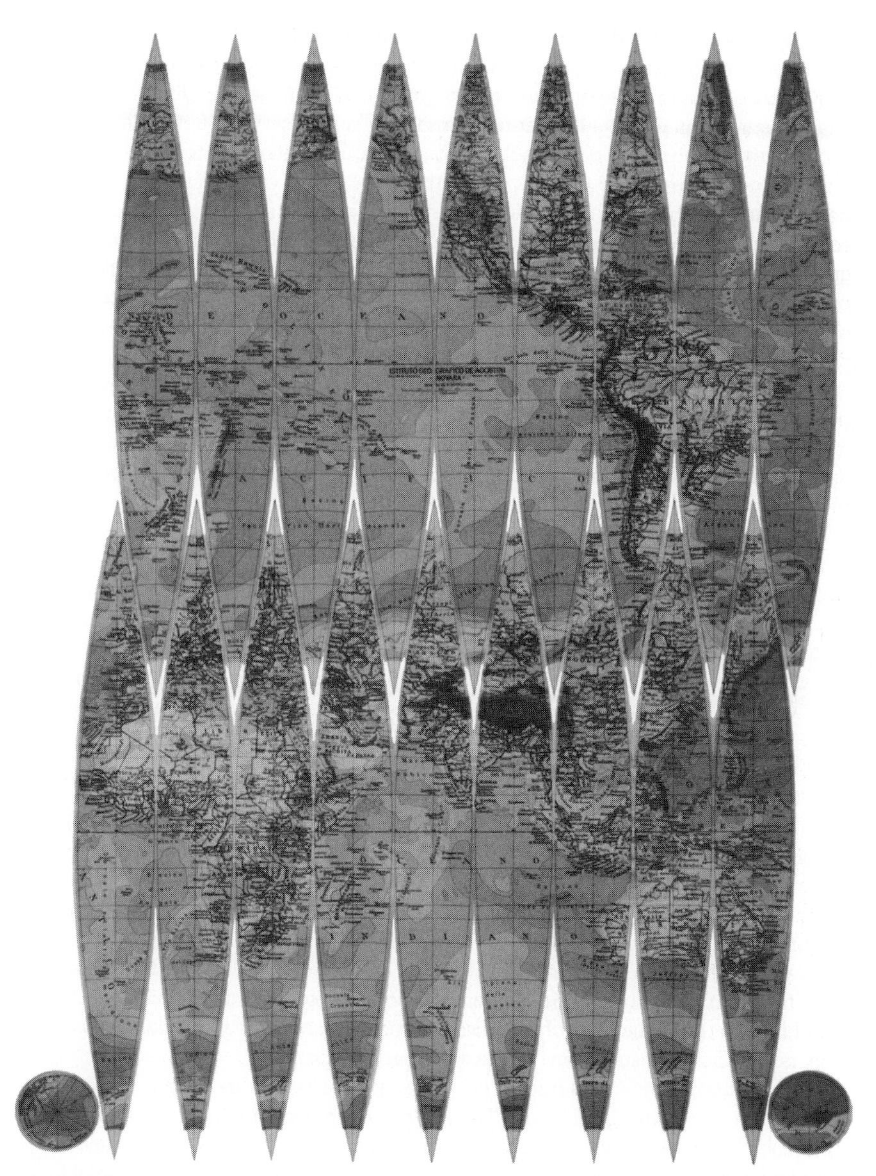

Globe gores from collections of Library of Congress. (*Istituto Geografico de Agostini, Novara, Italy*)

Globes are both artistically interesting and scientifically useful. Their principal value is in stimulating sound concepts of worldwide patterns and in rectifying errors induced by the limitations of flat maps. All flat maps distort the Earth's surface patterns, but carefully made globes constitute truer scale models of the Earth, with correct areas, shapes, and distances as well as continuity of surface. Globes have long been used as aids in navigation, in the teaching of earth sciences, and as room ornaments.

Many modern globes have special attachments to improve their utility. A meridian ring, extending from pole to pole, may be calibrated in degrees to measure latitude.

The longitude of points directly beneath that ring will be indicated at the intersection of the ring with the equatorial scale. A horizon ring at right angles to the meridian ring may be calibrated in miles or in meters, degrees, and hours to expedite distance and time measurement. A hinged horizon ring may be lifted to serve as a meridian ring, or placed in an oblique position to show great circle routes and distances. [A.C.G.]

Gonorrhea A common sexually transmitted disease caused by the bacterium *Neisseria gonorrhoeae*. Humans are the only natural hosts for *N. gonorrhoeae*, which directly infects the epithelium of the mucous membranes of the human genital tract, pharynx, rectum, or conjunctiva. Local epithelial cell destruction usually occurs, but the organisms may spread to adjacent organs or disseminate via the bloodstream. In women, local complications include inflammation of the uterine lining (endometritis), inflammation of the fallopian tube (salpingitis), inflammation of the abdominal wall (peritonitis), and inflammation of Bartholin's glands (bartholinitis); in men, periurethral abscess and inflammation of a duct connected to the testes (epididymitis). Systemic manifestations such as arthritis or dermatitis may develop, and rarely endocarditis or meningitis.

Women are disproportionately affected by the complications of gonorrhea. Acute pelvic inflammatory disease and salpingitis, the most serious complications of gonorrhea, result in ectopic pregnancy and infertility. Gonococcal infection during pregnancy may also predispose women to premature rupture of membranes, delivery in less than full term, and postpartum endometritis. During childbirth, the gonococcus may infect the conjunctiva of the infant and result in the infection ophthalmia neonatorum. This infection is a serious complication that remains common in less developed countries and can lead to permanent damage to the eye and blindness.

Gonorrhea continues to be the most commonly reported communicable disease in the United States, although incidence has declined since 1984. Risk factors that may influence the probability of infection include number of sexual partners, lack of barrier contraceptives, and young age.

Gonorrhea is an infection spread by physical contact with the mucosal surfaces of an infected person, usually a sexual partner. The risk of infection depends on the anatomic site, the amount of substance containing bacteria, and the number of exposures. Variations in host susceptibility have not been well defined. In a small but significant proportion of infections, there are no symptoms. These individuals are important in the epidemiology of this disease because gonorrhea is usually spread by carriers who have no symptoms or have ignored symptoms.

Control of gonorrhea depends on early diagnosis, effective treatment, and identification of asymptomatic individuals. The last has been accomplished, in part, through screening programs. However, complete control has not been possible because of the emergence and spread of strains that are resistant to less-expensive antimicrobial treatments such as penicillin and tetracycline.

There is no evidence that infected individuals develop long-lasting immunity to reinfection, and vaccination is not available. Thus, the prevention of gonorrhea relies on behavior modification and risk reduction, use of appropriate screening and diagnostic tests, routine use of highly effective antibiotics, early identification and treatment of sexual partners of individuals with gonorrhea, and the appropriate use of barrier methods such as condoms.

An increasing proportion of infections are due to antibiotic-resistant strains of *N. gonorrhoeae*. Chromosomally mediated resistance to multiple antibiotics as well as plasmid-mediated resistance to beta-lactam antibiotics and tetracycline occurs in strains from both developed and developing countries. Nevertheless, infections can be effectively treated with third-generation cephalosporins (for example, ceftriaxone) or fluoroquinolones (for example, ciprofloxacin or ofloxacin). *See* SEXUALLY TRANSMITTED DISEASES. [S.A.L.; S.A.Mo.]

Grain crops Crop plants that belong to the grass family (Gramineae), generally grown for their edible starchy seeds. They also are referred to as cereal crops and include wheat, rice, maize (corn), barley, rye, oats, sorghum (jowar), and millet. The grain of all these is used directly for human food and also for livestock, especially maize, barley, oats, and sorghum.

An important attribute of these grain crops is the easy manner in which they can be stored. The grain often dries naturally before harvest to a safe moisture content (10–12%), or can easily be dried with modern equipment. Grain placed in adequate storage facilities can then be protected against insect infestations and maintained in sound condition for years. *See* CORN; RICE; WHEAT. [E.G.H.]

Grass crops Members of the family Gramineae cultivated as forage and grain for consumption. The grasses are the most useful of all the plants that cover the Earth. The cereal grasses supply directly three-fourths of the energy and over half of the protein in food consumed by humans. Indirectly, these cereals together with the forage grasses supply most of the food for the domestic animals that provide milk, meat, eggs, and much of the draft power required to grow crops. *See* CEREAL; CORN; RICE; WHEAT.

The bamboos are of vast importance in the Indo-Malay region, where they are used in building houses, bridges, furniture, rafts, water pipes, vessels for holding water, and so forth. *See* BAMBOO.

The grasses protect soil from erosion and help conserve water resources. More than any other family of plants, the sod-forming grasses blanket golf courses, athletic fields, lawns, parks, and cemeteries with a protective covering that beautifies and enhances the environment. No other family of plants in the vast plant kingdom is so useful to humans. *See* EROSION; SOIL CONSERVATION.

Grass stems have solid joints (nodes) and leaves arranged in two rows, with one leaf at each joint (see illustration). The leaves consist of the sheath, which fits around the stem like a split tube, and the blade, which is commonly long and narrow. Seed heads are made up of minute flowers on tiny branchlets, often several crowded together, but always two-ranked like the leaves. The flowers are generally wind-pollinated. The seeds are enclosed between two bracts, or glumes, which remain on the seed when ripe.

The 600 genera grouped into 14 tribes that make up the grass family may be annual or perennial. Annuals and some perennial grasses are bunch grasses which spread only by seeding. Others, mostly perennials, also spread by creeping stems called stolons when above ground and rhizomes when below the soil surface. The creeping grasses form the best sods and surpass others for soil conservation; they are also the best turf grasses. All grasses have fine fibrous root systems that permeate the soil extensively to depths ranging from much less than 3 ft to more than 10 ft (1 to 3 m). The roots are short-lived, are continually being replaced, and in the process increase the organic matter content of the soil.

Grasses are distributed throughout the world. Annual species predominate in the adverse environments found in the deserts and the arctic areas. Temperature is the principal factor determining the distribution of perennial grasses. Perennial grasses are frequently classed as cool- or warm-season grasses depending upon the season in which they make their best growth.

Annuals and most perennial grasses reproduce sexually and are propagated by seed. Many of the grasses are cross-pollinated largely by the wind. In some species,

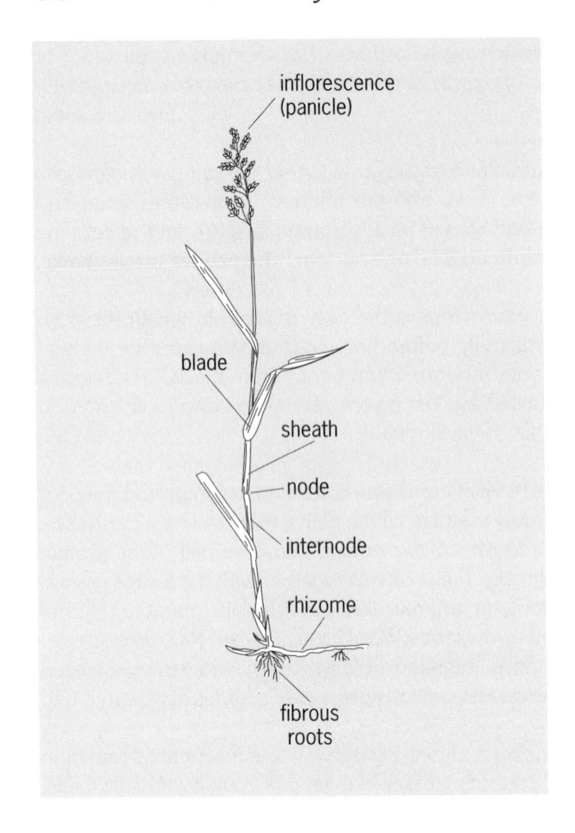

A typical grass plant. (*After P. D. Strausbaugh and E. L. Core, Flora of West Virginia, West Va. Univ. Bull., ser. 52, no. 12–2, pt. 1, p. 67, 1952*)

cross-pollination is facilitated by self-incompatibility that occurs at variable frequencies. Most grasses produce perfect flowers containing both male and female organs. The male organs (anthers) must be carefully removed before they shed pollen to make controlled hybrids. A few species, largely tropical perennials, reproduce by apomixis, simply defined as vegetative reproduction through the seed. Apomictic seeds produce the same genotype as the plant that produced them. *See* FLOWER; POLLINATION; REPRODUCTION (PLANT). [G.W.Bur.]

Grassland ecosystem A biological community that contains few trees or shrubs, is characterized by mixed herbaceous (nonwoody) vegetation cover, and is dominated by grasses or grasslike plants. Mixtures of trees and grasslands occur as savannas at transition zones with forests or where rainfall is marginal for trees. About 1.2×10^8 mi^2 (4.6×10^7 km^2) of the Earth's surface is covered with grasslands, which make up about 32% of the plant cover of the world. In North America, grasslands include the Great Plains, which extend from southern Texas into Canada. The European meadows cross the subcontinent, and the Eurasian steppe ranges from Hungary eastward through Russia to Mongolia; the pampas cover much of the interior of Argentina and Uruguay. Vast and varied savannas and velds can be found in central and southern Africa and throughout much of Australia. *See* SAVANNA.

Grasslands occur in regions that are too dry for forests but that have sufficient soil water to support a closed herbaceous plant canopy that is lacking in deserts. Thus, temperate grasslands usually develop in areas with 10–40 in. (25–100 cm) of annual precipitation, although tropical grasslands may receive up to 60 in. (150 cm). Grasslands are found primarily on plains or rolling topography in the interiors of great land masses, and from sea level to elevations of nearly 16,400 ft (5000 m) in the Andes. Because of their continental location they experience large differences in seasonal climate and wide ranges in diurnal conditions. In general, there is at least one dry season during the year, and drought conditions occur periodically. *See* DROUGHT; PLANT-WATER RELATIONS.

Significant portions of the world's grasslands have been modified by grazing or tillage or have been converted to other uses. The most fertile and productive soils in the world have developed under grassland, and in many cases the natural species have been replaced by cultivated grasses (cereals). *See* CEREAL; GRASS CROPS.

Different kinds of grasslands develop within continents, and their classification is based on similarity of dominant vegetation, presence or absence of specific dominant species, or prevailing climate conditions.

The climate of grasslands is one of daily and seasonal extremes. Deep winter cold does not preclude grasslands since they occur in some of the coldest regions of the world. However, the success of grasslands in the Mediterranean climate shows that marked summer drought is not prohibitive either. In North America, the rainfall gradient decreases from an annual precipitation of about 40 in. (100 cm) along the eastern border of the tallgrass prairie at the deciduous forest to only about 8 in. (20 cm) in the shortgrass prairies at the foothills of the Rocky Mountains. A similar pattern exists in Europe. Growing-season length is determined by temperature in the north latitudes and by available soil moisture in many regions, especially those adjacent to deserts. Plants are frequently subjected to hot and dry weather conditions, which are often exacerbated by windy conditions that increase transpirational water loss from the plant leaves.

Soils of mesic temperate grasslands are usually deep, about 3 ft (1 m), are neutral to basic, have high amounts of organic matter, contain large amounts of exchangeable bases, and are highly fertile, with well-developed profiles. The soils are rich because rainfall is inadequate for excessive leaching of minerals and because plant roots produce large amounts of organic material. With less rainfall, grassland soils are shallow, contain less organic matter, frequently are lighter colored, and may be more basic. Tropical and subtropical soils are highly leached, have lower amounts of organic material because of rapid decomposition and more leaching from higher rainfall, and are frequently red to yellow.

Grassland soils are dry throughout the profile for a portion of the year. Because of their dense fibrous root system in the upper layers of the soil, grasses are better adapted than trees to make use of light rainfall showers during the growing season. When compared with forest soils, grassland soils are generally subjected to higher temperatures, greater evaporation, periodic drought, and more transpiration per unit of total plant biomass. *See* BIOMASS; SOIL.

Throughout the year, flowering plants bloom in the grasslands with moderate precipitation, and flowers bloom after rainfall in the drier grasslands. With increasing aridity and temperature, grasslands tend to become less diverse in the number of species; they support more warm-season species; the complexity of the vegetation decreases; the total above-ground and below-ground production decreases; but the ratio of above-ground to below-ground biomass becomes smaller.

There are many more invertebrate species than any other taxonomic group in the grassland ecosystem. Invertebrates play several roles in the ecosystem. For example, many are herbivorous, and eat leaves and stems, whereas others feed on the roots of plants. Earthworms process organic matter into small fragments that decompose rapidly, scarab beetles process animal dung on the soil surface, flies feed on plants and are pests to cattle, and many species of invertebrates are predaceous and feed on other invertebrates. Soil nematodes, small nonarthropod invertebrates, include forms that are herbivorous, predaceous, or saprophagous, feeding on decaying organic matter. *See* SOIL ECOLOGY.

Most of the reptiles and amphibians in grassland ecosystems are predators. Relatively few bird species inhabit the grassland ecosystem, although many more species are found in the flooding pampas of Argentina than in the dry grasslands of the western United States. Their role in the grassland ecosystem involves consumption of seeds, invertebrates, and vertebrates; seed dispersal; and scavenging of dead animals.

Small mammals of the North American grassland include moles, shrews, gophers, ground squirrels, and various species of mice. Among intermediate-size animals are the opossum, fox, coyote, badger, skunk, rabbit, and prairie dog; large animals include various types of deer and elk. The most characteristic large mammal species of the North American grassland is the bison, although many of these animals were eliminated in the late 1800s. Mammals include both ruminant (pronghorns) and nonruminant (prairie dogs) herbivores, omnivores (opossum), and predators (wolves).

Except for large mammals and birds, the animals found in the grassland ecosystem undergo relatively large population variations from year to year. These variations, some of which are cyclical and others more episodic, are not entirely understood and may extend over several years. Many depend upon predator–prey relationships, parasite or disease dynamics, or weather conditions that influence the organisms themselves or the availability of food, water, and shelter. *See* POPULATION ECOLOGY.

Within the grassland ecosystem are enormous numbers of very small organisms, including bacteria, fungi, algae, and viruses. From a systems perspective, the hundreds of species of bacteria and fungi are particularly important because they decompose organic material, releasing carbon dioxide and other gases into the atmosphere and making nutrients available for recycling. Bacteria and some algae also capture nitrogen from the atmosphere and fix it into forms available to plants. *See* NITROGEN FIXATION; SOIL MICROBIOLOGY.

Much of the grassland ecosystem has been burned naturally, probably from fires sparked by lightning. Human inhabitants have also routinely started fires intentionally to remove predators and undesirable insects, to improve the condition of the rangeland, and to reduce cover for predators and enemies; or unintentionally. Thus, grasslands have evolved under the influences of grazing and periodic burning, and the species have adapted to withstand these conditions. If burning or grazing is coupled with drought, however, the grassland will sustain damage that may require long periods of time for recovery by successional processes. *See* AGROECOSYSTEM; ECOLOGICAL SUCCESSION; ECOSYSTEM. [P.G.Ri.]

Great circle, terrestrial A circle or near-circle representing a trace on the Earth's surface of a plane that passes through the center of the Earth and divides it into equal halves (see illustration). The Equator is a great circle, the trace of the plane that bisects and is perpendicular to the Earth's axis. Planes through the Poles cut the Earth along meridians. All meridians are great circles; actually, they are not quite circular because of the slightly flattened Earth. The equatorial diameter is 1.0034 times the size of the polar diameter. All parallels other than the Equator are called small circles, being smaller than a great circle.

Two common methods can be used to calculate the distance of a great circle arc. One method uses trigonometric functions: $\cos D = \sin a \sin b + \cos a \cos b \cos c$. Here D is the arc distance between points A and B in degrees, a is the latitude of A, b is the latitude of B, and c is the difference in longitude between A and B. After D is calculated,

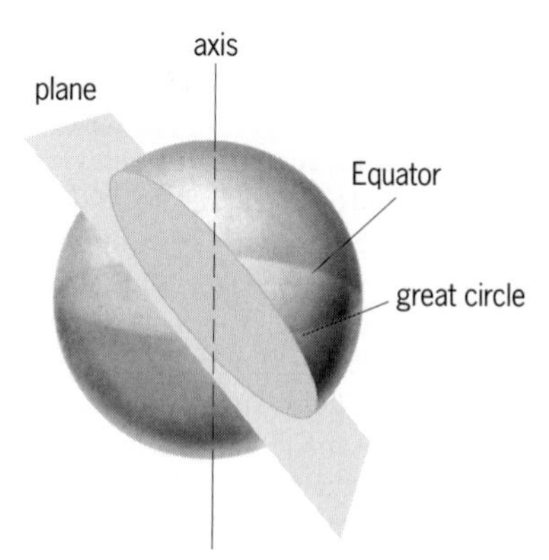

Diagram of a great circle described by a plane through the center of the Earth.

it can be converted to a linear distance measure by multiplying *D* by the length of one degree of the Equator, which is 111.32 km or 69.17 mi.

The other method uses the azimuthal equidistant projection. Unlike the gnomonic projection, the azimuthal equidistant projection can be centered at any point on the Earth's surface and can show the entire sphere. More importantly, a straight line from the center of the projection to any other point is a great circle route and the distances are at a comparable (consistent) scale between the two points. The azimuthal equidistant projection is therefore useful in showing any movement directed toward or away from a center, such as seismic waves, radio transmissions, missiles, and aircraft flights. [K.t.C.]

Greenhouse effect The ability of a planetary atmosphere to inhibit heat loss from the planet's surface, thereby enhancing the surface warming that is produced by the absorption of solar radiation. For the greenhouse effect to work efficiently, the planet's atmosphere must be relatively transparent to sunlight at visible wavelengths so that significant amounts of solar radiation can penetrate to the ground. Also, the atmosphere must be opaque at thermal wavelengths to prevent thermal radiation emitted by the ground from escaping directly to space. The principle is similar to a thermal blanket, which also limits heat loss by conduction and convection. In recent decades the term has also become associated with the issues of global warming and climate change induced by human activity. *See* ATMOSPHERE.

Basic understanding of the greenhouse effect dates back to the 1820s, when the French mathematician and physicist Joseph Fourier performed experiments on atmospheric heat flow and pondered the question of how the Earth stays warm enough for plant and animal life to thrive; and to the 1860s, when the Irish physicist John Tyndall demonstrated by means of quantitative spectroscopy that common atmospheric trace gases, such as water vapor, ozone, and carbon dioxide, are strong absorbers and emitters of thermal radiant energy but are transparent to visible sunlight. It was clear to Tyndall that water vapor was the strongest absorber of thermal radiation and, therefore, the most influential atmospheric gas controlling the Earth's surface temperature. The principal components of air, nitrogen and oxygen, were found to be radiatively inactive, serving instead as the atmospheric framework where water vapor and carbon dioxide can exert their influence.

The impact of water vapor behavior was noted by the American geologist Thomas Chamberlin who, in 1905, described the greenhouse contribution by water vapor as a positive feedback mechanism. Surface heating due to another agent, such as carbon dioxide or solar radiation, raises the surface temperature and evaporates more water vapor which, in turn, produces additional heating and further evaporation. When the heat source is taken away, excess water vapor precipitates from the atmosphere, reducing its contribution to the greenhouse effect to produce further cooling. This feedback interaction converges and, in the process, achieves a significantly larger temperature change than would be the case if the amount of atmospheric water vapor had remained constant. The net result is that carbon dioxide becomes the controlling factor of long-term change in the terrestrial greenhouse effect, but the resulting change in temperature is magnified by the positive feedback action of water vapor.

Besides water vapor, many other feedback mechanisms operate in the Earth's climate system and impact the sensitivity of the climate response to an applied radiative forcing. Determining the relative strengths of feedback interactions between clouds, aerosols, snow, ice, and vegetation, including the effects of energy exchange between the atmosphere and ocean, is an actively pursued research topic in current climate modeling. *See* CLIMATE MODIFICATION. [A.A.L.]

Greenhouse technology Along with low tunnels and high tunnels, greenhouses are structures used to grow plants under protected conditions. The progression of terms shows the level (low to high) of technical sophistication in the plant-growing systems. Low tunnels, also called row covers, primarily advance the growing season for outdoor crops (for example, tomatoes, melons, strawberries, and sweet corn). Low tunnels are created using long, narrow strips of transparent plastic material (often polyethylene) buried in the ground along their outer edges to cover one or several adjacent rows of plants grown directly in the soil. High tunnels are large versions of low tunnels, raised sufficiently above the ground that people can walk within them. Greenhouse (or glass-houses) are relatively permanent structures (usually glass or plastic with aluminum or steel frames) equipped with several means of environmental modification.

Construction. Free-standing greenhouses are the most basic structural type. Cross-sectional shapes can be classified as arch, hoop, or gable (see illustration). Multispan greenhouses are typically connected by a series of roof gutters to create a single air space. Large multispan greenhouses can cover several hectares under one roof, and they are the design of choice for larger commercial greenhouse operators. Floors are frequently made of concrete, although gravel floors with concrete walkways may be used to reduce cost.

Light transmittance is important when selecting a covering material. Glass provides the most light to the plants and retains its light transmittance; however, various rigid and film plastic glazing materials are used because of initial lower costs.

Environment control. Environment control typically encompasses air temperature, supplemental light, air movement (circulation and mixing), and carbon dioxide concentration. Some degree of relative humidity control may also be included.

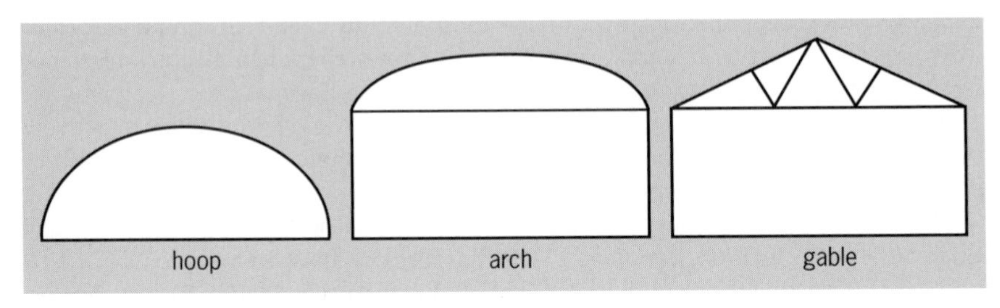

Common commercial greenhouse shapes.

Integrated control by computer, found in most modern greenhouses, provides the flexibility of zoned control of each environmental parameter without conflicting control signals (for example, ventilating while supplementing carbon dioxide).

Structural insulation opportunities in greenhouses are minimal and heat requirements are high in comparison to most other types of buildings. Efforts to conserve heat, such as insulating the north wall and part of the north roof, have often shown negative benefits by reducing natural light and degrading plant growth and quality. Movable, horizontal, indoor curtain systems (which also double as movable shade systems) can save approximately one-quarter of the yearly heat in a cold climate.

Greenhouses can be heated by oil or natural gas. Heat delivery is either hydronic (hot water) or by steam.

Solar loads in greenhouses are so great that mechanical cooling, as by air conditioning, is prohibitively expensive. Options for greenhouse cooling are thus limited. The most typical cooling mechanism is ventilation, either natural or mechanical. The next step of cooling is to use evaporative means. Cooling can be obtained by spraying a fine mist into the ventilation air or by pulling outside air through matrices or structures that are wetted to cool the air flowing past them.

Greenhouse lighting may be used for photoperiodic reasons, or for enhancing growth. Photoperiodic lighting is a very low intensity light during the night to break the darkness period and induce plant responses representative of summer (short nights and long days). Greenhouse supplemental lighting is usually provided by high-pressure sodium (HPS) lights because of their relatively high energy efficiency. *See* PHOTOPERIODISM; PHOTOSYNTHESIS.

Optimum concentrations of carbon dioxide are often in the range 800–1000 ppm, which can lead to 25% greater growth provided that other inputs are not limited. Carbon dioxide can be added through carefully controlled flue gases from the greenhouse heating system, or from tanks of liquid carbon dioxide.

Mechanization and automation. Many greenhouse operations that were formerly done by hand are now mechanized or automated. Root medium is mixed, fertilized, and placed directly into flats and pots by machine. Seeding and transplanting can be by machine. Plant watering and fertilizing (termed "fertigation" when combined) can now be automated. Automatic material movement at harvest, coordinated by a computer, is no longer unusual.

Nutrition management and hydroponics. Plant fertilizers are composed of a mix of salts that are electrically conducting when dissolved into water. This characteristic leads to the use of electrical conductivity as a measure of fertilizer concentration. Computer programs have been developed that are suitable for balancing a nutrient mix to achieve close approximations to the desired molar ratios of elements. *See* PLANT MINERAL NUTRITION.

Hydroponics is defined as growing plants without using soil. However, a root medium such as sand, gravel, or rockwool may be used. Two common hydroponics systems that use no root medium are the nutrient film technique (NFT) and deep flow troughs (DFT).

Economics. Modern greenhouse technologies have mirrored developments in most of agriculture in that increased labor efficiency, larger sizes of greenhouse operations, and mass production of a few crops, or even a single crop, have become the rule to be profitable. The current dynamic in the greenhouse industry in the United States is characterized by the entry of many growers in small, specialized operations, and consolidations and mergers of large operations. *See* FLORICULTURE; PLANT GROWTH; PLANT-WATER RELATIONS.

[L.D.A.]

Gregarinia A subclass of the class Telosporea. These protozoans occur principally as extracellular parasites in the digestive tracts and body cavities of invertebrates. There are three orders: the Archigregarinida, whose life cycle embraces both sexual and asexual phases; the Eugregarinida, which increases only by sporogony; and the Neogregarinida, whose life cycle involves schizogony and gamont formation. [E.R.B./N.D.L.]

Ground-water hydrology The occurrence, circulation, distribution, and properties of any liquid water residing beneath the surface of the earth. Generally ground water is that fraction of precipitation which infiltrates the land surface and subsequently moves, in response to various hydrodynamic forces, to reappear once again as seeps or in a more obvious fashion as springs. Most of ground-water discharge is not evident because it occurs through the bottoms of surface water bodies.

Ground water can be found, at least in theory, in any geological horizon containing interconnected pore space. Thus a ground-water reservoir (an analogy to an oil reservoir) can be a classical porous medium, such as sand or sandstone; a fractured, relatively impermeable rock, such as granite; or a cavernous geologic horizon, such as certain limestone beds. Ground-water reservoirs which readily yield water to wells are known as aquifers; in contrast, aquitards are formations which do not normally provide adequate water supplies, and aquicludes are considered, for all practical purposes, to be impermeable. These terms are, of course, subjective descriptions; the flow of water which constitutes an economically viable supply depends upon the intended use and the availability of alternative sources. *See* AQUIFER.

To effectively utilize ground water as a natural resource, it is necessary to be able to forecast the impact of exploitation on water availability. When ground water is used for water supply, a concern is the potential energy in the aquifer as reflected in the water levels in the producing well or neighboring wells. When a ground-water reservoir which does not readily transmit water is tapped, the energy loss associated with flow to the well can be such that the well must be drilled to prohibitively great depths to provide adequate supplies. On the other hand, in a formation able to transmit fluid easily, water levels may drop because the reservoir is being depleted of water. This is generally encountered in reservoirs of limited areal extent or those in which natural infiltration has been reduced either naturally or through human activities.

Problems involving ground-water quantity were once the primary concern of hydrologists; interest is now focused on ground-water quality. Ground-water contamination is a serious problem, particularly in the highly urbanized areas of the United States. *See* HYDROLOGY; WATER POLLUTION. [G.F.P.]

Guild A group of species that utilize the same kinds of resources, such as food, nesting sites, or places to live, in a similar manner. Emphasis is on ecologically associated groups that are most likely to compete because of similarity in ecological niches, even though species can be taxonomically unrelated. The term was derived from the guild in human society composed of people engaged in an activity or trade held in common.

The guild concept focuses attention on the ways in which ecologically related species differ enough to permit coexistence, or avoid competitive displacement. For example, new places to live for some plants are provided by badger mounds in dense tall-grass prairie vegetation.

The guild is also commonly used as the smallest unit in an ecosystem in studies relating to environmental impact, wildlife management, and habitat classification. A representative species of a guild may be selected for study involving the uncertain assumption that environmental impact will influence this species in the same way as other guild members. *See* ECOSYSTEM. [P.W.P.]

Gulf of California A young, elongate ocean basin on the west coast of Mexico. It is flanked on the west by the narrow mountainous peninsula and continental shelf of Baja California, while the eastern margin has a wide continental shelf and coastal plain. The floor of the gulf consists of a series of basins 3300–12,000 ft (1000–3600 m) deep, whereas the northern gulf is dominated by a broad shelf which is the result of deltaic deposition from the Colorado River. The structural depression of the gulf continues northward into the Imperial Valley of California, which is cut off from the ocean by the delta of the Colorado River. *See* CONTINENTAL MARGIN.

Most of the gulf lies within an arid climate, with 4–6 in. (10–15 cm) of annual rainfall over Baja California and ranging on the eastern side from 4 in. (10 cm) in the north to about 34 in. (85 cm) in the southeast. No year-round streams enter the gulf on the west; a series of intermediate-size rivers flow in on the east side; and the major source of fresh-water sediment came from the Colorado River at the north prior to damming it upstream in the United States.

Water circulation is driven by seasonal wind patterns. Surface water is blown into the gulf in the summer by the southwesterly wind regime. In the winter, surface water is driven out of the gulf by the northwesterly wind regime, and upwelling occurs along the eastern margin, resulting in high organic productivity. Bottom sediments of the gulf range from deltaic sediments of the Colorado River at the north and coalesced deltas of the intermediate-size rivers on the east. A strong oxygen minimum occurs between 990 and 3000 ft (300 and 900 m) water depth, where seasonal influx of terrigenous sediments and blooms of diatoms due to upwelling produce varved sediments consisting of alternating diatom-rich and clay-rich layers. Rates of sediment accumulation are high, and total sediment fill beneath the Colorado River delta at the north may attain thicknesses of greater than 6 mi (10 km), even though the structural depression and the underlying crust are geologically young. *See* DELTA; MARINE SEDIMENTS; OCEAN CIRCULATION; UPWELLING. [J.R.C.]

Gulf of Mexico A subtropical semienclosed sea bordering the western North Atlantic Ocean. It connects to the Caribbean Sea on the south through the Yucatan Channel and with the Atlantic on the east through the Straits of Florida. To the north, it is bounded by North America, to the west and south by Mexico and Central America, and on the east and southeast by Florida and Cuba respectively.

The continental shelves surrounding the gulf are very broad along the eastern (Florida), northern (Texas, Louisiana, Mississippi, Alabama), and southern (Campeche) area, averaging 125–186 mi (200–300 km) wide. The continental shelves along the western and southwestern (Mexico) and southeastern (Cuba) boundaries of the gulf are narrow, often being less than 12 mi (20 km) wide. Between the continental shelves and the Sigsbee Abyssal Plain are three steep continental slopes: the Florida Escarpment off west Florida, the Campeche Escarpment off Yucatan, and the Sigsbee Escarpment south of Texas and Louisiana. Two major submarine canyons crease the gulf's shelf areas: the De Soto Canyon near the Florida-Alabama border, and the Campeche Canyon west of the Yucatan Peninsula. *See* CONTINENTAL MARGIN; ESCARPMENT; MARINE GEOLOGY.

Compared with the North American rivers, the Mexican rivers are short, but they still provide approximately 20% of the fresh-water input to the gulf because of extensive orographic rainfall from the trade winds that dominate the southern flank of the basin. Meteorologically, the Gulf of Mexico is a transition zone between the tropical wind system (easterlies) and the westerly frontal-passage-dominated weather (in winter particularly) to the north, punctuated with intense tropical storms in summer/autumn called the West Indian Hurricane. Much of the atmospheric moisture supplied to the

North American heartland during spring and summer has its origin over the gulf, and thus it is a vital element in the so-called North American Monsoon. *See* HURRICANE; MONSOON METEOROLOGY; STORM SURGE; TROPICAL METEOROLOGY.

The Gulf Stream System dominates the oceanic circulation in the Gulf of Mexico. The Yucatan Current, flowing northward into the eastern Gulf of Mexico, is the first recognizable western boundary current in the Gulf Stream System. North of the Yucatan Peninsula, the flow penetrates into the eastern gulf (where it is called the Gulf Loop Current) at varying distances with a distinctive chronology, loops around clockwise, and finally exits through the Straits of Florida, where it is called the Florida Current. This intense current reaches to more than 3300 ft (1000 m) depth, and transports 1.1×10^9 ft^3/s (3×10^7 m^3/s) of water, an amount 1800 times that of the Mississippi River. *See* GULF STREAM; MEDITERRANEAN SEA; OCEAN CIRCULATION.

Surrounding the Gulf of Mexico are many population centers that exploit the numerous estuaries, lagoons, and oil and gas fields. Coral reefs off Yucatan, Cuba, and Florida provide important fishery and recreational activities. There are extensive wetlands along most coastal boundaries with ecological connections to many seagrass beds nearshore and coastal mangrove forests of Mexico, Cuba, and Florida. This biogeographic confluence creates one of the most productive marine areas on Earth, providing the food web for commercially important species such as lobster, demersal (bottom-dwelling) fish, and shrimp; this same ecology supports large populations of sea turtles and marine mammals. The coastal and nearshore waters also support large phytoplankton populations. The juxtaposition of these enormous marine resources and human activities has led to a distinctive anthropogenic impact on the health of the marine ecosystem. *See* BIOGEOGRAPHY; ESTUARINE OCEANOGRAPHY; FOOD WEB; MANGROVE; MARINE ECOLOGY; REEF; WETLANDS. [G.A.Ma.]

Gulf Stream A great ocean current transporting about 70,000,000 tons (63,000,000 metric tons) of water per second (1000 times the discharge of the Mississippi River) northward from the latitude of Florida to the Grand Banks off Newfoundland. The Gulf Stream is thought of as a portion of a great horizontal circulation in the ocean, where each particle of water executes a closed circuit, sometimes moving slowly in midocean regions and other times rapidly in strong currents like the Gulf Stream. Thus the beginning and end of the Stream have arbitrary geographical limits. *See* ATLANTIC OCEAN.

The Gulf Stream is a narrow (62 mi or 100 km) and swift (up to 5 knots or 250 cm/s) eastward-flowing current jet which is embedded in a weaker and broader mean westward flow and which is surrounded by intense eddies. As it leaves the coast at Cape Hatteras, the Stream meanders from side to side like a river.

The near-surface Gulf Stream transports warm water from southern latitudes eastward to the Grand Banks, where the flow becomes broader and weaker, separating into several branches and eddies. About half the near-surface flow continues eastward across the Mid-Atlantic Ridge, and half recirculates southwestward, with part of the recirculation consisting of a countercurrent located south of the Stream.

The Gulf Stream is predominantly driven by the large-scale wind pattern, the westerlies in the north and the trades in the south. The winds exert a torque on the ocean that, due to the shape and rotation of the Earth, causes a large western-intensified gyre. Cold, deep water is formed in northern seas and flows southward as a western boundary current; warm water flows northward and replaces it. *See* OCEAN CIRCULATION. [P.R.]

H

Hackberry A medium-sized to large tree, *Celtis occidentalis*. It occurs in the eastern half of the United States, except in the extreme south, and is characterized by corky or warty bark, by alternate, long-pointed serrate leaves unequal at the base, and by a small drupaceous fruit, with thin, sweet, edible flesh. Both species are used for furniture, boxes, and baskets, for shelterbelts, and as shade trees. Sugarberry (*C. laevigata*) is similar to hackberry and grows in the southeast United States. [A.H.G./K.P.D.]

Hail Precipitation composed of chunks or lumps of ice formed in strong updrafts in cumulonimbus clouds. Individual lumps are called hailstones. Most hailstones are spherical or oblong, some are conical, and some are bumpy and irregular. Diameters range from 0.2 to 6 in. (5 to 150 mm) or more. That is, the largest stones are grapefruit or softball size, and the smallest are pea size.

Very often hailstones are observed to be made of alternating rings of clear and white ice (see illustration). These rings indicate the growth processes of the hail. The milky or white portion of the growth occurs when small cloud droplets are collected

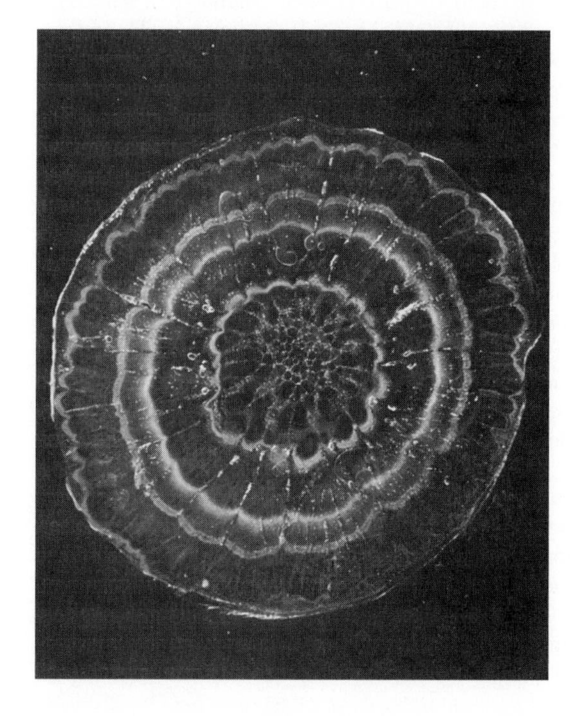

Cross section of a large hailstone showing the structure of alternating rings of clear and white ice. (*Alberta Research Council, Edmonton*)

by the hailstone and freeze almost instantaneously, trapping bubbles of air between the droplets and creating a milky appearance. The clear portion is formed when many droplets are collected so rapidly that a film of water spreads over the stone and freezes gradually, giving time for any trapped air bubbles to escape from the liquid.

The most favorable conditions for hail formation occur in the mountainous, high plains regions of the world. Hailstorms normally have relatively high, cool cloud bases and very strong updrafts within the clouds to carry the hailstones into the cooler regions of the cloud, where maximum growth occurs. Both small ice particles and supercooled liquid water (liquid water at temperatures below 32°F or 0°C) are needed for the ice particles to grow into hailstones. *See* CLOUD PHYSICS; PRECIPITATION (METEOROLOGY). [H.D.O.]

Hardy-Weinberg formula A basic mathematical relation used in population genetics. It gives the proportion of the various genotypes in a randomly mating population in terms of the frequencies of the genes. The formula was discovered independently in 1908 by G. H. Hardy, a British mathematician, and W. Weinberg, a German physician. *See* HUMAN GENETICS; POPULATION GENETICS.

In its simplest form the Hardy-Weinberg formula may be stated thus: If p is the proportion of gene A in the population and $q(= 1 - p)$ is the proportion of gene a, then after one generation of random mating the three genotypes AA, Aa, and aa will occur in the proportions p^2, $2pq$, and q^2. In other words the genotypes are given by the appropriate terms in the expansion of the binomial $(p + q)^2$. The extension to multiple alleles is direct.

The formula holds only for an infinite population and assumes random mating in the absence of significant mutation pressure or gene transfer between populations. However, it is an accurate approximation in many populations. *See* GENETICS. [J.F.Cr.]

Hazardous waste Any solid, liquid, or gaseous waste materials that, if improperly managed or disposed of, may pose substantial hazards to human health and the environment. Every industrial country in the world has had problems with managing hazardous wastes. Improper disposal of these waste streams in the past has created a need for very expensive cleanup operations. Efforts are under way internationally to remedy old problems caused by hazardous waste and to prevent the occurrence of other problems in the future.

A waste is considered hazardous if it exhibits one or more of the following characteristics: ignitability, corrosivity, reactivity, and toxicity. Ignitable wastes can create fires under certain conditions; examples include liquids, such as solvents, that readily catch fire, and friction-sensitive substances. Corrosive wastes include those that are acidic and those that are capable of corroding metal (such as tanks, containers, drums, and barrels). Reactive wastes are unstable under normal conditions. They can create explosions, toxic fumes, gases, or vapors when mixed with water. Toxic wastes are harmful or fatal when ingested or absorbed. When they are disposed of on land, contaminated liquid may drain (leach) from the waste and pollute groundwater. *See* GROUND-WATER HYDROLOGY; WATER POLLUTION.

Hazardous wastes may arise as by-products of industrial processes. They may also be generated by households when commercial products are discarded. These include drain openers, oven cleaners, wood and metal cleaners and polishes, pharmaceuticals, oil and fuel additives, grease and rust solvents, herbicides and pesticides, and paint thinners.

The predominant waste streams generated by industries in the United States are corrosive wastes, spent acids, and alkaline materials used in the chemical, metal-finishing, and petroleum-refining industries. Many of these waste streams contain heavy metals, rendering them toxic. Solvent wastes are generated in large volumes both by manufacturing industries and by a wide range of equipment maintenance industries that generate spent cleaning and degreasing solutions. Reactive wastes come primarily from the chemical industries and the metal-finishing industries. The chemical and primary-metals industries are the major sources of hazardous wastes.

There is a growing acceptance throughout the world of the desirability of using waste management hierarchies for solutions to problems of hazardous waste. A typical sequence involves source reduction, recycling, treatment, and disposal. Source reduction comprises the reduction or elimination of hazardous waste at the source, usually within a process. Recycling is the use or reuse of hazardous waste as an effective substitute for a commercial product or as an ingredient or feedstock in an industrial process.

Treatment is any method, technique, or process that changes the physical, chemical, or biological character of any hazardous waste so as to neutralize such waste; to recover energy or material resources from the waste; or to render such waste nonhazardous, less hazardous, safer to manage, amenable for recovery, amenable for storage, or reduced in volume. Disposal is the discharge, deposit, injection, dumping, spilling, leaking, or placing of hazardous waste into or on any land or body of water so that the waste or any constituents may enter the air or be discharged into any waters, including groundwater.

There are various alternative waste treatment technologies, for example, physical treatment, chemical treatment, biological treatment, incineration, and solidification or stabilization treatment. These processes are used to recycle and reuse waste materials, reduce the volume and toxicity of a waste stream, or produce a final residual material that is suitable for disposal. The selection of the most effective technology depends upon the wastes being treated.

There are abandoned disposal sites in many countries where hazardous waste has been disposed of improperly in the past and where cleanup operations are needed to restore the sites to their original state. Cleaning up such sites involves isolating and containing contaminated material, removal and redeposit of contaminated sediments, and in-place and direct treatment of the hazardous wastes involved. As the state of the art for remedial technology improves, there is a clear preference for processes that result in the permanent destruction of contaminants rather than the removal and storage of the contaminating materials. [H.M.F.]

Heartwater disease A rickettsial disease, also known as cowdriosis, which is caused by the microorganism *Cowdria ruminantium* and is transmitted by ticks of the genus *Amblyomma*. The disease occurs in wild and domestic ruminants, primarily cattle, sheep, and goats, in sub-Saharan Africa and some Caribbean islands (for example, Guadeloupe, Antigua, and Marie-Galante).

Heartwater disease is characterized by fluid in the pericardium of the heart, high fever, lung edema, and nervous symptoms that range from mild incoordination and exaggerated reflexes to convulsions seen in acute infections. The course of acute heartwater disease is 2–6 days, and recovery is rare. However, young animals have a high rate of natural resistance.

The organism is susceptible to tetracycline antibiotics. However, once marked nervous symptoms have developed, recovery usually does not occur.

Control and prevention of heartwater is achieved by tick control or immunization. [K.M.K.]

Heartworms Heartworm (*Dirofilaria immitus*) is a nematode parasite that resides within the host's large pulmonary arteries and right heart chambers. It primarily infests dogs but may also infest foxes, wolves, coyotes, ferrets, sea lions, horses, and cats. A dog can be infested with one to several hundred adult heartworms, which can grow to 12 in. (30 cm). During their 3–5-year life-span, heartworms can cause serious and often life-threatening damage to the heart and lungs. Endemic areas require a reservoir of infected animals (usually dogs) and the presence of mosquitoes, the intermediate host, which transmit the larval stages to a new host.

Adult female heartworms release their microscopic offspring called microfilariae into the bloodstream. A mosquito becomes infested with these circulating microfilariae while taking a blood meal from the dog. The microfilariae develop into mature larvae within the mosquito during the next 10–14 days. As the mosquito feeds again, the mature larvae are injected into the new host. Once in the dog, it takes approximately 6 months for these larvae to complete the cycle by migrating to the large arteries of the lung and right chambers of the heart.

Adult heartworms stimulate a progressive proliferation of the artery lining (endarteritis) that gradually restricts the blood flow to the lungs. The resulting increase in the pulmonary artery blood pressure (pulmonary hypertension) causes the right ventricle to pump harder, eventually leading to right heart failure. In advanced cases the lung fibrosis and heart changes may be permanent. When adult worms die either naturally or with treatment, their fragments become lodged distally in the smaller pulmonary arteries causing an exaggerated proliferation of the vessel lining, the formation of blood clots, and an intense local inflammatory reaction. Blood flow is severely restricted or totally blocked, resulting in severe coughing, coughing up blood, and difficulty in breathing (dyspnea).

Initial signs of disease include coughing, exercise intolerance, and weight loss. As the disease progresses, these symptoms become more pronounced. With advanced disease, dogs begin to exhibit progressive signs of pulmonary disease and associated heart failure, including fainting spells, collapse, difficulty in breathing, coughing up blood, and fluid accumulation around the lungs (hydrothorax) or within the abdominal cavity (ascites). Rarely, a rapidly fatal condition called vena caval syndrome may be observed in young dogs with massive heartworm infestation.

Heartworm infection can usually be determined by examining a blood sample for the presence of circulating microfilariae. Those infestations in which the adults produce no circulating microfilariae are termed occult infections, which can be accurately diagnosed by identifying specific circulating immunologic substances (uterine antigens) released into the blood by adult females.

All but the most advanced cases of heartworm disease can usually be treated successfully. Treatment involves multiple steps to eliminate both adult and microfilariae stages. Daily or monthly heartworm preventive medication is strongly recommended, especially during the mosquito season, in infested areas. The objective is killing the infectious larval stages before they develop into adults. Only dogs that test negative for heartworms should be placed on preventive medication because of the risk of serious reactions. [W.D.F.]

Heat balance, terrestrial atmospheric The balance of various types of energy in the atmosphere and at the Earth's surface. At the top of the atmosphere, the incoming solar radiation that is absorbed by the Earth-atmosphere system is approximately balanced by the terrestrial radiation emitted from this system over long periods of time. The flux of solar energy (energy per time) across a surface of unit area normal to the solar beam at the mean distance between the Sun and the Earth is referred to as the solar constant. Based on recent satellite measurements, a value of 1365 watts per square meter (W/m^2) for the solar constant has been suggested. Because

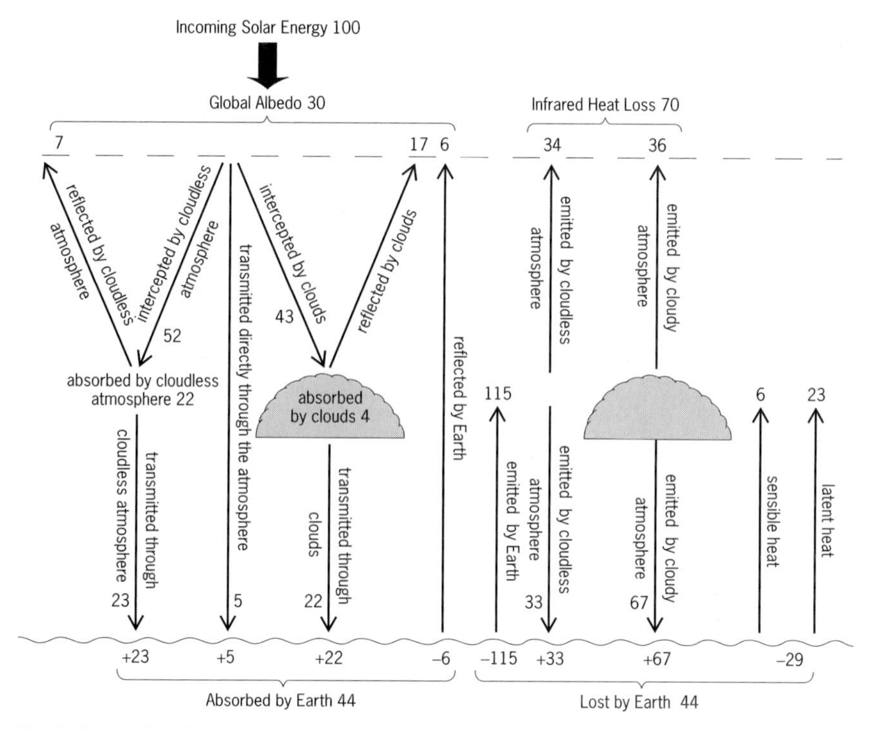

Heat balance of the Earth-atmosphere system. The incoming solar energy is taken to be 100 units. On a climatic scale, the incoming solar energy at the top of the atmosphere is approximately balanced by the reflected solar energy and thermal infrared heat loss. At the surface, the heat balance involves sensible and latent heat components, in addition to net radiative energy.

the area of the spherical Earth is four times that of its cross section facing the parallel solar beam, the top of the Earth's atmosphere receives an average of about 341 W/m^2. Based on measurements from satellite radiation budget experiments, about 30% of this is reflected back to space, and is referred to as the global albedo. The reflecting power of the Earth-atmosphere system includes the scattering of molecules, aerosols, and various types of clouds, as well as reflection by different surfaces. Thus, only about 70% of the incoming solar flux, that is, about 239 W/m^2, is absorbed within the Earth-atmosphere system. For this system to be in thermodynamic equilibrium or balance, it must emit the same amount of thermal infrared radiation. *See* SOLAR ENERGY.

For the presentation of internal heat balance components, the effective solar constant of 341 W/m^2 may be arbitrarily represented by 100 units (see illustration). Of these units, roughly 26 are absorbed within the atmosphere, including 22 by clear column and 4 by clouds. A total of 30 units are reflected back to space, including about 7 from clear column, 17 from cloudy atmospheres, and 6 directly from the Earth's surface. The remaining 44 units are absorbed by the surface. The Earth-atmosphere system emits terrestrial radiation according to its temperature and composition distributions. The upward flux from the warmer surface accounts for about 115 units. The colder troposphere emits both upward and downward fluxes, with about 70 and 100 units at the top and surface, respectively. The clear and cloudy portions are 34 and 36 at the top and 33 and 67 at the surface, respectively. The net upward flux at the surface, representing the difference between the flux emitted by the surface and the downward flux from the atmosphere reaching the surface, is about 15 units. *See* TERRESTRIAL RADIATION.

As a result of thermal emission, the atmosphere loses 55 units. With absorption of the incoming solar flux contributing only 26 units, the net radiative loss from the atmosphere amounts to about 29 units. This deficit is balanced by convective fluxes of sensible and latent heat associated with temperature gradient and evaporation. Based on statistical analyses, the average annual ratio of sensible to latent heat loss at the surface has a global value of about 0.27. It follows that the latent and sensible heat components are about 23 and 6 units, respectively, in order to produce an overall heat balance at the surface (see illustration). The atmosphere experiences a net radiative cooling that must be balanced by the latent heat of condensation released in precipitation processes and by the convection and conduction of sensible heat from the underlying surface. [K.N.L.]

Hemlock The genus *Tsuga* of the pine family, characterized by flattened needles with two white lines beneath the needlelike leaves, which have distinct short stalks. The cones are small and pendent.

Eastern hemlock (*T. canadensis*) occurs in eastern Canada, the Great Lakes states, and the Appalachians. Minutely toothed leaves are characteristic of this species. The wood is hard and strong, and is used for construction, boxes, crates, and paper pulp. The bark is one of the principal domestic sources of tannin. The eastern hemlock is a common ornamental tree.

Carolina hemlock (*T. caroliniana*), a species found in the southern Appalachians, has entire needles and is sometimes grown as an ornamental. The western hemlock (*T. heterophylla*) grows in the extreme Northwest and in Alaska. Its needles resemble those of the eastern hemlock, but the white lines beneath are not so distinct. It is an important lumber tree, with uses similar to these of the eastern species. [A.H.G./K.P.D.]

Hepatitis An inflammation of the liver caused by a number of etiologic agents, including viruses, bacteria, fungi, parasites, drugs, and chemicals. The most common infectious hepatitis is of viral etiology. All types of hepatitis are characterized by distortion of the normal hepatic lobular architecture due to varying degrees of necrosis of individual liver cells or groups of liver cells, acute and chronic inflammation, and Kupffer cell enlargement and proliferation. There is usually some degree of disruption of normal bile flow, which causes jaundice. The severity of the disease is highly variable and often unpredictable.

A frequently occurring form of hepatitis is caused by excessive ethyl alcohol intake and is referred to as alcoholic hepatitis. It usually occurs in chronic alcoholics and is characterized by fever, high white blood cell count, and jaundice. Some drugs are capable of damaging the liver and can occasionally cause enough damage to produce clinical signs and symptoms. Among these drugs are tetracycline, methotrexate, anabolic and contraceptive steroids, phenacetin, halothane, chlorpromazine, and phenylbutazone.

Clinical features of hepatitis include malaise, fever, jaundice, and serum chemical tests revealing evidence of abnormal liver function. In most mild cases of hepatitis, treatment consists of bedrest and analgesic drugs. In those individuals who develop a great deal of liver cell necrosis and subsequently progress into a condition known as hepatic encephalopathy, exchange blood transfusions are often used. This is done with the hope of removing or diluting the toxic chemicals thought to be the cause of this condition. Chronic hepatitis is a condition defined clinically by evidence of liver disease for at least 6 consecutive months. [S.P.H.]

Hepatitis C is a disease of the liver caused by the hepatitis C virus (HCV). The prevalence of HCV infection worldwide is 3% (170 million people), with infection rates in North America ranging from 1 to 2% of the population. A simulation analysis estimated that in the period from 1998 to 2008 there will be an increase of 92% in the incidence of cirrhosis of the liver, resulting in a 126% increase in the incidence of liver, failures and a 102% increase in the incidence of hepatocellular carcinoma (HCC), all attributed to HCV.

Hepatitis C virus can be transmitted only by blood-to-blood contact. With the institution of screening of blood, intravenous drug use has become the major source of transmission in North America. Approximately 89% of people who use intravenous drugs for one year become infected with HCV.

Management strategies can be divided into three main areas: surveillance of patients with chronic HCV infection who have not developed cirrhosis; surveillance of patients with established cirrhosis; and strategies to eradicate HCV. [N.Ar.; N.G.; G.L.]

Herbarium A collection of pressed and dried plant specimens, and a description of when, where, and by whom they were collected, arranged in a systematic manner, and serving as a permanent physical record of the occurrence of an individual plant at a specific place and time. Herbaria may contain specimens from the full range of organisms that have classically been considered plants: fungi, lichens, algae, bryophytes, ferns and their allies, gymnosperms, and angiosperms. Many herbaria also accumulate and manage special collections such as liquid-preserved parts for anatomical studies, wood, seeds, or specially preserved material suitable for extraction of deoxyribonucleic acid (DNA) or other chemical constituents. Many groups of plants, especially those with succulent or fleshy parts, are not suitable for preservation as dry, flat specimens because they lose many of their important features in the drying process. Consequently, these plants are often preserved in liquid. Specimens are used in taxonomic and ecological research, such as morphological studies, and for comparative identification and verification of unknown specimens. *See* PLANT GEOGRAPHY; PLANT TAXONOMY. [J.C.So.]

Herbicide Any chemical used to destroy or inhibit plant growth, especially of weeds or other undesirable vegetation. There are well over a hundred chemicals in common usage as herbicides. Many of these are available in several formulations or under several trade names. The variety of materials are conveniently classified according to the properties of the active ingredient as either selective or nonselective. Selective herbicides are those that kill some members of a plant population with little or no injury to others. Nonselective herbicides are those that kill all vegetation to which they are applied. Further subclassification is by method of application, such as preemergence (soil-applied before plant emergence) or postemergence (applied to plant foliage). Additional terminology sometimes applied to describe the mobility of post-emergence herbicides in the treated plant is contact (nonmobile) or translocated (mobile—that is, killing plants by systemic action).

A rapidly expanding use for nonselective herbicides is the destruction of vegetation before seeding in the practice of reduced tillage or no tillage. Some are also used to kill annual grasses in preparation for seeding perennial grasses in pastures. Additional uses are in fire prevention, elimination of highway hazards, destruction of plants that are hosts for insects and plant diseases, and killing of poisonous or allergen-bearing plants.

Preemergence or postemergence application methods derive naturally from the properties of the herbicidal chemical. The distinction between pre- and postemergence is not always clear-cut. For example, atrazine can exert its herbicidal action either following root absorption from a preemergence application or after leaf absorption from a postemergence treatment. [R.O.R.]

Herbivory The consumption of living plant tissue by animals. Herbivorous species occur in most of the major taxonomic groups of animals. Herbivorous insects alone may account for one-quarter of all species. The fraction of all plant biomass that is eaten by herbivores varies widely among plants and ecosystems, ranging from less than 1% to nearly 90%. In terms of both the number of species involved and the role that herbivory plays in the flow of energy and nutrients in ecosystems, herbivory is a key ecological interaction between species.

Herbivory usually does not kill the plant outright, although there are striking exceptions (such as bark beetle outbreaks that decimate conifer trees over thousands of square kilometers). Nevertheless, chronic attack by herbivores can have dramatic cumulative effects on the size, longevity, or reproductive output of individual plants. As a consequence, plants have evolved several means to reduce the level of damage from herbivores and to amellorate the impact of damage.

Many plants possess physical defenses that interfere mechanically with herbivore feeding on or attachment to the plant. In addition, plant tissues may contain chemical compounds that render them less digestible or even toxic to herbivores. Many plant compounds even can cause death if consumed by unadapted herbivores. While natural selection imposed by herbivores was the likely force driving the elaboration of these plant chemicals, humans have subsequently found many uses for the chemicals as active components of spices, stimulants, relaxants, hallucinogens, poisons, and drugs. An exciting recent finding is that some plants possess induced resistance, elevated levels of physical or chemical defenses that are brought on by herbivore damage and confer enhanced resistance to further damage.

Herbivores can either avoid or counteract plant defenses. Many herbivores avoid consuming the plant tissues that contain the highest concentrations of toxic or antinutritive chemicals. Herbivores have also evolved an elaborate array of enzymes to detoxify otherwise lethal plant chemicals. Because few herbivores have the ability to detoxify the chemical compounds produced by all the plant species they encounter, many herbivores have restricted diets; the larvae of more than half of all species of butterflies and moths include only a single genus of plants in their diets. Some insect species that have evolved the means to tolerate toxic plant chemicals have also evolved ways to use them in their own defense. Larvae of willow beetles store plant compounds in glands along their back. When the larvae are disturbed, the glands exude droplets of the foul-smelling compounds, which deter many potential predators.

If a plant evolved the ability to produce a novel chemical compound that its herbivores could not detoxify, the plant and its descendants would be freed for a time from the negative effects of herbivory. A herbivore that then evolved the means to detoxify the new compound would enjoy an abundance of food and would increase until the level of herbivory on the plant was once again high, favoring plants that acquire yet another novel antiherbivore compound. These repeated rounds of evolution of plant defenses and herbivore countermeasures (coevolution) over long periods of time help to explain similar patterns of evolutionary relatedness between groups of plant species and the herbivorous insect species that feed on them.

Plants and their herbivores seldom occur in isolation, and other species can influence the interaction between plants and herbivores. For example, mammalian herbivores often rely on gut microorganisms to digest cellulose in the plant material they consume.

Thus, herbivory occurs against a backdrop of multiple interactions involving the plants, the herbivores, and other species in the ecological community. [W.F.Mo.]

Herpes Any virus of the herpesvirus group, which comprises a family of 70 species, 5 of which are pathogenic to humans; the term also refers to any infection caused by these viruses. Since these pathogens are ubiquitous in nature, most individuals of all populations are exposed to and thus immunized to these viruses. The five pathogenic groups include herpes simplex I and II, varicella-zoster, cytomegalovirus, and the Epstein-Barr virus.

In nonimmunized hosts, the vast majority of all herpes infections present symptoms of nonspecific viral illnesses which resolve spontaneously. However, the infections that cause clinical disease in fact may cause serious morbidity and mortality in afflicted individuals. Reactivation of herpes infection, characteristic of the immunocompromised host, is an important cause of mortality in the treatment of patients with advanced cancer, and is a potential complication of an otherwise possibly curable systemic disease.

Herpesviruses have a deoxyribonucleic acid (DNA) core and are 150 to 200 nanometers in size with icosahedral symmetry, and are coated by a protein barrier, the capsid, derived from the infected host cells. The surface of the virions in general contains protein-carbohydrate structures which allow cellular attachment and thus cellular penetration. All viruses require living cells for their replication; the virus may replicate and destroy the cell, or replicate and allow cell survival, or incorporate its viral gene structure into the host gene structure. This incorporation phenomenon is designated as latency. For example, herpes simplex virus exhibits the phenomenon of latency within nerve cells in the area of previous infection. The Epstein-Barr virus characteristically causes latent infection in lymphocytes (white blood cells in the circulating blood), and the cytomegalic virus also causes latent infection within lymphocytes and possibly within nerve cells. Once the viral genome is incorporated into the host cell, antiviral drugs are of no use, since therapeutic agents cannot selectively destroy or inhibit the viral genome. Factors which are possibly involved in the reactivation of latent virus generally revolve around some depression of the host immune response system. Viral genome incorporation into host cells is of great interest as several herpesvirus types are implicated in the development of cancer.

The foundation of therapeutic intervention for all herpesviruses involves a series of chemicals with structures similar to the base pairs which compose the viral DNA structure. The base analogs compete with or inhibit viral enzymes necessary for the assembly of DNA. *See* VIRUS.

Herpes simplex I and II infections are spread by intimate contact of mucocutaneous surfaces during the period of virus shedding from active lesions. They usually affect the genitalia, but may affect the oral mucosa, causing painful ulcerations which crust and heal. Upon healing, the virus resides in latent form within local nerve cells. Viral reactivation is poorly understood, but may relate in part to the host immune system. The type II virus has been linked to the development of uterine cervical carcinoma, however its precise role remains a question.

Herpes simplex virus I (cold sores, fever blisters) afflicts 20–40% of the population in the United States and usually affects the oropharynx, causing pharyngitis, tonsillitis, gingivostomatitis, or keratitis (eye inflammation) as primary infections. Inflammation of the mouth, eye, or brain may occur as a secondary infection.

Primary infection (airborne) due to herpes varicella-zoster usually affects preschool children, causing chickenpox, with rare complications usually affecting the immunocompromised host. Secondary infection usually afflicts the elderly when latent viral reactivation occurs, presumably due to an immune imbalance in the host, and involves the spread of virus along the skin in the anatomic distribution of nerve (this disorder is known as shingles).

Cytomegalovirus is ubiquitous, with the majority of infections remaining subclinical. Adult syndromes include a mononucleosislike syndrome and hepatitis, both of which are self-limited diseases in the normal host. However, reactivation of latent infection is a major source of morbidity and especially mortality in the compromised host, for example, the patient being treated with chemoradiotherapy for advanced malignant disease. *See* CYTOMEGALOVIRUS INFECTION; HEPATITIS.

The characteristic clinical syndrome caused by Epstein-Barr virus infection includes generalized lymphadenopathy, hepatosplenomegaly, pharyngitis, tonsillitis, and general fatigue and fever. This disorder affects individuals of all ages, but predominantly adolescents. The majority of children are subclinically infected. This mononucleosis syndrome is usually a self-limited disorder, and investigational drugs in use for prophylaxis of high-risk individuals include interferons and acyclovir. Epstein-Barr virus is suspected to be of etiologic importance in Burkitt's African lymphoma. *See* ANIMAL VIRUS. [D.J.D.]

Hickory Any species of the genus *Carya*, formerly known botanically as *Hicoria*. Hickories are mostly tall forest trees characterized by strong, terminal, scaly winter buds, pinnately compound leaves (see illustration), solid pith (not chambered), and fruit with an outer husk or exocarp which splits more or less readily into four parts, revealing a nut with a hard shell or endocarp.

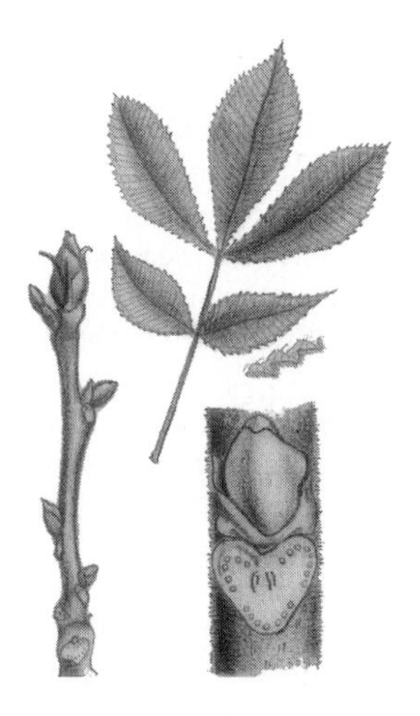

Twigs, buds, and leaves of shagbark hickory (*Carya ovata*).

The shagbark hickory (*C. ouata*) is found in the eastern half of the United States and adjacent Canada. It is the most important species because of the commercial value of its nuts, the hickory nuts of commerce, and of its wood. The pecan (*C. illinoensis*) is also a valuable species because of its commercially popular, thin-shelled, sweet nuts. Other species are the mockernut, shellbark, and pignut hickories. The remarkably tough and strong wood of all species makes it the world's best wood for tool handles. It is also used for parts of furniture, flooring, boxes, and crates, and for smoking meats. [A.H.G./K.P.D.]

Hill and mountain terrain Land surfaces characterized by roughness and strong relief. The distinction between hills and mountains is usually one of relative size or height, but the terms are loosely and inconsistently used.

Uplift of the Earth's crust is necessary to give mountain and hill lands their distinctive elevation and relief, but most of their characteristic features—peaks, ridges, valleys, and so on—have been carved out of the uplifted masses by streams and glaciers. Hill lands, with their lesser relief, indicate only lesser uplift, not a fundamentally different course of development. The features of hill and mountain lands are chiefly valleys and divides produced by sculpturing agents, especially running water and glacier ice. Local peculiarities in the form and pattern of these features reflect the arrangement and character of the rock materials within the upraised crustal mass that is being dissected.

Hill and mountain terrain occupies about 36% of the Earth's land area. The greater portion of that amount is concentrated in the great cordilleran belts that surround the Pacific Ocean, the Indian Ocean, and the Mediterranean Sea. Additional rough terrain, generally low mountains and hills, occurs outside the cordilleran systems in eastern North and South America, northwestern Europe, Africa, and western Australia. Eurasia is the roughest continent, more than half of its total area and most of its eastern portion being hilly or mountainous. Africa and Australia lack true cordilleran belts. The broad-scale pattern of crustal disturbance, and hence of rough lands, is now known to be related to the relative movements of a worldwide system of immense crustal plates. *See* MOUNTAIN. [E.H.Ha.]

Hirudinea A class of the annelid worms commonly known as leeches. These organisms are parasitic or predatory and have terminal suckers for attachment and locomotion. Most inhabit inland waters, but some are marine and a few live on land in damp places. The majority feed by sucking the blood of other animals, including humans.

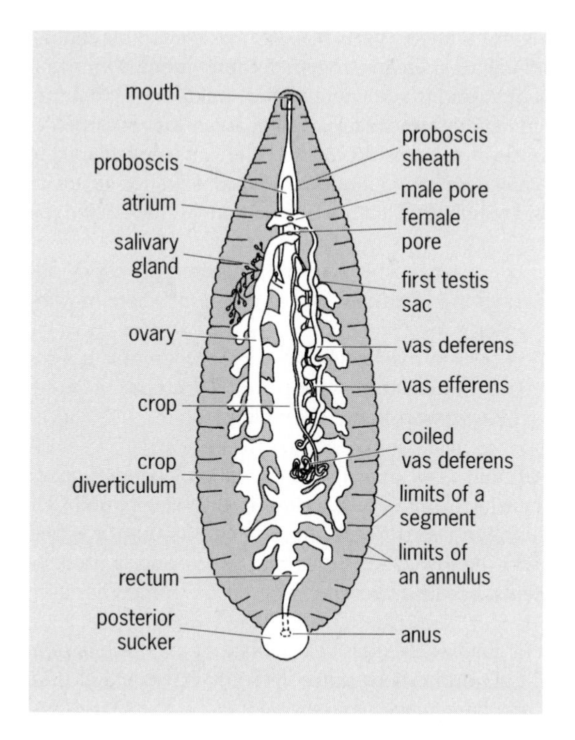

General structure of a leech. Male reproductive system Is shown on the right, the female on the left. (*After K. H. Mann, A key to the British freshwater leeches, Freshwater Blol. Ass. Scl. Publ., 14:3–21, 1954*)

Leeches differ from other annelids in having the number of segments in the body fixed at 34, chaetae or bristles lacking, and the coelomic space between the gut and the body wall filled with packing tissue (see illusration). In a typical leech the first six segments of the body are modified to form a head, bearing eyes, and a sucker, and the last seven segments are incorporated into a posterior sucker.

The mouth of a leech opens within the anterior sucker, and there are two main methods of piercing the skin of the host to obtain blood: an eversible proboscis or three jaws, each shaped like half a circular saw, placed just inside the mouth. The process of digestion is very slow, and a meal may last a leech for 9 months. The carnivorous forms have lost most or all of their gut diverticula and resemble earthworms in having a straight, tubular gut. Leeches are hermaphroditic, having a single pair of ovaries and several pairs of testes.

The importance of leeches as a means of making incisions for the letting of blood or the relief of inflammation is declining, and in developed countries the bloodsucking parasites of mammals are declining, because of lack of opportunity for contact with the hosts. In other countries they are still serious pests. *See* RHYNCHOBDELLAE. [K.H.M.]

Hog cholera A highly contagious epizootic disease of pigs, also known as classical (or European) swine fever. The causative agent is a virus in the genus *Pestivirus*. This disease is the subject of statutory controls in a majority of countries, and has been eradicated from many areas, including the United States, Canada, Australasia, and parts of Europe. Clinically and pathologically, it closely resembles African swine fever, which is caused by an unrelated virus. *See* ANIMAL VIRUS.

Hog cholera can occur in European wild boar, but among domestic species only pigs are affected. Humans are not susceptible. The primary mode of transmission is by contact or proximity. Infected animals shed virus in all bodily secretions, including aerosols of respiratory mucus. The virus survival time in aerosol is short, and airborne transmission over long distances is not a factor. Virus may also be spread by contact with contaminated equipment and vehicles. It can survive for many months in frozen or refrigerated meat from infected pigs, and is not inactivated by mild forms of curing.

Pigs of any age may be affected. There are typically a high fever, loss of appetite, and dullness. Other symptoms include blotchy discoloration of the skin (particularly the extremities), incoordination and weakness of the hindquarters, constipation followed by diarrhea, gummed-up eyes, and coughing. Death occurs within 4–7 days, and the mortality is usually high.

The chronic form of disease is characterized by dullness, unthriftiness, capricious appetite, and variable degrees of coughing, diarrhea, and emaciation. There may be joint swellings and ulceration of the skin.

Strains vary in virulence, and hog cholera may still be suspected when milder signs occur in epizootic form. Low-virulence strains may produce few signs apart from reproductive failure in sows or congenital tremors in their offspring. [S.E.]

Holly The American species of holly (*Ilex opaca*) has evergreen leaves. It grows naturally in the eastern and southeastern United States close to the Atlantic and Gulf coasts, in the Mississippi Valley, and westward to Oklahoma and Missouri. It is best known for its bright red berries. The heartwood takes a high polish and is used for cabinet work and musical instruments; because it resembles ivory, it is sometimes used for keys for pianos and organs.

The English holly (*I. aquifolium*) is cultivated extensively in the extreme northwestern United States, but is not hardy in the northeastern states. Its spiny leaves are glossier than those of the American holly and have wavier margins. [A.H.G./K.P.D.]

Hophornbeam The genus *Ostrya* of the birch family, represented in North America by two species. *Ostrya virginiana* is widely distributed in the eastern half of the United States and in the highlands of southern Mexico and Guatemala. It can be recognized by its fruit, which closely resembles that of the hop vine, and by its very scaly bark. The scales usually occur in narrow, more or less parallel, vertical strips. The leaves are sharply and doubly serrate. This is one of several trees known as ironwood because of its hard, strong wood; it is used for fence posts, tool handles, mallets, and other articles requiring hardness and strength. *Ostrya knowltonii* is a small rare tree of the southwestern United States. [A.H.G./K.P.D.]

Hornbeam The genus *Carpinus* of the birch family, represented in the United States by *C. caroliniana*, the American hornbeam or blue beech. Hornbeam is a small tree, and it has a smooth, steel-gray, fluted bark. It grows throughout the eastern half of the United States, especially in moist soil along banks of streams; it is sometimes called water beech. When mature, it is easily recognized by its peculiar bark, by the doubly serrate leaves resembling those of sweet birch, and by the small, pointed, angular winter buds with scales in four rows. The fruit is a small nutlet subtended by a three-lobed serrate bract. The wood is very hard, giving rise to the name iron-wood.

The European hornbeam (*C. betulus*) is often cultivated in parks and estates. It can be distinguished by its larger size, larger winter buds, and larger three-lobed, almost entire fruiting bracts. [A.H.G./K.P.D.]

Horticultural crops Intensively managed plants cultivated for food or for esthetic purposes. Plant agriculture is divided traditionally into the fields of agronomy (herbaceous field crops, mainly grains, forages, oilseeds, and fiber crops), forestry (forest trees and products), and horticulture (garden crops, particularly fruits, vegetables, spices and herbs, and all plants grown for ornamental use). Most horticultural plants are utilized in the living state, with water essential to quality; thus most horticultural plants and products are highly perishable. *See* AGRICULTURAL SCIENCE (PLANT); AGRONOMY; FLORICULTURE; FOREST AND FORESTRY.

Horticultural crops are usually classified as edibles or ornamentals. Edible crops which are used for direct human consumption are commonly subdivided into fruits or vegetables, but this classification is traditional and difficult to define precisely.

Fruit crops in the horticultural sense are cultivated for tissues associated with the botanical fruit, that is, seed-bearing structures derived from the flower, which are usually pulpy and tasteful. Trees or shrubs bearing nuts, characterized by a hard shell separated by a firm inner kernel (the seed), are often treated as a special category of fruit crops. *See* FRUIT.

Vegetable crops in the horticulture sense are commonly herbaceous plants grown as annuals or biennials and occasionally as perennials that have edible parts (including, confusingly, the botanical fruit). Examples of edible parts include the root (sweet potato), tuber (potato), young shoot (asparagus), leaf (spinach), flower buds (cauliflower), fruit (tomato), and seed (pea).

Plants grown for ornamental use, such as cut flowers, bedding plants, interior foliage plants, or landscape plants, represent an enormous group and include thousands of species. [J.J.]

Hospital infections Infections acquired during a hospital stay; also known as nosocomial infections. They may be recognized during or after hospitalization. They usually appear during hospitalization, but as many as 25% of infections related to surgery occur after discharge. Most nosocomial infections (93%) are caused by bacteria; fungi account for about 6%; and viruses, protozoa, and parasites account for the remaining 1%.

Nosocomial infections occur most frequently in the urinary tract, in surgical wounds, as a complication of pneumonia, and in association with bacteremia. *See* PNEUMONIA.

Viral infections, although sometimes difficult to recognize, are becoming an increasing concern in hospitals, particularly because of the possibility of transmitting the human immunodeficiency virus (HIV), the causative organism for acquired immune deficiency syndrome (AIDS). Since the virus is not transmitted by the respiratory route or by casual contact, no one is endangered simply by being in the same hospital with HIV-infected patients. *See* ACQUIRED IMMUNE DEFICIENCY SYNDROME (AIDS); INFECTION; MEDICAL BACTERIOLOGY. [B.B.D.]

Human ecology The study of how the distributions and numbers of humans are determined by interactions with conspecific individuals, with members of other species, and with the abiotic environment. Human ecology encompasses both the responses of humans to, and the effects of humans on, the environment. Human ecology today is the combined result of humans' evolutionary nature and cultural developments. *See* BIOSPHERE; ECOLOGICAL COMMUNITIES; ECOSYSTEM.

Humans' strong positive and negative emotional responses to components of the environment evolved because our ancestors' responses to environmental information affected survival and reproductive success. Early humans needed to interpret signals from other organisms and the abiotic environment, and they needed to evaluate and select habitats and the resources there. These choices were emotionally driven. For example, food is one of the most important resources provided by the environment. Gathering food requires decisions of where to forage and what items to select. Anthropologists often use the theory of optimal foraging to interpret how these decisions are made. The theory postulates that as long as foragers have other valuable ways to spend their time or there are risks associated with seeking food, efficient foraging will be favored even when food is not scarce. This approach has facilitated development of simple foraging models and more elaborate models of food sharing and gender division of labor, symbolic communication, long-term subsistence change, and cross-cultural variation in subsistence practices.

Significant modification of the environment by people was initiated by the domestication of fire, used to change vegetation structure and influence populations of food plants and animals. Vegetation burning is still common in the world, particularly in tropic regions. The arrival of humans with sophisticated tools precipitated the next major transformation of Earth, the extinction of large vertebrates. Agriculture drove the third major human modification of environments. Today about 35–40% of terrestrial primary production is appropriated by people, and the percentage is rising.

Humans will continue to exert powerful influences on the functioning of the Earth's ecological systems. The human population is destined to increase for many years. Rising affluence will be accompanied by increased consumption of resources and, hence, greater appropriation of the Earth's primary production. Nevertheless, many future human ecology scenarios are possible, depending on how much the human population grows and how growth is accommodated, the efficiency with which humans use and recycle resources, and the value that people give to preservation of biodiversity. *See* ECOLOGY; ENVIRONMENT. [G.H.O.]

Human genetics A discipline concerned with genetically determined resemblances and differences among human beings. Technological advances in the visualization of human chromosomes have shown that abnormalities of chromosome number or structure are surprisingly common and of many different kinds, and that they account for birth defects or mental impairment in many individuals as well as for numerous early spontaneous abortions. Progress in molecular biology has clarified the molecular structure of chromosomes and their constituent genes and the ways in which change in the molecular structure of a gene can lead to a disease. Concern about possible genetic damage through environmental agents and the possible harmful effects of hazardous substances in the environment on prenatal development has also stimulated research in human genetics. The medical aspects of human genetics have become prominent as nonhereditary causes of ill health or early death, such as infectious disease or nutritional deficiency, have declined, at least in developed countries.

In normal humans, the nucleus of each normal cell contains 46 chromosomes, which comprise 23 different pairs. Of each chromosome pair, one is paternal and the other maternal in origin. In turn, only one member of each pair is handed on through the reproductive cell (egg or sperm) to each child. Thus, each egg or sperm has only 23 chromosomes, the haploid number; fusion of egg and sperm at fertilization will restore the double, or diploid, chromosome number of 46.

The segregation of chromosome pairs during meiosis allows for a large amount of "shuffling" of genetic material as it is passed down the generation. Two parents can provide $2^{23} \times 2^{23}$ different chromosome combinations. This enormous source of variation is multiplied still further by the mechanism of crossing over, in which homologous chromosomes exchange segments during meiosis.

Twenty-two of the 23 chromosome pairs, the autosomes, are alike in both sexes; the other pair comprises the sex chromosomes. A female has a pair of X chromosomes; a male has a single X, paired with a Y chromosome which he has inherited from his father and will transmit to each of his sons. Sex is determined at fertilization, and depends on whether the egg (which has a single X chromosome) is fertilized by an X-bearing or a Y-bearing sperm.

Any gene occupies a specific chromosomal position, or locus. The alternative genes at a particular locus are said to be alleles. If a pair of alleles are identical, the individual is homozygous; if they are different, the individual is heterozygous.

Genetic variation has its origin in mutation. The term is usually applied to stable changes in DNA that alter the genetic code and thus lead to synthesis of an altered protein. The genetically significant mutations occur in reproductive cells and can therefore be transmitted to future generations. Natural selection acts upon the genetic diversity generated by mutation to preserve beneficial mutations and eliminate deleterious ones.

A very large amount of genetic variation exists in the human population. Everyone carries many mutations, some newly acquired but others inherited through innumerable generations. Though the exact number is unknown, it is likely that everyone is heterozygous at numerous loci, perhaps as many as 20%. *See* MUTATION.

The patterns of inheritance of characteristics determined by single genes or gene pairs depend on two conditions: (1) whether the gene concerned is on an autosome (autosomal) or on the X chromosome (X-linked); (2) whether the gene is dominant, that is, expressed in heterozygotes (when it is present on only one member of a chromosomal pair and has a normal allele) or is recessive (expressed only in homozygotes, when it is present at both chromosomes). [M.W.T.]

A quantitative trait is one that is under the control of many factors, both genetic and environmental, each of which contributes only a small amount to the total variability of the trait. The phenotype may show continuous variation (for example, height and skin color), quasicontinuous variation (taking only integer values—such as the number of ridges in a fingerprint), or it may be discontinuous (a presence/absence trait, such as diabetes or mental retardation). With discontinuous traits, it is assumed that there exists an underlying continuous variable and that individuals having a value of this variable above (or below) a threshold possess the trait.

A trait that "runs in families" is said to be familial. However, not all familial traits are hereditary because relatives tend to share common environments as well as common genes.

The variability of almost any trait is partly genetic and partly environmental. A rough measure of the relative importance of heredity and environment is an index called heritability. For example, in humans, the heritability of height is about 0.75. That is, about 75% of the total variance in height is due to variability in genes that affect height and 25% is due to exposure to different environments. [C.De.]

Hereditary diseases. Medical genetics has become an integral part of preventive medicine (that is, genetic counseling, including prenatal diagnostics). Hereditary diseases may be subdivided into three classes: chromosomal diseases; hereditary diseases with simple, mendelian modes of inheritance; and multifactorial diseases.

One out of 200 newborns suffers from an abnormality that is caused by a microscopically visible deviation in the number or structure of chromosomes. The most important clinical abnormality is Down syndrome—a condition due to trisomy of chromosome 21, one of the smallest human chromosomes. This chromosome is present not twice but three times; the entire chromosome complement therefore comprises 47, not 46, chromosomes. Down syndrome occurs one to two times in every 1000 births; its pattern of abnormalities derives from an imbalance of gene action during embryonic development. Down syndrome is a good example of a characteristic pattern of abnormalities that is produced by a single genetic defect.

Other autosomal aberrations observed in living newborns that lead to characteristic syndromes include trisomies 13 and 18 (both very rare), and a variety of structural aberrations such as translocations (exchanges of chromosomal segments between different chromosomes) and deletions (losses of chromosome segments). Translocations normally have no influence on the health status of the individual if there is no gain or loss of chromosomal material (these are called balanced translocations). However, carriers of balanced translocations usually run a high risk of having children in whom the same translocation causes gain or loss of genetic material, and who suffer from a characteristic malformation syndrome.

Clinical syndromes caused by specific aberrations vary, but certain clinical signs are common: low birth weights (small for date); a peculiar face; delayed general, and especially mental, development, often leading to severe mental deficiency; and multiple malformations, including abnormal development of limbs, heart, and kidneys.

Less severe signs than those caused by autosomal aberrations are found in individuals with abnormalities in number (and, sometimes, structure) of sex chromosomes. This is because in individuals having more than one X chromosome, the additional X chromosomes are inactivated early in pregnancy. For example, in women, one of the two X chromosomes is always inactivated. Inactivation occurs at random so that every normal woman is a mosaic of cells in which either one or the other X chromosome is active. Additional X chromosomes that an individual may have received will also be inactivated; in trisomies, genetic imbalance is thus avoided to a certain degree. However, inactivation is not complete; therefore, individuals with trisomies—for example, XXY (Klinefelter syndrome), XXX (triple-X syndrome), or XYY—or monosomies (XO; Turner syndrome) often show abnormal sexual development, intelligence, or behavior.

In contrast to chromosomal aberrations, the genetic defects in hereditary diseases with simple, mendelian modes of inheritance cannot be recognized by microscopic examination; as a rule, they must be inferred more indirectly from the phenotype and the pattern of inheritance in pedigrees. The defects are found in the molecular structure of the DNA. Often, one base pair only is altered, although sometimes more complex molecular changes, such as deletions of some bases or abnormal recombination, are involved. Approximately 1% of all newborns have, or will develop during their lives, a hereditary disease showing a simple mendelian mode of inheritance.

In medical genetics, a condition is called dominant if the heterozygotes deviate in a clearly recognizable way from the normal homozygotes, in most cases by showing an abnormality. Since such dominant mutations are usually rare, almost no homozygotes are observed.

In some dominant conditions, the harmful phenotype may not be expressed in a gene carrier (this is called incomplete penetrance), or clinical signs may vary in severeness between carriers (called variable expressivity). Penetrance and expressivity may be influenced by other genetic factors; sometimes, for example, by the sex of the affected person, whereas in other instances, the constitution of the "normal" allele has been implicated. Environmental conditions may occasionally be important. In most cases, however, the reasons are unknown.

X-linked modes of inheritance occur when the mutant allele is located on the X chromosome. The most important X-linked mode of inheritance is the recessive one. Here, the males (referred to as hemizygotes since they have only one allele) are affected, since they have no normal allele. The female heterozygotes, on the other hand, will be unaffected, since the one normal allele is sufficient for maintaining function. A classical example is hemophilia A, in which one of the serum factors necessary for normal blood clotting is inactive or lacking. (The disease can now be controlled by repeated substitution of the deficient blood factor—a good example for phenotypic therapy of a hereditary disease by substitution of a deficient gene product.) Male family members are affected whereas their sisters and daughters, while being unaffected themselves, transmit the mutant gene to half their sons. Only in very rare instances, when a hemophilic patient marries a heterozygous carrier, are homozygous females observed.

There are thousands of hereditary diseases with simple mendelian modes of inheritance, but most common anomalies and diseases are influenced by genetic variability at more than one gene locus. Most congenital malformations, such as congenital heart disease, cleft lip and palate, neural tube defects and many others, fall into this category, as do the constitutional diseases, such as diabetes mellitus, coronary heart disease, anomalies of the immune response and many mental diseases, such as schizophrenia or affective disorders. All of these conditions are common and often increase in frequency with advanced age. [F.V.]

Biochemical genetics. Biochemical genetics began with the study of inborn errors of metabolism. These are diseases of the body chemistry in which a small molecule such as a sugar or amino acid accumulates in body fluids because an enzyme responsible for its metabolic breakdown is deficient. This molecular defect is the result of mutation in the gene coding for the enzyme protein. The accumulated molecule, dependent on its nature, is responsible for the causation of a highly specific pattern of disease.

The field of biochemical genetics expanded with the recognition that similar heritable defective enzymes interfere with the breakdown of very large molecules, such as mucopolysaccharides and the complex lipids that are such prominent components of brain substance. The resultant storage disorders present with extreme alterations in morphology and bony structure and with neurodegenerative disease.

The majority of hereditary disorders of metabolism are inherited in an autosomal recessive fashion. In these families, each parent carries a single mutant gene on one chromosome and a normal gene on the other. Most of these mutations are rare. In populations with genetic diversity, most affected individuals carry two different mutations in the same gene. Some metabolic diseases are coded for by genes on the X chromosome. Most of these disorders are fully recessive, and so affected individuals are all males, while females carrying the gene are clinically normal. The disorders that result from mutations in the mitochondrial genome are inherited in nonmendelian fashion because mitochondrial DNA is inherited only from the mother. Those that carry a mutation are heteroplasmic; that is, each carries a mixed population of mitochondria, some with the mutation and some without.

Phenylketonuria (PKU) is a prototypic biochemical genetic disorder. It is an autosomally recessive disorder in which mutations demonstrated in a sizable number of families lead, when present in the genes on both chromosomes, to defective activity of the enzyme that catalyzes the first step in the metabolism of phenylalanine. This results in accumulation of phenylalanine and a recognizable clinical disease whose most prominent feature is severe retardation of mental development.

The diseases that result from mutation in mitochondrial DNA have been recognized as such only since the 1990s. They result from point mutations, deletions, and other rearrangements. A majority of these disorders express themselves chemically in elevated concentrations of lactic acid in the blood or cerebrospinal fluid. Many of the disorders are known as mitochondrial myopathies (diseases of muscles) because skeletal myopathy or cardiomyopathy are characteristic features. [W.L.Ny.]

Humidity Atmospheric water-vapor content, expressed in any of several measures, especially relative humidity, absolute humidity, humidity mixing ratio, and specific humidity.

Relative humidity is the ratio, in percent, of the moisture actually in the air to the moisture it would hold if it were saturated at the same temperature and pressure. It is a useful index of dryness or dampness for determining evaporation, or absorption of moisture.

Absolute humidity is the weight of water vapor in a unit volume of air expressed, for example, as grams per cubic meter or grains per cubic foot.

Humidity mixing ratio is the weight of water vapor mixed with unit mass of dry air, usually expressed as grams per kilogram. Specific humidity is the weight per unit mass of moist air and has nearly the same values as mixing ratio. [J.R.F.]

Humus The amorphous, ordinarily dark-colored, colloidal matter in soil, representing a complex of the fractions of organic matter of plant, animal, and microbial origin that are most resistant to decomposition.

Humus consists of the combined residues of organic materials which have lost their original structure following the rapid decomposition of the simpler ingredients and includes synthesized cell substance as well as by-products of microorganisms. It is not a definite substance and is in a continual state of flux, disappearing by slow decomposition, and being constantly renewed by incorporation of residual matter. With a balance between these processes, humus, though not static, remains relatively uniform in nature and amount in a given soil. It constitutes a reservoir of stabilizing material which imparts beneficial physical, chemical, and biological properties to soil. Fertile soils are rich in humus.

Humus improves the texture of soils. It exerts a binding effect on sandy soils, and loosens the harder, clayey soils, thus increasing their porosity and permeability. It increases the moisture-holding capacity and improves the granular structure by cementing mineral particles into stable crumbs. This helps soils resist the pulverizing and eroding action of wind, water, and cultivation. As a storehouse of elements important to plants, humus functions as a regulator of soil processes by liberating gradually nutrients that would otherwise drain away. A soil rich in humus provides optimum conditions for the development of beneficial microorganisms and constitutes the best medium for growth of plants.

Peat is a type of humus that results from the decomposition of plant material under conditions of excessive moisture or in areas submerged in water. It is an organic deposit formed in marshes and swamps by the partial decomposition of countless generations of a variety of plants. *See* BOG; PEAT. [A.G.L.]

Hurricane A tropical cyclone whose maximum sustained winds reach or exceed a threshold of 119 km/h (74 mi/h). In the western North Pacific ocean it is known as a typhoon. Many tropical cyclones do not reach this wind strength. *See* CYCLONE.

Maximum surface winds in hurricanes range up to about 200 mi (320 km) per hour. However, much greater losses of life and property are attributable to inundation from hurricane tidal surges and riverine or flash flooding than from the direct impact of winds on structures.

Tropical cyclones of hurricane strength occur in lower latitudes of all oceans except the South Atlantic and the eastern South Pacific, where combinations of cooler sea temperatures and prevailing winds whose velocities vary sharply with height prevent the establishment of a central warm core through a deep enough layer to sustain the hurricane wind system.

In the United States, property losses resulting from hurricanes have climbed steadily because of the increasing number of seashore structures. However, the loss of life, which has been huge in many storms, has decreased markedly. This is due mainly to the fact that warnings, aided by a more complete surveillance from aircraft and satellite, and extensive programs of public education, have become more accurate and more effective. Improvements in methodology for hurricane prediction have reduced the error in pinpointing hurricane landfall and have greatly reduced the probability of larger errors in prediction. *See* TROPICAL METEOROLOGY. [R.Sim.; J.Sim.]

Hydrography The measurement and description of the physical features and conditions of navigable waters and adjoining coastal areas, including oceans, rivers, and lakes. It involves geodesy, physical oceanography, marine geology, geophysics, photogrammetry (in coastal areas), remote sensing, and marine cartography. Basic parameters observed during a hydrographic survey are time, geographic position, depth of water, and bottom type. However, observation, analysis, and prediction of tides and currents area are also normally included in order to reduce depth measurements to a common vertical datum.

A principal objective of hydrography is to provide for safe navigation and protection of the marine environment through the production of up-to-date nautical charts and related publications. In addition, hydrographic data are essential to a multitude of other activities such as global studies, for example, shoreline erosion and sediment transport studies; coastal construction; delimitation of maritime boundaries; environmental protection and pollution control; exploration and exploitation of marine resources, both living and nonliving; and development of marine geographic information systems (GIS). *See* GEOGRAPHIC INFORMATION SYSTEMS.

Modern depth information is achieved with sonar measurements. Dual-frequency echo sounders are used, with a high-frequency, narrow beam to measure the depth below the vessel, and a lower-frequency, wider beam to obtain larger coverage of the terrain. Side-scan sonar, an instrument that transmits acoustic signals obliquely through the water, is normally towed behind the survey vessel and displays the returning echoes via an onboard graphic recorder. Although this technique does not allow exact determination of position and depth (both can be approximated), it provides excellent resolution with a depiction with what lies to either side of the vessel. Multibeam hydrographic survey systems consist of hull-mounted arrays such that a fan-shaped array of sound beams is transmitted perpendicular to the direction of the ship%s track. This provides for the possibility of 100% coverage of the sea floor.

Laser airborne systems mounted in fixed-wing aircraft or helicopters are also available for hydrographic surveys. The system emits a two-color laser beam, usually green and red, such that a return is received from the surface of the water by the red laser and from the bottom by the lower-frequency green laser, allowing the depth to be determined from the time difference. They can be operated in depths down to 165 ft (50 m), but more normally to 66 ft (20 m), depending on water clarity. Hydrographers use tide-coordinated aerial photography to delineate the high and low water lines for charting, which in turn is used for base-line determination of offshore boundaries. Satellite positioning of the aircraft using the Global Positioning System with carrier phase measurement and postprocessing of the data provides for determination of the position of the aircraft of the decimeter level. [C.An.; G.An.]

Hydrology The study of the waters of the Earth: their occurrence, circulation, and distribution; their chemical and physical properties; and their reaction with the environment, including their relation to living things. *See* TERRESTRIAL WATER.

Water in liquid and solid form covers most of the crust of the Earth. By a complex process powered by gravity and the action of solar energy, an endless exchange of water, in vapor, liquid, and solid forms, takes place between the atmosphere, the oceans, and the crust. This is known as the hydrologic cycle. Water circulates in the air and in the oceans, as well as over and below the surface of landmasses. The distribution of water in the planet is uneven. General patterns of circulation are present in the atmosphere, the oceans, and the landmasses, but regional features are very irregular and seemingly random in detail. Therefore, while causal relations underlie the overall process, it is believed that important elements of chance affect local hydrological events. *See* ATMOSPHERIC GENERAL CIRCULATION.

Whereas the global linkages of the hydrologic cycle are recognized, the science of hydrology has traditionally confined its direct concern to the detailed study of the portion of the cycle limited by the physical boundaries of the land; thus, it has generally excluded specialized investigations of the ocean (which is the subject of the science of oceanography) and the atmosphere (which is the subject of the science of meteorology). The heightened interest in anthropogenically induced environmental impacts has, however, underlined the critical role of the hydrologic cycle in the global transport and budgeting of mass, heat, and energy. Hydrology has become recognized as a science concerned with processes at the local, regional, and global scales. This enhanced status has strengthened its links to meteorology, climatology, and oceanography. *See* CLIMATOLOGY; METEOROLOGY; OCEANOGRAPHY.

A number of field measurements are performed for hydrologic studies. Among them are the amount and intensity of precipitation; the quantities of water stored as snow and ice, and their changes in time; discharge of streams; rates and quantities of infiltration into the soil, and movement of soil moisture; rates of production from wells and changes in their water levels as indicators of ground-water storage; concentration of chemical elements, compounds, and biological constituents in surface and ground waters; amounts of water transferred by evaporation and evapotranspiration to the atmosphere from snow, lakes, streams, soils, and vegetation; and sediment lost from the land and transported by streams.

In addition, hydrology is concerned with research on the phenomena and mechanisms involved in all physical and biological components of the hydrologic cycle, with the purpose of understanding them sufficiently to permit quantitative predictions and forecasting. The field investigations and measurements not only provide the data whereby the behavior of each component may be evaluated in detail, permitting formulation in quantitative terms, but also give a record of the historical performance of the entire system. Thus, two principal vehicles for hydrological forecasting and prediction become available: a set of elemental processes, whose operations are expressible in mathematical terms, linked to form deterministic models that permit the prediction of hydrologic events for given conditions; and a group of records or time series of measured hydrologic variables, such as precipitation or runoff, which can be analyzed by statistical methods to formulate stochastic models that permit inferences to be made on the future likelihood of hydrologic events. *See* GROUND-WATER HYDROLOGY; HYDROSPHERE. [M.A.Ma.]

Hydrometeorology The study of the occurrence, movement, and changes in the state of water in the atmosphere. The term is also used in a more restricted sense, especially by hydrologists, to mean the study of the exchange of water between the atmosphere and continental surfaces. This includes the processes of precipitation and direct condensation, and of evaporation and transpiration from natural surfaces. Considerable emphasis is placed on the statistics of precipitation as a function of area and time for given locations or geographic regions.

Water occurs in the atmosphere primarily in vapor or gaseous form. The average amount of vapor present tends to decrease with increasing elevation and latitude and also varies strongly with season and type of surface. Precipitable water, the mass of vapor per unit area contained in a column of air extending from the surface of the Earth to the outer extremity of the atmosphere, varies from almost zero in continental arctic air to about 6 g/cm^2 in very humid, tropical air.

Although a trivial proportion of the water of the globe is found in the atmosphere at any one instant, the rate of exchange of water between the atmosphere and the continents and oceans is high. Evaporation from the ocean surface and evaporation and transpiration from the land are the sources of water vapor for the atmosphere. Water vapor is removed from the atmosphere by condensation and subsequent precipitation in the form of rain, snow, sleet, and so on. The amount of water vapor removed by direct condensation at the Earth's surface (dew) is relatively small. *See* HYDROLOGY; METEOROLOGY; PRECIPITATION (METEOROLOGY). [E.M.R.]

Hydrosphere The water portion of the Earth as distinguished from the solid part and from the gaseous outer envelope (atmosphere). Approximately 74% of the Earth's surface is covered by water, in either the liquid or solid state. These waters, combined with minor contributions from ground waters, constitute the hydrosphere.

The oceans account for about 97% of the weight of the hydrosphere, while the amount of ice reflects the Earth's climate, being higher during periods of glaciation. There is a considerable amount of water vapor in the atmosphere. The circulation of the waters of the hydrosphere results in the weathering of the landmasses. The annual evaporation from the world oceans and from land areas results in an annual precipitation of 320,000 km^3 (76,000 mi^3) on the world oceans and 100,000 km^3 (24,000 mi^3) on land areas. The rainwater falling on the continents, partly taken up by the ground and partly by the streams, acts as an erosive agent before returning to the seas.

The unique chemical properties of water make it an effective solvent for many gases, salts, and organic compounds. Circulation of water and the dissolved material it contains is a highly dynamic process driven by energy from the Sun and the interior of the Earth. Each component has its own geochemical cycle or pathway through the hydrosphere, reflecting the component's relative abundance, chemical properties, and utilization by organisms. The introduction of materials by humans has significantly altered the composition and environmental properties of many natural waters. *See* GROUND-WATER HYDROLOGY; HYDROLOGY; LAKE; TERRESTRIAL WATER. [J.S.H.]

Hypermastigida An order of the Protozoa in the class Zoomastigophorea comprising the most complex flagellates, both structurally and in modes of division. All inhabit the alimentary canal of termites, cockroaches, and woodroaches. These organisms are multiflagellate. The nucleus is single and the organisms are plastic and slow-moving, generally ovoid to elongate. Flagella occur in spiral rows, in tufts, or over the enfire body. These flagellates vary from 15 to 350 micrometers in size. *See* PROTOZOA. [J.B.L.]

Hypodermis The outermost cell layer of the cortex, also called the exodermis, of plants. It forms a prominent layer immediately under the epidermis in many but not all plants. Like the endodermis, it develops Casparian strips, suberin deposits, and cellulose deposits impregnated with phenolic or quinoidal substances. The hypodermis may produce substances that act as a barrier to the entry of pathogens, and in some plants it may function in the absorption of water and the selection of ions that enter the plant. *See* CORTEX (PLANT); ENDODERMIS. [D.S.V.F.]

Hypoxia The failure of oxygen to gain access to, or to be utilized by, the body. Although the term anoxia is commonly used, a more precise term, hypoxia, is more often applicable because there is seldom a complete oxygen defect.

Oxygen deprivation may result from interference with some stage of the inspiration, lung diffusion, blood transport, cellular absorption, and final utilization by enzyme systems. A defect at any one or more of these major stages quite often induces a decreased ability of other related mechanisms to survive. This is seen most dramatically in any form of hypoxia in which the brain is deprived of the necessary oxygen for more than a few minutes. Nerve cell degeneration begins quickly, and although the original cause of hypoxia is removed, damage to the respiratory centers prevents resumption of breathing.

The term anoxia is used by many authorities to indicate an oxygen deficiency at the tissue level, and failure of cellular respiration may be designated histotoxic anoxia. There are other terms employed to differentiate the type of oxygen deficiency or the stage in the total respiratory process where defects occur. [E.G.St./N.K.M.]

I

Ice field A network of interconnected glaciers or ice streams, with common source area or areas, in contrast to ice sheets and ice caps. (An ice sheet is a broad, cakelike glacial mass with a relatively flat surface and gentle relief. Ice caps are properly defined as domelike glacial masses, usually at high elevation.) Being generally associated with terrane of substantial relief, ice-field glaciers are mostly of the broad-basin, cirque, and mountain-valley type. Thus, different sections of an ice field are often separated by linear ranges, bedrock ridges, and nunataks. [M.M.Mi.]

Iceberg A large mass of glacial ice broken off and drifted from parent glaciers or ice shelves along polar seas. Icebergs should be distinguished from polar pack ice which is sea ice, or frozen sea water, though rafted or hummocked fragments of the latter may resemble small bergs. *See* GLACIOLOGY; SEA ICE.

Icebergs are classified by shape and size. The terms used are arched, blocky, dome, pinnacled, tabular, valley, and weathered for berg description, and bergy-bit and growler for berg fragments ranging smaller than cottage size above water. The lifespan of an iceberg may be indefinite while the berg remains in cold polar waters, eroding only slightly during summer months. But under the influence of ocean currents, an iceberg that drifts into warmer water will disintegrate rapidly.

In the Arctic, icebergs (see illustration) originate chiefly from glaciers along Greenland coasts. It is estimated that a total of about 16,000 bergs are calved annually in the Northern Hemisphere, of which over 90% are of Greenland origin; but only about half of these have a size or source location to enable them to achieve any significant drift. No icebergs are discharged or drift into the North Pacific Ocean or its adjacent seas, except a few small bergs each year that calve from the piedmont glaciers along the Gulf of Alaska.

Arctic iceberg, eroded to form a valley or dry-dock type: grotesque shapes are common to the glacially produced icebergs of the North.

In the Southern Ocean, bergs originate from the giant ice shelves all along the Antarctic continent. These result in huge, tabular bergs or ice islands several hundred feet high and often over a hundred miles in length, which frequent the entire waters of the Antarctic seas. [R.P.D.]

Indian Ocean The smallest and geologically the most youthful of the three oceans. It differs from the Pacific and Atlantic oceans in two important aspects. First, it is landlocked in the north, does not extend into the cold climatic regions of the Northern Hemisphere, and consequently is asymmetrical with regard to its circulation. Second, the wind systems over its equatorial and northern portions change twice each year, causing an almost complete reversal of its circulation.

The eastern and western boundaries of the Indian Ocean are 147 and 20°E, respectively. In the southeastern Asian waters the boundary is usually placed across Torres Strait, and then from New Guinea along the Lesser Sunda Islands, across Sunda Strait and Singapore Strait.

The ocean floor is divided into a number of basins by a system of ridges. The largest is the Mid-Ocean Ridge, the greater part of which has a rather deep rift valley along its center. It lies like an inverted Y in the central portions of the ocean and ends in the Gulf of Aden. The Sunda Trench, stretching along Java and Sumatra, is the only deep-sea trench in the Indian Ocean. East of the Mid-Ocean Ridge, deep-sea sediments are chiefly red clay; in the western half of the ocean, globigerina ooze prevails and, near the Antarctic continent, diatom ooze.

Atmospheric circulation over the northern and equatorial Indian Ocean is characterized by the changing monsoons. In the southern Indian Ocean atmospheric circulation undergoes only a slight meridional shift during the year. The surface circulation is caused largely by winds and changes in response to the wind systems. In addition, strong boundary currents are formed, especially along the western coastline, as an effect of the Earth's rotation and of the boundaries created by the landmasses.

North of 10°S the changing monsoons cause a complete reversal of surface circulation twice a year. In February, during the Northeast Monsoon, flow north of the Equator is mostly to the west and the North Equatorial Current is well developed. Its water turns south along the coast of Somaliland and returns to the east as the Equatorial Countercurrent between about 2 and 10°S. In August, during the Southwest Monsoon, the South Equatorial Current extends to the north of 10°S; most of its water turns north along the coast of Somaliland, forming the strong Somali Current. North of the Equator flow is from west to east and is called the Monsoon Current. Parts of this current turn south along the coast of Sumatra and return to the South Equatorial Current. During the two transition periods between the Northeast and the Southwest monsoons in April–May and in October, a strong jetlike surface current flows along the Equator from west to east in response to the westerly winds during these months. *See* OCEAN CIRCULATION.

Both semidiurnal and diurnal tides occur in the Indian Ocean. The semidiurnal tides rotate around three amphidromic points situated in the Arabian Sea, southeast of Madagascar, and west of Perth. The diurnal tide also has three amphidromic points: south of India, in the Mozambique Channel, and between Africa and Antarctica. It has more the character of a standing wave, oscillating between the central portions of the Indian Ocean, the Arabian Sea, and the waters between Australia and Antarctica. [K.W.]

Industrial meteorology The application of meteorological information to industrial, business, or commercial problems. Generally, industrial meteorology is a branch of applied meteorology, which is the broad field where weather data, analyses, and forecasts are put to practical use. The term "private sector meteorology" has taken on the broader context of traditional industrial meteorology, expanding to include the provision of weather instrumentation/remote sensing devices, systems development and integration, and various consulting services to government and academia as well as value-added products and services to markets in industry (such as media, aviation, and utilities). Some areas in which industrial meteorology may be applied include environmental health and air-pollution control, weather modification, agricultural and forest management, and surface and air transportation. *See* AERONAUTICAL METEOROLOGY; AGRICULTURAL METEOROLOGY; METEOROLOGICAL INSTRUMENTATION.

Specific examples of the uses of industrial meteorology include many in the public sphere. For example, electric utilities need hourly predictions of temperature, humidity, and wind to estimate system load. In addition, they need to know when and where thunderstorms will impact their service area, so that crews can be deployed to minimize or correct disruptions to their transmission and distribution systems. Highway departments need to know when and where frozen precipitation will affect their service areas so that crews can be alerted, trucks loaded with sand and salt, and, if necessary, contractors hired to assist. Since a few degrees' change in temperature, or a slight change in intensity of snow or ice, determines the type of treatment required, early prediction and close monitoring of these parameters are critical.

Agricultural enterprises, from farmers to cooperatives to food manufacturers, rely on precise weather information and forecasts. Weather is the single most important factor in determining crop growth and production. Thus, monitoring and prediction of drought, floods, heat waves, and freezes are of extreme importance. *See* WEATHER FORECASTING AND PREDICTION.

Professionals involved with the meteorological aspects of air pollution are generally concerned with the atmospheric transport, distribution, transformation, and removal mechanisms of air pollutants. They are often called upon to evaluate the effectiveness of pollution control technologies or regulatory (policy) actions used to achieve and maintain air-quality goals. *See* AIR POLLUTION. [T.S.G.]

Industrial wastewater treatment A group of unit processes designed to separate, modify, remove, and destroy undesirable substances carried by wastewater from industrial sources. United States governmental regulations have been issued that involve volatile organic substances, designated priority pollutants; aquatic toxicity as defined by a bioassay; and in some cases nitrogen and phosphorus. As a result, sophisticated technology and process controls have been developed for industrial wastewater treatment.

Wastewater streams that are toxic or refractory should be treated at the source, and there are a number of technologies available. For example, wet air oxidation of organic materials at high temperature and pressure (2000 lb/in. or 14 kilopascals and 550°F or 288°C) is restricted to very high concentrations of these substances. Macroreticular (macroporous) resins are specific for the removal of particular organic materials, and the resin is regenerated and used again. Membrane processes, particularly reverse osmosis, are high-pressure operations in which water passes through a semipermeable membrane, leaving the contaminants in a concentrate. *See* HAZARDOUS WASTE.

Pretreatment and primary treatment processes address the problems of equalization, neutralization, removal of oil and grease, removal of suspended solids, and precipitation of heavy metals.

Aerobic biological treatment is employed for the removal of biodegradable organics. An aerated lagoon system is applicable (where large land areas are available) for treating nontoxic wastewaters, such as generated by pulp and paper mills. Fixed-film processes include the trickling filter and the rotating biological contactor. In these processes, a biofilm is generated on a surface, usually plastic. As the wastewater passes over the film, organics diffuse into the film, where they are biodegraded. Anaerobic processes are sometimes employed before aerobic processes for the treatment of high-strength, readily degradable wastewaters. The primary advantages of the anaerobic process is low sludge production and the generation of energy in the form of methane (CH_4) gas. *See* BIODEGRADATION; SEWAGE TREATMENT.

Biological processes can remove only degradable organics. Nondegradable organics can be present in the influent wastewater or be generated as oxidation by-products in the biological process. Many of these organics are toxic to aquatic life and must be removed from the effluent before discharge. The most common technology to achieve this objective is adsorption on activated carbon.

In some cases, toxic and refractory organics can be pretreated by chemical oxidation using ozone, catalyzed hydrogen peroxide, or advanced oxidation processes. In this case the objective is not mineralization of the organics but detoxification and enhanced biodegradability.

Biological nitrogen removal, both nitrification and denitrification, is employed for removal of ammonia from wastewaters. While this process is predictable in the case of municipal wastewaters, many industrial wastewaters are inhibitory to the nitrifying organisms.

Volatile organics can be removed by air or steam stripping. Air stripping is achieved by using packed or tray towers in which air and water counterflow through the tower. In steam stripping, the liquid effluent from the column is separated as an azeotropic mixture.

Virtually all of the processes employed for industrial wastewater treatment generate a sludge that requires some means of disposal. In general, the processes employed for thickening and dewatering are the same as those used in municipal wastewater treatment. Waste activated sludge is usually stabilized by aerobic digestion in which the degradable solids are oxidized by prolonged aeration.

Most landfill leachates have high and variable concentrations of organic and inorganic substances. All municipal and most industrial landfill leachates are amenable to biological treatment and can be treated anaerobically or aerobically, depending on the effluent quality desired. Activated carbon has been employed to remove nondegradable organics. In Europe, some plants employ reverse osmosis to produce a high-quality effluent. *See* WATER POLLUTION. [W.W.E.]

Infant diarrhea Diarrhea and its complications are the most important causes of infant death in most developing regions. The causes of the illness vary from dietary incompata-bilities to intestinal infection. The most important infectious causes are, in approximate order of importance: rotavirus, the bacteria *Shigella* (causing dysentery) and *Salmonella*, the parasite *Giardia lamblia*, and enteropathogenic *Escherichia coli* bacteria (a common cause of hospital nursery outbreaks). Breast-feeding is associated with a decreased occurrence of diarrhea and represents a major means of preventing infantile diarrhea in the developing world. *See* DIARRHEA. [H.L.D.]

Infection A term considered by some to mean the entrance, growth, and multiplication of a microorganism (pathogen) in the body of a host, resulting in the establishment of a disease process. Others define infection as the presence of a microorganism in host tissues whether or not it evolves into detectable pathologic effects. The host may be a bacterium, plant, animal, or human being, and the infecting agent may be viral, rickettsial, bacterial, fungal, or protozoan.

A differentiation is made between infection and infestation. Infestation is the invasion of a host by higher organisms such as parasitic worms. *See* EPIDEMIOLOGY; HOSPITAL INFECTIONS; MEDICAL BACTERIOLOGY; MEDICAL MYCOLOGY; MEDICAL PARASITOLOGY; OPPORTUNISTIC INFECTIONS; PATHOGEN; VIRUS. [D.N.La.]

Infectious disease A pathological condition spread among biological species. Infectious diseases, although varied in their effects, are always associated with viruses, bacteria, fungi, protozoa, multicellular parasites and aberrant proteins known as prions. A complex series of steps, mediated by factors contributed by both the infectious agent and the host, is required for microorganisms or prions to establish an infection or disease. Worldwide, infectious diseases are the third leading cause of human death.

The most common relationship between a host and a microorganism is a commensal one, in which advantages exist for both organisms. For example, hundreds of billions of bacteria of many genera live in the human gastrointestinal tract, coexisting in ecological balance without causing disease. These bacteria help prevent the invasion of the host by more virulent organisms. In exchange, the host provides an environment in which harmless bacteria can readily receive nutrients. There are very few microorganisms that cause disease every time they encounter a host. Instead, many factors of both host and microbial origin are involved in infectious disease. These factors include the general health of the host, previous exposure of the host to the microorganism, and the complement of molecules produced by the bacteria.

Spread of a pathogenic microorganism among individual hosts is the hallmark of an infectious disease. This process, known as transmission, may occur through four major pathways: contact with the microorganism, airborne inhalation, through a common vehicle such as blood, or by vector-borne spread.

The manner in which an infectious disease develops, or its pathogenesis, usually follows a consistent pattern. To initiate an infection, there must be a physical encounter as which the microorganism enters the host. The most frequent portals of entry are the respiratory, gastrointestinal, and genitourinary tracts as well as breaks in the skin. Surface components on the invading organism determine its ability to adhere and establish a primary site of infection. The cellular specificity of adherence of microorganisms often limits the range of susceptible hosts. For example, although measles and distemper viruses are closely related, dogs do not get measles and humans do not get distemper. From the initial site of infection, microorganisms may directly invade further into tissues or travel through the blood or lymphatic system to other organs.

Microorganisms produce toxins that can cause tissue destruction at the site of infection, can damage cells throughout the host, or can interfere with the normal metabolism of the host. The damage that microorganisms cause is directly related to the toxins they produce. Toxins are varied in their mechanism of action and host targets. *See* CHOLERA; STAPHYLOCOCCUS; TETANUS.

The host's reaction to an infecting organism is the inflammatory response, the body's most important internal defense mechanism. Although the inflammatory response is also seen as secondary to physical injury and nonspecific immune reactions, it is a reliable indicator of the presence of pathogenic microorganisms. Immune cells known as lymphocytes and granulocytes are carried by the blood to the site of infection. These cells either engulf and kill, or secrete substances which inhibit and neutralize, microorganisms. Other white blood cells, primarily monocytes, recognize foreign organisms and transmit chemical signals to other cells of the host's immune system, triggering the production of specific antibodies or specialized killer cells, both of which are lethal to the infecting microorganism. Any influence that reduces the immune system's ability to respond to foreign invasion, such as radiation therapy, chemotherapy, or destruction of immune cells by an immunodeficiency virus such as HIV, increases the likelihood that a organism will cause disease within the host.

Chemical compounds that are more toxic to microorganisms than to the host are commonly employed in the prevention and treatment of infectious disease; however, the emergence of drug-resistant organisms has led to increases in the morbidity and mortality associated with some infections. Other methods for controlling the spread of infectious diseases are accomplished by breaking a link in the chain of transmission between the host, microorganism, and mode of spread by altering the defensive capability of the host. Overall, the three most important advances to extend human life are clean water, vaccination, and antibiotics (in that order of importance).

Water-borne infections are controlled by filtration and chlorination of municipal water supplies. Checking food handlers for disease, refrigeration, proper cooking, and eliminating rodent and insect infestation have markedly reduced the level of food poisonings. The transmission of vector-borne diseases can be controlled by eradication of the vector. Blood-borne infections are reduced by screening donated blood for antibodies specific for HIV and other viruses and by rejecting donations from high-risk donors. For diseases such as tuberculosis, the airborne spread of the causative agent, *Mycobacterium tuberculosis*, can be reduced by quarantining infected individuals. The spread of sexually transmitted diseases, including AIDS, syphilis, and herpes simplex, can be prevented by inhibiting direct contact between the pathogenic microorganism and uninfected hosts. *See* ACQUIRED IMMUNE DEFICIENCY SYNDROME (AIDS); FOOD POISONING; WATER-BORNE DISEASE. [P.J.McN.]

Infectious mononucleosis A disease of children and young adults, characterized by fever and enlarged lymph nodes and spleen. EB (Epstein-Barr) herpesvirus is the causative agent.

Onset of the disease is slow and nonspecific with variable fever and malaise; later, cervical lymph nodes enlarge, and in about 50% of cases the spleen also becomes enlarged. The disease lasts 4–20 days or longer. Epidemics are common in institutions where young people live. EB virus infections occurring in early childhood are usually asymptomatic. In later childhood and adolescence, the disease more often accompanies infection—although even at these ages inapparent infections are common. [J.L.Me.]

Infectious myxomatosis A viral disease of European rabbits (*Oryctolagus*) and domestic rabbits, spread mainly by biting insects (mosquitoes and rabbit fleas) and characterized by edematous swellings of the skin, particularly on the head and anogenital area. The disease is caused by infection with myxoma virus, a pox virus, which occurs naturally in certain species of the genus *Sylvilagus* in North, Central, and South America. In these rabbits, infection results generally in localized, nonmalignant tumors that disappear in weeks or months. There is no effective treatment once clinical signs have appeared. Preventive measures include restriction of contact with insect vectors and vaccination with active or inactivated myxoma virus or with Shope fibroma virus. *See* ANIMAL VIRUS. [J.Ro.]

Infectious papillomatosis A nonfatal viral disease that occurs naturally in cottontail rabbits in states bordering the Mississippi River as well as in Oklahoma and Texas, and in brush rabbits in California. Cottontail rabbits from other parts of the United States and domestic rabbits (*Oryctolagus*) are also susceptible. The disease is caused by infection with Shope papilloma virus; it is spread naturally by contamination of broken skin or by the rabbit tick. The disease is characterized by cutaneous warts (papillomas) which can persist for months or years, sometimes becoming very prominent. Rabbits can be immunized against papillomatosis by injection of active or inactivated virus. *See* ANIMAL VIRUS. [J.Ro.]

Inflorescence A flower cluster segregated from any other flowers on the same plant, together with the stems and bracts (reduced leaves) associated with it. Certain plants produce inflorescences, whereas others produce only solitary flowers. *See* FLOWER. [G.J.W.]

Influenza An acute respiratory viral infection characterized by fever, chills, sore throat, headache, body aches, and severe cough. The term flu, which is frequently used incorrectly for various respiratory and even intestinal illnesses (such as stomach flu), should be used only for illness with these classic symptoms. The onset is typically abrupt, in contrast to common colds which begin slowly and progress over a period of days. Influenza is usually epidemic in occurrence. The first documented pandemic, or global epidemic, of influenza is considered to have been in 1580. The influenza pandemic of 1918, the most famous occurrence, was responsible for at least 20 million deaths worldwide.

The three types of influenza viruses, types A, B, and C, are classified in the virus family Orthomyxoviridae, and they are similar, but not identical, in structure and morphology. Types A and B are more similar in physical and biologic characteristics to each other than they are to type C. Influenza viruses may be spherical or filamentous in shape, and they are of medium size among common viruses of humans. *See* ANIMAL VIRUS.

When a cell is infected by two similar but different viruses of one type, especially type A, various combinations of the original parental viruses may be packaged or assembled into the new progeny; thus, a progeny virus may be a mixture of gene segments from each parental virus and therefore may gain a new characteristic, for example, a new surface protein. This phenomenon is called genetic reassortment, and the frequency with which it occurs and leads to viruses with new features is a significant cause of the constant appearance of new variants of the virus. In the laboratory, reassortment occurs between animal and human strains as well as between human strains. It probably occurs in nature also, and is thought to contribute to the appearance of new strains that infect humans. Generally, if a new variant is sufficiently different from the vaccine currently in use, the vaccine will provide limited or no protection.

The influenza virus has a short incubation period; that is, there is only a period of 1–3 days between infection and symptoms, and this leads to the abrupt development of symptoms that is a hallmark of influenza infections. The virus is typically shed in the throat for 5–7 days. Complete recovery from uncomplicated influenza usually takes several days, and the individual may feel weak and exhausted for a week or more after the major symptoms disappear. The two main complications of influenza are primary influenza virus pneumonia and secondary bacterial pneumonia. Primary influenza pneumonia is relatively infrequent, occurring in less than 1% of cases during an epidemic, although mortality may be 25–30%. The damage to epithelial cells and subsequent loss of the ability to clear particles from the respiratory tract can lead to secondary bacterial pneumonia. This problem commonly occurs in elderly individuals or those with underlying chronic lung disease or similar problems. Influenza-induced pneumonia may cause as many as 20,000 deaths in a typical influenza season. Another complication, known as Reye's syndrome, may follow influenza, and is more common in children. This disease of the brain develops within 2–12 days of a systemic viral infection, and can result in vomiting, liver damage, coma, and sometimes death.

All three types of influenza viruses can cause disease in humans, but there are significant differences in severity of the disease and the range of hosts. In contrast to the large number of animal species infected by type A virus, types B and C are only rarely isolated from animals and infect predominantly humans.

The presence or absence of antibodies is very important in the epidemiology of influenza. In individuals with no immunity, attack rates may reach 70% and severe illness may result. Even low levels of antibody may provide partial protection in an individual and decrease the severity of the illness to only coldlike symptoms.

During an epidemic, one strain of influenza is predominant, but it is not unusual for two or more other strains to be present as minor infections in a population. Outbreaks of influenza occur during cold-weather months in temperate climates, and typically most cases cluster on a period of 1–2 months, in contrast to broader periods of illness with many other respiratory viruses. An increased death rate due to primary pneumonia and bacterial superinfection is common and is one of the ways that public health authorities monitor an epidemic.

Control and prevention of influenza are attempted through the use of drugs and vaccines. Inactivated viral vaccines are used to prevent influenza, although use of attenuated live strains of the virus may better stimulate the cell-mediated immune response and provide higher-quality and longer-lasting immunity. The makeup of the vaccine is modified annually, based upon predictions of the expected prevalent strain for each flu season, but usually contains antigens of two type A viruses and one type B virus. These vaccines take advantage of the natural ability of the viral nucleic acid to reassort and form new strains. The vaccines utilize strains that are not virulent and will replicate at lower temperatures, as found in the nasopharnyx, but not at higher temperatures as found in the lower respiratory tract. The techniques of modern biotechnology are employed to clone copies of parts of the virus or to provide oligonucleotides corresponding to crucial functional areas of the virus, to obtain improved protection and reduced side effects. *See* BIOTECHNOLOGY. [J.M.Q.]

Inoculation The process of introducing a microorganism or suspension of microorganism into a culture medium. The medium may be (1) a solution of nutrients required by the organism or a solution of nutrients plus agar; (2) a cell suspension (tissue culture); (3) embryonated egg culture; or (4) animals, for example, rat, mouse, guinea pig, hamster, monkey, birds, or human being. When animals are used, the purpose usually is the activation of the immunological defenses against the organism. This is a form of vaccination, and quite often the two terms are used interchangeably. Both constitute a means of producing an artificial but active immunity against specific organisms, although the length of time given by such protection may vary widely with different organisms. [E.G.St./N.K.M.]

Insect control, biological The use of parasitoids, predators, and pathogens to reduce injurious pest insect populations and consequently the damage they cause. Viruses and bacteria are the most commonly used pathogens, but fungi, protozoa, and nematodes may also be important biological control agents.

Three ecological assumptions underlie biological control. First, natural enemies are among the prime factors responsible for the regulation, or control, of pest populations. Second, the influence exerted by parasitoids, predators, and pathogens is density-dependent. Density dependence means that the killing power of the natural enemy increases as the prey or host density increases. Conversely, the mortality induced by density-dependent natural enemies decreases as host density increases. In the dominant, or classical, form of biological control the third assumption is found: when an insect species escapes into a new area without its natural enemies, it reaches outbreak levels and becomes a pest. Biological-control practitioners believe regulation can be reestablished by importing the natural enemies of the pest from its area of origin.

In classical biological control, all efforts are typically directed toward establishing the natural enemies that were left behind in the area of origin. Classical biological control is by far the most frequently used form, assuming one excludes the use of resistant plant varieties as biological control.

Conservation involves manipulation of the environment in order to favor survival, reproduction, or any other aspect of the natural enemy's biology that affects its function as a biological control agent.

Aspects of research on and application of biological control may provide new or improved approaches. The improvement of biological control agents through selection, hybridization, or genetic engineering techniques may play an important role. A major strategy for control of pest insects may involve the use of genetic engineering to introduce traits into natural enemies that enhance their performance, or mortality-causing traits of natural enemies, such as insect pathogens, into plants. *See* BREEDING (PLANT); GENETIC ENGINEERING. [P.B.]

Insecta A class of the phylum Arthropoda, sometimes called the Hexapoda. In fact, Hexapoda is a superclass consisting of both Insecta and the related class Parainsecta (containing the springtails and proturans). Class Insecta is the most diverse group of organisms, containing about 900,000 described species, but there are possibly as many as 5 million to perhaps 20 million actual species of insects. Like other arthropods, they have an external, chitinous covering. Fossil insects dating as early as the Early Devonian have been found.

Classification. The class Insecta is divided into orders on the basis of the structure of the wings and the mouthparts, on the type of metamorphosis, and on various other characteristics. There are differences of opinion among entomologists as to the limits of some of the orders. The orders of insects (and their relatives the parainsects) are shown below.

Superclass Hexapoda
  Class Parainsecta
    Order: Protura; proturans
      Collembola; springtails
  Class Insecta
    Subclass Monocondylia
      Order: Diplura; telsontails
        Archaeognatha; bristletails
    Subclass Dicondylia
      Infraclass Apterygota
        Order: Zygentoma; silverfish, firebrats
      Infraclass Pterygota
        Section Palaeoptera
          Order: Ephemeroptera; mayflies
            Odonata; damselflies and dragonflies
        Section Neoptera
          Hemimetabola

Order: Plecoptera; stoneflies
Grylloblattodea; rockcrawlers
Orthoptera; grasshoppers, katydids, crickets
Phasmatodea; walkingsticks
Mantodea; mantises
Blattodea; cockroaches
Isoptera; termites
Dermaptera; earwigs
Embioptera; webspinners
Zoraptera; zorapterans
Psocoptera; psocids, booklice
Phthiraptera; lice
Thysanoptera; thrips
Hemiptera; cicadas, hoppers, aphids, whiteflies, scales

Holometabola
Order: Megaloptera; dobsonflies, alderflies
Raphidioidea; snakeflies
Neuroptera; lacewings, antlions
Coleoptera; beetles
Strepsiptera; twisted-wing parasites
Mecoptera; scorpionflies
Siphonaptera; fleas
Diptera; true flies
Trichoptera; caddisflies
Lepidoptera; moths, butterflies
Hymenoptera; sawflies, wasps, ants, bees

Morphology. Insects are usually elongate and cylindrical in form, and are bilaterally symmetrical. The body is segmented, and the ringlike segments are grouped into three distinct regions: the head, thorax, and abdomen. The head bears the eyes, antennae, and mouthparts; the thorax bears the legs and wings, when wings are present; the abdomen usually bears no locomotor appendages but often bears some appendages at its apex. Most of the appendages of an insect are segmented.

The skeleton is primarily on the outside of the body and is called an exoskeleton. However, important endoskeletal structures occur, particularly in the head. The body wall of an insect serves not only as a covering, but also as a supporting structure to which many important muscles are attached. The body wall of an insect is composed of three principal layers: the outer cuticula, which contains, among other chemicals, chitin; a cellular layer, the epidermis, which secretes the chitin; and a thin noncellular layer beneath the epidermis, the basement membrane. The surface of an insect's body consists of a number of hardened plates, or sclerites, separated by sutures or membranous areas, which permit bending or movement.

A pair of compound eyes usually cover a large part of the head surface. In addition most insects also possess two or three simple eyes, the ocelli, usually located on the upper part of the head between the compound eyes; each of these has a single lens.

Insect mouthparts typically consist of a labrum, or upper lip; a pair each of mandibles and maxillae; a labium, or lower lip; and a tonguelike structure, the hypopharynx. These structures are variously modified in different insect groups and are often used in classification and identification. The type of mouthparts an insect has determines how it feeds and what type of damage it is capable of causing.

Several forms of antennae are recognized, to which various names are applied; they are used extensively in classification. The antennae are usually located between or below the compound eyes and are often reduced to a very small size. They are sensory in function and act as tactile organs, organs of smell, and in some cases organs of hearing.

Insects are the only winged invertebrates, and their dominance as a group is probably due to their wings. Immature insects do not have fully developed wings, except in the mayflies. The wings may be likened to the two sides of a cellophane bag that have been pressed tightly together. The form and rigidity of the wing are due to the stiff chitinous veins which support and strengthen the membranous portion. At the base are small sclerites which serve as muscle attachments and produce consequent wing movement. The wings vary in number, placement, size, shape, texture, and venation, and in the position at which they are held at rest. Adult insects may be wingless or may have one pair of wings on the mesothorax, or, more often two pairs. There is a common basic pattern of wing venation in insects which is variously modified and in general quite specific for different large groups of insects. Much of insect classification depends upon these variations. A knowledge of fossil insects depends largely upon the wings, because they are among the more readily fossilized parts of the insect body.

Internal anatomy. The intake of oxygen, its distribution to the tissues, and the removal of carbon dioxide are accomplished by means of an intricate system of tubes called the tracheal system. The principal tubes of this system, the tracheae, open externally at the spiracles. Internally they branch extensively, extend to all parts of the body, and terminate in simple cells, the tracheoles. Many adaptations for carrying on respiration are known.

Insects possess an alimentary tract consisting of a tube, usually coiled, which extends from the mouth to the anus. It is differentiated into three main regions: the foregut, midgut, and hindgut. Valves between the three main divisions of the alimentary canal regulate the passage of food from one region to another.

The excretory system consists of a group of tubes with closed distal ends, the Malpighian tubules, which arise as evaginations of the anterior end of the hindgut. They vary in number from 1 to over 100, and extend into the body cavity. Various waste products are taken up from the blood by these tubules and passed out via the hindgut and anus.

The circulatory system of an insect is an open one. The only blood vessel is a tube located dorsal to the alimentary tract and extending through the thorax and abdomen. The posterior portion of this tube, the heart, is divided into a series of chambers, each of which has a pair of lateral openings called ostia. The anterior part of the tube is called the dorsal aorta.

The nervous system consists of a brain, often called the supraesophageal ganglion, located in the head above the esophagus; a subesophageal ganglion, connected to the brain by two commissures that extend around each side of the esophagus; and a ventral nerve cord, typically double, extending posteriorly through the thorax and abdomen from the subesophageal ganglion. In the nerve cords there are enlargements, called ganglia. Typically, there is a pair to each body segment. From each ganglion of the chain, nerves extend to each adjacent segment of the body, and also extend from the brain to part of the alimentary canal.

Reproduction in insects is nearly always sexual, and the sexes are separate. Variations from the usual reproductive pattern occur occasionally. In many social insects, such as the ants and bees, certain females, the workers, may be unable to reproduce because their sex organs are undeveloped; in some insects, individuals occasionally occur that have characters of both sexes, called gynandromorphs. Also, parthenogenesis—the process of females giving rise to females—is known in some species.

Metamorphosis. After insects hatch from an egg, they begin to increase in size and will also usually change, to some degree at least, in form and often in appearance. This developmental process is metamorphosis. The growth of an insect is accompanied by a series of molts, or ecdyses, in which the cuticle is shed and renewed.

The molt involves not only the external layers of the body wall, the cuticula, but also the cuticular linings of the tracheae, foregut, and hindgut; the cast skins often retain the shape of the insects from which they were shed. The shedding process begins with a splitting of the old cuticle. This split grows and the insect eventually wriggles out of the old cuticle. The new skin, remains soft and pliable long enough for the body to expand to its fullest capacity before hardening.

Insects differ regarding the number of molts during their growing period. Many have as few as four molts; a few species have 40 or more, and the latter continue to molt throughout life.

Insects have been grouped or classified on the basis of the type of metamorphosis they undergo. Although all entomologists do not agree upon the same classification, the following outline is presented:

1. Ametabolous or primitive: No distinct external changes are evident with an increase in size.

2. Hemimetabolous: Direct metamorphosis that is simple and gradual; immature forms resemble the adults except in size, wings, and genitalia. Immatures are referred to as nymphs or naiads if aquatic.

3. Holometabolous: Complete, or indirect, metamorphosis; stages in this developmental type are: egg→larva→pupa→adult (or imago). [D.M.DeL.]

Fossils. Insects and parainsects have a rich fossil record that extends to 415 million years, representing all taxonomic orders and 70% of all families that occur today, Insect deposits are characterized by an abundance of exceptionally well-preserved deposits known as Lagerstätten. Lagerstätten refer not only to the familiar amber deposits that entomb insects in hardened tree resin, but more importantly to a broad variety of typically laminar, sedimentary deposits. These deposits, formed in lake basins, are the most persistent of insect-bearing deposits and document the evolution of insect biotas during the past 300 million years. By contrast, the oldest amber is approximately 120 million years old and extends modern lineages and associated taxa to the Early Cretaceous. Other major types of insect deposits include terrestrial shales and fine-grained sandstones marginal to marine deposits during the Early and Middle Devonian, a proliferation of nodular ironstone-bearing strata of late Carboniferous age from the equatorial lowlands of the paleocontinent Euramerica, and distinctive lithographic limestones worldwide from the Middle Jurassic to Early Cretaceous. More modern deposits are Miocene to Recent sinter deposits created by hydrothermal zones with mineral-rich waters, and similarly aged asphaltum, representing the surface accumulation of tar. Lastly, insects are abundant in many Pleistocene glacial deposits of outwash and stranded lake sediments, formed by the waxing and waning of alpine and continental glaciers.

Various types of fossil documentation are important for understanding insect paleobiology, such as the body-fossil history of mouthparts. A recent study of insect mouthparts reveals a fivefold phase of increasing mouthpart disparity through time. This geochronologic deployment of the 34 basic types of modern insect mouthparts began during the Early Devonian with a few generalized types, was expanded during the late Carboniferous to early Permian to include major modifications of mandibulate and piercing-and-sucking types, and increased significantly again during the Late

Triassic to Early Jurassic to include filter-feeding mouthpart types and others involved in the ecologic penetration of aquatic ecosystems, and also intricate interactions with other animal and seed-plant hosts. During the Late Jurassic, there was expansion of mouthpart types involved in fluid-feeding on plant, fungal, and animal tissues and during the Early Cretaceous mouthpart innovation was completed by the addition of a few specialized mouthpart types involved in blood-feeding and other specialized associations. A comparison of taxonomic diversity and mouthpart disparity reveals that the generation of taxa has proceeded overall in a semilogarithmic increase reflected in a concave curve, whereas morphologic innovation, as revealed by mouthpart disparity, is a logistic process evidenced by a convex curve. This suggests that the deployment of basic morphologic types typically precedes taxonomic diversification in insect fossil history. [C.C.La]

Insecticide A material used to kill insects and related animals by disruption of vital processes through chemical action. Insecticides may be inorganic or organic chemicals. The principal source is from chemical manufacturing, although a few are derived from plants.

Insecticides are classified according to type of action as stomach poisons, contact poisons, residual poisons, systemic poisons, fumigants, repellents, attractants, insect growth regulators, or pheromones. Many act in more than one way. Stomach poisons are applied to plants so that they will be ingested as insects chew the leaves. Contact poisons are applied in a manner to contact insects directly, and are used principally to control species which obtain food by piercing leaf surfaces and withdrawing liquids. Residual insecticides are applied to surfaces so that insects touching them will pick up lethal dosages. Systemic insecticides are applied to plants or animals and are absorbed and translocated to all parts of the organisms, so that insects feeding upon them will obtain lethal doses. Fumigants are applied as gases, or in a form which will vaporize to a gas, so that they can enter the insects' respiratory systems. Repellents prevent insects from closely approaching their hosts. Attractants induce insects to come to specific locations in preference to normal food sources. Insect growth regulators are generally considered to act through disruption of biochemical systems or processes associated with growth or development, such as control of metamorphosis by the juvenile hormones, regulation of molting by the steroid molting hormones, or regulation of enzymes responsible for synthesis or deposition of chitin. Pheromones are chemicals which are emitted by one sex, usually the female, for perception by the other, and function to enhance mate location and identification; pheromones are generally highly species- specific.

Formulation of insecticides is extremely important in obtaining satisfactory control. Common formulations include dusts, water suspensions, emulsions, and solutions. Accessory agents, including dust carriers, solvents, emulsifiers, wetting and dispersing agents, stickers, deodorants or masking agents, synergists, and antioxidants, may be required to obtain a satisfactory product.

Proper timing of insecticide applications is important in obtaining satisfactory control. Whatever the technique used, the application of insecticides should be correlated with the occurrence of the most susceptible or accessible stage in the life cycle of the pest involved. By and large, treatments should be made only when economic damage by a pest appears to be imminent.

Among problems associated with insect control are the development of strains of insects resistant to insecticides; the assessment of the significance of small, widely distributed insecticide residues in and upon the environment; the development of better and more reliable methods for forecasting insect outbreaks; the evolvement of control programs integrating all methods—physical, physiological, chemical, biological, and cultural—for which practicality was demonstrated; the development of equipment and procedures to detect chemicals much below the part-per-million and microgram levels. As a consequence of the provisions of the Federal Insecticide, Fungicide, and Rodenticide Act as amended by the Federal Environmental Pesticide Control Act of 1972, there have been increased efforts to obtain data delineating mammalian toxicology, persistence in the environment, and immediate chronic impact of pesticides upon nontarget invertebrate and vertebrate organisms occupying aquatic, terrestrial, and arboreal segments of the environment. *See* INSECT CONTROL, BIOLOGICAL; INSECTA; PESTICIDE. [G.F.L.]

Insolation The incident radiant energy emitted by the Sun, which reaches a unit horizontal area of the Earth's surface. The term is a contraction of incoming solar radiation. About 99.9% of the Sun's energy is in the spectral range of 0.15–4.0 micrometers. About 95% of this energy is in the range of 0.3–2.4 μm; 1.2% is below 0.3 μm and 3.6% is above 2.4 μm. The bulk of the insolation (99%) is in the spectral region of 0.25–4.0 μm. About 40% is found in the visible region of 0.4–0.7 μm and only 10% is in wavelengths shorter than the visible. Energy of wavelengths shorter than 0.29 μm is absorbed high in the atmosphere by nitrogen, oxygen, and ozone.

Insolation depends on several factors: (1) the solar constant—that is, the amount of energy that in a unit time reaches a unit plane surface perpendicular to the Sun's rays outside the Earth's atmosphere, when the Earth is at its mean distance from the Sun; (2) the Sun's elevation in the sky; (3) the amount of solar radiation returned to space at the Earth-atmosphere boundary; and (4) the amount of solar radiation absorbed by the atmosphere and the amount of solar radiation reflected at the lower boundary of the Earth. Insolation is commonly expressed in units of watts per square meter, or calories per square centimeter per minute, also known as langley/min. For instance, the mean value of the solar constant has been estimated as 1368 W/m^2 (~1.96 ly/min), and the average insolation in summer for a midlatitude clear region could be 340 W/m^2 (700 ly/day), while for a cloudy region it is only about 120 W/m^2 (250 ly/day). *See* ATMOSPHERE; TERRESTRIAL RADIATION. [R.T.P.]

International Date Line The 180° meridian, where each day officially begins and ends. As a person travels eastward, against the apparent movement of the Sun, 1 h is gained for every 15° of longitude; traveling westward, time is lost at the same rate. Two people starting from any meridian and traveling around the world in opposite directions at the same speed would have the same time when they meet, but would be 1 day apart in date. When a traveler goes west across the line, a day is lost; if it is Monday to the east, it will be Tuesday immediately as the traveler crosses the International Date Line.

The 180° meridian is ideal for serving as the International Date Line (see illustration). It is exactly halfway around the world from the zero, or Greenwich, meridian, from which all longitude is reckoned. It also falls almost in the center of the largest ocean; consequently there is the least amount of inconvenience as regards population centers. A few deviations in the alignment have been made, such as swinging the line east around Siberia to keep that area all in the same day, and westward around the Aleutian

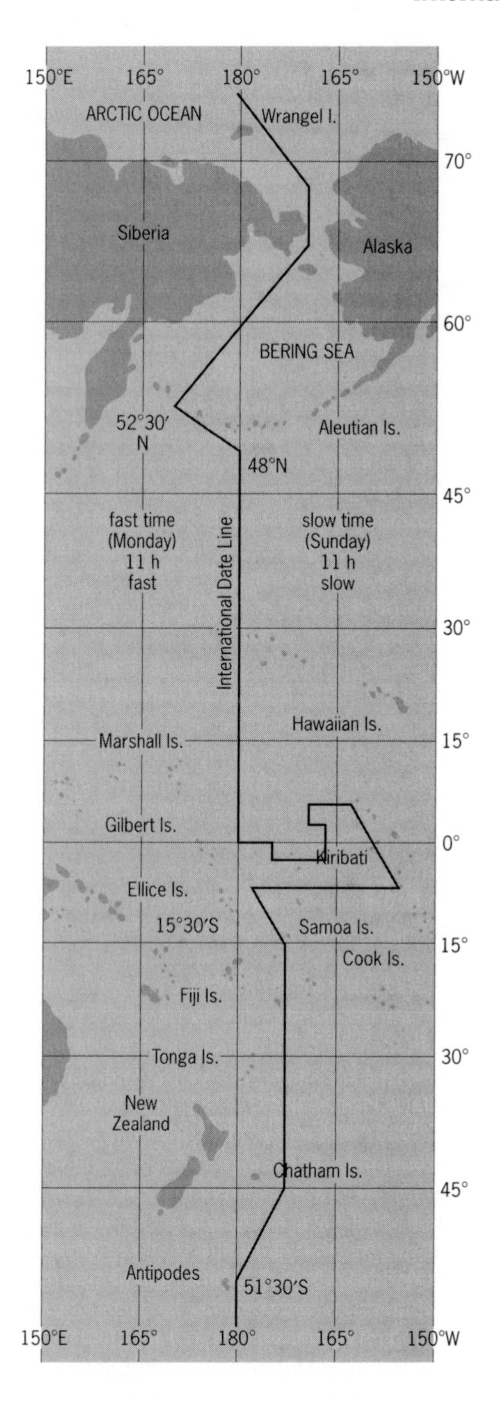

The International Date Line.

Islands so that they will be within the same day as the rest of Alaska. Other variations for the same purpose have been made near Kiribati, at the Equator, and the Fiji Islands, in the South Pacific. *See* MATHEMATICAL GEOGRAPHY. [V.H.E.]

Invasion ecology The study of the establishment, spread, and ecological impact of species translocated from one region or continent to another by humans. Biological invasions have gained attention as a tool for basic research, used to study the ecology and evolution of populations and of novel biotic interactions; and as a conservation issue tied to the preservation of biodiversity. The invasion of nonindigenous (also called exotic, alien, or nonnative) species is a serious concern for those charged with managing and protecting natural as well as managed ecosystems. *See* ECOLOGY; POPULATION ECOLOGY.

Ecologists make a distinction between introduced species, meaning any species growing outside its natural habitat including cultivated or domesticated organisms, and invasive species, meaning the subset of introduced species that establish free-living populations in the wild. The great majority of introduced species (approximately 90% as estimated from some studies) do not become invasive. While certain problem invaders, such as the zebra mussel (*Dreissena polymorpha*), exact enormous economic and ecological costs, other introduced species are generally accepted as beneficial additions, such as most major food crops.

Intentional plant introductions have been promoted primarily by the horticulture industry to satisfy the public's desire for novel landscaping. However, plants have also been introduced for agriculture, for silviculture, and for control of soil erosion. Intentional animal introductions include game species brought in for sport hunting or fishing. Unlike these examples, intentional introductions can also include species that are not necessarily intended to form self-sustaining populations, such as those promoted by the aquarium or pet trade.

Species introduced accidentally are "hitchhikers." Shipping ballast has been a major vector, first in the form of soil carrying terrestrial invertebrates and plant seeds or rhizomes, and more recently in the form of ballast water carrying planktonic larvae from foreign ports. While many species are introduced in ballast or by similar means, hitchhikers can also be unwanted parasites that bypass importation and quarantine precautions. For example, many nonindigenous agricultural weeds have been imported in contaminated seed lots.

Certain types of habitats seem to have higher numbers of established nonindigenous species than others. The characteristics that make a site open to invasion must be determined. For example, islands are notably vulnerable to invasions. Islands usually have fewer resident species to begin with, leading to the conjecture that simpler systems have less biotic resistance to invaders. That is, an introduced species is less likely to be met by a resident competitor, predator, or pathogen capable of excluding it. The idea of biotic resistance is also consistent with the idea that complexity confers stability in natural systems. *See* INVASION ECOLOGY.

A second generalization about invasibility is that ecosystems with high levels of anthropogenic disturbance, such as agricultural fields or roadsides, seem to be more invaded. Increased turnover of open space in these sites could provide more opportunities for the establishment of new species. An alternative explanation is that many species that adapted to anthropogenic habitats in Europe simply tagged along as humans re-created those habitats in new places. Those species would naturally have an advantage over native species at exploiting human disturbances. A final suggestion by proponents of ecosystem management is that disturbance (including, in this context, a disruption of natural disturbance regimes, for example, fire suppression) weakens the inherent resistance of ecosystems and promotes invasion.

Invasive species can have several different types of impacts. First, they can affect the traits and behavior of resident organisms (for example, causing a shift in diet, size, or shape of the native species they encounter). Second, impacts can occur at the level of the population, either by changing the abundance of a native population or by changing its genetic composition. Hybridization between an invader and a closely related native can result in introgression and genetic pollution. The endpoint can be the de facto extinction of the native species when the unique aspects of its genome are overwhelmed. Third, impacts can occur at the level of ecological communities. When individual populations are reduced or even driven extinct by competition or predation by an invasive species, the result is a decrease in the overall biodiversity of the invaded site. Finally, invaders can impact not only other species but the physical characteristics of an ecosystem as well.

There are two main contributing factors in determining which species have the biggest impacts: abundance and special characteristics. Invaders that reach extremely high density simply overwhelm all other organisms. Other species have special traits that result in an impact out of proportion to their numbers.

Because of the economic and conservation importance of nonindigenous species, much of invasion ecology focuses on the prevention, eradication, and control of invaders, and the restoration of sites after control. Research has emphasized the importance of early detection and eradication of problem species. Biological control has been touted as an environmentally friendly alternative to herbicides and pesticides. *See* ALLELOPATHY; ECOLOGICAL COMMUNITIES; ECOLOGICAL SUCCESSION; SPECIATION; SPECIES CONCEPT. [I.M.P.]

Ironwood The name given to any of at least 10 kinds of tree in the United States, including the American hornbeam (*Carpinus caroliniana*), eastern hophombeam (*Ostrya virginiana*), buckthorn bumelia (*Bumelia lycioides*), tough bumelia (*B. tenax*), buckwheat tree (*Cliftonia monophylla*), and swamp cyrilla or swamp ironwood (*Cyrilla racemiflora*). Leadwood (*Krugiodendron ferreum*), a native of southern Florida, has the highest specific gravity of all woods native to the United States and is also known as black ironwood. *See* HOPHORNBEAM; HORNBEAM. [A.H.G./K.P.D.]

Irrigation (agriculture) The artificial application of water to the soil to produce plant growth. Irrigation also cools the soil and atmosphere, making the environment favorable for plant growth. Water is applied to crops by surface, subsurface, sprinkler, and drip irrigation.

Surface irrigation includes furrow and flood methods. The furrow method is used for row crops. Corrugations or rills are small furrows used on close-growing crops. The flow, carried in furrows, percolates into the soil. Flow to the furrow is usually supplied by siphon tubes, spiles, gated pipe, or valves from buried pipe. In the flood method, controlled flooding is done with border strips, contour or bench borders, and basins.

Subirrigation is a type of irrigation accomplished by raising the water table to the root zone of the crop or by carrying moisture to the root zone by perforated underground pipe. Either method requires special soil conditions for successful operation.

A sprinkler system consists of pipelines which carry water under pressure from a pump or elevated source to lateral lines along which sprinkler heads are spaced at appropriate intervals. Laterals are moved from one location to another by hand or tractor, or they are moved automatically. The side-roll wheel system, which utilizes the lateral as an axle (see illustration), is very popular as a labor-saving method.

Drip irrigation is a method of providing water to plants almost continuously through small-diameter tubes and emitters. It has the advantage of maintaining high moisture levels at relatively low capital costs. It can be used on very steep, sandy, and rocky areas and can utilize saline waters better than most other systems. The system has been most

A side-roll sprinkler system which uses the main supply line (often more than 1000 ft, or 300 m, long) to carry the sprinkler heads and as the axle for wheels.

popular in orchards and vineyards, but is also used for vegetables, ornamentals, and for landscape plantings. *See* LAND DRAINAGE (AGRICULTURE); TERRACING (AGRICULTURE).

[M.A.H.]

Island biogeography The distribution of plants and animals on islands. Islands harbor the greatest number of endemic species. The relative isolation of many islands has allowed populations to evolve in the absence of competitors and predators, leading to the evolution of unique species that can differ dramatically from their mainland ancestors.

Plant species produce seeds, spores, and fruits that are carried by wind or water currents, or by the feet, feathers, and digestive tracts of birds and other animals. The dispersal of animal species is more improbable, but animals can also be carried long distances by wind and water currents, or rafted on vegetation and oceanic debris. Long-distance dispersal acts as a selective filter that determines the initial composition of an island community. Many species of continental origin may never reach islands unless humans accidentally or deliberately introduce them. Consequently, although islands harbor the greatest number of unique species, the density of species on islands (number of species per area) is typically lower than the density of species in mainland areas of comparable habitat. *See* POPULATION DISPERSAL.

Once a species reaches an island and establishes a viable population, it may undergo evolutionary change because of genetic drift, climatic differences between the mainland and the island, or the absence of predators and competitors from the mainland. Consequently, body size, coloration, and morphology of island species often evolve rapidly, producing forms unlike any related species elsewhere. Examples include the giant land tortoises of the Galápagos, and the Komodo dragon, a species of monitor lizard from Indonesia. *See* POPULATION GENETICS.

If enough morphological change occurs, the island population becomes reproductively isolated from its mainland ancestor, and it is recognized as a unique species. Because long-distance dispersal is relatively infrequent, repeated speciation may occur as populations of the same species successively colonize an island and differentiate. The most celebrated example is Darwin's finches, a group of related species that inhabit the Galápagos Islands and were derived from South American ancestors. The island species have evolved different body and bill sizes, and in some cases occupy unique ecological niches that are normally filled by mainland bird species. The morphology of these finches was first studied by Charles Darwin and constituted important evidence for his theory of natural selection. *See* ANIMAL EVOLUTION; SPECIATION.

Island biogeography theory has been extended to describe the persistence of single-species metapopulations. A metapopulation is a set of connected local populations in a fragmented landscape that does not include a persistent source pool region. Instead, the fragments themselves serve as stepping stones for local colonization and extinction. The most successful application of the metapopulation model has been to spotted owl populations of old-growth forest fragments in the northwestern United States. *See* BIOGEOGRAPHY; ECOLOGICAL COMMUNITIES; ECOSYSTEM. [N.J.Go.]

Isobar (meteorology) A curve along which pressure is constant. Leading examples of its uses are in weather forecasting and meteorology. The most common weather maps are charts of weather conditions at the Earth's surface and mean sea level, and they contain isobars as principal information. Areas of bad or unsettled weather are readily defined by roughly circular isobars around low-pressure centers at mean sea level. Likewise, closed isobars around high-pressure centers define areas of generally fair weather. *See* AIR PRESSURE.

A principal use of isobars stems from the so-called geostrophic wind, which approximates the actual wind on a large scale. The direction of the geostrophic wind is parallel to the isobars, in the sense that if an observer stands facing away from the wind, higher pressures are to the person's right if in the Northern Hemisphere and to the left if in the Southern. Thus, in the Northern Hemisphere, flow is counterclockwise about low-pressure centers and clockwise about high-pressure centers, with the direction of the flow reversed in the Southern Hemisphere. *See* WEATHER MAP. [F.B.Sh.]

J,K

Jet stream A relatively narrow, fast-moving wind current flanked by more slowly moving currents. Jet streams are observed principally in the zone of prevailing westerlies above the lower troposphere and in most cases reach maximum intensity, with regard both to speed and to concentration, near the tropopause. At a given time, the position and intensity of the jet stream may significantly influence aircraft operations because of the great speed of the wind at the jet core and the rapid spatial variation of wind speed in its vicinity. Lying in the zone of maximum temperature contrast between cold air masses to the north and warm air masses to the south, the position of the jet stream on a given day usually coincides in part with the regions of greatest storminess in the lower troposphere, though portions of the jet stream occur over regions which are entirely devoid of cloud. The jet stream is often called the polar jet, because of the importance of cold, polar air. The subtropical jet is not associated with surface temperature contrasts, like the polar jet. Maxima in wind speed within the jet stream are called jet streaks. *See* ATMOSPHERIC GENERAL CIRCULATION. [F.S.; H.B.B.]

Johne's disease Chronic inflammation of the mucosa of the ileocecal valve and adjacent tissues of the gastrointestinal tract of cattle, sheep, goats, and captive wild ruminants, caused by the bacterium *Mycobacterium paratuberculosis*. Transmission of *M. paratuberculosis* is primarily by ingestion of feces from animals shedding the organism. The incubation period varies from 2 to 3 years or more. Diseased animals in advanced stages have intermittent or persistent diarrhea without fever and often become emaciated.

Johne's disease is diagnosed by serologic tests and mycobacteriologic examinations conducted on feces or tissues collected by biopsy or at necropsy. To confirm a diagnosis of Johne's disease, it is necessary to isolate and identify the etiologic agent.

Therapeutic drugs are not available for routine treatment of animals. A killed whole-cell vaccine is available for use in calves 1–35 days of age. Live attenuated strains of *M. paratuberculosis* have been used for vaccinating cattle in a few countries, but are not approved for use in the United States or Canada. *Mycobacterium paratuberculosis* has not been shown to cause disease in humans; however, a *M. paratuberculosis*-like organism has been isolated from a few individuals with Crohn's disease. *See* MYCOBACTERIAL DISEASES. [C.O.T.]

Karst topography Distinctive associations of third-order, erosional landforms indented into second-order structural forms such as plains and plateaus. They are produced by aqueous dissolution, either acting alone or in conjunction with (and as the trigger for) other erosion processes. Karst is largely restricted to the most soluble rocks, which are salt, gypsum and anhydrite, and limestone and dolostone.

The essence of the karst dynamic system is that meteoric water (rain or snow) is routed underground, because the rocks are soluble, rather than flowing off in surface river channels. It follows that dissolutional caves develop in fracture systems, resurging as springs at the margins of the soluble rocks or in the lowest places. A consequence is that most karst topography is "swallowing topography," assemblages of landforms created to deliver meteoric water down to the caves.

Karst landforms develop at small, intermediate, and large scales. Karren is the general name given to small-scale forms—varieties of dissolutional pits, grooves, and runnels. Individuals are rarely greater than 10 m (30 ft) in length or depth, but assemblages of them can cover hundreds of square kilometers. On bare rock, karren display sharp edges; circular pits or runnels extending downslope predominate. Beneath soil, edges are rounded and forms more varied and intricate.

Sinkholes, also known as dolines or closed depressions, are the diagnostic karst (and pseudokarst) landform. They range from shallow, bowllike forms, through steep-sided funnels, to vertical-walled cylinders. Asymmetry is common. Individual sinkholes range from about 1 to 1000 m (3 to 3300 ft) in diameter and are up to 300 m (1000 ft) deep. Many may become partly or largely merged.

Dry valleys and gorges are carved by normal rivers, but progressively lose their water underground (via sinkholes) as the floors become entrenched into karst strata. Many gradations exist, from valleys that dry up only during dry seasons (initial stage) to those that are without any surface channel flow even in the greatest flood periods (paleovalleys). They are found in most plateau and mountain karst terrains and are greatest where river water can collect on insoluble rocks before penetrating the karst (allogenic rivers).

Poljes, a Serbo-Croatian term for a field, is the generic name adopted for the largest individual karst landform. This is a topographically closed depression with a floor of alluvium masking an underlying limestone floor beveled flat by planar corrosion.

Karst plains and towers are the end stage of karst topographic development in some regions, produced by long-sustained dissolution or by tectonic lowering. The plains are of alluvium, with residual hills (unconsumed intersinkhole limestone) protruding through. Where strata are massively bedded and the hills are vigorously undercut by seasonal floods or allogenic rivers, they may be steepened into vertical towers. [D.Fo.]

Kennel cough A common, highly contagious respiratory disease of dogs, also known as canine infectious tracheobronchitis. Several different bacteria and viruses are usually associated with the disease. Symptoms are generally mild but may vary widely depending on the agent, the host, and environmental factors. The main feature of the disease is sudden onset of violent coughing in dogs that had a recent exposure to other, infected dogs. The disease is easily transmitted between dogs by droplets in the air or direct contact, and often occurs as outbreaks or as a seasonal infection. Most dogs completely recover within 2 weeks; however, chronic and severe forms of the disease sometimes occur.

Infectious agents commonly associated with the disease are the bacterium *Bordetella bronchiseptica* and canine parainfluenza virus. Each agent is capable of producing a mild form of the disease; however, most single-agent infections probably show no symptoms of disease. Several species of mycoplasmas have been isolated from the lower respiratory tract of dogs with kennel cough, but always in combination with another agent (for example, bordetella or canine parainfluenza virus). These mycoplasmas are normally found in the upper respiratory tract of healthy dogs. *See* BORDETELLA.

Close contact with other dogs is usually required for transmission of kennel cough. Each of the viral agents of the disease, and possibly some of the mycoplasmas, has

host ranges restricted to dogs. Because infections that show no symptoms of disease are also common, it is sometimes difficult to determine the source of the infection. Canine parainfluenza virus and *B. bronchiseptica* do not usually persist longer than a few weeks or a few months, respectively, in an individual dog.

Treatment of kennel cough is often unwarranted. However, antitussives, bronchodilators, and corticosteroids are used to relieve coughing, and antimicrobials are used to treat or prevent bronchopneumonia. The risk of acquiring kennel cough can be reduced by minimizing exposure to infectious agents. [D.A.Be.]

Klebsiella A genus of gram-negative, nonmotile bacteria. Characteristic large mucoid colonies are due to production of a large amount of capsular material. Species of *Klebsiella* are commonly found in soil and water, on plants, and in animals and humans. Harmless strains of *Klebsiella* are beneficial because they fix nitrogen in soil. Pathogenic species include *K. pneumoniae, K. rhinoscleromatis*, and *K. ozaenae*, also known as *K. pneumoniae* subspecies *pneumoniae, rhinoscleromatis*, and *ozaenae*.

Klebsiella pneumoniae is the second most frequently isolated colon-related bacterium in clinical laboratories. The carbohydrate-containing capsule of *Klebsiella* promotes virulence by protecting the encased bacteria from ingestion by leukocytes; nonencapsulated variants of *Klebsiella* do not cause disease. Capsular types 1 and 2 cause pneumonia; types 8, 9, 10, and 24 are commonly associated with urinary tract infections. *See* ESCHERICHIA; PNEUMONIA.

Klebsiella accounts for a large percentage of hospital-acquired infections, mostly skin infections (in immunocompromised burn patients), bacteremia, and urinary tract infections. It is also the most common contaminant of intravenous fluids such as glucose solutions and other medical devices. *See* HOSPITAL INFECTIONS.

Klebsiella may produce *E. coli*-like enterotoxins and cause acute gastroenteritis in infants and young children. Enteric illnesses due to *Klebsiella* are more predominant where populations are more crowded and conditions less sanitary. Other virulence factors of *Klebsiella* include a relatively high ability to survive and multiply outside the host in a variety of environments, and its relatively simple growth requirements. *See* ENDOTOXIN.

Klebsiella rhinoscleromatis causes rhinoscleroma, a chronic destructive granulomatous disease of the upper respiratory tract that is most common in eastern Europe, central Africa, and tropical South America. *Klebsiella ozaenae* is one cause of chronic rhinitis (ozena), a destructive atrophy of the nasal mucosa, and is infrequently isolated from urinary tract infections and bacteremia. *See* MEDICAL BACTERIOLOGY. [D.J.Ev.]

L

Lake An inland body of water, small to moderately large in size, with its surface exposed to the atmosphere. Most lakes fill depressions below the zone of saturation in the surrounding soil and rock materials. Generically speaking, all bodies of water of this type are lakes, although small lakes usually are called ponds, tarns (in mountains), and less frequently pools or meres. The great majority of lakes have a surface area of less than 100 mi^2 (259 km^2). More than 30 well-known lakes, however, exceed 1500 mi^2 (3885 km^2) in extent, and the largest freshwater body, Lake Superior, North America, covers 31,180 mi^2 (80,756 km^2). Most lakes are relatively shallow features of the Earth's surface. Because of their shallowness, lakes in general may be considered evanescent features of the Earth's surface, with a relatively short life in geological time.

Lakes differ as to the salt content of the water and as to whether they are intermittent or permanent. Most lakes are composed of fresh water, but some are more salty than the oceans. Generally speaking, a number of water bodies which are called seas are actually salt lakes; examples are the Dead, Caspian, and Aral seas. All salt lakes are found under desert or semiarid climates, where the rate of evaporation is high enough to prevent an outflow and therefore a discharge of salts into the sea.

Lakes with fresh waters also differ greatly in the composition of their waters. Because of the balance between inflow and outflow, fresh lake water composition tends to assume the composite dissolved solids characteristics of the waters of the inflowing streams—with the lake's age having very little influence. Under a few special situations, as crater lakes in volcanic areas, sulfur or other gases may be present in lake water, influencing color, taste, and chemical reaction of the water. *See* FRESH-WATER ECOSYSTEM; HYDROSPHERE; MEROMICTIC LAKE; SURFACE WATER.

Both natural and artificial lakes are economically significant for their storage of water, regulation of stream flow, adaptability to navigation, and recreational attractiveness. A few salt lakes are significant sources of minerals. *See* EUTROPHICATION. [E.A.A.]

Land drainage (agriculture) The removal of water from the surface of the land and the control of the shallow groundwater table improves the soil as a medium for plant growth. The sources of excess water may be precipitation, snowmelt, irrigation water, overland flow or underground seepage from adjacent areas, artesian flow from deep aquifers, floodwater from channels, or water applied for such special purposes as leaching salts from the soil or for achieving temperature control.

The purpose of agricultural drainage can be summed up as the improvement of soil water conditions to enhance agricultural use of the land. Such enhancement may come about by direct effects on crop growth, by improving the efficiency of farming operations or, under irrigated conditions, by maintaining or establishing a favorable salt regime. Drainage systems are engineering structures that remove water according to the principles of soil physics and hydraulics. The consequences of drainage, however, may also include a change in the quality of the drainage water. Agricultural drainage is divided into two broad classes; surface and subsurface. Some installations serve both purposes. [J.N.Lu.]

Land-use planning The long-term development or conservation of an area and the establishment of a relationship between local objectives and regional goals. Land-use planning is often guided by laws and regulations. The major instrument for current land-use planning is the establishment of zones that divide an area into districts which are subject to specified regulations. Although land-use planning is sometimes done by private property owners, the term usually refers to permitting by government agencies. Land-use planning is conducted at a variety of scales, from plans by local city governments to regulations by federal agencies. The United States has never developed a national land-use plan because land use is considered a local concern.

A major part of local planning is zoning, the division of areas into districts. Zones cover most potential uses, such as residential, commercial, light industry, heavy industry, open space, or transportation infrastructure (such as rail lines or highways). Detailed regulations guide how each zone can be used. As a result of pressures from rapid growth, some cities have begun to write growth management plans that limit the pace of growth. Comprehensive city plans aimed to limit the pace of growth have been accepted by the courts.

Very few plans have been undertaken at a statewide scale. Each state plan differs by the needs and philosophy of the state. The state plans represent a balance of regional structures that address widespread growth with local powers that keep specific decision-making at the local level.

Environmental regulations are among the few national-level policies that have direct implications for land-use planning. Four of the major types of environmental laws that impact land-use planning are wetland laws, clean-air laws, clean-water laws, and laws for the protection of endangered species. *See* ENDANGERED SPECIES; WETLANDS.

Land-use planning, in large part, has focused on urban planning. Increasingly, land-use planning is done at larger scales and involves multiple issues. Awareness of environmental concerns, coupled with the wide availability of technical tools that include digital maps at all scales, has led to new approaches to land-use planning. These approaches often use ideas from landscape ecology, such as the concepts of patches; edges, boundaries, and fragmentation; buffer zones; and corridors and connectivity. *See* ECOLOGY, APPLIED; LANDSCAPE ECOLOGY. [C.Sc.]

Landscape ecology The study of the distribution and abundance of elements within landscapes, the origins of these elements, and their impacts on organisms and processes. A landscape may be thought of as a heterogeneous assemblage or mosaic of internally uniform elements or patches, such as blocks of forest, agricultural fields, and housing subdivisions. Biogeographers, land-use planners, hydrologists, and ecosystem ecologists are concerned with patterns and processes at large scale. Landscape ecologists bridge these disciplines in order to understand the interplay between the natural and human factors that influence the development of landscapes, and the impacts of landscape patterns on humans, other organisms, and the flows of materials and energy among patches. Much of landscape ecology is founded on the notion that many observations, such as the persistence of a small mammal population within a forest patch, may be fully understood only by accounting for regional as well as local factors.

Factors that lead to the development of a landscape pattern include a combination of human and nonhuman agents. The geology of a region, including the topography and soils along with the regional climate, is strongly linked to the distribution of surface

water and the types of vegetation that can exist on a site. These factors influence the pattern of human settlement and the array of past and present uses of land and water. One prevalent effect of humans is habitat fragmentation, which arises because humans tend to reduce the size and increase the isolation among patches of native habitat.

The pattern of patches on a landscape can in turn have direct effects on many different processes. The structure and arrangement of patches can affect the physical movement of materials such as nutrients or pollutants and the fate of populations of plants and animals. Many of these impacts can be traced to two factors, the role of patch edges and the connectedness among patches.

The boundary between two patches often act as filters or barriers to the transport of biological and physical elements. As an example, leaving buffer strips of native vegetation along stream courses during logging activities can greatly reduce the amount of sediment and nutrients that reach the stream from the logged area. Edge effects can result when forests are logged and there is a flux of light and wind into areas formerly located in the interior of a forest. In this example, edges can be a less suitable habitat for plants and animals not able to cope with drier, high-light conditions. When habitats are fragmented, patches eventually can become so small that they are all edge. When this happens, forest interior dwellers may become extinct. When patch boundaries act as barriers to movement, they can have pronounced effects on the dynamics of populations within and among patches. In the extreme, low connectivity can result in regional extinction even when a suitable habitat remains. This can occur if populations depend on dispersal from neighboring populations. When a population becomes extinct within a patch, there is no way for a colonist to reach the vacant habitat and reestablish the population. This process is repeated until all of the populations within a region disappear. Landscape ecologists have promoted the use of corridors of native habitat between patches to preserve connectivity despite the fragmentation of a landscape. [D.Sk.]

Landslide The perceptible downward sliding, falling, or flowing of masses of soil, rock, and debris (mixtures of soil and weathered rock fragments). Landslides range in size from a few cubic meters to over 10^9 m^3 (3.5×10^{10} ft^3), their velocities range from a few centimeters per day to over 100 m/s (330 ft/s), and their displacements may be several centimeters to several kilometers. *See* MASS WASTING.

The U.S. Highway Research Board classification divides landsliding of rock, soil, and debris, on the basis of the types of movement, into falls, slides, and flows. Other classifications consider flows, along with creep and other kinds of landslides, as general forms of mass wasting.

Falls occur when soil or rock masses free-fall through air. Falls are usually the result of collapse of cliff overhangs which result from undercutting by rivers or simply from differential erosion. Slides invariably involve shear displacement or failure along one or more narrow zones or planes. Internal deformation of the sliding mass after initial failure depends on the kinetic energy of the moving mass (size and velocity), the distance traveled, and the internal strength of the mass. Flows have internal displacement and a shape that resemble those of viscous fluids. Relatively weak and wet masses of shale, weathered rock, and soil may move in the form of debris flows and earthflows; water-soaked soils or weathered rock may displace as mudflows.

Mining and civil engineering works have induced myriads of landslides, a few of them of a catastrophic nature. Open-pitmines and road cuts create very high and steep slopes, often quite close to their stability limit. Local factors (weak joints, fault planes) or temporary ones (surges of water pressure inside the slopes, earthquake shocks) induce the failure of some of these slopes. The filling of reservoirs submerges the lower portion of natural, marginally stable slopes or old landslides. Water lowers slope stability by softening clays and by buoying the lowermost portion, or toe, of the slope.

Advances in soil and rock engineering have improved the knowledge of slope stability and the mechanics of landsliding. Small and medium-sized slopes in soil and rock can be made more stable. Remedial measures include lowering the slope angle, draining the slope, using retaining structures, compressing the slope with rock bolts or steel tendons, and grouting. *See* ENGINEERING GEOLOGY; EROSION; SOIL MECHANICS. [A.S.N.]

Larch A genus, *Larix*, of the pine family, with deciduous needles and short spurlike branches, which annually bear a crown of needles. The cones are small and persistent, varying by species in size, number, and form of the cone scales. The tamarack (*L. laricina*), also called hackmatack, is a native species. It has an erect, narrowly pyramidal habit, and grows in the northeastern United States, west to the Lake states, and across Canada to Alaska. The tough resinous wood is durable in contact with the soil and is used for railroad ties, posts, sills, and boats. Other uses include the manufacture of excelsior, cabinet work, interior finish, and utility poles.

The western larch (*L. occidentalis*), the most important and largest of all the species, grows in the northwestern United States and southeastern British Columbia. [A.H.G./K.P.D.]

Leaf A lateral appendage which is borne on a plant stem at a node (joint) and which usually has a bud in its axil. In most plants, leaves are flattened in form, although they may be nearly cylindrical with a sheathing base as in onion. Leaves usually contain chlorophyll and are the principal organs in which the important processes of photosynthesis and transpiration occur.

Morphology. A complete dicotyledon leaf consists of three parts: the expanded portion or blade; the petiole which supports the blades; and the leaf base. Stipules are small appendages that arise as outgrowths of the leaf base and are attached at the base of the petiole. The leaves of monocotyledons may have a petiole and a blade, or they may be linear in shape without differentiation into these parts; in either case the leaf base usually encircles the stem. The leaves of grasses consist of a linear blade attached to the stem by an encircling sheath.

Leaves are borne on a stem in a definite fixed order, or phyllotaxy, according to species (Fig. 1). For identification purposes, leaves are classified according to type (Fig. 2) and shape (Fig. 3), and types of margins (Fig. 4), tips, and bases (Fig. 5). The arrangement of the veins, or vascular bundles, of a leaf is called venation (Fig. 6). The main longitudinal veins are usually interconnected with small veins. Reticulate venation is most common in dicotyledons, parallel venation in monocotyledons.

Surfaces of leaves provide many characteristics that are used in identification. A surface is glabrous if it is smooth or free from hairs; glauçous if covered with a whitish, waxy material, or "bloom"; scabrous if rough or harsh to the touch; pubescent, a

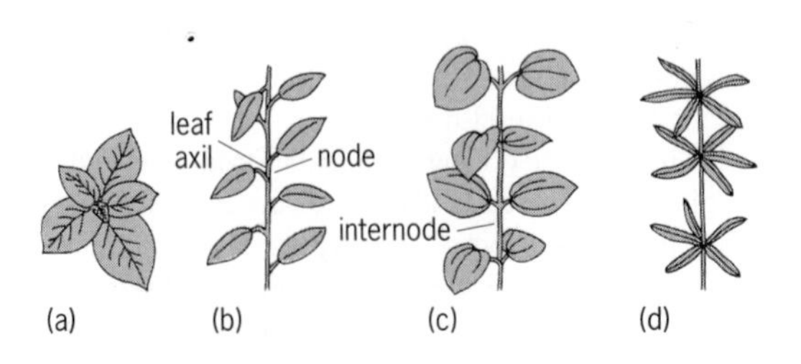

Fig. 1. Leaf arrangement. (*a*) Helical (top view). (*b*) Helical with elongated internodes (alternate). (*c*) Opposite (decussate). (*d*) Whorled (verticillate).

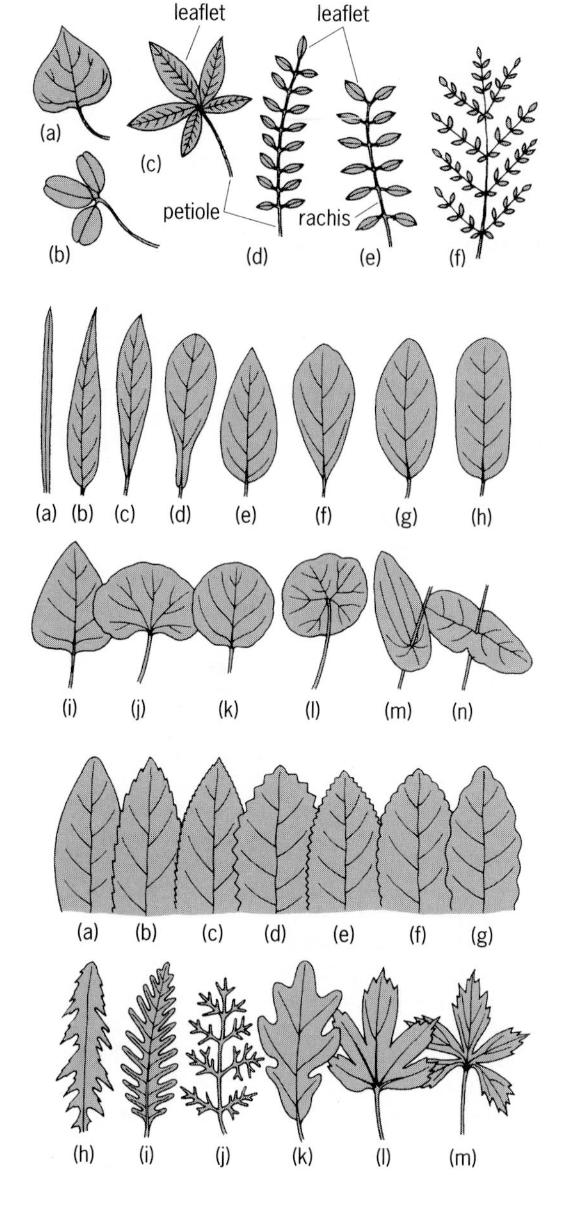

Fig. 2. Leaf types. (*a*) Simple. (*b*) Trifoliate. (*c*) Palmately compound. (*d*) Odd-pinnately compound. (*e*) Even-pinnately compound. (*f*) Decompound.

Fig. 3. Leaf shapes. (*a*) Linear. (*b*) Lanceolate. (*c*) Oblanceolate. (*d*) Spatulate. (*e*) Ovate. (*f*) Obovate. (*g*) Elliptic. (*h*) Oblong. (*i*) Deltoid. (*j*) Reniform. (*k*) Orbicular. (*l*) Peltate. (*m*) Perfoliate. (*n*) Connate.

Fig. 4. Leaf margins of various types. (*a*) Entire. (*b*) Serrate. (*c*) Serrulate. (*d*) Dentate. (*e*) Denticulate. (*f*) Crenate. (*g*) Undulate. (*h*) Incised. (*i*) Pinnatifid. (*j*) Dissected. (*k*) Lobed. (*l*) Cleft. (*m*) Parted.

general term for surfaces that are hairy; puberulent if covered with very fine, downlike hairs; villous if covered with long, soft, shaggy hairs; hirsute if the hairs are short, erect, and stiff; and hispid if they are dense, bristly, and harshly stiff.

The texture may be described as succulent when the leaf is fleshy and juicy; hyaline if it is thin and almost wholly transparent; chartaceous if papery and opaque but thin; scarious if thin and dry, appearing shriveled; and coriaceous if tough, thickish, and leathery.

Leaves may be fugacious, failing nearly as soon as formed; deciduous, failing at the end of the growing season; marcescent, withering at the end of the growing season but

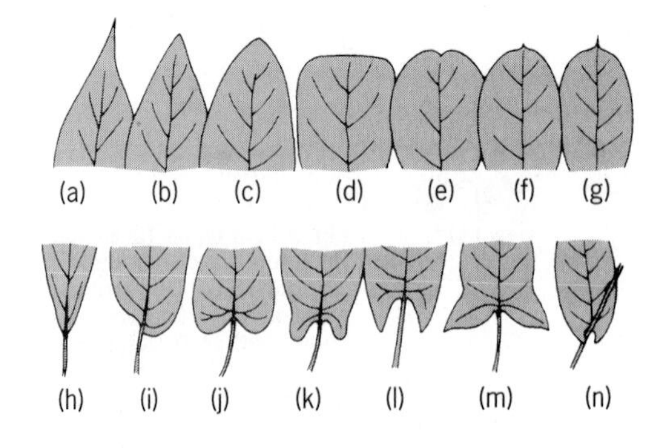

Fig. 5. Leaf tips and bases. (*a*) Acuminate. (*b*) Acute. (*c*) Obtuse. (*d*) Truncate. (*e*) Emarginate. (*f*) Mucronate. (*g*) Cuspidate. (*h*) Cuneate. (*i*) Oblique. (*j*) Cordate. (*k*) Auriculate. (*l*) Sagittate. (*m*) Hastate. (*n*) Clasping.

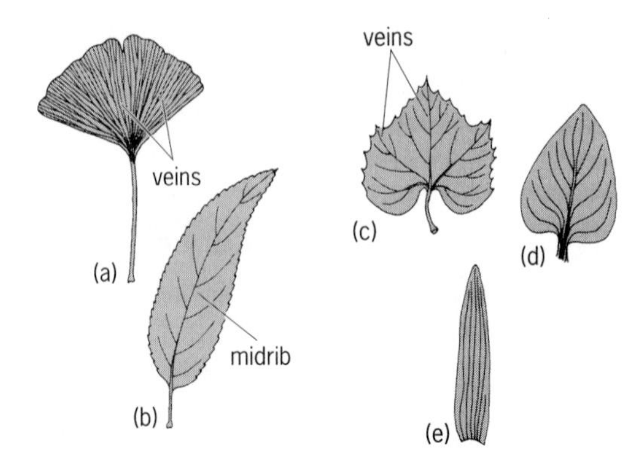

Fig. 6. Leaf venation. (*a*) Dichotomous. (*b*) Pinnate reticulate. (*c*) Palmate reticulate. (*d*) Parallel (expanded leaf). (*e*) Parallel (linear leaf).

not falling until toward spring; or persistent, remaining on the stem for more than one season, the plant thus being evergreen. *See* DECIDUOUS PLANTS; EVERGREEN PLANTS.

Anatomy. The foliage leaf is the chief photosynthetic organ of most vascular plants. Although leaves vary greatly in size and form, they share the same basic organization of internal tissues and have similar developmental pathways. Like the stem and root, leaves consist of three basic tissue systems: the dermal tissue system, the vascular tissue system, and the ground tissue system. However, unlike stems and roots which usually have radial symmetry, the leaf blade usually shows dorsiventral symmetry, with vascular and other tissues being arranged in a flat plane.

Stems and roots have apical meristems and are thus characterized by indeterminate growth; leaves lack apical meristems, and therefore have determinate growth. Because leaves are more or less ephemeral organs and do not function in the structural support of the plant, they usually lack secondary growth and are composed largely of primary tissue only. *See* ROOT (BOTANY); STEM.

The internal organization of the leaf is well adapted for its major functions of photosynthesis, gas exchange, and transpiration. The photosynthetic cells, or chlorenchyma tissue, are normally arranged in horizontal layers, which facilitates maximum interception of the Sun's radiation. The vascular tissues form an extensive network throughout the leaf so that no photosynthetic cell is far from a source of water, and carbohydrates produced by the chlorenchyma cells need travel only a short distance to reach the phloem in order to be transported out of the leaf (Fig. 7). The epidermal tissue forms a continuous covering over the leaf so that undue water loss is reduced, while at the

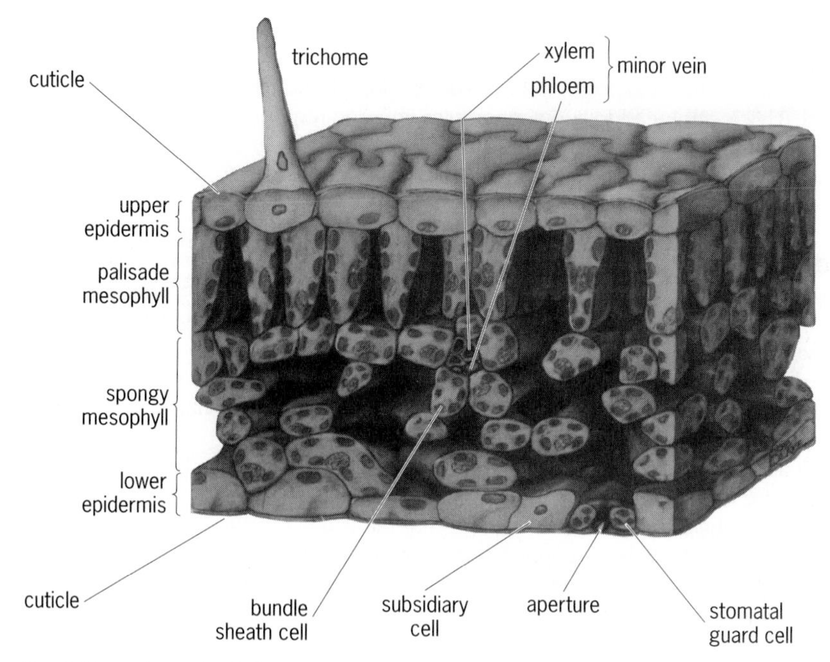

Fig. 7. Three-dimensional diagram of internal structure of a typical dicotyledon leaf.

same time the exchange of carbon dioxide and oxygen is controlled. *See* EPIDERMIS (PLANT); PARENCHYMA; PHLOEM; XYLEM. [N.G.D.]

Legionnaires' disease A type of pneumonia usually caused by infection with the bacterium *Legionella pneumophila*, but occasionally with a related species (such as *L. micdadei* or *L. dumoffii*). The disease was first observed in an epidemic among those attending an American Legion convention in Philadelphia, Pennsylvania, in 1976. The initial symptoms are headache, fever, muscle aches, and a generalized feeling of discomfort. The fever rises rapidly, reaching 102–105°F (32–41°C), and is usually accompanied by cough, shortness of breath, and chest pains. Abdominal pain and diarrhea are often present. The mortality rate can be as high as 15% in untreated or improperly diagnosed cases. Erythromycin, new-generation fluroquinolones, and rifampicin are considered highly effective medications, whereas the penicillins and cephalosporins are ineffective.

While epidemics of Legionnaires' disease (also referred to as legionellosis) can often be traced to a common source (cooling tower, potable water, or hot tub), most cases seem to occur sporadically. It is estimated that *Legionella* spp. account for approximately 4% of all community- and hospital-acquired pneumonia. Legionnaires' disease is most fequently associated with persons of impaired immune status. *Legionella* bacteria are commonly found in fresh water and moist soils worldwide and are often spread to humans through inhalation of aerosols containing the bacteria. Legionnaires' disease is not a communicable disease, indicating that human infection is not part of the survival strategy of these bacteria. Therefore, the legionellae are considered opportunistic pathogens of humans. It is technology (air conditioning) and the ability to extend life through medical advances (such as transplantation and treatments for terminal diseases) that have brought these bacteria into proximity with a susceptible population.

For most humans exposed to *L. pneumophila*, infection is asymptomatic or short-lived. This is attributed to a potent cellular immune response in healthy individuals. Recovery from Legionnaires' disease often affords immunity against future infection. However, no vaccine exists at the present time. *See* MEDICAL BACTERIOLOGY; PNEUMONIA.

[P.S.H.]

Legume forages Plants of the legume family used for livestock feed, grazing, hay, or silage. Legume forages are usually richer in protein, calcium, and phosphorus than other kinds of forages; such as grass. The production, preservation, and use of forage legumes require special skills on most soils. One important requirement is a supply of the needed symbiotic nitrogen-fixing bacteria if these are not already in the soil. Protection from weeds, injurious insects, diseases, and other harmful influences is often required. *See* NITROGEN FIXATION.

Alfalfa is the most important legume forage crop in the United States; it is used mainly for hay but is often grazed. White clover and the annual lespedezas are the most extensively grown legumes for grazing particularly in the southeastern United States. Red clover was an important crop prior to 1930 but is minor now. About a dozen other species of legumes are used for cultivated forage in the United States, and a large number are grown for range grazing. *See* COVER CROPS.

[P.T.]

Leptospirosis An acute febrile disease of humans produced by spirochetes of many species of *Leptospira*. The incubation period is 6–15 days. Among the prominent features of the disease are fever, jaundice, muscle pains, headaches, hepatitis, albuminuria, and multiple small hemorrhages in the conjunctiva or skin. Meningeal involvement often occurs. The febrile illness subsides after 3–10 days. Fatal cases show hemorrhagic lesions in the kidney, liver, skin, muscles, and central nervous system.

Wild rodents are the principal reservoirs, although natural infection occurs in swine, cattle, horses, and dogs and may be transmitted to humans through these animals. Humans are infected either through contact with the urine or flesh of diseased animals, or indirectly by way of contaminated water or soil, the organisms entering the body through small breaks in the skin or mucous membrane.

[T.B.T.]

Lightning An abrupt, high-current electric discharge that occurs in the atmospheres of the Earth and other planets and that has a path length ranging from hundreds of feet to tens of miles. Lightning occurs in thunderstorms because vertical air motions and interactions between cloud particles cause a separation of positive and negative charges.

The vast majority of lightning flashes between cloud and ground begin in the cloud with a process known as the preliminary breakdown. After perhaps a tenth of a second, a highly branched discharge, the stepped leader, appears below the cloud base and propagates downward in a succession of intermittent steps. The leader channel is usually negatively charged, and when the tip of a branch of the leader gets to within about 30 m (100 ft) of the ground, the electric field becomes large enough to initiate one or more upward connecting discharges, usually from the tallest objects in the local vicinity of the leader. When contact occurs between an upward discharge and the stepped leader, the first return stroke begins. The return stroke is basically a very intense, positive wave of ionization that propagates up the partially ionized leader channel into the cloud at a speed close to the speed of light. After a pause of 40–80 milliseconds, another leader, the dart leader, forms in the cloud and propagates down the previous

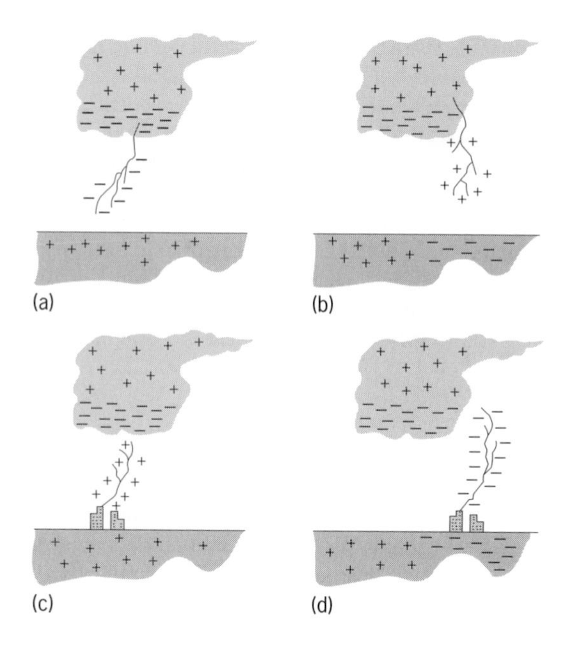

Sketches of the different types of lightning between an idealized cloud and the ground: (*a*) type 1, (*b*) type 2, (*c*) type 3, and (*d*) type 4. Channel development within the cloud is not shown. Type 1 is the most common form of cloud-to-ground lightning, and type 4 is very rare. (*After M. A. Uman, The Lightning Discharge, Academic Press, 1987*)

return-stroke channel without stepping. When the dart leader makes contact with the ground, a subsequent return stroke propagates back to the cloud. A typical cloud-to-ground flash lasts 0.2–0.3 s and contains about four return strokes; lightning often appears to flicker because the human eye is capable of just resolving the interval between these strokes.

Lightning between cloud and ground is usually classified according to the direction of propagation and polarity of the initial leader. For example, in the most frequent type of cloud-to-ground lightning a negative discharge is initiated by a downward propagating leader as described above (illus. *a*). In this case, the total discharge will effectively lower negative charge to ground or, equivalently, will deposit positive charge in the cloud.

A discharge can be initiated by a downward-propagating positive leader (illus. *b*). Positive discharges occur less frequently than negative ones, but positive discharges are often quite deleterious. Another type of lightning is a ground-to-cloud discharge that begins with a positive leader propagating upward (illus. *c*); this type is relatively rare and is usually initiated by a tall structure or a mountain peak. The rarest form of lightning is a discharge that begins with a negative leader propagating upward (illus. *d*).

The electric currents that flow in return strokes have been measured during direct strikes to instrumented towers. The peak current in a negative first stroke is typically 30 kiloamperes, with a zero-to-peak rise time of just a few microseconds. This current decreases to half-peak value in about 50 microseconds, and then low-level currents of hundreds of amperes may flow for a few to hundreds of milliseconds. The long-continuing currents produce charge transfers on the order of tens of coulombs and are frequently the cause of fires. Subsequent return strokes have peak currents that are typically 10–15 kA, and somewhat faster current rise times. Five percent of the negative discharges to ground generate peak currents that exceed 80 kA, and 5% of the positive discharges exceed 250 kA. Positive flashes frequently produce very large charge transfers, with 50% exceeding 80 coulombs and 5% exceeding 350 coulombs. [E.P.K.]

Red sprites, elves, and blue jets are upper atmospheric optical phenomena associated with thunderstorms and have only recently been documented using low-light-level television technology. Sprites are massive but weak luminous flashes appearing directly above active thunderstorms coincident with cloud-to-ground or intracloud lightning. They extend from the cloud tops to about 95 km (59 mi) and are predominantly red. High-speed photometer measurements show that the duration of sprites is only a few milliseconds. Their brightness is comparable to a moderately bright auroral arc. Elves are associated with sprites. They are optical emissions of approximately 1 millisecond, with a fast lateral, horizontal expansion that emits more red than blue light. They occur at altitudes of 75–95 km (47–59 mi). Blue jets are optical ejections from the top of the electrically active core regions of thunderstorms. Following the emergence from the top of the thundercloud, they typically propagate upward in narrow cones of about 15° full width at vertical speeds of roughly 100 km/s (60 mi/s), fanning out and disappearing at heights of about 40–50 km (25–30 mi). [R.E.O.]

Lignumvitae A tree, *Guaiacum sanctum*, also known as holywood lignumvitae, which is cultivated to some extent in southern California and tropical Florida. Lignumvitae is native in the Florida Keys, Bahamas, West Indies, and Central and South America. It is an evergreen tree of medium size with abruptly pinnate leaves. The tree yields a resin or gum known as gum guaiac or resin of guaiac which is used in medicine. The very heavy black heartwood is used in bowling balls, blocks and pulleys, and parts of instruments. [A.H.G./K.P.D.]

Limnology The study of lakes, ponds, rivers, streams, swamps, and reservoirs that make up inland water systems. Each of these inland aquatic environments is physically

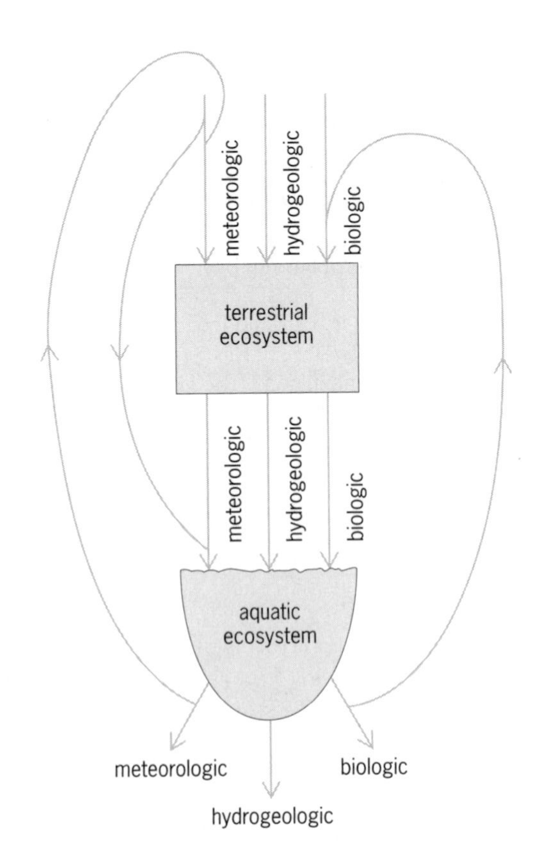

Diagrammatic model of the functional linkages between terrestrial and aquatic ecosystems. Vectors may be meteorologic, hydrogeologic, or biologic components moving nutrients or energy along the pathway shown.

and chemically connected with its surroundings by meteorologic and hydrogeologic processes (see illustration).

Aquatic systems with excellent physical conditions for production of organisms and high nutrient levels may show signs of eutrophication. Eutrophic lakes are generally identified by large numbers of phytoplankton and aquatic macrophytes and by low oxygen concentrations in the profundal zone. *See* ECOLOGY; EUTROPHICATION; FRESHWATER ECOSYSTEM; HYDROLOGY; LAKE; RIVER. [J.E.S.]

Listeriosis An infectious disease of humans and animals caused by the bacteria *Listeria monocytogenes* and *L. ivanovii*. Both humans and animals can be carriers, which excrete the bacterium in feces. Sheep, goats, and cattle can excrete the bacteria in milk, without clinical symptoms of mastitis. The most important pathway of infection is probably through food. *Listeria monocytogenes* has frequently been isolated from grass silage (fermented fodder), especially from silage of inferior quality, which is an important source of infection in ruminants. Relatively few animals develop clinical disease, but a high proportion can be latent carriers. Humans most likely ingest the bacteria with contaminated food, such as meat and meat products, raw milk and milk products, and unwashed vegetables.

Encephalitis is the most common form of disease in ruminants, and septicemia with involvement of several organs, including the pregnant uterus, occurs most commonly in monogastric animals, including very young sheep, goats, and calves.

Most human cases are sporadic, but food-borne epidemics occur. Abortions, perinatal disease, and disease in immunosuppressed individuals are most common. Perinatal disease is dominated by septicemia, widespread microscopic abscesses, and meningitis. In adults, meningitis is by far the most common manifestation, but in immunosuppressed individuals encephalitis occurs, possibly with a pathogenesis similar to encephalitis in ruminants.

Clinical diagnosis is based on symptoms, on isolation of *L. monocytogenes*, and on histopathological examination of affected tissue, especially brain tissue. *See* MEDICAL BACTERIOLOGY. [H.Gro.]

Locust (forestry) A name commonly applied to two trees, the black locust (*Robinia pseudoacacia*) and the honey locust (*Gleditsia triacanthos*). Both of these commercially important trees have podlike fruits similar to those of the pea or bean.

The black locust is native in the Appalachian and the Ozark regions and is now widely naturalized in the eastern United States, southern Canada, and Europe. The wood is hard and durable in the soil and is used for fence posts, mine timbers, poles, and railroad ties.

The honey locust is native in the Appalachian and the Mississippi Valley regions, but is also widely naturalized in the eastern United States and southern Canada. The reddish wood is hard, strong, and coarse-grained and takes a high polish. Because it is durable in contact with soil, it is used for fence posts and railroad ties; it is also used for construction, furniture, and interior finish. [A.H.G./K.P.D.]

Loess Silt-dominated sediment of eolian (windblown) origin. Loess is a common deposit in and near areas that were glaciated during the Quaternary Period, and most loess deposits are indirectly related to glaciation. *See* EOLIAN LANDFORMS; QUATERNARY.

Loess is a well-sorted clastic deposit which is unconsolidated, relatively homogeneous, seemingly nonstratified, and extremely porous. Colors range from buff to shades of pink, gray, yellow, or brown. Silt-sized particles, most of which are 0.0002–0.002 in. (0.005–0.05 mm) in diameter, usually make up 60–90% of the deposit, with small amounts of fine sand and small to moderate amounts of clay-sized material. The particles are generally angular to subangular.

Quartz is the dominant mineral, with subordinate amounts of feldspar, calcite, dolomite, clay minerals, and small amounts of other minerals. Clay minerals are primarily smectite, illite, and chlorite. They occur as silt-sized aggregates and, along with calcite, as coatings or fillings on silt grains, in interstices, and in vertical tubes left from the decay of grass roots. These latter characteristics partially bind the particles together and give loess with relatively large dry strength. As a result, many loess deposits maintain near-vertical slopes in both natural and artificial cuts.

Loess occurs as a relatively thin (generally <90 ft or 30 m), blanket-type deposit which drapes over an irregular landscape. It is common in many areas of the world, but is particularly thick near valleys that served as meltwater drainageways during Quaternary glaciation. Loess also may be derived from desert areas, in which case the particles must be produced by either weathering processes or eolian abrasion. *See* SEDIMENTOLOGY; SOIL. [W.H.J.]

Logging Those processes required to bring all or a portion of a tree from the stump to the mill facilities. Logging (tree harvesting) processes are clustered into tree conversion, woods transport (off-road transportation), landing operations (wood transfer), transport from landing to mill facility (truck, water, or rail), and unloading at the mill facility (wood transfer).

The start of harvesting is the cutting down of trees with hand tools, chain saws, or mechanized felling machines. The tree may be further cut into suitable lengths (bucking), or it may be transported whole or in tree-lengths. Tree products may be allocated during bucking with the aid of a computer on the felling and bucking machine or by a faller using a log order list or a hand-held computer to help decide the log products to make. The objective of the tree falling operation is to fell the tree with minimum damage, to avoid damaging surrounding trees, to minimize soil and water impacts, and to position the tree or logs for the next phase of harvesting. The goal of bucking is to produce the most valuable assortment of logs from the tree while considering the physical capability of the skidding (log-dragging), yarding (moving of logs to a landing), or forwarding (log-carrying) equipment.

Logs in lengths from about 1 to 10 m (3 to 33 ft) or other products must be transported from the stump to a place where they are further processed (often called a landing). In some cases, entire trees are pulled to the landing. Humans, animals, crawler tractors and wheeled skidders (machines that drag the logs), forwarders (machines that carry loads of logs), farm tractors with winches or trailers, cable logging systems, balloons, or even helicopters transport logs and tree products to landings.

At the landing, the logs or trees may be stored or directly processed for transport. They may be loaded onto trucks, trains, barges, or ships, or prepared for water transport. Whole trees brought to a landing may have limbs and bark removed, and then be chipped and loaded into chip vans for transport to a pulp mill. Tree-length segments may be delimbed and bucked into logs for different market destinations. Trees may be shredded, chunked, or processed through machines for use as fuel. The allocation process may include measurement by volume or weight of the products.

Because logs are heavy, they are normally loaded mechanically, although some regions still use manual or animal methods involving ramps. There are two general types of mechanical loaders used at roadside: swingboom loaders with grapples, and front-end loaders fitted with a log fork or grapple. Both are mobile, mounted on tracked or rubber-tired carriers. Forwarders usually unload themselves either into log decks for storage or onto setout trailers.

Trucks are most commonly used to transport log products to mill facilities. They vary from small vehicles hauling 5–8 tons on straight beds to large specialized off-highway vehicles hauling 50 tons or more. A variety of truck trailers are used depending upon the type of product. Trees or long logs may be loaded onto pole trailers. Short logs, 2.5 m (8 ft) and less, are often stacked sideways on flat-bed trailers with bunks. Chips or flakes are hauled in specially designed chip vans. Water transport in barges, as log rafts, and as free-floating logs is used in some areas.

Log products can be unloaded from truck trailers by lifting, rolling, or dumping over the side or end, depending upon the type of trailer. Trees and long logs are usually lifted from trailers by a grapple on mobile wheel loaders or overhead cranes (which can unload the entire truck in one pass). Shorter logs are often unloaded using slings. In some cases, short wood for pulp is swept directly off the trailer and fed into a debarking machine to eliminate rehandling. Chip trailers are often tilted and end-dumped on large hydraulic ramps. [J.Ses.; J.Ga.]

Louping ill A viral disease of sheep, capable of producing central nervous system manifestations. It occurs chiefly in the British Isles. Infections have been reported, although rarely, among persons working with sheep. In sheep the disease is usually biphasic with a systemic influenzalike phase, which is followed by encephalitic signs. In infected humans the first phase is generally the extent of the illness. The virus is a member of the Russian tick-borne complex of the group B arboviruses. Characteristics, diagnosis, and epidemiology are similar to those of other viruses of this complex. *See* ANIMAL VIRUS; ARBOVIRAL ENCEPHALITIDES. [J.L.Me.]

Lumber Timber sawed or split into planks, boards, and similar products. Lumber can come in many forms, species, and types from a wide variety of commercial sources. Because most lumber is manufactured similarly and graded by standardized rules, it is fairly uniform throughout the United States.

Lumber is manufactured from round logs primarily in rectangular shapes of different dimensions. Lumber length is recorded in actual dimensions. Width and thickness are traditionally recorded in nominal dimensions, which are somewhat more than actual dimensions. Lumber is classified by thickness into three categories: (1) board, lumber less than 38 mm (nominally 2 in.) thick; (2) dimension, lumber from 38 mm to, but not including, 114 mm (nominally 5 in.) thick; and (3) timber, lumber 114 mm (nominally 5 in.) or more in thickness in the least dimension. *See* LOGGING.

Lumber can be produced with either a rough or surfaced (dressed) finish. Rough-sawn lumber has surface imperfections caused by the primary sawing operations. Surfaced lumber is smoothed on either one or both sides and one or both edges.

In general, the grade of a piece of lumber is based on the number, character, and location of features that may lower the strength, durability, or utility value of the wood. Lumber grading can be divided into two main categories: remanufacture "shop grade" and structural "stress grade." Sorting of lumber for remanufacture is based on visual inspection. The wood is designated shop grade on the proportion of defect-free or clear cuttings of a certain size that can be made from a piece of lumber. The larger volume and more frequent number of clear cuttings, the higher the grade. Pieces of lumber graded for structural uses are put into classes with similar mechanical properties called stress grades. Stress grades are characterized by (1) one or more sorting criteria, (2) a set of allowable properties for engineering design, and (3) a unique grade name. The allowable properties are inferred through visual grading criteria or are determined nondestructively by machine-grading criteria.

Visual grading is the oldest stress-grading method. It is based on the premise that mechanical properties of lumber differ from mechanical properties of clear wood. Growth characteristics, which affect properties and can be seen and judged by eye, are used to sort the lumber into stress grades. Typical visual sorting criteria include density, decay, proportion of heartwood and sapwood, slope of grain, knots, shake, checks and splits, wane, and pitch pockets.

Machine-graded lumber is evaluated by a machine using a nondestructive test followed by visual grading to evaluate certain characteristics that the machine cannot or may not properly evaluate. Machine-stress-rated (MSR), machine-evaluated (MEL), and E-rated lumber are three types of machine-graded lumber. Machine-graded lumber allows for better sorting of material for specific applications in engineered structures.

Clear, straight-grained lumber can be about 50% stronger when dry than when wet. For lumber containing knots, the increase in strength with decreasing moisture content is dependent on lumber quality. For timber, often no adjustment for moisture content is made because properties are assigned on the basis of wood in the green condition. *See* WOOD PROPERTIES. [D.E.Kr.]

Lyme disease A multisystem illness caused by the tick-borne spirochete *Borrelia burgdorferi*. The disease, also known as Lyme borreliosis, generally begins with a unique expanding skin lesion, erythema migrans, which is often accompanied by symptoms resembling those of influenza or meningitis. During the weeks or months following the tick bite, some individuals may develop cardiac and neurological abnormalities, particularly meningitis or inflammation of the cranial or peripheral nerves. If the disease is untreated, intermittent or chronic arthritis and progressive encephalomyelitis may develop months or years after primary infection.

The causative agent, *B. burgdorferi*, is a helically shaped bacterium with dimensions of 0.18–0.25 by 4–30 micrometers. Once thought to be limited to the European continent, Lyme borreliosis and related disorders are now known to occur also in North America, Russia, Japan, China, Australia, and Africa, where *B. burgdorferi* is maintained and transmitted by ticks of the genus *Ixodes*, namely *I. dammini, I. pacificus*, and possibly *I. scapularis* in the United States, *I. ricinus* in Europe, and *I. persulcatus* in Asia. Reports of Lyme disease in areas where neither *I. dammini* nor *I. pacificus* is present suggest that other species of ticks or possibly other bloodsucking arthropods such as biting flies or fleas may be involved in maintaining and transmitting the spirochetes.

All stages of Lyme borreliosis may respond to antibiotic therapy. Early treatment with oral tetracycline, doxycycline, penicillin, amoxicillin, or erythromycin can shorten the duration of symptoms and prevent later disease.

Prevention and control of Lyme borreliosis must be directed toward reduction of the tick population. This can be accomplished through reducing the population of animals that serve as hosts for the adult ticks, elimination of rodents that are not only the preferred hosts but also the source for infecting immature ticks with *B. burgdorferi*, and application of tick-killing agents to vegetation in infested areas. Personal use of effective tick repellents and toxins is also recommended. *See* INFECTIOUS DISEASE; INSECTICIDE.

Lyme disease affects not only humans but also domestic animals such as dogs, horses, and cattle that serve as hosts for the tick vectors. Animals affected show migratory, intermittent arthritis in some joints similar to that observed in humans. [W.Ba.; J.J.Ka.]

Magnolia A genus of trees with large, chiefly white flowers, and simple, entire, usually large alternate leaves. In the winter the twigs may be recognized by their aromatic odor when bruised.

The most important species commercially is *Magnolia acuminata*, commonly called cucumber tree, which grows in the Appalachian and Ozark mountains. The fruit is red when ripe and resembles a small cucumber in shape.

The wood of the magnolia is similar to that of the tulip tree and is rather soft, but it is of such wide natural dimensions that it is valued for furniture, cabinetwork, flooring, and interior finish.

Magnolia species occur naturally in a broad belt in the eastern United States and Central America, with a similar region in eastern Asia and the Himalayas. [A.H.G./K.P.D.]

Mahogany A hard, red or yellow-brown wood which takes a high polish and is extensively used for furniture and cabinetwork. The West Indies mahogany tree (*Swietenia mahagoni*), a native of tropical regions in North and South America, is a large evergreen tree with smooth pinnate leaves. Together with other species it yields the world's most valuable cabinet wood. In the United States it occurs naturally only in the extreme southern tip of Florida, but it is planted elsewhere in the state as an ornamental and shade tree. [A.H.G./K.P.D.]

Malaria A disease caused by members of the protozoan genus *Plasmodium*, a widespread group of sporozoans that parasitize the human liver and red blood cells. Four species can infect humans: *P. vivax*, causing vivax or benign tertian malaria; *P. ovale,* a very similar form found chiefly in central Africa that causes ovale malaria; *P. malariae,* which causes malariae or quartan malaria; and *P. falciparum,* the highly pathogenic causative organism of falciparum or malignant tertian malaria. Malaria is characterized by periodic chills, fever, and sweats, often leading to severe anemia, an enlarged spleen, and other complications that may result in loss of life, especially among infants whose deaths are almost always attributed to falciparum malaria. The infective agents are inoculated into the human bloodstream by the bite of an infected female *Anopheles* mosquito, more than 60 species of which can carry the infection to humans. The disease is found in all tropical and some temperate regions, but it has been eradicated in North America, Europe, and Russia. Despite control efforts, malaria has probably been the greatest single killer disease throughout human history and continues to be a major infectious disease. *See* EPIDEMIC.

The vast reproductive capacity of *Plasmodium* parasites is illustrated by their life cycle, which begins as a series of asexual divisions in human liver and then red blood cells. Transfer of the parasites to the mosquito host depends on the rate of sexual multiplication that begins in the infected human red blood cells and is completed in the mosquito stomach, followed by asexual multiple division of the product of sexual fusion. Clinical malaria usually begins 7–18 days after infection with sporozoites. Red cell infections tend to follow a remarkably synchronous division cycle. The parasite progresses from merozoite to a vegetative phase (trophozoite) to a division stage (schizont), ending with the new generation of merozoites ready to break out in a burst of parasite releases and initiate the chills-fever-sweat phase of the disease. The sequence of chills, fever, and sweats is the result of simultaneous red cell destruction at 48- or 72-h intervals. *See* SPOROZOA.

Chloroquine remains the drug of choice for prevention as well as treatment of vivax, ovale, and malariae malaria. However, most strains of falciparum malaria have become strongly chloroquine-resistant. For prevention of chloroquine-resistant falciparum malaria (and in some areas vivax malaria is now chloroquine-resistant as well) a weekly dose of mefloquine beginning a week before, then during, and for 4 weeks after leaving the endemic area is recommended. Chloroquine-resistant malaria is chiefly treated with the oldest known malaricide, quinine, in the form of quinine sulfate, plus pyrimethamine-sulfadoxine. *See* DRUG RESISTANCE.

Failure of earlier efforts to eradicate malaria and the rapid spread of resistant strains of both parasites and their mosquito vectors necessitated renewed interest in prevention of exposure by avoidance of mosquito bites using pyrethrin-treated bednets, coverage of exposed skin during active mosquito periods (usually dawn, dusk, and evening hours), and use of insect-repellent lotions. A balance between epidemiological and immunological approaches to prevention, and the continued development of new drugs for prophylaxis and treatment are recognized as the most effective means to combat one of the most dangerous and widespread threats to humankind from an infectious agent. *See* MEDICAL PARASITOLOGY. [D.He.]

Mangrove A taxonomically diverse assemblage of trees and shrubs that form the dominant plant communities in tidal, saline wetlands along sheltered tropical and subtropical coasts. The development and composition of mangrove communities depend largely on temperature, soil type and salinity, duration and frequency of inundation, accretion of silt, tidal and wave energy, and cyclone or flood frequencies. Extensive mangrove communities seem to correlate with areas in which the water temperature of the warmest month exceeds 75°F (24°C), and they are absent from waters that never exceed 75°F (24°C) during the year. Intertidal, sheltered, low-energy, muddy sediments are the most suitable habitats for mangrove communities, and under optimal conditions, forests up to 148 ft (45 m) in height can develop. Where less favorable conditions are found, mangrove communities may reach maturity at heights of only 3 ft (1 m). *See* ECOSYSTEM.

Plants of the mangrove community belong to many different genera and families, many of which are not closely related to one another phylogenetically. However, they do share a variety of morphological, physiological, and reproductive adaptations that enable them to grow in an unstable, harsh, and salty environment. Approximately 80 species of plants belonging to about 30 genera in over 20 families are recognized throughout the world as being indigenous to mangroves. About 60 species occur on the east coasts of Africa and Australasia, whereas about 20 species are found in the Western Hemisphere. At the generic level, *Avicennia* and *Rhizophora* are the dominant plants of mangrove communities throughout the world, with each genus having several closely related species in both hemispheres. At the species level, however, only a few species, such as the portia tree (*Thespesia populnea*), the mangrove fern (*Acrostichum aureum*), and the swamp hibiscus (*Hibiscus tiliaceus*), occur in both hemispheres.

The mangrove community is often strikingly zoned parallel to the shoreline, with a sequence of different species dominating from open water to the landward margins. These zones are the response of individual species to gradients of inundation frequency, waterlogging, nutrient availability, and soil salt concentrations across the intertidal area, rather than a reflection of ecological succession, as earlier studies had suggested. *See* ECOLOGICAL SUCCESSION.

Most plants of the mangrove community are halophytes, well adapted to salt water and fluctuations of tide level. Many species show modified root structures such as stilt or prop roots, which offer support on the semiliquid or shifting sediments, whereas others have erect root structures (pneumatophores) that facilitate oxygen penetration to the roots in a hypoxic environment. Salt glands, which allow excess salt to be extruded through the leaves, occur in several species; others show a range of physiological mechanisms that either exclude salt from the plants or minimize the damage excess salts can cause by separating the salt from the sensitive enzyme systems of the plant. Several species have well-developed vivipary of their seeds, whereby the hypocotyl develops while the fruit is still attached to the tree. The seedlings are generally buoyant, able to float over long distances in the sea and rapidly establish themselves once stranded in a suitable habitat.

A mangrove may be considered either a sheltered, muddy, intertidal habitat or a forest community. The sediment surface of mangrove communities abounds with species that have marine affinities, including brightly colored fiddler crabs, mound-building mud lobsters, and a variety of mollusks and worms, as well as specialized gobiid fish (mudskippers). The waterways among the mangroves are important feeding and nursery areas for a variety of juvenile finfish as well as crustaceans. Animals with forest affinities that are associated with mangroves include snakes, lizards, deer, tigers, crab-eating monkeys, bats, and many species of birds.

Economically, mangroves are a major source of timber, poles, thatch, and fuel. The bark of some trees is used for tanning materials, whereas other species have food or medicinal value. *See* ECOLOGICAL COMMUNITIES; FOREST MANAGEMENT. [P.Sae.]

Maple A genus, *Acer*, of broad-leaved, deciduous trees including about 115 species in North America, Asia, Europe, and North Africa. This genus is characterized by simple, opposite, usually palmately lobed (rarely pinnate) leaves, generally inconspicuous flowers, and a fruit consisting of two longwinged samaras or keys (see illustration).

The most important commercial species is the sugar or rock maple (*A. saccharum*), called hard maple in the lumber market. This tree grows in the eastern half of the

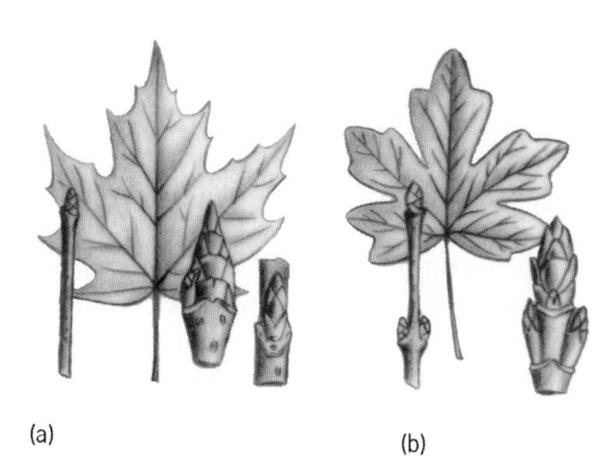

Characteristic maple leaves, twigs, and buds. (*a*) Sugar maple (*Acer saccharum*). (*b*) Hedge maple (*A. campestre*).

United States and adjacent Canada. It can be recognized by its gray furrowed bark, sharp-pointed scaly winter buds, and symmetrical oval outline of the crown.

Maples rank third in the production of hardwood lumber. Hard maple is used for flooring, furniture, boxes, crates, woodenware, spools, bobbins, motor vehicle parts, veneer, railroad ties, and pulpwood. It is the source of maple sugar and syrup and is planted as a shade tree. [A.H.G./K.P.D.]

Marine biological sampling The collection and observation of living organisms in the sea, including the quantitative determination of their abundance in time and space. The biological survey of the ocean depends to a large extent on specially equipped vessels. Sampling in intertidal regions at low tide is one of the few instances where it is possible to observe and collect marine organisms without special apparatus.

A primary aim of marine biology is to discover how ocean phenomena control the distribution of organisms. Sampling is the means by which this aim is accomplished. Traditional techniques employ the use of samplers attached to wires lowered over the side of a ship by means of hydraulic winches. These samplers include bottles designed for enclosing seawater samples from particular depths, fine-meshed nets that are towed behind the ship to sieve out plankton and fish, and grabs or dredges that are used to collect animals inhabiting the ocean bottom. These types of gear are relied upon in many circumstances; however, they illustrate some of the problems common to all methods by which the ocean is sampled. First, sampling is never synoptic, which means that it is not possible to sample an area of ocean so that conditions can be considered equivalent at each point. Usually, it is assumed that this is so. Second, there are marine organisms for which there exists no sampling methodology. For example, knowledge of the larger species of squid is confined to the few animals that have been washed ashore. A third problem concerns the representativeness of the samples collected. The open ocean has no easily definable boundaries, and organisms are not uniformly distributed. The actual sampling is regularly done out of view of the observer; thus, sampling effectiveness is often difficult to determine. Furthermore, navigational systems are not error-free, and therefore the position of the sample is never precisely known. All developments in methods for sampling the ocean try to resolve one or more of these difficulties by improving synopticity, devising more efficient sampling gear, or devising methods for observation such that more meaningful samples can be obtained.

Direct and remote observation methods provide valuable information on the undersea environment and thus on the representativeness of various sampling techniques. Personnel-operated deep-submergence research vessels (DSRVs) are increasingly being employed to observe ocean life at depth and on the bottom, and for determining appropriate sampling schemes. The deep-submergence research vessels are used with cameras and television recording equipment and are also fitted with coring devices, seawater samplers, and sensors of various types. Cameras are deployed from surface ships on a wire. Other cameras are operated unattended at the bottom for months at a time, recording changes occurring there. Scuba diving is playing a larger role, especially in open-ocean areas, and is used to observe marine organisms in their natural habitat as well as to collect the more fragile marine planktonic forms such as foraminifera, radiolaria, and jellyfish. Remotely piloted vehicles (RPVs) will continue to assume greater importance in sampling programs since they can go to greater depths than can divers, and they overcome a limitation in diving in that remotely piloted vehicles can be operated at night. Optical sensors carried aboard Earth-orbiting satellites can provide images of ocean color over wide areas. Ocean color is related to the turbidity and also to the amount of plant material in the seawater. This thus establishes a means by which sampling programs carried out from ships can be optimized. *See* DIVING; SEAWATER FERTILITY; UNDERWATER PHOTOGRAPHY; UNDERWATER VEHICLE. [J.Marr.]

Marine conservation The management of marine species and ecosystems to prevent their decline and extinction. As in terrestrial conservation, the goal of marine conservation is to preserve and protect biodiversity and ecosystem function through the preservation of species, populations, and habitats. The importance of conserving marine species and ecosystems is growing as a consequence of human activities. Negative impacts on marine biological systems are caused by such actions as overfishing; overutilization, degradation, and loss of coastal and marine habitats; introduction of nonnative species; and intensification of global climate change, which alters oceanic circulation and disrupts existing trophic relationships. Marine conservation biologists seek to reduce the negative effects of all these actions by conducting directed research and helping to develop management strategies for particular species, communities, habitats, or ecosystems.

A variety of approaches and tools are used in marine conservation. These include population assessment; mitigation, recovery, and restoration efforts; establishment of marine protected areas; and monitoring programs. Many of these approaches overlap with those in terrestrial conservation. However, fundamental differences between terrestrial and marine environments in spatial dimension, habitat type, and organismal life history require that basic conservation techniques be modified for application to the marine environment. *See* BIODIVERSITY; ECOSYSTEM; MARINE ECOLOGY; OCEANOGRAPHY.

Effective management requires knowledge of the size and status of populations. Trends in abundance can be detected through stock assessment methods first developed for marine fisheries and subsequently modified for application to other marine organisms. These methods use estimates of population size, reproduction, survivorship, and immigration to determine whether populations are increasing, decreasing, or stable. Population viability analysis is a specialized statistical assessment in which demographic and environmental information is used to determine the probability that a population will persist in a particular environment for a specified period of time. This method can be used to guide management decisions, and has been used in efforts to manage marine mammals, turtles, seabirds, and other species. *See* ECOLOGICAL COMMUNITIES; ECOLOGICAL METHODS; POPULATION ECOLOGY.

Depleted, threatened, or endangered populations are often subject to mitigation or recovery efforts. The purpose of these efforts is to reduce the immediate threat of extinction or extirpation. This is typically achieved by direct human intervention to increase the size of a population or to prevent further decline in population size. Methods used to achieve recovery for fish and marine invertebrates include reducing fishing quotas, restricting the use of certain types of fishing gear, restricting the seasonal or annual distribution of a fishery, or closing fisheries altogether.

Recovery efforts can be most successful if they are based on multispecies or ecosystem-level management strategies. These strategies take into account positive and negative interactions between species, such as facilitation, competition, and predation. They further take into account interactions between species and their environment. Key to the success of assessment and recovery programs is identification of the appropriate biological unit for conservation (for example, population, subspecies, stock, or evolutionarily significant unit). Maintaining genetic diversity is an important goal of conservation biology, because genetic diversity confers evolutionary potential. Thus, conservation efforts often are aimed at populations that are genetically distinct from other populations of the same species.

Restoration efforts are aimed at returning habitats to an ecologically functional condition, usually consistent with some previous, more pristine condition. *See* ENDANGERED SPECIES; FISHERIES ECOLOGY; MARINE FISHERIES.

Marine protected areas are set aside for the protection or recovery of species, habitats, or ecosystems. They include marine parks, marine reserves, marine sancturaries, harvest refugia, and voluntary or legislated no-take areas. Some marine protected areas allow for consumptive use (such as fishing) or extraction of resources (for example, oil drilling), while others are closed to most human activities.

Monitoring programs are necessary to determine the outcome of specific conservation actions and to guide future conservation decisions. Monitoring programs vary according to the objectives of specific conservation projects but typically include such activities as long-term surveys of population size and status, and the development of mathematical models to help predict specific outcomes. [T.Kl.]

Marine ecology An integrative science that studies the basic structural and functional relationships within and among living populations and their physical-chemical environments in marine ecosystems. Marine ecology draws on all the major fields within the biological sciences as well as oceanography, physics, geology, and chemistry. Emphasis has evolved toward understanding the rates and controls on ecological processes that govern both short- and long-term events, including population growth and survival, primary and secondary productivity, and community dynamics and stability. Marine ecology focuses on specific organisms as well as on particular environments or physical settings. *See* ENVIRONMENT.

Marine environments. Classification of marine environments for ecological purposes is based very generally on two criteria, the dominant community or ecosystem type and the physical-geological setting. Those ecosystems identified by their dominant community type include mangrove forests, coastal salt marshes, submersed seagrasses and seaweeds, and tropical coral reefs. Marine environments identified by their physical-geological setting include estuaries, coastal marine and nearshore zones, and open-ocean-deep-sea regions. *See* DEEP-SEA FAUNA; ECOLOGICAL COMMUNITIES; PHYTOPLANKTON; ZOOPLANKTON.

An estuary is a semienclosed area or basin with an open outlet to the sea where fresh water from the land mixes with seawater. The ecological consequences of freshwater input and mixing create strong gradients in physical-chemical characteristics, biological activity and diversity, and the potential for major adverse impacts associated with human activities. Because of the physical forces of tides, wind, waves, and freshwater input, estuaries are perhaps the most ecologically complex marine environment. They are also the most productive of all marine ecosystems on an area basis and contain within their physical boundaries many of the principal marine ecosystems defined by community type. *See* ESTUARINE OCEANOGRAPHY; MANGROVE; SALT MARSH.

Coastal and nearshore marine ecosystems are generally considered to be marine environments bounded by the coastal land margin (seashore) and the continental shelf 300–600 ft (100–200 m) below sea level. The continental shelf, which occupies the greater area of the two and varies in width from a few to several hundred kilometers, is strongly influenced by physical oceanographic processes that govern general patterns of circulation and the energy associated with waves and currents. Ecologically, the coastal and nearshore zones grade from shallow water depths, influenced by the adjacent landmass and input from coastal rivers and estuaries, to the continental shelf break, where oceanic processes predominate. Biological productivity and species diversity and abundance tend to decrease in an offshore direction as the food web becomes supported only by planktonic production. Among the unique marine ecosystems associated with coastal and nearshore water bodies are seaweed-dominated communities (for example, kelp "forests"), coral reefs, and upwellings. *See* CONTINENTAL MARGIN; REEF; UPWELLING.

Approximately 70% of the Earth's surface is covered by oceans, and more than 80% of the ocean's surface overlies water depths greater than 600 ft (200 m), making open-ocean–deep-sea environments the largest, yet the least ecologically studied and understood, of all marine environments. The major oceans of the world differ in their extent of landmass influence, circulation patterns, and other physical-chemical properties. Other major water bodies included in open-ocean–deep-sea environments are the areas of the oceans that are referred to as seas. A sea is a water body that is smaller than an ocean and has unique physical oceanographic features defined by basin morphology. Because of their circulation patterns and geomorphology, seas are more strongly influenced by the continental landmass and island chain structures than are oceanic environments.

Within the major oceans, as well as seas, various oceanographic environments can be defined. A simple classification would include water column depths receiving sufficient light to support photosynthesis (photic zone); water depths at which light penetration cannot support photosynthesis and which for all ecological purposes are without light (aphotic zone); and the benthos or bottom-dwelling organisms. Classical oceanography defines four depth zones; epipelagic, 0–450 ft (0–150 m), which is variable; mesopelagic, 450–3000 ft (150–1000 m); bathypelagic, 3000–12,000 ft (1000–4000 m); and abyssopelagic, greater than 12,000 ft (4000 m). These depth strata correspond approximately to the depth of sufficient light penetration to support photosynthesis; the zone in which all light is attenuated; the truly aphotic zone; and the deepest oceanic environments.

Marine ecological processes. Fundamental to marine ecology is the discovery and understanding of the principles that underlie the organization of marine communities and govern their behavior, such as controls on population growth and stability, quantifying interactions among populations that lead to persistent communities, and coupling of communities to form viable ecosystems. The basis of this organization is the flow of energy and cycling of materials, beginning with the capture of radiant solar energy through the processes of photosynthesis and ending with the remineralization of organic matter and nutrients.

Photosynthesis in seawater is carried out by various marine organisms that range in size from the microscopic, single-celled marine algae to multicellular vascular plants. The rate of photosynthesis, and thus the growth and primary production of marine plants, is dependent on a number of factors, the more important of which are availability and uptake of nutrients, temperature, and intensity and quality of light. Of these three, the last probably is the single most important in governing primary production and the distribution and abundance of marine plants. Considering the high attenuation of light in water and the relationships between light intensity and photosynthesis, net autotrophic production is confined to relatively shallow water depths. The major primary producers in marine environments are intertidal salt marshes and mangroves, submersed seagrasses and seaweeds, phytoplankton, benthic and attached microalgae, and—for coral reefs—symbiotic algae (zooxanthellae). On an areal basis, estuaries and nearshore marine ecosystems have the highest annual rates of primary production. From a global perspective, the open oceans are the greatest contributors to total marine primary production because of their overwhelming size.

The two other principal factors that influence photosynthesis and primary production are temperature and nutrient supply. Temperature affects the rate of metabolic reactions, and marine plants show specific optima and tolerance ranges relative to photosynthesis. Nutrients, particularly nitrogen, phosphorus, and silica, are essential for marine plants and influence both the rate of photosynthesis and plant growth. For many phytoplankton-based marine ecosystems, dissolved inorganic nitrogen is considered the principal limiting nutrient for autotrophic production, both in its limiting behavior and in its role in the eutrophication of estuarine and coastal waters. *See* PHOTOSYNTHESIS.

Marine food webs and the processes leading to secondary production of marine populations can be divided into plankton-based and detritus-based food webs. They approximate phytoplankton-based systems and macrophyte-based systems. For planktonic food webs, current evidence suggests that primary production is partitioned among groups of variously sized organisms, with small organisms, such as cyanobacteria, playing an equal if not dominant role at times in aquatic productivity. The smaller autotrophs—both through excretion of dissolved organic compounds to provide a substrate for bacterial growth and by direct grazing by protozoa (microflagellates and ciliates)—create a microbially based food web in aquatic ecosystems, the major portion of autotrophic production and secondary utilization in marine food webs may be controlled, not by the larger organisms typically described as supporting marine food webs, but by microscopic populations.

Macrophyte-based food webs, such as those associated with salt marsh, mangrove, and seagrass ecosystems, are not supported by direct grazing of the dominant vascular plant but by the production of detrital matter through plant mortality. The classic example is the detritus-based food webs of coastal salt marsh ecosystems. These ecosystems, which have very high rates of primary production, enter the marine food web as decomposed and fragmented particulate organics. The particulate organics of vascular plant origin support a diverse microbial community that includes bacteria, flagellates, ciliates, and other protozoa. These organisms in turn support higher-level consumers.

Both pelagic (water column) and benthic food webs in deep ocean environments depend on primary production in the overlying water column. For benthic communities, organic matter must reach the bottom by sinking through a deep water column, a process that further reduces its energy content. Thus, in the open ocean, high rates of secondary production, such as fish yields, are associated with areas in which physical-chemical conditions permit and sustain high rates of primary production over long periods of time, as is found in upwelling regions.

Regardless of specific marine environment, microbial processes provide fundamental links in marine food webs that directly or indirectly govern flows of organic matter and nutrients that in turn control ecosystem productivity and stability. *See* BIOLOGICAL PRODUCTIVITY; ECOLOGY; ECOSYSTEM; SEAWATER FERTILITY. [R.We.]

Marine fisheries The harvest of animals and plants from the ocean to provide food and recreation for people, food for animals, and a variety of organic materials for industry. It is now generally agreed that the world catch is approaching a maximum, which may be less than 100,000,000 metric tons per year. If methods can be devised to harvest smaller organisms not heretofore used because they have been too costly to catch and process, it has been estimated that the yield could perhaps be increased severalfold. Russia is said to have succeeded in developing an acceptable human food product from Antarctic krill. [J.L.McH.]

Marine geology The study of the portion of the Earth beneath the oceans. Approximately 70% of the Earth's surface is covered with water. Marine geology involves the study of the sea floor; of the sediments, rocks, and structures beneath the sea floor; and of the processes that are responsible for their formation. The average depth of the ocean is about 3800 m (12,500 ft), and the greatest depths are in excess of 11,000 m (36,000 ft; the Marianas Trench). Hence, the study of the sea floor necessitates employing a complex suite of techniques to measure the characteristic properties of the Earth's surface beneath the oceans. Contrary to popular views, only a minority of marine geological investigations involve the direct observation of the sea floor by scuba diving or in submersibles. Rather, most of the ocean floor has been investigated by surface ships using remote-sensing geophysical techniques, and more recently by the use of satellite observations.

The oceanic crust is relatively young, having been formed entirely within the last 200 million years (m.y.), a small fraction of the nearly 5-billion-year history of the Earth. The process of renewing or recycling the oceanic crust is the direct consequence of plate tectonics and sea-floor-spreading processes. It is therefore logical that the geologic history of the sea floor be outlined within the framework of plate tectonic tenets. Where plates move apart, molten lava reaches the surface to fill the voids, creating new oceanic crust. Where the plates come together, oceanic crust is thrust back within the interior of the Earth, creating the deep oceanic trenches. These trenches are located primarily around the rim of the Pacific Ocean. The material can be traced by using the distribution of earthquakes to depths of about 700 km (420 mi). At that level, the character of the subducted lithosphere is lost, and this material is presumably remelted and assimilated with the surrounding upper-mantle material. *See* EARTHQUAKE.

Mid-oceanic ridges. Most of the ocean floor can be classified into three broad physiographic regions, one grading into the other (see illustration). The approximate centers of the ocean basin are characterized by spectacular, globally encircling mountain ranges, the mid-oceanic ridge (MOR) system, which formed as the direct consequence of the splitting apart of oceanic lithosphere. The detailed morphologic characteristics of these mountain ranges depend somewhat upon the rate of separation of the plates involved. Abyssal hill relief, especially within 500 km (300 mi) of ridge crest, is noticeably rougher on the slow-spreading Mid-Atlantic Ridge than on the fast-spreading East Pacific Rise. The profile of the East Pacific Rise is also broader and shallower than for the Mid-Atlantic Ridge. If the entire mid-oceanic ridge system were spreading rapidly,

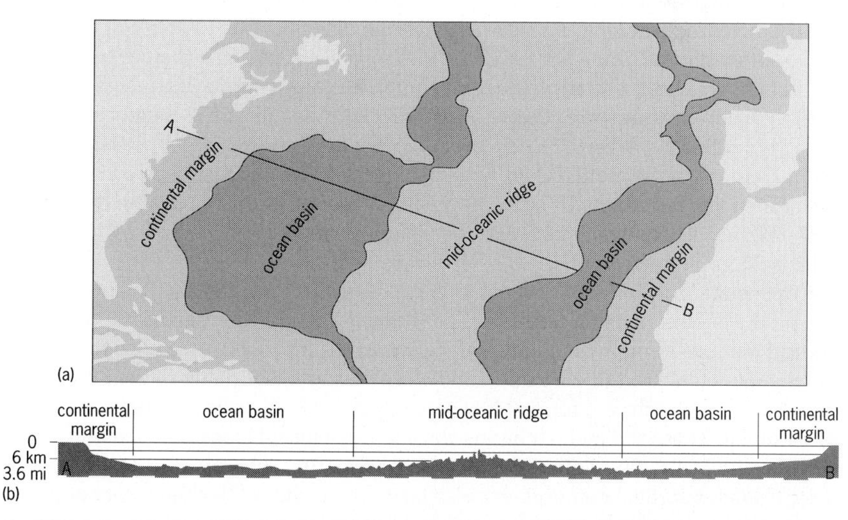

Geology of the North Atlantic Ocean. (*a*) Physiographic divisions of the ocean floor. (*b*) Principal morphologic features along the profile between North America and Africa.

the expanded volume of the ridge system would displace water from the ocean basins onto the continents.

The broad cross-sectional shape of this mid-ocean mountain range can be related directly and simply to its age. The depth of the mid-oceanic ridge at any place is a consequence of the steady conduction of heat to the surface and the associated cooling of the oceanic crust and lithosphere. As it cools, contracts, and becomes more dense, the oceanic crust plus the oceanic lithosphere sink isostatically (under its own weight) into the more fluid substrate (the asthenosphere). Hence, the depth to the top of the oceanic crust is a predictable function of the age of that crust; departures from such depth predictions represent oceanic depth anomalies. These depth anomalies are presumably formed as a consequence of processes other than lithospheric cooling, such as intraplate volcanism. The Hawaiian island chain and the Polynesian island groups are examples of this type of volcanism. *See* MID-OCEANIC RIDGE; OCEANIC ISLANDS; VOLCANOLOGY.

Basins. The deep ocean basins, which lie adjacent to the flanks of the mid-oceanic ridge, represent the older portions of the sea floor that were once the shallower flanks of the ridge (see illustration). The bulk of sediments found on the ocean floor can be broadly classified as terrigenous or biogenic. Terrigenous sediments are derived from adjacent landmasses and are brought to the sea through river systems. This sediment load is sometimes transported across the continental shelves, often utilizing, as pathways, submarine canyons that dissect the shelves, the continental slope, and the continental rise. Biogenic sediments are found in all parts of the ocean, either intermixed with terrigenous sediments or in near "pure form" in those areas inaccessible to terrigenous sedimentation.

Biogenic sediments are composed mostly of the undissolved tests of siliceous and calcareous microorganisms, which settle slowly to the sea floor. This steady so-called pelagic rain typically accumulates at rates of a few centimeters per thousand years. The composition and extent of the input to the biogenic sediment depend upon the composition and abundances of the organisms, which in turn are largely reflective of the water temperature and the available supply of nutrients. The Pacific equatorial zones and certain other regions of deep ocean upwelling are rich in nutrients and correspondingly rich in the microfauna and flora of the surface waters. Such regions are characterized by atypically high pelagic sediment rates. *See* UPWELLING.

Continental margins. The continental margins lie at the transition zone between the continents and the ocean basins and mark a major change from deep water to shallow water and from thin oceanic crust to thick continental crust. Rifted margins are found bounding the Atlantic Ocean (see illustration). These margins represent sections of the South American and North American continents that were once contiguous to west Africa and northwest Africa, respectively. These supercontinents were rifted apart 160–200 m.y. ago as the initial stages of sea-floor spreading and the birth of the present Atlantic Ocean sea floor.

Continental margins are proximal to large sources of terrestrial sediments that are the products of continental erosion. The margins are also the regions of very large vertical motions through time. This vertical motion is a consequence of cooling of the rifted continental lithosphere and subsidence. During initial rifting of the continents, fault-bound rift basins are formed that serve as sites of deposition for large quantities of sediment. These sedimented basins constitute significant loads onto the underlying crust, giving rise to an additional component of margin subsidence. The continental margins are of particular importance also because, as sites of thick sediment accumulations (including organic detritus), they hold considerable potential for the eventual formation and concentration of hydrocarbons. As relatively shallow areas, they are also accessible to offshore exploratory drilling and oil and gas production wells.

Many sedimentary aprons or submarine fans are found seaward of prominent submarine canyons that incise the continental margins. Studies of these sedimentary deposits have revealed a number of unusual surface features that include a complex system of submarine distributary channels, some with levees. The channel systems control and influence sediment distribution by depositional or erosional interchannel flows. Fans are also effected by major instantaneous sediment inputs caused by large submarine mass slumping and extrachannel turbidity flows.

In contrast to the rifted margins, the continental margins that typically surround the Pacific Ocean represent areas where plates are colliding. As a consequence of these collisions, the oceanic lithosphere is thrust back into the interior of the Earth; the loci of underthrusting are manifest as atypically deep ocean sites known as oceanic trenches. The processes of subducting the oceanic lithosphere give rise to a suite of tectonic and morphologic features characteristically found in association with the oceanic trenches. An upward bulge of the crust is created seaward of the trench that represents the flexing of the rigid oceanic crust as it is bent downward at the trench. The broad zone landward of most trenches is known as the accretionary prism and represents the accumulation of large quantities of sediment that was carried on the oceanic crust to the trench. Because the sediments have relatively little strength, they are not underthrust with the more rigid oceanic crust, but they are scraped off. In effect, they are plastered along the inner wall of the trench system, giving rise to a zone of highly deformed sediments. These sediments derived from the ocean floor are intermixed with sediments transported downslope from the adjacent landmass, thus creating a classic sedimentary melange. *See* CONTINENTAL MARGIN; SEDIMENTOLOGY.

Anomalous features. In addition to the major morphologic and sediment provinces, parts of the sea floor consist of anomalous features that obviously were not formed by fundamental processes of sea-floor spreading, plate collisions, or sedimentation. Examples are long, linear chains of seamounts and islands. Many of these chains are thought to reflect the motion of the oceanic plates over hot spots that are fixed within the mantle. *See* SEAMOUNT AND GUYOT.

The presence of large, anomalously shallow regions known as oceanic plateaus may also represent long periods of anomalous regional magmatic activity that may have occurred either near divergent plate boundaries or within the plate. Alternatively, many oceanic plateaus are thought to be small fragments of continental blocks that have been dispersed through the processes of rifting and spreading, and have subsequently subsided below sea level to become part of the submarine terrain.

Other important features of the ocean floor are the so-called scars represented by fracture zone traces that were formed as part of the mid-oceanic ridge system, where the ridge axis was initially offset. Oceanic crusts on opposite sides of such offsets have different ages and hence they have different crustal depths. A structural-tectonic discontinuity exists across this zone of ridge axis offset known as a transform zone. Although relative plate motion does not occur outside the transform zone, the contrasting properties represented by the crustal age differences create contrasting topographic and subsurface structural discontinuities, which can sometimes be traced for great distances. Fracture zone traces define the paths of relative motion between the two plates involved. Those mapped by conventional methods of marine survey have provided fundamental information that allows rough reconstructions of the relative positions of the continents and oceans throughout the last 150–200 m.y. The study that deals with the relative motions of the plates is known as plate kinematics.

Marginal seas. The sea-floor features described so far are representative of the main ocean basins and reflect their evolution mostly through processes of plate tectonics. Other, more complicated oceanic regions, typically found in the western Pacific, include a variety of small, marginal seas (back-arc basins) that were formed by the same general processes as the main ocean basins. These regions define a number of small plates whose interaction is also more or less governed by the normal tenets of plate tectonics. One difficulty in studying these small basins is that they are typified by only short-lived phases of evolution. Frequent changes in plate motions interrupt the process, creating tectonic overprints and a new suite of ocean-floor features. Furthermore, conventional methods of analyzing rock magnetism or depths of the sea floor to date the underlying crust do not work well in these small regions. The small dimensions of these seas bring into play relatively large effects of nearby tectonic boundaries and render invalid key assumptions of these analytical techniques. The number of small plates that actually behave as rigid pieces is not well known, but it is probably only 10–20 for the entire world. [D.E.H.]

Marine microbiology An independent discipline applying the principles and methods of general microbiology to research in marine biology and biogeochemistry. Marine microbiology focuses primarily on prokaryotic organisms, mainly bacteria. Because of their small size and easy dispersability, bacteria are virtually ubiquitous in the marine environment. Furthermore, natural populations of marine bacteria comprise a large variety of physiological types, can survive long periods of starvation, and are able to start their metabolic activity as soon as a substrate becomes available. As a result, the marine environment, similar to soil, possesses the potential of a large variety of microbial processes that degrade (heterotrophy) but also produce (autotrophy) organic matter. Considering the fact that the marine environment represents about 99% of the biosphere, marine microbial transformations are of tremendous global importance. *See* BIOSPHERE.

Heterotrophic transformations. Quantitatively, the most important role of microorganisms in the marine environment is heterotrophic decomposition and remineralization of organic matter. It is estimated that about 95% of the photosynthetically produced organic matter is recycled in the upper 300–400 m (1000–1300 ft) of water, while the remaining 5%, largely particulate matter, is further decomposed during sedimentation. Only about 1% of the total organic matter produced in surface waters arrives at the deep-sea floor in particulate form. In other words, the major source of energy and carbon for all marine heterotrophic organisms is distributed over the huge volume of pelagic water mass with an average depth of about 3800 m (2.5 mi). In this highly dilute medium, particulate organic matter is partly replenished from dissolved organic carbon by microbial growth, the so-called microbial loop.

Of the large variety of organic material decomposed by marine heterotrophic bacteria, oil and related hydrocarbons are of special interest. Other environmentally detrimental pollutants that are directly dumped or reach the ocean as the ultimate sink by land runoff are microbiologically degraded at varying rates. Techniques of molecular genetics are aimed at encoding genes of desirable enzymes into organisms for use as degraders of particular pollutants.

A specifically marine microbiological phenomenon is bacterial bioluminescence, which may function as a respiratory bypass of the electron transport chain. Free-living luminescent bacteria are distinguished from those that live in symbiotic fashion in light organelles of fishes or invertebrates.

Photoautotrophs and chemoautotrophs. The type of photosynthesis carried out by purple sulfur bacteria uses hydrogen sulfide (instead of water) as a source of electrons and thus produces sulfur, not oxygen. Photoautotrophic bacteria are therefore limited to environments where light and hydrogen sulfide occur simultaneously, mostly in lagoons and estuaries. In the presence of sufficient amounts of organic substrates, heterotrophic sulfate-reducing bacteria provide the necessary hydrogen sulfide where oxygen is depleted by decomposition processes. Anoxygenic photosynthesis is also carried out by some blue-green algae, which are now classified as cyanobacteria. *See* CYANOBACTERIA; PHOTOSYNTHESIS.

Chemoautotrophic bacteria are able to reduce inorganic carbon to organic carbon (chemosynthesis) by using the chemical energy liberated during the oxidation of inorganic compounds. Their occurrence, therefore, is not light-limited but depends on the availability of oxygen and the suitable inorganic electron source. Their role as producers of organic carbon is insignificant in comparison with that of photosynthetic producers (exempting the processes found at deep-sea hydrothermal vents). The oxidation of ammonia and nitrite to nitrate (nitrification) furnishes the chemically stable and biologically most available form of inorganic nitrogen for photosynthesis. *See* NITROGEN CYCLE.

The generation of methane and acetic acid from hydrogen and carbon dioxide stems from anaerobic bacterial chemosynthesis, and is common in anoxic marine sediments. *See* METHANOGENESIS (BACTERIA).

Marine microbial sulfur cycle. Sulfate is quantitatively the most prominent anion in seawater. Since it can be used by a number of heterotrophic bacteria as an electron acceptor in respiration following the depletion of dissolved oxygen, the resulting sulfate reduction and the further recycling of the reduced sulfur compounds make the marine environment microbiologically distinctly different from fresh water and most soils. The marine anaerobic, heterotrophic sulfate-reducing bacteria are classified in three genera; *Desulfovibrio, Desulfotomaculum*, and *Clostridium*.

The marine aerobic sulfur-oxidizing bacteria fall into two groups: the thiobacilli and the filamentous or unicellular organisms. While the former comprise a wide range from obligately to facultatively chemoautotrophic species (requiring none or some organic compounds), few of the latter have been isolated in pure culture, and chemoautotrophy has been demonstrated in only a few.

Hydrothermal vent bacteria. Two types of hydrothermal vents have been investigated: warm vents (8–25°C or 46–77°F) with flow rates of 1–2 cm (0.4–0.8 in.) per second, and hot vents (260–360°C or 500–600°F) with flow rates of 2 m (6.5 ft) per second. In their immediate vicinity, dense communities of benthic invertebrates are found with a biomass that is orders of magnitude higher than that normally found at these depths and dependent on photosynthetic food sources. This phenomenon has been explained by the bacterial primary production of organic carbon through the chemosynthetic oxidation of reduced inorganic compounds. The chemical energy required for this process is analogous to the light energy used in photosynthesis and is provided by the geothermal reduction of inorganic chemical species. The specific compounds contained in the emitted vent waters and suitable for bacterial chemosynthesis are mainly hydrogen sulfide, hydrogen, methane, and reduced iron and manganese. The extremely thermophilic microorganisms isolated from hydrothermal vents belong, with the exception of the genus *Thermotoga,* to the Archaebacteria. Of eight archaeal genera, growing within a temperature range of about 75–110°C (165–230°F), three are able to grow beyond the boiling point of water, if the necessary pressure is applied to prevent boiling. These organisms are strictly anaerobic. However, unlike mesophilic bacteria, hyperthermophilic marine isolates tolerate oxygen when cooled below their minimum growth temperature. *See* ARCHAEBACTERIA. [H.W.J.]

Marine sediments The accumulation of minerals and organic remains on the sea floor. Marine sediments vary widely in composition and physical characteristics as a function of water depth, distance from land, variations in sediment source, and the physical, chemical, and biological characteristics of their environments. The study of marine sediments is an important phase of oceanographic research and, together with the study of sediments and sedimentation processes on land, constitutes the subdivision of geology known as sedimentology. *See* OCEANOGRAPHY.

Traditionally, marine sediments are subdivided on the basis of their depth of deposition into littoral 0–66 ft (0–20 m), neritic 66–660 ft (20–200 m), and bathyal 660–6600 ft (200–2000 m) deposits. This division overemphasizes depth. More meaningful, although less rigorous, is a distinction between sediments mainly composed of materials derived from land, and sediments composed of biological and mineral material originating in the sea. Moreover, there are significant and general differences between deposits formed along the margins of the continents and large islands, which are influenced strongly by the nearness of land and occur mostly in fairly shallow water, and the pelagic sediments of the deep ocean far from land.

Sediments of continental margins. These include the deposits of the coastal zone, the sediments of the continental shelf, conventionally limited by a maximum depth of 330–660 ft (100–200 m), and those of the continental slope. Because of large differences in sedimentation processes, a useful distinction can be made between the coastal deposits on one hand (littoral), and the open shelf and slope sediments on the other (neritic and bathyal). Furthermore, significant differences in sediment characteristics and sedimentation patterns exist between areas receiving substantial detrital material from land, and areas where most of the sediment is organic or chemical in origin.

Coastal sediments include the deposits of deltas, lagoons, and bays, barrier islands and beaches, and the surf zone. The zone of coastal sediments is limited on the seaward side by the depth to which normal wave action can stir and transport sand, which depends on the exposure of the coast to waves and does not usually exceed 66–100 ft (20–30 m); the width of this zone is normally a few miles. The sediments in the coastal zone are usually land-derived. The material supplied by streams is sorted in the surf zone; the sand fraction is transported along the shore in the surf zone, often over long distances, while the silt and clay fractions are carried offshore into deeper water by currents. Consequently, the beaches and barrier islands are constructed by wave action mainly from material from fairly far away, although local erosion may make a contribution, while the lagoons and bays behind them receive their sediment from local rivers.

The types and patterns of distribution of the sediments are controlled by three factors and their interaction: (1) the rate of continental runoff and sediment supply; (2) the intensity and direction of marine transporting agents, such as waves, tidal currents, and wind; and (3) the rate and direction of sea level changes. The balance between these three determines the types of sediment to be found. *See* DELTA.

On most continental shelves, equilibrium has not yet been fully established and the sediments reflect to a large extent the recent rise of sea level. Only on narrow shelves with active sedimentation are present environmental conditions alone responsible for the sediment distribution. Sediments of the continental shelf and slope belong to one or more of the following types: (1) biogenic (derived from organisms and consisting mostly of calcareous material); (2) authigenic (precipitated from sea water or formed by chemical replacement of other particles, for example, glauconite, salt, and phosphorite); (3) residual (locally weathered from underlying rocks); (4) relict (remnants of earlier environments of deposition, for example, deposits formed during the transgression leading to the present high sea level stand); (5) detrital (products of weathering and erosion of land, supplied by streams and coastal erosion, such as gravels, sand, silt, and clay).

Much of the fine-grained sediment transported into the sea by rivers is not permanently deposited on the self but kept in suspension by waves. This material is slowly carried across the shelf by currents and by gravity flow down its gentle slope, and is finally deposited either on the continental slope or in the deep sea. If submarine canyons occur in the area, they may intercept these clouds, or suspended material, channel them, and transport them far into the deep ocean as turbidity currents. If the canyons intersect the nearshore zone where sand is transported, they can carry this material also out into deep water over great distances. *See* CONTINENTAL MARGIN; REEF. [T.H.V.A.]

Deep-sea sediments. In general, classifications are difficult to apply because so many deep-sea sediments are widely ranging mixtures of two or more end-member sediment types. However, they can be divided into biogenic and nonbiogenic sediments.

Biogenic sediments, those formed from the skeletal remains of various kinds of marine organisms, may be distinguished according to the composition of the skeletal material, principally either calcium carbonate or opaline silica. The most abundant contributors of calcium carbonate to the deep-sea sediments are the planktonic foraminiferids, coccolithophorids and pteropods. Organisms which extract silica from the sea water and whose hard parts eventually are added to the sediment are radiolaria, diatoms, and to a lesser degree, dilicoflagellates and sponges. The degree to which deep-sea sediments in any area are composed of one or more of these biogenic types depends on the organic productivity of the various organisms in the surface water, the degree to which the skeletal remains are redissolved by sea water while setting to the bottom, and the rate of sedimentation of other types of sediment material. Where sediments are composed largely of a single type of biogenic material, it is often referred to as an ooze, after its consistency in place on the ocean floor.

The nonbiogenic sediment constituents are principally silicate materials and, locally, certain oxides. These may be broadly divided into materials which originate on the continents and are transported to the deep sea (detrital constituents) and those which originate in place in the deep sea, either precipitating from solution (authigenic minerals) or forming from the alteration of volcanic or other materials. The coarser constituents of detrital sediments include quartz, feldspars, amphiboles, and a wide spectrum of other common rock-forming minerals. The finer-grained components also include some quartz and feldspars, but belong principally to a group of sheet-silicate minerals known as the clay minerals, the most common of which are illite, montmorillonite, kaolinite, and chlorite. The distributions of several of these clay minerals have yielded information about their origins on the continents and, in several cases, clues to their modes of transport to the oceans. [P.E.Bi.]

Maritime meteorology Those aspects of meteorology that occur over, or are influenced by, ocean areas. Maritime meteorology serves the practical needs of surface and air navigation over the oceans. Phenomena such as heavy weather, high seas, tropical storms, fog, ice accretion, sea ice, and icebergs are especially important because they seriously threaten the safety of ships and personnel. The weather and ocean conditions near the air-ocean interface are also influenced by the atmospheric planetary boundary layer, the ocean mixed layer, and ocean fronts and eddies.

To support the analysis and forecasting of many meteorological and oceanographic elements over the globe, observations are needed from a depth of roughly 1 km (0.6 mi) in the ocean to a height of 30 km (18 mi) in the atmosphere. In addition, the observations must be plentiful enough in space and time to keep track of the major features of interest, that is, tropical and extratropical weather systems in the atmosphere and fronts and eddies in the ocean. Over populated land areas, there is a fairly dense meteorological network; however, over oceans and uninhabited lands, meteorological observations are scarce and expensive to make, except over the major sea lanes and air routes. Direct observations in the ocean, especially below the sea surface, are insufficient to make a synoptic analysis of the ocean except in very limited regions. Fortunately, remotely sensed data from meteorological and oceanographic satellites are helping to fill in some of these gaps in data. Satellite data can provide useful information on the type and height of clouds, the temperature and humidity structure in the atmosphere, wind velocity at cloud level and at the sea surface, the ocean surface temperature, the height of the sea, and the location of sea ice. Although satellite-borne sensors cannot penetrate below the sea surface, the height of the sea can be used to infer useful information about the density structure of the ocean interior.

The motion of the atmosphere and the ocean is governed by the laws of fluid dynamics and thermodynamics. These laws can be expressed in terms of mathematical equations that can be put on a computer in the form of a numerical model and used to help analyze the present state of the fluid system and to forecast its future state. This is the science of numerical prediction, and it plays a very central role in marine meteorology and physical oceanography.

The first step in numerical prediction is known as data assimilation. This is the procedure by which observations are combined with the most recent numerical prediction valid at the time that the observations are taken. This combination produces an analysis of the present state of the atmosphere and ocean that is better than can be obtained from the observations alone. Data assimilation with a numerical model increases the value of a piece of data, because it spreads the influence of the data in space and time in a dynamically consistent way.

The second step is the numerical forecast itself, in which the model is integrated forward in time to predict the state of the atmosphere and ocean at a future time. Models of the global atmosphere and world ocean, as well as regional models with higher spatial resolution covering limited geographical areas, are used for this purpose. In meteorology and oceanography the success of numerical prediction depends on collecting sufficient data to keep track of meteorological and oceanographic features of interest (including those in the earliest stages of development), having access to physically complete and accurate numerical models of the atmosphere and ocean, and having computer systems powerful enough to run the models and make timely forecasts. *See* WEATHER FORECASTING AND PREDICTION. [R.L.Han.]

Mass wasting A generic term for downslope movement of soil and rock, primarily in response to gravitational body forces. Mass wasting is distinct from other erosive processes in which particles or fragments are carried down by the internal energy of wind, running water, or moving ice and snow.

The stability of slope-making materials is lost when their shear strength (or sometimes their tensile strength) is overcome by shear (or tensile) stresses, or when individual particles, fragments, and blocks are induced to topple or tumble. The shear and tensile strength of earth materials depends on their mineralogy and structure. Processes that generally decrease the strength of earth materials include one or more of the following: structural changes, weathering, groundwater, and meteorological changes. Stresses in slopes are increased by steepening, heightening, and external loading due to static and dynamic forces. Processes that increase stresses can be natural or result from human activities. Although other classifications exist, these movements can be conveniently classified according to their velocity into two types: creep and landsliding. *See* SOIL MECHANICS.

Geologically, creep is the imperceptible downslope movement at rates as slow as a fraction of millimeter per year; its cumulative effects are ubiquitously expressed in slopes as the downhill bending of bedded and foliated rock, bent tree trunks, broken retaining walls, and tilted structures. There are two varieties of geologic creep. Seasonal creep is the slow, episodic movement of the uppermost several centimeters of soil, or fractured and weathered rock. It is especially important in regions of permanently frozen ground. Rheologic creep, sometimes called continuous creep, is a time-dependent deformation at relatively constant shear stresses of masses of rock, soil, ice, and snow. This type of creep affects rock slopes down to depths of a few hundred meters, as well as the surficial layer disturbed by seasonal creep. Continuous creep is most conspicuous in weak rocks and in regions where high horizontal stresses (several tens of bars or several megapascals) are known to exist in rock masses at depths of 330 to 660 ft (100 to 200 m).

Landsliding includes all perceptible mass movements. Three types are generally recognized on the basis of the type of movement: falls, slides, and flows. Falls involve free-falling material; in slides the moving mass displaces along one or more narrow shear zones; and in flows the distribution of velocities within the moving mass resembles that of a viscous flow. *See* LANDSLIDE.

Mass wasting is an important consideration in the interaction between humans and the environment. Deforestation accelerates soil creep. Engineering activities such as damming and open-pit mining are known to increase landsliding. On the other hand, enormous natural rock avalanches have buried entire villages and claimed tens of thousands of lives. [A.S.N.]

Mastitis Inflammation of the mammary gland. This condition is most frequently caused by infection of the gland with bacteria that are pathogenic for this organ. It has been described in humans, cows, sheep, goats, pigs, horses, and rabbits. Mastitis causes lactating women to experience pain when nursing the child, it damages mammary tissue, and the formation of scar tissue in the breast may cause disfigurement.

The mammary gland is composed of a teat and a glandular portion. The gland has defensive mechanisms to prevent and overcome infection with bacteria. Nonspecific defense mechanisms include teat duct keratin, lactoferrin, lactoperoxidase, and complement. Specific defense mechanisms are mediated by antibodies and include opsonization, direct lysis of pathogens, and toxin neutralization. Milk contains epithelial cells, macrophages, neutrophils, and lymphocytes. To induce mastitis, a pathogen must first pass through the teat duct to enter the gland, survive the bacteriostatic and bactericidal mechanisms, and multiply. Bacteria possess virulence factors such as capsules and toxins which enable them to withstand these protective mechanisms.

When bacteria multiply within the gland, there is a release of inflammatory mediators and an influx of neutrophils. The severity of mammary infection is classified according to the clinical signs. In humans, infection occurs during lactation, with clinical episodes most frequent during the first 2 months of lactation. In acute puerperal mastitis the tissue becomes hot, swollen, red, and painful, and a fever may be present. In the absence of treatment, this may progress to a pus-forming mastitis, with the development of breast abscesses.

In acute puerperal mastitis of humans, suitable antibiotics are administered by the intravenous or intramuscular route, while in the abscess form surgical drainage is provided in addition to antibacterial therapy. Penicillins, cephalosporins, and erythromycin are administered locally into the infected gland after milking for 1–2 days. Additional antibiotics are given systemi-cally, that is, intravenously or intramuscularly, in severe cases of mastitis, and also to improve bacteriologic cure rates. [N.L.N.]

Mathematical ecology The application of mathematical theory and technique to ecology. The earliest studies in ecology were by naturalists interested in organisms and their relationships to the environment. Such investigations continue to this day as a central part of the subject, and have focused attention on understanding the ecological and evolutionary relationships among species. For the most part, such approaches are retrospective, designed to help in understanding how current ecological relationships developed, and to place that development within appropriate evolutionary context. The second major branch is applied ecology, and derives from the need to manage the environment and its resources. Here the necessity for rigorous mathematical treatments is obvious, but the goals are quite different from those in evolutionary ecology. Management and control are the objectives, and the relevant time horizon lies in the future. The focus is no longer simply to derive understanding and explanation; rather, one seeks methods for prediction and algorithms for control.

There has been a dramatic increase in mathematical activity concerning the modeling and control of epidemics, and an increasing recognition of the need to view such problems in their proper ecological context as host-parasite interactions. Researchers are using mathematical models to help to understand the factors underlying disease outbreaks, and to develop methods for control such as vaccination strategies. *See* EPIDEMIOLOGY.

Finally, the need for environmental protection in the face of threats from such competing stresses as toxic substances, acid precipitation, and power generation has led to the development of more sophisticated models that address the responses to stress of community and ecosystem characteristics, for example, succession, productivity, and nutrient cycling. *See* ECOLOGY; ECOLOGY, APPLIED; ENVIRONMENTAL ENGINEERING. [S.A.L.]

Mathematical geography The branch of geography that examines human and physical activities on the Earth's surface using models and statistical analysis. The primary areas in which mathematical methods are used include the analysis of spatial patterns, the processes that are responsible for creating and modifying these patterns, and the interactions among spatially separated entities.

What sets geographic methods apart from other quantitative disciplines is geography's focus on place and relative location. Latitude and longitude provide an absolute system of recording spatial data, but geographic databases also typically contain large amounts of relative and relational data about places. Thus, geographers have devoted much effort to accounting for spatial interrelations while maintaining consistency with the assumptions of mathematical models and statistical theory. *See* GEOGRAPHY.

Spatial pattern methodologies attempt to describe the arrangement of phenomena over space. In most cases these phenomena are either point or area features, though computers now allow for advanced three-dimensional modeling as well. Point and area analyses use randomness (or lack of pattern) as a dividing point between two opposite pattern types—dispersed or clustered.

An important innovation in geographic modeling has been the development of spatial autocorrelation techniques. Unlike conventional statistics, in which many tests assume that observations are independent and unrelated, very little spatial data can truly be considered independent. Soil moisture or acidity in one location, for example, is a function of many factors, including the moisture or acidity of nearby points. Because most physical and human phenomena exhibit some form of spatial interrelationships, several statistical methods, primarily based on the Moran Index, have been developed to measure this spatial autocorrelation. Once identified, the presence and extent of spatial autocorrelation can be built into the specification of geographical models to more accurately reflect the behavior of spatial phenomena. [J.C.Co.]

Measles An acute, highly infectious viral disease with cough, fever, and maculopapular rash. It is of worldwide endemicity.

The virus enters the body via the respiratory system, multiplies there, and circulates in the blood. Cough, sneezing, conjunctivitis, photophobia, and fever occur, with Koplik's spots (small red spots containing a bluish-white speck in the center) in the mouth.

A rash appears after 14 days' incubation and persists 5–10 days. Serious complications may occur in 1 out of 15 persons; these are mostly respiratory (bronchitis, pneumonia), but neurological complications are also found. Encephalomyelitis occurs rarely. Permanent disabilities may ensue for a significant number of persons. Measles is one of the leading causes of death among children in the world, particularly in the developing countries.

In unvaccinated populations, immunizing infections occur in early childhood during epidemics which recur after 2–3 years' accumulation of susceptible children. Transmission is by coughing or sneezing. Measles is infectious from the onset of symptoms until a few days after the rash has appeared. Second attacks of measles are very rare. Treatment is symptomatic.

Killed virus vaccine should not be used, as certain vaccinees become sensitized and develop local reactions when revaccinated with live attenuated virus, or develop a severe illness upon contracting natural measles. Live attenuated virus vaccine effectively prevents measles; vaccine-induced antibodies persist for years. [J.L.Me.]

Medical bacteriology The study of bacteria that cause human disease. The field encompasses the detection and identification of bacterial pathogens, determination of the sensitivity and mechanisms of resistance of bacteria to antibiotics, the mechanisms of virulence, and some aspects of immunity to infection. *See* VIRULENCE.

The clinical bacteriology laboratory identifies bacterial pathogens present in specimens such as sputum, pus, blood, and spinal fluid, or from swabs of skin, throat, rectal, or urogenital surfaces. Identification involves direct staining and microscopic examination of these materials, and isolation of bacteria present in the material by growth in appropriate media. The laboratory must differentiate bacterial pathogens from harmless bacteria that colonize humans. Species and virulent strains of bacteria can be identified on the basis of growth properties, metabolic and biochemical tests, and reactivity with specific antibodies.

Recent advances in the field of diagnostic bacteriology have involved automation of biochemical testing; the development of rapid antibody-based detection methods; and the application of molecular biology techniques. Once a bacterial pathogen has been identified, a major responsibility of the diagnostic bacteriology laboratory is the determination of the sensitivity of the pathogen to antibiotics. This involves observation of the growth of the bacteria in the presence of various concentrations of antibiotics. The process has been made more efficient by the development of automated instrumentation.

An increasingly serious problem in the therapy of infectious diseases is the emergence of antibiotic-resistant strains of bacteria. An important area of research is the mechanisms of acquisition of antibiotic resistance and the application of this knowledge to the development of more effective antibiotics. *See* BACTERIAL PHYSIOLOGY AND METABOLISM.

The study of bacterial pathogenesis involves the fields of molecular genetics, biochemistry, cell biology, and immunology. In cases where the disease is not serious and easily treated, research may involve the deliberate infection of human volunteers. Otherwise, various models of human disease must be utilized. These involve experimental infection of animals and the use of tissue cell culture systems. Modern molecular approaches to the study of bacterial pathogenesis frequently involve the specific mutation or elimination of a bacterial gene thought to encode a virulence property, followed by observation of the mutant bacteria in a model system of human disease. In this way, relative contributions of specific bacterial traits to different stages of the disease process can be determined. This knowledge permits the design of effective strategies for intervention that will prevent or cure the disease. *See* BACTERIAL GENETICS.

The presence of specific antibodies is frequently useful in the diagnosis of bacterial diseases in which the pathogen is otherwise difficult to detect. An example is the sexually transmitted disease syphilis; the diagnosis must be confirmed by the demonstration of antibodies specific for *T. pallidum*.

Immunity to some bacteria that survive intracellularly is not mediated by antibodies but by immune effector cells, known as T cells, that activate infected cells to kill the bacteria that they contain. An active area of research is how bacterial components are presented to the immune system in a way that will induce effective cell-mediated immunity. This research may lead to the development of T-cell vaccines effective against intracellular bacterial pathogens.

For disease entities caused by specific bacteria *see* ANTHRAX; BOTULISM; BRUCELLOSIS; CHOLERA; DIPHTHERIA; GLANDERS; GONORRHEA; JOHNE'S DISEASE; LISTERIOSIS; PLAGUE; TETANUS; TUBERCULOSIS; TULAREMIA. For disease entities caused by more than one microorganism *see* FOOD POISONING; INFANT DIARRHEA; MENINGITIS; PNEUMONIA. For groups of disease-producing bacteria *See* MEDICAL BACTERIOLOGY; PNEUMOCOCCUS; STREPTOCOCCUS. [S.L.M.]

Medical mycology The study of fungi (molds and yeasts) that cause human disease. Fungal infections are classified according to the site of infection on the body or whether an opportunistic setting is necessary to establish disease. Fungal infections that occur in an opportunistic setting have become more common due to conditions that compromise host defenses, especially cell-mediated immunity. Such conditions include acquired immunodeficiency syndrome (AIDS), cancer, and immunosuppressive therapy to prevent transplant rejection or to control inflammatory syndromes. Additionally, opportunistic fungal infections have become more significant as severely debilitated individuals live longer because of advances in modern medicine, and nosocomial (hospital-acquired) fungal infections are an increasing problem. Early diagnosis with treatment of the fungal infection and control of the predisposing cause are essential. *See* OPPORTUNISTIC INFECTIONS.

Antifungal drug therapy is extremely challenging since fungi are eukaryotes, as are their human hosts, leading to problems with toxicity or cross-reactivity with host molecules. Most antifungal drugs target the fungal cell membrane or wall. The "gold standard" for therapy of most severe fungal infections is amphotericin B, which binds to ergosterol, a membrane lipid found in most fungi and some other organisms but not in mammals. Unfortunately, minor cross-reactive binding of amphotericin B to cholesterol in mammalian cell membranes can lead to serious toxicity, especially in the kidney where the drug is concentrated. Recent advances in antifungal therapy include the use of liposomal amphotericin B and newer azoles such as fluconazole and itraconazole, which show reduced toxicity or greater specificity. Conversely, drug resistance in pathogenic fungi is an increasing problem, as it is in bacteria.

Candidiasis is the most common opportunistic fungal infection, and it has also become a major nosocomial infection in hospitalized patients. *Candida albicans* is a dimorphic fungus with a yeast form that is a member of the normal flora of the surface of mucous membranes. In an opportunistic setting, the fungus may proliferate and convert to a hyphal form that invades these tissues, the blood, and other organs. The disease may extend to the blood or other organs from various infected sites in patients who are suffering from a grave underlying disease or who are immunocompromised. Other important opportunistic fungal diseases include aspergillosis, mucormycosis, and cryptococcus.

Healthy persons can acquire disease from certain pathogenic fungi following inhalation of their fungal spores. The so-called deep or systemic mycoses are all caused by different species of soil molds; most infections are unrecognized and produce no or few symptoms. However, in some individuals infection may spread to all parts of the body from the lung, and treatment with amphotericin B or an antifungal azole drug is essential.

Other fungal infections develop when certain species of soil molds are inoculated deep into the subcutaneous tissue, such as by a deep thorn prick or other trauma. A specific type of lesion develops with each fungus as it grows within the tissue. Proper wound hygiene will prevent these infections.

Ringworm, also known as dermatophytosis or tinea, is the most common of all fungal infections. Some species of pathogenic molds can grow in the stratum corneum, the dead outermost layer of the skin. Disease results from host hypersensitivity to the metabolic products of the infecting mold as well as from the actual fungal invasion. Tinea corporis, ringworm of the body, appears as a lesion on smooth skin and has a red, circular margin that contains vesicles. The lesion heals with central clearing as the margin advances. On thick stratum corneum, such as the interdigital spaces of the feet, the red, itching lesions, known as athlete's foot or tinea pedis, become more serious if secondary bacterial infection develops. The ringworm fungi may also invade the hair shaft (tinea capitis) or the nail (onychomycosis). Many pharmaceutical agents are available to treat or arrest such infections, but control of transmission to others is important. *See* FUNGAL INFECTIONS; FUNGI; YEAST. [C.Ha.; J.P.W.]

Medical parasitology The study of diseases of humans caused by parasitic agents. It is commonly limited to parasitic worms (helminths) and the protozoa. Current usage places the various nonprotozoan microbes in distinct disciplines, such as virology, rickettsiology, and bacteriology.

Nematodes. The roundworms form an extremely large yet fairly homogeneous assemblage, most of which are free-living (nonparasitic). Some parasitic nematodes, however, may cause disease in humans (zoonosis), and others cause disease limited to human hosts (anthroponosis). Among the latter, several are enormously abundant and widespread. *See* NEMATA.

The giant roundworm (*Ascaris lumbricoides*) parasitizes the small intestine, probably affecting over a billion people; and the whipworm (*Trichuris trichiura*) infects the human colon, probably affecting a half billion people throughout the tropics. Similarly, the hookworms of humans, *Necator americanus* in the Americas and the tropical regions of Africa and Asia, and *Ancylostoma duodenale* in temperate Asia, the Mediterranean, and Middle East, suck blood from the small intestine and cause major debilitation, especially among the undernourished. The human pinworm (*Enterobius vermicularis*) infects the large intestine of millions of urban dwellers. Most intestinal nematodes, which require a period of egg maturation outside the human host before they are infective, are associated with fecal contamination of soil or food crops and are primarily rural in distribution.

The nonintestinal nematodes are spread by complex life cycles that usually involve bloodsucking insects. One exception is the guinea worm (*Dracunculus medinensis*), a skin-infecting 2–3-ft (0.6–1-m) worm transmitted by aquatic microcrustaceans that are ingested in drinking water that has been contaminated by larvae that escape from the skin sores of infected humans. Such bizarre life cycles are typical of many helminths. Other nematodes of humans include (1) the filarial worms, which are transmitted by mosquitoes and may induce enormously enlarged fibrous masses in legs, arms, or genitalia (elephantiasis), and (2) *Onchocerca volvulus*, which is transmitted by blackflies (genus *Simulium*) and forms microscopic embryos (microfilariae) in the eyes causing high incidence of blindness in Africa and parts of central and northern South America.

A more familiar tissue-infecting nematode of temperate regions is *Trichinella spiralis*, the pork or trichina worm, which is the agent of trichinosis. The tiny spiraled larvae encyst in muscle and can carry the infection to humans and other carnivorous mammals who eat raw or undercooked infected meat.

Trematodes. Parasites of the class Trematoda vary greatly in size, form, location in the human host, and disease produced, but all go through an initial developmental period in specific kinds of fresh-water snails, where they multiply as highly modified larvae of different types. Ultimately, an infective larval stage (cercaria) escapes in large numbers from the snail and continues the life cycle. Each trematode species follows a highly specific pathway from snail to human host, usually by means of another host or transport mechanism. These include the intestinal, liver, blood, and lung flukes. *See* SCHISTOSOMIASIS.

Cestodes. Tapeworms, the other great assemblage of parasitic flatworms, parasitize most vertebrates, with eight or more species found in humans. Their flat ribbonlike body form consists of a chain of hermaphroditic segments. Like the trematodes, their life cycles are complex, although not dependent on a snail host. The enormous beef tapeworm of humans, *Taenia saginata*, is transmitted by infected beef ("measly beef") from cattle that grazed where human feces containing egg-filled tapeworm segments contaminated the soil. Other tapeworms include the pork, dog, and broad (or fish) tapeworms.

Protozoa. Of the many protozoa that can reside in the human gut, only the invasive strain of *Entamoeba histolytica* causes serious disease. This parasite, ingested in water contaminated with human feces containing viable cysts of *E. histolytica*, can cause the disease amebiasis, which in its most severe form is known as amebic dysentery. Another common waterborne intestinal protozoon is the flagellate *Giardia lamblia*, which causes giardiasis, a mild to occasionally serious or long-lasting diarrhea. *See* GIARDIASIS; PROTOZOA.

Other flagellate parasites infect the human skin, bloodstream, brain, and viscera. The tsetse fly of Africa carries to humans the blood-infecting agents of trypanosomiasis, or African sleeping sickness, *Trypanosoma brucei gambiense* and *T. brucei rhodesiense*. The infection can be fatal if the parasites cross the blood-brain barrier. In Latin America, the flagellate *T. cruzi* is the agent of Chagas' disease, a major cause of debilitation and premature heart disease among those who are poorly housed. The infection is transmitted in the liquid feces of a conenose bug (genus *Triatoma*) and related insects. The infective material is thought to be scratched into the skin or rubbed in the eye, especially by sleeping children.

Another group of parasitic flagellates includes the macrophage-infecting members of the genus *Leishmania*, which are transmitted by blood-sucking midges or sand flies. Cutaneous leishmaniasis is characterized by masses of infected macrophages in the skin, which induce long-lasting dermal lesions of varying form and severity. The broad spectrum of host-parasite interactions is well exemplified by leishmaniases. The various manifestations of the disease are the result of the particular species of agent and vector, the immunological status of the host, the presence or absence of reservoir hosts, and the pattern of exposure.

Two remaining major groups of protozoa are the ciliates and the sporozoans. The former group is largely free-living, with only a single species, *Balantidium coli*, parasitic in humans (and pigs). This large protozoon is found in the large intestine, where it can cause balantidiasis, an ulcerative disease. The sporozoans, on the other hand, are all parasitic and include many parasites of humans. The most important are the agents of malaria. Other disease agents are included in the genera *Isospora, Sarcocystis, Cryptosporidium*, and *Toxoplasma*. *Pneumocystis*, a major cause of death among persons with acquired immune deficiency syndrome (AIDS), was formerly considered a protozoon of uncertain relationship, but now it is thought to be a member of the Fungi. *See* ACQUIRED IMMUNE DEFICIENCY SYNDROME (AIDS); MALARIA; SPOROZOA.

Toxoplasma gondii, the agent of toxoplasmosis, infects as many as 20% of the world's population. It can penetrate the placenta and infect the fetus if the mother has not been previously infected and has no antibodies. As with most medically important parasites, the great majority of *Toxoplasma* infections remain undetected and nonpathogenic. The parasite primarily affects individuals lacking immune competence—the very young, the very old, and the immunosuppressed. *See* MEDICAL BACTERIOLOGY; MEDICAL MYCOLOGY; ZOONOSES. [D.He.]

Medical waste Any solid waste that is generated in the diagnosis, treatment, or immunization of human beings or animals, in research pertaining thereto, or in the production or testing of biologicals. Since the development of disposable medical products in the early 1960s, the issue of medical waste has confronted hospitals and regulators. Previously, reusable products included items such as linen, syringes, and bandages; they were sterilized or disinfected prior to reuse, and the principal waste product was limited to human pathological tissue.

Most hazardous substances are described by their relevant properties, such as corrosive, poison, or flammable. Medical waste was originally defined in terms of its infectious properties, and thus it was called infectious waste. However, given the difficulty of identifying pathogenic organisms in waste that might cause disease, it has become standard practice to define medical waste by types or categories. While definitions differ somewhat under different regulations, in the United States the Centers for Disease Control and Prevention (CDC) cite four categories of infective wastes that should require special handling and treatment: laboratory cultures and stocks, pathology wastes, blood, and items that possess sharp points such as needles and syringes (sharps). These categories require that the generator of these wastes exercise judgment in identifying the material to be included.

The waste category that has generated a great deal of interest is sharps. Needles and syringes, in particular, pose risks, since the instruments can penetrate into the body, increasing the potential for disease transmission. Improper disposal of these items in the past has been the catalyst for increased regulation and tighter management control.

Treatment of medical waste constitutes a method for rendering it noninfectious prior to disposal in a landfill or other solid-waste site. The treatment technologies currently used for medical waste include incineration, sterilization, chemical disinfection, and microwave, as well as others under development. *See* HAZARDOUS WASTE. [R.A.Sp.]

Mediterranean Sea The Mediterranean Sea lies between Europe, Asia Minor, and Africa. It is completely landlocked except for the Strait of Gibraltar, the Bosporus, and the Suez Canal. The Mediterranean is conveniently divided into an eastern basin and a western basin, which are joined by the Strait of Sicily and the Strait of Messina.

The total water area of the Mediterranean is 965,900 mi^2 (2,501,000 km^2), and its average depth is 5040 ft (1536 m). The greatest depth in the western basin is 12,200 ft (3719 m), in the Tyrrhenian Sea. The eastern basin is deeper, with a greatest depth of 18,140 ft (5530 m) in the Ionian Sea about 34 mi (55 km) off the Greek mainland. The Atlantic tide disappears in the Strait of Gibraltar. The tides of the Mediterranean are predominantly semidiurnal. [J.Ly.]

Mendelism Fundamental principles governing the transmission of genetic traits, discovered by an Augustinian monk Gregor Mendel in 1856. Mendel performed his first set of hybridization experiments with pea plants. Although the pea plant is normally self-fertilizing, it can be easily crossbred, and grows to maturity in a single season. True breeding strains, each with distinct characteristics, were available from local seed merchants. For his experiments, Mendel chose seven sets of contrasting characters or traits. For stem height, the true breeding strains tall (7 ft or 2.1 m) and dwarf (18 in. or 45 cm) were used. He also selected six other sets of traits, involving the shape and color of seeds, pod shape and color, and the location of flowers on the plant stem.

The most simple crosses performed by Mendel involved only one pair of traits; each such experiment is known as a monohybrid cross. The plants used as parents in these crosses are known as the P_1 (first parental) generation. When tall and dwarf plants were crossed, the resulting offspring (called the F_1 or first filial generation) were all tall. When members of the F_1 generation were self-crossed, 787 of the resulting 1064 F_2 (second filial generation) plants were tall and 277 were dwarf. The tall trait is expressed in both the F_1 and F_2 generations, while the dwarf trait disappears in the F_1 and reappears in the F_2 generation. The trait expressed in the F_1 generation Mendel called the dominant trait, while the recessive trait is unexpressed in the F_1 but reappears in the F_2. In the F_2, about three-fourths of the offspring are tall and one-fourth are dwarf (a 3:1 ratio). Mendel made similar crosses with plants exhibiting each of the other pairs of traits, and in each case all of the F_1 offspring showed only one of the parental traits and, in the F_2, three-fourths of the plants showed the dominant trait and one-fourth exhibited the recessive trait. In subsequent experiments, Mendel found that the F_2 recessive plants bred true, while among the dominant plants one-third bred true and two-thirds behaved like the F_1 plants.

Law of segregation. To explain the results of his monohybrid crosses, Mendel derived several postulates. First, he proposed that each of the traits is controlled by a factor (now called a gene). Since the F_1 tall plants produce both tall and dwarf offspring, they must contain a factor for each, and thus he proposed that each plant contains a pair of factors for each trait. Second, the trait which is expressed in the F_1 generation is controlled by a dominant factor, while the unexpressed trait is controlled by a recessive factor. To prevent the number of factors from being doubled in each generation, Mendel postulated that factors must separate or segregate from each other during gamete formation. Therefore, the F_1 plants can produce two types of gametes, one type containing a factor for tall plants, the other a factor for dwarf plants. At fertilization, the random combination of these gametes can explain the types and ratios of offspring in the F_2 generation (see illustration). *See* GENE.

Independent assortment. Mendel extended his experiments to examine the inheritance of two characters simultaneously. Such a cross, involving two pairs of contrasting traits, is known as a dihybrid cross. For example, Mendel crossed plants with tall stems and round seeds with plants having dwarf stems and wrinkled seeds. The F_1 offspring were all tall and had round seeds. When the F_1 individuals were self-crossed, four types of offspring were produced in the following proportions: 9/16 were tall, round; 3/16 were tall, wrinkled; 3/16 were dwarf, round; and 1/16 were dwarf, wrinkled. On the

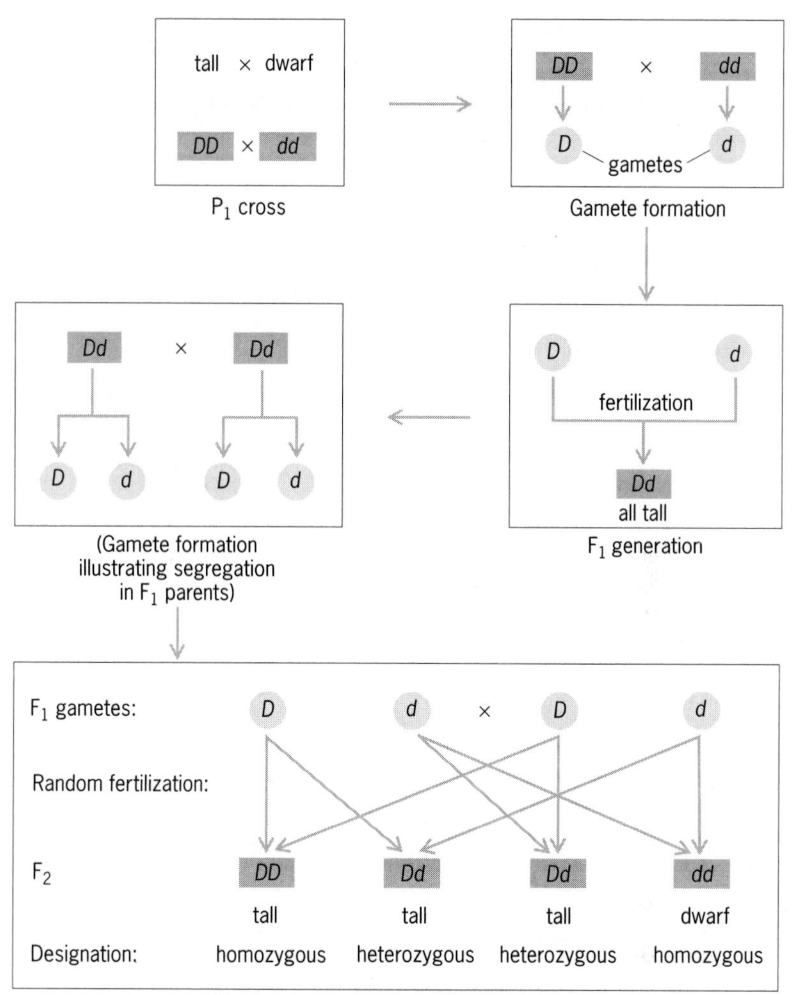

Schematic representation of a monohybrid cross. Pure-bred tall and dwarf strains are crossed, and yield typical 3:1 ratio in the F_2 generation. *D* and *d* represent the tall and dwarf factors (genes), respectively. (*After W. S. Klug and M. R. Cummings, Concepts of Genetics, Charles E. Merrill, 1983*)

basis of similar results in other dihybrid crosses, Mendel proposed that during gamete formation, segregating pairs of factors assort independently of one another. As a result of segregation, each gamete receives one member of every pair of factors [this assumes that the factors (genes) are located on different chromosomes]. As a result of independent assortment, all possible combinations of gametes will be found in equal frequency. In other words, during gamete formation, round and wrinkled factors segregate into gametes independently of whether they also contain tall or dwarf factors.

It might be useful to consider the dihybrid cross as two simultaneous and independent monohybrid crosses. In this case, the predicted F_2 results are 3/4 tall, 1/4 dwarf, and 3/4 round, 1/4 wrinkled. Since the two sets of traits are inherited independently, the number and frequency of phenotypes can be predicted by combining the two events:

$$3/4\,\text{tall}\begin{cases}3/4\,\text{round} & (3/4)(3/4) = 9/16\,\text{tall, round}\\ 1/4\,\text{wrinkled} & (3/4)(1/4) = 3/16\,\text{tall, wrinkled}\end{cases}$$

$$1/4\,\text{dwarf}\begin{cases}3/4\,\text{round} & (1/4)(3/4) = 3/16\,\text{dwarf, round}\\ 1/4\,\text{wrinkled} & (1/4)(1/4) = 1/16\,\text{dwarf, wrinkled}\end{cases}$$

This 9:3:3:1 ratio is known as a dihybrid ratio and is the result of segregation, independent assortment, and random fertilization. *See* GENETICS. [M.R.C.]

Meningitis Inflammation of the meninges. Certain types of meningitis are associated with distinctive abnormalities in the cerebrospinal fluid. With certain types of meningitis, especially bacterial, the causative organism can usually be recovered from the fluid.

Meningeal inflammation in most cases is caused by invasion of the cerebrospinal fluid by an infectious organism. Noninfectious causes also occur. For example, in immune-mediated disorders antigen-antibody reactions can cause meningeal inflammation. Other noninfectious causes of meningitis are the introduction into the cerebrospinal fluid of foreign substances such as alcohol, detergents, chemotherapeutic agents, or contrast agents used in some radiologic imaging procedures. Meningeal inflammation brought about by such foreign irritants is called chemical meningitis. Inflammation also can occur when cholesterol-containing fluid or lipid-laden material leaks into the cerebrospinal fluid from some intracranial tumors.

Bacterial meningitis is among the most feared of human infectious diseases because of its possible seriousness, its rapid progression, its potential for causing severe brain damage, and its frequency of occurrence. Most cases of bacterial meningitis have an acute onset. Common clinical manifestations are fever, headache, vomiting, stiffness of the neck, confusion, seizures, lethargy, and coma. Symptoms of brain dysfunction are caused by transmission of toxic materials from the infected cerebrospinal fluid into brain tissue and the disruption of arterial perfusion and venous drainage from the brain because of blood vessel inflammation. These factors also provoke cerebral swelling, which increases intracranial pressure. Before antibiotics became available, bacterial meningitis was almost invariably fatal.

Most types of acute bacterial meningitis are septic-borne in that they originate when bacteria in the bloodstream (bacteremia, septicemia) gain entrance into the cerebrospinal fluid. Meningitis arising by this route is called primary bacterial meningitis. Secondary meningitis is that which develops following direct entry of bacteria into the central nervous system, which can occur at the time of neurosurgery, in association with trauma, or through an abnormal communication between the external environment and the cerebrospinal fluid.

Many viruses can cause meningeal inflammation, a condition referred to as viral aseptic meningitis. The most common viral causes include the enteroviruses, the various herpesviruses, viruses transmitted by arthropods, the human immunodeficiency virus type I (HIV-1), and formerly, the mumps virus. If the virus attacks mainly the brain rather than the spinal cord, the disorder is termed viral encephalitis. *See* ANIMAL VIRUS; ARBOVIRAL ENCEPHALITIDES; ENTEROVIRUS; HERPES.

Fungal, parasitic, and rickettsial meningitis are less common in the United States than are bacterial and viral. These infections are more likely to be subacute or chronic than those caused by bacteria or viruses; in most cases, the meningeal inflammation is associated with brain involvement. An acute form of aseptic meningitis can occur in the spirochetal diseases, syphilis and Lyme disease. *See* LYME DISEASE; MEDICAL MYCOLOGY; MEDICAL PARASITOLOGY; RICKETTSIOSES; SYPHILIS. [W.E.B.]

Meningococcus A major human pathogen belonging to the bacterial genus *Neisseria*, and the cause of meningococcal meningitis and meningococcemia. The official designation is *N. meningitidis*. The meningococcus is a gram-negative, aerobic, nonmotile diplococcus. It is fastidious in its growth requirements and is very susceptible to adverse physical and chemical conditions.

Humans are the only known natural host of the meningococcus. Transmission occurs by droplets directly from person to person. Fomites and aerosols are probably unimportant in the spread of the organism. The most frequent form of host-parasite relationship is asymptomatic carriage in the nasopharynx.

The most common clinical syndrome caused by the meningococcus is meningitis, which is characterized by fever, headache, nausea, vomiting, and neck stiffness and has a fatality rate of 15% (higher in infants and adults over 60). Disturbance of the state of consciousness quickly occurs, leading to stupor and coma. Many cases also have a typical skin rash consisting of petechiae or purpura. *See* MENINGITIS. [R.Go.]

Meromictic lake A lake whose water is permanently stratified and therefore does not circulate completely throughout the basin at any time during the year. Normally lakes in the temperate zone mix completely during the spring and autumn when water temperatures are approximately the same from top to bottom. In meromictic lakes there are no periods of overturn or complete mixing because seasonal changes in the thermal gradient are either small or overridden by the stability of a chemical gradient, or the deeper waters are physically inaccessible to the mixing energy of the wind. Most commonly, the vertical stratification is stabilized by a chemical gradient in meromictic lakes.

The upper stratum of water in a meromictic lake is mixed by the wind and is called the mixolimnion. The bottom, denser stratum, which does not mix with the water above, is referred to as the monimolimnion. The transition layer between these strata is called the chemocline.

Of the hundreds of thousands of lakes on the Earth, only about 120 are known to be meromictic. In general, meromictic lakes in North America are restricted to: sheltered basins that are proportionally very small in relation to depth and that often contain colored water, basins in arid regions, and isolated basins in fiords. *See* FRESH-WATER ECOSYSTEM; LAKE; LIMNOLOGY. [G.E.Li.]

Mesosphere A layer within the Earth's atmosphere that extends from about 50 to 85 km (31 to 53 mi) above the surface. The mesosphere is predominantly characterized by its thermal structure. On average, mesospheric temperature decreases with increasing height.

Temperatures range from as high as 12°C (53°F) at the bottom of the mesosphere to as low as −133°C (−208°F) at its top. The top of the mesosphere, called the mesopause, is the coldest area of the Earth's atmosphere. Temperature increases with increasing altitude above the mesopause in the layer known as the thermosphere, which absorbs the Sun's extreme ultraviolet radiation. In the stratosphere, the atmospheric layer immediately below the mesosphere, the temperature also increases with height. The stratosphere is where ozone, which also absorbs ultraviolet radiation from the Sun, is most abundant. The transition zone between the mesosphere and the stratosphere is called the stratopause. Mesospheric temperatures are comparatively cold because very little solar radiation is absorbed in this layer. Meteorologists who predict weather conditions or study the lowest level of the Earth's atmosphere, the troposphere, often refer to the stratosphere, mesosphere, and thermosphere collectively as the upper atmosphere. However, scientists who study these layers distinguish between them; they also refer to the stratosphere and mesosphere as the middle atmosphere. *See* ATMOSPHERE; METEOROLOGY; STRATOSPHERE; THERMOSPHERE; TROPOSPHERE.

In the lower part of the mesosphere, the difference between the temperature at the summer and winter poles is of order 35°C (63°F). This large temperature gradient produces the north-south or meridional winds that blow from summer to winter. Temperatures in the upper mesosphere are colder in summer and warmer in winter, resulting in return meridional flow from the summer to the winter hemisphere. Although the temperature gradient in the upper part of the mesosphere remains large, additional complications result in wind speeds that are much slower than they are in the lower part of the mesosphere. Winds in the east-west or zonal direction are greatest at mesospheric middle latitudes. Zonal winds blow toward the west in summer and toward the east in winter. Like their meridional counterparts, zonal winds are comparatively strong near the bottom of the mesosphere and comparatively weak near the top. Thus, on average both temperature and wind speed decrease with increasing height in the mesosphere. *See* ATMOSPHERIC GENERAL CIRCULATION.

Meteors which enter the Earth's atmosphere vaporize in the upper mesosphere. These meteors contain significant amounts of metallic atoms and molecules which may ionize. Metallic ions combined with ionized water clusters make up a large part of the D-region ionosphere that is embedded in the upper mesosphere.

The upper mesosphere is also where iridescent blue clouds can be seen with the naked eye and photographed in twilight at high summer latitudes when the Sun lights them up in the otherwise darkening sky. These clouds are called noctilucent clouds (NLC). Noctilucent clouds are believed to be tiny ice crystals that grow on bits of meteoric dust.

Large-scale atmospheric circulation patterns transport tropospheric air containing methane and carbon dioxide from the lower atmosphere into the middle atmosphere. While carbon dioxide warms the lower atmosphere, it cools the middle and upper atmosphere by releasing heat to space. Methane breaks down and contributes to water formation when it reaches the middle atmosphere. If the air is sufficiently cold, the water can freeze and form noctilucent clouds. Temperatures must be below −129°C (−200°F) for noctilucent clouds to form. These conditions are common in the cold summer mesopause region at high latitudes. [M.Hag.]

Meteorological instrumentation Devices that measure or estimate properties of the Earth's atmosphere. Meteorological instruments take many forms, from simple mercury thermometers and barometers to complex observing systems that remotely sense winds, thermodynamic properties, and chemical constituents over large volumes of the atmosphere.

Weather station measurements provide a description of conditions near the ground. In addition to the average regional conditions, these measurements also provide local information on mesoscale phenomena such as cold fronts, sea breezes, and disturbed conditions resulting from nearby thunderstorms. Traditional thermodynamic instruments are mechanical or heat-conductive devices relying on the expansion and contraction of metallic and nonmetallic liquids or solid materials as a function of temperature, pressure, and humidity. Among these are the mercury, alcohol, and bimetallic thermometers for measurement of temperature, mercury and metallic bellow (aneroid) barometers for measurement of pressure, human hair hygrometers, and wet/dry-bulb thermometers (called psychrometers) for measurement of relative humidity. Mercury barometers are simply weighing devices that balance the mass of the atmospheric column against the mass of a mercury column. On average, a column of atmosphere weighs the same as 76 cm (29.92 in.) of mercury. Psychrometers measure humidity by means of the wet-bulb depression technique. A moist thermometer is cooled by evaporation when relative humidity is less than 100%. The temperature difference between wet and dry thermometers is referred to as the wet-bulb depression, a well-known function of relative humidity at standard airflow speeds. A related method of humidity measurement is the chilled mirror technique (dewpointer). A polished surface is cooled to the temperature of water vapor saturation, at which point the cooled surface becomes fogged. Dewpoint saturation uniquely defines humidity at a known temperature and pressure. *See* DEW POINT.

Precipitation measurement devices may be described as precision buckets, which measure the depth or weight of that which falls into them. These gages work best for rainfall, but they are also used in an electrically heated mode for weighing snow. Rulers are routinely used for measurement of snow depth. Time-resolved measurements of rainfall are traditionally made by counting quantum amounts (0.01 in. or 0.25 mm) of rain with a small, mechanically controlled tipping bucket located beneath a large collecting orifice. Modern rain measuring is sometimes performed along short paths via drop-induced scintillations of infrared radiation, which is emitted by a laser. When the raindrop size distribution is needed, optical-shadowing spectrometers are employed, as are momentum-measuring impact distrometers, devices that measure the number density versus the size distribution of raindrops or other hydrometeors. *See* SNOW GAGE.

Wind measurements are performed by anemometers, some of which use wind-driven spinning cups for wind speed determination. Vanes are used in conjunction with cups for indication of wind direction. Alternatively, three-axis propeller anemometers may be employed to provide orthogonal components of the three-dimensional wind vector. Many hybrids of these basic approaches continue to be successfully employed. Fast-response sonic anemometers employ ultrasound transmission, where the apparent propagation speed of sound is measured. The difference between this measured speed and the actual speed for a fluid at rest is the wind speed. Such measurements are made on a time scale of 0.01 s and are used to determine the fluxes of momentum, water vapor, sensible heat, and other scalars in the planetary boundary layer. *See* WIND MEASUREMENT.

Balloon-borne vertical profiles or soundings of temperature, humidity, and winds are central to computerized (numerical) weather prediction. Such observations are made simultaneously or synoptically worldwide on a daily basis. The temperature and humidity sensors are lightweight expendable versions of traditional surface station instruments. Balloon drift during ascent provides the wind measurement. The preferred method of tracking these rawinsondes is to use global navigation aid systems such as Omega, Loran-C, and the Global Positioning System. Parachute-borne dropsondes are often released from aircraft in data-sparse regions.

Remote sensing, principally via electromagnetic radiation, is a mainstay of modern meteorology. Such devices typically operate in the optical, infrared, millimeter-wave, microwave, and high-frequency radio regions of the electromagnetic spectrum. Passive radiometers typically operate at infrared and microwave frequencies; they are used for estimates of temperature, water vapor, cloud heights, cloud liquid water mass, and trace-gas concentrations. These observations are made from the ground, aircraft, and satellites, usually measuring naturally emitted radiation. Radarlike, active remote-sensing devices are among the most powerful tools available to meteorology. Collectively, these instruments are capable of measuring kinematic, microphysical, chemical, and thermodynamic properties of the troposphere at high spatial and temporal resolution. Active meteorological remote sensors are principally deployed on land, ships, and aircraft platforms, as well as aboard satellites. Unlike passive instruments, active remote sensors can precisely resolve the distance at which a measurement is located.

At optical frequencies, lidars measure conditions in relatively clear air. Capabilities include determining the properties of tenuous clouds; determining concentrations of aerosol, ozone, and water vapor; and measuring winds through the Doppler frequency-shift effect. Millimeter-wave radars are used to probe opaque, nonprecipitating clouds. Polarimetric and Doppler techniques reveal hydrometeor type, water mass, and air motions.

The best-known meteorological remote sensor is the microwave weather radar. In addition to measuring rainfall and tracking movement of storms, powerful and sensitive meteorological radars can measure detailed flow fields in and around storms by using hydrometeors, insects, and blobs of water vapor as reflective targets. These radars can also distinguish between rain, hail, and snow. When Doppler measurements are combined with the atmospheric equations of motion, thermodynamic perturbation fields, such as buoyancy, are revealed inside violent convective storms. At ultrahigh and very high radio frequencies, radars known as wind profilers measure the mean wind as a function of height in the clear and cloudy air. Superior to infrequent weather balloons, radio wind profiling methods permit continuous measurement of winds with regularity and high accuracy. When radio wind profilers are colocated with acoustic transponders, the speed of sound is easily measured through radar tracking of the acoustic wave. This permits the computation of atmospheric density and temperature profiles, on which the speed of sound is strongly dependent. *See* METEOROLOGY; RADAR METEOROLOGY. [R.E.Car.]

Meteorology A discipline involving the study of the atmosphere and its phenomena. Meteorology and climatology are rooted in different parent disciplines, the former in physics and the latter in physical geography. They have, in effect, become interwoven to form a single discipline known as the atmospheric sciences, which is devoted to the understanding and prediction of the evolution of planetary atmospheres and the broad range of phenomena that occur within them. The atmospheric sciences comprise a number of interrelated subdisciplines. *See* CLIMATOLOGY.

Atmospheric dynamics (or dynamic meteorology) is concerned with the analysis and interpretation of the three-dimensional, time-varying, macroscale motion field. It is a branch of fluid dynamics, specialized to deal with atmospheric motion systems on scales ranging from the dimensions of clouds up to the scale of the planet itself. The activity within dynamic meteorology that is focused on the description and interpretation of large-scale (greater than 1000 km or 600 mi) tropospheric motion systems such as extratropical cyclones has traditionally been referred to as synoptic meteorology, and that devoted to mesoscale (10–1000 km or 6–600 mi) weather systems such as severe thunderstorm complexes is referred to as mesometeorology. Both synoptic meteorology and mesometeorology are concerned with phenomena of interest in weather forecasting, the former on the day-to-day time scale and the latter on the time scale of minutes to hours.

The complementary field of atmospheric physics (or physical meteorology) is concerned with a wide range of processes that are capable of altering the physical properties and the chemical composition of air parcels as they move through the atmosphere. It may be viewed as a branch of physics or chemistry, specializing in processes that are of particular importance within planetary atmospheres. Overlapping subfields within atmospheric physics include cloud physics, which is concerned with the origins, morphology, growth, electrification, and the optical and chemical properties of the droplets within clouds; radiative transfer, which is concerned with the absorption, emission, and scattering of solar and terrestrial radiation by aerosols and radiatively active trace gases within planetary atmospheres; atmospheric chemistry, which deals with a wide range of gas-phase and heterogeneous (that is, involving aerosols or cloud droplets) chemical and photochemical reactions on space scales ranging from individual smokestacks to the global ozone layer; and boundary-layer meteorology or micrometeorology, which is concerned with the vertical transfer of water vapor and other trace constituents, as well as heat and momentum across the interface between the atmosphere and the underlying surfaces and their redistribution within the lowest kilometer of the atmosphere by motions on scales too small to resolve explicitly in global models. Aeronomy is concerned with physical processes in the upper atmosphere (above the 50-km or 30-mi level). *See* AERONOMY; ATMOSPHERIC CHEMISTRY; ATMOSPHERIC GENERAL CIRCULATION; CLOUD PHYSICS; MICROMETEOROLOGY; TERRESTRIAL RADIATION.

Although atmospheric dynamics and atmospheric physics in some circumstances can be successfully pursued as separate disciplines, important problems such as the development of numerical weather prediction models and the understanding of the global climate system require a synthesis. Physical processes such as radiative transfer and the condensation of water vapor onto cloud droplets are ultimately responsible for the temperature gradients that drive atmospheric motions, and the motion field, in turn, determines the evolving, three-dimensional setting in which the physical processes take place.

The atmospheric sciences cannot be completely isolated from related disciplines. On time scales longer than a month, the evolution of the state of the atmosphere is influenced by dynamic and thermodynamic interactions with the other elements of the climate system, that is, the oceans, the cryosphere, and the terrestrial biosphere. A notable example is the El Niño-Southern Oscillation phenomenon in the equatorial Pacific Ocean, in which changes in the distribution of surface winds force anomalous ocean currents; the currents can alter the distribution of sea-surface temperature, which in turn can alter the distribution of tropical rainfall, thereby inducing further changes in the surface wind field. On a time scale of decades or longer, the cycling of chemical species such as carbon, nitrogen, and sulfur between these same global reservoirs also influences the evolution of the climate system. Human activities represent an increasingly significant atmospheric source of some of the radiatively active trace gases that play a role in regulating the temperature of the Earth. *See* BIOSPHERE; MARITIME METEOROLOGY; TROPICAL METEOROLOGY.

Throughout the atmospheric sciences, prediction is a unifying theme that sets the direction for research and technological development. Prediction on the time scale of minutes to hours is concerned with severe weather events such as tornadoes, hail, and flash floods, which are manifestations of intense mesoscale weather systems, and with urban air-pollution episodes; day-to-day prediction is usually concerned with the more ordinary weather events and changes that attend the passage of synoptic-scale weather systems such as extratropical cyclones; and seasonal prediction is concerned with regional climate anomalies such as drought or recurrent and persistent cold air outbreaks. Prediction on still longer time scales involves issues such as the impact of human activity on the temperature of the Earth, regional climate, the ozone layer, and the chemical makeup of precipitation. *See* CLIMATE MODELING; DROUGHT; HAIL; TORNADO.

The evolution of the atmospheric sciences from a largely descriptive field to a mature, quantitative physical science discipline is apparent in the development of vastly improved predictive capabilities based upon the numerical integration of specialized versions of the Navier-Stokes equations, which include sophisticated parametrizations of physical processes such as radiative transfer, latent heat release, and microscale motions. The so-called numerical weather prediction models have largely replaced the subjective and statistical prediction methods that were widely used as a basis for day-to-day weather forecasting. The state-of-the-art numerical models exhibit significant skill for forecast intervals as long as about a week.

A distinction is often made between weather prediction, which is largely restricted to the consideration of dynamic and physical processes internal to the atmosphere, and climate prediction, in which interactions between the atmosphere and other elements of the climate system are taken into account. The importance and complexity of these interactions tend to increase with the time scale of the phenomena of interest in the forecast. Weather prediction involves shorter time frames (days to weeks), in which the information contained in the initial conditions is the dominant factor in determining the evolution of the state of the atmosphere; and climate prediction involves longer time frames (seasons and longer), for boundary forcing is the dominant factor in determining the state of the atmosphere.

Atmospheric prediction has benefited greatly from major advances in remote sensing. Geostationary and polar orbiting satellites provide continuous surveillance of the global distribution of cloudiness, as viewed with both visible and infrared imagery. These images are used in positioning of features such as cyclones and fronts on synoptic charts. Cloud motion vectors derived from consecutive images provide estimates of winds in regions that have no other data. Passive infrared and microwave sensors aboard satellites also provide information on the distribution of sea-surface temperature, sea state, land-surface vegetation, snow and ice cover, as well as vertical profiles of temperature and moisture in cloud-free regions. Improved ground-based radar imagery and vertical profiling devices provide detailed coverage of convective cells and other significant mesoscale features over land areas. Increasingly sophisticated data assimilation schemes are being developed to incorporate this variety of information into numerical weather prediction models on an operational basis. *See* ATMOSPHERE; CYCLONE; FRONT; RADAR METEOROLOGY; SATELLITE METEOROLOGY. [J.M.Wa.]

Methanogenesis (bacteria) The microbial formation of methane, which is confined to anaerobic habitats where occurs the production of hydrogen, carbon dioxide, formic acid, methanol, methylamines, or acetate—the major substrates used by methanogenic microbes (methanogens). In fresh-water or marine sediments, in the intestinal tracts of animals, or in habitats engineered by humans such as sewage sludge or biomass digesters, these substrates are the products of anaerobic bacterial metabolism. Methanogens are terminal organisms in the anaerobic microbial food chain—the final product, methane, being poorly soluble, anaerobically inert, and not in equilibrium with the reaction which produces it.

Two highly specialized digestive organs, the rumen and the cecum, have been evolved by herbivores to delay the passage of cellulose fibers so that microbial fermentation may be complete. In these organs, large quantities of methane are produced from hydrogen and carbon dioxide or formic acid by methanogens. From the rumen, an average cow may belch 26 gallons (100 liters) of methane per day.

Methanogens are the only living organisms that produce methane as a way of life. The biochemistry of their metabolism is unique and definitively delineates the group. Two reductive biochemical strategies are employed: an eight-electron reduction of carbon dioxide to methane or a two-electron reduction of a methyl group to methane. All methogens form methane by reducing a methyl group. The major energy-yielding reactions used by methanogens utilize substrates such as hydrogen, formic acid, methanol, acetic acid, and methylamine. Dimethyl sulfide, carbon monoxide, and alcohols such as ethanol and propanol are substrates that are used less frequently. *See* ARCHAEBACTERIA; BACTERIAL PHYSIOLOGY AND METABOLISM. [R.S.W.]

Microbial ecology The study of interrelationships between microorganisms and their living and nonliving environments. Microbial populations are able to tolerate and to grow under varying environmental conditions, including habitats with extreme environmental conditions such as hot springs and salt lakes. Understanding the environmental factors controlling microbial growth and survival offers insight into the distribution of microorganisms in nature, and many studies in microbial ecology are concerned with examining the adaptive features that permit particular microbial species to function in particular habitats.

Within habitats some microorganisms are autochthonous (indigenous), filling the functional niches of the ecosystem, and others are allochthonous (foreign), surviving in the habitat for a period of time but not filling the ecological niches. Because of their diversity and wide distribution, microorganisms are extremely important in ecological processes. The dynamic interactions between microbial populations and their surroundings and the metabolic activities of microorganisms are essential for supporting productivity and maintaining environmental quality of ecosystems. Microorganisms are crucial for the environmental degradation of liquid and solid wastes and various pollutants and for maintaining the ecological balance of ecosystems—essential for preventing environmental problems such as acid mine drainage and eutrophication. *See* ECOSYSTEM; EUTROPHICATION.

The various interactions among microbial populations and between microbes, plants, and animals provide stability within the biological community of a given habitat and ensure conservation of the available resources and ecological balance. Interactions between microbial populations can have positive or negative effects, either enhancing the ability of populations to survive or limiting population densities. Sometimes they result in the elimination of a population from a habitat. *See* RHIZOSPHERE.

The transfer of carbon and energy stored in organic compounds between the organisms in the community forms an integrated feeding structure called a food web. Microbial decomposition of dead plants and animals and partially digested organic matter in the decay portion of a food web is largely responsible for the conversion of organic matter to carbon dioxide. *See* BIOMASS; FOOD WEB.

Only a few bacterial species are capable of biological nitrogen fixation. In terrestrial habitats, the microbial fixation of atmospheric nitrogen is carried out by free-living bacteria, such as *Azotobacter*, and by bacteria living in symbiotic association with plants, such as *Rhizobium* or *Bradyrhizobium* living in mutualistic association within nodules on the roots of leguminous plants. In aquatic habitats, cyanobacteria, such as *Anabaena* and *Nostoc*, fix atmospheric nitrogen. The incorporation of the bacterial genes controlling nitrogen fixation into agricultural crops through genetic engineering may help improve yields. Microorganisms also carry out other processes essential for the biogeochemical cycling of nitrogen. *See* BIOGEOCHEMISTRY; NITROGEN CYCLE; NITROGEN FIXATION.

The biodegradation (microbial decomposition) of waste is a practical application of microbial metabolism for solving ecological problems. Solid wastes are decomposed by microorganisms in landfills and by composting. Liquid waste (sewage) treatment uses microbes to degrade organic matter, thereby reducing the biochemical oxygen demand (BOD). *See* ESCHERICHIA; SEWAGE TREATMENT. [R.M.A.]

Microbiology The multidisciplinary science of microorganisms. The prefix micro generally refers to an object sufficiently small that a microscope is required for visualization. In the seventeenth century, Anton van Leeuwenhoek first documented observations of bacteria by using finely ground lenses. Bacteriology, as a precursor science to microbiology, was based on Louis Pasteur's pioneering studies in the nineteenth century, when it was demonstrated that microbes as minute simple living organisms were an integral part of the biosphere involved in fermentation and disease. Microbiology matured into a scientific discipline when students of Pasteur, Robert Koch, and others sustained microbes on various organic substrates and determined that microbes caused chemical changes in the basal nutrients to derive energy for growth. Modern microbiology continued to evolve from bacteriology by encompassing the identification, classification, and study of the structure and function of a wide range of microorganisms including protozoa, algae, fungi, viruses, rickettsia, and parasites as well as bacteria. The comprehensive range of organisms is reflected in the major subdivisions of microbiology, which include medical, industrial, agricultural, food, and dairy. *See* ALGAE; BACTERIOLOGY; BIOTECHNOLOGY; FUNGI; MEDICAL BACTERIOLOGY; MEDICAL MYCOLOGY; MEDICAL PARASITOLOGY; PROTOZOA; RICKETTSIOSES; VIRUS. [E.W.V.]

Micrometeorology The study of small-scale meteorological processes associated with the interaction of the atmosphere and the Earth's surface. The lower boundary condition for the atmosphere and the upper boundary condition for the underlying soil or water are determined by interactions occurring in the lowest atmospheric layers. Momentum, heat, water vapor, various gases, and particulate matter are transported vertically by turbulence in the atmospheric boundary layer and thus establish the environment of plants and animals at the surface. These exchanges are important in supplying energy and water vapor to the atmosphere, which ultimately determine large-scale weather and climate patterns. Micrometeorology also includes the study of how air pollutants are diffused and transported within the boundary layer and the deposition of pollutants at the surface.

In many situations, atmospheric motions having time scales between 15 min and 1 h are quite weak. This represents a spectral gap that provides justification for distinguishing micrometeorology from other areas of meteorology. Micrometeorology studies phenomena with time scales shorter than the spectral gap (time scales less than 15 min to 1 h and horizontal length scales less than 2–10 km or 1–6 mi). Some phenomena studied by micrometeorology are dust devils, mirages, dew and frost formation, evaporation, and cloud streets. *See* AIR POLLUTION; ATMOSPHERE.

Much of the early understanding of micrometeorology was obtained by studying conditions in large, flat, uniform areas that are relatively simple situations. Micrometeorologists have turned their attention to more complex situations that represent conditions over more of the Earth's surface. The micrometeorology of complex terrain, that is, hills and mountains, is important for air pollution in many towns and cities and for visibility in national parks and for locating wind generators. Another interest is the study of micrometeorology in areas of widely varied surface conditions. For instance, several different crops, dry unirrigated lands, lakes, and rivers may be located near one another. In these cases it is important to understand how the micrometeorology associated with each of these surfaces interacts to produce the overall heat and moisture fluxes of the region so that these areas can be correctly included in weather and climate forecast computer programs. *See* CLIMATOLOGY; MOUNTAIN METEOROLOGY; WEATHER FORECASTING AND PREDICTION.

Microscale meteorological features are too small to be observed by the standard national and international weather observing network. Generally, micrometeorological phenomena must be studied during specific experiments by using specially designed instruments. Instruments used to study turbulent fluxes must be able to respond to very rapid fluctuations. Special cup anemometers are made from very light materials, and high-quality bearings are used to minimize drag. Other anemometers use the speed of sound waves or measure the temperature of heated wires to measure wind. Tiny thermometers are used, so that time constants are short. Instruments are usually placed on towers or in aircraft, or are suspended in packages from tethered balloons. Instruments have been developed that can measure turbulence remotely. Wind speed and boundary-layer convection can be measured with Doppler radar, lidar devices using lasers, and sodar (sound detection and ranging) using sound waves. *See* METEOROLOGICAL INSTRUMENTATION; METEOROLOGY. [S.A.St.]

Mid-Oceanic Ridge An interconnected system of broad submarine rises totaling about 60,000 km (36,000 mi) in length, the longest mountain range system on the planet. The origin of the Mid-Oceanic Ridge is intimately connected with plate tectonics. Wherever plates move apart sufficiently far and fast for oceanic crust to form in the void between them, a branch of the Mid-Oceanic Ridge will be created. In plan view the plate boundary of the Mid-Oceanic Ridge comprises an alternation of spreading centers (or axes or accreting plate boundaries) interrupted or offset by a range of different discontinuities, the most prominent of which are transform faults. As the plates move apart, new oceanic crust is formed along the spreading axes, and the ideal transform fault zones are lines along which plates slip past each other and where oceanic crust is neither created nor destroyed.

Separation of plates causes the hot upper mantle to rise along the spreading axes of the Mid-Oceanic Ridge; partial melting of this rising mantle generates magmas of basaltic composition that segregate from the mantle and rise in a narrow zone at the axis of the Mid-Oceanic Ridge to form the oceanic crust. The partially molten mantle "freezes" to the sides and bottoms of the diverging plates to form the mantle lithosphere that, together with the overlying "rind" of oceanic crust, comprises the lithospheric plate. At the axis of the Mid-Oceanic Ridge the underlying column of crust and mantle is hot and thermally expanded; this thermal expansion explains why the Mid-Oceanic Ridge is a ridge. With time, a column of crust plus mantle lithosphere cools and shrinks as it moves away from the ridge axis as part of the plate. The gentle regional slopes of the Mid-Oceanic Ridge therefore represent the combined effects of sea-floor spreading (divergent plate motion) and thermal contraction.

The height and thermal contraction rate of the ridge crest are relatively independent of the rate of sea-floor spreading; thus the width and regional slopes of the Mid-Oceanic Ridge depend primarily on the rate of plate separation (spreading rate). Where the plates are separating at 2 cm (0.8 in.) per year, the Mid-Oceanic Ridge has five times the regional slope but only one-fifth the width of a part of the ridge forming where the plates are separating at 10 cm (4 in.) per year. One consequence of the relation between the width and plate separation rate of the Mid-Oceanic Ridge is that more ocean water is displaced, thereby raising sea level, during times of globally faster plate motion.

Although the Mid-Oceanic Ridge exhibits little systematic depth variation along much of its length, there are several bulges (swells) of shallower sea floor. For reasons not well understood, the sea-floor bulges are more prominent along parts of the Mid-Oceanic Ridge where the rate of plate separation (spreading rate) is slower, for example, along the northern Mid-Atlantic Ridge and the Southwest Indian Ridge.

The axis of the Mid-Oceanic Ridge—that is, the active plate boundary between two separating plates—is a narrow zone only a few kilometers wide, characterized by frequent earthquakes, intermittent volcanism, and scattered clusters of hydrothermal vents where seawater, percolating downward and heated by proximity to hot rock, is expelled back into the ocean at temperatures as high as 350°C (660°F). Surrounding such vents are deposits of hydrothermal minerals rich in metals, as well as exotic animal communities including, in some vent fields, giant tubeworms and clams. *See* MARINE GEOLOGY; VOLCANO. [P.R.V.]

Migratory behavior Regularly occurring, oriented seasonal movements of individuals of many animal species. The term migration is used to refer to a diversity of animal movements, ranging from short-distance dispersal and one-way migration to round-trip migrations occurring on time scales from hours (the vertical movements of aquatic plankton) to years (the return of salmon to their natal streams following several years and thousands of kilometers of travel in the open sea).

Many temperate zone species, including many migrants, are known to respond physiologically to changes in the day length with season (photoperiodism). For example, many north temperate organisms are triggered to come into breeding condition by the interaction between the lengthening days in spring and their biological clocks (circadian rhythms). Similar processes, acting through the endocrine system, bring animals into migratory condition.

To perform regular oriented migrations, animals need some mechanism for determining and maintaining compass bearings. Animals use many environmental cues as sources of directional information. Work with birds has shown that species use several compasses.

Many species of vertebrates and invertebrates possess a time-compensated Sun compass. With such a system, the animal can determine absolute compass directions at any time of day; that is, its internal biological clock automatically compensates for the changing position of the Sun as the Earth rotates during the day. Many arthropods, fish, salamanders, and pigeons can perceive the plane of polarization of sunlight, and may use that information to help localize the Sun even on partly cloudy days.

Only birds that migrate at night have been shown to have a star compass. Unlike the Sun compass, it appears not to be linked to the internal clock. Rather, directions are determined by reference to star patterns which seem to be learned early in life.

Evidence indicates that several insects, fish, a salamander, certain bacteria, and birds may derive directional information from the weak magnetic field of the Earth.

Many kinds of animals show the ability to return to specific sites following a displacement. The phenomenon can usually be explained by familiarity with landmarks near "home" or sensory contact with the goal. For example, salmon are well known for their ability to return to their natal streams after spending several years at sea. Little is known about their orientation at sea, but they recognize the home stream by chemical (olfactory) cues in the water. The young salmon apparently imprint on the odor of the stream in which they were hatched. Current evidence indicates that birds imprint on or learn some feature of their birthplace, a prerequisite for them to be able to return to that area following migration. On its first migration, a young bird appears to fly in a given direction for a programmed distance. Upon settling in a wintering area, it will also imprint on that locale and will thereafter show a strong tendency to return to specific sites at both ends of the migratory route.

Only in birds can an unequivocal case be made for the existence of true navigation, that is, the ability to return to a goal from an unfamiliar locality in the absence of direct sensory contact with the goal. This process requires both a compass and the analog of a map. Present evidence suggests that the map is not based on information from the Sun, stars, landmarks, or magnetic field. Other possibilities such as olfactory, acoustic, or gravitational cues are being investigated, but the nature of the navigational component of bird homing remains the most intriguing mystery in this field. [K.P.A.]

Mononchida An order of nematodes having a full complement of cephalic sensilla on the lips in two circlets of 6 and 10. The amphids are small and cuplike, and are located just posterior to the lateral lips; the amphidial aperture is either slit-like or ellipsoidal. The stoma is globular and heavily cuticularized, and is derived primarily from the cheilostome. The stoma bears one or more massive teeth that may be opposed by denticles in either transverse or longitudinal rows. The esophagus is cylindrical conoid, with a heavily cuticularized luminal lining. The excretory system is atrophied. Males have ventromedial supplements and paired spicules. The gubernaculum may possess lateral accessory pieces. Females have one or two ovaries. Caudal glands and a spinneret are common; however, they may be degenerate or absent.

There are three mononchid superfamilies: The Mononchoidea contain some of the most common and easily recognized free-living nonparasitic nematodes that occur in soils and fresh waters throughout the world. The closely related Bathyodontoidea are inhabitants of soil or fresh water and prey on small microorganisms. The nonparasitic Monochuloidea comprise both soil and fresh-water species, all of which are predators of microfauna. *See* NEMATA. [A.R.M.]

Monsoon meteorology The study of the structure and behavior of the atmosphere in those areas of the world that have monsoon climates. In lay terminology, monsoon connotes the rains of the wet summer season that follows the dry winter. However, for mariners, the term monsoon has come to mean the seasonal wind reversals.

In true monsoon climates, both the wet summer season that follows the dry winter and the seasonal wind reversals should occur. Winds from cooler oceans blow toward heated continents in summer, bringing warm, unsettled, moisture-laden air and the season of rains, the summer monsoon. In winter, winds from the cold heartlands of the continents blow toward the oceans, bringing dry, cool, and sunny weather, the winter monsoon.

Based on these criteria, monsoon climates of the world include almost all of the Eastern Hemisphere tropics and subtropics, which is about 25% of the surface area of the Earth. The areas of maximum seasonal precipitation straddle or are adjacent to the Equator. Two of the world's areas of maximum precipitation (heavy rainfall) are within the domain of the monsoons: the central and south African region, and the larger south Asia-Australia region. The monsoon surface winds emanate from the cold continents of the winter hemisphere, cross the Equator, and flow toward and over the hot summer-hemisphere land masses.

India presents the classic example of a monsoon climate region, with an annual cycle that brings southwesterly winds and heavy rains in summer (the Indian southwest monsoon) and northeasterly winds and dry weather in winter (the northeast winter monsoon).

Like all weather systems on Earth, monsoons derive their primary source of energy from the Sun. About 30% of the Sun's energy that enters the top of the atmosphere is transmitted back to space by cloud and surface reflections. Little of the remainder is absorbed directly by the clear atmosphere; it is absorbed at the Earth's surface according to a seasonal cycle. The opposition of seasons in the Northern and Southern hemispheres leads to a slow movement of surface air across the Equator from winter hemisphere to summer hemisphere, forced by horizontal pressure gradients and vertical buoyancy forces resulting from differential seasonal heating. Such a seasonally reversing rhythm is most pronounced in the monsoon regions. *See* ATMOSPHERE; INSOLATION; METEOROLOGY; TROPICAL METEOROLOGY. [J.S.Fe.]

Moraine An accumulation of glacial debris, usually till, with distinct surface expression related to some former ice front position. End moraine, the most common form, is an uneven ridge of till built in front of or around the terminus of a glacier margin, and reflects some degree of equilibrium between rate of ice motion, supply of rock debris at the ice front, temperature of the glacier base, and shape and resistance of underlying bedrock (*see* illustration).

End moraine in Pennsylvania (*U.S. Geological Survey*).

If an end moraine represents the farthest forward position a glacier ever moved, it is a terminal moraine. It demonstrates a steady-state condition for a period of time within the ice body where constant forward motion is balanced by frontal melting; and a continual supply of debris, as on an endless conveyor belt, is brought forward to the glacier terminus. If the ice front then melts farther back than it moves forward, till is spread unevenly over the land as ground moraine. If a retreatal position of steady-state equilibrium is maintained again, a recessional moraine may be constructed.

Drumlins, produced by glacier streamlining of ground moraine, are probably the best-known moraine forms. [S.E.Wh.]

Moraxella A genus of bacteria that are parasites of mucous membranes. Subgenus *Moraxella* is characterized by gram-negative rods that are often very short and plump, frequently resembling a coccus, and usually occurring in pairs. Subgenus *Branhamella* has gram-negative cocci occurring as single cells or in pairs with the adjacent sides flattened. They are usually harmless parasites of humans and other warm-blooded animals and are generally considered not to be highly pathogenic. Most species may be opportunistic pathogens in predisposed or debilitated hosts.

There are presently six species in the subgenus *Moraxella*: *M.* (*M.*) *lacunata* (also known as *Diplobacillus moraxaxenfeld* and *liquefaciens*), *M.* (*M.*) *bovis*, *M.* (*M.*) *nonliquefaciens* (also known as *Bacillus duplex nonliquefaciens*), *M.* (*M.*) *atlantae*, *M.* (*M.*) *phenylpyruvica*, and *M.* (*M.*) *osloensis*. The different species are recognized on the basis of phenotypic properties, including liquefaction of coagulated serum, hemolysis of human blood in blood agar media, nitrate reduction, phenylalanine deaminase activity, urease activity, and growth on mineral salts medium with ammonium ion and acetate as the sole carbon source.

The subgenus *Branhamella* presently contains four species: *M.* (*B.*) *catarrhalis*, *M.* (*B.*) *caviae*, *M.* (*B.*) *ovis*, and *M.* (*B.*) *cuniculi*. The different species are recognized on the basis of hemolysis of human blood in blood agar media, and nitrate and nitrite reduction, among other properties.

Moraxella (*M.*) *lacunata*, the type species of the subgenus *Moraxella*, was a significant causative agent of human conjunctivitis and keratitis in the past but is only rarely isolated at present. Infectious keratoconjunctivitis in cattle, called pinkeye, is caused by *M.* (*M.*) *bovis*. *Moraxella* (*M.*) *nonliquefaciens* is considered to be a well-established parasite of humans and rarely causes disease, but it has been associated with endophthalmitis and pneumonitis with pulmonary abscess. *Moraxella* (*M.*) *osloensis*, usually a harmless parasite, has been frequently associated with such human infections as osteomyelitis, endocarditis, septicemia, meningitis, stomatitis, and septic arthritis.

Moraxella (*B.*) *catarrhalis*, the type species of *Branhamella*, is the only species of this subgenus recovered from humans. The organism is considered to be a well-adapted parasite but has been judged the etiologic agent of middle-ear infection, maxillary sinus infection, bronchitis, tracheitis, conjunctivitis, pneumonia, otitis media of infants, respiratory disease in the compromised host, septicemia, meningitis, and endocarditis. The remaining *Branhamella* species (*caviae*, *ovis*, and *cuniculi*) are parasites of guinea pigs, sheep, cattle, and rabbits.

Moraxella species are susceptible to most antimicrobial agents with the exception of the lincomycins. Their usually high susceptibility to the penicillins is a feature that separates them from most other gram-negative rods. *See* CLINICAL MICROBIOLOGY. [G.G.]

Mosquito Any member of the family Culicidae in the insect order Diptera. Mosquitoes are holometabolous insects and all larval stages are aquatic. Adults are recognized by their long proboscis for piercing and sucking, and characteristic scaled wing venation. This is a relatively large group of well-known flies with nearly 3000 species in 34 genera reported in the world. There are 13 genera and 167 recognized species of mosquitoes in North America north of Mexico. Almost 75% of these species belong to three genera: *Aedes* (78 species), *Culex* (29 species), and *Anopheles* (16 species).

Adult females lay their eggs on or near water. Most larvae, or wrigglers, feed on algae and organic debris that they filter from the water with their oral brushes, although certain genera may be predaceous and feed on other mosquito larvae. Larvae go through three molts and four instars before pupation. Pupae, or tumblers, are active but nonfeeding stages in which metamorphosis to the adult stage occurs. Both larvae and pupae usually breathe through air tubes at the surface of the water.

Adult male mosquitoes are relatively short-lived, and do not suck blood, but feed primarily on nectar and other plant juices. Females also feed on nectar as their primary energy source, but they require a blood meal for egg production in most species. Some mosquito species are very host-specific, blood-feeding only on humans, birds, mammals, or even reptiles and amphibians, although many species will feed on any available host.

Mosquitoes are of major importance in both human and veterinary medicine. They can cause severe annoyance and blood loss when they occur in dense populations, and they act as vectors of three important groups of disease-causing organisms: *Plasmodium*, the protozoan parasite that produces malaria; filarial worms, parasitic nematodes causing elephantiasis in humans and heartworm disease in canines; and arboviruses, which are the causative agents of yellow fever, dengue fever, LaCrosse encephalitis, St. Louis encephalitis, western equine encephalomyelitis, eastern and Venezuelan equine encephalitis, and several other viral diseases. Human malaria is transmitted exclusively by *Anopheles*, filariasis by *Culex*, *Anopheles*, and *Aedes*, and arboviruses primarily by *Culex* and *Aedes* species. *See* ARBOVIRAL ENCEPHALITIDES; HEARTWORMS; INSECTA; MALARIA; MEDICAL PARASITOLOGY; YELLOW FEVER. [B.M.Ch.]

Mountain A feature of the Earth's surface that rises high above its base and has generally steep slopes and a relatively small summit area. Commonly the features designated as mountains have local heights measurable in thousands of feet, lesser features of the same type being called hills, but there are many exceptions. *See* HILL AND MOUNTAIN TERRAIN.

Mountains rarely occur as isolated individuals. Instead they are usually found in roughly circular groups or massifs, such as the Olympic Mountains of northwestern Washington, or in elongated ranges, like the Sierra Nevada of California. An array of linked ranges and groups, such as the Rocky Mountains, the Alps, or the Himalayas, is a mountain system. North America, South America, and Eurasia possess extensive cordilleran belts, within which the bulk of their higher mountains occur. *See* MOUNTAIN SYSTEMS.

As a rule, mountains represent portions of the Earth's crust that have been raised above their surroundings by upwarping, folding, or buckling, and have been deeply carved by streams or glaciers into their present surface form. Some individual peaks and massifs have been constructed upon the surface by outpourings of lava or eruptions of volcanic ash. [E.H.Ha.]

Mountain meteorology The effects of mountains on the atmosphere, ranging over all scales of motion, including very small (such as turbulence), local (for instance, cloud formations over individual peaks or ridges), and global (such as the monsoons of Asia and North America).

The most readily perceived effects of a mountain, or even of a hill, are related to the blocking of air flow. When there is sufficient wind, the air either goes around the obstacle or over it, causing waves in the flow similar to those in a river washing over a boulder. Since ascending air cools by adiabatic expansion, the saturation point of water vapor may be reached in such waves as they form over an obstacle, and a cloud then forms in the ascending branch of the wave motion. Such a cloud dissipates in the descending branch where adiabatic warming takes place. The shapes and amplitudes of these lee waves (they form over and to the lee of mountains) depend not only on the thermal stability and on the vertical wind shear in the overlying atmosphere but also on the shape of the underlying terrain. *See* CLOUD; CLOUD PHYSICS.

On a grander scale, mountain ranges, such as the Sierras of North and South America, place an obstacle in the path of the westerly winds (that is, winds from the west), which generally prevail in middle latitudes. Such a blockage tends to generate a high-pressure region upwind from the mountains (this may be viewed as air piling up as it prepares to jump the hurdle), and a low-pressure area downwind. Thus, there is a stronger push against the mountains on the high-pressure western side than on the low-pressure eastern side. The net effect is the slowing down of the atmospheric flow (mountain torque).

Less subtle than mountain torque effects are the large-scale meanders that develop in the global flow patterns once they have been perturbed, mainly by the North and South American Andes and by the Plateau of Tibet and its Himalayan mountain ranges. These meanders in the large-scale flow are known as planetary waves. They appear prominently in the pressure patterns of hemispheric or global weather maps. *See* WEATHER MAP.

The major monsoon circulations interact with the global circulation, shaped in part by sea-surface temperature anomalies in the equatorial Pacific. The various aspects of mountain meteorology, therefore, have to be viewed within the larger picture. There is a continuous interaction between the weather effects on all space and time scales generated by the mountains and the weather patterns that prevail elsewhere on the Earth. *See* METEOROLOGY. [E.R.R.]

Mountain systems Long, broad, linear to arcuate belts in the Earth's crust where extreme mechanical deformation and thermal activity have been (or are being) concentrated.

Mountain systems in the general sense occur both on continents and in ocean basins, but the geological properties of the systems in continental as opposed to oceanic settings are distinctly different. The mechanical strain in classical, continental mountain systems is expressed in the presence of major folds, faults, and intensive fracturing and cleavage. Thermal effects are in the form of vast volcanic outpourings, intruded bodies of igneous magma, and metamorphism. Uplift and deformation in young mountain systems are conspicuously displayed in the physiographic forms of topographic relief. Where mountain building is presently taking place, the dynamics are partly expressed in warping of the land surface and significant shallow or deep earthquake activity. Locations of ancient mountain systems in continental regions now beveled flat by erosion are clearly disclosed by the presence of highly deformed, intruded, and metamorphosed rocks.

Two basic classes of oceanic mountain systems exist. A world-encircling oceanic rift mountain system has been built along the extensional tectonic boundary between plates diverging at rates of 0.8–2.4 in. or 2–6 cm per year from the mid-oceanic ridges. This rift mountain system is exposed to partial view in Iceland. The second type, island arc mountain systems, occur in oceanic basins where the crust dives downward at trench sites, thus underthrusting adjacent oceanic crust. *See* MARINE GEOLOGY; OCEANIC ISLANDS.

The classical, conspicuous mountain systems of the Earth occur at the continent/ocean interface, for this is the site where plate convergence has led to major sedimentation, subduction of oceanic crust under continents, collision of island arc mountain systems with continents, and head-on collision of continents. [G.H.D.]

Mulberry A genus (*Morus*) of trees characterized by milky sap and simple, often lobed, alternate leaves. White mulberry (*M. alba*) was introduced into the United States from China during the 19th century as a source of food for silkworms. The silkworm project was unsuccessful, but the trees remained and are common in cities and on the borders of forests. Red mulberry (*M. rubra*) grows in the eastern half of the United States and in southern Ontario. The wood is used for fence posts, furniture, interior finish, agricultural implements, and barrels. [A.H.G./K.P.D.]

Multiple cropping Planting two or more species in the same field in the same year. Preserved through history to maintain biological, economic, and nutritional diversity, multiple-species systems still are used by the majority of the world's farmers, especially in developing countries. Where farm size is small and the lack of capital has made it difficult to mechanize and expand, farm families that need a low-risk source of food and income often use multiple cropping. These systems maintain a green and growing crop canopy over the soil through much of the year, the total season depending on rainfall and temperature. Systems with more than one crop frequently make better use of total sunlight, water, and available nutrients than is possible with a single crop. The family has a more diverse supply of food and more than one source of income, with both spread over much of the year.

Multiple-cropping patterns are described by the number of crops per year and the intensity of crop overlap. Double cropping or triple cropping signifies systems with two or three crops planted sequentially with no overlap in growth cycle. Intercropping indicates that two or more crops are planted at the same time, or at least planted so that significant parts of their growth cycles overlap. Relay cropping describes the planting of a second crop after the first crop has flowered; in this system there still may be some competition for water or nutrients. When a crop is harvested and allowed to regrow from the crowns or root systems, the term ratoon cropping is used. Sugarcane, alfalfa, and sudangrass are commonly produced in this way, while the potential exists for such tropical cereals as sorghum and rice. Mixed cropping, strip cropping, associated cropping, and alternative cropping represent variations of these systems. *See* AGRICULTURAL SCIENCE (PLANT); AGRICULTURE; AGRONOMY. [C.A.F.]

Mumps An acute contagious viral disease, characterized chiefly by enlargement of the parotid glands (parotitis).

Besides fever, the chief signs and symptoms are the direct mechanical effect of swelling on glands or organs where the virus localizes. One or both parotids may swell rapidly, producing severe pain when the mouth is opened. In orchitis, the testicle is inflamed but is enclosed by an inelastic membrane and cannot swell; pressure necrosis produces atrophy, and if both testicles are affected, sterility may result. The ovary may enlarge, without sequelae.

An attenuated live virus vaccine can induce immunity without parotitis. It is recommended particularly for adults exposed to infected children, for students in boarding schools and colleges, and for military troops. [J.L.Me.]

Mushroom A macroscopic fungus with a fruiting body (also known as a sporocarp). Approximately 14% (10,000) described species of fungi are considered mushrooms. Mushrooms grow aboveground or underground. They have a fleshy or nonfleshy texture. Many are edible, and only a small percentage are poisonous.

Mushrooms reproduce via microscopic spheres (spores) that are roughly comparable to the seeds of higher plants. Spores are produced in large numbers on specialized structures in or on the fruiting body. Spores that land on a suitable medium absorb moisture, germinate, and produce hyphae that grow and absorb nutrients from the substratum. If suitable mating types are present and the mycelium (the threadlike filaments or hyphae that become interwoven) develops sufficiently to allow fruiting, the life cycle will continue. In nature, completion of the life cycle is dependent on many factors, including temperature, moisture and nutritional status of the substratum, and gas exchange capacity of the medium.

Fewer than 20 species of edible mushrooms are cultivated commercially. The most common cultivated mushroom is *Agaricus bisporus*, followed by the oyster mushroom (*Pleurotus* spp.). China is the leading mushroom-producing country; Japan leads the world in number of edible species cultivated commercially.

Mushrooms may be cultivated on a wide variety of substrates. They are grown from mycelium propagated on a base of steam-sterilized cereal grain. This grain and mycelium mixture is called spawn, which is used to seed mushroom substrata.

Mushrooms contain digestible crude protein, all essential amino acids, vitamins (especially provitamin D-2), and minerals; they are high in potassium and low in sodium, saturated fats, and calories. Although they cannot totally replace meat and other high-protein food in the diet, they can be considered an important dietary supplement and a health food.

Fungi have been used for their medicinal properties for over 2000 years. Although there remains an element of folklore in the use of mushrooms in health and medicine, several important drugs have been isolated from mushroom fruiting bodies and mycelium. The best-known drugs obtained are lentinan from *L. edodes*, grifolin from *Grifola frondosa*, and krestin from *Coriolus versicolor*. These compounds are protein-bound polysaccharides or long chains of glucose, found in the cell walls, and function as antitumor immunomodulatory drugs. *See* FUNGI; MEDICAL MYCOLOGY.
[D.M.C.]

Muskeg A term derived from Chippewan Indian for "grassy bog." In North America ecologists apply diverse usage, but most include peat bogs or tussock meadows, with variable woody vegetation such as spruce or tamarack. Plant remains accumulate when trapped in the water or in media such as sphagnum moss which inhibit decay. The peat might accrue indefinitely or might approach a steady state of raised bogs, blanket bogs, or forest if input became balanced by erosion or by loss as methane and CO_2.

Organic terrain occurs on every continent. The largest expanses of it are in Russia (especially western Siberia) and Canada. Typical bog or spruce-larch muskeg develops in cool temperature and subarctic to arctic lands. In the subtropics and the tropics there is considerable organic terrain, as in Paraguay, Uruguay, and Guyana in South America. Thus, despite differences in climate or floristics, the phenomenon of peat formation persists if local aeration or biochemical conditions hinder decay. Despite climatic and biotic differences, gross peat structure categories seem comparable the world over. *See* BOG; PEAT.
[N.W.R.]

Mutagens and carcinogens A mutagen is a substance or agent that induces heritable change in cells or organisms. A carcinogen is a substance that induces unregulated growth processes in cells or tissues of multicellular animals, leading to cancer. Although mutagen and carcinogen are not synonymous terms, the ability of a substance to induce mutations and its ability to induce cancer are strongly correlated. Mutagenesis refers to processes that result in genetic change, and carcinogenesis (the processes of tumor development) may result from mutagenic events. *See* MUTATION.

A mutation is any change in a cell or in an organism that is transmitted to subsequent generations. Mutations can occur spontaneously or be induced by chemical or physical agents. The cause of mutations is usually some form of damage to DNA or chromosomes that results in some change that can be seen or measured. However, damage can occur in a segment of DNA that is a noncoding region and thus will not result in a mutation. Mutations may or may not be harmful, depending upon which function is affected. They may occur in either somatic or germ cells. Mutations that occur in germ cells may be transmitted to subsequent generations, whereas mutations in somatic cells are generally of consequence only to the affected individual.

Not all heritable changes result from damage to DNA. For example, in growth and differentiation of normal cells, major changes in gene expression occur and are transmitted to progeny cells through changes in the signals that control genes that are transcribed into ribonucleic acid (RNA). It is possible that chemicals and radiation alter these processes as well. When such an effect is seen in newborns, it is called teratogenic and results in birth defects that are not transmitted to the next generation. However, if the change is transmissible to progeny, it is a mutation, even though it might have arisen from an effect on the way in which the gene is expressed. Thus, chemicals can have somatic effects involving genes regulating cell growth that could lead to the development of cancer, without damaging DNA.

Cancer arises because of the loss of growth control by dearrangement of regulatory signals. Included in the phenotypic consequences of mutations are alterations in gene regulation brought about by changes either in the regulatory region or in proteins involved with coordinated cellular functions. Altered proteins may exhibit novel interactions with target substrates and thereby lose the ability to provide a regulatory function for the cell or impose altered functions on associated molecules. Through such a complex series of molecular interactions, changes occur in the growth properties of normal cells leading to cancer cells that are not responsive to normal regulatory controls and can eventually give rise to a visible neoplasm or tumor. While mutagens can give rise to neoplasms by a process similar to that described above, not all mutagens induce cancer and not all mutational events result in tumors.

The identification of certain specific types of genes, termed oncogenes, that appear to be causally involved in the neoplastic process has helped to focus mechanistic studies on carcinogenesis. Oncogenes can be classified into a few functionally different groups, and specific mutations in some of the genes have been identified and are believed to be critical in tumorigenesis. Tumor suppressor genes or antioncogenes provide a normal regulatory function; by mutation or other events, the loss of the function of these genes may release cells from normal growth-control processes, allowing them to begin the neoplastic process. *See* ONCOLOGY.

There are a number of methods and systems for identifying chemical mutagens. Mutations can be detected at a variety of genetic loci in very diverse organisms, including bacteria, insects, cultured mammalian cells, rodents, and humans. Spontaneous and induced mutations occur very infrequently, the estimated rate being less than 1 in 10,000 per gene per cell generation. This low mutation rate is probably the result of a combination of factors that include the relative inaccessibility of DNA to damaging agents and the ability of cellular processes to repair damage to DNA.

Factors that contribute to the difficulty in recognizing substances that may be carcinogenic to humans include the prevalence of cancer, the diversity of types of cancer, the generally late-life onset of most cancers, and the multifactorial nature of the disease process. Approximately 50 substances have been identified as causes of cancer in humans, but they probably account for only a small portion of the disease incidence. *See* CANCER (MEDICINE); HUMAN GENETICS; MUTATION. [R.W.T.]

Mutation Any alteration capable of being replicated in the genetic material of an organism. When the alteration is in the nucleotide sequence of a single gene, it is referred to as gene mutation; when it involves the structures or number of the chromosomes, it is referred to as chromosome mutation, or rearrangement. Mutations may be recognizable by their effects on the phenotype of the organism (mutant).

Gene mutations. Two classes of gene mutations are recognized: point mutations and intragenic deletions. Two different types of point mutation have been described. In the first of these, one nucleic acid base is substituted for another. The second type of change results from the insertion of a base into, or its deletion from, the polynucleotide sequence. These mutations are all called sign mutations or frame-shift mutations because of their effect on the translation of the information of the gene.

More extensive deletions can occur within the gene which are sometimes difficult to distinguish from mutants which involve only one or two bases. In the most extreme case, all the informational material of the gene is lost.

A single-base alteration, whether a transition or a transversion, affects only the codon or triplet in which it occurs. Because of code redundancy, the altered triplet may still insert the same amino acid as before into the polypeptide chain, which in many cases is the product specified by the gene. Such DNA changes pass undetected. However, many base substitutions do lead to the insertion of a different amino acid, and the effect of this on the function of the gene product depends upon the amino acid and its importance in controlling the folding and shape of the enzyme molecule. Some substitutions have little or no effect, while others destroy the function of the molecule completely.

Single-base substitutions may sometimes lead not to a triplet which codes for a different amino acid but to the creation of a chain termination signal. Premature termination of translation at this point will lead to an incomplete and generally inactive polypeptide.

Sign mutations (adding or subtracting one or two bases to the nucleic acid base sequence of the gene) have a uniformly drastic effect on gene function. Because the bases of each triplet encode the information for each amino acid in the polypeptide product, and because they are read in sequence from one end of the gene to the other without any punctuation between triplets, insertion of an extra base or two bases will lead to translation out of register of the whole sequence distal to the insertion or deletion point. The polypeptide formed is at best drastically modified and usually fails to function at all. This sometimes is hard to distinguish from the effects of intragenic deletions. However, whereas extensive intragenic deletions cannot revert, the deletion of a single base can be compensated for by the insertion of another base at, or near, the site of the original change. *See* GENE; GENETIC CODE.

Chromosomal changes. Some chromosomal changes involve alterations in the quantity of genetic material in the cell nuclei, while others simply lead to the rearrangement of chromosomal material without altering its total amount.

Origins of mutations. Mutations can be induced by various physical and chemical agents or can occur spontaneously without any artificial treatment with known mutagenic agents.

Until the discovery of x-rays as mutagens, all the mutants studied were spontaneous in origin; that is, they were obtained without the deliberate application of any mutagen. Spontaneous mutations occur unpredictably, and among the possible factors responsible for them are tautomeric changes occurring in the DNA bases which alter their pairing characteristics, ionizing radiation from various natural sources, naturally occurring chemical mutagens, and errors in the action of the DNA-polymerizing and correcting enzymes.

Spontaneous chromosomal aberrations are also found infrequently. One way in which deficiencies and duplications may be generated is by way of the breakage-fusion-bridge cycle. During a cell division one divided chromosome suffers a break near its tip, and the sticky ends of the daughter chromatids fuse. When the centromere divides and the halves begin to move to opposite poles, a chromosome bridge is formed, and breakage may occur again along this strand. Since new broken ends are produced, this sequence of events can be repeated. Unequal crossing over is sometimes cited as a source of duplications and deficiencies, but it is probably less important than often suggested.

In the absence of mutagenic treatment, mutations are very rare. In 1927 H. J. Muller discovered that x-rays significantly increased the frequency of mutation in *Drosophila*. Subsequently, other forms of ionizing radiation, for example, gamma rays, beta particles, fast and thermal neutrons, and alpha particles, were also found to be effective. Ultraviolet light is also an effective mutagen. The wavelength most employed experimentally is 253.7 nm, which corresponds to the peak of absorption of nucleic acids.

Some of the chemicals which have been found to be effective as mutagens are the alkylating agents which attack guanine principally although not exclusively. The N7 portion appears to be a major target in the guanine molecule, although the O^6 alkylation product is probably more important mutagenically. Base analogs are incorporated into

DNA in place of normal bases and produce mutations probably because there is a higher chance that they will mispair at replication. Nitrous acid, on the other hand, alters DNA bases in place. Adenine becomes hypoxanthine and cytosine becomes uracil. In both cases the deaminated base pairs differently from the parent base. A third deamination product, xanthine, produced by the deamination of guanine, appears to be lethal in its effect and not mutagenic. Chemicals which react with DNA to generate mutations produce a range of chemical reaction products not all of which have significance for mutagenesis.

Significance of mutations. Mutations are the source of genetic variability, upon which natural selection has worked to produce organisms adapted to their present environments. It is likely, therefore, that most new mutations will now be disadvantageous, reducing the degree of adaptation. Harmful mutations will be eliminated after being made homozygous or because the heterozygous effects reduce the fitness of carriers. This may take some generations, depending on the severity of their effects. Chromosome alterations may also have great significance in evolutionary advance. Duplications are, for example, believed to permit the accumulation of new mutational changes, some of which may prove useful at a later stage in an altered environment.

Rarely, mutations may occur which are beneficial: Drug yields may be enhanced in microorganisms; the characteristics of cereals can be improved. However, for the few mutations which are beneficial, many deleterious mutations must be discarded. Evidence suggests that the metabolic conditions in the treated cell and the specific activities of repair enzymes may sometimes promote the expression of some types of mutation rather than others. [B.J.K.]

Mutualism An interaction between two species that benefits both. Individuals that interact with mutualists experience higher sucess than those that do not. Hence, behaving mutualistically is advantageous to the individual, and it does not require any concern for the well-being of the partner. At one time, mutualisms were thought to be rare curiosities primarily of interest to natural historians. However, it is now believed that every species is involved in one or more mutualisms. Mutualisms are thought to lie at the root of phenomena as diverse as the origin of the eukaryotic cell, the diversification of flowering plants, and the pattern of elevated species diversity in tropical forests.

Mutualisms generally involve an exchange of substances or services that organisms would find difficult or impossible to obtain for themselves. For instance, *Rhizobium* bacteria found in nodules on the roots of many legume (bean) species fix atmospheric nitrogen into a form (NH_3) that can be taken up by plants. The plant provides the bacteria with carbon in the form of dicarboxylic acids. The carbon is utilized by the bacteria as energy for nitrogen fixation. Consequently, leguminous plants often thrive in nitrogen-poor environments where other plants cannot persist. Another well-known example is lichens, in which fungi take up carbon fixed during photosynthesis of their algae associates. *See* NITROGEN FIXATION.

A second benefit offered within some mutualisms is transportation. Prominent among these mutualisms is biotic pollination, in which certain animals visit flowers to obtain resources and return a benefit by transporting pollen between the flowers they visit. A final benefit is protection from one's enemies. For example, ants attack the predators and parasites of certain aphids in exchange for access to the aphids' carbohydrate-rich excretions (honeydew).

Another consideration about mutualisms is whether they are symbiotic. Two species found in intimate physical association for most or all of their lifetimes are considered to be in symbiosis. Not all symbioses are mutualistic; symbioses may benefit both, one, or neither of the partners.

Mutualisms can also be characterized as obligate or facultative (depending on whether or not the partners can survive without each other), and as specialized or generalized (depending on how many species can confer the benefit in question).

Two features are common to most mutualisms. First, mutualisms are highly variable in time and space. Second, mutualisms are susceptible to cheating. Cheaters can be individuals of the mutualist species that profit from their partners' actions without offering anything in return, or else other species that invade the mutualism for their own gain.

Mutualism has considerable practical significance. Certain mutualisms play central roles in humans' ability to feed the growing population. It has been estimated that half the food consumed is the product of biotic pollination. *See* ECOLOGY; PLANT PATHOLOGY.

[J.L.Br.]

Mycobacterial diseases Diseases caused by mycobacteria, a diffuse group of acid-fast, rod-shaped bacteria in the genus *Mycobacterium*. Some mycobacteria are saprophytes, while others can cause disease in humans. The two most important species are *M. tuberculosis* (the cause of tuberculosis) and *M. leprae* (the cause of leprosy); other species have been called by several names, particularly the atypical mycobacteria or the nontuberculous mycobacteria. *See* TUBERCULOSIS.

These bacteria are classified according to their pigment formation, rate of growth, and colony morphology. The most commonly involved disease site is the lungs. Nontuberculous mycobacteria are transmitted from natural sources in the environment, rather than from person to person, and thus are not a public health hazard. Nontuberculous mycobacteria have been cultured from various environmental sources.

The diagnosis of disease caused by nontuberculous mycobacteria can be difficult, since colonization or contamination of specimens may be present rather than true infection.

Pulmonary disease resembling tuberculosis is a most important manifestation of disease caused by nontuberculous mycobacteria. The symptoms and chest x-ray findings are similar to those seen in tuberculosis. *Mycobacterium kansasii* and *M. avium intracellulare* are the most common pathogens. The disease usually occurs in middle-aged men and women with some type of chronic coexisting lung disease. The pathogenic mechanisms are obscure. Pulmonary infections due to *M. kansasii* can be treated successfully with chemotherapy. The treatment of pulmonary infections due to *M. avium intracellulare* complex is difficult.

Chronic infection involving joints and bones, bursae, synovia, and tendon sheaths can be caused by various species.

Localized abscesses due to *M. fortuitum* or *M. chelonei* can occur after trauma, after surgical incision, or at injection sites. The usual treatment is surgical incision. The most common soft tissue infection is caused by *M. marinum*, which may be introduced, following an abrasion or trauma, from handling fish or fish tanks, or around a swimming pool. Treatment is surgical. *Mycobacterium ulcerans* causes a destructive skin infection in tropical areas of the world. It is treated by wide excision and skin grafting.

Disseminated *M. avium intracellulare* is one of the opportunistic infections seen in the acquired immune deficiency syndrome (AIDS). In individuals with AIDS, the organism has been cultured from lung, brain, cerebrospinal fluid, liver, spleen, intestinal mucosa, and bone marrow. No treatment has yet been effective in this setting. *See* ACQUIRED IMMUNE DEFICIENCY SYNDROME (AIDS).

[G.M.L.; L.B.R.]

Mycology The study of organisms classified under the kingdom Fungi. Common names for some of these organisms are mushrooms, boletes, bracket or shelf fungi, powdery mildew, bread molds, yeasts, puffballs, morels, stinkhorns, truffles, smuts, and rusts. Fungi are found in every ecological niche. Mycologists estimate that there are 1.5 million species of fungi, with only 70,000 species now described. Fungi typically have a filamentous-branched somatic structure surrounded by thick cell walls known as hyphae. The phyla considered to be true fungi are Chytridiomycota, Zygomycota, Ascomycota, and Basidiomycota. Other phyla that sometimes are included as true fungi are Myxomycota, Dictyosteliomycota, Acrasiomycota, and Plasmodiophoromycota. *See* FUNGAL ECOLOGY; FUNGAL GENETICS; FUNGI; MEDICAL MYCOLOGY; PLANT PATHOLOGY.

[S.C.Jo.]

Mycorrhizae Dual organs of absorption that are formed when symbiotic fungi inhabit healthy absorbing organs (roots, rhizomes, or thalli) of most terrestrial plants and many aquatics and epiphytes.

Mycorrhizae appear in the earliest fossil record of terrestrial plant roots. Roughly 80% of the nearly 10,000 plant species that have been examined are mycorrhizal. Present-day plants that normally lack mycorrhizae are generally evolutionarily advanced. It has been inferred that primitive plants evolved with a symbiosis between fungi and rhizoids or roots as a means to extract nutrients and water from soil. The degree of dependence varies between species or groups of plants. In absolute dependence, characteristic of perennial, terrestrial plants, the host requires mycorrhizae to survive. Some plants are facultative; they may form mycorrhizae but do not always require them. This group includes many of the world's more troublesome weeds. A minority of plant species characteristically lack mycorrhizae, so far as is known, including many aquatics, epiphytes, and annual weeds.

The three major types of mycorrhizae differ in structural details but have many functions in common. The fungus colonizes the cortex of the host root and grows its filaments (hyphae) into surrounding soil from a few centimeters to a meter or more. The hyphae absorb nutrients and water and transport them to host roots. The fungi thus tap far greater volumes of soil at a relatively lower energy cost than the roots could on their own. Moreover, many, if not all, mycorrhizal fungi produce extracellular enzymes and organic acids that release immobile elements such as phosphorus and zinc from clay particles, or phosphorus and nitrogen bound in organic matter. The fungi are far more physiologically capable in extracting or recycling nutrients in this way than the rootlets themselves.

Mycorrhizal fungi are relatively poorly competent in extracting carbon from organic matter. They derive energy from host-photosynthesized carbohydrates. Hosts also provide vitamins and other growth regulators that the fungi need.

The major types are ectomycorrhizae, vesicular-arbuscular mycorrhizae, and ericoid mycorrhizae. Ectomycorrhizae are the most readily observed type. Ectomycorrhizal hosts strongly depend on mycorrhizae to survive. Relatively few in number of species, they nonetheless dominate most forests outside the tropics. Vesicular-arbuscular mycorrhizae (sometimes simply termed arbuscular mycorrhizae) form with the great majority of terrestrial herbaceous plant species plus nearly all woody perennials that are not ectomycorrhizal. Vesicular-arbuscular mycorrhizal hosts range from strongly mycorrhiza-dependent, especially the woody perennials, to faculative, as are many grasses.

Ericoid mycorrhizae are restricted to the Ericales, the heath order. The hosts are strongly mycorrhiza-dependent. Though relatively few in number, heath species dominate large areas around the world and are common understory plants in many forests. Other mycorrhiza types include those special for the Orchidaceae (orchids) and Gentianaceae (gentians). *See* ASCOMYCOTA.

The succession of plants from pioneering through seral to climax communities is governed by availability of mycorrhizal propagules. When catastrophic fire, erosion, or clearcutting reduce the availability of mycorrhizal fungi in the soil, plants dependent on those fungi will have difficulty becoming established. Each mycorrhizal fungus has its own array of physiological characteristics. Some are especially proficient at releasing nutrients bound in organic matter, some produce more effective antibiotics or growth regulators than others, and some are more active in cool, hot, wet, or dry times of year than others. Healthy plant communities or crops typically harbor diverse populations of mycorrhizal fungal species. This diversity, evolved over a great expanse of time, is a hallmark of thriving ecosystems. Factors that reduce this diversity also reduce the resilience of ecosystems.

Mycorrhizal inoculation of plants in nurseries, orchards, and fields has succeeded in many circumstances, resulting in improved survival and productivity of the inoculated plants. Inoculation with selected fungi is especially important for restoring degraded sites or introducing exotics. Because ectomycorrhizal fungi include many premier edibles such as truffles, seedlings can also be inoculated to establish orchards for production of edible fungi. *See* FUNGI. [J.M.Tr.]

Mycotoxin Any of the mold-produced substances that may be injurious to vertebrates upon ingestion, inhalation, or skin contact. The diseases they cause, known as mycotoxicoses, need not involve the toxin-producing fungus. Diagnostic features characterizing mycotoxicoses are the following: the disease is not transmissible; drug and antibiotic treatments have little or no effect; in field outbreaks the disease is often seasonal; the outbreak is usually associated with a specific foodstuff; and examination of the suspected food or foodstuff reveals signs of fungal activity.

The earliest recognized mycotoxicoses were human diseases. Ergotism, or St. Anthony's fire, results from eating rye infected with *Claviceps purpurea*. Yellow rice disease, a complex of human toxicoses, is caused by several *Penicillium islandicum* mycotoxins. World attention was directed toward the mycotoxin problem with the discovery of the aflatoxins in England in 1961. The aflatoxins, a family of mycotoxins produced by *Aspergillus flavus* and *A. parasiticus*, can induce both acute and chronic toxicological effects in vertebrates. Aflatoxin B_1, the most potent of the group, is toxic, carcinogenic, mutagenic, and teratogenic. Major agricultural commodities that are often contaminated by aflatoxins include corn, peanuts, rice, cottonseed, and various tree nuts. *See* AFLATOXIN; ERGOT AND ERGOTISM. [A.Ci.; M.Kl.]

Myiasis The infestation of vertebrates by the larvae, or maggots, of numerous species of flies. These larvae may invade different parts of the bodies of these animals or may appear externally. The Diptera of medical and veterinary importance are largely confined to the families Oestridae, Calliphoridae, and Sarcophagidae.

In cutaneous myiasis, the larvae are found in or under the skin. There may be a migration of some species of these larvae through host tissues, resulting in a swelling with intense itching. Such a condition is known as larva migrans, or creeping eruption, and may require surgical treatment.

Intestinal myiasis in humans is usually the result of accidentally swallowing the eggs or larvae of these flies. It occurs commonly in many herbivores who ingest the eggs when feeding on contaminated herbage. The larvae settle in the stomach or intestinal tract of the animal host.

Cavity, or wound, myiasis occurs when the larvae invade natural orifices, such as the nasopharynx, vulva, and sinuses, or artificial openings such as wounds. External myiasis includes infestation by those maggots which are blood feeders. *See* MEDICAL PARASITOLOGY. [C.B.C.]

Myzostomaria An aberrant group of Polychaeta recognized for four families, each with one genus. Myzostomidae with *Myzostomum* comprise about 130 species from worldwide areas. Mesomyzostomidae with *Mesomyzostomum* has two species, from Japan and the Aru Islands. Protomyzostomidae with *Protomyzostomum* is known for three species, from Japan and the Murman Sea, and Stelechopidae with *Stelechopus* has a single species, from Crozet Island, Antarctic Ocean. All are either external or internal parasites of echinoderms, chiefly crinoids, and hence from deep water. Most are greatly depressed, broad and very small, measuring at most a few millimeters. The separation of this group of polychaetes into families and species is based on external and internal characters. *See* ANNELIDA; POLYCHAETA. [O.H.]

N

Natural gas A combustible gas that occurs in porous rock of the Earth's crust and is found with or near accumulations of crude oil. Being in gaseous form, it may occur alone in separate reservoirs. More commonly it forms a gas cap, or mass of gas, entrapped between liquid petroleum and impervious capping rock layer in a petroleum reservoir. Under conditions of greater pressure it is intimately mixed with, or dissolved in, crude oil.

Typical natural gas consists of hydrocarbons having a very low boiling point. Methane (CH_4) makes up approximately 85% of the typical gas. Ethane (C_2H_6) may be present in amounts up to 10%; and propane (C_3H_8) up to 3%. Butane (C_4H_{10}); pentane (C_5H_{12}); hexane; heptane; and octane may also be present.

Whereas normal hydrocarbons having 5–10 carbon atoms are liquids at ordinary temperatures, they have a definite vapor pressure and therefore may be present in the vapor form in natural gas. Carbon dioxide, nitrogen, helium, and hydrogen sulfide may also be present.

Types of natural gas vary according to composition and can be dry or lean (mostly methane) gas, wet gas (considerable amounts of so-called higher hydrocarbons), sour gas (much hydrogen sulfide), sweet gas (little hydrogen sulfide), residue gas (higher paraffins having been extracted), and casinghead gas (derived from an oil well by extraction at the surface). Natural gas has no distinct odor. Its main use is for fuel, but it is also used to make carbon black, natural gasoline, certain chemicals, and liquefied petroleum gas. Propane and butane are obtained in processing natural gas.

Gas occurs on every continent. Wherever oil has been found, a certain amount of natural gas is also present. Successful exploitation of these resources involves drilling, producing, gathering, processing, transporting, and metering the use of the gas. Long before supplies of natural gas run out or become expensively scarce, it is expected that some process of coal gasification will produce a gas which is completely interchangeable with natural gas and at a competitive price. This is important because coal makes up a majority of the world's known fossil fuel reserves. But when energy consumers indicated in the marketplace their preference for fluid and gaseous fuels over the solid forms, coal gasification research, already well under way, was given additional impetus.

[M.A.A.; M.T.H.]

Nemata A phylum of unsegmented worms. A classification of nematodes follows:

Phylum Nemata
Class: Adenophorea
Subclass: Enoplia

Order: Enoplida
Oncholaimida
Tripylida
Isolaimida
Mononchida
Dorylaimida
Stichosomida
Subclass: Chromadoria
Order: Araeolaimida
Chromadorida
Desmoscolecida
Desmodorida
Monhysterida
Class: Secernentea
Subclass: Rhabditia
Order: Rhabditida
Strongylida
Subclass: Spiruria
Order: Spirurida
Ascaridida
Subclass: Diplogasteria
Order: Diplogasterida
Tylenchida

Diagnosis. The Nemata are unsegmented or pseudosegmented (any superficial annulation limited to the cuticle) bilaterally symmetrical worms with a basically circular cross section. The body is covered by a noncellular cuticle. The cylindrical body is usually bluntly rounded anteriorly and tapering posteriorly. The body cannot be easily divided into head, neck, and trunk or tail, although a region posterior to the anus is generally referred to as the tail. The oral opening is terminal (rarely subterminal) and followed by the stoma, esophagus, intestine, and rectum which opens through a subterminal anus. Females have separate genital and digestive tract openings. In males the tubular reproductive system joins posteriorly with the digestive tract to form a cloaca. The sexes are separate and the gonads may be paired or unpaired. Females may be oviparous or ovoviviporous.

Adult nematodes are extremely variable in size, ranging from less than 0.012 in. (0.3 mm) to over 26 ft (8 m). Nematodes are generally colorless except for food in the intestinal tract or for those few species which have eyespots.

Life cycle. Reproduction among nematodes is either amphimictic or parthenogenetic (rarely hermaphroditic). After the completion of oogenesis the chitinous egg shell is formed and a waxy vitelline membrane forms within the egg shell; in some nematodes the uterine cells deposit an additional outermost albuminoid coating. Upon deposition or within the female body, the egg proceeds through embryonation to the eellike first- or second-stage larva, but following eclosion the larva proceeds through four molts to adulthood. This represents a direct life cycle, but among parasites more diversity occurs.

Distribution. Nemata comprise the third largest phylum of invertebrates, being exceeded only by Mollusca and Arthropoda. In sheer numbers of individuals they exceed all other metazoa. As parasites of animals they exceed all other helminths combined. Nematodes have been recovered from the deepest ocean floors to the highest mountains, from the Arctic to the Antarctic, and in soils as deep as roots can penetrate. [A.R.M.]

Nematicide A type of chemical used to kill plant-parasitic nematodes. Nematicides may be classed as soil fumigants or soil amendments, space fumigants, surface sprays, or dips. Soil treatments are commonly used because most plant-pathogenic species spend part or all of their life cycle in the soil, in or about the roots of plants. Nematicides may be liquids, gases, or solids, but on a field scale, liquids are most practical. *See* NEMATA; PESTICIDE. [D.J.R.]

New Zealand A landmass in the Southern Hemisphere, bounded by the South Pacific Ocean to the north, east, and south and the Tasman Sea to the west, with a total land area of 103,883 mi^2 (269,057 km^2). The exposed landmass represents about one-quarter of a subcontinent, with three-quarters submerged. This long, narrow, mountainous country, oriented northeast to southwest, consists of two main islands, North Island and South Island, surrounded by a much greater area of crust submerged to depths reaching 1.2 mi (2 km).

South Island lowlands are either alluvial plains as in Otago, Southland, and Nelson, or glacial outwash fans as in Westland and Canterbury. North Island lowlands such as Hawke's Bay, Wairarapa, and Manawatu are alluvial; the Waikato, Hauraki, and Bay of Plenty lowlands occupy structural basins that contain large volumes of reworked volcanic debris from the central volcanic region. The alluvial lowlands of both main islands form the most agriculturally productive areas of the country. *See* PLAINS.

The climate of New Zealand is influenced by three main factors: a location in latitudes where the prevailing airflow is westerly; an oceanic environment; and the mountain chains, which modify the weather systems as they pass eastward, causing high rainfalls on windward slopes and sheltering effects to leeward.

Weather is determined mostly by series of anticyclones and troughs of low pressure that produce alternating periods of settled and variable conditions. Westerly air masses are occasionally replaced by southerly airstreams, which bring cold conditions with snow in winter and spring to areas south of 39°S, and northerly tropical maritime air, which brings warm humid weather to the north and east coasts. *See* METEOROLOGY.

Rainfall on land is 16–470 in. (400–12,000 mm) per year, with the highest rainfall being on the western windward slopes of the mountains, and the lowest on the eastern basins in the lee of the Southern Alps in Central Otago and south Canterbury. Annual rain days are at least 130 for most of North Island, but on South Island the totals are far more variable, with over 200 occurring in Fiordland, 180 on the west coast, and fewer than 80 in Central Otago. Summer droughts are relatively common in Northland, and in eastern regions of both islands. *See* DROUGHT; PRECIPITATION (METEOROLOGY).

Droughts, springtime air frosts, and hailstorms are the major common climatic hazards for the farming industry, but floods associated with prolonged intense rainstorms are the major general hazard.

The economy is heavily dependent on the natural resources soil, water, and plants. New Zealand has few exploitable minerals, but possesses a climate generally favorable for agriculture, pastoral farming, renewable forestry, and tourism. With a small population (3.4 million), much of its manufacturing is concerned with processing produce from the land and surrounding seas, and supplying the needs of those industries.

Because of its high relief and its location on an active crustal plate boundary in the zone of convergence between Antarctic air masses and tropical air masses, New Zealand is prone to high-intensity and high-frequency natural hazards—earthquakes, volcanic eruptions, large and small landslides, and floods. [M.J.Se.]

Newcastle disease A viral infection that affects the digestive, intestinal, and respiratory tracts and the neurological system of birds. The causative agent is an enveloped ribonucleic acid (RNA) virus that is classified as a paramyxovirus. *See* PARAMYXOVIRUS.

Newcastle disease occurs in five forms based on a virulence in chickens ranging from inapparent infection to severe disease and death. Viscerotropic-velogenic Newcastle disease causes a very severe infection, producing hemorrhagic lesions in the intestinal tract and high mortality. The neurotropic-velogenic type is also highly lethal and produces neurologic and respiratory signs in infected birds. The mesogenic form causes an acute respiratory or neurologic infection that may be lethal only in young birds. The lentogenic type is a mild or inapparent respiratory infection of chickens. The last group includes the viruses causing inapparent or asymptomatic infections of the digestive tract.

The wide susceptibility of avian species to infection with Newcastle disease has complicated control. Newcastle disease is spread worldwide by the international transportation of live birds disseminating the virus. Control of and protection from Newcastle disease can be achieved by the correct use of vaccines. Lentogenic and some mesogenic strains are used to produce vaccines that can be administered by aerosol, intranasal drops, or intramuscular injection, or as an additive to the drinking water. *See* ANIMAL VIRUS.

[M.L.V.]

Nitrogen cycle The collective term given to the natural biological and chemical processes through which inorganic and organic nitrogen are interconverted. It includes the process of ammonification, ammonia assimilation, nitrification, nitrate assimilation, nitrogen fixation, and denitrification.

Nitrogen exists in nature in several inorganic compounds, namely N_2, N_2O, NH_3, NO_2^-, and NO_3^-, and in several organic compounds such as amino acids, nucleotides, amino sugars, and vitamins. In the biosphere, biological and chemical reactions continually occur in which these nitrogenous compounds are converted from one form to another. These interconversions are of great importance in maintaining soil fertility and in preventing pollution of soil and water.

An outline showing the general interconversions of nitrogenous compounds in the soil-water pool is presented in the illustration. There are three primary reasons why organisms metabolize nitrogen compounds: (1) to use them as a nitrogen source, which means first converting them to NH_3, (2) to use certain nitrogen compounds as an energy source such as in the oxidation of NH_3 to NO_2^- and of NO_2^- to NO_3^- , and (3) to use certain nitrogen compounds (NO_3^-) as terminal electron acceptors under conditions where oxygen is either absent or in limited supply. The reactions and products involved in these three metabolically different pathways collectively make up the nitrogen cycle.

There are two ways in which organisms obtain ammonia. One is to use nitrogen already in a form easily metabolized to ammonia. Thus, nonviable plant, animal, and microbial residues in soil are enzymatically decomposed by a series of hydrolytic and other reactions to yield biosynthetic monomers such as amino acids and other small-molecular-weight nitrogenous compounds. These amino acids, purines, and pyrimidines are decomposed further to produce NH_3 which is then used by plants and bacteria for biosynthesis, or these biosynthetic monomers can be used directly by some microorganisms. The decomposition process is called ammonification.

The second way in which inorganic nitrogen is made available to biological agents is by nitrogen fixation (this term is maintained even though N_2 is now called dinitrogen), a process in which N_2 is reduced to NH_3. Since the vast majority of nitrogen is in the form of N_2, nitrogen fixation obviously is essential to life. The N_2-fixing process

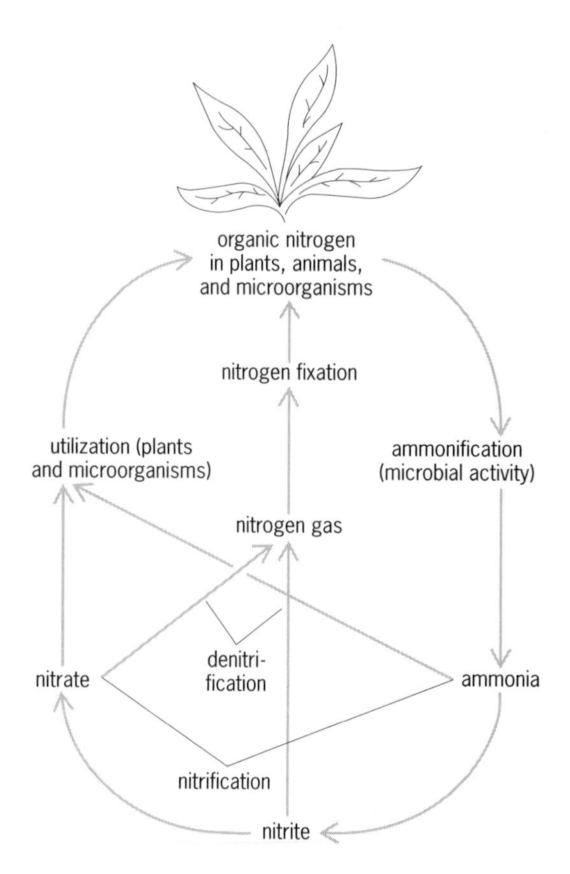

Diagram of the nitrogen cycle.

is confined to prokaryotes (certain photosynthetic and nonphotosynthetic bacteria). The major nitrogen fixers (called diazotrophs) are members of the genus *Rhizobium*, bacteria that are found in root nodules of leguminous plants, and of the cyanobacteria (originally called blue-green algae). *See* NITROGEN FIXATION. [L.E.Mo.]

Nitrogen fixation The chemical or biological conversion of atmospheric nitrogen (N_2) into compounds which can be used by plants, and thus become available to animals and humans. In the 1990s, chemical and biological processes together contributed about 260 million tons (230 million metric tons) of fixed nitrogen per year globally. Industrial production of nitrogen fertilizer accounted for about 85 million tons (80 million metric tons) of nitrogen per year, while spontaneous chemical processes, such as lightning, ultraviolet irradiation, and combustion, leading to the synthesis of nitrogen oxides from O_2 and N_2, may have accounted for 44 million tons (40 million metric tons) per year. The remainder, roughly half of the global input of newly fixed nitrogen, arose from biological processes. World agriculture, which is very dependent on nitrogen fixation, is increasingly reliant on chemical nitrogen sources.

Chemical fixation. Three chemical processes for fixing atmospheric nitrogen have been developed. All require considerable thermal or electrical energy and yield different products. In arc processes, which are now rarely used, air is passed through an electric arc and about 1% nitric oxide is formed, which can be chemically converted to nitrates. In the cyanamide process, which is now obsolete, heating calcium carbide in nitrogen generates calcium cyanamide, which when moistened hydrolyzes to urea and ammonia. In the widely used Haber process, hydrogen (generated by heating natural gas) is mixed with nitrogen (from air), and burned to yield a nitrogen-hydrogen mixture. The nitrogen-hydrogen mixture is compressed (10–80 megapascals) and heated (200–700°C or 390–1300°F) in the presence of a metal oxide catalyst to give ammonia. The Haber process is the major source of ammonia used for fertilizer. *See* FERTILIZER.

Biological fixation. Only prokaryotes—bacteria, archaea, and cyanobacteria (earlier called blue-green algae)—fix nitrogen. Nitrogen-fixing microbes, called diazotrophs, fall into two main groups, free-living and symbiotic. *See* ARCHAEBACTERIA; BACTERIA; CYANOBACTERIA; PROKARYOTAE.

The free-living diazotrophs are subclassified. Aerobic diazotrophs, of which there are over 50 genera, including *Azotobacter*, methane-oxidizing bacteria, and cyanobacteria, require oxygen for growth and fix nitrogen when oxygen is present. *Azotobacter*, some related bacteria, and some cyanobacteria fix nitrogen in ordinary air, but most members of this group fix nitrogen only when the oxygen concentration is low. Free-living diazotrophs, which fix nitrogen only when oxygen is absent or vanishingly low, are widespread. The genera *Bacillus* and *Klebsiella* include many strains of this type, and representatives of symbiotic diazotrophs behave in this way as well. *See* ALGAE; BACTERIAL PHYSIOLOGY AND METABOLISM.

The best-known symbiotic bacteria belong to the genus *Rhizobium*. Species of *Rhizobium*, or related genera, such as *Bradyrhizobium* and *Sinorhizobium*, colonize the roots of leguminous plants and stimulate the formation of nodules within which they fix nitrogen microaerobically. Both plants and bacteria show specificity; for example, certain types of plants require special strains of rhizobia. Some types of rhizobium, such as *Bradyrhizobium*, can fix nitrogen in the absence of plant tissue, but require low oxygen, though most rhizobia fix nitrogen only within the nodules. *See* SOIL MICROBIOLOGY.

The enzymes responsible for nitrogen fixation are called nitrogenases. The most common nitrogenase consists of two proteins, one large containing molybdenum, iron, and inorganic sulfur (the MoFe-protein or dinitrogenase), the other smaller containing iron and inorganic sulfur (the Fe-protein or dinitrogenase reductase). Nitrogenase reduces one molecule of N_2 to two of ammonia (NH_3), a reaction which is accompanied by the conversion of 16 molecules of adenosine triphosphate (ATP) to adenosine diphosphate (ADP) and the release of one molecule of H_2 as a by-product. Nitrogenase is irreversibly destroyed by air, so all aerobic diazotrophs have developed means of restricting access of oxygen to the active enzyme. [J.Po.]

North America The third largest continent, extending from the narrow isthmus of Central America to the Arctic Archipelago. The physical environments of North America, like the rest of the world, are a reflection of specific combinations of the natural factors such as climate, vegetation, soils, and landforms. *See* CONTINENT.

Location. North America covers 9,400,000 mi^2 (24,440,000 km^2) and extends north to south for 5000 mi (8000 km) from Central America to the Arctic. It is bounded by the Pacific Ocean on the west and the Atlantic Ocean on the east. The Gulf of Mexico is a source of moist tropical air, and the frozen Arctic Ocean is a source of polar air. With the major mountain ranges stretching north-south, North America is the only continent providing for direct contact of these polar and tropical air masses, leading to frequent climatically induced natural hazards such as violent spring tornadoes, extreme droughts, subcontinental floods, and winter blizzards, which are seldom found on other continents. *See* AIR MASS; ARCTIC OCEAN; ATLANTIC OCEAN; GULF OF MEXICO; PACIFIC OCEAN.

Geologic structure. The North American continent includes (1) a continuous, broad, north-south-trending western cordilleran belt stretching along the entire Pacific coast; (2) a northeast-southwest-trending belt of low Appalachian Mountains paralleling the Atlantic coast; (3) an extensive rolling region of old eroded crystalline rocks in the north-central and northeastern part of the continent called the Canadian Shield; (4) a large, level interior lowland covered by thick sedimentary rocks and extending from the Arctic Ocean to the Gulf of Mexico; and (5) a narrow coastal plain along the Atlantic Ocean and the Gulf of Mexico. These broad structural geologic regions provide the framework for the natural regions of this continent and affect the location and nature of landform, climatic, vegetation, and soil regions.

Canadian Shield. Properly referred to as the geological core of the continent, the exposed Canadian Shield extends about 2500 mi (4000 km) from north to south and almost as much from east to west. The rest of it dips under sedimentary rocks that overlap it on the south and west. The Canadian Shield consists of ancient Precambrian rocks, over 500 million years old, predominantly granite and gneiss, with very complex structures indicating several mountain-building episodes. It has been eroded into a rolling surface of low to moderate relief with elevations generally below 2000 ft (600 m). Its surface has been warped into low domes and basins, such as the Hudson Basin, in which lower Paleozoic rocks, including Ordovician limestones, have been preserved. Since the end of the Paleozoic Era, the Shield has been dominated by erosion. Parts of the higher surface remain at about 1500–2000 ft (450–600 m) above sea level, particularly in the Labrador area. The Shield remained as land throughout the Mesozoic Era, but its western margins were covered by a Cretaceous sea and by Tertiary terrestrial sediments derived from the Western Cordillera.

The entire exposed Shield was glaciated during the Pleistocene Epoch, and its surface was intensely eroded by ice and its meltwaters, erasing major surface irregularities and eastward-trending rivers that were there before. The surface is now covered by glacial till, outwash, moraines, eskers, and lake sediments, as well as drumlins formed by advancing ice. A deranged drainage pattern is evolving on this surface with thousands of lakes of various sizes. *See* DRUMLIN; ESKER; MORAINE.

The Canadian Shield extends into the United States as Adirondack Mountains in New York State, and Superior Upland west of Lake Superior.

Southeastern Coastal Plain. The Southeastern Coastal Plain is geologically the youngest part of the continent, and it is covered by the youngest marine sedimentary rocks. This flat plain, which parallels the Atlantic and Gulf coastline, extends for over 3000 mi (4800 km) from Cape Cod, Massachusetts, to the Yucatán Peninsula in Mexico. It is very narrow in the north but increases in width southward along the Atlantic coast and includes the entire peninsula of Florida. As it continues westward along the Gulf, it widens significantly and includes the lower Mississippi River valley. It is very wide in Texas, narrows again southward in coastal Mexico, and then widens in the Yucatán Peninsula and continues as a wide submerged plain, or a continental shelf, into the sea. *See* COASTAL PLAIN.

Extending from Cape Cod, Massachusetts, to Mexico and Central America, the Coastal Plain is affected by a variety of climates and associated vegetation. While a humid, cool climate with four seasons affects its northernmost part, subtropical air masses affect the southeastern part, including Florida, and hot and arid climate dominates Texas and northern Mexico; Central America has hot, tropical climates.

Varied soils characterize the Coastal Plain, including the fertile alluvial soils of the Mississippi Valley. Broadleaf forests are present in the northeast, citrus fruits grow in Florida, grasslands dominate the dry southwest, and tropical vegetation is present on Central American coastal plains.

Eastern Seaboard Highlands. Between the Southeastern Coastal Plain and the extensive interior provinces lies a belt of mountains that, by their height and pattern, create a significant barrier between the eastern seaboard and the interior of North America. These mountains consist of the Adirondack Mountains and the New England Highlands.

The Adirondack Mountains are a domal extension of the Canadian Shield, about 100 mi (160 km) in diameter, composed of complex Precambrian rocks. The New England Highlands consist of a north-south belt of mountains east of the Hudson Valley, including the Taconic mountains in the south and the Green mountains in the north, and continuing as the Notre Dame Mountains along the St. Lawrence Valley and the Chic-Choc Mountains of the Gaspé Peninsula. The large area of New England east of these mountains is an eroded surface of old crystalline rocks culminating in the center as the White Mountains, with their highest peak of the Presidential Range, Mount Washington, reaching over 6200 ft (1880 m). This area has been intensely glaciated, and it meets the sea in a rugged shoreline. Nova Scotia and Newfoundland have a similar terrain.

New England is a hilly to mountainous region carved out of ancient rocks, eroded by glaciers, and covered by glacial moraines, eskers, kames, erratics, and drumlins, with hundreds of lakes scattered everywhere. It has a cool and moist climate with four seasons, thin and acid soils, and mixed coniferous and broadleaf forests.

Appalachian Highlands. The Appalachian Highlands are traditionally considered to consist of four parts: the Piedmont, the Blue Ridge Mountains, the Ridge and Valley Section, and the Appalachian Plateau. These subregions are all characterized by different geologic structures and rock types, as well as different geomorphologies.

The northern boundary of the entire Appalachian System is an escarpment of Paleozoic rocks trending eastward along Lake Erie, Lake Ontario, and the Mohawk Valley. The boundary then swings south along Hudson River Valley and continues southwestward along the Fall Line to Montgomery, Alabama. The western boundary trends northeastward through Cumberland Plateau in Tennessee, and up to Cleveland, Ohio, where it joins the northern boundary. Together with New England, this region forms the largest mountainous province in eastern United States.

Interior Domes and Basins Province. The southwestern part of the Appalachian Plateau, overlain mainly by the Mississippian and Pennsylvanian sedimentary rocks, has been warped into two low structural domes called the Blue Grass and Nashville Basins, and a structural basin, drained by the Green River; its southern fringe is called the Pennyroyal Region. The Interior Dome and Basin Province is contained roughly between the Tennessee River in the south and west and the Ohio River in the north.

There is no boundary on the east, because the domes are part of the same surface as the Appalachian Plateau. However, erosional escarpments, forming a belt of hills called knobs, clearly mark the topographic domes and basins. The northern dome, called the Blue Grass Basin or Lexington Plain, has been eroded to form a basin surrounded by a series of inward-facing cuesta escarpments. The westernmost cuesta reaches about 600 ft (180 m) elevation while the central part of the basin lies about 1000 ft (300 m) above sea level, which is higher than the surrounding hills. This gently rolling surface with deep and fertile soils exhibits some solutional karst topography. *See* FLUVIAL EROSION LANDFORMS.

Ozark and Ouachita Highlands. The Paleozoic rocks of the Pennyroyal Region continue Westward across southern Illinois to form another dome of predominantly Ordovician rocks, called the Ozark Plateau. This dome, located mainly in Missouri and Arkansas, has an abrupt east side, and a gently sloping west side, called the Springfield Plateau. Its surface is stream eroded into hilly and often rugged topography that is developed mainly on limestones, although shales, sandstone, and chert are present. Much residual chert, eroded out of limestone, is present on the surface. There are some karst features, such as caverns and springs. In the northeast, Precambrian igneous rocks protrude to form the St. Francois Mountains, which reach an elevation of 1700 ft (515 m).

Central Lowlands. One of the largest subdivisions of North America is the Central Lowlands province which is located between the Appalachian Plateau on the east, the Interior Domes and Basins Province and the Ozark Plateau on the south, and the Great Plains on the west. It includes the Great Lakes section and the Manitoba Lowland in Canada. This huge lowland in the heart of the continent (whose elevations vary from about 900 ft or 270 m above sea level in the east and nearly 2000 ft or 600 m in the west) is underlain by Paleozoic rocks that continue from the Appalachian Plateau and dip south under the recent coastal plain sediments; meet the Cretaceous rocks on the west; and overlap the crystalline rocks of the Canadian Shield on the northeast.

The present surface of nearly the entire Central Lowlands, roughly north of the Ohio River and east of the Missouri River, is the creation of the Pleistocene ice sheets. When the ice formed and spread over Canada, and southward to the Ohio and Missouri rivers, it eroded much of the preexisting surface. During deglaciation, it left its deposits over the Canadian Shield and the Central Lowlands.

The Central Lowlands are drained by the third longest river system in the world, the Missouri-Mississippi, which is 3740 mi (6000 km) long. This mighty river system, together with the Ohio and the Tennessee, drains not only the Central Lowlands but also parts of the Appalachian Plateau and the Great Plains, before it crosses the Coastal Plain and ends in the huge delta of the Mississippi. The river carries an enormous amount of water and alluvium and continues to extend its delta into the Gulf. In 1993 it reached a catastrophic level of a hundred-year flood, claimed an enormous extent of land and many lives, and created an unprecedented destruction of property. This flood again alerted the population to the extreme risk of occupying a river floodplain. *See* FLOODPLAIN; RIVER.

Great Plains. The Great Plains, which lie west of the Central Lowlands, extend from the Rio Grande and the Balcones Escarpment in Texas to central Alberta in Canada. On the east, they are bounded by a series of escarpments, such as the Côteau du Missouri in the Dakotas. The dry climate with less than 20 in. (50 cm) of precipitation, and steppe grass vegetation growing on calcareous soils, help to determine the eastern boundary of the Great Plains. On the west, the Great Plains meet the abrupt front of the Rocky Mountains, except where the Colorado Piedmont and the lower Pecos River Valley separate them from the mountains.

The Great Plains region shows distinct differences between its subsections from south to north. The southernmost part, called the High Plains or Llano Estacado, and Edwards Plateaus are the flattest. While Edwards Plateau, underlain by limestones of the Cretaceous age, reveals solutional karst features, the High Plains have the typical Tertiary bare cap rock surface, devoid of relief and streams.

The central part of the Great Plains has a recent depositional surface of loess and sand. The Sand Hills of Nebraska form the most extensive sand dunes area in North America, covering about 24,000 mi^2 (62,400 km^2). They are overgrown by grass and have numerous small lakes. The loess region to the south provides spectacular small canyon topography. *See* DUNE.

The northern Great Plains, stretching north of Pine Ridge and called the Missouri Plateau, have been intensly eroded by the western tributaries of the Missouri River into river breaks and interfluves. In extreme cases, badlands were formed, such as those of the White River and the Little Missouri.

The terrain of the Canadian Great Plains consists of three surfaces rising from east to west: the Manitoba, Saskatchewan, and Alberta Prairies developed on level Creteceous and Tertiary rocks. Climatic differences between the arid and warm southern part and the cold and moist northern part have resulted in regional differences. The eastern boundary of the Saskatchewan Plain is the segmented Manitoba Escarpment, which extends for 500 mi (800 km) northwestward, and in places rises 1500 ft (455 m) above the Manitoba Lowland. Côteau du Missouri marks the eastern edge of the higher Alberta Plain.

Western Cordillera. The mighty and rugged Western Cordilleras stretch along the Pacific coast from Alaska to Mexico. There are three north-south-trending belts: (1) Brooks Range, Mackenzie Mountains, and the Rocky Mountains to the north and Sierra Madre Oriental in Mexico; (2) Interior Plateaus, including the Yukon Plains, Canadian Central Plateaus and Ranges, Columbia Plateau, Colorado Plateau, and Basin and Range Province stretching into central Mexico; and (3) Coastal Mountains from Alaska Range to California, Baja California, and Sierra Madre Occidental in Mexico.

This subcontinental-size mountain belt has the highest mountains, greatest relief, roughest terrain, and most beautiful scenery of the entire continent. It has been formed by earth movements resulting from the westward shift of the North American lithospheric plate. The present movements, and the resulting devastating earthquakes along the San Andreas fault system paralleling the Pacific Ocean, are part of this process.

This very high, deeply eroded and rugged Rocky Mountains region comprises several distinct parts: Southern, Middle, and Northern Rockies, plus the Wyoming Basin in the United States, and the Canadian Rockies. The Southern Rockies, extending from Wyoming to New Mexico, include the Laramie Range, the Front Range, and Spanish Peaks with radiating dikes on the east; Medicine Bow, Park, and Sangre de Cristo ranges in the center; and complex granite Sawatch Mountains and volcanic San Juan Mountains of Tertiary age on the west. Most of the ranges are elongated anticlines with exposed Precambrian granite core, and overlapping Paleozoic and younger sedimentary rocks which form spectacular hogbacks along the eastern front. There are about 50 peaks over 14,000 ft (4200 m) high, while the Front Range alone has about 300 peaks over 13,000 ft (3940 m) high. The southern Rocky Mountains, heavily glaciated into a beautiful and rugged scenery with permanent snow and small glaciers, form a major part of the Continental Divide.

The interior Plateaus and Ranges Province of the Western Cordillera lies between the Rocky Mountains and the Coastal Mountains. It is an extensive and complex region. It begins in the north with the wide Yukon Plains and Uplands; narrows into the Canadian Central Plateaus and Ranges; widens again into the Columbia Plateau, Basin and Range Province, and Colorado Plateau; and finally narrows into the Mexican Plateau and the Central American isthmus.

The coastal Lowlands and Ranges extend along the entire length of North America and include Alaskan Coast Ranges, Aleutian Islands, Alaska Range, Canadian Coast Ranges, and a double chain of the Cascade Mountains and Sierra Nevada on the east, and Coast Ranges on the west, separated by Puget Sound, Willamette Valley, and Great Valley of California. These ranges continue southward as Lower California Peninsula, Baja California, and Sierra Madre Occidental in Mexico.

The basin-and-range type of terrain of the southwest United States continues into northern Mexico and forms its largest physiographic region, the Mexican Plateau. This huge tilted block stands more than a mile above sea level—from about 4000 ft (1200 m) in the north, it rises to about 8000 ft (2400 m) in the south. The Mexican Plateau is separated from the Southern Mexican Highlands (Sierra Madre del Sur) by a low, hot and dry Balsas Lowland drained by the Balsas River. To the east of the Southern Highlands lies a lowland, the Isthmus of Tehuantepec, which is considered the divide between North and Central America. Here the Pacific and Gulf coasts are only 125 mi (200 km) apart. The lowlands of Mexico are the coastal plains. The Gulf Coastal Plain trends southward for 850 mi from the Rio Grande to the Yucatán Peninsula. It is about 100 mi (160 km) wide in the north, just a few miles wide in the center, and very wide in the Yucatan Peninsula. Barrier beaches, lagoons, and swamps occur along this coast. The Pacific Coastal Plains are much narrower and more hilly. North-south-trending ridges of granite characterize the northern part, and islands are present offshore. Toward the south, sandbars, lagoons, and deltaic deposits are common.

East of the Isthmus of Tehuantepec begins Central America with its complex physiographic and tectonic regions. This narrow, mountainous isthmus is geologically connected with the large, mountainous islands of the Greater Antilles in the Carribean. They are all characterized by east-west-trending rugged mountain ranges, with deep depressions between them. One such mountain system begins in Mexico and continues in southern Cuba, Puerto Rico, and the Virgin Islands. North of this system, called the Old Antillia, lies the Antillian Foreland, consisting of the Yucatán Peninsula and the Bahama Islands. Central American mountains are bordered on both sides by active volcanic belts. Along the Pacific, a belt of young volcanoes extends for 800 mi (1280 km) from Mexico to Costa Rica. Costa Rica and Panama are mainly a volcanic chain of mountains extending to South America. Nicaragua is dominated by a major crustal fracture trending northwest-southeast. [B.Z.B.]

North Pole That end of the Earth's axis which points toward the North Star, Polaris (Alpha Ursae Minoris). It is the geographical pole where all meridians converge, and should not be confused with the north magnetic pole, which is in the Canadian Archipelago. The North Pole's location falls near the center of the Arctic Sea. The North Pole has phenomena unlike any other place except the South Pole. For 6 months the Sun does not appear above the horizon, and for 6 months it does not go below the horizon. As there is a long period (about 7 weeks) of continuous twilight before March 21 and after September 23, the period of light is considerably longer than the period of darkness. [V.H.E.]

North Sea A flooded portion of the northwest continental margin of Europe occupying an area of over 200,000 mi^2 (500,000 km^2). The North Sea has extensive marine fisheries and important offshore oil and gas reserves. In the south, its depth is less than 150 ft (50 m), but north of 58° it deepens gradually to 600 ft (200 m) at the top of the continental slope. A band of deep water down to 1200 ft (400 m) extends around the south and west coast of Norway and is known as the Norwegian Trench.

The nontidal residual current circulation of the southern North Sea is mainly determined by wind velocity, but in the north, well-defined non-wind-driven currents have been identified, especially in the summer. Two of these currents bring in water from outside the North Sea; one flows through the channel between Orkney and Shetland (the Fair Isle current), and the other follows the continental slope north of Shetland and merges with the Fair Isle current southwest of Norway before entering the Skagerrak. The north-flowing Norwegian coastal current provides the exit route for North Sea waters, and is formed from the waters of these two major inflows and from other much smaller inputs such as river runoff, the English Channel, and the Baltic Sea.

There is a rich diversity of zooplankton within the North Sea. Copepods are of particular importance in the food web. There are a wide range of fish stocks in the North Sea and adjacent waters and, in terms of species exploited by commercial fisheries, they constitute the richest area in the northeast Atlantic. The commercially important stocks exploited for human consumption include cod, haddock, whiting, pollock, plaice, sole, herring, mackerel, lobster, prawn, and brown shrimp (*Crangon crangon*). A number of stocks are used for fishmeal and oil; these stocks include sand eel, Norway pout, blue whiting, and sprat. *See* MARINE ECOLOGY; ZOOPLANKTON. [H.D.D.]

Numerical taxonomy The grouping by numerical methods of taxonomic units based on their character states. The application of numerical methods to taxonomy, dating back to the rise of biometrics in the late nineteenth century, has received a great deal of attention with the development of the computer and computer technology. Numerical taxonomy provides methods that are objective, explicit, and repeatable, and is based on the ideas first put forward by M. Adanson in 1963. These ideas, or principles, are that the ideal taxonomy is composed of information-rich taxa based on as many features as possible, that *a priori* every character is of equal weight, that overall similarity between any two entities is a function of the similarity of the many characters on which the comparison is based, and that taxa are constructed on the basis of diverse character correlations in the groups studied.

In the early stages of development of numerical taxonomy, phylogenetic relationships were not considered. However, numerical methods have made possible exact measurement of evolutionary rates and phylogenetic analysis. Furthermore, rapid developments in the techniques of direct measurement of the homologies of deoxyribonucleic acid (DNA), and ribonucleic acid (RNA) between different organisms now provide an estimation of "hybridization" between the DNAs of different taxa and, therefore, possible evolutionary relationships. Thus, research in numerical taxonomy often includes analyses of the chemical and physical properties of the nucleic acids of the organisms the data from which are correlated with phenetic groupings established by numerical techniques. *See* PHYLOGENY; TAXONOMIC CATEGORIES. [R.R.C.]

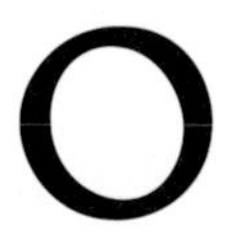

Oak A genus (*Quercus*) of trees, some of which are shrubby, with about 200 species, mainly in the Northern Hemisphere. About 50 species are native in the United States. All oaks have scaly winter buds, usually clustered at the ends of the twigs, and single at the nodes. The fruit is a nut (acorn). The leaves are simple and usually lobed.

Oaks furnish the most important hardwood lumber in the United States. Principal uses are for charcoal, barrels, building construction, flooring, railroad ties, mine timbers, boxes, crates, vehicle parts, ships, agricultural implements, caskets, woodenware, fence posts, piling, and veneer. Oak is also used for pulp and paper products. [A.H.G./K.P.D.]

Oasis An isolated fertile area, usually limited in extent and surrounded by desert. The term was initially applied to small areas in Africa and Asia typically supporting trees and cultivated crops with a water supply from springs and from seepage of water originating at some distance. However, the term has been expanded to include areas receiving moisture from intermittent streams or artificial irrigation systems. Thus the floodplains of the Nile and Colorado rivers can be considered vast oases, as can arid areas irrigated by humans. *See* DESERT.

Oases are restricted to climatic regions where precipitation is insufficient to support crop production. Such regions may be classified as extremely arid (annual rainfall less than 2 in. or 50 mm), arid (annual rainfall less than 10 in. or 250 mm), and semiarid (rainfall less than 20 in. or 500 mm). Many African and Asian oases are in extremely arid areas. Most oases are found in warm climates. Oasis soils are weakly developed, high in organic matter but often saline, and have been strongly affected by human occupation. [W.G.McG.]

Ocean One of the major subdivisions of the interconnected body of salt water that occupies almost three-quarters of the Earth's surface. Earth is the only planet in the solar system whose surface is covered with significant quantities of water. Of the nearly 1.4 billion cubic kilometers of water found either on the surface or in relatively accessible underground supplies, more than 97% is in the oceans. *See* OCEANOGRAPHY.

Oceans and the seas that connect them cover some 73% of the surface of the Earth, with a mean depth of 3729 m (12,234 ft) (see table). More than 70% of the oceans have a depth between 3000 and 6000 m (10,000 and 20,000 ft). Less than 0.2% of the oceans have depths as great as 7000 m (23,000 ft).

The oceans are cold and salty. Some 50% have a temperature between 0 and 2°C (32 and 36°F) and a salinity between 34.0 and 35.0. To a high degree of approximation, a salinity of 34 is the equivalent of 34 grams of salt in a kilogram of seawater. Water with a temperature above a few degrees Celsius is confined to a relatively thin surface layer of the ocean. *See* SEAWATER.

Ocean basin characteristics

	Area, km^2	Volume, km^3	Mean depth, m
Pacific	181,344,000	714,410,000	3940
Atlantic	94,314,000	337,210,000	3575
Indian	74,118,000	284,608,000	3840
Arctic	12,257,000	13,702,000	1117
Total	362,033,000	1,349,929,000	3729

Ocean salinity is primarily controlled by the balance of precipitation, river runoff, and evaporation of water at the sea surface. The highest salinities are found in major evaporation basins with little rainfall or river runoff, such as the Red Sea. The lowest salinities are found near the mouths of major rivers such as the Amazon. *See* RED SEA.

Nearly all elements known to humankind have been found dissolved in seawater, and those that have not are assumed to be present. However, all but a few are found in very small amounts. Sodium chloride accounts for some 85% of the dissolved salts, and an additional four ions (sulfate, magnesium, calcium, and potassium) bring the total to more than 99.3%. The ratio of ions is remarkably constant from one ocean to another and from top to bottom of each.

The oceans are continually transporting excess heat (warm water) from the tropics toward the Poles and returning colder water toward the tropics. This process of moving excess heat from lower (south of 40°) to higher (north of 40°) latitudes is shared approximately equally by the oceans and the atmosphere. A significant part of the ocean heat exchange process is carried out by the major ocean currents, the "named" currents such as the Gulf Stream, Brazil Current, California Current, and Kuroshio. These currents are primarily driven by the winds, and there is considerable similarity in their pattern from one ocean basin to another. *See* GULF STREAM.

The average winds over the North and South Atlantic as well as the North and South Pacific oceans come out of the west (westerlies) at the middle latitudes and from the east at the lower latitudes (trade winds). The frictional drag of these winds on the surface of the water imparts a spin or torque to the surface of the ocean, clockwise in the Northern Hemisphere and counterclockwise in the Southern Hemisphere. The major exception is the Indian Ocean north of the Equator, where the circulation is strongly influenced by the winds of the seasonal monsoon. *See* ATLANTIC OCEAN; EQUATORIAL CURRENTS; INDIAN OCEAN; OCEAN CIRCULATION; PACIFIC OCEAN. [J.A.K.]

Ocean circulation The general circulation of the ocean. The term is usually understood to include large-scale, nearly steady features, such as the Gulf Stream, as well as current systems that change seasonally but are persistent from one year to the next, such as the Davidson Current, off the northwestern United States coast and the equatorial currents in the Indian Ocean. A great number of energetic motions have periods of a month or two and horizontal scales of a few hundred kilometers—a very low-frequency turbulence, collectively called eddies. Energetic motions are also concentrated near the local inertial period (24 h, at 30° latitude) and at the periods associated with tides (primarily diurnal and semidiurnal).

The greatest single driving force for currents, as for waves, is the wind. Furthermore, the ocean absorbs heat at low latitudes and loses it at high latitudes. The resultant effect on the density distribution is coupled into the large-scale wind-driven circulation. Some subsurface flows are caused by the sinking of surface waters made dense by cooling or high evaporation. *See* OCEAN WAVES.

Except in western boundary currents, and in the Antarctic Circumpolar Current, the system of strong surface currents is restricted mainly to the upper 330–660 ft (100–200 m) of the sea. The mid-latitude anticyclonic gyres, however, are coherent in the mean well below 3300 ft (1000 m). The average speeds of the open-ocean surface currents remain mostly below 0.4 knot (20 cm/s). Exceptions to this are found in the western boundary currents, such as the Gulf Stream, and in the Equatorial Currents of the three oceans, all of which have velocities of 2–4 knots (1–2 m/s).

The deep circulation results in part from the wind stress and in part from the internal pressure forces which are maintained by the budgets of heat, salt, and water. Both groups of forces are dependent upon atmospheric influences. Apart from Coriolis and frictional forces, the topography of the sea bottom exercises a decisive influence on the course of deep circulation.

The deep circulation in marginal seas depends largely on the climate of the region, whether arid or humid. Under the influence of an arid climate, evaporation is greater than precipitation. The marginal sea is therefore filled with relatively salty water of a high density. Its surface lies at a lower level than that of the neighboring ocean. Examples of this type are the Mediterranean Sea, Red Sea, and Persian Gulf. The deep circulation of marginal seas in humid climates shows a different pattern. The level of the sea is higher than in the neighboring ocean. Therefore, the surface water with its lower density and accordingly its lower salinity flows outward, and the relatively salty ocean water of higher density flows over the sill into the marginal sea. Examples of this circulation are the Baltic Sea with the shallow Darsser and Drogden rises, the Norwegian and Greenland fiords, and the Black Sea with its entrance through the Bosporus. *See* BALTIC SEA; BLACK SEA; FIORD; MEDITERRANEAN SEA.

The deep circulation in the oceans is more difficult to perceive than the circulation in the marginal seas. In addition to the internal pressure forces, determined by the distribution of density and the piling up of water by the wind, there are also the influences of Coriolis forces and large-scale turbulence. There are areas in tropical latitudes in which the surface water, as a result of strong evaporation, has a relatively high density. In thermohaline convection, the water sinks while flowing horizontally until it reaches a density corresponding to its own, and then spreads out horizontally. In this way the colder and deeper levels of the oceans take on a layered structure consisting of the so-called bottom water, deep water, and intermediate water. *See* ATLANTIC OCEAN; PACIFIC OCEAN. [W.Stu.]

Wherever oceanographers have made long-term current and temperature measurements, they have found energetic fluctuations with periods of several weeks to several months. These low-frequency fluctuations (compared to tides) are caused by oceanic mesoscale eddies which are in many respects analogous to the atmospheric mesoscale pressure systems that form weather. Like the weather, mesoscale eddies often dominate the instantaneous current, and are thought to be an integral part of the ocean's general circulation.

Eddies occur in virtually all oceans and seas, but their amplitude varies greatly from place to place. The largest amplitudes are found on the western sides of the oceans in conjunction with the strongest ocean currents (the Gulf Stream in the North Atlantic, the Kuroshio in the North Pacific) and near the Equator. Much weaker eddies are found in the ocean interior, distant from major currents. This consistent pattern of eddy amplitude suggests that instabilities of western boundary currents are an important source of eddy energy. Atmospheric forcing by variable winds can also generate eddies, and is probably most important at low latitudes where the horizontal scales of the oceanic eddies best match the scales of the atmospheric forcing. [J.F.P.]

Ocean waves The irregular moving bumps and hollows on the ocean surface. Winds blowing over the ocean, in addition to producing currents, create surface water undulations called waves or a sea. The characteristics of these waves (or the state of the sea) depend on the speed of the wind, the length of time that it has blown, the distance over which it has blown, and the depth of the water. If the wind dies down, the waves that remain are called a dead sea.

Surface waves. Ocean surface waves are propagating disturbances at the atmosphere-ocean interface. They are the most familiar ocean waves. Surface waves are also seen on other bodies of water, including lakes and rivers.

A simple sinusoidal wave train is characterized by three attributes: wave height (H), the vertical distance from trough to crest; wavelength (L), the horizontal crest-to-crest distance; and wave period (T), the time between passage of successive crests past a fixed point. The phase velocity ($C = L/T$) is the speed of propagation of a crest. For a given ocean depth (h), wavelength increases with increasing period. The restoring force for these surface waves is predominantly gravitational. Therefore, they are known as surface gravity waves, unless their wavelength is shorter than 1.8 cm (0.7 in.), in which case surface tension provides the dominant restoring force.

Surface gravity waves may be classified according to the nature of the forces producing them. Tides are ocean waves induced by the varying gravitational influence of the Moon and Sun. They have long periods, usually 12.42 h for the strongest constituent. Storm surges are individual waves produced by the wind and dropping barometric pressure associated with storms; they characteristically last several hours. Earthquakes or other large, sudden movements of the Earth'scrust can cause waves, called tsunamis, which typically have periods of less than an hour. Wakes are waves resulting from relative motion of the water and a solid body, such as the motion of a ship through the sea or the rapid flow of water around a rock. Wind-generated waves, having periods from a fraction of a second to tens of seconds, are called wind waves. Like tides, they are ubiquitous in the ocean, and continue to travel well beyond their area of generation. The ocean is never completely calm. *See* STORM SURGE; TSUNAMI.

The growth of wind waves by the transfer of energy from the wind is not fully understood. At wind speeds less than 1.1 m/s (2.5 mi/h), a flat water surface remains unruffled by waves. Once generated, waves gain energy from the wind by wave-coupling of pressure fluctuations in the air just above the waves. For waves traveling slower than the wind, secondary, wave-induced airflows shift the wave-induced pressure disturbance downwind so the lowest pressure is ahead of the crests. This results in energy transfer from the wind to the wave, and hence growth of the wave.

If a constant wind blows over a sufficient length of ocean, called the fetch, for a sufficient length of time, a wave field develops whose statistical characteristics depend only on wind velocity. In particular, the spectrum of sea-surface elevation for such a fully-developed sea has the form of the equation shown, where f is frequency ($= 1/T$),

$$S(f) = A\frac{g^2}{f^5}e^{-1.25(f_m/f)^4}$$

$g = 9.8$ m/s^2 (32 ft/s^2) is gravitational acceleration, $f_m = 0.13\, g/U$ is the frequency of the spectral peak (U = wind speed at 10 m or 32.8 ft elevation), and $A = 5.2 \times 10^{-6}$ is a constant.

Because of viscosity, surface waves lose energy as they propagate, short-period waves being dampened more rapidly than long-period waves. Waves with long periods (typically 10 s or more) can travel thousands of kilometers with little energy loss. Such waves, generated by distant storms, are called swell.

When waves propagate into an opposing current, they grow in height. For example, when swell from a Weddell Sea storm propagates northeastward into the southwestward-flowing Agulhas Current off South Africa, high steep waves are formed. Many large ships in this region have been severely damaged by such waves.

Because actual ocean waves consist of many components with different periods, heights, and directions, occasionally a large number of these components can, by chance, come in phase with one another, creating a freak wave with a height several times the significant wave height of the surrounding sea. According to linear theory, waves with different periods propagate with different speeds in deep water, and hence the wave components remain in phase only briefly. But nonlinear effects are bound to be significant in a large wave. In such a wave, the effects of nonlinearity can compensate for those of dispersion, allowing a solitary wave to propagate almost unchanged. Consequently, a freak wave can have a lifetime of a minute or two. [M.Wi.]

Internal waves. Internal waves are wave motions of stably stratified fluids in which the maximum vertical motion takes place below the surface of the fluid. The restoring force is mainly due to gravity; when light fluid from upper layers is depressed into the heavy lower layers, buoyancy forces tend to return the layers to their equilibrium positions. In the oceans, internal oscillations have been observed wherever suitable measurements have been made. The observed oscillations can be analyzed into a spectrum with periods ranging from a few minutes to days. At a number of locations in the oceans, internal tides, or internal waves having the same periodicity as oceanic tides, are prominent.

Internal waves are important to the economy of the sea because they provide one of the few processes that can redistribute kinetic energy from near the surface to abyssal depths. When they break, they can cause turbulent mixing despite the normally stable density gradient in the ocean. Internal waves are known to cause time-varying refraction of acoustic waves because the sound velocity profile in the ocean is distorted by the vertical motions of internal waves. Internal waves have been found by recording fluctuating currents in middepths by moored current meters, by acoustic backscatter Doppler methods, and by studies of the fluctuations of the depths of isotherms as recorded by instruments repeatedly lowered from shipboard or by autonomous instruments floating deep in the water.

Internal waves are thought to be generated in the sea by variations of the wind pressure and stress at the sea surface, by the interaction of surface waves with each other, and by the interaction of tidal motions with the rough sea floor. [C.S.C.]

Oceanic islands Islands rising from the deep sea floor. Oceanic islands range in size from mere specks of rock or sand above the reach of tides to large masses such as Iceland (39,800 mi^2 or 103,000 km^2). Excluded are islands that have continental crust, such as the Seychelles, Norfolk, or Sardinia, even though surrounded by ocean; all oceanic islands surmount volcanic foundations. A few of these have active volcanoes, such as on Hawaii, the Galápagos islands, Iceland, and the Azores, but most islands are on extinct volcanoes. On some islands, the volcanic foundations have subsided beneath sea level, while coral reefs growing very close to sea level have kept pace with the subsidence, accumulating thicknesses of as much as 5000 ft (1500 m) of limestone deposits between the underlying volcanic rocks and the present-day coral islands. *See* REEF; VOLCANO.

Oceanic islands owe their existence to volcanism that began on the deep sea floor and built the volcanic edifices, flow on flow, up to sea level and above. The highest of the oceanic islands is Hawaii, where the peak of Mauna Kea volcano reaches 14,000 ft (4200 m). Most volcanic islands are probably built from scratch in less than 10^6 years, but minor recurrent volcanism may continue for millions of years after the main construction stage. *See* VOLCANOLOGY.

Islands in regions of high oceanic fertility are commonly host to colonies of sea birds, and the deposits of guano have been an important source of phosphate for fertilizer. On some islands, for example, Nauru in the western equatorial Pacific, the original guano has been dissolved and phosphate minerals reprecipitated in porous host limestone rocks. The principal crop on most tropical oceanic islands is coconuts, exploited for their oil content, but some larger volcanic islands, with rich soils and abundant water supplies, are sites of plantations of sugarcane and pineapple. Atoll and barrier-reef islands have very limited water supplies, depending on small lenses of ground water, augmented by collection of rainwater. *See* ATOLL; ISLAND BIOGEOGRAPHY; REEF. [E.L.Wi.]

Oceanography The science of the sea; including physical oceanography, marine chemistry, marine geology, and marine biology. The need to know more about the impact of marine pollution and possible effects of the exploitation of marine resources, together with the role of the ocean in possible global warming and climate change, means that oceanography is an important scientific discipline. Improved understanding of the sea has been essential in such diverse fields as fisheries conservation, the exploitation of underwater oil and gas reserves, and coastal protection policy, as well as in national defense strategies. The scientific benefits include not only improved understanding of the oceans and their inhabitants, but important information about the evolution of the Earth and its tectonic processes, and about the global environment and climate, past and present, as well as possible future changes. *See* CLIMATE HISTORY; MARINE SEDIMENTS; MARITIME METEOROLOGY.

The traditional basis of modern oceanography is the hydrographic station. Hydrographic studies are still carried out at regular intervals, with the research vessel in a specific position. Seawater temperature, depth, and salinity can be measured continuously by a probe towed behind the ship. The revolution in electronics has provided not only a new generation of instruments for studying the sea but also new ways of collecting and analyzing the data they produce. Computers are employed in gathering and processing data in all fields, and are also used in the creation of mathematical models to aid in understanding. Much information can also be gained by remote sensing using satellites, which are also a valuable navigational aid. These provide data on sea surface temperature and currents, and on marine productivity. Satellite altimetry gives information on wave height and winds and even bottom topography (because this affects sea level). Deep-sea cameras and submersibles now permit visual evidence of creatures in remote depths. *See* HYDROGRAPHY; MARINE ECOLOGY; SEAWATER.

Since the early 1900s, all recorded ocean depths have been incorporated in the General Bathymetric Chart of the Ocean. The amount of data available increased greatly with the introduction of continuous echo sounders; subsequently, side-scan sonar permitted very detailed topographical surveys to be made of the ocean floor. The features thus revealed, in particular the midocean ridges (spreading centers) and deep trenches (subduction zones), are integral to the theory of plate tectonics. An important discovery made toward the end of the twentieth century was the existence of hydrothermal vents, where hot mineral-rich water gushes from the Earth's interior. The deposition of minerals at these sites and the discovery of associated ecosystems make them of potential economic as well as great scientific interest. *See* MARINE GEOLOGY; MID-OCEANIC RIDGE. [M.De.]

Oil field waters Waters of varying mineral content which are found associated with petroleum and natural gas or have been encountered in the search for oil and gas. They are also called oil field brines, or brines. They include a variety of underground waters, usually deeply buried, and have a relatively high content of dissolved mineral matter. These waters may be (1) present in the pore space of the reservoir rock with the oil or gas, (2) separated by gravity from the oil or gas and thus lying below it, (3) at the edge of the oil or gas accumulation, or (4) in rock formations which are barren of oil and gas. Brines are commonly defined as water containing high concentrations of dissolved salts. Potable or fresh waters usually are not considered oil field waters but may be encountered, generally at shallow depths, in areas where oil and gas are produced.

Probably the most important geological use of oil field water analyses is their application to the quantitative interpretation of electrical and neutron well logs, particularly micrologs. *See* PETROLEUM GEOLOGY. [P.McG.]

Oligochaeta A class of the phylum Annelida including worms such as the earthworms. There are 21 families with over 3000 species. These animals exhibit both external and internal segmentation. They usually possess setae which are not borne on parapodia. Oligochaetes are hermaphroditic. The gonads are few in number and situated in the anterior part of the body, the male gonads being anterior to the female gonads. The gametes are discharged through special ducts, the oviducts and sperm ducts. A clitellum is present at maturity. There is no larval stage during development.

The oligochaetes are primarily fresh-water and burrowing terrestrial animals. A few are marine and several species occur in the intertidal zone.

Oligochaetes are cylindrical, elongated animals with the anterior mouth usually overhung by a fleshy lobe, the prostomium, and the anus terminal. The body plan is that of a tube within a tube. Externally, the segments are marked by furrows. The setae or bristles are borne on most segments. Other external features are the pores of the reproductive systems opening on certain segments, the openings of the nephridia, and in many earthworms dorsal pores which open externally from the coelom. Some aquatic species have extensions of the posterior part of the body which function as gills.

The oligochaetes have been used in studies of physiology, regeneration, and metabolic gradients. Some aquatic forms are important in studies of stream pollution as indicators of organic contamination. Earthworms are important in turning over the soil and reducing vegetable material into humus. It is likely that fertile soil furnishes a suitable habitat for earthworms, rather than being a result of their activity. *See* ANNELIDA. [P.C.H.]

Oncology The study of cancer. There are five major areas of oncology: etiology, prevention, biology, diagnosis, and treatment. As a clinical discipline, it draws upon a wide variety of medical specialties; as a research discipline, oncology also involves specialists in many areas of biology and in a variety of other scientific areas. Oncology has led to major progress in the understanding not only of cancer but also of normal biology.

Cancer defies simple definition. It is a disease that develops when the orderly relationship of cell division and cell differentiation becomes disordered. In cancer, dividing cells seem to lose the capacity to differentiate, and they acquire the ability to invade through basement membranes and spread (metastasize) to many areas of the body through the bloodstream or lymphatics. Cancer is usually clonal, that is, it develops initially in a single cell. That abnormal cell then produces progeny that may behave rather heterogeneously. Some progeny continue to divide, some develop the capacity to metastasize, and some develop resistance to therapeutic agents. This single cell and its progeny, if unchecked, typically lead to the death of the host. *See* CANCER (MEDICINE).

Causes of cancer. Cancer is generally thought to result from one or more permanent genetic changes in a cell. In some cells a single mutational event can lead to neoplastic transformation, but for most tumors it appears that carcinogenesis is a multistep process. Although some rare congenital conditions lead to cancer in infancy, the vast majority of human cancers arise as a result of the complex interplay between genetic and environmental factors. Without question, there are forms of cancer clearly related to particular environmental exposures; it is equally clear, however, that these factors act on a genetic substrate that may be either susceptible or resistant to the development of cancer.

The emergence of cancer appears to involve the accumulation of genetic damage in a target tissue. Such complex genetic changes specific to tissues appear to underlie the progression to cancer. Such multistep progression is quite complicated to study in experimental systems. Much work has focused on the identification, isolation, and characterization of oncogenes, which have the ability to transform normal cells into cancer cells. More than 50 bona fide or putative oncogenes have been characterized and mapped throughout the human genome. *See* HUMAN GENETICS.

Environmental factors involved in the development of cancers can be chemical, physical, or biological carcinogenic agents. At least three stages occur in the natural history of cancer development from environmental factors. The first stage is initiation, which is a specific alteration in the deoxyribonucleic acid (DNA) of a target cell; environmental agents may act by inducing expression of oncogenes. The second phase, promotion, involves the reversible stimulation of expansion of the initiated cell or the reversible alteration of gene expression in that cell or its progeny. Because promotion is thought to be reversible, it is a target for prevention. The final phase of carcinogenesis is progression. It is characterized by the development of aneuploidy and clonal variation in the tumor; these in turn result in invasiveness and metastasis. *See* MUTAGENS AND CARCINOGENS.

Cancer prevention. An obvious starting point for cancer prevention is avoidance of environmental agents that contribute to carcinogenesis.

The role of diet in cancer prevention is controversial. Epidemologic evidence suggests a particularly strong link between a high-fat, high-calorie, low-fiber diet and an increased risk of colon cancer. But a change to a low-fat, low-calorie, high-fiber diet may not alter the risk. The addition to the diet of carotenoids, selenium, vitamins A, D, and E, and some short-chain fatty acids may prevent cancers in high-risk populations, but there is no evidence that any dietary supplement will prevent cancer.

There are a variety of clinical settings in which surgery may prevent cancer. For example, surgical removal of the thyroid will prevent medullary carcinoma in individuals with certain types of multiple endocrine neoplasia, breast removal can be preventive in familial breast cancer, and removal of the ovaries can prevent cancer in familial ovarian cancer.

Cancer biology. The study of cancer biology picks up where cancer etiology leaves off, namely, at the point where the tumor has developed into a clonal cluster of autonomously proliferating cells. The pathological correlate of this stage of tumor development is carcinoma in situ; a condition in which no tissue destruction is evident, but atypical-appearing cancer cells are present at their site of origin. The transition from carcinoma in situ to locally invasive cancer is accompanied by dissolution of the basement membrane, penetration of tumor cells through the membrane and into the supportive tissues, and disruption of the supportive tissues. Expansion of the primary tumor in locally invasive cancer is always accompanied by the development of blood vessels. The tumor cells can also invade regional blood vessels and lymphatics and circulate throughout the body, attaching to endothelium in a distant organ site, inducing retraction of the endothelium, and becoming attached to the endothelial basement membrane. Once attached to the basement membrane, the tumor cells are covered over by the endothelial cells and effectively separated from the flow of blood. Local dissolution of the basement membrane then occurs, allowing the tumor to completely spread into the tissue and reestablish a blood flow in the breached vessel. As it grows, more blood vessel development nourishes the enlarging tumor.

During metastasis, tumor cells must overcome host defenses. They have various mechanisms to do so. For example, they produce new cell surface receptors to facilitate basement membrane and matrix binding; make new enzymes such as collagenases, serine proteases, metalloproteinases, cysteine proteinases, and endoglycosidases to facilitate their invasiveness; and secrete motility factors to enable them to move through the holes and pathways created by their enzymes. They avoid detection by the immune system through a variety of techniques. Unlike animal tumors, most human tumors are poorly immunogenic. Tumor cells often produce factors that are immunosuppressive. *See* TUMOR.

An unexplained feature of metastasis is the propensity of certain tumor types to spread to specific organs.

Tumor detection. There are two major strategies to detect tumors at the earliest possible stage in their history: responding to the seven warning signals of cancer and screening populations at high risk. The seven danger signals of cancer are (1) unusual bleeding or discharge, (2) a lump or thickening in the breast or elsewhere, (3) a sore that does not heal, (4) change in bowel or bladder habits, (5) persistent hoarseness or cough, (6) persistent indigestion or difficulty in swallowing, and (7) change in a wart or mole.

Diagnosis. The diagnosis of cancer depends on the careful examination of biopsy material. Cancers arising in tissues having ectodermal or endodermal origins are generally called carcinomas; those derived from glands are called adenocarcinomas. Cancers arising in tissues derived from mesoderm are called sarcomas; those of lymphohematopoietic origin are lymphomas and leukemias. The cardinal microscopic features of malignancy are anaplasia, invasion, and metastasis.

Once a diagnosis of cancer is made, it is critical to determine the extent to which the disease has spread. This is called staging. It is distinct from grading, which is an assessment of histologic atypia performed with a microscope. Staging entails performing a careful physical examination, various radiographic studies, and perhaps surgical procedures (biopsies, endoscopies) to examine those sites to which a particular tumor type is most likely to spread. For example, patients with breast cancer often undergo evaluation of the liver, brain, and bones to search for metastatic disease, whereas patients with lymphoma generally require assessment of lymph node groups, bone marrow, and liver. Often the results of such staging tests determine the nature and extent of therapy.

Treatment. There are four major approaches to cancer treatment: surgery, radiation therapy, chemotherapy, and biological therapy. These modalities are often used together with additive or synergistic effects. Surgery and radiation therapy are most effective in curing localized tumors and together result in the cure of about 40% of all newly diagnosed cases. Once the cancer has spread to regional nodes or distant sites, it is generally incurable with the use of local therapies alone. Systemic administration of a combination of chemotherapeutic agents may cure another 10–15% of all patients.

Relieving the symptoms of cancer and alleviating the side effects of agents used to treat it is another important aspect of treatment. Many agents and interventions are available for these purposes. Pharmacologic agents can control nausea and vomiting. Various strategies are available to control pain, improve appetite, and combat insomnia and mood changes. Surgical procedures and radiological techniques can palliate many of the complications of cancer that formerly were incapacitating. Even when the hope for a cure has dwindled, the oncologist can relieve much suffering. [D.L.L.]

Opportunistic infections Infections that cause a disease only when the host's immune system is impaired. The classic opportunistic infection never leads to disease in the normal host. The protozoon *Pneumocystis carinii* infects nearly everyone at some point in life but never causes disease unless the immune system is severely depressed. The most common immunologic defect associated with pneumocystosis is acquired immune deficiency syndrome (AIDS). *See* ACQUIRED IMMUNE DEFICIENCY SYNDROME (AIDS).

A compromised host is an individual with an abnormality or defect in any of the host defense mechanisms that predisposes that person to an infection. The altered defense mechanisms or immunity can be either congenital, that is, occurring at birth and genetically determined, or acquired. Congenital immune deficiencies are relatively rare. Acquired immunodeficiencies are associated with a wide variety of conditions such as (1) the concomitant presence of certain underlying diseases such as cancer, diabetes, cystic fibrosis, sickle cell anemia, chronic obstructive lung disease, severe burns, and cirrhosis of the liver; (2) side effects of certain medical therapies and drugs such as corticosteroids, prolonged antibiotic usage, anticancer agents, alcohol, and nonprescribed recreational drugs; (3) infection with immunity-destroying microorganisms such as the human immunodeficiency virus that leads to AIDS; (4) age, both old and young; and (5) foreign-body exposure, such as occurs in individuals with prosthetic heart valves, intravenous catheters, and other indwelling prosthetic devices.

Virtually any microorganism can become an opportunist. The typical ones fall into a number of categories and may be more likely to be associated with a specific immunologic defect. Examples include (1) gram-positive bacteria: both *Staphylococcus aureus* and the coagulase-negative *S. epidermidis* have a propensity for invading the skin and as well as catheters and other foreign implanted devices; (2) gram-negative bacteria: the most common is *Escherichia coli* and the most lethal is *Pseudomonas aeruginosa*; these pathogens are more likely to occur in cases of granulocytopenia (granulocyte deficiency, as occurs in leukemia or chemotheraphy; (3) acid-fast bacteria: *Mycotuberculum tuberculosis* is more likely to reactivate in the elderly and in those individuals with underlying malignancies and AIDS; (4) protozoa: defects in cell-mediated immunity, such as AIDS, are associated with reactivated infection with *Toxoplasma gondii* and *Cryptosporidium*; (5) fungi: *Cryptococcus neoformans* is a fungus that causes meningitis in individuals with impaired cell-mediated immunity such as AIDS, cancer, and diabetes; *Candida albicans* typically causes blood and organ infection in individuals with granulocytopenia. *See* ESCHERICHIA; MEDICAL MYCOLOGY; STAPHYLOCOCCUS; TUBERCULOSIS.

The first step in treatment of opportunistic infections involves making the correct diagnosis, which is often difficult as many of the pathogens can mistakenly be thought of as benign. The second step involves administration of appropriate antimicrobial agents. As a third step, if possible, the underlying immune defect needs to be corrected. *See* INFECTION; MEDICAL BACTERIOLOGY. [R.Mur.]

Organic evolution The modification of living organisms during their descent, generation by generation, from common ancestors. Organic, or biological, evolution is to be distinguished from other phenomena to which the term evolution is often applied, such as chemical evolution, cultural evolution, or the origin of life from nonliving matter. Organic evolution includes two major processes: anagenesis, the alteration of the genetic properties of a single lineage over time; and cladogenesis, or branching, whereby a single lineage splits into two or more distinct lineages that continue to change anagenetically.

Anagenesis consists of change in the genetic basis of the features of the organisms that constitute a single species. Populations in different geographic localities are considered members of the same species if they can exchange members at some rate and hence interbreed with each other, but unless the level of interchange (gene flow) is very high, some degree of genetic difference among different populations is likely to develop. The changes that transpire in a single population may be spread to other populations of the species by gene flow. *See* SPECIES CONCEPT.

Almost every population harbors several different alleles at each of a great many of the gene loci; hence many characteristics of a species are genetically variable. All genetic variations ultimately arise by mutation of the genetic material. Broadly defined, mutations include changes in the number or structure of the chromosomes and changes in individual genes, including substitutions of individual nucleotide pairs, insertion and deletion of nucleotides, and duplication of genes. Many such mutations alter the properties of the gene products (ribonucleic acid and proteins) or the timing or tissue localization of gene action, and consequently affect various aspects of the phenotype (that is, the morphological and physiological characteristics of an organism). Whether and how a mutation is phenotypically expressed often depends on developmental (epigenetic) events. *See* GENE; GENETIC CODE; MUTATION.

Natural selection is a consistent difference in the average rate at which genetically different entities leave descendants to subsequent generations; such a difference arises from differences in fitness (that is, in the rate of survival, reproduction, or both). In fact, a good approximate measure of the strength of natural selection is the difference between two such entities in their rate of increase. The entities referred to are usually different alleles at a locus, or phenotypically different classes of individuals in the population that differ in genotype. Thus selection may occur at the level of the gene, as in the phenomenon of meiotic drive, whereby one allele predominates among the gametes produced by a heterozygote. Selection at the level of the individual organism, the more usual case, entails a difference in the survival and reproductive success of phenotypes that may differ at one locus or at more than one locus. As a consequence of the difference in fitness, the proportion of one or the other allele increases in subsequent generations. The relative fitness of different genotypes usually depends on environmental conditions.

Different alleles of a gene that provides an important function do not necessarily differ in their effect on survival and reproduction; such alleles are said to be neutral. The proportion of two neutral alleles in a population fluctuates randomly from generation to generation by chance, because not all individuals in the population have the same number of surviving offspring. Random fluctuations of this kind are termed random genetic drift. If different alleles do indeed differ in their effects on fitness, both genetic drift and natural selection operate simultaneously. The deterministic force of natural selection drives allele frequencies toward an equilibrium, while the stochastic (random) force of genetic drift brings them away from that equilibrium. The outcome for any given population depends on the relative strength of natural selection (the magnitude of differences in fitness) and of genetic drift (which depends on population size).

The great diversity of organisms has come about because individual lineages (species) branch into separate species, which continue to diverge. This splitting process, speciation, occurs when genetic differences develop between two populations that prevent them from interbreeding and forming a common gene pool. The genetically based characteristics that cause such reproductive isolation are usually termed isolating mechanisms. Reproductive isolation seems to develop usually as a fortuitous by-product of genetic divergence that occurs for other reasons (either by natural selection or by genetic drift). *See* SPECIATION.

A frequent consequence of natural selection is that a species comes to be dominated by individuals whose features equip them better for the environment or way of life of the species. Such features are termed adaptations. Although many features of organisms are adaptive, not all are, and it is a serious error to suppose that species are capable of attaining ideal states of adaptation. Some characteristics are likely to have developed by genetic drift rather than natural selection, and so are not adaptations; others are side effects of adaptive features, which exist because of pleiotropy or developmental correlations.

Higher taxa are those above the species level, such as genera and families. A taxon such as a genus is typically a group of species, derived from a common ancestor, that share one or more features so distinctive that they merit recognition as a separate taxon. The degree of difference necessary for such recognition, however, is entirely arbitrary: there are often no sharp limits between related genera, families, or other higher taxa, and very often the diagnostic character exists in graded steps among a group of species that may be arbitrarily divided into different higher taxa. Moreover, a character that in some groups is used to distinguish higher taxa sometimes varies among closely related species or even within species. In addition, the fossil record of many groups shows that a trait that takes on very different forms in two living taxa has developed by intermediate steps along divergent lines from their common ancestor; thus the inner ear bones of mammals may be traced to jaw elements in reptiles that in turn are homologous to gill arch elements in Paleozoic fishes.

The characteristics of a species evolve individually or in concert with certain other traits that are developmentally or functionally correlated. Because of this mosaic pattern of evolution, it is meaningful to speak of the rate of evolution of characters, but not of species or lineages as total entities. Thus in some lineages, such as the so-called living fossils, many aspects of morphology have evolved slowly since the groups first came into existence, but evolution of their deoxyribonucleic acid and amino acid sequences has proceeded at much the same rate as in other lineages. Every species, including the living fossils, is a mixture of traits that have changed little since the species' remote ancestors, and traits that have undergone some evolutionary change in the recent past. The history of life is not one of progress in any one direction, but of adaptive radiation on a grand scale: the descendants of any one lineage diverge as they adapt to different resources, habitats, or ways of life, acquiring their own specialized features as they do so. There is no evidence that evolution has any goal, nor does the mechanistic theory of evolutionary processes admit of any way in which genetic change can have a goal or be directed toward the future. However, for life taken as a whole, the only clearly discernible trend is toward ever-increasing diversity. [D.J.Fu.]

Organic geochemistry The study of the abundance and composition of naturally occurring organic substances, their origins and fate, and the processes that affect their distributions on Earth and in extraterrestrial materials. These activities share the common need for identification, measurement, and assessment of organic matter in its myriad forms.

Organic geochemistry was born from a curiosity about the organic pigments extractable from petroleum and black shales. It developed with extensive investigations of the chemical characteristics of petroleum and petroleum source rocks as clues to their occurrence and formation, and now encompasses a broad scope of activities within interdisciplinary areas of earth and environmental science. This range of studies recognizes the potential of geological records of organic matter to help characterize sedimentary depositional environments and to provide evidence of ancient life and indications of evolutionary developments through the Earth's history. Organic geochemistry includes determinations of anthropogenic contaminants amid the natural background of organic molecules and the assessment of their environmental impact and fate. Marine organic geochemistry addresses and interprets aquatic processes involving carbon species. It involves investigations of the chemical character of particulate and dissolved organic matter, evaluation of oceanic primary production including the factors (light, temperature, nutrient availability) that influence the uptake of carbon dioxide (CO_2), the composition of marine organisms, and the subsequent processing of organic constituents through the food web. Organic geochemistry extends to broader biogeochemical issues, such as the carbon cycle, and the effects of changing carbon dioxide levels, especially efforts to use geochemical data and proxies to help constrain global climate models. Examination of the organic chemistry of meteorites and lunar materials also falls within its compass, and as a critical part of the quest for remnants of life on Mars, such extraterrestrial studies are now regaining the prominence they held in the 1970s during lunar exploration. *See* GEOCHEMISTRY.

Global inventories of carbon. Carbon naturally exists as oxidized and reduced forms in carbonate carbon and organic matter. The major reservoir of both forms of carbon on Earth is the geosphere. It contains carbonate minerals deposited as sediments and organic matter accumulated from the remains of dead organisms. Estimates of the size of the geological reservoir of carbon vary within the range of 5 to 7×10^{22} g, of which 75% is carbonate carbon and 25% is organic carbon. The amounts of carbon contained in living biota (5×10^{17} g), dissolved in the ocean (4×10^{19} g), and present in atmospheric gases (7×10^{17} g) are miniscule compared to the quantity of organic carbon buried in the rock record. The importance of buried organic matter extends beyond its sheer magnitude; it includes the fossil fuels—coal, natural gas, and petroleum—that supply 85% of the world's energy. *See* BIOGEOCHEMISTRY; FOSSIL FUEL; NATURAL GAS; PETROLEUM; SEDIMENTARY ROCKS.

Sedimentary organic matter. The vast amounts of organic matter contained in geological materials represent the accumulated vestiges of organisms amassed over the expanse of geological time. Yet, survival of organic cellular constituents of biota into the rock record is the exception rather than the norm. Only a small portion of the carbon fixed by organisms during primary production, especially by photosynthesis, escapes degradation as it settles through the water column and eludes microbial alteration during subsequent incorporation and assimilation into sedimentary detritus. *See* BIODEGRADATION.

Sedimentary organic matter can be divided operationally into solvent-extractable bitumen and insoluble kerogen. Bitumens contain a myriad of structurally distinct molecules, especially hydrocarbons, which can be individually identified (such as by gas chromatography-mass spectrometry) although they may be present in only minute quantities (nanograms or picograms). The range of components includes many biomarkers that retain structural remnants inherited from their source organisms, which attest to their biological origins and subsequent geological fate.

Biomarkers are individual compounds whose chemical structures carry evidence of their origins and history. Recognition of the specificity of biomarker structures initially helped confirm that petroleum was derived from organic matter produced by biological processes. Of the thousands of individual petroleum components, hundreds reflect precise biological sources of organic matter, which distinguish and differentiate their disparate origins. The diagnostic suites of components may derive from individual families of organisms, but contributions at a species level can occasionally be recognized. Biomarker abundances and distributions help to elucidate sedimentary environments, providing evidence of depositional settings and conditions. They also reflect sediment maturity, attesting to the progress of the successive, sequential transformations that convert biological precursors into geologically occurring products. Thus, specific biomarker characteristics permit assessment of the thermal history of individual rocks or entire sedimentary basins. *See* BASIN.

Carbon isotopes. Carbon naturally occurs as three isotopes: carbon-12 (^{12}C), carbon-13 (^{13}C), and radiocarbon (^{14}C). Temporal excursions in the ^{13}C values of sediment sequences can reflect perturbations of the global carbon cycle. Radiocarbon is widely employed to date archeological artifacts, but the sensitivity of its measurement also permits its use in exploration of the rates of biogeochemical cycling in the oceans. This approach permits assessment of the ages of components in sediments, demonstrating that bacterial organic matter is of greater antiquity than components derived from phytoplankton sources. *See* MARINE SEDIMENTS; RADIOCARBON DATING. [S.C.B.]

P

Pacific islands A geographic designation that includes thousands of mainly small coral and volcanic islands scattered across the Pacific Ocean from Palau in the west to Easter Island in the east. Island archipelagos off the coast of the Asian mainland, such as Japan, Philippines, and Indonesia, are not included even though they are located within the Pacific Basin. The large island constituting the mainland of Papua New Guinea and Irian Jaya is also excluded, along with the continent of Australia and the islands that make up Aotearoa or New Zealand. The latter, together with the Asian Pacific archipelagos, contain much larger landmasses, with a greater diversity of resources and ecosystems, than the oceanic islands, commonly labelled Melanesia, Micronesia, and Polynesia. *See* AUSTRALIA; NEW ZEALAND; OCEANIC ISLANDS.

The great majority of these islands are between 4 and 4000 mi^2 (10 and 10,000 km^2) in land surface area. The three largest islands include the main island of New Caledonia (6220 mi^2 or 16,100 km^2), Viti Levu (4053 mi^2 or 10,497 km^2) in Fiji, and Hawaii (4031 mi^2 or 10,440 km^2) the big island in the Hawaiian chain. When the 80-mi (200-km) Exclusive Economic Zones are included in the calculation of surface area, some Pacific island states have very large territories. These land and sea domains, far more than the small, fragmented land areas per se, capture the essence of the island world that has meaning for Pacific peoples.

Oceanic islands are often classified on the basis of the nature of their surface lithologies. A distinction is commonly made between the larger continental islands of the western Pacific, the volcanic basalt island chains and clusters of the eastern Pacific, and the scattered coral limestone atolls and reef islands of the central and northern Pacific.

It has been suggested that a more useful distinction can be drawn between plate boundary islands and intraplate islands. The former are associated with movements along the boundaries of the great tectonic plates that make up the Earth's surface. Islands of the plate boundary type form along the convergent, divergent, or tranverse plate boundaries, and they characterize most of the larger island groups in the western Pacific. These islands are often volcanically and tectonically active and form part of the Pacific so-called Ring of Fire, which extends from Antarctica in a sweeping arc through New Zealand, Vanuatu, Bougainville, and the Philippines to Japan.

The intraplate islands comprise the linear groups and clusters of islands that are thought to be associated with volcanism, either at a fixed point or along a linear fissure. Volcanic island chains such as the Hawaii, Marquesas, and Tuamotu groups are classic examples. Others, which have their volcanic origins covered by great thickness of coral, include the atoll territories of Kiribati, Tuvalu, and the Marshall Islands. Another type of intraplate island is isolated Easter Island, possibly a detached piece of a mid-ocean ridge. The various types of small islands in the Pacific are all linked geologically to much larger structures that lie below the surface of the sea. These structures contain the answers to some puzzles about island origins and locations, especially when considered in terms of the plate tectonic theory of crustal evolution. *See* MARINE GEOLOGY; MID-OCEANIC RIDGE; SEAMOUNT AND GUYOT; VOLCANO.

The climate of most islands in the Pacific is dominated by two main forces: ocean circulation and atmospheric circulation. Oceanic island climates are fundamentally distinct from those of continents and islands close to continents, because of the small size of the island relative to the vastness of the ocean surrounding it. Because of oceanic influences, the climates of most small, tropical Pacific islands are characterized by little variation through the year compared with climates in continental areas.

The major natural hazards in the Pacific are associated either with seasonal climatic variability (especially cyclones and droughts) or with volcanic and tectonic activity. *See* CLIMATE HISTORY. [R.D.Be.]

Pacific Ocean The Pacific Ocean has an area of 6.37×10^7 mi^2 (1.65×10^8 km^2) and a mean depth of 14,000 ft (4280 m). It covers 32% of the Earth's surface and 46% of the surface of all oceans and seas, and its area is greater than that of all land areas combined. Its mean depth is the greatest of the three oceans and its volume is 53% of the total of all oceans. Its greatest depths in the Marianas and Japan trenches are the world's deepest, more than 6 mi (10 km).

The two major wind systems driving the waters of the ocean are the westerlies which lie about 40–50° lat in both hemispheres (the "roaring forties") and the trade winds from the east which dominate in the region between 20°N and 20°S. These give momentum directly to the west wind drift (flow to the east) in high latitudes and to the equatorial currents which flow to the west. At the continents there is flow of water from one system to the other and huge circulatory systems result. *See* OCEAN CIRCULATION; SOUTHEAST ASIAN WATERS.

The swiftest flow (greater than 2 knots) is found in the Kuroshio Current near Japan. It forms the northwestern part of a huge clockwise gyre whose north edge lies in the west wind drift centered at about 40°N, whose eastern part is the south-flowing California Current, and whose southern part is the North Equatorial Current.

Equatorward of 30° lat heat received from the Sun exceeds that lost by reflection and back radiation, and surface waters flowing into these latitudes from higher latitudes (California and Peru currents) increase in temperature as they flow equatorward and turn west with the Equatorial Current System. They carry heat poleward and transfer part of it to the high-latitude cyclones along the west wind drift. The temperature of the equatorward currents along the eastern boundaries of the subtropical anticyclones is thus much lower than that of the currents of their western boundaries at the same latitudes. The highest temperatures (more than 82°F or 28°C) are found at the western end of the equatorial region. Along the Equator itself somewhat lower temperatures are found. The cold Peru Current contributes to its eastern end, and there is apparent upwelling of deeper, colder water at the Equator.

Upwelling also occurs at the edge of the eastern boundary currents of the subtropical anticyclones. When the winds blow strongly equatorward (in summer) the surface waters are driven offshore, and the deeper colder waters rise to the surface and further reduce the low temperatures of these equatorward-flowing currents. *See* UPWELLING.

The limiting temperature in high latitudes is that of freezing. Ice is formed at the surface at temperatures slightly less than 30°F (−1°C) depending upon the salinity; further loss of heat is retarded by its insulating effect. The ice field covers the northern and eastern parts of the Bering Sea in winter, and most of the Sea of Okhotsk, including that part adjacent to Hokkaido (the north island of Japan). Summer temperatures, however, reach as high as 43°F (6°C) in the northern Bering Sea and as high as 50°F (10°C) in the northern part of the Sea of Okhotsk. *See* BERING SEA.

Pack ice reaches to about 62°S from Antarctica in October and to about 70°S in March, with icebergs reaching as far as 50°S. *See* ICEBERG; SEA ICE.

Surface waters in high latitudes are colder and heavier than those in low latitudes. As a result, some of the high-latitude waters sink below the surface and spread equatorward, mixing mostly with water of their own density as they move, and eventually become the dominant water type in terms of salinity and temperature of that density over vast regions.

The most conspicuous water masses formed in the Pacific are the Intermediate Waters of the North and of the South Pacific, which on the vertical sections include the two huge tongues of low salinity extending equatorward beneath the surface from about 55°S and from about 45°N. The southern tongue is higher in salinity and density and lies at a greater depth. [J.L.Re.]

Paleoclimatology The study of ancient climates. Climate is the long-term expression of weather; in the modern world, climate is most noticeably expressed in vegetation and soil types and characteristics of the land surface. To study ancient climates, paleoclimatologists must be familiar with various disciplines of geology, such as sedimentology and paleontology, and with climate dynamics, which includes aspects of geography and atmospheric and oceanic physics. Understanding the history of the Earth's climate system greatly enhances the ability to predict how it might behave in the future. *See* CLIMATOLOGY.

Information about ancient climates comes principally from three sources: sedimentary deposits, including ancient soils; the past distribution of plants and animals; and the chemical composition of certain marine fossils. These are all known as proxy indicators of climate (as opposed to direct indicators, such as temperature, which cannot be measured in the past). In addition, paleoclimatologists use computer models of climate that have been modified for application to ancient conditions.

Like modern climatologists, paleoclimatologists are concerned with boundary conditions, forcing, and response. Boundary conditions are the limits within which the climate system operates. The boundary conditions considered by paleoclimatologists depend on the part of Earth history that is being studied. For the recent past, that is, the last few million years, boundary conditions that can change on short time scales are considered, for example, atmospheric chemistry. For the more distant past, paleoclimatologists must also consider boundary conditions that change on long time scales. Geographic features—that is, the positions of the continents, the location and orientation of major mountain ranges, the positions of shorelines, and the presence or absence of epicontinental seaways—are important for understanding paleoclimatic patterns. Forcing is a change in boundary conditions, such as continental drift, and response is how forcing changes the climate system. Forcing and response are cause and effect in paleoclimatic change.

Proxy indicators of paleoclimate are abundant in the geologic record. Important sedimentary indicators forming on land are coal, eolian sandstone (ancient sand dunes), evaporites (salt), tillites (ancient glacial deposits), and various types of paleosols (ancient soils), such as bauxite (aluminum ore) and laterite (some iron ores). Coals may form where conditions are favorable for growth of plants and accumulation and preservation of peat, conditions that are partly controlled by climate, especially seasonality of rainfall.

Fossil indicators provide information about climate mostly by their distribution (paleobiogeography), although a few specific types of fossils may be indicative of certain climatic conditions. The latter are usually fossils from the younger part of the geologic record and are closely related to modern species that have narrow environmental tolerances. Another type of information available for documenting paleoclimatic patterns and change is stable isotope geochemistry of fossils and certain types of sedimentary rock. Many elements that are used by organisms to make shells, teeth, and stems occur naturally in several different forms, known as isotopes. The most climatically useful isotopes are those of oxygen (O). Although the effects of temperature change and ice volume change can be difficult to distinguish, the analysis of oxygen isotopes has provided a powerful quantitative tool for the study of both long-term temperature change and the history of the polar ice caps.

A great deal of research in paleoclimatology has been devoted to understanding the causes of climatic change, and the overriding conclusion is that any given shift in the paleoclimatic history of the Earth was brought about by multiple factors operating in concert. The most important forcing factors for paleoclimatic variation are changes in paleogeography and atmospheric chemistry and variations in the Earth's orbital parameters. *See* ATMOSPHERIC CHEMISTRY; BIOGEOCHEMISTRY. [J.T.P.; E.J.Ba.]

Parainfluenza virus A member of the genus *Paramyxovirus* of the family Paramyxoviridae which is associated with a variety of respiratory illnesses. The virus particles range in size from 90 to 200 nanometers, agglutinate red blood cells, and (like the influenza viruses) contain a receptor-destroying enzyme. They differ from the influenza viruses in their large size, their possession of the larger ribonucleoprotein helix characteristic of the paramyxoviruses, their tendency to lyse as well as agglutinate erythrocytes, and their generally poor growth in eggs. *See* PARAMYXOVIRUS.

Four subgroups are known, designated parainfluenza 1, 2, 3, and 4. Types 1, 2, and 3 are distributed throughout the world, but thus far type 4 has been found only in the United States. Parainfluenza 1 and 3 are ubiquitous endemic agents producing infections all through the year. Types 2 and 4 occur more sporadically. With all of the parainfluenza viruses, most primary infections take place early in life. About half of the first infections with parainfluenza 1, about two-thirds of those with parainfluenza 2, and three-fourths of those with parainfluenza 3 produce febrile illnesses. The target organ of type 3 is the lower respiratory tract, with first infections frequently resulting in bronchial pneumonia, bronchiolitis, or bronchitis. Type 1 is the chief cause of croup, but the other types have also been incriminated, to the extent that one-half of all cases of croup can be shown to be caused by parainfluenza viruses. *See* ANIMAL VIRUS; INFLUENZA. [J.L.Me.; M.E.Re.]

Paramo A biological community, essentially a grassland, covering extensive high areas in equatorial mountains of the Western Hemisphere. Geographically, paramos are limited to the Northern Andes and adjacent mountains. Paramos occur in alpine regions above timberline and are controlled by a complex of climatic and soil factors peculiar to mountains near the Equator. The richly diverse flora and the fauna of the paramos are adapted to severely cold, mostly wet conditions. Humans have found some paramos suitable for living and use. [H.G.B.]

Paramyxovirus A subgroup of myxoviruses that includes the viruses of mumps, measles, parainfluenza, respiratory syncytial (RS) disease, and Newcastle disease. Like influenza viruses, the paramyxoviruses are ribonucleic acid (RNA)-containing viruses and possess an ether-sensitive lipoprotein envelope. *See* ANIMAL VIRUS; MEASLES; MUMPS; NEWCASTLE DISEASE; PARAINFLUENZA VIRUS. [J.L.Me.]

Parenchyma A ground tissue chiefly concerned with the manufacture and storage of food. The primary functions of plants, such as photosynthesis, assimilation, respiration, storage, secretion, and excretion—those associated with living protoplasm—proceed mainly in parenchymal cells. Parenchyma is frequently found as a homogeneous tissue in stems, roots, leaves, and flower parts. Other tissues, such as sclerenchyma, xylem, and phloem, seem to be embedded in a matrix of parenchyma; hence the use of the term ground tissue with regard to parenchyma is derived. The parenchymal cell is one of the most frequently occurring cell types in the plant kingdom. *See* PLANT ANATOMY; PLANT PHYSIOLOGY.

Typical parenchyma occurs in pith and cortex of roots and stems as a relatively undifferentiated tissue composed of polyhedral cells that may be more or less compactly arranged and show little variation in size or shape. The mesophyll, that is, the tissue located between the upper and lower epidermis of leaves, is a specially differentiated parenchyma called chlorenchyma because its cells contain chlorophyll in distinct chloroplastids.

This chlorenchymatous tissue is the major locus of photosynthetic activity and consequently is one of the more important variants of parenchyma. Specialized secretory parenchymal cells are found lining resin ducts and other secretory structures. *See* PHOTOSYNTHESIS. [R.L.Hu.]

Pasteurellosis A variety of infectious diseases caused by the coccobacilli *Pasteurella multocida* and *P. haemolytica*; the term also applies to diseases caused by any *Pasteurella* species. All *Pasteurella* species occur as commensals in the upper respiratory and alimentary tracts of their various hosts. Although varieties of some species cause primary disease, many of the infections are secondary to other infections or result from various environmental stresses. *Pasteurella* species are generally extracellular parasites that elicit mainly a humoral immune response. Several virulence factors have been identified. *See* VIRULENCE.

Pasteurella multocida is the most prevalent species of the genus causing a wide variety of infections in many domestic and wild animals, and humans. It is a primary or, more frequently, a secondary pathogen of cattle, swine, sheep, goats, and other animals. As a secondary invader, it is often involved in pneumonic pasteurellosis of cattle (shipping fever) and in enzootic or mycoplasmal pneumonia of swine. It is responsible for a variety of sporadic infections in many animals, including abortion, encephalitis, and meningitis. It produces severe mastitis in cattle and sheep, and toxin-producing strains are involved in atrophic rhinitis, an economically important disease of swine. Hemorrhagic septicemia, caused by capsular type B strains, has been reported in elk and deer in the United States.

All strains of *P. haemolytica* produce a soluble cytotoxin (leukotoxin) that kills various leukocytes of ruminants, thus lowering the primary pulmonary defense. It is the principal cause of the widespread pneumonic pasteurellosis of cattle. Other important diseases caused by certain serotypes of *P. haemolytica* are mastitis of ewes and septicemia of lambs.

All of the *Pasteurella* species can be isolated by culturing appropriate clinical specimens on blood agar. Multiple drug resistance is frequently encountered. Treatment is effective if initiated early. Among the drugs used are penicillin and streptomycin, tetracyclines, chloramphenicol, sulphonamides, and some cephalosporins. Sound sanitary practices and segregation of affected animals may help limit the spread of the major pasteurelloses. Live vaccines and bacterins (killed bacteria) are used for the prevention of some. [G.R.Ca.]

Pathogen Any agent capable of causing disease. The term pathogen is usually restricted to living agents, which include viruses, rickettsia, bacteria, fungi, yeasts, protozoa, helminths, and certain insect larval stages. *See* DISEASE.

Pathogenicity is the ability of an organism to enter a host and cause disease. The degree of pathogenicity, that is, the comparative ability to cause disease, is known as virulence. The terms pathogenic and nonpathogenic refer to the relative virulence of the organism or its ability to cause disease under certain conditions. This ability depends not only upon the properties of the organism but also upon the ability of the host to defend itself (its immunity) and prevent injury. The concept of pathogenicity and virulence has no meaning without reference to a specific host. For example, gonococcus is capable of causing gonorrhea in humans but not in lower animals. *See* MEDICAL MYCOLOGY; MEDICAL PARASITOLOGY; PLANT PATHOLOGY; PLANT VIRUSES AND VIROIDS; VIRULENCE. [D.N.La.]

Pathotoxin A chemical of biological origin, other than an enzyme, that plays an important causal role in a plant disease. Most pathotoxins are produced by plant pathogenic fungi or bacteria, but some are produced by higher plants, and one has been reported to be the product of an interaction between a plant and a bacterial pathogen. Some pathogen-produced pathotoxins are highly selective in that they cause severe damage and typical disease symptoms only on plants susceptible to the pathogens that produce them. Others are nonselective and are equally toxic to plants susceptible or resistant to the pathogen involved. A few pathotoxins are species-selective, and are damaging to many but not all plant species. In these instances, some plants resistant to the pathogen are sensitive to its toxic product. *See* PLANT PATHOLOGY. [H.Wh.]

Peat A dark-brown or black residuum produced by the partial decomposition and disintegration of mosses, sedges, trees, and other plants that grow in marshes and other wet places. Forest-type peat, when buried and subjected to geological influences of pressure and heat, is the natural forerunner of most coal. Moor peat is formed in relatively elevated, poorly drained moss-covered areas, as in parts of Northern Europe. *See* HUMUS. [G.H.C.]

Pediculosis Human infestation with lice. There are two biological varieties of the human louse, *Pediculus humanus*, var. *capitis* and var.*corporis*, each showing a strong preference for a specific location on the human body. *Pediculus humanus capitis* colonizes the head and *P. h. corporis* lives in the body-trunk region.

These lice are wingless insects which are ectoparasites. Their mouthparts are modified for piercing skin and sucking blood. The terminal segments of their legs are modified into clawlike structures which are utilized to grasp hairs and clothing fibers.

Lice are important vectors of human diseases. Their habit of sucking blood and their ability to crawl rapidly from one human to another transmit such diseases as typhus (rickettsial) and epidemic relapsing fever (spirochetal). The body fluids and feces of infected lice transmit these diseases. [R.Su.]

Pedology Defined narrowly, a science that is concerned with the nature and arrangement of horizons in soil profiles; the physical constitution and chemical composition of soils; the occurrence of soils in relation to one another and to other elements of the environment such as climate, natural vegetation, topography, and rocks; and the modes of origin of soils. Pedology so defined does not include soil technology, which is concerned with uses of soils.

Broadly, pedology is the science of the nature, properties, formation, distribution, and function of soils, and of their response to use, management, and manipulation. The first definition is widely used in the United States and less so in other countries. The second definition is worldwide. *See* SOIL; SOIL MECHANICS. [R.W.S.]

Pericycle As commonly defined, the outer boundary of the stele of plants. Originally it was interpreted as a band of cells between the phloem and the innermost layer (endodermis) of the cortex. Such pericycle is commonly found in roots and, in lower vascular plants, also in stems. In higher vascular plants, however, a distinct layer of cells may not be present between the phloem and the cortex. The pericycle, if present, may be composed of parenchyma or sclerenchyma cells with relatively thin or heavily thickened walls. It may be one to several layers in radial dimensions.

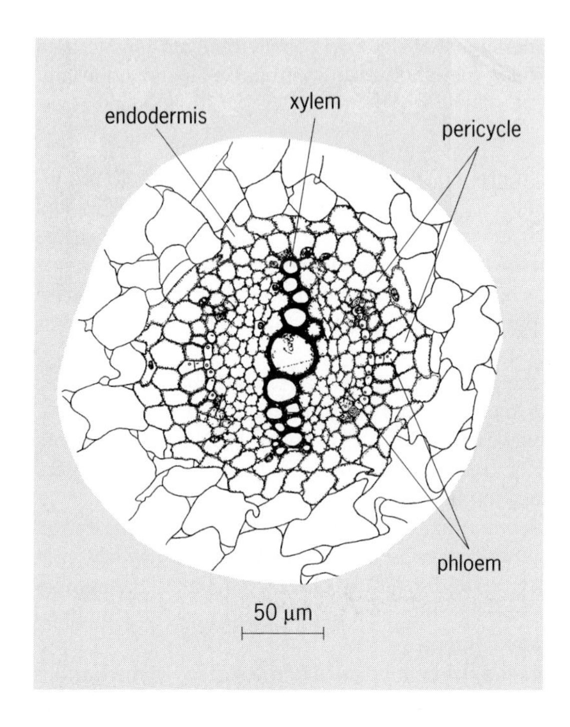

Transection of central part of sugarbeet. (*From K. Esau, Hilgardia, 9(8), 1935*)

Primordia of branch roots commonly arise in the pericycle in seed plants, most frequently outside the xylem ridges (see illustration). The first cork cambium may also arise in the pericycle of those roots that have secondary vascular tissues. In roots, a part of the vascular cambium itself (that outside the primary xylem ridges) originates from pericycle cells. *See* CORTEX (PLANT); ENDODERMIS; PARENCHYMA; PHLOEM; ROOT (BOTANY); SCLERENCHYMA; STEM; XYLEM. [V.I.C.]

Periderm A group of tissues which replaces the epidermis in the plant body. Its main function is to protect the underlying tissues from desiccation, freezing, heat injury, mechanical destruction, and disease. Although periderm may develop in leaves and fruits, its main function is to protect stems and roots. The fundamental tissues which compose the periderm are the phellogen, phelloderm, and phellem.

The phellogen is the meristematic portion of the periderm and consists of one layer of initials. These exhibit little variation in form, appearing rectangular and somewhat flat in cross and radial sections, and polygonal in tangential sections.

The phelloderm cells are phellogen derivatives formed inward. The number of phelloderm layers varies with species, season, and age of the periderm. In some species, the periderm lacks the phelloderm altogether. The phelloderm consists of living cells with photosynthesizing chloroplasts and cellulosic walls.

The phellem, or cork, cells are phellogen derivatives formed outward. These cells are arranged in tiers with almost no intercellular spaces except in the lenticel regions. After completion of their differentiation, the phellem cells die and their protoplasts disintegrate. The cell lumens remain empty, excluding a few species in which various crystals can be found. The remarkable impermeability of the suberized cell walls is largely due to their impregnation with waxes, tannins, cerin, friedelin, and phellonic and phellogenic acids.

Lenticels are loose-structured openings that develop usually beneath the stomata and that facilitate gas transport through the otherwise impermeable layers of phellem. *See* BARK; SCLERENCHYMA. [Y.W.; H.Wi.]

Permafrost Perennially frozen ground, occurring wherever the temperature remains below 32°F (0°C) for several years, whether the ground is actually consolidated by ice or not and regardless of the nature of the rock and soil particles of which the earth is composed. Perhaps 25% of the total land area of the Earth contains permafrost; it is continuous in the polar regions and becomes discontinuous and sporadic toward the Equator. During glacial times permafrost extended hundreds of miles south of its present limits in the Northern Hemisphere.

Temperature of permafrost at the depth of no annual change, about 30–100 ft (10–30 m), crudely approximates mean annual air temperature. It is below 23°F (−5°C) in the continuous zone, between 23–30°F (−5 and −1°C) in the discontinuous zone, and above 30°F (−1°C) in the sporadic zone. Temperature gradients vary horizontally and vertically from place to place and from time to time.

Ice is one of the most important components of permafrost, being especially important where it exceeds pore space. Physical properties of permafrost vary widely from those of ice to those of normal rock types and soil. The cold reserve, that is, the number of calories required to bring the material to the melting point and melt the contained ice, is determined largely by moisture content.

Permafrost develops today where the net heat balance of the surface of the Earth is negative for several years. Much permafrost was formed thousands of years ago but remains in equilibrium with present climates. Permafrost eliminates most groundwater movement, preserves organic remains, restricts or inhibits plant growth, and aids frost action. It is one of the primary factors in engineering and transportation in the polar regions. [R.F.B.]

Pesticide A material useful for the mitigation, control, or elimination of plants or animals detrimental to human health or economy. Algicides, defoliants, desiccants, herbicides, plant growth regulators, and fungicides are used to regulate populations of undesirable plants which compete with or parasitize crop or ornamental plants. Attractants, insecticides, miticides, acaricides, molluscicides, nematocides, repellants, and rodenticides are used principally to reduce parasitism and disease transmission in domestic animals, the loss of crop plants, the destruction of processed food, textile, and wood products, and parasitism and disease transmission in humans.

Some pesticides are obtained from plants and minerals. Examples include the insecticides cryolite, a mineral, and nicotine, rotenone, and the pyrethrins which are extracted from plants. A few pesticides are obtained by the mass culture of microorganisms. Two examples are the toxin produced by *Bacillus thuringiensis*, which is active against moth and butterfly larvae, and the so-called milky disease of the Japanese beetle produced by the spores of *B. popilliae*. Most pesticides, however, are products which are chemically manufactured. Two outstanding examples are the insecticide DDT and the herbicide 2,4-D.

Concern over the undesirable effects of pesticides on nonpest organisms culminated in laws to prevent exposure of either humans or the environment to unreasonable hazard from pesticides through rigorous registration procedures. The purpose of regulations are to classify pesticides for general or restricted use as a function of acute toxicity, to certify the qualifications of users of restricted pesticides, to identify accurately and label pesticide products, and to ensure proper and safe use of pesticides. Recommendations as to the product and method of choice for control of any pest problem—weed, insect, or varmint—are best obtained from county or state agricultural extension specialists. [G.F.L.]

Sophisticated methods of pest control are continually being developed. Highly specific synthetic insect hormones are being developed. In an increasing number of pest situations, a natural predator of an insect has been introduced, or conditions are maintained that favor the propagation of the predator. The numbers of the potential pest species are thereby maintained below a critical threshold. An insect control program in which use of insecticides is only one aspect of a strategy based on ecologically sound measures is known as integrated pest management. *See* AGRICULTURAL CHEMISTRY; CHEMICAL ECOLOGY; FUNGISTAT AND FUNGICIDE; HERBICIDE; INSECT CONTROL, BIOLOGICAL; INSECTICIDE. [R.W.Ri.]

Petroleum Unrefined, or crude, oil is found underground and under the sea floor, in the interstices between grains of sandstone and limestone or dolomite (not in caves). Petroleum is a mixture of liquids varying in color from nearly colorless to jet black, in viscosity from thinner than water to thicker than molasses, and in density from light gases to asphalts heavier than water. It can be separated by distillation into fractions that range from light color, low density, and low viscosity to the opposite extreme. In places where it has oozed from the ground, its volatile fractions have vaporized, leaving the dense, black parts of the oil as a pool of tar or asphalt (such as the Brea Tar Pits in California). Much of the world's crude oil is today produced from drilled wells.

Petroleum consists mostly of hydrocarbon molecules. The four main classes of hydrocarbons are paraffins (also called alkanes), olefins (alkenes), cycloparaffins (cycloalkanes), and aromatics. Olefins are absent in crude oil but can be formed in certain refining processes. The simplest hydrocarbon is one carbon atom bonded to four hydrogen atoms (chemical formula CH_4), and is called methane.

Petroleum usually contains all of the possible hydrocarbon structures except alkenes, with the number of carbon atoms per molecule going up to a hundred or more. These fractions include compounds that contain sulfur, nitrogen, oxygen, and metal atoms. The proportion of compounds containing these atoms increases with increasing size of the molecule.

Asphaltic molecules contain many cyclic compounds in which the rings contain sulfur, nitrogen, or oxygen atoms; these are called heterocyclic compounds. An example is pyridine.

It is generally agreed that petroleum formed by processes similar to those which yielded coal, but was derived from small animals rather than from plants. Dead organisms have been buried in mud over millions of years. Further layers deposited over these mud layers have in some cases reached a thickness of thousands of feet, and compacted the layers beneath them, until the mud has become shale rock. The mud layers were heated and compressed by the layers above. The bodies of the organisms in the mud were decomposed and converted into fatty liquids and solids. Heating these fatty materials over a very long time caused their molecules to break into smaller fragments and combine into larger ones, so the original range of molecular size was spread greatly into the range found in crude oil. Bacteria were usually present, and helped remove oxygen from the molecules and turned them into hydrocarbon compounds. The great pressure of the overlying rock layers helped to force the oil out of the compacted mud (shale) layers into less compacted limestone, dolomite, or sandstone layers next to the shale layers. *See* ORGANIC GEOCHEMISTRY; PETROLEUM GEOLOGY; SEDIMENTOLOGY.

At depths greater than about 25,000 ft (7620 m), the temperature is so high that the oil conversion processes go all the way to natural gas and soot. Natural gas formed by the conversion processes is now also found over a variety of depths which do not indicate the depth and temperature of their origin. *See* NATURAL GAS.

The oil formed by the natural thermal and bacterial processes was squeezed out of the compacting mud layers into sandstone or limestone layers and migrated upward in tilted layers. Tectonic processes caused such uptilting and bulging of layers to form ridges and domes. When the ridges and domes were covered by shale already formed, the pores of the shale were too tiny to let the oil through, so the shale acted as a sealing cap. When the oil could not rise farther, it was trapped. Porous rock in such a structure that contains oil or gas is called an oil or gas reservoir.

The recovery from typical reservoirs is not as high as might be thought. Multiple-layer reservoirs will typically contain oil-bearing layers with a wide range of permeability. When recovery from the highest-permeability layers is as complete as it can be, the low-permeability layers will usually have been only slightly depleted, despite all efforts to improve the recovery. Despite recovery efforts, half or more of the oil originally present in oil reservoirs is still in them. [E.L.Cl.]

Heavy oil and tar sand oil (bitumen) are petroleum hydrocarbons found in sedimentary rocks. They are formed by the oxidation and biodegradation of crude oil, and occur in the liquid or semiliquid state in limestones, sandstones, or sands.

These oils are characterized by their viscosity; however, density (or API gravity) is also used when viscosity measurements are not available. Heavy oils contain 3 wt% or more sulfur and as much as 200 ppm vanadium. Titanium, zinc, zirconium, magnesium, manganese, copper, iron, and aluminum are other trace elements that can be found in these deposits. Their high naphthenic acid content makes refinery processing equipment vulnerable to corrosion. [E.Ok.]

Petroleum geology The practice of utilizing geological principles and applying geological concepts to the discovery and recovery of petroleum. Related fields in petroleum discovery include geochemistry and geophysics. The related areas in petroleum recovery are petroleum and chemical engineering. *See* CHEMICAL ENGINEERING; GEOCHEMISTRY.

Petroleum occurs in a liquid phase as crude oil and condensate, and in a gaseous phase as natural gas. The phase is dependent on the kind of source rock from which the petroleum was formed and the physical and thermal environment in which it exists. Most petroleum occurs at varying depths below the ground surface, but generally petroleum existing as a liquid (crude oil) is found at depths of less than 20,000 ft (6100 m) while natural gas is found both at shallow depths and at depths exceeding 30,000 ft (9200 m). In some cases, oil may seep to the surface, forming massive deposits of oil or tar sands. Natural gas also seeps to the surface but escapes into the atmosphere, leaving little or no surface trace. *See* NATURAL GAS; PETROLEUM.

Most petroleum is found in sedimentary basins in sedimentary rocks, although many of the 700 or so sedimentary basins of the world contain no known significant accumulations. Several conditions must exist for the accumulation of petroleum: (1) There must be a source rock, usually high in organic matter, from which petroleum can be generated. (2) There must be a mechanism for the petroleum to move, or migrate. (3) A reservoir rock with voids to hold petroleum fluids must exist. (4) The reservoir must be in a configuration to constitute a trap and be covered by a seal—any kind of low-permeability or dense rock formation that prevents further migration. If any of these conditions do not exist, petroleum either will not form or will not accumulate in commercially extractable form. *See* BASIN; SEDIMENTARY ROCKS.

The aim of petroleum geologists is to find traps or accumulations of petroleum. The trap not only must be defined but must exist where other conditions such as source and reservoir rocks occur.

To locate these traps, the geologist must rely on subsurface information and data gathered by drilling exploratory wells and data obtained by geophysical surveying. These data, once interpreted, are used to construct maps, cross sections, and models that are used to infer or to actually depict subsurface configurations that might contain petroleum. Such depictions are prospects for drilling.

Oil and gas must be trapped in an individual reservoir in sufficient quantities to be commercially producible. Worldwide, 25% of all oil discovered so far is contained in only ten fields, seven of which are in the Middle East. Fifty percent of all oil discovered to date is found in only 50 fields.

Most of the large and fairly obvious fields in the United States have been discovered, except those possibly existing in frontier or lightly explored areas such as Alaska and the deep waters offshore. Few areas of the world remain entirely untested, but many areas outside the United States are only partly explored, and advanced techniques have yet to be deployed in the recovery of oil and gas found so far.

Greater efforts in petroleum geology along with petroleum engineering are being made to increase recovery from existing fields. Of all oil discovered so far, it is estimated that there will be recovery of only 35% on the average. Recovering some part of this huge oil resource will require geological reconstruction of reservoirs, a kind of very detailed and small-scale exploration. These reconstructions and models have allowed additional recovery of oil that is naturally movable in the reservoir. If the remaining oil is immobile because it is too viscous or because it is locked in very small pores or is held by capillary forces, techniques must be used by the petroleum geologist and the petroleum engineer to render the oil movable. [W.L.Fi.]

Petroleum microbiology Those aspects of microbiology in which crude oil, refined petroleum products, or pure hydrocarbons serve as nutrients for the growth of microorganisms or are altered as a result of their activities. Applications of petroleum microbiology include oil pollution control, enhanced oil recovery, microbial contamination of petroleum fuels and oil emulsions, and conversion of petroleum hydrocarbons into microbial products.

Many species of bacteria, fungi, and algae have the enzymatic capability to use petroleum hydrocarbons as food. Biodegradation of petroleum requires an appropriate mixture of microorganisms, contact with oxygen gas, and large quantities of utilizable nitrogen and phosphorus compounds and smaller amounts of other elements essential for the growth of all microorganisms. Part of the hydrocarbons are converted into carbon dioxide and water and part into cellular materials, such as proteins and nucleic acids. The requirement for a mixture of different microorganisms arises from the fact that petroleum is composed of a wide variety of different groups of hydrocarbons, whereas any specific microorganism is highly specialized with regard to the type of hydrocarbon it can digest. The bacterial genera that contain the most frequently isolated hydrocarbon degraders are *Pseudomonas*, *Acinetobacter*, *Flavobacterium*, *Brevibacterium*, *Corynebacterium*, *Arthrobacter*, *Mycobacterium*, and *Nocardia*. The fungal genera that contain oil utilizers include *Candida*, *Cladosporium*, *Rhodotorula*, *Torulopsis*, and *Trichosporium*. *See* BIODEGRADATION.

Oil pollution results from natural hydrocarbon seeps, accidental spills, and intentional discharge of oily materials into the environment. Once the oil is released and comes into contact with water, air, and the necessary salts, microorganisms present in the environment begin the natural process of petroleum biodegradation. If this process did not occur, the world's oceans would soon become completely covered with a layer of oil. The reason that oil spills become a pollution problem is that the natural microbial systems for degrading the oil become temporarily overwhelmed. *See* WATER POLLUTION.

The largest potential application of petroleum microbiology is in the field of enhanced oil recovery. Microbial products, as well as viable microorganisms, have been used as stimulation agents to enhance oil recovery from petroleum reservoirs. Xanthan, a polysaccharide produced by *Xanthomonas campestris*, is used as a waterflood thickening agent in oil recovery. Emulsan, a lipopolysaccharide produced by a strain of *Acinetobacter calcoaceticus*, stabilizes oil-in-water emulsions. A number of other microbial products are being tested for potential application in enhanced oil recovery processes. Field tests have indicated that injection of viable microorganisms with their nutrients into petroleum reservoirs can lead to enhanced oil recovery, presumably due to production of carbon dioxide gas, acids, and surfactants.

A variety of valuable materials, such as amino acids, carbohydrates, nucleotides, vitamins, enzymes, antibiotics, citric acid, long-chain dicarboxylic acids, and biomass, can be produced by microbial processes using petroleum hydrocarbons as substrates. The main advantage of using hydrocarbons as substrates is their lower cost. Also, certain products, such as tetradecane-1,14-dicarboxylic acid, a raw material for preparing perfumes, are synthesized in higher yields on hydrocarbon than on carbohydrate substrates.

The most active area of research and development in petroleum microbiology since the mid-1960s has been in the large-scale production and concentration of microorganisms for animal feed and human food. Dried microbial cells are collectively referred to as single-cell protein. In spite of its advantages, single-cell protein has not yet played a significant role in providing protein for animal feed or human consumption. However, many scientists are optimistic about its potential. The ability of microorganisms to utilize petroleum also has its detrimental aspects, particularly with respect to the deterioration of petroleum fuels, asphalt coatings, and oil emulsions used with cutting machinery. All hydrocarbons become contaminated if they come into contact with water during storage. *See* BACTERIAL PHYSIOLOGY AND METABOLISM . [E.Ros.]

Phloem The principal food-conducting tissue in vascular plants. Its conducting cells are known as sieve elements, but phloem may also include companion cells, parenchyma cells, fibers, sclereids, rays, and certain other cells. As a vascular tissue, phloem is spatially associated with xylem, and the two together form the vascular system. *See* XYLEM.

Sieve elements differ from phloem parenchyma cells in the structure of their walls and to some extent in the character of their protoplasts. Sieve areas, distinctive structures in sieve element walls, are specialized primary pit fields in which there may be numerous modified plasmodesmata. Plasmodesmata are strands of cytoplasm connecting the protoplasts of two contiguous cells. These strands are often surrounded by callose, a carbohydrate material, that appears to form rapidly in plants when they are placed under stress.

Typical sieve cells are long elements in which all the sieve areas are of equal specialization, though sieve areas may be more numerous in some walls than in others. In contrast, a sieve-tube member has some sieve areas more specialized than others; that is, the pores, or modified plasmodesmata, are larger in some sieve areas. Parts of the walls containing such sieve areas are called sieve plates.

Companion cells are specialized parenchyma cells that occur in close ontogenetic and physiologic association with sieve tube members. Some sieve-tube members lack companion cells. The precise functional relationship between these two kinds of cells is unknown.

Parenchyma cells in the phloem occur singly or in strands of two or more cells. They store starch, frequently contain tannins or crystals, commonly enlarge as the sieve elements become obliterated, or may be transformed into sclereids or cork cambium cells.

Phloem fibers vary greatly in length (from less than 0.04 in. or 1 mm in some plants to 20 in. or 50 cm in the ramie plant). The secondary walls are commonly thick and typically have simple pits, but may or may not be lignified. [M.A.W.]

Phoresy A relationship between two different species of organisms in which the larger, or host, organism transports a smaller organism, the guest. It is regarded as a type of commensalism in which the relationship is limited to transportation of the guest. [C.B.C.]

Photomorphogenesis The regulatory effect of light on plant form, involving growth, development, and differentiation of cells, tissues, and organs. Morphogenic influences of light on plant form are quite different from light effects that nourish the plant through photosynthesis, since the former usually occur at much lower energy levels than are necessary for photosynthesis. Light serves as a trigger in photomorphogenesis, frequently resulting in energy expenditure orders of magnitude larger than the amount required to induce a given response. Photomorphogenic processes determine the nature and direction of a plant's growth and thus play a key role in its ecological adaptations to various environmental changes. *See* PHOTOSYNTHESIS.

Morphogenically active radiation is known to control seed and spore germination, growth and development of stems and leaves, lateral root initiation, opening of the hypocotyl or epicotyl hook in seedlings, differentiation of the epidermis, formation of epidermal hairs, onset of flowering, formation of tracheary elements in the stem, and form changes in the gametophytic phase of ferns, to mention but a few of such known phenomena. Many nonmorphogenic processes in plants are also basically controlled by light independent of photosynthesis. Among these are chloroplast movement, biochemical reactions involved in the synthesis of flavonoids, anthocyanins, chlorophyll, and carotenoids, and leaf movements in certain legumes. [W.R.Br.]

Photoperiodism The growth, development, or other responses of organisms to the length of night or day or both. Photoperiodism has been observed in plants and animals, but not in bacteria (prokaryotic organisms), other single-celled organisms, or fungi.

A true photoperiodism response is a response to the changing day or night. Some species respond to increasing day lengths and decreasing night lengths (for example, by forming flowers or developing larger gonads); this is called a long-day response. Other species may exhibit the same response, or the same species may respond in some different way, to decreasing days and increasing nights; this is a short-day response. Sometimes a response is independent or nearly independent of day length, and is said to be day-neutral. There are many plant responses to photoperiod. These include development of reproductive structures in lower plants (mosses) and in flowering plants; rate of flower and fruit development; stem elongation in many herbaceous species as well as coniferous and deciduous trees (usually a long-day response and possibly the most widespread photoperiodism response in higher plants); autumn leaf drop and formation of winter dormant buds (short days); development of frost hardiness (short days); formation of roots on cuttings; formation of many underground storage organs such as bulbs (onions, long days), tubers (potato, short days), and storage roots (radish, short days); runner development (strawberry, short day); balance of male to female flowers or flower parts (especially in cucumbers); aging of leaves and other plant parts; and even such obscure responses as the formation of foliar plantlets (such as the minute plants formed on edges of *Bryophyllum* leaves), and the quality and quantity of essential oils (such as those produced by jasmine plants). Note that a single plant, for example, the strawberry, might be a short-day plant for one response and a long-day plant for another response.

Animal responses. There are also many responses to photoperiod in animals, including control of several stages in the life cycle of insects (for example, diapause) and the long-day promotion in birds of molting, development of gonads, deposition of body fat, and migratory behavior. Even feather color may be influenced by photoperiod (as in the ptarmigan). In several mammals the induction of estrus and spermatogenic activity is controlled by photoperiod (sheep, goat, snowshoe hare), as is fur color in certain species (snowshoe hare). Growth of antlers in American elk and deer can be controlled by controlling day length. Increasing day length causes antlers to grow, whereas decreasing day length causes them to fall off. By changing day lengths rapidly, a cycle of antler growth can be completed in as little as 4 months; slow changes can extend the cycle to as long as 2 years. When attempts are made to shorten or extend these limits even more, the cycle slips out of photoperiodic control and reverts to a 10–12-month cycle, apparently controlled by an internal annual "clock."

Seasonal responses. Response to photoperiod means that a given manifestation will occur at some specific time during the year. Response to long days (shortening nights) normally occurs during the spring, and response to short days (lengthening nights) usually occurs in late summer or autumn. Since day length is accurately determined by the Earth's rotation on its tilted axis as it revolves in its orbit around the Sun, detection of day length provides an extremely accurate means of determining the season at a given latitude. Such other environmental factors as temperature and light levels also vary with the seasons but are clearly much less dependable from year to year.

Mechanisms. It has long been the goal of researchers on photoperiodism to understand the plant or animal mechanisms that account for the responses. Light must be detected, the duration of light or darkness must be measured, and this time measurement must be metabolically translated into the observed response: flowering, stem elongation, gonad development, fur color, and so forth. Basic mechanisms differ not only between plants and animals but among different species as well. The roles (synchronization, anticipation, and so on) are similar in all organisms that exhibit photoperiodism, but the mechanisms through which these roles are achieved are apparently quite varied.

Strongest inhibition of flowering in short-day plants comes when the light interruption occurs around the time of the critical night (about 7–9 h for cocklebur plants), but actual effectiveness also depends on the length of the dark period. With short-day cockleburs, the shorter the night, the less the flowering and the longer the time that light inhibits flowering.

Orange-red wavelengths used as a night interruption are by far the most effective part of the spectrum in inhibition of short-day responses and promotion of long-day responses (flowering in most studies), and effects of orange-red light can be completely reversed by subsequent exposure of plants to light of somewhat longer wavelengths, called far-red light. These observations led in the early 1950s to discovery of the phytochrome pigment system, which is apparently the molecular machinery that detects the light effective in photoperiodism of higher plants. *See* PHYTOCHROME.

In photoperiodism of short-day plants, an optimum response is usually obtained when phytochrome is in the far-red receptive form during the day and the red-receptive form during the night. Although normal daylight contains a balance of red and far-red wavelengths, the red-receptive form is most sensitive, so the pigment under normal daylight conditions is driven mostly to the far-red receptive form. At dusk this form is changed metabolically, and the red-receptive form builds up. It is apparently this shift in the form of phytochrome that initiates measurement of the dark period. This is how a plant "sees": when the far-red-sensitive form of the pigment is abundant, the plant "knows" it is in the light; the red-sensitive form (or lack of far-red form) indicates to the plant's biochemistry that it is in the dark.

The measurement of time—the durations of the day or night—is the very essence of photoperiodism. The discovery of a biological clock in living organisms was made in the late 1920s. It was shown that the movement of leaves on a bean plant (from horizontal at noon to vertical at midnight) continued uninterruptedly for several days, even when plants were placed in total darkness and at a constant temperature, and that the time between given points in the cycle (such as the most vertical leaf position) was almost but not exactly 24 h. In the case of bean leaves, it was about 25.4 h. Many other cycles have now been found with similar characteristics in virtually all groups of plants and animals. There is strong evidence that the clocks are internal and not driven by some daily change in the environment. Such rhythms are called circadian.

Circadian rhythms usually have period lengths that are remarkably temperature-insensitive, which is also true of time measurement in photoperiodism. Furthermore, the rhythms are normally highly sensitive to light, which may shift the cycle to some extent. Thus, daily rhythms in nature are normally synchronized with the daily cycle as the Sun rises and sets each day. Their circadian nature appears only when they are allowed to manifest themselves under constant conditions of light (or darkness) and temperature, so that their free-running periods can appear. [F.B.S.]

Photorespiration Light-dependent carbon dioxide release and oxygen uptake in photosynthetic organisms caused by the fixation of oxygen instead of carbon dioxide during photosynthesis. This oxygenation reaction forms phosphoglycolate, which represents carbon lost from the photosynthetic pathway. Phosphoglycolate also inhibits photosynthesis if it is allowed to accumulate in the plant. The reactions of photorespiration break down phosphoglycolate and recover 75% of the carbon to the photosynthetic reaction sequence. The remaining 25% of the carbon is released as carbon dioxide. Photorespiration reduces the rate of photosynthesis in plants in three ways: carbon dioxide is released; energy is diverted from photosynthetic reactions to photorespiratory reactions; and competition between oxygen and carbon dioxide reduces the efficiency of the important photosynthetic enzyme ribulose-bisphosphate (RuBP) carboxylase. There is no known function of the oxygenation reaction; most scientists believe it is an unavoidable side reaction of photosynthesis. *See* PHOTOSYNTHESIS.

The rate of photosynthesis can be stimulated as much as 50% by reducing photorespiration. Since photosynthesis provides the material necessary for plant growth, photorespiration inhibits plant growth by reducing the net rate of carbon dioxide assimilation (photosynthesis). Plants grow faster and larger under nonphotorespiratory conditions, in either low oxygen or high carbon dioxide atmospheres. Most of the beneficial effects on plant growth achieved by increasing CO_2 may result from the reduced rate of photorespiration. *See* PLANT GROWTH.

There are some plants that avoid photorespiration under certain conditions by actively accumulating carbon dioxide inside the cells that have ribulose-bisphosphate carboxylase/oxygenase. Many cacti do this by taking up carbon dioxide at night and then releasing it during the day to allow normal photosynthesis. These plants are said to have crassulacean acid metabolism (CAM). Another group of plants, including corn (*Zea mays*), take up carbon dioxide by a special accumulating mechanism in one part of the leaf, then transport it to another part of the leaf for release and fixation by normal photosynthesis. The compound used to transport the carbon dioxide has four carbon atoms, and so these plants are called C_4 plants. Plants that have no mechanism for accumulating carbon dioxide produce the three-carbon compound phosphoglycerate directly and are therefore called C_3 plants. Most species of plants are C_3 plants. *See* PLANT RESPIRATION. [T.D.S.]

Photosynthesis The manufacture in light of organic compounds (primarily certain carbohydrates) from inorganic materials by chlorophyll- or bacteriochlorophyll-containing cells. This process requires a supply of energy in the form of light. In chlorophyll-containing plant cells and in cyanobacteria, photosynthesis involves oxidation of water (H_2O) to oxygen molecules, which are released into the environment. In contrast, bacterial photosynthesis does not involve O_2 evolution—instead of H_2O, other electron donors, such as H_2S, are used. This article will focus on photosynthesis in plants. *See* BACTERIAL PHYSIOLOGY AND METABOLISM; CHLOROPHYLL; PLANT RESPIRATION.

The light energy absorbed by the pigments of photosynthesizing cells, especially by the pigment chlorophyll or bacteriochlorophyll, is efficiently converted into stored chemical energy. Together, the two aspects of photosynthesis—the conversion of inorganic into organic matter, and the conversion of light energy into chemical energy—make it the fundamental process of life on Earth: it is the ultimate source of all living matter and of all life energy.

The net overall chemical reaction of plant photosynthesis is shown in the equation below, where $\{CH_2O\}$ stands for a carbohydrate (sugar).

$$H_2O + CO_2 + \text{light energy} \xrightarrow[\text{enzymes}]{\text{chlorophyll}} \{CH_2O\} + O_2$$

The photochemical reaction in photosynthesis belongs to the type known as oxidation-reduction, with CO_2 acting as the oxidant (hydrogen or electron acceptor) and water as the reductant (hydrogen or electron donor). The unique characteristic of this particular oxidation-reduction is that it goes "in the wrong direction" energetically; that is, it converts chemically stable materials into chemically unstable products. Light energy is used to make this "uphill" reaction possible.

Photosynthesis is a complex, multistage process. Its main parts are (1) the primary photochemical process in which light energy absorbed by chlorophyll is converted into chemical energy, in the form of some energy-rich intermediate products; and

(2) the enzyme-catalyzed "dark" (that is, not photochemical) reactions by which these intermediates are converted into the final products—carbohydrates and free oxygen.

Experiments suggest that plants contain two pigment systems. One (called photosystem I, or PS I, sensitizing reaction I) contains the major part of chlorophyll *a*; the other (called photosystem II, or PS II, sensitizing reaction II) contains some chlorophyll *a* and the major part of chlorophyll *b* or other auxiliary pigments (for example, the red and blue pigments, called phycobilins, in red and blue-green algae, and the brown pigment fucoxanthol in brown algae and diatoms). It appears that efficient photosynthesis requires the absorption of an equal number of light quanta in PS I and in PS II; and that within both systems excitation energy undergoes resonance migration from one pigment to another until it ends in special molecules of chlorophyll *a* called the reaction centers. The latter molecules then enter into a series of chemical reactions that result in the oxidation of water to produce O_2 and the reduction of nicotinamide adenine dinucleotide phosphate ($NADP^+$). Chromatophores from photosynthetic bacteria and chloroplasts from green plants, when illuminated in the presence of adenosine diphosphate (ADP) and inorganic phosphate, also use light energy to synthesize adenosine triphosphate (ATP); this photophosphorylation could be associated with some energy-releasing step in photosynthesis. [G.; R.Gov.]

The light-dependent conversion of radiant energy into chemical energy as ATP and reduced nicotinamide adenine dinucleotide phosphate (NADPH) serves as a prelude to the utilization of these compounds for the reductive fixation of CO_2 into organic molecules. Such molecules, broadly designated as photosynthates, are usually but not invariably in the form of carbohydrates such as glucose polymers or sucrose, and form the base for the nutrition of all living things. Collectively, the biochemical processes by which CO_2 is assimilated into organic molecules are known as the photosynthetic dark reactions, not because they must occur in darkness, but because light—in contrast to the photosynthetic light reactions—is not required.

C_3 photosynthesis. The essential details of C_3 photosynthesis can be seen in Fig. 1. Three molecules of CO_2 combine with three molecules of the five-carbon compound

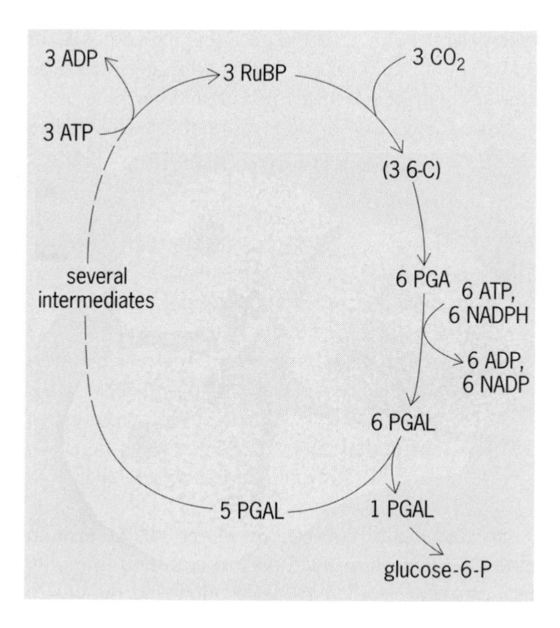

Fig. 1. Schematic outline of the Calvin (C_3) carbon dioxide assimilation cycle.

ribulose bisphosphate (RuBP) in a reaction catalyzed by RuBP carboxylase to form three molecules of an enzyme-bound six-carbon compound. These are hydrolyzed into six molecules of the three-carbon compound phosphoglyceric acid (PGA), which are phosphorylated by the conversion of six molecules of ATP (releasing ADP for photophosphorylation via the light reactions). The resulting compounds are reduced by the NADPH formed in photosynthetic light reactions to form six molecules of the three-carbon compound phosphoglyceraldehyde (PGAL). One molecule of PGAL is made available for combination with another three-carbon compound, dihydroxyacetone phosphate, which is isomerized from a second PGAL (requiring a second "turn" of the Calvin-cycle wheel) to form a six-carbon sugar. The other five PGAL molecules, through a complex series of enzymatic reactions, are rearranged into three molecules of RuBP, which can again be carboxylated with CO_2 to start the cycle turning again. The net product of two "turns" of the cycle, a six-carbon sugar (glucose-6-phosphate) is formed either within the chloroplast in a pathway leading to starch (a polymer of many glucose molecules), or externally in the cytoplasm in a pathway leading to sucrose (condensed from two six-carbon sugars, glucose and fructose).

C_4 photosynthesis. Initially, the C_3 cycle was thought to be the only route for CO_2 assimilation, although it was recognized by plant anatomists that some rapidly growing plants (such as maize, sugarcane, and sorghum) possessed an unusual organization of the photosynthetic tissues in their leaves (Kranz morphology). It was then demonstrated that plants having the Kranz anatomy utilized an additional CO_2 assimilation route now known as the C_4-dicarboxylic acid pathway (Fig. 2). Carbon dioxide enters a mesophyll cell, where it combines with the three-carbon compound phosphoenolpyruvate (PEP) to form a four-carbon acid, oxaloacetic acid, which is reduced to malic acid or transaminated to aspartic acid. The four-carbon acid moves into bundle sheath cells, where the acid is decarboxylated, the CO_2 assimilated via the C_3 cycle, and the resulting three-carbon compound, pyruvic acid, moves back into the mesophyll cell and is transformed into PEP, which can be carboxylated again. The two cell types, mesophyll and bundle sheath, are not necessarily adjacent, but in all documented cases of C_4 photosynthesis the organism had two distinct types of green cells. C_4 metabolism is classified into three types, depending on the decarboxylation reaction used with the four-carbon acid in the bundle sheath cells:

1. NADP-ME type (sorghum):

$$\text{NADP}^+ + \text{malic acid} \xrightarrow{\text{NADP-malic enzyme}} \text{pyruvic acid} + CO_2 + \text{NADPH}$$

2. NAD-ME type (*Atriplex* species):

$$\text{NAD}^+ + \text{malic acid} \xrightarrow{\text{NAD-malic enzyme}} \text{pyruvic acid} + CO_2 + \text{NADH}$$

3. PCK type (*Panicum* species):

$$\text{Oxaloacetic acid} + \text{ATP} \xrightarrow{\text{phosphoenol pyruvate carboxykinase}} \text{PEP} + CO_2 + \text{ADP}$$

CAM photosynthesis. Under arid and desert conditions, where soil water is in short supply, transpiration during the day when temperatures are high and humidity is low may rapidly deplete the plant of water, leading to desiccation and death. By

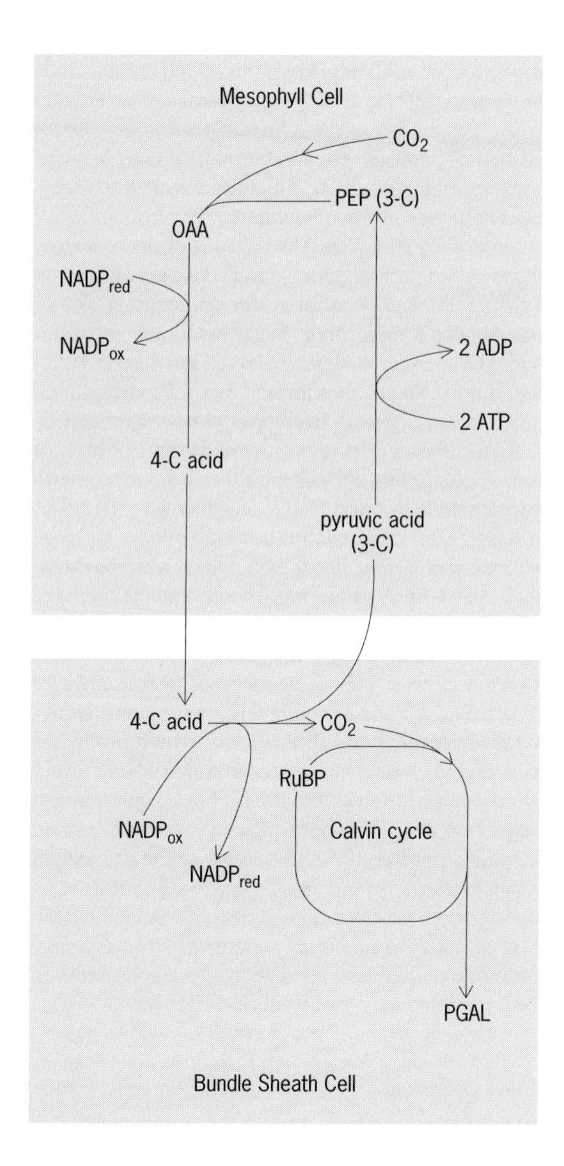

Fig. 2. Schematic outline of the Hatch-Slack (C_4) carbon dioxide assimilation route in two cell types of a NADP-ME-type plant.

keeping stomata closed during the day, water can be conserved, but the uptake of CO_2, which occurs entirely through the stomata, is prevented. Desert plants in the Crassulaceae, Cactaceae, Euphorbiaceae, and 15 other families evolved, apparently independently of C_4 plants, an almost identical strategy of assimilating CO_2 by which the CO_2 is taken in at night when the stomata open; water loss is low because of the reduced temperatures and correspondingly higher humidities. First studied in plants of the Crassulaceae, the process has been called crassulacean acid metabolism (CAM).

In contrast to C_4, where two cell types cooperate, the entire process occurs within an individual cell; the separation of C_4 and C_3 is thus temporal rather than spatial. At night, CO_2 combines with PEP through the action of PEP carboxylase, resulting in the formation of oxaloacetic acid and its conversion into malic acid. The PEP is formed from starch or sugar via the glycolytic route of respiration. Thus, there is a daily reciprocal relationship between starch (a storage product of C_3 photosynthesis) and the accumulation of malic acid (the terminal product of nighttime CO_2 assimilation). [M.G.; G.A.B.]

Phycobilin Any member of a class of intensely colored pigments found in some algae that absorb light for photosynthesis. Phycobilins are structurally related to mammalian bile pigments, and they are unique among photosynthetic pigments in being covalently bound to proteins (phycobiliproteins). In at least two groups of algae, phycobiliproteins are aggregated in a highly ordered protein complex called a phycobilisome.

Phycobilins occur only in three groups of algae: cyanobacteria (blue-green algae), Rhodophyta (red algae), and Cryptophyceae (cryptophytes), and are largely responsible for their distinctive colors, including blue-green, yellow, and red. Five different phycobilins have been identified to date, but the two most common are phycocyanobilin, a blue pigment, and phycoerythrobilin, a red pigment. In the cell, these pigments absorb light maximally in the orange (620-nanometers) and green (550-nm) portion of the visible light spectrum, respectively. A blue-green light (495-nm) absorbing pigment, phycourobilin, is found in some cyanobacteria and red algae. A yellow light (575-nm) absorbing pigment, phycobiliviolin (also called cryptoviolin) is apparently found in all cryptophytes but in only a few cyanobacteria. A fifth phycobilin, which absorbs deep-red light (697 nm), has been identified spectrally in some cryptophytes, but its chemical properties are unknown.

Phycobilins are associated with the photosynthetic light-harvesting system in chloroplasts of red algae and cryptophytes and with the photosynthetic membranes of cyanobacteria, which lack chloroplasts. Phycobilins are covalently bound to a water-soluble protein that aggregates on the surface of the photosynthetic membrane. All other photosynthetic pigments (for example, chlorophylls and carotenoids) are bound to photosynthetic membrane proteins by hydrophobic attraction. Phycobiliprotein can constitute a major fraction of an alga. In some cyanobacteria, phycobiliproteins can account for more than 50% of the soluble protein and one-quarter of the dry weight of the cell. *See* CELL PLASTIDS.

Phycobilins are photosynthetic accessory pigments that absorb light efficiently in the yellow, green, orange, or red portion of the light spectrum, where chlorophyll *a* only weakly absorbs. Light energy absorbed by phycobilins is transferred with greater than 90% efficiency to chlorophyll *a*, where it is used for photosynthesis. *See* CHLOROPHYLL; PHOTOSYNTHESIS. [T.M.K.]

Phylogeny The genealogical history of organisms, both living and extinct. Phylogeny represents the historical pattern of relationships among organisms which has resulted from the actions of many different evolutionary processes. Phylogenetic relationships are depicted by branching diagrams called cladograms, or phylogenetic trees. Cladograms show relative affinities of groups of organisms called taxa. Such groups of organisms have some genealogical unity, and are given a taxonomic rank such as species, genera, families, or orders. For example, two species of cats—say, the lion (*Panthera leo*) and the tiger (*Panthera tigris*)—are more closely related to each other than either is to the gray wolf (*Canis latrans*). The family including all cats, Felidae, is more closely related to the family including all dogs, Canidae, than either is to the family that includes giraffes, Giraffidae. The lion and tiger, and the Felidae and Canidae, are called sister taxa because of their close relationship relative to the gray wolf, or to the Giraffidae, respectively.

Cladograms thus depict a hierarchy of relationships among a group of taxa (illus. *a*). Branch points, or nodes, of a cladogram represent hypothetical common ancestors (not specific real ancestors), and the branches connect descendant sister taxa. If the taxa

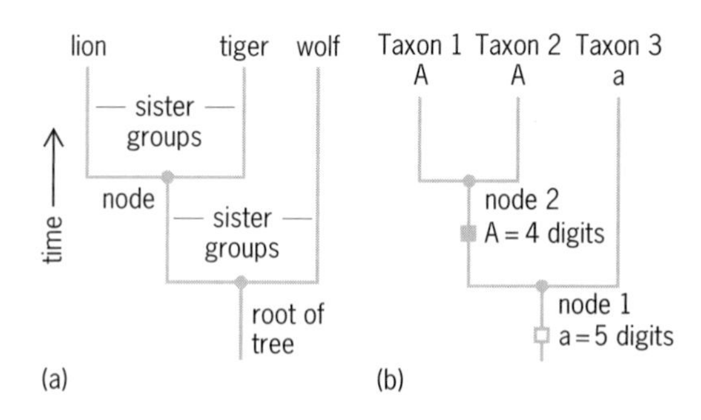

Phylogenetic trees. (*a*) A tree representing a hierarchical pattern of sister-group relationships (those taxa descended from a common ancestor). (*b*) Relationships are determined by identifying derived characters; in this case the condition of five digits is primitive, whereas the loss of a digit is derived and unites taxa 1 and 2.

being considered are species, nodes are taken to signify speciation events. The goal of the science of cladistics, or phylogenetic analysis, is to discover these sister-group (cladistic) relationships and to identify what are termed monophyletic groups—two or more taxa postulated to have a single, common origin.

The acceptance of a cladogram depends on the empirical evidence that supports it relative to alternative hypotheses of relationship for those same taxa. Evidence for or against alternative phylogenetic hypotheses comes from the comparative study of the characteristics of those taxa. Similarities and differences are determined by comparison of the anatomical, behavioral, physiological, or molecular [such as deoxyribonucleic acid (DNA) sequences] attributes among the taxa. A statement that two features in two or more taxa are similar and thus constitute a shared character is, in essence, a preliminary hypothesis that they are homologous; that is, the taxa inherited the specific form of the feature from their common ancestor. However, not all similarities are homologs; some are developed independently through convergent or parallel evolution, and although they may be similar in appearance, they had different histories and thus are not really the same feature. In cladistic theory, shared homologous similarities are either primitive (plesiomorphic condition) or derived (apomorphic condition), whereas nonhomologous similarities are termed homoplasies (or sometimes, parallelisms or convergences). This distinction over concepts and terminology is important because only derived characters constitute evidence that groups are actually related.

As evolutionary lineages diversify, some characters will become modified. Examples include the enlargement of forelimbs or the loss of digits on the hand. Thus, during evolution the foot of a mammal might transform from a primitive condition of having five digits to a derived form with only four digits (illus. *b*). Following branching at node 1, the foot in one lineage undergoes an evolutionary modification involving the loss of a digit (expressed as character state A). A subsequent branching event then produced taxa 1 and 2, which inherited that derived character. The lineage leading to taxon 3, however, retained the primitive condition of five digits (character state a). The presence of the shared derived character, A, is called a synapomorphy, and identifies taxa 1 and 2 as being more closely related to each other than either is to taxon 3. Distinguishing between the primitive and derived conditions of a character within a group of taxa (the ingroup) is usually accomplished by comparisons to groups postulated to have more distant relationships (outgroups). Character states that are present in ingroups but not outgroups are postulated to be derived. Systematists have developed computer programs that attempt to identify shared derived characters (synapomorphies) and, at the same time, use them to construct the best phylogenetic trees for the available data.

Knowledge of phylogenetic relationships provides the basis for classifying organisms. A major task of the science of systematics is to search for monophyletic groups. Some groups, such as birds and mammals, are monophyletic; that is, phylogenetic analysis suggests they are all more closely related to each other than to other vertebrates. However, other traditional groups, such as reptiles, have been demonstrated to be nonmonophyletic (some so-called reptiles, such as dinosaurs and their relatives, are more closely related to birds than they are to other reptiles such as snakes). Classifications based on monophyletic groups are termed natural classifications. Phylogenies are also essential for understanding the distributional history, or biogeography, of organisms. Knowing how organisms are related to one another helps the biogeographer to decipher relationships among areas and to reconstruct the spatial histories of groups and their biotas. *See* ANIMAL EVOLUTION; BIOGEOGRAPHY; TAXONOMIC CATEGORIES. [J.Cr.]

Physical geography The study of the Earth's surface features and associated processes. Physical geography aims to explain the geographic patterns of climate, vegetation, soils, hydrology, and landforms, and the physical environments that result from their interactions. Physical geography merges with human geography to provide a synthesis of the complex interactions between nature and society.

The basic content of physical geography comprises a number of areas of specialization. Climatology, the scientific study of climates, concerns the total complex of weather conditions at a given location over an extended time period; it deals not only with average conditions but with extremes and variations. Geomorphology is the interpretive description and explanation of landforms and the fluvial, glacial, coastal, and eolian process that operate on them. The forms, processes, and patterns within the biosphere, including vegetation and animal distributions, are studied as biogeography. With strong ties to fluvial geomorphology, geographic hydrology concerns the scientific study of water from the aspects of distribution, movement, and utilization. Soil geography, with emphasis on the origin, characteristics, classification, and utilization potential of soils, provides an area of specialization with links to land use. Ultimately, the physical geography of a region is understood through an integration of the multiple aspects. [J.E.O.]

Physiological ecology (animal) A discipline that combines the study of physiological processes, the functions of living organisms and their parts, with ecological processes that connect the individual organism with population dynamics and community structure. *See* POPULATION ECOLOGY.

Physiological ecologists focus on whole-animal function and adjustments to ever-changing environments, in both laboratory and field. Short-term behavioral adjustments and longer-term physiological adjustments tend to maximize the fitness of animals, that is, their capacity to survive and reproduce successfully. Among the processes that physiological ecologists study are temperature regulation, energy metabolism and energetics, nutrition, respiratory gas exchange, water and osmotic balance, and responses to environmental stresses. These environmental stresses may include climate variation, nutrition, disease, and toxic exposure. For instance, climate affects animal heat and mass balances, and such changes affect body temperature regulation. Behavioral temperature regulation (typically, avoidance of temperature extremes) modifies mass and energy intake and expenditure, and the difference between intake and expenditure provides the discretionary mass for growth and reproduction. Mortality risk (survivorship) also depends on temperature-dependent behavior, which determines daily activity. Activity time constrains the time for foraging and habitat selection, which in turn influence not only mortality risk but also community composition. Animals are similarly constrained in their discretionary mass and energy by reduction in nutrition, which decreases absorbed food, and by disease and toxins, which may elevate the costs to maintain a higher body temperature (fever). *See* BEHAVIORAL ECOLOGY. [W.Po.]

Physiological ecology (plant) The branch of plant science that seeks physiological (mechanistic) explanations for ecological observations. Emphasis is placed on understanding how plants cope with environmental variation at the physiological level, and on the influence of resource limitations on growth, metabolism, and reproduction of individuals within and among plant populations, along environmental gradients, and across different communities and ecosystems. The responses of plants to natural, controlled, or manipulated conditions above and below ground provide a basis for understanding how the features of plants enable their survival, persistence, and spread. Information gathered is often used to identify the physiological and morphological features of a plant that permit adaptation to different sets of environmental conditions.

The environments that plants occupy are often subject to variation or change. The ecophysiological characteristics of these plants must be able to accommodate this or the plants face extinction. Given the right conditions, ample time, and genetic variation among a group of interbreeding individuals, plant populations and species can evolve to accommodate marked ecological change or habitat heterogeneity. If evolutionary changes in physiology or morphology occur on a local or regional scale, populations within a single species may diverge in their characteristics. Separate ecological races (ecotypes) arise in response to an identifiable, set of environmental conditions. Ecotypes are genetically distinct and are particularly well suited to the local or regional environment they occupy. Such ecotypes can often increase the geographical range and amplitude of environmental conditions that the species occupies or tolerates. Ecotypes may also occur as a series of populations arrayed over a well-defined environmental gradient called an ecocline. In contrast, if ecotypes are not present, some plant species may still be able to accommodate a wide range of growth conditions through morphological and physiological adjustments, by acclimation to a single factor (such as light) or acclimatization to a complex suite of factors which define the entire habitat. Acclimatization can occur when individuals from several different regions or populations are grown in a common location and adjust, physiologically or morphologically, to this location. Acclimation and acclimatization can therefore be defined as the ability of a single genotype (individual) to express multiple phenotypes (outward appearances) in response to variable growing conditions. Neither requires underlying genetic changes, though some genetic change might occur which could mean that the response seen may itself evolve. Acclimation and acclimatization may also be called phenotypic plasticity. *See* PLANT EVOLUTION.

Studies of metabolic rates in relation to environmental conditions within populations, ecotypes, or species provide a way to measure the tolerance limits expressed at different scales. These data in turn help identify the scales at which different adaptations are expressed, and enhance an understanding of the evolution of physiological processes. Combining observations and measurements from the field with those obtained in laboratory and controlled environment experiments can help identify which conditions may be most influential on plant processes and therefore what may have shaped the physiological responses seen. Laboratory and controlled environment (common garden) experiments also assist in helping identify how much of the variation expressed in a particular metabolic process can be assigned to a particular environmental factor and how much to the plants themselves and the genetic and developmental plasticity they possess. *See* ECOLOGY; ECOSYSTEM; PLANT PHYSIOLOGY. [T.E.D.]

Phytoalexin Any antibiotic produced by plants in response to microorganisms. Plants use physical and chemical barriers as a first line of defense. When these barriers are breached, however, the plant must actively protect itself by employing a variety of strategies. Plant cell walls are strengthened, and special cell layers are produced to block further penetration of the pathogen. These defenses can permanently stop a pathogen when fully implemented, but the pathogen must be slowed to gain time.

The rapid defenses available to plants include phytoalexin accumulation, which takes a few hours, and the hypersensitive reaction, which can occur in minutes. The hypersensitive reaction is the rapid death of plant cells in the immediate vicinity of the pathogen. Death of these cells is thought to create a toxic environment of released plant components that may in themselves interfere with pathogen growth, but more importantly, damaged cells probably release signals to surrounding cells and trigger a more comprehensive defense effort. Thus, phytoalexin accumulation is just one part of an integrated series of plant responses leading from early detection to eventual neutralization of a potentially lethal invading microorganism.

The tremendous capacity of plants to produce complex chemical compounds is reflected in the structural diversity of phytoalexins. Each plant species produces one or several phytoalexins, and the types of phytoalexins produced are similar in related species. The diversity, complexity, and toxicity of phytoalexins may provide clues about their function. The diversity of phytoalexins may reflect a plant survival strategy. That is, if a plant produces different phytoalexins from its neighbors, it is less likely to be successfully attacked by pathogens adapted to its neighbor's phytoalexins. Diversity and complexity, therefore, may reflect the benefits of using different deterrents from those found in other plants. *See* PLANT PATHOLOGY. [A.R.A.]

Phytochrome A pigment that controls most photomorphogenic responses in higher plants. Mechanisms have evolved in plants that allow them to adapt their growth and development to more efficiently seek and capture light and to tailor their life cycle to the climatic seasons. These mechanisms enable the plant to sense not only the presence of light but also its intensity, direction, duration, and spectral quality. Plants thus regulate important developmental processes such as seed germination, growth direction, growth rate, chloroplast development, pigmentation, flowering, and senescence, collectively termed photomorphogenesis.

To perceive light signals, plants use several receptor systems that convert light absorbed by specific pigments into chemical or electrical signals to which the plants respond. This signal conversion is called photosensory transduction. Pigments used include cryptochrome, a blue light-absorbing pigment; an ultraviolet light-absorbing pigment; and phytochrome, a red/far-red light-absorbing pigment.

Phytochrome consists of a compound that absorbs visible light (chromophore) bound to a protein. The chromophore is an open-chain tetrapyrrole closely related to the photosynthetic pigments found in the cyanobacteria and similar in structure to the circular tetrapyrroles of chlorophyll and hemoglobin. Phytochrome is one of the most intensely colored pigments found in nature, enabling phytochrome in seeds to sense even the dim light present well beneath the surface of the soil and allowing leaves to perceive moonlight. *See* CHLOROPHYLL.

Phytochrome can exist in two stable photointerconvertible forms, P_r or P_{fr}, with only P_{fr} being biologically active. Absorption of red light (near 666 nanometers) by inactive P_r converts it to active P_{fr}, while absorption of far-red light (near 730 nm) by active P_{fr} converts phytochrome back to inactive P_r. Plants frequently respond quantitatively to light by detecting the amount of P_{fr} produced. As a result, the amount of P_{fr} must be strictly regulated nonphotochemically by precisely controlling both the synthesis and degradation of the pigment.

Phytochrome has a variety of functions in plants. Initially, production of P_{fr} is required for many seeds to begin germination. This requirement prevents germination of seeds that are buried too deep in the soil to successfully reach the surface. In etiolated (dark-grown) seedlings, phytochrome can measure an increase in light intensity and duration through the increased formation of P_{fr}. Light direction also can be deduced from the asymmetry of P_{fr} levels from one side of the plant to the other. Different phytochrome responses vary in their sensitivity to P_{fr}; some require very low levels of P_{fr} (less than 1% of total phytochrome) to elicit a maximal response, while others require almost all of the pigment to be converted to P_{fr}. Thus, as the seedling grows toward the soil surface, a cascade of photomorphogenic responses are induced, with the more sensitive responses occurring first. This chain of events produces a plant that is mature and photosynthetically competent by the time it finally reaches the surface. Production of P_{fr} also makes the plant aware of gravity, inducing shoots to grow up and roots to grow down into the soil. *See* PLANT MOVEMENTS; SEED.

In light-grown plants, phytochrome allows for the perception of daylight intensity, day length, and spectral quality. Intensity is detected through a measurement of phytochrome shuttling between P_r and P_{fr}; the more intense the light, the more interconversion. This signal initiates changes in chloroplast morphology to allow shaded leaves to capture light more efficiently. If the light is too intense, phytochrome will also elicit the production of pigments to protect plants from photodamage.

Temperate plants use day length to tailor their development, a process called photoperiodism. How the plant measures day length is unknown, but it involves phytochrome and actually measures the length of night. *See* PHOTOPERIODISM.

Finally, phytochrome allows plants to detect the spectral quality of light, a form of color vision, by measuring the ratio of P_r to P_{fr}. When a plant is grown under direct sun, the amounts of red and far-red light are approximately equal, and the ratio of P_r to P_{fr} in the plant is about 1:1. Should the plant become shaded by another plant, the P_r/P_{fr} ratio changes dramatically to 5:1 or greater. This is because the shading plant's chlorophyll absorbs much of the red light needed to produce P_{fr} and absorbs almost none of the far-red light used to produce P_r. For a shade-intolerant plant, this change in P_r/P_{fr} ratio induces the plant to grow taller, allowing it to grow above the canopy.

It is not known how phytochrome elicits the diverse array of photomorphogenic responses, but the regulatory action must result from discrete changes in the molecule following photoconversion of P_r to P_{fr}. These changes must then start a chain of events in the photosensory transduction chain leading to the photomorphogenic response. Many photosensory transduction chains probably begin by responding to P_{fr} or the P_r/P_{fr} ratio and branch off toward discrete end points. *See* PHOTOMORPHOGENESIS. [R.D.V.]

Phytoplankton Mostly autotrophic microscopic algae which inhabit the illuminated surface waters of the sea, estuaries, lakes, and ponds. Many are motile. Some perform diel (diurnal) vertical migrations, others do not. Some nonmotile forms regulate their buoyancy. However, their locomotor abilities are limited, and they are largely transported by horizontal and vertical water motions.

A great variety of algae make up the phytoplankton. Diatoms (class Bacillariophyceae) are often conspicuous members of marine, estuarine, and fresh-water plankton. Dinoflagellates (class Dinophyceae) occur in both marine and fresh-water environments and are important primary producers in marine and estuarine environments. Coccolithophorids (class Haptophyceae) are also marine primary producers of some importance. They do not occur in fresh water.

Even though marine and fresh-water phytoplankton communities contain a number of algal classes in common, phytoplankton samples from these two environments will appear quite different. These habitats support different genera and species and groups of higher rank in these classes. Furthermore, fresh-water plankton contains algae belonging to additional algal classes either absent or rarely common in open ocean environments. These include the green algae (class Chlorophyceae), the euglenoid flagellates (class Euglenophyceae), and members of the Prasinophyceae.

The phytoplankton in aquatic environments which have not been too drastically affected by human activity exhibit rather regular and predictable seasonal cycles. Coastal upwelling and divergences, zones where deeper water rises to the surface, are examples of naturally occurring phenomena which enrich the mixed layer with needed nutrients and greatly increase phytoplankton production. In the ocean these are the sites of the world's most productive fisheries. *See* EUTROPHICATION. [R.W.H.]

Phytotronics Research using whole plants and conducted under controlled environmental conditions to determine responses to a single or known combination of environmental elements. Originally, the term phytotronics was used to identify research conducted specifically in phytotrons where controlled plant growth units are available for simultaneous use. The name phytotronics is also often applied to any research conducted with whole plants in a controlled environment plant growth chamber or room. [H.H.]

Picornaviridae A viral family made up of the small (18–30 nanometer) ether-sensitive viruses that lack an envelope and have a ribonucleic acid (RNA) genome. The name is derived from "pico" meaning very small, and RNA for the nucleic acid type. Picornaviruses of human origin include the following subgroups: enteroviruses (polioviruses, coxsackieviruses, and echoviruses) and rhinoviruses. There are also picornaviruses of lower animals (for example, bovine foot-and-mouth disease, a rhinovirus). *See* ANIMAL VIRUS; COXSACKIEVIRUS; ECHOVIRUS; ENTEROVIRUS; FOOT-AND-MOUTH DISEASE; POLIOMYELITIS; RHINOVIRUS. [J.L.Me.]

Pine The genus *Pinus*, of the pine family, characterized by evergreen leaves, usually in tight clusters (fascicles) of two to five, rarely single. There are about 80 known species distributed throughout the Northern Hemisphere. Botanically the leaves are of two kinds: (1) a scalelike form, the primary leaf, which subtends a much shortened and eventually deciduous shoot bearing (2) the secondary leaves or needles. The wood of pines is easily recognized by the numerous resin ducts and by the characteristic resinous odor. [A.H.G./K.P.D.]

Pith The central zone of tissue of an axis in which the vascular tissue is arranged as a hollow cylinder. Pith is present in most stems and in some roots. Stems without pith rarely occur in angiosperms but are characteristic of psilopsids, lycopsids, *Sphenophyllum*, and some ferns. Roots of some ferns, many monocotyledons, and some dicotyledons include a pith, although most roots have xylem tissue in the center.

Pith is composed usually of parenchyma cells often arranged in longitudinal files. This arrangement results from predominantly transverse division of pith mother cells near the apical meristem. *See* PARENCHYMA; ROOT (BOTANY); STEM. [H.W.Bl.]

Plague An infectious disease of humans and rodents caused by the bacterium *Yersinia pestis*. The sylvatic (wild-animal) form persists today in more than 200 species of rodents throughout the world. The explosive urban epidemics of the Middle Ages, known as the Black Death, resulted when the infection of dense populations of city rats living closely with humans introduced disease from the Near East. The disease then was spread both by rat fleas and by transmission between humans. During these outbreaks, as much as 50% of the European population died. At present, contact with wild rodents and their fleas, sometimes via domestic cats and dogs, leads to sporadic human disease. *See* INFECTIOUS DISEASE.

After infection by *Y. pestis*, fleas develop obstruction of the foregut, causing regurgitation of plague bacilli during the next blood meal. The rat flea, *Xenopsylla cheopsis*, is an especially efficient plague vector, both between rats and from rats to humans. Human (bubonic) plague is transmitted by the bite of an infected flea; after several days, a painful swelling (the bubo) of local lymph nodes occurs. Bacteria can then spread to other organ systems, especially the lung; fever, chills, prostration, and death may occur. Plague pneumonia develops in 10–20% of all bubonic infections. In some individuals, the skin may develop hemorrhages and necrosis (tissue death), probably the origin of the ancient name, the Black Death. The last primary pneumonic plague outbreak in the United States occurred in 1919, when 13 cases resulting in 12 deaths developed before the disease was recognized and halted by isolation of cases.

Bubonic plaque is suspected when the characteristic painful, swollen glands develop in the groin, armpit, or neck of an individual who has possibly been exposed to wild-animal fleas in an area where the disease is endemic. Immediate identification is possible by microscopic evaluation of bubo aspirate stained with fluorescent-tagged antibody. Antibiotics should be given if plague is suspected or confirmed. Such treatment is very effective if started early. The current overall death rate, approximately 15%, is reduced to less than 5% among patients treated at the onset of symptoms. *See* MEDICAL BACTERIOLOGY. [D.L.Pa.]

Plains The relatively smooth sections of the continental surfaces, occupied largely by gentle rather than steep slopes and exhibiting only small local differences in elevation. Because of their smoothness, plains lands, if other conditions are favorable, are especially amenable to many human activities. Thus it is not surprising that the majority of the world's principal agricultural regions, close-meshed transportation networks, and concentrations of population are found on plains. Large parts of the Earth's plains, however, are hindered for human use by dryness, shortness of frost-free season, infertile soils, or poor drainage. Because of the absence of major differences in elevation or exposure or of obstacles to the free movement of air masses, extensive plains usually exhibit broad uniformity or gradual transition of climatic characteristics.

Somewhat more than one-third of the Earth's land area is occupied by plains. With the exception of ice-sheathed Antarctica, each continent contains at least one major expanse of smooth land in addition to numerous smaller areas. The largest plains of North America, South America, and Eurasia lie in the continental interiors, with broad extensions reaching to the Atlantic (and Arctic) Coast. The most extensive plains of Africa occupy much of the Sahara and reach south into the Congo and Kalahari basins. Much of Australia is smooth, with only the eastern margin lacking extensive plains. *See* TERRAIN AREAS.

Surfaces that approach true flatness, while not rare, constitute a minor portion of the world's plains. Most commonly they occur along low-lying coastal margins, the lower sections of major river systems, or the floors of inland basins. Nearly all are the products of extensive deposition by streams or in lakes or shallow seas. The majority of plains, however, are distinctly irregular in surface form, as a result of valley-cutting by streams or of irregular erosion and deposition by continental glaciers. [E.H.Ha.]

Plant An organism that belongs to the Kingdom Plantae (plant kingdom) in biological classification. The study of plants is called botany. *See* BOTANY; CLASSIFICATION, BIOLOGICAL.

The Plantae share the characteristics of multicellularity, cellulose cell walls, and photosynthesis using chlorophylls *a* and *b* (except for a few plants that are secondarily heterotrophic). Most plants are also structurally differentiated, usually having organs specialized for anchorage, support, and photosynthesis. Tissue specialization for photosynthetic, conducting, and covering functions is also characteristic. Plants have a sporic (rather than gametic or zygotic) life cycle that involves both sporophytic and gametophytic phases, although the latter is evolutionarily reduced in the majority of species. Reproduction is sexual, but diversification of breeding systems is a prominent feature of many plant groups. *See* PHOTOSYNTHESIS; REPRODUCTION (PLANT).

A conservative estimate of the number of described species of plants is 250,000. There are possibly two or three times that many species as yet undiscovered, primarily in the Southern Hemisphere. Plants are categorized into nonvascular and vascular groups, and the latter into seedless vascular plants and seed plants. The nonvascular plants include the liverworts, hornworts, and mosses. The vascular plants without seeds are the ground pines, horsetails, ferns, and whisk ferns; seed plants include cycads, ginkgos, conifers, gnetophytes, and flowering plants. Each of these groups constitutes a division in botanical nomenclature, which is equivalent to a phylum in the zoological system. *See* PLANT TAXONOMY. [M.La.]

Plant anatomy The area of plant science concerned with the internal structure of plants. It deals both with mature structures and with their origin and development.

The plant anatomist dissects the plant and studies it from different planes and at various levels of magnification. At the level of the cell, anatomy overlaps plant cytology, which deals exclusively with the cell and its contents. Sometimes the name plant histology is applied to the area of plant anatomy directed toward the study of cellular details of tissues. *See* PLANT CELL. [K.E.]

Plant-animal interactions The examination of the ecology of interacting plants and animals by using an evolutionary, holistic perspective. For example, the chemistry of defensive compounds of a plant species may have been altered by natural-selection pressures resulting from the long-term impacts of herbivores. Also, the physiology of modern herbivores may be modified from that of thousands of years ago as adaptations for the detoxification or avoidance of plant defensive chemicals have arisen.

The application of the theories based on an understanding of plant-animal interactions provides an understanding of problems in modern agricultural ecosystems. In addition, plant-animal interactions have practical applications in medicine. For example, a number of plant chemicals, such as digitalin from the foxglove plant, that evolved as herbivore-defensive compounds have useful therapeutic effects on humans. *See* AGROECOSYSTEM.

Effects of interaction types for each species*

Interaction	Effect on species A	Effect on species B
Mutualism	+	+
Commensalism	+	0
Antagonism	+	−
Competition	−	−
Amensalism	0	−
Neutralism	0	0

*+ = beneficial, − = harmful, 0 = neutral.

The evolutionary consequences of plant-animal interactions vary, depending on the effects on each participant. Interaction types range from mutualisms, that is, relationships which are beneficial to both participating species, to antagonisms, in which the interaction benefits only one of the participating species and negatively impacts the other. Interaction types are defined on the basis of whether the impacts of the interaction are beneficial, harmful, or neutral for each interacting species (see table). [W.G.A.]

Fossil record. Plants and animals interact in a variety of ways within modern ecosystems. These interactions may range from simple examples of herbivory (animals eating plants) to more complex interactions such as pollination or seed and fruit dispersal. Animals also rely on plants for food and shelter. The complex interactions between these organisms over geologic time not only have resulted in an abundance and diversity of organisms in time and space but also have contributed to many of the evolutionary adaptations found in the biological world.

Paleobiologists have attempted to decipher some of the interrelationships that existed between plants and animals throughout geologic time. The ecological setting in which the organisms lived in the geologic past is being analyzed in association with the fossils. Thus, as paleobiologists have increased their understanding of certain fossil organisms, it has become possible to consider some aspects of the ecosystems in which they lived, and in turn, how various types of organisms interacted.

Herbivory. Perhaps the most widespread interaction between plants and animals is herbivory, in which plants are utilized as food. One method of determining the extent of herbivory in the fossil record is by analyzing the plant material that has passed through the digestive gut of the herbivore.

The stems of some fossil plants show tissue disruption similar to various types of wounds occurring in plant parts that have been pierced by animal feeding structures. As plants developed defense systems in the form of fibrous layers covering inner, succulent tissues, some animals evolved piercing mouthparts that allowed them to penetrate these thick-walled layers. In some fossil plants, it is also possible to see evidence of wound tissue that has grown over these penetration sites. *See* HERBIVORY.

Mimicry. Another example of the interactions between plants and animals that can be determined from the fossil record is mimicry. Certain fossil insects have wings that are morphologically identical to plant leaves, thus providing camouflage from predators as the insect rested on a seed fern frond.

Pollination. The transfer of pollen from the pollen sacs to the receptive stigma in angiosperms or to the seed in gymnosperms is an example of an ancient interaction between plants and animals. It has been suggested that pollination in some groups initially occurred as a result of indiscriminate foraging behavior by certain animals, and later evolved specifically as a method to effect pollination. The size, shape, and organization of fossil pollen grains provide insight into potential pollination vectors. *See* POLLINATION. [T.N.T.]

Plant cell The basic unit of structure and function in nearly all plants. Although plant cells are variously modified in structure and function, they have many common features. The most distinctive feature of all plant cells is the rigid cell wall, which is absent in animal cells. The range of specialization and the character of association of plant cells is very wide. In the simplest plant forms a single cell constitutes a whole organism and carries out all the life functions. In just slightly more complex forms, cells are associated structurally, but each cell appears to carry out the fundamental life functions, although certain ones may be specialized for participation in reproductive processes. In the most advanced plants, cells are associated in functionally specialized tissues, and associated tissues make up organs such as the leaves, stem, and root.

Plant and animal cells are composed of the same fundamental constituents—nucleic acids, proteins, carbohydrates, lipids, and various inorganic substances—and are organized in the same fundamental manner. A characteristic of their organization is the presence of unit membranes composed of phospholipids and associated proteins and in some instances nucleic acids.

Perhaps the most conspicuous and certainly the most studied of the features peculiar to plant cells is the presence of plastids. The plastids are membrane-bound organelles with an inner membrane system. Chlorophylls and other pigments are associated with the inner membrane system. *See* CELL PLASTIDS; CHLOROPHYLL. [W.G.W.]

Plant communication Movement of signals or cues, presumably chemical, among individual plants or plant parts. These chemical cues are a consequence of damage to plant tissues and stimulate physiological changes in the undamaged "receiving" plant or tissue. There are very few studies of this phenomenon, and so theories of its action and significance are fairly speculative.

Plants produce a wealth of secondary metabolites that do not function in the main, or primary, metabolism of the plant, which includes photosynthesis, nutrient acquisition, and growth. Since many of these chemicals have very specific negative effects on animals or pathogens, ecologists speculate that they may be produced by plants as defenses. Plant chemical defenses either may be present all of the time (constitutive) or may be stimulated in response to attack (induced). Those produced in response to attack by pathogens are called phytoalexins. In order to demonstrate the presence of an induced defense, the chemistry of plant tissues or their suitability to some "enemy" (via a bioassay) must be compared before and after real or simulated attack. Changes found in the chemistry and suitability of control or unattacked plants when nearby experimental plants are damaged imply that some signal or cue has passed from damaged to undamaged plants. Controlled studies have shown that responses in undamaged plants are related to the proximity of a damaged neighbor. *See* ALLELOPATHY; PLANT METABOLISM. [J.C.Sc.]

Plant evolution The process of biological and organic change within the plant kingdom by which the characteristics of plants differ from generation to generation. The main levels (grades) of evolution have long been clear from comparisons among living plants, but the fossil record has been critical in dating evolutionary events and revealing extinct intermediates between modern groups, which are separated from each other by great morphological gaps. Plant evolution has been clarified by cladistic methods for estimating relationships among both living and fossil groups. These methods attempt to reconstruct the branching of evolutionary lines (phylogeny) by using shared evolutionary innovations (for example, presence of a structure not found in other groups) as evidence that particular organisms are descendants of the same ancestral lineage (a monophyletic group, or clade). Many traditional groups are actually grades rather than clades; these are indicated below by names in quotes.

Most botanists restrict the term plants to land plants, which invaded the land after 90% of Earth history. There is abundant evidence of photosynthetic life extending back 3.5 billion years to the early Precambrian, in the form of microfossils resembling cyanobacteria (prokaryotic blue-green algae) and limestone reefs (stromatolites) made by these organisms. Larger cells representing eukaryotic "algae" appear in the late Precambrian, followed by macroscopic "algae" and animals just before the Cambrian. *See* ALGAE; PROKARYOTAE.

Origin of land plants. Cellular, biochemical, and molecular data place the land plants among the "green algae," specifically the "charophytes," which resemble land plants in their mode of cell division and differentiated male and female gametes (oogamy). Land plants themselves are united by a series of innovations not seen in "charophytes," many of them key adaptations required for life on land. They have an alternation of generations, with a haploid, gamete-forming (gametophyte) and diploid, spore-forming (sporophyte) phase. Their reproductive organs (egg-producing archegonia, sperm-producing antheridia, and spore-producing sporangia) have a protective layer of sterile cells. The sporophyte, which develops from the zygote, begins its life inside the archegonium. The spores, produced in fours by meiosis, are air-dispersed, with a resistant outer wall that prevents desiccation. *See* REPRODUCTION (PLANT).

Land plants have been traditionally divided into "bryophytes" and vascular plants (tracheophytes). These differ in the relative role of the sporophyte, which is subordinate and permanently attached to the gametophyte in "bryophytes" but dominant and independent in vascular plants. In vascular plants, tissues are differentiated into an epidermis with a waxy cuticle that retards water loss and stomates for gas exchange, parenchyma for photosynthesis and storage, and water- and nutrient-conducting cells (xylem, phloem). However, cladistic analyses imply that some "bryophytes" are closer to vascular plants than others. This implies that the land-plant life cycle originated before the full suite of vegetative adaptations to land life, and that the sporophyte began small and underwent a trend toward elaboration and tissue specialization. *See* EPIDERMIS (PLANT); PHOTOSYNTHESIS.

In the fossil record, the first recognizable macroscopic remains of land plants are Middle Silurian vascular forms with a branched sporophyte, known as "rhyniophytes." These differed from modern plants in having no leaves or roots, only dichotomously branching stems with terminal sporangia. However, spore tetrads formed by meiosis are known from older beds (Middle Ordovician); these may represent more primitive, bryophytic plants.

In one of the most spectacular adaptive radiations in the history of life, vascular plants diversified through the Devonian. At the beginning of this period, vegetation was low and probably confined to wet areas, but by the Late Devonian, size had increased in many lines, resulting in large trees and forests with shaded understory habitats. Of the living groups of primitive vascular plants, the lycopsids (club mosses) branched off first, along with the extinct "zosterophyllopsids." A second line, the "trimerophytes," gave rise to sphenopsids (horsetails) and ferns (filicopsids). This radiation culminated in the coal swamp forests of the Late Carboniferous, with tree lycopsids (Lepidodendrales), sphenopsids (*Calamites*), and ferns (Marattiales). Remains of these plants make up much of the coal of Europe and eastern North America, which were then located on the Equator.

Seed plants. Perhaps the most significant event after the origin of land plants was evolution of the seed. Primitive seed plants ("gymnosperms") differ from earlier groups in their reproduction, which is heterosporous (producing two sizes of spores), with separate male and female gametophytes packaged inside the pollen grain (microspore), and the ovule (a sporangium with one functional megaspore, surrounded by an integument, which develops into the seed). The transfer of sperm (two per pollen grain) from one sporophyte to another through the air, rather than by swimming, represents a step toward independence from water for reproduction. This step must have helped plants invade drier areas than they had previously occupied. In addition, seed plants have new vegetative features, particularly secondary growth, which allows production of a thick trunk made up of secondary xylem (wood) surrounded by secondary phloem and periderm (bark). Together, these innovations have made seed plants the dominant organisms in most terrestrial ecosystems ever since the disappearance of the Carboniferous coal swamps. *See* ECOSYSTEM; SEED.

A major breakthrough in understanding the origin of seed plants was recognition of the "progymnosperms" in the Middle and Late Devonian. These plants, which were the first forest-forming trees, had secondary xylem, phloem, and periderm, but they still reproduced by spores, implying that the anatomical advances of seed plants arose before the seed. Like sphenopsids and ferns, they were apparently derived from "trimerophytes." The earliest seed plants of the Late Devonian and Carboniferous, called "seed ferns" because of their frondlike leaves, show steps in origin of the seed. Origin of the typical mode of branching in seed plants, from buds in the axils of the leaves, occurred at about the same time.

Seed plants became dominant in the Permian during a shift to drier climate and extinction of the coal swamp flora in the European-American tropical belt, and glaciation in the Southern Hemisphere Gondwana continents. Early conifers predominated in the tropics; extinct glossopterids inhabited Gondwana. Moderation of climate in the Triassic coincided with the appearance of new seed plant groups as well as more modern ferns. Many Mesozoic groups show adaptations for protection of seeds against animal predation, while flowers of the Bennettitales constitute the first evidence for attraction of insects for cross-pollination, rather than transport of pollen by wind.

Angiosperms. The last major event in plant evolution was the origin of angiosperms (flowering plants), the seed plant group that dominates the modern flora. The flower, typically made up of protective sepals, attractive petals, pollen-producing stamens, and ovule-producing carpels (all considered modified leaves), favors more efficient pollen transfer by insects. The ovules are enclosed in the carpel, so that pollen germinates on the sticky stigma of the carpel rather than in the pollen chamber of the ovule. The carpels (separate or fused) develop into fruits, which often show special adaptations for seed dispersal. Other advances include an extreme reduction of the gametophytes, and double fertilization whereby one sperm fuses with the egg, and the second sperm with two other gametophyte nuclei to produce a triploid, nourishing tissue called the endosperm. Angiosperms also developed improved vegetative features, such as more efficient water-conducting vessels in the wood and leaves with several orders of reticulate venation. These features may have contributed to their present dominance in tropical forests, previously occupied by conifers with scale leaves. *See* RAINFOREST.

Most botanists believed that the most primitive living angiosperms are "magnoliid dicots," based on their "gymnosperm"-like pollen, wood anatomy, and flower structure. Studies of Cretaceous fossil pollen, leaves, and flowers confirm this view by showing a rapid but orderly radiation beginning with "magnoliid"-like and monocotlike types, followed by primitive eudicots (with three pollen apertures), some related to sycamores and lotuses.

Both morphological and molecular data imply that angiosperms are monophyletic and most closely related to Bennettitales and Gnetales, a seed plant group that also radiated in the Early Cretaceous but later declined to three living genera. Since all three groups have flowerlike structures, suggesting that the flower and insect pollination arose before the closed carpel, they have been called anthophytes. These relationships, plus problematical Triassic pollen grains and macrofossils with a mixture of angiospermlike and more primitive features, suggest that the angiosperm line goes back to the Triassic, although perhaps not as fully developed angiosperms. Within angiosperms, it is believed that "magnoliids" are relatively primitive, monocots and eudicots are derived clades, and wind-pollinated temperate trees such as oaks, birches, and walnuts (Amentiferae) are advanced eudicots. However, "magnoliids" include both woody plants, and herbs, and their flowers range from large, complex, and insect-pollinated to minute, simple, and wind-pollinated. These extremes are present among the earliest Cretaceous angiosperms, and cladistic analyses disagree on which is most primitive.

Although plant extinctions at the end of the Cretaceous have been linked with radiation of deciduous trees and proliferation of fruits dispersed by mammals and birds, they were less dramatic than extinctions in the animal kingdom. Mid-Tertiary cooling led to contraction of the tropical belt and expansion of seasonal temperate and arid zones. These changes led to the diversification of herbaceous angiosperms and the origin of open grassland vegetation, which stimulated the radiation of hoofed mammals, and ultimately the invention of human agriculture. *See* AGRICULTURE; FLOWER; PLANT KINGDOM. [J.A.Do.]

Plant geography The study of the spatial distributions of plants and vegetation and of the environmental relationships which may influence these distributions. Plant geography (or certain aspects of it) is also known as phytogeography, phytochorology, geobotany, geographical botany, or vegetation science.

A flora is the collection of all plant species in an area, or in a period of time, independent of their relative abundances and relationships to one another. The species can be grouped and regrouped into various kinds of floral elements based on some common feature. For example, a genetic element is a group of species with a common evolutionary origin; a migration element has a common route of entry into the territory; a historical element is distinct in terms of some past event; and an ecological element is related to an environmental preference. An endemic species is restricted to a particular area, which is usually small and of some special interest. The collection of all interacting individuals of a given species, in an area, is called a population.

An area is the entire region of distribution or occurrence of any species, element, or even an entire flora. The description of areas is the subject of areography, while chorology studies their development. The local distribution within the area as a whole, as that of a swamp shrub, is the topography of that area. Areas are of interest in regard to their general size and shape, the nature of their margin, whether they are continuous or disjunct, and their relationships to other areas. Closely related plants that are mutually exclusive are said to be vicarious (areas containing such plants are also called vicarious). A relict area is one surviving from an earlier and more extensive occurrence. On the basis of areas and their floristic relationships, the Earth's surface is divided into floristic regions, each with a distinctive flora.

Floras and their distribution have been interpreted mainly in terms of their history and ecology. Historical factors, in addition to the evolution of the species themselves, include consideration of theories of shifting continental masses, changing sea levels, and orographic and climatic variations in geologic time, as well as theories of island biogeography, all of which have affected migration and perpetuation of floras. The main ecological factors include the immediate and contemporary roles played by climate, soil, animals, and humans. *See* ISLAND BIOGEOGRAPHY.

Vegetation refers to the mosaic of plant life found on the landscape. The vegetation of a region has developed from the numerous elements of the local flora but is shaped also by nonfloristic physiological and environmental influences. Vegetation is an organized whole, at a higher level of integration than the separate species, composed of those species and their populations. Vegetation may possess emergent properties not necessarily found in the species themselves. Sometimes vegetation is very weakly integrated, as pioneer plants of an abandoned field. Sometimes it is highly integrated, as in an undisturbed tropical rainforest. Vegetation provides the main structural and functional framework of ecosystems. *See* ECOSYSTEM.

Plant communities are an important part of vegetation. No definition has gained universal acceptance, in part because of the high degree of independence of the species themselves. Thus, the community is often only a relative social continuity in nature, bounded by a relative discontinuity, as judged by competent botanists. *See* ECOLOGICAL COMMUNITIES.

In looking at vegetation patterns over larger areas, it is the basic physiognomic distinctions between grassland, forest, and desert, with such variants as woodland (open forest), savanna (scattered trees in grassland), and scrubland (dominantly shrubs), which are most often emphasized. These general classes of vegetation structure can be broken down further by reference to leaf types and seasonal habits (such as evergreen or deciduous). Geographic considerations may complete the names of the main vegetation formation types, also called biomes (such as tropical rainforest, boreal coniferous forest, or temperate grasslands). Such natural vegetation regions are most closely related to climatic patterns and secondarily to soil or other environmental factors. *See* ALTITUDINAL VEGETATION ZONES.

Vegetational plant geography has emphasized the mapping of such vegetation regions and the interpretation of these in terms of environmental (ecological) influences. Distinction has been made between potential and actual vegetation, the latter becoming more important due to human influence. *See* VEGETATION AND ECOSYSTEM MAPPING.

Some plant geographers point to the effects of ancient human populations, natural disturbances, and the large-herbivore extinctions and climatic shifts of the Pleistocene on the species composition and dynamics of so-called virgin vegetation. On the other hand, it has been shown that the site occurrence and geographic distributions of plant and vegetation types can be predicted surprisingly well from general climatic and other environmental patterns. Unlike floristic botany, where evolution provides a single unifying principle for taxonomic classification, vegetation structure and dynamics have no single dominant influence.

Basic plant growth forms (such as broad-leaved trees, stem-succulents, or forbs) have long represented convenient groups of species based on obvious similarities. When these forms are interpreted as ecologically significant adaptations to environmental factors, they are generally called life forms and may be interpreted as basic ecological types.

In general, basic plant types may be seen as groups of plant taxa with similar form and ecological requirements, resulting from similar morphological responses to similar environmental conditions. When similar morphological or physiognomic responses occur in unrelated taxa in similar but widely separated environments, they may be called convergent characteristics. *See* PLANTS, LIFE FORMS OF.

As human populations alter or destroy more and more of the world's natural vegetation, problems of species preservation, substitute vegetation, and succession have increased in importance. This is especially true in the tropics, where deforestation is proceeding rapidly. Probably over half the species in tropical rainforests have not yet even been identified. Because nutrients are quickly washed out of tropical rainforest soils, cleared areas can be used for only a few years before they must be abandoned to erosion and much degraded substitute vegetation. Perhaps the greatest current challenge in plant geography is to understand tropical vegetation and succession sufficiently well to design self-sustaining preserves of the great diversity of tropical vegetation. *See* BIOGEOGRAPHY; ECOLOGY; RAINFOREST. [E.O.B.]

Plant growth An irreversible increase in the size of the plant. As plants, like other organisms, are made up of cells, growth involves an increase in cell numbers by cell division and an increase in cell size. Cell division itself is not growth, as each new cell is exactly half the size of the cell from which it was formed. Only when it grows to the same size as its progenitor has growth been realized. Nonetheless, as each cell has a maximum size, cell division is considered as providing the potential for growth.

While growth in plants consists of an increase in both cell number and cell size, animal growth is almost wholly the result of an increase in cell numbers. Another important difference in growth between plants and animals is that animals are determinate in growth and reach a final size before they are mature and start to reproduce. Plants have indeterminate growth and, as long as they live, continue to add new organs and tissues. In a plant new cells are produced all the time, and some parts such as leaves and flowers may die, while the main body of the plant persists and continues to grow. The basic processes of cell division are similar in plants and animals, though the presence of a cell wall and vacuole in plant cells means that there are certain important differences. This is particularly true in plant cell enlargement, as plant cells, being restrained in size by a cellulose cell wall, cannot grow without an increase in the wall. Plant cell growth is thus largely a property of the cell wall.

Sites of cell division. Cell division in plants takes place in discrete zones called meristems. The stem and root apical meristems produce all the primary (or initial) tissues of the stem and root. The cylindrical vascular cambium produces more conducting cells at the time when secondary thickening (the acquisition of a woody nature) begins. The vascular cambium is a sheet of elongated cells which divide to produce xylem or water-conducting cells on the inside, and phloem or sugar-conducting cells on the outside. Unlike the apical meristems whose cell division eventually leads to an increase in length of the stem and root, divisions of the vascular cambium occur when that part of the plant has reached a fixed length, and lead only to an increase in girth, not in length. The final meristematic zone, the cork cambium, is another cylindrical sheet of cells on the outer edge of older stems and roots of woody plants. It produces new outer cells only, and these cells differentiate into the corky layers of the bark so that new protective layers are produced as the tree increases in circumference. *See* BUD; PERIDERM; ROOT (BOTANY); STEM.

Controls. Plant growth is affected by internal and external factors. The internal controls are all the product of the genetic instructions carried in the plant. These influence the extent and timing of growth and are mediated by signals of various types transmitted within the cell, between cells, or all around the plant. Intercellular communication in plants may take place via hormones (or chemical messengers) or by other forms of communication not well understood. There are several hormones (or groups of hormones), each of which may be produced in a different location, that have a different target tissue and act in a different manner.

The external environments of the root and shoot place constraints on the extent to which the internal controls can permit the plant to grow and develop. Prime among these are the water and nutrient supplies available in the soil. Because cell expansion is controlled by cell turgor, which depends on water, any deficit in the water supply of the plant reduces cell turgor and limits cell elongation, resulting in a smaller plant. *See* PLANT-WATER RELATIONS.

Mineral nutrients are needed for the biochemical processes of the plant. When these are in insufficient supply, growth will be less vigorous, or in extreme cases it will cease altogether. *See* PLANT MINERAL NUTRITION.

An optimal temperature is needed for plant growth. The actual temperature range depends on the species. In general, metabolic reactions and growth increase with temperature, though high temperature becomes damaging. Most plants grow slowly at low temperatures, 32–50°F (0–10°C), and some tropical plants are damaged or even killed at low but above-freezing temperatures.

Light is important in the control of plant growth. It drives the process of photosynthesis which produces the carbohydrates that are needed to osmotically retain water in the cell for growth. *See* PHOTOSYNTHESIS.

Fruits and seeds. Fruits and seeds are rich sources of hormones. Initial hormone production starts upon pollination and is further promoted by ovule fertilization. These hormones promote the growth of both seed and fruit tissue. Fruits grow initially by cell division, then by cell enlargement, and finally sometimes by an increase in air spaces.

The growth of a seed starts at fertilization. A small undifferentiated cell mass is produced from the single-celled zygote. This proceeds to form a small embryo consisting of a stem tip bearing two or more leaf primordia at one end and a root primordium at the other. Either the endosperm or the cotyledons enlarge as a food store. *See* FRUIT; SEED.

Flowering. At a certain time a vegetative plant ceases producing leaves and instead produces flowers. This often occurs at a particular season of the year. The determining factor for this event is day length (or photoperiod). Different species of plants respond to different photoperiods. *See* FLOWER; PHOTOPERIODISM.

The light signal for flowering is received by the leaves, but it is the stem apex that responds. Exposing even a single leaf to the correct photoperiod can induce flowering. Clearly, then, a signal must travel from the leaf to the apex. Grafting a plant that has been photoinduced to flower to one not so induced can cause the noninduced plant to flower. It has been proposed that a flower-inducing hormone travels from the leaf to the stem apex and there induces changes in the development of the cells such that the floral morphology results.

Dormancy. At certain stages of the life cycle, most perennial plants cease growth and become dormant. Plants may cease growth at any time if the environmental conditions are unfavorable. When dormant, however, a plant will not grow even if the conditions are favorable. *See* DORMANCY.

Leaf abscission. As a perennial plant grows, new leaves are continuously or seasonally produced. At the same time the older leaves are shed because newer leaves are metabolically more efficient in the production of photosynthates. A total shedding of tender leaves may enable the plant to withstand a cold period or drought. In temperate deciduous trees, leaf abscission is brought about by declining photoperiods and temperatures. *See* ABSCISSION; PLANT MORPHOGENESIS. [P.J.D.]

Plant keys Artificial analytical constructs for identifying plants. The identification, nomenclature, and classification of plants are the domain of plant taxonomy, and one basic responsibility of taxonomists is determining if the plant at hand is identical to a known plant. The dichotomous key provides a shortcut for identifying plants that eliminates searching through numerous descriptions to find one that fits the unknown plant.

A key consists of series of pairs (couplets) of contradictory statements (leads). Each statement of a couplet must “lead” to another couplet or to a plant name. Each couplet provides an either-or proposition wherein the user must accept one lead as including the unknown plant in question and must simultaneously reject the opposing lead. The user then proceeds from acceptable lead to acceptable lead of successive couplets until a name for the unknown plant is obtained. For confirmation, the newly identified plant should then be compared with other known specimens or with detailed descriptions of that species.

Keys are included in monographic or revisionary treatments of groups of plants, most often for a genus or family. Books containing extended keys, coupled with detailed descriptions of each kind of plant (taxon), for a given geographic region are called manuals or floras, though the latter technically refers to a simple listing of names of plants for a given region. *See* PLANT KINGDOM; PLANT TAXONOMY. [D.J.Pi.]

Plant kingdom The worldwide array of plant life, including plants that have roots in the soil, plants that live on or within other plants and animals, plants that float on or swim in water, and plants that are carried in the air. Fungi used to be included in the plant kindom because they looked more like plants than animals and did not move about. It is now known that fungi are probably closer to animals in terms of their evolutionary relationships. Also once included in plants were the "blue-green algae," which are now clearly seen to be bacteria, although they are photosynthetic (and presumably the group of organisms from which the chloroplasts present in true plants were derived). The advent of modern methods of phylogenetic DNA analysis has allowed such distinctions, but even so, what remains of the plantlike organisms is still remarkably divergent and difficult to classify.

Plants range in size from unicellular algae to giant redwoods. Some plants complete their life cycles in a matter of hours, whereas the bristlecone pines are known to be over 4000 years old. Plants collectively are among the most poorly understood of all forms of life, with even their most basic functions still inadequately known, including how they sense gravity and protect themselves from infection by bacteria, viruses, and fungi. Furthermore, new species are being recorded every year.

Within the land plants, a great deal of progress has been made in sorting out phylogenetic (evolutionary) relationships of extant taxa based on DNA studies, and the system of classification listed below includes these changes. The angiosperms or flowering plants (Division Magnoliophyta) have recently been reclassified based on phylogenetic studies of DNA sequences. Within the angiosperms, several informal names are indicated in parentheses; these names may at some future point be formalized, but for the present they are indicated in lowercase letters because they have not been formally recognized under the Code of Botanical Nomenclature.

It is known that the bryophytes (Division Bryophyta) are not closely related to each other, but which of the three major groups is closest to the other land plants is not yet clear. Among the extant vascular plants, Lycophyta are the sister group to all the rest, with all of the fernlike groups forming a single monophyletic (natural) group, which is reflected here in the classification by putting them all under Polypodiophyta. This group is the sister to the extant seed plants, within which all gymnosperms form a group that is sister to the angiosperms. Therefore, if Division is taken as the highest category within Embryobionta (the embryo-forming plants), then the following scheme would reflect the present state of knowledge of relationships (an asterisk indicates that a group is known only from fossils).

Subkingdom Thallobionta (thallophytes)
- Division Rhodophycota (red algae)
  - Class Rhodophyceae
- Division Chromophycota
  - Class: Chrysophyceae (golden or golden-brown algae)
    - Prymnesiophyceae
    - Xanthophyceae (yellow-green algae)
    - Eustigmatophyceae
    - Bacillariophyceae (diatoms)
    - Dinophyceae (dinoflagellates)
    - Phaeophyceae (brown algae)
    - Raphidophyceae (chloromonads)
    - Cryptophyceae (cryptomonads)
- Division Euglenophycota (euglenoids)
  - Class Euglenophyceae
- Division Chlorophycota (green algae)
  - Class: Chlorophyceae
    - Charophyceae
    - Prasinophyceae

Subkingdom Embryobionta (embryophytes)
- Division Rhyniophyta*
  - Class Rhyniopsida
- Division Bryophyta
  - Class Hepaticopsida (liverworts)
    - Subclass Jungermanniidae

Order: Takakiales
Calobryales
Jungermanniales
Metzgeriales
Subclass Marchantiidae
Order: Sphaerocarpales
Monocleales
Marchantiales
Class: Anthocerotopsida (hornworts)
Sphagnopsida (peatmosses)
Andreaeopsida (granite mosses)
Bryopsida (mosses)
Subclass: Archidiidae
Bryidae
Order: Fissidentales
Bryoxiphiales
Schistostegales
Dicranales
Pottiales
Grimmiales
Seligeriales
Encalyptales
Funariales
Splachnales
Order: Bryales
Mitteniales
Orthotrichales
Isobryales
Hookeriales
Hypnales
Subclass: Buxbaumiidae
Tetraphididae
Dawsoniidae
Polytrichidae
Division Lycophyta
Class Lycopsida
Order: Lycopodiales
Asteroxylales*
Protolepidodendrales*
Selaginellales*
Lepidedendrales*
Isoetales
Class Zosterophyllopsida*
Division Polypodiophyta
Class Polypodopsida
Order: Equisetales
Marattiales
Sphenophyllales*
Pseudoborniales*
Psilotales
Ophioglossales
Noeggerathiales*
Protopteridales*
Polypodiales
Class Progymnospermopsida*
Division Pinopsida
Class Ginkgoopsida
Order: Calamopityales*
Callistophytales*
Peltaspermales*
Ginkgoales
Leptostrobales*
Caytoniales
Arberiales*
Pentoxylales*
Class Cycadopsida
Order: Lagenostomales*
Trigonocarpales*
Cycadales
Bennettiales*
Class Pinopsida
Order: Cordaitales*
Pinales
Podocarpales
Gnetales
Division Magnoliophyta (angiosperms, flowering plants)
Class Magnoliopsida
unplaced groups: Amborellaceae, Ceratophyllaceae, Chloranthaceae, Nymphaeaceae, etc.
eumagnoliids
Order: Magnoliales
Laurales
Piperales
Winterales
monocotyledons
Order: Acorales
Alismatales
Asparagales
Dioscoreales
Liliales
Pandanales
commelinids
Arecales
Commelinales
Poales
Zingiberales
eudicotyledons
(basal eudicots)

Order: Ranunculales
Proteales
Buxales
Trochodendrales
(core eudicots)
Order: Berberidopsidales
Gunnerales
Dilleniales
Santalales
Caryophyllales
Saxifragales
Rosidae
Order: Vitales
Myrtales
Geraniales
Crossosomatales
(eurosid I)
Order: Celastrales
Cucurbitales
Fabales
Fagales
Malpighiales
Oxalidales
Rosales
Zygophyllales
(eurosid II)
Order: Brassicales
Malvales
Sapindales
Asteridae
Order: Cornales
Ericales
(euasterid I)
Order: Garryales
Gentianales
Lamiales
Solanales
(euasterid II)
Order: Apiales
Aquifoliales
Asterales
Dipsacales

See DEOXYRIBONUCLEIC ACID (DNA); PLANT EVOLUTION; PLANT PHYLOGENY; PLANT TAXONOMY. [M.W.C.; M.F.F.]

Plant metabolism The complex of physical and chemical events of photosynthesis, respiration, and the synthesis and degradation of organic compounds. Photosynthesis produces the substrates for respiration and the starting organic compounds used as building blocks for subsequent biosyntheses of nucleic acids, amino acids, and proteins, carbohydrates and organic acids, lipids, and natural products. *See* PHOTORESPIRATION; PHOTOSYNTHESIS; PLANT RESPIRATION. [I.P.T.]

Plant mineral nutrition The relationship between plants and all chemical elements other than carbon, hydrogen, and oxygen in the environment. Plants obtain most of their mineral nutrients by extracting them from solution in the soil or the aquatic environment. Mineral nutrients are so called because most have been derived from the weathering of minerals of the Earth's crust. Nitrogen is exceptional in that little occurs in minerals: the primary source is gaseous nitrogen of the atmosphere.

Some of the mineral nutrients are essential for plant growth; others are toxic, and some absorbed by plants may play no role in metabolism. Many are also essential or toxic for the health and growth of animals using plants as food. Six basic facts have been established: (1) plants do not need any of the solid materials in the soil—they cannot even take them up; (2) plants do not need soil microorganisms; (3) plant roots must have a supply of oxygen; (4) all plants require at least 14 mineral nutrients; (5) all of the essential mineral nutrients may be supplied to plants as simple ions of inorganic salts in solution; and (6) all of the essential nutrients must be supplied in adequate but nontoxic quantities. These facts provide a conceptually simple definition of and test for an essential mineral nutrient. A mineral nutrient is regarded as essential if, in its absence, a plant cannot complete its life cycle.

Nutrients which plants require in relatively large amounts, that is, the essential macronutrients, are nitrogen, sulfur, phosphorus, calcium, potassium, and magnesium.

Iron is not required in large amounts and hence is regarded as an essential micronutrient or trace element. With the progressive development of better techniques for purifying water and salts, the list of essential nutrients for all plants has expanded to include boron, manganese, zinc, copper, molybdenum, and chlorine. Evidence has accumulated in support of nickel being essential. In addition, sodium and silicon have been shown to be essential for some plants, beneficial to some, and possibly of no benefit to others. Cobalt has also been shown to be essential for the growth of legumes when relying upon atmospheric nitrogen. Claims that two other chemical elements (vanadium and selenium) may be essential micronutrients have still to be firmly established.

Mineral nutrients may be toxic to plants either because the specific nutrient interferes with plant metabolism or because its concentration in combination with others in solution is excessive and interferes with the plant's water relations. Other chemical elements in the environment may also be toxic. High concentrations of salts in soil solutions or aquatic environments may depress their water potential to such an extent that plants cannot obtain sufficient water to germinate or grow. Some desert plants growing in saline soils can accumulate salt concentrations of 20–50% dry weight in their leaves without damage, but salt concentrations of only 1–2% can damage the leaves of many species. *See* PLANT-WATER RELATIONS.

A number of elements interfere directly with other aspects of plant metabolism. Sodium is thought to become toxic when it reaches concentrations in the cytoplasm that depress enzyme activity or damage the structure of organelles, while the toxicity of selenium is probably due to its interference in metabolism of amino acids and proteins. The ions of the heavy metals, cobalt, nickel, chromium, manganese, copper, and zinc are particularly toxic in low concentrations, especially when the concentration of calcium in solution is low; increasing calcium increases the plant's tolerance. Aluminum is toxic only in acid soils. Boron may be toxic in soils over a wide pH range, and is a serious problem for sensitive crops in regions where irrigation waters contain excessive boron or where the soils contain unusually high levels of boron.

All plants grow poorly on very acid soils ($pH \leq 3.5$); some plants may grow reasonably well on somewhat less acid soils. Several factors may be involved, and their interactions with plant species are complex. The harmful effects of soil acidity in some areas have been exacerbated by industrial emissions resulting in acid rain and in deposition of substances which increase the acidity on further reaction in the soil, with consequent damage to plants and animals in these ecosystems. *See* ACID RAIN.

The elemental composition of plants is important to the health and productivity of animals which graze them. With the exception of boron, all elements which are essential for plant growth are also essential for herbivorous mammals. Animals also require sodium, iodine, and selenium and, in the case of ruminant herbivores, cobalt. As a result, animals may suffer deficiencies of any one of this latter group of elements when ingesting plants which are quite healthy but contain low concentrations of these elements. In addition, nutrients in forage may be rendered unavailable to animals through a variety of factors that prevent their absorption from the gut. Plants and animals differ also in their tolerance of high levels of nutrients, sometimes with deleterious results for grazing animals. For example, the toxicity of high concentrations of selenium in plants to animals grazing them, known as selenosis, was recognized when the puzzling and long-known "alkali disease" and "blind staggers" in grazing livestock in parts of the Great Plains of North America were shown to be symptoms of chronic and acute selenium toxicity. *See* NITROGEN CYCLE; PLANT TRANSPORT OF SOLUTES; RHIZOSPHERE; ROOT (BOTANY); SOIL CHEMISTRY. [J.F.L.]

Plant morphogenesis The origin and development of plant form and structure. Morphogenesis may be concerned with the whole plant, with a plant part, or with the subcomponents of a structure.

The establishment of differences at the two ends of a structure is called polarity. In plants, polar differences can be recognized very early in development. In the zygote, cytological differences at the two ends of the cell establish the position of the first cell division, and thus the fate of structures produced from the two newly formed cells. During the development of a plant, polarity is also exhibited in the plant axis (in the shoot and root tips). If a portion of a shoot or root is excised and allowed to regenerate, the end toward the shoot tip always regenerates shoots whereas the opposite end forms roots. Polarity is also evident on the two sides of a plant organ, such as the upper and lower surface of a leaf, sepal, or petal.

The diversity in plant form is produced mainly because different parts of the plant grow at different rates. Furthermore, the growth of an individual structure is different in various dimensions. Thus the rate of cell division and cell elongation as well as the orientation of the plane of division and of the axis of cell elongation ultimately establish the form of a structure. Such differential growth rates are very well orchestrated by genetic factors. Although the absolute growth rates of various parts of a plant may be different, their relative growth rates, or the ratio of their growth rates, are always constant. This phenomenon is called allometry (or heterogony), and it supports the concept that there is an interrelationship between the growth of various organs of a plant body. *See* PLANT GROWTH.

During development, either the removal of or changes in one part of the plant may drastically affect the morphogenesis of one or more other parts of the plant. This phenomenon is called correlation and is mediated primarily through chemical substances, such as nutrients and hormones.

The ultimate factors controlling the form of a plant and its various organs are the genes. In general, several genes interact during the development of a structure, although each gene plays a significant role. Thus, a mutation in a single gene may affect the shape or size of a leaf, flower, or fruit, or the color of flower petals, or the type of hairs produced on stems and leaves. There are at least two classes of genes involved in plant morphogenesis: regulatory genes that control the activity of other genes, and effector genes that are directly involved in a developmental process. The effector genes may affect morphogenesis through a network of processes, including the synthesis and activity of proteins and enzymes, the metabolism of plant growth substances, changes in the cytoskeleton and the rates and planes of cell division, and cell enlargement.

Plant form is also known to be affected by nutritional factors, such as sugars or nitrogen levels. For example, leaf shape can be affected by different concentrations of sucrose, and the sexuality of flowers is related to the nitrogen levels in the soil in some species. Inorganic ions (such as silver and cobalt) have also been known to affect the type of flower produced. *See* PLANT MINERAL NUTRITION.

Although genes are the ultimate controlling factors, they do not act alone, but interact with the existing environmental factors during plant development. Environmental factors, including light, temperature, moisture, and pressure, affect plant form. *See* PHYSIOLOGICAL ECOLOGY (PLANT); PLANT-WATER RELATIONS. [V.K.S.]

Plant movements The wide range of movements that allow plants to reorient themselves in relation to changed surroundings, to facilitate spore or seed dispersal, or, in the case of small free-floating aquatic plants, to migrate to regions optimal for their activities. There are two types of plant movement: abiogenic movements, which arise purely from the physical properties of the cells and therefore take place in nonliving tissues or organs; and biogenic movements, which occur in living cells or organs and require an energy input from metabolism.

Abiogenic movements. Drying or moistening of certain structures causes differential contractions or expansions on the two sides of cells and hence causes movements of curvature. Such movements are called hygroscopic and are usually associated with seed and spore liberation and dispersal. Examples of such movement occur in the "parachute" hairs of the fruit of dandelion (*Taraxacum officinale*), which are closed when damp but open when the air is dry to induce release from the heads and give buoyancy for wind dispersal.

Another type of abiogenic movement is due to changes in volume of dead water-containing cells. In the absence of a gas phase, water will adhere to lignocellulose cell walls. As water is lost by evaporation from the surface of these cells, considerable tensions can build up inside, causing them to decrease in volume while remaining full of water. The effect is most commonly seen in some grasses of dry habitats, such as sand dunes, where longitudinal rows of cells on one side of the leaf act as spring hinges, contracting in a dry atmosphere and causing the leaf to roll up into a tight cylinder, thus minimizing water loss by transpiration.

Biogenic movements. There are two types of biogenic movement. One of these is locomotion of the whole organism and is thus confined to small, simply organized units in an aqueous environment. The other involves the change in shape and orientation of whole organs of complex plants, usually in response to specific stimuli.

Locomotion. In most live plant cells the cytoplasm can move by a streaming process known as cyclosis. Energy for cyclosis is derived from the respiratory metabolism of the cell. The mechanism probably involves contractile proteins very similar to the actomyosin of animal muscles.

Cell locomotion is a characteristic of many simple plants and of the gametes of more highly organized ones. Motility in such cells is produced by cilia anchored in the peripheral layers of the cell and projecting into the surrounding medium.

Cell locomotion is usually not random but is directed by some environmental gradient. Thus locomotion may be in response to specific chemicals, in which case it is called chemotaxis. Light gradients induce phototaxis; temperature gradients induce thermotaxis; and gravity induces geotaxis. One or more of these environmental factors may operate to control movement to optimal living conditions.

Movement of organs. In higher plants, organs may change shape and position in relation to the plant body. When bending or twisting of the organ is evoked spontaneously by some internal stimulus, it is termed autonomous movement. The most common movements, however, are those initiated by external stimuli such as light and the force of gravity. Of these there are two kinds. In nastic movements (nasties), the stimulus usually has no directional qualities (such as a change in temperature), and the movement is therefore not related to the direction from which the stimulus comes. In tropisms, the stimulus has a direction (for instance, gravitational pull), and the plant movement direction is related to it.

The most common autonomous movement is circumnutation, a slow, circular, sometimes waving movement of the tips of shoots, roots, and tendrils as they grow; one complete cycle usually takes from 1 to 3 h. These movements are due to differential growth, but some may be caused by turgor changes in the cells of special hinge organs and are thus reversible.

1. *Nastic movements*. There are two kinds of nastic movements, due either to differential growth or to differential changes in the turgidity of cells. They can be triggered by a wide variety of external stimuli.

Photonastic (light/dark trigger) movements are characteristic of many flowers and inflorescences, which usually open in the light and close in the dark. Thermonasty (temperature-change trigger) is seen in the tulip and crocus flowers, which open in a warm room and close again when cooled. The most striking nastic movements are seen in the sensitive plant (*Mimosa pudica*). Its multipinnate leaves are very sensitive to touch or slight injury. Leaflets fold together, pinnae collapse downward, and the whole leaf sinks to hang limply.

Epinasty and hyponasty occur in leaves as upward and downward curvatures respectively. They arise either spontaneously or as the result of an external stimulus, such as exposure to the gas ethylene in the case of epinasty; they are not induced by gravity.

2. *Tropisms.* Of these the most universal and important are geotropism (or more properly gravitropism) and phototropism; others include thigmotropism and chemotropism.

In geotropism, the stimulus is gravity. The main axes of most plants grow in the direction of the plumb line with shoots upward (negative geotropism) and roots downward (positive geotropism).

In phototropism the stimulus is a light gradient, and unilateral light induces similar curvatures; those toward the source are positively phototropic; those away from the source are negatively phototropic. Main axes of shoots are usually positively phototropic, while the vast majority of roots are insensitive.

In thigmotropism (sometimes called haptotropism), the stimulus is touch; it occurs in climbing organs and is responsible for tendrils curling around a support. In many tendrils the response may spread from the contact area, causing the tight coiling of the basal part of the tendril into an elaborate and elastic spring.

Chemotropism is induced by a chemical substance. Examples are the incurling of the stalked digestive glands of the insectivorous plant *Drosera* and incurling of the whole leaf of *Pinguicula* in response to the nitrogenous compounds in the insect prey. A special case of chemotropism concerns response to moisture gradients; for example, under artificial conditions in air, the primary roots of some plants will curve toward and grow along a moist surface. This is called hydrotropism and may be of importance under natural soil conditions in directing roots toward water sources. *See* PLANT PHYSIOLOGY.

[L.J.A.]

Plant pathology The study of disease in plants; it is an integration of many biological disciplines and bridges the basic and applied sciences. As a science, plant pathology encompasses the theory and general concepts of the nature and cause of disease, and yet it also involves disease control strategies, with the ultimate goal being reduction of damage to the quantity and quality of food and fiber essential for human existence.

Kinds of plant diseases. Diseases were first classified on the basis of symptoms. Three major categories of symptoms were recognized long before the causes of disease were known; necroses, destruction of cell protoplasts (rots, spots, wilts); hypoplases, failure in plant development (chlorosis, stunting); and hyperplases, overdevelopment in cell number and size (witches'-brooms, galls). This scheme remains useful for recognition and diagnosis.

When fungi, and then bacteria, nematodes, and viruses, were recognized as causes of disease, it became convenient to classify diseases according to the responsible agent. If the agents were infectious (biotic), the diseases were classified as being "caused by bacteria," "caused by nematodes," or "caused by viruses." To this list were added phanerogams and protozoans, and later mollicutes (mycoplasmas, spiroplasmas), rickettsias, and viroids. In a second group were those diseases caused by such noninfectious (abiotic) agents as air pollutants, inadequate oxygen, and nutrient excesses and deficiencies.

Other classifications of disease have been proposed, such as diseases of specific plant organs, diseases involving physiological processes, and diseases of specific crops or crop groups (for example, field crops, fruit crops, vegetable crops).

Symptoms of plant diseases. Symptoms are expressions of pathological activity in plants. They are visible manifestations of changes in color, form, and structure: leaves may become spotted, turn yellow, and die; fruits may rot on the plants or in storage; cankers may form on stems; and plants may blight and wilt. Diagnosticians learn how to associate certain symptoms with specific diseases, and they use this knowledge in the identification and control of pathogens responsible for the diseases.

Those symptoms that are external and readily visible are considered morphological. Others are internal and primarily histological, for example, vascular discoloration of the xylem of wilting plants. Microscopic examination of diseased plants may reveal additional symptoms at the cytological level, such as the formation of tyloses (extrusion of living parenchyma cells of the xylem of wilted tissues into vessel elements).

It is important to make a distinction between the visible expression of the diseased condition in the plant, the symptom, and the visible manifestation of the agent which is responsible for that condition, the sign. The sign is the structure of the pathogen, and when present it is most helpful in diagnosis of the disease.

All symptoms may be conveniently classified into three major types because of the manner in which pathogens affect plants. Most pathogens produce dead and dying tissues, and the symptoms expressed are categorized as necroses. Early stages of necrosis are evident in such conditions as hydrosis, wilting, and yellowing. As cells and tissues die, the appearance of the plant or plant part is changed, and is recognizable in such common conditions as blight, canker, rot, and spot.

Many pathogens do not cause necrosis, but interfere with cell growth or development. Plants thus affected may eventually become necrotic, but the activity of the pathogen is primarily inhibitory or stimulatory. If there is a decrease in cell number or size, the expressions of pathological activity are classified as hypoplases; if cell number or size is increased, the symptoms are grouped as hyperplases. These activities are very specific and most helpful in diagnosis. In the former group are such symptoms as mosaic, rosetting, and stunting, with obvious reduction in plant color, structure, and size. In the latter group are gall, scab, and witches'-broom, all visible evidence of stimulation of growth and development of plant tissues. *See* CROWN GALL. [C.W.B.]

The primary agents of plant disease are fungi, bacteria, viruses and viroids, nematodes, parasitic seed plants, and a variety of noninfectious agents.

Fungi. More plant diseases are caused by fungi than by any other agent. The fungi that cause plant disease derive their food from the plant (host) and are called parasites. Those that can live and grow only in association with living plant tissues are obligate parasites. Some fungi obtain their food from dead organic matter and are known as saprobes or saprophytes. Still others can utilize food from either dead organic matter or from living plant cells, and are referred to as either facultative parasites or facultative saprophytes. The classes with plant disease-causing fungi are Plasmodiophoromycetes, Chytridiomycetes, Zygomycetes, Oomycetes, Ascomycetes, Basidiomycetes, and Deuteromycetes. *See* FUNGI. [C.W.E.]

Bacteria. Over 100 species of bacteria mainly in five genera cause disease in hundreds of different species of flowering plants. Destructive bacterial diseases affect the major cereal, vegetable, and fruit crops. None of the bacterial pathogens of plants causes serious diseases of humans or animals, and certain groups of green plants (mosses, ferns, conifers, and hardwood trees) have few or no major bacterial diseases. *See* BACTERIA.

Each species of bacteria produces a distinctive pattern of symptoms on those hosts that it attacks.

With a few exceptions, most of the bacteria that cause disease in plants are non-spore-forming, rod-shaped, gram-negative cells. It is not possible to separate plant pathogenic bacteria into species on the basis of colony characteristics, cell morphology, or staining characteristics. Therefore, biochemical and physical tests used to differentiate bacteria in general are also used in studies on plant pathogens. Helpful techniques include serological tests, DNA hybridization, sensitivity to phages, and gel electrophoresis of proteins. However, one of the most important of the tests for identification is the demonstration of pathogenicity to a specific plant.

Many foliage pathogens are dependent upon wind-driven splashing rain as the primary means of spread from plant to plant. Bacteria also may be spread from plant to plant by insects, in irrigation water, and by various cultural operations during the growing season. Bacteria can also survive the winter in insects such as flea beetles. Most bacterial plant pathogens survive adverse conditions in host plants. Only a small number of species survive for long periods of time in the soil in the absence of host plants. Bacterial plant pathogens are not capable of forcing their way through the cuticle of a leaf or bark of a stem. They must enter through wounds or natural openings.

The exact manner by which bacteria induce disease in plants is not fully understood. The surface area of hundreds of thousands of cells in intimate contact with the surrounding plant cells is very large, and enzymes, as well as toxic materials, can be released readily into the host tissue.

Certain vascular parasites produce gumlike substances or polysaccharides. Masses of bacterial cells embedded in the polysaccharide material and host responses to hormone imbalances reduce the rate of flow of water when such bacteria as the one causing bacterial wilt of tomato invade water-conducting tissue. The genetic control for tumor induction in the crown gall bacterium has been shown to reside in a plasmid, a nonchromosomal DNA element. The presence of the tumor-inducing plasmid ensures that genetic information in the bacterial cell is transferred and maintained in the transformed cells of the plant cancer resulting from infection.

Since primary sources of inoculum for new infections are often in undecomposed plant debris or seed, crop rotation and the use of pathogen-free seed or seed treatments are, in general, effective in reducing losses from a large number of bacterial diseases affecting foliage. Resistant varieties have been developed for a number of foliage diseases because spraying or dusting, in general, has been relatively ineffective. Certain antibiotics are effective, but not widely used because of the concern for the development of resistant strains. An insect-transmitted bacterium can be controlled if populations of the insect vector are reduced. [A.K.]

Viruses and viroids. Viruses and viroids are the simplest of the various causative agents of plant disease. The essential element of each of these two pathogens is an infective nucleic acid. The nucleic acid of viruses is covered by an exterior shell (coat) of protein, but that of viroids is not. *See* PLANT VIRUSES AND VIROIDS.

Approximately 400 plant viruses and about 10 viroids are known. The nucleic acid of most plant viruses is a single-stranded RNA; a number of isometric viruses have a double-stranded RNA. A few viruses contain double-stranded DNA, and several containing single-stranded DNA have been reported. The nucleic acid of viroids is a single-stranded RNA, but its molecular weight is much lower than that of viruses.

Some viruses, such as tobacco mosaic virus (TMV) and cucumber mosaic virus, are found in many plant species; others, such as wheat streak mosaic virus, occur only in a few grasses. Viruses are transmitted from plant to plant in several ways. The majority are transmitted by vectors such as insects, mites, nematodes, and fungi which acquire viruses during feeding upon infected plants. Some viruses are transmitted to succeeding generations by infected seed. Viroids are spread mainly by contact between healthy and diseased plants or by the use of contaminated cutting tools.

The control or prevention of virus diseases involves breeding for resistance, propagation of virus-free plants, use of virus-free seed, practices designed to reduce the spread by vectors, and, in some cases, the deliberate inoculation of plants with mild strains of a virus to protect them from the deleterious effects of severe strains. *See* PLANT VIRUSES AND VIROIDS. [R.I.H.]

Nematodes. All soils that support plant life contain nematodes living in the water films that surround soil particles. Most nematodes feed primarily on microscopic plants, animals, and bacteria, but a few are parasites of animals; another relatively small group of nematodes parasitize plants. *See* NEMATA.

Plant-parasitic nematodes are distinguished by their small size, about 1 mm average length, and mouthparts that are modified to form a hollow stylet which is inserted into plant cells. All of the plant parasites are placed into two orders, Tylenchida and Dorylaimida.

Plant injury is of three general types and is related to feeding habits. Migratory endoparasites destroy tissues as they feed, producing necrotic lesions in the root cortex. Other migratory endoparasites invade leaf tissues and produce extensive brown spots. Sedentary endoparasites do not kill host cells, but induce changes in host tissues, which lead to an elaborate feeding site or gall. The third general type of symptom is produced by certain migratory ectoparasites, where root tips are devitalized and cease to grow without any associated swelling or necrosis.

In addition to the plant injury that they cause directly, nematodes are important factors in disease complexes. Lesions and galls provide entrance courts for soil fungi and bacteria, and many diseases caused by soil-borne pathogens are more severe when nematodes are present. Important viruses are transmitted by nematodes of the order Dorylaimida.

Control of plant-parasitic nematodes often is based on selection of nonhosts for crop rotations or nematode-resistant varieties. Some plant species release compounds into the soil that are toxic to nematodes. Animal manures, compost, and other organic amendments enhance the buildup of natural enemies of nematodes. [R.A.R.]

Many parasitic seed plants (estimated at nearly 3000) attack other higher plants. In some families (for example, the mistletoes Loranthaceae and Viscaceae) all members are parasitic; in others, only a single genus is parasitic in an otherwise autotrophic family.

Most parasitic plants are terrestrial; that is, the parasitic connection with the host plant is through the roots. Other parasitic plants grow on the above-ground parts of the host. Some plants are classed as semiparasites because they can live in the soil as independent plants for a time, but are not vigorous or may not flower if they do not become attached to a suitable host. The nutritional status of parasitic plants ranges from total parasites with no chlorophyll (for example, the broomrapes, Orobanchaceae) to plants that are well supplied with chlorophyll and obtain primarily water and minerals from their hosts (many mistletoes; see illustration). [F.G.H.]

Noninfectious agents of disease. Plants with symptoms caused by noninfectious agents cannot serve as sources of further spread of the same disorder. Such noninfectious agents may be deficiencies or excesses of nutrients, anthropogenic pollutants, or biological effects by organisms external to the affected plants. On the farm, plant-damaging pollution may be caused by careless use of pesticides. Mishandled herbicides

Many small plants of a dwarf mistletoe (*Arceuthobium vaginatum*) parasitizing a ponderosa pine branch. This is the most damaging disease of ponderosa pine in many parts of the West.

are by far the most damaging to plants. Off the farm, anthropogenic air pollutants are generated by industrial processes, and by any heating or transportation method that uses fossil fuels. The most common air pollutants that damage plants are sulfur oxides and ozone. Sulfur oxides are produced when sulfur-containing fossil fuels are burned or metallic sulfides are refined. Human-generated ozone is produced by sunlight acting on clouds of nitrogen oxides and hydrocarbons that come primarily from automobile exhausts. *See* AIR POLLUTION; WATER POLLUTION. [G.N.A.]

Epidemiology of plant disease. Epidemiology is the study of the intensification of disease over time and the spread of disease in space. The botanical epidemiologist is concerned with the interrelationships of the host plant (suscept), the pathogen, and the environment, which are the components of the disease triangle. With a thorough knowledge of these components, the outbreak of disease may be forecast in advance, the speed at which the epidemic will intensify may be determined, control measures can be applied at critical periods, and any yield loss to disease can be projected. The maximum amount of disease occurs when the host plant is susceptible, the pathogen is aggressive, and the environment is favorable.

Epidemiologically, there are two main types of diseases: monocyclic, those that have but a single infection cycle (with the rare possibility of a second or even third cycle) per crop season; and polycyclic, those that have many, overlapping, concatenated cycles of infection per crop season. For both epidemiological types, the increase of disease slows as the proportion of disease approaches saturation or 100%. [R.D.Ber.]

Control of plant disease is defined as the maintenance of disease severity below a certain threshold, which is determined by economic losses. Diseases may be high in incidence but low in severity, or low in incidence but high in severity, and are kept in check by preventing the development of epidemics. The principles of plant disease control form the basis for preventing epidemics. However, the practicing agriculturist uses three approaches to the control of plant disease: cultural practices affecting the environmental requirement of the suscept-pathogen-environment triangle necessary for disease development, disease resistance, and chemical pesticides. [R.E.St.]

Plant phylogeny The evolutionary chronicle of plant life on the Earth. Understanding of this history is largely based on knowledge of extant plants, but the fossil record is playing an increasingly important role in refining and illuminating this picture. Study of deoxyribonucleic acid (DNA) sequences has also been revolutionizing this process in recent years. The molecular data (largely in the form of DNA sequences from several genes) have been demonstrated to be highly correlated with other information. *See* PHYLOGENY.

"Algae" was once a taxonomic designation uniting the lower photosynthetic organisms, but ultrastructural and molecular data have uncovered a bewildering diversity of species. Algae are now recognized as 10 divergent lineages on the tree of life that join organisms as distinct as bacteria and eukaryotic protozoans, ciliates, fungi, and embryophytes (including the land plants). In a biochemical context, the term "algae" defines species characterized by chlorophyll *a* photosynthesis (except Embryophyta); some of their descendants are heterotrophic (secondary chloroplast loss). Despite the variety of species it encompasses, the term "algae" also retains phylogenetic relevance. *See* ALGAE; CHLOROPHYLL; PHOTOSYNTHESIS. [G.W.Sa.]

Embryobionta, or embryophytes, are largely composed of the land plants that appear to have emerged 475 million years ago. The evidence indicates that land plants have not evolved from different groups of green algae (Chlorophyta) as suggested in the past, but instead share a common ancestor, which was a green alga. Land plants all have adaptations to the terrestrial environment, including an alternation of generations (sporophyte or diploid and gametophyte or haploid) with the sporophyte generation producing haploid spores that are capable of resisting desiccation and dispersing widely, a cuticle covering their outside surfaces, and separate male and female reproductive organs in the gametophyte stage. The life history strategies of land plants fall into two categories that do not reflect their phylogenetic relationships. The mosses, hornworts, and liverworts represent the first type, and they have expanded the haploid generation, upon which the sporophyte is dependent. Several recent analyses of DNA data as well as evidence from mitochondrial DNA structure have demonstrated that the liverworts alone are the remnants of the earliest land plants and that the mosses and hornworts are closer to the vascular plants (tracheophytes). The tracheophytes include a large number of extinct and relatively simple taxa, such as the rhiniophytes and horneophytes known only as Silurian and Devonian fossils. All tracheophytes are of the second category, and they have expanded the sporophyte generation. Among extant tracheophytes, the earliest branching are the lycopods or club mosses (*Lycopodium* and *Selaginella*), and there are still a diversity of other forms, including sphenophytes (horsetails, *Equisetum*) and ferns (a large and diverse group in which the positions of several families still are not clear).

All seeds plants take the reduction of the gametophyte generation a step further and make it dependent on the sporophyte, typically hiding it within reproductive structures, which are either cones or flowers. The first seed plants originated at least by the Devonian, and they are known to have a great diversity of extinct forms, including the seed ferns. There are two groups of extant seed-bearing plants, gymnosperms and angiosperms. In the gymnosperms, the seeds are not enclosed within tissue derived from the parent plant. There are four distinct groups of extant gymnosperms, often recognized as classes: Cycadopsida, Gnetopsida, Ginkgoopsida, and Pinopsida.

The angiosperms (also flowering plants or Magnoliopsida) are the dominant terrestrial plants, although the algae collectively must still be acknowledged as the most important in the maintenance of the Earth's ecological balance (fixation of carbon dioxide and production of oxygen). In angiosperms the seeds are covered by protective tissues derived from the parental plant. There are no generally accepted angiospermous fossils older than 120 million years, but the lineage is clearly much older based on DNA clocks and other circumstantial lines of evidence such as their current geographic distributions.

Traditionally the angiosperms have been divided into two groups, monocotyledons (monocots) and dicotyledons (dicots), based on the number of seed leaves. However, DNA sequence data have demonstrated that, although there are two groups, these are characterized by fundamentally different pollen organization, such that the monocots share with a group of dicots pollen with one pore whereas the rest of the dicots have pollen with three (or more) pores. [M.W.C.]

Plant physiology That branch of plant sciences that aims to understand how plants live and function. Its ultimate objective is to explain all life processes of plants by a minimal number of comprehensive principles founded in chemistry, physics, and mathematics.

Plant physiology seeks to understand all the aspects and manifestations of plant life. In agreement with the major characteristics of organisms, it is usually divided into three major parts: (1) the physiology of nutrition and metabolism, which deals with the uptake, transformations, and release of materials, and also their movement within and between the cells and organs of the plant; (2) the physiology of growth, development, and reproduction, which is concerned with these aspects of plant function; and (3) environmental physiology, which seeks to understand the manifold responses of plants to the environment. The part of environmental physiology which deals with effects of and adaptations to adverse conditions—and which is receiving increasing attention—is called stress physiology.

Plant physiological research is carried out at various levels of organization and by using various methods. The main organizational levels are the molecular or subcellular, the cellular, the organismal or whole-plant, and the population level. Work at the molecular level is aimed at understanding metabolic processes and their regulation, and also the localization of molecules in particular structures of the cell but with little if any consideration of other processes and other structures of the same cell. Work at the cellular level often deals with the same processes but is concerned with their integration in the cell as a whole. Research at the organismal level is concerned with the function of the plant as a whole and its different organs, and with the relationships between the latter.

Research at the population level, which merges with experimental ecology, deals with physiological phenomena in plant associations which may consist either of one dominant species (like a field of corn) or of numerous diverse species (like a forest). Work at the organismal and to some extent the population level is carried out in facilities permitting maintenance of controlled environmental conditions (light, temperature, water and nutrient supply, and so on). *See* PLANT METABOLISM; PLANT RESPIRATION; PHYSIOLOGICAL ECOLOGY (PLANT); PHYTOTRONICS. [A.L.]

Plant pigment A substance in a plant that imparts coloration. The photosynthetic pigments are involved in light harvesting and energy transfer in photosynthesis. This group of pigments comprises the tetrapyrroles, which include chlorophylls (chl) and phycobilins, and the carotenoids. The light-absorbing groups of these molecules, the chromophores, contain conjugated double bonds (alternating single and double bonds), which make them effective photoreceptors. The sum of the absorption spectra of the chlorphylls and the carotenoids, evident in the absorption spectrum of a green leaf, is equivalent to the action spectrum of photosynthesis. *See* CHLOROPHYLL; PHOTOSYNTHESIS; PHYCOBILIN.

The second major group comprises the anthocyanins, intensely colored plant pigments responsible for most scarlet, crimson, purple, mauve, and blue colors in higher plants. About 100 different anthocyanins are known. Unlike the chlorophylls and carotenoids, which are lipid-soluble chloroplast pigments, the anthocyanins are water-soluble and are located in the cell vacuole. Chemically, they are a class of flavonoids and are particularly closely related, both structurally and biosynthetically, to the flavonols. Their value to the plant lies in the contrasting colors they provide in flower and fruit, against the green background of the leaf, to attract insects and animals for purposes of pollination and seed dispersal. [S.Ra.]

Plant propagation The deliberate, directed reproduction of plants using plant cells, tissues, or organs. Asexual propagation, also called vegetative propagation, is accomplished by taking cuttings, by grafting or budding, by layering, by division of plants, or by separation of specialized structures such as tubers, rhizomes, or bulbs. This method of propagation is used in agriculture, in scientific research, and in professional and recreational gardening. It has a number of advantages over seed propagation: it retains the genetic constitution of the plant type almost completely; it is faster than seed propagation; it may allow elimination of the nonfruiting, juvenile phase of the plant's life; it preserves unique, especially productive, or esthetically desirable plant forms; and it allows plants with roots well adapted for growth on poor soils to be combined with tops that produce superior fruits, nuts, or other products. *See* BREEDING (PLANT); REPRODUCTION (PLANT). [C.E.LaM.]

Tissue cultures and protoplast cultures are among the techniques that have been investigated for plant propagation; the success of a specific technique depends on a number of factors. Practical applications of such methods include the clonal propagation of desirable phenotypes and the commercial production of virus-free plants.

Plant tissue cultures are initiated by excising tissue containing nucleated cells and placing it on an enriched sterile culture medium. The response of a plant tissue to a culture medium depends on a number of factors: plant species, source of tissue, chronological age and physiological state of the tissue, ingredients of the culture medium, and physical culturing conditions, such as temperature, photoperiod, and aeration.

Though technically more demanding, successful culture of plant protoplasts involves the same basic principles as plant tissue culture. Empirical methods are used to determine detailed techniques for individual species; such factors as plant species, tissue source, age, culture medium, and physical culture conditions have to be considered. *See* PLANT CELL. [K.G.F.]

Plant respiration A biochemical process whereby specific substrates are oxidized with a subsequent release of carbon dioxide, CO_2. There is usually conservation of energy accompanying the oxidation which is coupled to the synthesis of energy-rich compounds, such as adenosine triphosphate (ATP), whose free energy is then used to drive otherwise unfavorable reactions that are essential for physiological processes such as growth. Respiration is carried out by specific proteins, called enzymes, and it is necessary for the synthesis of essential metabolites, including carbohydrates, amino acids, and fatty acids, and for the transport of minerals and other solutes between cells. Thus respiration is an essential characteristic of life itself in plants as well as in other organisms.

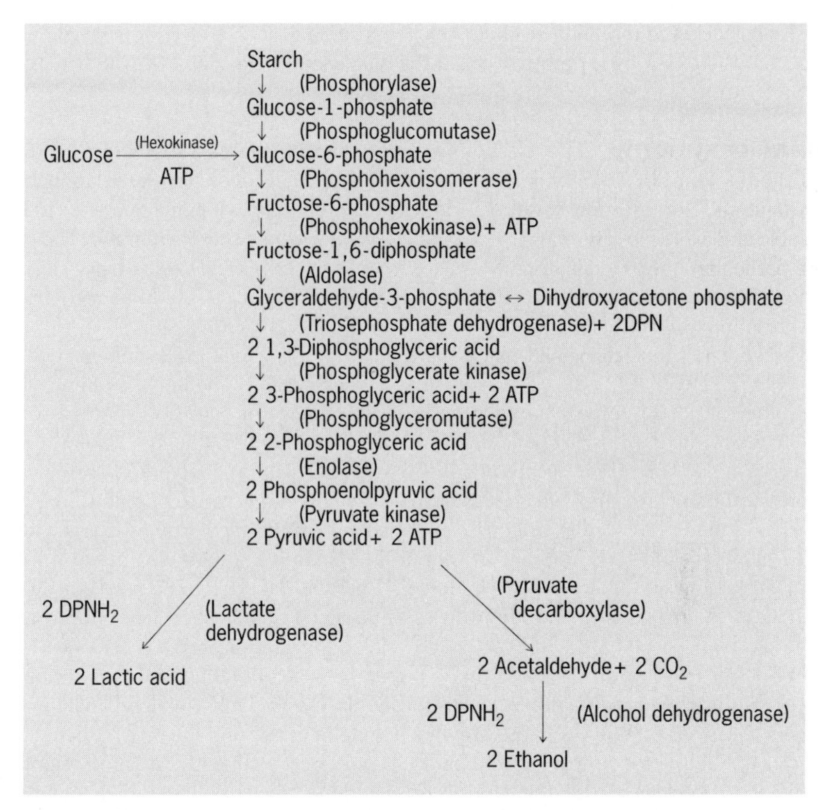

Reaction sequence for anaerobic glycolysis. The soluble enzymes are shown in parentheses.

Overall aerobic respiration is the end result of a sequence of many biochemical reactions that ultimately lead to O_2 uptake and CO_2 evolution. In the absence of O_2, as may occur in bulky plant tissues such as the potato tuber and carrot root and in submerged plants such as germinating rice seedlings, the breakdown of hexose does not go to completion. The end products are either lactic acid or ethanol, which are produced by anaerobic glycolysis or fermentation.

The sequence of reactions of anaerobic glycolysis or fermentation is shown in the illustration. The enzymes associated with anerobic glycolysis have been isolated from many plant tissues, but more often ethanol and not lactic acid is the final product.

In aerobic tissues, pyruvic acid produced during glycolysis is completely oxidized with the accompanying synthesis of much more ATP than in anaerobic glycolysis. Pyruvic acid oxidation takes place in the mitochondria by means of a cyclic sequence of reactions, the Krebs cycle (also known as the citric acid cycle) which begins when the first product of pyruvate oxidation, acetyl coenzyme A, reacts with oxaloacetic acid to produce citric acid. Oxaloacetic acid is eventually regenerated. Thus the cycle can be repeated. In terms of conservation of chemical energy, the Krebs cycle is about 12 times more efficient than anaerobic glycolysis per mole of glucose oxidized.

In addition to anaerobic glycolysis and the Krebs cycle, there are two other sequences of biochemical reactions related to respiration that are important in plant tissues: (1) The pentose phosphate pathway permits an alternate mechanism for converting hexose phosphate to pyruvate, and (2) in germinating fatty seeds the reactions of the Krebs cycle are modified so that acetyl coenzyme A is converted to succinic acid and then to hexose by a pathway called the glyoxylate cycle. *See* PHOTORESPIRATION; PHOTOSYNTHESIS; PLANT GROWTH; PLANT METABOLISM. [I.Z.]

Plant taxonomy The area of study focusing on the development of a classification system, or taxonomy, for plants based on their evolutionary relationships (phylogeny). The assumption is that if classification reflects phylogeny, reference to the classification will help researchers focus their work in a more accurate manner. The task is to make phylogeny reconstruction as accurate as possible. The basic unit of classification is generally accepted to be the species, but how a species should be recognized has been intensely debated. *See* PLANT KINGDOM; PLANT PHYLOGENY.

The earliest classifications of plants were those of the Greek philosophers such as Aristotle (384–322 B.C.) and Theophrastus (372–287 B.C.). The latter is often called the father of botany largely because he listed the names of over 500 species, some of which are still used as scientific names today. In the next 1600 years little progress occurred in plant taxonomy. It was not until the fifteenth century that there was renewed interest in botany, much of which was propelled by the medical use of plants. In 1753 Carolus Linnaeus, a Swedish botanist, published his *Species Plantarum*, a classification of all plants known to Europeans at that time. Linnaeus's system was based on the arrangement and numbers of parts in flowers, and was intended to be used strictly for identification (a system now referred to as an artificial classification as opposed to a natural classification, based on how closely related the species are).

In *Species Plantarum*, Linnaeus made popular a system of binomial nomenclature developed by the French botanist Gaspard Bauhin (1560–1624), which is still in use. Each species has a two-part name, the first being the genus and the second being the species epithet. For example, *Rosa alba* (italicized because it is Latin) is the scientific name of one species of rose; the genus is *Rosa* and the species epithet is *alba*, meaning white (it is not a requirement that scientific names be similar to common names or have real meaning, although such relevance is often the case). The genus name *Rosa* is shared by all species of roses, reflecting that they are thought to be more closely related to each other than to species in any other group.

Today, we understand that the best classification system is one that reflects the patterns of the evolutionary processes that produced these plants. The rules of botanical nomenclature (and those of zoology as well, although they are not identical) are part of an internationally accepted Code that is revised (minimally) at an international congress every 5 years. *See* PLANT EVOLUTION.

Use of common names in science and horticulture is not practical. Scientific names are internationally agreed upon so that a consistent taxonomic name is used everywhere for a given organism. In addition to genus and species, plants are classified by belonging to a family; related families are grouped into orders, and these are typically grouped into a number of yet higher and more encompassing categories. In general, higher categories are composed of many members of lower types—for example, a family may contain 350 genera, but some may be composed of a single genus with perhaps a single species if that species is distantly related to all others.

Many botanists use a number of intermediate categories between the level of genus and family, such as subfamilies, tribes, and subtribes, as well as some between species and genus, such as subgenera and sections, but none of these categories is formally mandated. They are useful nonetheless to reflect intermediate levels of relatedness, particularly in large families (composed of several hundreds or even thousands of species). Below the level of species, some botanists use the concept of subspecies (which is generally taken to mean a geographically distinct form of a species) and variety (which is often a genetic form or genotype, for example a white-flowered form of a typically blue-flowered species, or a form that is ecologically distinct).

The basic idea that plant classification should reflect evolutionary (genetic) relationships has been well accepted for some time, but the degree to which this could be assessed by the various means available differed. It has only recently become possible to assess genetic patterns of relatedness directly by analyzing DNA sequences. In the 1990s, DNA technology became much more efficient and less costly, resulting in a dramatic upsurge in the availability of DNA sequence data for various genes from each of the three genetic compartments present in plants (nuclear, mitochondrial, and plastid or chloroplast). In 1998 a number of botanists collectively proposed the first DNA-based classification of a major group of organisms, the angiosperms or flowering plants. For the first time, a classification was directly founded on assessments of the degree of relatedness made with objective, computerized methods of phylogeny reconstruction. Other data, such as chemistry and morphology, were also incorporated into these analyses, but by far the largest percentage of information came from DNA sequences—that is, relatedness was determined mostly on the basis of similarities in plants' genetic codes. The advantages of such a classification were immediately obvious: (1) it was not based on intuition about which category of information best reflected natural relationships; (2) it ended competition between systems based on differing emphases; (3) the analysis could be repeated by other researchers using either the same or different data (other genes or categories of information); and (4) it could be updated as new data emerged, particularly from studies of how chromosomes are organized and how morphology and other traits are determined by the genes that code for them.

At the same time that DNA data became more widely available as the basis for establishing a classification, a more explicit methodology for turning the results of a phylogenetic analysis into a formal classification became popular. This methodology, called cladistics, allowed a large number of botanists to share ideas of how the various taxonomic categories could be better defined. Although there remain a number of dissenting opinions about some minor matters of classification, it is now impossible for scientists to propose alternative ideas based solely on opinion. *See* PHYLOGENY; TAXONOMY. [M.W.C.; M.F.F.]

Plant transport of solutes The movement of organic and inorganic compounds through plant vascular tissues. Transport can take place over considerable distances; in tree species transport distances are often 100–300 ft (30–100 m).

This long-distance transport is necessary for survival in higher land plants in which specialized organs of uptake or synthesis are separated by a considerable distance from the organs of utilization. Diffusion is not rapid enough to account for the amount of material moved over such long distances. Rather, transport depends on a flowing stream of liquid in vascular tissues (phloem and xylem) that are highly developed structurally.

The movement of organic solutes occurs mainly in the phloem, where it is also known as translocation and where the direction of transport is from places of production, such as mature leaves, to places of utilization or storage, such as the shoot apex or developing storage roots. Organic materials translocated in the phloem include the direct products of photosynthesis (sugars) as well as compounds derived from them (nitrogenous compounds and plant hormones, for example). Some movement of organic solutes does occur in the xylem of certain species. Inorganic solutes or mineral elements, however, generally move with water in the xylem from sites of uptake in the roots to sites where water is lost from the plant, primarily the leaves. Some redistribution of the ions throughout the plant may then occur in the phloem.

The mechanism of phloem translocation is not known with certainty. Proposed mechanisms fall into two classes: one stresses the role of the conducting tissues in generating the moving force, and the other views the regions of supply and utilization as the source of this force. In the former group are mechanisms that depend on cytoplasmic streaming, electroosmosis, and activated diffusion in the sieve elements. The second group of theories, which has received more general acceptance in spite of a number of admitted limitations, includes a variety of mass-flow mechanisms. Theories of translocation must account for the important observations: polarity, bidirectional movement, velocity, energy requirement, turgor pressure, and phloem structure.

The model for the ascent of sap in the xylem which is correct according to all present evidence is called the cohesion hypothesis. According to this hypothesis, water is lost in the leaves by evaporation from cell-wall surfaces; water vapor then diffuses into the atmosphere by way of small pores between two specialized cells (guard cells). The guard cells and the pore are collectively called a stomate. This loss of water from the leaf causes movement of water out of the xylem in the leaf to the surfaces where evaporation is occurring. Water has a high internal cohesive force, especially in small tubes with wettable walls. In addition, the xylem elements and the cell walls provide a continuous water-filled system in the plant. Thus the loss of water from the xylem elements in the leaves causes a tension or negative pressure in the xylem sap. This tension is transmitted all the way down the stem to the roots, so that a flow of water occurs up the plant from the roots and eventually from the soil. The velocity of this sap flow in tree species ranges from 3 to approximately 165 ft/h (1 to 50 m/h), depending on the diameter of the xylem vessels. [S.S.D.]

Plant viruses and viroids Plant viruses are pathogens which are composed mainly of a nucleic acid (genome) normally surrounded by a protein shell (coat); they replicate only in compatible cells, usually with the induction of symptoms in the affected plant. Viroids are among the smallest infections agents known. Their circular, single-stranded riboncleic acid (RNA) molecule is less than one-tenth the size of the smallest viruses.

Viruses. Viruses can be seen only with an electron microscope (see illustration). Isometric (spherical) viruses range from 25 to 50 nanometers in diameter, whereas most anisometric (tubular) viruses are 12 to 25 nm in diameter and of various lengths (200–2000 nm), depending on the virus. The coat of a few viruses is covered by a membrane which is derived from its host.

Over 800 plant viruses have been recognized and characterized. The genomes of most of them, such as the tobacco mosaic virus (TMV), are infective single-stranded RNAs; some RNA viruses have double-stranded RNA genomes. Cauliflower mosaic virus and bean golden mosaic virus are examples of viruses having double-stranded and single-stranded deoxyribonucleic acid (DNA), respectively. The genome of many plant viruses is a single polynucleotide and is contained in a single particle, whereas the genomes of brome mosaic and some other viruses are segmented and distributed between several particles. There are also several low-molecular-weight RNAs (satellite RNAs) which depend on helper viruses for their replication.

The natural hosts of plant viruses are widely distributed throughout the higher-plant kingdom. Some viruses (TMV and cucumber mosaic virus) are capable of infecting over a hundred species in many families, whereas others, such as wheat streak mosaic virus, are restricted to a few species in the grass family. The replication of single-stranded RNA viruses involves release of the virus genome from the coat protein; the association of the RNA with the ribosomes of the cell; translation of the genetic information of the RNA into specific proteins, including subunits of the coat protein and possibly viral

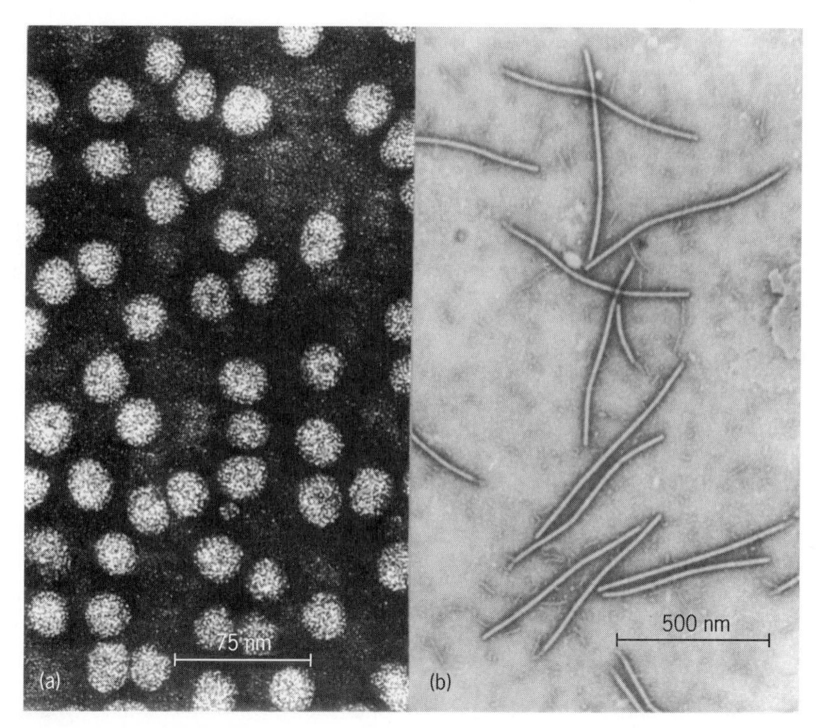

Representative plant viruses in purified virus preparation obtained from infected leaves. (*a*) Tobacco streak virus (isometric). (*b*) Pea seed-borne mosaic virus (anisometric).

RNA-synthesizing enzymes (replicases); transmission by vectors and diseases induction; synthesis of noninfective RNA using parental RNA as the template; and assembly of the protein subunits and viral RNA to form complete virus particles.

In other RNA viruses, such as lettuce necrotic yellows virus, an enzyme which is contained in the virus must first make a complementary (infective) copy of the RNA; this is then translated into enzymes and coat protein subunits. The replication of double-stranded RNA viruses is similar to that of lettuce necrotic yellows virus.

With double-stranded DNA viruses, viral DNA is uncoated in a newly infected cell and transported to the nucleus, where it associates with histones to form a closed circular minichromosome. Two major RNA species (35S and 19S) are transcribed from the minichromosome by a host-encoded enzyme and are translated in the cytoplasm to produce virus-associated proteins. The 35S RNA serves as the template for a viral enzyme which transcribes it to viral DNA, which is then encapsidated to form virus particles.

Symptoms are the result of an alteration in cellular metabolism and are most obvious in newly developing tissues. In some plants, depending on the virus, the initial infection does not spread because cells surrounding the infected cells die, resulting in the formation of necrotic lesions. Such plants are termed hypersensitive. The size and shape of leaves and fruit may be adversely affected, and in some instances plants may be killed. Not all virus infections produce distinctive symptoms.

The most common mode of transmission for many viruses is by means of vectors, mainly insects (predominantly aphids and leafhoppers), and to a lesser extent mites, soil-inhabiting fungi, and nematodes which acquire viruses by feeding on infected plants. Viruses transmitted by one class of vector are rarely transmitted by another, and there is often considerable specificity between strains of a virus and their vectors.

Some viruses are transmitted to succeeding generations mainly by embryos in seeds produced by infected plants; over 200 viruses are transmitted in this way.

Viroids. Only about 30 viroids are known, but they cause very serious diseases in such diverse plants as chrysanthemum, citrus, coconut, and potato. They can also be isolated from plants that do not exhibit symptoms. Viroids are mainly transmitted by vegetative propagation, but some, such as potato spindle tuber viroid, are transmitted by seed or by contact between infected and healthy plants. Tomato planta macho viroid is efficiently transmitted by aphids. *See* PLANT PATHOLOGY; VIRUS. [R.I.H.]

Plant-water relations Water is the most abundant constituent of all physiologically active plant cells. Leaves, for example, have water contents which lie mostly within a range of 55–85% of their fresh weight. Other relatively succulent parts of plants contain approximately the same proportion of water, and even such largely nonliving tissues as wood may be 30–60% water on a fresh-weight basis. The smallest water contents in living parts of plants occur mostly in dormant structures, such as mature seeds and spores. The great bulk of the water in any plant constitutes a unit system. This water is not in a static condition. Rather it is part of a hydrodynamic system, which in terrestrial plants involves absorption of water from the soil, its translocation throughout the plant, and its loss to the environment, principally in the process known as transpiration.

Cellular water relations. The typical mature, vacuolate plant cell constitutes a tiny osmotic system, and this idea is central to any concept of cellular water dynamics. Although the cell walls of most living plant cells are quite freely permeable to water and solutes, the cytoplasmic layer that lines the cell wall is more permeable to some substances than to others.

If a plant cell in a flaccid condition—one in which the cell sap exerts no pressure against the encompassing cytoplasm and cell wall—is immersed in pure water, inward osmosis of water into the cell sap ensues. This gain of water results in the exertion of a turgor pressure against the protoplasm, which in turn is transmitted to the cell wall. This pressure also prevails throughout the mass of solution within the cell. If the cell wall is elastic, some expansion in the volume of the cell occurs as a result of this pressure, although in many kinds of cells this is relatively small.

If a turgid or partially turgid plant cell is immersed in a solution with a greater osmotic pressure than the cell sap, a gradual shrinkage in the volume of the cell ensues; the amount of shrinkage depends upon the kind of cell and its initial degree of turgidity. When the lower limit of cell wall elasticity is reached and there is continued loss of water from the cell sap, the protoplasmic layer begins to recede from the inner surface of the cell wall. Retreat of the protoplasm from the cell wall often continues until it has shrunk toward the center of the cell, the space between the protoplasm and the cell wall becoming occupied by the bathing solution. This phenomenon is called plasmolysis.

In some kinds of plant cells movement of water occurs principally by the process of imbibition rather than osmosis. The swelling of dry seeds when immersed in water is a familiar example of this process.

Stomatal mechanism. Various gases diffuse into and out of physiologically active plants. Those gases of greatest physiological significance are carbon dioxide, oxygen, and water vapor. The great bulk of the gaseous exchanges between a plant and its environment occurs through tiny pores in the epidermis that are called stomates. Although stomates occur on many aerial parts of plants, they are most characteristic of, and occur in greatest abundance in, leaves. *See* EPIDERMIS (PLANT); LEAF.

Transpiration process. The term transpiration is used to designate the process whereby water vapor is lost from plants. Although basically an evaporation process, transpiration is complicated by other physical and physiological conditions prevailing in the plant. Whereas loss of water vapor can occur from any part of the plant which is exposed to the atmosphere, the great bulk of all transpiration occurs from the leaves. There are two kinds of foliar transpiration: (1) stomatal transpiration, in which water vapor loss occurs through the stomates, and (2) cuticular transpiration, which occurs directly from the outside surface of epidermal walls through the cuticle. In most species 90% or more of all foliar transpiration is of the stomatal type.

Transpiration is a necessary consequence of the relation of water to the anatomy of the plant, and especially to the anatomy of the leaves. Terrestrial green plants are dependent upon atmospheric carbon dioxide for their survival. In terrestrial vascular plants the principal carbon dioxide–absorbing surfaces are the moist mesophyll cells walls which bound the intercellular spaces in leaves. Ingress of carbon dioxide into these spaces occurs mostly by diffusion through open stomates. When the stomates are open, outward diffusion of water vapor unavoidably occurs, and such stomatal transpiration accounts for most of the water vapor loss from plants. Although transpiration is thus, in effect, an incidental phenomenon, it frequently has marked indirect effects on other physiological processes which occur in the plant because of its effects on the internal water relations of the plant.

Water translocation. In terrestrial rooted plants practically all of the water which enters a plant is absorbed from the soil by the roots. The water thus absorbed is translocated to all parts of the plant. The mechanism of the "ascent of sap" (all translocated water contains at least traces of solutes) in plants, especially tall trees, was one of the first processes to excite the interest of plant physiologists.

The upward movement of water in plants occurs in the xylem, which, in the larger roots, trunks, and branches of trees and shrubs, is identical with the wood. In the trunks or larger branches of most kinds of trees, however, sap movement is restricted to a few of the outermost annual layers of wood. *See* XYLEM.

Root pressure is generally considered to be one of the mechanisms of upward transport of water in plants. While it is undoubtedly true that root pressure does account for some upward movement of water in certain species of plants at some seasons, various considerations indicate that it can be only a secondary mechanism of water transport.

Upward translocation of water (actually a very dilute sap) is engendered by an increase in the negativity of water potential in the cells of apical organs of plants. Such increases in the negativity of water potentials occur most commonly in the mesophyll cells of leaves as a result of transpiration.

Water absorption. The successively smaller branches of the root system of any plant terminate ultimately in the root tips, of which there may be thousands and often millions on a single plant. Most absorption of water occurs in the root tip regions, and especially in the root hair zone. Older portions of most roots become covered with cutinized or suberized layers through which only very limited quantities of water can pass. *See* ROOT (BOTANY).

Whenever the water potential in the peripheral root cells is less than that of the soil water, movement of water from the soil into the root cells occurs. There is some evidence that, under conditions of marked internal water stress, the tension generated in the xylem ducts will be propagated across the root to the peripheral cells. If this occurs, water potentials of greater negativity could develop in peripheral root cells than would otherwise be possible. The absorption mechanism would operate in fundamentally the same way whether or not the water in the root cells passed into a state of tension.

The process just described, often called passive absorption, accounts for most of the absorption of water by terrestrial plants.

The phenomenon of root pressure represents another mechanism of the absorption of water. This mechanism is localized in the roots and is often called active absorption. Water absorption of this type only occurs when the rate of transpiration is low and the soil is relatively moist. Although the xylem sap is a relatively dilute solution, its osmotic pressure is usually great enough to engender a more negative water potential than usually exists in the soil water when the soil is relatively moist. A gradient of water potentials can thus be established, increasing in negativity across the epidermis, cortex, and other root tissues, along which the water can move laterally from the soil to the xylem. *See* PLANT MINERAL NUTRITION. [B.S.M.]

Plants, life forms of A term for the vegetative (morphological) form of the plant body. Life-form systems are based on differences in gross morphological features, and the categories bear no necessary relationship to reproductive structures, which form the basis for taxonomic classification. Features used in establishing life-form classes include deciduous versus evergreen leaves, broad versus needle leaves, size of leaves, degree of protection afforded the perennating tissue, succulence, and duration of life cycle (annual, biennial, or perennial).

There is a clear correlation between life forms and climates. For example, broad-leaved evergreen trees clearly dominate in the hot humid tropics, whereas broad-leaved deciduous trees prevail in temperature climates with cold winters and warm summers, and succulent cacti dominate American deserts. Although cacti are virtually absent from African deserts, members of the family Euphorbiaceae have evolved similar succulent life forms. Such adaptations are genetic, having arisen by natural selection.

Many life-form systems have been developed. The most successful and widely used system is that of C. Raunkiaer, proposed in 1905. Reasoning that it was the perennating buds (the tips of shoots which renew growth after a dormant season, either of cold or drought) which permit a plant to survive in a specific climate, Raunkiaer's classes were based on the degree of protection afforded the bud and the position of the bud relative to the soil surface. They applied to autotrophic, vascular, self-supporting plants. Raunkiaer's classificatory system is:

Phanerophytes: bud-bearing shoots in the air, predominantly woody trees and shrubs; subclasses based on height and on presence or absence of bud scales

Chamaephytes: bud within 10 in. (25 cm) of the surface, mostly prostrate or creeping shrubs

Hemicryptophytes: buds at the soil surface, protected by scales, snow, and litter

Cryptophytes: buds underneath the soil surface or under water

Therophytes: annuals, the seed representing the only perennating tissue

By determining the life forms of a sample of 1000 species from the world's floras, Raunkiaer showed a correlation between the percentage of species in each life-form class present in an area and the climate of the area. Raunkiaer concluded that there were four main phytoclimates: phanerophyte-dominated flora of the hot humid tropics, hemicryptophyte-dominated flora in moist to humid temperate areas, therophyte-dominated flora in arid areas, and a chamaephyte-dominated flora of high latitudes and altitudes.

Subsequent studies modified Raunkiaer's views. (1) Phanerophytes dominate, to the virtual exclusion of other life forms, in true tropical rainforest floras, whereas other life forms become proportionately more important in tropical climates with a dry season. (2) Therophytes are most abundant in arid climates and are prominent in temperate areas with an extended dry season, such as regions with Mediterranean climate.

(3) Other temperate floras have a predominance of hemicryptophytes with the percentage of phanerophytes decreasing from summer-green deciduous forest to grassland. (4) Arctic and alpine tundra are characterized by a flora which is often more than three-quarters chamaephytes and hemicryptophytes, the percentage of chamaephytes increasing with latitude and altitude. *See* PLANT GEOGRAPHY.

There has been interest in developing systems which describe important morphologic features of plants and which permit mapping and diagramming vegetation. Descriptive systems incorporate essential structural features of plants, such as stem architecture and height; deciduousness; leaf texture, shape, and size; and mechanisms for dispersal. These systems are important in mapping vegetation because structural features generally provide the best criteria for recognition of major vegetation units. *See* ALTITUDINAL VEGETATION ZONES; VEGETATION AND ECOSYSTEM MAPPING. [A.W.C.]

Plateau Any elevated area of relatively smooth land. Usually the term is used more specifically to denote an upland of subdued relief that on at least one side drops off abruptly to adjacent lower lands. In most instances the upland is cut by deep but widely separated valleys or canyons. Small plateaus that stand above their surroundings on all sides are often called tables, tablelands, or mesas. The abrupt edge of a plateau is an escarpment or, especially in the western United States, a rim. [E.H.Ha.]

Playa A nearly level, generally dry surface in the lowest part of a desert basin with internal drainage (see illustration). When its surface is covered by a shallow sheet of water, it is a playa lake. Playas and playa lakes are also called dry lakes, alkali flats, mud flats, saline lakes, salt pans, inland sabkhas, ephemeral lakes, salinas, and sinks.

A playa surface is built up by sandy mud that settles from floodwater when a playa is inundated by downslope runoff during a rainstorm. A smooth, hard playa occurs where ground-water discharge is small or lacking and the surface is flooded frequently. These mud surfaces are cut by extensive desiccation polygons caused by shrinkage of the drying clay. Puffy-ground playas form by crystallization of minerals as ground water evaporates in muds near the surface.

Subsurface brine is present beneath many playas. The type of brine depends on the original composition of the surface water and reflects the lithology of the rocks weathered in the surrounding mountains. *See* GROUND-WATER HYDROLOGY.

Light-colored playa in lowest part of Sarcobatus Flat in southern Nevada.

Numerous playas in the southwestern United States yield commercial quantities of evaporite minerals, commonly at shallow depths. Important are salt ($NaCl$) and the borates, particularly borax ($Na_2B_4O_7 \cdot 10H_2O$), kernite ($Na_2B_4O_7 \cdot 4H_2O$), ulexite ($NaCaB_5O_9 \cdot 8H_2O$), probertite ($NaCaB_5O_9 \cdot 5H_2O$), and colemanite ($Ca_2B_6O_{11} \cdot 5H_2O$). Soda ash (sodium carbonate; Na_2CO_3) is obtained from trona ($Na_3H(CO_3)_2 \cdot 2H_2O$) and gaylussite ($Na_2Ca(CO_3)_2 \cdot 5H_2O$). Lithium and bromine are produced from brine waters. *See* DESERT EROSION FEATURES; SALINE EVAPORITES. [J.F.Hu.]

Pneumococcus The major causative microorganism (*Streptococcus pneumoniae*) of lobar pneumonia. Pneumococci occur singly or as pairs or short chains of oval or lancet-shaped cocci, 0.05–1.25 micrometers each, flattened at proximal sides and pointed at distal ends. A capsule of polysaccharide envelops each cell or pair of cells. The organism is nonmotile and stains gram-positive unless degenerating.

Pneumococci have been isolated from the upper respiratory tract of healthy humans, monkeys, calves, horses, and dogs. Epizootics of pneumococcal infection have been described in monkeys, guinea pigs, and rats but are not the source of human infection. In humans, pneumococci may be found in the upper respiratory tract of nearly all individuals at one time or another. Following damage to the epithelium lining the respiratory tract, pneumococci may invade the lungs. They are the principal cause of lobar pneumonia in humans and may cause also pleural empyema, pericarditis, endocarditis, meningitis, arthritis, peritonitis, and infection of the middle ear. Approximately one of four cases of pneumococcal pneumonia is accompanied by invasion of the bloodstream by pneumococci, producing bacteremia. Although the high mortality of untreated pneumococcal infection has been reduced significantly by treatment with antibiotics, one of every six patients with bacteremic lobar pneumonia still succumbs despite optimal therapy. In addition, the number of isolates of pneumococci resistant to one or more antimicrobial drugs has been gradually but steadily increasing. For these reasons, prophylactic vaccination is recommended, especially for those segments of the population that are at high risk for fatal infection. The polyvalent vaccine contains the purified capsular polysaccharides of the 23 types that are responsible for 85% of bacteremic pneumococcal infection and has an aggregate efficacy of 65–70% in preventing infection with any of the types represented in it. [R.Au.]

Pneumonia An acute or chronic inflammatory disease of the lungs. More specifically when inflammation is caused by an infectious agent, the condition is called pneumonia; when the inflammatory process in the lung is not related to an infectious organism, it is called pneumonitis.

An estimated 45 million cases of infectious pneumonia occur annually in the United States, with up to 50,000 deaths directly attributable to it. Pneumonia is a common immediate cause of death in persons with a variety of underlying diseases. With the use of immunosuppressive and chemotherapeutic agents for treating transplant and cancer patients, pneumonia caused by infectious agents that usually do not cause infections in healthy persons (that is, pneumonia as an opportunistic infection) has become commonplace. Moreover, individuals with acquired immune deficiency syndrome (AIDS) usually die from an opportunistic infection, such as pneumocystis pneumonia or cytomegalovirus pneumonia. Concurrent with the variable and expanding etiology of pneumonia and the more frequent occurrence of opportunistic infections is the development of new antibiotics and other drugs used in the treatment of pneumonia. *See* ACQUIRED IMMUNE DEFICIENCY SYNDROME (AIDS); OPPORTUNISTIC INFECTIONS.

Bacteria, as a group, are the most common cause of infectious pneumonia, although influenza virus has replaced *Streptococcus pneumoniae* (*Diplococcus pneumoniae*) as the most common single agent. Some of the bacteria are normal inhabitants of the body and proliferate to cause disease only under certain conditions. Other bacteria are contaminants of food or water.

Most bacteria cause one of two main morphologic forms of inflammation in the lung. *Streptococcus pneumoniae* causes lobar pneumonia, in which an entire lobe of a lung or a large portion of a lobe becomes consolidated (firm, dense) and nonfunctional secondary to an influx of fluid and acute inflammatory cells that represent a reaction to the bacteria. This type of pneumonia is uncommon today, usually occurring in people who have poor hygiene and are debilitated. If lobar pneumonia is treated adequately, the inflammatory process may entirely disappear, although in some instances it undergoes a process called organization, in which the inflammatory tissue changes into fibrous tissue, usually rendering that portion of the lung nonfunctional.

The other morphologic form of pneumonia, which is caused by the majority of bacteria, is called bronchopneumonia. In this form there is patchy consolidation of lung tissue, usually around the small bronchi and bronchioles, again most frequently in the lower lobes. This type of pneumonia may also undergo complete resolution if there is adequate treatment, although rarely it organizes.

Viral pneumonia is usually a diffuse process throughout the lung and produces a different type of inflammatory reaction than is seen in bronchopneumonia or lobar pneumonia. Mycoplasma pneumonia, caused by *Mycoplasma pneumoniae*, is referred to as primary atypical pneumonia and causes an inflammatory reaction similar to that of viral pneumonia.

Pneumonia can be caused by a variety of other fungal organisms, especially in debilitated persons such as those with cancer or AIDS. *Mycobacterium tuberculosis*, the causative agent of pulmonary tuberculosis, produces an inflammatory reaction similar to fungal organisms. *See* MYCOBACTERIAL DISEASES; TUBERCULOSIS.

Legionella pneumonia, initially called Legionnaire's disease, is caused by bacteria of the genus *Legionella*. The condition is frequently referred to under the broader name of legionellosis. *See* LEGIONNAIRES' DISEASE.

The signs and symptoms of pneumonia and pneumonitis are usually nonspecific, consisting of fever, chills, shortness of breath, and chest pain. Fever and chills are more frequently associated with infectious pneumonias but may also be seen in pneumonitis. The physical examination of a person with pneumonia or pneumonitis may reveal abnormal lung sounds indicative of regions of consolidation of lung tissue. A chest x-ray also shows the consolidation, which appears as an area of increased opacity (white area). Cultures of sputum or bronchial secretions may identify an infectious organism capable of causing the pneumonia.

The treatment of pneumonia and pneumonitis depends on the cause. Bacterial pneumonias are treated with antimicrobial agents. If the organisms can be cultured, the sensitivity of the organism to a specific antibiotic can be determined. Viral pneumonia is difficult to treat, as most drugs only help control the symptoms. The treatment of pneumonitis depends on identifying its cause; many cases are treated with cortisone-type medicines. [S.P.H.]

Poison A substance which by chemical action and at low dosage can kill or injure living organisms. Broadly defined, poisons include chemicals toxic for any living form: microbes, plants, or animals. In common usage the word is limited to substances toxic for humans and mammals, particularly where toxicity is a substance's major property of medical interest. Because of their diversity in origin, chemistry, and toxic action, poisons defy any simple classification. Almost all chemicals with recognized physiological effects are toxic at sufficient dosage.

Origin and chemistry. Many poisons are of natural origin. Some bacteria secrete toxic proteins (for example, botulinus, diphtheria, and tetanus toxins) that are among the most poisonous compounds known. Lower plants notorious for poisonous properties are ergot (*Claviceps purpurea*) and a variety of toxic mushrooms. *See* ERGOT AND ERGOTISM; MUSHROOM; TOXIN.

Higher plants, which constitute the major natural source of drugs, contain a great variety of poisonous substances. Many of the plant alkaloids double as drugs or poisons, depending on dose. These include curare, quinine, atropine, mescaline, morphine, nicotine, cocaine, picrotoxin, strychnine, lysergic acid, and many others.

Poisons of animal origin (venoms) are similarly diverse. Toxic marine animals alone include examples of every phylum. Insects and snakes represent the best-known venomous land animals, but on land, too, all phyla include poison-producing species. Among mammalian examples are certain shrews with poison-producing salivary glands. *See* POISON GLAND.

Poisons of nonliving origin vary in chemical complexity from the toxic elements, for example, the heavy metals, to complex synthetic organic molecules. Most of the heavy metals (gold, silver, mercury, arsenic, and lead) are poisons of high potency in the form of their soluble salts. Strong acids or bases are toxic largely because of corrosive local tissue injury.

The chemically reactive gases hydrogen sulfide, hydrocyanic acid, chlorine, bromine, and ammonia are also toxic, even at low concentration, both because of their corrosiveness and because of more subtle chemical interaction with enzymes or other cell constituents.

Many organic substances of synthetic origin are highly toxic and represent a major source of industrial hazard. Most organic solvents are more or less toxic on ingestion or inhalation. Many alcohols, such as methanol, are much more toxic. Many solvents (for example, carbon tetrachloride, tetrachloroethane, dioxane, and ethylene glycol) produce severe chemical injury to the liver and other viscera, sometimes from rather low dosage.

Physiological actions. The action of poisons is generally described by the physiological or biochemical changes which they produce. For most poisons, a descriptive account can be given which indicates what organic system (for example, heart, kidney, liver, brain, and bone marrow) appears to be most critically involved and contributes most to seriously disordered body function or death. In many cases, however, organ effects are multiple, or functional derangements so generalized that a cause of death cannot be localized.

More precise understanding of the mechanism of poisons requires detailed knowledge of their action in chemical terms. Information of this kind is available for only a few compounds, and then in only fragmentary detail. Poisons that inhibit acetylcholinesterase have toxic actions traceable to a single blocked enzyme reaction, hydrolysis of normally secreted acetylcholine. Detailed understanding of the mechanism of chemical inhibition of cholinesterase is not complete, but allows some prediction of chemical structures likely to act as inhibitors.

Carbon monoxide toxicity is also partly understood in chemical terms, since formation of carboxyhemoglobin, a form incapable of oxygen transport, is sufficient to explain the anoxic features of toxicity.

Heavy metal poisoning in many cases is thought to involve inhibition of enzymes by formation of metal mercaptides with enzyme sulfhydryl groups, the unsubstituted form of which is necessary for enzyme action. This is a general reaction that may occur with a variety of sulfhydryl-containing enzymes in the body. Specific susceptible enzymes whose inhibition explains toxicity have not yet been well documented.

Metabolic antagonists active as poisons function by competitive blocking of normal metabolic reactions. Some antagonists may act directly as enzyme inhibitors, others may be enzymatically altered to form derivatives which are even more potent inhibitors at a later metabolic step.

Where poison mechanisms are relatively well understood, it has sometimes been possible to employ rationally selected antidotes.

Potency. The strength or potency of poisons is most frequently measured by the lethal dose, potency being inversely proportional to lethal dose. From statistically treated dose-response data, the dose killing 50% of the sample population can be determined, and is usually designated the MLD (median lethal dose) or LD_{50}. This is the commonest measure of toxic potency. *See* TOXICOLOGY. [E.A.]

Poison gland The specialized gland of certain fishes, as well as the granular glands and some mucous glands of many aquatic and terrestrial Amphibia. The poison glands of fishes are simple or slightly branched acinous structures which use the holocrine method of secreting a mucuslike substance. The poison glands of snakes are modified oral or salivary glands. Amphibian glands are simple, acinous, holocrine, with granular secretion. In some cases these amphibian poison glands produce mucus by a merocrine method of secretion. These glands function as protective devices. [O.E.N.]

Poisonous plants More than 700 species of seed plants, ferns, horsetails, and fungi that cause toxic, though rarely fatal, reactions in humans and animals. Human allergic responses, including hay fever, asthma, and dermatitis, are widespread. Allergic responses are produced by many different plant species, but most common are poison ivy, poison sumac, and Pacific poison oak (all are species of *Toxicodendron*). Internal injury by toxic plants is less common but can be detrimental or lethal. *See* ASTHMA.

Glycoside-containing plants. Glycosides are common compounds in plants. They decompose to form one or more sugars, but sometimes the remaining compounds, aglycones, can be quite poisonous. Cyanogenic glycosides, which produce hydrocyanic acid, are found worldwide in many plant families; the best known are in the rose family (Rosaceae) and in the pea family (Fabaceae). Leaves, bark, and seeds of stone fruits such as cultivated and wild cherries, plums, peaches, bitter almonds, and apricots contain the glycoside amygdalin, which hydrolyzes to form hydrocyanic acid that can be fatally toxic to humans or animals. The same toxic substance is found in apple and pear seeds. Cardiac glycosides are found in many unrelated species of plants. Those of the foxglove (*Digitalis purpurea*) contain a number of these glycosides used medicinally to slow and strengthen the heartbeat. Oleander (*Nerium oleander*), which is cultivated in the warmer parts of the United States, contains a toxic glycoside that has an action similar to that of digitalis.

Alkaloid-containing plants. Alkaloids, compounds containing a nitrogen atom, have specific pharmacological effects on both humans and animals. Found in many different plant families, they have been used in drug therapy since ancient times, but misuse of these plants can produce poisonings. The potato family (Solanaceae) has many species that contain a number of alkaloids. Hyoscyamine and atropine are the alkaloids occurring in belladonna or deadly nightshade (*Atropa belladonna*), black henbane (*Hyoscyamus niger*), thornapples and jimsonweed (*Datura*), and tree daturas or angel's-trumpets (*Brugmansia*). The black nightshades (*Solanum*) contain glycoalkaloids. Plants of tobacco, *Nicotiana*, contain numerous alkaloids, principally the very toxic nicotine or its isomer anabasine. Plants of poison hemlock (*Conium macalatum*) have several alkaloids similar to nicotine, which affect the central nervous system. Plants of rattlebox (*Crotalaria*), groundsel (*Senecio*), and fiddleneck (*Amsinckia*) have alkaloids of similar molecular structure. Anagyrine is a toxic alkaloid found in several species of lupine (*Lupinus*) in the western United States. Alkaloids present in species of monkshood (*Aconitum*) are extremely toxic. Larkspur plants (*Delphinium*) have similar toxic alkaloids affecting the central nervous system, causing excitability and muscular spasms. Plants of false hellebores (*Veratrum*) and death camas (*Zigadenus*) have complex alkaloids of similar structure and cause livestock deaths in the western United States.

Heath plants. Toxic resins, andromedotoxins, formed by members of the heath family (Ericaceae) are derived from diterpenes. The most toxic species are mountain laurel, sheep laurel, and bog laurel, all in the genus *Kalmia*.

Pokeweed. Pokeweed (*Phytolacca americana*) is a garden weed throughout the United States. but it is native to the eastern and central areas. The entire plant, especially the seeds and the large root, is poisonous. Human poisonings have resulted from inadvertently including parts of the root along with the shoots.

Waterhemlock. The waterhemlocks (*Cicuta*) are widespread in North America. The large underground tubers, mistakenly considered edible, have caused human poisonings and death. Livestock usually do not eat waterhemlock, but have been fatally poisoned. The highly toxic principle is an unsaturated aliphatic alcohol which acts directly on the central nervous system.

Oxalate poisoning. Oxalic acid, as oxalate salts, accumulates in large amounts in some species of plants such as those of the genera *Halogeton*, *Bassia*, *Rumex*, and *Oxalis*.

Nitrate poisoning. Nitrate poisoning is widespread and results in many cattle deaths yearly. Any disruption of the normal synthesis of nitrates into amino acids and proteins causes large accumulations of nitrates in various species of plants, particularly in the goosefoot family (Chenopodiaceae).

Allelopathic toxins. Allelopathic phytotoxins are chemical compounds produced by vascular plants that inhibit the growth of other vascular plants. Residues of grain sorghum (*Sorghum bicolor*) can markedly reduce the following year's growth of wheat and some weedy grass seedlings. Locoweeds, belonging to the genera *Oxytropis* and *Astragalus*, produce an unknown toxin that causes loss of livestock that become addicted to eating these unpalatable plants. Other species of *Astragalus* produce toxic aliphatic nitro compounds. Still other species of *Astragalus* accumulate toxic quantities of selenium which usually causes chronic poisoning of livestock. *See* ALLELOPATHY.

Fungi. Every year human fatalities occur from ingestion of wild poisonous mushrooms. Those gathered in the wild require individual identification since toxic species may grow alongside edible ones. Ergot fungus, infecting many species of grasses, causes widespread poisoning of livestock. *See* ERGOT AND ERGOTISM; TOXICOLOGY. [T.C.F.]

Polar meteorology The science of weather and climate in the high latitudes of the Earth. In the polar regions the Sun never rises far above the horizon and remains below it continuously for part of the year, so that snow and ice can persist for long periods even at low elevations. The meteorological processes that result have distinctive local and large-scale characteristics in both polar regions. [U.R.]

Poliomyelitis An acute infectious viral disease which in its serious form affects the central nervous system and, by destruction of motor neurons in the spinal cord, produces flaccid paralysis. However, about 99% of infections are either inapparent or very mild. *See* ANIMAL VIRUS.

The virus probably enters the body through the mouth; primary multiplication occurs in the throat and intestine. Transitory viremia occurs; the blood seems to be the most likely route to the central nervous system. The severity of the infection may range from a completely inapparent through minor influenzalike illness, or an aseptic meningitis syndrome (nonparalytic poliomyelitis) with stiff and painful back and neck, to the severe forms of paralytic and bulbar poliomyelitis. In all clinical types, virus is regularly present in the enteric tract. In paralytic poliomyelitis the usual course begins as a minor illness but progresses, sometimes with an intervening recession of symptoms (hence biphasic), to flaccid paralysis of varying degree and persistence. When the motor neurons affected are those of the diaphragm or of the intercostal muscles, respiratory paralysis occurs. Bulbar poliomyelitis results from viral attack on the medulla (bulb of the brain) or higher brain centers, with respiratory, vasomotor, facial, palatal, or pharyngeal disturbances.

Poliomyelitis occurs throughout the world. In temperate zones it appears chiefly in summer and fall, although winter outbreaks have been known. It occurs in all age groups, but less frequently in adults because of their acquired immunity. The virus is spread by human contact; the nature of the contact is not clear, but it appears to be associated with familial contact and with interfamily contact among young children. The virus may be present in flies.

Inactivated poliovirus vaccine (Salk; IPV), prepared from virus grown in monkey kidney cultures, was developed and first used in the United States, but oral poliovirus vaccine (Sabin; OPV) is now generally used throughout the world. The oral vaccine is a living, attenuated virus. [J.L.Me.]

Pollen The small male reproductive bodies produced in the pollen sacs of seed plants (gymnosperms and angiosperms).

On maturation in the pollen sac, a pollen grain may reach 0.00007 mg as in spruce, or less than 1/20 of this weight. A grain usually has two waxy, durable outer walls, the exine, and an inner fragile wall, the intine. These walls surround the contents with their nuclei and reserves of starch and oil.

Pollen identification depends on interpretation of morphological features. Exine and aperture patterns are especially varied in the more highly evolved dicots, so that recognition at family, genus, or even species level may be possible despite the small surface area available on a grain (Figs. 1 and 2). Since the morphological characters are con-

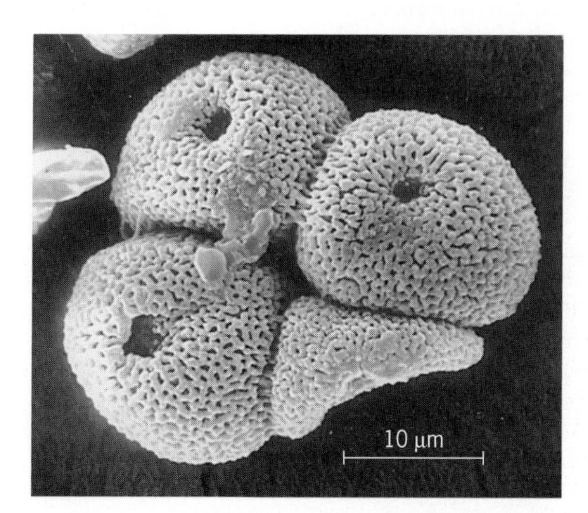

Fig. 1. Tetrad of cattail (*Typha latifolia*), with grains cohering, one-pored, exine reticulate. (*Scanning electron micrograph by C. M. Drew, U.S. Naval Weapons Center, Cina Lake, California*).

Fig. 2. Young tetrad of *Lavatera* (mallow family), with grains free in callose of pollen mother cell. (*Photomicrograph by Luc Waterkeyn*)

servative in the extreme, usually changing very slowly through geologic time, studies of fine detail serve to establish the lineal descent of many plants living today.

Extreme variations in size may occur within a family, but pollen grains range mainly from 24 to 50 micrometers, with the dicot range being from 2 μm in *Myosotis* to 250 μm in *Mirabilis*; the monocots range from about 15 to 150 μm or more in the ginger family, with eelgrass (*Zostera*) having pollen measuring 2550 $\times$ 3.7 μm in a class of its own; living gymnosperms range from 15 μm in *Gnetum* to about 180 μm in *Abies* (including sacs), while fossil types range from about 11 μm (one-furrowed) to 300 μm.

Most grains are free (monads) though often loosely grouped because of the spines or sticky oils and viscin threads. Compound grains (polyads) are richly developed in some angiosperm families, commonly occurring in four (tetrads), or in multiples of four up to 64 or more.

In most microspores (pollen and spore) the polar axis runs from the inner (proximal) face to the outer (distal) face, as oriented during tetrad formation. The equator crosses it at right angles. Bilateral grains dominate in the gymnosperms and monocots, the polar axis usually being the shorter one, with the single aperture on the distal side. On the other hand, almost all dicot pollen is symmetrical around the polar axis (usually the long axis), with shapes ranging mainly from spheroidal to ellipsoidal, with rounded equatorial outlines, and sometimes "waisted" as in the Umbelliferae. Three-pored grains often have strikingly triangular outlines in polar view.

Rarely lacking, the flexible membranes of apertures are sometimes covered only by endexine. They allow for sudden volume changes, as for the emergence of the germ tube. They are classified as furrows, with elongate outlines, and pores usually more or less circular in shape. Short slitlike intermediate forms occur. A few families have no apertures, some as a result of reduction of exine as an adaptation to wind- or water-pollination. Gymnosperms may also lack openings, or have one small papilla; most have one furrow, or a long weak area.

Few pollen grains are completely smooth (psilate) at ordinary magnifications; most have sculpture both on the surface and the structure below it. *See* FLOWER; POLLINATION; REPRODUCTION (PLANT). [L.M.C.]

Pollination The transport of pollen grains from the plant parts that produce them to the ovule-bearing organs, or to the ovules (seed precursors) themselves. In gymnosperms, the pollen, usually dispersed by the wind, is simply caught by a drop of fluid excreted by each freely exposed ovule. In angiosperms, where the ovules are contained in the pistil, the pollen is deposited on the pistil's receptive end (the stigma), where it germinates. *See* FLOWER.

Without pollination, there would be no fertilization; it is thus of crucial importance for the production of fruit crops and seed crops. Pollination also plays an important part in plant breeding experiments aimed at increasing crop production through the creation of genetically superior types. *See* BREEDING (PLANT); REPRODUCTION (PLANT).

Self- and cross-pollination. In most plants, self-pollination is difficult or impossible, and there are various mechanisms which are responsible. For example, in dichogamous flowers, the pistils and stamens reach maturity at different times; in protogyny, the pistils mature first, and in protandry, the stamens mature before the pistils. Selfing is also impossible in dioecious species, where some plants bear flowers that have only pistils (pistillate or female flowers), while other individuals have flowers that produce only pollen (staminate or male flowers). In monoecious species, where pistillate and staminate flowers are found in the same plant, self-breeding is at least reduced. Heterostyly is another device that promotes outbreeding. Here some flowers (pins) possess a long pistil and short stamens, while others (thrums) exhibit the reverse condition; each plant individual bears only pins or only thrums.

Flower attractants. As immobile organisms, plants normally need external agents for pollen transport. These can be insects, wind, birds, mammals, or water, roughly in that order of importance. In some plants the pollinators are simply trapped; in the large majority of cases, however, the flowers offer one or more rewards, such as sugary nectar, oil, solid food bodies, perfume, sex, an opportunity to breed, a place to sleep, or some of the pollen itself. For the attraction of pollinators, flowers provide either visual or olfactory signals. Color includes ultraviolet, which is perceived as a color by most insects and at least some hummingbird species. Fragrance is characteristic of flowers pollinated by bees, butterflies, or hawkmoths, while carrion or dung odors are produced by flowers catering to certain beetles and flies. A few orchids, using a combination of olfactory and visual signals, mimic the females of certain bees or wasps so successfully that the corresponding male insects will try to mate with them, thus achieving pollination (pseudocopulation).

While some flowers are "generalists," catering to a whole array of different animals, others are highly specialized, being pollinated by a single species of insect only. Extreme pollinator specificity is an important factor in maintaining the purity of plant species in the field, even in those cases where hybridization can easily be achieved artificially in a greenhouse or laboratory, as in most orchids. The almost incredible mutual adaptation between pollinating animal and flower which can frequently be observed exemplifies the idea of coevolution. *See* POLLEN. [B.J.D.M.]

Polychaeta The largest class of the phylum Annelida, containing 68–70 families. About 1600 genera and 10,000 species have been named from worldwide areas; about one-fourth of this number may be synonymous. Polychaeta (meaning "many setae") is conveniently though not clearly divisible into the Errantia, or free-moving annelids, and Sedentaria, or tubicolous families.

The body may be long, cylindrical, and multisegmented, or short and compact, with a limited number of segments. It consists of prostomium (Fig. 1), or head; peristomium, or first segment around the mouth; trunk, or body proper; and tail region, or pygidium. Most segments have highly diagnostic paired, lateral fleshy appendages called

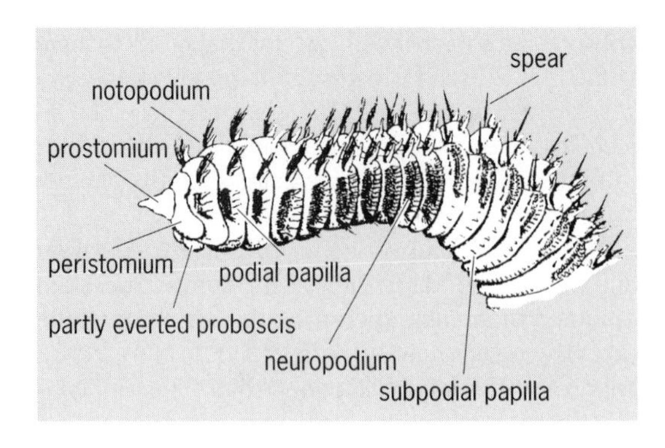

Fig. 1. Terminology of the anterior parts of the body, based on *Phylo* (Orbiniidae).

parapodia. These are provided with secreted supporting rods and spreading fascicles of setae, or hooks, which display remarkable specificity.

The anterior end, or prostomium, may be a simple lobe derived from the larval trochophore, modified as a pseudoannulated cone, or covered by peristomial structures so as to be invisible. Oral tentacles for food gathering may be eversible from the buccal cavity; they may be long, slender, or thick and their surface smooth or papillated.

The anterior, preoral end may be developed as a thick, fleshy papillated, nonretractile proboscis (Fig. 2), or the prostomium may be completely retractile into the first several segments and protected by a cage formed of setae directed forward, or concealed by a compact operculum formed of setae of the first several segments. The anterior end of the alimentary tract is muscular or epithelial; it may be covered with soft papillae or hard structures. These structures function for secretion, food gathering, and maintaining traction; they are named for their form or function.

The trunk is the main body region and is composed of metameres numbering few to many. They may be similar to one another (homonomous) as in Errantia, or different (heteronomous) resulting in anterior thoracic and posterior abdominal regions.

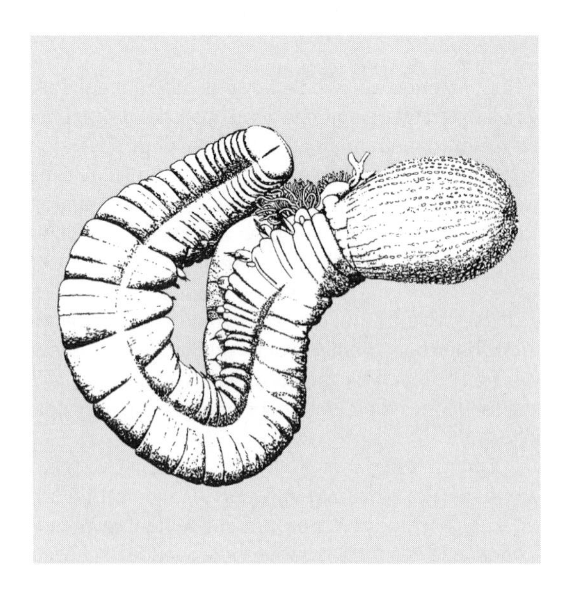

Fig. 2. Nonretractile proboscis organ preceding prostomium in *Artacama* (Terebellidae).

Reproduction is highly evolved and diversified; it can be sexual or asexual. Sexual reproduction is usually dioecious, with the two sexes similar. In rare cases it is dimorphic.

Polychaetes range in length from a fraction of 0.04 in. (1 mm) to more than 144 in. (360 cm). Colors and patterns are varied and specific, due to pigment and refraction of light. Littoral, warm-water species may be brilliant and multicolored, whereas polar and deep-water species tend to be drab or sometimes melanistic to almost black.

Most polychaetes are free-living; some of the remaining members are commensal with another animal for attachment surface, for food, or for protection. Polychaetes are distributed in all marine habitats and show remarkable specificity according to latitude, depth, and kinds of substrata. Most of the families tend to be represented in any major geographic area, although taxa may differ with place. *See* ANNELIDA. [O.H.]

Polyoma virus A papovavirus that infects rodents. The name derives from the capability of this virus to induce a wide variety of tumors when inoculated into newborn animals. The icosahedral viral particle consists of deoxyribonucleic acid (DNA) and protein only. The genome is a small, double-stranded, closed circular DNA molecule approximately 5300 base pairs in length. It encodes the T antigens expressed early in the productive cycle and in transformed cells, and the viral capsid proteins, expressed late in the productive cycle.

Polyoma virus is endemic in most wild populations of mice, but causes no disease. Tumors produced by this virus are unknown in the wild. Inoculation of large quantities of virus into newborn rodents, however, induces a number of tumor types, particularly many sarcomas and carcinomas. These tumors contain the viral genome but produce few infectious viral particles. Infected animals produce neutralizing antibodies directed against structural components of the virus; tumor-bearing animals produce antibodies to antigens (T antigens) which are present in tumor cells but not in the virus particle.

The virus is easily propagated to high titer in mouse embryo tissue culture, resulting in the lysis of the cells and the production of a hemagglutinin. This "productive infection" occurs only in mouse cells. Small numbers of "abortively infected" cells retain the viral genome and become transformed. Transformed cells are tumorigenic when injected into syngeneic animals, and contain the T antigens but produce no virus. The expression of the T antigens has been shown to be necessary and sufficient to induce cell transformation. *See* ONCOLOGY; TUMOR VIRUSES; VIRUS. [S.M.D.]

Poplar Any tree of the genus *Populus*, family Salicaceae, marked by simple, alternate leaves which are usually broader than those of the willow, the other American representative of this family. Poplars have scaly buds, bitter bark, flowers and fruit in catkins, and a five-angled pith. *See* WILLOW.

Some species are commonly called cottonwood because of the cottony hairs attached to the seeds. Other species, called aspens, have weak, flattened leaf stalks which cause the leaves to flutter in the slightest breeze. One of the important species in the United States is the quaking, or trembling, aspen (*P. tremuloides*). The soft wood of this species is used for paper pulp. The European aspen (*P. nigra*), which is similar to the quaking aspen, is sometimes planted, and its variety, *italica*, the Lombardy poplar of erect columnar habit, is used in landscape planting. The black cottonwood (*P. trichocarpa*) is the largest American poplar and is also the largest broad-leaved tree in the forests of the Pacific Northwest. The cottonwood or necklace poplar (*P. deltoides*) is native in the eastern half of the United States. In the balsam or tacamahac poplar (*P. balsamifera*), the resin is used in medicine as an expectorant. The wood is used for veneer, boxes, crates, furniture, paper pulp, and excelsior. [A.H.G./K.P.D.]

Population dispersal The process by which groups of living organisms expand the space or range within which they live. Dispersal operates when individual organisms leave the space that they have occupied previously, or in which they were born, and settle in new areas. Natal dispersal is the first movement of an organism from its birth site to the site in which it first attempts to breed. Adult dispersal is a subsequent movement when an adult organism changes its location in space. As individuals move across space and settle into new locations, the population to which they belong expands or contracts its overall distribution. Thus, dispersal is the process by which populations change the area they occupy.

Migration is the regular movement of organisms during different seasons. Many species migrate between wintering and breeding ranges. Such migratory movement is marked by a regular return in future seasons to previously occupied regions, and so usually does not involve an expansion of population range. Some migratory species show astounding abilities to return to the exact locations used in previous seasons. Other species show no regular movements, but wander aimlessly without settling permanently into a new space. Wandering (called nomadism) is typical of species in regions where the availibility of food resources are unpredictable from year to year. Neither migration nor nomadism is considered an example of true dispersal. *See* MIGRATORY BEHAVIOR.

Virtually all forms of animals and plants disperse. In most higher vertebrates, the dispersal unit is an entire organism, often a juvenile or a member of another young age class. In other vertebrates and many plants, especially those that are sessile (permanently attached to a surface), the dispersal unit is a specialized structure (disseminule). Seeds, spores, and fruits are disseminules of plants and fungi; trochophores and planula larvae are disseminules of sea worms and corals, respectively. Many disseminules are highly evolved structures specialized for movement by specific dispersal agents such as wind, water, or other animals.

A special case of zoochory (dispersal using animal agents) involves transport by humans. The movement of people and cargo by cart, car, train, plane, and boat has increased the potential dispersal of weedy species worldwide. Many foreign aquatic species have been introduced to coastal areas by accidental dispersal of disseminules in ship ballast water. The zebra mussel is one exotic species that arrive in this manner and is now a major economic problem throughout the Great Lakes region of North America. Some organisms have been deliberately introduced by humans into new areas. Domestic animals and plants have been released throughout the world by farmers. A few pest species were deliberately released by humans; European starlings, for example.

Some of the most highly coevolved dispersal systems are those in which the disseminule must be eaten by an animal. Such systems have often evolved a complex series of signals and investments by both the plant and the animal to ensure that the seeds are dispersed at an appropriate time and that the animal is a dependable dispersal agent. Such highly evolved systems are common in fruiting plants and their dispersal agents, which are animals called frugivores. Fruiting plants cover their seeds with an attractive, edible package (the fruit) to get the frugivore to eat the seed. To ensure that fruits are not eaten until the seeds are mature, plants change the color of their fruits as a signal to show that the fruits are ready for eating.

Many plants in the tropical rainforests are coevolved to have their seeds dispersed by specific animal vectors, including birds, mammals, and ants. Many tropical trees, shrubs, and herbaceous plants are specialized to have their seeds dispersed by a single animal species. Temperate forest trees, in contrast, often depend on wind dispersal of both pollen and seeds.

Dispersal barriers are physical structures that prevent organisms from crossing into new space. Oceans, rivers, roads, and mountains are examples of barriers for species whose disseminules cannot cross such features. It is believed that the creation of physical barriers is the primary factor responsible for the evolution of new species. A widespread species can be broken into isolated fragments by the creation of a new physical barrier. With no dispersal linking the newly isolated populations, genetic differences that evolve in each population cannot be shared between populations. Eventually, the populations may become so different that no interbreeding occurs even if dispersal pathways are reconnected. The populations are then considered separate species. *See* SPECIATION.

Dispersal is of major concern for scientists who work with rare and endangered animals. Extinction is known to be more prevalent in small, isolated populations. Conservation biologists believe that many species exist as a metapopulation, that is, a group of populations interconnected by the dispersal of individuals or disseminules between subpopulations. The interruption of dispersal in this system of isolated populations can increase the possibility of extinction of the whole metapopulation. Conservation plans sometimes propose the creation of corridors to link isolated patches of habitat as a way of increasing the probability of successful dispersal. *See* EXTINCTION (BIOLOGY); POPULATION DISPERSION. [J.B.D.]

Population dispersion The spatial distribution at any particular moment of the individuals of a species of plant or animal. Under natural conditions organisms are distributed either by active movements, or migrations, or by passive transport by wind, water, or other organisms. The act or process of dissemination is usually termed dispersal, while the resulting pattern of distribution is best referred to as dispersion. Dispersion is a basic characteristic of populations, controlling various features of their structure and organization. It determines population density, that is, the number of individuals per unit of area, or volume, and its reciprocal relationship, mean area, or the average area per individual. It also determines the frequency, or chance of encountering one or more individuals of the population in a particular sample unit of area, or volume. The ecologist therefore studies not only the fluctuations in numbers of individuals in a population but also the changes in their distribution in space. *See* POPULATION DISPERSAL.

Principal types of dispersion. The dispersion pattern of individuals in a population may conform to any one of several broad types, such as random, uniform, or contagious (clumped). Any pattern is relative to the space being examined; a population may appear clumped when a large area is considered, but may prove to be distributed at random with respect to a much smaller area.

Random or haphazard implies that the individuals have been distributed by chance. In such a distribution, the probability of finding an individual at any point in the area is the same for all points. Hence a truly random pattern will develop only if each individual has had an equal and independent opportunity to establish itself at any given point. Examples of approximately random dispersions can be found in the patterns of settlement by free-floating marine larvae and of colonization of bare ground by airborne disseminules of plants. Nevertheless, true randomness appears to be relatively rare in nature.

Uniform distribution implies a regularity of distance between and among the individuals of a population. Perfect uniformity exists when the distance from one individual to its nearest neighbor is the same for all individuals. Patterns approaching uniformity are most obvious in the dispersion of orchard trees and in other artificial plantings, but the tendency to a regular distribution is also found in nature, as for example in the relatively even spacing of trees in forest canopies, the arrangement of shrubs in deserts, and the distribution of territorial animals.

The most frequent type of distribution encountered is contagious or clumped, indicating the existence of aggregations or groups in the population. Clusters and clones of plants, and families, flocks, and herds of animals are common phenomena. The formation of groups introduces a higher order of complexity in the dispersion pattern, since the several aggregations may themselves be distributed at random, evenly, or in clumps. An adequate description of dispersion, therefore, must include not only the determination of the type of distribution, but also an assessment of the extent of aggregation if the latter is present.

Factors affecting dispersion. The principal factors that determine patterns of population dispersion include (1) the action of environmental agencies of transport, (2) the distribution of soil types and other physical features of the habitat, (3) the influence of temporal changes in weather and climate, (4) the behavior pattern of the population in regard to reproductive processes and dispersal of the young, (5) the intensity of intra- and interspecific competition, and (6) the various social and antisocial forces that may develop among the members of the population. Although in certain cases the dispersion pattern may be due to the overriding effects of one factor, in general populations are subject to the collective and simultaneous action of numerous distributional forces and the dispersion pattern reflects their combined influence. When many small factors act together on the population, a more or less random distribution is to be expected, whereas the domination of a few major factors tends to produce departure from randomness.

Optimal population density. The degree of aggregation which promotes optimum population growth and survival varies according to the species and the circumstances. Groups or organisms often flourish best if neither too few nor too many individuals are present; they have an optimal population density at some intermediate level. The concept of an intermediate optimal population density is sometimes known as Allee's principle. *See* ECOLOGICAL COMMUNITIES; POPULATION ECOLOGY; POPULATION GENETICS. [F.C.E.]

Population ecology The study of spatial and temporal patterns in the abundance and distribution of organisms and of the mechanisms that produce those patterns. Species differ dramatically in their average abundance and geographical distributions, and they display a remarkable range of dynamical patterns of abundance over time, including relative constancy, cycles, irregular fluctuations, violent outbreaks, and extinctions. The aims of population ecology are threefold: (1) to elucidate general principles explaining these dynamic patterns; (2) to integrate these principles with mechanistic models and evolutionary interpretations of individual life-history tactics, physiology, and behavior as well as with theories of community and ecosystem dynamics; and (3) to apply these principles to the management and conservation of natural populations.

In addition to its intrinsic conceptual appeal, population ecology has great practical utility. Control programs for agricultural pests or human diseases ideally attempt to reduce the intrinsic rate of increase of those organisms to very low values. Analyses of the population dynamics of infectious diseases have successfully guided the development of vaccination programs. In the exploitation of renewable resources, such as in forestry or fisheries biology, population models are required in order to devise sensible harvesting strategies that maximize the sustainable yield extracted from exploited populations. Conservation biology is increasingly concerned with the consequences of habitat fragmentation for species preservation. Population models can help characterize minimum viable population sizes below which a species is vulnerable to rapid extinction, and can help guide the development of interventionist policies to save endangered species. Finally, population ecology must be an integral part of any attempt to bring the world's burgeoning human population into harmonious balance with the environment. *See* ECOLOGY; MATHEMATICAL ECOLOGY; THEORETICAL ECOLOGY. [R.Hol.]

Population genetics The study of both experimental and theoretical consequences of mendelian heredity on the population level, in contradistinction to classical genetics which deals with the offspring of specified parents on the familial level. The genetics of populations studies the frequencies of genes, genotypes, and phenotypes, and the mating systems. It also studies the forces that may alter the genetic composition of a population in time, such as recurrent mutation, migration, and intermixture between groups, selection resulting from genotypic differential fertility, and the random changes incurred by the sampling process in reproduction from generation to generation. This type of study contributes to an understanding of the elementary step in biological evolution. The principles of population genetics may be applied to plants and to other animals as well as humans. *See* GENETICS; MENDELISM. [C.C.L.]

Population viability The ability of a population to persist and to avoid extinction. The viability of a population will increase or decrease in response to changes in the rates of birth, death, and growth of individuals. In natural populations, these rates are not stable, but undergo fluctuations due to external forces such as hurricanes and introduced species, and internal forces such as competition and genetic composition. Such factors can drive populations to extinction if they are severe or if several detrimental events occur before the population can recover. *See* ECOLOGY; POPULATION ECOLOGY.

One of the most important uses of population viability models comes from modern conservation biology, which uses these models to determine whether a population is in danger of extinction. This is called population viability analysis (PVA) and consists of demographic and genetic models that are used to make decisions on how to manage populations of threatened or endangered species. The National Research Council has called population viability analysis "the cornerstone, the obligatory tool by which recovery objectives and criteria [for endangered species] are identified." *See* ECOLOGICAL MODELING. [G.LeB.; T.E.M.]

Precipitation (meteorology) The fallout of water drops or frozen particles from the atmosphere. Liquid types are rain or drizzle, and frozen types are snow, hail, small hail, ice pellets (also called ice grains; in the United States, sleet), snow pellets (graupel, soft hail), snow grains, ice needles, and ice crystals. In England sleet is defined as a mixture of rain and snow, or melting snow. Deposits of dew, frost, or rime, and moisture collected from fog are occasionally also classed as precipitation. *See* HAIL; SNOW.

All precipitation types are called hydrometeors, of which additional forms are clouds, fog, wet haze, mist, blowing snow, and spray. Whenever rain or drizzle freezes on contact with the ground to form a solid coating of ice, it is called freezing rain, freezing drizzle, or glazed frost; it is also called an ice storm or a glaze storm, and sometimes is popularly known as silver thaw or erroneously as a sleet storm. *See* CLOUD; FOG.

Rain, snow, or ice pellets may fall steadily or in showers. Steady precipitation may be intermittent though lacking sudden bursts of intensity. Hail, small hail, and snow pellets occur only in showers; drizzle, snow grains, and ice crystals occur as steady precipitation. Showers originate from instability clouds of the cumulus family, whereas steady precipitation originates from stratiform clouds.

The amount of precipitation, often referred to as precipitation or simply as rainfall, is measured in a collection gage. It is the actual depth of liquid water which has fallen on the ground, after frozen forms have been melted, and is recorded in millimeters or inches and hundredths. A separate measurement is made of the depth of unmelted snow, hail, or other frozen forms. *See* SNOW GAGE. For discussions of other topics related to precipitation; CLOUD PHYSICS; DEW; DEW POINT; HUMIDITY; HYDROLOGY; HYDROMETEOROLOGY; RAIN SHADOW; WEATHER MODIFICATION. [J.R.F.]

Precipitation measurement Instruments used to measure the amount of rain or snow that falls on a level surface. Such measurements are made with instruments known as precipitation gages. A precipitation gage can be as simple as an open container on the ground to collect rain, snow, and hail; it is usually more complex, however, because of the need to avoid wind effects, enhance accuracy and resolution, and make a measurement representative of a large area. Precipitation is measured as the depth to which a flat horizontal surface would have been covered per unit time if no water were lost by runoff, evaporation, or percolation. Depth is expressed in inches or millimeters, typically per day. The unit of time is often understood and not stated explicitly. Snow and hail are converted to equivalent depth of liquid water. *See* METEOROLOGICAL INSTRUMENTATION; PRECIPITATION (METEOROLOGY); SNOW SURVEYING. [F.V.B.]

Accurate quantitative precipitation measurement is probably the most important weather radar application. It is extremely valuable for hydrological applications such as watershed management and flash flood warnings. Radar can make rapid and spatially contiguous measurements over vast areas of a watershed at relatively low cost. *See* PRECIPITATION (METEOROLOGY); RADAR METEOROLOGY. [R.J.Do.]

Precision agriculture The application of technologies and agronomic principles to manage spatial and temporal variability associated with all aspects of agricultural production for the purpose of improving crop performance and environmental quality. The intent of precision agriculture is to match agricultural inputs and practices to localized conditions within a field (site-specific management) and to improve the accuracy of their application. The finer-scale management of precision agriculture is in contrast to whole-field or whole-farm management strategies, where management decisions and practices are uniformly applied throughout a field or farmstead.

Successful implementation of precision agriculture requires three basic steps. First, farmers must obtain accurate maps of the spatial variability of factors (soils, plants, and pests) that determine crop yield and quality and/or factors that cause environmental degradation. Second, once known, variability can be managed using site-specific management recommendations and accurate input control technologies. Third, precision agriculture requires an evaluation component to understand the economic, environmental, and social impacts on the farm and adjacent ecosystems and to provide feedback on cropping system performance.

Precision agriculture is technology-enabled, information-based, and decision-focused, because it relies on an increasing level of detail in information acquired with technology to improve decision making in crop production. Consequently, precision agriculture will evolve as technology, information management, and decision tools emerge in this era of rapid technological advancement. *See* AGRICULTURE; AGRONOMY. [F.J.P.; P.N.]

Predator-prey interactions Predation occurs when one animal (the predator) eats another living animal (the prey) to utilize the energy and nutrients from the body of the prey for growth, maintenance, or reproduction. In the special case in which both predator and prey are from the same species, predation is called cannibalism. Sometimes the prey is actually consumed by the predator's offspring. This is particularly prevalent in the insect world. Insect predators that follow this type of lifestyle are called parasitoids, since the offspring grow parasitically on the prey provided by their mother.

Predation is often distinguished from herbivory by requiring that the prey be an animal rather than a plant or other type of organism (bacteria). To distinguish predation from decomposition, the prey animal must be killed by the predator. Some organisms occupy a gray area between predator and parasite. Finally, the requirement that both energy and nutrients be assimilated by the predator excludes carnivorous plants from being predators, since they assimilate only nutrients from the animals they consume. *See* FOOD WEB.

Population dynamics refers to changes in the sizes of populations of organisms through time, and predator-prey interactions may play an important role in explaining the population dynamics of many species. They are a type of antagonistic interaction, in which the population of one species (predators) has a negative effect on the population of a second (prey), while the second has a positive effect on the first. For population dynamics, predator-prey interactions are similar to other types of antagonistic interactions, such as pathogen-host and herbivore-plant interactions.

Community structure refers generally to how species within an ecological community interact. The simplest conception of a community is as a food chain, with plants or other photosynthetic organisms at the bottom, followed by herbivores, predators that eat herbivores, and predators that eat other predators. This simple conception works well for some communities. Nonetheless, the role of predator species in communities is often not clear. Many predators change their ecological roles over their lifetime. Many insect predators that share the same prey species are also quite likely to kill and devour each other. This is called intraguild predation, since it is predation within the guild of predators. Furthermore, many species are omnivores, feeding at different times as either predators or herbivores. Therefore, the role of particular predator species in a community is often complex.

Predator-prey interactions may have a large impact on the overall properties of a community. For example, most terrestrial communities are green, suggesting that predation on herbivores is great enough to stop them from consuming the majority of plant material. In contrast, the biomass of herbivorous zooplankton in many aquatic communities is greater than the biomass of the photosynthetic phytoplankton, suggesting that predation on zooplankton is not enough to keep these communities green. *See* POPULATION ECOLOGY. [A.R.I.]

Prion disease Transmissible spongiform encephalopathies in both humans and animals. Scrapie is the most common form in animals, while in humans the most prevalent form is Creutzfeldt-Jakob disease. This group of disorders is characterized at a neuropathological level by vacuolation of the brain's gray matter (spongiform change). They were initially considered to be examples of slow virus infections. Experimental work has consistently failed to demonstrate detectable nucleic acids—both ribonucleic acid (RNA) and deoxyribonucleic acid (DNA)—as constituting part of the infectious agent. Contemporary understanding suggests that the infectious particles are composed predominantly, or perhaps even solely, of protein, and from this concept was derived the acronym prion (proteinaceous infectious particles). Also of interest is the apparent paradox of how these disorders can be simultaneously infectious and yet inherited in an autosomal dominant fashion (from a gene on a chromosome other than a sex chromosome).

Disorders. Scrapie, which occurs naturally in sheep and goats, was the first of the spongiform encephalopathies to be described. An increasing range of animal species have been recognized as occasional natural hosts of this type of disease. Bovine spongioform encephalopathy, commonly known as mad cow disease, has been epidemic in British cattle. The first confirmed cases were reported in late 1986. By early 1995 it had been identified in almost 150,000 cattle and more than half of all British herds. Its exact origin is not known, but claims that it came from sheep are now discredited.

So far, animal models have indicated that only central nervous system tissue has been shown to transmit the disease after oral ingestion—a diverse range of other organs, including udder, skeletal muscle, lymph nodes, liver, and buffy coat of blood (white blood cells) proving noninfectious.

The currently recognized spectrum of human disorders encompasses kuru, Creutzfeldt-Jakob disease, Gerstmann-Straussler-Scheinker disease, and fatal familial insomnia. All, including familial cases, have been shown to be transmissible to animals and hence potentially infectious; all are invariably fatal with no effective treatments currently available.

Human-to-human transmission. A variety of mechanisms of human-to-human transmission have been described. Transmission is due in part to the ineffectiveness of conventional sterilization and disinfection procedures to control the infectivity of transmissible spongiform encephalopathies. Numerically, pituitary hormone–related Creutzfeldt-Jakob disease is the most important form of human-to-human transmission of disease. However, epidemiological evidence suggests that there is no increased risk of contracting Creutzfeldt-Jakob disease from exposure in the form of close personal contact during domestic and occupational activities. Incubation periods in cases involving human-to-human transmission appear to vary enormously, depending upon the mechanism of inoculation. Current evidence suggests that transmission of Creutzfeldt-Jakob disease from mother to child does not occur. Two important factors pertaining to transmissibility are the method of inoculation and the dose of infectious material administered. A high dose of infectious material administered by direct intracerebral inoculation is clearly the most effective method of transmissibility and generally provides the shortest incubation time. *See* MUTATION; SCRAPIE. [C.L.Mas.; S.J.Co.]

Prokaryotae A group of predominantly unicellular microorganisms or infectious agents of cells (the viruses), lacking nuclei, and having asexual and chromonemal reproduction and unidirectional recombination. The Prokaryotae may be considered to include a kingdom for viruses, although such "organisms" are considered acellular or noncellular (even nonliving) by many authorities, and one for the typical moneran forms, the many kinds of bacteria plus the cyanobacteria (the blue-green algae) and the Prochlorophycota. *See* VIRUS.

Bacteria, viruses, and blue-green algae possess little in common, besides such superficial characters as microscopic (or ultramicroscopic) size and frequent involvement in causing diseases in other organisms, including human beings, and such negative characters as not being eukaryotic, not possessing any mouth opening, and not being multicellular (or, in the case of viruses, even cellular) in their organization. In the smallest dimension, prokaryotes measure from 0.2 to 10 micrometers; viruses show a diameter of 10 to 300 nanometers, the largest, therefore, just barely overlapping in width with the smallest bacterium.

Monerans (above the virus level) are generally solitary, unicellular forms; but some species are filamentous, colonial, or mycelial. Some are also motile, either by gliding or by the action of bacterial flagella containing the protein flagellin. Modes of nutrition are diverse: absorptive, chemosynthetic, photoheterotrophic, and photoautotrophic.

Respiration is anaerobic or aerobic, or facultatively either one. Respiratory and photosynthetic functions are both generally associated with the plasma membrane system: there are no specialized organelles such as mitochondria or plastids, although thylakoids are present in cyanobacteria and in Prochlorophycota.

Virus particles can survive in a dried, crystalline, metabolically inert state. Bacteria may produce endospores of great resistance to a variety of environmental stresses; the trophic forms occur ubiquitously in aquatic or moist habitats, including cells and tissues of hosts belonging to all other groups of organisms. Complex viruses have an envelope surrounding the nucleocapsid; many bacteria possess rigid cell walls, and some produce outer sheaths. *See* ALGAE; BACTERIA; CYANOBACTERIA . [J.O.C.]

Protista The kingdom comprising all single-celled forms of living organisms in both the five-kingdom and six-kingdom systems of classification. Kingdom Protista encompasses both Protozoa and Protophyta, allowing considerable integration in the classification of both these animallike and plantlike organisms, all of whose living functions as individuals are carried out within a single cell membrane. Among the kingdoms of cellular organisms, this definition can be used to distinguish the Protista from the Metazoa (sometimes named Animalia) for many-celled animals, or from the Fungi and from the Metaphyta (or Plantae) for many-celled green plants.

The most significant biological distinction is that which separates the bacteria and certain other simply organized organisms, including blue-green algae (collectively, often designated Kingdom Monera), from both Protista and all many-celled organisms. The bacteria are described as prokaryotic; both the Protista and the cells of higher plants and animals are eukaryotic. Structurally, a distinguishing feature is the presence of a membrane, closely similar to the bounding cell membrane, surrounding the nuclear material in eukaryotic cells, but not in prokaryotic ones. *See* PROKARYOTAE; PROTOZOA.

The definition that can separate the Protista from many-celled animals is that the protistan body never has any specialized parts of the cytoplasm under the sole control of a nucleus. In some protozoa, there can be two, a few, or even many nuclei, rather than one, but no single nucleus ever has separate control over any part of the protistan cytoplasm which is specialized for a particular function. In contrast, in metazoans there are always many cases of nuclei, each in control of cells of specialized function.

Most authorities would agree that the higher plants, the Metazoa, and the Parazoa (or sponges) almost certainly evolved (each independently) from certain flagellate stocks of protistans. [W.D.R.H.]

Protozoa A group of eukaryotic microorganisms traditionally classified in the animal kingdom. Although the name signifies primitive animals, some Protozoa (phytoflagellates and slime molds) show enough plantlike characteristics to justify claims that they are plants.

Protozoa are almost as widely distributed as bacteria. Free-living types occur in soil, wet sand, and in fresh, brackish, and salt waters. Protozoa of the soil and sand live in films of moisture on the particles. Habitats of endoparasites vary. Some are intracellular, such as malarial parasites in vertebrates, which are typical Coccidia in most of the cycle. Other parasites, such as *Entamoeba histolytica*, invade tissues but not individual cells. Most trypanosomes live in the blood plasma of vertebrate hosts. Many other parasites live in the lumen of the digestive tract or sometimes in coelomic cavities of invertebrates, as do certain gregarines. *See* GREGARINIA.

Many Protozoa are uninucleate, others are binucleate or multinucleate, and the number of nuclei also may vary at different stages in a life cycle. Protozoa range in size from 1 to 10^6 micrometers. Colonies are known in flagellates, ciliates, and Sarcodina.

Although marked differentiation of the reproductive and somatic zooids characterizes certain colonies, such as *Volvox*, Protozoa have not developed tissues and organs.

Morphology. A protozoan may be a plastic organism (ameboid type), but changes in form are often restricted by the pellicle. A protective layer is often secreted outside the pellicle, although the pellicle itself may be strengthened by incorporation of minerals. Secreted coverings may fit closely, for example, the cellulose-containing theca of Phytomonadida and Dinoflagellida, analogous to the cell wall in higher plants. The dinoflagellate theca (Fig. 1*a*) may be composed of plates arranged in a specific pattern. Tests, as seen in Rhizopodea (Arcellinida, Gromiida, Foraminiferida), may be composed mostly of inorganic material, although organic (chitinous) tests occur in certain species. Siliceous skeletons, often elaborate, characterize the Radiolaria (Figs. 1*d* and 2*c*). A vase-shaped lorica, from which the anterior part of the organism or its appendages may be extended, occurs in certain flagellates (Fig. 1*b*) and ciliates (Fig. 1*c*). Certain marine ciliates (Tintinnida) are actively swimming loricate forms.

Flagella occur in active stages of Mastigophora and flagellated stages of certain Sarcodina and Sporozoa. A flagellum consists of a sheath enclosing a matrix in which an axoneme extends from the cytoplasm to the flagellar tip. In certain groups the sheath shows lateral fibrils (mastigonemes) which increase the surface area and also

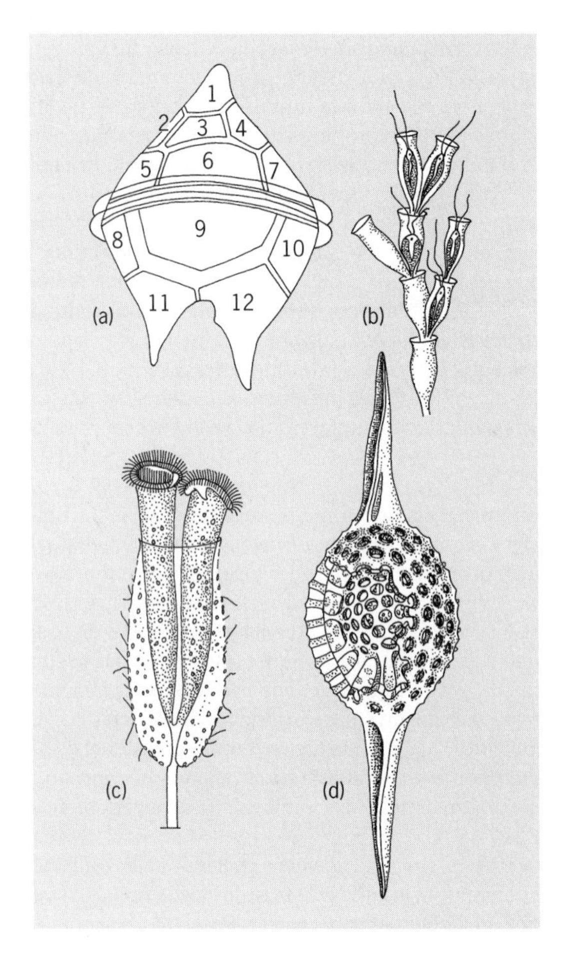

Fig. 1. External coverings of Protozoa. (*a*) Theca of dinoflagellate (*Peridinium*), showing separate plates. (*b*) Lorica of a colonial chrysomonad, *Dinobryon*. (*c*) Two zooids within a lorica of a peritrich, *Cothurnia*. (*d*) A radiolarian skeleton, siliceous type. (*After L. H. Hyman, The Invertebrates, vol. 1, McGraw-Hill, 1940*)

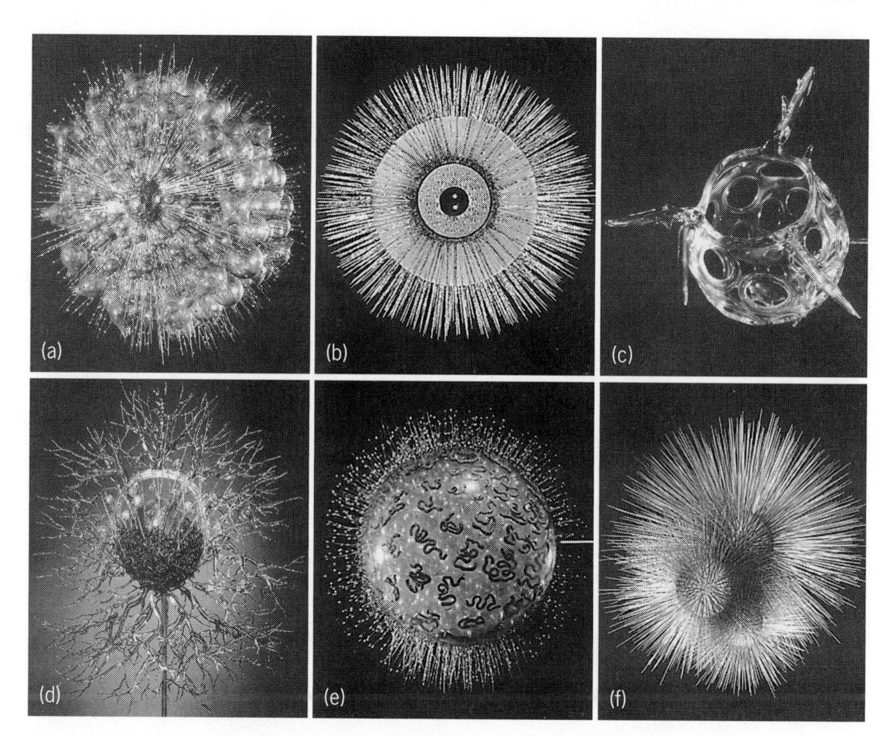

Fig. 2. Glass models of marine Protozoa. Radiolarian types: (*a*) *Trypanosphaera transformata* Haeckel (Indian Ocean); (*b*) *Actissa princeps* Haeckel (Indian and Pacific oceans); (*c*) *Peridium spinipes* Haeckel (Pacific Ocean); (*d*) *Lithocircus magnificus* Haeckel (Atlantic Ocean); (*e*) *Collozoum serpentinum* Haeckel (Atlantic Ocean). (*f*) Foraminiferan type: the pelagic *Globigerina bulloides* d'Orbigny (which is found in all seas). (*American Museum of Natural History*)

may modify direction of the thrust effecting locomotion. Although typically shorter than flagella, cilia are similar in structure.

Two major types of pseudopodia have been described, the contraction-hydraulic and the two-way flow types. The first are lobopodia with rounded tips and ectoplasm denser than endoplasm. The larger ones commonly contain granular endoplasm and clear ectoplasm. Two-way flow pseudopodia include reticulopodia of Foraminiferida and related types, filoreticulopodia of Radiolaria, and axopodia of certain Heliozoia.

In addition to nuclei, food vacuoles (gastrioles) in phagotrophs, chromatophores and stigma in many phytoflagellates, water-elimination vesicles in many Protozoa, and sometimes other organelles, the cytoplasm may contain mitochondria, Golgi material, pinocytotic vacuoles, stored food materials, endoplasmic reticulum, and sometimes pigments of various kinds.

Nutrition. In protozoan feeding, either phagotrophic (holozoic) or saprozoic (osmotrophic) methods predominate in particular species. In addition, chlorophyll-bearing flagellates profit from photosynthesis; in fact, certain species have not been grown in darkness and may be obligate phototrophs.

Phagotrophic ingestion of food, followed by digestion in vacuoles, is characteristic of Sarcodina, ciliates, and many flagellates. Digestion follows synthesis of appropriate enzymes and their transportation to the food vacuole. Details of ingestion vary. Formation of food cups, or gulletlike invaginations to enclose prey, is common in more or less ameboid organisms, such as various Sarcodina, many flagellates, and at least a few Sporozoa. Entrapment in a sticky reticulopodial net occurs in Foraminiferida and certain other Sarcodina. A persistent cytostome and gullet are involved in phagotrophic ciliates and a few flagellates. Many ciliates have buccal organelles (membranes, membranelies, and closely set rows of cilia) arranged to drive particles to the cytostome. Particles pass through the cytostome into the cytopharynx (gullet), at the base of which food vacuoles (gastrioles) are formed. Digestion occurs in such vacuoles.

By definition saprozoic feeding involves passage of dissolved foods through the cortex. It is uncertain to what extent diffusion is responsible, but enzymatic activities presumably are involved in uptake of various simple sugars, acetate and butyrate. In addition, external factors, for example, the pH of the medium, may strongly influence uptake of fatty acids and phosphates.

Reproduction. Reproduction occurs after a period of growth which ranges, in different species, from less than half a day to several months (certain Foraminiferida). General methods include binary fission, budding, plasmotomy, and schizogony. Fission, involving nuclear division and replication of organelles, yields two organisms similar in size. Budding produces two organisms, one smaller than the other. In plasmotomy, a multinucleate organism divides into several, each containing a number of nuclei. Schizogony, characteristic of Sporozoa, follows repeated nuclear division, yielding many uninucleate buds.

Simple life cycles include a cyst and an active (trophic) stage undergoing growth and reproduction. In certain free-living and parasitic species, no cyst is developed. Dimorphic cycles show two active stages; polymorphic show several. The former include adult and larva (Suctoria); flagellate and ameba (certain Mastigophora and Sarcodina); flagellate and palmella (nonflagellated; certain Phytomonadida); and ameba and plasmodium (Mycetozoia especially).

Parasitic protozoa. Parasites occur in all major groups. Sporozoa are exclusively parasitic, as are some flagellate orders (Trichomonadida, Hypermastigida, and Oxymonadida), the Opalinata, Piroplasmea, and several ciliate orders (Apostomatida, Astomatida, and Entodiniomorphida). Various other groups contain both parasitic and free-living types. Protozoa also serve as hosts of other protozoa, certain bacteria, fungi, and algae.

Relatively few parasites are distinctly pathogenic, causing amebiasis, visceral leishmaniasis (kala azar), sleeping sickness, Chagas' disease, malaria, tick fever of cattle, dourine of horses, and other diseases. *See* MALARIA; SPOROZOA. [R.P.H.]

Pseudomonas A genus of gram-negative, nonsporeforming, rod-shaped bacteria. Motile species possess polar flagella. They are strictly aerobic, but some members do respire anaerobically in the presence of nitrate. Some species produce acids oxidatively from carbohydrates; none is fermentative and none photosynthetic.

Members of the genus *Pseudomonas* cause a variety of infective diseases; some species cause disease of plants. One species, *P. mallei*, is a mammalian parasite, and is the causative agent of glanders, an infectious disease of horses that occasionally is transmitted to humans by direct contact. *Pseudomonas aeruginosa* is the most significant cause of hospital-acquired infections, particularly in predisposed patients with metabolic, hematologic, and malignant diseases. The spectrum of clinical disease ranges from urinary tract infections to septicemia, pneumonia, meningitis, and infections of postsurgical and posttraumatic wounds. *See* GLANDERS; HOSPITAL INFECTIONS; MENINGITIS; PNEUMONIA. [G.L.Gi.]

Public health An effort organized by society to protect, promote, and restore the people's health. It is the combination of sciences, skills, and beliefs that is directed to the maintenance and improvement of health through collective or social actions. The programs, services, and institutions of public health emphasize the prevention of disease and the health needs of the population as a whole. Additional goals include the reduction of the amount of disease, premature death, disability, and discomfort in the population.

The basic sciences of public health include epidemiology and vital statistics, which measure health status and assess health trends in the population. Epidemiology is also a powerful research method, used to identify causes and calculate risks of acquiring or dying of many conditions. Many sciences, including toxicology and microbiology, are applied to detect, monitor, and correct physical, chemical, and biological hazards in the environment. Such applications are being used to address concerns about a deteriorating global environment. The social and behavioral sciences have become more prominent in public health since the recognition that such factors as indolence, loneliness, personality type, and addiction to tobacco contribute to the risk of premature death and chronic disabling diseases. *See* EPIDEMIOLOGY.

In most industrial nations, public health services are organized nationally, regionally, and locally. National public health services are usually responsible for setting, monitoring, and maintaining health standards, for promoting good health, for collecting and compiling national health statistics, and for supporting and performing research on diseases important to public health. Regional (for example, state) public health services deal mainly with major health protection activities such as ensuring safe water and food supplies; they may also operate screening programs for early detection of disease and are responsible for health care of certain groups such as chronic mentally ill persons. Local public health services (in cities, large towns, and some rural communities) conduct a variety of personal public health services, such as immunization programs, health education, health surveillance and advice for mothers and newborn babies, and personal care of vulnerable groups such as the elderly and housebound long-term sick. Local health services also investigate and control epidemics and other communicable conditions such as sexually transmitted diseases.

National public health services communicate with each other in efforts to control diseases of international importance, and they collaborate worldwide under the auspices of the World Health Organization (WHO). While much of the work of WHO has been concentrated in the developing nations, it has also been involved in global efforts to control major epidemic diseases and to set standards for hazardous environmental and occupational exposures. [J.M.La.]

Puna An alpine biological community in the central portion of the Andes Mountains of South America. Sparsely vegetated, treeless stretches cover high plateau country (altiplano) and slopes of central and southern Peru, Bolivia, northern Chile, and northwestern Argentina. The poor vegetative cover and the puna animals are limited by short seasonal precipitation as well as by the low temperatures of high altitudes.

Like the paramos of the Northern Andes, punas occur above timberline, and extend upward, in modified form, to perpetual snow. Due to greater heights of the Central Andean peaks, aeolian regions, that is, regions supplied with airborne nutrients above the upper limit of vascular plants, generally the snowline, are more extensive here than above the paramos. *See* PARAMO. [H.G.B.]

Quaternary A period that encompasses at least the last 3,000,000 years of the Cenozoic Era, and is concerned with major worldwide glaciations and their effect on land and sea, on worldwide climate, and on the plants and animals that lived then. The Quaternary is divided into the Pleistocene Epoch and Holocene. The universal

Era	Period	Subperiod
CENOZOIC	QUATERNARY	
	TERTIARY	
MESOZOIC	CRETACEOUS	
	JURASSIC	
	TRIASSIC	
PALEOZOIC	PERMIAN	
	CARBONIFEROUS	Pennsylvanian
		Mississippian
	DEVONIAN	
	SILURIAN	
	ORDOVICIAN	
	CAMBRIAN	
PRECAMBRIAN		

term Pleistocene is gradually re-placing Quaternary; Holocene involves the last 7000 years since the Pleistocene. [S.E.Wh.]

Rabies An acute, encephalitic viral infection. Human beings are infected from the bite of a rabid animal, usually a dog. Canine rabies can infect all warm-blooded animals, and death usually results. *See* ANIMAL VIRUS.

The virus is believed to move from the saliva-infected wound through sensory nerves to the central nervous system, multiply there with destruction of brain cells, and thus produce encephalitis, with severe excitement, throat spasm upon swallowing (hence hydrophobia, or fear of water), convulsions, and death—with paralysis sometimes intervening before death.

All bites should immediately be cleaned thoroughly with soap and water, and a tetanus shot should be considered. The decision to administer rabies antibody, rabies vaccine, or both depends on four factors: the nature of the biting animal; the existence of rabies in the area; the manner of attack (provoked or unprovoked) and the severity of the bite and contamination by saliva of the animal; and recommendations by local public health officials. *See* PUBLIC HEALTH; TETANUS.

Diagnosis in the human is made by observation of Negri bodies (cytoplasmic inclusion bodies) in brains of animals inoculated with the person's saliva, or in the person's brain after death. A dog which has bitten a person is isolated and watched for 10 days for signs of rabies; if none occur, rabies was absent. If signs do appear, the animal is killed and the brain examined for Negri bodies, or for rabies antigen by testing with fluorescent antibodies.

Individuals at high risk, such as veterinarians, must receive preventive immunization. If exposure is believed to have been dangerous, postexposure prophylaxis should be undertaken. If antibody or immunogenic vaccine is administered promptly, the virus can be prevented from invading the central nervous system. An inactivated rabies virus vaccine is available in the United States. It is made from virus grown in human or monkey cell cultures and is free from brain proteins that were present in earlier Pasteur-type vaccines. This material is sufficiently antigenic that only four to six doses of virus need be given to obtain a substantial antibody response. [J.L.Me.]

Radar meteorology The application of radar to the study of the atmosphere and to the observation and forecasting of weather. Meteorological radars transmit electromagnetic waves at microwave and radio-wave frequencies. Water and ice particles, inhomogeneities in the radio refractive index associated with atmospheric turbulence and humidity variations, insects, and birds scatter radar waves. The backscattered energy received at the radar constitutes the returned signal. Meteorologists use the amplitude, phase, and polarization state of the backscattered energy to deduce the location and intensity of precipitation, the wind speed along the direction of the radar beam, and precipitation type (for example, rain or hail). *See* WEATHER FORECASTING AND PREDICTION.

Much of the understanding of the structure of storms derives from measurements made with networks of Doppler radars. They are used to investigate the complete three-dimensional wind fields associated with storms, fronts, and other meteorological phenomena. *See* PRECIPITATION (METEOROLOGY); THUNDERSTORM. [R.M.Ra.]

Radiocarbon dating A method of obtaining age estimates on organic materials which has been used to date samples as old as 75,000 years. The method has provided age determinations in archeology, geology, geophysics, and other branches of science.

Radiocarbon (^{14}C) determinations can be obtained on wood; charcoal; marine and fresh-water shell; bone and antler; peat and organic-bearing sediments; carbonate deposits such as tufa, caliche, and marl; and dissolved carbon dioxide (CO_2) and carbonates in ocean, lake, and ground-water sources. Each sample type has specific problems associated with its use for dating purposes, including contamination and special environmental effects. While the impact of ^{14}C dating has been most profound in archeological research and particularly in prehistoric studies, extremely significant contributions have also been made in hydrology and oceanography. In addition, beginning in the 1950s the testing of thermonuclear weapons injected large amounts of artificial ^{14}C ("bomb ^{14}C") into the atmosphere, permitting it to be used as a geochemical tracer.

Carbon (C) has three naturally occurring isotopes. Both ^{12}C and ^{13}C are stable, but ^{14}C decays by very weak beta decay (electron emission) to nitrogen-14 (^{14}N) with a half-life of approximately 5700 years. Naturally occurring ^{14}C is produced as a secondary effect of cosmic-ray bombardment of the upper atmosphere. As $^{14}CO_2$, it is distributed on a worldwide basis into various atmospheric, biospheric, and hydrospheric reservoirs on a time scale much shorter than its half-life. Metabolic processes in living organisms and relatively rapid turnover of carbonates in surface ocean waters maintain ^{14}C levels at approximately constant levels in most of the biosphere. The natural ^{14}C activity in the geologically recent contemporary "prebomb" biosphere was approximately 13.5 disintegrations per minute per gram of carbon.

To the degree that ^{14}C production has proceeded long enough without significant variation to produce an equilibrium or steady-state condition, ^{14}C levels observed in contemporary materials may be used to characterize the original ^{14}C activity in the corresponding carbon reservoirs. Once a sample has been removed from exchange with its reservoir, as at the death of an organism, the amount of ^{14}C begins to decrease as a function of its half-life. A ^{14}C age determination is based on a measurement of the residual ^{14}C activity in a sample compared to the activity of a sample of assumed zero age (a contemporary standard) from the same reservoir. The relationship between the ^{14}C age and the ^{14}C activity of a sample is given by the equation below, where t is

$$t = \frac{1}{\lambda} \ln \frac{A_o}{A_s}$$

radiocarbon years B.P. (before the present), λ is the decay constant of ^{14}C (related to the half-life $t_{1/2}$ by the expression $t_{1/2} = 0.693/\lambda$), A_o is the activity of the contemporary standards, and A_s is the activity of the unknown age samples. Conventional radiocarbon dates are calculated by using this formula, an internationally agreed half-life value of 5568 ± 30 years, and a specific contemporary standard.

The naturally occurring isotopes of carbon occur in the proportion of approximately 98.9% ^{12}C, 1.1% ^{13}C, and 10^{-10}% ^{14}C. The extremely small amount of radiocarbon in natural materials was one reason why ^{14}C was one of the isotopes which had been produced artificially in the laboratory before being detected in natural concentrations. A measurement of the ^{14}C content of an organic sample will provide an accurate determination of the sample's age if it is assumed that (1) the production of ^{14}C by cosmic rays has remained essentially constant long enough to establish a steady state in the $^{14}C/^{12}C$ ratio in the atmosphere, (2) there has been a complete and rapid mixing of ^{14}C throughout the various carbon reservoirs, (3) the carbon isotope ratio in the sample has not been altered except by ^{14}C decay, and (4) the total amount of carbon in any reservoir has not been altered. In addition, the half-life of ^{14}C must be known with sufficient accuracy, and it must be possible to measure natural levels of ^{14}C to appropriate levels of accuracy and precision. [R.E.T.]

Radioecology The study of the fate and effects of radioactive materials in the environment. It derives its principles from basic ecology and radiation biology.

Responses to radiation stress have consequences for both the individual organism and for the population, community, or ecosystem of which it is a part. When populations or individuals of different species differ in their sensitivities to radiation stress, for example, the species composition of the entire biotic community may be altered as the more radiation-sensitive species are removed or reduced in abundance and are replaced in turn by more resistant species. Such changes have been documented by studies in which natural ecological systems, including grasslands, deserts, and forests, were exposed to varying levels of controlled gamma radiation stress. *See* POPULATION ECOLOGY.

Techniques of laboratory toxicology are also available for assessing the responses of free-living animals to exposure to low levels of radioactive contamination in natural environments. This approach uses sentinel animals, which are either tamed, imprinted on the investigator, or equipped with miniature radio transmitters, to permit their periodic relocation and recapture as they forage freely in the food chains of contaminated habitats. When the animals are brought back to the laboratory, their level of radioisotope uptake can be determined and blood or tissue samples taken for analysis. In this way, even subtle changes in deoxyribonucleic acid (DNA) structure can be evaluated over time. These changes may be suggestive of genetic damage by radiation exposure. In some cases, damage caused by a radioactive contaminant may be worsened by the synergestic effects of other forms of environmental contaminants such as heavy metals.

Because of the ease with which they can be detected and quantified in living organisms and their tissues, radioactive materials are often used as tracers. Radioactive tracers can be used to trace food chain pathways or determine the rates at which various processes take place in natural ecological systems. Although most tracer experiments were performed in the past by deliberately introducing a small amount of radioactive tracer into the organism or ecological system to be studied, they now take advantage of naturally tagged environments where trace amounts of various radioactive contaminants were inadvertently released from operating nuclear facilities such as power or production reactors or waste burial grounds.

An important component of radioecology, and one that is closely related to the study of radioactive tracers, is concerned with the assessment and prediction of the movement and concentration of radioactive contaminants in the environment in general, and particularly in food chains that may lead to humans. *See* ECOLOGY; ENVIRONMENTAL TOXICOLOGY; FOOD WEB. [I.L.B.]

Radioisotope geochemistry A branch of environmental geochemistry and isotope geology concerned with the occurrence of radioactive nuclides in sediment, water, air, biological tissues, and rocks. The nuclides have relatively short half-lives ranging from a few days to about 10^6 years, and occur only because they are being produced by natural or anthropogenic nuclear reactions or because they are the intermediate unstable daughters of long-lived naturally occurring radioactive isotopes of uranium and thorium. The nuclear radiation, consisting of alpha particles, beta particles, and gamma rays, emitted by these nuclides constitutes a potential health hazard to humans. However, their presence also provides opportunities for measurements of the rates of natural processes in the atmosphere and on the surface of the Earth.

The unstable daughters of uranium and thorium consist of a group of 43 radioactive isotopes of 13 chemical elements, including all of the naturally occurring isotopes of the chemical elements radium, radon, polonium, and several others. A second group of radionuclides is produced by the interaction of cosmic rays with the chemical elements of the Earth's surface and atmosphere. This group includes hydrogen-3 (tritium), beryllium-10, carbon-14, aluminum-26, silicon-32, chlorine-36, iron-55, and others. A third group of radionuclides is produced artificially by the explosion of nuclear devices, by the operation of nuclear reactors, and by various particle accelerators used for research in nuclear physics. Some of the radionuclides produced in nuclear reactors decay sufficiently slowly to be useful for geochemical research, including strontium-90, cesium-137, iodine-129, and isotopes of plutonium. The explosion of nuclear devices in the atmosphere has also contributed to the abundances of certain radionuclides that are produced by cosmic rays such as tritium and carbon-14. *See* DATING METHODS. [G.Fau.]

Rain shadow An area of diminished precipitation on the lee side of mountains. There are marked rain shadows, for example, east of the coastal ranges of Washington, Oregon, and California, and over a larger region, much of it arid, east of the Cascade Range and Sierra Nevadas. All mountains decrease precipitation on their lee; but rain shadows are sometimes not marked if moist air often comes from different directions, as in the Appalachian region.

The causes of rain shadow are (1) precipitation of much of the moisture when air is forced upward on the windward side of the mountains, (2) deflection or damming of moist air flow, and (3) downward flow on the lee slopes, which warms the air and lowers its relative humidity. [J.R.F.]

Rainforest Forests that occur in continually wet climates with no dry season. There are relatively small areas of temperate rainforests in the Americas and Australasia, but most occur in the tropics and subtropics.

The most extensive tropical rainforests are in the Americas. These were originally 1.54×10^6 mi^2 (4×10^6 km^2) in extent, about half the global total, and mainly in the Amazon basin. A narrow belt also occurs along the Atlantic coast of Brazil, and a third block lies on the Pacific coast of South America, extending from northern Peru to southern Mexico.

Tropical rainforests have a continuous canopy (commonly 100–120 ft or 30–36 m tall) above which stand huge emergent trees, reaching 200 ft (60 m) or taller. Within the rainforest canopy are trees of many different sizes, including pygmies, that reach only a few feet. Trees are the main life form and are often, for purposes of description and analysis, divided into strata or layers. Trees form the framework of the forest and support an abundance of climbers, orchids, and other epiphytes, adapted to the microclimatic conditions of the different zones of the canopy, from shade lovers in the gloomy, humid lower levels, to sun lovers in the brightly lit, hotter, and drier upper levels. Most trees have evergreen leaves, many of which are pinnate or palmate. These features of forest structure and appearance are found throughout the world's lowland tropical rainforests. There are other equally distinctive kinds of rainforest in the lower and upper parts of perhumid tropical mountains, and additional types on wetlands.

Rainforests occur where the monthly rainfall exceeds 4 in. (100 mm) for 9–12 months. They merge into other seasonal or monsoon forests where there is a stronger dry season (3 months or more with 2.5 in. or 60 mm of rainfall). The annual mean temperature in the lowlands is approximately 64°F (18°C). There is no season unfavorable for growth.

Primary rainforests are exceedingly rich in species of both plants and animals. There are usually over 100 species of trees 2.5 in. (10 cm) in diameter or bigger per 2.4 acres (1 ha). There are also numerous species of climbers and epiphytes. Flowering and fruiting occur year-round, but commonly there is a peak season; animal breeding may be linked to this. Secondary rainforests are much simpler. There are fewer tree species, less variety from location to location, and fewer epiphytes and climbers; the animals are also somewhat different. *See* ECOLOGICAL SUCCESSION.

Tropical rainforests are a source of resins, dyes, drugs, latex, wild meat, honey, rattan canes, and innumerable other products essential to rural life and trade. Modern technology for extraction and for processing has given timber of numerous species monetary value, and timber has come to eclipse other forest products in importance. The industrial nations use much tropical hardwood for furniture, construction, and plywood. Rainforest timbers, however, represent only 11% of world annual industrial wood usage, a proportion that has doubled since 1950. West Africa was the first main modern source, but by the 1960s was eclipsed by Asia, where Indonesia and Malaysia are the main producers of internationally traded tropical hardwoods. Substantial logging has also developed in the neotropics. *See* FOREST ECOSYSTEM. [T.C.Wh.]

Red Sea A body of water that separates northeastern Africa from the Arabian Peninsula. The Red Sea forms part of the African Rift System, which also includes the Gulf of Aden and a complex series of continental rifts in East Africa extending as far south as Malawi. The Red Sea extends for 1920 km (1190 mi) from Ras (Cape) Muhammed at the southern tip of the Sinai Peninsula to the Straits of Bab el Mandab at the entrance to the Gulf of Aden. At Sinai the Red Sea splits into the Gulf of Suez, which extends for an additional 300 km (180 mi) along the northwest trend of the Red Sea and the nearly northward-trending Gulf of Aqaba. The 175-km-long (109-mi) Gulf of Aqaba forms the southern end of the Levant transform, a primarily strike-slip fault system extending north into southern Turkey. The Levant transform also includes the Dead Sea and Sea of Galilee and forms the northwestern boundary of the Arabian plate. *See* ESCARPMENT; FAULT AND FAULT STRUCTURES.

The Red Sea consists of narrow marginal shelves and coastal plains and a broad main trough with depths ranging from about 400 to 1200 m (1300 to 3900 ft). The main trough is bisected by a narrow (<60 km or 37 mi wide) axial trough with a very rough bottom morphology and depths of greater than 2000 m (6600 ft). The maximum recorded depth is 2920 m (9580 ft). *See* REEF.

Water circulation in the Red Sea is driven by monsoonal wind patterns and changes in water density due to evaporation. Evaporation in the Red Sea is sufficient to lower the sea level by over 2 m (6.6 ft) per year. No permanent rivers flow into the sea, and there is very little rainfall. As a result, there must be a net inflow of water from the Gulf of Aden to compensate for evaporative losses. During the winter monsoon, prevailing winds in the Red Sea are from the south, and there is a surface current from the Gulf of Aden into the Red Sea. During the summer monsoon, the wind in the Red Sea blows strongly from the north, causing a surface current out of the Red Sea. *See* MONSOON METEOROLOGY; SEAWATER. [J.R.Co.]

Redwood A member of the pine family, *Sequoia semper-virens*, is the tallest tree in the Americas, attaining a height of 350 ft (107 m) and a diameter of 27 ft (8.2 m). Its present range is limited to a strip along the Pacific Coast, extending from southwest Oregon to south of San Francisco. The leaves are evergreen, sharply pointed, small, disposed in two vertical rows on short branches, and scalelike on the main stem. The cones are egg-shaped. The bark is a dull red-brown, on old trees sometimes 1 ft (0.3 m) thick, densely fibrous, and highly resistant to fire. The tree gets its common name from the color of the bark as well as that of the heartwood.

The wood holds paint well and is used for bridge timbers, tanks, flumes, silos, posts, shingles, paneling, doors, caskets, furniture, siding, and many other building purposes. *See* PINE. [A.H.G./K.P.D.]

Reef A mass or ridge of rock or rock-forming organisms in a water body, a rock trend on land or in a mine, or a rocky trend in soil. Usually the term reef means a rocky menace to navigation, within 6 fathoms (11 m) of the water surface. Various kinds of calcium carbonate–secreting animals and plants create biogenic, or organic, reefs throughout the warmer seas. Most biogenic reefs are made of corals and associated organisms, but some entire reefs and important parts of others consist mainly of lime-secreting algae, hydrozoans, annelids, oysters, or sponges. *See* ALGAE; SCLERACTINIA.

The term fringing reef refers to a coral or other biogenic reef that fringes the edge of the land. A barrier reef ordinarily made of corals or other organisms parallels the shore at the seaward side of a natural lagoon. An atoll is an annular coral reef that surrounds a lagoon. *See* ATOLL. [P.Cl.]

Reforestation The reestablishment of forest cover either naturally or artificially. Given enough time, natural regeneration will usually occur in areas where temperatures and rainfall are adequate and when grazing and wildfires are not too frequent.

Reforestation occurs on land where trees have been recently removed due to harvesting or to natural disasters such as a fire, landslide, flooding, or volcanic eruption. When abandoned cropland, pastureland, or grasslands are converted to tree cover, the practice is termed afforestation (where no forest has existed in recent memory). Afforestation is common in countries such as Australia, South Africa, Brazil, India, and New Zealand. Although natural regeneration can occur on abandoned cropland, planting trees will decrease the length of time required until the first harvest of wood. Planting also has an advantage in that both tree spacing and tree species can be prescribed. The selection of tree species can be very important since it affects both wood quality and growth rates. Direct seeding is also used for both afforestation and reforestation, although it often is less successful and requires more seed than tree planting. Unprotected seed are often eaten by birds and rodents, and weeds can suppress growth of newly germinated seed. For these reasons, direct seeding accounts for only about 5% and 1% of artificial reforestation in Canada and the United States, respectively. [D.B.So.]

Relapsing fever An acute infectious disease characterized by recurring fever. It is caused by spirochetes of the genus *Borrelia* and transmitted by the body louse (*Pediculus humanus humanus*) and by ticks of the genus *Ornithodoros*.

Louse-borne relapsing fever, caused by *Borrelia recurrentis*, is typically epidemic. Epidemics, once widespread on all continents, are rare but still occur in certain parts of South America, Africa, and Asia. Tick-borne relapsing fevers are endemic. They are more widely distributed throughout the Eastern and Western hemispheres. At least 15 species of *Borrelia* have been recognized as causative agents.

After incubation of 2–10 days, the initial attack begins abruptly with chills, high fever, headache, and pains in muscles and joints, and lasts 2–8 days, ending by crisis. A remission period of 3– 10 days is followed by a relapse similar to the initial attack but milder. There may be 4–5 relapses, although occasionally 10 or more have been recorded. Mortality varies from 2 to 5% but may be considerably higher during epidemics.

Chlortetracycline is the most effective antibiotic drug, but penicillin, oxytetracycline, and streptomycin also have therapeutic value.

The best way to prevent relapsing fever is to control louse and tick populations with effective insecticides and acaricides. *See* MEDICAL BACTERIOLOGY; PEDICULOSIS. [W.Bu.]

Renewable resources Agricultural materials used as feedstocks for industrial processes. For many centuries agricultural products were the main sources of raw material for the manufacturing of soap, paint, ink, lubricants, grease, paper, cloth, drugs, and a host of other nonfood products. During the early 1900s, the advances in organic synthesis in western Europe and the United States led to the use of coal as an alternative resource; in the 1940s, oil and natural gas were added as starting materials as a result of great advances in catalysis and polymer sciences. Since then the petrochemical industry has grown rapidly as the result of the abundance and low price of the starting materials as well as the development of new products. However, with the rapidly increasing economies of the nations of the world, these developments did not ever result in reduction in the utilization of agricultural products as industrial materials.

Animal fats, marine and vegetable oils, and their fatty acid derivatives have always played a major role in the manufacturing of many industrial products. Some of these commodities are produced solely for industrial end uses; examples are linseed, tung, castor (not counting minor amounts used for medicinal purposes), and sperm whale oils. Others, such as tallow and soybean oil, are used for both edible and industrial products.

Starch, cellulosics, and gums also have been used for many centuries as industrial materials, whereas sugar crops, such as sugarcane and sugarbeet, have mainly satisfied world food requirements.

Natural rubber and turpentine are excellent examples of plant-derived hydrocarbons. The development of synthetic rubbers during and after World War II has never threatened the demand for natural rubber; there is generally a world shortage. Turpentine is a product of the wood and paper pulp industry and is used as a solvent and thinner in paints and varnishes.

The threat that industrial nations might be separated from part or all of their traditional sources of raw materials through political and economic upheavals or natural calamities has resulted in a renewed effort to develop additional crops for local agriculture. In the United States, research has provided a number of candidate species that either are now in commercial development or are ready for the time when circumstances warrant such development. Examples are jojoba (liquid wax ester to replace sperm whale oil), guayule (alternate source of natural rubber), kenaf (paper fiber with annual yields much higher than available from trees), and crambe and meadow-foam (long-chain fatty acids, since erucic acid is no longer available from rapeseed oil). There is also active research involving *Cuphea* species (alternate source of lauric and other medium-chain fatty acids, to augment coconut oil), *Vernonia* (source of epoxy oil), and several other promising plants. For example, the Chinese tallow tree has the potential of producing 2.2 tons per acre (5 metric tons per hectare) of seed oil that could be used for manufacturing fuel and other chemicals. [L.H.P.]

Reproduction (plant) The formation by a plant of offspring that are either exact copies or reasonable likenesses. When the process is accomplished by a single individual without fusion of cells, it is referred to as asexual; when fusion of cells is involved, whether from an individual or from different donors, the process is sexual.

Asexual reproduction. Using the technique of tissue culture, higher green plants can be regenerated from a single cell and can usually flower and set seed normally when removed and placed in soil. This experiment shows that each cell of the plant body carries all the information required for formation of the entire organism. The culture of isolated cells or bits of tissue thus constitutes a means of vegetative propagation of the plant and can provide unlimited copies identical to the organism from which the cells were derived.

All other vegetative reproductive devices of higher plants are elaborations of this basic ability and tendency of plant cells to produce tissue masses that can organize into growing points (meristems) to yield the typical patterns of differentiated plant organs. For example, a stem severed at ground level may produce adventitious roots. Similarly, the lateral buds formed along stems can, if excised, give rise to entire plants. The "eyes" of the potato tuber, a specialized fleshy stem, are simply buds used in vegetative propagation of the crop. In many plants, cuttings made from fleshy roots can similarly form organized buds and reconstitute the plant by vegetative propagation. Thus, each of the vegetative organs of the plant (leaf, stem, and root) can give rise to new plants by asexual reproduction. *See* PLANT PROPAGATION.

Sexual reproduction. While in asexual reproduction, the genetic makeup of the progeny rarely differs greatly from that of the parent, the fusion of cells in sexual reproduction can give rise to new genetic combinations, resulting in new types of plants. The life cycle of higher green plants consists of two distinct generations, based on the chromosomal complement of their cells. The sporophyte generation is independent and dominant in the flowering plants and ferns, but small, nongreen, and dependent in the mosses, and contains the $2n$ number of chromosomes. The diploidy results in each case from the fusion of sperm and egg to form the zygote, which then develops into an embryo and finally into the mature sporophyte. The sporophyte generation ends with the formation of $1n$ spores by reduction division, or meiosis, in a spore mother cell. The spore then develops into the gametophyte generation, which in turn produces the sex cells, or gametes. The gametophyte generation ends when gametes fuse to form the zygote, restoring the $2n$ situation typical of sporophytes.

In flowering plants, the gametophyte or $1n$ generation is reduced to just a few cells (generally three for the male and eight for the female). The male gametophyte is formed after meiosis occurs in the microspore mother cells of the anther, yielding a tetrad of $1n$ microspores. Each of these microspores then divides mitotically at least twice. The first division produces the tube nucleus and the generative nucleus. The generative nucleus then divides again to produce two sperms. These nuclei are generally not separated by cell walls, but at this stage the outer wall of the spore becomes thickened and distinctively patterned—a stage typical of the mature male gametophyte, the pollen grain. *See* FLOWER; POLLEN; POLLINATION.

Each pollen grain has a weak pore in its wall, through which the pollen tube emerges at the time of germination. Pollen germinates preferentially in the viscous secretion on the surface of the stigma, and its progress down the style to the ovary is guided through specific cell-to-cell recognition processes. Throughout its growth, which occurs through the deposition of new cell wall material at the advancing tip, the pollen tube is controlled by the tube nucleus, usually found at or near the tip. When the pollen tube, responding to chemical signals, enters the micropyle of the ovule, its growth ceases and the tip bursts, discharging the two sperms into the embryo sac, the female gametophyte of the ovary.

The female gametophyte generation, like the male, arises through meiotic division of a $2n$ megaspore mother cell. This division forms four $1n$ megaspores, of which three usually disintegrate, the fourth developing into an eight-nucleate embryo sac by means of three successive mitotic divisions. The eight nuclei arrange themselves into two groups of four, one at each pole of the embryo sac. Then one nucleus from each pole moves to the center of the embryo sac. One of the three nuclei at the micropylar end of the embryo sac is the female gamete, the egg, which fuses with one of the sperm nuclei to form the zygote, the first cell of the sporophyte generation, which produces the embryo. The second sperm fuses with the two polar nuclei at the center of the embryo sac to form a $3n$ cell that gives rise to the endosperm of the seed, the tissue in which food is stored. The entire ovule ripens into the seed, with the integuments forming the protective seed coat. The entire ovary ripens into a fruit, whose color, odor, and taste are attractive to animals, leading to dispersal of the seeds. The life cycle is completed when the seed germinates and grows into a mature sporophyte with flowers, in which meiotic divisions will once again produce $1n$ microspores and megaspores.

Nonflowering higher plants such as the ferns and mosses also show a distinct alternation of generations. The familiar fern plant of the field is the sporophyte generation. Meiosis occurs in sporangia located in special places on the leaves, generally the undersides or margins. A spore mother cell produces a tetrad of $1n$ spores, each of which can germinate to produce a free-living, green gametophyte called a prothallus. On the prothallus are produced male and female sex organs called antheridia and archegonia, which give rise to sperms and eggs, respectively. Sperms, motile because of their whiplike flagella, swim to the archegonium, where they fertilize the egg to produce the zygote that gives rise to the sporophyte generation again.

In mosses, by contrast, the dominant green generation is the gametophyte. Antheridia or archegonia are borne at the tips of these gametophytes, where they produce sperms and eggs, respectively. When suitably wetted, sperms leave the antheridium, swim to a nearby archegonium, and fertilize the egg to produce a $2n$ zygote that gives rise to a nongreen, simple, dependent sporophyte. The moss sporophyte consists mainly of a sporangium at the end of a long stalk, at the base of which is a mass of tissue called the foot, which absorbs nutrients from the green, photosynthetic gametophyte. Meiosis occurs in the sporangium when a spore mother cell gives rise to four reduced spores. Each spore can germinate, giving rise to a filamentous structure from which leafy gametophytic branches arise, completing the life cycle.

Various members of the algae that reproduce sexually also display alternation of generations, producing sperms and eggs in antheridia and oogonia. Sporophyte and gametophyte generations may each be free-living and independent, or one may be partially or totally dependent on the other. *See* FRUIT; PLANT PHYSIOLOGY; POPULATION DISPERSAL; SEED. [A.W.G.]

Reservoir A place or containment area where water is stored. Where large volumes of water are to be stored, reservoirs usually are created by the construction of a dam across a flowing stream. When water occurs naturally in streams, it is sometimes not available when needed. Reservoirs solve this problem by capturing water and making it available at later times.

In addition to large reservoirs, many small reservoirs are in service. These include varieties of farm ponds, regulating lakes, and small industrial or recreational facilities. In some regions, small ponds are called tanks. Small reservoirs can have important cumulative effects in rural regions

Reservoirs can be developed for single or multiple purposes, such as to supply water for people and cities, to provide irrigation water, to lift water levels to make navigation possible on streams, and to generate electricity.

Another purpose of reservoirs is to control floods by providing empty spaces for flood waters to fill, thereby diminishing the rate of flow and water depth downstream of the reservoir.

Reservoirs also provide for environmental uses of water by providing water to sustain fisheries and meet other fish and wildlife needs, or to improve water quality by providing dilution water when it is needed in downstream sections of rivers. Reservoirs may also have esthetic and recreational value, providing boating, swimming, fishing, rafting, hiking, viewing, photography, and general enjoyment of nature. *See* RIVER ENGINEERING. [N.S.Gr.]

Respirator A device designed to protect the wearer from noxious gases, vapors, and aerosols or to supply oxygen or doses of medication to the wearer. Respirators are used widely in industry to protect workers against harmful atmospheres, and in the military to protect personnel against chemical, biological, or radioactive warfare agents. Respirators are classified according to whether they are atmosphere-supplying or air-purifying.

Atmosphere-supplying respirators are used in atmospheres deficient in oxygen or extremely hazardous to the health of the wearer. Such atmospheres can occur in unventilated cellars, wells, mines, burning buildings, and enclosures containing inert gas. The self-contained breathing apparatus (SCBA) is a completely self-contained unit with the air supply or the oxygen-generating material being carried by the wearer. Air-supplied respirators are equipped with the same variety of facepieces as the SCBA, however these respirators can have the air supplied to the facepiece by means of a hose and a blower—the hose mask—or from a compressed-air source equipped with proper airflow and pressure-regulating equipment—the air-line mask.

In an air-purifying respirator, ambient air is passed through a purifying medium to remove the contaminants. However, these devices do not provide oxygen or protect against oxygen-deficient atmospheres. A widely used air-purifying respirator is the nonpowered, or negative-pressure, respirator (see illustration). Ambient air is inhaled

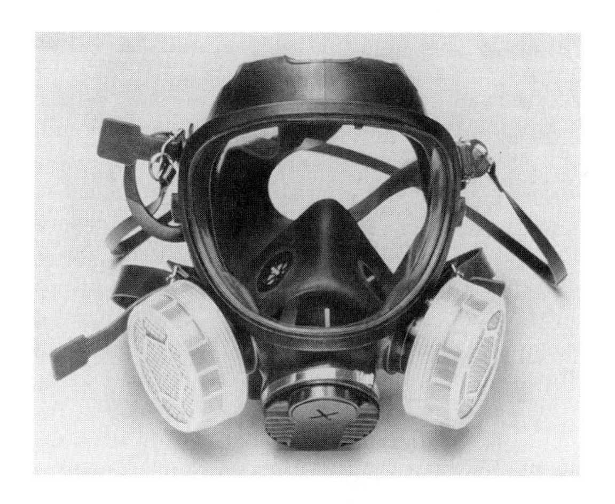

Negative-pressure air-purifying respirator.

through the purifying medium in the replaceable cartridges and exhaled through an exhaust valve. In the case of the powered air-purifying respirator, an external blower, usually powered by a belt or helmet-mounted battery pack, forces air through the purifying medium and supplies it to the wearer under positive pressure, thus minimizing the problem of face-seal leakage. [B.Y.H.L.; D.A.Ja.]

Respiratory syncytial virus A virus belonging to the Paramyxoviridae, genus *Pneumovirus*. This virus, although unrelated to any other known respiratory disease agent and differing from the parainfluenza viruses in a number of important characteristics, has been associated with, a large proportion of respiratory illnesses in very young children, particularly bron-chiolitis and pneumonia. It appears to be one of the major causes of these serious illnesses of infants. It is the only respiratory virus that occurs with its greatest frequency in infants in their first 6 months of life. In older infants and children, a milder illness is produced.

The clinical disease in young infants may be the result of an antigen-antibody reaction that occurs when the infecting virus meets antibody transmitted from the mother. For this reason respiratory syncytial vaccines that stimulate production of antibodies in the serum, but not in the nasal secretions, may do more harm than good. *See* ANIMAL VIRUS. [J.I.L.; M.E.Re.]

Restoration ecology A field in the science of conservation that is concerned with the application of ecological principles to restoring degraded, derelict, or fragmented ecosystems. The primary goal of restoration ecology (also known as ecological restoration) is to return a community or ecosystem to a condition similar in ecological structure, function, or both, to that existing prior to site disturbance or degradation.

A reference framework is needed to guide any restoration attempt—that is, to form the basis of the design (for example, desired species composition and density) and monitoring plan (for example, setting milestones and success criteria for restoration projects). Such a reference system is derived from ecological data collected from a suite of similar ecosystems in similar geomorphic settings within an appropriate biogeographic region. Typically, many sites representing a range of conditions (for example, pristine to highly degraded) are sampled, and statistical analyses of these data reveal what is possible given the initial conditions at the restoration site. *See* ECOLOGY, APPLIED; ECOSYSTEM. [P.L.Fi.]

Retrovirus A family of viruses distinguished by three characteristics: (1) genetic information in ribonucleic acid (RNA); (2) virions possess the enzyme reverse transcriptase; and (3) virion morphology consists of two proteinaceous structures, a dense core and an envelope that surrounds the core. Some viruses outside the retrovirus family have some of these characteristics, but none has all three. Numerous retroviruses have been described; they are found in all families of vertebrates. *See* ANIMAL VIRUS .

The genome is composed of two identical molecules of single-stranded RNA, which are similar in structure and function to cellular messenger RNA. Deoxyribonucleic acid (DNA) is not present in the virions of retroviruses. The reverse transcriptase in each virus makes a DNA copy of the RNA genome shortly after entry of the virus into the host cell. The discovery of this enzyme changed thinking in biology. Previously, the only known direction for the flow of genetic information was from DNA to RNA, yet retroviruses make DNA copies of their genome by using an RNA template. This reversal of genetic information was considered backward and hence the family name retrovirus, meaning backward virus.

Once the DNA copy of the RNA genome is made, it is inserted directly into one of the chromosomes of the host cell. This results in new genetic information being acquired by the host species. The study of reverse transcriptase has led to other discoveries of how retroviruses add a variety of new genetic information into the host. One such class of genes carried by retroviruses is oncogenes, meaning tumor genes. Retroviral oncogenes appear to be responsible for tumors in animals.

Two distinct retroviruses have been discovered in humans. One is human T-cell lymphotropic virus type 1 (HTLV-1), a type C-like virus associated with adult T-cell leukemia. The other is the human acquired immune deficiency syndrome (AIDS) virus, a type E lentivirus. *See* ACQUIRED IMMUNE DEFICIENCY SYNDROME (AIDS). [P.A.Ma.]

Rhabditida An order of nematodes in which the number of labia varies from a full complement of six to three or two or none. The tubular stoma may be composed of five or more sections called rhabdions. The three-part esophagus always ends in a muscular bulb that is invariably valved. The excretory tube is cuticularly lined, and paired lateral collecting tubes generally run posteriorly from the excretory cell; some taxa have anterior tubules also. Females have one or two ovaries; when only one is present, the vulva shifts posteriorly. The cells of the intestine may be uninucleate, binucleate, or tetranucleate, and the hypodermal cells may also be multinucleate.

There are eight superfamilies in the order: Rhabditoidea, Allionematoidea; Bunonematoidea, Cephaloboidea, Panagrolaimoidea, Robertioidea, Chambersiellolidea, and Elaphonematoidea. The Rhabditoidea are one of the largest nematode superfamilies and contain many important parasites of humans and domestic animals. This superfamily is distinguished by the well-developed cylindrical stoma and three-part esophagus that ends in a valved terminal bulb. In parasitic species, adult stages and some larval stages lack the valved terminal bulb. Though most species of Rhabditoidea are free-living feeders on terrestrial bacteria, others are important in the biological control of insects or as parasites of mammals. *See* MEDICAL PARASITOLOGY. [A.R.M.]

Rheumatic fever An illness that follows an upper respiratory infection with the group A streptococcus (*Streptococcus pyogenes*) and is characterized by inflammation of the joints (arthritis) and the heart (carditis). Arthritis typically involves multiple joints and may migrate from one joint to another. The carditis may involve the outer lining of the heart, the heart muscle itself, or the inner lining of the heart. A minority of affected individuals also develop a rash (erythema marginatum), nodules under the skin, or Sydenham'schorea (a neurologic disorder characterized by involuntary, uncoordinated movements of the legs, arms, and face). Damage to heart valves may be permanent and progressive, leading to severe disability or death from rheumatic heart disease years after the initial attack. The disease occurs an average of 19 days after the infection and is thought to be the result of an abnormal immunologic reaction to the group A streptococcus. Initial attacks of rheumatic fever generally occur among individuals aged 5 to 15. Those who have had one attack are highly susceptible to recurrences after future streptococcal infections.

Initial attacks of rheumatic fever can be prevented by treatment of strep throat with penicillin for at least 10 days. Patients who have had an episode of rheumatic fever should continue taking antibiotics for many years to prevent group A streptococcal infections that may trigger a recurrence of rheumatic fever. *See* STREPTOCOCCUS.

[A.L.Bi.; J.B.Da.]

Rhinovirus A genus of the family Picornaviridae. Members of the human rhinovirus group include at least 113 antigenically distinct types. Like the enteroviruses, the rhinoviruses are small (17–30 nanometers), contain ribonucleic acid (RNA), and are not inactivated by ether. Unlike the enteroviruses, they are isolated from the nose and throat rather than from the enteric tract, and are unstable if kept under acid conditions (pH 3–5) for 1–3 h. Rhinoviruses have been recovered chiefly from adults with colds and only rarely from patients with more severe respiratory diseases. *See* COMMON COLD; ENTEROVIRUS.

In a single community, different rhinovirus types predominate during different seasons and during different outbreaks in a single season, but more than one type may be present at the same time.

Although efforts have been made to develop vaccines, none is available. Problems that hinder development of a useful rhinovirus vaccine include the short duration of natural immunity even to the specific infecting type, the large number of different antigenic types of rhinovirus, and the variation of types present in a community from one year to the next. *See* ANIMAL VIRUS; PICORNAVIRIDAE. [J.L.Me.; M.E.Re.]

Rhizosphere The soil region subject to the influence of plant roots. It is characterized by a zone of increased microbiological activity and is an example of the relationship of soil microbes to higher plants.

A sharp boundary cannot be drawn between the rhizosphere and the soil unaffected by the plant (edaphosphere). At the root surface the rhizosphere effect is most intense, falling off sharply with increasing distance.

Growth of a plant markedly changes the microbial population of soil within its influence. In the rhizosphere there are more microorganisms than in soil distant from the plant. This increase is most pronounced with bacteria but is evident with other groups. The rhizosphere effect is seen in seedling plants; it increases with the age of the plant and usually reaches a maximum at the stage of greatest vegetative growth. Upon death of the plant the microbial population reverts to the level of the surrounding soil. Leguminous plants support higher rhizosphere populations than nonlegumes. The stimulation of microorganism growth in the rhizosphere results chiefly from the liberation of readily available organic substances by the growing plant. *See* SOIL MICROBIOLOGY. [A.G.L.]

Rhynchobdellae An order of the class Hirudinea. These leeches possess an eversible proboscis and lack hemoglobin in the blood. They may be divided into two families, the Glossi-phoniidae and the Ichthyobdellidae. Glossiphoniidae are flattened, mostly small leeches occurring chiefly in fresh water. Ichthyobdellidae typically have cylindrical bodies with conspicuous, powerful suckers used to attach themselves to passing fish. They frequently have lateral appendages which aid in respiration. *See* HIRUDINEA. [K.H.M.]

Rice The plant *Oryza sativa* is the major source of food for nearly one-half of the world's population. The most important rice-producing countries are mainland China, India, and Indonesia, but in many smaller countries rice is the leading food crop. In the United States, rice production is largely concentrated in selected areas of Arkansas, California, Louisiana, and Texas. *See* WHEAT.

Over 95% of the world rice crop is used for human food. Although most rice is boiled, a considerable amount is consumed as breakfast cereals. Rice starch also has many uses. Broken rice is used as a livestock feed and for the production of alcoholic beverages. The bran from polished rice is used for livestock feed; the hulls are used for fuel and cellulose. The straw is used for thatching roofs in the Orient and for making paper, mats, hats, and baskets. Rice straw is also woven into rope and used as cordage for bags. This crop serves a multitude of purposes in countries where agriculture is dependent largely upon rice.

Rice is unlike many other cereal grains in that all cultivated varieties belong to the same species and have 12 pairs of chromosomes, as do most wild types. The extent of variation in morphological and physiological characteristics within this single species is greater than for any other cereal crop. *See* GENETICS.

Rice is an annual grass plant varying in height from 2 to 6 ft (0.6 to 1.8 m). Plants tiller, that is, develop new shoots freely, the number depending upon spacing and soil fertility. The inflorescence is an open panicle. Flowers are perfect and normally self-pollinated, with natural crossing seldom exceeding 3–4%. A distinct characteristic of the flower is the six anthers rather than the customary three of other grasses. Spikelets have a single floret, lemma and palea completely enclosing the caryopsis or fruit, which may be yellow, red, brown, or black. Lemmas may be awnless, partly awned, or fully awned. Threshed rice, which retains its lemma and palea, is called rough rice or paddy. *See* FLOWER; FRUIT; GRASS CROPS; INFLORESCENCE; REPRODUCTION (PLANT).

In the United States, only about 25 varieties are in commercial production. Cultivated rices are classified as upland and lowland. Upland types, which can be grown in high-rainfall areas without irrigation, produce relatively low yields. The lowland types, which are grown submerged in water for the greater part of the season, produce higher yields. In contrast to most plants, rice can thrive when submerged because oxygen is transported from the leaves to the roots. All rice in the United States is produced under lowland or flooded conditions. Rice varieties are also classified as long- or short-grain. Most long-grain rices have high amylose content and are dry or fluffy when cooked, while most short-grain rices have lower amylose content and are sticky when cooked. In the United States a third grain length is recognized: medium-grain. The medium-grain rices have cooking qualities similar to those of short-grain varieties. *See* AGRICULTURAL SCIENCE (PLANT); GRAIN CROPS. [J.N.R.]

The rice kernel has four primary components: the hull or husk, the seedcoat or bran, the embryo or germ, and the endosperm. The main objective of milling rice is to remove the indigestible hull and additional portions of bran to yield whole unbroken endosperm. Rice milling involves relatively uncomplicated abrasive and separatory procedures which provide a variety of products dependent on the degree of bran removal or the extent of endosperm breakage.

Instant rice is made from whole grain rice by pretreating under controlled cooking, cooling, and drying conditions to impart the quick-cooking characteristic. Ready-to-eat breakfast rice cereals are prepared from milled rice as flakes or puffs. Rice bran oil was developed as a result of increased extraction of lipids from rice bran. It is utilized as an edible-grade oil in a variety of applications as well as an industrial feedstock for soap and resin manufacture. *See* CEREAL. [M.A.U.]

Rickettsioses Often severe infectious diseases caused by several diverse and specialized bacteria, the rickettsiae and rickettsia-like organisms. The best-known rickettsial diseases infect humans and are usually transmitted by parasitic arthropod vectors.

Rickettsiae and rickettsia-like organisms are some of the smallest microorganisms visible under a light microscope. Although originally confused with viruses, in part because of their small size and requirements for intracellular replication, rickettsiae and rickettsia-like organisms are characterized by basic bacterial (gram-negative) morphologic features. Their key metabolic enzymes are variations of typical bacterial enzymes. The genetic material of rickettsiae and rickettsia-like organisms likewise seems to conform to basic bacterial patterns. The genome of all rickettsia-like organisms consists of double-stranded deoxyribonucleic acid (DNA).

Rickettsiae enter host cells by phagocytosis and reproduce by simple binary fission. The site of growth and reproduction varies among the various genera.

Clinically, the rickettsial diseases of humans are most commonly characterized by fever, headache, and some form of cutaneous eruption, often including diffuse rash, as in epidemic and murine typhus and Rocky Mountain spotted fever, or a primary ulcer or eschar at the site of vector attachment, as in Mediterranean spotted fever and scrub typhus. Signs of disease may vary significantly between individual cases of rickettsial disease. Q fever is clinically exceptional in several respects, including the frequent absence of skin lesions.

All of the human rickettsial diseases, if diagnosed early enough in the infection, can usually be effectively treated with the appropriate antibiotics. Tetracycline and chloramphenicol are among the most effective antibiotics used; they halt the progression of the disease activity, but do so without actually killing the rickettsial organisms. Presumably, the immune system is ultimately responsible for ridding the body of infectious organisms. Penicillin and related compounds are not considered effective.

Most rickettsial diseases are maintained in nature as diseases of nonhuman vertebrate animals and their parasites. Human infection may usually be regarded as peripheral to the normal natural infection cycles, and human-to-human transmission is not the rule. However, the organism responsible for epidemic typhus (*Rickettsia prowazekii*) and the agent responsible for trench fever (*Rochalimaea quintana*) have the potential to spread rapidly within louse-ridden human populations. *See* ZOONOSES.

All known spotted fever group organisms are transmitted by ticks. Despite a global distribution in the form of various diseases, nearly all spotted fever group organisms share close genetic, antigenic, and certain pathologic features. Examples of human diseases include Rocky Mountain spotted fever (in North and South America), fièvre boutonneuse or Mediterranean spotted fever (southern Europe), South African tick-bite fever (Africa), Indian tick typhus (Indian subcontinent), and Siberian tick typhus (northeastern Europe and northern Asia). If appropriate antibiotics are not administered, Rocky Mountain spotted fever, for example, is a life-threatening disease. *See* INFECTIOUS DISEASE. [R.L.Re.]

Rift valley One of the geomorphological expressions between two tectonic plates that are opening relative to each other or sliding past each other. The term originally was used to describe the central graben structures of such classic continental rift zones as the Rhinegraben and the East African Rift, but the definition now encompasses mid-oceanic ridge systems with central valleys such as the Mid-Atlantic Ridge. *See* MID-OCEANIC RIDGE.

Continental and oceanic rift valleys are end members in what many consider to be an evolutionary continuum. In the case of continental rift valleys, plate separation is incomplete, and the orientation of the stress field relative to the rift valley can range from nearly orthogonal to subparallel. Strongly oblique relationships are probably the norm. In contrast, oceanic rift valleys mark the place where the trailing edges of two distinctly different plates are separating. The separation is complete, and the spreading is organized and focused, resulting in rift valleys that tend to be oriented orthogonal or suborthogonal to the spreading directions.

The basic cross-sectional form of rift valleys consists of a central graben surrounded by elevated flanks. It is almost universally accepted that the central grabens of continental rift valleys are subsidence features. The crystalline basement floors of some parts of the Tanganyika and Malawi (Nyasa) rift valleys in East Africa lie more than 5 mi (9 km) below elevated flanks.

In continental rift valleys the true cross-sectional form is typically asymmetric, with the rift floors tilted toward the most elevated flank. Most of the subsidence is controlled by one border fault system, and most of the internal faults parallel the dip of the border faults. *See* FAULT AND FAULT STRUCTURES.

Oceanic rift valleys are also distinctly separated into segments by structures known as transform faults. The cross-sectional form of oceanic rift valleys can be markedly asymmetric. It is unlikely that the cross-sectional form of oceanic rift valleys is related genetically to that of continental rift valleys, except in the broadest possible terms. [B.R.R.]

Rift Valley fever An arthropod-borne (primarily mosquito), acute, febrile, viral disease of humans and numerous species of animals. Rift Valley fever is caused by a ribonucleic acid (RNA) virus in the genus *Phlebovirus* of the family Bunyaviridae. In sheep and cattle, it is also known as infectious enzootic hepatitis. First described in the Rift Valley of Africa, the disease presently occurs in west, east, and south Africa and has extended as far north as Egypt. Historically, outbreaks of Rift Valley fever have occurred at 10–15-year intervals in normally dry areas of Africa subsequent to a period of heavy rainfall.

In humans, clinical signs of Rift Valley fever are influenzalike, and include fever, headache, muscular pain, weakness, nausea, epigastric pain, and photophobia. Most people recover within 4–7 days, but some individuals may have impaired vision or blindness in one or both eyes; a small percentage of infected individuals develop a hemorrhagic syndrome and die.

Rift Valley fever should be suspected when high abortion rates, high mortality, or extensive liver lesions occur in newborn animals. The diagnosis is confirmed by isolating the virus from tissues of the infected animal or human. Control of the disease is best accomplished by widespread vaccination of susceptible animals to prevent amplification of the virus and, thus, infection of vectors. Any individual that works with infected animals or live virus in a laboratory should be vaccinated. *See* ANIMAL VIRUS. [C.A.Me.]

Rinderpest An acute or subacute, contagious viral disease of ruminants and swine, manifested by high fever, lacrimal discharge, profuse diarrhea, erosion of the epithelium of the mouth and of the digestive tract, and high mortality.

Rinderpest (also known as cattle plague) is caused by a ribonucleic acid (RNA) virus classified in the genus *Morbillivirus* within the Paramyxoviridae family. This virus is closely associated with the viruses of human measles, peste des petits ruminants of sheep and goats, canine distemper, and phocine distemper, and with the dolphin morbillivirus. Although there are significant differences in virulence, only one serotype of rinderpest virus is known. The rinderpest virus is easily inactivated by heat and survives outside the host for a short time. Therefore, transmission of rinderpest is by direct contact between animals.

Although all cloven-hoofed animals are considered susceptible to rinderpest, clinical cases are mostly seen in cattle and water buffalo. Rinderpest is characterized by the development of high fever that lasts for several days until just prior to death. The morbidity in susceptible cattle and buffalo is greater than 90%, with death of almost all clinically affected animals. Rinderpest has been controlled in most African nations, but persists in regions of eastern Africa. *See* ANIMAL VIRUS; PARAMYXOVIRUS. [A.To.]

River A natural, fresh-water surface stream that has considerable volume compared with its smaller tributaries. The tributaries are known as brooks, creeks, branches, or forks. Rivers are usually the main stems and larger tributaries of the drainage systems that convey surface runoff from the land. Rivers flow from headwater areas of small tributaries to their mouths, where they may discharge into the ocean, a major lake, or a desert basin.

Rivers flowing to the ocean drain about 68% of the Earth's land surface. Regions draining to the sea are termed exoreic, while those draining to interior closed basins are endoreic. Areic regions are those which lack surface streams because of low rainfall or lithoogic conditions.

Sixteen of the largest rivers account for nearly half of the total world river flow of water. The Amazon River alone carries nearly 20% of all the water annually discharged by the world's rivers. Rivers also carry large loads of sediment. The total sediment load for all the world's rivers averages about 22×10^9 tons (20×10^9 metric tons) brought to the sea each year. Sediment loads for individual rivers vary considerably. The Yellow River of northern China is the most prolific transporter of sediment. Draining an agricultural region of easily eroded loess, this river averages about 2×10^9 tons (1.8×10^9 metric tons) of sediment per year, one-tenth of the world average. *See* DEPOSITIONAL SYSTEMS AND ENVIRONMENTS; LOESS.

River discharge varies over a broad range, depending on many climatic and geologic factors. The low flows of the river influence water supply and navigation. The high flows are a concern as threats to life and property. However, floods are also beneficial. The ancient Egyptian civilization was dependent upon the Nile River floods to provide new soil and moisture for crops. Floods are but one attribute of rivers that affect human society. Means of counteracting the vagaries of river flow have concerned engineers for centuries. In modern times many of the world's rivers are managed to conserve the natural flow for release at times required by human activity, to confine flood flows to the channel and to planned areas of floodwater storage, and to maintain water quality at optimum levels. *See* FLOODPLAIN; RIVER ENGINEERING. [V.R.B.]

River engineering A branch of civil engineering that involves the control and utilization of rivers for the benefit of humankind. Its scope includes river training, channel design, flood control, water supply, navigation improvement, hydraulic structure design, hazard mitigation, and environmental enhancement. River engineering is also necessary to provide protection against floods and other river disasters. The emphasis is often on river responses, long-term and short-term, to changes in nature, and stabilization and utilization, such as damming, channelization, diversion, bridge construction, and sand or gravel mining. Evaluation of river responses is essential at the conceptual, planning, and design phases of a project and requires the use of fundamental principles of river and sedimentation engineering. *See* RIVER; STREAM TRANSPORT AND DEPOSITION. [H.H.C.]

River tides Tides that occur in rivers emptying directly into tidal seas. These fides show three characteristic modifications of ocean tides. (1) The speed at which the tide travels upstream depends on the depth of the channel. (2) The further upstream, the longer the duration of the falling tide and the shorter the duration of the rising tide. (3) The range of the tide decreases with distance upstream.

In a river the difference between the depths of water at high and low tides may be relatively large, leading to a marked difference between the speeds at which high and low tides move. The difference in depth between various points on the river also partially explains the second modification, or duration of fall and rise. In addition, the river flow, which may fluctuate widely, helps a failing tide but hinders a rising tide, increasing the difference in duration.

The third modification or decrease in tidal range upstream may be accounted for by loss of energy of the water through friction with the sides and bottom of the channel. Although friction always saps energy from the tide, if the channel becomes constricted within a short distance, the water may be forced into a smaller space, thus producing a larger tidal range. [B.K.]

Rock mechanics Application of the principles of mechanics and geology to quantify the response of rock when it is acted upon by environmental forces, particularly when human-induced factors alter the original ambient conditions. Rock mechanics is an interdisciplinary engineering science that requires interaction between physics, mathematics, and geology, and civil, petroleum, and mining engineering. The present state of knowledge permits only limited correlations between theoretical predictions and empirical results. Therefore, the most useful principles are based upon data obtained from laboratory and in-place measurements and from prototype behavior (behavior of the completed engineering works). Increasing emphasis is upon in-place measurements because rock properties are regarded as site-specific; that is, the properties of the rock system at one site probably will be significantly different from those at another site, even if geologic environments are similar. *See* ENGINEERING GEOLOGY.

Because of the interdisciplinary aspects, there is no standardization of rock mechanics terminology. However, the following terms and definitions are useful.

Environmental factors are the natural factors and human influences that require consideration in engineering problems in rock mechanics. The major natural factors are geology, ambient stresses, and hydrology. The human influences derive from the application of chemical, electrical, mechanical, or thermal energy during construction (or destruction) processes.

The ambient stress field is the distribution and numerical value of the stresses in the environment prior to its disturbance by humans.

The term rock system includes the complete environment that can influence the behavior of that portion of the Earth's crust that will become part of an engineering structure. Generally, all natural environmental factors are included.

A rock element is the coherent, intact piece of rock that is the basic constituent of the rock system and which has physical, mechanical, and petrographic properties that can be described or measured by laboratory tests on each such element. The concepts of rock system and rock element enable the concomitant engineering design to be optimized according to the principles of system engineering.

"Rock failure" occurs when a rock system or element no longer can perform its intended engineering function. Failure may be evidenced by fractures, distortion of shape, or reduction in strength. "Failure mechanism" includes the causes for the manner of rock failure. *See* SOIL MECHANICS. [W.R.J.]

Rodenticide A toxic chemical that is used to kill pest rodents and sometimes other pest mammals, including moles, rabbits, and hares. Most rodenticides are used to control rats and house mice.

Rodenticides are generally combined with some rodent-preferred food item such as grain (corn, wheat, oats) or a combination of grains in low yet effective amounts. Bait formulations may be in pellet forms or incorporated in paraffin blocks of varying sizes. As a safeguard against accidental ingestion by nontarget species, baits are placed either where they are inaccessible to children, domestic animals, or wildlife, or within tamper-resistant bait boxes designed to exclude all but rodent-size animals.

As a group, anticoagulant rodenticides dominate the market and are sold under a wide variety of trade names. In order of their development, they are warfarin, pindone (Pival), diphacinone, and chlorophacinone. When small amounts of these anticoagulants are consumed over several days, death results from internal bleeding. The newer, second-generation anticoagulant rodenticides, such as brodifacoum, bromadiolone, and difethialone, were developed to counteract the growing genetic resistance in rats and house mice to the earlier anticoagulants, especially warfarin. The second-generation compounds are more potent and capable of being lethal following a single night's feeding, although death is generally delayed by several days. Rodenticides that do not belong to the anticoagulant group include zinc phosphide, bromethalin, and cholecalciferol (vitamin D_3). The feeding and lethal characteristics differ among them. Acutely toxic strychnine baits are also available but are restricted to underground application, primarily for pocket gophers and moles. Several lethal fumigants or materials that produce poisonous gases are used to kill rodents in burrows and within other confined areas such as unoccupied railway cars or buildings. Lethal fumigants include aluminum phosphide, carbon dioxide, chloropicrin, and smoke or gas cartridges, which are ignited to produce carbon monoxide and other asphyxiating gases.

Because of their high toxicity, rodenticides are inherently hazardous to people, domestic animals, and wildlife. They are highly regulated, as are certain other types of pesticides. Some rodenticides can be purchased and used only by trained certified or licensed pest control operators, while others with a greater safety margin can be used by the general public. All rodenticides must be used in accordance with the label directions and may be prohibited where they may jeopardize certain endangered species. *See* PESTICIDE. [R.E.Ma.]

Root (botany) The absorbing and anchoring organ of vascular plants. Roots are simple axial organs that produce lateral roots, and sometimes buds, but bear neither leaves nor flowers. Elongation occurs in the root tip. The older portion of the root, behind the root tip, may thicken through cambial activity. Some roots, grass for example, scarcely thicken, but tree roots can become 4 in. (10 cm) or more in diameter near the stem. Roots may be very long. The longest maple (*Acer*) roots are usually as long as the tree is tall, but the majority of roots are only a few inches long. The longest roots may live for many years, while small roots may live for only a few weeks or months.

Root tips and the root hairs on their surface take up water and minerals from the soil. They also synthesize amino acids and growth regulators (gibberellins and cytokinins). These materials move up through the woody, basal portion of the root to the stem. The thickened, basal portion of the root anchors the plant in the soil. Thickened roots, such as carrots, can store food that is later used in stem growth.

Roots usually grow in soil where: it is not too dense to stop root tip elongation; there is enough water and oxygen for root growth; and temperatures are high enough (above 39°F or 4°C) to permit root growth, but not so high that the roots are killed (above 104°F or 40°C). In temperate zones most roots are in the uppermost 4 in. (10 cm) of the soil; root numbers decrease so rapidly with increasing depth that few roots are found more than 6 ft (2 m) below the surface. Roots grow deeper in areas where the soil is hot and dry; roots from desert shrubs have been found in mines more than 230 ft (70 m) below the surface. In swamps with high water tables the lack of oxygen restricts roots to the uppermost soil layers. Roots may also grow in the air. Poison ivy vines form many small aerial roots that anchor them to bark or other surfaces.

The primary root originates in the seed as part of the embryo, normally being the first organ to grow. It grows downward into the soil and produces lateral second-order roots that emerge at right angles behind the root tip. Sometimes it persists and thickens to form a taproot. The second-order laterals produce third-order laterals and so on until there are millions of roots in a mature tree root system. In contrast to the primary root, most lateral roots grow horizontally or even upward. In many plants a few horizontal lateral roots thicken more than the primary, so no taproot is present in the mature root system.

Adventitious roots originate from stems or leaves rather than the embryo or other roots. They may form at the base of cut stems, as seen in the horticultural practice of rooting cuttings. [B.F.W.]

Rous sarcoma The first filterable agent (virus) known to cause a solid tumor in chickens. It was discovered in 1911 by P. F. Rous, who won the Nobel prize in 1967 for his discovery. It is a ribonucleic acid virus and belongs to the avian leukosis group. Certain strains of the virus cause tumors in hamsters, rabbits, monkeys, and other species. The Rous virus is known as a "defective" virus in that it is incapable of producing tumors by itself but requires another closely related virus of the avian leukosis group to act as a "helper" for the production of the foci. *See* ANIMAL VIRUS; TUMOR VIRUSES. [A.E.Mo.]

Rubber tree *Hevea brasiliensis*, a member of the spurge family (Euphorbiaceae) and a native of the Amazon valley. It is the natural source of commercial rubber.

It has been introduced into all the tropical countries supporting the rainforest type of vegetation, and is grown extensively in established plantations, especially in Malaysia. The latex from the trees is collected and coagulated. The coagulated latex is treated in different ways to produce the kind of rubber desired. Rubber is made from the latex of a number of other plants, but *Hevea* is the rubber plant of major importance. [P.D.St./E.L.C.]

Rubella A benign, infectious virus disease of humans characterized by cold-like symptoms and transient, generalized rash. This disease, also known as German measles, is primarily a disease of childhood. However, maternal infection during early pregnancy may result in infection of the fetus, giving rise to serious abnormalities and malformations. The congenital infection persists in the infant, who harbors and sheds virus for many months after birth.

In rubella infection acquired by ordinary person-to-person contact, the virus is believed to enter the body through respiratory pathways. Antibodies against the virus develop as the rash fades, increase rapidly over a 2–3-week period, and then fall during the following months to levels that are maintained for life. One attack confers life-long immunity, since only one antigenic type of the virus exists. Immune mothers transfer antibodies to their offspring, who are then protected for approximately 4–6 months after birth.

Live attenuated rubella vaccines have been available since 1969. The vaccine induces high antibody titers and an enduring and solid immunity. It may also induce secretory immunoglobulin (IgA) antibody in the respiratory tract and thus interfere with establishment of infection by wild virus. This vaccine is available as a single antigen or combined with measles and mumps vaccines (MMR vaccine). The vaccine induces immunity in at least 95% of recipients, and that immunity endures for at least 10 years. [J.L.Me.]

Rust (microbiology) Plant diseases caused by fungi of the order Uredinales and characterized by the powdery and usually reddish spores produced. There are more than 4000 species of rust fungi. All are obligate parasites (require a living host) in nature, and each species attacks only plants of particular genera or species. Morphologically identical species that attack different host genera are further classified as special forms (*formae speciales*); for example, *Puccinia graminis* f. sp. *tritici* attacks wheat and *P. graminis* f. sp. *hordei* attacks barley. Each species or special form can have many physiological races that differ in their ability to attack different cultivars (varieties) of a host species. Rusts are among the most destructive plant diseases. Economically important examples include wheat stem rust, white pine blister rust, and coffee rust. *See* UREDINALES.

Rust fungi have complex life cycles, producing up to five different fruiting structures with distinct spore types that appear in a definite sequence. Macrocyclic (long-cycled) rust fungi produce all five spore types, whereas microcyclic (short-cycled) rust fungi produce only teliospores and basidiospores. Some macrocyclic rust fungi complete their life cycle on a single host and are called autoecious, whereas others require two different or alternate hosts and are called heteroecious. *See* FUNGI; PLANT PATHOLOGY. [E.A.M.]

S

Saline evaporites Deposits of bedded sedimentary rocks composed of salts precipitated during solar evaporation of surface or near-surface brines derived from seawater or continental waters. Dominant minerals in ancient evaporite beds are anhydrite (along with varying amounts of gypsum) and halite, which make up more than 85% of the total sedimentary evaporite salts. Many other salts make up the remaining 15%; their varying proportions in particular beds can be diagnostic of the original source of the mother brine. *See* SEAWATER; SEDIMENTARY ROCKS.

Today, brines deposit their salts within continental playas or coastal salt lakes and lay down beds a few meters thick and tens of kilometers across. In contrast, ancient, now-buried evaporite beds are often much thicker and wider; they can be up to hundreds of meters thick and hundreds of kilometers wide. Most ancient evaporites were formed by the evaporation of saline waters within hyperarid areas of huge seaways typically located within arid continental interiors. The inflow brines in such seaways were combinations of varying proportions of marine and continental ground waters and surface waters. There are few modern depositional counterparts to these ancient evaporites, and none to those beds laid down when whole oceanic basins dried up, for example, the Mediterranean some 5.5 million years ago. *See* BASIN; DEPOSITIONAL SYSTEMS AND ENVIRONMENTS; MEDITERRANEAN SEA; PLAYA.

Evaporite salts precipitate by the solar concentration of seawater, continental water, or hybrids of the two. The chemical makeup, salinity (35‰), and the proportions of the major ions in modern seawater are near-constant in all the world's oceans, with sodium (Na) and chloride (Cl) as the dominant ions and calcium (Ca) and sulfate (SO_4) ions present in smaller quantities [$Na(Ca)SO_4Cl$ brine]. Halite and gypsum anhydrites have been the major products of seawater evaporation for at least the past 2 billion years, but the proportions of the more saline minerals, such as sylvite/magnesium sulfate ($MgSO_4$) salts, appear to have been more variable. [J.Wa.]

Salmonelloses Diseases caused by *Salmonella*. These include enteritis and septicemia with or without enteritis. *Salmonella typhi*, *S. paratyphi A*, *B*, and *C*, and occasionally *S. cholerae suis* cause particular types of septicemia called typhoid and paratyphoid fever, respectively; while all other types may cause enteritis or septicemia, or both together.

Typhoid fever has an incubation period of 5–14 days. It is typified by a slow onset with initial bronchitis, diarrhea or constipation, a characteristic fever pattern (increase for 1 week, plateau for 2 weeks, and decrease for 2–3 weeks), a slow pulse rate, development of rose spots, swelling of the spleen, and often an altered consciousness; complications include perforation of the bowel and osteomyelitis. Typhoid fever leaves the individual with a high degree of immunity. Vaccination with an oral vaccine gives an individual considerable protection for about 3 years. The only effective antibiotic is chloramphenicol.

Paratyphoid fever has a shorter course and is generally less severe than typhoid fever. Vaccination is an ineffective protective measure.

Enteric fevers, that is, septicemias due to types of *Salmonella* other than those previously mentioned, are more frequent in the United States than typhoid and paratyphoid fever but much less frequent than *Salmonella* enteritis. In children and in previously healthy adults, enteric fevers are most often combined with enteritis and have a favorable outlook. The organisms involved are the same as those causing *Salmonella* enteritis. Chloramphenicol or ampicillin are used in treatment. However, resistant strains have been observed. *See* DRUG RESISTANCE.

Inflammation of the small bowel due to *Salmonella* is one of the most important bacterial zoonoses. The most frequent agents are *S. typhimurium*, *S. enteritidis*, *S. newport*, *S. heidelberg*, *S. infantis*, and *S. derby*. The incubation period varies from 6 h to several days. Diarrhea and fever are the main symptoms; the intestinal epithelium is invaded, and early bacteremia is probable. Predisposed are persons with certain preexisting diseases (the same as for enteric fevers), very old and very young individuals, and postoperative patients. Antimicrobial treatment serves only to prolong the carrier state and has no effect on the disease. [A.W.C.V.G.]

Salt dome An upwelling of crystalline rock salt and its aureole of deformed sediments. A salt pillow is an immature salt dome comprising a broad salt swell draped by concordant strata. A salt stock is a more mature, pluglike diapir of salt that has pierced, or appears to have pierced, overlying strata. Most salt stocks are 0.6–6 mi (1–10 km) wide and high. Salt domes are closely related to other salt upwellings, some of which are much larger. Salt canopies, which form by coalescence of salt domes and tongues, can be more than 200 mi (300 km) wide.

Exploration for oil and gas has revealed salt domes in more than 100 sedimentary basins that contain rock salt layers several hundred meters or more thick. The salt was precipitated from evaporating lakes in rift valleys, intermontaine basins, and especially along divergent continental margins. Salt domes are known in every ocean and continent. *See* BASIN.

Salt domes consist largely of halite (NaCl, common table salt). Other evaporites, such as anhydrite ($CaSO_4$) and gypsum ($CaSO_4 \cdot 2H_2O$), form thinner layers within the rock salt. *See* SALINE EVAPORITES.

Salt domes supply industrial commodities, including fuel, minerals, chemical feedstock, and storage caverns. Giant oil or gas fields are associated with salt domes in many basins around the world, especially in the Middle East, North Sea, and South Atlantic regions. Salt domes are also used to store crude oil, natural gas (methane), liquefied petroleum gas, and radioactive or toxic wastes. [M.P.A.J.]

Salt gland A specialized gland located around the eyes and nasal passages in marine turtles, snakes, and lizards, and in birds such as the petrels, gulls, and albatrosses, which spend much time at sea. In the marine turtle it is an accessory lacrimal gland which opens into the conjunctival sac. In seagoing birds and in marine lizards it opens into the nasal passageway. Salt glands copiously secrete a watery fluid containing a high percentage of salt, higher than the salt content of urine in these species. As a consequence, these animals are able to drink salt-laden sea water without experiencing the dehydration necessary to eliminate the excess salt via the kidney route. [O.E.N.]

Salt marsh A maritime habitat characterized by grasses, sedges, and other plants that have adapted to continual, periodic flooding. Salt marshes are found primarily throughout the temperate and subarctic regions.

The tide is the dominating characteristic of a salt marsh. The salinity of the tide defines the plants and animals that can survive in the marsh area. The vertical range of the tide determines flooding depths and thus the height of the vegetation, and the tidal cycle controls how often and how long vegetation is submerged. Two areas are delineated by the tide: the low marsh and the high marsh. The low marsh generally floods and drains twice daily with the rise and fall of the tide; the high marsh, which is at a slightly higher elevation, floods less frequently. *See* MANGROVE.

Salt marshes usually are developed on a sinking coastline, originating as mud flats in the shallow water of sheltered bays, lagoons, and estuaries, or behind sandbars. They are formed where salinity is high, ranging from 20 to 30 parts per thousand of sodium chloride. Proceeding up the estuary, there is a transitional zone where salinity ranges from 20 to less than 5 ppt. In the upper estuary, where river input dominates, the water has only a trace of salt. This varying salinity produces changes in the marsh—in the kinds of species and also in their number. Typically, the fewest species are found in the salt marsh and the greatest number in the fresh-water tidal marsh. *See* ESTUARINE OCEANOGRAPHY.

The salt marsh is one of the most productive ecosystems in nature. In addition to the solar energy that drives the photosynthetic process of higher rooted plants and the algae growing on the surface muds, tidal energy repeatedly spreads nutrient-enriched waters over the marsh surface. Some of this enormous supply of live plant material may be consumed by marsh animals, but the most significant values are realized when the vegetation dies and is decomposed by microorganisms to form detritus. Dissolved organic materials are released, providing an essential energy source for bacteria that mediate wetland biogeochemical cycles (carbon, nitrogen, and sulfur cycles). *See* BIOGEOCHEMISTRY; BIOLOGICAL PRODUCTIVITY.

The salt marsh serves as a sediment sink, a nursery habitat for fishes and crustaceans, a feeding and nesting site for waterfowl and shorebirds, a habitat for numerous unique plants and animals, a nutrient source, a reservoir for storm water, an erosion control mechanism, and a site for esthetic pleasures. Appreciation for the importance of salt marshes has led to federal and state legislation aimed at their protection. [F.C.D.]

Sand Unconsolidated granular material consisting of mineral, rock, or biological fragments between 63 micrometers and 2 mm in diameter. Finer material is referred to as silt and clay; coarser material is known as gravel. Sand is usually produced primarily by the chemical or mechanical breakdown of older source rocks, but may also be formed by the direct chemical precipitation of mineral grains or by biological processes. Accumulations of sand result from hydrodynamic sorting of sediment during transport and deposition. *See* DEPOSITIONAL SYSTEMS AND ENVIRONMENTS; SEDIMENTARY ROCKS.

Most sand originates from the chemical and mechanical breakdown, or weathering, of bedrock. Since chemical weathering is most efficient in soils, most sand grains originate within soils. Rocks may also be broken into sand-size fragments by mechanical processes, including diurnal temperature changes, freeze-thaw cycles, wedging by salt crystals or plant roots, and ice gouging beneath glaciers. *See* WEATHERING PROCESSES.

Because sand is largely a residual product left behind by incomplete chemical and mechanical weathering, it is usually enriched in minerals that are resistant to these processes. Quartz not only is extremely resistant to chemical and mechanical weathering but is also one of the most abundant minerals in the Earth's crust. Many sands dominantly consist of quartz. Other common constituents include feldspar, and fragments of igneous or metamorphic rock. Direct chemical precipitation or hydrodynamic processes can result in sand that consists almost entirely of calcite, glauconite, or dense dark-colored minerals such as magnetite and ilmenite.

Although sand and gravel has one of the lowest average per ton values of all mineral commodities, the vast demand makes it among the most economically important of all mineral resources. Sand and gravel is used primarily for construction purposes, mostly as concrete aggregate. Pure quartz sand is used in the production of glass, and some sand is enriched in rare commodities such as ilmenite (a source of titanium) and in gold. [M.J.J.]

Sandalwood The name applied to any species of the genus *Santalum* of the sandalwood family (Santalaceae). However, the true sandalwood is the hard, close-grained, aromatic heartwood of a parasitic tree, *S. album*, of the Indo-Malayan region. This fragrant wood is used in ornamental carving, cabinet work, and as a source of certain perfumes. The odor of the wood is an insect repellent, and on this account the wood is much used in making boxes and chests. The fragrant wood of a number of species in other families bears the same name, but none of these is the real sandalwood. [P.D.St./E.L.C.]

Sassafras A medium-sized tree, *Sassafras albidum*, of the eastern United States, extending north as far as southern Maine. Sometimes it is only a shrub in the north, but from Pennsylvania southward heights of 90 ft (27 m) or more with diameters of 4–7 ft (1.2–2.1 m) have been reported for this plant. Sassafras is said to live from 700 to 1000 years. It can be recognized by the bright-green color and aromatic odor of the twigs and leaves. The leaves are simple or mitten-shaped (hence a common name "mitten-tree"), or they may have lobes on both sides of the leaf blade. [A.H.G./K.P.D.]

Satellite meteorology The branch of meteorological science that uses meteorological sensing elements on satellites to define the past and present state of the atmosphere. Meteorological satellites can measure a wide spectrum of electromagnetic radiation in real time, providing the meteorologist with a supplemental source of data.

Modern satellites are sent aloft with multichannel high-resolution radiometers covering an extensive range of infrared and microwave wavelengths. Radiometers sense cloudy and clear-air atmospheric radiation at various vertical levels, atmospheric moisture content, ground and sea surface temperatures, and ocean winds, and provide visual imagery as well.

There are two satellite platforms used for satellite meteorology: geostationary and polar. Geostationary (geo) satellites orbit the Earth at a distance that allows them to make one orbit every 24 hours. By establishing the orbit over the Equator, the satellite appears to remain stationary in the sky. This is important for continuous scanning of a region on the Earth for mesoscale (approximately 10–1000 km horizontal) forecasting.

Polar satellites orbit the Earth in any range of orbital distances with a high inclination angle that causes part of the orbit to fly over polar regions. The orbital distance of 100–200 mi (160–320 km) is selected for meteorological applications, enabling the satellite to fly over a part of the Earth at about the same time every day. With orbital distances of a few hundred miles, the easiest way to visualize the Earth-satellite relationship is to think of a satellite orbiting the Earth pole-to-pole while the Earth rotates independently beneath the orbiting satellite. The advantage of polar platforms is that they eventually fly over most of the Earth. This is important for climate studies since one set of instruments with known properties will view the entire world.

The enormous aerial coverage by satellite sensors bridges many of the observational gaps over the Earth's surface. Satellite data instantaneously give meteorologists up-to-the minute views of current weather phenomena.

Images derived from the visual channels are presented as black and white photographs. The brightness is solely due to the reflected solar light illuminating the Earth. Visible images are useful for determining general cloud patterns and detailed cloud structure. In addition to clouds, visible imagery shows snowcover, which is useful for diagnosing snow amount by observing how fast the snow melts following a storm. Cloud patterns defined by visual imagery can give the meteorologist detailed information about the strength and location of weather systems, which is important for determining storm motion and provides a first guess or forecast as to when a storm will move into a region. *See* CLOUD.

More quantitative information is available from infrared sensors, which measure radiation at longer wavelengths (from infrared to microwave). By analyzing the infrared data, the ground surface, cloud top, and even intermediate clear air temperatures can be determined 24 hours a day. By relating the cloud top temperature in the infrared radiation to an atmospheric temperature profile from balloon data, cloud top height can be estimated. This is a very useful indicator of convective storm intensity since more vigorous convection will generally extend higher in the atmosphere and appear colder.

The advent of geosynchronous satellites allowed the position of cloud elements to be traced over time. These cloud movements can be converted to winds, which can provide an additional source of data in an otherwise unobserved region. These techniques are most valuable for determination of mid- and high-level winds, particularly over tropical ocean areas. Other applications have shown that low-level winds can be determined in more spatially limited environments, such as those near thunderstorms, but those winds become more uncertain when the cloud elements grow vertically into air with a different speed and direction (a sheared environment). *See* WIND.

By using a wide variety of sensors, satellite data provide measurements of phenomena from the largest-scale global heat and energy budgets down to details of individual thunderstorms. Having both polar orbiting and geosynchronous satellites allows coverage over most Earth locations at time intervals from 3 minutes to 3 hours.

The greatest gain with the introduction of weather satellites was in early detection, positioning, and monitoring of the strength of tropical storms (hurricanes, typhoons). Lack of conventional meteorological data over the tropics (particularly the oceanic areas) makes satellite data indispensable for this task. The hurricane is one of the most spectacular satellite images. The exact position, estimates of winds, and qualitative determination of strength are possible with continuous monitoring of satellite imagery in the visible channels. In addition, infrared sensors provide information on cloud top height, important for locating rain bands. Microwave sensors can penetrate the storm to provide an indication of the interior core's relative warmth, closely related to the strength of the hurricane, and sea surface temperature to assess its development potential. *See* HURRICANE; TROPICAL METEOROLOGY.

Most significant weather events experienced by society—heavy rain or snow, severe thunderstorms, or high winds—are organized by systems that have horizontal dimensions of about 60 mi (100 km). These weather systems, known as mesoscale convective systems, often fall between stations of conventional observing networks. Hence, meteorologists might miss them were it not for satellite sensing. *See* HAIL; METEOROLOGY; PRECIPITATION (METEOROLOGY); THUNDERSTORM; TORNADO; WEATHER FORECASTING AND PREDICTION. [D.L.B.; J.A.McG.]

Savanna The term savanna was originally used to describe a tropical grassland with more or less scattered dense tree areas. This vegetation type is very abundant in tropical and subtropical areas, primarily because of climatic factors. The modern definition of savanna includes a variety of physiognomically or environmentally similar vegetation types in tropical and extratropical regions. The physiognomically savannalike extratropical vegetation types (forest tundra, forest steppe, and everglades) differ greatly in environment and species composition.

In the widest sense savanna includes a range of vegetation zones from tropical savannas with vegetation types such as the savanna woodlands to tropical grassland and thornbush. In the extratropical regions it includes the "temperate" and "cold savanna" vegetation types known under such names as taiga, forest tundra, or glades. *See* GRASSLAND ECOSYSTEM; TAIGA; TUNDRA. [H.Li.]

Scarlet fever An acute contagious disease that results from infection with *Streptococcus pyogenes* (group A streptococci). It most often accompanies pharyngeal (throat) infections with this organism but is occasionally associated with wound infection or septicemia. Scarlet fever is characterized by the appearance, about 2 days after development of pharyngitis, of a red rash that blanches under pressure and has a sandpaper texture. Usually the rash appears first on the trunk and neck and spreads to the extremities. The rash fades after a week, with desquamation, or peeling, generally occurring during convalescence. The disease is usually self-limiting, although severe forms are occasionally seen with high fever and systemic toxicity. Appropriate antibiotic therapy is recommended to prevent the onset in susceptible individuals of rheumatic fever and rheumatic heart disease. *See* MEDICAL BACTERIOLOGY; RHEUMATIC FEVER; STREPTOCOCCUS. [E.D.Gr.]

Schistosomiasis A disease in which humans are parasitized by any of three species of blood flukes: *Schistosoma mansoni*, *S. haematobium*, and *S. japonicum*. Adult *S. mansoni* prefer the veins of the hemorrhoidal plexus, *S. haematobium* those of the vesical plexus, and *S. japonicum* those of the small intestine. The disease is also known as bilharziasis.

An embryonated egg passed in feces or urine hatches in fresh water, liberating a miracidium larva which penetrates into specific gastropod snails. The larval cycle in the snail lasts for about 1 month. The cercaria emerges from the mollusk, swims in the water, and penetrates the skin of the final host upon coming in contact with it.

Schistosomiasis is an agricultural hazard for all ages in irrigated lands or swamps. Elsewhere fluvial waters are the main source of infection, in which case incidence is marked in human beings who are less than 15 years old and is higher among boys than among girls. [J.F.M.]

Scleractinia An order of the subclass Zoantharia which comprises the true or stony corals (see illustration). These are solitary or colonial anthozoans which attach to a firm substrate. They are profuse in tropical and subtropical waters and contribute to the formation of coral reefs or islands. Some species are free and unattached.

Most of the polyp is impregnated with a hard calcareous skeleton secreted from ectodermal calcioblasts. The solitary corals form cylindrical, discoidal, or cuneiform skeletons, whereas colonial skeletons are multifarious. The polyps increase rapidly by intra- or extratentacular budding, and the skeletons of polyps which settle in groups may fuse to form a colony. The pyriform, ciliated planula swims with its aboral extremity, which is composed of an ectodermal sensory layer, directed anteriorly. Planulation occurs periodically in conformity with lunar phases in many tropical species. [K.At.]

Solitary coral polyps, *Oulangia* sp.

Scleractinian corals possess robust skeletons, so they have a rich fossil record. Because they are restricted mainly to tropical belts, they help indicate the position of the continents throughout the Mesozoic and Cenozoic periods. They are also important for understanding the evolution of corals and the origin and maintenance of reef diversity through time. Pleistocene corals shows persistent reef coral communities throughout the last several hundred thousand years. Environmental degradation has led to the dramatic alteration of living coral communities during the past several decades. [J.M.Pan.]

Sclerenchyma Single cells or aggregates of cells whose principal function is thought to be mechanical support of plants or plant parts. Sclerenchyma cells have thick secondary walls and may or may not remain alive when mature. They vary greatly in form and are of widespread occurrence in vascular plants. Two general types, sclereids and fibers, are widely recognized, but since these intergrade, the distinction is sometimes arbitrary. [N.H.B.]

Scrapie A transmissible, usually fatal disease of adult sheep characterized by degeneration of the central nervous system. The disease is known in Great Britain, France, Belgium, Iceland, the United States, Canada, and northern India. Scrapie has certain similarities with kuru, a human disease in New Guinea, and mink encephalopathy.

Scrapie affects both sexes and is insidious in its onset, starting with hyperexcitability and progressive itch. Later, loss of wool occurs when the animal rubs against fixed objects or bites and nibbles its skin. Some animals do not rub but are either nervous and tremble when approached or appear sleepy. Incoordination of gait is constant and usually more evident in the hindquarters. In the final stages the sheep, being unable to stand, lie down, become emaciated, and die. *See* PRION DISEASE. [I.Zl.]

Sea breeze A diurnal, thermally driven circulation in which a surface convergence zone often exists between airstreams having over-water versus over-land histories. The sea breeze is one of the most frequently occurring small-scale (mesoscale) weather systems. It results from the unequal sensible heat flux of the lower atmosphere over adjacent solar-heated land and water masses. Because of the large thermal inertia of a water body, during daytime the air temperature changes little over the water while over land the air mass warms. Occurring during periods of fair skies and generally weak large-scale winds, the sea breeze is recognizable by a wind shift to onshore, generally several hours after sunrise. On many tropical coastlines the sea breeze is an almost daily occurrence. It also occurs with regularity during the warm season along mid-latitude coastlines and even occasionally on Arctic shores. Especially during periods of very light winds, similar though sometimes weaker wind systems occur over the shores of large lakes and even wide rivers and estuaries (lake breezes, river breezes). At night, colder air from the land often will move offshore as a land breeze. Typically the land breeze circulation is much weaker and shallower than its daytime counterpart. *See* ATMOSPHERIC GENERAL CIRCULATION; METEOROLOGY.

The occurrence and strength of the sea breeze is controlled by a variety of factors, including land-sea surface temperature differences; latitude and day of the year; the synoptic wind and its orientation with respect to the shoreline; the thermal stability of the lower atmosphere; surface solar radiation as affected by haze, smoke, and stratiform and convective cloudiness; and the geometry of the shoreline and the complexity of the surrounding terrain. *See* WIND. [W.A.Ly.]

Sea-floor imaging The process whereby mapping technologies are used to produce highly detailed images of the sea floor. High-resolution images of the sea floor are used to locate and manage marine resources such as fisheries and oil and gas reserves, identify offshore faults and the potential for coastal damage due to earthquakes, and map out and monitor marine pollution, in addition to providing information on what processes are affecting the sea floor, where these processes occur, and how they interact. *See* MARINE GEOLOGY.

Side-scan sonar provides a high-resolution view of the sea floor. In general, a side-scan sonar consists of two sonar units attached to the sides of a sled tethered to the back of a ship. Each sonar emits a burst of sound that insonifies a long, narrow corridor of the sea floor extending away from the sled. Sound reflections from the corridor that echo back to the sled are then recorded by the sonar in their arrival sequence, with echoes from points farther away arriving successively later. The sonars repeat this sequence of "talking" and listening every few seconds as the sled is pulled through the water so that consecutive recordings build up a continuous swath of sea-floor reflections, which provide information about the texture of the sea floor.

The best technology for mapping sea-floor depths or bathymetry is multibeam sonar. These systems employ a series of sound sources and listening devices that are mounted on the hull of a survey ship. As with side-scan sonar, every few seconds the sound sources emit a burst that insonifies a long, slim strip of the sea floor aligned perpendicular to the ship's direction. The listening devices then begin recording sounds from within a fan of narrow sea-floor corridors that are aligned parallel to the ship and that cross the insonified strip. By running the survey the same way that one mows a lawn, adjacent swaths are collected parallel to one another to produce a complete sea-floor map of an area.

The most accurate and detailed view of the sea floor is provided by direct visual imaging through bottom cameras, submersibles, remotely operated vehicles, or if the waters are not too deep, scuba diving. Because light is scattered and absorbed in waters greater than about 33 ft (10 m) deep, the sea-floor area that bottom cameras can image is no more than a few meters. This limitation has been partly overcome by deep-sea submersibles and remotely operated vehicles, which provide researchers with the opportunity to explore the sea floor close-up for hours to weeks at a time. But even the sea-floor coverage that can be achieved with these devices is greatly restricted relative to side-scan sonar, multibeam sonar, and satellite altimetry.

The technology that provides the broadest perspective but the lowest resolution is satellite altimetry. A laser altimeter is mounted on a satellite and, in combination with land-based radars that track the satellite's altitude, is used to measure variations in sea-surface elevation to within 2 in. (5 cm). Removing elevation changes due to waves and currents, sea-surface height can vary up to 660 ft (200 m). These variations are caused by minute differences in the Earth's gravity field, which in turn result from heterogeneities in the Earth's mass. These heterogeneities are often associated with sea-floor topography. By using a mathematical function that equates sea-surface height to bottom elevations, global areas of the sea floor can be mapped within a matter of weeks. However, this approach has limitations. Sea-floor features less than 6–9 mi (10–15 km) in length are generally not massive enough to deflect the ocean surface, and thus go undetected. Furthermore, sea-floor density also affects the gravity field; and where different-density rocks are found, such as along the margins of continents, the correlation between Earth's gravity field and sea-floor topography breaks down. [L.F.Pr.]

Sea ice Ice formed by the freezing of seawater. Ice in the sea includes sea ice, river ice, and land ice. Land ice is principally icebergs. River ice is carried into the sea during spring breakup and is important only near river mouths. The greatest part, probably 99% of ice in the sea, is sea ice. *See* ICEBERG.

The freezing point temperature and the temperature of maximum density of seawater vary with salinity. When freezing occurs, small flat plates of pure ice freeze out of solution to form a network which entraps brine in layers of cells. As the temperature decreases more water freezes out of the brine cells, further concentrating the remaining brine so that the freezing point of the brine equals the temperature of the surrounding pure ice structure. The brine is a complex solution of many ions.

The brine cells migrate and change size with changes in temperature and pressure. The general downward migration of brine cells through the ice sheet leads to freshening of the top layers to near zero salinity by late summer. During winter the top surface temperature closely follows the air temperature, whereas the temperature of the underside remains at freezing point, corresponding to the salinity of water in contact.

The sea ice in any locality is commonly a mixture of recently formed ice and old ice which has survived one or more summers. Except in sheltered bays, sea ice is continually in motion because of wind and current. [W.Ly.]

Seamount and guyot A seamount is a mountain that rises from the ocean floor; a submerged flat-topped seamount is termed a guyot. By arbitrary definition, seamounts must be at least 3000 ft (about 900 m) high, but in fact there is a continuum of smaller undersea mounts, down to heights of only about 300 ft (100 m). Some seamounts are high enough temporarily to form oceanic islands, which ultimately subside beneath sea level. There are on the order of 10,000 seamounts in the world ocean, arranged in chains (for example, the Hawaiian chain in the North Pacific) or as isolated features. In some chains, seamounts are packed closely to form ridges. Very large oceanic volcanic constructions, hundreds of kilometers across, are called oceanic plateaus. *See* MARINE GEOLOGY; OCEANIC ISLANDS; VOLCANO.

Almost all seamounts are the result of submarine volcanism, and most are built within less than about 1 million years. Seamounts are made by extrusion of lavas piped upward in stages from sources within the Earth's mantle to vents on the seafloor. Seamounts provide data on movements of tectonic plates on which they ride, and on the rheology of the underlying lithosphere. The trend of a seamount chain traces the direction of motion of the lithospheric plate over a more or less fixed heat source in the underlying asthenosphere part of the Earth's mantle. [E.L.Wi.]

Seawater An aqueous solution of salts of a rather constant composition of elements whose presence determines the climate and makes life possible on the Earth and which constitutes the oceans, the mediterranean seas, and their embayments. The

Major constituents of seawater (salinity 35 psu)*

Positive ions	Amount, g/kg	Negative ions	Amount, g/kg
Sodium (Na^+)	10.752	Chloride (Cl^-)	19.345
Magnesium (Mg^{2+})	1.295	Bromide (Br^-)	0.066
Potassium (K^+)	0.390	Fluoride (F^-)	0.0013
Calcium (Ca^{2+})	0.416	Sulfate (SO_4^-)	2.701
Strontium (Sr^{2+})	0.013	Bicarbonate (HCO_3^-)	0.145
		Boron hydroxide ($B(OH)_3^-$)	0.027

*Water, 965 psu; dissolved materials, 35 psu.

physical, chemical, biological, and geological events therein are the studies that are grouped as oceanography. Water is most often found in nature as seawater (about 98%). The rest is ice, water vapor, and fresh water. The basic properties of seawater, their distribution, the interchange of properties between sea and atmosphere or land, the transmission of energy within the sea, and the geochemical laws governing the composition of seawater and sediments are the fundamentals of oceanography. *See* HYDROSPHERE; OCEANOGRAPHY.

The major chemical constituents of seawater are cations (positive ions) and anions (negative ions) [see table]. In addition, seawater contains the suspended solids, organic substances, and dissolved gases found in all natural waters. A standard salinity of 35 practical salinity units (psu; formerly parts per thousand, or ‰) has been assumed. While salinity does vary appreciably in oceanic waters, the fractional composition of salts is remarkably constant throughout the world's oceans. In addition to the dissolved salts, natural seawater contains particulates in the form of plankton and their detritus, sediments, and dissolved organic matter, all of which lend additional coloration beyond the blue coming from Rayleigh scattering by the water molecules. Almost every known natural substance is found in the ocean, mostly in minute concentrations. [J.L.Re.]

Seawater fertility A measure of the potential ability of seawater to support life. Fertility is distinguished from productivity, which is the actual production of living material by various trophic levels of the food web. Fertility is a broader and more general description of the biological activity of a region of the sea, while primary production, secondary production, and so on, is a quantitative description of the biological growth at a specified time and place by a certain trophic level. Primary production that uses recently recycled nutrients such as ammonium, urea, or amino acids is called regenerated production to distinguish it from the new production that is dependent on nitrate being transported by mixing or circulation into the upper layer where primary production occurs. New production is organic matter, in the form of fish or sinking organic matter, that can be exported from the ecosystem without damaging the productive capacity of the system. *See* BIOLOGICAL PRODUCTIVITY.

The potential of the sea to support growth of living organisms is determined by the fertilizer elements that marine plants need for growth. Fertilizers, or inorganic nutrients as they are called in oceanography, are required only by the first trophic level in the food web, the primary producers; but the supply of inorganic nutrients is a fertility-regulating process whose effect reaches throughout the food web. When there is an abundant supply to the surface layer of the ocean that is taken up by marine plants and converted into organic matter through photosynthesis, the entire food web is enriched, including zooplankton, fish, birds, whales, benthic invertebrates, protozoa, and bacteria. *See* DEEP-SEA FAUNA; FOOD WEB; MARINE FISHERIES.

The elements needed by marine plants for growth are divided into two categories depending on the quantities required: The major nutrient elements that appear to determine variations in ocean fertility are nitrogen, phosphorus, and silicon. The micronutrients are elements required in extremely small, or trace, quantities including essential metals such as iron, manganese, zinc, cobalt, magnesium, and copper, as well as vitamins and specific organic growth factors such as chelators. Knowledge of the fertility consequences of variations in the distribution of micronutrients is incomplete, but consensus among oceanographers is that the overall pattern of ocean fertility is set by the major fertilizer elements—nitrogen, phosphorus, and silicon—and not by micronutrients.

Two types of marine plants carry out primary production in the ocean: microscopic planktonic algae collectively called phytoplankton, and benthic algae and sea grasses attached to hard and soft substrates in shallow coastal waters.

The benthic and planktonic primary producers are a diverse assemblage of plants adapted to exploit a wide variety of marine niches; however, they have in common two basic requirements for the photosynthetic production of new organic matter: light energy and the essential elements of carbon, hydrogen, nitrogen, oxygen, phosphorus, sulfur, and silicon for the synthesis of new organic molecules. These two requirements are the first-order determinants of photosynthetic growth for all marine plants and, hence, for primary productivity everywhere in the ocean.

The regions of the world's oceans differ dramatically in overall fertility. In the richest areas, the water is brown with diatom blooms, fish schools are abundant, birds darken the horizon, and the sediments are fine-grained black mud with a high organic content. In areas of low fertility, the water is blue and clear, fish are rare, and the bottom sediments are well-oxidized carbonate or clay. These extremes exist because the overall pattern of fertility is determined by the processes that transport nutrients to the sunlit upper layer of the ocean where there is energy for photosynthesis. *See* SEAWATER.

[R.T.B.]

Sedimentary rocks Rocks that accumulate at the surface of the Earth, under ambient temperatures. Together with extruded hot lavas, sedimentary rocks form a thin cover of stratified material (the stratisphere) over the deep-seated igneous and metamorphic rocks that constitute the bulk of the Earth's crust. Sediments cover about three-quarters of the land and of the ocean floor. The thickness of the stratisphere is generally measured in kilometers, and locally reaches about 15 km (50,000 ft). *See* EARTH CRUST.

Most sediments accumulate as sand and dust or mud. Being deposited from fluids (air, water) under the influence of gravity, they tend to assume level surfaces (though locally steep slopes may be developed, as in dunes and reefs). Changes in supply of sediment and in depositing agencies change the nature of the deposits from day to day and from millennium to millennium, and commonly interrupt the process altogether. As a result, the accumulated mantle of sediment has a layered structure, divided into beds or strata. Sediments become compacted as waters are squeezed out of them during burial and tectonism, and become cemented as remaining pore space becomes filled by newly growing minerals, mainly calcite or quartz. Bacterial degradation of organic matter, invasion by other fluids, and changes in temperature continue to alter the chemical environment, and lead to alteration of unstable mineral phases. Such processes are included in the term diagenesis. Soft sediment thus becomes converted to rock, but the geologist includes both in the concept of sedimentary rocks.

When sediments are carried to greater depths or are otherwise subjected to high heat or pressure, growth of new minerals and plastic deformation destroy sedimentary structures and metamorphose the rock. Alternatively, the sediment melts in transition to igneous rock. Thus, sedimentary rocks are recycled through geologic time. Most of the crust under the continents, consisting of igneous and metamorphic rocks, has probably passed through the sedimentary state at some point. Despite such losses, sedimentary rocks have locally survived from very early (Archean) times, nearly 4 billion years ago.

Sediments are almost entirely derived from transfer of materials within the Earth's crust. First in importance is gradation, the wearing away of the highlands and the deposition of the products in the low spots: subsiding basins and the oceans. Second is crustal volcanism, which produces large ash falls from explosive volcanoes, and recycles ions to the surface in hot springs. Small amounts are contributed from the mantle underlying the crust: mainly pumice produced when mantle-derived oceanic basalts interact with water. A small fraction of sediment consists of organic matter created by organisms from carbon dioxide and water. Water frozen in the atmosphere transiently covers parts of the stratisphere with ice, while traces of extraterrestrial matter continue to be added from meteorites. *See* WEATHERING PROCESSES.

Though sediments contain such a large range of diverse constituents occurring in a wide variety of mixtures, such mixtures are generally dominated by one or two constituents, and thus may be grouped into a number of classes, each of which can be divided into families.

Detrital sediments are alternately transported and deposited, reeroded, and redeposited on their way to a more permanent resting place, so that their constituents may carry the imprints of a complex history, while the structure of the deposit testifies to the last depositional episode. *See* DEPOSITIONAL SYSTEMS AND ENVIRONMENTS.

Pyroclastic sediments originate from volcanic vents. Submarine eruptions form pumice, or frothy glass, much of which floats widely. The important contributions are great eruptions of glass droplets are ejected into atmosphere and stratosphere to fall as a rain of pumice, sand, and silt, in some cases mixed with crystals. Pyroclastic rocks, largely composed of glass, are readily altered to clay minerals (montmorillonite) in weathering. They produce excellent soils. Beds of montmorillonite (bentonites) are mined for preparation of artificial muds such as those used in well drilling. *See* VOLCANO; VOLCANOLOGY.

Chemical sediments represent the precipitation of materials carried in solution, either by simple chemical precipitation or by the activity of organisms.

Carbonate rocks form about 20% of all sediments. In natural waters, calcium and magnesium are mainly held in solution by virtue of carbon dioxide. In many fresh waters and in the surficial ocean, withdrawal of carbon dioxide—by warming of the water or by the consumption of carbon dioxide in green-plant photosynthesis—leads to supersaturation and to the deposition of calcium carbonate. This normally yields a lime mud of microscopic crystals. Even more important is the secretion of calcium carbonate skeletons, ultimately deposited on ocean floors, by some algae and by a large variety of animals, ranging from microscopic foraminifera to corals and molluscan shells. Carbonate rocks are a major ingredient of portland cement. They are crushed in large quantities for use in road building, agriculture, and smelting, and in the chemical industry. They also furnish building and ornamental stone. Carbonate rocks contain a large share of the world's petroleum resources.

Evaporites are formed in bays, estuaries, and lakes of arid regions. On progressive evaporation, seawater first forms deposits of calcium sulfate as gypsum or anhydrite, followed by halite (NaCl) and ultimately potash and magnesium salts. Evaporation of lake water may yield different precipitates such as trona, borax, and silicates. Much of what is sold as table salt is mined from evaporite deposits, as is potash fertilizer. Plaster of paris is produced from gypsum or anhydrite, and the chemical industry relies on evaporite deposits of various types. *See* FERTILIZER; SALINE EVAPORITES.

Nondetrital siliceous rocks such as silicon dioxide (silica) is second only to carbonate in the dissolved load of most streams. Organisms take up nearly all silica supplied, covering much of the deep-sea floor with radiolarian and diatomaceous ooze. Over geological time spans, diagenetic alteration converts these into dull white opal-ct or quartz porcellanites, or into the solid, waxy-looking mosaics of fine quartz grains known as chert or flint. Diatom ooze is mined for abrasives and filters, as well as for insulation.

Carbonaceous sediments are the result of organic activity, and are of two sorts: the peat-coal series and the kerogens. Peat is used for local fuel in boggy parts of the world. Lignite and bituminous coals continue to be important fuels. *See* PEAT. [A.G.F.]

Sedimentology The study of natural sediments, both lithified (sedimentary rocks) and unlithified, and of the processes by which they are formed. Sedimentology includes all those processes that give rise to sediment or modify it after deposition: weathering, which breaks up or dissolves preexisting rocks so that sediment may form from them; mechanical transportation; deposition; and diagenesis, which modifies sediment after deposition and burial within a sedimentary basin and converts it into sedimentary rock. Sediments deposited by mechanical processes (gravels, sands, muds) are known as clastic sediments, and those deposited predominantly by chemical or biological processes (limestones, dolomites, rock salt, chert) are known as chemical sediments. *See* SEDIMENTARY ROCKS; WEATHERING PROCESSES.

The raw materials of sedimentation are the products of weathering of previously formed igneous, metamorphic, or sedimentary rocks. In the present geological era, 66% of the continents and almost all of the ocean basins are covered by sedimentary rocks. Therefore, most of the sediment now forming has been derived by recycling previously formed sediment. Identification of the oldest rocks in the Earth's crust, formed more than 3×10^9 years ago, has shown that this process has been going on at least since then. Old sedimentary rocks tend to be eroded away or converted into metamorphic rocks, so that very ancient sedimentary rocks are seen at only a few places on Earth. *See* EARTH CRUST.

Major controls. The major controls on the sedimentary cycle are tectonics, climate, worldwide (eustatic) changes in sea level, the evolution of environments with geological time, and the effect of rare events.

Tectonics are the large-scale motions (both horizontal and vertical) of the Earth's crust. Tectonics are driven by forces within the interior of the Earth but have a large effect on sedimentation. These crustal movements largely determine which areas of the Earth's crust undergo uplift and erosion, thus acting as sources of sediment, and which areas undergo prolonged subsidence, thus acting as sedimentary basins. Rates of uplift may be very high (over 10 m or 33 ft per 1000 years) locally, but probably such rates prevail only for short periods of time. Over millions of years, uplift even in mountainous regions is about 1 m (3.3 ft) per 1000 years, and it is closely balanced by rates of erosion. Rates of erosion, estimated from measured rates of sediment transport in rivers and from various other techniques, range from a few meters per 1000 years in mountainous areas to a few millimeters per 1000 years averaged over entire continents. *See* BASIN.

Climate plays a secondary but important role in controlling the rate of weathering and sediment production. The more humid the climate, the higher these rates are. A combination of hot, humid climate and low relief permits extensive chemical weathering, so that a larger percentage of source rocks goes into solution, and the clastic sediment produced consists mainly of those minerals that are chemically inert (such as quartz) or that are produced by weathering itself (clays). Cold climates and high relief favor physical over chemical processes. *See* CLIMATOLOGY.

Tectonics and climate together control the relative level of the sea. In cold periods, water is stored as ice at the poles, which can produce a worldwide (eustatic) lowering of sea level by more than 100 m (330 ft). Changes in sea level, whether local or worldwide, strongly influence sedimentation in shallow seas and along coastlines; sea-level changes also affect sedimentation in rivers by changing the base level below which a stream cannot erode its bed.

One of the major conclusions from the study of ancient sediments has been that the general nature and rates of sedimentation have been essentially unchanged during the last billion years of geological history. However, this conclusion, uniformitarianism, must be qualified to take into account progressive changes in the Earth's environment through geological time, and the operation of rare but locally or even globally important catastrophic events. The most important progressive changes have been in tectonics and atmospheric chemistry early in Precambrian times, and in the nature of life on the Earth, particularly since the beginning of the Cambrian.

Throughout geological time, events that are rare by human standards but common on a geological time scale, such as earthquakes, volcanic eruptions, and storms, produced widespread sediment deposits. There is increasing evidence for a few truly rare but significant events, such as the rapid drying up of large seas (parts of the Mediterranean) and collisions between the Earth and large meteoric or cometary bodies (bolides).

Sediment is moved either by gravity acting on the sediment particles or by the motions of fluids (air, water, flowing ice), which are themselves produced by gravity. Deposition takes place when the rate of sediment movement decreases in the direction of sediment movement; deposition may be so abrupt that an entire moving mass of sediment and fluid comes to a halt (mass deposition, for example, by a debris flow), or so slow that the moving fluid (which may contain only a few parts per thousand of sediment) leaves only a few grains of sediment behind. The settling velocity depends on the density and viscosity of the fluid, as well as on the size, shape, and density of the grains. *See* STREAM TRANSPORT AND DEPOSITION.

Chemical sedimentation. Chemical weathering dissolves rock materials and delivers ions in solution to lakes and the ocean. The concentrations of ions in river and ocean water are quite different, showing that some ions must be removed by sedimentation. Comparison of the modern rate of delivery of ions to the ocean, with their concentration in the oceans, shows that some are removed very rapidly (residence times of only a few thousand years) whereas others, such as chlorine and sodium, are removed very slowly (residence times of hundreds of millions of years).

Biological effects. Many so-called chemical sediments are actually produced by biochemical action. Much is then reworked by waves and currents, so that the chemical sediment shows clastic textures and consists of grains rounded and sorted by transport. Depositional and diagenetic processes, however, are often strongly affected by organic action, no matter what the origin of the sediment. Plants in both terrestrial and marine environments tend to trap sediment, enhancing deposition and slowing erosion.

Sedimentary environments and facies. Sedimentary rocks preserve the main direct evidence about the nature of the surface environments of the ancient Earth and the way they have changed through geological time. Thus, besides trying to understand the basic principles of sedimentation, sedimentologists have studied modern and ancient sediments as records of ancient environments. For this purpose, fossils and primary sedimentary structures are the best guide. These structures are those formed at the time of deposition, as opposed to those formed after deposition by diagenesis, or by deformation. In describing sequences of sedimentary rocks in the field (stratigraphic sections), sedimentologists recognize compositional, structural, and organic aspects of rocks that can be used to distinguish one unit of rocks from another. Such units are known as sedimentary facies, and they can generally be interpreted as having formed in different environments of deposition. Though there are a large number of different sedimentary environments, they can be classified in a number of general classes, and their characteristic facies are known from studies of modern environments. *See* FACIES (GEOLOGY). [G.V.M.]

Seed A fertilized ovule containing an embryo which forms a new plant upon germination. Seed-bearing characterizes the higher plants—the gymnosperms (conifers and allies) and the angiosperms (flowering plants). Gymnosperm (naked) seeds arise on the surface of a structure, as on a seed scale of a pine cone. Angiosperm (covered) seeds develop within a fruit, as the peas in a pod. *See* FLOWER; FRUIT.

Structure. One or two tissue envelopes, or integuments, form the seed coat which encloses the seed except for a tiny pore, the micropyle (see illustration). The micropyle is near the funiculus (seed stalk) in angiosperm seeds. The hilum is the scar left when the seed is detached from the funiculus. Some seeds have a raphe, a ridge near the hilum opposite the micropyle, and a bulbous strophiole. Others such as nutmeg possess arils, outgrowths of the funiculus, or a fleshy caruncle developed from the seed coat near the hilum, as in the castor bean. The embryo consists of an axis and attached cotyledons (seed leaves). The part of the axis above the cotyledons is the epicotyl (plumule); that below, the hypocotyl, the lower end of which bears a more or less developed primordium of the root (radicle). The epicotyl, essentially a terminal bud, possesses an apical meristem (growing point) and, sometimes, leaf primordia. The seedling stem develops from the epicotyl. An apical meristem of the radicle produces the primary root of the seedling, and transition between root and stem occurs in the hypocotyl. *See* ROOT (BOTANY); STEM.

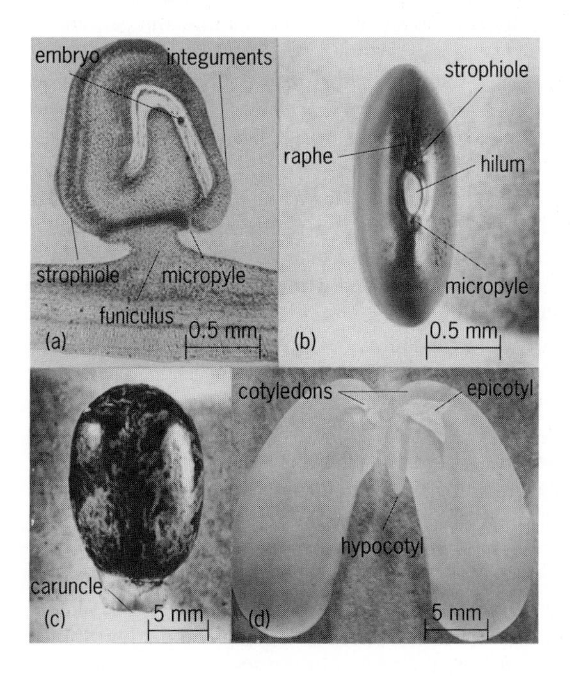

Seed structures. (*a*) Median longitudinal section of pea ovule shortly after fertilization, showing attachment to pod tissues. (*b*) Mature kidney bean. (*c*) Mature castor bean. (*d*) Opened embryo of mature kidney bean.

Two to many cotyledons occur in different gymnosperms. The angiosperms are divided into two major groups according to number of cotyledons: the monocotyledons and the dicotyledons. Mature gymnosperm seeds contain an endosperm (albumen or nutritive tissue) which surrounds the embryo. In some mature dicotyledon seeds the endosperm persists, the cotyledons are flat and leaflike, and the epicotyl is simply an apical meristem. In other seeds, such as the bean, the growing embryo absorbs the endosperm, and food reserve for germination is stored in fleshy cotyledons. The endosperm persists in common monocotyledons, for example, corn and wheat; and the cotyledon, known as the scutellum, functions as an absorbing organ during germination. Grain embryos also possess a coleoptile and a coleorhiza sheathing the epicotyl and the radicle, respectively. The apical meristems of lateral seed roots also may be differentiated in the embryonic axis near the scutellum of some grains.

Many so-called seeds consist of hardened parts of the fruit enclosing the true seed which has a thin, papery seed coat. Among these are the achenes, as in the sunflower, dandelion, and strawberry, and the pits of stone fruits such as the cherry, peach, and raspberry. Many common nuts also have this structure. Mechanisms for seed dispersal include parts of both fruit and seed. *See* POPULATION DISPERSAL.

Economic importance. Propagation of plants by seed and technological use of seed and seed products are among the most important activities of modern society. Specializations of seed structure and composition provide rich sources for industrial exploitation apart from direct use as food. Common products include starches and glutens from grains, hemicelluloses from guar and locust beans, and proteins and oils from soybeans and cotton seed. Drugs, enzymes, vitamins, spices, and condiments are obtained from embryos, endosperms, and entire seeds, often including the fruit coat. Most of the oils of palm, olive, and pine seeds are in the endosperm. Safflower seed oil is obtained mainly from the embryo, whereas both the seed coat and embryo of cotton seed are rich in oils. *See* PLANT ANATOMY; REPRODUCTION (PLANT). [R.M.R.]

Physiology. Physical and biochemical processes of seed growth and germination are controlled by genetic and environmental factors. Conditions of light, temperature, moisture, and oxygen affect the timing and ability of a seed to mature and germinate. Seed development (embryogenesis) is concerned with the synthesis and storage of carbohydrate, protein, and oil to supply nutrients to the germinating seedling prior to soil emergence. Seed development occurs in several stages: rapid cell division, seed fill, and desiccation. The timing of each stage is species-specific and environmentally influenced.

Dormancy. Seed dormancy is the inability of a living seed to germinate under favorable conditions of temperature, moisture, and oxygen. Dormancy does not occur in all seeds, but typically occurs in plant species from temperate and colder habitats. This process allows for a delay in seed germination until environmental conditions are adequate for seedling survival. At least three types of seed dormancy are recognized: primary, secondary (induced), and enforced. Primary dormancy occurs during seed maturation, and the seed does not germinate readily upon being shed. Secondary and enforced dormancy occur after the seed is shed and may be caused by adverse environmental factors such as high or low temperature, absence of oxygen or light, low soil moisture, and presence of chemical inhibitors. Seeds with secondary dormancy will not germinate spontaneously when environmental conditions improve, and need additional environmental stimuli. Seeds with enforced dormancy germinate readily upon removal of the environmental limitation. Regulation of dormancy may be partly controlled by hormones. *See* DORMANCY.

Dormancy is terminated in a large number of species when an imbibed seed is illuminated with white light. Biochemical control of this process is related to the functioning of a single pigment, phytochrome, frequently located in the seed coat or embryonic axis. Phytochrome imparts to the seed the ability to interpret light quality, such as that under an existing vegetative canopy, and to distinguish light from dark with respect to its position in the soil. Phytochrome also is affected by temperature and is involved in the seasonal control of the ending of dormancy. Hormones that promote germination of dormant seeds include gibberellins, cytokinins, ethylene, and auxins.

Germination. Germination is the process whereby a viable seed takes up water and the radicle (primary root) or hypocotyl emerges from the seed under species-specific conditions of moisture, oxygen, and temperature. Dormant seeds must undergo additional environmental stimuli to germinate. The germinating seed undergoes cell expansion, as well as increases in respiration, protein synthesis, and other metabolic activities prior to emergence of the growing seedling. [C.A.L.]

Seiche A short-period oscillation in an enclosed or semienclosed body of water, analogous to the free oscillation of water in a dish. The initial displacement of water from a level surface can arise from a variety of causes, and the restoring force is gravity, which always tends to maintain a level surface. Once formed, the oscillations are characteristic only of the geometry of the basin itself and may persist for many cycles before decaying under the influence of friction. The term "seiche" appears to have been first used to describe the rhythmic oscillation of the water surface in Lake Geneva, which occasionally exposed large areas of the lake bed that are normally submerged.

Seiches can be generated when the water is subject to changes in wind or atmospheric pressure gradients or, in the case of semi-enclosed basins, by the oscillation of adjacent connected water bodies having a periodicity close to that of the seiche or of one of its harmonics. Other, less frequent causes of seiches include heavy precipitation over a portion of the lake, flood discharge from rivers, seismic disturbances, submarine mudslides or slumps, and tides. The most dramatic seiches have been observed after earthquakes. [A.Wu.; D.M.F.]

Seismic stratigraphy Determination of the nature of sedimentary rocks and their fluid content from analysis of seismic data. Seismic stratigraphy is divided into seismic-sequence (facies) analysis and reflection-character analysis.

In seismic-sequence analysis the first step is to separate seismic-sequence units, also called seismic-facies units. This is usually done by mapping unconformities where they are shown by angularity. Angularity below an unconformity may be produced by erosion at an angle across the former bedding surfaces or by toplap (offlap), and angularity above an unconformity may be produced by onlap or downlap, the latter distinction being based on geometry. The unconformities are then followed along reflections from the points where they cannot be so identified, advantage being taken of the fact that the unconformity reflection is often relatively strong. The procedure often followed is to mark angularities in reflections by small arrows before drawing in the boundaries.

Seismic-facies units are three-dimensional, and many of the conclusions from them are based on their three-dimensional shape. The appearance on seismic lines in the dip and strike directions is often very different. For example, a fan-shaped unit might show a progradational pattern in the dip direction and discontinuous, overlapping arcuate reflections in the strike direction. *See* FACIES (GEOLOGY).

Reflection-character analysis may be based on information from boreholes which suggests that a particular interval may change nearby in a manner which increases its likelihood to contain hydrocarbon accumulations. Lateral changes in the wave shape of individual reflection events may suggest where the stratigraphic changes or hydrocarbon accumulations may be located.

Where sufficient information is available to develop a reliable model, expected changes are postulated and their effects are calculated and compared to observed seismic data. The procedure is called synthetic seismogram manufacture; it usually involves calculating seismic data based on sonic and density logs from boreholes, sometimes based on a model derived in some other way. The sonic and density data are then changed in the manner expected for a postulated stratigraphic change, and if the synthetic seismogram matches the actual seismic data sufficiently well, it implies that the changes in earth layering are similar to those in the model. [R.E.Sh.]

Sendai virus A member of the viruses in the type species Parainfluenza 1, genus *Paramyxovirus*, family Paramyxoviridae; it is also called hemagglutinating virus of Japan (HVJ). Sendai virus was originally recovered in Sendai, Japan, from mice inoculated with autopsy specimens from newborns who died of fatal pneumonitis in an epidemic in 1952. Subsequent attempts to isolate this virus from humans were, however, mostly unsuccessful, although mice are commonly infected with Sendai virus along with rats, guinea pigs, hamsters, and pigs. It is believed that the natural host of Sendai virus is the mouse and that the virus is usually nonpathogenic for humans. *See* ANIMAL VIRUS; PARAINFLUENZA VIRUS. (N.I.)

Septic tank A single-story, watertight, on-site treatment system for domestic sewage, consisting of one or more compartments, in which the sanitary flow is detained to permit concurrent sedimentation and sludge digestion. The septic tank is constructed of materials not subject to decay, corrosion, or decomposition, such as precast concrete, reinforced concrete, concrete block, or reinforced resin and fiberglass. The tank must be structurally capable of supporting imposed soil and liquid loads. Septic tanks are used primarily for individual residences, isolated institutions, and commercial complexes such as schools, prisons, malls, fairgrounds, summer theaters, parks, or recreational facilities. Septic tanks have limited use in urban areas where sewers and municipal treatment plants exist.

Septic tanks do not treat sewage; they merely remove some solids and condition the sanitary flow so that it can be safely disposed of to a subsurface facility such as a tile field, leaching pools, or buried sand filter. The organic solids retained in the tank undergo a process of liquefaction and anaerobic decomposition by bacterial organisms. The clarified septic tank effluent is highly odorous, contains finely divided solids, and may contain enteric pathogenic organisms. The small amounts of gases produced by the anaerobic bacterial action are usually vented and dispersed to the atmosphere without noticeable odor or ill effects. *See* SEWAGE; SEWAGE TREATMENT. [G.Pa.]

Sequoia The giant sequoia or big tree (*Sequoia gigantea*) occupies a limited area in California and is said to be the oldest and most massive of all living things. The leaves are evergreen, scalelike, and overlapping on the branches. In height sequoia is a close second to the redwood (300–330 ft or 90–100 m) but the trunk is more massive. Sequoia trees may be 27–30 ft (8–9 m) in diameter 10 ft (3 m) from the ground. The stump of one tree showed 3400 annual rings. The red-brown bark is 1–2 ft (0.3–0.6 m) thick and spongy. Vertical grooves in the trunk give it a fluted appearance. The heartwood is dull purplish-brown and lighter and more brittle than that of the redwood. The wood and bark contain much tannin, which is probably the cause of the great resistance to insect and fungus attack. The most magnificent trees are within the General Grant and Sequoia National Parks. *See* REDWOOD. [A.H.G./K.P.D.]

Sewage Water-carried wastes, in either solution or suspension, that flow away from a community. Also known as wastewater flows, sewage is the used water supply of the community. It is more than 99.9% pure water and is characterized by its volume or rate of flow, its physical condition, its chemical constituents, and the bacteriological organisms that it contains. Depending on their origin, wastewaters can be classed as sanitary, commercial, industrial, or surface runoff.

The spent water from residences and institutions, carrying body wastes, ablution water, food preparation wastes, laundry wastes, and other waste products of normal living, are classed as domestic or sanitary sewage. Liquid-carried wastes from stores and service establishments serving the immediate community, termed commercial wastes, are included in the sanitary or domestic sewage category if their characteristics are similar to household flows. Wastes that result from an industrial process or the production or manufacture of goods are classed as industrial wastes. Their flows and strengths are usually more varied, intense, and concentrated than those of sanitary sewage. Surface runoff, also known as storm flow or overland flow, is that portion of precipitation that runs rapidly over the ground surface to a defined channel. Precipitation absorbs gases and particulates from the atmosphere, dissolves and leaches materials from vegetation and soil, suspends matter from the land, washes spills and debris from urban streets and highways, and carries all these pollutants as wastes in its flow to a collection point. Discharges are classified as point-source when they emanate from a pipe outfall, or non-point-source when they are diffused and come from agriculture or unchanneled urban land drainage runoff. *See* HYDROLOGY; PRECIPITATION (METEOROLOGY).

Wastewaters from all of these sources may carry pathogenic organisms that can transmit disease to humans and other animals; contain organic matter that can cause odor and nuisance problems; hold nutrients that may cause eutrophication of receiving water bodies; and may contain hazardous or toxic materials. Proper collection and safe, nuisance-free disposal of the liquid wastes of a community are legally recognized as a necessity in an urbanized, industrialized society. *See* PUBLIC HEALTH; SEWAGE TREATMENT; TOXICOLOGY. [G.Pa.]

Sewage treatment Unit processes used to separate, modify, remove, and destroy objectionable, hazardous, and pathogenic substances carried by wastewater in solution or suspension in order to render the water fit and safe for intended uses. Treatment removes unwanted constituents without affecting or altering the water molecules themselves, so that wastewater containing contaminants can be converted to safe drinking water. Stringent water quality and effluent standards have been developed that require reduction of suspended solids (turbidity), biochemical oxygen demand (related to degradable organics), and coliform organisms (indicators of fecal pollution); control of pH as well as the concentration of certain organic chemicals and heavy metals; and use of bioassays to guarantee safety of treated discharges to the environment.

In all cases, the impurities, contaminants, and solids removed from all wastewater treatment processes must ultimately be collected, handled, and disposed of safely, without damage to humans or the environment.

Treatment processes are chosen on the basis of composition, characteristics, and concentration of materials present in solution or suspension. The processes are classified as pretreatment, preliminary, primary, secondary, or tertiary treatment, depending on type, sequence, and method of removal of the harmful and unacceptable constituents. Pretreatment processes equalize flows and loadings, and precondition wastewaters to neutralize or remove toxics and industrial wastes that could adversely affect sewers or inhibit operations of publicly owned treatment works. Preliminary treatment processes protect plant mechanical equipment; remove extraneous matter such as grit, trash, and debris; reduce odors; and render incoming sewage more amenable to subsequent treatment and handling. Primary treatment employs mechanical and physical unit processes to separate and remove floatables and suspended solids and to prepare wastewater for biological treatment. Secondary treatment utilizes aerobic microorganisms in biological reactors to feed on dissolved and colloidal organic matter. As these microorganisms reduce biochemical oxygen demand and turbidity (suspended solids), they grow, multiply, and form an organic floc, which must be captured and removed in final settling tanks. Tertiary treatment, or advanced treatment, removes specific residual substances, trace organic materials, nutrients, and other constituents that are not removed by biological processes. Most advanced wastewater treatment systems include denitrification and ammonia stripping, carbon adsorption of trace organics, and chemical precipitation. Evaporation, distillation, electrodialysis, ultrafiltration, reverse osmosis, freeze drying, freeze-thaw, floatation, and land application, with particular emphasis on the increased use of natural and constructed wetlands, are being studied and utilized as methods for advanced wastewater treatment to improve the quality of the treated discharge to reduce unwanted effects on the receiving environment. *See* SEWAGE; WETLANDS.

On-site sewage treatment for individual homes or small institutions uses septic tanks, which provide separation of solids in a closed, buried unit. Effluent is discharged to subsurface absorption systems. *See* SEPTIC TANK; WATER TREATMENT. [G.Pa.]

Sexually transmitted diseases Infections that are acquired and transmitted by sexual contact. Although virtually any infection may be transmitted during intimate contact, the term sexually transmitted disease is restricted to conditions that are largely dependent on sexual contact for their transmission and propagation in a population. The term venereal disease is literally synonymous with sexually transmitted disease but traditionally is associated with only five long-recognized diseases (syphilis, gonorrhea, chancroid, lymphogranuloma venereum, and donovanosis). Sexually transmitted diseases occasionally are acquired nonsexually (for example, by newborn infants from their mothers, or by clinical or laboratory personnel handling pathogenic organisms or infected secretions), but in adults they are virtually never acquired by contact with contaminated intermediaries such as towels, toilet seats, or bathing facilities. However, some sexually transmitted infections (such as human immunodeficiency virus infection, viral hepatitis, and cytomegalovirus infection) are transmitted primarily by sexual contact in some settings and by nonsexual means in others. *See* GONORRHEA; SYPHILIS.

The sexually transmitted diseases may be classified in the traditional fashion, according to the causative pathogenic organisms, as follows:

- Bacteria
  - *Chlamydia trachomatis*
  - *Neisseria gonorrhoeae*
  - *Treponema pallidum*
  - *Mycoplasma genitalium*
  - *Mycoplasma hominis*
  - *Ureaplasma urealyticum*
  - *Haemophilis ducreyi*
  - *Calymmatobacterium granulomatis*
  - *Salmonella* species
  - *Shigella* species
  - *Campylobacter* species
- Viruses
  - Human immunodeficiency viruses (types 1 and 2)
  - Herpes simplex viruses (types 1 and 2)
  - Hepatitis viruses B, C, D
  - Cytomegalovirus
  - Human papillomaviruses
  - Molluscum contagiosum virus
  - Kaposi sarcoma virus
- Protozoa
  - *Trichomonas vaginalis*
  - *Entamoeba histolytica*
  - *Giardia lamblia*
  - *Cryptosporidium* and related species
- Ectoparasites
  - *Phthirus pubis* (pubic louse)
  - *Sarcoptes scabiei* (scabies mite)

Sexually transmitted diseases may also be classified according to clinical syndromes and complications that are caused by one or more pathogens as follows:

1. Acquired immunodeficiency syndrome (AIDS) and related conditions
2. Pelvic inflammatory disease
3. Female infertility
4. Ectopic pregnancy
5. Fetal and neonatal infections
6. Complications of pregnancy
7. Neoplasia
8. Human papillomavirus and genital warts
9. Genital ulcer-inguinal lymphadenopathy syndromes
10. Lower genital tract infection in women
11. Viral hepatitis and cirrhosis
12. Urethritis in men
13. Late syphilis
14. Epididymitis
15. Gastrointestinal infections
16. Acute arthritis
17. Mononucleosis syndromes
18. Molluscum contagiosum
19. Ectoparasite infestation

See ACQUIRED IMMUNE DEFICIENCY SYNDROME (AIDS); CANCER (MEDICINE); DRUG RESISTANCE; HEPATITIS; PUBLIC HEALTH.

Most of these syndromes may be caused by more than one organism, often in conjunction with nonsexually transmitted pathogens. They are listed in the approximate order of their public health impact. [H.H.Ha.]

Shipping fever A severe inflammation of the lungs (pneumonia) commonly seen in North American cattle after experiencing the stress of transport. This disease occurs mainly in 6–9-month-old beef calves transported to feedlots. The characteristic pneumonia is caused primarily by the bacterium *Pasteurella haemolytica* serotype A1; thus a synonym for shipping fever is bovine pneumonic pasteurellosis.

Pasteurella haemolytica replicates rapidly in the upper respiratory tract and is inhaled into the lungs, where pneumonia develops in the deepest region of the lower respiratory tract (the pulmonary alveoli). Viruses can also concurrently damage pulmonary alveoli and enhance the bacterial pneumonia.

Symptoms initially include reduced appetite, high fever, rapid and shallow respiration, depression, and a moist cough. During the later stages of disease, cattle lose weight and have labored breathing. The lesion causing these symptoms is a severe pneumonia accompanied by inflammation of the lining of the chest cavity, and is called a fibrinous pleuropneumonia. Without vigorous treatment, shipping fever can cause death in 5–30% of affected cattle.

Treatment is aimed at eliminating bacteria by using antibiotics, reducing inflammation by using nonsteroidal anti-inflammatory drugs, and separating affected cattle. Vaccines can stimulate immunity to the viruses and bacteria associated with shipping fever. *See* ANIMAL VIRUS; PNEUMONIA. [A.W.Co.]

Siderophores Low-molecular-mass molecules that have a high specificity for chelating or binding iron. Siderophores are produced by many microorganisms, including bacteria, yeast, and fungi, to obtain iron from the environment. More than 500 different siderophores have been identified from microorganisms. Some bacteria produce more than one type of siderophore. *See* BACTERIA; FUNGI; YEAST.

Iron is required by aerobic bacteria and other living organisms for a variety of biochemical reactions in the cell. Although iron is the fourth most abundant element in the Earth's crust, it is not readily available to bacteria. Iron is found in nature mostly as insoluble precipitates that are part of hydroxide polymers. Bacteria living in the soil or water must have a mechanism to solubilize iron from these precipitates in order to assimilate iron from the environment. Iron is also not freely available in humans and other mammals. Most iron is found intracellularly in heme proteins and ferritin, an iron storage compound. Iron outside cells is tightly bound to proteins. Therefore, bacteria that grow in humans or other animals and cause infections must have a mechanism to remove iron from these proteins and use it for their own energy and growth needs. Siderophores have a very high affinity for iron and are able to solubilize and transport ferric iron (Fe^{3+}) in the environment and also to compete for iron with mammalian proteins such as transferrin and lactoferrin. The majority of bacteria and fungi use siderophores to solubilize and transport iron. Microorganisms can use either siderophores produced by themselves or siderophores produced by other microorganisms.

The many different types of siderophores can generally be classified into two structural groups, hydroxamates and catecholate compounds. Despite their structural differences, all form an octahedral complex with six binding coordinates for Fe^{3+}.

Siderophores have potential applications in the treatment of some human diseases and infections. Some siderophores are used therapeutically to treat chronic or acute iron overload conditions in order to prevent iron toxicity in humans. Individuals who have defects in blood cell production or who receive multiple transfusions can sometimes have too much free iron in the body. However, in order to prevent infection during treatment for iron overload, it is important to use siderophores that cannot be used by bacterial pathogens.

A second clinical application of siderophores is in antibiotic delivery to bacteria. Some gram-negative bacteria are resistant to antibiotics because they are too big to diffuse through the outer-membrane porins. However, siderophore-antibiotic combination compounds have been synthesized that can be transported into the cell using the siderophore receptor. *See* DRUG RESISTANCE. [P.A.So.]

Siliceous sediment Fine-grained sediment and sedimentary rock dominantly composed of the microscopic remains of the unicellular, silica-secreting plankton diatoms and radiolarians. Minor constituents include extremely small shards of sponge spicules and other microorganisms such as silicoflagellates. Siliceous sedimentary rock sequences are often highly porous and can form excellent petroleum source and reservoir rocks. *See* SEDIMENTARY ROCKS.

Given their biologic composition, siliceous sediments provide some of the best geologic records of the ancient oceans. Diatoms did not evolve until the late Mesozoic; thus the majority of siliceous rocks older than approximately 150 million years are formed by radiolarians. Geologists map the distribution of ancient siliceous sediments now pushed up onto land by plate tectonic processes, and can thus determine which portions of the ancient seas were biologically productive; this knowledge in turn can give great insight into regions of the Earth's crust that may be economically productive (for example oil-containing regions). The vast oil reserves of coastal California are predominantly found in the Monterey Formation, a highly porous diatomaceous siliceous sedimentary sequence distributed along the western seaboard of the United States. [R.W.Mu.]

Silviculture The theory and practice of controlling the establishment, composition, and growth of stands of trees for any of the goods (including timber, pulp, energy, fruits, and fodder) and benefits (water, wildlife habitat, microclimate amelioration, and carbon sequestration) that they may be called upon to produce. In practicing silviculture, the forester draws upon knowledge of all natural factors that affect trees growing upon a particular site, and guides the development of the vegetation, which is either essentially natural or only slightly domesticated, to best meet the demands of society in general and ownership in particular. Based on the principles of forest ecology and ecosystem management, silviculture is more the imitation of natural processes of forest growth and development than a substitution for them.

The spatial patterns in which old trees are removed and the species that replace them determine the structure and developmental processes of the new stands. If all the trees are replaced at once with a single species, the result is so-called pure stand or monoculture in which all of the trees form a single canopy of foliage that is lifted ever higher as the stand develops. If several species start together from seeds, from small trees already present, or from sprouts, as a single cohort, the different species usually tend to grow at different rates in height. Some are adapted to develop in the shade of their sun-loving neighbors. The result is a stratified mixture. Such stands grow best on soils or in climates, such as in tropical moist forests, where water is not a limiting factor and the vegetation collectively uses most of the photosynthetically active light. If trees are replaced in patches or strips, the result is an uneven-aged stand which may be of one species or many.

These different spatial and temporal patterns of stand structure are created by different methods of reproduction. The simplest is the clear-cutting method, in which virtually all of the vegetation is removed. Although it is sometimes possible to rely on adjacent uncut stands as sources of seed, it is usually necessary to reestablish the new stand by artificial seeding or planting after clear-cutting. The seed tree method differs only in that a limited number of trees are temporarily left on the area to provide seed.

In the shelterwood method, enough trees are left on the cutting area to reduce the degree of exposure significantly and to provide a substantial source of seed. In this method the growth of a major portion of the preexisting crop continues, and the old trees are not entirely removed until the new stand is well established. The three methods just described lead to the creation of essentially even-aged stands.

The choice of methods of regeneration cutting depends on the ecological status of the species and stands desired. Species that characterize the early stages of succession, the so-called pioneers, will endure, and usually require the kind of exposure to sunlight resulting from heavy cutting or, in nature, severe fire, catastrophic windstorms, floods, and landslides. These pioneer species usually grow rapidly in youth but are short-lived and seldom attain large size. The species that attain greatest age and largest size are ordinarily those which are intermediate in successional position and tolerance of shade. Some of them will become established after the severe exposure of clear-cutting, but they often start best with light initial shade such as that created by shelterwood cutting. Their longevity and large size result from the fact that they are naturally adapted to reproduce after disturbances occurring at relatively long intervals. The shade-tolerant species representing late or climax stages in the succession are adapted to reestablish themselves in their own shade. These species represent natural adaptations to the kinds of fatal disturbance caused by insects, disease, and atmospheric agencies rather than to the more complete disturbance caused by fire. *See* ECOLOGICAL SUCCESSION; FOREST ECOSYSTEM.

Much silvicultural practice is aimed at the creation and maintenance of pure, even-aged stands of single species. This approach is analogous to that of agriculture and simplifies administration, harvesting, and other operations. The analogy is often carried to the extent of clear-cutting and planting of large tracts with intensive site preparation, especially with species representative of the early or intermediate stages of succession. Mixed stands are more difficult to handle from the operational standpoint but are more resistant to injury from insects, fungi, and other damaging agencies which usually tend to attack single species. They are also more attractive than pure stands and make more complete use of the light, water, and nutrients available on the site. They usually do not develop unless these site factors are comparatively favorable; where soil moisture or some other factor is limiting, it may be possible for only a single well-adapted species to grow. If the site is highly favorable, as in tropical rainforest, river floodplains, or moist ravines, it is so difficult to maintain pure stands that mixed stands are inevitable.

The application of silviculture involves a number of accessory practices other than cutting. In localities of high fire risks, it may be desirable to burn the slash (logging debris) after cutting. Not only does this reduce the potential fuel, but it may also help the establishment of seedlings by baring the mineral soil or reducing the physical barrier represented by the slash. Slash disposal is most often necessary where the cutting has been very heavy or where the climate is so cold or dry that decay is slow. Deliberate prescribed burning of the litter beneath existing stands of fire-resistant species is sometimes carried out even in the absence of cutting to reduce the fuel for wild fires, to kill undesirable understory species, to enhance the production of forage for wild and domestic animals, and to improve seedbed conditions. *See* FOREST FIRE.

Integrated schedules of treatment for stands are called silvicultural systems. They cover both intermediate and reproduction treatments but are classified and named in terms of the general method of reproduction cutting contemplated. Such programs are evolved for particular situations and kinds of stands with due regard for all the significant biological and economic considerations involved. These considerations include the desired uses of the land, kinds of products and services sought, prospective costs and returns of the enterprise presented by management of the stand, funds available for long-term investment in stand treatments, harvesting techniques and equipment employed, reduction of losses from damaging agencies, and the natural requirements that must be met in reproducing the stand and fostering its growth. *See* FOREST AND FORESTRY; FOREST MANAGEMENT. [D.M.S.; M.S.A.]

Sirocco A southerly or southeasterly wind current from the Sahara or from the deserts of Saudi Arabia which occurs in advance of cyclones moving eastward through the Mediterranean Sea. The sirocco is most pronounced in the spring, when the deserts are hot and the Mediterranean cyclones are vigorous. It is observed along the southern and eastern coasts of the Mediterranean Sea from Morocco to Syria as a hot, dry wind capable of carrying sand and dust great distances from the desert source. The sirocco is cooled and moistened in crossing the Mediterranean and produces an oppressive, muggy atmosphere when it extends to the southern coast of Europe. *See* AIR MASS; WIND. [F.S.]

Smallpox An acute infectious viral disease characterized by severe systemic involvement and a single crop of skin lesions that proceeds through macular, papular, vesicular, and pustular stages. Smallpox is caused by variola virus, a brick-shaped, deoxyribonucleic acid-containing member of the Poxviridae family. Strains of variola virus are indistinguishable antigenically, but have differed in the clinical severity of the disease caused. Following a 13-year worldwide campaign coordinated by the World Health Organization (WHO), smallpox was declared eradicated by the World Health Assembly in May 1980. Smallpox is the first human disease to be eradicated.

Humans were the only reservoir and vector of smallpox. The disease was spread by transfer of the virus in respiratory droplets during face-to-face contact. Before vaccination, persons of all ages were susceptible. It was a winter-spring disease; there was a peak incidence in the drier spring months in the Southern Hemisphere and in the winter months in temperate climates. The spread of smallpox was relatively slow. The incubation period was an average of 10–12 days, with a range of 7–17 days. Fifteen to forty percent of susceptible persons in close contact with an infected individual developed the disease.

There were two main clinically distinct forms of smallpox, variola major and variola minor. Variola major, prominent in Asia and west Africa, was the more severe form, with widespread lesions and case fatality rates of 15–25% in unvaccinated persons, exceeding 40% in children under 1 year. From the early 1960s to 1977, variola minor was prevalent in South America and south and east Africa; manifestations were milder, with a case fatality rate of less than 1%.

There is no specific treatment for the diseases caused by poxviruses. Supportive care for smallpox often included the systemic use of penicillins to minimize secondary bacterial infection of the skin. When lesions occurred on the cornea, an antiviral agent (idoxuridine) was advised.

Edward Jenner, a British general medical practitioner who used cowpox to prevent smallpox in 1796, is credited with the discovery of smallpox vaccine (vaccinia virus). However, the global smallpox eradication program did not rely only on vaccination. Although the strategy for eradication first followed a mass vaccination approach, experience showed that intensive efforts to identify areas of epidemiologic importance, to detect outbreaks and cases, and to contain them would have the greatest effect on interrupting transmission. In 1978, WHO established an International Commission to confirm the absence of smallpox worldwide. The recommendations made by the commission included abandoning routine vaccination except for laboratory workers at special risk. *See* ANIMAL VIRUS. [J.Bre.]

Smog The noxious mixture of gases and particles commonly associated with air pollution in urban areas. Harold Antoine des Voeux is credited with coining the term in 1905 to describe the air pollution in British towns. *See* AIR POLLUTION.

The constituents of smog affect the human cardio-respiratory system and pose a health threat. Individuals exposed to smog can experience acute symptoms ranging from eye irritation and shortness of breath to serious asthmatic attacks. Under extreme conditions, smog can cause mortality, especially in the case of the infirm and elderly. Smog can also harm vegetation and likely leads to significant losses in the yields from forests and agricultural crops in affected areas.

The only characteristic of smog that is readily apparent to the unaided observer is the low visibility or haziness that it produces, due to tiny particles suspended within the smog. Observation of the more insidious properties of smog—the concentrations of toxic constituents—requires sensitive analytical instrumentation. Technological advances in these types of instruments, along with the advent of high-speed computers to simulate smog formation, have led to an increasing understanding of smog and its causes.

Smog is an episodic phenomenon because specific meteorological conditions are required for it to accumulate near the ground. These conditions include calm or stagnant winds which limit the horizontal transport of the pollutants from their sources, and a temperature inversion which prevents vertical mixing of the pollutants from the boundary layer into the free troposphere. *See* METEOROLOGY; STRATOSPHERE; TROPOSPHERE.

Smog can be classified into three types: London smog, photochemical smog, and smog from biomass burning.

London smog arises from the by-products of coal burning. These by-products include soot particles and sulfur oxides. During cool damp periods (often in the winter), the soot and sulfur oxides can combine with fog droplets to form a dark acidic fog. As nations switch from coal to cleaner-burning fossil fuels such as oil and gas as well as alternate energy sources such as hydroelectric and nuclear, London smogs cease. *See* ACID RAIN.

Photochemical smog is a more of a haze than a fog and is produced by chemical reactions in the atmosphere that are triggered by sunlight. A. J. Hagen-Smit first unraveled the chemical mechanism that produces photochemical smog. He irradiated mixtures of volatile organic compounds (VOC) and nitrogen oxides (NO_x) in a reaction chamber. After a few hours, Hagen-Smit observed the appearance of cracks in rubber bands stretched across the chamber. Knowing that ozone (O_3) can harden and crack rubber, Hagen-Smit correctly reasoned that photochemical smog was caused by photochemical reactions involving VOC and NO_x, and that one of the major oxidants produced in this smog was O_3. *See* ATMOSPHERIC CHEMISTRY.

While generally not as dangerous as London smog, photochemical smog contains a number of noxious constituents. Ozone, a strong oxidant that can react with living tissue, is one of these noxious compounds. Another is peroxyacetyl nitrate (PAN), an eye irritant that is produced by reactions between NO_2 and the breakdown products of carbonyls. Particulate matter having diameters of about 10 micrometers or less are of concern because they can penetrate into the human respiratory tract during breathing and have been implicated in a variety of respiratory ailments.

Probably the oldest type of smog known to humankind is produced from the burning of biomass or wood. It combines aspects of both London smog and photochemical smog since the burning of biomass can produce copious quantities of smoke as well as VOC and NO_x. [W.L.Ch.]

Smoke Fumes and smoke are dispersions of finely divided solids or liquids in a gaseous medium. The particle-size range is 0.01–5.0 micrometers. Typical dispersions are smokes from incomplete combustion of organic matter such as tobacco, wood, and coal; soot or carbon black; oil-vapor mists; chemical fumes such as sulfur trioxide (SO_3) and phosphorus pentoxide (P_2O_5) mists, ammonium chloride (NH_4Cl), and metal oxides; and the products of hydrolysis of metal chlorides by moist air. Oil-vapor and P_2O_5 mists (formed by burning phosphorus in moist air) have been extensively used in military operations to produce screening smokes. *See* AIR POLLUTION. [H.H.St./A.J.T.]

Smut (microbiology) The dark powdery masses of “smut spores” (teliospores) that develop in living plant tissues infected by species of *Ustilago*, *Tilletia*, and similar plant parasitic fungi. Molecular and ultrastructural data show that smut fungi comprise two phylogenetically distinct lines within the Basidiomycota; recent classification places them in two different classes, Ustilaginomycetes and Urediniomycetes. Ustilaginomycetes contains most of the smut fungi, and several groups of morphologically distinct, nonsmut plant parasites, including *Exobasidium*, *Graphiola*, and *Microstroma*. Smut fungi belonging to the genus *Microbotryum*, best known as anther smuts of Caryophyllaceae, are now placed within the Urediniomycetes with rust fungi and allied taxa. Ustilaginomycetes and Urediniomycetes also contain a number of saprotrophic, yeastlike fungi that are related to the plant parasitic smuts. Yeastlike saprotrophs that produce teliospores include *Tilletiaria* in the Ustilaginomycetes, and *Leucosporidium*, *Rhodosporidium*, and *Sporobolomyces* in the Urediniomycetes. Yeastlike saprotrophs in the Ustilaginomycetes that do not form teliospores but reproduce in another manner similar to *Ustilago* and *Tilletia* include *Pseudozyma* and *Tilletiopsis*, respectively. *See* FUNGI.

Teliospores of plant parasitic smut fungi form in a fruiting structure called a sorus. Sori are commonly produced in the inflorescence, leaves, or stems of the host, although the root is the site of sorus formation in smuts belonging to the genus *Entorrhiza*. Teliospores develop within the sorus by conversion of dikaryotic mycelial cells into thick-walled resistant spores within which paired nuclei fuse. Meiosis also occurs in teliospores of some smut fungi, but meiotic division is more characteristic of the tubular basidium that develops at germination.

There are 1200 species and 50 different genera of known smut fungi that infect over 4000 species of angiosperms. Smut fungi occur on both monocotyledonous and dicotyledonous hosts but are most economically important as pathogens of barley, corn, oats, onions, rice, sorghum, sugarcane, and wheat. Control of smut diseases varies with species and includes fungicidal seed treatments and use of resistant crop varieties. *See* BASIDIOMYCOTA; PLANT PATHOLOGY. [J.R.Ba.; L.M.Ca.]

Snow Frozen precipitation resulting from the growth of ice crystals from water vapor in the Earth's atmosphere.

As ice particles fall out in the atmosphere, they melt to raindrops when the air temperature is a few degrees above 32°F (0°C), or accumulate on the ground at colder temperatures. At temperatures above −40°F (−40°C), individual crystals begin growth on icelike aerosols (often clay particles 0.1 micrometer in diameter), or grow from cloud droplets (10 μm in diameter) frozen by similar particles. At lower temperatures, snow crystals grow on cloud droplets frozen by random molecular motion. At temperatures near 25°F (−4°C), crystals sometimes grow on ice fragments produced during soft hail (graupel) growth. Snow crystals often grow in the supersaturated environment provided by a cloud of supercooled droplets; this is known as the Bergeron-Findeisen process for formation of precipitation. When crystals are present in high concentrations (100 particles per liter) they grow in supersaturations lowered by mutual competition for available vapor.

Ice crystals growing under most atmospheric conditions (air pressure down to 0.2 atm or 20 kilopascals and temperatures 32 to −58°F or 0 to −50°C) have a hexagonal crystal structure, consistent with the arrangement of water molecules in the ice lattice, which leads to striking hexagonal shapes during vapor growth. The crystal habit (ratio of growth along and perpendicular to the hexagonal axis) changes dramatically with temperature. Both field and laboratory studies of crystals grown under known or controlled conditions show that the crystals are platelike above 27°F (−3°C) and between 18 and −13°F (−8 and −25°C), and columnlike between 27 and 18°F (−3 and −8°C) and below −13°F (−25°C).

Individual crystals fall in the atmosphere at velocity up to 0.5 m s^{-1} (1.6 ft s^{-1}). As crystals grow, they fall at higher velocity, which leads, in combination with the high moisture availability in a supercooled droplet cloud, to sprouting of the corners to form needle or dendrite skeletal crystals.

Under some conditions crystals aggregate to give snowflakes. This happens for the dendritic crystals that grow near 5°F (−15°C), which readily interlock if they collide with each other, and for all crystals near 32°F (0°C). Snowflakes typically contain several hundred individual crystals.

When snow reaches the ground, changes take place in the crystals. At temperatures near 32°F (0°C) the crystals rapidly lose the delicate structure acquired during growth, sharp edges evaporate, and the crystals take on a rounded shape, some 1–2 mm (0.04–0.08 in.) in diameter. These grains sinter together at their contact points to give snow some structural rigidity. The specific gravity varies from ~0.05 for freshly fallen "powder" snow to ~0.4 for an old snowpack. *See* PRECIPITATION (METEOROLOGY).

[J.Hal.]

Snow gage An instrument for measuring the amount of water equivalent in snow; more commonly known as a snow sampler. Frequently snow samplers are made of a lightweight seamless aluminum tube consisting of easily coupled lengths. Other snow samplers have been developed from material such as fiber glass and plastic for use in shallow snow, deep dense snow, and so forth.

To obtain a measurement the sampler is pushed vertically through the snow to the ground surface. The sampler, together with its snow core, is withdrawn and weighed. The water equivalent of the snow layer is obtained by subtracting the weight of the sampler from the total. In addition to any error introduced by the scale, there is usually a 6–8% error in the weight of samples taken in this manner.

Automatic devices that permit the remote observation of the water equivalent of the snow have been developed. These devices also permit telemetering the data to a central location, eliminating the need for travel to the snowfields. *See* SNOW; SNOW SURVEYING. [R.T.Be.]

Snow line A term generally used to refer to the elevation of the lower edge of a snow field. In mountainous areas, it is not truly a line but rather an irregular, commonly patchy border zone, the position of which in any one sector has been determined by the amount of snowfall and ablation. These factors may vary considerably from one part to another. On glacier surfaces the snow line is sometimes referred to as the glacier snow line or névé line (the outer limit of retained winter snow cover on a glacier).

Year-to-year variation in the position of the orographical snow line is great. The mean position over many decades, however, is important as a factor in the development of nivation hollows and protalus ramparts in deglaciated cirque beds. *See* GLACIOLOGY. [M.M.Mi.]

Snow surveying A technique for providing an inventory of the total amount of snow covering a drainage basin or a given region. Most of the usable water in western North America originates as mountain snowfall that accumulates during the winter and spring and appears several months later as streamflow. Snow surveys were established to provide an estimate of the snow water equivalent (that is, the depth of water produced from melting the snow) for use in predicting the volume of spring runoff. They are also extremely useful for flood forecasting, reservoir regulation, determining hydropower requirements, municipal and irrigation water supplies, agricultural productivity, wildlife survival, and building design, and for assessing transportation and recreation conditions.

Conventional snow surveys are made at designated sites, known as snow courses, at regular intervals each year throughout the winter period. A snow sampler is used to measure the snow depth and water equivalent at a series of points along a snow course. Average depth and water equivalent are calculated for each snow course. Satellite remote sensing and data relay are technologies used to obtain information on snow cover in more remote regions. *See* SNOW GAGE; SURFACE WATER. [B.E.Go.]

Snowfield and névé The term snowfield is usually applied to mountain and glacial regions to refer to an area of snow-covered terrain with definable geographic margins. Where the connotation is very general and without regard to geographical limits, the term snow cover is more appropriate; but glaciology requires more precise terms with respect to snowfield areas.

These terms differentiate according to the physical character and age of the snow cover. Technically, a snowfield can embrace only new or old snow (material from the current accumulation year). Anything older is categorized as firn or ice. The term névé is a descriptive phrase used to refer to consolidated granular snow not yet changed to glacier ice. Because of the need for simple terms, however, it has become acceptable to use the term névé when specifically referring to a geographical area of snowfields on mountain slopes or glaciers (that is, an area covered with perennial "snow" and embracing the entire zone of annually retained accumulation). *See* GLACIOLOGY. [M.M.Mi.]

Social hierarchy A fundamental aspect of social organization that is established by fighting or display behavior and results in a ranking of the animals in a group. Social, or dominance, hierarchies are observed in many different animals, including insects, crustaceans, mammals, and birds. In many species, size, age, or sex determines dominance rank. Dominance hierarchies often determine first or best access to food, social interactions, or mating within animal groups.

When two animals fight, several different behavioral patterns can be observed. Aggressive acts and submissive acts are both parts of a fight. Aggression and submission, together, are known as agonistic behavior. An agonistic relationship in which one animal is dominant and the other is submissive is the simplest type of dominance hierarchy. In nature, most hierarchies involve more than two animals and are composed of paired dominant-subordinate relationships. The simplest dominance hierarchies are linear and are known as pecking orders. In such a hierarchy the top individual (alpha) dominates all others. The second-ranked individual (beta) is submissive to the dominant alpha but dominates the remaining animals. The third animal (gamma) is submissive to alpha and beta but dominates all others. This pattern is repeated down to the lowest animal in the hierarchy, which cannot dominate any other group member.

Other types of hierarchies result from variations in these patterns. If alpha dominates beta, beta dominates gamma, but gamma dominates alpha, a dominance loop is formed. In some species a single individual dominates all members of the social group, but no consistent relationships are formed among the other animals. In newly formed hierarchies, loops or other nonlinear relationships are common, but these are often resolved over time so that a stable linear hierarchy is eventually observed.

Males often fight over access to females and to mating with them. Male dominance hierarchies are seen in many hooved mammals (ungulates). Herds of females use dominance hierarchies to determine access to food. Agonistic interactions among females are often not as overtly aggressive as those among males, but the effects of the dominance hierarchy can easily be observed. In female dairy cattle, the order of entry into the milking barn is determined by dominance hierarchy, with the alpha female entering first.

Because dominant animals may have advantages in activities such as feeding and mating, they will have more offspring than subordinate animals. If this is the case, then natural selection will favor genes for enhanced fighting ability. Heightened aggressive behavior may be counterselected by the necessity for amicable social interactions in certain circumstances. Many higher primates live in large groups of mixed sex and exhibit complex social hierarchies. In these groups, intra- and intersexual dominance relationships determine many aspects of group life, including feeding, grooming, sleeping sites, and mating. Macaque, baboon, and chimpanzee societies are characterized by cooperative alliances among individuals that are more important than individual fighting ability in maintaining rank.

Social hierarchies provide a means by which animals can live in groups and exploit resources in an orderly manner. In particular, food can be distributed among group members with little ongoing conflict. Another motivation for group living is mutual defense. Even though subordinates receive less food or have fewer opportunities to mate, they may have greatly increased chances of escaping predation. *See* BEHAVIORAL ECOLOGY; POPULATION ECOLOGY; TERRITORIALITY. [M.D.Bre.]

Social insects Insects that share resources and reproduce cooperatively. The shared resources are shelter, defense, and food (collection or production). After a period of population growth, the insects reproduce in several ways. As social insect groups grow, they evolve more differentiation between members but reintegrate into a more closely organized system known as eusocial. These are the most advanced societies with individual polymorphism, and they contain insects of various ages, sizes, and shapes. All the eusocial insects are included in the orders Isoptera (termites) and Hymenoptera (wasps, bees, and ants). *See* INSECTA.

The social insects have evolved in various patterns. In the Hymenoptera, the society is composed of only females; males are produced periodically for their sperm. They usually congregate and attract females, or they visit colonies with virgin females and copulate there. In the Hymenoptera, sex is determined largely by whether the individual has one or two sets of chromosomes. Thus the queen has the power to determine the sex of her offspring: if she lets any of her stored sperm reach the egg, a female is produced; if not, a male results. In the more primitive bees and wasps, social role (caste) is influenced by interaction with like but not necessarily related individuals. The female that can dominate the others assumes the role of queen, even if only temporarily. Domination is achieved by aggression, real or feigned, or merely by a ritual that is followed by some form of salutation by the subordinates. This inhibits the yolk-stimulating glands and prevents the subordinates from contributing to egg production; if it fails to work, the queen tries to destroy any eggs that are laid. Subordinate females take on more and more of the work of the group for as long as the queen is present and well. At first, all the eggs are fertilized and females develop, with the result that virgin females inhabit the nest for the first batches. They are often undernourished, and this, together with their infertility, reduces their urge to leave the nest and start another one. Such workers are said to be produced by maternal manipulation.

Reproductive ants, like termites, engage in a massive nuptial flight, after which the females, replete with sperm, go off to start a new nest. At some stage after the nuptials, the reproductives break off their wings, which have no further use. Workers, however, never have wings because they develop quickly and pass right through the wing-forming stages; their ovaries and genitalia are also reduced. Ant queens can prevent the formation of more queens; as with the honeybee, they do this behaviorally by using pheromones. They also force the workers to feed all larvae the same diet. To this trophogenic caste control is added a blastogenic control; eggs that are laid have a developmental bias toward one caste or another. This is not genetic; bias is affected by the age of the queen and the season: more worker-biased eggs are laid by young queens and by queens in spring. In some ants, workers mature in various sizes. Since they have disproportionately large heads, the biggest workers are used mainly for defense; they also help with jobs that call for strength, like cutting vegetation or cracking nuts.

Social insects make remarkable nests that protect the brood as well as regulate the microclimate. The simplest nests are cavities dug in soil or soft wood, with walls smoothed and plastered with feces that set hard. Chambers at different levels in the soil are frequently connected by vertical shafts so that the inhabitants can choose the chamber with the best microclimate. Termites and ants also make many different types of arboreal nests. These nests are usually made of fecal material, but one species of ant (*Oecophylla*) binds leaves together with silk produced in the salivary glands of their larvae that the workers hold in their jaws and spin across leaves. A whole group of ants (for example, *Pseudomyrmex*) inhabit the pith of plants.

Social bees use wax secreted by their cuticular glands and frequently blended with gums from tree exudates for their nest construction. Cells are made cooperatively by a curtain of young bees that scrape wax from their abdomen, chew it with saliva, and mold it into the correct shape; later it is planed and polished. With honeybees the hexagonal comb reaches perfection as a set of back-to-back cells, each sloping slightly upward to prevent honey from running out. The same cells are used repeatedly for brood and for storage; or they may be made a size larger for rearing males. Only the queen cell is pendant, with a circular cross section and an opening below.

The ubiquity and ecological power of social insects depend as much on their ability to evolve mutualistic relations with other organisms as on the coherence of their social organization. Wood and the cellulose it contains is normally available as a source of energy only to bacteria and fungi. However, it is used as a basic resource by both termites and ants that have evolved a technique of culturing these organisms. Though lower termites have unusual protozoa as intestinal symbionts, higher termites have bacteria in pouches in their hind gut. Many have a fungus that they culture in special chambers in their nests. The termites feed on woody debris, leaves, and grass cuttings; the fungus digests these materials with the aid of termite feces and produces soft protein-rich bodies that the termites share with their juveniles and reproductives, neither of which are able to feed themselves. Protected from the weather, the fungus can remain active throughout the dry seasons—an inestimable advantage in the subtropics.

Many ants collect and store seeds that they mash and feed directly to their larvae. Provided they are collected when dry and stored in well-ventilated chambers, these seeds can remain viable and edible for an entire season. The plants benefit because not all the seeds are eaten; some that start to germinate are thrown out with the rubbish of the colony—in effect a way of planting them. Others are left behind by the ants when they change nests. In this way, grass seeds can extend their range into dry areas that they could not reach alone.

The dispersal of plant pollen by bees is a well-known symbiosis, and it has led to the evolution of many strange shapes, colors, and scents in flowers. Quite specific flower-bee relationships may exist in which one plant may use very few species of bees for the transfer of pollen. *See* POLLINATION; POPULATION ECOLOGY. [M.V.B.]

Soil Finely divided rock-derived material containing an admixture of organic matter and capable of supporting vegetation. Soils are independent natural bodies, each with a unique morphology resulting from a particular combination of climate, living plants and animals, parent rock materials, relief, the groundwaters, and age. Soils support plants, occupy large portions of the Earth's surface, and have shape, area, breadth, width, and depth. Soil, as used here, differs in meaning from the term as used by engineers, where the meaning is unconsolidated rock material. *See* PEDOLOGY.

Origin and classification. Soil covers most of the land surface as a continuum. Each soil grades into the rock material below and into other soils at its margins, where changes occur in relief, groundwater, vegetation, kinds of rock, or other factors which influence the development of soils. Soils have horizons, or layers, more or less parallel to the surface and differing from those above and below in one or more properties, such as color, texture, structure, consistency, porosity, and reaction (see illustration). The succession of horizons is called the soil profile.

Soil formation proceeds in stages, but these stages may grade indistinctly from one into another. The first stage is the accumulation of unconsolidated rock fragments, the parent material. Parent material may be accumulated by deposition of rock fragments moved by glaciers, wind, gravity, or water, or it may accumulate more or less in place from physical and chemical weathering of hard rocks. The second stage is the formation

Photograph of a soil profile showing horizons. The dark crescent-shaped spots at the soil surface are the result of plowing. The dark horizon is the principal horizon of accumulation of organic matter that has been washed down from the surface. The thin wavy lines were formed in the same manner. 1 in. = 2.5 cm.

of horizons. This stage may follow or go on simultaneously with the accumulation of parent material. Soil horizons are a result of dominance of one or more processes over others, producing a layer which differs from the layers above and below. *See* WEATHERING PROCESSES.

Systems of soil classification are influenced by concepts prevalent at the time a system is developed. The earliest classifications were based on relative suitability for different crops, such as rice soils, wheat soils, and vineyard soils. Over the years, many systems of classification have been attempted but none has been found markedly superior. Two bases for classification have been tried. One basis has been the presumed genesis of the soil; climate and native vegetation were given major emphasis. The other basis has been the observable or measurable properties of the soil.

The Soil Survey staff of the U.S. Department of Agriculture and the land-grant colleges adopted the current classification scheme in 1965. This system differs from earlier systems in that it may be applied to either cultivated or virgin soils. Previous systems have been based on virgin profiles, and cultivated soils were classified on the presumed characteristics or genesis of the virgin soils. The new system has six categories, based on both physical and chemical properties. These categories are the order, suborder, great group, subgroup, family, and series, in decreasing rank. The orders and the general nature of the included soils are given in the table. The suborder

Soil orders

Order	Formative element in name	General nature of soils
Alfisols	alf	Gray to brown surface horizons, medium to high base supply, with horizons of clay accumulation; usually moist, but may be dry during summer
Aridisols	id	Pedogenic horizons, low in organic matter, and usually dry
Entisols	ent	Pedogenic horizons lacking
Histosols	ist	Organic (peats and mucks)
Inceptisols	ept	Usually moist, with pedogenic horizons of alteration of parent materials but not of illuviation
Mollisols	oil	Nearly black organic-rich surface horizons and high base supply
Oxisols	ox	Residual accumulations of inactive clays, free oxides, kaolin, and quartz; mostly tropical
Spodosols	od	Accumulations of amorphous materials in subsurface horizons
Ultisols	ult	Usually moist, with horizons of clay accumulation and a low supply of bases
Vertisols	ert	High content of swelling clays and wide deep cracks during some season

narrows the ranges in soil moisture and temperature regimes, kinds of horizons, and composition, according to which of these is most important. The taxa (classes) in the great group category group soils that have the same kinds of horizons in the same sequence and have similar moisture and temperature regimes. The great groups are divided into subgroups that show the central properties of the great group, intergrade subgroups that show properties of more than one great group, and other subgroups for soils with atypical properties that are not characteristic of any great group. The families are defined largely on the basis of physical and mineralogic properties of importance to plant growth. The soil series is a group of soils having horizons similar in differentiating characteristics and arrangement in the soil profile, except for texture of the surface portion, and developed in a particular type of parent material.

Surveys. Soil surveys include those researches necessary (1) to determine the important characteristics of soils, (2) to classify them into defined series and other units, (3) to establish and map the boundaries between kinds of soil, and (4) to correlate and predict adaptability of soils to various crops, grasses, and trees; behavior and productivity of soils under different management systems; and yields of adapted crops on soils under defined sets of management practices. Although the primary purpose of soil surveys has been to aid in agricultural interpretations, many other purposes have become important, ranging from suburban planning, rural zoning, and highway location, to tax assessment and location of pipelines and radio transmitters. This has happened because the soil properties important to the growth of plants are also important to its engineering uses.

Two kinds of soil maps are made. The common map is a detailed soil map, on which soil boundaries are plotted from direct observations throughout the surveyed area. Reconnaissance soil maps are made by plotting soil boundaries from observations made at intervals. The maps show soil and other differences that are of significance for present or foreseeable uses.

[G.D.S.]

Physical properties. Physical properties of soil have critical importance to growth of plants and to the stability of cultural structures such as roads and buildings. Such properties commonly are considered to be: size and size distribution of primary particles and of secondary particles, or aggregates, and the consequent size, distribution, quantity, and continuity of pores; the relative stability of the soil matrix against disruptive forces, both natural and cultural; color and textural properties, which affect absorption and radiation of energy; and the conductivity of the soil for water, gases, and heat. These usually would be considered as fixed properties of the soil matrix, but actually some are not fixed because of influence of water content. The additional property, water content—and its inverse, gas content—ordinarily is transient and is not thought of as a property in the same way as the others. However, water is an important constituent, despite its transient nature, and the degree to which it occupies the pore space generally dominates the dynamic properties of soil. Additionally, the properties listed above suggest a macroscopic homogeneity for soil which it may not necessarily have. In a broad sense, a soil may consist of layers or horizons of roughly homogeneous soil materials of various types that impart dynamic properties which are highly dependent upon the nature of the layering. Thus, a discussion of dynamic soil properties must include a description of the intrinsic properties of small increments as well as properties it imparts to the system.

From a physical point of view it is primarily the dynamic properties of soil which affect plant growth and the strength of soil beneath roads and buildings. While these depend upon the chemical and mineralogical properties of particles, particle coatings, and other factors discussed above, water content usually is the dominant factor. Water content depends upon flow and retention properties, so that the relationship between water content and retentive forces associated with the matrix becomes a key physical property of a soil. *See* EROSION; GROUND-WATER HYDROLOGY; SOIL MECHANICS. [W.H.G.]

Soil chemistry The study of the composition and chemical properties of soil. Soil chemistry involves the detailed investigation of the nature of the solid matter from which soil is constituted and of the chemical processes that occur as a result of the action of hydrological, geological, and biological agents on the solid matter. Because of the broad diversity among soil components and the complexity of soil chemical processes, the application of a wide variety of concepts and methods employed in the chemistry of aqueous solutions, of amorphous and crystalline solids, and of solid surfaces is required.

Elemental composition. The elemental composition of soil varies over a wide range, permitting only a few general statements to be made. Those soils that contain less than 12–20% organic carbon are termed mineral. All other soils are termed organic. Carbon, oxygen, hydrogen, nitrogen, phosphorus, and sulfur are the most important constituents of organic soils and of soil organic matter in general. Carbon, oxygen, and hydrogen are most abundant; the content of nitrogen is often about one-tenth that of carbon, while the content of phosphorus or sulfur is usually less than one-fifth that of nitrogen (Table 1).

Besides oxygen, the most abundant elements found in mineral soils are silicon, aluminum, and iron. The distribution of chemical elements will vary considerably from soil to soil and, in general, will be different in a specific soil from the distribution of elements in the crustal rocks of the Earth. The most important micro or trace elements in soil are boron, copper, manganese, molybdenum, and zinc, since these elements are essential in the nutrition of green plants. Also important are cobalt, selenium, cadmium, and nickel. The average distribution of trace elements in soil is not greatly different from that in crustal rocks (Table 2).

Table 1. Average percentages of total carbon, total nitrogen, and organic phosphorus in selected soils

Soil	% C	% N	% P
Sand	2.5	.23	.04
Fine sandy loam	3.3	.23	.06
Medium loam	2.3	.22	.05
Clay loam, well drained	4.6	.36	.10
Clay loam, poorly drained	8.0	.43	.05
Peat	46.1	1.32	.03

Table 2. Average amounts of trace elements commonly found in soils and crustal rocks

Trace element	Soil, ppm*	Crustal rocks, ppm
As	6	1.8
B	10	10
Cd	.06	.2
Co	8	25
Cr	100	100
Cu	20	55
Mo	2	1.5
Ni	40	75
Pb	10	13
Se	.2	.05
V	100	135
Zn	50	70

*ppm = parts per million.

The elemental composition of soil varies with depth below the surface because of pedochemical weathering. The principal processes of this type that result in the removal of chemical elements from a given soil horizon are: (1) soluviation (ordinary dissolution in water), (2) cheluviation (complexation by organic or inorganic ligands), (3) reduction, and (4) suspension. The principal effect of these four processes is the appearance of alluvial horizons in which compounds such as aluminum and iron oxides, aluminosilicates, or calcium carbonate have been precipitated from solution or deposited from suspension. *See* WEATHERING PROCESSES.

Minerals. The minerals in soils are the products of physical, geochemical, and pedochemical weathering. Soil minerals may be either amorphous or crystalline. They may be classified further, approximately, as primary or secondary minerals, depending on whether they are inherited from parent rock or are produced by chemical weathering, respectively.

The bulk of the primary minerals that occur in soil are found in the silicate minerals. Chemical weathering of the silicate minerals is responsible for producing the most important secondary minerals in soil. These are found in the clay fraction and include aluminum and iron hydrous oxides (usually in the form of coatings on other minerals), carbonates, and aluminosilicates.

Ion exchange. A portion of the chemical elements in soil is in the form of cations that are not components of inorganic salts but that can be replaced reversibly by the cations of leaching salt solutions or acids. These cations are said to be exchangeable, and their total quantity is termed the cation exchange capacity (CEC) of the soil. The

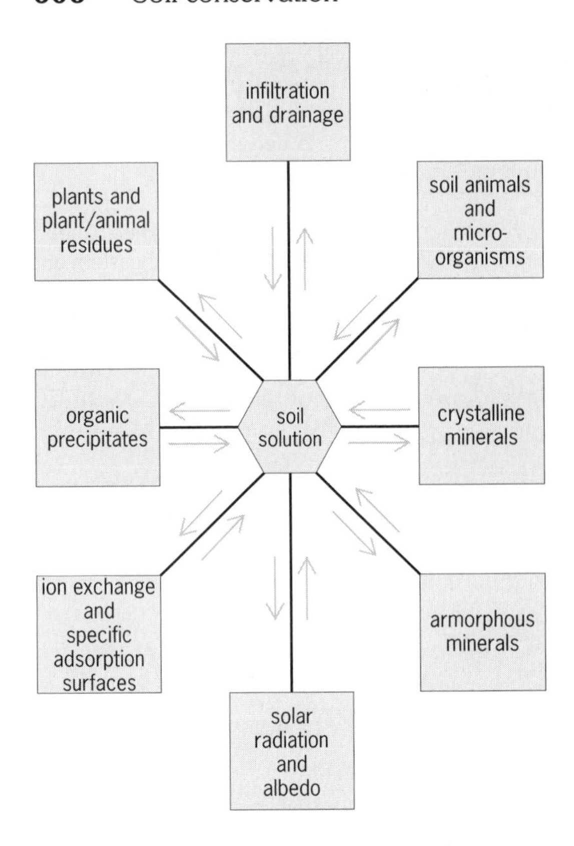

Factors influencing the chemistry of soil solution. (*Modified from J. F. Hodgson, Chemistry of the micronutrients in soil, Adv. Agron., 15:141, 1963*)

CEC of a soil generally will vary directly with the amounts of clay and organic matter present and with the distribution of clay minerals.

The stoichiometric exchange of the anions in soil for those in a leaching salt solution is a phenomenon of relatively small importance in the general scheme of anion reactions with soils. Under acid conditions ($pH < 5$) the exposed hydroxyl groups at the edges of the structural sheets or on the surfaces of clay-sized particles become protonated and thereby acquire a positive charge. The degree of protonation is a sensitive function of pH, the ionic strength of the leaching solution, and the nature of the clay-sized particle.

Soil solution. The solution in the pore space of soil acquires its chemical properties through time-varying inputs and outputs of matter and energy that are mediated by the several parts of the hydrologic cycle and by processes originating in the biosphere (see illustration). The soil solution thus is a dynamic and open natural water system whose composition reflects the many reactions that can occur simultaneously between an aqueous solution and an assembly of mineral and organic solid phases that varies with both time and space. *See* SOIL. [G.Sp.]

Soil conservation The practice of arresting and minimizing artificially accelerated soil deterioration. Its importance has grown because cultivation of soils for agricultural production, deforestation and forest cutting, grazing of natural range, and other disturbances of the natural cover and position of the soil have increased greatly in the last 100 years.

Erosion and deterioration. The exact extent of accelerated soil erosion in the world today is not known, particularly as far as the rate of soil movement is concerned. However, it may be said that nearly every semiarid area with cultivation or long-continued grazing, every hill land with moderate to dense settlement in humid temperate and subtropical climates, and all cultivated or grazed hill lands in the Mediterranean climate areas suffer to some degree from such erosion. Recognized problems of erosion are found in such culturally diverse areas as southern China, the Indian plateau, south Australia, the South African native reserves, Russia, Spain, the southeastern and midwestern United States, and Central America.

Within the United States the most critical areas have been the hill lands of the Piedmont and the interior Southeast, the Great Plains, the Palouse area hills of the Pacific Northwest, southern California hills, and slope lands of the Midwest. The high-intensity rainstorms of the Southeast, and the cyclical droughts of the Plains have predisposed the two larger areas to erosion. *See* EROSION.

Soil may deteriorate either by physical movement of soil particles from a given site or by depletion of the water-soluble elements in the soil which contribute to the nourishment of crop plants, grasses, trees, and other economically usable vegetation. The physical movement generally is referred to as erosion. Wind, water, glacial ice, animals, and tools used by humans may be agents of erosion. For purposes of soil conservation, the two most important agents of erosion are wind and water, especially as their effects are intensified by the disturbance of natural cover or soil position.

Depletion of soil nutrients obviously is a part of soil erosion. However, such depletion may take place in the absence of any noticeable amount of erosion. The disappearance of naturally stored nitrogen, potash, phosphate, and some trace elements from the soil also affects the usability of the soil for human purposes. The natural fertility of virgin soils always is depleted over time as cultivation continues, but the rate of depletion is highly dependent on management practices. *See* PLANT MINERAL NUTRITION; SOIL.

Accelerated erosion may be induced by any land use practice which denudes the soil surfaces of vegetative cover. For example, cultivation of any row crop on a slope without soil-conserving practices is an invitation to accelerated erosion. Cultivation of other crops, like the small grains, also may induce accelerated erosion, especially where fields are kept bare between crops to store moisture. Forest cutting, overgrazing, grading for highway use, urban land use, or preparation for other large-scale engineering works also may speed the erosion of soil.

Causes of soil mismanagement. One of the chief causes of erosion-inducing agricultural practices in the United States has been ignorance of their consequences. The cultivation methods of the settlers of western European stock who set the pattern of land use in this country came from a physical environment which was far less susceptible to erosion than North America, because of the mild nature of rainstorms and the prevailing soil textures in Europe. Corn, cotton, and tobacco, moreover, were crops unfamiliar to European agriculture. In eastern North America the combination of European cultivation methods and American interfilled crops resulted in generations of soil mismanagement.

On the Plains and in other susceptible western areas, small grain monoculture, particularly of wheat, encouraged the exposure of the uncovered soil surface so much of the time that water and wind inevitably took their toll. On rangelands, lack of knowledge as to the precipitation cycle and range capacity, and the urge to maximize profits every year contributed to a slower, but equally sure denudation of cover.

Finally, the United States has experienced extensive erosion in mountain areas because of forest mismanagement. Clear-cutting of steep slopes, forest burning for grazing purposes, inadequate fire protection, and shifting cultivation of forest lands have allowed vast quantities of soil to wash out of the slope sites where they could have produced timber and other forest values indefinitely.

Effects on other resources. Accelerated erosion may have consequences which reach far beyond the lands on which the erosion takes place and the community associated with them. During periods of heavy wind erosion, for example, the dust fall may be of economic importance over a wide area beyond that from which the soil cover has been removed. The most pervasive and widespread effects, however, are those associated with water erosion. Removal of upstream cover changes the regimen of streams below the eroding area.

A long chain of other effects also ensues. Because of the extremes of low water in denuded areas during dry seasons, water transportation is made difficult or impossible without regulation, fish and wildlife support is endangered or disappears, the capacity of streams to carry sewage and other wastes safely may be seriously reduced, recreational values are destroyed, and run-of-the-river hydroelectric generation reaches a very low level. *See* WATER CONSERVATION. [E.A.A.; D.J.P.]

Soil degradation Loss in the quality or productivity of soil as a result of human activities. Degradation is attributed to changes in soil nutrient status, loss of soil organic matter, deterioration of soil structure, and toxicity due to accumulations of naturally occurring or anthropogenic materials. The effects of soil degradation include loss of agricultural productivity, negative impacts on the environment and economic stability, and increased clearing of virgin forest and exploitation of marginally suitable land. *See* SOIL.

Soil degradation may impair the function of soil organisms and thus result in further problems. The soil microbial community is essential to cycling of nutrients and decomposition of wastes; thus hindrance of microbial activities may have serious ecological results. Degradative forces such as erosion that result in loss of organic matter may also result in long-term losses of microbial activity. *See* EROSION; SOIL ECOLOGY.

Soil degradation often affects productivity through depletion of plant nutrients. Excessive leaching of cations involved in buffering soil pH may result in soil acidification, changing the solubility, and thus availability, of certain nutrients to plants. Misuse can also result in concentrations of chemicals that are toxic to plants. *See* SOIL CHEMISTRY; SOIL FERTILITY.

Historically, mining of coal and various metal ores has been among the most devastating forms of land use. Mining often exposes large amounts of reduced (decreased oxidation state) minerals in the form of mine tailings and rubble. The acidic mine tailings are extremely difficult to revegetate, and without intervention they remain exposed, continuing to produce acidic drainage until oxidizable material is depleted. Left untreated, surface (strip) mines may require 50–150 years to recover.

In some cases, compounds are toxic to soil organisms when present at significant concentrations, and thus produce large shifts in microbial community structure. Eventually, microorganisms present in the soil degrade most organic contaminants, thus alleviating toxic effects.

Unlike organic contaminants, which are degraded to nontoxic forms, toxic metals usually become essentially permanent features of soils. Fortunately, adsorption of metals to soil colloids decreases the availability of the metals for movement in the environment or uptake by plants, animals, and microorganisms. *See* AGRICULTURAL CHEMISTRY. [G.K.S.]

Soil ecology The study of the interactions among soil organisms, and between biotic and abiotic aspects of the soil environment. Soil is made up of a multitude of physical, chemical, and biological entities, with many interactions occurring among them. Soil is a variable mixture of broken and weathered minerals and decaying organic matter. Together with the proper amounts of air and water, it supplies, in part, sustenance for plants as well as mechanical support. *See* SOIL.

Abiotic and biotic factors lead to certain chemical changes in the top few decimeters (8–10 in.) of soil. The work of the soil ecologist is made easier by the fact that the surface 10–15 cm (4–6 in.) of the A horizon has the majority of plant roots, microorganisms, and fauna. A majority of the biological-chemical activities occur in this surface layer.

The biological aspects of soil range from major organic inputs, decomposition by primary decomposers (bacteria, fungi, and actinomycetes), and interactions between microorganisms and fauna (secondary decomposers) which feed on them. The detritus decomposition pathway occurs on or within the soil after plant materials (litter, roots, sloughed cells, and soluble compounds) become available through death or senescence. Plant products are used by microorganisms (primary decomposers). These are eaten by the fauna which thus affect flows of nutrients, particularly nitrogen, phosphorus, and sulfur. The immobilization of nutrients into plants or microorganisms and their subsequent mineralization are critical pathways. The labile inorganic pool is the principal one that permits subsequent microorganism and plant existence. Scarcity of some nutrient often limits production. Most importantly, it is the rates of flux into and out of these labile inorganic pools which enable ecosystems to successfully function. *See* ECOLOGY; ECOSYSTEM; GUILD; SOIL; SYSTEMS ECOLOGY. [D.C.C.]

Soil fertility The ability of a soil to supply plant nutrients. Sixteen chemical elements are required for the growth of all plants: carbon, oxygen, and hydrogen (these three are obtained from carbon dioxide and water), plus the elements nitrogen, phosphorus, potassium, calcium, magnesium, sulfur, iron, manganese, zinc, copper, boron, molybdenum, and chlorine. Some plant species also require one or more of the elements cobalt, sodium, vanadium, and silicon. *See* PLANT MINERAL NUTRITION.

While carbon and oxygen are supplied to plants from carbon dioxide in the air, the other essential elements are supplied primarily by the soil. Of the latter, all except hydrogen from water are called mineral nutrients. Only part of the 13 essential mineral nutrients in soil are in a chemical form that can be immediately used by plants. The unusable (unavailable) parts, which eventually do become available to plants, are of two kind: they may be in organic combination (such as nitrogen in soil humus) or in solid inorganic soil particles (such as potassium in soil clays). The time for complete decomposition and dissolution of these compounds varies widely, from days to hundreds of years.

Soils exhibit a variable ability to supply the mineral nutrients needed by plants. This characteristic allows soils to be classified according to their level of fertility. This can vary from a deficiency to a sufficiency, or even toxicity (too much), of one or more nutrients. A serious deficiency of only one essential nutrient can still greatly reduce crop yields. Several soil properties are important in determining a soil's inherent fertility. One property is the adsorption and storage of nutrients on the surfaces of soil particles. Such adsorption of a number of nutrients is caused by an attraction of positively charged nutrients to negatively charged soil particles. This adsorption is called cation exchange (adsorbed cations can be exchanged with other cations in solution), and the quantity of nutrient cations a soil can adsorb is called its cation-exchange capacity. *See* SOIL CHEMISTRY.

The negative charge in soils is associated with clay particles, but some of the soil's cation-exchange capacity may arise from organic matter (humus) in the soil. Negative charges of organic matter arise largely from carboxylic and phenolic acid functional groups. Since these functional groups are weak acids, the negative charge from organic matter increases as the soil pH increases. The negative charges of soil organic matter can adsorb the same cations as described for the soil clays. The proportion of cation-exchange capacity arising from mineral clays and from organic matter depends on the proportions of each in the soil and on the kinds of clays. In most mineral soils, the soil clays comprise the greater proportion of cation-exchange capacity. Within the class of mineral soils, those soils with more clay and less sand and silt have the greatest cation-exchange capacity.

The amount and kind of acids on the cation exchange sites can have a substantial influence on a soil's perceived fertility. Two factors cause soils to become acid: when crops are harvested, exchangeable bases are removed as part of the crop; and exchangeable bases move with drainage water below the crop's root zone (leaching). Since much of the nitrogen fertilizer supplied to crops contains ammonium nitrogen (this is true of both manure and chemical fertilizers), the addition of high rates of nitrogen also enhances soil acidification. *See* FERTILIZER; NITROGEN CYCLE; SOIL. [D.E.Ki.]

Soil mechanics The study of the response of masses composed of soil, water, and air to imposed loads. Because both water and air are able to move through the soil pores, the discipline also involves the prediction of these transport processes. Soil mechanics provides the analytical tools required for foundation engineering, retaining wall design, highway and railway subbase design, tunneling, earth dam design, mine excavation design, and so on. Because the discipline relates to rock as well as soils, it is also known as geotechnical engineering. *See* ENGINEERING GEOLOGY.

Soil consists of a multiphase aggregation of solid particles, water, and air. This fundamental composition gives rise to unique engineering properties, and the description of the mechanical behavior of soils requires some of the most sophisticated principles of engineering mechanics. The terms multiphase and aggregation both imply unique properties. As a multiphase material, soil exhibits mechanical properties that show the combined attributes of solids, liquids, and gases. Individual soil particles behave as solids, and show relatively little deformation when subjected to either normal or shearing stresses. Water behaves as a liquid, exhibiting little deformation under normal stresses, but deforming greatly when subjected to shear. Being a viscous liquid, however, water exhibits a shear strain rate that is proportional to the shearing stress. Air in the soil behaves as a gas, showing appreciable deformation under both normal and shear stresses. When the three phases are combined to form a soil mass, characteristics that are an outgrowth of the interaction of the phases are manifest. Moreover, the particulate nature of the solid particles contributes other unique attributes.

When dry soil is subjected to a compressive normal stress, the volume decreases nonlinearly; that is, the more the soil is compressed, the less compressible the mass becomes. Thus, the more tightly packed the particulate mass becomes, the more it resists compression. The process, however, is only partially reversible, and when the compressive stress is removed the soil does not expand back to its initial state.

When this dry particulate mass is subjected to shear stress, an especially interesting behavior owing to the particulate nature of the soil solids results. If the soil is initially dense (tightly packed), the mass will expand because the particles must roll up and over each other in order for shear deformation to occur. Conversely, if the mass is initially loose, it will compress when subjected to a shear stress. Clearly, there must also exist a specific initial density (the critical density) at which the material will display zero volume change when subjected to shear stress. The term dilatancy has been applied to the relationship between shear stress and volume change in particulate materials. Soil is capable of resisting shear stress up to a certain maximum value. Beyond this value, however, the material undergoes large, uncontrolled shear deformation.

The other limiting case is saturated soil, that is, a soil whose voids are entirely filled with water. When such a mass is initially loose and is subjected to compressive normal stress, it tends to decrease in volume; however, in order for this volume decrease to occur, water must be squeezed from the soil pores. Because water exhibits a viscous resistance to flow in the microscopic pores of fine-grained soils, this process can require considerable time, during which the pore water is under increased pressure. This excess pore pressure is at a minimum near the drainage face of the soil mass and at a maximum near the center of the soil sample. It is this gradient (or change in pore water pressure with change in position within the soil mass) that causes the outflow of water and the corresponding decrease in volume of the soil mass.

Conversely, if an initially dense soil mass is subjected to shear stress, it tends to expand. The expansion, however, may be time-dependent because of the viscous resistance to water being drawn into the soil pores. During this time the pore water will be under decreased pressure. Thus, in saturated soil masses, changes in pore water pressure and time-dependent volume change can be induced by either changes in normal stress or by changes in shear stress. *See* ROCK MECHANICS; SOIL. [C.A.Mo.]

Soil microbiology The study of biota that inhabit the soil and the processes that they mediate. The soil is a complex environment colonized by an immense diversity of microorganisms. Soil microbiology focuses on the soil viruses, bacteria, actinomycetes, fungi, and protozoa, but it has traditionally also included investigations of the soil animals such as the nematodes, mites, and other microarthropods. These organisms, collectively referred to as the soil biota, function in a belowground ecosystem based on plant roots and litter as food sources. Modern soil microbiology represents an integration of microbiology with the concepts of soil science, chemistry, and ecology to understand the functions of microorganisms in the soil environment.

The surface layers of soil contain the highest numbers and variety of microorganisms, because these layers receive the largest amounts of potential food sources from plants and animals. The soil biota form a belowground system based on the energy and nutrients that they receive from the decomposition of plant and animal tissues. The primary decomposers are the bacteria and fungi.

Microorganisms, especially algae and lichen, are pioneering colonizers of barren rock surfaces. Colonization by these organisms begins the process of soil formation necessary for the growth of higher plants. After plants have been established, decomposition by microorganisms recycles the energy, carbon, and nutrients in dead plant and animal tissues into forms usable by plants. Therefore, microorganisms have a key role in the processing of materials that maintain life on the Earth. The transformations of elements between forms are described conceptually as the elemental cycles.

In the carbon cycle, microorganisms transform plant and animal residues into carbon dioxide and the soil organic matter known as humus. Humus improves the water-holding capacity of soil, supplies plant nutrients, and contributes to soil aggregation. Microorganisms may also directly affect soil aggregation. The extent of soil aggregation determines the workability or tilth of the soil. A soil with good tilth is suitable for plant growth because it is permeable to water, air, and roots. *See* HUMUS.

Soil microorganisms play key roles in the nitrogen cycle. The atmosphere is approximately 80% nitrogen gas (N_2), a form of nitrogen that is available to plants only when it is transformed to ammonia (NH_3) by either soil bacteria (N_2 fixation) or by humans (manufacture of fertilizers). Soil bacteria also mediate denitrification, which returns nitrogen to the atmosphere by transforming NO_3^- to N_2 or nitrous oxide (N_2O)

gas. Microorganisms are crucial to the cycling of sulfur, phosphorus, iron, and many micronutrient trace elements.

In addition to the elemental cycles, there are several interactions between plants and microbes which are detrimental or beneficial to plant growth. Some soil microorganisms are pathogenic to plants and cause plant diseases such as root rots and wilts. Many plants form symbiotic relationships with fungi called mycorrhizae (literally fungus-root). Mycorrhizae increase the ability of plants to take up nutrients and water. The region of soil surrounding plant roots, the rhizosphere, may contain beneficial microorganisms which protect the plant root from pathogens or supply stimulating growth factors. The interactions between plant roots and soil microorganisms is an area of active research in soil microbiology. *See* BIOGEOCHEMISTRY; MYCORRHIZAE; NITROGEN CYCLE; NITROGEN FIXATION; RHIZOSPHERE.

The incredible diversity of soil microorganisms is a vast reserve of potentially useful organisms. Many of the medically important antibiotics are produced by filamentous bacteria known as actinomycetes. The soil is the largest reservoir of these medically important microorganisms.

The numerous natural substances that are used by microorganisms indicate that soil microorganisms have diverse mechanisms for degrading a variety of compounds. Human activity has polluted the environment with a wide variety of synthetic or processed compounds. Many of these hazardous or toxic substances can be degraded by soil microorganisms. This is the basis for the treatment of contaminated soils by bioremediation, the use of microorganisms or microbial processes to detoxify and degrade environmental contaminants. Soil microbiologists study the microorganisms, the metabolic pathways, and the controlling environmental conditions that can be used to eliminate pollutants from the soil environment. *See* HAZARDOUS WASTE.

Microbiologists traditionally isolate pure strains of microorganisms by using culture methods. Methods that do not rely on culturing microorganisms include microscopic observation and biochemical or genetic analysis of specific cell constituents. The rates or controlling factors for microbial processes are studied by using methods from chemistry, biology, and ecology. Typically, these studies involve measuring the rate of production and consumption of a compound of interest. The results of these studies are commonly analyzed by using mathematical models. Models allow the information from one system to be generalized for different environmental conditions. *See* MICROBIOLOGY; SOIL; SOIL CHEMISTRY; SOIL ECOLOGY. [J.M.N.]

Soil sterilization A chemical or physical process that results in the death of soil organisms. This control method affects many organisms, even though the elimination of only specific weeds, fungi, bacteria, viruses, nematodes, or pests is desirable. Even if complete sterilization is achieved, it is short lived since organisms will recolonize this biological vacuum quite rapidly. Soil sterilization can be achieved through both physical and chemical means. Physical control measures include steam and solar energy. Chemical control methods include herbicides and fumigants. Dielectric heating and gamma irradiation are used less frequently as soil sterilization methods. Composting can be used to sterilize organic materials mixed with soil, but it is not used for the sterilization of soil alone. Soil sterilization is used in greenhouse operations, the production of high-value or specialty crops, and the control of weeds. [C.A.St.]

Solar energy The energy transmitted from the Sun. The upper atmosphere of Earth receives about 1.5×10^{21} watt-hours (thermal) of solar radiation annually. This vast amount of energy is more than 23,000 times that used by the human population of this planet, but it is only about one two-billionth of the Sun's massive outpouring—about 3.9×10^{20} MW.

The power density of solar radiation measured just outside Earth's atmosphere and over the entire solar spectrum is called the solar constant. According to the World Meteorological Organization, the most reliable (1981) value for the solar constant is 1370 ± 6 W/m^2.

Solar radiation is attenuated before reaching Earth's surface by an atmosphere that removes or alters part of the incident energy by reflection, scattering, and absorption. In particular, nearly all ultraviolet radiation and certain wavelengths in the infrared region are removed. However, the solar radiation striking Earth's surface each year is still more than 10,000 times the world's energy use. Radiation scattered by striking gas molecules, water vapor, or dust particles is known as diffuse radiation. Clouds are a particularly important scattering and reflecting agent, capable of reducing direct radiation by as much as 80 to 90%. The radiation arriving at the ground directly from the Sun is called direct or beam radiation. Global radiation is all solar radiation incident on the surface, including direct and diffuse.

Solar research and technology development aim at finding the most efficient ways of capturing low-density solar energy and developing systems to convert captured energy to useful purposes. Also of significant potential as power sources are the indirect forms of solar energy: wind, biomass, hydropower, and the tropical ocean surfaces. With the exception of hydropower, these energy resources remain largely untapped. *See* ENERGY SOURCES.

Five major technologies using solar energy are being developed. (1) The heat content of solar radiation is used to provide moderate-temperature heat for space comfort conditioning of buildings, moderate- and high-temperature heat for industrial processes, and high-temperature heat for generating electricity. (2) Photovoltaics convert solar energy directly into electricity. (3) Biomass technologies exploit the chemical energy produced through photosynthesis (a reaction energized by solar radiation) to produce energy-rich fuels and chemicals and to provide direct heat for many uses. (4) Wind energy systems generate mechanical energy, primarily for conversion to electric power. (5) Finally, a number of ocean energy applications are being pursued; the most advanced is ocean thermal energy conversion, which uses temperature differences between warm ocean surface water and cooler deep water to produce electricity. *See* BIOMASS; WIND.

Solar energy can be converted to useful work or heat by using a collector to absorb solar radiation, allowing much of the Sun's radiant energy to be converted to heat. This heat can be used directly in residential, industrial, and agricultural operations; converted to mechanical or electrical power; or applied in chemical reactions for production of fuels and chemicals.

A solar energy system is normally designed to be able to deliver useful heat for 6 to 10 h a day, depending on the season and weather. Storage capacity in the solar thermal system is one way to increase a plant's operating capacity.

There are four primary ways to store solar thermal energy: (1) sensible-heat-storage systems, which store thermal energy in materials with good heat-retention qualities; (2) latent-heat-storage systems, which store solar thermal energy in the latent heat of fusion or vaporization of certain materials undergoing a change of phase; (3) chemical energy storage, which uses reversible reactions (for example, the dissociation-association reaction of sulfuric acid and water); and (4) electrical or mechanical storage, particularly through the use of storage batteries (electrical) or compressed air (mechanical).

Photovoltaic systems convert light energy directly to electrical energy. Using one of the most versatile solar technologies, photovoltaic systems can, because of their modularity, be designed for power needs ranging from milliwatts to megawatts. They can be used to provide power for applications as small as a wristwatch to as large as an entire community. They can be used in centralized systems, such as a generator in a power plant, or in dispersed applications, such as in remote areas not readily accessible to utility grid lines.

Biomass energy is solar energy stored in plant and animal matter. Through photosynthesis in plants, energy from the Sun transforms simple elements from air, water, and soil into complex carbohydrates. These carbohydrates can be used directly as fuel (for example, burning wood) or processed into liquids and gases (for example, ethanol or methane). Biomass is a renewable energy resource because it can be harvested periodically and converted to fuel. *See* PHOTOSYNTHESIS.

Wind is a source of energy derived primarily from unequal heating of Earth's surface by the Sun. Energy from the wind has been used for centuries to propel ships, to grind grain, and to lift water. Wind turbines extract energy from the wind to perform mechanical work or to generate electricity.

Ocean thermal energy conversion uses the temperature difference between surface water heated by the Sun and deep cold water pumped from depths of 2000 to 3000 ft (600 to 900 m). This temperature difference makes it possible to produce electricity from the heat engine concept. Since the ocean acts as an enormous solar energy storage facility with little fluctuation of temperature over time, ocean thermal energy conversion, unlike most other renewable energy technologies, can provide electricity 24 h a day. [R.L.S.M.]

Sourwood A deciduous tree, *Oxydendrum arboreum*, of the heath family, indigenous to the southeastern section of the United States, and found from Pennsylvania to Florida and west to Indiana and Louisiana. It is usually a small or medium-sized tree. The wood is not used commercially. Sourwood is also known as sorrel tree, and it is widely planted as an ornamental. [A.H.G./K.P.D.]

South America The southernmost of the New World or Western Hemisphere continents, with three-fourths of it lying within the tropics. South America is approximately 4500 mi (7200 km) long and at its greatest width 3000 mi (4800 km). Its area is estimated to be about 7,000,000 mi^2 (18,000,000 km^2). South America has many unique physical features, such as the Earth's longest north-south mountain range (the Andes), highest waterfall (Angel Falls), highest navigable fresh-water lake (Lake Titicaca), and largest expanse of tropical rainforest (Amazonia). The western side of the continent has a deep subduction trench offshore, whereas the eastern continental shelf is more gently sloping and relatively shallow. *See* CONTINENT.

South America has three distinct regions: the relatively young Andes Mountains located parallel to the western coastline, the older Guiana and Brazilian Highlands located near the eastern margins of the continent, and an extensive lowland plains, which occupies the central portion of the continent. The regions have distinct physiographic and biotic features.

The Andes altitudes often exceed 20,000 ft (6000 m) and perpetual snow tops many of the peaks, even along the Equator. So high are the Andes in the northern half of the continent that few passes lie below 12,000 ft (3600 m). Because of the vast extent of the Andes, a greater proportion of South America than of any other continent lies above 10,000 ft (3000 m). The young, rugged, folded Andean peaks stand in sharp contrast to the old, worn-down mountains of the eastern highlands. Although the Andes appear to be continuous, most geologists believe that they consist of several structural units, more or less joined. They are a single range in southern Chile, two ranges in Bolivia, and dominantly three ranges in Peru, Ecuador, and Colombia.

Except in Bolivia, where they attain their maximum width of 400 mi (640 km), the Andes are seldom more than 200 mi (320 km) wide. The average height of the Andes is estimated to be 13,000 ft (3900 m). However, it is only north of latitude 35°S that the mountains exceed elevations of 10,000 ft (3000 m).

From the southern tip of Cape Horn north to 41°S latitude, the western coastal zone consists of a broad chain of islands where a mountainous strip subsided and the ocean invaded its valleys. This is one of the world's finest examples of a fiorded coast. Nowhere along the Pacific coast is there a true coastal plain. South of Arica, Chile, the bold, precipitous coast is broken by only a few deep streams, the majority of which carry no water for years at a time. Between Arica and Caldera, Chile, there are no natural harbors and almost no protected anchorages. In fact, South America's coastline is the least indented of all the continents except Africa's. *See* FIORD.

The Caribbean coast of Colombia is a lowland formed largely of alluvium, deposited by the Magdalena and Cauca rivers, and bounded by mountains on three sides. In Venezuela, the Central Highlands rise abruptly from the Caribbean, with lowlands around Lake Maracaibo, west of Puerto Cabello, and around the mouth of the Río Tuy of the Port of Guanta. The coastal region of Guyana, Suriname, and French Guiana is a low, swampy alluvial plain 10–30 mi (16–48 km) wide, and as much as 60 mi (96 km) wide along the larger rivers. This coastal plain is being built up by sediments carried by the Amazon to the Atlantic and then deflected westward by the equatorial current and cast upon the shore by the trade winds.

There is no broad coastal plain south of the Amazon and east of the Brazilian Highlands to afford easy access to the interior. The rise from the coastal strip to the interior is quite gradual in northeastern Brazil; but southward, between Bahia and Río Grande do Sul, the steep Serra do Mar is a formidable obstacle to transportation.

Along coastal Uruguay there is a transition between the hilly uplands and plateaus of Brazil and the flat Pampas of Argentina, whereas coastal Argentina as far south as the Río Colorado, in Patagonia, is an almost featureless plain. In Patagonia, steep cliffs rise from the water's edge. Behind these cliffs lies a succession of dry, flat-topped plateaus, surmounted occasionally by hilly land composed of resistant crystalline rocks. Separating southern Patagonia from Tierra del Fuego is the Strait of Magellan, which is 350 mi (560 km) long and 2–20 mi (3–32 km) wide. Threading through numerous islands, the strait is lined on each side with fiords and mountains.

There are three great river systems in South America and a number of important rivers that are not a part of these systems. The largest river system is the Amazon which, with its many tributaries, drains a basin covering 2,700,000 mi^2 (7,000,000 km^2), or about 40% of the continent. The next largest is the system composed of the Paraguay, Paraná, and La Plata rivers, the last being a huge estuary. The third largest river system, located in southern Venezuela, is the Orinoco, which drains 365,000 mi^2 (945,000 km^2) of land, emptying into the Atlantic Ocean along the northeast edge of the continent.

The plants and animals of the South American tropics are classified as Neotropical, defined by the separation of the South American and African continents during the Middle Cretaceous (95 million years ago). The Paraná basalt flow, which caps the Brazilian shield in southern Brazil and adjacent parts of Uruguay and Argentina, as well as western Africa, indicates the previous linkage between the South American and African continents. South America has many biotic environments, including the constantly moist tropical rainforest, seasonally dry deciduous forests and savannas, and high-altitude tundra and glaciated environments.

Amazonia contains the largest extent of tropical rainforest on Earth. It is estimated to encompass up to 20% of the Earth's higher plant species and is a critically important source of fresh water and oxygen. Structurally complex, the rainforest is composed of up to four distinct vertical layers of plants and their associated fauna. The layers often cluster at 10, 20, 98, and 164 ft (3, 6, 30, and 50 m) in height. The lower canopy and forest floor are usually open spaces because of the low intensity of light (around 1%) that reaches the forest floor. Over 75% of Amazonian soils are classified as infertile, acidic, or poorly drained, making them undesirable for agriculture because of nutrient deficiencies. Most of the nutrients in the tropical rainforest are quickly absorbed and stored in plant biomass because the high annual rainfall and associated leaching make it impossible to maintain nutrients in the soils. In addition to the high structural complexity of the tropical rainforest, there is considerable horizontal diversity or patchiness. As many as 300 separate species of trees can be found in a square mile (2.6 km^2) sample tract of rainforest in Brazil. The high complexity and species diversity of the rainforest are the result of long periods of relative stability in these regions. *See* RAINFOREST.

Deciduous forest are found in areas where there is seasonal drought and the trees lose their leaves in order to slow transpiration. The lower slopes of the Andes, central Venezuela, and central Brazil are areas where these formations are found. Conifer forests occur in the higher elevations of the Andes and the higher latitudes of Chile and Argentina. *See* DECIDUOUS PLANTS.

Tropical savannas occupy an extensive range in northern South America through southeastern Venezuela and eastern Colombia. Temperate savannas are found in Paraguay, Uruguay, the Pampas of Argentina, and to the south, Patagonia. Savannas are composed of a combination of grass and tree species. The climate in these areas is often quite hot with high rates of evapotranspiration and a pronounced dry season. Most of the plants and animals of these zones are drought-adapted and fire-adapted. Tall grasses up to 12 ft (3.5 m) are common as are thorny trees of the Acacia (Fabaceae) family. Many birds and mammals are found in these zones, including anteater, armadillo, capybara (the largest rodent on Earth), deer, jaguar, and numerous species of venomous snake, including rattlesnake and bushmaster (*mapanare*). *See* SAVANNA.

South America is unique in having a west-coast desert (the Atacama) that extends almost to the Equator, probably receiving less rain than any on Earth, and an east coast desert located poleward from latitude 40°S (the Patagonian). *See* DESERT.

In Bolivia and Peru the zone from 10,000 to 13,000 ft (3000 to 3900 m), though occasionally to 15,000–16,000 ft (4500 to 4800 m), is known as the *puna*. Here the hot days contrast sharply with the cold nights. Above the *puna*, from timberline to snowline, is the *paramo*, a region of broadleaf herbs and grasses found in the highest elevations of Venezuela, Colombia, and Ecuador. Many of the plant species in these environments are similar to those found at lower elevations; however, they grow closer to the ground in order to conserve heat and moisture. *See* PARAMO; PUNA. [D.A.Sa.; C.L.W.]

South Pole That end of the Earth's rotational axis opposite the North Pole. It is the southernmost point on the Earth and one of the two points through which all meridians pass (the North Pole being the other point). This is the geographic pole and is not the same as the south magnetic pole. The South Pole lies inland from the Ross Sea, within the land mass of Antarctica, at an elevation of about 9200 ft (2800 m).

There is no natural way to determine local Sun time because there is no noon position of the Sun, and shadows point north at all times, there being no other direction from the South Pole. *See* MATHEMATICAL GEOGRAPHY; NORTH POLE. [V.H.E.]

Southeast Asian waters All the seas between Asia and Australia and the Pacific and the Indian oceans. They form a geographical and oceanographical unit because of their special structure and position, and make up an area of 3,450,000 mi^2 (8,940,000 km^2), or about 2.5% of the surface of all oceans.

The surface circulation is completely reversed twice a year by the changing monsoon winds. The subsurface circulation carries chiefly the outrunners of the intermediate waters of the Pacific Ocean into these seas. The tides are mostly of the mixed type. Diurnal tides are found in the Java Sea, in the Gulf of Tonkin, and in the Gulf of Thailand. Semidiurnal tides with high amplitudes occur in the Malacca Straits. *See* INDIAN OCEAN; PACIFIC OCEAN. [K.W.]

The Southeast Asian seas are characterized by the presence of numerous major plate boundaries. In Southeast Asia the plate boundaries are identified, respectively, by young, small ocean basins with their spreading systems and associated high heat flow; deep-sea trenches and their associated earthquake zones and volcanic chains; and major strike-slip faults such as the Philippine Fault (similar to the San Andreas Fault in California). The Southeast Asian seas are thus composed of a mosaic of about 10 small ocean basins whose boundaries are defined mainly by trenches and volcanic arcs. The dimensions of these basins are much smaller than the basins of the major oceans. The major topographic features of the region are believed to represent the surface expression of plate interactions, the scars left behind on the sea floor. [D.E.H.]

Speciation The process by which new species of organisms evolve from preexisting species. It is part of the whole process of organic evolution. The modern period of its study began with the publication of Charles Darwin's and Alfred Russell Wallace's *Theory of Evolution by Natural Selection* in 1858, and Darwin's *On the Origin of Species* in 1859.

Belief in the fixity of species was almost universal before the middle of the nineteenth century. Then it was gradually realized that all species continuously change, or evolve; however, the causative mechanism remained to be discovered. Darwin proposed a mechanism. He argued that (1) within any species population there is always some heritable variation; the individuals differ among themselves in structure, physiology, and behavior; and (2) natural selection acts upon this variation by eliminating the less fit. Thus if two members of an animal population differ from each other in their ability to find a mate, obtain food, escape from predators, resist the ravages of parasites and pathogens, or survive the rigors of the climate, the more successful will be more likely than the less successful to leave descendants. The more successful is said to have greater fitness, to be better adapted, or to be selectively favored. Likewise among plants: one plant individual is fitter than another if its heritable characteristics make it more successful than the other in obtaining light, water, and nutrients, in protecting itself from herbivores and disease organisms, or in surviving adverse climatic conditions. Over the course of time, as the fitter members of a population leave more descendants than the less fit, their characteristics become more common.

This is the process of natural selection, which tends to preserve the well adapted at the expense of the ill adapted in a variable population. The genetic variability that must exist if natural selection is to act is generated by genetic mutations in the broad sense, including chromosomal rearrangements together with point mutations. *See* GENETICS; MUTATION.

If two separate populations of a species live in separate regions, exposed to different environments, natural selection will cause each population to accumulate characters adapting it to its own environment. The two populations will thus diverge from each other and, given time, will become so different that they are no longer interfertile. At this point, speciation has occurred: two species have come into existence in the place of one. This mode of speciation, speciation by splitting, is probably the most common mode. Two other modes are hybrid speciation and phyletic speciation; many biologists do not regard the latter as true speciation.

Many students of evolution are of the opinion that most groups of organisms evolve in accordance with the punctuated equilibrium model rather than by phyletic gradualism. There are two chief arguments for this view. First, it is clear from the fossil record that many species persist without perceptible change over long stretches of time and then suddenly make large quantum jumps to radically new forms. Second, phyletic gradualism seems to be too slow a process to account for the tremendous proliferation of species needed to supply the vast array of living forms that have come into existence since life first appeared on Earth. *See* ANIMAL EVOLUTION; POPULATION GENETICS; SPECIES CONCEPT. [E.C.P.]

Species concept The species is the fundamental unit of organization of the taxonomic system; of interactions between organisms as described by geneticists and ecologists; and of evolution as studied by phylogeneticists. As a category the term species resists definition; thus, a species concept is adopted as a framework within which biologists of various persuasions delineate the taxa with which they work at the species level. However, no universal concept has been accepted by all biologists for two fundamental reasons: (1) Different groups of organisms in nature are organized differently in terms of reproductive mechanisms and patterns; in degrees of differentiation among species in morphological, genetic, physiological, behavioral, biochemical, and other types of characters; and in the modes of speciation that have given rise to the members of the group. (2) The philosophy, training, working methods, and goals of different of biologists affect the manner in which each perceives the coherence or diversity of the biological world in general and that of the group of organisms in question in particular.

According to the taxonomic concept, a species consists of groups of individuals (populations) that are morphologically similar to one another, and differ morphologically from other such groups. There are several important ideas expressed in this concept. First, there is internal cohesiveness; that is, the members of the species share certain characteristics. Second, there is external distinction because other species have different characteristics, and thus species may be distinguished from one another. Third, the characteristics that a species possesses may be easily observed because they are phenotypic; that is, a species may be identified by its appearance.

Difficulty in applying the taxonomic concept arises with certain groups of organisms. Bacteria are often identified by physiological and biochemical tests requiring sophisticated laboratories and equipment; in addition, the mutation rate in bacteria is so high that the various traits used to identify them can change rapidly. In insects, the morphological differences between species may be very slight and easily overlooked. In certain groups of plants, hybridization and polyploidy have led to a continuous range of variation of characters, in which no discontinuities sufficient to distinguish species can be discerned. Critics claim that the purely phenetic approach of the taxonomic concept may not reflect real genetic or breeding relationships. However, this concept provides guidelines by which species may be recognized by ordinary (nonexperimental) means. The composition of a species so recognized can then be subjected to hypothesis testing within the framework of other concepts. *See* TAXONOMIC CATEGORIES.

According to the biological concept, a species is composed of groups of individuals (populations) that normally interbreed with one another. The fundamental ideas expressed by this concept are that the internal cohesiveness of a species is maintained by the exchange of genes through sexual reproduction (gene flow) and that the distinctness of the species is maintained by reproductive isolation (barriers to gene flow) from other groups of populations. If two populations do not exchange genes, they belong to separate species regardless of their morphological similarity.

This concept works well in those groups of organisms that are exclusively outbreeding, such as birds and mammals. However, it is difficult to apply to plants, in which interbreeding between morphologically very distinct species and even genera is common. Also, those organisms that do not reproduce sexually present problems of classification. Even in sexually reproducing organisms, populations that are morphologically identical but reproductively separated by geographic distance (disjuncts) present problems of classification within the framework of the concept. The populations might interbreed if they were in contact, but this can be determined only under artificial conditions and not in nature. However, the development of the biological species concept has contributed greatly to making taxonomy an evolutionary science because of its emphasis on the identification of genetic, rather than the very possibly superficial phenetic, relationship among organisms.

According to the evolutionary concept, a species is a lineage of ancestor-descendant populations that maintains its identity from other such lineages and that has its own evolutionary tendencies and historical fate. The important ideas expressed in this concept are the following. (1) All organisms, regardless of their mode of reproduction, belong to some evolutionary species. (2) Species need be reproductively isolated from one another to the extent that they maintain their distinction from other species. (3) There may or may not be a morphological discontinuity between species but, if there is, it is reasonable to hypothesize that more than one species is present. If there is not, other data such as that on breeding relationships may be used to recognize species.

The evolutionary concept encompasses the taxonomic concept, the biological concept, and other more narrowly defined concepts—for example, the ecological species, the genetic species, and the paleospecies. It is operational in that it provides guidelines for the recognition of species and for testing of hypotheses concerning membership in each species; it also is compatible with the Linnaean taxonomic hierarchy. As it becomes more widely used by working systematists, problems and difficulties with the concept may appear that will require its refinement. However, the evolutionary concept may in the long run be more acceptable to a wider group of biologists than any other yet proposed. *See* TAXONOMY. [M.A.La.]

Sphaerularoidea A superfamily of parasitic nematodes in the order Tylenchida. Adult females are hemocoel parasites of insects and mites; a few taxa contain both plant and insect parasites. In general, nematodes belonging to this group have three distinct phases in their life cycles: two free-living and one parasitic. In the free-living phase the female gonad is single, anteriorly directed, and with few developing oocytes and a prominent uterus filled with sperm. In the parasitic phase either the body becomes grossly enlarged and degenerates to a reproductive sac, or the uterus prolapses and gonadal development takes place outside the body. The males are always free-living and not infective. The most interesting taxon is *Sphaerularia bombi*, which prolapses the uterus. When totally prolapsed, the gonad becomes the parasite and the original female a useless appendage. It is not unusual for the prolapsed gonad to attain a volume 15,000 times that of the original female. *See* NEMATA. [A.R.M.]

Spirurida An order of nematodes in which the labial region is usually provided with two lateral labia or pseudolabia; in some taxa there are four or more lips; rarely lips are absent. Because of the variability in lip number, there is variation in the shape of the oral opening, which may be surrounded by teeth. The amphids are most often laterally located; however, in some taxa they may be located immediately posterior to the labia or pseudolabia. The stoma may be cylindrical and elongate or rudimentary. The esophagus is generally divisible into an anterior muscular portion and an elongate swollen posterior glandular region, where the multinucleate glands are located. Eclosion larvae are usually provided with a cephalic spine or hook and a porelike phasmid on the tail.

All known spirurid nematodes utilize an invertebrate in their life cycle; the definitive hosts are mammals, birds, reptiles, and rarely amphibians. The order contains four superfamilies: Spiruroidea, Physalopteroidea, Filarioidea, and Drilone-matoidea.

Spiruroidea. The Spiruroidea comprise parasitic nematodes whose life cycle always requires an intermediate host for larvae to the third stage. The definitive hosts are mammals, birds, fishes, reptiles, and rarely amphibians; and spiruroids may be located in the host's digestive tract, eye, or nasal cavity or in the female reproductive system. Morphologically, the lip region is variable in Spiruroidea, ranging from four lips to none. When lips are present, the lateral lips are well developed and are referred to as pseudolabia. The cephalic and cervical region may be ornamented with cordons, collarettes, or cuticular rings. The stoma is always well developed and is often provided with teeth just inside the oral opening. In birds the nematodes are often associated with the gizzard, and the damage caused results in death, generally by starvation. When the muscles of the gizzard are destroyed, seeds pass intact and cannot be digested.

This superfamily contains the largest of all known nematodes, *Placentonema gigantissima*, parasitic in the placenta of sperm whales. Mature females attain a length of 26 ft (8 m) and a diameter of 1 in. (2.5 cm). The adult female has 32 ovaries, which produce great numbers of eggs. [A.R.M.]

Filarioidea. The Filarioidea contain highly specialized parasites of most groups of vertebrates. They are particularly common in amphibians, birds, and mammals. While they cannot be classified as completely harmless, most of the many hundreds of known species are not associated with any recognized disease. A limited number of species produce serious diseases in humans, and a few others produce serious diseases in domestic or wild animals. The filarial parasites of humans are found almost exclusively in the tropics, with some extension into the subtropics.

There are no conspicuous divisions into distinct body regions. Sexual dimorphism is the rule; in common with other nematodes, the female filarioid is at least twice as long as the male, and often the difference is much greater. The adult worms are found in a wide variety of places in the body of the vertebrate host, but each species has its preferred host and preferred location within that host.

All the known filariae require a bloodsucking arthropod intermediate host, usually an insect and commonly a dipteron, in which to complete embryonation. The microfilariae are ingested as the arthropod feeds. After embryonation is completed, the resulting infective larvae gain entrance into the definitive vertebrate in association with the next feeding of the arthropod. *See* NEMATA.

Filariasis. Filariasis is a disease caused by Filarioidea in humans or lower animals. The term is loosely used to indicate mere infection by such organisms. In human medicine, filariasis commonly refers to the disease caused by, or to infection with, one of the mosquito-borne, elephantoid-producing filarioids—most frequently *Wuchereria bancrofti*, less frequently *Brugia malayi*, and more recently *B. timori*. The only specific laboratory aid to diagnosis is the detection and identification of the microfilariae.

Onchocerciasis. This disease is caused by *Onchocerca volvulus* in the subcutaneous lymphatics. It is characterized by subcutaneous nodules which are most conspicuous where the skin lies close over bony structures, via cranium, pelvic girdle, joints, and shoulder blades. When they are on the head, the microfilariae reach the eyes. Ocular disturbances vary from mild transient bleary vision to total and permanent blindness.

Loa loa. The African eye worm, *Loa loa*, is the filarioidean worm most commonly acquired by Caucasian immigrants, including missionaries, in Africa. Transmission is by daytime-feeding sylvan deer-flies, genus *Chrysops*. The only preventive measures are protective clothing, including head nets. Repellents have some value. Fortunately, serious damage is rare even when the worm gets into the eye. The areas of pitting edema known as calabar swellings are painful and diagnostic. They commonly occur on the wrists, hands, arms, or orbital tissues. [G.F.O.]

Sporozoa A subphylum of Protozoa, typically with spores. The spores are simple and have no polar filaments. There is a single type of nucleus. There are no cilia or flagella except for flagellated microgametes in some groups. In most Sporozoa there is an alternation of sexual and asexual stages in the life cycle. In the sexual stage, fertilization is by syngamy, that is, the union of male and female gametes. All Sporozoa are parasitic. The subphylum is divided into three classes—Telosporea, Toxoplasmea, and Haplosporea. *See* PROTOZOA. [N.D.L.]

Spruce Evergreen tree belonging to the genus *Picea* of the pine family. The needles are single, usually four-sided, and borne on little peglike projections; the cones are pendulous. Resin ducts in the wood may be seen with a magnifying lens, but they are fewer than in *Pinus*.

The white spruce (*P. glauca*), ranging from northern New England to the Lake States and Montana and northward into Alaska, is distinguished by the somewhat bluish cast of its needles, small cylindrical cones, and gray or pale-brown twigs without pubescence (hairs). Red spruce (*P. rubens*) is a similar tree but with greener foliage; smaller, more oval cones; and more or less pubescent twigs. Occurring naturally with white spruce in the northeastern United States and adjacent Canada, red spruce extends southward along the Appalachians into North Carolina. Black spruce (*P. mariana*) ranges from northern New England and Newfoundland to Alaska. However, it occurs sparingly in the Appalachians to West Virginia. The cones are smaller than in the white and red species and are egg-shaped or nearly spherical and persistent. The twigs are pubescent.

Blue spruce (*P. pungens*), also known as Colorado blue spruce, is probably the best known of the western species because of its wide use as an ornamental tree. The twigs are glabrous (without pubescence). Engelmann spruce (*P. engelmanni*) has needles usually of a deep blue-green color, sometimes much like those of the blue spruce but the young twigs are slightly hairy. The cones, although cylindrical, are smaller than in blue spruce. This species is also a Rocky Mountain tree like the blue spruce, but it is more widely distributed from British Columbia to Arizona and also in the mountains of Oregon and Washington. Sitka spruce (*P. sitchensis*) is the largest spruce in the Northern Hemisphere. The leaves have a pungent odor, are considerably flattened, and stand out from the twig in all directions. It ranges from Alaska to northern California. The Norway spruce (*P. abies*), the common spruce of Europe, is much planted in the United States for timber, as well as for ornamental purposes. It can be recognized by the dark-green color of the leaves; glabrous, pendent, short branchlets; and large cones, usually near the top of the tree. [A.H.G./K.P.D.]

Stalactites and stalagmites Stalactites, stalagmites, dripstone, and flowstone are travertine deposits in limestone caverns, formed by the evaporation of waters bearing calcium carbonate. Stalactites grow down from the roofs of caves and tend to be long and thin, with hollow cores. The water moves down the core and precipitates at the bottom, slowly extending the length while keeping the core open for more water to move down.

Stalagmites grow from the floor up and are commonly found beneath stalactites; they are formed from the evaporation of the same drip of water that forms the stalactite. Stalagmites are thicker and shorter than stalactites and have no central hollow core. *See* CAVE. [R.Si.]

Staphylococcus A genus of bacteria containing at least 28 species that are collectively referred to as staphylococci. Their usual habitat is animal skin and mucosal surfaces. Although the genus is known for the ability of some species to cause infectious diseases, many species rarely cause infections. Pathogenic staphylococci are usually opportunists and cause illness only in compromised hosts. *Staphylococcus aureus*, the most pathogenic species, is usually identified by its ability to produce coagulase (proteins that affect fibrinogen of the blood-clotting cascade). Since most other species of staphylococci do not produce coagulase, it is useful to divide staphylococci into coagulase-positive and coagulase-negative species. Coagulase-negative staphylococci are not highly virulent but are an important cause of infections in certain high-risk groups. Although *Staphylococcus* infections were once readily treatable with antibiotics, some strains have acquired genes making them resistant to multiple antimicrobial agents. *See* BACTERIA; DRUG RESISTANCE; MEDICAL BACTERIOLOGY.

Staphylococcus cells are spherical with a diameter of 0.5–1.5 micrometers. Clumps of staphylococci resemble bunches of grapes when viewed with a microscope, owing to cell division in multiple planes. The staphylococci have a gram-positive cell composition, with a unique peptidoglycan structure that is highly cross-linked with bridges of amino acids.

Most species are facultative anaerobes. Within a single species, there is a high degree of strain variation in nutritional requirements. Staphylococci are quite resistant to desiccation and high-osmotic conditions. These properties facilitate their survival in the environment, growth in food, and communicability.

In addition to genetic information on the chromosome, pathogenic staphylococci often contain accessory elements such as plasmids, bacteriophages, pathogenicity islands (DNA clusters containing genes associated with pathogenesis), and transposons. These elements harbor genes that encode toxins or resistance to antimicrobial agents and may be transferred to other strains. Genes involved in virulence, especially those coding for exotoxins and surface-binding proteins, are coordinately or simultaneously regulated by loci on the chromosome. *See* BACTERIAL GENETICS; BACTERIOPHAGE.

Most *Staphylococcus aureus* infections develop into a pyogenic (pus-forming) lesion caused by acute inflammation. Inflammation helps eliminate the bacteria but also damages tissue at the site of infection. Typical pyogenic lesions are abscesses with purulent centers containing leukocytes, fluid, and bacteria. Pyogenic infections can occur anywhere in the body. Blood infections (septicemia) can disseminate the organism throughout the body and abscesses can form internally.

Certain strains of *S. aureus* produce exotoxins that mediate two illnesses, toxic shock syndrome and staphylococcal scalded skin syndrome. In both diseases, exotoxins are produced during an infection, diffuse from the site of infection, and are carried by the blood (toxemia) to other sites of the body, causing symptoms to develop at sites distant from the infection. Toxic shock syndrome is an acute life-threatening illness mediated by staphylococcal superantigen exotoxins. Staphylococcal scalded skin syndrome, also known as Ritter's disease, refers to several staphylococcal toxigenic infections. It is characterized by dermatologic abnormalities caused by two related exotoxins, the type A and B exfoliative (epidermolytic) toxins.

Staphylococcal food poisoning is not an infection, but an intoxication that results from ingestion of staphylococcal enterotoxins in food. The enterotoxins are produced when food contaminated with *S. aureus* is improperly stored under conditions that allow the bacteria to grow. Although contamination can originate from animals or the environment, food preparers with poor hygiene are the usual source. Effective methods for preventing staphylococcal food poisoning are aimed at eliminating contamination through common hygiene practices, such as wearing gloves, and proper food storage to minimize toxin production. *See* FOOD POISONING.

Coagulase-positive staphylococci are the most important *Staphylococcus* pathogens for animals. Certain diseases of pets and farm animals are very prominent. *Staphylococcus aureus* is the leading cause of infectious mastitis in dairy animals. [G.Boh.]

Stem The organ of vascular plants that usually develops branches and bears leaves and flowers. On woody stems a branch that is the current season's growth from a bud is called a twig. The stems of some species produce adventitious roots. *See* ROOT (BOTANY).

General characteristics. While most stems are erect, aerial structures, some remain underground, others creep over or lie prostrate on the surface of the ground, and still others are so short and inconspicuous that the plants are said to be stemless, or acaulescent. When stems lie flattened immediately above but not on the ground, with tips curved upward, they are said to be decumbent, as in juniper. If stems lie flat on the ground but do not root at the nodes (joints), the stem is called procumbent or prostrate, as in purslane. If a stem creeps along the ground, rooting at the nodes, it is said to be repent or creeping, as in ground ivy.

Most stems are cylindrical and tapering, appearing circular in cross section; others may be quadrangular or triangular.

Herbaceous stems (annuals and herbaceous perennials) die to the ground after blooming or at the end of the growing season. They usually contain little woody tissue. Woody stems (perennials) have considerable woody supporting tissue and live from year to year. A woody plant with no main stem or trunk, but usually with several stems developed from a common base at or near the ground, is known as a shrub. [N.A.]

External features. A shoot or branch usually consists of a stem, or axis, and leafy appendages. Stems have several distinguishing features. They arise either from the epicotyl of the embryo in a seed or from buds. The stem bears both leaves and buds at nodes, which are separated by leafless regions or internodes, and sometimes roots and flowers (see illustration).

The nodes are the regions of the primary stem where leaves and buds arise. The number of leaves at a node is usually specific for each plant species. In deciduous plants which are leafless during winter, the place of former attachment of a leaf is marked by the leaf scar. The scar is formed in part by the abscission zone formed at the base of the leaf petiole. The stem regions between nodes are called internodes. Internode length varies greatly among species, in different parts of the same stem, and under different growing conditions.

Lenticels are small, slightly raised or ridged regions of the stem surface that are composed of loosely arranged masses of cells in the bark. Their intercellular spaces are continuous with those in the interior of the stem, therefore permitting gas exchange similar to the stomata that are present before bark initiation.

There are three major types of stem branching: dichoto-mous, monopodial, and sympodial. Dichotomy occurs by a division of the apical meristem to form two new axes. If the terminal bud of an axis continues to grow and lateral buds grow out as branches, the branching is called monopodial. If the apical bud terminates growth in

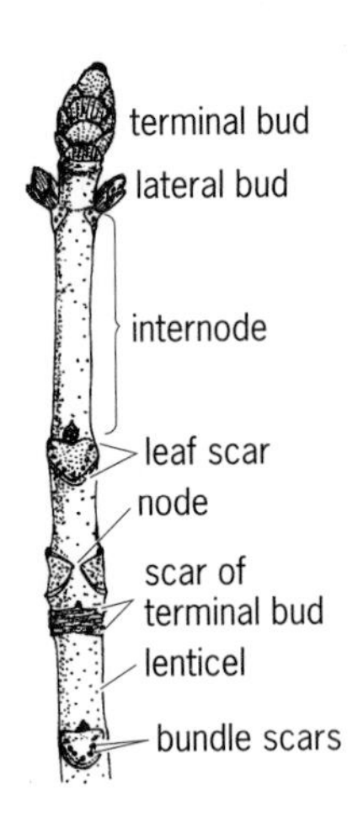

Winter woody twig (horse chestnut) showing apical dominance. (*After E. W. Sinnott and K. S. Wilson, Botany: Principles and Problems, 5th ed., McGraw-Hill, 1955*)

a flower or dies back and one or more axillary buds grow out, the branching is called sympodial. Often only one bud develops so that what appears to be single axis is in fact composed of a series of lateral branches arranged in linear sequence.

The large and conspicuous stems of trees and shrubs assume a wide variety of distinctive forms. Columnar stems are basically unbranched and form a terminal leaf cluster, as in palms, or lack obvious leaves, as in cacti. Branching stems have been classified either as excurrent, when there is a central trunk and a conical leaf crown, as in firs and other conifers, or as decur-rent (or deliquescent), when the trunk quickly divides up into many separate axes so that the crown lacks a central trunk, as in elm. *See* TREE. [J.B.F.]

Internal features. The stem is composed of the three fundamental tissue systems that are found also in all other plant organs: the dermal (skin) system, consisting of epidermis in young stems and peridem in older stems of many species; the vascular (conducting) system, consisting of xylem (water conduction) and phloem (food conduction); and the fundamental or ground tissue system, consisting of parenchyma and sclerenchyma tissues in which the vascular tissues are embedded. The arrangement of the vascular tissues varies in stems of different groups of plants, but frequently these tissues form a hollow cylinder enclosing a region of ground tissue called pith and separated from the dermal tissue by another region of ground tissue called cortex. *See* CORTEX (PLANT); EPIDERMIS (PLANT); PARENCHYMA; PHLOEM; PITH; SCLERENCHYMA; XYLEM.

Part of the growth of the stem results from the activity of the apical meristem located at the tip of the shoot. The derivatives of this meristem are the primary tissues; epidermis, primary vascular tissues, and the ground tissues of the cortex and pith. In many species, especially those having woody stems, secondary tissues are added to the primary. These tissues are derived from the lateral meristems, oriented parallel with the sides of the stem: cork cambium (phellogen), which gives rise to the secondary protective tissue periderm, which consists of phellum (cork), phellogen (cork cambium), and phelloderm (secondary cortex) and which replaces the epidermis; and vascular cambium, which is inserted between the primary xylem and phloem and forms secondary xylem (wood) and phloem.

The vascular tissues and the closely associated ground tissues—pericycle (on the outer boundary of vascular region), interfascicular regions (medullary or pith rays), and frequently also the pith—may be treated as a unit called the stele. The variations in the arrangement of the vascular tissues serve as a basis for distinguishing the stelar types. The word stele means column and thus characterizes the system of vascular and associated ground tissues as a column. This column is enclosed within the cortex, which is not part of the stele. *See* PERICYCLE. [J.E.Gu.]

Sterilization An act of destroying all forms of life on and in an object. A substance is sterile, from a microbiological point of view, when it is free of all living microorganisms. Sterilization is used principally to prevent spoilage of food and other substances and to prevent the transmission of diseases by destroying microbes that may cause them in humans and animals. Microorganisms can be killed either by physical agents, such as heat and irradiation, or by chemical substances.

Heat sterilization is the most common method of sterilizing bacteriological media, foods, hospital supplies, and many other substances. Either moist heat (hot water or steam) or dry heat can be employed, depending upon the nature of the substance to be sterilized. Moist heat is also used in pasteurization, which is not considered a true sterilization technique because all microorganisms are not killed; only certain pathogenic organisms and other undesirable bacteria are destroyed.

Many kinds of radiations are lethal, not only to microorganisms but to other forms of life. These radiations include both high-energy particles as well as portions of the electromagnetic spectrum.

Filtration sterilization is the physical removal of microorganisms from liquids by filtering through materials having relatively small pores. Sterilization by filtration is employed with liquid that may be destroyed by heat, such as blood serum, enzyme solutions, antibiotics, and some bacteriological media and medium constituents. Examples of such filters are the Berkefeld filter (diatomaceous earth), Pasteur-Chamberland filter (porcelain), Seitz filter (asbestos pad), and the sintered glass filter.

Chemicals are used to sterilize solutions, air, or the surfaces of solids. Such chemicals are called bactericidal substances. In lower concentrations they become bacteriostatic rather than bactericidal; that is, they prevent the growth of bacteria but may not kill them. Other terms having similar meanings are employed. A disinfectant is a chemical that kills the vegetative cells of pathogenic microorganisms but not necessarily the endospores of spore-forming pathogens. An antiseptic is a chemical applied to living tissue that prevents or retards the growth of microorganisms, especially pathogenic bacteria, but which does not necessarily kill them.

The desirable features sought in a chemical sterilizer are toxi-city to microorganisms but nontoxicity to humans and animals, stability, solubility, inability to react with extraneous organic materials, penetrative capacity, detergent capacity, noncorrosiveness, and minimal undesirable staining effects. Rarely does one chemical combine all these desirable features. Among chemicals that have been found useful as sterilizing agents are the phenols, alcohols, chlorine compounds, iodine, heavy metals and metal complexes, dyes, and synthetic detergents, including the quaternary ammonium compounds. [C.F.N.]

Stichosomida An order of nematodes composed of taxa that are parasitic in either vertebrates or invertebrates. The most distinguishing characteristic is the modification of the posterior esophagus into a stichosome, a series of glands exterior to the esophagus proper. The stichosome may be in one or two rows. The early larval stages possess a protrusible stylet that is absent in the adults. Amphids are postlabial.

There are three stichosomid superfamilies: Trichocephaloidea parasitize vertebrates. In Mermithoidea, adult worms are free living, but their larvae are parasitic in a variety of insects, arachnids, and pulmonate snails. Echinomermelloidea parasitize marine invertebrates; the superfamily comprises three families: Echinomermellidae, Marimermithidae, and Benthimermithidae. *See* NEMATA. [A.R.M.]

Storm surge An anomalous rise in water elevations caused by severe storms approaching the coast. A storm surge can be succinctly described as a large wave that moves with the storm that caused it. The surge is intensified in the nearshore, shallower regions where the surface stress caused by the strong onshore winds pile up water against the coast, generating an opposing pressure head in the offshore direction. However, there are so many other forces at play in the dynamics of the storm surge phenomenon, such as bottom friction, Earth's rotation, inertia, and interaction with the coastal geometry, that a simple static model cannot explain all the complexities involved. Scientists and engineers have dedicated many years in the development and application of sophisticated computer models to accurately predict the effects of storm surges.

The intensity and dimension of the storm causing a surge, and thus the severity of the ensuing surge elevations, depend on the origin and atmospheric characteristics of the storm itself. Hurricanes and severe extratropical storms are the cause of most significant surges. In general, hurricanes are more frequent in low to middle latitudes, and extratropical storms are more frequent in middle to high latitudes. *See* HURRICANE.
[S.R.S.]

Strangles A highly contagious disease of the upper respiratory tract of horses and other members of the family Equidae, characterized by inflammation of the pharynx and abscess formation in lymph nodes. This disease occurs in horses of all ages throughout the world. The causative agent is *Streptococcus equi*, a clonal pathogen apparently derived from an ancestral strain of *S. zooepidemicus*. It is an obligate parasite of horses, donkeys, and mules. *See* STREPTOCOCCUS.

Strangles is most common and most severe in young horses, and is very prevalent on breeding farms. The causative agent has been reported to survive for 7 weeks in pus but dies in a week or two on pasture. Transmission is either direct by nose or mouth contact or aerosol, or indirect by flies, drinking buckets, pasture, and feed. The disease is highly contagious under conditions of crowding, exposure to severe climatic conditions such as rain and cold, and prolonged transportation. Carrier animals, although of rare occurrence, are critical in maintenance of the streptococcus and in initiation of outbreaks.

The mean incubation period is about 10 days, with a range of 3–14 days. The animal becomes quieter, has fever of 39–40.5°C (102–105°F), nasal discharge, loss of appetite, and swelling of one or more lymph nodes of the mouth. Pressure of a lymph node on the airway may cause respiratory difficulty. Abscesses in affected lymph nodes rupture in 7–14 days, and rapid clinical improvement and recovery then ensues. Recovery is associated with formation of protective antibodies in the nasopharynx and in the serum.

Streptococcus equi is easily demonstrated in smears of pus from abscesses and in culture of pus or nasal swabs on colistin–nalidixic acid blood agar. Acutely affected animals also show elevated white blood cell counts and elevated fibrinogen.

Commercially available vaccines are injected in a schedule of two or three primary inoculations followed by annual boosters. However, the clinical attack rate may be reduced by only 50%, a level of protection much lower than that following the naturally occurring disease.

Procaine penicillin G is the antibiotic of choice and quickly brings about reduction of fever and lymph node enlargement. [J.F.T.]

Stratosphere The atmospheric layer that is immediately above the troposphere and contains most of the Earth's ozone. Here temperature increases upward because of absorption of solar ultraviolet light by ozone. Since ozone is created in sunlight from oxygen, a by-product of photosynthesis, the stratosphere exists because of life on Earth. In turn, the ozone layer allows life to thrive by absorbing harmful solar ultraviolet radiation. The mixing ratio of ozone is largest (10 parts per million by volume) near an altitude of 30 km (18 mi) over the Equator. The distribution of ozone is controlled by solar radiation, temperature, wind, reactive trace chemicals, and volcanic aerosols. *See* ATMOSPHERE; TROPOSPHERE.

The heating that results from absorption of ultraviolet radiation by ozone causes temperatures generally to increase from the bottom of the stratosphere (tropopause) to the top (stratopause) near 50 km (30 mi), reaching 280 K (45°F) over the summer pole. This temperature inversion limits vertical mixing, so that air typically spends months to years in the stratosphere. *See* TEMPERATURE INVERSION; TROPOPAUSE.

The lower stratosphere contains a layer of small liquid droplets. Typically less than 1 micrometer in diameter, they are made primarily of sulfuric acid and water. Occasional large volcanic eruptions maintain this aerosol layer by injecting sulfur dioxide into the stratosphere, which is converted to sulfuric acid and incorporated into droplets. Enhanced aerosol amounts from an eruption can last several years. By reflecting sunlight, the aerosol layer can alter the climate at the Earth's surface. By absorbing upwelling infrared radiation from the Earth's surface, the aerosol layer can warm the stratosphere. The aerosols also provide surfaces for a special set of chemical reactions that affect the ozone layer. Liquid droplets and frozen particles generally convert chlorine-bearing compounds to forms that can destroy ozone. They also tend to take up nitric acid and water and to fall slowly, thereby removing nitrogen and water from the stratosphere. The eruption of Mount Pinatubo (Philippines) in June 1991 is believed to have disturbed the Earth system for several years, raising stratospheric temperatures by more than 1 K (1.8°F) and reducing global surface temperatures by about 0.5 K (0.9°F).

Ozone production is balanced by losses due to reactions with chemicals in the nitrogen, chlorine, hydrogen, and bromine families. Reaction rates are governed by temperature, which depends on amounts of radiatively important species such as carbon dioxide. Human activities are increasing the amounts of these molecules and are thereby affecting the ozone layer. Evidence for anthropogenic ozone loss has been found in the Antarctic lower stratosphere. Near polar stratospheric clouds, chlorine and bromine compounds are converted to species that, when the Sun comes up in the southern spring, are broken apart by ultraviolet radiation and rapidly destroy ozone. This sudden loss of ozone is known as the anthropogenic Antarctic ozone hole. *See* STRATOSPHERIC OZONE. [M.H.H.]

Stratospheric ozone While ozone is found in trace quantities throughout the atmosphere, the largest concentrations are located in the lower stratosphere in a layer between 9 and 18 mi (15 and 30 km). Atmospheric ozone plays a critical role for the biosphere by absorbing the ultraviolet radiation with wavelength (λ) 240–320 nanometers. This radiation is lethal to simple unicellular organisms (algae, bacteria, protozoa) and to the surface cells of higher plants and animals. It also damages the genetic material of cells and is responsible for sunburn in human skin. The incidence of skin cancer has been statistically correlated with the observed surface intensity of ultraviolet wavelength 290–320 nm, which is not totally absorbed by the ozone layer. *See* STRATOSPHERE.

Ozone also plays an important role in photochemical smog and in the purging of trace species from the lower atmosphere. Furthermore, it heats the upper atmosphere by absorbing solar ultraviolet and visible radiation ($\lambda < 710$ nm) and thermal infrared radiation ($\lambda \simeq 9.6$ micrometers). As a consequence, the temperature increases steadily from about −60°F (220 K) at the tropopause (5–10 mi or 8–16 km altitude) to about 45°F (280 K) at the stratopause (30 mi or 50 km altitude). This ozone heating provides the major energy source for driving the circulation of the upper stratosphere and mesosphere. *See* ATMOSPHERIC GENERAL CIRCULATION; TROPOPAUSE.

Above about 19 mi (30 km), molecular oxygen (O_2) is dissociated to free oxygen atoms (O) during the daytime by ultraviolet photons, ($h\nu$), as shown in reaction (1). The oxygen atoms produced then form ozone (O_3) by reaction (2), where M is an arbitrary

$$O_2 + h\nu \rightarrow O + O \qquad \lambda < 242 \text{ nm} \tag{1}$$

$$O + O_2 + M \rightarrow O_3 + M \tag{2}$$

molecule required to conserve energy and momentum in the reaction. Ozone has a short lifetime during the day because of photodissociation, as shown in reaction (3).

$$O_3 + h\nu \rightarrow O_2 + O \qquad \lambda < 710 \text{ nm} \tag{3}$$

However, except above 54 mi (90 km), where O_2 begins to become a minor component of the atmosphere, reaction (3) does not lead to a net destruction of ozone. Instead the O is almost exclusively converted back to O_3 by reaction (2). If the odd oxygen concentration is defined as the sum of the O_3 and O concentrations, then odd oxygen is produced by reaction (1). It can be seen that reactions (2) and (3) do not affect the odd oxygen concentrations but merely define the ratio of O to O_3. Because the rate of reaction (2) decreases with altitude while that for reaction (3) increases, most of the odd oxygen below 36 mi (60 km) is in the form of O_3 while above 36 mi (60 km) it is in the form of O. The reaction that is responsible for a small fraction of the odd oxygen removal rate is shown as reaction (4). A significant fraction of the removal is

$$O + O_3 \rightarrow O_2 + O_2 \tag{4}$$

caused by the presence of chemical radicals [such as nitric oxide (NO), chlorine (Cl), bromine (Br), hydrogen (H), or hydroxyl (OH)], which serve to catalyze reaction (4) (see illustration).

The discovery in the mid-1980s of an ozone hole over Antarctica, which could not be explained by the classic theory of ozone and had not been predicted by earlier chemical models, led to many speculations concerning the causes of this event, which can be observed each year in September and October. As suggested by experimental and observational evidence, heterogeneous reactions on the surface of liquid or solid particles that produce Cl_2, HOCl, and $ClNO_2$ gas, and the subsequent rapid photolysis of these molecules, produces chlorine radicals (Cl, ClO) which in turn lead to the destruction of ozone in the lower stratosphere by a catalytic cycle [reactions (5)–(7)].

$$Cl + O_3 \rightarrow ClO + O_2 \tag{5}$$

$$ClO + ClO \rightarrow Cl_2O_2 \tag{6}$$

$$Cl_2O_2 + h\nu \rightarrow 2Cl + O_2 \tag{7}$$

Solar radiation is needed for these processes to occur.

Sites on which the reactions producing Cl_2, HOCl, and $ClNO_2$ can occur are provided by the surface of ice crystals in polar stratospheric clouds (PSCs). These clouds are formed between 8 and 14 mi (12 and 22 km) when the temperature drops below approximately −123°F (187 K). Other types of particles are observed at temperatures above the frost point of −123°F (187 K). These particles provide additional surface area

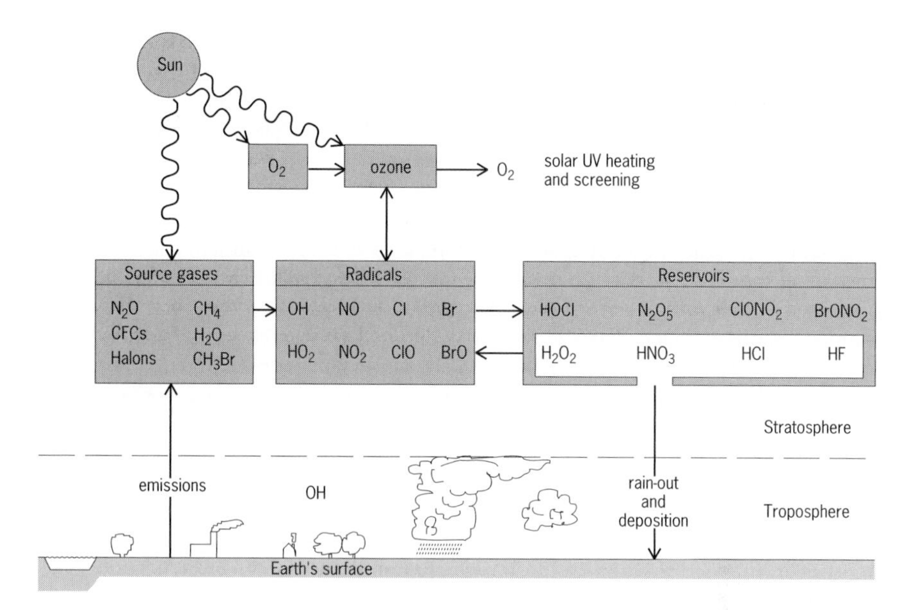

Principal chemical cycles in the stratosphere. The destruction of ozone is affected by the presence of radicals which are produced by photolysis or oxidation of source gases. Chemical reservoirs are relatively stable but are removed from the stratosphere by transport toward the troposphere and rain-out.

for these reactions to occur. Clouds are observed at high latitudes in winter. Because the winter temperatures are typically 20–30°F (10–15 K) colder in the Antarctic than in the Arctic, their frequency of occurrence is highest in the Southern Hemisphere. Thus, the formation of the springtime ozone hole over Antarctica is explained by the activation of chlorine and the catalytic destruction of O_3 which takes place during September, when the polar regions are sunlit but the air is still cold and isolated from midlatitude air by a strong polar vortex. Satellite observations made since the 1970s suggest that total ozone in the Arctic has been abnormally low during the 1990s, probably in relation to the exceptionally cold winter tempratures in the Arctic lower stratosphere recorded during that decade. [G.P.B.; R.G.Pr.]

Stream gaging The measurement of water discharge in streams. Discharge is the rate of movement of the stream's water volume. It is the product of water velocity times cross-sectional area of the stream channel. Several techniques have been developed for measuring stream discharge; selection of the gaging method usually depends on the size of the stream. The most accurate methods for measuring stream discharge make use of in-stream structures through which the water can be routed, such as flumes and weirs.

A flume is a constructed channel that constricts the flow through a control section, the exact dimensions of which are known. Through careful hydraulic design and calibration by laboratory experiments, stream discharge through a flume can be determined by simply measuring the water depth (stage) in the inlet or constricted sections. Appropriate formulas relate stage to discharge for the type of flume used.

A weir is used in conjunction with a dam in the streambed. The weir itself is usually a steel plate attached to the dam that has a triangular, rectangular, or trapezoidal notch over which the water flows. Hydraulic design and experimentation has led to calibration curves and appropriate formulas for many different weir designs. To calculate stream discharge through a weir, only the water stage in the reservoir created by the dam needs to be measured. Stream discharge can be calculated by using the appropriate formula that relates stage to discharge for the type of weir used. *See* HYDROLOGY; SURFACE WATER. [T.C.Wi.]

Stream transport and deposition The sediment debris load of streams is a natural corollary to the degradation of the landscape by weathering and erosion. Eroded material reaches stream channels through rills and minor tributaries, being carried by the transporting power of running water and by mass movement, that is, by slippage, slides, or creep. The size represented may vary from clay to boulders. At any place in the stream system the material furnished from places upstream either is carried away or, if there is insufficient transporting ability, is accumulated as a depositional feature. The accumulation of deposited debris tends toward increased ease of movement, and this tends eventually to bring into balance the transporting ability of the stream and the debris load to be transported. [L.B.L.]

Streptococcus A large genus of spherical or ovoid bacteria that are characteristically arranged in pairs or in chains resembling strings of beads. Many of the streptococci that constitute part of the normal flora of the mouth, throat, intestine, and skin are harmless commensal forms; other streptococci are highly pathogenic. The cells are gram-positive and can grow either anaerobically or aerobically, although they cannot utilize oxygen for metabolic reactions. Glucose and other carbohydrates serve as sources of carbon and energy for growth. All members of the genus lack the enzyme catalase. Streptococci can be isolated from humans and other animals.

Streptococcus pyogenes is well known for its participation in many serious infections. It is a common cause of throat infection, which may be followed by more serious complications such as rheumatic fever, glomerulonephritis, and scarlet fever. Other beta-hemolytic streptococci participate in similar types of infection, but they are usually not associated with rheumatic fever and glomerulonephritis. Group B streptococci, which are usually beta-hemolytic, cause serious infections in newborns (such as meningitis) as well as in adults. Among the alpha-hemolytic and nonhemolytic streptococci, *S. pneumoniae* is an important cause of pneumonia and other respiratory infections. Vaccines that protect against infection by the most prevalent capsular serotypes are available. The viridans streptococci comprise a number of species commonly isolated from the mouth and throat. Although normally of low virulence, these streptococci are capable of causing serious infections (endocarditis, abcesses). *See* PNEUMONIA . [K.Ru.]

Streptothricosis An acute or chronic infection of the epidermis, caused by the bacterium *Dermatophilus congolensis*. which results in an oozing dermatitis with scab formation. Streptothricosis includes dermatophilosis, mycotic dermatitis, lumpy wool, strawberry foot-rot, and cutaneous streptothricosis—diseases having a worldwide distribution and affecting a wide variety of species, including humans.

The infectious form of the bacterium is a coccoid, motile zoospore that is released when the skin becomes wet. Thus, the disease is closely associated with rainy seasons and wet summers. Zoospores lodge on the skin of susceptible animals and germinate by producing filaments which penetrate to the living epidermis, where the organism proliferates by branching mycelial growth.

Early cutaneous lesion of dermatophilosis in cattle reveals small vesicles, papules, and pus formation under hair plaques. An oozing dermatitis then appears as the disease progresses and exudates coalesce to form scabs, which change to hard crusts firmly adherent to the skin. The crusts enlarge and harden, and are often devoid of hair. Lesions occur on most areas of sheep, but the characteristic lesions in the wooled areas occur as numerous hard masses of crust or scab scattered irregularly over the back, flanks, and upper surface of the neck. Lesion resolution has been found to correlate with the presence of immunoglobulin A–containing plasma cells in the dermis and with the antibody levels to *D. congolensis* at the skin surface of infected sheep and cattle.

No single treatment is considered specific for dermatophilosis. Some observers claim that topical agents are successful for sheep and cattle. A number of systemic antibiotics are effective in treating the disease. A combination of streptomycin and penicillin has given good therapeutic results in both bovine and equine infections. [S.S.S.]

Strongylida An order of nematodes in which the cephalic region may be adorned with three or six labia or the labia may be replaced by a corona radiata. All strongylid nematodes are parasitic. The order embraces eight superfamilies: Strongyloidea, Diaphanocephaloidea, Ancylostomatoidea, Trichostrongyloidea, Metastrongyloidea, Cosmocercoidea, Oxyuroidea, and Heterakoidea.

Strongyloidea. The Strongyloidea contain important parasites of reptiles, birds, and mammals. The early larval stages may be free-living microbivores, but the adults are always parasitic. Three species are important parasites of horses, *Strongylus vulgaris, S. equinus*, and *S. edentatus*. All three undergo direct life cycles; that is, infestations are acquired by ingestion of contaminated food.

Trichostrongyloidea. The Trichostrongyloidea comprise obligate parasites of all vertebrates but fishes. Normally they are intestinal parasites, but some are found in the lungs. The species are important parasites of sheep, cattle, and goats. The adult females lay eggs in the intestinal tract, which are passed out with the feces. In the presence of oxygen the eggs hatch in a few days. When the larvae are ingested by an appropriate host, their protective sheath is lost, and they proceed through the fourth larval stage to adulthood in the intestinal tract, where they may enter the mucosa. No migration takes place outside the gastrointestinal tract.

Metastrongyloidea. The Metastrongyloidea comprise obligate parasites of terrestrial and marine mammals, found commonly in the respiratory tract. In their life cycle they utilize both paratenic and intermediate hosts, among them a variety of invertebrates, including earthworms and mollusks. Two important species are *Metastrongylus apri* (swine lungworm) and *Angiostrongylus cantonensis* (rodent lungworm).

Heterakoidea. The Heterakoidea are capable of parasitizing almost any warm-blooded vertebrate as well as reptiles and amphibians. The species *Ascaridia galli* is the largest known nematode parasite of poultry; males are 2–3 in. (50–76 mm) long, and females 3–4.5 in. (75–116 mm). [A.R.M.]

Oxyuroidea. The Oxyuroidea constitute a large group of the phylum Nemata. Hosts include terrestrial mammals, birds, reptiles, amphibians, fishes, insects, and other arthropods.

The species are small to medium sized and thin bodied. With one exception, known life cycles are direct. Typically the eggs pass out of the host's alimentary tract onto the ground, where they become fully embryonated and infective. Normally the infective egg does not hatch until a susceptible animal ingests it. The cecum and colon of the host are the typical locations of these parasites. Larvae in all stages of development and adults occur in the gut.

The human pinworm, *Enterobius vermicularis*, is probably the most contagious of all helminthic diseases. It is estimated that 10% of the world's population suffer from this parasite, the majority being children. Indeed, incidence among schoolchildren in the cool regions of the world often approaches 100%. Infection occurs when eggs are inhaled or ingested. The most common method of transmission is from anus to mouth. Because of the aerial transmission, this disease is highly contagious. Though the infection is seldom serious, the behavioral symptoms are disturbing: nail biting, teeth grinding, anal scratching, insomnia, nightmares, and even convulsions. Several medical treatments are available, but there is often the danger of reinfestation from contaminated objects within the household or institution. *See* NEMATA. [J.T.L.]

Surface water A term commonly used to designate the water flowing in stream channels. The term is sometimes used in a broader sense as opposed to "subsurface water." In this sense, surface water includes water in lakes, marshes, glaciers, and reservoirs as well as that flowing in streams. In the broadest sense, surface water is all the water on the surface of the Earth and thus includes the water of the oceans. Subsurface water includes water in the root zone of the soil and ground water flowing or stored in the rock mantle of the Earth. Subsurface water differs from surface water in the mechanics of its movement as well as in its location. Surface and subsurface water are two stages of the movement of the Earth's water through the hydrologic cycle. The world's ocean and atmospheric moisture are two other main stages of the grand water cycle of the Earth. *See* HYDROLOGY.

The table gives estimates of the amounts of water in various parts of the hydrologic cycle and their detention periods. It may be noted that surface water on the continents is but a small part of the world's water and that the bulk of that is in fresh-water lakes. However, the detention period is also short. This means that the surface-water part, and especially the water in the streams, is rapidly discharged and replenished. That is why surface water, as well as the shallower ground water, is called a renewable resource.

Distribution of the world's supply of water

Location		Volume of water, 10^9 acre-ft*	Percentage total	Detention period, years
World's oceans		1,060,000	97.39	5,000
Surface water on the continents				
Glaciers and polar ice caps	20,000		1.83	2,000
Fresh-water lakes	100		0.0093	100
Saline lakes and inland seas	68		0.0063	50
Average in stream channels	0.25		0.00002	0.05
Total surface water		20,200		700 av
Subsurface water on the continents				
Root zone of the soil	10		0.00094	0.25
Ground water above				
Ground water above 2500 ft	3,700		0.339	5
Ground water below 2500 ft	4,600		0.425	100
Total subsurface water		8,300		
Atmospheric water		115	0.0011	0.03
Total world water (rounded)		1,088,000	100	3,000

*109 acre-ft = 1.233×10^8 ha · m = 1.233×10^{12} m^3.
†2500 ft = 750 m.

Water that has a detention period of more than a generation is not renewed within sufficient time to be so considered. *See* GROUND-WATER HYDROLOGY; RIVER.

Precipitation that reaches the Earth is subdivided by processes of evaporation and infiltration into various routes of subsequent travel. Evaporation from wet land surfaces and from vegetation returns some of the water to the atmosphere immediately. Precipitation that falls at rates less than the local rate of infiltration enters the soil. Some of the infiltrated water is retained in the soil, sustaining plant life, and some reaches the ground water. The precipitation that exceeds the capacity of the soil to absorb water flows overland in the direction of the steepest slope and concentrates in rills and minor channels. During storms most of the water in surface streams is derived from that portion of the precipitation which fails to infiltrate the soil. *See* PRECIPITATION (METEOROLOGY).

The distinction between surface and subsurface water, though useful, should not obscure the fact that water on the surface and water underground is physically connected through pores, cracks, and joints in rock and soil material. In many areas, particularly in humid regions, surface water in stream channels is the visible part of a reservoir, which is partly underground; the water surface of a river is the visible extension of the surface of the ground water. [L.B.L.]

Swamp, marsh, and bog Wet flatlands, where mesophytic vegetation is a really more important than open water, which are commonly developed in filled lakes, glacial pits and potholes, or poorly drained coastal plains or floodplains. Swamp is a term usually applied to a wetland where trees and shrubs are an important part of the vegetative association, and bog implies lack of solid foundation. Some bogs consist of a thick zone of vegetation floating on water.

Unique plant associations characterize wetlands in various climates and exhibit marked zonation characteristics around the edge in response to different thicknesses of the saturated zone above the firm base of soil material. Coastal marshes covered with vegetation adapted to saline water are common on all continents. Presumably many of these had their origin in recent inundation due to post-Pleistocene rise in sea level. *See* MANGROVE. [L.B.L.]

Sweetgum The tree *Liquidambar styraciflua*, also called redgum, a deciduous tree of the southeastern United States. It is found northward as far as southwestern Connecticut, and also grows in Central America. Sweetgum is readily distinguished by its five-lobed, or star-shaped, leaves and by the corky wings or ridges usually developed on the twigs. The erect trunk is a dark gray, but the branches are lighter in color. In winter the persistent, spiny seedballs are an excellent diagnostic feature.

Sweetgum is used for furniture, interior trim, railroad ties, cigar boxes, crates, flooring, barrels, woodenware, and wood pulp, and it is one of the most important materials for plywood manufacture. Sweetgum is one of the most desirable ornamental trees, chiefly because of its brilliant autumn coloration. [A.H.G./K.P.D.]

Sycamore American sycamore (*Platanus occidentalis*) a member of the plane tree family, known also as American plane tree, buttonball, or buttonwood, and ranging from southern Maine to Nebraska and south into Texas and northern Florida. It has the most massive trunk of any American hardwood. Characteristic are the white patches which are exposed when outer layers of the bark slough off; the simple, large, lobed leaves whose stalks completely cover the conical winter buds; and the spherical fruit heads that are always borne singly in the American species and persist throughout the winter. The tough, coarse-grained wood is difficult to work, but is useful for butchers' blocks, saddle trees, vehicles, tobacco and cigar boxes, crates, and slack cooperage. [A.H.G./K.P.D.]

Syphilis A sexually transmitted infection of humans caused by *Treponema pallidum* ssp. *pallidum*, a corkscrew-shaped motile bacterium (spirochete). Due to its narrow width, *T. pallidum* cannot be seen by light microscopy but can be observed with staining procedures (silver stain or immunofluorescence) and with dark-field, phase-contrast, or electron microscopy. The organism is very sensitive to environmental conditions and to physical and chemical agents. The complete genome sequence of the *T. pallidum* Nichols strain has been determined. The nucleotide sequence of the small, circular treponemal chromosome indicates that *T. pallidum* lacks the genetic information for many of the metabolic activities found in other bacteria. Thus, this spirochete is dependent upon the host for most of its nutritional requirements. *See* BACTERIAL GENETICS.

Syphilis is usually transmitted through direct sexual contact with active lesions and can also be transmitted by contact with infected blood and tissues. If untreated, syphilis progresses through various stages (primary, secondary, latent, and tertiary). Infection begins as an ulcer (chancre) and may eventually involve the cardiovascular and central nervous systems, bones, and joints. Congenital syphilis results from maternal transmission of *T. pallidum* across the placenta to the fetus. *See* SEXUALLY TRANSMITTED DISEASES.

Treponema pallidum is an obligate parasite of humans and does not have a reservoir in animals or the environment. Syphilis has a worldwide distribution. Its incidence varies widely according to geographical location, socioeconomic status, and age group. Although syphilis is controlled in most developed countries, it remains a public health problem in many developing countries. Studies have shown that syphilis is a risk factor for infection with the human immunodeficiency virus (HIV) since syphilitic lesions may act as portals of entry for the virus. There is little natural immunity to syphilis infection or reinfection.

Parenteral penicillin G is the preferred antibiotic for treatment of all stages of syphilis. Alternative antibiotics for syphilis treatment include erythromycin and tetracycline. There is currently no vaccine to prevent syphilis. However, it is anticipated that information obtained from the *T. pallidum* genome sequence will lead to further improvements in diagnostic tests for syphilis and to the eventual development of a vaccine that would prevent infection. *See* PUBLIC HEALTH. [L.V.S.]

Systems ecology The analysis of how ecosystem function is determined by the components of an ecosystem and how those components cycle, retain, or exchange energy and nutrients. Systems ecology typically involves the application of computer models that track the flow of energy and materials and predict the responses of systems to perturbations that range from fires to climate change to species extinctions. Systems ecology is closely related to mathematical ecology, with the major difference stemming from systems ecology's focus on energy and nutrient flow and its borrowing of ideas from engineering. Systems ecology is one of the few theoretical tools that can simultaneously examine a system from the level of individuals all the way up to the level of ecosystem dynamics. It is an especially valuable approach for investigating systems so large and complicated that experiments are impossible, and even observations of the entire system are impractical. In these overwhelming settings, the only approach is to break down the research into measurements of components and then assemble a system model that pieces together all components. An important contribution of ecosystem science is the recognition that there are critical ecosystem services such as cleansing of water, recycling of waste materials, production of food and fiber, and mitigation of pestilence and plagues. *See* ECOLOGICAL COMMUNITIES; ECOLOGICAL ENERGETICS; ECOLOGY; ECOSYSTEM; GLOBAL CLIMATE CHANGE; THEORETICAL ECOLOGY. [P.M.Ka.]

T

Taiga A zone of forest vegetation encircling the Northern Hemisphere between the arctic-subarctic tundras in the north and the steppes, hardwood forests, and prairies in the south. The chief characteristic of the taiga is the prevalence of forests dominated by conifers. The dominant trees are particular species of spruce, pine, fir, and larch. Other conifers, such as hemlock, white cedar, and juniper, occur locally, and the broad-leaved deciduous trees, birch and poplar, are common associates in the southern taiga regions. Taiga is a Siberian word, equivalent to "boreal forest." *See* TUNDRA.

The northern and southern boundaries of the taiga are determined by climatic factors, of which temperature is most important. However, aridity controls the forest-steppe boundary in central Canada and western Siberia. In the taiga the average temperature in the warmest month, July, is greater than 50°F (10°C), distinguishing it from the forest-tundra and tundra to the north; however, less than four of the summer months have averages above 50°F (10°C), in contrast to the summers of the deciduous forest further south, which are longer and warmer. Taiga winters are long, snowy, and cold—the coldest month has an average temperature below 32°F (0°C). Permafrost occurs in the northern taiga. It is important to note that climate is as significant as vegetation in defining taiga. Thus, many of the world's conifer forests, such as those of the American Pacific Northwest, are excluded from the taiga by their high precipitation and mild winters. [J.C.Ri.]

Taxonomic categories Any one of a number of formal ranks used for organisms in a traditional Linnaean classification. Biological classifications are orderly arrangements of organisms in which the order specifies some relationship. Taxonomic classifications are usually hierarchical and comprise nested groups of organisms. The actual groups are termed taxa. In the hierarchy, a higher taxon may include one or more lower taxa, and as a result the relationships among taxa are expressed as a divergent hierarchy that is formally represented by tree diagrams. In Linnaean classifications, taxonomic categories are devices that provide structure to the hierarchy of taxa without the use of tree diagrams. By agreement, there is a hierarchy of categorical ranks for each major group of organisms, beginning with the categories of highest rank and ending with categories of lowest rank, and while it is not necessary to use all the available categories, they must be used in the correct order (see table).

Conceptually, the hierarchy of categories is different than the hierarchy of taxa. For example, the taxon Cnidaria, which is ranked as a phylum, includes the classes Anthozoa (anemones), Scyphozoa (jellyfishes), and Hydrozoa (hydras). Cnidaria is a particular and concrete group that is composed of parts. Anthozoa is part of, and included in, Cnidaria. However, categorical ranks are quite different. The category "class" is not part of, nor included in, the category "phylum." Rather, the category "class" is a shelf in the hierarchy, a roadmark of relative position. There are many

Categories commonly used in botanical and zoological classifications, from highest to lowest rank

Botanical categories	Zoological categories
Divisio	Phylum
Classis	Class
Ordo	Order
Familia	Family
Genus	Genus
Species	Species

animal taxa ranked as classes, but there is only one "class" in the Linnaean hierarchy. This is an important strength of the system because it provides a way to navigate through a classification while keeping track of relative hierarchical levels with only a few ranks for a great number of organisms.

When Linnaeus invented his categories, there were only class, order, family, genus, and species. These were sufficient to serve the needs of biological diversity in the late eighteenth century, but were quite insufficient to classify the increasing number of species discovered since 1758. As a result, additional categorical levels have been created. These categories may use prefixes, such as super- and sub-, as well as new basic levels such as tribe. An example of a modern expanded botanical hierarchy of ranks between family and species is:

Familia
Subfamilia
Tribus
Subtribus
Genus
Subgenus
Sectio
Subsectio
Series
Subseries
Species

Linnaean categories are the traditional devices used to navigate the hierarchy of taxa. But categories are only conventions, and alternative logical systems, such as those used by phylogenetic systematists (cladists), are frequently used. *See* CLASSIFICATION, BIOLOGICAL; PHYLOGENY; PLANT TAXONOMY; ZOOLOGICAL NOMENCLATURE. [E.O.W.]

Taxonomy The arrangement or classification of objects according to certain criteria. Systematics is a broader term applied to all comparative biology, including taxonomy. For classifying plants and animals, where the term taxonomy is most often applied, the criteria are characters of structure and function.

A given character usually has two or more states. These variations are used as the basis of biological classification, grouping together like species (in which the majority of the character states are alike) and separating unlike species (in which many of the character states are different). Since the acceptance by biologists of the concept of organic evolution, more and more effort has been made to produce systems of classification that conform to phylogenetic (that is, evolutionary) relationships. Taxonomy is thus concerned with classification, but ultimately classification itself depends upon phylogeny—the amount, direction, and sequence of genetic changes. Scientists try to classify lines, or clusters of lines, of descent. This has not always been the case, and in the past various other criteria have been used, such as whether organisms were edible (ancient times) and whether flowers had five stamens or four or some other number (Linnaean times). Modern taxonomists generally agree that the patterns or clusters of diversity they observe in nature, such as the groups of primates, the rodents, and the bats, are the objective results of purely biological processes acting at different times and places in the past. At the least, animal and plant taxonomy provides a method of communication, a system of naming; at the most, taxonomy provides a framework for the embodiment of all comparative biological knowledge. *See* CLASSIFICATION, BIOLOGICAL; NUMERICAL TAXONOMY; ORGANIC EVOLUTION; PHYLOGENY; PLANT TAXONOMY. [W.H.Wa.]

Temperature inversion The increase of air temperature with height; an atmospheric layer in which the upper portion is warmer than the lower. Such an increase is opposite, or inverse, to the usual decrease of temperature with height, or lapse rate, in the troposphere. However, above the tropopause, temperature increases with height throughout the stratosphere, decreases in the mesosphere, and increases again in the thermosphere. Thus inversion conditions prevail throughout much of the atmosphere much or all of the time, and are not unusual or abnormal. *See* AIR TEMPERATURE; ATMOSPHERE.

Inversions are created by radiative cooling of a lower layer, by subsidence heating of an upper layer, or by advection of warm air over cooler air or of cool air under warmer air. Outgoing radiation, especially at night, cools the Earth's surface, which in turn cools the lowermost air layers, creating a nocturnal surface inversion a few inches to several hundred feet thick.

Inversions effectively suppress vertical air movement, so that smokes and other atmospheric contaminants cannot rise out of the lower layer of air. California smog is trapped under an extensive subsidence inversion; surface radiation inversions, intensified by warm air advection aloft, can create serious pollution problems in valleys throughout the world; radiation and subsidence inversions, when horizontal air motion is sluggish, create widespread pollution potential, especially in autumn over North America and Europe. *See* AIR POLLUTION; SMOG. [A.Cou.]

Terracing (agriculture) A method of shaping land to control erosion on slopes of rolling land used for cropping and other purposes. In early practice the land was shaped into a series of nearly level benches or steplike formations. Modern practice in terracing, however, consists of the construction of low-graded channels or levees to carry the excess rainfall from the land at nonerosive velocities. The physical principle involved is that, when water is spread in a shallow stream, its flow is retarded by the roughness of the bottom of the channel and its carrying, or erosive, power is reduced. Since direct impact of rainfall on bare land churns up the soil and the stirring effect keeps it in suspension in overland flow and rills, terracing does not prevent sheet erosion. It serves only to prevent destruction of agricultural land by gullying and must be supplemented by other erosion-control practices, such as grass rotation, cover crops, mulching, contour farming, strip cropping, and increased organic matter content. *See* EROSION; SOIL CONSERVATION.

The two major types of terraces are the bench and the broadbase (see illustration). The bench terrace is essentially a steep-land terrace and consists of an almost vertical retaining wall, called a riser, or a steep vegetative slope to hold the nearly level surface of the soil for cultivation, orchards, vineyards, or landscaping. The broadbase terrace has the distinguishing characteristic of farmability; that is, crops can be grown on this

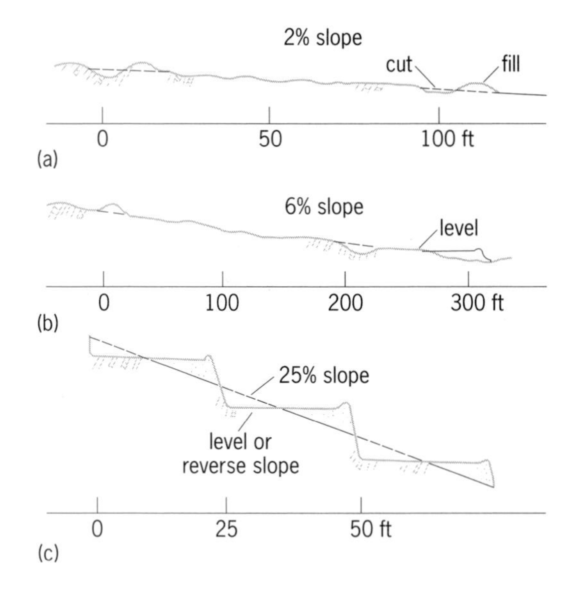

Types of terraces. (*a*) Broadbase. (*b*) Conservation bench. (*c*) Bench. 1 ft = 0.3 m. (*After Soil and Water Conservation Engineering, 2d ed., The Ferguson Foundation Agricultural Engineering Series, John Wiley and Sons, Inc., 1966*)

terrace and worked with modern-day machinery. These terraces are constructed either to remove or retain water and, based on their primary function, are classified either as graded or level. *See* LAND DRAINAGE (AGRICULTURE). [C.B.O.]

Terrain areas Subdivisions of the continental surfaces distinguished from one another on the basis of the form, roughness, and surface composition of the land. The pattern of landform differences is strongly reflected in the arrangement of such other features of the natural environment as climate, soils, and vegetation. These regional associations must be carefully considered in planning of activities as diverse as agriculture, transportation, city development, and military operations.

Eight classes of terrain are distinguished on the basis of steepness of slopes, local relief (the maximum local differences in elevation), cross-sectional form of valleys and divides, and nature of the surface material. Approximate definitions of terms used and percentage figures indicating the fraction of the world's land area occupied by each class are as follows; (1) flat plains: nearly level land, slight relief, 4%; (2) rolling and irregular plains: mostly gently sloping, low relief, 30%; (3) tablelands: upland plains broken at intervals by deep valleys or escarpments, moderate to high relief, 5%; (4) plains with hills or mountains: plains surmounted at intervals by hills or mountains of limited extent, 15%; (5) hills: mostly moderate to steeply sloping land of low to moderate relief, 8%; (6) low mountains: mostly steeply sloping, high relief, 14%; (7) high mountains: mostly steeply sloping, very high relief, 13%; and (8) ice caps: surface material, glacier ice, 11%. [E.H.Ha.]

Terrestrial coordinate system The perpendicular intersection of two curves or two lines, one relatively horizontal and the other relatively vertical, is the basis for finding and describing terrestrial location. The Earth's graticule, consisting of an imaginary grid of east-to-west-bearing lines of latitude and north-to-south bearing lines of longitude, is derived from the Earth's shape and rotation, and is rooted in spherical geometry. Plane coordinate systems, equivalent to horizontal X and vertical

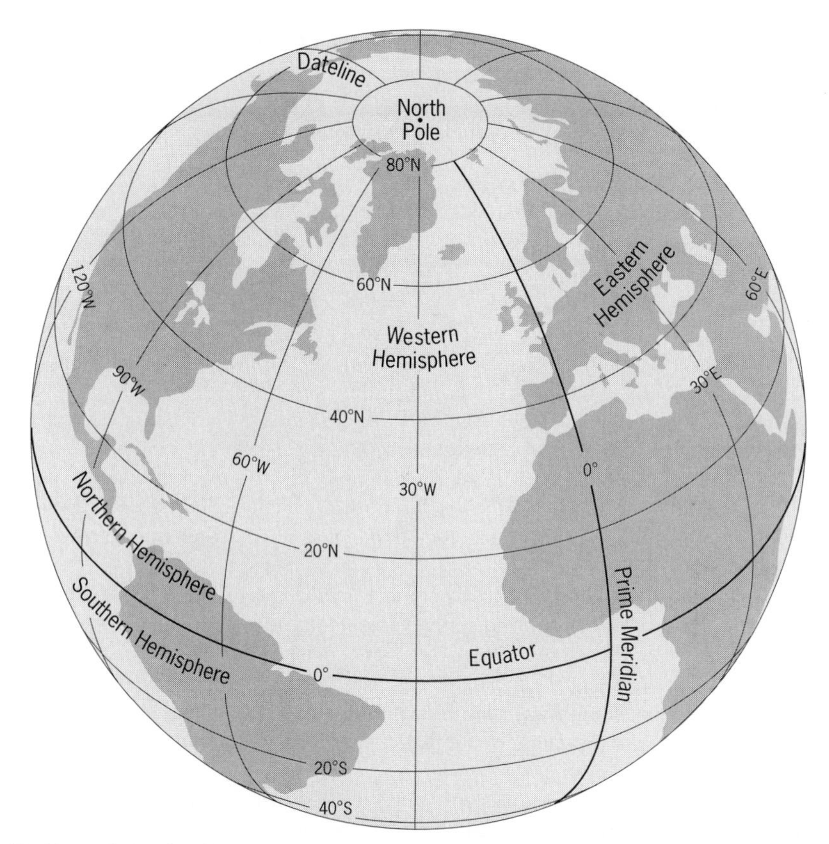

Earth's graticule. Meridians of longitude run from north to south, but are measured east or west of the Prime Meridian. Parallels run from east to west, but are measured north or south of the Equator.

Y coordinates, are based upon cartesian geometry and differ from the graticule in that they have no natural origin or beginning for their grids.

The Earth, which is essentially a sphere, rotates about an axis that defines the geographic North and South poles. The poles serve as the reference points on which the system of latitude and longitude is based (see illustration).

Latitude is arc distance (angular difference) from the Equator and is defined by a system of parallels, or lines that run east to west, each fully encompassing the Earth. The Equator is the parallel that bisects the Earth into the Northern and Southern hemispheres, and lies a constant 90° arc distance from both poles. As the only parallel to bisect the Earth, the Equator is considered a great circle. All other parallels are small circles (do not bisect the Earth), and are labeled by their arc distance north or south from the Equator and by the hemisphere in which they fall. Parallels are numbered from 0° at the Equator to 90° at the poles. For example, 42°S describes the parallel 42 degrees arc distance from the Equator in the Southern Hemisphere. For increased location precision, degrees of latitude and longitude are further subdivided into minutes ($1^\circ = 60'$) and seconds ($1' = 60''$). *See* EQUATOR; GREAT CIRCLE, TERRESTRIAL.

Longitude is defined by a set of imaginary curves extending between the two poles, spanning the Earth. These curves, called meridians, always point to true geographical north (or south) and converge at the poles. In the present-day system of longitude, meridians are numbered by degrees east or west of the beginning meridian, called the Prime Meridian or the Greenwich Meridian, which passes through the Royal Observatory in Greenwich, England. The Prime Meridian was assigned a longitude of 0°.

Since the Earth is fundamentally a sphere, its circumference describes a circle containing 360°, the arc distance through which the Earth rotates in 24 hours. The arc distance from the Prime Meridian describes the location of any meridian (see illustration). The 180° meridian is commonly referred to as the International Dateline. Together, the Prime Meridian and the International Dateline describe a great circle that bisects the Earth, as do all other meridian circles. The west half of the Earth, located between the Prime Meridian and the International Dateline, comprises the Western Hemisphere, and the east half on the opposite side forms the Eastern Hemisphere. Meridians within the Western Hemisphere are labeled with a W, and meridians within the Eastern Hemisphere are labeled with an E. A complete description of longitude includes an angular measurement and a hemispheric label. For example, 78°W is the meridian 78° west of the Prime Meridian. Neither the 0° meridian (Prime) nor the 180° meridian (Dateline) is given a hemispheric suffix because they divide the two hemispheres, and therefore do not belong to either one.

Coordinate system alternatives to the graticule evolved in the early twentieth century because of the complexity of using spherical geometry in determining latitude, longitude, and direction. Plane (two-dimensional) or cartesian coordinate systems presume that a relatively nonspherical Earth exists in smaller areas. Plane coordinates are superimposed upon these small areas, with coordinates being determined by the equivalent of a grid composed of a number of parallel vertical lines (X) and a complementary set of parallel horizontal lines (Y).

The State Plane Coordinate system (SPC) is used only in the United States and partitions each state into zones. Each zone has its own coordinate system. The number of zones designated in each state is determined by the size of the state. Zone boundaries follow either meridians or parallels depending on the shape of the state. All measurements are made in feet.

The Universal Transverse Mercator (UTM) system is a worldwide coordinate system in which locations are expressed using metric units. The basis for the UTM system is the Universal Transverse Mercator map projection. This projection becomes vastly distorted in polar areas above 80°, and for this reason the UTM system is confined to extend from 84°N to 80°S. The UTM system partitions the Earth into 60 north-south elongated zones, each having a width of 6° of longitude.

A number of other coordinate systems are in use today. Foremost among these are the U.S. Public Land Survey System, the Universal Polar Stereographic (UPS) system, and the World Geographic Reference (GEOREF) system. [S.Lav.]

Terrestrial ecosystem A community of organisms and their environment that occurs on the land masses of continents and islands. Terrestrial ecosystems are distinguished from aquatic ecosystems by the lower availability of water and the consequent importance of water as a limiting factor. Terrestrial ecosystems are characterized by greater temperature fluctuations on both a diurnal and seasonal basis than occur in aquatic ecosystems in similar climates. The availability of light is greater in terrestrial ecosystems than in aquatic ecosystems because the atmosphere is more transparent than water. Gases are more available in terrestrial ecosystems than in aquatic ecosystems. Those gases include carbon dioxide that serves as a substrate for photosynthesis, oxygen that serves as a substrate in aerobic respiration, and nitrogen that serves as a substrate for nitrogen fixation. Terrestrial environments are segmented into a subterranean portion from which most water and ions are obtained, and an atmospheric portion from which gases are obtained and where the physical energy of light is transformed into the organic energy of carbon-carbon bonds through the process of photosynthesis.

Terrestrial ecosystems occupy 55,660,000 mi^2 (144,150,000 km^2), or 28.2%, of Earth's surface. Although they are comparatively recent in the history of life (the first terrestrial organisms appeared in the Silurian Period, about 425 million years ago) and occupy a much smaller portion of Earth's surface than marine ecosystems, terrestrial ecosystems have been a major site of adaptive radiation of both plants and animals. Major plant taxa in terrestrial ecosystems are members of the division Magnoliophyta (flowering plants), of which there are about 275,000 species, and the division Pinophyta (conifers), of which there are about 500 species. Members of the division Bryophyta (mosses and liverworts), of which there are about 24,000 species, are also important in some terrestrial ecosystems. Major animal taxa in terrestrial ecosystems include the classes Insecta (insects) with about 900,000 species, Aves (birds) with 8500 species, and Mammalia (mammals) with approximately 4100 species. *See* PLANT TAXONOMY; TAXONOMY.

Organisms in terrestrial ecosystems have adaptations that allow them to obtain water when the entire body is no longer bathed in that fluid, means of transporting the water from limited sites of acquisition to the rest of the body, and means of preventing the evaporation of water from body surfaces. They also have traits that provide body support in the atmosphere, a much less buoyant medium than water, and other traits that render them capable of withstanding the extremes of temperature, wind, and humidity that characterize terrestrial ecosystems. Finally, the organisms in terrestrial ecosystems have evolved many methods of transporting gametes in environments where fluid flow is much less effective as a transport medium.

The organisms in terrestrial ecosystems are integrated into a functional unit by specific, dynamic relationships due to the coupled processes of energy and chemical flow. Those relationships can be summarized by schematic diagrams of trophic webs, which place organisms according to their feeding relationships. The base of the food web is occupied by green plants, which are the only organisms capable of utilizing the energy of the Sun and inorganic nutrients obtained from the soil to produce organic molecules. Terrestrial food webs can be broken into two segments based on the status of the plant material that enters them. Grazing food webs are associated with the consumption of living plant material by herbivores. Detritus food webs are associated with the consumption of dead plant material by detritivores. The relative importance of those two types of food webs varies considerably in different types of terrestrial ecosystems. Grazing food webs are more important in grasslands, where over half of net primary productivity may be consumed by herbivores. Detritus food webs are more important in forests, where less than 5% of net primary productivity may be consumed by herbivores. *See* FOOD WEB; SOIL ECOLOGY.

There is one type of extensive terrestrial ecosystem due solely to human activities and eight types that are natural ecosystems. Those natural ecosystems reflect the variation of precipitation and temperature over Earth's surface. The smallest land areas are occupied by tundra and temperate grassland ecosystems, and the largest land area is occupied by tropical forest. The most productive ecosystems are temperate and tropical forests, and the least productive are deserts and tundras. Cultivated lands, which together with grasslands and savannas utilized for grazing are referred to as agroecosystems, are of intermediate extent and productivity. Because of both their areal extent and their high average productivity, tropical forests are the most productive of all terrestrial ecosystems, contributing 45% of total estimated net primary productivity on land. *See* DESERT; ECOLOGICAL COMMUNITIES; ECOSYSTEM; FOREST AND FORESTRY; GRASSLAND ECOSYSTEM; SAVANNA; TUNDRA. [S.J.McN.]

Terrestrial radiation Electromagnetic radiation emitted from the Earth and its atmosphere. Terrestrial radiation, also called thermal infrared radiation or outgoing longwave radiation, is determined by the temperature and composition of the Earth's atmosphere and surface. The temperature structure of the Earth and the atmosphere is a result of numerous physical, chemical, and dynamic processes. In a one-dimensional context, the temperature structure is determined by the balance between radiative and convective processes.

The Earth's surface emits electromagnetic radiation according to the laws that govern a blackbody or a graybody. A blackbody absorbs the maximum radiation and at the same time emits that same amount of radiation so that thermodynamic equilibrium is achieved as to define a uniform temperature. A graybody is characterized by incomplete absorption and emission and is said to have emissivity less than unity. The thermal infrared emissivities from water and land surfaces are normally between 90 and 95%. It is usually assumed that the Earth's surfaces are approximately black in the analysis of infrared radiative transfer. Exceptions include snow and some sand surfaces whose emissivities are wavelength-dependent and could be less than 90%. Absorption and emission of radiation by atmospheric molecules are more complex and require a fundamental understanding of quantum mechanics. *See* ATMOSPHERE; HEAT BALANCE, TERRESTRIAL ATMOSPHERIC.

The radiant energy emitted from a number of temperatures covering the Earth and the atmosphere is measured as a function of wavenumber and wavelength. This energy is called Planck intensity (or radiance), and the units that are commonly used are denoted as watt per square meter per solid angle per wavenumber ($W/m^2 \cdot sr \cdot cm^{-1}$). Terrestrial radiation originating from the Earth-atmosphere-ocean system, as well as solar radiation reflected and scattered back to space, is measured on a daily basis by meteorological satellites. Instruments on meteorological satellites measure visible, ultraviolet, infrared, and microwave radiation.

Each spectral region provides meteorologists and other Earth system scientists with information about atmospheric ozone, water vapor, temperature, aerosols, clouds, precipitation, lightning, and many other parameters. Measuring atmospheric radiation allows the detection of sea and land temperature, snow and ice cover, and winds at the surface of the ocean. By tracking the movement of clouds and other atmospheric features, such as aerosols and water vapor, it is possible to obtain estimates of winds above the surface. *See* SATELLITE METEOROLOGY. [K.N.L.; T.H.V.H.]

Terrestrial water The total inventory of water on the Earth. Water is unevenly distributed over the Earth's surface in oceans, rivers, and lakes. In addition, the world's water is distributed throughout the atmosphere and also occurs as soil moisture, groundwater, ice caps, and glaciers. *See* ATMOSPHERE; GLACIOLOGY; GROUND-WATER HYDROLOGY; HYDROLOGY; LAKE; SURFACE WATER. [R.L.N.]

Territoriality Behavior patterns in which an animal actively defends a space or some other resource. One major advantage of territoriality is that it gives the territory holder exclusive access to the defended resource, which is generally associated with feeding, breeding, or shelter from predators or climatic forces. Feeding and breeding territories can be mobile, such as when an animal defends a newly obtained food source or a temporarily receptive mate. Stationary territories often serve multiple functions and include access to food, a place to rear young, and a refuge site from predators and the elements.

Territoriality can be understood in terms of the benefits and costs accrued to territory holders. Benefits include time saved by foraging in a known area, energy acquired through feeding on territorial resources, reduction in time spent on the lookout for predators, or increase in number of mates attracted and offspring raised. Costs usually involve time and energy expended in patrolling and defending the territorial site, and increased risk of being captured by a predator when engaged in territorial defense.

Because territories usually include resources that are in limited supply, active defense is often necessary. Such defense frequently involves a graded series of behaviors called displays that include threatening gestures such as vocalizations, spreading of wings or gill covers, lifting and presentation of claws, head bobbing, tail and body beating, and finally, direct attack. Direct confrontation can usually be avoided by advertising the location of a territory in a way that allows potential intruders to recognize the boundaries and avoid interactions with the defender. Such advertising may involve odors that are spread with metabolic by-products, such as urine or feces in dogs, cats, or beavers, or produced specifically as territory markers, as in ants. Longer-lasting territorial marks can involve visual signals such as scrapes and rubs, as in deer and bear. *See* CHEMICAL ECOLOGY; ETHOLOGY; POPULATION ECOLOGY. [G.S.H.]

Tetanus An infectious disease, also known as lockjaw, which is caused by the toxin of *Clostridium tetani*. The bacterium may be isolated from fertile soil and the intestinal tract or fecal material of humans and other animals. Infection commonly follows dirt contamination of deep wounds or other injured tissues.

The incubation period of tetanus is usually 5–10 days, and the disease is characterized by convulsive tonic contraction of voluntary muscles. Prevention of tetanus rests on the proper, prompt surgical care of contaminated wounds and prophylactic use of antitoxin if the individual has not been protected by active immunization with toxoid. [L.S.McC.]

Thallobionta One of the two commonly recognized subkingdoms of plants, encompassing the euglenoids and various classes of algae. In contrast to the more closely knit subkingdom Embryobionta, the Thallobionta (often also called Thallophyta) are diverse in pigmentation, food reserves, cell-wall structure, and flagellar structure. The Thallobionta are united more by the absence of certain specialized tissues or organs than by positive resemblances. They do not have the multicellular sex organs commonly found in most divisions of Embryobionta. Many of the Thallobionta are unicellular, and those which are multicellular seldom have much differentiation of tissues. None of them has tissues comparable to the xylem found in most divisions of the Embryobionta, and only some of the brown algae have tissues comparable to the phloem found in most divisions of the Embryobionta.

A large proportion of the Thallobionta are aquatic, and those which grow on dry land seldom reach appreciable size. The Thallobionta thus consist of all those plants which have not developed the special features that mark the progressive adaptation of the Embryobionta to life on dry land. *See* PLANT KINGDOM. [A.Cr.]

Theoretical ecology The use of models to explain patterns, suggest experiments, or make predictions in ecology. Because ecological systems are idiosyncratic, extremely complex, and variable, ecological theory faces special challenges. Unlike physics or genetics, which use fundamental laws of gravity or of inheritance, ecology has no widely accepted first-principle laws. Instead, different theories must be invoked for different questions, and the theoretical approaches are enormously varied. A central problem in ecological theory is determining what type of model to use and what to leave out of a model. The traditional approaches have relied on analytical models based on differential or difference equations; but recently the use of computer simulation has greatly increased. *See* ECOLOGICAL MODELING; ECOLOGY; ECOLOGY, APPLIED.

The nature of ecological theory varies depending on the level of ecological organization on which the theory focuses. The primary levels of ecological organization are (1) physiological and biomechanical, (2) evolutionary (especially applied to behavior), (3) population, and (4) community.

At the physiological and biomechanical level, the goals of ecological theory are to understand why particular structures are present and how they work. The approaches of fluid dynamics and even civil engineering have been applied to understanding the structures of organisms, ranging from structures that allow marine organisms to feed, to physical constraints on the stems of plants.

At the behavioral evolutionary level, the goals of ecological theory are to explain and predict the different choices that individual organisms make. Underlying much of this theory is an assumption of optimality: the theories assume that evolution produces an optimal behavior, and they attempt to determine the characteristics of the optimal behavior so it can be compared with observed behavior. One area with well-developed theory is foraging behavior (where and how animals choose to feed). Another example is the use of game theory to understand the evolution of behaviors that are apparently not optimal for an individual but may instead be better for a group. *See* BEHAVIORAL ECOLOGY .

The population level has the longest history of ecological theory and perhaps the broadest application. The simplest models of single-species populations ignore differences among individuals and assume that the birth rates and death rates are proportional to the number of individuals in the population. If this is the case, the rate of growth is exponential, a result that goes back at least as far as Malthus's work in the 1700s. As Malthus recognized, this result produces a dilemma: exponential growth cannot continue unabated. Thus, one of the central goals of population ecology theory is to determine the forces and ecological factors that prevent exponential growth and to understand the consequences for the dynamics of ecological populations. *See* POPULATION ECOLOGY.

Modifications and extensions of theoretical approaches like the logistic model (which uses differential equations to explain the stability of populations) have also been used to guide the management of renewable natural resources. Here, the most basic concept is that of the maximum sustainable yield, which is the greatest level of harvest at which a population can continue to persist. *See* ADAPTIVE MANAGEMENT; MATHEMATICAL ECOLOGY.

The primary goal of ecological theory at the community level is to understand diversity at local and regional scales. Recent work has emphasized that a great deal of diversity in communities may depend on trade-offs. For example, a trade-off between competitive prowess and colonization ability is capable of explaining why so many plants persist in North American prairies. Another major concept in community theory is the role of disturbance. Understanding how disturbances (such as fires, hurricanes, or wind storms) impacts communities is crucial because humans typically alter disturbance. *See* BIODIVERSITY; ECOLOGICAL COMMUNITIES. [A.Has.]

Thermal ecology The examination of the independent and interactive biotic and abiotic components of naturally heated environments. Geothermal habitats are present from sea level to the tops of volcanoes and occur as fumaroles, geysers, and hot springs. Hot springs typically possess source pools with overflow, or thermal, streams (rheotherms) or without such streams (limnotherms). Hot spring habitats have existed since life began on Earth, permitting the gradual introduction and evolution of species and communities adapted to each other and to high temperatures. Other geothermal habitats do not have distinct communities.

Hot-spring pools and streams, typified by temperatures higher than the mean annual temperature of the air at the same locality and by benthic mats of various colors, are found on all continents except Antarctica. They are located in regions of geologic activity where meteoric water circulates deep enough to become heated. The greatest densities occur in Yellowstone National Park (Northwest United States), Iceland, and New Zealand. Source waters range from 40°C (104°F) to boiling (around 100°C or 212°F depending on elevation), and may even be superheated at the point of emergence. Few hot springs have pH 5–6; most are either acidic (pH 2–4) or alkaline (pH 7–9). [C.Wi.]

Thermal wind The difference in the geostrophic wind between two heights in the atmosphere over a given position on Earth. It approximates the variation of the actual winds with height for large-scale and slowly changing motions of the atmosphere. Such structure in the wind field is of fundamental importance to the description of the atmosphere and to processes causing its day-to-day changes. The thermal wind embodies a basic relationship between vertical fluctuations of the horizontal wind and horizontal temperature gradients in the atmosphere. This relationship arises from the combination of the geostrophic wind law, the hydrostatic equation, and the gas law.

The geostrophic wind law applies directly to steady, straight, and unaccelerated horizontal motion and is a good approximation for large-scale and slowly changing motions in the atmosphere. The hydrostatic equation combined with the gas law relates the atmospheric pressure and temperature fields. The relationship is accurate for most atmospheric situations but not for small-scale and rapidly changing conditions such as in turbulence and thunderstorms. The equation gives the change of pressure in the vertical direction as a function of pressure and temperature. The key conclusion is that at a given level in the atmosphere the pressure change (decrease) with height is more rapid in cold air than in warm air. *See* ATMOSPHERE; TROPOSPHERE. [D.D.H.]

Thermosphere A rarefied portion of the atmosphere, lying in a spherical shell between 50 and 300 mi (80 and 500 km) above the Earth's surface, where the temperature increases dramatically with altitude. The thermosphere responds to the variable outputs of the Sun, the ultraviolet radiation at wavelengths less than 200 nanometers, and the solar wind plasma that flows outward from the Sun and interacts with the Earth's geomagnetic field. This interaction energizes the plasma, accelerates charged particles into the thermosphere, and produces the aurora borealis and aurora australis, which are nearly circular-shaped regions of luminosity that surround the magnetic north and south poles respectively. Embedded within the thermosphere is the ionosphere, a weakly ionized plasma.

In the thermosphere, these molecular species are subjected to intense solar ultraviolet radiation and photodissociation that gradually turns the molecular species into the atomic species oxygen, nitrogen, and hydrogen. Up to above 60 mi (100 km), atmospheric turbulence keeps the atmosphere well mixed, with the molecular concentrations dominating in the lower atmosphere. Above 60 mi, solar ultraviolet radiation most strongly dissociates molecular oxygen, and there is less mixing from atmospheric turbulence. The result is a transition area where molecular diffusion dominates and atmospheric species settle according to their molecular and atomic weights. Above 60 mi, atomic oxygen is the dominant species. *See* ATMOSPHERE.

About 60% of the solar ultraviolet energy absorbed in the thermosphere and ionosphere heats the ambient neutral gas and ionospheric plasma; 20% is radiated out of the thermosphere as airglow from excited atoms and molecules; and 20% is stored as chemical energy of the dissociated oxygen and nitrogen molecules, which is released later when recombination of the atomic species occurs. Most of the neutral gas heating that establishes the basic temperature structure of the thermosphere is derived from excess energy released by the products of ion-neutral and neutral chemical reactions occurring in the thermosphere and ionosphere.

The average vertical temperature profile is determined by a balance of local solar heating by the downward conduction of molecular thermal product to the region of minimum temperature near 50 mi (80 km). For heat to be conducted downward within the thermosphere, the temperature of the thermosphere must increase with altitude. The global mean temperature increases from about 200 K (−100°F) near 50 mi to 700–1400 K (800–2100°F) above 180 mi (300 km), depending upon the intensity of solar ultraviolet radiation reaching the Earth. Above 180 mi, molecular thermal conduction occurs so fast that vertical temperature differences are largely eliminated; the isothermal temperature in the upper thermosphere is called the exosphere temperature.

As the Earth rotates, absorption of solar energy in the thermosphere undergoes a daily variation. Dayside heating causes the atmosphere to expand, and the loss of heat at night causes it to contract. This heating pattern creates pressure differences that drive a global circulation, transporting heat from the warm dayside to the cool nightside. [R.G.R.]

Thunderstorm A convective storm accompanied by lightning and thunder and a variety of weather such as locally heavy rainshowers, hail, high winds, sudden temperature changes, and occasionally tornadoes. The characteristic cloud is the cumulonimbus or thunderhead, a towering cloud, generally with an anvil-shaped top. A host of accessory clouds, some attached and some detached from the main cloud, are often observed in conjunction with cumulonimbus. *See* LIGHTNING.

Thunderstorms are manifestations of convective overturning of deep layers in the atmosphere and occur in environments in which the decrease of temperature with height (lapse rate) is sufficiently large to be conditionally unstable and the air at low levels is moist. In such an atmosphere, a rising air parcel, given sufficient lift, becomes saturated and cools less rapidly than it would if it remained unsaturated because the released latent heat of condensation partly counteracts the expansional cooling. The rising parcel reaches levels where it is warmer (by perhaps as much as 18°F or 10°C over continents) and less dense than its surroundings, and buoyancy forces accelerate the parcel upward. The rising parcel is decelerated and its vertical ascent arrested at altitudes where the lapse rate is stable, and the parcel becomes denser than its environment. The forecasting of thunderstorms thus hinges on the identification of regions where the lapse rate is unstable, low-level air parcels contain adequate moisture, and surface heating or uplift of the air is expected to be sufficient to initiate convection. *See* FRONT.

Thunderstorms are most frequent in the tropics, and rare poleward of 60° latitude. Thunderstorms are most common during late afternoon because of the diurnal influence of surface heating.

Radar is used to detect thunderstorms at ranges up to 250 mi (400 km) from the observing site. Much of present-day knowledge of thunderstorm structure has been deduced from radar studies, supplemented by visual observations from the ground and satellites, and in-place measurements from aircraft, surface observing stations, and weather balloons. *See* METEOROLOGICAL INSTRUMENTATION; RADAR METEOROLOGY; SATELLITE METEOROLOGY.

Thunderstorms are considered severe when they produce winds greater than 58 mi/h (26 m/s or 50 knots), hail larger than 3/4 in. (19 mm) in diameter, or tornadoes. While thunderstorms are generally beneficial because of their needed rains (except for occasional flash floods), severe storms have the capacity of inflicting utter devastation over narrow swaths of the countryside. Severe storms are most frequently supercells which form in environments with high convective instability and moderate-to-large vertical wind shears. The supercell may be an isolated storm or part of a squall line. *See* HAIL; TORNADO. [R.D.J.]

Tick paralysis A loss of muscle function or sensation in humans or certain animals following the prolonged feeding of female ticks. Paralysis, of Landry's type, usually begins in the legs and spreads upward to involve the arms and other parts of the body. Evidence suggests that paralysis is due to a neurotoxin formed by the feeding ticks rather than the result of infection with microorganisms.

The disease has been reported in North America, Australia, South Africa, and occasionally in some European countries and is caused by appropriate species of indigenous ticks. In Australia, *Ixodes holocyclus* causes frequent cases in dogs, and occasional cases in humans, and paralysis has been known to progress even after removal of ticks. *Ixodes cubicundus* is associated with the disease in South Africa. [C.B.P.]

Tidal bore A part of a tidal rise in a river which is so rapid that water advances as a wall often several feet high. The phenomenon is favored by a substantial tidal range and a channel which shoals and narrows rapidly upstream, but the conditions are so critical that it is not common. Although the bore is a very striking feature, the tide continues to rise after the passage of the bore. Bores may be eliminated by changing channel depth or shape. *See* RIVER TIDES.

In North America three bores have been observed: at the head of the Bay of Fundy (see illustration), at the head of the Gulf of California, and at the head of Cook Inlet,

Tidal bore of the Petitcodiac River, Bay of Fundy, New Brunswick, Canada. Rise of water is about 4 ft (1.2 m). (*New Brunswick Travel Bureau*)

Alaska. The largest known bore occurs in the Tsientang Kiang, China. At spring tides this bore is a wall of water 15 ft (4.5 m) high moving upstream at 25 ft/s (7.5 m/s). [B.K.]

Tidal datum A reference elevation of the sea surface from which vertical measurements are made, such as depths of the ocean and heights of the land. The intersection of the elevation of a tidal datum with the sloping shore forms a line used as a horizontal boundary. In turn, this line is also a reference from which horizontal measurements are made for the construction of additional coastal and marine boundaries.

Since the sea surface moves up and down from infinitely small amounts to hundreds of feet over periods of less than a second to millions of years, it is necessary to stop the vertical motion in order to have a practical reference. This is accomplished by hydraulic filtering, numerical averaging, and segment definition of the record obtained from a tide gage affixed to the adjacent shore. Waves of periods up through wind waves are effectively damped by a restricting hole in the measurement well. Recorded hourly heights are averaged to determine the mean of the higher (or only) high tide of each tidal day (24.84 h), all the high tides, all the hourly heights, all the low tides, and the lower (or only) low tide. The length of the averaging segment is a specific 19 year, which averages all the tidal cycles through the regression of the Moon's nodes and the metonic cycle. [The metonic cycle is a time period of 235 lunar months (19 years); after this period the phases of the Moon occur on the same days of the same months.] But most of all, the 19-year segment is meaningful in terms of measurement capability, averaging meteorological events, and for engineering and legal interests. However, the 19-year segment must be specified and updated because of sea-level changes occurring over decades. The present tidal datum epoch is 1983 through 2001.

Tidal datums are legal entities. Because of variations in gravity, semistationary meteorological conditions, semipermanent ocean currents, changes in tidal characteristics, ocean density differences, and so forth, the sea surface (at any datum elevation) does not conform to a mathematically defined spheroid. [S.D.H.]

Tidal power Tidal-electric power is obtained by utilizing the recurring rise and fall of coastal waters. Marginal marine basins are enclosed with dams, making it possible to create differences in the water level between the ocean and the basins. The oscillatory flow of water filling or emptying the basins is used to drive hydraulic turbines which propel electric generators.

Large amounts of electric power could be developed in the world's coastal regions having tides of sufficient range, although even if fully developed this would amount to only a small percentage of the world's potential water (hydroelectric) power. [G.G.A.]

Tornado A violently rotating, tall, narrow column of air (vortex), typically about 300 ft (100 m) in diameter, that extends to the ground from a cumulonimbus cloud. The vast majority of tornadoes rotate cyclonically (counterclockwise in the Northern Hemisphere). Of all atmospheric storms, tornadoes are the most violent. *See* CLOUD; CYCLONE.

Tornadoes are made visible by a generally sharp-edged, funnel-shaped cloud pendant from the cloud base, and a swirling cloud of dust and debris rising from the ground (see illustration). The funnel consists of small waterdroplets that form as moist air entering the tornado's partial vacuum expands and cools. The condensation funnel may not extend all the way to the ground and may be obscured by dust. Many condensation funnels exist aloft without tangible signs that the vortex is in contact with the ground; these are known as funnel clouds. Tornado funnels assume various forms: a slender smooth rope, a cone (often truncated by the ground), a thick turbulent black cloud on the ground, or multiple funnels (vortices) that revolve around the axis of the overall tornado.

Many tornadoes evolve as follows: The tornado begins outside the precipitation region as a dust whirl on the ground and a short funnel pendant from a wall cloud on the southwest side of the thunderstorm; it intensifies as the funnel lengthens downward, and attains its greatest power as the funnel reaches its greatest width and is almost vertical; then it shrinks and becomes more tilted, and finally becomes contorted and ropelike as it decays. A downdraft and curtain of rain and large hail gradually spiral

The Cordell, Oklahoma, tornado of May 22, 1981, in its decay stage. (*National Severe Storms Laboratory/University of Mississippi Tornado Intercept Project*)

from the northeast cyclonically around the tornado, which often ends its life in rain. *See* HAIL; PRECIPITATION (METEOROLOGY); THUNDERSTORM.

Most tornadoes and practically all violent ones develop from a larger-scale circulation, the mesocyclone, which is 2–6 mi (3–9 km) in diameter and forms in a particularly virulent variety of thunderstorm, the supercell. The mesocyclone forms first at midaltitudes of the storm and in time develops at low levels and may extend to high altitudes as well. The tornado forms on the southwest side (Northern Hemisphere) of the storm's main updraft, close to the downdraft, after the development of the mesocyclone at low levels. Some supercells develop up to six mesocyclones and tornadoes repeatedly over great distances at roughly 45-min intervals. Tornadoes associated with supercells are generally of the stronger variety and have larger parent cyclones. Hurricanes during and after landfall may spawn numerous tornadoes from small supercells located in their rainbands. *See* HURRICANE.

Tornadoes are classified as weak, strong, or violent, or from F0 to F5 on the Fujita (F) scale of damage intensity. Sixty-two percent of tornadoes are weak (F0 to F1). These tornadoes have maximum windspeeds less than about 50 m/s (110 mi/h) and inflict only minor damage, such as peeling back roofs, overturning mobile homes, and pushing cars into ditches. Thirty-six percent of tornadoes are strong (F2 to F3) with maximum windspeeds estimated to be 50–90 m/s (110–200 mi/h). Strong tornadoes extensively damage the roofs and walls of houses but leave some walls partially standing. They demolish mobile homes, and lift and throw cars. The remaining 2% are violent (F4 to F5), with windspeeds in excess of about 90 m/s (200 mi/h). They level houses to their foundations, strew heavy debris over hundreds of yards, and make missiles out of heavy objects such as roof sections, vehicles, utility poles, and large, nearly empty storage tanks. *See* WIND.

Tornadoes occur most often at latitudes between 20° and 60°, and they are relatively frequent in the United States, Russia, Europe, Japan, India, South Africa, Argentina, New Zealand, and parts of Australia. Violent tornadoes are confined mainly to the United States, east of the Rocky Mountains.

Essentially, there are five atmospheric conditions that set the stage for wide-spread tornado development: (1) a surface-based layer, at least 3000 ft (1 km) deep, of warm, moist air, overlain by dry air at midlevels; (2) an inversion separating the two layers, preventing deep convection until the potential for explosive overturning is established; (3) rapid decrease of temperature with height above the inversion; (4) a combination of mechanisms, such as surface heating and lifting of the air mass by a front or upper-level disturbance, to eliminate the inversion locally; (5) pronounced vertical wind shear (variation of the horizontal wind with height). Specifically, storm-relative winds in the lowest 6000 ft (2 km) should exceed 20 knots (10 m/s) and veer (turn anticyclonically) with height at a rate of more than 10°/1000 ft (30°/km). Such conditions are prevalent in the vicinity of the jet stream and the low-level jet.

The first three conditions above indicate that the atmosphere is in a highly metastable state. There is a strong potential for thunderstorms with intense updrafts and downdrafts. The fourth condition is the existence of a trigger to release the instability and initiate the thunderstorms. The fifth is the ingredient for updraft rotation. *See* AIR MASS; FRONT; JET STREAM; TEMPERATURE INVERSION. [R.D.J.]

Toxicology The study of the adverse effects of chemical and physical agents on living organisms. Toxicology has also been referred to as the science of poisons. *See* ENVIRONMENTAL TOXICOLOGY; POISON.

The most important factor that influences the toxic effect of a specific chemical is the dose. All chemicals, including essential substances such as oxygen and water, produce toxic effects when administered in large enough doses. Another significant factor is the route of exposure. Living organisms may be exposed to a chemical by inhalation (into the lungs), ingestion (into the stomach), penetration through the skin, or, in special circumstances, injection into the body. In general, substances are absorbed into the body most efficiently through the lungs so that inhalation is often the most serious route of exposure.

A third factor is the fate of the chemical after the organism is exposed. The chemical may not be absorbed at all, limiting its possible adverse effects to the site of exposure. If it is absorbed, then it may travel throughout the body and has the potential to cause toxic effects at one or more sites remote from the site of entry. The remote sites where these adverse effects occur are called target organs.

Another significant variable is the time course of the exposure. A quantity of chemical administered at one time may have an effect even though the same quantity administered in small doses over time has no effect.

In view of the importance of timing in producing adverse effects, toxicologists distinguish between two broad classes of toxicity, acute and chronic. Acute toxicity refers to effects that occur shortly after a single exposure or small number of closely spaced exposures. Chronic toxicity refers to delayed effects that occur after long-term repeated exposures.

Traditionally, the effect of most concern for acute toxicants (such as cyanide) is death. Acute toxicity is generally measured by using an assay to determine the lethal dose; rodents are given single doses and the number that have died 14 days later is recorded. The data are plotted for each dose, and the dose that is lethal for 50% of the animals (lethal dose 50 or LD_{50}) is used as the criterion for acute toxicity.

Some synthetic chemicals, such as polychlorinated biphenyls (PCBs) and dichlorodiphenyltrichloroethane (DDT), exhibit their effects only after a number of repeated exposures and are considered chronic hazards, with cancer and reproductive effects being of greatest concern. To determine the dose at which chronic effects occur, rodents are exposed to daily doses of the chemical under study for long periods of time—from a few months to a lifetime. The highest dose at which no effects can be observed, the no observed effect level (NOEL), is used as a measure of chronic toxicity. *See* MUTAGENS AND CARCINOGENS. [M.Kam.; R.W.L.]

Toxin Properly, a poisonous protein, especially of bacterial origin. However, non-proteinaceous poisons, such as fungal aflatoxins and plant alkaloids, are often called toxins. *See* AFLATOXIN.

Bacterial exotoxins are proteins of disease-causing bacteria that are usually secreted and have deleterious effects. Several hundred are known. In some extreme cases a single toxin accounts for the principal symptoms of a disease, such as diphtheria, tetanus, and cholera. Bacteria that cause local infections with pus often produce many toxins that affect the tissues around the infection site or are distributed to remote organs by the blood. *See* CHOLERA; DIPHTHERIA; STAPHYLOCOCCUS; TETANUS.

Toxins may assist the parent bacteria to combat host defense systems, to increase the supply of certain nutrients such as iron, to invade cells or tissues, or to spread between hosts. Sometimes the damage suffered by the host organism has no obvious benefit to the bacteria. For example, botulinal neurotoxin in spoiled food may kill the person or animal that eats it long after the parent bacteria have died. In such situations it is assumed that the bacteria benefit from the toxin in some other habitat and that the damage to vertebrates is accidental. *See* FOOD POISONING.

Certain bacterial and plant toxins have the unusual ability to catalyze chemical reactions inside animal cells. Such toxins are always composed of two functionally distinct parts termed A and B, and they are often called A-B toxins. The B part binds to receptor molecules on the animal cell surface and positions the toxin upon the cell membrane. Subsequently, the enzymically active A portion of the toxin crosses the animal cell membrane and catalyzes some intracellular chemical reaction that disrupts the cell physiology or causes cell death.

A large group of toxins breach the normal barrier to free movement of molecules across cell membranes. In sufficient concentration such cytolytic toxins cause cytolysis, a process by which soluble molecules leak out of cells, but in lower concentration they may cause less obvious damage to the cell's plasma membrane or to its internal membranes.

Tetanus and botulinal neurotoxins block the transmission of nerve impulses across synapses. Tetanus toxin blockage results in spastic paralysis, in which opposing muscles contract simultaneously. The botulinal neurotoxins principally paralyze neuromuscular junctions and cause flaccid paralysis.

Gram-negative bacteria, such as *Salmonella* and *Hemophilus*, have a toxic component in their cell walls known as endotoxin or lipopolysaccharide. Among other detrimental effects, endotoxins cause white blood cells to produce interleukin-1, a hormone responsible for fever, malaise, headache, muscle aches, and other nonspecific consequences of infection. The exotoxins of toxic shock syndrome and of scarlet fever induce interleukin-1 and also tumor necrosis factor, which has similar effects. *See* ENDOTOXIN; SCARLET FEVER.

Toxoids are toxins that have been exposed to formaldehyde or other chemicals that destroy their toxicities without impairing immunogenicity. When injected into humans, toxoids elicit specific antibodies known as antitoxins that neutralize circulating toxins.

Such immunization (vaccination) is very effective for systemic toxinoses, such as diphtheria and tetanus. [D.M.Gi.]

Tree A perennial woody plant at least 20 ft (6 m) in height at maturity, having an erect stem or trunk and a well-defined crown or leaf canopy. However, no sharp lines can be drawn between trees, shrubs, and lianas (woody vines). The essence of the tree form is relatively large size, long life, and a slow approach to reproductive maturity. The difficulty of transporting water, nutrients, and storage products over long distances and high into the air against the force of gravity is a common problem of large treelike plants and one that is not shared by shrubs or herbs.

Classification. Almost all existing trees belong to the seed plants (Spermatophyta). An exception are the giant tree ferns which were more prominent in the forests of the Devonian Period and today exist only in the moist tropical regions. The Spermatophyta are divided into the Pinophyta (gymnosperms) and the flowering plants, Magnoliophyta (angiosperms). The gymnosperms bear their seed naked on modified leaves, called scales, which are usually clustered into structures called cones—for example, pine cones. By contrast the seed of angiosperms is enclosed in a ripened ovary, the fruit.

The orders Cycadales, Ginkgoales, and Pinales of the Pinophyta contain trees. *Ginkgo biloba*, the ancient maidenhair tree, is the single present-day member of the Ginkgoales. The Cycadales, characteristic of dry tropical areas, contain many species which are small trees. The Pinales, found throughout the world, supply much of the wood, paper, and building products of commerce. They populate at least one-third of all existing forest and include the pines (*Pinus*), hemlocks (*Tsuga*), cedars (*Cedrus*), spruces (*Picea*), firs (*Abies*), cypress (*Cupressus*), larches (*Larix*), Douglas-fir (*Pseudotsuga*), sequoia (*Sequoia*), and other important genera. The Pinales are known in the lumber trade as softwoods and are popularly thought of as evergreens, although some (for example, larch and bald cypress) shed their leaves in the winter. *See* CEDAR; CYPRESS; DOUGLAS-FIR; FIR; HEMLOCK; LARCH; PINE; SEQUOIA; SPRUCE.

In contrast to the major orders of gymnosperms which contain only trees, many angiosperm families are herbaceous and include trees only as an exception. Only a few are exclusively arborescent. The major classes of the angiosperms are the Liliopsida (monocotyledons) and the Magnoliopsida (dicotyledons). The angiosperm trees, commonly thought of as broad-leaved and known as hardwoods in the lumber market, are dicotyledons. Examples of important genera are the oaks (*Quercus*), elms (*Ulmus*), maples (*Acer*), and poplars (*Populus*). *See* ELM; MAPLE; OAK; POPLAR.

The Liliopsida contain few tree species, and these are never used for wood products, except in the round as posts. Examples of monocotyledonous families are the palms (Palmae), yucca (Liliaceae), bamboos (Bambusoideae), and bananas (Musaceae). *See* BAMBOO.

Morphology. The morphology of a tree is similar to that of other higher plants. Its major organs are the stem, or trunk and branches; the leaves; the roots; and the reproductive structures. Almost the entire bulk of a tree is nonliving. Of the trunk, branches, and roots, only the tips and a thin layer of cells just under the bark are alive. Growth occurs only in these meristematic tissues. Meristematic cells are undifferentiated and capable of repeated division. *See* FLOWER; LEAF; PLANT GROWTH; ROOT (BOTANY); STEM.

Growth. Height is a result of growth only in apical meristems at the very tips of the twigs. A nail driven into a tree will always remain at the same height, and a branch which originates from a bud at a given height will never rise higher. The crown of a tree ascends as a tree ages only by the production of new branches at the top and by the death and abscission of lower, older branches as they become progressively more shaded. New growing points originate from the division of the apical meristem and appear as buds in the axils of leaves. *See* BUD; PLANT GROWTH.

In the gymnosperms and the dicotyledonous angiosperms, growth in diameter occurs by division in only a single microscopic layer, three or four cells wide, which completely encircles and sheaths the tree. This lateral meristem is the cambium. It divides to produce xylem cells (wood) on the inside toward the core of the tree and phloem cells on the outside toward the bark. In trees of the temperate regions the growth of each year is seen in cross section as a ring. *See* BARK; PHLOEM; XYLEM.

Xylem elements become rigid through the thickening and modification of their cell wall material. The tubelike xylem cells transport water and nutrients from the root through the stem to the leaves. In time the xylem toward the center of the trunk becomes impregnated with various mineral and metabolic products, and it is no longer capable of conduction. This nonfunctional xylem is called heartwood and is recognizable in some stems by its dark color. The light-colored, functional outer layer of the xylem is the sapwood. *See* WOOD ANATOMY.

The phloem tissue transports dissolved carbohydrates and other metabolic products manufactured by the leaves throughout the stem and the roots. Most of the phloem cells are thin-walled and are eventually crushed between the bark and the cambium by the pressures generated in growth. The outer bark is dead and inelastic but the inner bark contains patches of cork cambium which produce new bark. As a tree increases in circumference, the old outer bark splits and fissures develop, resulting in the rough appearance characteristic of the trunks of most large trees.

In the monocotyledons the lateral cambium does not encircle a central core, and the vascular or conducting tissue is organized in bundles scattered throughout the stem. The trunk is not wood as generally conceived although it does in fact have secondary xylem. *See* DENDROLOGY; FOREST AND FORESTRY; PLANT PHYSIOLOGY; PLANT TAXONOMY.

[F.T.L.]

Tree diseases Diseases of both shade and forest trees have the same pathogens, but the trees differ in value, esthetics, and utility. In forests, disease is significant only when large numbers of trees are seriously affected. Diseases with such visible symptoms as leaf spots may be alarming on shade trees but hardly noticed on forest trees. Shade trees with substantial rot may be ornamentals with high value, whereas these trees would be worthless in the forest. Emphasis on disease control for the same tree species thus requires a different approach, depending on location of the tree.

Forest trees. From seed to maturity, forest trees are subject to many diseases. Annual losses of net sawtimber growth from disease (45%) are greater than from insects and fire combined. Young, succulent seedlings, especially conifers, are killed by certain soil-inhabiting fungi (damping-off). Root systems of older seedlings may be destroyed by combinations of nematodes and such fungi as *Cylindrocladium, Sclerotium*, and *Fusarium*. Chemical treatment of seed or soil with formulations containing nematicides and fungicides, and cultural practices unfavorable to root pathogens help to avoid these diseases.

Roots rots are caused by such fungi as *Heterobasidion* (=*Fomes*) *annosus* (mostly in conifers) and *Armillariella* (= *Armillaria*) *mellea* (mostly in hardwoods). These fungi cause heart rot in the roots and stems of large trees and also invade and kill young, vigorous ones.

In natural forests, leaf diseases are negligible, but in nurseries and plantations, fungal infections cause severe defoliation, retardation of height growth, or death. *Scirrhia acicola* causes brown spot needle blight and prevents early height growth of longleaf pine in the South; it defoliates Christmas tree plantations of Scotch pine (*Pinus sylvestris*) in northern states. Fungicides and prescribed burning are used successfully for control.

Oak wilt is a systemic disease, with the entire tree affected through its water-conducting system. The causal fungus, *Ceratocystis fagacearum*, spreads to nearby healthy trees by root grafts and to trees at longer distances by unrelated insects. The sporulating mats of the fungus develop between bark and wood, producing asexual and, sometimes, sexual spores, which are disseminated by insects. Control is possible by eradicating infected trees and by disruption of root grafts by trenching or by chemicals.

Stem rust diseases occur as cankers or galls on coniferous hosts and as minor lesions on other ones. A few, such as white pine blister rust and southern fusiform rust, are epidemic, lethal, and economically important. Others of less immediate importance (such as western gall rust) are capable of serious, widespread infection. Resistant varieties are favored for control. Other control measures include pruning out early infections and spraying nursery trees with chemicals during periods favoring needle infection.

Stem infections by numerous fungi, resulting in localized death of cambium and inner bark, range from lesions killing small stems in a year (annual) to gross stem deformities (perennial), where cankers enlarge with stem growth. Chestnut blight, first known in the United States in 1904, destroyed the American chestnut as a commercial species; and is an example of the annual lesion type. The less dramatic or devastating *Nectria* canker destroys stems of timber value, and is an example of the perennial lesion type.

All tree species, including decay-resistant ones such as redwood, are subject to ultimate disintegration by fungi. Decay fungi (Hymenomycetes) are associated with non-decay fungi (Deuteromycetes) and bacteria. These microflora enter the tree through wounds, branch stubs, and roots, and are confined to limited zones of wood by anatomical and wound-stimulated tissue barriers. The extent of decay is limited by compartmentalization of decay in trees. Trees aged beyond maturity are most often invaded by wood-rotting fungi; losses can be minimized by avoiding wounds and by shortening cutting rotations. Losses from rot are especially serious in overmature coniferous stands in the western United States, Canada, and Alaska. *See* FOREST PEST CONTROL; WOOD DEGRADATION.

Shade trees. Many shade trees are grown under conditions for which they are poorly adapted, and are subject to environmental stresses not common to forest trees. Both native and exotic trees planted out of natural habitats are predisposed to secondary pathogens following environmental stress of noninfectious origins. They are also susceptible to the same infectious diseases as forest trees. Appearance is more important than the wood produced, and individual value is higher per tree than for forest trees. Thus, disease control methods differ from those recommended for forest trees.

The most important and destructive shade tree disease known is Dutch elm disease, introduced from Europe to North America before 1930. The causal fungus, *Ophiostoma ulmi* (*Ceratocytis ulmi*), is introduced to the water-conducting system of healthy elms by the smaller European elm bark beetle (*Scolytus multistriatus*) or the American elm bark beetle (*Hylurgopinus rufipes*). One or more new and more aggressive strains of the fungus have arisen since 1970. More devastating than the original ones, they are destroying the elms in North America and Europe that survived earlier epidemics. Effective means of prevention are sanitation (destroying diseased and dying and dead elm wood); insecticidal sprays; disruption of root systems; and early pruning of new branch infections. Of much promise are resistant varieties of elm, systemic fungicides, and insect pheromones.

The most common bacterial disease of shade trees is wetwood of elm and certain other species. It is reported to be caused by a single bacterial species (*Erwinia nimipressuralis*), although causal associations of other bacteria are now suspected. The bacteria are normally present in the heartwood of mature elms and cause no disease unless they colonize sapwood by exterior wounds. Fermented sap under pressure bleeds from wounds and flows down the side of the tree. Sustained bleeding kills underlying cambial tissue. Internal gas pressure and forced spread of bacterial toxins inside living tissues of the tree can be reduced by strategic bleeding to avoid seepage into bark and cambium.

A second bacterial disease of elm, elm yellows, is caused by a mycoplasmalike organism, considered to be a unique kind of bacterium. Elm yellows is as lethal as Dutch elm disease but is more limited in distribution. The pathogen is carried by the elm leafhopper (*Scaphoideus luteolus*), which sucks phloem juice from leaf veins. Spread of disease also occurs through grafted root systems. Control measures include early destruction of infected trees, disruption of root systems, and insecticidal sprays. Injection with tetracycline and other antibiotics helps to slow the progress of the bacterium.

The most common and complex diseases of shade trees are diebacks and declines (such as maple decline). Many species show similar patterns of symptoms caused by multiple factors, but no single causal factor is known to cause any one of these diseases. Noninfectious agents of shade tree diseases are drought, soil compaction, mineral deficiency, soil pollution from waste or salt, air pollution, and so on. Trees affected experience chlorosis, premature fall coloration and abscission, tufting of new growth, dwarfing and sparseness of foliage, progressive death of terminal twigs and branches, and decline in growth. Such trees are often infested with borers and bark beetles, and infected by branch canker and root rot fungi. Noninfectious stress predisposes trees to infectious disease that is caused by different kinds of weakly parasitic fungi as secondary pathogens. *See* PLANT PATHOLOGY. [R.J.C.]

Tree physiology The study of how trees grow and develop in terms of genetics; biochemistry; cellular, tissue, and organ functions; and interaction with environmental factors. While many physiological processes are similar in trees and other plants, trees possess unique physiologies that help determine their outward appearance. These physiological processes include carbon relations (photosynthesis, carbohydrate allocation), cold and drought resistance, water relations, and mineral nutrition.

Three characteristics of trees that define their physiology are longevity, height, and simultaneous reproductive and vegetative growth. Trees have physiological processes that are more adaptable than those in the more specialized annual and biennial plants. Height allows trees to successfully compete for light, but at the same time this advantage creates transport and support problems. These problems were solved by the evolution of the woody stem which combines structure and function into a very strong transport system. Simultaneous vegetative and reproductive growth in adult trees causes significant competition for carbohydrates and nutrients, resulting in decreased vegetative growth. Trees accommodate both types of growth by having cyclical reproduction: one year many flowers and seeds are produced, followed by a year or more in which few or no flowers are produced.

Carbon relations. While biochemical processes of photosynthesis and carbon assimilation and allocation are the same in trees and other plants, the conditions under which these processes occur in trees are more variable and extreme. In evergreen species, photosynthesis can occur year round as long as the air temperature remains above freezing, while some deciduous species can photosynthesize in the bark of twigs and stem during the winter.

Carbon dioxide fixed into sugars moves through the tree in the phloem and xylem to tissues of high metabolism which vary with season and development. At the onset of growth in the spring, sugars are first mobilized from storage sites, primarily in the secondary xylem (wood) and phloem (inner bark) of the woody twigs, branches, stem, and roots. The sugars, stored as starch, are used to build new leaves and twigs, and if present, flowers. Once the new leaves expand, photosynthesis begins and sugars are produced, leading to additional leaf growth. Activation of the vascular cambium occurs at the same time, producing new secondary xylem and phloem. In late spring, the leaves begin photosynthesizing at their maximum rates, creating excess sugars which are translocated down the stem to support further branch, stem, and root growth. From midsummer through fall until leaf abscission (in deciduous trees) or until temperatures drop to freezing (in evergreen trees), sugars replenish the starch used in spring growth. Root growth may be stimulated at this time by sugar availability and warm soil temperatures. Throughout the winter, starch is used for maintenance respiration, but sparingly since low temperatures keep respiration rates low. *See* PHLOEM; PHOTOSYNTHESIS; XYLEM.

In adult trees, reproductive structures (flowers in angiosperms or strobili in gymnosperms) develop along with new leaves and represent large carbohydrate sinks. Sugars are preferentially utilized at the expense of leaf, stem, and root growth. This reduces the leaf area produced, affecting the amount of sugars produced during that year, thereby reducing vegetative growth even further. The reproductive structures are present throughout the growing season until seed dispersal and continually utilize sugars that would normally go to stem and root growth.

Cold resistance. The perennial nature of trees requires them to withstand low temperatures during the winter. Trees develop resistance to freezing through a process of physiological changes beginning in late summer. A tree goes through three sequential stages to become fully cold resistant. The process involves reduced cell hydration along with increased membrane permeability. The first stage is initiated by shortening days and results in shoot growth cessation, bud formation, and metabolic changes. Trees in this stage can survive temperatures down to 23°F (−5°C). The second stage requires freezing temperatures which alter cellular molecules. Starch breakdown is stimulated, causing sugar accumulation. Trees can survive temperatures as low as −13°F (−25°C) at this stage. The last stage occurs after exposure to very low temperatures (−22 to −58°F or −30 to −50°C), which increases soluble protein concentrations that bind cellular water, preventing ice crystallization. Trees can survive temperatures below −112°F (−80°C) in this stage. A few days of warmer temperatures, however, causes trees to revert to the second stage.

Water relations. Unlike annual plants that survive drought as seeds, trees have evolved traits that allow them to avoid desiccation. These traits include using water stored in the stem, stomatal closure, and shedding of leaves to reduce transpirational area. All the leaves can be shed and the tree survives on stored starch. Another trait of some species is to produce a long tap root that reaches the water table, sometimes tens of meters from the soil surface. On a daily basis, trees must supply water to the leaves for normal physiological function. If the water potential of the leaves drops too low, the stomata close, reducing photosynthesis. To maintain high water potential, trees use water stored in their stems during the morning which is recharged during the night. *See* PLANT-WATER RELATIONS.

Transport and support. Trees have evolved a means of combining long-distance transport between the roots and foliage with support through the production of secondary xylem (wood) by the vascular cambium. In older trees the stem represents 60–85% of the aboveground biomass. However, 90% of the wood consists of dead cells. These dead cells function in transport and support of the tree. As these cells develop and mature, they lay down thick secondary walls of cellulose and lignin that provide support, and then they die with the cell lumen becoming an empty tube. The interconnecting cells provide an efficient transport system, capable of moving 106 gal (400 liters) of water per day. The living cells in the wood (ray parenchyma) are the site of starch storage in woody stems and roots. *See* PLANT TRANSPORT OF SOLUTES.

Mineral nutrition. Nutrient deficiencies are similar in trees and other plants because of the functions of these nutrients in physiological processes. Tree nutrition is unique because trees require lower concentrations, and they are able to recycle nutrients within various tissues. Trees adapt to areas which are low in nutrients by lowering physiological functions and slowing growth rates. In addition, trees allocate more carbohydrates to root production, allowing them to exploit large volumes of soil in search of limiting nutrients. Proliferation of fine roots at the organic matter-mineral soil interface where many nutrients are released from decomposing organic matter allows trees to recapture nutrients lost by leaf fall. *See* PLANT MINERAL NUTRITION; PLANT PHYSIOLOGY; TREE.

[J.D.Jo.]

Trophic ecology The study of the structure of feeding relationships among organisms in an ecosystem. Researchers focus on the interplay between feeding relationships and ecosystem attributes such as nutrient cycling, physical disturbance, or the rate of tissue production by plants and the accrual of detritus (dead organic material). Feeding or trophic relationships can be represented as a food web or as a food chain. Food webs depict trophic links between all species sampled in a habitat, whereas food chains simplify this complexity into linear arrays of interactions among trophic levels. Thus, trophic levels (for example, plants, herbivores, detritivores, and carnivores) are amalgamations of species that have similar feeding habits. (However, not all species consume prey on a single trophic level. Omnivores are species that feed on more than one trophic level.) *See* ECOLOGY; ECOSYSTEM; FOOD WEB.

The three fundamental questions in the field of trophic ecology are: (1) What is the relationship between the length of food chains and plant biomass (the relative total amount of plants at the bottom of the food chain)? (2) How do resource supply to producers (plants) and resource demand by predators determine the relative abundance of organisms at each trophic level in a food chain? (3) How long are real food chains, and what factors limit food chain length?

A central theory in ecology is that "the world is green" because carnivores prevent herbivores from grazing green plant biomass to very low levels. Trophic structure (the number of trophic levels) determines trophic dynamics (as measured by the impact of herbivores on the abundance of plants). Indirect control of plant biomass by a top predator is called a trophic cascade. Cascades have been demonstrated to varying degrees in a wide variety of systems, including lakes, streams, subtidal kelp forests, coastal shrub habitats, and old fields. In all of these systems, the removal of a top predator has been shown to precipitate dramatic reductions in the abundance of species at lower trophic levels. Food chain theory predicts a green world when food chains have odd numbers of trophic levels, but a barren world (plants suppressed by herbivores) in systems with even numbers of trophic levels.

Although predators often have strong indirect effects on plant biomass as a result of trophic cascades, both predation (a top-down force) and resource supply to producers (a bottom-up force) play strong roles in the regulation of plant biomass. The supply of inorganic nutrients at the bottom of a food chain is an important determinant of the rate at which the plant trophic level produces tissue (primary production, or productivity). However, the degree to which nutrient supply enhances plant biomass accrual depends on two factors: (1) how many herbivores are present (which in turn depends on how many trophic levels there are in the system) and (2) the degree to which the herbivores can respond to increases in plant productivity and control plant biomass. The relative importance of top-down (demand) versus bottom-up (supply) forces is well illustrated by lake systems, in which the supply of phosphorus (bottom-up force) and the presence of piscivorous (fish-eating) fish (top-down force) have significant effects on the standing stock of phytoplankton, the plant trophic level in lake water columns. *See* BIOLOGICAL PRODUCTIVITY; BIOMASS; FRESH-WATER ECOSYSTEM; LAKE; PHYTOPLANKTON.

Increases in productivity may act to lengthen food chains. However, food chain length may be limited by the efficiency at which members of each trophic level assimilate energy as it moves up the food chain; the resilience of the chain (measured as the inverse of the time required for all trophic levels to return to previous abunance levels after a disturbance); and the size of the ecosystem—small habitats are simply not large enough to support the home range or provide ample habitat for larger carnivorous species. *See* ECOLOGICAL ENERGETICS; SYSTEMS ECOLOGY; THEORETICAL ECOLOGY. [J.L.S.; L.R.G.]

Tropic of Cancer The parallel of latitude about $23^1/_2{}^\circ$ (23.45°) north of the Equator. The importance of this line lies in the fact that its degree of angle from the Equator is the same as the inclination of the Earth's axis from the vertical to the plane of the ecliptic. Because of this inclination of the axis and the revolution of the Earth in its orbit, the vertical overhead rays of the Sun may progress as far north as $23^1/_2{}^\circ$. At no place north of the Tropic of Cancer will the Sun, at noon, be 90° overhead.

On June 21, the summer solstice (Northern Hemisphere), the Sun is vertical above the Tropic of Cancer. On this same day the Sun is 47° above the horizon at noon at the Arctic Circle, and at the Tropic of Capricorn, only 43° above the horizon. The Tropic of Cancer is the northern boundary of the equatorial zone called the tropics, which lies between the Tropic of Cancer and Tropic of Capricorn. *See* MATHEMATICAL GEOGRAPHY. [V.H.E.]

Tropic of Capricorn The parallel of latitude approximately $23^1/_2{}^\circ$ (23.45°)south of the Equator. It was named for the constellation Capricornus (the goat), for astronomical reasons which no longer prevail.

Because the Earth, in its revolution around the Sun, has its axis inclined $23^1/_2{}^\circ$ from the vertical to the plane of the ecliptic, the Tropic of Capricorn marks the southern limit of the zenithal position of the Sun. Thus, on December 22 (Southern Hemisphere summer, but northern winter solstice) the Sun, at noon, is 90° above the horizon.

The Tropic of Capricorn is the southern boundary of the equatorial zone referred to as the tropics, which lies between the Tropic of Capricorn and the Tropic of Cancer. *See* MATHEMATICAL GEOGRAPHY; TROPIC OF CANCER. [V.H.E.]

Tropical meteorology The study of atmospheric structure and behavior in the areas astride the Equator, roughly between 30° north and south latitude. The weather and climate of the tropics involve phenomena such as trade winds, hurricanes, intertropical convergence zones, jet streams, monsoons, and the El Niño Southern Oscillation. More energy is received from the Sun over the tropical latitudes than is lost to outer space (infrared radiation). The reverse is true at higher latitudes, poleward of 30°. The excess energy from the tropics is transported by winds to the higher latitudes, largely by vertical circulations that span roughly 30° in latitudinal extent. These circulations are known as Hadley cells.

For the most part, the oceanic tropics (the islands) experience very little change of day-to-day weather except when severe events occur. Tropical weather can be more adverse during the summer seasons of the respective hemispheres. The near equatorial belt between 5°S and 5°N is nearly always free from hurricanes and typhoons: the active belt lies outside this region over the tropics. The land areas experience considerable heating of the Earth's surface, and the summer-to-winter contrasts are somewhat larger there. For instance, the land areas of northern India experience air temperatures as high as 108°F (42°C) in the summer (near the Earth's surface), while in the winter season the temperatures remain 72°F (22°C) for many days. The diurnal range of temperature is also quite large over land areas on clear days during the summer (32°F or 18°C) as compared to winter (18°F or 10°C).

The steady northeast surface winds over the oceans of the Northern Hemisphere between 5° and 20°N and southeast winds over the corresponding latitudes of the southern oceans constitute the trade winds. Trade winds have intensities of around 5–10 knots (2.5–5 m/s). They are the equatorial branches of the anticyclonic circulation (known as the subtropical high pressure). The steadiness of wind direction is quite high in the trades. *See* WIND.

Hurricanes are also known as typhoons in the west Pacific and tropical cyclones in Indian Ocean and south Pacific. If the wind speed exceeds 65 knots (33 m/s) in a tropical storm, the storm is labeled a hurricane. A hurricane usually forms over the tropical oceans, north or south of 5° latitude from the Equator. *See* HURRICANE.

Intertropical convergence zones are located usually between 5 and 10°N latitude. They are usually oriented west to east and contain cloud clusters with rainfall of the order of 1.2–2 in. (30–50 mm) per day. The trade winds of the two hemispheres supply moisture to this precipitating system. *See* CLOUD PHYSICS; PRECIPITATION (METEOROLOGY).

A number of fast-moving air currents, known as jets, are important elements of the tropical general circulation. With speeds in excess of 30 knots (15 m/s), they are found over several regions of the troposphere. *See* ATMOSPHERIC GENERAL CIRCULATION; JET STREAM; TROPOSPHERE.

Basically the entire landmass from the west coast of Africa to Asia and extending to the date line experiences a phenomenon known as the monsoon. Monsoon circulations are driven by differential heating between relatively cold oceans and relatively warm landmasses. *See* MONSOON METEOROLOGY.

Every 2–6 years the eastern equatorial Pacific Ocean experiences a rise in sea surface temperature of about 5–9°F (3–5°). This phenomenon is known as El Niño, which is part of a larger cycle referred to as the El Niño Southern Oscillation (ENSO). The other extreme in the cycle is referred to as La Niña. El Niño has been known to affect global-scale weather. *See* MARITIME METEOROLOGY. [T.N.Kr.]

Tropopause The boundary between the troposphere and the stratosphere in the atmosphere. The tropopause is broadly defined as the lowest level above which the lapse rate (decrease) of temperature with height becomes less than 5.8°F mi^{-1} (2°C km^{-1}). In low latitudes the tropical tropopause is at a height of 9.3–11 mi at about −135°F (15–17 km at about 180 K), and the polar tropopause between tropics and poles is at about 6.2 mi at about −63°F (10 km at about 220 K). There is a well-marked "tropopause gap" or break where the tropical and polar tropopauses overlap at 30–40° latitude. The break is in the region of the subtropical jet stream and is of major importance for the transfer of air and tracers (humidity, ozone, radioactivity) between stratosphere and troposphere. The height of the tropopause varies seasonally and also daily with the weather systems, being higher and colder over anticyclones than over depressions. *See* AIR TEMPERATURE; ATMOSPHERE; STRATOSPHERE; TROPOSPHERE. [R.J.Mu.]

Troposphere The lowest major layer of the atmosphere. The troposphere extends from the Earth's surface to a height of 6–10 mi (10–16 km), the base of the stratosphere. It contains about four-fifths of the mass of the whole atmosphere. *See* ATMOSPHERE.

On the average, the temperature decreases steadily with height throughout this layer, with a lapse rate of about 19°F mi^{-1} (6.5°C km^{-1}), although shallow inversions (temperature increases with height) and greater lapse rates occur, particularly in the boundary layer near the Earth's surface. Appreciable water-vapor contents and clouds are almost entirely confined to the troposphere. Hence it is the seat of all important weather processes and the region where interchange by evaporation and precipitation (rain, snow, and so forth) of water substance between the surface and the atmosphere takes place. *See* CLIMATOLOGY; CLOUD PHYSICS; METEOROLOGY; WEATHER. [R.J.Mu.]

Trypanosomiasis A potentially fatal infection caused by parasites of the genus *Trypanosoma*.

The African trypanosomes, the cause of African trypanoso-miasis or African sleeping sickness, are flagellated protozoan parasites. *Trypanosoma brucei rhodesiense* and *T. b. gambiense* cause disease in humans. *Trypanosoma brucei* is restricted to domestic and wild animals. The trypanosomes are transmitted by the tsetse fly (*Glossina*), which is restricted to the African continent. The trypanosomes are taken up in a blood meal and grow and multiply within the tsetse gut. After 2–3 weeks, depending upon environmental conditions, they migrate into the salivary glands, where they become mature infective forms, and are then transmitted by the injection of infected saliva into a new host during a blood meal. The survival of the tsetse fly is dependent upon temperature and humidity, and the fly is confined by the Sahara to the north and by the colder drier areas to the south, an area about the size of the United States. Approximately 50 million people live within this endemic area, and 15,000–20,000 new human cases of African trypanosomiasis are reported annually.

In humans and other mammals the trypanosomes are extracellular. During the early stages of infection, the trypanosomes are found in the blood and lymph but not in cerebrospinal fluid. There are fever, malaise, and enlarged lymph nodes. In the absence of treatment the disease becomes chronic and the trypanosomes penetrate into the cerebrospinal fluid and the brain. The symptoms are headaches, behavioral changes, and finally the characteristic sleeping stage. Without treatment the individual sleeps more and more and finally enters a comatose stage which leads to death. Treatment is more difficult if the infection is not diagnosed until the late neurological stage.

Trypanosoma cruzi, the cause of American trypanosomiasis (Chagas' disease), is predominantly an intracellular parasite in the mammalian host. During the intracellular stage, *T. cruzi* loses its flagellum and grows predominantly in cells of the spleen, liver, lymphatic system, and cardiac, smooth, and skeletal muscle. The cells of the autonomic nervous system are also frequently invaded. *See* MEDICAL PARASITOLOGY. [J.R.See.]

Tsunami A set of ocean waves caused by any large, abrupt disturbance of the sea surface. If the disturbance is close to the coastline, tsunamis can demolish local coastal communities within minutes. A very large disturbance can both cause local devastation and export tsunami destruction thousands of miles away. Since 1850, tsunamis have been responsible for the loss of over 120,000 lives and billions of dollars of damage to coastal structures and habitats. Methods for predicting when and where the next tsunami will strike have not been developed; but once the tsunami is generated, forecasting its arrival and impact is possible through wave theory and measurement technology. *See* OCEAN WAVES.

Tsunamis are most commonly generated by earthquakes in marine and coastal regions. Major tsunamis are produced by large (greater than 7 on the Richter scale), shallow-focus (<30-km or 19-mi depth in the Earth) earthquakes associated with the movement of oceanic and continental plates. They frequently occur in the Pacific, where dense oceanic plates slide under the lighter continental plates. When these plates fracture, they cause a vertical movement of the sea floor that allows a quick and efficient transfer of energy from the solid earth to the ocean. The resulting tsunami propagates as a set of waves whose energy is concentrated at wavelengths corresponding to the earth movements (~100 km or 60 mi), at wave heights determined by vertical displacement (~1 m or 3 ft), and at wave directions determined by the adjacent coastline geometry. Because each earthquake is unique, every tsunami has unique wavelengths, wave heights, and directionality. From a warning perspective, this makes the problem of forecasting tsunamis in real time daunting. *See* EARTHQUAKE.

Other large-scale disturbances of the sea surface that can generate tsunamis are explosive volcanoes and asteroid impacts. The eruption of the volcano Krakatoa in the East Indies on August 27, 1883, produced a 30-m (100-ft) tsunami that killed over 36,000 people. *See* VOLCANO. [E.N.B.]

Tuberculosis An infectious disease caused by the bacillus *Mycobacterium tuberculosis*. It is primarily an infection of the lungs, but any organ system is susceptible, so its manifestations may be varied. Effective therapy and methods of control and prevention of tuberculosis have been developed, but the disease remains a major cause of mortality and morbidity throughout the world. The treatment of tuberculosis has been complicated by the emergence of drug-resistant organisms, including multiple-drug-resistant tuberculosis, especially in those with HIV infection. *See* ACQUIRED IMMUNE DEFICIENCY SYNDROME (AIDS).

Mycobacterium tuberculosis is transmitted by airborne droplet nuclei produced when an individual with active disease coughs, speaks, or sneezes. When inhaled, the droplet nuclei reach the alveoli of the lung. In susceptible individuals the organisms may then multiply and spread through lymphatics to the lymph nodes, and through the bloodstream to other sites such as the lung apices, bone marrow, kidneys, and meninges.

The development of acquired immunity in 2 to 10 weeks results in a halt to bacterial multiplication. Lesions heal and the individual remains asymptomatic. Such an individual is said to have tuberculous infection without disease, and will show a positive tuberculin test. The risk of developing active disease with clinical symptoms and positive cultures for the tubercle bacillus diminishes with time and may never occur, but is a lifelong risk. Only 5% of individuals with tuberculous infection progress to active disease. Progression occurs mainly in the first 2 years after infection; household contacts and the newly infected are thus at risk.

Many of the symptoms of tuberculosis, whether pulmonary disease or extrapulmonary disease, are nonspecific. Fatigue or tiredness, weight loss, fever, and loss of appetite may be present for months. A fever of unknown origin may be the sole indication of tuberculosis, or an individual may have an acute influenzalike illness. Erythema nodosum, a skin lesion, is occasionally associated with the disease.

The lung is the most common location for a focus of infection to flare into active disease with the acceleration of the growth of organisms. There may be complaints of cough, which can produce sputum containing mucus, pus- and, rarely, blood. Listening to the lungs may disclose rales or crackles and signs of pleural effusion (the escape of fluid into the lungs) or consolidation if present. In many, especially those with small infiltration, the physical examination of the chest reveals no abnormalities.

Miliary tuberculosis is a variant that results from the blood-borne dissemination of a great number of organisms resulting in the simultaneous seeding of many organ systems. The meninges, liver, bone marrow, spleen, and genitourinary system are usually involved. The term miliary refers to the lung lesions being the size of millet seeds (about 0.08 in. or 2 mm). These lung lesions are present bilaterally. Symptoms are variable.

Extrapulmonary tuberculosis is much less common than pulmonary disease. However, in individuals with AIDS, extrapulmonary tuberculosis predominates, particularly with lymph node involvement. Fluid in the lungs and lung lesions are other common manifestations of tuberculosis in AIDS. The lung is the portal of entry, and an extrapulmonary focus, seeded at the time of infection, breaks down with disease occurring.

Development of renal tuberculosis can result in symptoms of burning on urination, and blood and white cells in the urine; or the individual may be asymptomatic. The symptoms of tuberculous meningitis are nonspecific, with acute or chronic fever, headache, irritability, and malaise.

A tuberculous pleural effusion can occur without obvious lung involvement. Fever and chest pain upon breathing are common symptoms.

Bone and joint involvement results in pain and fever at the joint site. The most common complaint is a chronic arthritis usually localized to one joint. Osteomyelitis is also usually present.

Pericardial inflammation with fluid accumulation or constriction of the heart chambers secondary to pericardial scarring are two other forms of extrapulmonary disease.

The principal methods of diagnosis for pulmonary tuberculosis are the tuberculin skin test (an intracutaneous injection of purified protein derivative tuberculin is performed, and the injection site examined for reactivity), sputum smear and culture, and the chest x-ray. Culture and biopsy are important in making the diagnosis in extrapulmonary disease.

A combination of two or more drugs is used in the initial therapy of tuberculous disease. Drug combinations are used to lessen the chance of drug-resistant organisms surviving. The preferred treatment regimen for both pulmonary and extrapulmonary tuberculosis is a 6-month regimen of the antibiotics isoniazid, rifampin, and pyrazinamide given for 2 months, followed by isoniazid and rifampin for 4 months. Because of the problem of drug-resistant cases, ethambutol can be included in the initial regimen until the results of drug susceptibility studies are known. Once treatment is started, improvement occurs in almost all individuals. Any treatment failure or individual relapse is usually due to drug-resistant organisms. *See* DRUG RESISTANCE.

The community control of tuberculosis depends on the reporting of all new suspected cases so case contacts can be evaluated and treated appropriately as indicated. Individual compliance with medication is essential. Furthermore, measures to enhance compliance, such as directly observed therapy, may be necessary. *See* MYCOBACTERIAL DISEASES.

[G.Lo.]

Tubeworms The name given to marine polychaete worms (particularly to many species in the family Serpulidae) which construct permanent calcareous tubes on rocks, seaweeds, dock pilings, and ship bottoms. The individual tubes with hard walls of calcite-aragonite are firmly cemented to any hard substrate and to each other. Economically they are among the most important fouling organisms both on ship hulls (where they are second only to barnacles) and inside sea-water cooling pipes of power stations.

About 340 valid species of serpulid tubeworms have been described. The majority are truly marine, but several species thrive in brackish waters of low salinity, and one species occurs in fresh waters in Karst limestone caves. The wide geographical distribution of certain abundant species owes much to human transport on the bottoms of relatively fast ships and occurred within the last 120 years. *See* ANNELIDA; POLYCHAETA.

[W.D.R.-H.]

Tularemia A worldwide disease caused by infection with the bacterium *Francisella tularensis*, which affects multiple animal species. Infection in humans occurs frequently from skinning infected animals bare-handed or from the bites of infected animals, ticks, deer flies. The mortality rate varies by species, but with treatment it is low. Ungulates are frequently infected but generally suffer low mortality.

Tularemia can be difficult to differentiate from other diseases because it can have multiple clinical manifestations. Nonspecific signs frequently include fever, lethargy, anorexia, and increased pulse and respiration rates. The disease can overlap geographically with plague, and both may lead to enlarged lymph nodes (buboes). However, with tularemia, the buboes are more likely to ulcerate. If tularemic infection results from inhalation of dust from contaminated soil, hay, or grain, either pneumonia or a typhoidal syndrome can occur. Rarely, the route of entry for the bacteria is the eyes, leading to the oculoglandular type of tularemia. If organisms are ingested, the oropharyngeal form can develop, characterized by abdominal pain, diarrhea, vomiting, and ulcers. *See* PLAGUE.

Tularemia is not transmitted directly from individual to individual. If the infected person or animal is untreated, blood remains infectious for 2 weeks and ulcerated lesions are infectious for a month. Deer flies (*Chrysops discalis*) are infective for 2 weeks, and ticks are infective throughout their lifetime (usually 2 years).

A number of antibacterial agents are effective against *F. tularensis*, the most effective being streptomycin. Penicillin and the sulfonamides have no therapeutic effect. [M.Ei.]

Tulip tree A tree *Liriodendron tulipfera*, also known in forestry as yellow poplar, belonging to the magnolia family, Magnoliaceae. One of the largest and most valuable hardwoods of eastern North America, it is native from southern New England and New York westward to southern Michigan, and south to Louisiana and northern Florida.

Botanically, this tree is distinguished by leaves which are squarish at the tip as if cut off, true terminal buds flattened and covered by two valvate scales, an aromatic odor resembling that of magnolia, and cone-shaped fruit which is persistent in winter. The name tulip refers to the large greenish-yellow and orange-colored flowers.

The wood of the tulip tree is light yellow to brown, hence the common name yellow poplar, which is a misnomer. It is a soft and easily worked wood, used for construction, interior finish, containers (boxes, crates, baskets), woodenware, excelsior, veneer, and sometimes for paper pulp. [A.H.G./K.P.D.]

Tumor Literally, a swelling; in the past the term has been used in reference to any swelling of the body, no matter what the cause. However, the word is now being used almost exclusively to refer to a neoplastic mass, and the more general usage is being discarded.

A neoplastic mass or neoplasm is a pathological lesion characterized by the progressive or uncontrolled proliferation of cells. The cells involved in the neoplastic growth have an intrinsic heritable abnormality such that they are not regulated properly by normal methods. The stimulus which elicits this growth is not usually known.

It is common to divide tumors into benign or malignant. The decision as to which category a tumor should be assigned is usually based on information gained from gross or microscopic examination, or both. Benign neoplasms usually grow slowly, remain so localized that they cause little harm, and generally can be successfully and permanently removed. Malignant or cancerous neoplasms tend to grow rapidly, spread throughout the body, and recur if removed. Not all tumors which have been classified as benign are harmless to the host, and some can cause serious problems. Difficulties may occur as a result of mechanical pressure.

The cells of benign tumors are well differentiated. This means that the cells are very like the normal tissue in size, structure, and spatial relationship. The cells forming the tumor usually function normally. Cell proliferation usually is slow enough so that there is not a large number of immature cells. As the cellular mass increases in size, most benign tumors develop a fibrous capsule around them which separates them from the normal tissue. The cells of a benign tumor remain at the site of origin and do not spread throughout the body. Anaplasia (loss of differentiation) is not seen in benign tumors.

The cells of malignant tumors may be well differentiated, but most have some degree of anaplasia. Anaplastic cells tend to be larger than normal and are abnormal, even bizarre, in shape. The nuclei tend to be very large, and irregular, and they often stain darkly. Malignant tumors may be partially but never completely encapsulated. The cells of the cancer infiltrate and destroy surrounding tissue. They have the ability to metastasize; that is, cells from the primary tumor are disseminated to other regions of the body where they are able to produce secondary tumors called metastases.

In most cases the formation of a neoplasm is irreversible. It results from a permanent cellular defect which is passed on to daughter cells. Tumors should undergo medical appraisal to determine what treatment, if any, is needed. [N.K.M.; C.Qu.]

Tumor suppressor genes are a class of genes which, when mutated, predispose an individual to cancer. The mutations result in the loss of function of the particular tumor suppressor protein encoded by the gene. Although this class of genes was named for its link to human cancer, it is now clear that these genes play a critical role in the normal development, growth, and proliferation of cells and organs within the human body. The protein product of many tumor suppressor genes constrains cell growth and proliferation so that these events occur in a controlled manner. Thus, these genes appear to act in a manner antagonistic to that of oncogenes, which promote cell growth and proliferation.

The retinoblastoma (RB), p53, and p16 genes are the best-understood tumor suppressors . Inactivating mutations in the RB gene have been observed in retinoblastomas, osteosarcomas (cancer of the bone), as well as cancers of the lung, breast, and bladder. The p16 mutations have been observed in cancers of the skin, lung, breast, brain, bone, bladder, kidney, esophagus, and pancreas. The tumor suppressor p53 is the most frequently mutated gene associated with the development of many different types of human cancer, including those of the breast, lung, and colon. It is also associated with the rare inherited disease, Li-Fraumeni syndrome. Affected individuals manifest an increased likelihood of breast carcinomas, soft tissue sarcoma, brain tumors, osteosarcoma, leukemia, and adrenocortical carcinoma. Like RB and p16, p53 has a role in cell cycle regulation. In addition, p53 functions in the cell's decision on whether to undergo programmed cell death (apoptosis). Deregulaed cell proliferation and escape from apoptosis appear to be two common pathways leading to tumor formation. *See* CANCER (MEDICINE); GENE; MUTATION; ONCOLOGY; TUMOR VIRUSES. [M.Ew.]

Tumor viruses Viruses associated with tumors can be classified in two broad categories depending on the nucleic acid in the viral genome and the type of strategy to induce malignant transformation.

RNA viruses. The ribonucleic acid (RNA) tumor viruses are retroviruses. When they infect cells, the viral RNA is copied into deoxyribonucleic acid (DNA) by reverse transcription and the DNA is inserted into the host genome, where it persists and can be inherited by subsequent generation of cells. Transformation of the infected cells can be traced to oncogenes that are carried by the viruses but are not necessary for viral replication. The viral oncogenes are closely similar to cellular genes, the proto-oncogenes, which code for components of the cellular machinery that regulates cell proliferation, differentiation, and death. Incorporation into a retrovirus may convert proto-oncogenes into oncogenes in two ways: the gene sequence may be altered or truncated so that it codes for proteins with abnormal activity; or the gene may be brought under the control of powerful viral regulators that cause its product to be made in excess or in inappropriate circumstances. Retroviruses may also exert similar oncogenic effects by insertional mutation when DNA copies of the viral RNA are integrated into the host-cell genome at a site close to or even within proto-oncogenes. *See* RETROVIRUS.

RNA tumor viruses cause leukemias, lymphomas, sarcomas, and carcinomas in fowl, rodents, primates, and other species. The human T-cell leukemia virus (HTLV) types I and II are endemic in Southeast Asian populations and cause adult T-cell leukemia and hairy-cell leukemia. *See* AVIAN LEUKOSIS; ROUS SARCOMA.

DNA viruses. DNA viruses replicate lytically and kill the infected cells. Transformation occurs in nonpermissive cells where the infection cannot proceed to viral replication. The transforming ability of DNA tumor viruses has been traced to several viral proteins that cooperate to stimulate cell proliferation, overriding some of the normal growth control mechanisms in the infected cell and its progeny. Unlike retroviral oncogenes, DNA virus oncogenes are essential components of the viral genome and have no counterpart in the normal host cells. Some of these viral proteins bind to the protein products of two key tumor suppressor genes of the host cells, the retinoblastoma gene and the p53 gene, deactivating them and thereby permitting the cell to replicate its DNA and divide. Other DNA virus oncogenes interfere with the expression of cellular genes either directly or via interaction with regulatory factors. There is often a delay of several years between initial viral infection in the natural host species and the development of cancer, indicating that, in addition to virus-induced transformation, other environmental factors and genetic accidents are involved. A specific or general impairment of the host immune responses often plays an important role.

DNA tumor viruses belong to the families of papilloma, polyoma, adeno, hepadna, and herpes viruses and produce tumors of different types in various species. DNA tumor viruses are thought to play a role in the pathogenesis of about 15–20% of human cancers. These include Burkitt's lymphoma, nasopharyngeal carcinoma, immunoblastic lymphomas in immunosuppressed individuals and a proportion of Hodgkin's lymphomas that are all associated with the Epstein-Barr virus of the herpes family; and liver carcinoma associated with chronic hepatitis B virus infection. *See* ANIMAL VIRUS; CANCER (MEDICINE); INFECTIOUS PAPILLOMATOSIS; MUTATION; ONCOLOGY. [M.G.Ma.]

Tundra An area supporting some vegetation beyond the northern limit of trees, between the upper limit of trees and the lower limit of perennial snow on mountains, and on the fringes of the Antarctic continent and its neighboring islands. The term is of Lapp or Russian origin, signifying treeless plains of northern regions. Biologists, and particularly plant ecologists, sometimes use the term tundra in the sense of the vegetation of the tundra landscape. Tundra has distinctive characteristics as a kind of landscape and as a biotic community, but these are expressed with great differences according to the geographic region.

Characteristically tundra has gentle topographic relief, and the cover consists of perennial plants a few inches to a few feet or a little more in height. The general appearance during the growing season is that of a grassy sward in the wetter areas, a matted spongy turf on mesic sites, and a thin or sparsely tufted lawn or lichen heath on dry sites. In winter, snow mantles most of the surface. By far, most tundra occurs where the mean annual temperature is below the freezing point of water, and perennial frost (permafrost) accumulates in the ground below the depth of annual thaw and to depths at least as great as 1600 ft (500 m). *See* PERMAFROST. [W.S.B.]

Tung tree The plant *Aleurites fordii*, a species of the spurge family (Euphorbiaceae). The tree, native to central and western China, is the source of tung oil. It has been grown successfully in the southern United States. The globular fruit has three to seven large, hard, rough-coated seeds containing the oil, which is expressed after the seeds have been roasted. Tung oil is used to produce a hard, quick-drying, superior varnish, which is less apt to crack than other kinds. The foliage, sap, fruit, and commercial tung meal contain a toxic saponin, which causes gastroenteritis in animals that eat it. [P.D.St./E.L.C.]

Tupelo A tree belonging to the genus *Nyssa* of the sour gum family, Nyssaceae. The most common species is *N. sylvatica*, variously called pepperidge, black gum, or sour gum, the authorized name being black tupelo. Tupelo grows in the easternmost third of the United States; southern Ontario, Canada; and Mexico.

The tree can be identified by the comparatively small, obovate, shiny leaves and by branches that develop at a wide angle from the axis. The fruit is a small blue-black drupe, a popular food for birds. The wood is yellow to light-brown and hard to split because of the twisted grain. Tupelo is used for boxes, baskets, and berry crates, and as backing on which veneers of rarer and more expensive woods are glued. It is also used for flooring, rollers in glass factories, hatters' blocks, and gunstocks. [A.H.G./K.P.D.]

Tylenchida An order of nematodes in which the labial region is variable and may be distinctly set off or smoothly rounded and well developed; the hexaradiate symmetry is most often retained or discernible. The hollow stylet is the product of the cheilostome (conus, guiding apparatus, and framework) and the esophastome (shaft and knobs). Throughout the order the stylet may be present or absent and may be adorned with knobs. The variable esophagus is most often divisible into the corpus, isthmus, and glandular posterior bulb. The corpus is divisible into the procorpus and metacorpus. The metacorpus is generally valved but may not occur in some females and males, and the absence is characteristic of some taxa. The orifice of the dorsal esophageal gland opens either into the anterior procorpus or just anterior to the metacorporal valve. The excretory system is asymmetrical, and there is but one longitudinal collecting tubule. Females have one or two genital branches; when only one branch is present, it is anteriorly directed. Except for sex-reversed males, there is only one genital branch. Males may have one (=phasmid) or more caudal papillae. The spicules are always paired and variable in shape; they may be accompanied by a gubernaculum. The order comprises five superfamilies: Tylenchoidea, Criconematoidea, Saphaerularoidea, Aphelenchoidea, and Aphelenchoidoidea. *See* NEMATA; PLANT PATHOLOGY. [A.R.M.]

U

Underwater photography The techniques involved in using photographic equipment underwater. By far the greatest percentage of underwater photography is done within sport-diving limits in the tropical oceans.

Underwater photographers are faced with specific technical challenges. Water is 600 times denser than air and is predominantly blue in color. Depth affects light and creates physiological considerations for the photographer. As a result, underwater photography requires an understanding of certain principles of light beneath the sea.

As in all photography, consideration of the variables of light transmission is crucial to underwater photography. When sunlight strikes the surface of the sea, its quality and quantity change in several ways. As light travels from air to a denser medium, such as water, the light rays are bent (refracted); one result is magnification of underwater objects by one-third as compared to viewing them in air. The magnification effect must be considered when estimating distances underwater, which is critical for both focus and exposure. Light is absorbed when it propagates through water. Variables affecting the level of light penetration include the time of day (affects the angle at which the sunlight strikes the surface of the water); cloud cover; clarity of the water; depth (light is increasingly absorbed with increasing depth); and surface conditions (if the sea is choppy, more light will be reflected off the surface and less light transmitted to the underwater scene).

Depth affects not only the quantity of light but also the quality of light. Once light passes from air to water, different wavelengths of its spectrum are absorbed as a function of the color of the water and depth. Even in the clearest tropical sea, water serves as a powerful cyan (blue-green) filter. Natural full-spectrum photographs can be taken only with available light in very shallow depths. In ideal daylight conditions and clear ocean water, photographic film fails to record red at about 15 ft (4.5 m) in depth. Orange disappears at 30 ft (9 m), yellow at 60 ft (18 m), green at 80 ft (24 m), and at greater depth only blue and black are recorded on film. To restore color, underwater photographers must use artificial light. *See* SEAWATER.

The water column between photographer and subject degrades both the resolution of the image and the transmission of artificial light (necessary to restore color). Therefore, the most effective underwater photos are taken as close as possible to the subject, thereby creating the need for a variety of optical tools to capture subjects of various sizes within this narrow distance limitation.

There are two types of underwater cameras—amphibious and housed. Amphibious cameras may be used either underwater or topside, although some lenses are for underwater use only (known as water contact lenses). A housed camera is a conventional above-water camera that has been protected from the damaging effects of seawater by a waterproof enclosure. The amphibious camera is protected by a series of O-rings, primarily located at the lens mount, film loading door, shutter release, and other places where controls are necessary. The O-rings make the system not only resistant to leaks but also impervious to dust or inclement weather when used above water. [S.Fri.]

Deep-sea underwater photography—approximately 150 ft (35 m)—requires the design and use of special camera and lighting equipment. Watertight cases are required for both camera and light source, and they must be able to withstand the pressure generated by the sea. For each 33 ft (10 m) of depth, approximately one additional atmosphere ($\sim 10^2$ kilopascals) of pressure is exerted. At the greatest ocean depths, about 40,000 ft (12,000 m), a case must be able to withstand 17,600 lb/in.^{2} (1200 kg/cm^2). The windows for the lens and electrical seals must also be designed for such pressure to prevent water intrusion.

Auxiliary lighting is required, since daylight is absorbed in both intensity and hue. The camera must be positioned and triggered to render the desired photograph, and the great depths preclude a free-swimming human operator. Operation is often from a cable via sonar sensing equipment or from deep-diving underwater vehicles. Bottom-sensing switches can operate deep-sea cameras for photographing the sea floor, and remotely operated vehicles (ROVs) can incorporate both video and still cameras. *See* UNDERWATER VEHICLE.

When an observer descends to great depths in a diving vehicle, the camera can assist in documentation by recording what is seen. Furthermore, the visual data will assist in accurate description of the observed phenomena. Elapsed-time photography with a motion picture camera in the sea is important in studying sedimentation deposits caused by tides, currents, and storms. Similarly, the observation of biological activity taken with the elapsed-time camera and then speeded up for viewing may reveal processes that cannot ordinarily be observed. *See* DIVING. [H.E.E.; S.Fri.]

Underwater vehicle A submersible work platform designed to be operated either remotely or directly. Underwater vehicles are grouped into three categories: deep submersible vehicles (DSVs), remotely operated vehicles (ROVs), and autonomous underwater vehicles (AUVs). There are also hybrid vehicles which combine two or three categories on board a single platform. Within each category of submersible there are specially adaped vehicles for specific work tasks. These can be purpose-built or modifications of standard submersibles.

There are five types of DSVs: one-atmosphere untethered vehicles; one-atmosphere tethered vehicles, including observation/work bells; atmospheric diving suits; diver lock-out vehicles; and wet submersibles. While they differ mainly in configuration, source of power, and number of crew members, all carry a crew at 1-atm (10^2-kilopascal) pressure within a dry chamber. An exception is the wet submersible, where the crew is exposed to full depth pressure. The purpose of the DSV is to put the trained mind and eye to work inside the ocean. The earliest submersibles had very small viewing ports fitted into thick-walled steel hulls. In the mid-1960s, experimental work began on use of massive plastics (acrylics) as pressure hull materials. Today, submersibles with depth capabilities to 3300 ft (1000 m) are being manufactured with pressure hulls made entirely of acrylic. Essentially the hull is now one huge window.

The first ROVs were developed in the late 1950s for naval use. By the mid-1970s, they were used in the civil sector. The rapid acceptance of these submersibles is due to their relatively low cost and the fact that they do not put human life at risk when undertaking hazardous missions. However, their most important attribute is that they are less complex. By virtue of their surface-connecting umbilical cable, they can operate almost indefinitely since there is no human inside requiring life support and no batteries to be recharged. There are four types of ROV: tethered free-swimming vehicles, towed vehicles, bottom reliant vehicles, and structurally reliant vehicles.

AUVs are crewless and untethered submersibles which operate independent of direct human control. Their operations are controlled by a preprogrammed, on-board computer. They were first developed in the military where applications include such tasks as minefield location and mapping, minefield installation, submarine decoys, and covert intelligence collection. Civilian tasks include site monitoring, basic oceanographic data gathering, under-ice mapping, offshore structure and pipeline inspection, and bottom mapping. These submersibles are particularly useful where long-duration measurements and observations are to be made and where human presence is not required.

AUVs span a wide range of sizes and capabilities, related to their intended missions. Each is a mobile instrumentation platform with propulsion, sensors, and on-board intelligence designed to complete sampling tasks autonomously. At the large end of the scale, transport-class platforms in the order of 10 m (33 ft) length and 10 metric ton (11 tons) weight in air have been designed for missions requiring long endurance, high speed, large payloads, or high-power sensors. At the small end of the scale, network-class platforms in the order of 1 m (3.3 ft) length and 100 kg (220 lb) weight in air address missions requiring portability, multiple platforms, adaptive spatial sampling, and sustained presence in a specific region. Vehicles can also be categorized in terms of propulsion method (propeller-driven or buoyancy-driven) or in terms of their maximum operating depth.

Hybrid vehicles are those that combine crewed vehicles, remotely operated vehicles, and divers. For example, the hybrid *DUPLUS II* can operate either as a tethered free-swimming ROV or as a 1-atm tethered crewed vehicle. This evolved to provide capability for remotely conducting those tasks for which human skills are not needed, and then to put the human at the place where those skills are required. Other hybrid examples include ROVs that can be controlled remotely from the surface or at the work site by a diver performing maintenance and repair tasks. [D.Wal.; T.B.C.]

Upwelling The phenomenon or process involving the ascending motion of water in the ocean. Vertical motions are an integral part of ocean circulation, but they are a thousand to a million times smaller than the horizontal currents. Vertical motions are inhibited by the density stratification of the ocean because with increasing depth, as the temperature decreases, the density increases, and energy must be expended to displace water vertically. The ocean is also stratified in other properties; for example, nutrient concentration generally increases with depth. Thus even weak vertical flow may cause a significant effect by advecting nutrients to a new level. *See* GEOPHYSICAL FLUID DYNAMICS.

There are two important upwelling processes. One is the slow upwelling of cold abyssal water, occurring over large areas of the world ocean, to compensate for the formation and sinking of this deep water in limited polar regions. The other is the upwelling of subsurface water into the euphotic (sunlit) zone to compensate for a horizontal divergence of the flow in the surface layer, usually caused by winds. *See* OCEAN CIRCULATION. [R.L.Sm.]

Urban climatology The branch of climatology concerned with urban areas. These locales produce significant changes in the surface of the Earth and the quality of the air. In turn, surface climate in the vicinity of urban sites is altered. The era of urbanization on a worldwide scale has been accompanied by unintentional, measurable changes in city climate. *See* CLIMATOLOGY.

The process of urbanization changes the physical surroundings and induces alterations in the energy, moisture, and motion regime near the surface. Most of these alterations may be traced to causal factors such as air pollution; anthropogenic heat; surface waterproofing; thermal properties of the surface materials; and morphology of the surface and its specific three-dimensional geometry—building spacing, height, orientation, vegetative layering, and the overall dimensions and geography of these elements. Other factors that must be considered are relief, nearness to water bodies, size of the city, population density, and land-use distributions.

In general, cities are warmer than their surroundings, as documented over a century ago. They are islands or spots on the broader, more rural surrounding land. Thus, cities produce a heat island effect on the spatial distribution of temperatures. The timing of a maximum heat island is followed by a lag shortly after sundown, as urban surfaces, which absorbed and stored daytime heat, retain heat and affect the overlying air. Meantime, rural areas cool at a rapid rate.

A number of energy processes are altered to create warming, and various features lead to those alterations. City size, the morphology of the city, land-use configuration, and the geographic setting (such as relief, elevation, and regional climate) dictate the intensity of the heat island, its geographic extent, its orientation, and its persistence through time. Individual causes for heat island formation are related to city geometry, air pollution, surface materials, and anthropogenic heat emission. There are two atmosphere layers in an urban environment, besides the planetary boundary layer outside and extending well above the city: (1) The urban boundary layer is due to the spatially integrated heat and moisture exchanges between the city and its overlying air. (2) The surface of the city corresponds to the level of the urban canopy layer. Fluxes across this plane comprise those from individual units, such as roofs, canyon tops, trees, lawns, and roads, integrated over larger land-use divisions (for example, suburbs). [A.J.B.]

Uredinales An order of fungi known as plant rusts that belong to the division (phylum) Basidiomycota. In nature, all 7000 species are obligate parasites of many vascular plant species. They cause diseases known as rust on numerous cultivated crops, and many trees are also attacked. Each rust species infects one or just a few closely related plant host species. Rust fungi occur on all continents except Antarctica.

The body of a rust fungus consists of numerous microscopic, threadlike, branching hyphae that grow inside the tissues and between the cells of the host plant. Specialized feeding structures, haustoria, grow from these hyphae into the host cells of the plant and provide nourishment for the hyphae. Hyphal cells are binucleate. Depending upon the rust species and environmental conditions, up to six different kinds of spore-producing structures, the sori, may be produced by one rust species. These sori may be powdery, waxy, or crustlike, and whitish, yellow, orange, brown, or blackish.

Rust teliospores that are produced in sori called telia are essential elements for classification. Teliospores consist of one or more specialized probasidial cells. During the development of each of these cells, nuclei fuse and meiosis occurs, resulting in a septate metabasidium from which four haploid meiospores (basidiospores) are formed. Basidiospores are forcibly ejected from the tips of their tiny stalks and are wind-disseminated. After landing on a susceptible host, under proper conditions basidiospores produce new infections. In cold regions, a rust may survive the winter as dormant, thick-walled, dark-colored teliospores. In many rust species, teliospores mature and produce basidiospores with no or little dormancy, especially in the tropics.

Although most rust species require only one host species to complete their life cycles (autoecious rusts), some of the best-known rusts require two taxonomically unrelated hosts (heteroecious rusts).

Rust diseases are controlled most effectively by breeding resistant host varieties. Fungicides are used for some rusts. In the case of some heteroecious rusts, eradication of one of the hosts (the noneconomic one) has been successful. Some rust species are used to aid in biological control of weeds. *See* BASIDIOMYCOTA; EUMYCOTA; FUNGI; PLANT PATHOLOGY. [J.F.Hen.]

V

Vegetable ivory The seed of the tagua palm (*Phytelephas macrocarpa*) of tropical America. Each drupelike fruit contains six to nine bony seeds. The extremely hard endosperm of the seed is used as a substitute for ivory. Vegetable ivory can be carved and tooled to make buttons, chessmen, knobs, inlays, and various ornamental articles. [P.D.St./E.L.C.]

Vegetation and ecosystem mapping The graphic portrayal of spatial distributions of vegetation, ecosystems, or their characteristics. Vegetation is one of the most conspicuous and characteristic features of the landscape and has long been a convenient way to distinguish different regions; maps of ecosystems and biomes have been mainly vegetation maps. As pressure on the Earth's natural resources grows and as natural ecosystems are increasingly disturbed, degraded, and in some cases replaced completely, the mapping of vegetation and ecosystems, at all scales and by various methods, has become more important. *See* BIOME; ECOSYSTEM.

Three approaches have arisen for mapping general vegetation patterns, (1) based on vegetation structure or gross physiognomy, (2) based on correlated environmental patterns, and (3) based on important floristic taxa. The environmental approach provides the least information about the actual vegetation but succeeds in covering regions where the vegetation is poorly understood. Most modern classification systems use a combination of physiognomic and floristic characters. *See* PLANT GEOGRAPHY.

Mapping has expanded to involve other aspects of vegetation and ecosystems as well as new methodologies for map production. Functional processes such as primary production, decomposition rates, and climatic correlates (such as evapotranspiration) have been estimated for enough sites so that world maps can be generated. Structural aspects of ecosystems, such as total standing biomass or potential litter accumulations, are also being estimated and mapped. Quantitative maps of these processes or accumulations can be analyzed geographically to provide first estimates of important aspects of world biogeochemical budgets and resource potentials.

Computer-produced maps, using Geographic Information Systems (GIS), often coupled directly with predictive models, remote-sensing capabilities, and other techniques, have also revolutionized vegetation and ecosystem mapping. This gives scientists a powerful tool for modeling and predicting the outcome from global climate change, in that feedback from the world's vegetation can be accounted for. Before computer technology exploded in the early 1980s, the spatial scale and related resolution or grain of vegetation and ecosystem mapping was limited by the static nature of hard-copy maps. The advent of GIS technology enabled the analysis of digital maps at any spatial scale, and the only limitation was the resolution at which the data were originally mapped. In addition, GIS software is used for sophisticated spatial analyses on maps, and this was virtually impossible before. *See* CLIMATE MODELING; GEOGRAPHIC INFORMATION SYSTEMS. [B.E.F.; E.O.B.]

Vernalization The induction in plants of the competence or ripeness to flower by the influence of cold, that is, temperatures below the optimal temperature for growth. Vernalization thus concerns the first of the three phases of flower formation in plants. In the second stage, for which a certain photoperiod frequently is required, flowers are initiated. In the third stage flowers are unfolded. *See* FLOWER; PHOTOPERIODISM; PLANT GROWTH. [K.N.Z.]

Vestimentifera A phylum of benthonic marine worms that is restricted to habitats rich in sulfide (for example, hydrothermal vents and sulfide seeps) and that as adults lack a mouth, gut, and anus, so they are nourished by internal symbionts. All vestimentiferans live in tubes of varying hardness and rigidity. Tube material is secreted by internal glands and is a mixture of chitin and protein.

The phylum has two classes based on the arrangement and orientation of lamellae formed by fusion of branchial filaments: Axonobranchia with one order, Riftiida, and one monogeneric family with one species; and Basibranchia with two orders, Lamellibrachiida and Tevniida.

Vestimentiferans have four regions along their length. The most anterior, obturacular region is provided with a mass of well-vascularized filaments supported by a paired structure, the obturaculum. Normally these branchial filaments protrude from the opening of the tube and act as a gill, allowing for the exchange of dissolved substances between the worm's body and the seawater. The second region, the vestimentum, is quite muscular and serves to maintain the plume of branchial filaments in the open seawater by pressure on the inner tube surface at its opening. The vestimentum is the site of many glands that contribute material for lengthening the tube and thickening it near the opening. The third region, the trunk, is a single segment and constitutes about 75% of the total length of the worm. It is provided with a pair of large longitudinal blood vessels. The remains of the larval gut containing sulfide-oxidizing bacteria and the gonads are also present. The fourth and most posterior region, the opisthosome, is made up of many segments, each segment in the anterior portion bearing a row of small hooks that can be set into the inner surface of the tube. This provides an anchor against which the body of the worm retracts when the longitudinal muscles of the trunk contract during withdrawal into the tube. [M.L.J.]

Virulence The ability of a microorganism to cause disease. Virulence and pathogenicity are often used interchangeably, but virulence may also be used to indicate the degree of pathogenicity. Scientific understanding of the underlying mechanisms of virulence has increased rapidly due to the application of the techniques of biochemistry, genetics, molecular biology, and immunology. Bacterial virulence is better understood than that of other infectious agents.

Virulence is often multifactorial, involving a complex interplay between the parasite and the host. Various host factors, including age, sex, nutritional status, genetic constitution, and the status of the immune system, affect the outcome of the parasite-host interaction. Hosts with depressed immune systems, such as transplant and cancer patients, are susceptible to microorganisms not normally pathogenic in healthy hosts. Such microorganisms are referred to as opportunistic pathogens. The attribute of virulence is present in only a small portion of the total population of microorganisms, most of which are harmless or even beneficial to humans and other animals. *See* OPPORTUNISTIC INFECTIONS.

The spread of an infectious disease usually involves the adherence of the invading pathogen to a body surface. Next, the pathogen multiplies in host tissues, resisting or evading various nonspecific host defense systems. Actual disease symptoms are from damage to host tissues caused either directly or indirectly by the microorganism's components or products.

Most genetic information in bacteria is carried in the chromosome. However, genetic information is also carried on plasmids, which are independently replicating structures much smaller than the chromosome. Plasmids may provide bacteria with additional virulence-related capabilities (such as pilus formation, iron transport systems, toxin production, and antibiotic resistance). In some bacteria, several virulence determinants are regulated by a single genetic locus. *See* BACTERIA; VIRUS. [B.Wi.]

Virus Any of a heterogeneous class of agents that share three characteristics: (1) They consist of a nucleic acid genome surrounded by a protective protein shell, which may itself be enclosed within an envelope that includes a membrane; (2) they multiply only inside living cells, and are absolutely dependent on the host cells' synthetic and energy-yielding apparatus; (3) the initial step in multiplication is the physical separation of the viral genome from its protective shell, a process known as uncoating, which differentiates viruses from all other obligatorily intracellular parasites. In essence, viruses are nucleic acid molecules, that is, genomes that can enter cells, replicate in them, and encode proteins capable of forming protective shells around them. Terms such as "organism" and "living" are not applicable to viruses. It is preferable to refer to them as functionally active or inactive rather than living or dead.

The primary significance of viruses lies in two areas. First, viruses destroy or modify the cells in which they multiply; they are potential pathogens capable of causing disease. Many of the most important diseases that afflict humankind, including rabies, smallpox, poliomyelitis, hepatitis, influenza, the common cold, measles, mumps, chickenpox, herpes, rubella, hemorrhagic fevers, and the acquired immunodeficiency syndrome (AIDS) are caused by viruses. Viruses also cause diseases in livestock and plants that are of great economic importance. *See* ACQUIRED IMMUNE DEFICIENCY SYNDROME (AIDS); PLANT PATHOLOGY.

Second, viruses provide the simplest model systems for many basic problems in biology. Their genomes are often no more than one-millionth the size of, for example, the human genome; yet the principles that govern the behavior of viral genes are the same as those that control the behavior of human genes. Viruses thus afford unrivaled opportunities for studying mechanisms that control the replication and expression of genetic material.

Although viruses differ widely in shape and size (see illustration), they are constructed according to certain common principles. Basically, viruses consist of nucleic acid and protein. The nucleic acid is the genome which contains the information necessary for virus multiplication and survival, the protein is arranged around the genome in the form of a layer or shell that is termed the capsid, and the structure consisting of shell plus nucleic acid is the nucleocapsid. Some viruses are naked nucleocapsids. In others, the nucleocapsid is surrounded by a lipid bilayer to the outside of which "spikes" composed of glycoproteins are attached; this is termed the envelope. The complete virus particle is known as the virion, a term that denotes both intactness of structure and the property of infectiousness.

Viral genomes are astonishingly diverse. Some are DNA, others RNA; some are double-stranded, others single-stranded; some are linear, others circular; some have plus polarity, other minus (or negative) polarity; some consist of one molecule, others

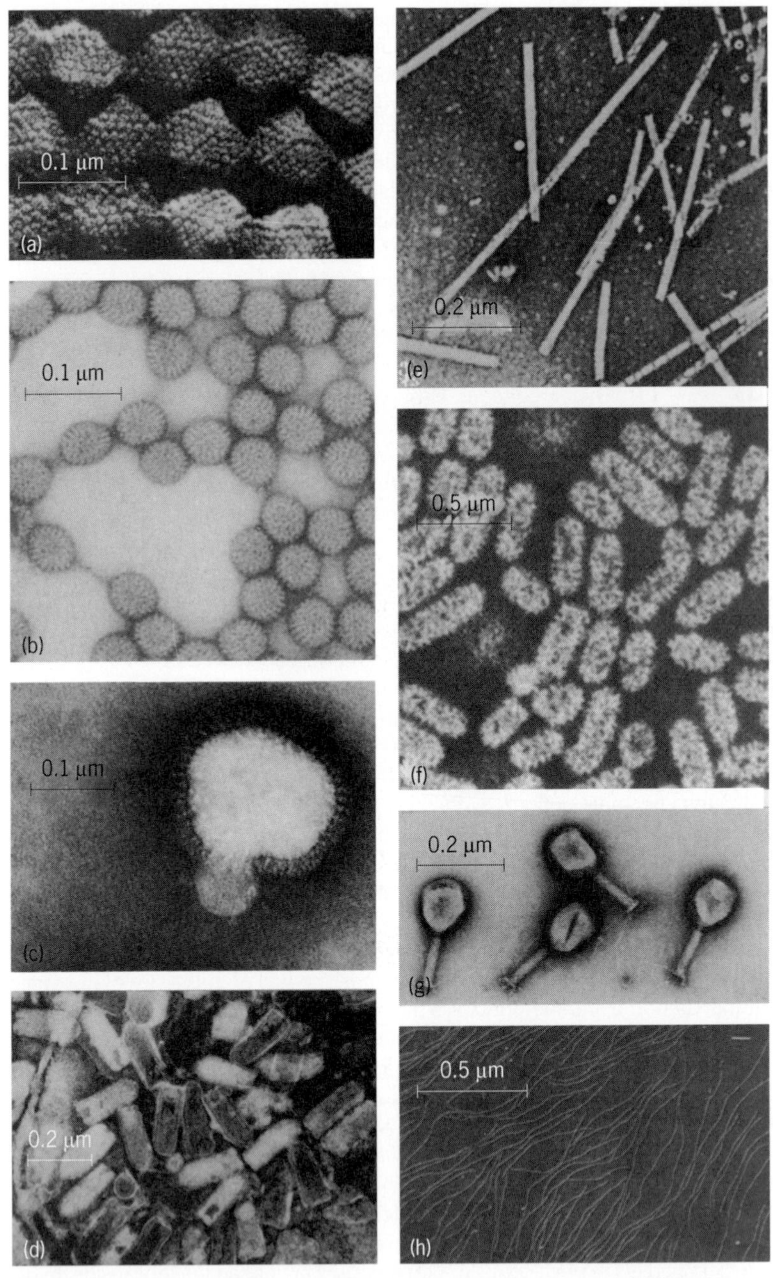

Electron micrographs of highly purified preparations of some viruses. (*a*) Adenovirus. (*b*) Rotavirus. (*c*) Influenza virus (*courtesy of George Leser*). (*d*) Vesicular stomatitis virus. (*e*) Tobacco mosaic virus. (*f*) Alfalfa mosaic virus. (*g*) T4 bacteriophage. (*h*) M13 bacteriophage.

of several (up to 12). They range from 3000 to 280,000 base pairs if double-stranded, and from 5000 to 27,000 nucleotides if single-stranded.

Viral genomes encode three types of genetic information. First, they encode the structural proteins of virus particles. Second, most viruses encode enzymes capable of transcribing their genomes into messenger RNA molecules that are then translated by host-cell ribosomes, as well as nucleic acid polymerases capable of replicating their genomes; many viruses also encode nonstructural proteins with catalytic and other functions necessary for virus particle maturation and morphogenesis. Third, many viruses encode proteins that interact with components of host-cell defense mechanisms against invading infectious agents. The more successful these proteins are in neutralizing these defenses, the more virulent viruses are.

The two most commonly observed virus-cell interactions are the lytic interaction, which results in virus multiplication and lysis of the host cell; and the transforming interaction, which results in the integration of the viral genome into the host genome and the permanent transformation or alteration of the host cell with respect to morphology, growth habit, and the manner in which it interacts with other cells. Transformed animal and plant cells are also capable of multiplying; they often grow into tumors, and the viruses that cause such transformation are known as tumor viruses. *See* CANCER (MEDICINE); ONCOLOGY; RETROVIRUS; TUMOR VIRUSES.

There is little that can be done to interfere with the growth of viruses, since they multiply within cells, using the cells' synthetic capabilities. The process, interruption of which has met with the most success in preventing virus multiplication, is the replication of viral genomes, which is almost always carried out by virus-encoded enzymes that do not exist in uninfected cells and are therefore excellent targets for antiviral chemotherapy. Another viral function that has been targeted is the cleavage of polyproteins, precursors of structural proteins, to their functional components by virus-encoded proteases; this strategy is being used with some success in AIDS patients. *See* CYTOMEGALOVIRUS INFECTION; HERPES; INFLUENZA; RESPIRATORY SYNCYTIAL VIRUS.

Antiviral agents on which much interest is focused are the interferons. Interferons are cytokines or lymphokines that regulate cellular genes concerned with cell division and the functioning of the immune system. Their formation is strongly induced by virus infection; they provide the first line of defense against viral infections until antibodies begin to form. Interferons interfere with the multiplication of viruses by preventing the translation of early viral messenger RNAs. As a result, viral capsid proteins cannot be formed and no viral progeny results.

By far the most effective means of preventing viral diseases is by means of vaccines. There are two types of antiviral vaccines, inactivated virus vaccines and attenuated active virus vaccines. Most of the antiviral vaccines currently in use are of the latter kind. The principle of antiviral vaccines is that inactivated virulent or active attenuated virus particles cause the formation of antibodies that neutralize a virulent virus when it invades the body. *See* ANIMAL VIRUS; PLANT VIRUSES AND VIROIDS. [W.K.J.]

Volcano A mountain or hill, generally steep-sided, formed by accumulation of magma (molten rock with associated gas and crystals) erupted through openings or volcanic vents in the Earth's crust; the term volcano also refers to the vent itself. During the evolution of a long-lived volcano, a permanent shift in the locus of principal vent activity can produce a satellitic volcanic accumulation as large as or larger than the parent volcano, in effect forming a new volcano on the flanks of the old.

Planetary exploration has revealed dramatic evidence of volcanoes and their products on the Earth's Moon, Mars, Mercury, Venus, and the moons of Jupiter, Neptune, and Uranus on a scale much more vast than on Earth. However, only the products and landforms of terrestrial volcanic activity are described here. *See* VOLCANOLOGY.

Volcanic vents. Volcanic vents, channelways for magma to ascend toward the surface, can be grouped into two general types: fissure and central (pipelike). Magma consolidating below the surface in fissures or pipes forms a variety of igneous bodies, but magma breaking the surface produces fissure or pipe eruptions. Fissures, most of them less than 10 ft (3 m) wide, may form in the summit region of a volcano, on its flanks, or near its base; central vents tend to be restricted to the summit area of a volcano. For some volcanoes or volcanic regions, swarms of fissure vents are clustered in swaths called rift zones.

Volcanic products. Magma erupted onto the Earth's surface is called lava. If the lava is chilled and solidifies quickly, it forms volcanic glass; slower rates of chilling result in greater crystallization before complete solidification. Lava may accrete near the vent to form various minor structures or may pour out in streams called lava flows, which may travel many tens of miles from the vents. During more violent eruption, lava torn into fragments and hurled into the air is called pyroclastic (fire-broken materials).

Volcanic gases. Violent volcanic explosions may throw dust and aerosols high into the stratosphere, where it may drift across the surface of the globe for many thousands of miles. Most of the solid particles in the volcanic cloud settle out within a few days, and nearly all settle out within a few weeks, but the gaseous aerosols (principally sulfuric acid droplets) may remain suspended in the stratosphere for several years. Such stratospheric clouds of volcanic aerosols, if sufficiently voluminous and long-lived, can have an impact on global climate. *See* ACID RAIN; AIR POLLUTION.

In general, water vapor is the most abundant constituent in volcanic gases; the water is mostly of meteoric (atmospheric) origin, but in some volcanoes can have a significant magmatic or juvenile component. Excluding water vapor, the most abundant gases are the various species of carbon, sulfur, hydrogen, chlorine, and fluorine.

Mudflows are common on steep-side volcanoes where poorly indurated or nonwelded pyroclastic material is abundant. Probably by far the most common cause, however, is simply heavy rain saturating a thick cover of loose, unstable pyroclastic material on the steep slope of the volcano, transforming the material into a mobile, water-saturated "mud," which can rush downslope at a speed as great as 50–55 mi (80–90 km) per hour. Such a dense, fast-moving mass can be highly destructive, sweeping up everything loose in its path.

Volcanic landforms. Much of the Earth's solid surface, on land and below the sea, has been shaped by volcanic activity. Landscape features of volcanic origin may be either positive (constructional) forms, the result of accumulation of volcanic materials, or negative forms, the result of the lack of accumulation or collapse.

Not all volcanoes show a graceful, symmetrical cone shape, such as that exemplified by Mount Fuji, Japan. Most volcanoes, especially those near tectonic plate boundaries, are more irregular, though of grossly conical shape. Such volcanoes, called stratovolcanoes or composite volcanoes, typically erupt explosively and are composed dominantly of andesitic, relatively viscous and short lava flows, interlayered with beds of ash and cinder that thin away from the principal vents. Volcanoes constructed primarily of fluid basaltic lava flows, which may spread great distances from the vents, typically are gentle-sloped, broadly upward convex structures. Such shield volcanoes, classic examples of which are Mauna Loa volcano, Hawaii, tend to form in oceanic intraplate regions and are associated with hot-spot volcanism. The shape and size of a volcano can vary widely between the simple forms of composite and shield volcanoes, depending on magma viscosity, eruptive style (explosive versus nonexplosive), migration of vent locations, duration and complexity of eruptive history, and posteruption modifications.

Some of the largest volcanic edifices are not shaped like the composite or shield volcanoes. In certain regions of the world, voluminous extrusions of very fluid basaltic lava from dispersed fissure swarms have built broad, nearly flat-topped accumulations. These voluminous outpourings of lava are known as flood basalts or plateau basalts.

Submarine volcanism. Deep submarine volcanism occurs along the spreading ridges that zigzag for thousands of miles across the ocean floor, and it is exposed above sea level only in Iceland. Because of the logistical difficulties in making direct observations posed by the great ocean depths, no deep submarine volcanic activity has been actually observed during eruption. However, evidence that deep-sea eruptions are happening is clearly indicated by (1) seismic and acoustic monitoring networks; (2) the presence of deep-ocean floor hydrothermal vents; (3) episodic hydrothermal discharges, measured and mapped as thermal and geochemical anomalies in the ocean water; and (4) the detection of new lava flows in certain segments of the oceanic ridge system. *See* MID-OCEANIC RIDGE.

Volcanic eruptions in shallow water are very similar in character to those on land but, on average, are probably somewhat more explosive, owing to heating of water and resultant violent generation of supercritical steam. Much of the ocean basin appears to be floored by basaltic lava. *See* OCEANIC ISLANDS.

Fumaroles and hot springs. Vents at which volcanic gases issue without lava or after the eruption are known as fumaroles. They are found on active volcanoes during and between eruptions and on dormant volcanoes, persisting long after the volcano itself has become inactive. Fumaroles grade into hot springs and geysers. The water of most, if not all, hot springs is predominantly of meteoric origin, and is not water liberated from magma. Some hot springs are of volcanic origin and the water may contain volcanic gases. *See* GEYSER.

Distribution of volcanoes. Over 500 active volcanoes are known on the Earth, mostly along or near the boundaries of the dozen or so lithospheric plates that compose the Earth's solid surface. Lithospheric plates show three distinct types of boundaries: divergent or spreading margins—adjacent plates are pulling apart; convergent margins (subduction zones)—plates are moving toward each other and one is being destroyed; and transform margins—one plate is sliding horizontally past another. All these types of plate motion are well demonstrated in the Circum-Pacific region, in which many active volcanoes form the so-called Ring of Fire. Some volcanoes, however, are not associated with plate boundaries, and many of these so-called intraplate volcanoes form roughly linear chains in the interior parts of the oceanic plates, for example, the Hawaiian-Emperor, Austral, Society, and Line archipelagoes in the Pacific Basin. Intraplate volcanism also has resulted in voluminous outpourings of fluid lava to form extensive plateau basalts, or of more viscous and siliceous pyroclastic products to form ash flow plains. [R.I.T.]

Volcanology The scientific study of volcanic phenomena, especially the processes, products, and hazards associated with active or potentially active volcanoes. It focuses on eruptive activity that has occurred within the past 10,000 years of the Earth's history, particularly eruptions during recorded history. Strictly speaking, it emphasizes the surface eruption of magmas and related gases, and the structures, deposits, and other effects produced thereby. Broadly speaking, however, volcanology includes all studies germane to the generation, storage, and transport of magma, because the surface eruption of magma represents the culmination of diverse physicochemical processes at depth. This article considers the activity of erupting volcanoes and the nature of erupting lavas. For a discussion of the distribution of volcanoes and the surface structures and deposits produced by them. *See* VOLCANO.

On average, about 50 to 60 volcanoes worldwide are active each year. About half of these constitute continuing activity that began the previous year, and the remainder are new eruptions. Analysis of historic records indicates that eruptions comparable in size to that of Mount St. Helens or El Chichón tend to occur about once or twice per decade, and larger eruptions such as Pinatubo about once per one or two centuries.

On a global basis, eruptions the size of that at Nevado del Ruiz in November 1985 are orders of magnitude more frequent.

Modern volcanology perhaps began with the founding of well-instrumented observations at Asama Volcano (Japan) in 1911 and at Kilauea Volcano (Hawaii) in 1912. The Hawaiian Volcano Observatory, located on Kilauea's caldera rim, began to conduct systematic and continuous monitoring of seismic activity preceding, accompanying, and following eruptions, as well as other geological, geophysical, and geochemical observations and investigations.

The eruptive characteristics, products, and resulting landforms of a volcano are determined predominantly by the composition and physical properties of the magmas involved in the volcanic processes (see table). Formed by partial melting of existing solid rock in the Earth's lower crust or upper mantle, the discrete blebs of magma consist of liquid rock (silicate melt) and dissolved gases. Driven by buoyancy, the magma blebs, which are lighter than the surrounding rock, coalesce as they rise toward the surface to form larger masses.

Generalized relationships between magma composition, relative viscosity, and common eruptive characteristics

Magma composition	Relative viscosity	Common eruptive characteristics
Basaltic	Fluidal	Lava fountains, flows, and pools
Andesitic	Less fluidal	Lava flows, explosive ejecta, ashfalls, and pyroclastic flows
Dacitic-rhyolitic	Viscous	Explosive ejecta, ashfalls, pyroclastic flows, and lava domes

Magma consists of three phases: liquid, solid, and gas. Volcanic gases generally are predominantly water; other gases include various compounds of carbon, sulfur, hydrogen, chlorine, and fluorine. All volcanic gases also contain minor amounts of nitrogen, argon, and other inert gases, largely the result of atmospheric contamination at or near the surface.

Temperatures of erupting magmas have been measured in lava flows and lakes, pyroclastic deposits, and volcanic vents by means of infrared sensors, optical pyrometers, and thermocouples. Reasonably good and consistent measurements have been obtained for basaltic magmas erupted from Kilauea and Mauna Loa volcanoes, Hawaii, and a few other volcanoes. Measured temperatures typically range between 2100 and 2200°F (1150 and 1200°C), and many measurements in cooling Hawaiian lava lakes indicate that the basalt becomes completely solid at about 1800°F (980°C). *See* GEOLOGIC THERMOMETRY.

The character of a volcanic eruption is determined largely by the viscosity of the liquid phase of the erupting magma and the abundance and condition of the gas it contains. Viscosity is in turn affected by such factors as the chemical composition and temperature of the liquid, the load of suspended solid crystals and xenoliths, the abundance of gas, and the degree of vesiculation. The subsequent violent expansion during eruption shreds the frothy liquid into tiny fragments, generating explosive showers of volcanic ash and dust, accompanied by some larger blocks (volcanic "bombs"); or it may produce an outpouring of a fluidized slurry of gas, semisolid bits of magma froth, and entrained blocks to form high-velocity pyroclastic flows, surges, and glowing avalanches.

Types of eruptions customarily are designated by the name of a volcano or volcanic area that is characterized by that sort of activity, even though all volcanoes show different modes of eruptive activity on occasion and even at different times during a single eruption.

Eruptions of the most fluid lava, in which relatively small amounts of gas escape freely with little explosion, are designated Hawaiian eruptions. Most of the lava is extruded as successive, thin flows that travel many miles from their vents. An occasional feature of Hawaiian activity is the lava lake, a pool of liquid lava with convectional circulation that occupies a preexisting shallow depression or pit crater.

Strombolian eruptions are somewhat more explosive eruptions of lava, with greater viscosity, and produce a larger proportion of pyroclastic material. Many of the volcanic bombs and lapilli assume rounded or drawn-out forms during flight, but commonly are sufficiently solid to retain these shapes on impact.

Generally still more explosive are the vulcanian type of eruptions. Angular blocks of viscous or solid lava are hurled out, commonly accompanied by voluminous clouds of ash but with little or no lava flow.

Peléean eruptions are characterized by the heaping up of viscous lava over and around the vent to form a steep-sided hill or volcanic dome. Explosions, or collapses of portions of the dome, may result in glowing avalanches (nuées ardentes).

Plinian eruptions are paroxysmal eruptions of great violence—named after Pliny the Elder, who was killed in A.D. 79 while observing the eruption of Vesuvius—and are characterized by voluminous explosive ejections of pumice and by ash flows. The copious expulsion of viscous siliceous magma commonly is accompanied by collapse of the summit of the volcano, forming a caldera, or by collapse of the broader region, forming a volcano-tectonic depression.

A major component of the science of volcanology is the systematic and, preferably, continuous monitoring of active and potentially active volcanoes. Scientific observations and measurements—of the visible and invisible changes in a volcano and its surroundings—between eruptions are as important, perhaps even more crucial, than during eruptions. Measurable phenomena important in volcano monitoring include earthquakes; ground movements; variations in gas compositions; and deviations in local gravity, electrical, and magnetic fields. These phenomena reflect pressure and stresses induced by subsurface magma movements and or pressurization of the hydrothermal envelope surrounding the magma reservoir. The monitoring of volcanic seismicity and ground deformations before, during, and following eruptions has provided the most useful and reliable information. *See* EARTHQUAKE.

Volcanoes are in effect windows into the Earth's interior; thus research in volcanology, in contributing to an improved understanding of volcanic phenomena, provides special insights into the chemical and physical processes operative at depth. However, volcanology also serves an immediate role in the mitigation of volcanic and related hydrologic hazards (mudflows, floods, and so on). Progress toward hazards mitigation can best be advanced by a combined approach. One aspect is the preparation of comprehensive volcanic hazards assessments of all active and potentially active volcanoes, including a volcanic risk map for use by government officials in regional and local land-use planning to avoid high-density development in high-risk areas. The other component involves improvement of predictive capability by upgrading volcano-monitoring methods and facilities to adequately study more of the most dangerous volcanoes. An improved capability for eruption forecasts and predictions would permit timely warnings of impending activity, and give emergency-response officials more lead time for preparation of contingency plans and orderly evacuation, if necessary. [R.I.T.]

Water-borne disease Disease acquired by drinking water contaminated at its source or in the distribution system, or by direct contact with environmental and recreational waters. Water-borne disease results from infection with pathogenic microorganisms or chemical poisoning.

These pathogenic microorganisms include viruses, bacteria, protozoans, and helminths. A number of microbial pathogens transmitted by the fecal-oral route are commonly acquired from water in developing countries where sanitation is poor. Viral pathogens transmitted via fecally contaminated water include hepatitis viruses A and E. Important bacterial pathogens transmitted via fecally contaminated water in the developing world are *Vibrio cholerae*, enterotoxigenic *Escherichia coli*, *Shigella*, and *Salmonella enterica* serotype Typhi. Water-borne protozoan pathogens in the developing world include *Giardia lamblia* and *Entamoeba histolytica*. The major water-borne helminthic infection is schistosomiasis; however, transmission is not fecal-oral. Another water-borne helmenthic infection is dracunculiasis (guinea worm infection).

In developed countries, fecal contamination of drinking water supplies is less likely. However, there have been outbreaks of diseases such as shigellosis and giardiasis associated with lapses in proper water treatment, such as cross-contamination of wastewater systems and potable water supplies. Animals are therefore more likely to play a role in water-borne disease in developed countries. Bacterial pathogens acquired from animal feces such as nontyphoid *S. enterica*, *Campylobacter jejuni*, and *E. coli* serotype O157:H7 have caused outbreaks of water-borne disease in developed countries where water is not properly chlorinated. Hikers frequently acquire *G. lamblia* infections from drinking untreated lake and stream water. *Giardia lamblia* may have animal reservoirs and can persist in the environment. A recently recognized pathogen apparently resistant to standard chlorination and filtration practices is the protozoan *Cryptosporidium parvum*. This organism is found in the feces of farm animals and may enter water supplies through agricultural runoff.

Chemical poisoning of drinking water supplies causes disease in both developing and developed countries. Lead, copper, and cadmium have been frequently involved. *See* CHOLERA; ESCHERICHIA; MEDICAL PARASITOLOGY; SCHISTOSOMIASIS. [S.L.M.]

Water conservation The protection, development, and efficient management of water resources for beneficial purposes. Nearly every human activity—from agriculture to transportation to daily living—relies on water resources and affects the availability and quality of those resources. Water resource development has played a role in flood control, agricultural production, industrial and energy development, fish and wildlife resource management, navigation, and a host of other activities. As a result of these impacts, natural hydrologic features have changed through time, pollution has decreased the quality of remaining water resources, and global climate change may affect the distribution of water in the future. *See* HYDROLOGY.

Water availability varies substantially between geographic regions, but it is also affected strongly by the population of the region. Asia, for example, has an extremely large total runoff but the lowest per-capita water availability. In addition, nearly 40% of the world's population lives in areas that experience severe to moderate water stress. Thus the combination of water and population distribution has resulted in a large difference in per-capita water use between countries.

Worldwide, nearly 4000 km^3 of water is withdrawn every year from surface and ground waters. This is a sixfold increase from the levels withdrawn in 1900 (since which time population has increased four times). Agriculture accounts for the greatest proportion of water use, with about two-thirds of water withdrawals and 85% of water consumption. It also accounts for a great proportion of the increase in water use, with irrigated cropland more than doubling globally since 1960. However, in Europe and North America particularly, industry consumes a large proportion of available water; industrial uses for water are anticipated to grow on other continents as well.

Land development has substantially affected the distribution of water resources. It is estimated that one-half of the natural wetlands in the world have been lost in the last century. In some areas, such as California, wetland loss is estimated to be greater than 90%. The vast majority of wetlands loss has been associated with agricultural development, but urban and industrial changes have reduced wetlands as well. River channels have also been altered to enhance irrigation, navigation, power production, and a variety of other human activities.

Ground-water resources have been depleted in the last century, with many aquifers or artesian sources being depleted more rapidly than they can be recharged. This is called ground-water overdraft. In the United States, ground-water overdraft is a serious problem in the High Plains from Nebraska to Texas and in parts of California and Arizona. *See* GROUND-WATER HYDROLOGY.

Streams have traditionally served for waste disposal. Towns and cities, industries, and mines provide thousands of pollution sources. Pollution dilution requires large amounts of water. Treatment at the source is safer and less wasteful than flushing untreated or poorly treated wastes downstream. However, sufficient flows must be released to permit the streams to dilute, assimilate, and carry away the treated effluents. *See* WATER POLLUTION.

The availability of fresh water is also likely to be affected by global climate change. There is substantial evidence that global temperatures have risen and will continue to rise. Although the precise effects of this temperature risk on water distribution are challenging to predict, most models of climate change do anticipate increased global precipitation. It is likely that some areas, particularly those at mid to high latitudes, will become wetter, but the increased precipitation will be more seasonal than current patterns. Other areas are likely to receive less precipitation than they do currently. In addition, many models predict increases in the intensity and frequency of severe droughts and floods in at least some regions. These changes will affect natural stream flow patterns, soil moisture, ground-water recharge, and thus the timing and intensity of human demands for fresh-water supplies. *See* GLOBAL CLIMATE CHANGE.

Land management vitally influences the distribution and character of runoff. Inadequate vegetation or surface organic matter; compaction of farm, ranch, or forest soils by heavy vehicles; frequent crop-harvesting operations; repeated burning; or excessive trampling by livestock or wild ungulates all expose the soil to the destructive energy of rainfall or rapid snowmelt. On such lands little water enters the soil, soil particles are dislodged and quickly washed into watercourses, and gullies may form. *See* LAND-USE PLANNING; SOIL CONSERVATION.

There are a variety of measures that can be taken to reduce water consumption. In the United States, for example, per-capita water usage dropped 20% from 1980 to 1995. In many cases, improvements to existing systems would contribute to additional water savings. In the United States, an average of 15% of the water in public supply systems (for cities with populations greater than 10,000) is unaccounted for, and presumably lost.

Improvements can also be achieved by changing industrial and agricultural practices. Agricultural water consumption has an estimated overall water use efficiency of 40%. More effective use of water in agricultural systems can be achieved, for example, with more efficient delivery methods such as drip irrigation. More accurate assessment of soil and plant moisture can allow targeted delivery of water at appropriate times. In industrial settings, recycling and more efficient water use has tremendous potential to reduce water consumption. Overall, industrial water usage dropped by 30% in California between 1980 and 1990, with some sectors achieving even greater reductions. Japan has achieved a 25% reduction in industrial water use since the 1970s. Additional potential to reduce this usage still exists even in locations where many conservation measures are already in place.

Residential water consumption can also be reduced through conservation measures. High-efficiency, low-flow toilets can reduce the water required to flush by 70% or more. Additional savings are possible with efficient faucet fixtures and appliances.

Water conservation in the United States faces a number of institutional as well as technological challenges. States must administer the regulatory provisions of their pollution-control laws, develop water quality standards and waste-treatment requirements, and supervise construction and maintenance standards of public service water systems. Some states can also regulate ground-water use to prevent serious overdrafts. Artesian wells may have to be capped, permits may be required for drilling new wells, or reasonable use may have to be demonstrated. Federal responsibilities consist largely of financial support or other stimulation of state and local water management. Federal legislation permits court action on suits involving interstate streams where states fail to take corrective action following persistent failure of a community or industry to comply with minimum waste-treatment requirements.

The watershed control approach to planning, development, and management rests on the established interdependence of water, land, and people. Coordination of structures and land-use practices is sought to prevent erosion, promote infiltration, and retard high flows (to prevent flooding). The Natural Resources Conservation and Forest Services of the Department of Agriculture administer the program. The Natural Resources Conservation Service cooperates with other federal and state agencies and operates primarily through the more than 2000 soil conservation districts.

Because watersheds often span political boundaries, many efforts to conserve and manage water require cooperation between states and countries. Many countries currently have international treaties addressing water allocation and utilization. In 1997, the United Nations adopted the Convention on the Law of the Non-navigational Uses of International Watercourses, which includes an obligation not to cause significant harm to other watercourse states, as well as provisions for dispute resolution. In addition, in 1996 the Global Water Partnership and the World Water Council were formed for the purpose of addressing ongoing international water concerns. [B.F.; M.McC.; G.Pr.]

The increasing utilization of the continental shelf for oil drilling and transport, siting of nuclear power plants, and various types of planned and inadvertent waste disposal, as well as for food and recreation, requires careful management of human activities in this ecosystem. Nearshore waters are presently subject to both atmospheric and coastal input of pollutants in the form of heavy metals, synthetic chemicals, petroleum hydrocarbons, radionuclides, and other urban wastes. Overfishing is an additional human-induced stress. Physical transport of pollutants, their modification by the coastal food web, and demonstration of transfer to humans are sequential problems of increasing complexity on the continental shelf.

One approach to quantitatively assess the above pollutant impacts is to construct simulation models of the coastal food web in a systems analysis of the continental shelf. Models of physical transport of pollutants have been the most successful, for example, as in studies of beach fouling by oil. Incorporation of additional biological and chemical terms in a simulation model, however, requires dosage response functions of the natural organisms to each class of pollutants, as well as a quantitative description of the "normal" food web interactions of the continental shelf. *See* ECOLOGICAL MODELING; FOOD WEB.

In addition to toxic materials introduced by oil spills, sewage, and agricultural and industrial run-off, coastal waters are vulnerable to thermal pollution. Thermal pollution is caused by the discharge of hot water from power plants or factories and from desalination plants. A large power installation may pump in 10^6 gal/min (63 m^3/s) of seawater to act as a coolant and discharge it at a temperature approximately 18°F (10°C) above that of the ambient water. In a shallow bay with restricted tidal flow, the rise in temperature can cause gross alterations to the natural ecology. Federal standards prohibit heating of coastal waters by more than 0.9°F (0.5°C). *See* THERMAL ECOLOGY.

Finally, dredging waters to fill wetlands for house lots, parking lots, or industrial sites destroys the marshes that provide sanctuary for waterfowl and for the young of estuarine fishes. As the bay bottom is torn up, the loosened sediments shift about with the current and settle in thick masses on the bottom, suffocating animals and plants. In this way, the marshes are eliminated and the adjoining bays are degraded as aquatic life zones. The northeast Atlantic states have lost 45,000 acres (182 km^2) of coastal wetlands in only 10 years, and San Francisco Bay has been nearly half obliterated by filling. Dredging to remove sand and gravel has the same disruptive effects as dredging for landfill or other purposes, whether the sand and gravel are sold for profit or used to replenish beach sand eroded away by storms. The dredging of boat channels adds to the siltation problem, and disposal of dredge spoils is being regulated in coastal areas.

[J.J.W.]

Water desalination The removal of dissolved minerals (including salts) from seawater or brackish water. This may occur naturally as part of the hydrologic cycle, or as an engineered process. Engineered water desalination processes, which produce potable water from seawater or brackish water, have become important because many regions throughout the world suffer from water shortages caused by the uneven distribution of the natural water supply and by human use.

Seawater, brackish water, and fresh water have different levels of salinity, which is often expressed by the total dissolved solids (TDS) concentration. Seawater has a TDS concentration of about 35,000 mg/L, and brackish water has a TDS concentration of 1000–10,000 mg/L. Water is considered fresh when its TDS concentration is below 500 mg/L, which is the secondary (voluntary) drinking water standard for the United States. Salinity is also expressed by the water's chloride concentration, which is about half of its TDS concentration. *See* SEAWATER.

Water desalination processes separate feed water into two streams: a fresh-water stream with a TDS concentration much less than that of the feed water, and a brine stream with a TDS concentration higher than that of the feed water.

Distillation is a process that turns seawater into vapor by boiling, and then condenses the vapor to produce fresh water. Boiling water is an energy-intensive operation, requiring about 4.2 kilojoules of energy (or latent heat) to raise the temperature of 1 kg of water by 1°C. After water reaches its boiling point, another 2257 kJ of energy (or the heat of vaporization) is required to convert it to vapor. The boiling point depends on ambient atmospheric pressure—at lower pressure, the boiling point of water is lower. Therefore, keeping water boiling can be accomplished either by providing a constant energy supply or by reducing the ambient atmospheric pressure.

Reverse osmosis, the process that causes water in a salt solution to move through a semipermeable membrane to the fresh-water side, is accomplished by applying pressure in excess of the natural osmotic pressure to the salt solution. The operational pressure of reverse osmosis for seawater desalination is much higher than that for brackish water, as the osmotic pressure of seawater at a TDS concentration of 35,000 mg/L is about 2700 kJ while the osmotic pressure of brackish water at a TDS concentration of 3000 mg/L is only about 230 kJ.

Salts dissociate into positively and negatively charged ions in water. The electrodialysis process uses semipermeable and ion-specific membranes, which allow the passage of either positively or negatively charged ions while blocking the passage of the oppositely charged ions. An electrodialysis membrane unit consists of a number of cell pairs bound together with electrodes on the outside. These cells contain an anion exchange membrane and cation exchange membrane. Feed water passes simultaneously in parallel paths through all of the cells, separating the product (water) and ion concentrate.

[C.C.K.L.; J.W.P.]

Water pollution A change in the chemical, physical, biological, and radiological quality of water that is injurious to its existing, intended, or potential uses (for example, boating, waterskiing, swimming, the consumption of fish, and the health of aquatic organisms and ecosystems). The term "water pollution" generally refers to human-induced (anthropogenic) changes to water quality. Thus, the discharge of toxic chemicals from a pipe or the release of livestock waste into a nearby water body is considered pollution. Conversely, nutrients that originate from animals in the wild or toxins that originate from natural processes are not considered pollution.

The contamination of ground water, rivers, lakes, wetlands, estuaries, and oceans can threaten the health of humans and aquatic life. Sources of water pollution are generally divided into two categories. The first is point-source pollution, in which contaminants are discharged from a discrete location. Sewage outfalls and oil spills are examples of point-source pollution. The second category is non-point-source or diffuse pollution, referring to all of the other discharges that deliver contaminants to water bodies. Acid rain and unconfined runoff from agricultural or urban areas are examples of non-point-source pollution. The principal contaminants of water include toxic chemicals, nutrients and biodegradable organics, and bacterial and viral pathogens.

Water pollution can threaten human health when pollutants enter the body via skin exposure or through the direct consumption of contaminated food or drinking water. Priority pollutants, including dichlorodiphenyl trichloroethane (DDT) and polychlorinated biphenyls (PCBs), persist in the natural environment and bioaccumulate in the tissues of aquatic organisms. These persistent organic pollutants are transferred up the food chain (in a process called biomagnification), and they can reach levels of concern in fish species that are eaten by humans. Finally, bacteria and viral pathogens can pose a public health risk for those who drink contaminated water or eat raw shellfish from polluted water bodies. *See* ENVIRONMENTAL TOXICOLOGY; FOOD WEB.

Contaminants have a significant impact on aquatic ecosystems. for example, enrichment of water bodies with nutrients (principally nitrogen and phosphorus) can result in the growth of algae and other aquatic plants that shade or clog streams. If wastewater containing biodegradable organic matter is discharged into a stream with inadequate dissolved oxygen, the water downstream of the point of discharge will become anaerobic and will be turbid and dark. Settleable solids, if present, will be deposited on the streambed, and anaerobic decomposition will occur. Over the reach of stream where the dissolved-oxygen concentration is zero, a zone of putrefaction will occur with the production of hydrogen sulfide, ammonia, and other odorous gases. Because many fish species require a minimum of 4–5 mg of dissolved oxygen per liter of water, they will be unable to survive in this portion of the stream.

Direct exposures to toxic chemicals is also a health concern for individual aquatic plants and animals. Chemicals (e.g., pesticides) are frequently transported to lakes and rivers via runoff, and they can have unintended and harmful effects on aquatic life. Toxic chemicals have been shown to reduce the growth, survival, reproductive output, and disease resistance of exposed organisms. These effects can have important consequences for the viability of aquatic populations and communities. *See* INSECTICIDE.

Wastewater discharges are most commonly controlled through effluent standards and discharge permits. Under this system, discharge permits are issued with limits on the quantity and quality of effluents. Water-quality standards are sets of qualitative and quantitative criteria designed to maintain or enhance the quality of receiving waters. Receiving waters are divided into several classes depending on their uses, existing or intended, with different sets of criteria designed to protect uses such as drinking water supply, bathing, boating, fresh-water and shellfish harvesting, and outdoor sports for seawater. For toxic compounds, chemical-specific or whole-effluent toxicity studies are used to develop standards and criteria. In the chemical-specific approach, individual criteria are used for each toxic chemical detected in the wastewater. Criteria can be developed to protect aquatic life against acute and chronic effects and to safeguard humans against deleterious health effects, including cancer. In the whole-effluent approach, toxicity or bioassay tests are used to determine the concentration at which the wastewater induces acute or chronic toxicity effects. *See* HAZARDOUS WASTE; SEWAGE TREATMENT. [N.Sc.; G.Tc.]

Water softening The process of removing divalent cations, usually calcium or magnesium, from water. When a sample of water contains more than 120 mg of these ions per liter (0.016 oz/gal), expressed in terms of calcium carbonate ($CaCO_3$), it is generally classified as a hard water. Hard waters are frequently unsuitable for many industrial and domestic purposes because of their soap-destroying power and tendency to form scale in equipment such as boilers, pipelines, and engine jackets. Therefore it is necessary to treat the water either to remove or to alter the constituents for it to be fit for the proposed use.

The principal water-softening processes are precipitation, cation exchange, electrical methods, or combinations of these. The factors to be considered in the choice of a softening process include the raw-water quality, the end use of softened water, the cost of softening chemicals, and the ways and costs of disposing of waste streams. *See* WATER TREATMENT. [Y.H.M.]

Water table The upper surface of the zone of saturation in permeable rocks not confined by impermeable rocks. It may also be defined as the surface underground at which the water is at atmospheric pressure. Saturated rock may extend a little above this level, but the water in it is held up above the water table by capillarity and is under less than atmospheric pressure; therefore, it is the lower part of the capillary fringe and is not free to flow into a well by gravity. Below the water table, water is free to move under the influence of gravity.

The position of the water table is shown by the level at which water stands in wells penetrating an unconfined water-bearing formation. Where a well penetrates only impermeable material, there is no water table and the well is dry. But if the well passes through impermeable rock into water-bearing material whose hydrostatic head is higher than the level of the bottom of the impermeable rock, water will rise approximately to the level it would have assumed if the whole column of rock penetrated had been permeable. This is called artesian water. *See* ARTESIAN SYSTEMS; GROUND-WATER HYDROLOGY.

[A.N.S./R.K.L.]

Water treatment Physical and chemical processes for making water suitable for human consumption and other purposes. Drinking water must be bacteriologically safe, free from toxic or harmful chemicals or substances, and comparatively free of turbidity, color, and taste-producing substances. Excessive hardness and high concentration of dissolved solids are also undesirable, particularly for boiler feed and industrial purposes. The treatment processes of greatest importance are sedimentation, coagulation, filtration, disinfection, softening, and aeration.

Sedimentation occurs naturally in reservoirs and is accomplished in treatment plants by basins or settling tanks. Plain sedimentation will not remove extremely fine or colloidal material within a reasonable time, and the process is used principally as a preliminary to other treatment methods.

Fine particles and colloidal material are combined into masses by coagulation. These masses, called floc, are large enough to settle in basins and to be caught on the surface of filters.

Suspended solids, colloidal material, bacteria, and other organisms are filtered out by passing the water through a bed of sand or pulverized coal, or through a matrix of fibrous material supported on a perforated core. Soluble materials such as salts and metals in ionic form are not removed by filtration. *See* FILTRATION.

There are several methods of treatment of water to kill living organisms, particularly pathogenic bacteria; the application of chlorine or chlorine compounds is the most common. Less frequently used methods include the use of ultraviolet light, ozone, or silver ions. Boiling is the favored household emergency measure.

Municipal water softening is common where the natural water has a hardness in excess of 150 parts per million. Two methods are used: (1) The water is treated with lime and soda ash to precipitate the calcium and magnesium as carbonate and hydroxide, after which the water is filtered; (2) the water is passed through a porous cation exchanger which has the ability of substituting sodium ions in the exchange medium for calcium and magnesium in the water. For high-pressure steam boilers or some other industrial processes, almost complete deionization of water is needed, and treatment includes both cation and anion exchangers.

Aeration is a process of exposing water to air by dividing the water into small drops, by forcing air through the water, or by a combination of both. Aeration is used to add oxygen to water and to remove carbon dioxide, hydrogen sulfide, and taste-producing gases or vapors. *See* WATER POLLUTION.

[R.H.]

Waterspout An intense columnar vortex (not necessarily containing a funnel-shaped cloud) of small horizontal extent, over water. Typical visible vortex diameters are of the order of 33 ft (10 m), but a few large waterspouts may exceed 330 ft (100 m) across. In the case of Florida waterspouts, only rarely does the visible funnel extend

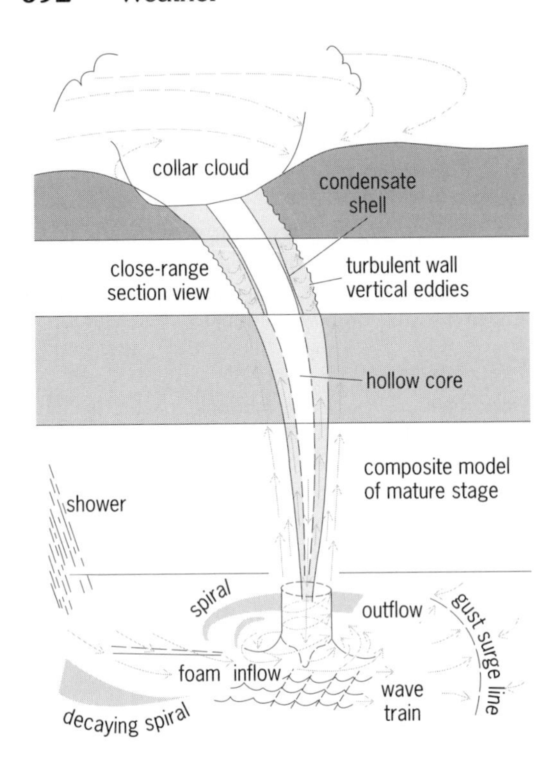

Composite schematic model of a mature waterspout. For scaling reference, the maximum funnel diameters in this stage, just below the collar cloud, range from 10 to 460 ft (3 to 140 m).

from parent cloudbase to sea surface. Like the tornado, most of the visible funnel is condensate. Therefore, the extension of the funnel cloud downward depends upon the distribution of ambient water vapor, ambient temperature, and pressure drop due to the vortex circulation strength. These vortices are most frequently observed during the warm season in the oceanic tropics and subtropics.

All waterspouts undergo a regular life cycle composed of five discrete but overlapping stages. (1) The dark-spot stage signifies a complete vortex column extending from cloud-base to sea surface. (2) The spiral-pattern stage is characterized by development of alternating dark- and light-colored bands spiraling around the dark spot on the sea surface. (3) The spray ring (incipient spray vortex) stage is characterized by a concentrated spray ring around the dark spot, with a lengthening funnel cloud above. (4) The mature waterspout stage (see illustration) is characterized by a spray vortex of maximum intensity and organization. (5) The decay stage occurs when the waterspout dissipates (often abruptly).

Waterspouts and tornadoes are qualitatively similar, differing only in certain quantitative aspects: tornadoes are usually more intense, move faster, and have longer lifetimes—especially maxi-tornadoes. Tornadoes are associated with intense, baroclinic (frontal), synoptic-scale disturbances with attendant strong vertical wind shear, while waterspouts are associated with weak, quasibarotropic disturbances (weak thermal gradients) and consequent weak vertical wind shear. *See* TORNADO; WIND. [J.H.G.]

Weather The state of the atmosphere, as determined by the simultaneous occurrence of several meteorological phenomena at a geographical locality or over broad areas of the Earth. When such a collection of weather elements is part of an interrelated physical structure of the atmosphere, it is termed a weather system, and includes phenomena at all elevations above the ground. More popularly, weather refers to a certain state of the atmosphere as it affects humans' activities on the Earth's surface. In this sense, it is often taken to include such related phenomena as waves at sea and floods on land.

A weather element is any individual physical feature of the atmosphere. At a given locality, at least seven such elements may be observed at any one time. These are clouds, precipitation, temperature, humidity, wind, pressure, and visibility. Each of these principal elements is divided into many subtypes. *See* WEATHER MAP.

The various forms of precipitation are included by international agreement among the hydrometeors, which comprise all the visible features in the atmosphere, besides clouds, that are due to water in its various forms. For convenience in processing weather data and information, this definition is made to include some phenomena not due to water, such as dust and smoke. Some of the more common hydrometeors include rain, snow, fog, hail, dew, and frost. *See* PRECIPITATION (METEOROLOGY).

Certain optical and electrical phenomena have long been observed among the weather elements. These include lightning, aurora, solar or lunar corona, and halo. *See* AIR MASS; ATMOSPHERE; CLOUD; FRONT; LIGHTNING; METEOROLOGY; WEATHER OBSERVATIONS; WIND. [P.F.C.]

Weather forecasting and prediction Processes for formulating and disseminating information about future weather conditions based upon the collection and analysis of meteorological observations. Weather forecasts may be classified according to the space and time scale of the predicted phenomena. Atmospheric fluctuations with a length of less than 100 m (330 ft) and a period of less than 100 s are considered to be turbulent. The study of atmospheric turbulence is called micrometeorology; it is of importance for understanding the diffusion of air pollutants and other aspects of the climate near the ground. Standard meteorological observations are made with sampling techniques that filter out the influence of turbulence. Common terminology distinguishes among three classes of phenomena with a scale that is larger than the turbulent microscale: the mesoscale, synoptic scale, and planetary scale. *See* MICROMETEOROLOGY.

The mesoscale includes all moist convection phenomena, ranging from individual cloud cells up to the convective cloud complexes associated with prefrontal squall lines, tropical storms, and the intertropical convergence zone. Also included among mesoscale phenomena are the sea breeze, mountain valley circulations, and the detailed structure of frontal inversions. Most mesoscale phenomena have time periods less than 12 h. The prediction of mesoscale phenomena is an area of active research. Most forecasting methods depend upon empirical rules or the short-range extrapolation of current observations, particularly those provided by radar and geostationary satellites. Forecasts are usually couched in probabilistic terms to reflect the sporadic character of the phenomena. Since many mesoscale phenomena pose serious threats to life and property, it is the practice to issue advisories of potential occurrence significantly in advance. These "watch" advisories encourage the public to attain a degree of readiness appropriate to the potential hazard. Once the phenomenon is considered to be imminent, the advisory is changed to a "warning," with the expectation that the public will take immediate action to prevent the loss of life.

The next-largest scale of weather events is called the synoptic scale, because the network of meteorological stations making simultaneous, or synoptic, observations serves to define the phenomena. The migratory storm systems of the extratropics are synoptic-scale events, as are the undulating wind currents of the upper-air circulation which accompany the storms. The storms are associated with barometric minima, variously called lows, depressions, or cyclones. The synoptic method of forecasting consists of the simultaneous collection of weather observations, and the plotting and analysis of these data on geographical maps. An experienced analyst, having studied several of these maps in chronological succession, can follow the movement and intensification of weather systems and forecast their positions. This forecasting technique requires the regular and frequent use of large networks of data. *See* WEATHER MAP.

Planetary-scale phenomena are persistent, quasistationary perturbations of the global circulation of the air with horizontal dimensions comparable to the radius of the Earth. These dominant features of the general circulation appear to be correlated with the major orographic features of the globe and with the latent and sensible heat sources provided by the oceans. They tend to control the paths followed by the synoptic-scale storms, and to draw upon the synoptic transients for an additional source of heat and momentum. *See* ATMOSPHERE; METEOROLOGICAL INSTRUMENTATION; WEATHER OBSERVATIONS.

Numerical weather prediction is the prediction of weather phenomena by the numerical solution of the equations governing the motion and changes of condition of the atmosphere. Numerical weather prediction techniques, in addition to being applied to short-range weather prediction, are used in such research studies as air-pollutant transport and the effects of greenhouse gases on global climate change. *See* AIR POLLUTION; GREENHOUSE EFFECT; JET STREAM. [J.P.G.; J.R.G.]

The first operational numerical weather prediction model consisted of only one layer, and therefore it could model only the temporal variation of the mean vertical structure of the atmosphere. Computers now permit the development of multilevel (usually about 10–20) models that could resolve the vertical variation of the wind, temperature, and moisture. These multilevel models predict the fundamental meteorological variables for large scales of motion. Global models with horizontal resolutions as fine as 125 mi (200 km) are being used by weather services in several countries. Global numerical weather prediction models require the most powerful computers to complete a 10-day forecast in a reasonable amount of time.

Research models similar to global models could be applied for climate studies by running for much longer time periods. The extension of numerical predictions to long time intervals (many years) requires a more accurate numerical representation of the energy transfer and turbulent dissipative processes within the atmosphere and at the air-earth boundary, as well as greatly augmented computing-machine speeds and capacities.

Long-term simulations of climate models have yielded simulations of mean circulations that strongly resemble those of the atmosphere. These simulations have been useful in explaining the principal features of the Earth's climate, even though it is impossible to predict the daily fluctuations of weather for extended periods. Climate models have also been used successfully to explain paleoclimatic variations, and are being applied to predict future changes in the climate induced by changes in the atmospheric composition or characteristics of the Earth's surface due to human activities. *See* CLIMATE HISTORY; CLIMATE MODIFICATION; PALEOCLIMATOLOGY. [R.A.An.]

Surface meteorological observations are routinely collected from a vast continental data network, with the majority of these observations obtained from the middle latitudes of both hemispheres. Commercial ships of opportunity, military vessels, and moored and drifting buoys provide similar in-place measurements from oceanic regions. Information on winds, pressure, temperature, and moisture throughout the troposphere and into the stratosphere is routinely collected from (1) balloon-borne instrumentation packages (radiosonde observations) and commercial and military aircraft which sample the free atmosphere directly; (2) ground-based remote-sensing instrumentation such as wind profilers (vertically pointing Doppler radars), the National Weather Service

Doppler radar network, and lidars; and (3) special sensors deployed on board polar orbiting or geostationary satellites. The remotely sensed observations obtained from meteorological satellites have been especially helpful in providing crucial measurements of areally and vertically averaged temperature, moisture, and winds in data-sparse (mostly oceanic) regions of the world. Such measurements are necessary to accommodate modern numerical weather prediction practices and to enable forecasters to continuously monitor global storm (such as hurricane) activity. *See* METEOROLOGICAL INSTRUMENTATION; RADAR METEOROLOGY.

Forecast products and forecast skill are classified as longer term (greater than 2 weeks) and shorter term. These varying skill levels reflect the fact that existing numerical prediction models such as the medium-range forecast have become very good at making large-scale circulation and temperature forecasts, but are less successful in making weather forecasts. An example is the prediction of precipitation amount and type given the occurrence of precipitation and convection. Each of these forecasts is progressively more difficult because of the increasing importance of mesoscale processes to the overall skill of the forecast. *See* PRECIPITATION (METEOROLOGY). [L.F.B.]

Nowcasting is a form of very short range weather forecasting. The term nowcasting is sometimes used loosely to refer to any area-specific forecast for the period up to 12 h ahead that is based on very detailed observational data. However, nowcasting should probably be defined more restrictively as the detailed description of the current weather along with forecasts obtained by extrapolation up to about 2 h ahead. Useful extrapolation forecasts can be obtained for longer periods in many situations, but in some weather situations the accuracy of extrapolation forecasts diminishes quickly with time as a result of the development or decay of the weather systems. *See* WEATHER. [K.A.B.]

Forecasts of time averages of atmospheric variables, for example, sea surface temperature, where the lead time for the prediction is more than 2 weeks, are termed long-range or extended-range climate predictions. Extended-range predictions of monthly and seasonal average temperature and precipitation are known as climate outlooks. The accuracy of long-range outlooks has always been modest because the predictions must encompass a large number of possible outcomes, while the observed single event against which the outlook is verified includes the noise created by the specific synoptic disturbances that actually occur and that are unpredictable on monthly and seasonal time scales. According to some estimates of potential predictability, the noise is generally larger than the signal in middle latitudes. [E.A.O'L.]

Weather map A map or a series of maps that is used to depict the evolution and life cycle of atmospheric phenomena at selected times at the surface and in the free atmosphere. Weather maps are used for the analysis and display of in-place observational measurements and computer-generated analysis and forecast fields derived from weather and climate prediction models by research and operational meteorologists, government research laboratories, and commercial firms. Similar analyses derived from sophisticated computer forecast models are displayed in map form for forecast periods of 10–14 days in advance to provide guidance for human weather forecasters. *See* METEOROLOGICAL INSTRUMENTATION; WEATHER OBSERVATIONS.

Rapid advances in computer technology and visualization techniques, as well as the continued explosive growth of the Internet distribution of global weather observations, satellite and radar imagery, and model analysis and forecast fields, have revolutionized how weather, climate, and forecast data and information can be conveyed to both the general public and sophisticated users in the public and commercial sectors. People and organizations with access to the Internet can access weather and climate information in a variety of digital or map forms in support of a wind range of professional and personal activities. *See* CLIMATOLOGY; RADAR METEOROLOGY; WEATHER FORECASTING AND PREDICTION. [L.F.B.]

Weather modification Human influence on the weather and, ultimately, climate. This can be either intentional, as with cloud seeding to clear fog from airports or to increase precipitation, or unintentional, as with air pollution, which increases aerosol concentrations and reduces sunlight. Weather is considered to be the day-to-day variations of the environment—temperature, cloudiness, relative humidity, wind-speed, visibility, and precipitation. Climate, on the other hand, reflects the average and extremes of these variables, changing on a seasonal basis. Weather change may lead to climate change, which is assessed over a period of years. *See* AIR POLLUTION; CLIMATE HISTORY; CLOUD PHYSICS.

Specific processes of weather modification are as follows: (1) Change of precipitation intensity and distribution result from changes in the colloidal stability of clouds. For example, seeding of supercooled water clouds with dry ice (solid carbon dioxide, CO_2) or silver iodide (AgI) leads to ice crystal growth and fall-out; layer clouds may dissipate, convective clouds may grow. (2) Radiation change results from changes of aerosol or clouds (deliberately with a smoke screen, or unintentionally with air pollution from combustion), from changes in the gaseous constituents of the atmosphere (as with carbon dioxide from fossil fuel combustion), and from changes in the ability of surfaces to reflect or scatter back sunlight (as replacing farmland by houses.) (3) Change of wind regime results from change in surface roughness and heat input, for example, replacing forests with farmland. *See* PRECIPITATION (METEOROLOGY). [J.Hal.]

Weather observations The measuring, recording, and transmitting of data of the variable elements of weather. In the United States the National Weather Service (NWS), a division of the National Oceanic and Atmospheric Administration (NOAA), has as one of its primary responsibilities the acquisition of meteorological information. The data are sent by various communication methods to the National Meteorological Center.

At the Center, the raw data are fed into large computers that are programmed to plot, analyze, and process the data and also to make prognostic weather charts. The processed data and the forecast guidance are then distributed by special National Weather Service systems and conventional telecommunications to field offices, other government agencies, and private meteorologists. They in turn prepare forecasts and warnings based on both processed and raw data. *See* WEATHER MAP.

A wide variety of meteorological data are required to satisfy the needs of meteorologists, climatologists, and users in marine activities, forestry, agriculture, aviation, and other fields. This has led to a dual surface-observation program: the Synoptic Weather Program and the Basic Observations Program. *See* AERONAUTICAL METEOROLOGY; AGRICULTURAL METEOROLOGY; INDUSTRIAL METEOROLOGY.

The Synoptic Weather Program is designed to assist in the preparation of forecasts and to provide data for international exchange. Worldwide surface observations are taken at standard times [0000, 0600, 1200, and 1800 Universal Time Coordinated (UTC)] and sent in synoptic code.

The Basic Observations Program routinely provides meteorological data every hour. Special observations are taken at any intervening time to report significant weather events or changes. Observation sites are located primarily at airports; a few are in urban centers. At these sites, human observers report the weather elements.

Present weather consists of a number of hydrometers, such as liquid or frozen precipitation, fog, thunderstorms, showers, and tornadoes, and of lithometers, such as haze, dust, smog, dust devils, and blowing sand. The amount of cloudiness is also reported. *See* FOG; PRECIPITATION (METEOROLOGY); SMOG; THUNDERSTORM; TORNADO.

Pressure measurements are read from either a mercury or precision aneroid barometer located at the station. A microbarograph provides a continuous record of the pressure, from which changes in specific intervals of time are reported. Pressure changes are frequently quite helpful in short-range prediction of weather events. *See* AIR PRESSURE.

Temperature and humidity are measured by a hygrothermometer, located near the center of the runway complex at many airport stations. The readings are transmitted to the observation site. The temperature dial indicator is equipped with pointers to determine maximum and minimum temperature extremes. *See* HUMIDITY.

Wind speed and direction measurements are telemetered into most airport stations. The equipment, consisting of an anemometer and a wind vane, is located near the center of the runway complex at participating airports; elsewhere it is placed in an unsheltered area. *See* WIND MEASUREMENT.

Various types of clouds and their heights are reported. The lowest height of opaque clouds covering half or more of the sky is known as the ceiling, and is normally measured by a ceilometer at first-order stations. *See* CLOUD.

Upper-air observations have been made by the National Weather Service with radiosondes. The radiosonde is a small, expendable instrument package that is suspended below a 6-ft-diameter (2-m) balloon filled with hydrogen or helium. As the radiosonde is carried aloft, sensors on it measure profiles of pressure, temperature, and relative humidity. By tracking the position of the radiosonde in flight with a radio direction finder or radio navigation system, such as Loran or the Global Positioning System (GPS), information on wind speed and direction aloft is also obtained.

Understanding and accurately predicting changes in the atmosphere requires adequate observations of the upper atmosphere. Radiosonde observations, plus routine aircraft reports, radar, and satellite observations, provide meteorologists with a three-dimensional picture of the atmosphere. *See* METEOROLOGICAL INSTRUMENTATION; WEATHER OBSERVATIONS; WIND MEASUREMENT.

Weather radars distributed throughout the United States are used to observe precipitation within a radius of about 250 nmi (460 km), and associated wind fields (utilizing the Doppler principle) within about 125 nmi (230 km). The primary component of this set of weather radars is known as NEXRAD (Next Generation Weather Radar). These radars provide information on rainfall intensity, likelihood of tornadoes or severe thunderstorms, projected paths of individual storms (both ambient and within-storm wind fields), and heights of storms for short-range (up to 3 h) forecasts and warnings. *See* RADAR METEOROLOGY.

Geostationary weather satellites near 22,000 mi (36,000 km) above the Earth transmit pictures depicting the cloud cover over vast expanses of the hemisphere. Using still photographs and animated images, the meteorologist can determine, among other things, areas of potentially severe weather and the motion of clouds and fog. In addition, the satellite does an outstanding job of tracking hurricanes over the ocean where few other observations are taken. *See* HURRICANE.

Ground-based lightning detection systems detect the electromagnetic wave that emanates from the lightning path as the lightning strikes the ground. Lightning information has proven to be operationally valuable to a wide variety of users and as a supplement to other observing systems, particularly radar and satellites. *See* LIGHTNING; METEOROLOGY; WEATHER FORECASTING AND PREDICTION. [F.S.Z.; R.L.L.]

Weathering processes The response of geologic materials to the environment (physical, chemical, and biological) at or near the Earth's surface. This response typically results in a reduction in size of the weathering materials; some may become as tiny as ions in solution.

The agents and energies that activate weathering processes and the products resulting therefrom have been classified traditionally as physical and chemical in type. In classic physical weathering, rock materials are broken by action of mechanical forces into smaller fragments without change in chemical composition, whereas in chemical weathering the process is characterized by change in chemical composition. In practice, however, the two processes commonly overlap.

Specific agents of weathering may be recognized and correlated with the types of effects they produce. Important agents of weathering are water in all surface occurrences (rain, soil and ground water, streams, and ocean); the atmosphere (H_2O, O_2, CO_2, wind); temperature (ambient and changing, especially at the freezing point of water); insolation (on large bare surfaces); ice (in soil and glaciers); gravity; plants (bacteria and macroforms); animals (micro and macro, including humans). Human modifications of otherwise geologic weathering that have increased exponentially during recent centuries include construction, tillage, lumbering, use of fire, chemically active industry (fumes, liquid, and solid effluents), and manipulation of geologic water systems.

Products of physical weathering include jointed (horizontal and vertical) rock masses, disintegrated granules, frost-riven soil and surface rock, and rock and soil flows. Products of chemical weathering include the soil, and the clays used in making ceramic structural products, whitewares, refractories, various fillers and coating of paper, portland cement, absorbents, and vanadium. These are the relatively insoluble products of weathering; characteristically they occur in clays, siltstones, and shales. Sand-size particles resulting from both physical and chemical weathering may accumulate as sandstones.

After precipitation, the relatively soluble products of chemical weathering give rise to products and rocks such as limestone, gypsum, rock salt, silica, and phosphate and potassium compounds useful as fertilizers. [W.D.K.]

Weeds Unwanted plants or plants whose negative values outweigh the positive values in a given situation. Weeds impact growers each year in reduced yield and quality of agricultural products. Especially in tropical areas, irrigation systems have become unusable because of clogging with weeds. Weeds can harbor deleterious disease organisms and insects that harm crops and livestock. In addition, weeds can cause allergic reactions and serious skin problems (poison ivy), break up pavement, slow or stop water flow in municipal water supplies, interfere with power lines, cause fire hazards around buildings and along railroad tracks, and produce poisonous plant parts. *See* ALLERGY.

The most serious weeds are those that succeed in invading new areas and surviving at the expense of other plants by monopolizing light, nutrients, and water or by releasing chemicals detrimental to the growth of surrounding vegetation (allelopathy). In the plant kingdom, dozens of species have been shown to release allelopathic chemicals from roots, leaves, and stems. *See* ALLELOPATHY.

Classification. Weeds can be classified as summer annuals, which germinate in the spring, set seed, and die in the fall (crabgrass); winter annuals, which germinate in the fall, set seed, and die in the spring (common chickweek); biennials, which germinate one year, overwinter, set seed, and die the following summer (wild carrot); simple perennials, which live for several years but spread only by seed (dandelion);

and creeping perennials, which live for several years and can spread both by seed and by underground roots or rhizomes (field bindweed).

Control methods. Hand pulling, fire, flooding, and tillage are useful for controlling weeds. Insects and pathogens have also been introduced to control certain weed species. Techniques such as herbicides, computerization of spray and tillage technology, remote sensing for weed mapping and identification, and laser treatment have also been explored. Herbicides have resulted in large improvements in the availability and quality of food. They have increased the feasibility of no-till agriculture, leading to significant reductions in soil erosion. They commonly kill weeds by disrupting a physiological process that is not present in animals. Some, however, are moderately high in toxicity and must be used carefully. Rare individual weeds that are genetically resistant to the herbicide have flourished and reproduced, leading to populations of weeds resistant to that herbicide. *See* HERBICIDE. [A.P.A.]

Biological control. Biological control involves the use of natural enemies (parasites, pathogens, and predators) to control pest populations. In the case of weed pests, the primary natural enemy groups utilized are arthropods, fungal pathogens, and vertebrates. The two major approaches are classical and inundative biological control. Biocontrol can be a highly effective and cost-efficient means of controlling weeds without the use of chemical herbicides.

Classical biocontrol (also termed the inoculative or importation method) is based on the principle of population regulation by natural enemies. Most naturalized weeds leave behind their natural enemies when they colonize new areas, and so can increase to significant densities. Classical biocontrol involves the importation of natural enemies, usually from the area of origin of the weed (and preferably from a part of its native range that is a good climatic match with the intended control area), and their field release. Imported biocontrol agents must be host specific to the target weed.

Inundative control is the mass production and periodic release of large numbers of biocontrol agents to achieve controlling densities. It can be used where existing populations of agents are lacking or where existing populations that are not self-sustaining at high, controlling densities can be augmented. A chief advantage of this method is that it can be integrated with conventional fanning practices on cultivated croplands.

Arthropods, especially insects, are heavily utilized as imported biocontrol agents in uncultivated environments such as grasslands and aquatic systems.

Rusts (Uredinales) are the fungal group most frequently employed as imported agents in classical biocontrol programs. The rust *Puccinia chondrillina*, imported from Italy, controlled the narrow-leaf form of skeletonweed (*Chondrilla juncea*) in Australia. Formulations of spores of foreign and endemic pathogens can be used as mycoherbicides in inundative applications.

Vertebrate animal agents, such as goats, typically do not possess a high degree of host-plant specificity, and their feeding has to be carefully managed to focus it on the target weeds. *See* ARTHROPODA; FUNGI. [C.E.Tu.]

Wetlands Ecosystems that form transitional areas between terrestrial and aquatic components of a landscape. Typically they are shallow-water to intermittently flooded ecosystems, which results in their unique combination of hydrology, soils, and vegetation. Examples of wetlands include swamps, fresh- and salt-water marshes, bogs, fens, playas, vernal pools and ponds, floodplains, organic and mineral soil flats, and tundra. As transitional elements in the landscape, wetlands often develop at the interface between drier uplands such as forests and farmlands, and deep-water aquatic systems such as lakes, rivers, estuaries, and oceans. Thus, wetland ecosystems are characterized by the presence of water that flows over, ponds on the surface of, or saturates the soil for at least some portion of the year.

Wetland soils can be either mineral (composed of varying percentages of sand, silt, or clay) or organic (containing 12–20% organic matter). Through their texture, structure, and landscape position, soils control the rate of water movement into and through the soil profile (the vertical succession of soil layers). Retention of water and organic carbon in the soil environment controls biogeochemical reactions that facilitate the functioning of wetland soils. *See* BIOGEOCHEMISTRY.

Vegetated wetlands are dominated by plant species, called hydrophytes, that are adapted to live in water or under saturated soil conditions. Adaptations that allow plants to survive in a water-logged environment include morphological features, such as pneumatophores (the "knees," or exposed roots, of the bald cypress), buttressed tree trunks, shallow root systems, floating leaves, hypertrophied lenticels, inflated plant parts, and adventitious roots. Physiological adaptations also allow plants to survive in a wetland environment. These include the ability of plants to transfer oxygen from the root system into the soil immediately surrounding the root (rhizosphere oxidation); the reduction or elimination of ethanol accumulation due to low concentrations of alcohol dehydrogenase; and the ability to concentrate malate (a nontoxic metabolite) instead of ethanol in the root system. *See* ROOT (BOTANY).

Wetlands differ with respect to their origin, position in the landscape, and hydrologic and biotic characteristics. For example, work has focused on the hydrology as well as the geomorphic position of wetlands in the landscape. This hydrogeomorphic approach recognizes and uses the fundamental physical properties that define wetland ecosystems to distinguish among classes of wetlands that occur in riverine, depressional, estuarine or lake fringe, mineral or organic soil flats, and slope environments.

The extent of wetlands in the world is estimated to be $2\text{–}3 \times 10^6$ mi^2 ($5\text{–}8 \times 10^6$ km^2), or about 4–6% of the Earth's land surface. Wetlands are found on every continent except Antarctica and in every clime from the tropics to the frozen tundra. Rice paddies, which comprise another 500,000–600,000 mi^2 ($1.3\text{–}1.5 \times 10^6$ km^2), can be considered as a type of domesticated wetland of great value to human societies worldwide. *See* BOG; MANGROVE; MUSKEG; PLAYA; SALT MARSH; TUNDRA.

Wetlands are often an extremely productive part of the landscape. They support a rich variety of waterfowl and aquatic organisms, and represent one of the highest levels of species diversity and richness of any ecosystem. Wetlands are an extremely important habitat for rare and endangered species.

Wetlands often serve as natural filters for human and naturally generated nutrients, organic materials, and contaminants. The ability to retain, process, or transform these substances is called assimilative capacity, and is strongly related to wetland soil texture and vegetation. The assimilative capacity of wetlands has led to many projects that use wetland ecosystems for wastewater treatment and for improving water quality. Wetlands also have been shown to prevent downstream flooding and, in some cases, to prevent ground-water depletion as well as to protect shorelines from storm damage. The best wetland management practices enhance the natural processes of wetlands by maintaining conditions as close to the natural hydrology of the wetland as possible. *See* GROUND-WATER HYDROLOGY.

The world's wetlands are becoming a threatened landscape. Loss of wetlands worldwide currently is estimated at 50%. Wetland loss results primarily from habitat destruction, alteration of wetland hydrology, and landscape fragmentation. Global warming may soon be added to this list, although the exact loss of coastal wetlands due to sea-level rise is not well documented. Worldwide, destruction of wetland ecosystems primarily has been through the conversion of wetlands to agricultural land.

Hydrologic modifications that destroy, alter, and degrade wetland systems include the construction of dams and water diversions, ground-water extraction, and the artificial manipulation of the amount, timing, and periodicity of water delivery. The primary impact of landscape fragmentation on wetland ecosystems is the disruption and degradation of wildlife migratory corridors, reducing the connectivity of wildlife habitats and rendering wetland habitats too small, too degraded, or otherwise irreversibly altered to support the critical life stages of plants and animals.

The heavy losses of wetlands in the world, coupled with the recognized values of these systems, have led to a number of policy initiatives at both the national and international levels.

Wetland restoration usually refers to the rehabilitation of degraded or hydrologically altered wetlands, often involving the reestablishment of vegetation. Wetland enhancement generally refers to the targeted restoration of one or a set of ecosystem functions over others, for example, the focused restoration of a breeding habitat for rare, threatened, or endangered amphibians. Wetland creation refers to the construction of wetlands where they did not exist before. Created wetlands are also called constructed or artificial wetlands. Restoring, enhancing, or creating a wetland requires a comprehensive understanding of hydrology and ecology, as well as engineering skills. *See* ECOSYSTEM; ESTUARINE OCEANOGRAPHY; HYDROLOGY; RIVER ENGINEERING; RESTORATION ECOLOGY. [W.J.M.; P.L.Fi.; L.C.Le.; S.R.St.]

Wheat A food grain crop. Wheat is the most widely grown food crop in the world, and is increasing in production. It ranks first in world crop production and is the national food staple of 43 countries. At least one-third of the world's population depends on wheat as its main staple. The principal food use of wheat is as bread, either leavened or unleavened. The United States is second to Russia in total production, but the average yield per acre in the United States is about twice that of Russia. Other major wheat-producing countries in the world are Canada, China, India, France, Argentina, and Australia.

Wheat is best adapted to a cool dry climate, but is grown in a wide range of soils and climates. Much of the world's wheat is seeded in the fall season and, after being dormant or growing very slowly during winter, it makes rapid growth in the spring and develops grain for harvest in early summer.

Wheat for milling is classified according to hardness, color, and best use. In the United States, there are seven official market classes of which the following five are the most important: (1) hard red winter, for bread; (2) hard red spring, for bread and rolls; (3) soft red winter, for cake and pastries; (4) white, for bread, breakfast foods, and pastries; and (5) durum, for macaroni products.

The wheat inflorescence is a spike bearing sessile spikelets arranged alternately on a zigzag rachis. Two, three, or more florets may develop in each spikelet and bear grains. The grain may be white, red (brown), or purple, and it may be hard or soft in texture. Size of the grain or caryopsis may be large, as in durum, or very small, as in shot wheat (*Triticum sphaerococcum*). Wheats vary in plant height and in the ability to produce tillers. The stems are usually hollow. The wheat grain is composed of the endosperm and embryo enclosed by bran layers. The endosperm portion is principally starch and is therefore used as energy food. Wheat is also an important protein source, especially for those people who use wheat as their main staple. *See* SEED.

Botanically, wheat is a member of the grass family to which rice, barley, corn, and several other cereal grain crops also belong. The *Triticum* genus includes a wide range of wheat forms. Taxonomic studies place the goat grasses (*Aegilops*) and wheat (*Triticum*) in one genus, *Triticum*. Wheat has been crossed with rye (*Secale*) and with *Agropyron* (a grass). New forms, called *Triticale*, have been derived from crossing rye and wheat followed by doubling the chromosomes in the hybrid.

Most countries in which wheat is grown have wheat breeding programs in which the objective is to develop more productive and more stable varieties (cultivars). Many methods are combined in these programs, but in nearly all of them specially selected parent types are crossbred followed by pure-line selection among the progeny to develop new combinations of merit. Varieties and genetic types from all over the world become candidate parents to provide the desired recombinations of good quality, winter and drought hardiness, straw strength, yield, and disease resistance. Wheats must be bred for specific milling processes and to provide quality end-use products. Many new varieties have complex pedigrees. *See* BREEDING (PLANT); GRAIN CROPS. [L.P.R.]

Milling of wheat has evolved from rudimentary crushing or cracking to sophisticated separation and refining. The main purpose of milling is isolation of the starch-protein matrix, that is, separation of the endosperm from the high-fiber bran and high-lipid germ. Under optimal conditions, milling yields a high-quality, uniformly colored flour with a relatively stable shelf-life. The flours of hard wheats (11 to 13% protein) develop strong gluten complexes during mixing and are therefore suitable for making bread. Whole soft wheats (9 to 11% protein) yield flours that are used primarily for cakes, cookies, and pastries. Durum wheat is used to produce a relatively coarse flour, semolina, used for manufacture of pasta products. [M.A.U.]

Whooping cough An acute infection of the tracheobronchial tree caused by *Bordetella pertussis*, a bacteria species exclusive to infected humans. The disease (also known as pertussis) follows a prolonged course beginning with a runny nose, and finally develops into violent coughing, followed by a slow period of recovery. The coughing stage can last 2–4 weeks, with a whooping sound created by an exhausted individual rapidly breathing in through a narrowed glottis after a series of wrenching coughs. The classical disease occurs in children 1–5 years of age, but in immunized populations infants are at greatest risk and adults with attenuated (and unrecognized) disease constitute a major source of transmission to others. *Bordetella pertussis* is highly infectious, particularly following face-to-face contact with an individual who is coughing. The disease is caused by structural components and extracellular toxins elaborated by *B. pertussis*. Multiple virulence factors produced by the organism play important roles at various stages of pertussis.

A vaccine produced from whole *B. pertussis* cells and combined with diphtheria and tetanus toxoids has been used throughout the world for routine childhood immunization. Concern over vaccine morbidity has caused immunization rates to decline in some developed countries. These drops in immunization rates have often been followed by widespread outbreaks of disease, including deaths. Considerable effort has been directed toward the development of a vaccine which would minimize side effects but maintain efficacy. A new acellular vaccine is available and has fewer side effects than the whole-cell vaccine. *See* DIPHTHERIA; TETANUS.

Although *B. pertussis* is susceptible to many antibiotics, their use has little effect once the disease reaches the coughing stage. Erythromycin is effective in preventing spread to close contacts and in the early stage. [K.J.R.]

Willow A deciduous tree and shrub of the genus *Salix*, order Salicales, common along streams and in wet places in the United States, Europe, and China. The twigs are often yellow-green and bear alternate leaves which are characteristically long, narrow, and pointed, usually with fine teeth along the margins. Flowers occur in catkins. The fruit contains several silky seeds.

Willow lumber is used for fuel and in making charcoal, excelsior, ball bats, boxes, crates, boats, waterwheels, and wicker furniture. The tough, pliable shoots of many species are used to make baskets; the bark of other species is used for tanning. Willows are of great value in checking soil erosion. A few species are ornamental shade trees. [J.F.F.]

Wind The motion of air relative to the Earth's surface. The term usually refers to horizontal air motion, as distinguished from vertical motion, and to air motion averaged over a chosen period of 1–3 min. Micrometeorological circulations (air motion over periods of the order of a few seconds) and others small enough in extent to be obscured by this averaging are thereby eliminated.

The direct effects of wind near the surface of the Earth are manifested by soil erosion, the character of vegetation, damage to structures, and the production of waves on water surfaces. At higher levels wind directly affects aircraft, missile and rocket operations, and dispersion of industrial pollutants, radioactive products of nuclear explosions, dust, volcanic debris, and other material. Directly or indirectly, wind is responsible for the production and transport of clouds and precipitation and for the transport of cold and warm air masses from one region to another. *See* ATMOSPHERIC GENERAL CIRCULATION; WIND MEASUREMENT.

Cyclonic and anticyclonic circulation are each a portion of the pattern of airflow within which the streamlines (which indicate the pattern of wind direction at any instant) are curved so as to indicate rotation of air about some central point of the cyclone or anticyclone. The rotation is considered cyclonic if it is in the same sense as the rotation of the surface of the Earth about the local vertical, and is considered anticyclonic if in the opposite sense. Thus, in a cyclonic circulation, the streamlines indicate counterclockwise (clockwise for anticylonic) rotation of air about a central point on the Northern Hemisphere or clockwise (counterclockwise for anticyclonic) rotation about a point on the Southern Hemisphere. When the streamlines close completely about the central point, the pattern is denoted respectively a cyclone or an anticyclone. Since the gradient wind represents a good approximation to the actual wind, the center of a cyclone tends strongly to be a point of minimum atmospheric pressure on a horizontal surface. Thus the terms cyclone, low-pressure area, or low are often used to denote essentially the same phenomenon.

Convergent or divergent patterns are said to occur in areas in which the (horizontal) wind flow and distribution of air density is such as to produce a net accumulation or depletion, respectively, of mass of air. The horizontal mass divergence or convergence is intimately related to the vertical component of motion. For example, since local temporal rates of change of air density are relatively small, there must be a net vertical export of mass from a volume in which horizontal mass convergence is taking place. Only thus can the total mass of air within the volume remain approximately constant.

The horizontal mass divergence or convergence is closely related to the circulation. In a convergent wind pattern the circulation of the air tends to become more cyclonic; in a divergent wind pattern the circulation of the air tends to become more anticyclonic. A convergent surface wind field is typical of fronts. As the warm and cold currents impinge at the front, the warm air tends to rise over the cold air, producing the typical frontal band of cloudiness and precipitation. *See* FRONT.

Zonal surface winds patterns result from a longitudinal averaging of the surface circulation. This averaging typically reveals a zone of weak variable winds near the Equator (the doldrums) flanked by northeasterly trade winds in the Northern Hemisphere and southeasterly trade winds in the Southern Hemisphere, extending poleward in each instance to about latitude 30°. The doldrum belt, particularly at places and times at which it is so narrow that the trade winds from the two hemispheres impinge upon it quite sharply, is designated the intertropical convergence zone, or ITCZ. The resulting convergent wind field is associated with abundant cloudiness and locally heavy rainfall. *See* MONSOON METEOROLOGY.

Local winds commonly represent modifications by local topography of a circulation of large scale. They are often capricious and violent in nature and are sometimes characterized by extremely low relative humidity. Examples are the mistral which blows down the Rhone Valley in the south of France, the bora which blows down the gorges leading to the coast of the Adriatic Sea, the foehn winds which blow down the Alpine valleys, the williwaws which are characteristic of the fiords of the Alaskan coast and the Aleutian Islands, and the chinook which is observed on the eastern slopes of the Rocky Mountains. *See* CHINOOK. [F.S.; H.B.B.]

Wind measurement The determination of three parameters: the size of an air sample, its speed, and its direction of motion. Air movement or wind is a vector that is specified by speed and direction; meteorological convention indicates wind direction is the direction from which the wind blows (for example, a southeast wind blows toward the northwest). Anemometers measure wind speed, while wind vanes indicate direction. On average, the wind blows horizontally over flat terrain; however, gusts, thermals, cloud outflows, and many other conditions have associated with them significant short-term vertical wind components. While research wind instruments typically measure both horizontal and vertical air movement, operational and personal wind sensors measure only the horizontal component.

There are many types of wind measurement instruments. In situ devices measure characteristics of air in contact with the instrument; often they are referred to as immersion sensors because they are immersed in the fluid (air) they measure. Remote wind sensors make measurements without physical contact with the portion of the atmosphere measured. Active remote sensors emit electromagnetic (for example, light or radio waves) or sound waves into the atmosphere and measure the amount and nature of the electromagnetic or acoustic power returned from the atmosphere. *See* METEOROLOGICAL INSTRUMENTATION; WIND. [W.F.D.]

Wind power The extraction of kinetic energy from the wind and conversion of it into a useful type of energy: thermal, mechanical, or electrical. Wind power has been used for centuries.

It has been estimated that the total wind power in the atmosphere averages about 3.6×10^{12} kW, which is an annual energy of about 107,000 quads (1 quad $= 2.931 \times 10^{11}$ kWh). Only a fraction of this wind energy can be extracted, estimated to be a maximum of 4000 quads per year. According to what is commonly known as the Betz limit, a maximum of 59% of this power can be extracted by a wind machine. Practical machines actually extract from 5 to 45% of the available power. Because the available wind power varies with the cube of wind speed, it is very important to find areas with high average wind speeds to locate wind machines. *See* WIND.

Most research on wind power has been concerned with producing electricity. Wind power is a renewable energy source that has virtually no environmental problems. However, wind power has limitations. Wind machines are expensive and can be located only where there is adequate wind. These high-wind areas may not be easily accessible or near existing high-voltage lines for transmitting the wind-generated energy. Another disadvantage occurs because the demand for electricity varies with time, and electricity production must follow the demand cycle. Since wind power varies randomly, it may not be available when needed. The storage of electrical energy is difficult and expensive, so that wind power must be used in parallel with some other type of generator or with nonelectrical storage. Wind power teamed with hydroelectric generators is attractive because the water can be used for energy storage, and operation with underground compressed-air storage is another option. *See* ENERGY SOURCES.

The most common type of wind turbine for producing electricity has a horizontal axis, with two or more aerodynamic blades mounted on the horizontal shaft. With a horizontal-axis machine, the blade tips can travel at several times the wind speed, which results in a high efficiency. The blade shape is designed by using the same aerodynamic theory as for aircraft. [G.Th.]

Wood anatomy Wood is composed mostly of hollow, elongated, spindle-shaped cells that are arranged parallel to each other along the trunk of a tree. The characteristics of these fibrous cells and their arrangement affect strength properties, appearance, resistance to penetration by water and chemicals, resistance to decay, and many other properties.

Just under the bark of a tree is a thin layer of cells, not visible to the naked eye, called the cambium. Here cells divide and eventually differentiate to form bark tissue to the outside of the cambium and wood or xylem tissue to the inside. This newly formed wood (termed sapwood) contains many living cells and conducts sap upward in the tree. Eventually, the inner sapwood cells become inactive and are transformed into heartwood. This transformation is often accompanied by the formation of extractives that darken the wood, make it less porous, and sometimes provide more resistance to decay. The center of the trunk is the pith, the soft tissue about which the first wood growth takes place in the newly formed twigs. *See* STEM.

In temperate climates, trees often produce distinct growth layers. These increments are called growth rings or annual rings when associated with yearly growth; many tropical trees, however, lack growth rings. These rings vary in width according to environmental conditions.

Many mechanical properties of wood, such as bending strength, crushing strength, and hardness, depend upon the density of wood; the heavier woods are generally stronger. Wood density is determined largely by the relative thickness of the cell wall and the proportions of thick- and thin-walled cells present. *See* WOOD PROPERTIES.

In hardwoods (for example, oak or maple), these three major planes along which wood may be cut are known commonly as end-grain, quarter-sawed (edge-grain) and plain-sawed (flat-grain) surfaces (see illustration).

Hardwoods have specialized structures called vessels for conducting sap upward. Vessels are a series of relatively large cells with open ends, set one above the other and continuing as open passages for long distances. In most hardwoods, the ends of the individual cells are entirely open; in others, they are separated by a grating. On the end grain, vessels appear as holes and are termed pores. The size, shape, and arrangement of pores vary considerably between species, but are relatively constant within a species.

Most smaller cells on the end grain are wood fibers which are the strength-giving elements of hardwoods. They usually have small cavities and relatively thick walls. Thin places or pits in the walls of the wood fibers and vessels allow sap to pass from one cavity to another. Wood rays are strips of short horizontal cells that extend in a radial direction. Their function is food storage and lateral conduction. *See* PARENCHYMA.

The rectangular units that make up the end grain of softwood are sections through long vertical cells called tracheids or fibers. Because softwoods do not contain vessel cells, the tracheids serve the dual function of transporting sap vertically and giving strength to the wood. The wood rays store and distribute sap horizontally.

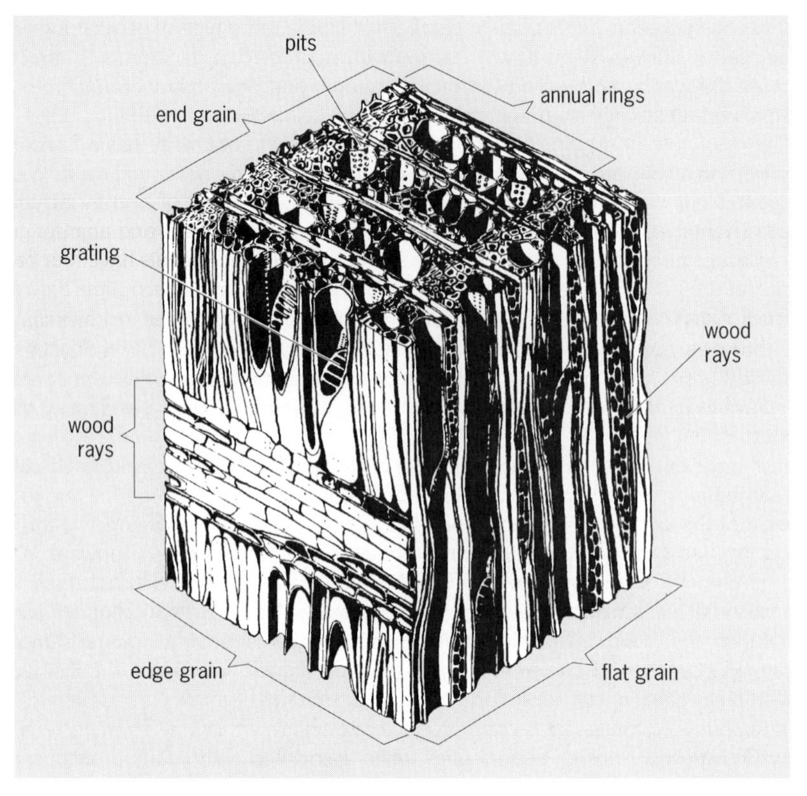

Structure of a typical hardwood. (*USDA*)

The principal compound in mature wood cells is cellulose, a polysaccharide of repeating glucose molecules which may reach 4 μm in length. These cellulose molecules are arranged in an orderly manner into structures about 10–25 nm wide called microfibrils. The microfibrils wind together like strands in a cable to form macrofibrils that measure about 0.5 μm in width and may reach 4 μm in length. These cables are as strong as an equivalent thickness of steel.

This framework of cellulose macrofibrils is cross-linked with hemicelluloses, pectins, and lignin. Lignin, the second most abundant polymer found in plants, gives the cell wall rigidity and the substance that cements the cells together. *See* PLANT ANATOMY; TREE. [R.B.M.]

Wood degradation Decay of the components of wood. Despite its highly integrated matrix of cellulose, hemicellulose, and lignin, which gives wood superior strength properties and a marked resistance to chemical and microbial attack, a variety of organisms and processes are capable of degrading wood. The decay process is a continuum, often involving a number of organisms over many years. Wood degrading agents are both biotic and abiotic, and include heat, strong acids or bases, organic chemicals, mechanical wear, and sunlight (uv degradation).

Abiotic degradation. Heat degrades both cellulose and hemicellulose, reducing strength and causing the wood to darken. At temperatures above 451°F (219°C), combustion occurs. Strong acids eventually degrade the carbohydrate portion of wood, reducing its strength. Strong bases attack the lignin, leaving the wood appearance bleached and white. Other chemicals, such as concentrated organics or salt solutions, can also disrupt the lignocellulosic matrix, reducing material properties of the wood. Sunlight, primarily through the action of ultraviolet light, also degrades wood through the creation of free radicals which then degrade the wood polymers. Mechanical wear of wood can occur in a variety of environments.

Biotic degradation. Biotic damage can occur from a variety of agents, including bacteria, fungi, insects, marine borers, and birds and animals. Birds and animals generally cause mechanical damage in isolated instances. All biotic agents have four basic requirements: adequate temperature (32–104°F or 0–40°C) with most optima between 77–90°F (25–32°C), oxygen (or other suitable terminal election acceptor), water, and a food source. Water is a critical element for biotic decay agents: it serves as reactant in degradative reactions, a medium for diffusion of enzymes into wood and degradative products back to the organism, and a wood swelling agent.

Bacteria are not major degraders of wood products, but they can damage pit membranes, thereby increasing permeability, and some are capable of cell wall degradation. *See* BACTERIA.

Fungi are among the most important wood-degrading organisms because they play an important role in terrestrial carbon cycling. Wood-degrading fungi can be classified as molds, stainers, soft rotters, brown rotters, and white rotters on the basis of the attack patterns. Molds, stainers, and soft rotters are members of the ascomycetes and the deuteromycetes (Fungi Imperfecti). *See* ASCOMYCOTA; FUNGI.

A number of insects have evolved to attack wood, including termites (Isoptera), beetles (Coleoptera), and bees and ants (Hymenoptera). Termites are the most important wood-degrading insects in most environments, and their activity causes severe economic losses.

In saline environments, marine borers can cause significant wood losses. Three groups of marine borers—shipworms, pholads, and gribbles (*Limnoria*)—cause most wood damage in these areas.

Wood protection. Protecting wood from degradation can take a number of forms. By far the simplest method is to employ designs which limit wood exposure to moisture. In some cases, however, water exclusion is not possible and alternative methods must be employed. The simplest of these methods is the use of heartwood from naturally durable species. Decay- or insect-resistant species include redwood (*Sequoia sempervirens*), western red cedar (*Thuja plicata*), and ekki (*Lophira alata*), while marine-borer-resistant heartwoods include greenheart (*Ocotea rodiaei*) and ekki. Most marine-borer-resistant woods contain high levels of silica which discourages marine borer attack, while species resistant to terrestrial decay agents often contain toxic phenolics. Wood can also be protected from degradation by spraying, dipping, soaking, or pressure treatment with preservatives. *See* WOOD PROPERTIES. [J.J.Mo.]

Wood properties Physical and mechanical characteristics of wood which are controlled by specific anatomy, moisture content, and to a lesser extent, mineral and extractive content. The properties are also influenced by wood's directional nature, which results in markedly different properties in the longitudinal, tangential, and radial directions or axes. Wood properties within a species vary greatly from tree to tree and within a single axis.

Physical properties. The physical properties (other than appearance) are moisture content, shrinkage, density, permeability, and thermal and electrical properties.

Moisture content is a major factor in the processing of wood because it influences all physical and mechanical properties, and durability and performance during use.

Normal in-use moisture content of processed wood that has been dried ranges 8–13%. Moisture content for wood is expressed on either a fractional or percentage basis. Moisture content is defined as the ratio of the mass of water contained in the wood to the mass of the same sample of dry wood.

Shrinkage occurs when wood loses moisture below the fiber saturation point. Above that point, wood is dimensionally stable. The amount of the shrinkage depends on its direction relative to grain orientation and the amount of moisture lost below the fiber saturation point. Wood shrinks significantly more in the radial and tangential directions than in the longitudinal direction.

The density of wood is determined by the amount of cell wall substance and the volume of voids caused by the cell cavities (lumens) of the fibers. Density can vary widely across a growth or annual ring. The percentage of earlywood and latewood in each growth ring determines the overall density of a wood sample.

Permeability is a measure of the flow characteristics of a liquid or gas through wood as a result of the total pressure gradient. Permeability is influenced by the anatomy of the wood cells, the direction of flow (radial, tangential, and longitudinal), and the properties of the fluid being measured. Permeability is also affected by the species, by whether the wood is sapwood or heartwood, and by the chemical and physical properties of the fluid.

The primary thermal properties of wood are conductivity, specific heat, and coefficient of thermal expansion. The conductivity of wood is determined by density, moisture content, and direction of conduction. Thermal conductivity in the transverse directions (radial and tangential) is approximately equal. Conductivity in the longitudinal direction is greater than in the transverse directions. For most processing operations, the dominant heating direction is transverse. Thermal conductivity is important to wood processing because heating—whether for drying, curing, pressing, or conditioning—is an integral step. Specific heat of wood is dependent on moisture content and, to less extent, on temperature.

Dry wood is an excellent insulator. By measuring wood's electrical resistance, electrical moisture meters accurately determine the moisture content of wood in the 5–25% range. Two other electrical properties of interest are the dielectric constant and the dielectric power factor for alternating current. These dielectric properties are dependent on density, moisture content, frequency of current, grain orientation, and temperature. The power factor is a measure of the stored energy that is converted to heat.

Mechanical properties. The mechanical properties of wood include elastic, strength, and vibration characteristics. These properties are dependent upon species, grain orientation, moisture content, loading rate, and size and location of natural characteristics such as knots.

Wood is both an elastic and plastic material. Elasticity manifests itself during loading and at moisture contents and temperatures that occur in most service uses of wood. The elastic stiffness or modulus of elasticity of wood is dependent on grain orientation, moisture content, species, temperature, and rate of loading. The stiffness of wood in the longitudinal (fiber) direction is utilized in the manufacture of composite products such as oriented strand board, in which the grain or fiber direction is controlled.

The strength of wood, like its elastic properties, is dependent upon rate of loading, species, moisture content, orientation, temperature, size and location of natural characteristics such as knots, and specimen size. The strength of individual wood fibers in the longitudinal direction can be significantly greater than that of larger samples with their complex anatomy and many defects. As with stiffness, the excellent strength characteristics of wood in the direction of the fiber can be maximized during the manufacture of wood composites by controlling fiber alignment.

Damping and sound velocity are two primary vibration phenomena of interest in structural applications. Damping occurs when internal friction dissipates mechanical energy as heat. The velocity of a sound wave through wood can be used to estimate mechanical stiffness and strength: the higher the velocity, the higher the stiffness and strength. Like other properties of wood, the velocity of sound along the three principal axes differs. Sound velocity in the longitudinal direction is two to four times greater than in the transverse directions. *See* WOOD ANATOMY. [J.B.Wi.]

X,Y

Xylem The principal water-conducting tissue and the chief supporting system of higher plants. This tissue and the associated phloem constitute the vascular system of vascular plants. Xylem is composed of various kinds of cells, living or nonliving. The structure of these cells differs in their functions, but characteristically all have a rigid and enduring cell wall that is well preserved in fossils.

In terms of their functions, the kinds of cells in xylem are those related principally to conduction and support, tracheids; to conduction, vessel members; to support, fibers; and to food storage, parenchyma. Vessel members and tracheids are often called tracheary elements. The cells in each of the four categories vary widely in structure. *See* PARENCHYMA.

Xylem tissues arise in later stages of embryo development of a given plant and are added to by differentiation of cells derived from the apical meristems of roots and stems. Growth and differentiation of tissues derived from the apical meristem provide the primary body of the plant, and the xylem tissues formed in it are called primary. Secondary xylem, when present, is produced by the vascular cambium.

In the trade, softwood is a name for xylem of gymnosperms (conifers) and hardwood for xylem of angiosperms. The terms do not refer to actual hardness of the wood. Woods of gymnosperms are generally composed only of tracheids, wood parenchyma, and small rays, but differ in detail. Resin ducts are present in many softwoods. Woods of angiosperms show extreme variation in both vertical and horizontal systems, but with few exceptions have vessels. [V.I.C.; K.E.]

Yaws An infectious disease of humans caused by the spirochete *Treponema pertenue*. It is also known as frambesia and is largely confined to the tropics. Usually yaws is contracted in childhood by direct contact or from small flies feeding in succession on infected lesions and open wounds. No race or age possesses natural immunity. [T.B.T.]

Yeast A collective name for those fungi which possess, under normal conditions of growth, a vegetative body (thallus) consisting, at least in part, of simple, single cells. The cells making up the thallus occur in pairs, in groups of three, or in straight or branched chains consisting of as many as 12 or more cells. Vegetative reproduction is characterized by budding or fission. Sexual reproduction also occurs in yeast, and is differentiated from that of other fungi by sexual states that are not enclosed in a fruiting body. Yeasts are a phylogenetically diverse group of organisms that occur in two divisions of fungi (Ascomycotina and Basidiomycotina) and 100 genera. The 700 or more species that have been described possibly represent only 1% of the species in nature, so the majority of the yeasts have yet to be discovered. Yeast plays a large part in industrial fermentation processes such as the production of industrial enzymes and chemicals, food products, industrial ethanol, and malt beverage and wine; in diseases of humans, animals and plants; in food spoilage; and as a model of molecular genetics. *See* GENETIC ENGINEERING; MEDICAL MYCOLOGY.

The shape and size of the individual cells of some species vary slightly, but in other species the cell morphology is extremely heterogeneous. The shape of yeast cells may be spherical, globose, ellipsoidal, elongate to cylindrical with rounded ends, more or less rectangular, pear-shaped, apiculate or lemon-shaped, ogival or pointed at one end, or tetrahedral. The diameter of a spherical cell may vary from 2 to 10 micrometers. The length of cylindrical cells is often 20–30 μm and, in some cases, even greater.

The asexual multiplication of yeast cells occurs by a budding process, by the formation of cross walls or fission, and sometimes by a combination of these two processes. Yeast buds are sometimes called blastospores or blastoconidia. When yeast reproduces by a fission mechanism, the resulting cells are termed arthrospores or arthroconidia.

Yeasts are categorized into two groups, based on their methods of sexual reproduction: the ascomycetous (Division Ascomycotina) and basidiomycetous (Division Basidiomycotina) yeasts.

The sexual spores of the ascomycetous yeasts are termed ascospores, which are formed in simple structures, often a vegetative cell. Such asci are called naked asci because of the absence of an ascocarp, which is a more complex fruiting body found in the higher Ascomycetes. If the vegetative cells are diploid, a cell may transform directly into an ascus after the $2n$ nucleus undergoes a reduction or meiotic division. *See* ASCOMYCOTA.

Certain yeasts have been shown to be heterothallic; that is, sporulation occurs when strains of opposite mating type (usually indicated by “a” and α) are mixed on sporulation media. However, some strains may be homothallic (self-fertile), and reduction division and karyogamy (fusion of two haploid nuclei) take place during formation of the sexual spore. Yeasts that produce sporogenous cells represent the teleomorphic form of the life cycle. In cases, in which sexual cycles are unknown, the yeast represents the asexual or anamorphic form. A species of yeast may be originally discovered in the anamorphic form and named accordingly; subsequently, the sexual state may be found and a name applied to represent the teleomorph. Consequently, the anamorphic and teleomorphic names will differ.

Basidiospores and teliospores are the sexual spores that are produced in the three classes of basidiomycetous yeasts: Urediniomycetes, Hymenomycetes, and Ustilaginomycetes. Sexual reproduction and life cycle in these yeasts is typical of other basidiomycetes in that it can include both unifactorial (bipolar) and bifactorial (tetrapolar) mating systems. *See* BASIDIOMYCOTA.

Some yeasts have the ability to carry out an alcoholic fermentation. Other yeasts lack this property. In addition to the fermentative type of metabolism, fermentative yeasts as a rule have a respiratory type of metabolism, whereas nonfermentative yeasts have only a respiratory, or oxidative, metabolism. Both reactions produce energy, with respiration producing by far the most, which is used in part for synthetic reactions, such as assimilation and growth. Part is lost as heat. In addition, small or sometimes large amounts of by-products are formed, including organic acids, esters, aldehydes, glycerol, and higher alcohols. When a fermenting yeast culture is aerated, fermentation is suppressed and respiration increases. This phenomenon is called the Pasteur effect. *See* FERMENTATION.

Yeasts are ubiquitous in nature. They exist on plants and animals; in waters, sediments, and soils; and in terrestrial, aquatic, and marine habitats. Yeasts require oxygen for growth and reproduction; therefore they do not inhabit anaerobic environments such as anoxic sediments. Many species have highly specific habitats, whereas others are found on a variety of substrates in nature. [J.W.Fe.; H.J.P.]

Yeast infection An infection mainly caused by fungi of the genus *Candida*. Although members of the genus *Candida* continue to be the most common agents of yeast infections, numerous nosocomial (related to medical treatment) factors have altered this etiologic pattern over the last 20 years. Reports in professional publications have described over 200 species in 25 yeast genera as being associated with human infections. *See* FUNGAL INFECTIONS; FUNGI; YEAST.

Etiologic agents. *Candida*, species particularly *C. albicans*, cause almost 70% of all yeast infections. However, recent changes in medical practices such as the use of broad-spectrum antibiotics, steroids, and immunosuppressive drugs, along with the recognition of new and highly debilitating diseases [for example, acquired immune deficiency syndrome (AIDS)], have increased the diversity of the agents of these infections. Species that were once rarely encountered in patient specimens (for example, *C. parapsilosis*, *C. krusei*, and *C. guilliermondii*) are being isolated with increased frequency. Equally important is the fact that yeasts that have never been described as the cause of human infections, (for example, *Blastoschizomyces capitatus*, *C. dubliniensis*, and *Trichosporon cutaneum*) are now being associated, albeit rarely, with human disease. *See* ACQUIRED IMMUNE DEFICIENCY SYNDROME (AIDS); MEDICAL MYCOLOGY; OPPORTUNISTIC INFECTIONS.

Predisposing factors. Since yeasts are not as adept as other microbial pathogens at evading or overwhelming the body's immune defenses, they generally require some disruption in the natural protective mechanisms of humans to initiate an infection. Minor breaks in the skin (such as those caused by a cut or scrape) to more significant disruptions in the integrity of the skin (as created by the delivery of medications or nutrients through intravenous catheters) can provide the means by which the yeasts on human skin may enter the body and potentially initiate an infection. The use of broad-spectrum antibacterial agents eliminates a large portion of the normal bacterial flora, allowing the yeasts in the mouth and intestines to grow rapidly and seed the blood to create a temporary benign infection or a chronic, potentially life-threatening infection. Natural hormonal imbalances as created by pregnancy or diabetes mellitus, as well as those caused by the use of medications such as corticosteroids, depress the immune system and predispose the individual to possible yeast infections. The common use of immunosuppressive drugs, and the depressed immunity associated with newer human diseases such as AIDS cause a decrease in the immune system's ability to prevent and eliminate yeast infections. Finally, the longer human life expectancy created by modern medical practices has contributed to an ever-increasing population of senior citizens with naturally lowered resistance to all forms of microbial infections.

Clinical symptomology. Although yeast infections usually affect the skin and mucous membranes, the illness can take several different forms, each with different symptoms. For example, in newborns and infants, candidiasis can appear as reddening and blisterlike lesions of skin infections (diaper rash), or it can present as white-gray lesions on the mucosal tissue lining the oral cavity (thrush). The symptoms observed in the vast majority of yeast infections mimic those of other microbial pathogens and thus are generally of little value in establishing their etiology. The diagnosis of most yeast infections involves obtaining detailed medical histories from patients, conducting extensive physical examinations, analyzing clinical laboratory data, and utilizing the education, training, and experience of the attending physicians and infectious disease specialists. *See* CLINICAL MICROBIOLOGY.

Treatment. The introduction of the first clinically effective antifungal antibiotic (nystatin) closely followed the use of antibacterial drugs (penicillin). However, nystatin was found to be effective only when brought into direct contact with the infectious agent. The broad-spectrum and fungicidal (killing fungi) activity of amphotericin B, another member of the same chemical family, maintains its use as the drug of choice for several yeast infections and as the antifungal of last resort when all other antibiotics have failed. Two closely related families of drugs, the azoles and triazoles, were introduced in the early 1970s and quickly established themselves as the first-line antibiotics for yeast infections. However, the newer triazoles are fungistatic (limiting growth) rather than fungicidal. As a result, recurrences of infection happen frequently once therapy is discontinued. Although new antifungal antibiotics continue to be introduced, the perfect drug with broad-spectrum and fungicidal activity, easy delivery, and limited side effects has yet to be found. *See* ANTIMICROBIAL AGENTS; FUNGISTAT AND FUNGICIDE. [I.F.S.]

Yellow fever An acute, febrile, mosquito-borne viral disease characterized in severe cases by jaundice, albuminuria, and hemorrhage. Inapparent infections also occur.

The agent is a flavivirus, an arbovirus of group B. The virus multiplies in mosquitoes, which remain infectious for life. After the mosquito ingests a virus-containing blood meal, an interval of 12–18 days (called the extrinsic incubation period) is required for it to become infectious. *See* ANIMAL VIRUS; ARBOVIRAL ENCEPHALITIDES.

The virus enters the body through a mosquito bite and multiplies in lymph nodes, circulates in the blood, and localizes in the liver, spleen, kidney, bone marrow, and lymph glands. The severity of the disease and the major signs and symptoms which appear depend upon where the virus localizes and how much cell destruction occurs. The incubation period is 3–6 days. At the onset, the individual has fever, chills, headache, and backache, followed by nausea and vomiting. A short period of remission often follows. On about the fourth day, the period of intoxication begins with a slow pulse relative to a high fever and moderate jaundice. In severe cases, there are high levels of protein in the urine, and manifestations of bleeding appear; the vomit may be black with altered blood; and there is an abnormally low number of lymphocytes in the blood. When the disease progresses to the severe stage (black vomit and jaundice), the mortality rate is high. However, the infection may be mild and go unrecognized. Diagnosis is made by isolation of the virus from the serum obtained from an individual as early as possible in the disease, or by the rise in serum antibody.

There are two major epidemiological cycles of yellow fever: classical or urban epidemic yellow fever, and sylvan or jungle yellow fever. Urban yellow fever involves person-to-person transmission by *Aedes aegypti* mosquitoes in the Western Hemisphere and West Africa. This mosquito breeds in the accumulations of water that accompany human settlement. Jungle yellow fever is primarily a disease of monkeys. In South America and Africa, it is transmitted from monkey to monkey by arboreal mosquitoes (*Haemagogus* and *Aedes* species) that inhabit the moist forest canopy. The infection in animals ranges from severe to inapparent. Persons who come in contact with these mosquitoes in the forest can become infected. Jungle yellow fever may also occur when an infected monkey visits a human habitation and is bitten by *A. aegypti*, which then transmits the virus to a human.

Vigorous mosquito abatement programs have virtually eliminated urban yellow fever. However, with the speed of modern air travel, the threat of a yellow fever outbreak exists where *A. aegypti* is present. An excellent attenuated live-virus vaccine is available. [J.L.Me.]

Yersinia A genus of bacteria in the Enterobacteriaceae family. The bacteria appear as gram-negative rods and share many physiological properties with related *Escherichia coli*. Of the 11 species of *Yersinia, Y. pestis, Y. enterocolitica*, and *Y. pseudotuberculosis* are etiological agents of human disease. *Yersinia pestis* causes flea-borne bubonic plague (the black death), an extraordinarily acute process believed to have killed over 200 million people during human history. Enteropathogenic *Y. pseudotuberculosis* and *Y. enterocolitica* typically cause mild chronic enteric infections. The remaining species either promote primary infection of fish (*Y. ruckeri*) or exist as secondary invaders or inhabitants of natural environments (*Y. aldovae, Y. bercovieri, Y. frederiksenii, Y. intermedia, Y. kristensenii, Y. mollaretii*, and *Y. rohdei*). *See* MEDICAL BACTERIOLOGY; PLAGUE. [R.R.B.]

Yew A genus of evergreen trees and shrubs, *Taxus*, with a fruit containing a single seed surrounded by a scarlet, fleshy, cuplike envelope (aril). The leaves are flat and acicular (needle-shaped), green below, with stalks extending downward on the stem. The only native American species of commercial importance is the Pacific yew (*T. breuifolia*), a medium-sized tree of the Pacific Coast and northern Rocky Mountain regions. Although it is not a common tree, its wood is sometimes used for poles, paddles, bows, and small cabinetwork.

The English yew (*T. baccata*), native in Europe, North Africa, and northern Asia, and the Japanese yew (*T. cuspidate*) are much cultivated in the United States as evergreen ornamentals. [A.H.G./K.P.D.]

Z

Zoogeography The subdivision of the science of biogeography that is concerned with the detailed description of the distribution of animals and how their past distribution has produced present-day patterns. Scientists in this field attempt to formulate theories that explain the present distributions as elucidated by geography, physiography, climate, ecological correlates (especially vegetation), geological history, the canons of evolutionary theory, and an understanding of the evolutionary relationships of the particular animals under study.

The field of zoogeography is based upon five observations and two conclusions. The observations are as follows. (1) Each species and higher group of animals has a discrete nonrandom distribution in space and time (for example, the gorilla occurs only in two forest areas in Africa). (2) Different geographical regions have an assemblage of distinctive animals that coexist (for example, the fauna of Africa south of the Sahara with its monkeys, pigs, and antelopes is totally different from the fauna of Australia with its platypuses, kangaroos, and wombats). (3) These differences (and similarities) cannot be explained by the amount of distance between the regions or by the area of the region alone [for example, the fauna of Europe and eastern Asia is strikingly similar although separated by 6900 mi (11,500 km) of land, while the faunas of Borneo and New Guinea are extremely different although separated by a tenth of that distance across land and water]. (4) Faunas strikingly different from those found today previously occurred in all geographical regions (for example, dinosaurs existed over much of the world in the Cretaceous). (5) Faunas resembling those found today or their antecedents previously occurred, sometimes at sites far distant from their current range (for example, the subtropical-warm temperate fauna of Eocene Wyoming, including many fresh-water fishes, salamander, and turtle groups, is now restricted to the southeastern United States).

The conclusions are as follows. (1) There are recognizible recurrent patterns of animal distribution. (2) These patterns represent faunas composed of species and higher groups that have evolved through time in association with one another.

Two rather different approaches have dominated the study of zoogeography since the beginning of the nineteenth century: ecological and historical. Ecological zoogeography attempts to explain current distribution patterns principally in terms of the ecological requirements of animals, with particular emphasis on environmental parameters, physiological tolerances, ecological roles, and adaptations. The space and time scales in this approach are narrow, and emphasis is upon the statics and dynamics of current or very recent events. Historical zoogeography recognizes that each major geographical area has a different assemblage of species, that certain systematic groups of organisms tend to cluster geographically, and that the interaction of geography, climate, and evolutionary processes over a long time span is responsible for the patterns or general tracks. Emphasis in this approach is upon the statics and dynamics of major geographical and geological events ranging across vast areas and substantial time intervals of up to millions of years. The approach is based on concordant evolutionary association of diverse groups through time. *See* ECOLOGY. [J.M.S.]

Zoological nomenclature The system of naming animals that was adopted by zoologists and detailed in the International Code of Zoological Nomenclature, which applies to both living and extinct animals. The present system is founded on the 10th edition of C. Linnaeus's *Systema Naturae* (1758) and has evolved through international agreements culminating in the Code adopted in 1985. The primary objective of the Code is to promote the stability of the names of taxa (groups of organisms) by providing rules concerning name usage and the activity of naming new taxa. The rules are binding for taxa ranked at certain levels and nonbinding on taxa ranked at other levels.

Zoological nomenclature is built around four basic features. (1) The correct names of certain taxa are either unique or unique combinations. (2) These names are formed and treated as Latin names and are universally applicable, regardless of the native language of the zoologist. (3) The Code for animals is separate and independent from similar codes for plants and bacteria. (4) No provisions of the Code are meant to restrict the intellectual freedom of individual scientists to pursue their own research.

There are four common reasons why nomenclature may change. (1) New species are found that were once considered parts of other species. (2) Taxonomic revisions may uncover older names or mistakes in identification of types. (3) Taxa may be combined, creating homonyms that require replacement. (4) Concepts of the relationships of animals change. Stability is subservient to progress in understanding animal diversity.

The articles in the Code are directed toward the names of taxa at three levels. The family group includes taxa ranked as at the family and tribe levels (including super- and subfamilies). The genus group includes taxa below subtribe and above species. The species group includes taxa ranked as species or subspecies. Taxa above the family group level are not specifically treated, and their formation and use are not strictly regulated. For each group, provisions are made that are either binding or recommended.

Binominal nomenclature. The basis for naming animals is binominal nomenclature, that is, a system of two-part names. The first name of each species is formed from the generic name, and the second is a trivial name, or species epithet. The two names agree in gender unless the specific epithet is a patronym (named for a person). The combination must be unique; no other animal can have the same binominal. The formal name of a species also includes the author, so the formal name for humans is *Homo sapiens* Linnaeus. The genus, as a higher taxon, may have one or many species, each with a different epithet. *Homo* includes *H. sapiens, H. erectus, H. habilis*, and so forth. One feature of the Linnaean binominal system is that species epithets can be used over and over again, so long as they are used in different genera. *Tyrannosaurus rex* is a large dinosaur, and *Percina rex* is a small fresh-water fish. The epithet *rex* is not the species name of *Percina rex* because all species names are binominal in form. It is recommended that names of genera and species be set in a different typeface from normal text; italics is conventional. Different names for the same species are termed synonyms, and the senior synonym is usually correct (principle of priority). Modern species descriptions are accompanied by a description that attempts to show how the species is different from others, and the designation of one or more type specimens.

Higher taxa. All higher taxonomic names have one part (uninominal) and are plural. Names of taxa of the family and genus groups must be unique. The names of genera are in Latin or latinized, are displayed in italics, and may be used alone. The names of the family group are formed by a root and an ending specific to a particular hierarchical level: family Hominidae (root + idae), subfamily Homininae (root + inae). The endings of superfamilies (root + oidea) and tribes (root + ini) are recommended but not mandated. The endings of taxa higher than the family group vary. For example, orders of fishes are formed by adding -iformes to the root (Salmoniformes), while in insects the ending is usually -ptera (Coleoptera). *See* CLASSIFICATION, BIOLOGICAL; TAXONOMIC CATEGORIES. [E.O.W.]

Zoology The science that deals with knowledge of animal life. With the great growth of information about animals, zoology has been much subdivided. Some major fields are anatomy, which deals with gross and microscopic structure; physiology, with living processes in animals; embryology, with development of new individuals; genetics, with heredity and variation; parasitology, with animals living in or on others; natural history, with life and behavior in nature; ecology, with the relation of animals to their environments; evolution, with the origin and differentiation of animal life; and taxonomy, with the classification of animals. *See* GENETICS; PHYLOGENY; PLANT EVOLUTION; TAXONOMY. [T.I.S.]

Zoonoses Infections of humans caused by the transmission of disease agents that naturally live in animals. People become infected when they unwittingly intrude into the life cycle of the disease agent and become unnatural hosts. Zoonotic helminthic diseases, caused by parasitic worms, involve many species of helminths, including nematodes (roundworms), trematodes (flukes), cestodes (tapeworms), and acanthocephalans (thorny-headed worms). Helminthic zoonoses may be contracted from domestic animals such as pets, from edible animals such as seafood, or from wild animals. Fortunately, most kinds of zoonotic helminthic infections are caused by rare human parasites.

The best-recognized example of a food-borne zoonotic helminthic disease is trichinosis, caused by the trinchina worm, *Trichinella spiralis*, a tiny nematode. People commonly become infected by eating inadequately prepared pork, but a sizable proportion of victims now contract the worms by eating the meat of wild carnivores, such as bear. Trichinosis is usually a mild disease, manifested by symptoms and signs of intestinal and muscular inflammation, but in heavy infections damage done by the larvae to the heart and central nervous system can be life threatening. Because of public awareness about properly cooking pork and federal regulations about feeding pigs, trichinosis has become uncommon in the United States. People who eat inadequately prepared marine fish may become infected with larval nematodes. Of the many potential (and rare) helminthic zoonoses from wild animals in the United States, *Baylisascaris procyonis* is particularly dangerous. The nematode is highly prevalent in raccoons, the definitive host. *See* MEDICAL PARASITOLOGY; NEMATA. [D.E.No.]

Zooplankton Animals that inhabit the water column of oceans and lakes and lack the means to counteract transport currents. Zooplankton inhabit all layers of these water bodies to the greatest depths sampled, and constitute a major link between primary production and higher trophic levels in aquatic ecosystems. Many zooplankton are capable of strong swimming movements and may migrate vertically from tens to hundreds of meters; others have limited mobility and depend more on water turbulence to stay afloat. All zooplankton, however, lack the ability to maintain their position against the movement of large water masses.

Zooplankton can be divided into various operational categories. One means of classification is based on developmental stages and divides animals into meroplankton and holoplankton. Meroplanktonic forms spend only part of their life cycles as plankton and include larvae of benthic worms, mollusks, crustaceans, echinoderms, coral, and even insects, as well as the eggs and larvae of many fishes. Holoplankton spend essentially their whole existence in the water column. Examples are chaetognaths, pteropods, larvaceans, siphonophores, and many copepods. Nearly every major taxonomic group of animals has either meroplanktonic or holoplanktonic members.

Size is another basis of grouping all plankton. A commonly accepted size classification scheme includes the groupings: picoplankton (<2 micrometers), nanoplankton (2–20 μm), microplankton (20–200 μm), mesoplankton (0.2–20 mm), macroplankton (20–200 mm), and megaplankton (>200 mm).

The classic description of the trophic dynamics of plankton is a food chain consisting of algae grazed by crustacean zooplankton which are in turn ingested by fishes. This model may hold true to a degree in some environments such as upwelling areas, but it masks the complexity of most natural food webs. Zooplankton have an essential role in linking trophic levels, but several intermediate zooplankton consumers can exist between the primary producers (phytoplankton) and fish. Thus, food webs with multiple links to different organisms indicate the versatility of food choice and energy transfer and are a more realistic description of the planktonic trophic interactions.

Size is of major importance in planktonic food webs. Most zooplankton tend to feed on organisms that have a body size smaller than their own. However, factors other than size also modify feeding interactions. Some phytoplankton are noxious and are avoided by zooplankton, and others are ingested but not digested. Furthermore, zooplankton frequently assume different feeding habits as they grow from larval to adult form. They may ingest bacteria or phytoplankton at one stage of their life cycle and become raptorial feeders later. Other zooplankton are primarily herbivorous but also ingest heterotrophic protists and can opportunistically become carnivorous. Consequently, omnivory, which is considered rare in terrestrial systems, is a relatively common trophic strategy in the plankton. In all food webs, some individuals die without being consumed and are utilized by scavengers and ultimately by decomposers (bacteria and fungi). *See* ECOLOGY; ECOSYSTEM; MARINE ECOLOGY; PHYTOPLANKTON. [R.W.Sa.]

1
Appendix

2
Contributors

3
Index

BIBLIOGRAPHIES

AGRICULTURE

Adams, C.R., K. Bamdford, and M.P. Early, *Principles of Horticulture*, 3d ed., 1998.
Barrick, R.K., and H. Harmon, *Animal Science*, 1984.
Blakely, J., and D.H. Bade, *The Science of Animal Husbandry*, 5th ed., 1989.
Brady, N.C. (ed.), *Advances in Agronomy*, vols. 28–45, 1976–1991.
Ensminger, M.E. (ed.), *Animal Science*, 9th ed., 1991.
Heath, E., et al., *Forages: The Science of Grassland Agriculture*, 4th ed., 1985.
Lockhart, J.A., and A.J. Wiseman, *Introduction to Crop Husbandry: Including Grassland*, 6th ed., 1988.
Martin, J.H., et al., *Principles of Field Crop Production*, 4th ed., 1986.
Miller, D.A., *Forage Crops*, 1984.
Rechcigl, M., Jr. (ed.), *Handbook of Agricultural Productivity*, vol. 1: *Plant Productivity*, vol. 2: *Animal Productivity*, 1982.
Sparks, D.L. (ed.), *Advances in Agronomy*, vols. 47–49, 1992–1993.

Journals:

Agronomy Journal, American Society of Agronomy, bimonthly.
Journal of Agricultural Science (England), bimonthly.
Journal of Animal Science, American Society of Animal Science, monthly.

CLIMATOLOGY AND METEOROLOGY

Ahrens, C.D., *Meteorology Today*, 7th ed., 2002.
Berger, A.L., S.H. Schneider, and J.C. Duplessy (eds.), *Climate and Geo-Sciences*, 1989.
Bryant, E., *Climate Process and Change*, 1997.
Danielson, E.W., J. Levin, and E. Abrams, *Meteorology*, 2d ed., 2002.
Finlayson-Pitts, B.J., and J.N. Pitts, Jr., *Chemistry of the Upper and Lower Atmosphere*, 2000.
Glantz, M.H. (ed.), *The Role of Regional Organizations in Context of Climate Change*, 1993.
Hornak, K.A., *Dictionary of Meteorology and Climatology, English/Spanish*, 1997.
Holton, J.R., *An Introduction to Dynamic Meteorology*, 3d ed., 1993.
Liou, K.N., *An Introduction to Atmospheric Radiation*, 2d ed., 2002.
Lutgens, F., E. Tarbuck, and D. Tasa, *The Atmosphere: An Introduction to Meteorology*, 9th ed., 2003.
Oliver, J.E., and J.J. Hidore, *Climatology: An Atmospheric Science*, 2d ed., 2001.
Pruppacher, H.R., and J.D. Klett, *Microphysics of Clouds and Precipitation*, 1997.
Scorer, R.S., *Dynamics of Meteorology and Climate*, 1997.
Wang, P.K., *Ice Microdynamics*, 2002.

Journals:

Bulletin of the American Meteorological Society, American Meteorological Society, monthly.
Journal of the Atmospheric Sciences, American Meteorological Society, semimonthly.
Journal of Climate, American Meteorological Society, monthly.
Journal of Applied Meteorology, American Meteorological Society, monthly.

CONSERVATION

Dasmann, R.F., *Environmental Conservation*, 6th ed., 2004.
Frankham, R., J.D. Ballou, and D.A. Briscoe, *A Primer of Conservation Genetics*, 2004.

Hambler, C., *Conservation* (Studies in Biology), 2004.
Meffe, G.K., and C.R. Carroll, *Principles of Conservation Biology*, 3d ed., 2005.
Kircher, H.B., *Our Natural Resources and Their Conservation*, 7th rev. ed., 1992.
Owen, O., and D.D. Chiras, *Natural Resources Conservation: An Ecological Approach*, 5th ed., 1990.
Primack, R.B., *Essentials of Conservation Biology*, 2004.

Journals:

American Institute for Conservation Journal, semiannually.
Audubon Magazine, Audubon Society, bimonthly.
Conservation, bimonthly.
New York State Conservationist, New York State Department of Environmental Conservation, bimonthly.

ECOLOGY

Bolen, E.G., and W.L. Robinson, *Wildlife Ecology and Management*, 5th ed., 2002.
Conner, J.K., and D.L. Hartl, *A Primer of Ecological Genetics*, 2004.
Gotelli, N.J., *A Primer of Ecology*, 3d ed., 2001.
Molles, Jr., M.C., *Ecology: Concepts and Applications*, 2d ed., 2002.
Pianka, E.R., *Evolutionary Ecology*, 6th ed., 1999.
Ricklefs, R.E., and G. Miller, *Ecology*, 4th ed., 1999.
Roughgarden, J., et al. (eds.), *Perspectives in Ecological Theory*, 1989.
Smith, R.L., and T. Smith, *Ecology and Field Biology*, 6th ed., 2001.
Walker, L.R., and R. del Moral, *Primary Succession and Ecosystem Rehabilitation* (Cambridge Studies in Ecology), 2003.
Westman, W.E., *Ecology Impact, Assessment and Environmental Planning*, 1985.

Journals:

Annual Review of Ecology and Systematics, annually.
Ecological Monographs, Ecological Society of America, quarterly.
Ecology, Ecological Society of America, 6 issues per year.
Journal of Environmental Sciences, Institute of Environmental Sciences, bimonthly.

ENVIRONMENTAL CHEMISTRY

Armour, M., *Hazardous Laboratory Chemicals Disposals Guide*, 3d spiral ed., 2003.
Burke, R., *Hazardous Materials Chemistry for Emergency Responders*, 2d ed., 2002.
Furr, K.A., *CRC Handbook of Laboratory Safety*, 5th ed., 2000.
Hathaway, G.J., N.H. Proctor, and J.P. Hughes, *Proctor and Hughes' Chemical Hazards of the Workplace*, 4th ed., 1996.
Lund, H.F. (ed.), *The McGraw-Hill Recycling Handbook*, 1993.
Manahan, S.E., *Environmental Chemistry*, 8th ed., 2004.
Parkin, G., *Chemistry for Environmental Engineering and Science*, 2002.
Pawlowski, L., et al. (eds.), *Chemistry for Protection of the Environment*, 1998.
Quigley, D.R., *The Essential Pocket Book of Emergency Chemical Management*, spiral ed., 1996.

Journals:

Environmental Science and Technology, American Chemical Society, semimonthly.
Journal of Environmental Engineering, American Society of Civil Engineers, monthly.

ENVIRONMENTAL ENGINEERING

Banham, R., *The Architecture of the Well-Tempered Environment*, 2d rev. ed., 1984.
Barrett, G.W., and R. Rosenberg (eds.), *Stress Effects on Natural Ecosystems*, 1982.

Holmes, J.R., *Practical Waste Management*, 1990.
Kiely, G., *Environmental Engineering*, 1996.
Lin, S.D., *Handbook of Environmental Engineering Calculations*, 2000.
Linaweaver, F.P. (ed.), *Environmental Engineering*, 1992.
Revelle, C.S., E.E. Whitlatch, and J.R. Wright, *Civil Engineering Systems*, 2d ed., 2003.
Salvato, J.A., N.L. Nemerow, and F.J. Agardy, *Environmental Engineering and Sanitation*, 5th ed., 2003.
White, I.D., D. Mottershead, and S.J. Harrison, *Environmental Systems: An Introductory Text*, 2d ed., 1993.

Journals:

The Diplomate, American Academy of Environmental Engineers, quarterly.
Environmental Engineering Science, monthly.
Journal of Environmental Engineering, monthly.
Journal of Environmental Sciences, Institute of Environmental Engineers, bimonthly.

FORESTRY

Anderson, D., and I.I. Holland (eds.), *Forests and Forestry*, 4th ed., 1997.
Gholz, H.L. (ed.), *Agroforestry: Realities, Possibilities and Potentials*, 1987.
Landsberg, J.J., and S.T. Gower, *Applications of Physiological Ecology to Forest Management*, 1996.
Leuschner, W.A., *Introduction to Forest Resource Management*, 1992.
Matthews, J.D., *Silvicultural Systems*, 1991.
Sharpe, G.W., and C. Hendes, *Introduction to Forestry*, 6th ed., 1995.
Shugart, H.H., *A Theory of Forest Dynamics: The Ecological Implications of Forest Succession Models*, 2003.
Young, A., *Agroforestry for Soil Conservation* (Science and Practice of Agroforestry), 1989.

Journals:

Forest Products Journal, Forest Products Research Society, monthly.
Forest Science, Society of American Foresters, quarterly.
Journal of Forestry, Society of American Foresters, monthly.

GEOCHEMISTRY

Brownlow, A.H., *Geochemistry*, 2d ed., 1995.
Chester, R., *Marine Geochemistry*, 2d ed., 2003.
Engel, M., and S. Mako (eds.), *Organic Chemistry, Principles and Applications* (Topics in Geobiology), 1993.
Faure, G., *Principles and Applications of Inorganic Geochemistry*, 2d ed., 1997.
Hoefs, J., *Stable Isotope Geochemistry*, 5th ed., 2004.

Journals:

Chemical Geology, Elsevier, monthly.
Geochimica et Cosmochimica Acta, Elsevier, semimonthly.

GEOGRAPHY

Gaile, G.L., and C.J. Willmott (eds.), *Geography in America at the Dawn of the 21st Century*, 2004.
Gould, P., *Becoming a Geographer (Space, Place, and Society)*, 2000.
McKnight, T.L., and D. Hess, *Physical Geography: A Landscape Appreciation*, 8th ed., 2004.
Muehrcke, P.C., *Map Use: Reading, Analysis and Interpretation*, 1998.
Robinson, A.H., et al., *Elements of Cartography*, 1995.

Strahler, A.H., and A.N. Strahler, *Introducing Physical Geography*, 3d ed., 2002.
Wieden, F.T., *Physical Geography*, 1995.

Journals:

Focus, American Geographical Society, quarterly.
Geographical Review, American Geographical Society, quarterly.
The Professional Geographer, Association of American Geographers, quarterly.

GEOLOGY

Ager, D.V., *Nature of the Stratigraphical Record*, 3d ed., 1993.
Condie, K.C., and R.E. Sloon, *Origin and Evolution of Earth: Principles of Historical Geology*, 1998.
Hallam, A., *Great Geological Controversies*, 2d ed., 1990.
Lutgens, F.K., E.J. Tarbuck, and D. Tasa, *Essentials of Geology*, 8th ed., 2002.
Montgomery, C.W., *Physical Geology*, 3d ed., 1992.
Press, F., and R. Siever, *Understanding Earth*, 3d ed., 2000.
Skinner, B.J., and S.C. Porter, *Dynamic Earth: An Introduction to Physical Geology*, 5th ed., 2003.
Tarbuck, E.J., F.K. Lutgens, and D. Tasa, *The Earth: An Introduction to Physical Geology*, 8th ed., 2004.

Journals:

Annual Reviews of Earth and Planetary Sciences, annually.
Episodes, International Union of Geological Sciences, quarterly.
Geotimes, American Geological Institute, monthly.
GSA Bulletin, Geological Society of America, monthly.
Journal of Geology, University of Chicago Press, bimonthly.

HYDROLOGY

Barcelona, M., *Contamination of Groundwater: Prevention Assessment, Restoration* (Pollution Technology Review), 1990.
Black, P.E., *Watershed Hydrology*, 2d ed., 1996.
Dunne, T., and L.B. Leopold, *Water in Environmental Planning*, 1995.
Gupta, R.S.S., *Hydrology and Hydraulic Systems*, 2001.
Linsley, R.K., et al., *Hydrology for Engineers*, 3d ed. (McGraw-Hill Series in Water Resources & Environmental Engineering), 1982.
Ponce, V.M., *Engineering Hydrology: Principles and Practices*, 1989.
Wilson, E.M., *Engineering Hydrology*, 4th ed., 1990.

Journals:

Hydrological Sciences Journal, International Association of Hydrological Sciences (U.K.), quarterly.
Journal of Hydrology, Elsevier, monthly.

MEDICAL MICROBIOLOGY

Brooks, G.F., et al., *Jawetz, Melnick & Adelberg's Medical Microbiology*, 23d ed., 2004.
Levinson, W., *Medical Microbiology and Immunology*, 8th ed., 2004.
Mims, C.A., H.M. Dockrell, and R.V. Goering, *Medical Microbiology*, 3d ed., 2004.
Ryan, K.J., and C.G. Ray, *Sherris Medical Microbiology: An Introduction to Infectious Diseases*, 4th ed., 2003.

Journals:

Annual Review of Microbiology, annually.

Antimicrobial Agents and Chemotherapy, monthly.
Journal of Clinical Microbiology, monthly.
Journal of Medical Microbiology, monthly.

MYCOLOGY

Ainsworth, G.C., et al., *Ainsworth and Bisby's Dictionary of Fungi*, 9th ed., 2001.
Burnett, J.H., *Fungal Populations and Species* (Life Science), 2003.
Carlile, M.J., et al.,*The Fungi*, 2d ed., 2001.
Khachatourians, G.G., et al., *Applied Mycology and Biotechnology: Fungal Genomics*, 2003.
Ulloa, M., and R.T. Hanlin, *Illustrated Dictionary of Mycology*, 2000.
Watanabe, T., *Pictorial Atlas of Soil and Seed Fungi: Morphologies of Cultured Fungi and Key to Species*, 2d ed., 2002.

Journals:

FEMS Yeast Research, Elsevier, quarterly.
Fungal Genetics and Biology, Elsevier, 9 issues per year.
Medical Mycology, Taylor & Francies, bimonthly.
Mycological Research, British Mycological Society, monthly.
Mycologist, British Mycological Society, monthly.

OCEANOGRAPHY

Colling, A., *Ocean Circulation*, 2d ed., 2001.
Garrison, T.S., *Oceanography: An Invitation to Marine Science*, 5th ed., 2005.
Knauss, J.A., *An Introduction to Physical Oceanography*, 2d ed., 1996.
Lalli, C.M., *Biological Oceanography*, 2d ed., 1997.
Pilson, M.E.Q., *An Introduction to the Chemistry of the Sea*, 1998.
Thurman, H.V., and A.P. Trujillo, *Essentials of Oceanography*, 8th ed., 2004.

Journals:

Journal of Physical Oceanography, American Meteorological Society, monthly.
Oceanus, Woods Hole Oceanographic Institute, semiannually.

PLANT ANATOMY

Bowes, B.G., *A Colour Atlas of Plant Structure*, 1996.
Dickison, W.C., *Integrative Plant Anatomy*, 2000.
Fahn, A., *Plant Anatomy*, 4th ed., 1990.
Mauseth, J.D., *Plant Anatomy*, 1988.
Nabors, M., *Introduction to Botany*, 2004.
Northington, D.K., et al., *The Botanical World*, 2d ed., 1995.
Pearson, L.C., *The Diversity of Evolution of Plants*, 1995.
Raven, P.H., R.F. Evert, and S.E. Eichhorn, *Biology of Plants*, 1999.
Rudall, P., *Anatomy of Flowering Plants: An Introduction to Structure and Development*, 1987.

PLANT PATHOLOGY

Agrios, G.N., *Plant Pathology*, 4th ed., 1997.
Dickinson, M., and J. Beynon (eds.), *Molecular Plant Pathology*, 2000.
Lucas, G.B., C.L. Campbell, and L.T. Lucas, *Introduction to Plant Diseases: Identification and Management*, 2d ed., 1992.
Lucas, J.A., *Plant Pathology and Plant Pathogens*, 1998.
Matthews, R.E., et al., *Plant Virology*, 4th ed., 2001.
Sinclair, W.A., H. Lyon, and W.T. Johnson, *Diseases of Trees and Shrubs*, 1987.

Strange, R.N., *Introduction of Plant Pathology*, 2003.
Talbot, N., *Plant-Pathogen Interactions*, 2004.
Vidhyasekaran, P., *Concise Encyclopedia of Plant Pathology*, 2004.
Zamir, K. P. (ed.), *Fungal Disease Resistance in Plants: Biochemistry, Molecular Biology, and Genetic Engineering*, 2004.

Journals:

Annual Review of Phytopathology, annually.
Plant Disease, American Phytopathological Society, monthly.

PLANT PHYSIOLOGY

Baker, N.R., *Photosynthesis and the Environment* (Advances in Photosynthesis), 1996.
Bewley, J.D., and M. Black, *Seeds: Physiology Development and Germination* (The Language of Science), 2d ed., 1994.
Blankenship, R.E., *Molecular Mechanisms of Photosynthesis*, 2002.
Chrispeels, M.J., and D.E. Sadava, *Plants, Genes, and Crop Biotechnology*, 2d ed., 2003.
Davies, P.J., *Plant Hormones: Physiology, Biochemistry and Molecular Biology*, 1995.
Epstrin, E., and A.J. Bloom, *Mineral Nutrition of Plants: Principles and Perspectives*, 2d ed., 2004.
Kramer, P.J., and J.S. Boyer, *Water Relations of Plants and Soils*, 1995.
Leyser, O., and S. Day, *Mechanisms in Plant Development*, 2003.
Raven, P.H., et al., *Biology of Plants*, 6th ed., 1998.
Taiz, L., and E. Zeiger, *Plant Physiology*, 3d ed., 2002.

Journals:

Current Opinion in Plant Biology, Elsevier, bimonthly.
The Plant Cell, American Society of Plant Biologists, monthly.
Plant Cell and Environment, monthly.
Plant Journal, bimonthly.
Plant Physiology, American Society of Plant Biologists, monthly.
Planta, Springer-Verlag, monthly.
Trends in Plant Science, monthly.

SOIL SCIENCE

Bowles, J.E., *Engineering Properties of Soils and Their Measurement*, 4th ed., 1992.
Brady, N.C., and R.R. Weil, *The Nature and Properties of Soils*, 13th ed., 2001.
Crunkilton, J.R., et al., *The Earth and Agriscience*, 1995.
Daniels, R.B., and R.D. Hammer, *Soil Geomorphology*, 1992.
Dowdy, R.H. (ed.), *Chemistry in the Soil Environment*, 1981.
Foster, A.B., and D.A. Bosworth, *Approved Practices in Soil Conservation*, 5th ed., 1982.
Foth, H.D., *Fundamentals of Soil Science*, 8th ed., 1990.
Hanks, R.J., *Applied Soil Physics: Soil Water and Temperature Applications*, 2d ed., 1992.
Morgan, R.P.C., *Soil Erosion and Conservation*, 2d ed., 1995.
Plaster, E.J., *Soil Science and Management*, 3d ed., 1997.

Journals:

Agronomy Journal, American Society of Agronomy, bimonthly.

SYSTEMATICS

Avise, J.C., *Phylogeography: The History and Formation of Species*, 2000.
Brooks, D.R., and D.A. McLennan, *The Nature of Diversity: An Evolutionary Voyage of Discovery*, 2002.

Ereshefsky, M., *The Poverty of the Linnaean Hierarchy: A Philosophical Study of Biological Taxonomy*, 2001.
Felsenstein, J., *Inferring Phylogenies*, 2004.
Ghiselin, M.T., *Metaphysics and the Origin of Species*, 1997.
Hillis, D.M., C. Moritz, and B.K. Mable (eds.), *Molecular Systematics*, 2d ed., 1996.
Schuh, R.T., *Biological Systematics: Principles and Applications*, 2000.
Wilson, R.A. (ed.), *Species: New Interdisciplinary Essays*, 1999.

Journals:

Cladistics, Willi Hennig Society, bimonthly.
Systematic Biology, Society of Systematic Zoology, bimonthly.

VETERINARY MEDICINE

Aiello, S.E. (ed.), *The Merck Veterinary Manual*, 8th ed., 1998.
Ettinger, S.J., and E.C. Feldman, *Textbook of Veterinary Internal Medicine*, 4th ed., 1995.
Fenner, W.R., *Quick Reference to Veterinary Medicine*, 3d ed., 2000.
Howard, J.L., and R.A. Smith, *Current Veterinary Therapy 4: Food Animal Practice*, 1999.
Timoney, J.F., *Hagan and Bruner's Microbiology and Infectious Diseases of Domestic Animals*, 8th ed., 1988.

VIROLOGY

Belshe, R.B., *Textbook of Human Virology*, 2d ed., 1990.
Diener, T.O. (ed.), *The Viroids* (The Viruses), 1987.
Dimmock, N.J., P.D. Griffiths, and C.R. Madeley (eds.), *Control of Virus Diseases*, 1990.
Fields, B.N., *Virology*, 3d ed., 1995.
Flint, S.J., et al. (eds.), *Principles of Virology: Molecular Biology, Pathogenesis, and Control*, 2d ed., 2000.
ICTVdB: The Universal Virus Database of the International Committee on Taxonomy of Viruses, maintained by C. Büchen-Osmond [http://www.ncbi.nlm.nih.gov/ICTVdB/ ICTVdBintro.htm].
Knipe, D.M. *Fundamental Virology*, 4th ed., 2001.
Matthews, R.E., et al., *Matthew's Plant Virology*, 4th ed., 2001.
Nathanson, N. (ed.), *Viral Pathogenesis*, 1997.
Voyles, B.A., *Biology of the Viruses*, 2d ed., 2002.
Wagner, E.K., and M.J. Hewlett, *Basic Virology*, 1999.

Journals:

Advances in Virus Research, irregularly.
Excerpta Medica, Section 47: Virology, 10 issues per year.
Journal of Virology, American Society for Microbiology, bimonthly.

ZOOLOGY

Alcock, J., *Animal Behavior: An Evolutionary Approach*, 7th ed., 2001.
Barnes, R.D., R.S. Fox, and E.E. Ruppert, *Invertebrate Zoology*, 7th ed., 2003.
Bond, C.E., *Biology of Fishes*, 2d ed., 1996.
Dorit, R., W.F. Walker, and R.D. Barnes, *Zoology*, 1991.
Feduccia, A., and E. McCrady, *Torrey's Morphogenesis of the Vertebrates*, 5th ed., 1991.
Grzimek, B. (ed.), *Grzimek's Encyclopedia of Mammals*, 2d ed., 1989.
Hickman, C.P., L.S. Roberts, and A. Larson, *Integrated Principles of Zoology*, 11th ed., 2000.
Hildebrand, M., and G. Goslow, *Analysis of Vertebrate Structure*, 5th ed., 2001.
Kardong, K.V., *Vertebrates*, 2d ed., 1998.
Linzey, D.W., *Vertebrate Biology*, 2001.
McFarland, D., *Animal Behavior, Psychobiology, Ethology, and Evolution*, 3d ed., 1998.

Meglitsch, P.A., and F.R. Schram, *Invertebrate Zoology*, 3d ed., 1991.
Miller, S.A., and J.P. Harley, *Zoology*, 6th ed., 2004.
Pearse, V., et al., *Living Invertebrates*, 1987.
Pough, F., et al., *Vertebrate Life*, 6th ed., 2001.

Journals:

Integrative and Comparative Biology, Society for Integrative and Comparative Biology, bimonthly.
Journal of Zoology, monthly.

Equivalents of commonly used units for the U.S. Customary System and the metric system

1 inch = 2.5 centimeters (25 millimeters)	1 centimeter = 0.4 inch	1 inch = 0.083 foot
1 foot = 0.3 meter (30 centimeters)	1 meter = 3.3 feet	1 foot = 0.33 yard (12 inches)
1 yard = 0.9 meter	1 meter = 1.1 yards	1 yard = 3 feet (36 inches)
1 mile = 1.6 kilometers	1 kilometer = 0.62 mile	1 mile = 5280 feet (1760 yards)
1 acre = 0.4 hectare	1 hectare = 2.47 acres	
1 acre = 4047 square meters	1 square meter = 0.00025 acre	
1 gallon = 3.8 liters	1 liter = 1.06 quarts = 0.26 gallon	1 quart = 0.25 gallon (32 ounces; 2 pints)
1 fluid ounce = 29.6 milliliters	1 milliliter = 0.034 fluid ounce	1 pint = 0.125 gallon (16 ounces)
32 fluid ounces = 946.4 milliliters		1 gallon = 4 quarts (8 pints)
1 quart = 0.95 liter	1 gram = 0.035 ounce	1 ounce = 0.0625 pound
1 ounce = 28.35 grams	1 kilogram = 2.2 pounds	1 pound = 16 ounces
1 pound = 0.45 kilogram	1 kilogram = 1.1×10^{-3} ton	1 ton = 2000 pounds
1 ton = 907.18 kilograms		
$°F = (1.8 \times °C) + 32$	$°C = (°F - 32) \div 1.8$	

Conversion factors for the U.S. Customary System, metric system, and International System

A. Units of length

Units	*cm*	*m*	*in.*	*ft*	*yd*	*mi*
1 cm	= 1	0.01	0.3937008	0.03280840	0.01093613	6.213712×10^{-6}
1 m	= 100.	1	39.37008	3.280840	1.093613	6.213712×10^{-4}
1 in.	= 2.54	0.0254	1	0.08333333 . . .	0.02777777 . . .	1.578283×10^{-5}
1 ft	= 30.48	0.3048	12.	1	0.3333333 . . .	$1.893939\ldots \times 10^{-4}$
1 yd	= 91.44	0.9144	36.	3.	1	$5.681818\ldots \times 10^{-4}$
1 mi	$= 1.609344 \times 10^{5}$	1.609344×10^{3}	6.336×10^{4}	5280.	1760.	1

B. Units of area

Units	*cm²*	*m²*	*in.²*	*ft²*	*yd²*	*mi²*
1 cm^2	= 1	10^{-4}	0.1550003	1.076391×10^{-3}	1.195990×10^{-4}	3.861022×10^{-11}
1 m^2	$= 10^{4}$	1	1550.003	10.76391	1.195990	3.861022×10^{-7}
1 $in.^2$	= 6.4516	6.4516×10^{-4}	1	$6.944444\ldots \times 10^{-3}$	7.716049×10^{-4}	2.490977×10^{-10}
1 ft^2	= 929.0304	0.09290304	144.	1	0.1111111 . . .	3.587007×10^{-8}
1 yd^2	= 8361.273	0.8361273	1296.	9.	1	3.228306×10^{-7}
1 mi^2	$= 2.589988 \times 10^{10}$	2.589988×10^{6}	4.014490×10^{9}	2.78784×10^{7}	3.0976×10^{6}	1

C. Units of volume

Units	m^3	cm^3	liter	$in.^3$	ft^3	qt	gal
1 m^3	= 1	10^6	10^3	6.102374×10^4	35.31467×10^{-3}	1.056688	264.1721
1 cm^3	= 10^{-6}	1	10^{-3}	0.06102374	3.531467×10^{-5}	1.056688×10^{-3}	2.641721×10^{-4}
1 liter	= 10^{-3}	1000.	1	61.02374	0.03531467	1.056688	0.2641721
1 $in.^3$	= 1.638706×10^{-5}	16.38706	0.01638706	1	5.787037×10^{-4}	0.01731602	4.329004×10^{-3}
1 ft^3	= 2.831685×10^{-2}	28316.85	28.31685	1728.	1	2.992208	7.480520
1 qt	= 9.463529×10^{-4}	946.3529	0.9463529	57.75	0.03342014	1	0.25
1 gal (U.S.)	= 3.785412×10^{-3}	3785.412	3.785412	231.	0.1336806	4.	1

D. Units of mass

Units	g	kg	oz	lb	metric ton	ton
1 g	= 1	10^{-3}	0.03527396	2.204623×10^{-3}	10^{-6}	1.102311×10^{-6}
1 kg	= 1000.	1	35.27396	2.204623	10^{-3}	1.102311×10^{-3}
1 oz (avdp)	= 28.34952	0.02834952	1	0.0625	2.834952×10^{-5}	3.125×10^{-5}
1 lb (avdp)	= 453.5924	0.4535924	16.	1	4.535924×10^{-4}	$5. \times 10^{-4}$
1 metric ton	= 10^6	1000.	35273.96	2204.623	1	1.102311
1 ton	= 907184.7	907.1847	32000.	2000.	0.9071847	1

Conversion factors for the U.S. Customary System, metric system, and International System (*cont.*)

E. Units of density

Units	$g \cdot cm^{-3}$	$g \cdot L^{-1}$, $kg \cdot m^{-3}$	$oz \cdot in.^{-3}$	$lb \cdot in.^{-3}$	$lb \cdot ft^{-3}$	$lb \cdot gal^{-1}$
1 $g \cdot cm^{-3}$	= 1	1000.	0.5780365	0.03612728	62.42795	8.345403
1 $g \cdot L^{-1}$, $kg \cdot m^{-3}$	= 10^{-3}	1	5.780365×10^{-4}	3.612728×10^{-5}	0.06242795	8.345403×10^{-3}
1 $oz \cdot in.^{-3}$	= 1.729994	1729.994	1	0.0625	108.	14.4375
1 $lb \cdot in.^{-3}$	= 27.67991	27679.91	16.	1	1728.	231.
1 $lb \cdot ft^{-3}$	= 0.01601847	16.01847	9.259259×10^{-3}	5.787037×10^{-4}	1	0.1336806
1 $lb \cdot gal^{-1}$	= 0.1198264	119.8264	4.749536×10^{-3}	4.329004×10^{-3}	7.480519	1

F. Units of pressure

Units	Pa, $N \cdot m^{-2}$	$dyn \cdot cm^{-2}$	*bar*	*atm*	$kgf \cdot cm^{-2}$	*mmHg (torr)*	*in. Hg*	$lbf \cdot in.^{-2}$
1 Pa, 1 $N \cdot m^{-2}$	= 1	10	10^{-5}	9.869233×10^{-6}	1.019716×10^{-5}	7.500617×10^{-3}	2.952999×10^{-4}	1.450377×10^{-4}
1 $dyn \cdot cm^{-2}$	= 0.1	1	10^{-6}	9.869233×10^{-7}	1.019716×10^{-6}	7.500617×10^{-4}	2.952999×10^{-5}	1.450377×10^{-5}
1 bar	= 10^5	10^6	1.	0.9869233	1.019716	750.0617	29.52999	14.50377
1 atm	= 101325	1013250	1.01325	1	1.033227	760.	29.92126	14.69595
1 $kgf \cdot cm^{-2}$	= 98066.5	980665	0.980665	0.9678411	1	735.5592	28.95903	14.22334
1 mmHg (torr)	= 133.3224	1333.224	1.333224×10^3	1.315789×10^{-3}	1.359510×10^{-3}	1	0.03937008	0.01933678
1 in. Hg	= 3386.388	33863.88	0.03386388	0.03342105	0.03453155	25.4	1	0.4911541
1 $lbf \cdot in.^{-2}$	= 6894.757	68947.57	0.06894757	0.06804596	0.07030696	51.71493	2.036021	1

G. Units of energy

Units	g mass (energy equiv)	J	eV	cal	cal_{IT}	Btu_{IT}	kWh	hp-h	ft-lbf	$ft^3 \cdot lbf \cdot in.^{-2}$	liter-atm
1 g mass (energy equiv)	= 1	8.987552×10^{13}	5.609589×10^{32}	2.148076×10^{3}	2.146640×10^{13}	8.518555×10^{10}	2.496542×10^{7}	3.347918×10^{7}	6.628878×10^{13}	4.603388×10^{11}	8.870024×10^{11}
1 J	$= 1.112650 \times 10^{-14}$	1	6.241509×10^{18}	0.2390057	0.2388459	9.478172×10^{-4}	$2.777777\ldots \times 10^{-7}$	3.725062	0.7375622	5.121960×10^{-3}	9.869233×10^{-3}
1 eV	$= 1.782662 \times 10^{-33}$	1.602177×10^{-19}	1	3.829294×10^{-20}	3.826733×10^{-20}	1.518570×10^{-22}	4.450490×10^{-26}	5.968206×10^{-26}	1.181705×10^{-19}	8.206283×10^{-22}	1.581225×10^{-21}
1 cal	$= 4.655328 \times 10^{-14}$	4.184	2.611448×10^{19}	1	0.9993312	3.965667×10^{-3}	$1.1622222\ldots \times 10^{-6}$	1.558562×10^{-6}	3.085960	2.143028×10^{-2}	0.04129287
1 cal_{IT}	$= 4.658443 \times 10^{-14}$	4.1868	2.613195×10^{19}	1.000669	1	3.968321×10^{-3}	1.163×10^{-6}	1.559609×10^{-6}	3.088025	2.144462×10^{-2}	0.04132050
1 Btu_{IT}	$= 1.173908 \times 10^{-11}$	1055.056	6.585141×10^{21}	252.1644	251.9958	1	2.930711×10^{-4}	3.930148×10^{-4}	778.1693	5.403953	10.41259
1 kWh	$= 4.005540 \times 10^{-8}$	3600000.	2.246943×10^{25}	860420.7	859845.2	3412.142	1	1.341022	2655224.	18349.06	35529.24
1 hp-h	$= 2.986931 \times 10^{-8}$	2384519.	1.675545×10^{25}	641615.6	641186.5	2544.33	0.7456998	1	1980000.	13750.	26494.15
1 ft-lbf	$= 1.508551 \times 10^{-14}$	1.355818	8.462351×10^{18}	0.3240483	0.3238315	1.285067×10^{-3}	3.766161×10^{-7}	$5.050505\ldots \times 10^{-7}$	1	$6.944444\ldots \times 10^{-3}$	0.01338088
1 $ft^3 \cdot lbf \cdot in.^{-2}$	$= 2.172313 \times 10^{-12}$	195.2378	1.218578×10^{21}	46.66295.	46.63174	0.1850497	5.423272×10^{-5}	$7.272727\ldots \times 10^{-5}$	144.	1	1.926847
1 liter-atm	$= 1.127393 \times 10^{-12}$	101.325	6.324209×10^{20}	24.21726	24.20106	0.09603757	2.814583×10^{-5}	3.774419×10^{-5}	74.73349	0.5189825	1

The chemical elements

Name	Symbol	At. no.	Name	Symbol	At. no.
Actinium	Ac	89	Meitnerium	Mt	109
Aluminum	Al	13	Mendelevium	Md	101
Americium	Am	95	Mercury	Hg	80
Antimony	Sb	51	Molybdenum	Mo	42
Argon	Ar	18	Neodymium	Nd	60
Arsenic	As	33	Neon	Ne	10
Astatine	At	85	Neptunium	Np	93
Barium	Ba	56	Nickel	Ni	28
Berkelium	Bk	97	Niobium	Nb	41
Beryllium	Be	4	Nitrogen	N	7
Bismuth	Bi	83	Nobelium	No	102
Bohrium	Bh	107	Osmium	Os	76
Boron	B	5	Oxygen	O	8
Bromine	Br	35	Palladium	Pd	46
Cadmium	Cd	48	Phosphorus	P	15
Calcium	Ca	20	Platinum	Pt	78
Californium	Cf	98	Plutonium	Pu	94
Carbon	C	6	Polonium	Po	84
Cerium	Ce	58	Potassium	K	19
Cesium	Cs	55	Praseodymium	Pr	59
Chlorine	Cl	17	Promethium	Pm	61
Chromium	Cr	24	Protactinium	Pa	91
Cobalt	Co	27	Radium	Ra	88
Copper	Cu	29	Radon	Rn	86
Curium	Cm	96	Rhenium	Re	75
Darmstadtium	Ds	110	Rhodium	Rh	45
Dubnium	Db	105	Roentgenium	Rg	111
Dysprosium	Dy	66	Rubidium	Rb	37
Einsteinium	Es	99	Ruthenium	Ru	44
Element 112*		112	Rutherfordium	Rf	104
Erbium	Er	68	Samarium	Sm	62
Europium	Eu	63	Scandium	Sc	21
Fermium	Fm	100	Seaborgium	Sg	106
Fluorine	F	9	Selenium	Se	34
Francium	Fr	87	Silicon	Si	14
Gadolinium	Gd	64	Silver	Ag	47
Gallium	Ga	31	Sodium	Na	11
Germanium	Ge	32	Strontium	Sr	38
Gold	Au	79	Sulfur	S	16
Hafnium	Hf	72	Tantalum	Ta	73
Hassium	Hs	108	Technetium	Tc	43
Helium	He	2	Tellurium	Te	52
Holmium	Ho	67	Terbium	Tb	65
Hydrogen	H	1	Thallium	Tl	81
Indium	In	49	Thorium	Th	90
Iodine	I	53	Thulium	Tm	69
Iridium	Ir	77	Tin	Sn	50
Iron	Fe	26	Titanium	Ti	22
Krypton	Kr	36	Tungsten	W	74
Lanthanum	La	57	Uranium	U	92
Lawrencium	Lr	103	Vanadium	V	23
Lead	Pb	82	Xenon	Xe	54
Lithium	Li	3	Ytterbium	Yb	70
Lutetium	Lu	71	Yttrium	Y	39
Magnesium	Mg	12	Zinc	Zn	30
Manganese	Mn	25	Zirconium	Zr	40

*This element does not have an official name or symbol.

Periodic table

(The atomic numbers are listed above the symbols identifying the elements. The heavy line separates metals from nonmetals.)

	1	2	3	4	5	6	7	8	9	10	11	12	13	14	15	16	17	18
1s																	1 H Hydrogen	2 He Helium
s / p	3 Li Lithium	4 Be Beryllium											5 B Boron	6 C Carbon	7 N Nitrogen	8 O Oxygen	9 F Fluorine	10 Ne Neon
	11 Na Sodium	12 Mg Magnesium											13 Al Aluminum	14 Si Silicon	15 P Phosphorus	16 S Sulfur	17 Cl Chlorine	18 Ar Argon
d	19 K Potassium	20 Ca Calcium	21 Sc Scandium	22 Ti Titanium	23 V Vanadium	24 Cr Chromium	25 Mn Manganese	26 Fe Iron	27 Co Cobalt	28 Ni Nickel	29 Cu Copper	30 Zn Zinc	31 Ga Gallium	32 Ge Germanium	33 As Arsenic	34 Se Selenium	35 Br Bromine	36 Kr Krypton
	37 Rb Rubidium	38 Sr Strontium	39 Y Yttrium	40 Zr Zirconium	41 Nb Niobium	42 Mo Molybdenum	43 Tc Technetium	44 Ru Ruthenium	45 Rh Rhodium	46 Pd Palladium	47 Ag Silver	48 Cd Cadmium	49 In Indium	50 Sn Tin	51 Sb Antimony	52 Te Tellurium	53 I Iodine	54 Xe Xenon
	55 Cs Cesium	56 Ba Barium	71 Lu Lutetium	72 Hf Hafnium	73 Ta Tantalum	74 W Tungsten	75 Re Rhenium	76 Os Osmium	77 Ir Iridium	78 Pt Platinum	79 Au Gold	80 Hg Mercury	81 Tl Thallium	82 Pb Lead	83 Bi Bismuth	84 Po Polonium	85 At Astatine	86 Rn Radon
	87 Fr Francium	88 Ra Radium	103 Lr Lawrencium	104 Rf Rutherfordium	105 Db Dubnium	106 Sg Seaborgium	107 Bh Bohrium	108 Hs Hassium	109 Mt Meitnerium	110 Ds Darmstadtium	111 Rg Roentgenium	112	113	114	115	116	117	118

f														
	57 La Lanthanum	58 Ce Cerium	59 Pr Praseodymium	60 Nd Neodymium	61 Pm Promethium	62 Sm Samarium	63 Eu Europium	64 Gd Gadolinium	65 Tb Terbium	66 Dy Dysprosium	67 Ho Holmium	68 Er Erbium	69 Tm Thulium	70 Yb Ytterbium
	89 Ac Actinium	90 Th Thorium	91 Pa Protactinium	92 U Uranium	93 Np Neptunium	94 Pu Plutonium	95 Am Americium	96 Cm Curium	97 Bk Berkelium	98 Cf Californium	99 Es Einsteinium	100 Fm Fermium	101 Md Mendelevium	102 No Nobelium

Classification of living organisms

- Domain Archaea[a]
  - Phylum Crenarchaeota
    - Class Thermoprotel
      - Order Thermoproteales
      - Order Desulfurococcales
      - Order Sulfolobales
  - Phylum Euryarchaeota
    - Class Methanobacteria
      - Order Methanobacteriales
    - Class Methanoccocci
      - Order Methanococcales
      - Order Methanomicrobiales
      - Order Methanosarcinales
    - Class Halobacteria
      - Order Halobacteriales
    - Class Thermoplasmata
      - Order Thermoplasmatales
    - Class Thermococci
      - Order Thermococcales
    - Class Archaeoglobi
    - Class Methanopyrl
      - Order Methanopyrales

- Domain Bacteria
  - Phylium Aquificae
    - Class Aquificae
      - Order Aquificales
  - Phylum Thermotogae
    - Class Thermotogae
      - Order Thermotogales
  - Phylum Thermodesulfobacteria
    - Class Thermodesulfobacteria
      - Order Thermodesulfobacteriales
  - Phylum Deinococcus-Thermus
    - Class Deinococci
      - Order Deinococcales
      - Order Thermales
  - Phylum Chryslogenetes
    - Class Chrysiogenetes
      - Order Chrysiogenales
  - Phylum Chloroflexi
    - Class Chloroflexi
      - Order Chloroflexales
      - Order Herpetosiphonales
  - Phylum Thermomicrobia
    - Class Thermomicrobia
      - Order Thermomicrobiales
  - Phylum Nitrospira
    - Class Nitrospira
      - Order Nitrospirales
  - Phylum Deferribacteres
    - Class Deferribacteres
      - Order Deferribacterales
  - Phylum Cyanobacteria
    - Class Cyanobacteria
  - Phylum Chlorobi
    - Class Chlorobia
      - Order Chlorobiales
  - Phylum Proteobacteria
    - Class Alphaproteobacteria
      - Order Rhodospirillales
      - Order Rickettsiales
      - Order Rhodobacterales
      - Order Sphingomonadales
      - Order Caulobacterales
      - Order Rhizobiales
    - Class Betaproteobacteria
      - Order Burkholderiales
      - Order Hydrogenophilales
      - Order Methylophilales
      - Order Neisseriales
      - Order Nitrosomonadales
      - Order Rhodocyclales
    - Class Cammaproteobacteria
      - Order Chromatiales
      - Order Acidithiobacillales
      - Order Xanthomonadales
      - Order Cardiobacteriales
      - Order Thiotrichales
      - Order Legionellals
      - Order Methylococcales
      - Order Oceanospirillales
      - Order Pseudomonadales
      - Order Alteromonadales
      - Order Vibrionales
      - Order Aeromonadales
      - Order Enterobacteriales
      - Order Pasteurellales
    - Class Deltaproteobacteria
      - Order Desulfurellales
      - Order Desulfovibrionales
      - Order Desulfobacterales
      - Order Desulfuromonadales
      - Order Syntrophobacterales
      - Order Bdellovibrionales
      - Order Myxococcales
    - Class Epsilonproteobacteria
      - Order Campylobacterales
  - Phylum Firmicutes
    - Class Clostridia
      - Order Clostridiales
      - Order Thermoanaerobacteriales
      - Order Haloanaerobiales
    - Class Mollicutes
      - Order Mycoplasmatales
      - Order Entomoplasmatales
      - Order Acholeplasmatales
      - Order Anaeroplasmatales
    - Class Bacilli
      - Order Bacillales
      - Order Lactobacillales
  - Phylum Actinobacteria
    - Class Actinobacteria
    - Subclass Acidimicrobidae
      - Order Acidimicrobiales
      - Suborder Acidimicrobineae
    - Subclass Rubrobacteridae
      - Order Rubrobacterles
      - Suborder Rubrobacterineae
    - Subclass Coriobacteridae
      - Order Coriobacteriales
      - Suborder Cariobacterineae
    - Subclass Sphaerobacteridae
      - Order Sphaeriobacteriales
      - Suborder Sphaerobacterineae
    - Subclass Actinobacteridae
      - Order Actinomyietales
      - Suborder Actiomycineae
      - Suborder Micrococcineae
      - Suborder Corynebacterineae
      - Suborder Micromonosporineae
      - Suborder Propionibacterineae
      - Suborder Pseudonocardineae
      - Suborder Streptomycineae
      - Suborder Streptosporangineae
      - Suborder Frankineae
      - Suborder Glycomycineae
      - Order Bifidobacteriales
  - Phylum Planctomycetes

Classification of living organisms (*cont.*)

Class Planctomycetacia
Order Planctomycetales
Phylum Chlamydiae
Class Chlamydiae
Order Chlamydiales
Phylum Spirochaetes
Class Spirochaetes
Order Spirochaetales
Phylum Fibrobacteres
Class Fibrobacteres
Order Fibrobacterales
Phylum Acidobacteria
Class Acidobacteria
Order Acidobacteriales
Phylum Bacteroidetes
Class Bacteroidetes
Order Bacteroidales
Class Flavobacteria
Order Flavobacteriales
Class Sphingobacteria
Order Sphingobacteriales
Phylum Fusobacteria
Class Fusobacteria
Order Fusobacteriales
Phylum Verrucomicrobia
Class Verrucomicrobiae
Order Verrucomicrobiales
Phylum Dictyoglomus
Class Dictyoglomi
Order Dictyoglomales

Domain Eukarya[b]

Kingdom Protista
Phylum Metamonada
Phylum Trichozoa
Subphylum Parabasala
Class Trichomonadea
Class Hypermastigotea

Subkingdom Neozoa
Phylum Choanozoa
Phylum Amoebozoa
Subphylum Lobosa
Subphylum Conosa
Class Archamoebae
Class Mycetozoa
Phylum Foraminifera
Phylum Percolozoa
Phylum Euglenozoa
Class Euglenoidea
Class Saccostomae
Phylum Sporozoa
Subphylum Gregarinae
Subphylum Coccidiomorpha
Subphylum Perkinsida
Subphylum Manubrispora
Phylum Ciliophora
Phylum Radiozoa
Phylum Heliozoa
Phylum Rhodophyta
Class Rhodophyceae
Subclass Banglophycidae
Order Bangiales
Order Compsopogonales
Order Porphyridiales
Order Rhodochaetales
Subclass Florideophycidae
Order Acrochaetiales
Order Ahnfeltiales
Order Balbianiales
Order Balliales
Order Batrachospermales
Order Bonnemaisoniales
Order Ceramiales
Order Colaconematales
Order Corallinales
Order Gelidiales
Order Gigartinales
Order Gracilarlales
Order Halymeniales
Order Hildenbrandiales
Order Nemaliales
Order Palmariales
Order Plocamiales
Order Rhodogorgonales
Order Rhodymeniales
Order Thoreales
Phylum Chrysophyta
Class Bacillariophyceae
Subclass Bacillariophycidae
Order Achnanthales
Order Bacillariales
Order Cymbellales
Order Dictyoneidales
Order Lyrellales
Order Mastogloiales
Order Naviculales
Order Rhopalodiales
Order Surirellales
Order Thallassiophysales
Subclass Biddulphiophycidae
Order Anaulales
Order Biddulphiales
Order Hemlaulales
Order Triceratiales
Subclass Chaetocerotophycidae
Order Chaetocerotales
Order Leptocylindrales
Subclass Corethrophycidae
Order Cymatosirales
Subclass Coscinodiscophycidae
Order Arachnoidiscales
Order Asterolamprales
Order Aulacoseirales
Order Chrysaanthemodiscales
Order Coscinodiscales
Order Ethmodiscales
Order Melosirales
Order Orthoseirales
Order Parallales
Order Stictocyclales
Order Stictodiscales
Subclass Cymatosirophycidae
Order Cymatosirales
Subclass Eunotiophycidae
Order Eunotiales
Subclass Fragilariophycidae
Order Ardissoneales
Order Cyclophorales
Order Climacospheniales
Order Fragllariales
Order Licmorphorales
Order Protoraphidales
Order Rhabdonematales
Order Rhaphoneidales
Order Striatellales
Order Tabellariales
Order Thalassionematales
Order Toxariales
Subclass Lithodesmiophycidae
Order Lithodesmialescidae
Subclass Rhizosoleniophycidae
Order Rhizosoleniales
Subclass Thalassiosirophycidae
Order Thalassiosirales
Class Bolidophyceae
Order Bolidomonadales
Class Chrysomerophyceae
Order Chrysomeridales
nom. nud.
Class Chrysophyceae
Order Chromulinales

Classification of living organisms (*cont.*)

        - Order Hibberdiales
    - Class Dictyochophyceae
        - Order Dictyochales
        - Order Pedinellales
        - Order Rhizochromulinales
    - Class Eustigmatophyceae
        - Order Eustigmatales
    - Class Pelagophyceae
        - Order Pelagomonadales
        - Order Sarcinochrysidales
    - Class Phaeophyceae
        - Order Ascoseirales
        - Order Chordariales
        - Order Cutleriales
        - Order Desmarestiales
        - Order Dictysiphonales
        - Order Dictyotales
        - Order Durvillaeales
        - Order Ectocarpales
        - Order Fucales
        - Order Laminariales
        - Order Scytosiphonales
        - Order Sphacelariales
        - Order Sporochnales
        - Order Tilopteridiales
    - Class Phaeothamniophyceae
        - Order Phaeothamniales
        - Order Pleurochloridellales
    - Class Pinguiophyceae
        - Order Pinguiochrysidales
    - Class Raphidophyceae
        - Order Rhaphidomonadales
    - Class Synurophyceae
        - Order Synurales
    - Class Xanthophyceae (=Tribophyceae)
        - Order Botrydiales
        - Order Chloramoebales
        - Order Heterogloeales
        - Order Mischococcales
        - Order Rhizochloridales
        - Order Tribonematales
        - Order Vaucheriales
- Phylum Cryptophyta
    - Class Cryptophyceae
        - Order Cryptomonadales
        - Order Cryptococcales
- Phylum Glaucocystophyta
    - Class Glaucocystophycaae
        - Order Cyanophorales
        - Order Glaucocystales
        - Order Gloeochaetales
- Phylum Prymnesiophyta (=Haptophyta)
    - Class Pavlovophyceae
        - Order Pavlovales
    - Class Prymnesiophyceae
        - Order Coccolithales
        - Order Isochrysidales
        - Order Phaeocystales
        - Order Prymneslales
- Phylum Dinophyta
    - Class Dinophyceae
        - Order Actiniscales
        - Order Blastodiniales
        - Order Chytriodiniales
        - Order Desmocapsales
        - Order Desmomonadales
        - Order Dinophysales
        - Order Gonyaulacales
        - Order Gymnodiniales
        - Order Kokwitziellaless
        - Order Nannoceratopslales
        - Order Noctilucales
        - Order Oxyrthinales
        - Order Peridiniales
        - Order Phytodiniales
        - Order Prorocentrales
        - Order Ptychodiscales
        - Order Pyrocysales
        - Order Suessiales
        - Order Syndiniales
        - Order Thoracosphaerales
- Phylum Chlorophyta
    - Class Charophyceae
        - Order Charales
        - Order Chlorokybaees
        - Order Coleochaetales
        - Order Klebsormidiales
        - Order Zygnematales
    - Class Chlorophyceae
        - Order Chaetophorales
        - Order Chlorococcales
        - Order Cladophorales
        - Order Odeogoniales
        - Order Sphaeropleales
        - Order Volvocales
        - Order Pleurastrales
    - Class Prasinophyceae
        - Order Chlorodendrales
        - Order Mamiellales
        - Order Pseudoscourfeldiales
        - Order Pyramimonidales
    - Class Trebouxiophyceae
        - Order Trebouxiales
    - Class Ulvophyceae
        - Order Bryopsidales
        - Order Caulerpales
        - Order Codiolales
        - Order Dasycladales
        - Order Halimedales
        - Order Prasioeales
        - Order Siphonocladales
        - Order Trentepohliales
        - Order Ulotrichales
        - Order Ulvales
- Phylum Euglenophyta
    - Class Euglenophyceae
        - Order Euglenales
        - Order Euglenamorphales
        - Order Eutreptiales
        - Order Heteronematales
        - Order Rhabdomonadales
        - Order Sphenomonadales
- Phylum Acrasiomycota
    - Class Acrasiomycetes
        - Order Acrasiales
- Phylum Dictyosteliomycota
    - Class Dictyosteliomycetes
        - Order Dictyosteliales
- Phylum Myxomycota
    - Class Myxomycetes
        - Order Liceales
        - Order Echinosteliales
        - Order Trichiales
        - Order Physarales
        - Order Stemonitales
        - Order Ceratiomyxales
    - Class Protosteliomycetes
        - Order Protosteliales
- Phylum Plasmodiophoromycota
    - Class Plasmodiophoromycetes
        - Order Plasmodiophorales
- Phylum Oomycota
    - Class Oomycetes
        - Order Saprolegniales
        - Order Salilagenidiales
        - Order Leptomitales
        - Order Myzocytiopsidales
        - Order Rhipidiales
        - Order Pythiales
        - Order Peronosporales

Classification of living organisms (*cont.*)

Phylum Hyphochytriomycota
Class Hyphochytriomycetes
Order Hyphochytriales
Phylum Labyrinthulomycota
Class Labyrinthulomycetes
Order Labyrinthulales
Phylum Chytridiomycota
Class Chytridiomycetes
Order Blastocladiales
Order Chytridiales
Order Monoblepharidales
Order Neocallimastigales
Order Spizellomycetales
Phylum Zygomycota
Class Trichomycetes
Order Amoebidiales
Order Asellariales
Order Eccrinales
Order Harpellales
Class Zygomycetes
Order Mucorales
Order Dimargaritales
Order Kickxellales
Order Endogonales
Order Glomales
Order Entomophthorales
Order Zoopagales
Phylum Ascomycota
Class Archiascomycetes
Order Taphrinales
Order Schizosaccharomyce-tales
Class Saccharomycetes
Order Saccharomycetales
Class Plectomycetes
Order Eurotiales
Order Ascosphaerales
Order Onygenales
Class Laboulbeniomycetes
Order Laboulbeniales
Order Spathulosporales
Class Pyrenomycetes
Order Hypocreales
Order Melanosporales
Order Microascales
Order Phylachorales
Order Ophiostomatales
Order Diaporthales
Order Calosphaceriales
Order Xylariales
Order Sordariales
Order Meliolales
Order Halosphaeriales
Class Discomycetes
Order Medeolarlales
Order Rhytismatales
Order Ostropales
Order Cyttariales.
Order Helotiales
Order Neolectales
Order Gyalectales
Order Lecanorales
Order Lichinales
Order Peltigerales
Order Pertusariales
Order Teloschistales
Order Caliciales
Order Pezizales
Class Loculoascomycetes
Order Coryneliales
Order Dothideales
Order Myriangiales
Order Arthoniales
Order Pyrenulales
Order Asterinales
Order Capnodiales
Order Chaetothyriales
Order Patellariales
Order Pleosporales
Order Melanommatales
Order Trichotheliales
Order Verrucariales
Phylum Basidiomycota
Class Basidiomycetes
Subclass Heterobasidiomy-cetes
Order Agricostibales
Order Atractiellales
Order Auriculariales
Order Heterogastridiales
Order Tremellales
Subclass Homobasidiomycetes
Order Agaricales
Order Boletales
Order Bondarzewiales
Order Cantharellales
Order Ceratobasidiaes
Order Cortinariales
Order Dacrymycetales
Order Fistulinales
Order Ganodermatales
Order Gautieriales
Order Gomphales
Order Hericiales
Order Hymenoghaetales
Order Hymenogastrales
Order Lachnocladiales
Order Lycoperdales
Order Melanogastrales
Order Nidulariales
Order Phallales
Order Poriales
Order Russulales
Order Schizophyllales
Order Sclerodermatales
Order Stereales
Order Thelephorales
Order Tulasnellales
Order Tulostomatales
Class Ustomycetes
Order Cryptobasidiales
Order Cryptomycocola-cales
Order Exobasidiales
Order Graphiolales
Order Platyglocales
Order Sporidiales
Order Ustilaginales
Class Tellomycetes
Order Septobasidiales
Order Uredinales
Phylum Deuteromycetes (Asexual Ascomycetes and Basidiomycetes)
Class Hyphomycetes
Order Hyphomycetales
Order Stibeilales
Order Tuberculariales
Class Agonomycetes
Order Agonomycetales
Class Coelomycetes
Order Melanconiales
Order Sphaeropsidales
Order Pycnothyriales

Kingdom Plantae

Subkingdom Embryobionta
Division Hepaticophyta
Class Junermanniopsida
Order Calobryales
Order Jungermanniales
Order Metzgeriales

Classification of living organisms (*cont.*)

Class Marchantiopsida
Order Sphaerocarpales
Order Monocleales
Order Marchantiales
Division Anthocerotophyta
Class Anthocerotopsida
Order Anthocerotales
Division Bryophyta
Class Sphagnicopsida
Order Sphagnicales
Class Andreaeopsida
Order Andreaeles
Class Bryopsida
Order Archidiales
Order Bryales
Order Buxbaumiales
Order Dicranales
Order Encalyptales
Order Fissidentales
Order Funariales
Order Grimmiales
Order Hookeriales
Order Hypnobryales
Order Isobryales
Order Orthotrichales
Order Pottiales
Order Orthotrichales
Order Seligerales
Order Splachnales
Division Lycophyta
Class Lycopsida
Order Isoetales
Order Lycopodiales
Order Selaginellales
Division Polypodiophyta
Class Polypodopsida
Order Equisetales
Order Marattiales
Order Ophioglossales
Order Polypodiales
Order Psilotales
Division Pinophyta
Class Ginkgopsida
Order Ginkgoales
Class Cycadopsida
Order Cycadales
Class Pinopsida
Order Pinales
Order Podocarpales
Order Gnetales
Division Magnoliophyta
|unplaced orders|
Order Ceratophyllales
Order Chloranthales
Class Amborellopsida
Order Amborellales
Class Austrobaileyales
Order Austrobaileyales
Class Liliopsida
Order Acorales
Order Alismatales
Order Arecales
Order Asparagales
Order Commelinales
Order Dioscoreales
Order Liliales
Order Pandanales
Order Poales
Order Zingiberales
Class Magnoliopsida
Order Magnoliales
Order Laurales
Order Piperales
Order Canellales
Class Nymphaeopsida
Order Nymphaeales
Class Rosopsida
|unplaced orders|
Order Berberidopsidales
Order Buxales
Order Gunnerales
Order Proteales
Order Saxifragales
Order Santalales
Order Trochodendrales
Subclass Caryophyllidae
Order Caryophyliales
Order Dilleniales
Subclass Ranunculidae
Order Ranunculales
Subclass Rosidae
|unplaced orders|
Order Crossosomatales
Order Geraniales
Order Myrtales
Order Vitales
Superorder Rosanae
Order Celastrales
Order Cucurbitales
Order Fabales
Order Fagales
Order Malpighiales
Order Oxalidales
Order Rosales
Order Zygophyllales
Superorder Malvanae
Order Brassicales
Order Malvales
Order Sapindales
Subclass Asteridae
|unplaced order|
Order Boraginales
Superorder Cornanae
Order Cornales
Superorder Ericanae
Order Ericles
Superorder Lamianae
Order Garryales
Order Gentianales
Order Lamiales
Order Solanales
Superorder Asteranae
Order Apiales
Order Aquifoliales
Order Asterales
Order Dipsacales

Kingdom Animalia

Subkingdom Parazoa
Phylum Porifera
Subphylum Cellularia
Class Demospongiae
Class Calcarea
Subphylum Symplasma
Class Hexactinellida
Phylum Placozoa

Subkingdom Eumetazoa
Phylum Cnidaria
(=Coelenterata)
Class Scyphozoa
Order Stauromedusae
Order Coronatae
Order Semaeostomeae
Order Rhizostomeae
Class Cubozoa
Order Cubomedusae
Class Hydrozoa
Order Hydroida
Order Milleporina
Order Stylasterina
Order Trachylina

Classification of living organisms (*cont.*)

Order Siphonophora
Order Chondrophora
Order Actinulida
Class Anthozoa
Subclass Alcyonaria
(=Octocorallia)
Order Stolonifera
Order Gorgonacea
Order Alcyonacea
Order Pennatulacea
Subclass Zoantharia
(=Hexacorallia)
Order Actinaria
Order Corallimorpharia
Order Scleractinia
Order Zoanthinaria
(=Zoanthidea)
Order Ceriantharia
Order Ptychodactiaria
Order Antipatharia
Phylum Ctenophora
Class Tentaculata
Order Cydippida
Order Platyctenida
Order Lobata
Order Cestida
Order Ganeshida
Order Thalassocalycida
Class Nuda
Order Beroida
Phylum Platyhelminthes
Class Turbellaria
Order Acoela
Order Rhabdocoela
Order Catenullda
Order Macrostomida
Order Nemertodermatida
Order Lecithoepitheliata
Order Polycladida
Order Prolecithophora
(=Holocoela)
Order Proseriata
Order Tricladida
Order Neorhabdocoela
Class Cestoda
Subclass Cestodaria
Subclass Eucestoda
Order Caryophyllidea
Order Spathebothriidea
Order Trypanorhyncha
Order Pseudophyllidea
Order Tetraphyllidea
Order Cyclophyllidea
Class Monogenea
Class Trematoda
Subclass Digenea
Order Strigeidida
Order Azygiida
Order Echinostomida
Order Plagiorchiida
Order Opisthorchiida
Subclass Aspidogastrea
(=Aspidobothrea)
Phylum Mesozoa
Class Orthonectida
Class Rhombozoa
Order Dicyemida
Order Heterocyemida
Phylum Myxozoa
(=Myxospora)
Phylum Nemertea
(=Rhynchocoela,
Nemertinea)
Class Anopia
Order Palaeonemertea
(=Palaeonemertini)
Order Heteronemertea
Class Enopia
Order Hoplonemertea
(=Hoplonemertini)
Order Bdelionemertea
Phylum Gnathostomuilda
Order Filospermoidea
Order Bursovaginoidea
Phylum Gastrotricha
Order Chaetonotida
Order Macrodasyida
Phylum Cycliophora
Phylum Rotifera
Class Monogononta
Order Ploima
Order Flosculariaceae
Order Collothecaceae
Class Bdelloidea
Class Seisonidea
Phylum Acanthocephaia
Class Archiacanthocephaia
Class Eoacanthocephaia
Class Palaeacanthocephala
Phylum Nematoda (=Nemata)
Class Adenophorea
Subclass Enoplia
Order Enoplida
Order Dorylaimida
Order Trichocephalida
Order Mermithida
Subclass Chromadoria
Class Secernentea
Subclass Rhabditia
Order Rhabditida
Order Ascaridida
Order Strongylida
Subclass Spiruria
Order Spirurida
Order Camallanida
Subclass Diplogasteria
Phylum Nematomorpha
Class Nectonematoida
Class Gordioida
Phylum Priapulida
Phylum Kinorhyncha
(=Echinoderida)
Class Cyclorhagida
Class Homalorhagida
Phylum Loricifera
Phylum Mollusca
Subphylum Aculifera
Class Polyplacophora
Class Aplacophora
Subclass Neomeniophora
(=Solenogastres)
Subclass Chaetodermomor-
pha
(=Caudofoveata)
Subphylum Conchifera
Class Monoplacophora
Class Gastropoda
Subclass Prosobranchia
Order Archaeogastropoda
Order Mesogastropoda
(=Taenioglossa)
Order Neogastropoda
Subclass Opisthobranchia
Order Cephalaspidea
Order Runcinoidea
Order Acochlidioidea
Order Sacoglossa
(=Ascoglossa)
Order Anaspidea
(=Aplysiacea)
Order Notaspidea
Order Thecosomata
Order Gymnosomata

Classification of living organisms (*cont.*)

Order Nudibranchia
Subclass Pulmonata
Order Archaeopulmonata
Order Basommatophora
Order Stylommatophora
Order Systellommato-
phora
Class Bivalvia
(=Pelecypoda)
Subclass Protobranchia
(=Palaeotaxodonta,
Cryptodonta)
Subclass Pteriomorphia
Subclass Paleoheterodonta
Subclass Heterodonta
Subclass Anomalodesmata
Class Scaphopoda
Class Cephalopoda
Subclass Nautiloidea
Subclass Coleoidea
(=Dibranchiata)
Order Sepioidea
Order Teuthoidea
(=Decapoda)
Order Vampyromorpha
Order Octopoda
Phylum Annelida
Class Polychaeta
Order Phyllodocida
Order Spintherida
Order Eunicida
Order Spionida
Order Chaetopterida
Order Magelonida
Order Psammodrilida
Order Cirratulida
Order Flabelligerida
Order Ophelilda
Order Capitellida
Order Owenilda
Order Terebellida
Order Sabellida
Order Protodrilida
Order Myzostomida
Class Clitellata
Subclass Oligochaeta
Order Lumbriculida
Order Haplotaxida
Subclass Hirudinea
Order Rhynchobdeilae
Order Arhynchobdellae
Order Branchiobdellida
Order Acanthobdellida
Class Pogonophora
(=Siboglinidae)
Subclass Perviata
(=Frenulata)
Subclass Obturata
(=Vestimentifera)
Class Echiura
Order Echiura
Order Xenopneusta
Order Heteromyota
Phylum Sipuncula
Phylum Arthropoda
Subphylum Chelicerata
Class Merostomata
Order Xiphosura
Class Arachnida
Order Scorpiones
Order Uropygi
Order Amblypygi
Order Araneae
Order Ricinulei
Order Pseudoscorpiones
Order Solifugae
(=Solpugida)
Order Opiliones
Order Acari
Class Pycnogonida
(=Pantopoda)
Subphylum Mandibulata
Class Myriapoda
Order Chilopoda
Order Dipiopoda
Order Symphyia
Order Pauropoda
Class insecta (=Hexapoda)
Subclass Apterygota
Order Thysanura
Order Collembola
Subclass Pterygota
Superorder Hemime-
tabola
Order Ephemeroptera
Order Odonata
Order Blattaria
Order Mantodea
Order isoptera
Order Grylioblattaria
Order Orthoptera
Order Phasmida
(=Phasmatoptera)
Order Dermaptera
Order Embiidina
Order Plecoptera
Order Psocoptera
Order Anoplura
Order Mallophaga
Order Thysanoptera
Order Hemiptera
Order Homoptera
Superorder Holometabola
Order Neuroptera
Order Coleoptera
Order Strepsiptera
Order Mecoptera
Order Siphonaptera
Order Diptera
Order Trichoptera
Order Lepidoptera
Order Hymenoptera
Class Crustacea
Subclass Cephalocarida
Subclass Malacostraca
Superorder Syncarida
Superorder Hoplocarida
Order Stomatopoda
Superorder Peracarida
Order Thermosbaenacea
Order Mysidacea
Order Cumacea
Order Tanaidacea
Order Isopoda
Order Amphipoda
Superorder Eucarida
Order Euphausiacea
Order Decapoda
Subclass Branchiopoda
Order Notostraca
Order Cladocera
Order Conchostraca
Order Anostraca
Subclass Ostracoda
Order Myodocopa
Order Podocopa
Subclass Mystacocarida
Subclass Copepoda
Order Calanoida
Order Harpacticoida
Order Cyclopoida
Order Monstrilloida
Order Siphonostomatoida

Classification of living organisms (*cont.*)

Order Poecilostomatoida
Subclass Branchiura
Subclass Pentastomida
Order Cephalobaenida
Order Porocephalida
Subclass Tantulocarida
Subclass Remipedia
Subclass Cirripedia
Order Acrothoracica
Order Ascothoracica
Order Thoracica
Order Rhizocephala
Phylum Tardigrada
Class Heterotardigrada
Class Mesotardigrada
Class Eutardigrada
Order Parachela
Order Apochela
Phylum Onychophora
Phylum Phoronida
Phylum Brachiopoda
Class inarticulata
Order Lingulida
Order Acrotretida
Class Articulata
Order Rhynchonellida
Order Terebratulida
Phylum Bryozoa
(=Ectoprocta, polyzoa)
Class Phylactolaemata
Class Stenolaemata
Class Gymnolaemata
Order Ctenostomata
Order Chellostomata
Phylum Entoprocta
(=Kamptozoa)
Phylum Chaetognatha
Class Sagittoidea
Order Phragmophora
Order Aphragmophora
Phylum Echinodermata
Subphylum Crinozoa
Class Crinoidea
Order Millericrinida
Order Cyrtocrinida
Order Bourgueticrinida
Order Isocrinida
Order Comatulida
Subphylum Asterozoa
Class Stelleroidea
Subclass Somasteroidea
Subclass Ophiuroidea
Order Phrynophiurida
Order Ophiurida
Subclass Asteroidea
Order Platyasterida
Order Paxillosida
Order Valvatida
Order Spinulosida
Order Forcipulata
Order Brisingida
Class Concentricycloidea
Subphylum Echinozoa
Class Echinoidea
Order Cidaroida
Order Echinothuroida
Order Diadematoida
Order Arbacioida
Order Temnopleuroida
Order Echinoida
Order Holectypoida
Order Clypeasteroida
Order Spatangoida
Class Holothuroidea
Order Dendrochirotida
Order Aspidochirotida
Order Elasipodida
Order Apodida
Order Molpadiida
Phylum Hemichordata
Class Enteropneusta
Class Pterobranchia
Phylum Chordata
Subphylum Urochordata
(=Tunicata)
Class Ascidiacea
Order Aspiousobranchia
Order Phlebobranchia
Order Stolidobranchia
Class Larvacea
(=Appendicularia)
Class Thaliacea
Order Pyrosomida
Order Doliolida
Order Salpida
Subphylum Cephalochordata
(=Acrania)
Phylum Chordata[c]
Subphylum Vertebrata
Superclass Agnatha
Class Myxini
Order Myxiniformes
Class Cephalaspidomorphi
Order Petromyzontiformes
Superclass Gnathostomata
Class Chondrichthyes
Subclass Holocephali
Order Chimaeriformes
Subclass Elasmobranchii
Order Hexanchiformes
Order Squaliformes
Order Pristiophoriformes
Order Squatiniformes
Order Pristiformes
Order Rhinobatiformes
Order Torpediniformes
Order Myliobatiformes
Order Heterodontiformes
Order Orectolobiformes
Order Lamniformes
Order Carchiniformes
Class Sarcopterygii
Subclass Coelacanthimor-
morpha
Order Coelacanthiformes
Subclass Porolepimorpha
and Dipnol
Order Ceratodontiformes
Order Lepidosireniformes
Class Actinopterygii
Subclass Chondrostei
Order Polypteriformes
Order Acipenseriformes
Subclass Neopterygli
Order Semionotiformes
Order Amiiformes
Division Teiestei
Subdivision Osteoglosso-
morpha
Order Osteoglossiformes
Subdivision Elopomorpha
Order Elopiformes
Order Albuliformes
Order Anguilliformes
Order Saccopharyngi-
formes
Subdivision Clupeomorpha
Order Clupeiformes
Subdivision Euteleostei
Superorder Ostariophysi
Order Gonorthynchiformes
Order Cypriniformes
Order Characiformes

Classification of living organisms (*cont.*)

Order Siluriformes
Order Gymnotiformes
Superorder Protacanthopterygii
Order Esociformes
Order Osmeriformes
Order Salmoniformes
Superorder Stenopterygii
Order Stomiformes
Order Ateleopodiformes
Superorder Cyclosquamata
Order Aulopiformes
Superoder Scopelomorpha
Order Myctophiformes
Superorder Lampridiomorpha
Order Lampridiformes
Superorder Polymixiomorpha
Order Polymixiiformes
Superorder Paracanthopterygii
Order Percopsiformes
Order Ophidiiformes
Order Gadiformes
Order Batrachoidiformes
Order Lophiiformes
Superorder Acanthopterygil
Order Mugiliformes
Order Atherinomorpha
Order Beloniformes
Order Cyprinodontiformes
Order Stephanoberyciformes
Order Beryciformes
Order Zeiformes
Order Gasterosteiformes
Order Synbranchiformes
Order Scorpeaniformes
Order Perciformes
Order Pleurnectiformes
Order Tetraodontiformes
Class Amphibia
Subclass Lissamphibia
Order Gymnophiona
Order Caudata (Urodela)
Order Anura–frogs and toads
Class Reptilia
Subclass Anapsida
Order Testudines
Subclass Diapsida
Order Sphenodonta
Order Squamata
Suborder Lacertilia
Suborder Serpentes
Order Crocodylia
Infraclass Eoaves
Order Struthioniformes
Order Tinamiformes
Infraclass Neoaves
Order Craciformes
Order Galliformes
Order Anseriformes
Order Turniciformes
Order Piciformes
Order Galbuliformes
Order Bucerotiformes
Order Upupiformes
Order Trogoniformes
Order Coraciiformes
Order Coliiformes
Order Cuculiformes
Order Psittaciformes
Order Apodiformes
Order Trochiliformes
Order Musophagiformes
Order Strigiformes
Order Columbiformes
Order Grulformes
Order Ciconliformes
Suborder Charadrii
Suborder Ciconii
Order Passeriformes
Class Mammalia (Synapsida)
Order Monotremata
Order Didelophimorphia
Order Paucituberculata
Order Microbiotheria
Order Dasyuromorphia
Order Peramelemorphia
Order Notoryctemorphia
Order Diprotodontia
Order Xenarthra
Order insectivora
Order Scandentia
Order Dermoptera
Order Chiroptera
Order Primates
Order Carnivora
Order Cetacea
Order Sirenia
Order Proboscidea
Order Perissodactyla
Order Hyracoidea
Order Tubulidentata
Order Artiodactyla
Order Pholidota
Order Rodentia
Suborder Sciurognathi
Suborder Hystricognathi
Order Lagomorpha
Order Macroscelidea

[a]Derived from G. M. Garrity et. al., *Taxonomic Outline of the Prokaryotes*, Release 2, January 2002, Springer-Verlag. New York. http://dx.doi.org/10.1007/bergeysouthline. Readers interested in determning taxonomic composition of lower taxa may obtain this document free of charge.
[b]Condensed from Jan A. Pechenik, *Biology of the Invertebrates*, 4th ed., McGraw-Hill, 2000.
[c]Condensed from Donald Linzey. *Vertebrate Biology*, Appendix I: Classification of Living Vertebrates, McGraw-Hill, 2001.

Note: The contributions of the following to the updating of this classification scheme are gratefully acknowledged: Dr. Craig Balley, Dr. Mark Chase, Dr. George M. Garrity, Dr. S.C. Jong, Dr. Robert Knowlton, Dr. Donald Linzey.

Soil orders

Order	Formative element in name	General nature
Alfisols	alf	Soils with gray to brown surface horizons, medium to high base supply, with horizons of clay accumulation; usually moist, but may be dry during summer
Aridisols	id	Soils with pedogenic horizons, low in organic matter, and usually dry
Entisols	ent	Soils without pedogenic horizons
Histosols	ist	Organic soils (peats and mucks)
Inceptisols	ept	Soils that are usually moist, with pedogenic horizons of alteration of parent materials but not of illuviation
Mollisols	oll	Soils with nearly black, organic-rich surface horizons and high base supply
Oxisols	ox	Soils with residual accumulations of inactive clays, free oxides, kaolin, and quartz; mostly tropical
Spodosols	od	Soils with accumulation of amorphous materials in subsurface horizons
Ultisols	ult	Soils that are usually moist, with horizons of clay accumulation and a low supply of bases
Vertisols	ert	Soils with high content of swelling clays and wide deep cracks during some seasons

Carbon cycle

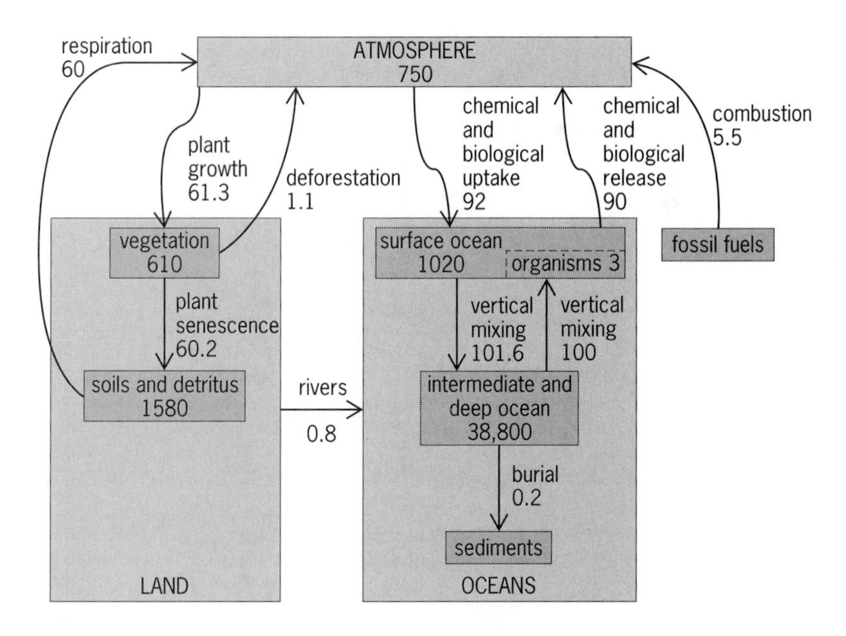

The storage of carbon in the atmosphere and terrestrial and oceanic ecosystems is in picograms (1 Pg=10^{15} g), and fluxes of carbon between boxes are in Pg y^{-1}.

Nitrogen cycle

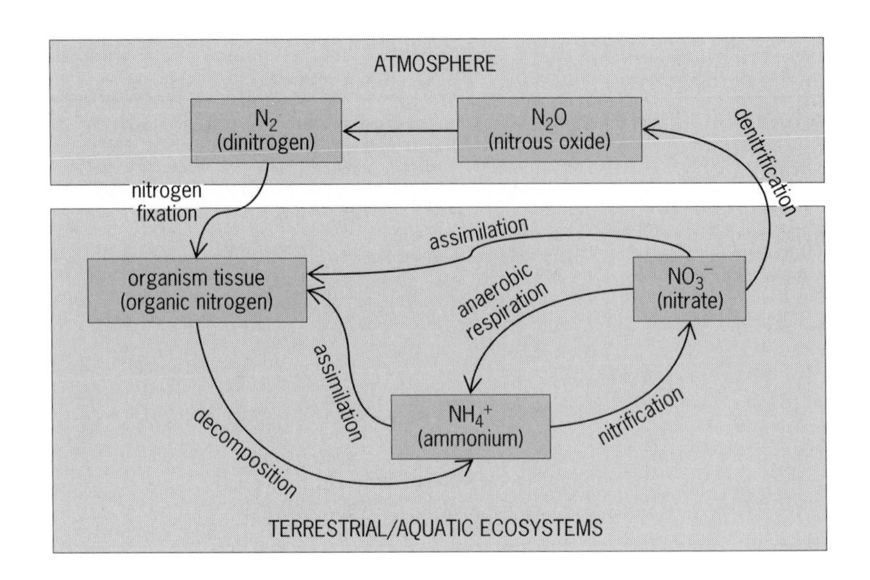

Structure of the atmosphere

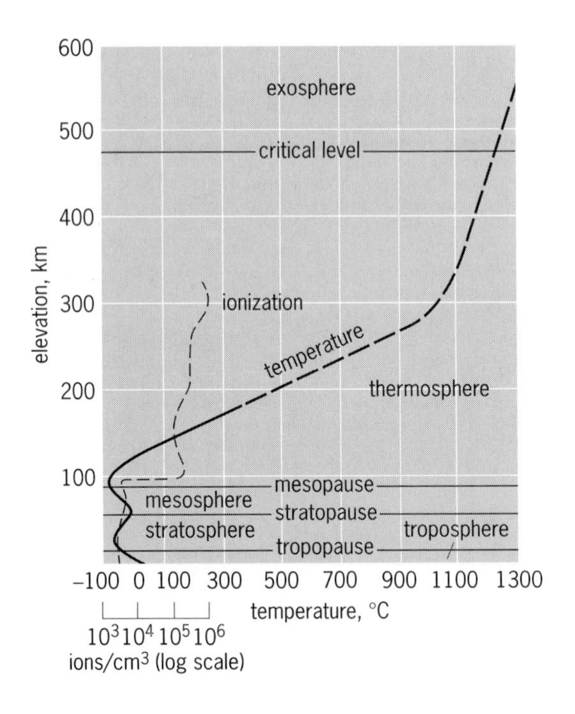

The log scale applies only to the ionization curve. 1 km = 0.6 mi; °F = (°C × 1.8) + 32.

Major sources and types of indoor air pollutants

Sources	*Pollutants*
Combustion with appliances using fossil fuels or wood	Particulate matter Nitrogen oxides Carbon monoxide Carbon dioxide Lead and trace metals Hydrocarbons Volatile organic compounds
Tobacco smoking	Particulate matter Carbon monoxide Carbon dioxide Nitrogen oxides Hydrocarbons Volatile organic compounds Radon progeny
Building and furnishing materials	Hydrocarbons (especially aldehydes) Volatile organic compounds Particulate matter Radon progeny Molds and other allergens
Water reservoirs (fixtures for air conditioning, cleaning, or treating)	Molds Bacilli and other bacteria
Consumer products	Halogenated hydrocarbons Volatile organic compounds Trace metals
Animals (pets and opportunistic dwellers) and plants	Allergens Carbon dioxide
Infiltration	Particulate matter Nitrogen oxides Sulfur oxides Pollen Molds

Major categories of water pollutants*

Category	Examples	Sources
A. Causes of health problems		
1. Infectious agents	Bacteria, viruses, parasites	Human and animal excreta
2. Organic chemicals	Pesticides, plastics, detergents, oil, and gasoline	Industrial, household, and farm use
3. Inorganic chemicals	Acids, caustics, salts, metals	Industrial effluents, household cleansers, surface runoff
4. Radioactive materials	Uranium, thorium, cesium, iodine, radon	Mining and processing of ores, power plants, weapons production, natural sources
B. Causes of ecosystem disruption		
1. Sediment	Soil, silt	Land erosion
2. Plant nutrients	Nitrates, phosphates, ammonium	Agricultural and urban fertilizers, sewage, manure
3. Oxygen-demanding wastes	Animal manure and plant residues	Sewage, agricultural runoff, paper mills, food processing
4. Thermal	Heat	Power plants, industrial cooling

*Reproduced with permission from W. P. Cunningham et al., *Environmental Science: A Global Concern*, 7th ed., McGraw-Hill, 2003.

Top fifteen hazardous substances, 2003

Substance	*Source*	*Toxic effects*
Arsenic	From elevated levels in soil or water	Multiple organ systems affected. Heart and blood vessel abnormalities, liver and kidney damage, impaired nervous system function.
Lead	Lead-based paint Lead additives in gasoline	Neurological damage. Affects brain development in children. Large doses affect brain and kidneys in adults and children.
Metallic mercury	Air or water at contaminated sites	Permanent damage to brain, kidneys, developing fetus.
Vinyl chloride	Plastics manufacturing Air or water at contaminated sites	Acute effects: dizziness, headache, unconsciousness, death. Chronic effects: liver, lung, and circulatory damage.
Polychlorinated biphenyls (PCBs)	Eating contaminated fish Industrial exposure	Probable carcinogens. Acne and skin lesions.
Benzene	Industrial exposure Glues, cleaning products, gasoline	Acute effects: drowsiness, headache, death at high levels. Chronic effects: damages blood-forming tissues and immune system; also carcinogenic.
Cadmium	Released during combustion Living near a smelter or power plant Picked up in food	Probable carcinogen, kidney damage, lung damage, high blood pressure.
Polycyclic aromatic hydrocarbons	Tobacco smoke and charbroiled meats	Probable carcinogens. Difficulty reproducing and possible birth defects.
Benzo[a]pyrene	Product of combustion of gasoline or other fuels In smoke and soot	Probable carcinogen, possible birth defects.
Benzo[b]fluoranthene	Product of combustion of gasoline and other fuels Inhaled in smoke	Probable carcinogen.
Chloroform	Contaminated air and water Many kinds of industrial settings	Affects central nervous system, liver, and kidneys; probable carcinogen.
DDT	From food with low levels of contamination Still used as pesticide in parts of world	Probable carcinogen; possible long-term effect on liver; possible reproductive problems.
Aroclor 1254 (a mixture of PCBs)	From food and air	Probable carcinogens Acne and skin lesions.
Aroclor 1260 (a mixture of PCBs)	From food and air	Probable carcinogens Acne and skin lesions.
Dibenz[a,h]anthracene	Incomplete burning of coal, oil, gas, wood, tobacco	Known animal carcinogen; probable human carcinogen.

SOURCE: Data from Agency for Toxic Substances and Disease Registry. (Adapted from E. D. Enger and B. F. Smith, *Environmental Science: A Study of Interrelationships*, 8th ed., McGraw-Hill, 2002)

BIOGRAPHICAL LISTING

Adanson, Michel (1727–1806), French naturalist. Classified plants in his *Les Familles Naturelles des Plantes.*

Afzelius, Adam (1750–1837), Swedish botanist. Founded the Linnaean Institute.

Agassiz, Jean Louis Rudolphe (1807–1873), Swiss-born American naturalist. Wrote books on ichthyology, especially relating to classification.

Andrews, Roy Chapman (1884–1960), American naturalist. Discovered many plant and animal fossils.

Audubon, John James (1785–1851), Haitian-born American ornithologist and artist. Made drawings and paintings of birds and animals.

Bang, Bernhard Laurits Frederik (1848–1932), Danish veterinarian. Discovered method of eradicating bovine tuberculosis; discovered *Brucella abortus*, the agent of contagious abortion (Bang's disease) and brucellosis.

Bassham, James Alan (1922–), American chemist. Helped to elucidate basic photosynthetic carbon cycle.

Bates, Henry Walter (1825–1892), English naturalist. Discovered Batesian mimicry among butterflies and moths.

Beebe, Charles William (1877–1962), American naturalist. Pioneer in deep-sea exploration; made ornithological collections.

Behring, Emil Adolph von (1854–1917), German bacteriologist. Produced diphtheria and tetanus antitoxins; Nobel Prize, 1901.

Bentham, George (1800–1884), English botanist. With J. Hooker, wrote *Genera Plantarum.*

Bergey, David Hendricks (1860–1937), American bacteriologist. Authority on classification of bacteria.

Blumberg, Baruch Samuel (1925–), American physician and biologist. Research leading to a test for hepatitis viruses in blood and to an experimental hepatitis vaccine; Nobel Prize, 1976.

Bose, Jagadis Chandra (1858–1937), Indian plant physiologist and physicist. Founded the Bose Research Institute in Calcutta; investigated photosynthesis, "nervous mechanism" of plants, and other plant subjects.

Buffon, Georges Louis Leclerc, Comte de (1707–1788), French naturalist. Compiled *Histoire Narurelle*, a monumental work on natural history.

Burbank, Luther (1849–1926), American horticulturist. Experimented on crossing and in-breeding of plant varieties.

Calvin, Melvin (1911–1997), American chemist. With J. A. Bassham, traced the path of carbon in photosynthesis; Nobel Prize, 1961.

Carver, George Washington (1864–1943), American botanist. Did research on industrial uses of the peanut.

Cesalpino, Andrea (1519–1603), Italian physician and botanist. First to attempt to classify plants according to characteristics of fruit and seed in his *De Plantis.*

Cohn, Ferdinand Julius (1828–1898), German botanist. A founder of bacteriology; did research in plant pathology; first to classify bacteria according to genus and species.

Crutzen, Paul J. (1933–), Dutch chemist. Demonstrated that nitrogen oxides react catalytically with ozone, thus accelerating the rate of reduction of the ozone layer; Nobel Prize, 1995.

Cuvier, Georges Léopold Chrétien Frédéric Dagobert, Baron (1769–1838), French naturalist. Made a detailed classification of the animal kingdom; wrote on comparative anatomy.

Dam, Carl Peter Henrik (1895–1976), Danish biochemist and nutritionist. Discovered vitamin K and studied its role in human hemorrhagic disease; Nobel Prize, 1943.

Darwin, Charles Robert (1809–1882), English naturalist. Proposed far-reaching theory of evolution of species and theory of natural selection in his *Origin of Species.*

De Vries, Hugo (1848–1935), Dutch botanist. Formulated the mutation theory of evolution.

Dick, George Frederick (1881–1967), American physician and bacteriologist. Isolated scarlet fever streptococci, developed scarlet fever streptococcus antitoxin, and developed Dick test.

Dobzhansky, Theodosius (1900–1975), Russian-born American biologist. Elucidated the mechanisms of heredity and variation through studies of *Drosophila.*

Fanning, John Thomas (1837–1911), American civil engineer. Designed water works and water supply systems.

Frisch, Karl von (1886–1982), Austrian zoologist. Discovered means by which bees communicate information about the distance and direction of food; Nobel Prize, 1973.

Gajdusek, Daniel Carleton (1923–), American physician and virologist. Discovered causal virus and transmission mechanism of kuru; Nobel Prize, 1976.

Gesner, Konrad von (1516–1565), Swiss naturalist. Wrote *Historia Animalium*, beginning zoology as a science.

Haldane, John Bourdon Sanderson (1892–1964), British geneticist and physiologist. Pioneered in mathematical treatment of population genetics; studied respiration in humans; wrote about enzymes.

Hooker, Joseph Dalton (1817–1911), English botanist. With G. Bentham, wrote *Genera Plantarum*; prepared works on the flora of New Zealand, Antarctica, and India.

Humboldt, Friedrich Heinrich Alexander, Baron von (1769–1859), German naturalist. Founder of physical geography; made scientific explorations of South America and Central Asia.

Ingenhousz, Jan (1730–1799), Dutch physician and naturalist. Demonstrated the cycle of photosynthesis in plants.

Jussieu, Antoine Laurent de (1748–1836), French botanist. Wrote *Genera Plantarum*, the basis of modern natural botanical classification.

Kitasato, Shibasaburo (1852–1931), Japanese bacteriologist. Independently of A. E. J. Yersin, discovered the bacillus of bubonic plague; isolated bacilli of symptomatic anthrax, dysentery, and tetanus.

Klebs, Edwin (1834–1913), German pathologist. Described diphtheria (Klebs-Löffler) bacillus; studied bacteriology of malaria, anthrax, and tuberculosis.

Koch, Robert (1843–1910), German physician and bacteriologist. Studied cholera, tuberculosis, and bubonic plague; showed a specific bacillus to be the cause of anthrax; discovered the tubercle bacillus; Nobel Prize, 1905.

Lamarck, Jean Baptiste Pierre Antoine de Monet, Chevailer de (1744–1829), French naturalist. Proposed theory that changes in animal and plant structure are caused by changes in environment; classified animals into vertebrates and invertebrates.

Laveran, Charles Louis Alphonse (1845–1922), French physician. Discovered the malaria parasite; researched sleeping sickness; Nobel Prize, 1907.

Linnaeus, Carolus, real name Carl von Linné (1707–1778), Swedish botanist. Developed the Linnaean system of biological classification.

Löffler, Friedrich August Johannes (1852–1915). German bacteriologist. Isolated the diphtheria (Klebs-Löffler) bacillus; developed protective serum against foot-and-mouth disease.

Lorenz, Konrad Zacharias (1903–1989), Austrian zoologist. Pioneered in study of animal behavior patterns; discovered imprinting in birds; Nobel Prize, 1973.

Lyell, Charles (1797–1875), British geologist. Wrote *Principles of Geology*, refuting catastrophic theory of geological changes.

Mendel, Gregor Johann (1822–1884), Austrian botanist. Formulated Mendel's laws of heredity, the foundation of genetics.

Metchnikoff, Élie (1845–1916), Russian-born French zoologist and bacteriologist. Work on cholera and immunology; Nobel Prize, 1908.

Mohl, Hugo von (1805–1872), German botanist. Worked on the anatomy and physiology of higher plant forms; discovered protoplasm.

Molina, Mario J. (1943–), American chemist. Demonstrated that chemically inert chlorofluorocarbon (CFC) could be transported up to the ozone layer and could react with ultraviolet light and deplete the ozone layer; Nobel Prize, 1995.

Muller, Hermann Joseph (1890–1967), American geneticist. Studied genetic mutation rates under natural and artificial conditions; discovered the effect of x-rays on mutation rate; Nobel Prize, 1946.

Müller, Paul Hermann (1899–1965), Swiss chemist. Discovered the insecticidal properties of DDT; Nobel Prize, 1948.

Nicolle, Charles Jules Henri (1866–1936), French physician. Discovered the louse to be the transmission vector of typhus; Nobel Prize, 1928.

Noguchi, Hideyo (1876–1928), Japanese bacteriologist. First to produce pure cultures of syphilis spirochetes; discovered the parasite of yellow fever.

Orr, John Boyd, Baron (1880–1971). Scottish physiologist and nutritionist. Work on animal nutrition; pioneer in science of human nutrition; Nobel Peace Prize, 1949.

Pearson, Karl (1857–1936), English applied mathematician, statistician, and biometrician. Pioneered in application of statistics to biology; introduced chi-square test.

Pfeiffer, Richard Friedrich Johann (1858–1945), German bacteriologist. Discovered Pfeiffer's bacillus in influenza; described Pfeiffer's reaction for determination of cholera.

Piccard, Auguste (1884–1962), Swiss physicist. Conducted a data-collecting exploration of the stratosphere in an airtight gondola of a balloon; constructed and tested a bathysphere for deep-sea exploration.

Prusiner, Stanley B. (1942–), American neurologist. Discovered prions, a new biological agent of infection; Nobel Prize, 1997.

Ray or Wray, John (1627?–1706), English naturalist. Identified the difference between mono- and dicotyledons; arranged plants according to their natural form, the foundation of the natural system of classification.

Ross, Ronald (1857–1932), British physician. Proved that malaria is transmitted by the female *Anopheles* mosquito; Nobel Prize, 1902.

Roux, Pierre Paul Emile (1853–1933), French physician and bacteriologist. Helped develop modern serum therapeutics, especially concerning diphtheria.

Sachs, Julius von (1832–1897), German botanist. Studied the connection between sunlight and chlorophyll; worked on heliotropism and geotropism.

Theiler, Max (1899–1972), South African physician and virologist. Developed a vaccine to prevent human yellow fever; Nobel Prize, 1951.

Tinbergen, Nikolaas (1907–1988), Dutch-born British zoologist. Pioneered in study of social behavior of animals and their responses to complex stimuli; conducted experimental studies of the effects of selection pressures and evolutionary response to them; Nobel Prize, 1973.

Wallace, Alfred Russel (1823–1913), English naturalist. Originated, independently of C. Darwin, theory of natural selection; postulated Wallace's line regarding geographical distribution of animals.

Wegener, Alfred Lothar (1880–1930), German geologist. Presented the idea of continental drift.

Weller, Thomas Huckle (1915–), American virologist and parasitologist. Isolated the virus of chickenpox and herpes zoster and proved the common etiology of the two diseases; first to propagate German measles virus; Nobel Prize, 1954.

Williamson, William Crawford (1816–1895), English naturalist. Laid the foundation for paleobotany and showed the importance of plant life forms in coal.

Wright, Almroth Edward (1861–1947), British physician and pathologist. Studied parasitic disease; introduced inoculation against typhoid.

Yersin, Alexandre Émile Jean (1863–1943), Swiss bacteriologist. Discovered the bubonic plague bacillus in Hong Kong, working independently of S. Kitasato, and developed a serum for it.

Zinsser, Hans (1878–1940), American bacteriologist. Developed methods of immunization against typhus.

Contributor Initials

Each article in the Encyclopedia is signed with the contributor's initials. This section gives all such initials. The contributor's name is provided. The contributor's affiliation can then be found by turning to the next section.

A

A.A.L. Andrew A. Lacis
A.B.McG. Ann B. McGuire
A.Ci. Alex Ciegler
A.Cou. Arnold Court
A.Cr. Arthur Cronquist
A.C.G. Arch C. Gerlach
A.C.L. Alcinda C. Lewis
A.D.H. Arthur D. Hasler
A.D.M. Andrew D. Miall
A.E.Fr. A. E. Freeman
A.E.Mo. Alice E. Moore
A.G.F. Alfred G. Fischer
A.G.L. Allan G. Lochhead
A.G.M. Antonios G. Mikos
A.Has. Alan Hastings
A.H.G. Arthur H. Graves
A.H.St. Alfred H. Sturtevant
A.I.D. Alex I. Donaldson
A.J.F.G. A. J. F. Griffiths
A.J.T. Aaron J. Teller
A.Ko. Allan Konopka
A.L. Anton Lang
A.Lo. Alan Longacre
A.L.Bi. A. L. Bisno
A.L.deW. A. L. de Week
A.L.G. Arnold L. Gordon
A.L.Ko. Arthur L. Koch
A.L.She. Albert L. Sheffer
A.M.M. Alexander M. Mood
A.M.We. Adrian M. Wenner
A.N.S. Albert N. Sayre
A.P.A. Arnold P. Appleby
A.R.A. Arthur R. Ayers
A.R.M. Armand R. Maggenti
A.To. Alfonso Torres
A.We. Alison Weiss
A.Wu. Alfred Wüest
A.W.C. Arthur W. Cooper
A.W.Co. Anthony W. Confer
A.W.C.V.G. Alexander W. C. von Graevenitz
A.W.G. Arthur W. Galston
A.W.H.D. Antoni W. H. Damman

B

B.A.M. Brian A. Mauer
B.A.S. B. A. Summers
B.B.Ch. Bruno B. Chromel
B.B.D. Bruce B. Dan
B.C.S. B. C. Sutton
B.E. Baltus Erasmus
B.E.F. Blake E. Feist
B.E.Go. Barry E. Goodison
B.E.S.G. Brian E. S. Gunning
B.F. Bernard Frank
B.F.W. B. F. Wilson
B.Hi. Bruce Hicks
B.J.D.M. Bastiaan J. D. Meeuse
B.J.K. Brian J. Kilbey
B.J.M. Basil J. Mason
B.J.S. Brian J. Skinner
B.K. Blair Kinsman
B.L.B. Blaine L. Blad
B.L.I. B. Lynn Ingram
B.M.Ch. Bruce M. Christiansen
B.R.R. Brian R. Rust
B.S.M. Bernard S. Meyer
B.T.B. Bernard T. Bormann
B.Wi. Brian Wilkinson
B.W.C. Bruce W. Calnek
B.Y.H.L. Benjamin Y. H. Liu
B.Y.Ta. Bernard Y. Tao
B.Z.B. Barbara Zakrewska Borowieka

C

C.An. Christian Andreasen
C.A.F. Charles A. Francis
C.A.H. Carol A. Hoffman
C.A.Me. Charles A. Mebus
C.A.Mo. Charles A. Moore
C.A.St. Carl A. Strausbaugh
C.B.C. Charles B. Curtin
C.B.O. Charles B. Ogburn
C.B.P. Cornelius B. Philip
C.B.V.N. Cornelis B. Van Niel
C.C.K.L. Clark C. K. Liu
C.C.L. Ching Chun Li
C.E.LaM. C. E. LaMotte
C.F.N. Charles F. Niven, Jr.
C.G.J. Clive G. Jones
C.Ha. Carlyn Halde
C.H.S. Christopher H. Scholz
C.In. Chuck Ingels
C.J.I. Charles J. Issel
C.J.Ka. Clarence J. Kado
C.L.W. C. Langdon White
C.Qu. Carol Quaife
C.R.Ca. C. Ronald Carroll
C.Sc. Cheryl Schultz
C.S.M. Craighton S. Mauk
C.V. Charles Vitek
C.V.C. Charles V. Crittenden
C.Wi. Conrad Wickstrom
C.W.B. Carl W. Boothroyd
C.W.N. Chester W. Newton

D

D.A.Be. David A. Bemis
D.A.Ja. Daniel A. Japuntich
D.A.Sa. Deborah A. Salazar
D.B. David Barr
D.B.Lo. D. B. Loope
D.B.So. David B. South
D.C.C. David C. Coleman
D.C.F. Derek C. Ford
D.C.M. David C. Morrison
D.D.E. Dennis D. Eberl
D.D.Fo. Dennis D. Focht
D.D.H. David D. Houghton
D.D.J. Dennis D. Juranek
D.E.G. David E. Giannasi
D.E.H. Dennis E. Hayes
D.E.Ki. David E. Kissel
D.E.No. Donald E. Norris
D.E.Te. Dennis E. Teeguarden
D.Fo. Derek Ford
D.F.Ow. Denis F. Owen
D.Ge. Dennis Genito
D.He. Donald Heyneman
D.J.D. Daniel J. Diekema: Drug resistance
D.J.D. David J. Dabbs: Herpes
D.J.Dep. Donald J. DePaulo
D.J.E. Don J. Easterbrook
D.J.Ev. Doyle J. Evans, Jr.
D.J.Fu. Douglas J. Futuyma
D.J.Hor. David J. Horn
D.J.J. Daniel J. Jones
D.J.P. Donald J. Patton
D.J.Pi. Donald J. Pinkava
D.J.R. Dewey J. Raski
D.L.B. Daniel L. Birkenheuer
D.L.Ha. Dennis L. Hartmann
D.L.J. Donald Lee Johnson
D.L.L. Dan L. Longo
D.L.Pa. Darwin L. Palmer
D.Lam. Dennis Lamb
D.M.C. D. Martinez-Carrera
D.M.DeL. Dwight M. DeLong
D.M.F. David M. Farmer
D.M.Gi. D. Michael Gill
D.M.S. David M. Smith
D.M.Se. Denise M. Seliskar
D.N.La. Daniel N. Lapedes
D.Sim. Daniel Simberloff
D.Sk. David Skelly

E

E.A. Elihu Abrahams
E.A.A. Edward A. Ackerman
E.A.M. Eugene A. Milus
E.C.P. E. C. Pielou
E.D.Gr. Ernest D. Gray
E.F.B. Edmour F. Blouin
E.Gry. Ellen Gryj
E.G.H. E. G. Heyne
E.G.N. Elton G. Nelson
E.G.St. Edward G. Stuart
E.H.Ha. Edwin H. Hammond
E.I. Earl Ingerson
E.J.Ba. Eric J. Barron

E.Ke. Edwin Kessler
E.K.B. Elizabeth K. Berner
E.L.C. Earl L. Core
E.L.Cl. Elmond L. Claridge
E.L.L. Elbert L. Little, Jr.
E.L.Wi. Edward L. Winterer
E.M.R. Eugene M. Rasmusson
E.N.B. Eddie N. Bernard
E.O.B. Elgene O. Box
E.O.W. Edward O. Wiley
E.P.Go. Eric P. Goosby
E.P.K. E. Philip Krider
E.Ros. Eugene Rosenberg
E.R.B. Elery R. Becker
E.R.R. Elmar R. Reiter
E.S.S. Edward S. Sarachik
E.W.V. Edward W. Voss, Jr.

F

F.A.M. Frederick A. Murphy
F.B.Go. Frank B. Golley
F.B.S. Frank B. Salisbury
F.B.Sh. Frederick B. Shuman
F.C.D. Franklin C. Daiber
F.C.E. Francis C. Evans
F.H.Lu. Frank H. Ludlam
F.H.M. F. H. May
F.L.Sc. Frederick L. Schuster
F.M.D. Frank Matthews Dugan
F.S. Frederick Sanders
F.S.B. Francis H. Brown
F.S.Z. Frederick S. Zbar
F.T.A. Fredrich T. Addicott
F.T.L. F. Thomas Ledig
F.V.B. Ferdinando V. Boero

G

G. Govindjee
G.An. Giuseppe Angrisano
G.A.Ma. George A. Maul
G.A.Mo. Gary A. Moll
G.Bo. Gerard Bond
G.Boh. Greg Bohach
G.B.L. Glen B. Lesins
G.C.G. Gerard C. Gambs
G.D.S. Guy D. Smith
G.E.Li. Gene E. Likens
G.Fau. Gunter Faure
G.F.L. George F. Ludvik
G.F.P. George F. Pinder
G.F.S. G. F. Sprague
G.G. Gerald Gilardi
G.G.A. George G. Adkins
G.H.C. Gilbert H. Cady
G.H.D. George H. Davis
G.H.O. Gordon H. Orians
G.J. Giles Johnson
G.J.Ga. Gerald J. Gray
G.J.R. Gerald J. Romick
G.J.W. George J. Wilder
G.K.S. Gerald K. Sims
G.Lo. George Lordi
G.L.Gi. Gerald L. Gilardi
G.M.L. George M. Lordi
G.M.S. George M. Savage
G.P. George Papageorgiou
G.Pa. Gerald Palevsky
G.Pr. George Press
G.R.Ca. G. R. Carter
G.R.N. Gerald R. North
G.Sp. Garrison Sposito
G.S.H. Gene S. Helfman
G.T. Geza Teleki
G.Tc. George Tchobanoglous
G.Th. Gary Thomann
G.V.M. Gerard V. Middleton
G.We. Gunter Weller
G.W.Bur. Glenn W. Burton
G.W.Sa. Gary W. Saunders
G.Z. Geoffrey Zubay

H

H.A.Fr. Henry A. Fribourg
H.B.B. Howard B. Bluestein
H.B.Pi. Harry B. Pionke
H.B.S. Howard B. Sprague
H.C.Wi. Kurd C. Willett
H.Di. Hugh Dingle
H.D.D. H. D. Dooley
H.D.O. Harold D. Orville
H.E.B. Harold E. Burkhart
H.E.L. H. E. Landsberg
H.E.We. H. E. Wendler
H.Gro. Hallstein Gronstol
H.G.B. Harriet G. Barclay
H.H. Henry Hellmers
H.H.C. Howard H. Chang
H.H.Ha. H. Hunter Handsfield
H.H.St. Henry H. Storch
H.J.P. Herman J. Phaff
H.J.Wi. Herold J. Wiens
H.Li. Helmut Lieth
H.L.D. Herbert L. Dupont
H.L.G.S. Henry L. G. Stroyan
H.L.Sc. Hilary Lappin-Scott
H.M.F. Harry M. Freeman
H.P.W. Hilda P. Willett
H.S.L. Harry S. Ladd
H.Wh. Harry Wheeler
H.Wi. Hugh Wilcox
H.W.J. Holger W. Jannasch

I

I.L.B. I. Lehr Brisbin, Jr.
I.P.T. Irwin P. Ting
I.Z. Israel Zelitch
I.Zl. I. Zlotnik

J

J.A.Do. James A. Doyle
J.A.McG. John A. McGinley
J.Bre. Joel Breman
J.B.D. John B. Dunning, Jr.
J.B.Da. James B. Dale
J.B.Jo. Jeremy B. Jones
J.B.L. James B. Lackey
J.B.Wi. James B. Wilson
J.Cr. Joel Cracraft
J.C.Co. Jonathan C. Comer
J.C.Ri. J. C. Ritchie
J.C.Ro. Jerome C. Rozen, Jr.
J.C.Sc. Jack C. Schultz
J.C.So. James C. Solomon
J.D.Jo. Jon D. Johnson
J.D.MacD. J. D. MacDougall
J.E.Ba. John E. Banks
J.E.F. James E. Ferrell
J.E.K. John E. Kutzbach
J.E.O. John E. Oliver
J.E.O'L. John E. O'Leary
J.E.S. James E. Schindler
J.F.Cr. James F. Crow
J.F.F. James F. Ferry
J.F.Hu. John F. Hubert
J.F.L. J. F. Loneragan
J.F.M. Jose F. Maldonado-Moll
J.F.P. James F. Price
J.F.T. John F. Timoney
J.Ga. John Garland
J.G.F. J. G. Futral
J.G.H. James G. Horsfall
J.G.S. James G. Speight
J.Hal. John Hallett
J.Hu. Jack Huebler
J.H.Car. Jack H. Carlson
J.H.Fi. Joseph H. Field
J.H.G. Joseph H. Golden
J.J. Jules Janick
J.J.Ka. James J. Kasmierczak
J.J.Mo. Jeffrey J. Morrell
J.K.St. Jeffrey K. Stone
J.Ly. John Lyman
J.L.Ba. Jeffrey L. Bada
J.L.K. J. Laurence Kulp
J.L.McH. J. L. McHugh
J.L.Me. Joseph L. Melnick
J.L.Re. Joseph L. Reid
J.L.S. John L. Sabo
J.L.Sto. Jeffrey L. Stott
J.L.Wi. James L. Wilson
J.M. Jonathan Marks
J.Ma. Jack Major
J.Marr. John Marr
J.M.La. John M. Last
J.M.N. Jeanette M. Norton
J.M.Q. John M. Quarles
J.M.S. Jay M. Savage
J.M.Tr. James M. Trappe
J.M.Wa. John M. Wallace
J.N. Jerome Namias
J.N.Lu. James N. Luthin
J.N.R. J. N. Rutger
J.O.C. John O. Corliss
J.Po. John Postgate
J.P.G. J. P. Gerrity
J.P.M. J. Patrick Muffler
J.P.M. Joseph P. Mascarenhas: Gametogenesis
J.P.W. Jon P. Woods
J.Ro. John Ross
J.R.Ba. J. R. Bandoni
J.R.C. Joseph R. Curray
J.R.Co. James R. Cochran
J.R.F. J. R. Fulks
J.R.G. John R. Gyakum
J.R.H. James R. Holton
J.R.See. John Richard Seed
J.S. Julius Schachter
J.Ses. John Sessions

J.Sim. Joanne Simpson
J.S.C. John S. Campbell
J.S.Fe. Jay S. Fein
J.S.H. Jeffrey S. Hanor
J.T.Le. J. T. Lee
J.T.O. James T. Oris
J.T.P. Judith Totman Parrish
J.Wa. Johm Warren
J.W.Be. J. W. Bennett
J.W.C. J. W. Costerton
J.W.Ez. John W. Ezzell
J.W.Fe. Jack W. Fell
J.W.Gei. John W. Geissman
J.W.M. James W. Murray
J.W.P. Jae-Woo Park
J.W.V. James W. Valentine

K

K.A. Knut Aagaard
K.At. Kenji Atoda
K.Bu. Kevin Burke
K.B.C. Kenneth B. Cumberland
K.E. Katherine Esau
K.G. K. Grasshoff
K.H.M. Kenneth H. Mann
K.H.N. Kenneth H. Nealson
K.J.H. Katherine J. Hansen
K.J.R. Kenneth J. Ryan
K.M.K. Katherine M. Kocan
K.M.M. Kenneth M. Marsh
K.M.S. Kaye M. Shedlock
K.N.L. K. N. Liou
K.N.Z. Klaus Napp-Zinn
K.P.A. Kenneth P. Able
K.P.D. Kenneth P. Davis
K.Ru. Kathryn Ruoff
K.R.D. K. R. Dyer
K.T.C. Kang-tsung Chang
K.V.S. Karlene V. Schwartz
K.W. Klaus Wyrtki
K.W.F. Karl W. Flessa

L

L.B. Lane Barksdale
L.B.L. Luna B. Leopold
L.B.R. Lee B. Reichman
L.Co. Lee Conch
L.C.Le. Lyndon C. Lee
L.E.Mo. Leonard E. Mortenson
L.F.B. Lance F. Bosart
L.H.P. L.H. Princen
L.J.A. Leslie J. Audus
L.M.C. Lucy M. Cranwell
L.M.Ca. Lori M. Carris
L.P.R. Louis P. Reitz
L.R.G. Leah R. Gerber
L.S.McC. Leland S. McClung
L.V.M. Larry V. McIntire
L.V.S. Lola V. Stamm

M

M.A.A. Michael A. Adewumi
M.A.H. Mel A. Hagood
M.A.La. Meredith A. Lane
M.A.Ma. Michelle A. Marvier
M.A.W. Michael A. Walsh
M.Ch. Martha Christensen
M.De. Margaret Deacon
M.Do. Mike Doyle
M.D.Bre. Michael D. Breen
M.Ei. Millicent Eidson
M.E.Re. M. E. Reichmann
M.F.F. Michael F. Fay
M.G.Ma. Maria G. Masucci
Mo.H. Monto Ho
M.Hag. Maura Hagan
M.H.G. Michael H. Glantz
M.H.H. Matthew H. Hitchman
M.J.Cou. Mary J. Coulombe
M.J.G.A. M. J. G. Appel
M.J.J. Mark J. Johnson
M.J.Se. M. J. Selby
M.Kam. Michael Kamrin
M.Kl. Maren Klich
M.Ko. Michelle Kominz
M.La. Meredith Lane
M.L.J. Meredith L. Jones
M.L.V. Mary Lynne Vickers
M.McC. Michelle McClure
M.M.Mi. Maynard M. Miller
M.Pa. Miguel Pascual
M.P.A.J. M. P. A. Jackson
M.Ru. Manfred Ruddat
M.R.C. Michael R. Cummings
M.R.F. Mark Roy Fuller
M.R.J.S. M. R. J. Salton
M.S.A. Mark S. Ashton
M.T.H. Michel T. Halbouty
M.V.B. M. V. Brian
M.Wi. M. Wilkening
M.W.C. Mark W. Chase
M.W.T. Margaret W. Thompson
M.Z. Milton Zaitlin

N

N.A. Nelle Ammons
N.B.M. Norman B. Marshall
N.C.P. Niels C. Pedersen
N.D.L. Norman D. Levine
N.G.D. Nancy G. Dengler
N.H.B. Norman H. Boke
N.I. Nakao Ishida
N.J.Go. Nicholas J. Gottelli
N.K.M. N. Karle Mottet
N.L.N. Neil L. Norcross
N.Sc. Nathaniel Scholz
N.S.Gr. Neil S. Grigg
N.W.R. Norman W. Radforth

O

O.E.N. Olin E. Nelsen
O.H. Olga Hartman

P

P.A.Ma. Preston A. Marx
P.B. Pedro Barbosa
P.B.C. Philip B. Cowles
P.Cl. Preston Cloud
P.C.H. Perry C. Holt
P.C.Si. P. C. Silva
P.D. Patricia Dell'Arciprete
P.D.St. Perry D. Strausbaugh
P.D.W. P. D. Walton
P.Fr. Peter Frumhoff
P.F.C. Philip F. Clapp
P.G.Ri. Paul G. Risser
P.J.D. Peter J. Davies
P.J.McN. Peter J. McNamara
P.J.S. Perry J. Sampson
P.J.Z. Paul J. Zwerman
P.Ka. Peter Kareiva
P.K.M. Pamela K. Mulligan
P.L.Fi. Peggy L. Fiedler
P.L.S. Peter L. Steponkus
P.McG. Prestor McGain
P.M.Ka. P. M. Kareiva
P.M.R. Patricia M. Rodier
P.R. Philip Richardson
P.Ran. Peter Randerson
P.Ri. Paul Risser
P.Ro. Patricia Rosa
P.R.M. Peter R. Marker
P.R.V. Peter R. Vogt
P.Sae. Peter Saenger
P.Sc. Paul Schimmel
P.S.H. Paul S. Hoffman
P.T. Paul Tabor
P.W.P. Peter W. Price

R

R.Au. Robert Austrian
R.A.Ber. Robert A. Berner
R.A.Cor. Robert A. Corbitt
R.A.D. Richard A. Davis, Jr.
R.A.Fin. Richard A. Finkelstein
R.A.Go. Richard A. Goodman
R.A.L. Roy A. Larson
R.B.C. Richard B. Couch
R.B.M. Regis B. Miller
R.Ch. Robert Chen
R.D.J. Robert Davies-Jones
R.D.V. Richard D. Vierstra
R.E.Car. Richard E. Carbone
R.E.D. Robert E. Dickinson
R.E.Ga. Roger E. Garrett
R.E.H. Robert E. Hungate
R.E.Ma. Rex E. Marx
R.E.Sh. Robert E. Sheriff
R.E.T. R. E. Taylor
R.F.B. Robert F. Black
R.F.Bo. R. F. Bozarth
R.F.Co. R. Frank Cook
R.F.Di. Robert F. Dill
R.F.Fl. Richard F. Flint
R.F.H. R. F. Horsnail
R.G. Raymond Giese
R.Go. Ronald Gold
R.Gov. R. Govindjee
R.H. Richard Hazen
R.Hol. Robert Holt
R.H.Do. Robert H. Dodds
R.H.E. Richard H. Eyde
R.I.H. R. I. Hamilton

R.I.T. Robert I. Tilling
R.J.C. Richard J. Campana
R.J.D. Raymond J. Deland
R.J.Mu. R. J. Murgatroyd
R.K.L. Ronald K. Linde
R.K.Li. Ray K. Linsley
R.LeB.H. Roger LeB. Hooke
R.L.A. Richard L. Armstrong
R.L.E. Robert L. Edmonds
R.L.Fi. Robert L. Fisher
R.L.Han. Robert L. Haney
R.L.Hu. Robert L. Hulbary
R.L.L. Ronald L. Lavoie
R.L.M. R. L. Mower
R.L.Moe. Richard L. Moe
R.L.N. Raymond L. Nace
R.L.Re. Russell L. Regnery
R.L.S.M. Robert L. San Martin
R.L.Sm. Robert Leroy Smith
R.Mur. Robert Murphy
R.M.A. Ronald M. Atlas
R.M.Al. R. McNeill Alexander
R.M.Ke. Robert M. Kelly
R.M.M. Richard M. Mitterer
R.M.R. Roger M. Reeve
R.M.Ra. Robert M. Rauber
R.N.H. Rodney N. Hader
R.O.R. Rodney O. Radke
R.P.D. Robertson P. Dinsmore
R.P.H. Richard P. Hall
R.Ri. Ralph Riley
R.R.B. Robert R. Brubaker
R.R.C. Rita R. Colwell
R.R.Sc. Ronald R. Schnabel
R.S. Rudolf Schmid
R.Si. Raymond Siever
R.Sim. Robert Simpson
R.Su. Richard Sudds
R.S.W. Ralph S. Wolfe
R.T.B. Richard T. Barber
R.T.Be. Robert T. Beaumont
R.T.Ha. Richard T. Hanlin
R.T.P. Rachel T. Pinker
R.U. R. Ullrich
R.We. Richard Wetzel
R.W.H. Robert W. Holmes
R.W.Ha. Richard W. Harris
R.W.L. Robert W. Leader
R.W.S. Roy W. Simonson
R.W.Sa. Robert W. Sanders
R.W.T. Raymond W. Tennant

S

S.A.L. Simon A. Levin
S.A.M. Shelby A. Miller
S.A.Mo. Stephen A. Morse
S.A.R. Scott A. Redhead
S.A.St. Steven A. Stage
S.C.B. Simon C. Brassell
S.C.Jo. S. C. Jong
S.D.H. Steacy D. Hicks
S.E. Steven Edwards
S.E.Wh. Sidney E. White
S.Gr. S. Granick
S.G.P. S. George Philander
S.J.McN. Samuel J. McNaughton
S.Lav. Stephen Lavin
S.L.A. Sandra L. Anagnostakis
S.L.M. Steve L. Moseley
S.M.D. S. M. Dilworth
S.M.F. Sydney M. Finegold
S.Pi. Stuart Pimm
S.P.H. Samuel P. Hammer
S.Ra. Shirley Raps
S.Ri. Saul Rich
S.R.Bo. Steven R. Bolin
S.R.E. Steven R. Emerson
S.R.H. Steven R. Hanna
S.R.S. Sergio R. Signorini
S.R.St. Scott R. Stewart
S.S.D. S. Sovonick-Dunford
S.S.S. Shelley S. Sutherland

T

T.B.T. Thomas B. Turner
T.C.F. Thomas C. Fuller
T.C.Wh. T. C. Whitmore
T.C.Wi. Thomas C. Winter
T.D.S. Thomas D. Sharkey
T.E.D. T. E. Dawson
T.H.V.A. Tjeerd H. Van Andel
T.H.V.H. Thomas H. Vender Haar
T.I.S. Tracy I. Storer
T.J.T. T. J. Toy
T.L.Ha. Thomas L. Hale
T.L.M. Tom L. McKnight
T.L.W. Timothy L. White
T.M.F. Thomas M. Frost
T.M.K. Todd M. Kana
T.N.Kr. T. N. Krishnamurti
T.Sc. Tim Schowalter

U, V

U.R. Uwe Radok
V.A. Vera Alexander
V.H.E. Van H. English
V.I.C. Vernon I. Cheadle
V.K.S. V. K. Sawhney
V.L.E. Virginia L. Ernster
V.R.B. Victor R. Baker

W

W.A.Ly. Walter A. Lyons
W.Bu. Willy Burgdorfer
W.C.Wo. William C. Wonders
W.D.K. Walter D. Keller
W.D.Mo. Walter D. Mooney
W.D.R.H. W. D. Russell-Hunter
W.E.B. William E. Bell
W.F.D. Walter F. Dabberdt
W.F.F. W. F. Furter
W.F.Mo. William F. Morris
W.G.A. Warren G. Abrahamson
W.G.McG. William G. McGinnies
W.G.W. W. Gordon Whaley
W.H.C Wallis H. Clark, Jr.
W.H.J. W. Hilton Johnson
W.H.Wa. Warren H. Wagner, Jr.
W.J.M. William J. Mitsch
W.K.J. W. K. Joklik
W.Ly. Waldo Lyon
W.L.Ch. William L. Chameides
W.L.Fi. William L. Fisher
W.Ma. Walter Mannheim
W.Po. William Porter
W.Re. Walter Reid
W.R.Br. Winslow R. Briggs
W.R.J. William R. Judd
W.Stu. Wilton Sturges
W.Swe. Will Swearingen
W.S.B. William S. Benninghof
W.S.Br. Willard S. Bromley
W.W.E. W. W. Epstein
W.W.Sp. Wesley W. Spink

Y

Y.H.M. Yi Hua Ma
Y.W. Y. Waisel

Contributor Affiliations

This list comprises all contributors to the Encyclopedia. A brief affiliation is provided for each author. This list may be used in conjunction with the previous section to fully identify the contributor of each article.

A

Aagaard, Prof. Knut. Department of Oceanography, University of Washington.
Able, Dr. Kenneth P. Department of Biological Sciences, State University of New York, Albany.
Abrahams, Prof. Elihu. Department of Physics, Rutgers University.
Abrahamson, Prof. Warren G., II. Department of Biology, Bucknell University, Lewisburg, Pennsylvania.
Ackerman, Dr. Edward A. Deceased; formerly, Carnegie Institution of Washington.
Addicott, Prof. Fredrick T. Department of Botany, University of California, Davis.
Adewumi, Prof. Michael A. Petroleum and Natural Gas Engineering, Pennsylvania State University.
Adkins, George G. Retired; formerly, Chief, Department of River Basins, Bureau of Power, Federal Power Commission.
Alexander, Prof. R. McNeill. Department of Biology, University of Leeds, England.
Alexander, Dr. Vera. Institute of Marine Science, University of Alaska.
Ammons, Dr. Nelle. Retired; Department of Biology, West Virginia University.
Anagnostakis, Dr. Sandra L. Department of Pathology & Ecology, Connecticut Agriculture Experimental Station, New Haven.
Andreasen, Dr. Christian. Bureau Hydrographique International, Monaco.
Angrisano, Rear Admiral Guiseppe. Bureau Hydrographique International, Monaco.
Appel, Dr. Max J. G. James A. Baker Institute for Animal Health, Cornell University, Ithaca, New York.
Appleby, Dr. Arnold P. Professor Emeritus of Crop Science, Oregon State University, Corvallis.
Armstrong, Dr. Richard L. Institute of Arctic and Alpine Research, University of Colorado.
Ashton, Dr. Mark S. Yale School of Forestry and Environmental Studies, New Haven, Connecticut.
Atoda, Dr. Kenji. Professor of Zoology, Tohoku University, Sendai, Japan.
Atwood, Dr. Ronald L. Foote Mineral Company, Kings Mountain, North Carolina.
Audus, Dr. Leslie J. Professor of Botany (retired), Bedford College, University of London, England.
Austrian, Dr. Robert. Department of Research Medicine, School of Medicine, University of Pennsylvania.
Ayers, Arthur R. Assistant Professor, Cellular and Development Biology, Harvard University.

B

Bada, Dr. Jeffrey L. Scripps Institution of Oceanography, University of California, San Diego-La Jolla.
Baker, Dr. Victor R. Department of Geosciences, University of Arizona.
Barber, Dr. Richard T. Duke University Marine Laboratory.
Barbosa, Dr. Pedro. Department of Entomology, College of Life Sciences, University of Maryland.
Barclay, Dr. Harriet G. Professor of Botany, Department of Life Sciences, University of Tulsa.
Barksdale, Dr. Lane. Department of Microbiology, School of Medicine, New York University Medical Center.
Barr, Dr. David. Department of Entomology, Royal Ontario Museum, Toronto, Ontario, Canada.
Barron, Prof. Eric. Director, Earth System Science Center, Pennsylvania State University, University Park.
Beaumont, Robert T. Director of Meteorological Operations, E. G. G., Inc., Boulder, Colorado.
Becker, Dr. Elery R. Deceased; formerly, Division of Life Sciences, Arizona State University.
Bell, Dr. William E. Department of Pediatrics, University of Iowa.
Bemis, Dr. David A. Director, Clinical Bacteriology and Mycology Services, University of Tennessee, Knoxville.
Bennett, Prof. Joan W. Department of Cell and Molecular Biology, Tulane University, New Orleans, Louisiana.
Benninghoff, Dr. William S. Department of Botany, University of Michigan.
Bernard, Dr. E. N. National Oceanic and Atmospheric Administration, Pacific Marine Environmental Laboratory, Seattle, Washington.
Berner, Dr. Elizabeth K. Department of Geology and Geophysics, Yale University.
Berner, Dr. Robert A. Department of Geology and Geophysics, Yale University.
Birkenheuer, Dr. Daniel L. Lakewood, Colorado.
Bisno, Dr. A. L. Veterans Administration Medical Center, Miami, Florida.
Black, Dr. Robert F. Department of Geology, University of Connecticut.
Blad, Dr. Blaine L. Agricultural Meteorology Section, University of Nebraska.
Blouin, Dr. Edmour F. College of Veterinary Medicine, Oklahoma State University, Stillwater.
Bluestein, Prof. Howard B. Department of Meteorology, University of Oklahoma.
Boero, Dr. Ferdinando. Professor of Zoology, Università di Lecce, Dipartmento di Biologia, Stazione di Biologia Marina, Lecce, Italy.
Bohach, Dr. Greg. Department of Microbiology, Molecular Biology, and Biochemistry, University of Idaho, Moscow.
Boke, Dr. Norman H. George Lynn Cross Research Professor of Botany, University of Oklahoma.
Bolin, Dr. Steven R. Research Leader, Virology Cattle Research, Agricultural Research Service, U.S. Department of Agriculture, Ames, Iowa.

Bond, Dr. Gerard. Lamont-Doherty Geological Observatory, Palisades, New York.
Boothroyd, Dr. Carl W. Department of Plant Pathology, Cornell University.
Bormann, Dr. Bernard T. Department of Forest Sciences, Oregon State University, Corvallis.
Borowiecki, Prof. Barbara Z. Department of Geography, University of Wisconsin, Milwaukee.
Bosart, Prof. Lance F. Department of Atmospheric Sciences, State University of New York, Albany.
Box, Dr. Elgene O. Department of Geography, University of Georgia.
Bozarth, Dr. Robert F. Department of Life Science, Indiana State University, Terre Haute.
Brassell, Prof. Simon C. Professor of Geological Sciences, Biogeochemical Laboratories, Indiana University, Bloomington.
Breed, Dr. Michael D. Environmental Population and Organismic Biology, University of Colorado.
Breman, Dr. Joel D. Division of Parasitic Diseases, Centers for Disease Control, Department of Health and Human Services, Atlanta, Georgia.
Brian, Dr. M. V. Institute of Terrestrial Ecology, Furzebrook Research Station, Wareham, England.
Briggs, Dr. Winslow R. Department of Plant Biology, Carnegie Institution of Washington, Stanford, California.
Brisbin, Dr. I. Lehr. Savannah River Ecology Laboratory, University of Georgia, Aiken, South Carolina.
Bromley, Willard S. Consulting Forester and Association Consultant, New Rochelle, New York.
Brubaker, Prof. Robert R. Department of Microbiology, Michigan State University, East Lansing.
Burgdorfer, Dr. Willy. Scientist Emeritus, Laboratory of Pathology, Rocky Mountain Laboratories, National Institutes of Health, Department of Health and Human Services, Hamilton, Montana.
Burke, Prof. Kevin C. Department of Geosciences, University of Houston.
Burkhart, Dr. Harold E. Department of Forestry, College of Forestry and Wildlife Resources, Virginia Polytechnic Institute and State University, Blacksburg.

C

Cady, Dr. Gilbert H. Deceased; formerly, Consulting Coal Geologist, Urbana, Illinois.
Calnek, Dr. Bruce W. Chairperson, Department of Avian and Aquatic Animal Medicine, College of Veterinary Medicine, Cornell University.
Campana, Dr. Richard J. Department of Botany and Plant Pathology, University of Maine.
Campbell, Dr. John S. Department of Biological Sciences, University of Lethbridge, Alberta, Canada.
Carbone, Dr. Richard E. Director, Atmospheric Technology Division, National Center for Atmospheric Research, Boulder, Colorado.
Carlson, Dr. Jack H. Rhone Meriux, Inc., Athens, Georgia.
Carris, Dr. Lori M. Department of Plant Pathology, Washington State University, Pullman.
Carroll, Dr. C. Ronald. Institute of Ecology, University of Georgia, Athens.
Carter, Dr. G. R. Professor Emeritus, Department of Patho-biology, Virginia-Maryland Regional College of Veterinary Medicine, Virginia Polytechnic Institute and State University, Blackburg.
Chameides, Prof. William L. School of Earth and Atmospheric Sciences, Georgia Institute of Technology, Atlanta.
Chang, Prof. Howard H. Department of Civil and Environmental Engineering, San Diego State University, San Diego, California.
Chang, Prof. Kang-tsung. Department of Geography, University of Idaho, Moscow.
Cheadle, Prof. Vernon I. Chancellor, University of California, Santa Barbara.
Chen, Dr. Robert T. Medical Epidemiologist, Infant Immunization Section, Centers for Disease Control, Surveillance Investigations and Research Branch, Atlanta, Georgia.
Christensen, Dr. Bruce M. Assistant Professor, Department of Veterinary Science, University of Wisconsin, Madison.
Christensen, Dr. Martha. Professor Emeritus, Department of Botany, University of Wyoming, Laramie.
Ciegler, Dr. Alex. Southern Regional Research Center, USDA Science and Education Administration, New Orleans.
Clapp, Philip F. National Weather Service, National Oceanic and Atmospheric Administration, Washington, D.C.
Claridge, Dr. Elmond L. Retired; formerly, Director of Graduate Program in Petroleum Engineering, Chemical Engineering Department, University of Houston Central Campus, Houston.
Clark, Prof. Wallis H., Jr. Director, Aquaculture Program, University of California, Davis.
Cloud, Dr. Preston E., Jr. Department of Geology, University of California, Santa Barbara.
Cochran, Dr. James R. Lamont-Doherty Earth Observatory, Columbia University, Palisades, New York.
Coleman, Dr. David C. Department of Entomology, University of Georgia.
Colwell, Dr. Rita R. Department of Microbiology, Division of Agricultural and Life Sciences, University of Maryland.
Comer, Prof. Jonathan. Deparment of Geography, Oklahoma State University, Stillwater.
Confer, Prof. Anthony W. Head, Department of Veterinary, Pathology, Oklahoma State University, Stillwater.
Cook, Dr. Frank. Research Assistant Professor, Department of Veterinary Science, Gluck Equine Research Center, University of Kentucky, Lexington.
Cooper, Dr. Arthur W. Department of Botany, North Carolina State University.
Corbitt, Robert A. Associate, Metcalf & Eddy, Inc., Atlanta, Georgia.
Core, Prof. Earl L. Professor of Botany, Department of Biology, West Virginia University.
Corliss, Dr. John O. Department of Zoology, University of Maryland.
Costerton, Dr. J. William F. Department of Biology, University of Calgary, Alberta, Canada.

Couch, Dr. Richard B. Professor of Naval Architecture and Marine Engineering, Ship Hydrodynamics Laboratory, University of Michigan.

Coulombe, Dr. Mary J. Director, Timber Access and Supply, American Forest and Paper Association, Washington, D.C.

Court, Dr. Arnold. Retired; formerly, Department of Climatology, California State University.

Cowles, Prof. Philip B. Professor Emeritus of Microbiology, School of Medicine, Yale University.

Cracraft, Dr. Joel. Department of Ornithology, American Museum of Natural History, New York.

Cranwell, Dr. Lucy M. Department of Geosciences, University of Arizona.

Crittenden, Dr. Charles V. Geographer, Economic Development Administration, U.S. Department of Commerce.

Cronquist, Dr. Arthur. Director of Botany, New York Botanical Gardens, Bronx, New York.

Crow, Dr. James F. Department of Genetics, University of Wisconsin.

Cummings, Dr. Michael R. Department of Biological Sciences, University of Illinois.

Curray, Dr. Joseph R. Scripps Institution of Oceanography, University of California, La Jolla.

Curtin, Dr. Charles B. Department of Biology, Creighton University.

D

Dabberdt, Dr. Walter F. Atmospheric Technology Division, National Center for Atmospheric Research, Boulder, Colorado.

Dabbs, Dr. David J. Pathologist, Department of Pathology, University of Washington.

Daiber, Dr. Franklin C. College of Marine Studies, University of Delaware.

Dale, Dr. James B. Veteran Affairs Medical Center, Memphis, Tennessee.

Damman, Prof. Antoni W. H. Department of Biology, University of Connecticut.

Davies, Dr. Peter J. Department of Botany, Cornell University.

Davies-Jones, Dr. Robert P. Meteorologist, National Severe Storms Laboratory, U.S. Department of Commerce, Norman, Oklahoma.

Davis, Dr. George H. Department of Geological Sciences, University of Arizona.

Davis, Prof. Kenneth P. Deceased; formerly, School of Forestry, Yale University.

Davis, Dr. Richard A., Jr. Department of Geology, University of South Florida.

Dawson, Dr. Todd E. Ecology and Systematics, Cornell University, Carson Hall, Ithaca, New York.

Deacon, Margaret. Department of Oceanography, University of Southampton, United Kingdom.

Deland, Dr. Raymond J. Retired; formerly, Department of Meteorology and Oceanography, New York University.

Dell'Arciprete, Dr. Patricia. Central Nacional Patagónico (CONICET), Chubut, Argentina.

DeLong, Dr. Dwight M. College of Biological Sciences, Ohio State University.

Dengler, Dr. Nancy G. Department of Botany, University of Toronto, Ontario, Canada.

DePaolo, Dr. Donald J. Department of Geology and Geophysics, University of California, Berkeley.

de Week, Dr. A. L. Institut für Klinische Immunologic Inselspital, Universitat Bern, Switzerland.

Dickinson, Prof. Robert E. Department of Atmospheric Physics, University of Arizona, Tucson.

Diekema, Dr. Daniel James. University of Iowa College of Medicine, Department of Pathology, Division of Medical Microbiology, Iowa City.

Dill, Dr. Robert F. Ocean Mining Administration, U.S. Department of the Interior.

Dingle, Prof. Hugh. Department of Entomology, College of Agricultural and Environmental Sciences, University of California, Davis.

Dinsmore, Robertson P. International Ice Patrol, Woods Hole Oceanographic Institution, Woods Hole, Massachusetts.

Dodds, Robert H. Deceased; formerly, Personnel Manager, Gibbs and Hill, Inc., New York.

Donaldson, Dr. Alex I. Pirbright Laboratory, Pirbright, Surrey, United Kingdom.

Dooley, H. D. Marine Laboratory, Department of Agriculture and Fisheries, Aberdeen, Scotland.

Doyle, Prof. James A. Department of Biological Sciences, University of California, Davis.

Doyle, Prof. Michael P. Director, Center for Food Safety and Quality Enhancement, University of Georgia, Griffin.

Dugan, Dr. Frank M. American Type Culture Collection, Manassas, Virginia.

Dunning, Dr. John B., Jr. Department of Forestry and Natural Resources, Purdue University.

DuPont, Dr. Herbert L. Medical School, University of Texas, Houston.

Dyer, Dr. K. R. Institute of Oceanographic Sciences, Somerset, England.

E

Easterbrook, Don J. Department of Geology, Western Washington University, Bellingham.

Eberl, Dr. Dennis D. U.S. Geological Survey, Denver Federal Center, Denver, Colorado.

Edmonds, Dr. Robert L. College of Forest Resources, University of Washington, Seattle.

Edwards, Dr. Steven. Department of Virology, Central Veterinary Laboratory-Weybridge, United Kingdom.

Eidson, Dr. Millicent. Veterinarian, Santa Fe, New Mexico.

Emerson, Dr. Steven. Department of Oceanography, University of Washington.

English, Prof. Van H. Department of Geography, Dartmouth College.

Epstein, Dr. W. W. Department of Chemistry, University of Utah.

Erasmus, Dr. Baltus J. Head, Onderstepoort Vaccine Factory, Republic of South Africa.

Ernster, Dr. Virginia L. Department of Epidemiology and International Health, School of Medicine, University of California, San Francisco.

Esau, Prof. Katherine. Department of Botany, University of California, Santa Barbara.

Evans, Dr. Doyle J. Chief, Bacterial Enteropathogens Laboratory, Veterans Affairs Medical Center, Houston, Texas.
Evans, Prof. Francis C. Division of Biological Sciences, University of Michigan.
Eyde, Dr. Richard H. Department of Botany, Smithsonian Institution.
Ezzell, Dr. John W. Chief, Special Pathogens Branch, U.S. Army Medical Research Institute, Fort Detrick, Maryland.

F

Farmer, Dr. David M. Department of the Environment, Pacific Region, Institute of Ocean Sciences, Sidney, British Columbia, Canada.
Faure, Dr. Gunter. Department of Geological Sciences, Ohio State University.
Fein, Dr. Jay S. Division of Atmospheric Sciences, National Science Foundation, Washington, D.C.
Feist, Dr. Blake E. Northwest Fisheries Science Center, Environmental Conservation Division–Watershed Program, Seattle, Washington.
Fell, Dr. Jack W. University of Miami, Rosenstiel School of Marine and Atmospheric Science, Marine Biology and Fisheries, Key Biscayne, Florida.
Ferrell, Dr. James E., Jr. Department of Molecular Pharmacology, Stanford University, School of Medicine.
Ferry, Dr. James F. Department of Biology, Madison College.
Fiedler, Dr. Peggy L. Piedmont, California.
Finegold, Dr. Sydney M. Chief, Infectious Disease Section, Veterans Administration Wadsworth Hospital Center, Los Angeles, California.
Finkelstein, Dr. Richard. Department of Microbiology, University of Missouri School of Medicine, Columbia.
Fischer, Prof. Alfred G. San Pedro, California.
Fisher, Dr. Robert L. Geological Research Division, Scripps Institution of Oceanography, La Jolla, California.
Fisher, Dr. William L. Department of Geological Science, University of Texas, Austin.
Flessa, Dr. Karl Walter. Division of Earth Sciences, National Science Foundation, Washington, D.C.
Flint, Dr. Richard F. Department of Geology and Geophysics, Yale University.
Focht, Prof. Dennis D. Department of Soil and Environmental Sciences, University of California, Riverside.
Ford, Prof. Derek C. Department of Geography, McMaster University, Hamilton, Ontario, Canada.
Frank, Prof. Bernard. Deceased; formerly, Professor of Watershed Management, Colorado State University.
Freeman, Dr. A. E. Department of Animal Science, Iowa State University.
Freeman, Harry M. Chief, Waste Minimization Branch, Risk Reduction Engineering Laboratory, U.S. Environmental Protection Agency, Cincinnati, Ohio.
Fribourg, Prof. Henry A. Department of Plant and Soil Science, University of Tennessee, Knoxville.
Frost, Dr. Thomas M. Center for Limnology, University of Wisconsin.
Frumhoff, Dr. Peter. Department of Entomology, University of California, Davis.
Fulks, J. R. Retired; formerly, National Weather Service, Chicago, Illinois.
Fuller, Dr. Mark Roy. Patuxent Wildlife Reserve Center, Laurel, Maryland.
Fuller, Dr. Thomas C. Retired; formerly, Supervisor, Botany and Seed Laboratories, California Department of Food and Agriculture, Sacramento.
Furter, Dr. W. F. Dean of Graduate Studies and Research, Royal Military College of Canada.
Futral, Prof. J. G. Department of Agricultural Engineering, Georgia Agricultural Experiment Station, Experiment, Georgia.
Futuyma, Dr. Douglas J. Section of Ecology and Systematics, Cornell University.

G

Galston, Dr. Arthur W. Department of Biology, Yale University.
Gambs, Gerard C. Vice President (retired), Ford, Bacon, & Davis, Inc., New York.
Garland, Dr. John. Professor, Department of Forest Engineering, Oregon State University, Corvallis.
Garrett, Dr. Roger E. Department of Agricultural Engineering, University of California, Davis.
Genito, Dennis. Pasture Systems and Watershed Management Research Laboratory, U.S. Department of Agriculture–Agricultural Research Service, University Park, Pennsylvania.
Gerber, Leah R. National Center for Ecological Analysis and Synthesis, Santa Barbara, California.
Gerlach, Dr. Arch C. Chief Geographer, Geological Survey, U.S. Department of the Interior.
Gerrity, Dr. Joseph P. National Meteorological Center, National Oceanic and Atmospheric Administration, Camp Springs, Maryland.
Giannasi, Dr. David E. Department of Botany, University of Georgia.
Giese, Raymond. Department of Mechanical Engineering, University of Minnesota.
Gilardi, Dr. Gerald L. Microbiology Laboratory, North General Hospital, New York.
Gill, Dr. D. Michael. Department of Molecular Biology and Microbiology, Tufts University.
Glantz, Dr. Michael. National Center for Atmospheric Research, Boulder, Colorado.
Gold, Dr. Ronald. Hospital for Sick Children, Toronto, Ontario, Canada.
Golley, Dr. Frank B. Institute of Ecology, Athens, Georgia.
Goodison, Dr. Barry E. Superintendent, Hydrometeorological Impact and Development Section, Canadian Climate Centre, Atmospheric Environment Service, Downsview, Ontario.
Goodman, Dr. Richard A. Division of Public Health, Emory University.
Goosby, Dr. Eric P. AIDS Education and Training Centers Program, U.S. Public Health Service, Rockville, Maryland.
Gordon, Dr. Arnold L. Lamont-Doherty Geological Observatory, Palisades, New York.

Gotelli, Dr. Nick. Department of Biology, University of Vermont, Burlington.
Govindjee, Dr. Department of Botany and Department of Physiology and Biophysics, University of Illinois, Urbana.
Govindjee, Dr. Rajni. Retired; formerly, Department of Botany, University of Illinois, Urbana.
Granick, Dr. S. Department of Biochemistry, Rockefeller University.
Grasshoff, Dr. K. Institut für Meereskunde an der Universitat Kiel, Germany.
Graves, Dr. Arthur H. Deceased; formerly, Consultant in Genetics, Connecticut Agricultural Experiment Station.
Gray, Dr. Ernest. Variety Club Children's Hospital, Department of Pediatrics, University of Minnesota.
Gray, Dr. Gerald J. Vice President, Forest Policy Center, American Forests, Washington, D.C.
Griffiths, Dr. A. J. F. Department of Botany, University of British Columbia, Vancouver, Canada.
Grigg, Prof. Neil S. Head, Department of Civil Engineering, Colorado State University, Fort Collins.
Gronstol, Prof. Hallstein. Department of Large Animal Clinical Sciences, Norwegian College of Veterinary Medicine, Oslo, Norway.
Gunning, Dr. Brian E. S. Department of Botany, Queen's University, Belfast, Ireland.
Gyakum, Dr. John R. Department of Meteorology, McGill University, Montreal, Quebec, Canada.

H

Hader, Rodney N. Secretary, American Chemical Society, Washington, D.C.
Hagan, Dr. Maura. National Center for Atmospheric Research, Boulder, Colorado.
Hagood, Mel A. Irrigated Agriculture Research and Extension Center, Washington State University.
Halbouty, Dr. Michael T. Consulting Geologist and Petroleum Engineer, Houston, Texas.
Halde, Dr. Carlyn. Department of Microbiology and Immunology, School of Medicine, University of California, San Francisco.
Hall, Dr. Richard P. Deceased; formerly, Professor of Zoology, University of California, Los Angeles.
Hallet, Dr. John. Director, Atmospheric Ice Physics Laboratory, Desert Research Institute, Atmospheric Sciences Center, University of Nevada.
Hamilton, Dr. R. I. Research Branch, Agriculture Canada, Vancouver, British Columbia, Canada.
Hammar, Dr. Samuel P. The Diagnostic Specialties Laboratory, Washington.
Hammer, Edward E. Retired; formerly, Fluorescent and High Intensity Systems Department, General Electric Company, Cleveland, Ohio.
Handsfield, Dr. H. Hunter. Director, Sexually Transmitted Disease Control Program, Seattle-King Country Public Health Department, Harborview Medical Center, Seattle, Washington.
Haney, Dr. Robert L. Department of Meteorology, Naval Postgraduate School, Monterey, California.
Hanlin, Dr. Richard T. Department of Plant Pathology, University of Georgia, Athens.
Hanna, Dr. Steven R. Sigma Research Corporation, Concord, Massachusetts.
Hanor, Jeffrey S. Department of Geology, Louisiana State University.
Hansen, Dr. Katherine J. Department of Earth Sciences, Montana State University, Bozeman.
Harris, Prof. Richard W. Department of Environmental Horticulture, University of California, Davis.
Hartman, Dr. Olga. Deceased; formerly, Allan Hancock Foundation, University of Southern California.
Hartmann, Dr. Dennis L. Department of Atmospheric Sciences, University of Washington, Seattle.
Hasler, Dr. Arthur D. Laboratory of Limnology, University of Wisconsin.
Hastings, Dr. Alan. Division of Environmental Studies, University of California at Davis.
Hayes, Dr. Dennis E. Lamont-Doherty Geological Observatory, Columbia University, Palisades, New York.
Hazen, Richard. Hazen and Sawyer, Consulting Engineers, New York.
Helfman, Dr. Gene S. Department of Zoology, University of Georgia.
Hellmers, Dr. Henry. Botany Department, Duke University.
Heyne, Prof. E. G. Professor of Plant Breeding, Department of Agronomy, Kansas State University.
Heyneman, Dr. Donald. Chair, Health and Medical Sciences Program, University of California, Berkeley.
Hicks, Dr. Bruce B. Atmospheric Turbulence and Diffusion Division, National Oceanic and Atmospheric Administration, Oak Ridge, Tennessee.
Hicks, Steacy D. Sterling, Virginia.
Hitchman, Dr. Matthew H. Department of Meteorology, University of Wisconsin, Madison.
Hoffman, Dr. Paul. Department of Microbiology and Immunology, Faculty of Medicine, Dalhousie University, Halifax, Nova Scotia, Canada.
Holmes, Dr. Robert W. Department of Biological Sciences, University of California, Santa Barbara.
Holt, Dr. Perry C. Department of Biology, Virginia Polytechnic Institute.
Holton, Prof. James R. Department of Atmospheric Sciences, University of Washington.
Hooke, Dr. Roger LeB. Deer Isle, Maine.
Horn, Dr. David J. Department of Entomology, College of Agriculture, Ohio State University.
Horsfall, Dr. James G. Director, Connecticut Agricultural Experiment Station, New Haven.
Horsnail, Dr. R. F. Amax Exploration, Inc., Denver, Colorado.
Houghton, Dr. David D. Department of Meteorology, University of Wisconsin, Madison.
Hubert, Prof. John F. Department of Geosciences, University of Massachusetts, Amherst.
Huebler, Dr. Jack. Senior Vice President, Institute of Gas Technology, IIT Center, Chicago, Illinois.

Hulbary, Dr. Robert L. Department of Botany, University of Iowa.
Hungate, Prof. Robert E. Department of Bacteriology, University of California, Davis.

I

Ingels, Chuck A. University of California Cooperative Extension, Sacramento.
Ingerson, Dr. Earl. Department of Geology, University of Texas, Austin.
Ingram, Dr. B. Lynn. Department of Geology and Geophysics, University of California, Berkeley.
Ishida, Dr. Nakao. Department of Bacteriology, Tohoku University School of Medicine, Sendai, Japan.
Issel, Dr. Charles J. College of Agriculture, Veterinary Science, Gluck Equine Research Center, University of Kentucky, Lexington.

J

Jackson, Dr. M. P. A. Senior Research Scientist, Bureau of Economic Geology, University of Texas, Austin.
Janick, Prof. Jules. Department of Horticulture, Purdue University.
Jannasch, Dr. Holger W. Senior Scientist, Woods Hole Oceanographic Institution, Woods Hole, Massachusetts.
Japuntich, Daniel A. Occupational Health and Safety Products Division, 3M Company, St. Paul, Minnesota.
Johnson, Dr. Donald Lee. Department of Geography, University of Illinois, Urbana.
Johnson, Dr. Giles. School of Biological Sciences, University of Manchester, United Kingdom.
Johnson, Dr. Jon D. Department of Forestry, University of Florida, Gainesville.
Johnson, Dr. W. Hilton. Department of Geology, University of Illinois, Urbana-Champaign.
Joklik, Dr. W. K. Department of Microbiology and Immunology, Duke University Medical Center.
Jones, Dr. Clive G. Institute of Ecosystem Studies, New York Botanical Garden, Millbrook, New York.
Jones, Dr. Daniel J. Professor and Chairman, Department of Earth Sciences, California State College, Bakersfield.
Jones, Dr. Jeremy B. Department of Biological Sciences, University of Nevada, Las Vegas.
Jones, Dr. Meredith L. National Museum of Natural History, Smithsonian Institution, Washington, D.C.
Judd, Dr. William R. School of Civil Engineering, Purdue University.
Juranek, Dr. Dennis D. Center for Disease Control, Atlanta, Georgia.

K

Kamrin, Dr. Michael A. Center for Environmental Toxicology, Department of Pathology, Michigan State University.
Kana, Dr. Todd M. Horn Point Laboratory, Cambridge, Maryland.
Kareiva, Prof. Peter. Department of Zoology, University of Washington, Seattle.
Karner, Dr. Garry D. Lamont-Doherty Earth Observatory, Palisades, New York.
Keller, Prof. Walter D. Department of Geology, University of Missouri.
Kessler, Prof. Edwin. Director, National Severe Storms Laboratory, Norman, Oklahoma.
Kilbey, Dr. Brian J. Department of Genetics, Institute of Animal Genetics, University of Edinburgh, Scotland.
Kinsman, Prof. Blair. College of Marine Studies, University of Delaware.
Kissel, Dr. David E. Department of Agronomy, Kansas State University.
Klich, Dr. Maren. Microbiologist, Food and Feed Safety Research, U.S. Department of Agriculture–Agricultural Research Service, New Orleans, Louisiana.
Kocan, Prof. Katherine M. Department of Veterinary Pathology, Oklahoma State University, Stillwater.
Koch, Prof. Arthur L. Department of Biology, Indiana University.
Kominz, Michelle. Lamont-Doherty Geological Observatory, Palisades, New York.
Konopka, Dr. Allan E. Department of Biological Sciences, Purdue University.
Krider, Prof. E. Philip. Director, Institute of Atmospheric Physics, University of Arizona.
Krishnamurti, Dr. T. N. Department of Meteorology, Florida State University.
Kulp, J. Laurence. President, Teledyne Isotopes, Westwood, New Jersey.
Kutzbach, Prof. John E. Department of Meteorology, University of Wisconsin.

L

Lacis, Dr. Andrew A. NASA Goddard Institute for Space Studies, New York.
Ladd, Dr. Harry S. (Retired) National Museum, Smithsonian Institution.
Lamb, Prof. Dennis. Department of Meteorology, Pennsylvania State University.
LaMotte, Dr. Clifford E. Botany Department, Iowa State University.
Landsberg, Dr. H. E. Deceased; formerly, Institute for Physical Science and Technology, University of Maryland.
Lane, Dr. Meredith A. Director, University of Kansas Herbarium, Lawrence.
Lang, Dr. Anton. Department of Botany, MUS-DOE Plant Research Laboratory, Michigan State University.
Lapedes, Daniel N. Deceased; formerly, Editor in Chief, "McGraw-Hill Encyclopedia of Science and Technology," McGraw-Hill, Inc., New York.
Lappin-Scott, Hilary. Department of Biology, University of Calgary, Alberta, Canada.
Larson, Dr. Roy A. Department of Horticulture, North Carolina State University.
Last, Prof. John M. Faculty of Health Sciences, School of Medicine/Epidemiology and Community Medicine, University of Ottawa, Ontario, Canada.
Lavin, Prof. Stephen. Department of Geography, University of Nebraska, Lincoln.
Lavoie, Dr. Ronald L. Chief, Program Requirements and Development Division, U.S. Department of Commerce, National Oceanic

and Atmospheric Administration, Silver Spring, Maryland.

Leader, Dr. Robert W. Center for Environmental Toxicology, Department of Pathology, Michigan State University.

Ledig, Dr. F. Thomas. School of Forestry, Yale University.

Lee, Dr. J. T. Cooperative Institute for Mesoscale Meterological Studies, National Oceanic and Atmospheric Administration, University of Oklahoma.

Lee, Lyndon C. L. C. Lee & Associates, Inc., Seattle, Washington.

Leopold, Dr. Luna B. Department of Geology and Geophysics, University of California, Berkeley.

Lesins, Dr. Glen. Department of Oceanography, Dalhousie University, Halifax, Nova Scotia, Canada.

Levine, Dr. Norman D. College of Veterinary Medicine, University of Illinois, Urbana.

Lewis, Dr. Alcinda C. Institute of Ecosystem Studies, New York Botanical Garden, Millbrook.

Li, Dr. Ching Chun. Graduate School of Public Health, University of Pittsburgh.

Lieth, Dr. Helmut. Department of Biochemistry, University of North Carolina.

Likens, Prof. Gene E. Director, Institute of Ecosystem Studies, The New York Botanical Garden, Millbrook, New York.

Linde, Dr. Ronald K. President, Envirodyne, Inc., Los Angeles, California.

Linsley, Prof. Ray K. Department of Civil Engineering, Stanford University.

Liou, Prof. Kuo-Nan. Department Chair, Department of Atmospheric Sciences, University of California, Los Angeles.

Little, Elbert L., Jr. Dendrologist, Forest Service, U.S. Department of Agriculture.

Liu, Benjamin Y. H. Department of Mechanical Engineering, University of Minnesota.

Liu, Prof. Clark C. K. Department of Civil Engineering, University of Hawaii at Manoa.

Lochhead, Dr. John H. London, England.

Loneragan, Prof. J. F. School of Environmental and Life Science, Murdoch University, Perth, Australia.

Longacre, Alan. (Retired) Fluor Engineers and Constructors, Irvine, California.

Longo, Dr. Dan L. Director, National Cancer Institute, Frederick Cancer Research Facility, Maryland.

Loope, Dr. David B. Department of Geology, University of Nebraska Medical Center.

Lordi, George M. Department of Medicine, College of Medicine and Dentistry, Newark, New Jersey.

Ludlam, Prof. Frank H. Deceased; formerly, Department of Meteorology, Imperial College, London, England.

Ludvik, Dr. George F. Insecticide Application Research, Agricultural Research and Development Department, Monsanto Company, St.

Luthin, Prof. James N. Department of Civil Engineering, University of California, Davis.

Lyman, Dr. John. Department of Oceanography, University of North Carolina.

Lyon, Dr. Waldo. Arctic Submarine Research Laboratory, Naval Undersea Warfare Center, San Diego, California.

Lyons, Dr. Walter A. Certified Consulting Meteorologist, Forensic Meteorology Associates, Fort Collins, Colorado.

M

Ma, Dr. Yi Hua. Department of Chemical Engineering, Worcester Polytechnic Institute.

Macdoughall, Prof. J. Douglas. Geological Research Division, Scripps Institution of Oceanography, La Jolla, California.

Maggenti, Dr. Armand R. Division of Nematology, University of California, Davis.

Major, Dr. Jack. Department of Botany, University of California, Davis.

Maldonado-Moll, Dr. Jose F. Special Assistant for Science and Technology, Sistema Universitario de la Fundacion Educative Ana G. Mendez Rio Piedras, Puerto Rico.

Mann, Dr. Kenneth H. Fisheries Research Board of Canada, Dartmouth, Nova Scotia.

Mannheim, Prof. Walter. Med. Zentrum filr Hygiene, Universitat Marburg, Germany.

Marks, Dr. Jonathan. Department of Sociology and Anthropology, University of North Carolina at Charlotte.

Marsh, Dr. Kenneth N. Thermodynamic Research Center, Texas A & M University.

Marshall, Dr. Norman B. Retired; formerly, Senior Principal Scientific Officer, British Museum of Natural History, London, England.

Martínez-Carrera, Dr. D. College of Postgraduates in Agricultural Sciences, Mushroom Biotechnology, Puebla, Mexico.

Marvier, Dr. Michelle A. Department of Biology, Santa Clara University, Santa Clara, California.

Marx, Dr. Preston A. California Primate Research Center, University of California, Davis.

Mason, Dr. Basil J. Program Director, Center for Environmental Technology, Imperial College of Science and Technology, London, England.

Masucci, Dr. Maria G. Microbiology and Tumor Biology Center, Karolinska Institute, Stockholm, Sweden.

Mauk, Dr. Craighton S. Research Associate, Department of Horticultural Science, Mountain Horticultural Crops Research and Extension Center, School of Agriculture and Life Sciences, North Carolina State University, Fletcher.

Maul, Prof. George A. Director, Division of Marine and Environmental Systems, Florida Institute of Technology, Melbourne.

Maurer, Dr. Brian A. Department of Fisheries and Wildlife, Michigan State University, East Lansing.

May, F. H. Consultant, Kerr-McGee Corporation, Whittier, California.

McClung, Dr. Leland S. Professor Emeritus, Department of Microbiology, Indiana University.

McClure, Dr. Michelle. Northwest Fisheries Science Center, Seattle, Washington.

McGinley, Dr. John A. Chief, Forecast Research Group, National Oceanic and Atmospheric Administration, Environmental Research Laboratory, Boulder, Colorado.

McGinnies, Dr. William G. Professor of Dendrochronology and Arid Land Ecologist, Office of Arid Land Studies, University of Arizona.

McGuire, Ann B. Aquaculture Program, University of California, Davis.
McHugh, Dr. J. L. Marine Sciences Research Center, State University of New York, Stony Brook.
McIntyre, Dr. Andrew. Lamont-Doherty Geological Observatory, Palisades, New York; Queens College of the City University of New York.
McKnight, Prof. Tom. Department of Geography, University of Caifornia, Los Angeles, California.
McNamara, Dr. Peter. Department of Medical Microbiology, University of Wisconsin, Madison.
McNaughton, Prof. Samuel J. Department of Biology, Syracuse University.
Mebus, Dr. Charles A. Laboratory Chief, Foreign Animal Disease Diagnostic Laboratory, U.S. Department of Agriculture, Greenport, New York.
Meeuse, Dr. Bastiaan J. D. Department of Botany, University of Washington.
Melnick, Dr. Joseph L. Department of Virology and Epidemiology, Baylor College of Medicine.
Meyer, Dr. Bernard S. Department of Botany, Ohio State University.
Middleton, Dr. Gerard V. Department of Geology, McMaster University, Hamilton, Ontario, Canada.
Mikos, Prof. Antonios G. Director of John W. Cox Laboratory of Biomedical Engineering, Department of Chemical Engineering, Rice University, Houston, Texas.
Miller, Prof. Maynard M. Department of Geology, Michigan State University; Director, Foundation for Glacial and Environmental Research, Seattle, Washington.
Miller, Dr. Regis B. Project Leader, Center for Wood Anatomy Research, Forest Products Laboratory, U.S. Department of Agriculture, Forest Service, Madison, Wisconsin.
Miller, Dr. Shelby A. Argonne National Laboratory, Argonne, Illinois.
Milus, Dr. E. A. Department of Plant Pathology, University of Arkansas, Fayetteville.
Mitsch, Prof. William J. Graduate Program Environmental Science, School of Natural Resources, Ohio State University, Columbus.
Mitterer, Dr. Richard M. Department of Geosciences, University of Texas, Dallas.
Moll, Dr. Gary A. Director, Urban and Community Forestry, American Forestry Association, Washington, D.C.
Mood, Dr. Alexander M. Director, Public Policy Research Organization, University of California, Irvine.
Moore, Dr. Alice E. Sloan-Kettering Institute of Cancer Research, New York.
Moore, Prof. Charles A. Department of Civil Engineering, Ohio State University.
Morrell, Dr. Jeffrey J. Department of Forest Products, Oregon State University, Corvallis.
Morris, Dr. William F. Department of Zoology, Duke University, Durham, North Carolina.
Morrison, Dr. David C. Associate Director, Basic Research Programs, Department of Microbiology, University of Kansas Medical Center, Kansas City.
Morse, Dr. Stephen A. Director, Division of Sexually Transmitted Diseases Laboratory Research, Centers for Disease Control and Protection, Atlanta, Georgia.
Mortenson, Leonard E. Department of Biological Sciences, Purdue University.
Moseley, Dr. Steve L. Department of Microbiology, University of Washington School of Medicine, Seattle.
Mottet, Dr. N. Karle. Professor of Pathology and Director of Hospital Pathology, University Hospital, University of Washington.
Mulligan, Dr. Pamela K. Formerly, Department of Biochemistry, University of North Carolina.
Murgatroyd, Dr. R. J. Meteorological Office, Bracknell, England.
Murphy, Dr. Frederick A. Centers for Disease Control, Department of Health and Human Services, Tucker, Georgia.
Murphy, Dr. Robert L. Division of Infectious Diseases, Northwestern University Medical School, Chicago, Illinois.
Murray, Dr. James W. School of Oceanography, University of Washington, Seattle.

N

Nace, Dr. Raymond L. Geological Survey, U.S. Department of the Interior, Raleigh, North Carolina.
Namias, Jerome. Climate Research Group, Scripps Institution of Oceanography, La Jolla, California.
Napp-Zinn, Prof. Klaus. Botanical Institute, Cologne, Germany.
Nealson, Dr. Kenneth H. Center for Great Lakes Studies, University of Wisconsin.
Nelson, Elton G. Collaborator (WOC), Crops Research Division, U.S. Department of Agriculture, Beltsville, Maryland.
Nelsen, Dr. Olin E. Department of Biology, University of Pennsylvania.
Newton, Dr. Chester W. National Center for Atmospheric Research, Boulder, Colorado.
Norcross, Prof. Neil L. College of Veterinary Medicine, Cornell University, Ithaca, New York.
North, Dr. Gerald R. Director, Climate System Research Program, Department of Meteorology, Texas A&M University.
Norton, Dr. Jeanette. Department of Plants, Soils and Biometerology, Utah State University, Logan.

O

Ogburn, Charles B. Cooperative Extension Service, Auburn University.
O'Leary, Dr. John E. Department of Forest Engineering, Oregon State University.
Oliver, Prof. John E. Department of Geography, Geology, and Anthropology, Indiana State University, Terre Haute, Indiana.
Orians, Dr. Gordon. Department of Zoology, University of Washington, Seattle.
Oris, Dr. James T. Department of Zoology, Miami University, Oxford, Ohio.
Orville, Dr. Harold D. Institute for Atmospheric Science, South Dakota School of Mines, Rapid City, South Dakota.
Owen, Dr. Denis F. Department of Biology, Oxford Polytechnic.

P

Palevsky, Dr. Gerald. Consulting Professional Engineer, Hastings-on-Hudson, New York.
Palmer, Prof. Darwin L. Chief, Veterans Administration Hospital, Medical Center, Department of Medicine, Infectious Disease Division, University of New Mexico School of Medicine.
Papageorgiou, George. Research Associate, Department of Botany, University of Illinois, Urbana.
Park, Prof. Jae-Woo. Department of Civil Engineering, University of Hawaii at Manoa.
Parrish, Dr. Judith Totman. Department of Geosciences, University of Arizona.
Pascual, Dr. Miguel. Centro Nacional Patagónico (CONICET), Chubut, Argentina.
Patterson, Dr. David J. Department of Zoology, University of Bristol, England.
Pedersen, Prof. Niels C. School of Veterinary Medicine, University of California, Davis.
Phaff, Dr. Herman J. Department of Food Science and Technology, College of Agriculture and Environmental Science, University of California, Davis.
Philander, Prof. S. George. Program in Atmospheric and Oceanic Sciences, Princeton University.
Philip, Dr. Cornelius B. Department of Entomology, California Academy of Sciences, San Francisco.
Pielou, Dr. E. C. Department of Biology, University of Lethbridge, Alberta, Canada.
Pimm, Dr. Stuart. Department of Zoology, University of Tennessee, Knoxville.
Pinder, Dr. George F. Department of Civil and Geological Engineering, Princeton University.
Pinkava, Dr. Donald J. Director, ASU Herbarium, Department of Botany, Arizona State University.
Pinker, Prof. Rachel T. Department of Meteorology, University of Maryland.
Pionke, Dr. Harry B. Pasture Systems and Watershed Management Research Laboratory, U.S. Department of Agriculture–Agricultural Research Service, University Park, Pennsylvania.
Postgate, Prof. John R. Department of Microbiology, University of Sussex, England.
Price, Dr. James F. Department of Physical Oceanography, Woods Hole Oceanographic Institution, Massachusetts.
Price, Dr. Peter W. Department of Entomology, Northern Arizona University.
Princen, L. H. Associate Center Director, North Regional Research Center, Department of Agriculture, Peoria, Illinois.

Q

Quaife, Carol. Department of Pathology, University of Washington.
Quarles, Dr. John M. Department of Medical Microbiology & Immunology, Texas A & M University, College Station.

R

Radforth, Dr. Norman W. Muskeg Research Institute, University of New Brunswick, Canada.
Radke, Dr. Rodney O. Agricultural Product Research Laboratory, Monsanto Company, St. Louis, Missouri.
Radok, Dr. Uwe. Environmental Research (CIRES), University of Colorado.
Randerson, Dr. Peter. Department of Applied Biology, University of Wales Institute of Science and Technology, Cardiff.
Raps, Dr. Shirley. Department of Biological Sciences, Hunter College, City University of New York.
Raski, Dr. Dewey J. Department of Nematology, University of California, Davis.
Rasmusson, Dr. Eugene M. Geophysical Fluid Dynamics Laboratory, Environmental Science Services Administration, Princeton, New Jersey.
Rauber, Prof. Robert M. Department of Atmospheric Science, University of Illinois at Urbana-Champaign, Urbana.
Redhead, Dr. Scott A. Mycologist, Biosystematics Research Centre, Agriculture Canada, Research Branch, Central Experimental Farm, Ottawa, Ontario, Canada.
Reeve, Dr. Roger M. Western Regional Research Laboratory, U.S. Department of Agriculture, Albany, California.
Regnery, Dr. Russell. Supervisory Research Microbiologist, Viral and Rickettsial Zoonoses Branch, Department of Health and Human Services, Centers for Disease Control and Prevention, Atlanta, Georgia.
Reichman, Lee B. Department of Medicine, College of Medicine and Dentistry, Newark, New Jersey.
Reichmann, Dr. M. E. Department of Microbiology, University of Illinois.
Reid, Joseph L. Scripps Institution of Oceanography, La Jolla, California.
Reid, Dr. Walter V. World Resources Institute, Washington, D.C.
Reiter, Dr. Elmar R. Department of Atmospheric Science, Colorado State University.
Reitz, Dr. Louis P. (Retired) Corps Research Division, Agricultural Research Service, U.S. Department of Agriculture, Beltsville, Maryland.
Rich, Dr. Saul. Connecticut Agricultural Experiment Station, New Haven, Connecticut.
Richardson, Dr. Philip. Department of Physical Oceanography, Woods Hole Oceanographic Institution, Woods Hole, Massachusetts.
Riley, Dr. Ralph. Plant Breeding Institute, Cambridge, England.
Risser, Dr. Paul G. Vice President for Research, University of New Mexico.
Ritchie, Dr. J. C. Department of Biology, Trent University, Peterborough, Ontario, Canada.
Rodier, Dr. Patricia M. Department of Anatomy, School of Medicine, University of Virginia.
Romick, Dr. Gerald J. Applied Physics Laboratory, Johns Hopkins University, Laurel, Maryland.
Rosa, Dr. Patricia. National Institutes of Health and Allergies, Rocky Mountain Laboratories, Hamilton, Montana.
Rosenberg, Dr. Eugene. Department of Microbiology, George S. Wise Faculty of Life Sciences, Tel Aviv University, Israel.
Ross, Dr. J. Central Science Laboratory, Surrey, United Kingdom.

Ruddat, Dr. Manfred. Department of Molecular Genetics and Cell Biology, University of Chicago, Barnes Laboratory.
Ruouff, Dr. Kathryn L. Francis Blake Bacteriology Laboratories, Department of Microbiology and Molecular Genetics, Massachusetts General Hospital, Harvard Medical School, Boston, Massachusetts.
Russell-Hunter, Prof. W. D. Professor of Zoology, Department of Biology, Syracuse University.
Rust, Dr. Brian R. Department of Geology, University of Ottawa, Ontario, Canada.
Rutger, J. N. Department of Agronomy, University of California, Davis.
Ryan, Dr. Kenneth J. Department of Pathology, University of Arizona Medical Center, Tucson.

S

Sabo, Dr. John L. National Center for Ecological Analysis and Synthesis, Santa Barbara, California.
Saenger, Dr. Peter. Centre for Coastal Management, University of New England, Northern Rivers, Lismore, Australia.
Salazar, Prof. Deborah A. Department of Geography, Oklahoma State University, Stillwater.
Salisbury, Dr. Frank B. Plant Science Department, Utah State University.
San Martin, Robert L. Deputy Assistant Secretary for Renewable Energy, Department of Energy, Washington, D.C.
Sanders, Dr. Frederick. Department of Meteorology, Massachusetts Institute of Technology.
Sanders, Dr. Robert W. Department of Zoology, University of Georgia.
Sarachik, Prof. Edward S. Department of Atmospheric Sciences, University of Washington, Seattle.
Saunders, Dr. Gary W. Department of Biology, University of New Brunswick, Fredericton, New Brunswick, Canada.
Savage, Dr. George M. Director, Product Research and Development, Upjohn International Inc., Kalamazoo, Michigan.
Savage, Dr. Jay M. Department of Biological Sciences, University of Southern California.
Sawhney, Dr. V. K. Department of Biology, University of Saskatchewan, Saskatoon, Canada.
Sayre, Dr. Albert N. Deceased; formerly, Consulting Groundwater Geologist, Behre Dolbear and Company.
Schachter, Dr. Julius. Department of Laboratory Medicine, University of California, San Francisco.
Schimmel, Dr. Curt. Silicon Graphics, Inc., Mountain View, California.
Schindler, Dr. James E. Department of Biological Sciences, College of Sciences, Clemson University.
Schmid, Dr. Rudolf. Department of Botany, University of California, Berkeley.
Schnabel, Dr. Ronald. Pasture Systems and Watershed Management Research Laboratory, U.S. Department of Agriculture–Agricultural Research Service, University Park, Pennsylvania.
Scholz, Christopher H. Lamont-Doherty Geological Observatory, Palisades, New York.
Scholz, Dr. Nathaniel. Northwest Fisheries Science Center, Environmental Conservation Division, Seattle, Washington.
Schowalter, Prof. Timothy D. Department of Entomology, Oregon State University, Corvallis.
Schultz, Dr. Cheryl. National Center for Ecological Analysis and Synthesis, University of California, Santa Barbara.
Schultz, Dr. Jack. Gypsy Moth Research Center, Department of Entomology, College of Agriculture, Pesticide Research Laboratory and Graduate Study Center, Pennsylvania State University.
Schuster, Prof. Frederick L. Department of Biology, Brooklyn College.
Schwartz, Dr. Karlene V. Department of Biology, University of Massachusetts, Boston.
Seed, Prof. John Richard. Department of Epidemiology, School of Public Health, University of North Carolina, Chapel Hill.
Selby, Prof. Michael J. Department of Earth Science, University of Waikato, New Zealand.
Seliskar, Dr. Denise M. Associate Research Scientist, College of Marine Studies, University of Delaware.
Sessions, Dr. John. Department of Forest Engineering, Oregon State University, Corvallis.
Sharkey, Dr. Thomas D. Department of Botany, University of Wisconsin.
Shedlock, Dr. Kaye M. U.S. Geological Survey, Administrative Officer, Denver, Colorado.
Sheriff, Dr. Robert E. Department of Geology, University of Houston.
Shuman, Dr. Frederick B. Director, National Meteorological Center, National Oceanic and Atmospheric Administration, Washington, D.C.
Siever, Dr. Raymond. Department of Geological Sciences, Harvard University.
Signorini, Dr. Sergio R. Division of Ocean Sciences, National Science Foundation, Arlington, Virginia.
Silva, Dr. Paul C. Department of Botany, University of California, Berkeley.
Simberloff, Prof. Daniel. Department of Biological Science, Florida State University.
Simonson, Dr. Roy W. Director (retired), Soil Classification and Correlation, U.S. Department of Agriculture, Hyattsville, Maryland.
Simpson, Dr. Joanne. Head, Severe Storms Branch, Goddard Space Flight Center, NASA, Greenbelt, Maryland.
Simpson, Prof. Robert H. Retired; formerly, Director, Experimental Meteorology Laboratory, National Weather Service, Miami, Florida.
Sims, Dr. G. K. Agricultural Research Service, U.S. Department of Agriculture, Urbana, Illinois.
Skelly, Dr. David. School of Forestry and Environmental Studies, Yale University, New Haven, Connecticut.
Skinner, Prof. Brian J. Department of Geology and Geophysics, Yale University, New Haven, Connecticut.

Smith, Prof. David M. Morris K. Jeesup Professor of Silviculture, School of Forestry, Yale University.
Smith, Dr. Robert L. College of Oceanic and Atmospheric Sciences, Oregon State University, Corvallis.
Solomon, Dr. James C. Associate Curator, Missouri Botanical Garden, St. Louis.
Sovonick-Dunford, Dr. S. Department of Biological Sciences, University of Cincinnati.
Speight, Dr. James G. Western Research Institute, Laramie, Wyoming.
Spink, Dr. Wesley W. Regents, Professor of Medicine, School of Medicine, University of Minnesota.
Sposito, Dr. Garrison. Department of Plant and Soil Biology, University of California, Berkeley.
Sprague, Prof. G. F. Department of Agronomy, University of Illinois, Urbana.
Sprague, Dr. Howard B. Agricultural Consultant, Washington, D.C.
Stage, Dr. Steven A. Baton Rouge, Louisiana.
Stamm, Dr. Lola. Department of Epidemiology, School of Public Health, University of North Carolina, Chapel Hill.
Steponkus, Prof. Peter L. Department of Agronomy, Cornell University.
Stewart, Scott R. L. C. Lee & Associates, Inc., Seattle, Washington.
Stone, Dr. Jeffrey K. Department of Botany and Plant Pathology, Oregon State University, Corvallis.
Storch, Dr. Henry H. Deceased; formerly, Assistant Professor of Chemistry, New York University.
Storer, Dr. Tracy I. Deceased; formerly, Department of Zoology, University of California, Davis.
Stott, Dr. Jeffrey L. Department of Veterinary Microbiology and Immunology, School of Veterinary Medicine, University of California, Davis.
Strausbaugh, Prof. Carl A. Kimberly Research and Extension Center, University of Idaho, Kimberly.
Stroyan, Dr. Henry L. G. Entomology Department, Ministry of Agriculture, Fisheries and Food, Harpenden, England.
Stuart, Dr. Edward G. Deceased; formerly, West Virginia School of Medicine.
Sturges, Dr. Wilton. Department of Oceanography, Florida State University.
Sturtevant, Prof. Alfred H. Thomas Hunt Morgan Professor of Biology, Emeritus, California Institute of Technology.
Sudds, Dr. Richard. Department of Biological Sciences, State University of New York, Pittsburgh.
Summers, Dr. B. A. James A. Baker Institute for Animal Health, Cornell University, Ithaca, New York.
Sutherland, Dr. S. S. Senior Research Officer, Animal Health Laboratories, Department of Agriculture, Western Australia.
Sutton, Dr. B. C. Head of Taxonomic and Identification Services, Mycological Institute, Ferry Lane, Surrey, United Kingdom.
Swearingen, Dr. Will. NASA-Montana State University TechLink, Bozeman, Montana.

T

Tabor, Paul. Soil Conservation Service, Athens, Georgia.
Tao, Dr. B. Y. School of Agricultural and Biological Engineering, Purdue University, West Lafayette, Indiana.
Taylor, Dr. R. E. Associate Professor and Director, Radiocarbon Laboratory, University of California, Riverside.
Tchobanoglous, Prof. George. Department of Civil Engineering, University of California, Davis.
Teaguarden, Dr. Dennis E. Professor Emeritus of Forestry, University of California, Berkeley.
Teleki, Dr. Geza. Deceased; formerly, Department of Geology, George Washington University.
Teller, Dr. Aaron J. Teller Environmental Systems, Worcester, Massachusetts.
Thomann, Dr. Gary C. Power Technologies, Inc., Schenectady, New York.
Thompson, Dr. Margaret W. Departments of Medical Genetics and Pediatrics, University of Tronoto and the Hospital for Sick Children, Toronto, Ontario, Canada.
Timoney, Dr. John F. Department of Veterinary Science, University of Kentucky, Lexington.
Ting, Dr. Irwin P. Department of Botany and Plant Sciences, University of California, Riverside.
Torres, Dr. Alfonso. Head, Diagnostic Services Section, National Veterinary Services Laboratory, U.S. Department of Agriculture, Greenport, New York.
Toy, Dr. T. J. Department of Geography, University of Denver, Colorado.
Trappe, Dr. James M. Department of Forest Science, Oregon State University, Corvallis.
Turner, Dr. Thomas B. Professor of Microbiology and Dean Emeritus, School of Medicine, Johns Hopkins University.

U

Ullrich, Dr. Robert C. Department of Botany, University of Vermont, Burlington.

V

Valentine, Dr. J. W. Department of Geology, University of California, Santa Barbara.
Van Andel, Dr. Tjeerd H. Department of Geology, Stanford University.
Van Niel, Dr. Cornelius B. Deceased; formerly, Hopkins Marine Station, Pacific Grove, California.
Vickers, Dr. Mary Lynne. Department of Veterinary Science, Livestock Disease Diagnostic Center, University of Kentucky, Lexington.
Vierstra, Dr. Richard. Horticulture Department, University of Wisconsin.
Vitek, Dr. Charles. Medical Epidemiologist, HIV Vaccine Section, Division of HIV/AIDS Prevention, Centers for Disease Control and Prevention, Atlanta, Georgia.
Vogt, Dr. Peter. Department of the Navy, Naval Research Laboratory, Washington, D.C.
Von Graevenitz, Alexander W. C. Department of Medical Microbiology, University of Zurich, Switzerland.

Voss, Prof. Edward W., Jr. Department of Microbiology, University of Illinois, Urbana.

W

Wagner, Dr. Warren H., Jr. Department of Botany, University of Michigan.
Waisel, Prof. Y. Department of Botany, Tel Aviv University, Israel.
Wallace, Dr. John M. Department of Atmospheric Sciences, University of Washington, Seattle.
Walsh, Dr. Michael A. Department of Biology, Utah State University.
Walton, Dr. Peter D. Department of Plant Science, University of Alberta, Edmonton, Canada.
Warren, Dr. John J. K. Resources Pty. Ltd., Mitcham, Australia.
Weiss, Dr. Allison. Department of Molecular Genetics, Biochemistry, and Microbiology, University of Cincinnati, Ohio.
Weller, Prof. Gunter. Global Change and Arctic Systems, University of Alaska, Fairbanks.
Wendler, Dr. Helen E. Department of Biology, Massachusetts Institute of Technology.
Wenner, Prof. Adrian M. Department of Biological Sciences, University of California, Santa Barbara.
Wetzel, Dr. Richard. College of William and Mary, Virginia Institute of Marine Science, School of Marine Science, Gloucester Point, Virginia.
Whaley, Dr. W. Gordon. Cell Research Institute, University of Texas, Austin.
Wheeler, Dr. Harry E. Department of Plant Pathology, University of Kentucky.
White, Dr. C. Langdon. Professor of Geography, Stanford University.
White, Dr. Sidney E. Department of Geology and Mineralogy, Ohio State University.
White, Dr. Tim. School of Forest Resources, University of Florida, Gainesville.
Whitmore, Dr. T. C. Department of Geography, University of Cambridge, United Kingdom.
Wickstrom, Dr. Conrad. Department of Biological Sciences, Kent State University.
Wiens, Prof. Herold J. Deceased; formerly, Department of Geography, Yale University.
Wilcox, Dr. Hugh. State University College of Forestry, Syracuse University.
Wilder, Dr. George J. Department of Biological Sciences, University of Illinois, Chicago.
Wiley, Prof. Edward O., III. Curator, Natural History Museum, University of Kansas, Lawrence.
Wilkening, Dr. M. Department of Physics, New Mexico Institute of Mining and Technology.
Wilkinson, Dr. Brian J. Department of Biological Sciences, Illinois State University.
Willett, Prof. Hilda P. Director of Graduate Studies, Department of Microbiology, Duke University Medical Center, Durham, North Carolina.
Willett, Prof. Kurd C. Department of Meteorology, Massachusetts Institute of Technology.
Wilson, Dr. B. F. Department of Forestry and Wildlife Management, University of Massachusetts.
Wilson, Dr. James B. Department of Forest Products, Oregon State University, Corvallis.
Wilson, Dr. James L. Department of Geology and Mineralogy, University of Michigan.
Winter, Dr. Thomas C. U.S. Geological Survey, Department of the Interior, Denver, Colorado.
Winterer, Dr. Edward L. Geological Research Division, Scripps Institution of Oceanography, La Jolla, California.
Wolfe, Dr. Ralph S. Marine Biological Laboratory, Woods Hole, Massachusetts.
Wonders, Dr. William C. Department of Geography, University of Alberta, Canada.
Woods, Dr. Jon P. Assistant Professor, Department of Medical Microbiology and Immunology, University of Wisconsin Medical School, Madison.
Wyrtki, Dr. Klaus. Department of Oceanography, University of Hawaii, Manoa.
Wüest, Dr. Alfred Johny. Applied Aquatic Ecology (APEC), Kastanienbaum, Switzerland.

Z

Zaitlin, Dr. Milton. Associate Director, Biotechnology Program, Plant Pathology, Cornell University.
Zbar, Dr. Frederick. Chevy Chase, Maryland.
Zelitch, Dr. Israel. Department of Biochemistry and Genetics, Connecticut Agricultural Experiment Station, New Haven.
Zlotnik, Dr. I. Microbiological Research Establishment, Wilts, England.
Zubay, Dr. Geoffrey. Sherman Fairchild Center for the Life Sciences, Department of Biological Sciences, Columbia University.
Zwerman, Prof. Paul J. Department of Agronomy, Cornell University.

Index

The asterisk indicates page numbers of an article title.

B

C

D

E

F

G

H

I

J

K

L

M

N

O

P

Q

R

S

U

V

W

X

Y

Z